Aids to Understanding Wo

iris, rainbow: *iris*
iso-, equal: *iso*tonic
kerat-, horn: *kerat*in
labi-, lip: *labi*a
labyrinth-, maze: *labyrinth*
lacri-, tears: *lacri*mal gland
lacun-, pool: *lacun*a
laten-, hidden: *laten*t
-lemm, rind or peel: neuri*lemm*a
leuko-, white: *leuko*cyte
lingu-, tongue: *lingu*al tonsil
lip-, fat: *lip*ids
-logy, study of: physio*logy*
-lyte, dissolvable: electro*lyte*
macro-, large: *macro*phage
macula, spot: *macula* lutea
meat-, passage: auditory *meat*us
melan-, black: *melan*in
mening-, membrane: *mening*es
mens-, month: *mens*trual
meta-, change: *meta*bolism
mict-, to pass urine: *mict*urition
mit-, thread: *mit*osis
mono-, one: *mono*saccharide
mons-, mountain: *mons* pubis
morul-, mulberry: *morul*a
moto-, moving: *moto*r
mut-, change: *mut*ation
myo-, muscle: *myo*fibril
nat-, to be born: pre*nat*al
nephr-, kidney: *nephr*on
neutr-, neither one nor the other: *neutr*al
nod-, knot: *nod*ule
nutri-, nourish: *nutri*ent
odont-, tooth: *odont*oid process
olfact-, to smell: *olfact*ory
-osis, abnormal increase in production: leukocyt*osis*
oss-, bone: *oss*eous tissue
papill-, nipple: *papill*ary muscle
para-, beside: *para*thyroid glands
pariet-, wall: *pariet*al membrane
patho-, disease: *patho*gen
pelv-, basin: *pelv*ic cavity
peri-, around: *peri*cardial membrane
phag-, to eat: *phag*ocytosis
pino-, to drink: *pino*cytosis
pleur-, rib: *pleur*al membrane
plex-, interweaving: choroid *plex*us
poie-, to make: hemoto*poie*sis
poly-, many: *poly*unsaturated
pseudo-, false: *pseudo*stratified epithelium
puber-, adult: *puber*ty
pylor-, gatekeeper: *pylor*ic sphincter
sacchar-, sugar: mono*sacchar*ide
sarco-, flesh: *sa*
scler-, hard: *scler*a
seb-, grease: *seb*aceous gland
sens-, feeling: *sens*ory neuron
-som, body: ribo*som*e
squam-, scale: *squam*ous epithelium
stasis-, standing still: homeo*stasis*
strat-, layer: *strat*ified
syn-, together: *syn*thesis
systol-, contraction: *systol*e
tachy-, rapid: *tachy*cardia
tetan-, stiff: *tetan*ic
thromb-, clot: *thromb*ocyte
toc-, birth: oxy*toc*in
-tomy, cutting: ana*tomy*
trigon-, triangle: *trigon*e
troph-, well fed: muscular hyper*trophy*
tropic-, influencing: adrenocortico*tropic*
tympan-, drum: *tympan*ic membrane
umbil-, navel: *umbil*ical cord
ventr-, belly or stomach: *ventr*icle
vill-, hair: *vill*i
vitre-, glass: *vitre*ous humor
zym-, ferment: en*zym*e

Essentials of Human Anatomy and Physiology

Sixth Edition

David Shier
Washtenaw Community College

◆

Jackie Butler
Grayson County Community College

◆

Ricki Lewis
The University at Albany

Boston, Massachusetts Burr Ridge, Illinois Dubuque, Iowa
Madison, Wisconsin New York, New York San Francisco, California St. Louis, Missouri

WCB/McGraw-Hill

A Division of The ***McGraw·Hill*** *Companies*

HOLE'S ESSENTIALS OF HUMAN ANATOMY AND PHYSIOLOGY, SIXTH EDITION

This book is printed on recycled, acid-free paper containing 10% postconsumer waste.

1 2 3 4 5 7 8 9 0 QPH/QPH 0 9 8 7

ISBN 0-697-28252-X (paper)
0-697-28251-1 (case)

Publisher: *Michael D. Lange*
Sponsoring editor: *Kristine Noel Tibbetts*
Developmental editor: *Kelly A. Drapeau*
Marketing manager: *Keri L. Witman*
Project manager: *Marla K. Irion*
Production supervisor: *Mary E. Haas*
Designer: *K. Wayne Harms*
Photo research coordinator: *John C. Leland*
Art editor: *Brenda A. Ernzen*
Compositor: *Shepherd, Inc.*
Typeface: *10.5/12 Garamond*
Printer: *Quebecor Printing Book Group/Hawkins*

Cover and interior design: *Lisa Delgado*

The credits section for this book begins on page 594 and is considered an extension of the copyright page.

Library of Congress Catalog Card Number: 96-78736

http://www.mhcollege.com

QuickStudy:

Computerized Study Guide for *Hole's Essentials of Human Anatomy and Physiology*

by Nancy A. Sickles Corbett

Focus study time only in the areas where you need the most help.

Available on diskette:
IBM: 0-697-32925-9
Macintosh: 0-697-32926-7

How does it work?

- Take the computerized test for each chapter.
- The program presents you with feedback for each answer.
- A page-referenced study plan is then created for all incorrect answers.
- Look up the answers to the questions you missed.
- This easy-to-use *QuickStudy* is not a duplicate of the printed study guide that accompanies your text. It is full of new study aids.

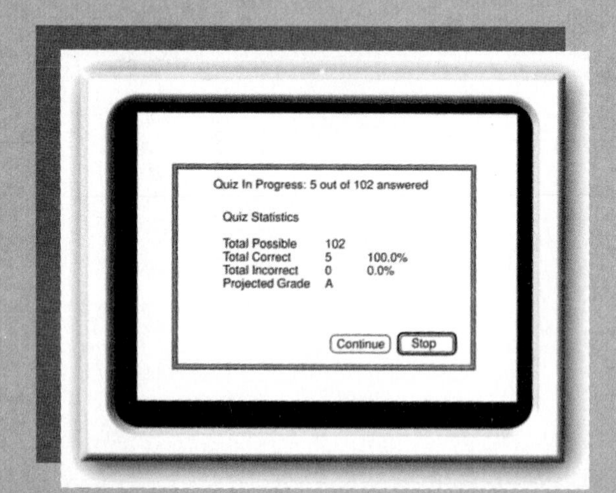

The program provides instant feedback on quiz results.

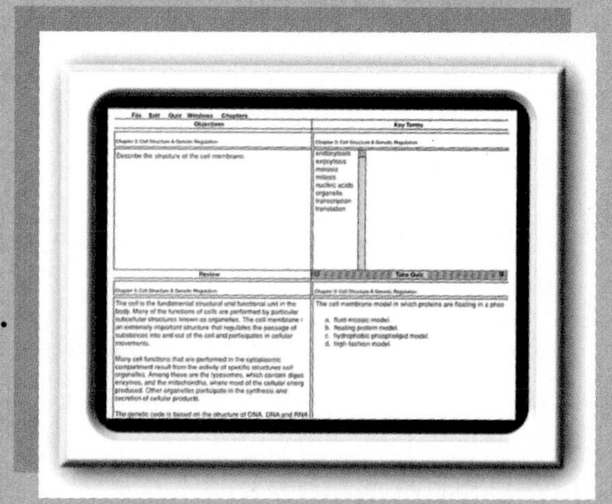

Four different review methods are available.

- **Learning Objectives** outline the major points of each text chapter.
- **Key Terms** in each chapter are listed and defined.
- **Review** examines important concepts and facts in detail.
- **Take Quiz** allows testing on any combination of—or all—chapters.

To order any of these products, contact your bookstore manager or call our Customer Service Department at 800-338-3987.

Brief

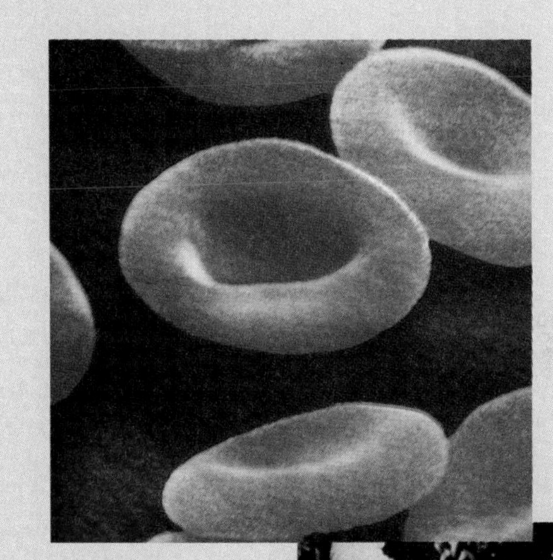

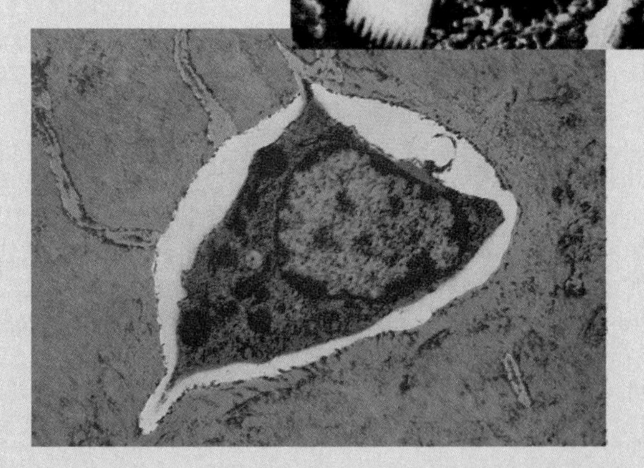

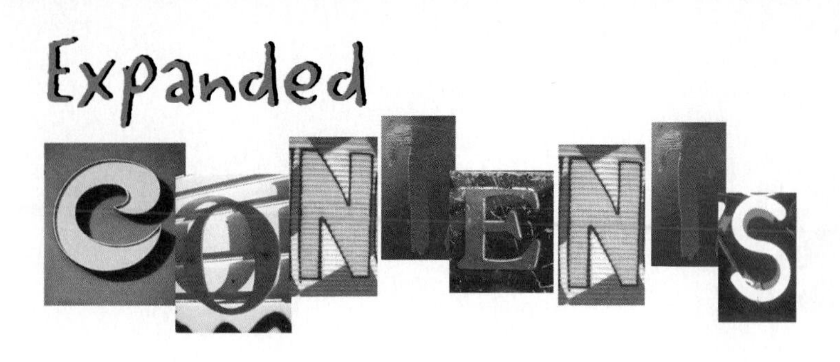
Expanded
CONTENTS

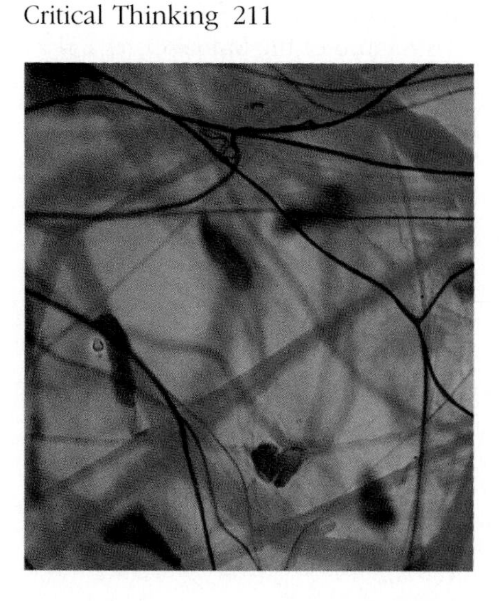

UNIT 3
Integration and Coordination

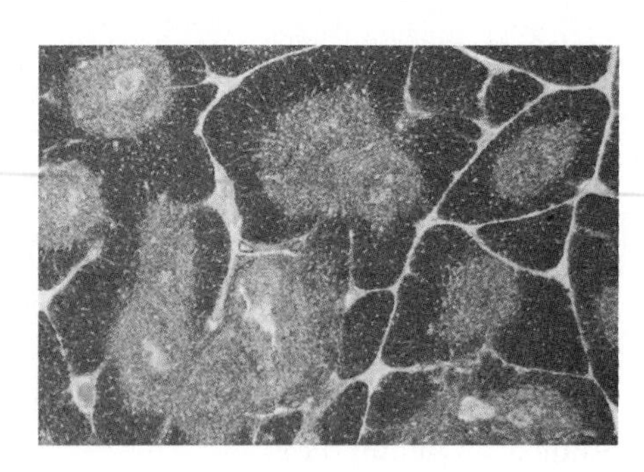

UNIT 4
TRANSPORT

UNIT 5
ABSORPTION AND EXCRETION

UNIT 6
THE HUMAN LIFE CYCLE

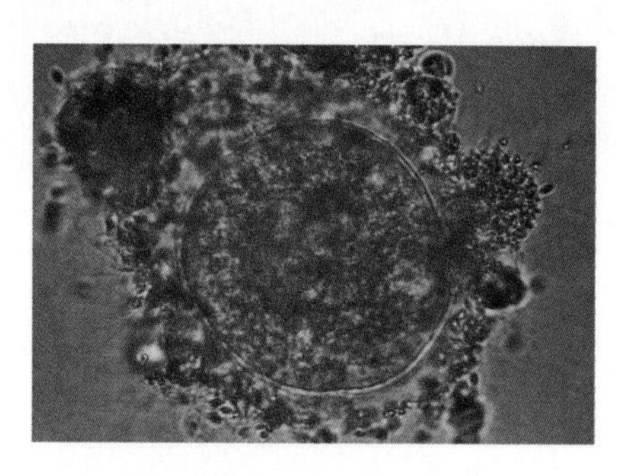

On the Importance of Studying Anatomy and Physiology

Imagine losing a major part of a primary organ system. When WCB/McGraw-Hill sales representative Amy Porter traveled around Texas telling professors about *Hole's Essentials of Human Anatomy and Physiology,* she had no idea that she would soon become intimately involved in comprehending several of her body's organ systems—and how they were turning against her.

In October 1995, ten days after taking a common antibiotic, Amy broke out in a rash that turned to blisters and rapidly covered her body. Amy also had a raging fever. In the emergency room, a dermatologist diagnosed Stevens-Johnson syndrome, which quickly worsened into its most severe form, called toxic epidermal necrolysis. In short, Amy's entire epidermis peeled off, as if she had suffered devastating burns everywhere. The terror and pain were unimaginable.

Fortunately, the skin bank at the University of Texas at Galveston was able to cover Amy's body with donor skin, anchored with a thousand staples, which protected her for several weeks. Amy, dulled by drugs, endured painful daily baths. Gradually, the skin grafts wore off as new skin from beneath replaced them.

Only one in 2 million people develop Stevens-Johnson syndrome; only 10% survive. Because of her severe case, Amy's odds were even worse. But she made it.

Amy Porter learned more than she ever wanted to know about the integumentary, immune, and nervous systems, and the body's control of temperature and fluids. Her story dramatically illustrates the importance of a knowledge of anatomy and physiology. Yet, even for those of us lucky enough to enjoy good health, understanding how the human body functions can be fascinating and valuable.

The goal of this sixth edition of *Hole's Essentials of Human Anatomy and Physiology* is to introduce you to the structure and function of the human body in an interesting and highly readable manner. The text is especially written for students of one-semester courses in anatomy and physiology who are pursuing careers in allied health fields and whose backgrounds in physical and biological sciences are minimal.

A New Team

We began where most revisions begin—addressing reviewers' concerns. We made corrections, updated terminology, and adjusted the writing style, while carefully retaining the flavor and breadth of coverage of the original. We added exciting opening vignettes to the chapters, sprinkled in many brand-new boxes, and placed greater emphasis and updated information on physiology throughout the text.

The three of us—a physiologist, a microbiologist, and a geneticist—were strangers when brought together just four summers ago. Immersed in this project, we became great friends, and a true collaboration was born. Our hope is that the result not only retains the comprehensiveness of John Hole's work, but also embraces our diverse training, our many teaching experiences, and our shared enthusiasm.

What's New

Chapter Opening Vignettes

Just as the true "story" of Amy Porter opens this preface, each chapter begins with a true vignette and accompanying photograph that immediately reveal the relevance of chapter material. Some tales may be familiar from news reports; others introduce new individuals. The chapter opening vignettes may suggest possible careers, as they vividly apply chapter principles.

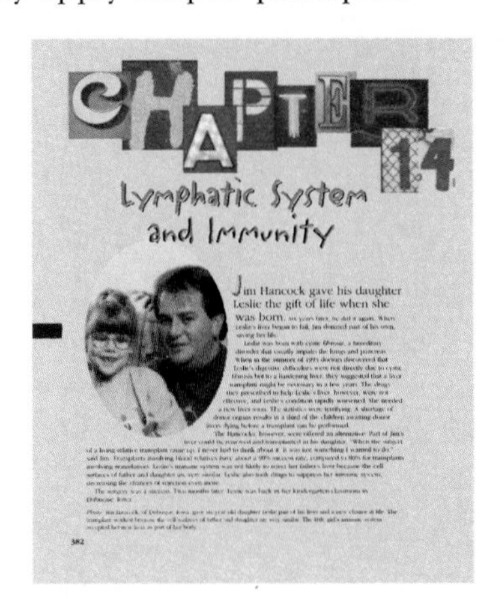
CHAPTER 14

Lymphatic System and Immunity

Jim Hancock gave his daughter Leslie the gift of life when she was born.

382

Style

The sixth edition is easier to read and flows more smoothly. It has been meticulously edited to change passive voice to active voice, to remove wordiness, to define all terms as they are used, and to include more transitions. The changes are subtle, but the reading is easier and more meaningful.

ORGANization Illustrations

Each of the eleven body system chapters contains a full-page illustration entitled ORGANization that describes the interrelationships of organ systems, a general theme of this edition. This integrated approach is central to understanding anatomy and physiology, but all too often, students study each chapter in an isolated manner and miss how one system relates to other systems. The ORGANization illustrations give you a quick review of system interactions. ORGANization illustrations can be springboards for class discussion, provide ideas for further study or term papers, review chapter concepts, and reinforce the "big picture" in learning and applying the principles of anatomy and physiology.

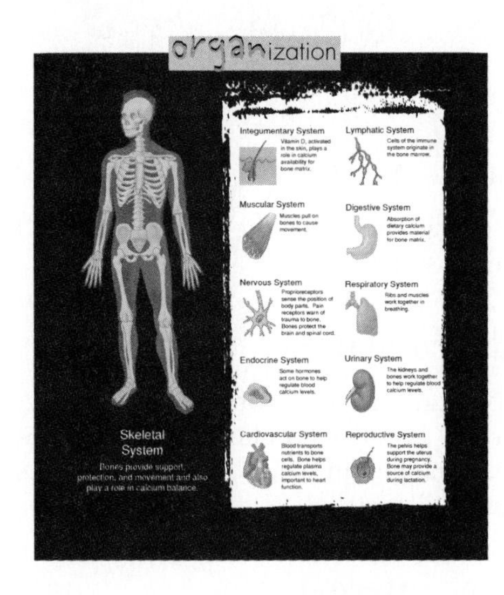

Chapter Reorganization

Chapter sequence is reorganized to meet the needs of adopters. It also follows the standard format of competing books and is more logical. The following compares fifth and sixth edition chapter orders:

Fifth Edition	*Sixth Edition*
Chapter 12 Digestive System	Chapter 12 Blood
Chapter 13 Respiratory System	Chapter 13 Cardiovascular System
Chapter 14 Blood	Chapter 14 Lymphatic System and Immunity
Chapter 15 Cardiovascular System	Chapter 15 Digestion and Nutrition
Chapter 16 Lymphatic System and Immunity	Chapter 16 Respiratory System

The complete chapter list in sequence is:

Unit 1	**Levels of Organization**
Chapter 1	Introduction to Human Anatomy and Physiology
Chapter 2	Chemical Basis of Life
Chapter 3	Cells
Chapter 4	Cellular Metabolism
Chapter 5	Tissues
Unit 2	**Support and Movement**
Chapter 6	Skin and the Integumentary System
Chapter 7	Skeletal System
Chapter 8	Muscular System
Unit 3	**Integration and Coordination**
Chapter 9	Nervous System
Chapter 10	Somatic and Special Senses
Chapter 11	Endocrine System
Unit 4	**Transport**
Chapter 12	Blood
Chapter 13	Cardiovascular System
Chapter 14	Lymphatic System and Immunity
Unit 5	**Absorption and Excretion**
Chapter 15	Digestion and Nutrition
Chapter 16	Respiratory System
Chapter 17	Urinary System
Chapter 18	Water, Electrolyte, and Acid-Base Balance
Unit 6	**The Human Life Cycle**
Chapter 19	Reproductive Systems
Chapter 20	Pregnancy, Growth, and Development

Content and Emphasis Changes

- In chapter 5, "Tissues," new micrographs and corresponding artwork, fresh examples, and icons help you locate tissues in the human body.
- At the request of reviewers, nutrition is incorporated back into the digestive system chapter (chapter 15).

- Chapter 19, "Reproductive Systems," and chapter 20, "Pregnancy, Growth, and Development" are substantially updated and include new material on current and developing medical technology.
- Physiology is emphasized more throughout the text, especially in the cellular, muscular, nervous, endocrine, cardiovascular, lymphatic and immune, and digestive system chapters, and in the discussions of cellular metabolism, genetics, and membrane transport.

Aids to Learning

Hole's Essentials of Human Anatomy and Physiology is written with the student in mind. Several features make learning difficult concepts and terminology easier.

Chapter Objectives

Before you begin to study a chapter, carefully read the chapter objectives. These indicate what you should be able to do after mastering the information within the narrative. The review exercises at the end of each chapter are phrased as detailed objectives, and reading them before beginning your study is also helpful. Both the chapter objectives and the review exercises are guides that identify important chapter topics.

Aids to Understanding Words

Aids to Understanding Words, found at the beginning of each chapter and in the appendix, helps build your vocabulary. These sections define root words, stems, prefixes, and suffixes that help you discover word meanings, and then apply these definitions in example words. Studying these lists will help you discover and remember scientific word meanings.

Key Terms

A list of key terms and their phonetic pronunciations at the beginning of each chapter helps build your science vocabulary. The key terms are boldfaced and defined within the chapter and are likely to be found in subsequent chapters as well. The glossary at the end of the book explains phonetic pronunciations.

Review Questions within the Narrative

This edition continues to have review questions at the ends of major sections in each chapter to test your understanding of the material just covered.

Illustrations

The revamped illustration program is perhaps the most obvious change in the new edition. Detail, clarity, accuracy, and consistency prevail, with frequent icons for orientation and to establish a sense of scale. Color is consistent from chapter to chapter—a cell nucleus is not orange in one chapter, purple in another. More than half of the illustrations are new to this edition, many from the highly successful *Hole's Human Anatomy and Physiology,* 7th edition.

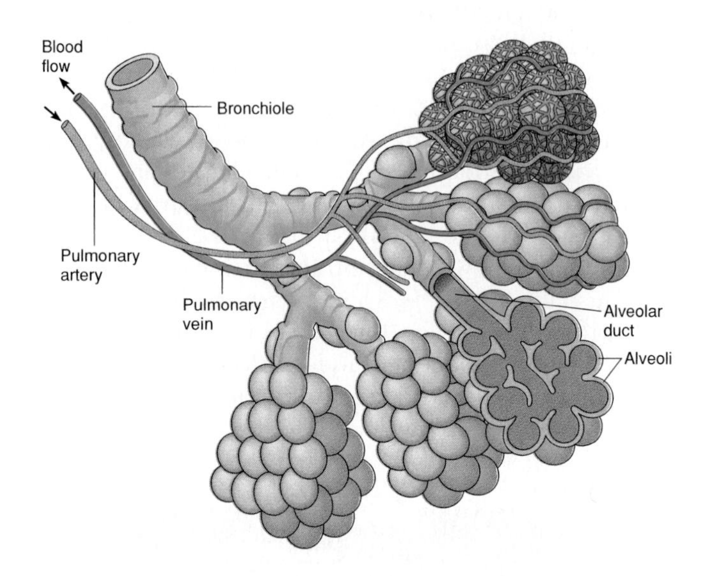

Design

A new design improves the readability and flow of the text and illustrations.

Boxed Information

Short paragraphs in colored boxes apply ideas and facts in the narrative to clinical situations. Some boxes describe changes in body structure and function that result from disease.

Topics of Interest

Longer boxes, entitled "Topic of Interest," are new or rewritten. They discuss disorders, physiological responses to environmental factors, and other topics of general interest.

ORGANization Illustrations

ORGANization illustrations, found at the end of selected chapters, conceptually link the highlighted body system to every other system and reinforce the dynamic interplays between groups of organs. These graphic representations stress the "big picture" in learning and applying anatomy and physiology.

Clinical Terms

At the end of many chapters are lists of related terms often used in clinical situations. These terms, along with their phonetic pronunciations and brief definitions, are helpful to your understanding of medical terminology.

Chapter Summaries

An end-of-chapter summary in outline form helps you review the chapter's main ideas.

Review Exercises and Critical Thinking

End-of-chapter review exercises check understanding of the chapter's major ideas. Critical thinking questions, many of them new, coax you to apply information to clinical situations.

Reference Plates

Reference plates continue to be a strong feature in this edition. One set illustrates the structure and location of major internal organs and comes from *Hole's Human Anatomy and Physiology,* 7th edition. Its white background offers better contrast, making the art more vibrant and detail easier to see. Another set of reference plates depicts the structural detail of the human skull. All reference plates have a colored tab down the length of the page for easy access.

Multimedia Tie-ins

A dancing man icon also appears in many of the figure legends throughout the chapters. This icon represents the Dynamic Human CD-ROM and signals the reader that information relating to this figure can be found on the CD-ROM. *The Dynamic Human* CD-ROM can be packaged with *Hole's Essentials of Human Anatomy and Physiology* for minimal fee or can be bought separately. A correlation guide can be found at the end of the preface.

A videotape icon appears in many of the figure legends throughout the chapters. This icon alerts the reader that an animation related to the figure is available from the WCB Life Science Animations videotape series. Students can purchase the videotapes, and the tapes are available to qualified adopters. A correlation guide is included in this preface. The five tape titles are:

1. Chemistry, The Cell, and Energetics (1–11)
2. Cell Division, Heredity, Genetics, Reproduction, and Development (12–21)
3. Animal Biology #1 (22–31)
4. Animal Biology #2 (32–45)
5. Plant Biology, Evolution, and Ecology (46–53)

Unit and Chapter Tabs

Colored tabs on page edges highlight unit and chapter openers and provide easier reference.

Supplementary Materials

The following supplementary materials help the instructor plan class work and presentations, and aid students in their learning activities:

1. *Laboratory Manual for Hole's Essentials of Human Anatomy and Physiology* by Terry R. Martin has been heavily revised and is now more closely correlated with the sixth edition of the text. Correlations to *The Virtual Physiology* CD-ROM by WCB/McGraw-Hill are included in an appendix. An Intelitool lab supplement is also offered with the revised edition. ISBN 0-697-32918-6
2. *Instructor's Manual and Test Item File* by Jeffrey and Karianne Prince, with contributions from Connie Schoepske and Kermit Harless, includes for each text chapter a purpose, a chapter summary, topics for discussion or individual research, possible demonstrations, related transparencies, related films and software, and suggestions for additional reading. The appendixes include a list of transparencies, answers to review exercises and critical-thinking questions, a correlation with the laboratory manual, film, videocassette, and software distributors, and biological supply houses. A test item file with test questions for each chapter and a visuals testbank of transparency masters provide material for student quizzing or practice. The instructor's manual also provides more detailed information on subjects that instructors may want to talk about in class, but not necessarily test students on. This information is called "In Case You Wondered" and is formatted for distribution as handouts. ISBN 0-697-32917-8
3. *Student Study Guide* by Nancy A. Sickles Corbett contains chapter overviews, chapter objectives, focus questions, mastery tests, study activities, and answer keys that correspond to the text chapters. ISBN 0-697-32920-8
4. *Microtest* is a computerized testing service offered free upon request to qualified text adopters. A complete test item file is available on computer diskette for use with Windows and Macintosh computers. Windows version 0-697-32930-5 Mac version 0-697-32931-3

5. *Transparencies* consist of 250 acetate sheets of full-color illustrations from the text that complement classroom lectures or can be used for short quizzes. ISBN 0-697-32921-6
6. *Student Study Art Notebook* contains the 250 illustrations from the transparency set. With this notebook, students no longer have to worry about whether they will be able to see leader lines and labels in a large lecture hall. ISBN 0-697-32919-4
7. *Instructor's Manual for the Laboratory Manual* by Terry R. Martin correlates the laboratory manual with text chapters, provides a suggested time schedule for each laboratory exercise, gives instructional suggestions, and answers the questions in the lab reports. A Frame Number Directory in the appendix correlates appropriate sections of the laboratory manual to *The Dynamic Human* videodisc. ISBN 0-697-32916-X
8. *QuickStudy* is a computerized study guide containing true-false, multiple-choice, and fill-in-the-blank test questions on a diskette. This is available for use with IBM and Macintosh computers. IBM 0-697-32925-9 Mac 0-697-32926-7
9. The WCB/McGraw-Hill Visual Resource Library for *Hole's Essentials of Human Anatomy and Physiology* is on a CD-ROM containing virtually all of the illustrations from the textbook. An easy-to-use program enables you to move quickly among the images, show or hide labels, and create your own multimedia presentation. ISBN 0-697-32927-5
10. *Histology Color Slides* include a set of seventy micrographs of tissues, organs, and other body features described in the textbook. ISBN 0-697-28931-1
11. *Intelitool Supplement to Accompany Anatomy and Physiology Laboratory Manuals* by Terry R. Martin contains four Intelitool laboratory exercises on muscle physiology, reflex physiology, electrocardiography, and spirometry. ISBN 0-697-33368-X
12. *World Wide Web Home Page* for *Hole's Essentials of Human Anatomy and Physiology*. The address is www.mhhe.com/sciencemath/biology/holeessentials/. The page links to relevant anatomy and physiology sites, as well as to appropriate WCB/McGraw-Hill multimedia products.

Other learning aids available from WCB/McGraw-Hill include:

13. *The Dynamic Human CD-ROM* illustrates the important relationships between anatomical structures and their functions in the human body. Realistic computer visualization and three-dimensional visualizations are the premier features of this CD-ROM, which is available for use with Macintosh and Windows computers. Mac version 0-697-37909-4 Windows version 0-697-37910-8
14. *The Dynamic Human Videodisc* contains all of the animations (200+) from *The Dynamic Human* CD-ROM. A bar code directory is also available. ISBN 0-697-38937-5
15. *WCB Life Science Animations Videotape Series* consists of five videotapes containing fifty-three animations of key physiological processes.
#1: 0-697-25068-7 #2: 0-697-25069-5
#3: 0-697-25070-9 #4: 0-697-25071-7
#5: 0-697-26600-1
Another videotape, entitled *Physiological Concepts of Life Science,* contains similar animations. ISBN 0-697-21512-1
16. *Explorations in Human Biology CD-ROM* consists of sixteen interactive modules that stress human physiology. Students actively investigate vital processes as they explore each module filled with color, sound, and movement. The CD-ROM is available for use with Macintosh and Windows computers. Mac version 0-697-22964-5 Windows version 0-697-22963-7
17. *Explorations in Cell Biology and Genetics CD-ROM* contains interactive concepts related to key anatomy and physiology topics. An instructor can use the CD-ROM in lecture and/or place it in a lab or resource center for students. The CD-ROM is available for use with Macintosh and Windows computers. ISBN 0-697-29214-2
18. *Life Science Living Lexicon CD-ROM* contains a comprehensive collection of life science terms, including definitions of their roots, prefixes, and suffixes, as well as audio pronunciations and illustrations. The lexicon is student-interactive, featuring quizzing and note-taking capabilities. ISBN 0-697-29266-5
19. *The Virtual Physiology Lab CD-ROM* contains ten dry labs of the most common and important physiology experiments. ISBN 0-697-37994-9
20. *WCB Anatomy and Physiology Videodisc* is a four-sided videodisc with more than thirty animations of physiological processes, as well as line art and micrographs. A bar code directory is also available. ISBN 0-697-27716-X
21. *WCB Anatomy and Physiology Videotape Series* includes:
    - Introduction to Cat Dissection: Musculature ISBN 0-697-11630-1
    - Blood Cell Counting, Identification, and Grouping ISBN 0-697-11629-8

- Internal Organs and the Circulatory System of the Cat ISBN 0-697-13922-0
- Introduction to the Human Cadaver and Prosection ISBN 0-697-11177-6

22. *Human Anatomy and Physiology Study Cards* by Van De Graaff et al. is a boxed set of 3-by-5-inch cards. The 300 study cards are a well-organized and illustrated synopsis of the structure and function of the human body, and offer a quick and effective review of human anatomy and physiology. ISBN 0-697-26447-5
23. *Coloring Guide to Anatomy and Physiology* by Robert and Judith Stone emphasizes learning through color association and thoroughly reviews anatomical and physiological concepts. ISBN 0-697-17109-4
24. *Atlas of Skeletal Muscles* by Robert and Judith Stone is a guide to the structures, functions, locations, and actions of human skeletal muscles. ISBN 0-697-13790-2
25. *Laboratory Atlas of Anatomy and Physiology* by Eder et al. is a full-color atlas containing histology, human skeletal anatomy, human muscular anatomy, dissections, and reference tables. ISBN 0-697-39480-8
26. *Case Histories in Human Physiology* by Donna Van Wynsberghe and Gregory Cooley provides opportunities for integrating thinking and for problem solving. An answer key is also available. ISBN 0-697-13791-0
27. *Survey of Infectious and Parasitic Diseases* by Kent M. Van De Graaff presents essential information on 100 of the most common and clinically significant diseases. ISBN 0-697-27535-3

Acknowledgments

Any textbook is the result of hard work by a large team. Although we directed the revision, many "behind-the-scenes" people at WCB/McGraw-Hill were indispensable to the project. We would like to thank our editorial team of Kevin Kane, Michael Lange, Kris Noel Tibbetts, Kelly Drapeau, and Darlene Schueller; our production team, which included Marla Irion, Brenda Ernzen, and John Leland; the copyeditor Mary Monner; and most of all, John Hole, for giving us the opportunity and freedom to continue his classic work.

We especially thank our wonderful and patient families for their support.

David Shier
Jackie Butler
Ricki Lewis

Reviewers

We would like to acknowledge the valuable contributions of the reviewers for the sixth edition who read either portions or all of the manuscript as it was being prepared, and who provided detailed criticisms and ideas for improving the narrative and the illustrations. They include the following:

Regina Almeida
San Bernardino Valley College

Frederick D. Arguello
Yuba Community College

Elizabeth Bastio
Villa Julie College

Peter Biesemeyer
North Country Community College

Suzanne E. Blank
Kirkwood Community College

Elaine M. Blasko
West Liberty State College

Cynthia A. Bottrell
Scott Community College

Tom Carson
Bossier Parish Community College

Diane K. Chaddock
Southwestern Michigan College

Suzzette F. Chopin
Texas A&M University-Corpus Christi

Denis A. Coble
University of Connecticut

Janet Colwell
St. Francis School of Practical Nursing

Judith D. Cunningham
Montgomery County Community College

Thomas A. Davis
Loras College

Tom Dudley
Angelina College

Phillip Eichman
University of Rio Grande

George Fortunato
Suffolk Community College

Eugenia M. Fulcher
Swainsboro Technical Institute

Rita J. Fulton
Belmont Technical College

Louis A. Giacinti
Milwaukee Area Technical College

Canzater Gillespie
Milwaukee Area Technical College

Carol A. Goff
Warren County Area Vocational Technical School

Linda A. Hall
Ozarka Technical College

Kathleen Hayes
State of Connecticut, Practical Nurse Education

Ann L. Henninger
Wartburg College

Shawn Henry
Dakota State University

Dawn Holtzmeier
Hocking College

Sandra Hsu
Merritt College

Mark W. Huntington
Manchester College

Drusilla Beal Jolly
Forsyth Technical Community College

J. Michael Jones
Culver-Stockton College

Inez Devlin-Kelly
Bakersfield College

Glenn E. Kietzmann
Wayne State College

Roger L. Lane
Kent State University

Mary E. Larscheid
Iowa Lakes Community College

Margot Lowe
Los Angeles Harbor College

Dorothy G. May
Park College

Juan E. Mejia
South Texas Community College

Christine Miller
Bay Mills Community College

Lewis M. Milner
North Central Technical College

Patricia T. Mote
Oglethorpe University
DeKalb College
Pace Academy College

Norrine E. Nolan
Milwaukee Area Technical College

Diane B. Pelletier
Brenneke School of Massage

Michael Postula
Parkland College

Susan Puglisi
Norwalk Community Technical College

Jackie McLeod-Roberts
Nene College of Higher Education

Anna Rovid
Shippensburg University

Norma H. Rubin
The University of Texas Medical Branch

David K. Saunders
Emporia State University

Kandie D. Semmelman
Delaware Technical and Community College

John E. Sheard
Eastern Michigan University

Andrew M. Smith
Butler University

Barbara Stout
Stark State College of Technology

Lauren I. Strader
North Georgia Technical Institute

Michael A. Vitale
Daytona Beach Community College

John Wakeman
Louisiana Tech University

Nancy A. Wall
Lawrence University

Steve Weber
Vici Public Schools

Leesa Whicker
Davidson County Community College

Lucy White
Ivy Tech State College

Claudia M. Williams
Campbell University

Clarence C. Wolfe
Northern Virginia Community College

Paul H. Yancey
Whitman College

Terry L. Zeman
Clark State Community College

Correlation of the *WCB Life Science Animations* (*LSA*)

Figure Number	LSA Number	Title	Figure Number	LSA Number	Title
2.4	1	Formation of an Ionic Bond	13.1	37	Blood Circulation
3.2	2	Journey into a Cell	13.7	37	Blood Circulation
3.3	2	Journey into a Cell	13.10	37	Blood Circulation
3.4	2	Journey into a Cell	13.12	38	Production of Electrocardiogram
3.5	2, 4	Journey into a Cell, Cellular Secretion	13.15	38	Production of Electrocardiogram
3.6	2	Journey into a Cell	13.22	37	Blood Circulation
3.7	2	Journey into a Cell	13.31	37	Blood Circulation
3.9	2	Journey into a Cell	13.36	37	Blood Circulation
3.10	2	Journey into a Cell	14.3	41	B-Cell Immune Response
3.17	3	Endocytosis	14.6	41	B-Cell Immune Response
3.18	12	Mitosis	14.14	42	Structure and Function of Antibodies
4.2	36	Digestion of Lipids	14.15	41	B-Cell Immune Response
4.6	5, 6, 11	Glycolysis, Oxidative Respiration including Krebs Cycle, ATP as an Energy Carrier	14.16	42	Structure and Function of Antibodies
4.8	5, 6	Glycolysis, Oxidative Respiration including Krebs Cycle	15.4	33	Peristalsis
4.15	16	Transcription of a Gene	15.27	36	Digestion of Lipids
4.17	16, 17	Transcription of a Gene, Protein Synthesis	19.4	19	Spermatogenesis
4.18	17	Protein Synthesis	19.5	13, 19	Meiosis, Spermatogenesis
4.19	15	DNA Replication	19.8	13, 20	Meiosis, Oogenesis
8.1	29	Levels of Muscle Structure	20.3	21	Human Embryonic Development
8.7	30	Sliding Filament Model of Muscle Contraction	20.5	21	Human Embryonic Development
9.3	22	Formation of Myelin Sheath	20.7	21	Human Embryonic Development
9.13	24	Signal Integration	20.8	21	Human Embryonic Development
11.3	28	Peptide Hormone Action (cAMP)	20.10	21	Human Embryonic Development
12.16	40	A, B, O Blood Types	20.13	21	Human Embryonic Development

Correlation of *The Dynamic Human*

Chapter 1
1.8 Anatomical Orientation/ Visible Human/Thorax
1.9 Anatomical Orientation/ Visible Human/Abdomen
1.10 Anatomical Orientation/ Planes
TA1 Anatomical Orientation/ Visible Human/Head
Anatomical Orientation/ Visible Human/Abdomen

Chapter 5
5.1 Respiratory System/Histology/Alveoli
5.2 Urinary System/Anatomy/ Nephron Anatomy
Urinary System/Anatomy/ Kidney Anatomy
5.3 Digestive System/Histology/Duodenal Villi
Digestive System/Histology/ Fundic Stomach
5.4 Respiratory System/Histology/Trachea
Respiratory System/Histology/Bronchiole
5.5 Digestive System/Histology/Esophagus
5.6 Urinary System/Histology/ Bladder
5.9 Immune and Lymphatic System/Anatomy/Microscopic Components/Macrophage
5.10 Immune and Lymphatic System/Anatomy/Microscopic Components/Mast Cell
5.14 Skeletal System/Histology/ Hyaline Cartilage
5.15 Skeletal System/Histology/ Elastic Cartilage
5.16 Skeletal System/Histology/Fibrocartilage
5.17 Skeletal System/Histology/ Compact Bone
5.19 Muscular System/Histology/ Skeletal Muscle (longitudinal)
Muscular System/Histology/ Skeletal Muscle (cross sectional)
5.20 Muscular System/Histology/ Smooth Muscle
5.21 Muscular System/Histology/ Cardiac Muscle
Cardiovascular System/Histology/Cardiac Muscle
5.22 Nervous System/Histology/ Spinal Neuron

Chapter 7
7.1 Skeletal Muscle/Explorations/ Cross Section of a Long Bone
7.3 Skeletal Muscle/Histology/ Compact Bone
7.8 Skeletal Muscle/Anatomy/ Gross Anatomy
7.9 Skeletal Muscle/Anatomy/ Cranial Anatomy
7.11 Skeletal Muscle/Anatomy/ Cranial Anatomy
7.12 Skeletal Muscle/Anatomy/ Cranial Anatomy
7.34 Skeletal Muscle/Explorations/ Types of Joints/Fibrous Joints
7.35 Skeletal Muscle/Explorations/ Types of Joints/Synovial Joints
7.36 Skeletal Muscle/Explorations/ Types of Joints/Synovial Joints
7.37 Skeletal Muscle/Explorations/ Types of Joints/Synovial Joints/Synovial Joint Motion
7.38 Skeletal Muscle/Explorations/ Types of Joints/Synovial Joints/Synovial Joint Motion
7.39 Skeletal Muscle/Explorations/ Types of Joints/Synovial Joints/ Synovial Joint Motion

Chapter 8
8.1 Muscular System/Anatomy/ Skeletal Muscle Anatomy
8.2 Muscular System/Anatomy/ Skeletal Muscle Anatomy
8.5 Muscular System/Explorations/ Neuromuscular Junction
8.7 Muscular System/Explorations/ Sliding Filament Theory
8.13 Muscular System/Explorations/ Muscle Action Around Joints

Chapter 9
9.1 Nervous System/Histology/ Spinal Neuron
9.13 Nervous System/Explorations/ Neural Network
9.15 Nervous System/Explorations/ Reflex Arc
9.20 Nervous System/Anatomy/ Spinal Cord Anatomy
9.23 Nervous System/Anatomy/ Gross Anatomy of Bone
9.24 Nervous System/Anatomy/ Gross Anatomy of Bone
9.27 Nervous System/Anatomy/ Gross Anatomy of Bone

Chapter 10
10.4 Nervous System/Explorations/ Olfaction
10.5 Nervous System/Explorations/Taste
10.6 Nervous System/Explorations/Zones of Taste
10.7 Nervous System/Explorations/ Hearing
10.9 Nervous System/Explorations/ Hearing
10.10 Nervous System/Explorations/ Hearing
Nervous System/Histology/ Organ of Corti
10.12 Nervous System/Explorations/ Static Equilibrium
10.13 Nervous System/Explorations/ Dynamic Equilibrium
10.14 Nervous System/Explorations/ Dynamic Equilibrium
10.18 Nervous System/Explorations/Vision
Nervous System/Histology/Eye
10.23 Nervous System/Histology/Retina

Chapter 11
11.1 Endocrine System/Anatomy
11.4 Endocrine System/Explorations/Endocrine Function
11.6 Endocrine System/Anatomy/ Hypothalamus and Pituitary Gland
11.7 Endocrine System/Anatomy/ Hypothalamus and Pituitary Gland
11.8 Endocrine System/Explorations/ Hypothalamo-Hypophyseal Portal Axis
11.9 Endocrine System/Anatomy/ Thyroid Gland
Endocrine System/Histology/ Thyroid Gland
11.10 Endocrine System/Histology/Thyroid Gland
11.12 Endocrine System/Anatomy/Adrenal Gland
Endocrine System/Histology/ Adrenal Cortex
Endocrine System/Histology/ Adrenal Medulla
11.14 Endocrine System/Anatomy/ Pancreas

Chapter 12
12.16 Immune and Lymphatic System/Clinical Concepts/ Blood Type

Chapter 13
13.1 Cardiovascular System/Explorations/ Heart Dynamics/Blood Flow
13.2 Cardiovascular System/Anatomy/ 3D Viewer: Thoracic Anatomy
Cardiovascular System/Anatomy/ Gross Anatomy of the Heart
13.3 Cardiovascular System/Anatomy/ Gross Anatomy of the Heart
13.4 Cardiovascular System/Anatomy/ Gross Anatomy of the Heart
13.9 Cardiovascular System/Anatomy/ Gross Anatomy of the Heart

Correlation of *The Dynamic Human*—*continued*

13.10 Cardiovascular System/Explorations/ Heart Dynamics/Blood Flow
13.11 Cardiovascular System/Explorations/ Heart Dynamics/Cardiac Cycle
13.12 Cardiovascular System/Explorations/ Heart Dynamics/Conduction System
Cardiovascular System/Explorations/ Heart Dynamics/ECG
13.18 Cardiovascular System/Explorations/ Generic Vasculature
Cardiovascular System/Histology/ Vasculature
13.19 Cardiovascular System/Explorations/ Generic Vasculature
13A Cardiovascular System/Histology/ Vasculature
13.23 Cardiovascular System/Explorations/ Generic Vasculature
Cardiovascular System/Histology/ Vasculature

Chapter 14
14.1 Immune and Lymphatic System/ Explorations/Lymph Formation and Movement
14.2 Immune and Lymphatic System/ Anatomy/Gross Anatomy/ Lymphatic Vessels
Immune and Lymphatic System/ Anatomy/Gross Anatomy/Lymph Node
14.6 Immune and Lymphatic System/ Histology/Lymph Node
Immune and Lymphatic System/ Explorations/Lymph Formation and Movement
14.9 Immune and Lymphatic System/ Anatomy/Gross Anatomy/Thymus
Immune and Lymphatic System/Histology/Thymus
14.10 Immune and Lymphatic System/ Anatomy/Gross Anatomy/Spleen
Immune and Lymphatic System/ Histology/Spleen
14.15 Immune and Lymphatic System/ Explorations/Specific Immunity
14.16 Immune and Lymphatic System/ Explorations/Specific Immunity

Chapter 15
15.1 Digestive System/Anatomy/ Gross Anatomy
15.3 Digestive System/Histology/Duodenum
15.4 Digestive System/Explorations/ Esophagus
15.5 Digestive System/Explorations/ Oral Cavity
Digestive System/Histology/ Submandibular Gland
15.9 Digestive System/Histology/Tooth
15.10 Digestive System/Explorations/ Oral Cavity
Digestive System/Histology/ Submandibular Gland
15.11 Digestive System/Explorations/ Stomach
15.12 Digestive System/Histology/ Fundic Stomach
15.13 Digestive System/Histology/ Fundic Stomach
15.15 Digestive System/Anatomy/ Gross Anatomy
Digestive System/Explorations/ Pancreas
15.17 Digestive System/Anatomy/ 3D Viewer
Digestive System/Anatomy/ Gross Anatomy
15.19 Digestive System/Histology/Liver
15.20 Digestive System/Clinical Concepts/Gallstones
15.21 Digestive System/Explorations/ Gallbladder and Liver
15.25 Digestive System/Histology/ Duodenal Villi
Digestive System/Histology/ Duodenum
Digestive System/Explorations/ Duodenum
15.26 Digestive System/Histology/ Duodenal Villi
Digestive System/Histology/ Duodenum
15.27 Digestive System/Explorations/ Duodenum

Chapter 16
16.1 Respiratory System/Anatomy/ Gross Anatomy
Respiratory System/Anatomy/ 3D Viewer
16.2 Respiratory System/Anatomy/ Gross Anatomy
16.3 Respiratory System/Histology/ Nasal Cavity
16.6 Respiratory System/Anatomy/ Gross Anatomy
16.8 Respiratory System/Histology/Lung
16.9 Respiratory System/Histology/Alveoli
16.10 Respiratory System/Explorations/ Gas Exchange
16.11 Respiratory System/Explorations/ Mechanics of Breathing
16.12 Respiratory System/Explorations/ Mechanics of Breathing
16.13 Respiratory System/Explorations/ Mechanics of Breathing
16.14 Respiratory System/Clinical Concepts/Spirometry
16.18 Respiratory System/Histology/Lung
16.19 Respiratory System/Explorations/ Gas Exchange
16.20 Respiratory System/Explorations/ Oxygen Transport

Chapter 17
17.1 Urinary System/Anatomy/ Gross Anatomy
17.2 Urinary System/Anatomy/ Kidney Anatomy
Urinary System/Anatomy/ Nephron Anatomy
17.5 Urinary System/Anatomy/ Nephron Anatomy
Urinary System/Histology/ Renal Cortex
17.6 Urinary System/Anatomy/3D Viewer: Nephron
17.8 Urinary System/Explorations/ Urine Formation
17.10 Urinary System/Explorations/ Urine Formation
17.11 Urinary System/Explorations/ Urine Formation
17.15 Urinary System/Histology/Bladder
17.16 Urinary System/Anatomy/ Gross Anatomy

Chapter 19
19.1 Reproductive System/Anatomy/ Male Reproductive Anatomy
Reproductive System/Anatomy/3D Viewer: Male Reproductive Anatomy
19.2 Reproductive System/Histology/Testis
19.4 Reproductive System/Explorations/ Male/Spermatogenesis
19.5 Reproductive System/Explorations/ Male/Spermatogenesis
19.7 Reproductive System/Anatomy/ Female Reproductive Anatomy
Reproductive System/Anatomy/ 3D Viewer: Female Reproductive Anatomy
19.8 Reproductive System/Explorations/ Female/Oogenesis
19.9 Reproductive System/Explorations/ Female/Ovarian Cycle
19.10 Reproductive System/Histology/ Ovarian Follicle
19.11 Reproductive System/Anatomy/ Female Reproductive Anatomy
19.13 Reproductive System/Explorations/ Female/Menstrual Cycle
Reproductive System/Explorations/ Female/Ovarian Cycle
19.16 Reproductive System/ Clinical Concepts/Tubal Ligation
Reproductive System/ Clinical Concepts/Vasectomy

CHAPTER 1

Introduction to Human Anatomy and Physiology

Sarah Sabrina was dying. Something was blocking her throat, making it increasingly difficult for her to breathe. Happily, a tonsillectomy saved her life—and made history. Sarah Sabrina became the first monkey to have her tonsils removed. An ear, nose, and throat specialist whose typical patients are children performed the surgery at a hospital for humans.

After a veterinarian was unable to identify the cause of Sarah's choking, the golden, long-haired spider monkey—one of only fifty remaining in the world—found herself in the office of Dr. John Angerosa, an otolaryngologist. The doctor jokes that he had to anesthetize her for the initial exam because she would not open her mouth and say "ah." When he realized that it was her tonsils causing the blockage, and not a tumor, Dr. Angerosa arranged for a tonsillectomy. A nearby hospital admitted this patient of another species (without insurance) because of her endangered and life-threatened status.

Dr. Angerosa consulted with primatologist Jane Goodall, who told him that tonsillectomies on monkeys simply were not done. Veterinarians' knowledge of cat and dog anatomy was not much help either. So the doctor had to rely on his memory of human anatomy and physiology, recalling diagrams and explanations in his textbooks while probing the monkey's throat, comparing and contrasting it to that of a human.

The successful surgery will help Sarah Sabrina to reach her full life span of forty-five years. For Dr. Angerosa, the experience of saving Sarah Sabrina's life made him appreciate the value of knowing human anatomy and physiology, which, fortunately, is similar to that of a monkey. He sums up the situation: "You never know how you're going to use what you learned in anatomy and physiology!"

Photo: A physician drew on his knowledge of human anatomy and physiology to perform a tonsillectomy on a monkey.

Although humans are a particular kind of living organism, we share traits with other organisms. We carry on life processes and thus demonstrate the characteristics of life. We have requirements that must be met if we are to survive, and our lives depend on maintaining a stable internal environment.

A discussion of the traits that humans have in common with other organisms and of the way the complex human body is organized provides a beginning for the study of anatomy and physiology in chapter 1. This chapter also introduces a set of special terms used to describe body parts.

Chapter Objectives

After studying this chapter, you should be able to:

1. Define *anatomy* and *physiology,* and explain how they are related. (p. 4)
2. List and describe the major characteristics of life. (p. 4)
3. List and describe the major requirements of organisms. (p. 5)
4. Define *homeostasis,* and explain its importance to survival. (p. 6)
5. Describe a homeostatic mechanism. (p. 6)
6. Explain biological levels of organization. (p. 7)
7. Describe the locations of the major body cavities. (p. 9)
8. List the organs located in each major body cavity. (p. 9)
9. Name the membranes associated with the thoracic and abdominopelvic cavities. (p. 11)
10. Name the major organ systems, and list the organs associated with each. (p. 12)
11. Describe the general functions of each organ system. (p. 12)
12. Properly use the terms that describe relative positions, body sections, and body regions. (p. 14)

Aids to Understanding Words

append- [to hang something] *append*icular: Pertaining to the limbs.

cardi- [heart] peri*cardi*um: A membrane that surrounds the heart.

cran- [helmet] *cran*ial: Pertaining to the portion of the skull that surrounds the brain.

dors- [back] *dors*al: A position toward the back.

homeo- [same] *homeo*stasis: The maintenance of a stable internal environment.

-logy [study of] physio*logy*: The study of body functions.

meta- [change] *meta*bolism: The chemical changes that occur within the body.

pariet- [wall] *pariet*al membrane: A membrane that lines the wall of a cavity.

pelv- [basin] *pelv*ic cavity: A basin-shaped cavity enclosed by the pelvic bones.

peri- [around] *peri*cardial membrane: A membrane that surrounds the heart.

pleur- [rib] *pleur*al membrane: A membrane that encloses the lungs and lines the thoracic cavity.

-stasis [standing still] homeo*stasis*: The maintenance of a stable internal environment.

-tomy [cutting] ana*tomy*: The study of structure, which often involves cutting or removing body parts.

Key Terms

abdominopelvic (ab-dom″ĭ-no-pel′vik)
absorption (ab-sorp′shun)
anatomy (ah-nat′o-me)
appendicular (ap″en-dik′u-lar)
assimilation (ah-sim″ĭ-la′shun)
axial (ak′se-al)
circulation (ser-ku-la′shun)
digestion (di-jest′yun)
excretion (ek-skre′shun)
homeostasis (ho″me-ō-sta′sis)
metabolism (mĕ-tab′o-lizm)
negative feedback (neg′ah-tiv fēd′bak)
organelle (or″gan-el′)
organism (or′gah-nizm)
parietal (pah-ri′ĕ-tal)
pericardial (per″ĭ-kar′de-al)
peritoneal (per″ĭ-to-ne′al)
physiology (fiz″e-ol′o-je)
pleural (ploo′ral)
reproduction (re″pro-duk′shun)
respiration (res″pĭ-ra′shun)
thoracic (tho-ras′ik)
visceral (vis′er-al)

The accent marks used in the pronunciation guides are derived from a simplified system of phonetics that is standard in medical usage. The single accent (′) denotes the major stress and identifies the most heavily pronounced syllable in the word. The double accent (″) indicates the secondary stress. A syllable marked with a double accent receives less emphasis than the syllable that carries the main stress, but more emphasis than neighboring unstressed syllables.

Introduction

The study of the human body probably began with our earliest ancestors, who must have been curious about how their bodies worked, as we are today. At first, their interests most likely concerned injuries and illnesses because healthy bodies demand little attention from their owners. Their healers relied heavily on superstitions and notions about magic. However, as healers tried to help the sick, they began to discover useful ways of examining and treating the human body. They observed the effects of injuries, noticed how wounds healed, and examined dead bodies to determine causes of death. They also found that certain herbs and potions could sometimes be used to treat coughs, headaches, and other common problems.

Over time, people began to believe that humans could understand forces that caused natural events. They began observing the world around them more closely, asking questions and seeking answers. This set the stage for the development of modern science.

As techniques for making accurate observations and performing careful experiments evolved, knowledge of the human body expanded rapidly (fig. 1.1). At the same time, early medical providers coined many new terms to name body parts, to describe their locations, and to explain their functions. These terms, most of which originated from Greek and Latin words, formed the basis for the language of anatomy and physiology. (The names of some modern medical and applied sciences are listed on page 18.)

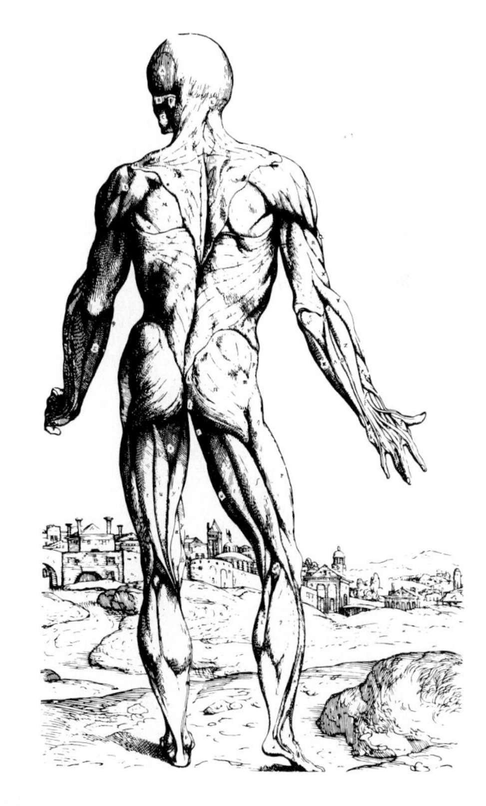

***Figure* 1.1**

The study of the human body has a long history, as this illustration from the second book of *De Humani Corporis Fabrica* by Andreas Vesalius, issued in 1543, illustrates. (Note the similarity to the anatomical position, described later in the chapter.)

1. What factors probably stimulated an early interest in the human body?
2. What kinds of activities helped promote the development of modern science?

Anatomy and Physiology

Anatomy is the branch of science that deals with the structure (morphology) of body parts—their forms and how they are organized. **Physiology,** on the other hand, concerns the functions of body parts—what they do and how they do it.

The topics of anatomy and physiology are difficult to separate because the structures of body parts are so closely associated with their functions. Body parts form a well-organized unit—the **human organism**—and each part functions in the unit's operation. A particular body part's function depends on the way the part is constructed—that is, how its subparts are organized. For example, the organization of parts in the human hand with its long, jointed fingers makes it easy to grasp objects; the hollow chambers of the heart are adapted to pump blood through tubular blood vessels; the shape of the mouth enables it to receive food; and teeth are shaped to break solid foods into small pieces (fig. 1.2).

Anatomy and physiology are ongoing as well as ancient fields. Researchers frequently discover new information about physiology, particularly at the molecular level. Although unusual, new parts of human anatomy are discovered today, too. Recently, researchers identified a small piece of connective tissue between the upper part of the spinal cord and a muscle at the back of the head. This connective tissue bridge may be the trigger for pain impulses in certain types of tension headaches.

1. Why is it difficult to separate the topics of anatomy and physiology?
2. List several examples that illustrate how the structure of a body part makes possible its function.

Characteristics of Life

Before beginning a more detailed study of anatomy and physiology, it is helpful to consider some of the traits humans share with other organisms. These characteristics of life include the following:

1. **Movement** often refers to a self-initiated change in an organism's position or to its traveling from one place to another. However, the term also applies to motion of internal parts, such as the beating of the heart.
2. **Responsiveness** is an organism's ability to sense changes taking place inside or outside its body and to react to these changes. Seeking water to quench thirst is a response to water loss from body tissues; moving away from a hot fire is another example of responsiveness.

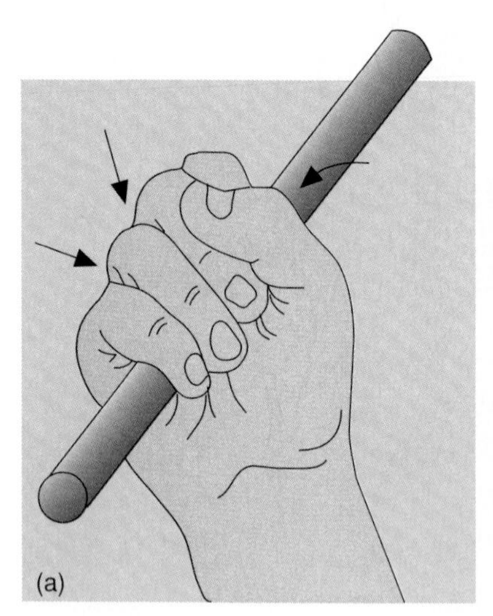

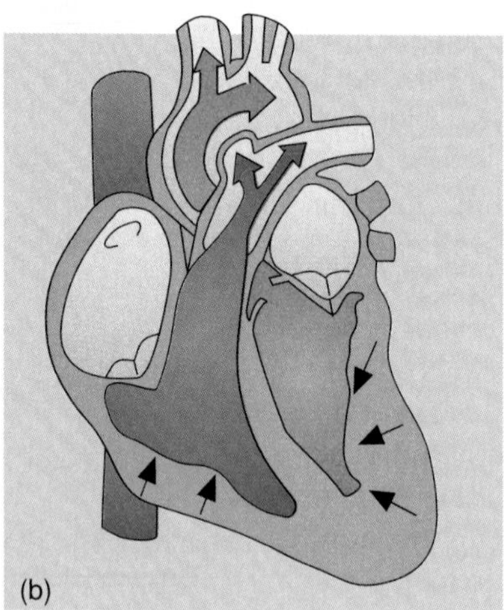

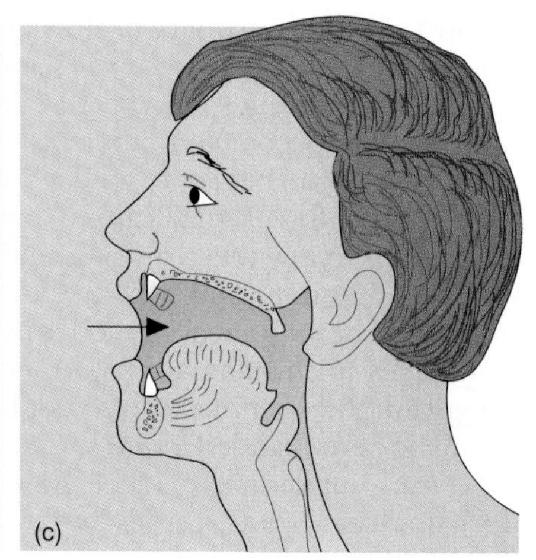

Figure 1.2

The structures of body parts are closely related to their functions. (*a*) The hand is adapted for grasping, (*b*) the heart for pumping blood, and (*c*) the mouth for receiving food.

Health-care workers repeatedly monitor patients' *vital signs*—observable body functions that reflect essential metabolic activities. Vital signs indicate that a person is alive. Assessment of vital signs includes measuring body temperature and blood pressure and monitoring rates and types of pulse and breathing movements. Absence of vital signs signifies death. A person who has died displays no spontaneous muscular movements (including those of the breathing muscles and beating heart), does not respond to stimuli (even the most painful that can be ethically applied), exhibits no reflexes (such as the knee-jerk reflex and the pupillary reflexes of the eye), and generates no brain waves (demonstrated by a flat encephalogram, which reflects a lack of metabolic activity in the brain). ■

3. **Growth** is an increase in body size, usually without any important change in shape. It occurs whenever an organism produces new body materials faster than old ones are worn out or depleted.
4. **Reproduction** is the process of making a new individual, as when parents produce an offspring. Cells reproduce, too, usually to repair injured tissues.
5. **Respiration** is the process of obtaining oxygen, using oxygen to release energy from food substances, and removing the resultant gaseous wastes.
6. **Digestion** chemically and mechanically breaks down food substances into simpler forms that cells can absorb and use.
7. **Absorption** is the passage of substances through certain membranes, as when digestive products pass through the membrane that lines the intestine and enter body fluids.
8. **Circulation** is the movement of substances within the body in body fluids.
9. **Assimilation** is the changing of absorbed substances into forms that are chemically different from those that entered body fluids.
10. **Excretion** is the removal of wastes that body parts produce as a result of their activities.

Each of these characteristics of life—in fact, everything an organism does—depends upon physical and chemical changes that occur within body parts. Taken together, these changes are called **metabolism.**

1. What are the characteristics of life?
2. How are the characteristics of life dependent on metabolism?

Maintenance of Life

The structures and functions of almost all body parts help maintain the life of the organism. The only exceptions are an organism's reproductive structures, which ensure that its species will continue into the future.

Requirements of Organisms

Life requires certain environmental factors, including the following:

1. **Water** is the most abundant chemical in the body. It is required for many metabolic processes and provides the environment in which most of them take place. Water also transports substances within the organism and is important in regulating body temperature.
2. **Foods** are substances that provide the body with necessary chemicals (nutrients) in addition to water. Some of these chemicals are used as energy sources, others supply raw materials for building new living matter, and still others help regulate vital chemical reactions.
3. **Oxygen** is a gas that makes up about one-fifth of ordinary air. It is used to release energy from food substances. This energy, in turn, drives metabolic processes.
4. **Heat** is a form of energy. It is a product of metabolic reactions, and the amount of heat present partly determines the rate at which these reactions occur. Generally, the more heat, the more rapidly chemical reactions take place. (*Temperature* is a measurement of the amount of heat.)
5. **Pressure** is an application of force to something. For example, the force on the outside of the body due to the weight of air above it is called *atmospheric pressure.* In humans, this pressure is important in breathing. Similarly, organisms living under water are subjected to *hydrostatic pressure*—a pressure a liquid exerts—due to the weight of water above them. In humans, heart action

produces blood pressure (another form of hydrostatic pressure), which forces blood through blood vessels.

Although organisms require water, food, oxygen, heat, and pressure, these factors alone are not enough to ensure survival. Both the quantities and the qualities of such factors are also important. For example, the amount of water entering and leaving an organism must be regulated, as must the concentration of oxygen in body fluids. Similarly, survival depends on the quality as well as the quantity of food available—that is, food must supply the correct nutrients in adequate amounts.

Homeostasis

As an organism moves, factors in its external environment may change. If the organism is to survive, conditions within the fluids surrounding its body cells (its **internal environment**) must remain relatively stable. In other words, body parts function efficiently only when the concentrations of water, nutrients, and oxygen and the conditions of heat and pressure remain within certain narrow limits. This maintenance of a stable internal environment is called **homeostasis.**

To better understand this idea of maintaining a stable environment, imagine a room equipped with a furnace and an air conditioner (fig. 1.3). Suppose the room temperature is to remain near 20° C (68° F), so the thermostat is adjusted to a **set point** of 20° C. Because a thermostat is sensitive to temperature changes, it will signal the furnace to start and the air conditioner to stop whenever the room temperature drops below the set point. If the temperature rises above the set point, the thermostat will stop the furnace and start the air conditioner. As a result, the room will maintain a relatively constant temperature.

A similar **homeostatic mechanism** regulates body temperature in humans. The "thermostat" is a temperature-sensitive region in a temperature control center of the brain. In healthy persons, the set point of the brain's thermostat is at or near 37° C (98.6° F).

If a person is exposed to a cold environment and body temperature begins to drop, the brain senses this change and triggers heat-generating and heat-conserving activities. For example, small groups of muscles are stimulated to contract involuntarily, an action called *shivering*. Such muscular contractions produce heat, which helps warm the body. At the same time, blood vessels in the skin are signaled to constrict so that less warm blood flows through them. In this way, heat that might otherwise be lost is retained in deeper tissues.

If a person is becoming overheated, the brain's temperature control center triggers a series of changes that promote loss of body heat. For example, it stimulates sweat glands in the skin to secrete perspiration. As this water evaporates from the surface, heat is carried away and the skin is cooled. At the same time, the brain center causes blood vessels in the skin to

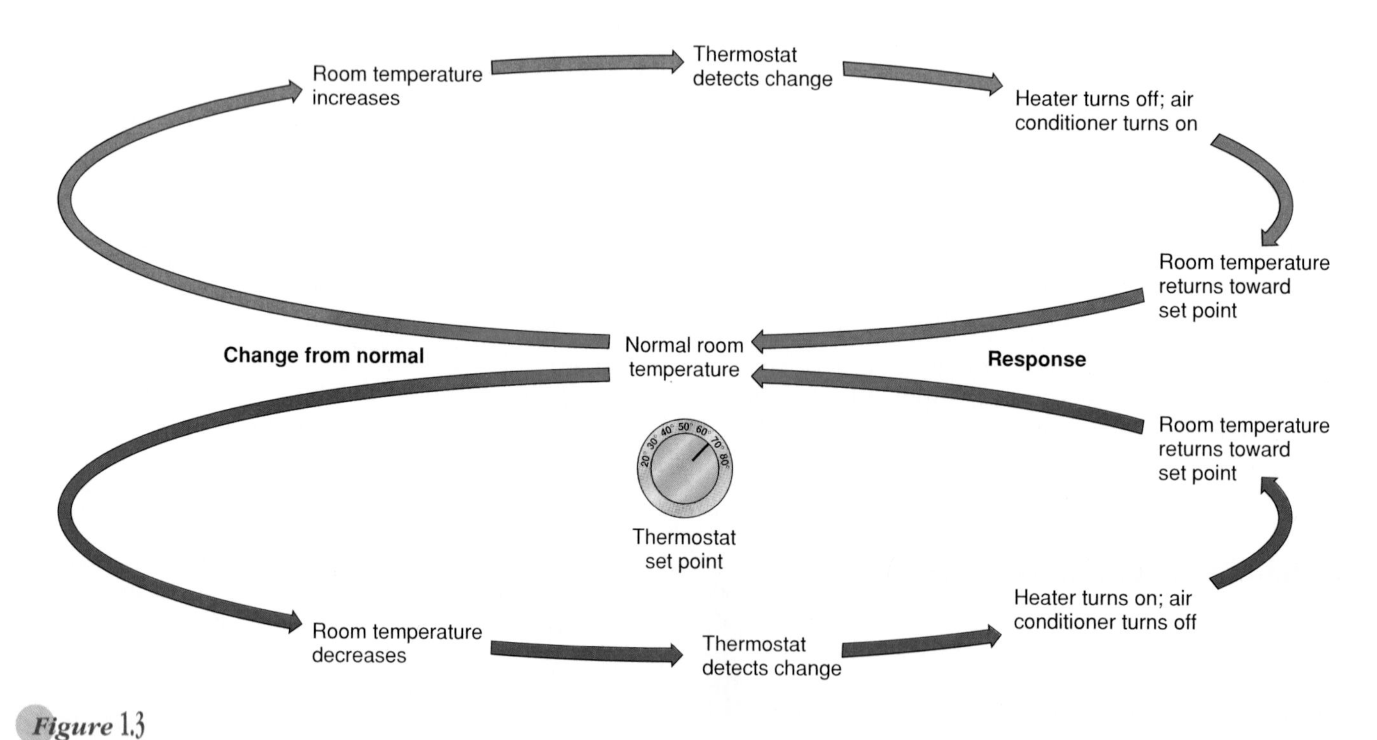

Figure 1.3

A thermostat signals an air conditioner and a furnace to turn on or off to maintain a relatively stable room temperature. This system is an example of a homeostatic mechanism.

dilate. This allows the blood carrying heat from deeper tissues to reach the surface, where heat is lost to the outside. The brain stimulates an increase in heart rate, which sends a greater volume of blood into surface vessels. The brain also stimulates an increase in breathing rate, which allows more heat-carrying air to be expelled from the lungs (fig. 1.4). Body temperature regulation is discussed further in chapter 6.

Another homeostatic mechanism regulates the blood pressure in the blood vessels (arteries) leading away from the heart. In this instance, pressure sensitive parts (receptors) in the walls of these vessels sense changes in blood pressure and signal a pressure control center of the brain. If blood pressure is above the set point, the brain signals the heart, causing its chambers to contract more slowly and with less force. Because of this decreased heart action, less blood enters the blood vessels, decreasing the pressure inside them. If blood pressure is below the set point, the brain center signals the heart to contract more rapidly and with greater force so that the pressure in the vessels increases. Chapter 13 discusses regulation of blood pressure in more detail.

In the previous examples, homeostasis is maintained by a self-regulating control mechanism that receives signals (or feedback) about changes away from the normal set point and causes reactions that tend to return conditions to normal. Since the changes away from the normal state stimulate responses in the opposite direction, the responses are called *negative,* and the homeostatic control mechanism is said to act by a process of **negative feedback.** Examples of negative feedback are presented throughout the book.

Homeostatic mechanisms maintain a relatively constant internal environment, yet physiological values may vary slightly in a person from time to time or from one person to the next. Therefore, both normal values for an individual and the idea of a **normal range** for the general population are clinically important.

1. What requirements of organisms does the external environment provide?
2. Why is homeostasis important to survival?
3. Describe two homeostatic mechanisms.

Levels of Organization

Early investigators focused their attention on larger body structures, since they were limited in their ability to observe small parts. Studies of small parts had to

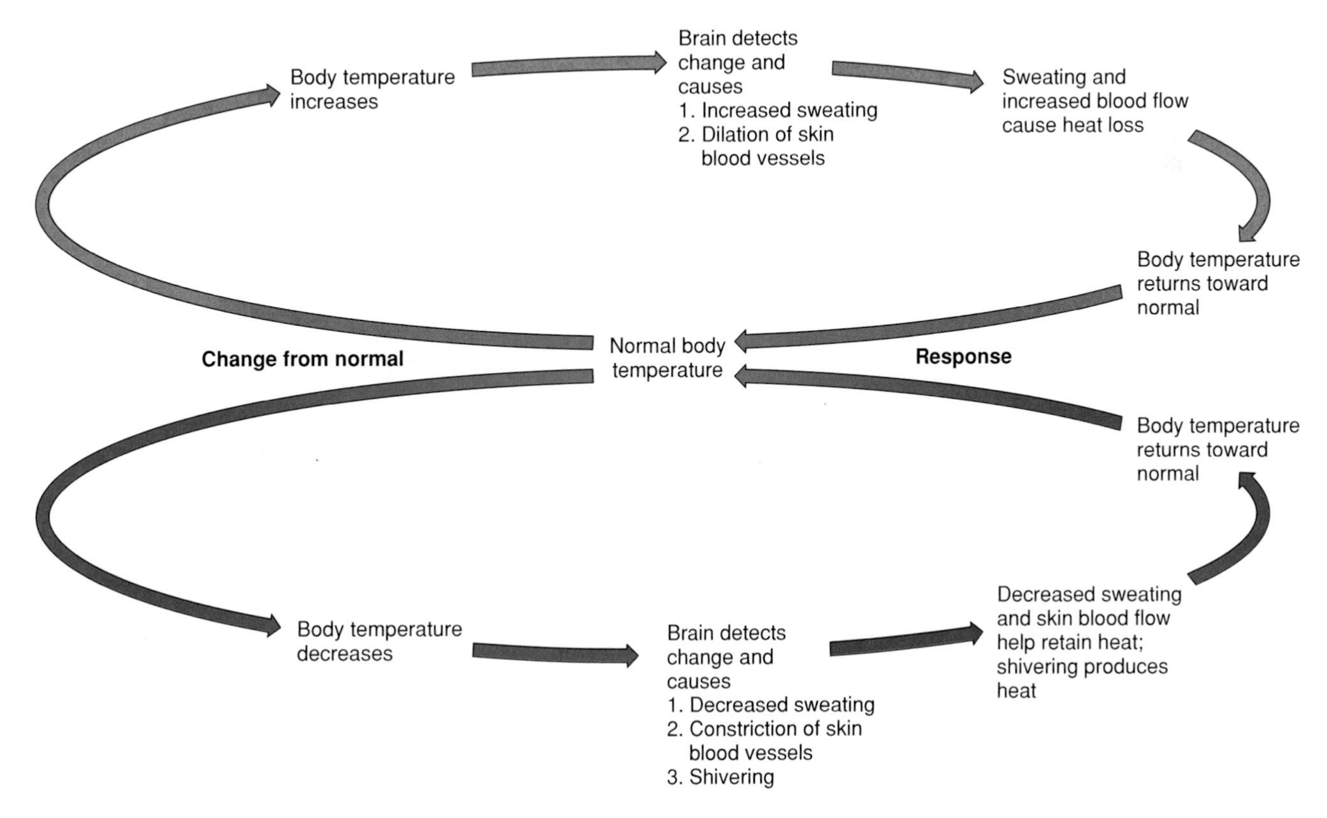

Figure 1.4

The homeostatic mechanism that regulates body temperature.

wait for the invention of magnifying lenses and microscopes, which came into use about 400 years ago. With these tools, investigators discovered that larger body structures were made up of smaller parts, which, in turn, were composed of even smaller ones.

Today, scientists recognize that all materials, including those that make up the human body, are composed of chemicals (fig. 1.5). Chemicals consist of invisible particles called **atoms**, which join to form **molecules.** Small molecules can combine in complex ways to form larger molecules called **macromolecules.**

Within the human organism, the basic unit of structure and function is a **cell**, which is microscopic. Although individual cells vary in size, shape, and specialized functions, all have certain characteristics in common. For instance, all human cells contain structures called **organelles** that carry on specific activities. Organelles are composed of aggregates of large molecules, such as proteins, carbohydrates, lipids, and nucleic acids.

Cells may be organized into layers or masses that have common functions. Such a group of cells forms a **tissue.** Groups of different tissues form **organs**—complex structures with specialized functions—and groups of organs that function closely together comprise **organ systems.** Organ systems make up an **organism.**

Body parts can be thought of as having different *levels of organization,* such as the *atomic level, molecular level,* or *cellular level.* Furthermore, body parts vary in complexity from one level to the next. That is, atoms are less complex than molecules, molecules are less complex than organelles, tissues are less complex than organs, and so forth.

Chapters 2–6 discuss these levels of organization in more detail. Chapter 2 describes the atomic and molecular levels. Chapter 3 deals with organelles and cellular structures and functions, and chapter 4 explores cellular metabolism. Chapter 5 describes tissues. Chapter 6 presents membranes (linings) as examples of organs, and the skin and its accessory organs as an example of an organ system. Beginning with chapter 6, the structures and functions of each of the organ systems are described in detail.

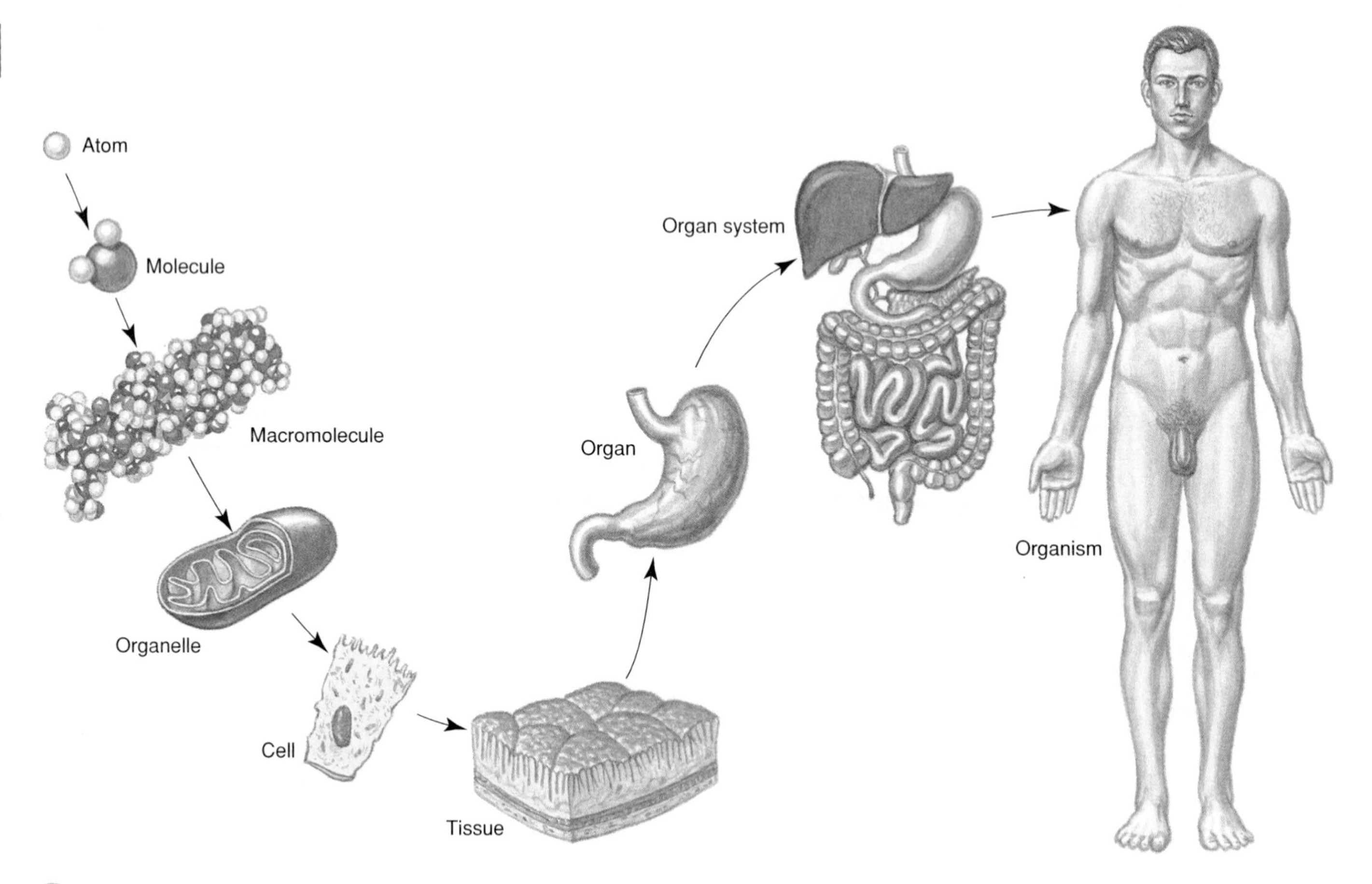

Figure 1.5

A human body is composed of parts within parts, which vary in complexity.

1. How does the human body illustrate levels of organization?
2. What is an organism?
3. How do body parts at different levels of organization vary in complexity?

Organization of the Human Body

The human organism is a complex structure composed of many parts. Its major features include several body cavities, layers of membranes within these cavities, and a variety of organ systems.

Body Cavities

The human organism can be divided into an **axial portion**, which includes the head, neck, and trunk, and an **appendicular portion**, which includes the upper and lower limbs. Within the axial portion are two major cavities: a **dorsal cavity** and a larger **ventral cavity** (fig. 1.6*a*). The organs within such a cavity are called visceral organs or **viscera.** The dorsal cavity can be subdivided into two parts: the **cranial cavity** within the skull, which houses the brain; and the **vertebral canal,** which contains the spinal cord within sections of the backbone (vertebrae). The ventral cavity consists of a **thoracic cavity** and an **abdominopelvic cavity.**

The thoracic cavity is separated from the lower abdominopelvic cavity by a broad, thin muscle called the **diaphragm.** The thoracic cavity wall is composed of skin, skeletal (voluntary) muscles, and various bones.

A region called the **mediastinum** separates the thoracic cavity into two compartments, which contain the right and left lungs. The remaining thoracic viscera—heart, esophagus, trachea, and thymus gland—are located within the mediastinum (fig. 1.6*b*).

The abdominopelvic cavity, which includes an upper abdominal portion and a lower pelvic portion, extends from the diaphragm to the floor of the pelvis. Its wall consists primarily of skin, skeletal muscles, and bones. The viscera within the **abdominal cavity** include the stomach, liver, spleen, gallbladder, kidneys, and most of the small and large intestines.

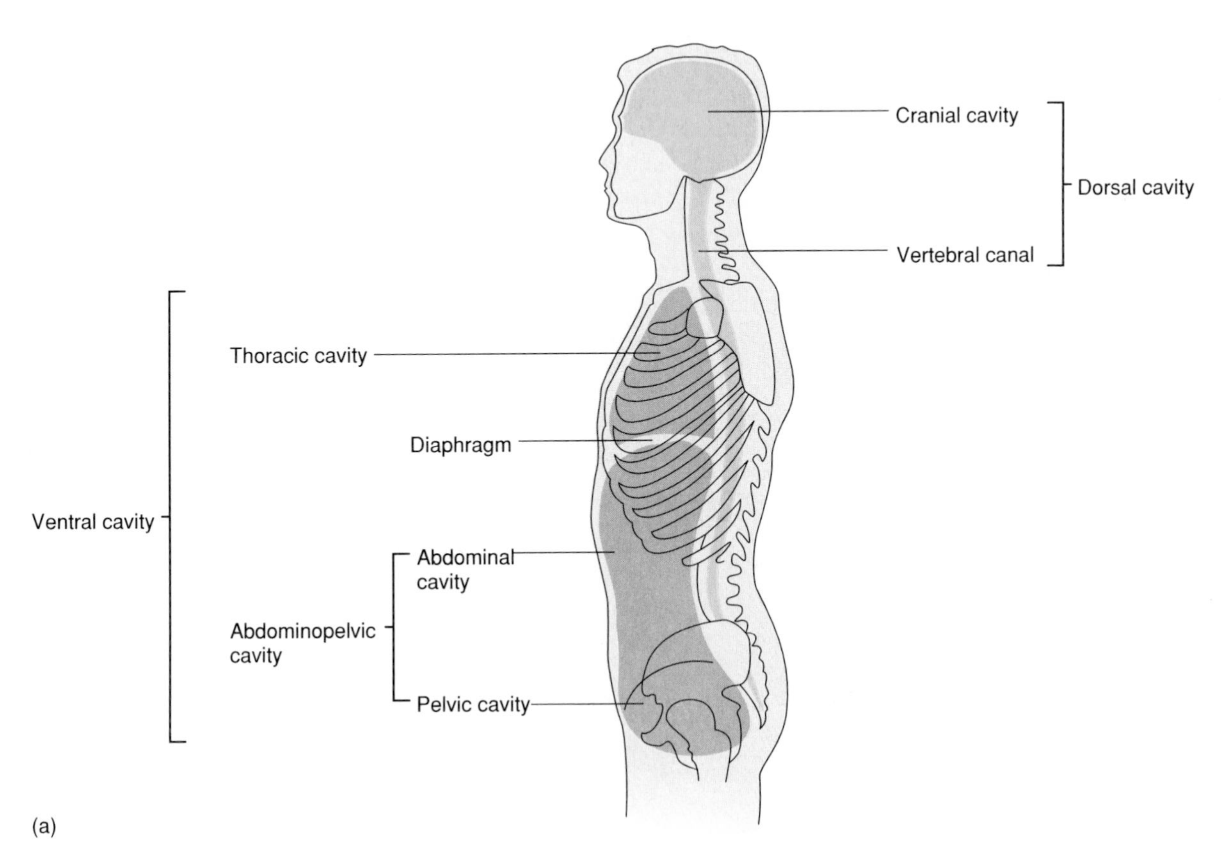

Figure 1.6

Major body cavities. (*a*) Lateral view. (*b*) Coronal view.

Continued

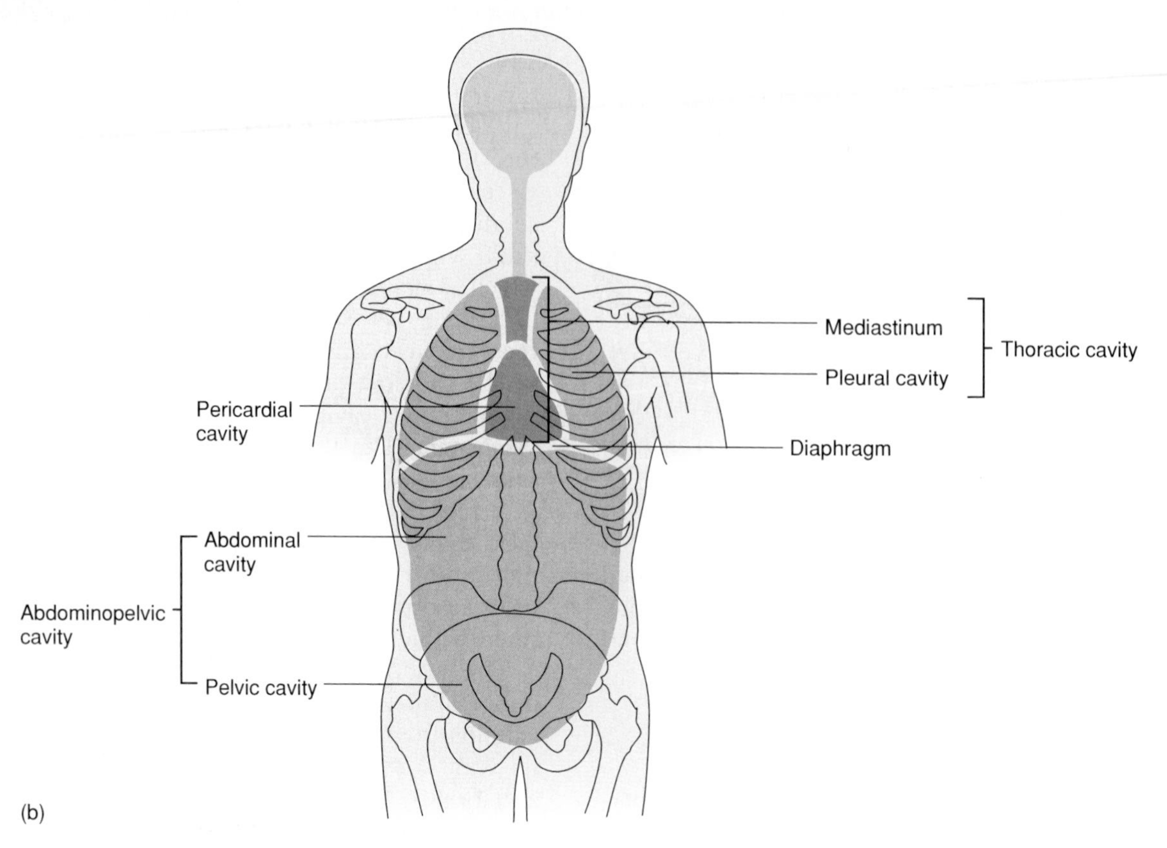

Figure 1.6—*Continued*

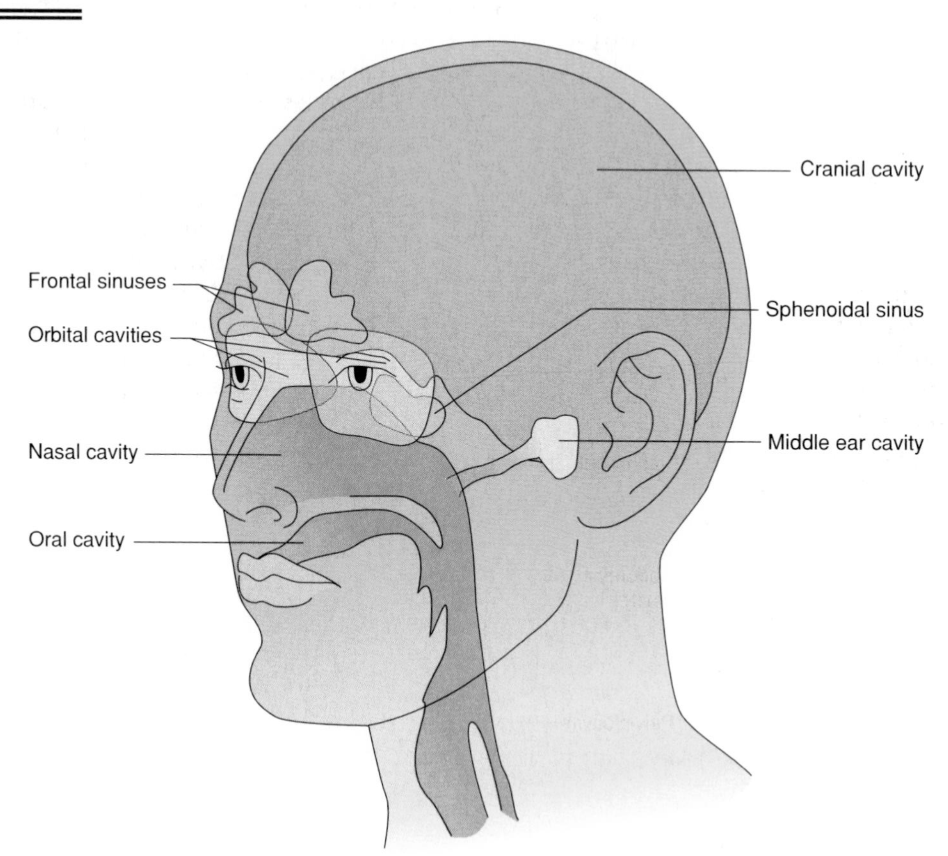

Figure 1.7

The cavities within the head include the oral, nasal, orbital, cranial, and middle ear cavities, as well as several sinuses.

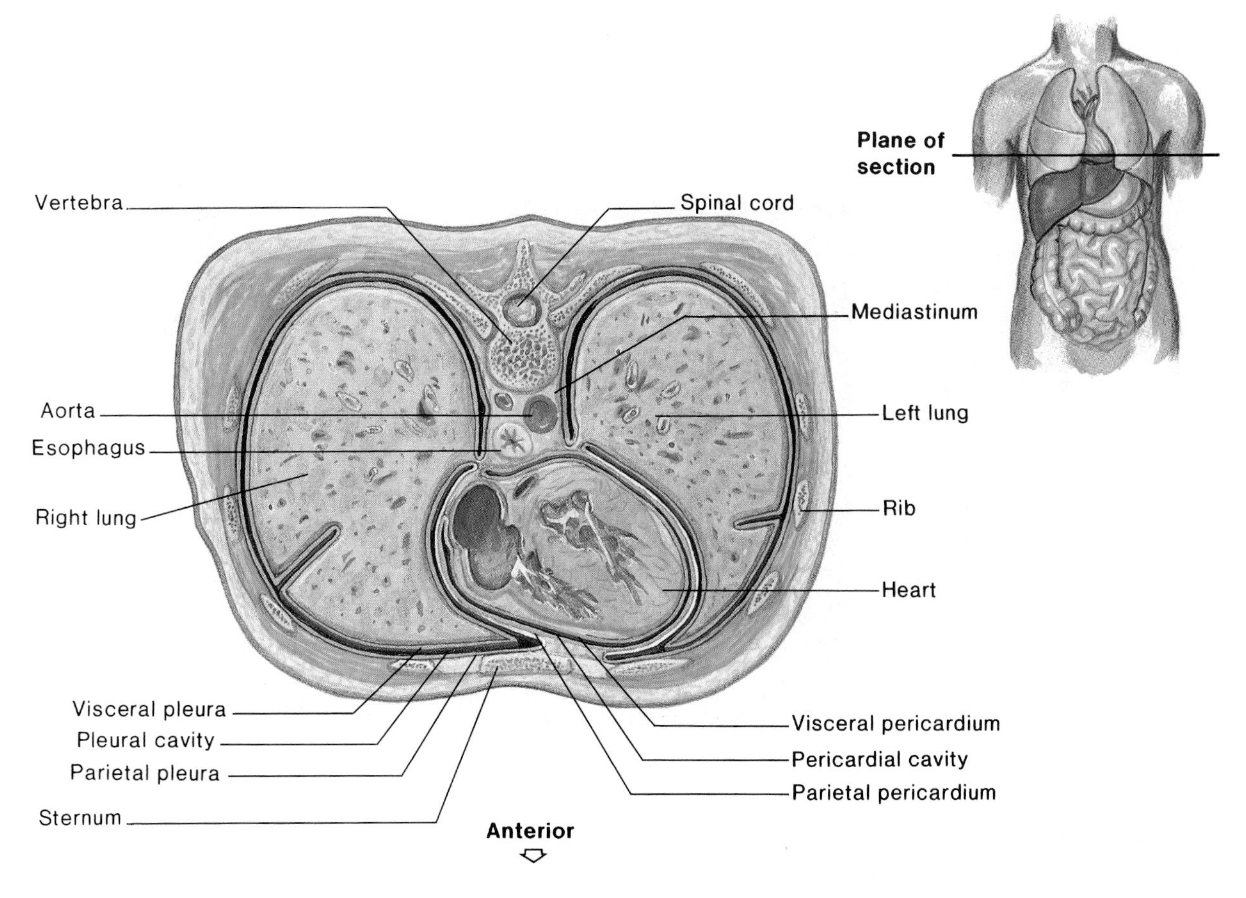

Figure 1.8

A transverse section through the thorax reveals the serous membranes associated with the heart and lungs (*superior view*).

The **pelvic cavity** is the portion of the abdominopelvic cavity enclosed by the hip bones (see chapter 7). It contains the terminal portion of the large intestine, the urinary bladder, and the internal reproductive organs.

Smaller cavities within the head include (fig. 1.7):

1. **Oral cavity,** containing the teeth and tongue.
2. **Nasal cavity,** located within the nose and divided into right and left portions by a nasal septum. Several air-filled *sinuses* connect to the nasal cavity (see chapter 7). These include the frontal and sphenoidal sinuses shown in figure 1.7.
3. **Orbital cavities,** containing the eyes and associated skeletal muscles and nerves.
4. **Middle ear cavities,** containing the middle ear bones.

Thoracic and Abdominopelvic Membranes

The walls of the right and left thoracic compartments, which contain the lungs, are lined with a membrane called the *parietal pleura* (fig. 1.8). A similar membrane, called the *visceral pleura,* covers the lungs themselves. (Note: **parietal** refers to the membrane attached to the wall of a cavity; **visceral** refers to the membrane that is deeper—toward the interior—and covers an internal organ, such as a lung.)

The parietal and visceral **pleural membranes** are separated by a thin film of watery fluid (serous fluid) that they secrete. While no actual space normally exists between these membranes, the potential space between them is called the *pleural cavity* (see fig. 1.6*b*).

The heart, which is located in the broadest portion of the mediastinum, is surrounded by **pericardial membranes.** A thin *visceral pericardium* covers the heart's surface and is separated from a thicker *parietal pericardium* by a small amount of fluid. The *pericardial cavity* (see fig. 1.6*b*) is the potential space between these membranes.

In the abdominopelvic cavity, the lining membranes are called **peritoneal membranes** (fig. 1.9). A *parietal peritoneum* lines the wall, and a *visceral peritoneum* covers each organ in the abdominal cavity. The *peritoneal cavity* is the potential space between these membranes.

***Figure* 1.9**

Transverse section through the abdomen (*superior view*).

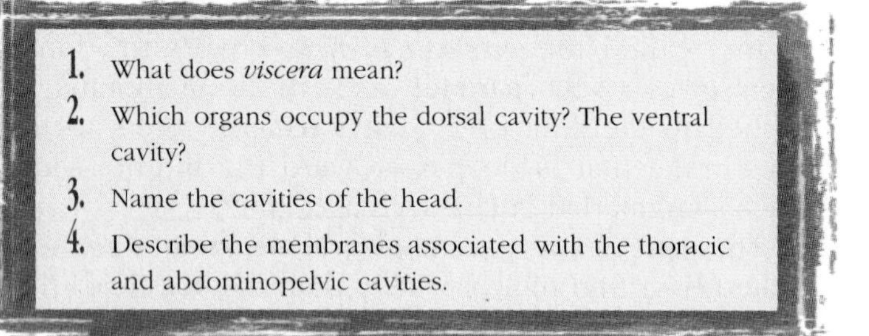

1. What does *viscera* mean?
2. Which organs occupy the dorsal cavity? The ventral cavity?
3. Name the cavities of the head.
4. Describe the membranes associated with the thoracic and abdominopelvic cavities.

Organ Systems

The human organism consists of several organ systems. Each system includes a set of interrelated organs that work together to provide specialized functions. As you read about each system, you may want to consult the illustrations of the human torso and locate some of the organs listed in the description (see Reference Plates, pp. 22–29).

Body Covering

Organs of the **integumentary system** (see chapter 6) include the skin and various accessory organs, such as the hair, nails, sweat glands, and sebaceous glands. These parts protect underlying tissues, help regulate body temperature, house a variety of sensory receptors, and synthesize certain products.

Support and Movement

The organs of the skeletal and muscular systems (see chapters 7 and 8) support and move body parts. The **skeletal system** consists of bones, as well as ligaments and cartilages that bind bones together. These parts provide frameworks and protective shields for softer tissues, are attachments for muscles, and act with muscles when body parts move. Tissues within bones also produce blood cells and store inorganic salts.

Muscles are the organs of the **muscular system.** By contracting and pulling their ends closer together, muscles provide forces that cause body movements. They also maintain posture and are the main source of body heat.

Integration and Coordination

For the body to act as a unit, its parts must be integrated and coordinated. The nervous and endocrine systems control and adjust various organ functions, which maintains homeostasis.

The **nervous system** (see chapter 9) consists of the brain, spinal cord, nerves, and sense organs (see chapter 10). Nerve cells within these organs use electrochemical signals called *nerve impulses* to communicate with one another and with muscles and glands. Each impulse produces a relatively short-term effect on its target. Some nerve cells act as specialized sensory receptors that can detect changes inside and outside the body. Other nerve cells receive the impulses transmitted from these sensory receptors and interpret and act on the information received. Still other nerve cells carry impulses from the brain or spinal cord to muscles or glands, stimulating them to contract or to secrete products.

The **endocrine system** (see chapter 11) includes all the glands that secrete chemical messengers called *hormones*. The hormones, in turn, move away from the glands in body fluids, such as blood or tissue fluid (fluid from the spaces within tissues). Usually, a particular hormone affects only a particular group of cells, which is called its *target tissue*. The effect of a hormone is to alter the metabolism of the target tissue. Compared to nerve impulses, hormonal effects occur over a relatively long time period. Organs of the endocrine system include the pituitary, thyroid, parathyroid, and adrenal glands, as well as the pancreas, ovaries, testes, pineal gland, and thymus gland.

Transport

Two organ systems transport substances throughout the internal environment. The **cardiovascular system** (see chapters 12 and 13) includes the heart, arteries, veins, capillaries, and blood. The heart is a muscular pump that helps force blood through the blood vessels. Blood transports gases, nutrients, hormones, and wastes. It carries oxygen from the lungs and nutrients from the digestive organs to all body cells, where these biochemicals are used in metabolic processes. Blood also transports hormones from endocrine glands to their target tissues and carries wastes from body cells to the excretory organs, where the wastes are removed from the blood and released to the outside.

The **lymphatic system** (see chapter 14) is sometimes considered part of the cardiovascular system. It is composed of the lymphatic vessels, lymph fluid, lymph nodes, thymus gland, and spleen. This system transports some of the tissue fluid back to the bloodstream and carries certain fatty substances away from the digestive organs. Cells of the lymphatic system are called lymphocytes, and they defend the body against infections by removing disease-causing microorganisms and viruses from tissue fluid.

Absorption and Excretion

Organs in several systems absorb nutrients and oxygen and excrete various wastes. The organs of the **digestive system** (see chapter 15), for example, receive foods from the outside. Then they break down food molecules into simpler forms that can pass through cell membranes and thus be absorbed. Materials that are not absorbed are transported back to the outside and eliminated. Certain digestive organs also produce hormones and thus function as parts of the endocrine system. The digestive system includes the mouth, tongue, teeth, salivary glands, pharynx, esophagus, stomach, liver, gallbladder, pancreas, small intestine, and large intestine. Chapter 15 also discusses nutrition.

The organs of the **respiratory system** (see chapter 16) take air in and out and exchange gases between the blood and the air. More specifically, oxygen passes from air within the lungs into the blood, and carbon dioxide leaves the blood and enters the air. The nasal cavity, pharynx, larynx, trachea, bronchi, and lungs are parts of this system.

The **urinary system** (see chapter 17) consists of the kidneys, ureters, urinary bladder, and urethra. The kidneys remove wastes from blood and help maintain the body's water and electrolyte balance. The product of these activities is urine. Other portions of the urinary system store urine and transport it outside the body. Chapter 18 discusses the urinary system's role in maintaining water, electrolyte, and acid-base balance.

Reproduction

Reproduction is the process of producing offspring (progeny). Cells reproduce when they divide and give rise to new cells. The **reproductive system** of an organism, however, produces whole new organisms like itself (see chapter 19).

The male reproductive system includes the scrotum, testes, epididymides, vasa deferentia, seminal vesicles, prostate gland, bulbourethral glands, penis, and urethra. These parts produce and maintain sperm cells (spermatozoa). Components of the male reproductive system also transfer sperm cells into the female reproductive tract.

The female reproductive system consists of the ovaries, uterine tubes, uterus, vagina, clitoris, and vulva. These organs produce and maintain female sex cells (the egg cells [ova]), receive sperm cells, and transport the sperm cells and egg cells within the female system. The female reproductive system also supports the development of prenatal humans, such as embryos and fetuses, and enables the birth process to proceed.

1. Name each of the major organ systems, and list the organs of each system.
2. Describe the general functions of each organ system.

Anatomical Terminology

To communicate effectively with one another, investigators have developed a set of terms to describe anatomy that have precise meanings. Some of these terms concern the relative positions of body parts, others relate to imaginary planes along which cuts may be made, and still others describe body regions.

Use of such terms assumes that the body is in the **anatomical position.** This means that the body is standing erect, face forward, with upper limbs at the sides and with the palms forward.

Relative Positions

Terms of relative position describe the location of one body part with respect to another. They include the following:

1. **Superior** means that a body part is above another part or is closer to the head. (The thoracic cavity is superior to the abdominopelvic cavity.)
2. **Inferior** means that a body part is below another body part or is toward the feet. (The neck is inferior to the head.)
3. **Anterior** (or *ventral*) means toward the front. (The eyes are anterior to the brain.)
4. **Posterior** (or *dorsal*) is the opposite of anterior; it means toward the back. (The pharynx is posterior to the oral cavity.)
5. **Medial** relates to an imaginary midline dividing the body into equal right and left halves. A body part is medial if it is closer to this line than another part. (The nose is medial to the eyes.)
6. **Lateral** means toward the side with respect to the imaginary midline. (The ears are lateral to the eyes.)
7. **Proximal** describes a body part that is closer to a point of attachment or closer to the trunk of the body than another part. (The elbow is proximal to the wrist.)
8. **Distal** is the opposite of proximal. It means that a particular body part is farther from the point of attachment or farther from the trunk than another part. (The fingers are distal to the wrist.)
9. **Superficial** means situated near the surface. (The epidermis is the superficial layer of the skin.) *Peripheral* also means outward or near the surface. It describes the location of certain blood vessels and nerves. (The nerves that branch from the brain and spinal cord are peripheral nerves.)
10. **Deep** describes parts that are more internal. (The dermis is the deep layer of the skin.)

Body Sections

Observing the relative locations and organization of internal body parts requires cutting or sectioning the body along various planes (fig. 1.10). The following terms describe such planes and sections:

1. **Sagittal** refers to a lengthwise cut that divides the body into right and left portions. If a sagittal section passes along the midline and divides the body into equal parts, it is called *median* (midsagittal).
2. **Transverse** (or *horizontal*) refers to a cut that divides the body into superior and inferior portions.
3. **Coronal** (or *frontal*) refers to a section that divides the body into anterior and posterior portions.

Sometimes, a cylindrical organ such as a blood vessel is sectioned. In this case, a cut across the structure is called a *cross section,* an angular cut is an *oblique section,* and a lengthwise cut is a *longitudinal section* (fig. 1.11).

Body Regions

A number of terms designate body regions. The abdominal area, for example, is subdivided into the following nine regions, as figure 1.12 shows:

1. **Epigastric region** The upper middle portion.
2. **Left** and **right hypochondriac regions** On each side of the epigastric region.
3. **Umbilical region** The middle portion.
4. **Left** and **right lumbar regions** On each side of the umbilical region.
5. **Hypogastric region** The lower middle portion.
6. **Left** and **right iliac regions** (left and right inguinal regions) On each side of the hypogastric region.

The following terms are commonly used to refer to various body regions (fig. 1.13 illustrates some of these regions):

Abdominal (ab-dom′ĭ-nal) The region between the thorax and pelvis.

Acromial (ah-kro′me-al) The point of the shoulder.

Antebrachial (an″te-bra′ke-al) The forearm.

Antecubital (an″te-ku′bĭ-tal) The space in front of the elbow.

Axillary (ak′sĭ-ler″e) The armpit.

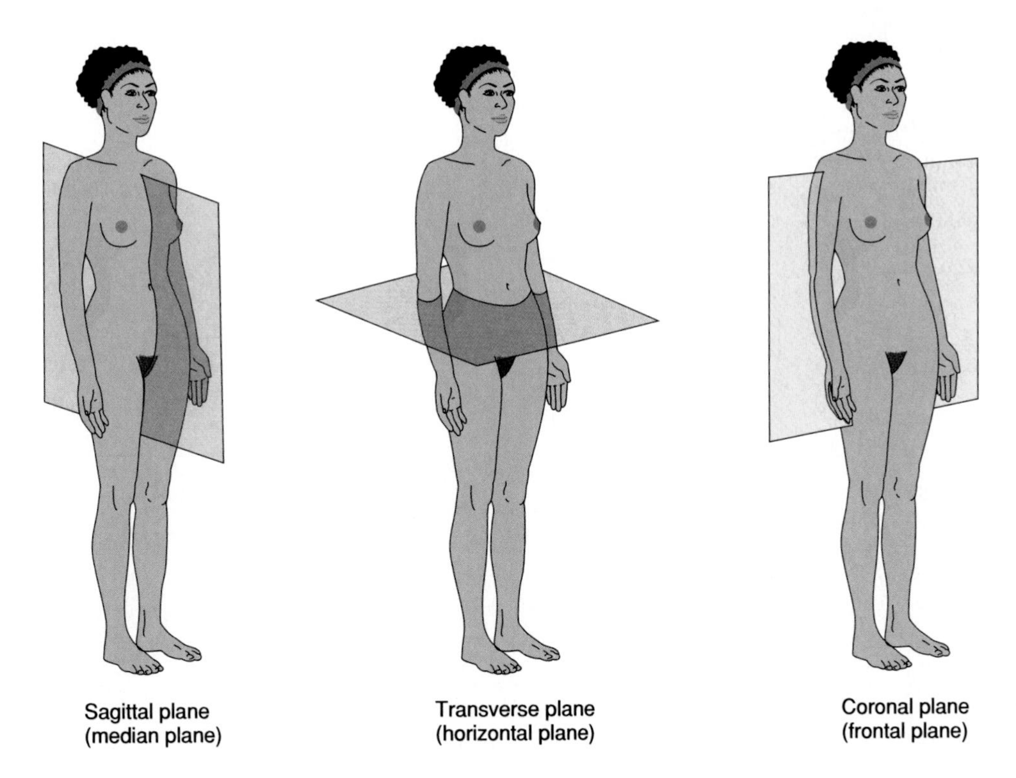

***Figure* 1.10**

Observation of internal parts requires sectioning the body along various planes.

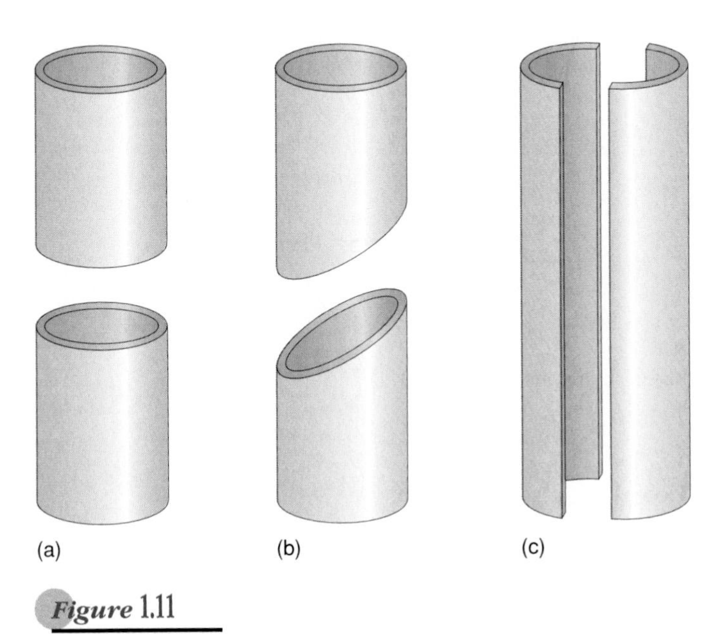

***Figure* 1.11**

Cylindrical parts may be cut in (*a*) cross section, (*b*) oblique section, or (*c*) longitudinal section.

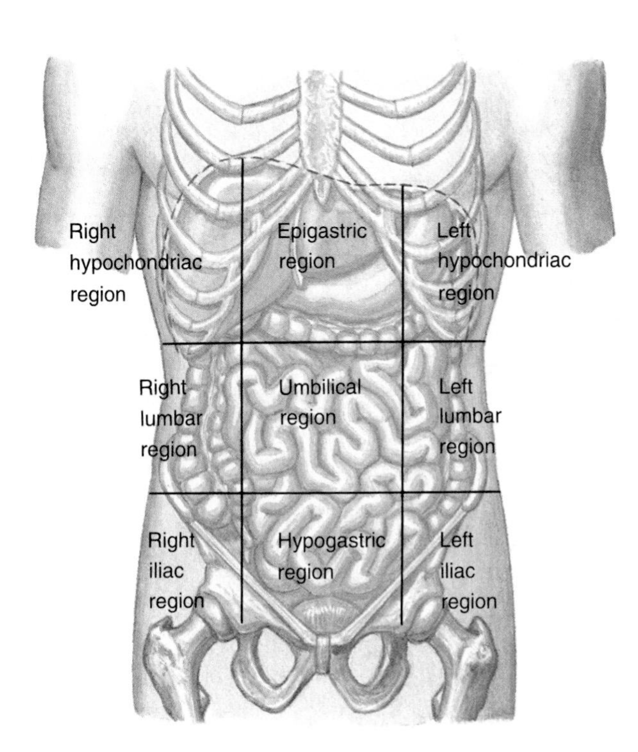

***Figure* 1.12**

The abdominal area is subdivided into nine regions.

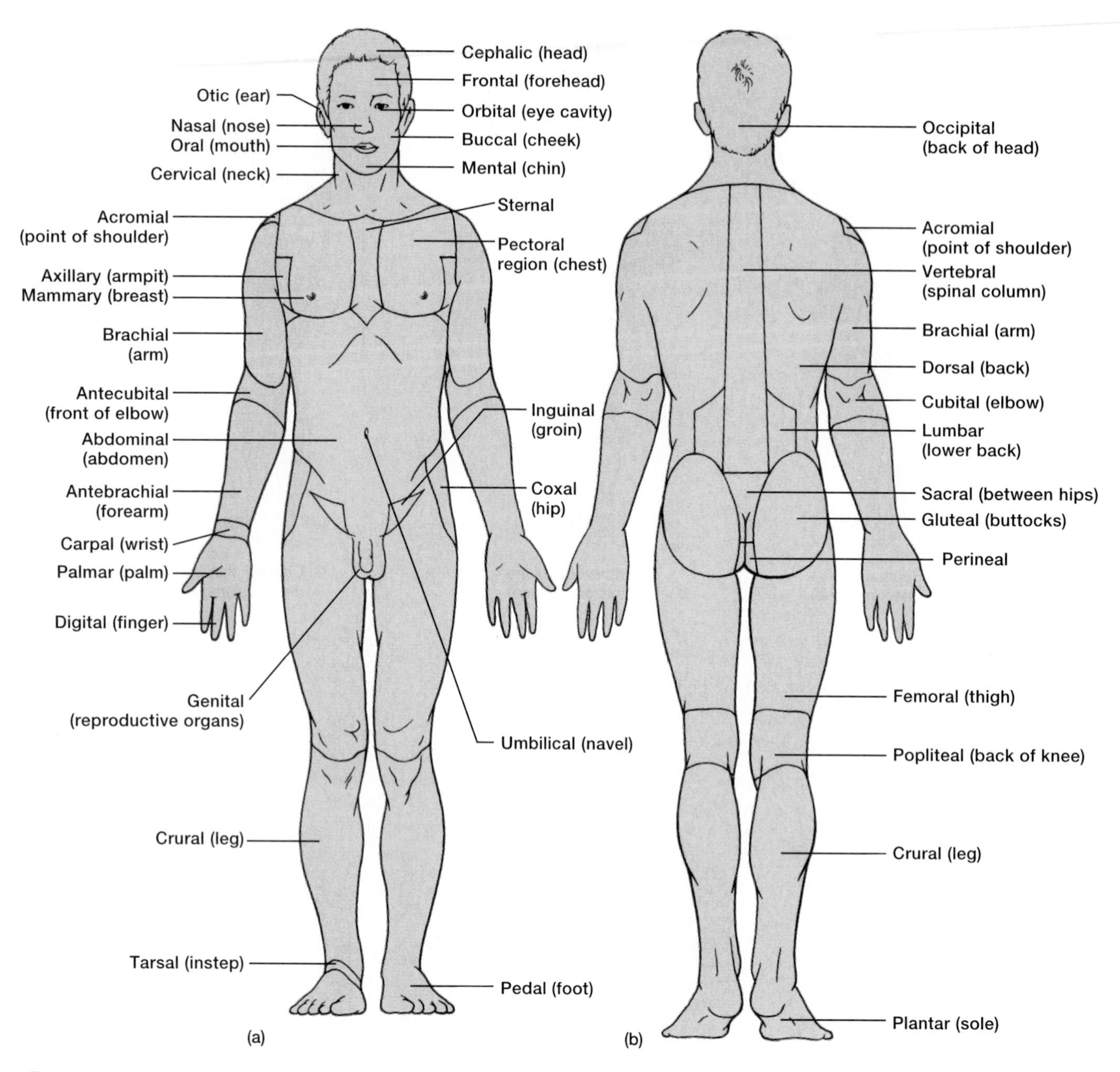

Figure 1.13

Some terms used to describe body regions. (*a*) Anterior regions. (*b*) Posterior regions.

Brachial (bra′ke-al) The arm.
Buccal (buk′al) The cheek.
Carpal (kar′pal) The wrist.
Celiac (se′le-ak) The abdomen.
Cephalic (sĕ-fal′ik) The head.
Cervical (ser′vĭ-kal) The neck.
Costal (kos′tal) The ribs.
Coxal (kok′sal) The hip.
Crural (kro͞o r′al) The leg.
Cubital (ku′bĭ-tal) The elbow.
Digital (dij′ĭ-tal) The finger.
Dorsal (dor′sal) The back.
Femoral (fem′or-al) The thigh.

Radiologists use a procedure called *computerized tomography,* or CT scanning, to visualize internal organ sections (fig. 1A). In this procedure, an X-ray-emitting device moves around the body region being examined. At the same time, an X-ray detector moves in the opposite direction on the other side. As the devices move, an X-ray beam passes through the body from hundreds of different angles. Since tissues and organs of varying composition within the body absorb X rays differently, the amount of X ray reaching the detector varies from position to position.

Measurements that the X-ray detector makes are recorded in the memory of a computer, which later combines them mathematically and creates a sectional image of the internal body parts on a viewing screen. ■

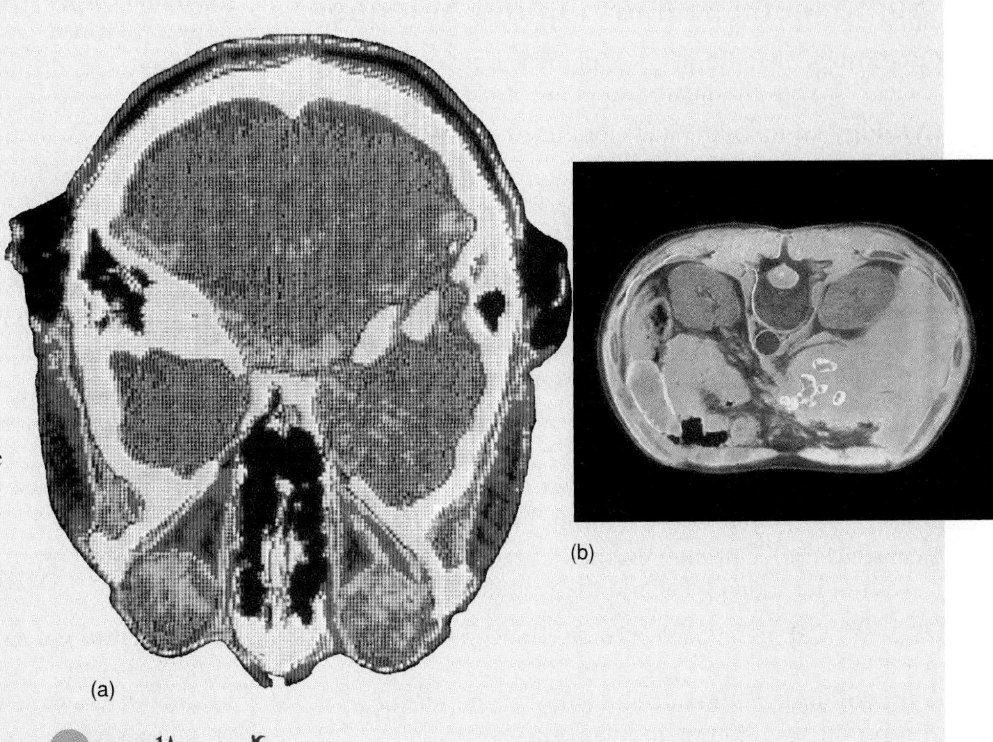

Figure 1A

Falsely colored CT (computerized tomography) scans of the (*a*) head and (*b*) abdomen. Note: These are not shown in correct relative size.

Frontal (frun′tal) The forehead.

Genital (jen′ĭ-tal) The reproductive organs.

Gluteal (gloo′te-al) The buttocks.

Inguinal (ing′gwĭ-nal) The depressed area of the abdominal wall near the thigh (groin).

Lumbar (lum′bar) The region of the lower back between the ribs and the pelvis (loin).

Mammary (mam′er-e) The breast.

Mental (men′tal) The chin.

Nasal (na′zal) The nose.

Occipital (ok-sip′ĭ-tal) The lower posterior region of the head.

Oral (o′ral) The mouth.

Orbital (or′bi-tal) The eye cavity.

Otic (o′tik) The ear.

Palmar (pahl′mar) The palm of the hand.

Patellar (pah-tel′ar) The front of the knee.

Pectoral (pek′tor-al) The chest.

Pedal (ped′al) The foot.

Pelvic (pel′vik) The pelvis.

Perineal (per″ĭ-ne′al) The region between the anus and the external reproductive organs (perineum).

Plantar (plan′tar) The sole of the foot.

Popliteal (pop″lĭ-te′al) The area behind the knee.

Sacral (sa′kral) The posterior region between the hipbones.

Sternal (ster′nal) The middle of the thorax, anteriorly.

Tarsal (tahr′sal) The instep of the foot.

Umbilical (um-bil′ĭ-kal) The navel.

Vertebral (ver′te-bral) The spinal column.

1. Describe the anatomical position.
2. Using the appropriate terms, describe the relative positions of several body parts.
3. Describe three types of body sections.
4. Name the nine regions of the abdomen.

Some Medical and Applied Sciences

Cardiology (kar″de-ol′o-je) Branch of medical science dealing with the heart and heart diseases.

Cytology (si-tol′o-je) Study of the structure, function, and abnormalities of cells.

Dermatology (der″mah-tol′o-je) Study of the skin and its diseases.

Endocrinology (en″do-krĭ-nol′o-je) Study of hormones, hormone-secreting glands, and their diseases.

Epidemiology (ep″ĭ-de″me-ol′o-je) Study of the factors determining the distribution and frequency of the occurrence of health-related conditions within a defined human population.

Gastroenterology (gas″tro-en″ter-ol′o-je) Study of the stomach and intestines and their diseases.

Geriatrics (jer″e-at′riks) Branch of medicine dealing with older individuals and their medical problems.

Gerontology (jer″on-tol′o-je) Study of the aging process.

Gynecology (gi″nĕ-kol′o-je) Study of the female reproductive system and its diseases.

Hematology (hēm″ah-tol′o-je) Study of the blood and blood diseases.

Histology (his-tol′o-je) Study of the structure and function of tissues, also called microscopic anatomy.

Immunology (im″u-nol′o-je) Study of the body's resistance to disease.

Neonatology (ne″o-na-tol′o-je) Study of newborns and the treatment of their disorders.

Nephrology (nĕ-frol′o-je) Study of the structure, function, and diseases of the kidneys.

Neurology (nu-rol′o-je) Study of the nervous system and its disorders.

Obstetrics (ob-stet′riks) Branch of medicine dealing with pregnancy and childbirth.

Oncology (ong-kol′o-je) Study of cancers.

Ophthalmology (of″thal-mol′o-je) Study of the eye and eye diseases.

Orthopedics (or″tho-pe′diks) Branch of medicine dealing with the muscular and skeletal systems and their problems.

Otolaryngology (o″to-lar″in-gol′o-je) Study of the ear, throat, and larynx, and their diseases.

Pathology (pah-thol′o-je) Study of structural and functional changes that disease produces.

Pediatrics (pe″de-at′riks) Branch of medicine dealing with children and their diseases.

Pharmacology (fahr″mah-kol′o-je) Study of drugs and their uses in the treatment of disease.

Podiatry (po-di′ah-tre) Study of the care and treatment of feet.

Psychiatry (si-ki′ah-tre) Branch of medicine dealing with the mind and its disorders.

Radiology (ra″de-ol′o-je) Study of X rays and radioactive substances and their uses in the diagnosis and treatment of diseases.

Toxicology (tok″sĭ-kol′o-je) Study of poisonous substances and their effects upon body parts.

Urology (u-rol′o-je) Branch of medicine dealing with the urinary system, apart from the kidneys (nephrology), and the male reproductive system, and their diseases.

Summary Outline

Introduction (p. 3)

1. Early interest in the human body probably developed as people became concerned about injuries and illnesses.
2. Primitive doctors began to learn how certain herbs and potions affected body functions.
3. The belief that humans could understand forces that caused natural events led to the development of modern science.
4. A set of terms originating from Greek and Latin words is the basis for the language of anatomy and physiology.

Anatomy and Physiology (p. 4)

1. Anatomy describes the form and organization of body parts.
2. Physiology considers the functions of anatomical parts.
3. The function of an anatomical part depends on the way it is constructed.

Characteristics of Life (p. 4)

Characteristics of life are traits all organisms share.

1. These characteristics include:
   a. Movement—changing body position or moving internal parts.
   b. Responsiveness—sensing and reacting to internal or external changes.
   c. Growth—increasing size without changing shape.
   d. Reproduction—producing offspring.
   e. Respiration—obtaining oxygen, using oxygen to release energy from foods, and removing gaseous wastes.
   f. Digestion—breaking down food substances into component nutrients that the intestine can absorb.
   g. Absorption—moving substances through membranes and into body fluids.
   h. Circulation—moving substances through the body in body fluids.
   i. Assimilation—changing substances into chemically different forms.
   j. Excretion—removing body wastes.
2. These activities constitute metabolism.

Maintenance of Life (p. 5)

The structures and functions of body parts maintain the life of the organism.

1. Requirements of organisms
   a. Water is used in many metabolic processes, provides the environment for metabolic reactions, and transports substances.
   b. Food supplies energy, raw materials for building new living matter, and chemicals necessary in vital reactions.
   c. Oxygen releases energy from food materials. This energy drives metabolic reactions.
   d. Heat is a product of metabolic reactions and helps govern the rates of these reactions.
   e. Pressure is an application of force to something. In humans, atmospheric and hydrostatic pressures help breathing and blood movements, respectively.
2. Homeostasis
   a. If an organism is to survive, the conditions within its body fluids must remain relatively stable.
   b. Maintenance of a stable internal environment is called *homeostasis.*
   c. Homeostatic mechanisms help regulate body temperature and blood pressure.
   d. Homeostatic mechanisms act through negative feedback.

Levels of Organization (p. 7)

The body is composed of parts with different levels of organization.

1. Matter is composed of atoms.
2. Atoms join to form molecules.
3. Organelles are built of groups of large molecules.
4. Cells, which contain organelles, are the basic units of structure and function that form the body.
5. Cells are organized into tissues.
6. Tissues are organized into organs.
7. Organs that function closely together comprise organ systems.
8. Organ systems constitute the organism.
9. These levels of organization vary in complexity from one level to the next.

Organization of the Human Body (p. 9)

1. Body cavities
   a. The axial portion of the body contains the dorsal and ventral cavities.
      (1) The dorsal cavity includes the cranial cavity and the vertebral canal.
      (2) The ventral cavity includes the thoracic and abdominopelvic cavities, which are separated by the diaphragm.
   b. The organs within a body cavity are called *viscera.*
   c. The mediastinum separates the thoracic cavity into right and left compartments.
   d. Other body cavities include the oral, nasal, orbital, and middle ear cavities.
2. Thoracic and abdominopelvic membranes
   a. Thoracic membranes
      (1) Pleural membranes line the thoracic cavity (parietal pleura) and cover the lungs (visceral pleura).
      (2) Pericardial membranes surround the heart (parietal pericardium) and cover its surface (visceral pericardium).
      (3) The pleural and pericardial cavities are the potential spaces between the respective parietal and visceral membranes.
   b. Abdominopelvic membranes
      (1) Peritoneal membranes line the abdominopelvic cavity (parietal peritoneum) and cover the organs inside (visceral peritoneum).
      (2) The peritoneal cavity is the potential space between the parietal and visceral membranes.
3. Organ systems
   The human organism consists of several organ systems. Each system includes a set of interrelated organs.
   a. Body covering
      (1) The integumentary system includes the skin, hair, nails, sweat glands, and sebaceous glands.
      (2) It protects underlying tissues, regulates body temperature, houses sensory receptors, and synthesizes various substances.
   b. Support and movement
      (1) Skeletal system
         (a) The skeletal system is composed of bones, as well as cartilages and ligaments that bind bones together.
         (b) It provides a framework, protective shields, and attachments for muscles. It also produces blood cells and stores inorganic salts.
      (2) Muscular system
         (a) The muscular system includes the muscles of the body.
         (b) It moves body parts, maintains posture, and produces body heat.
   c. Integration and coordination
      (1) Nervous system
         (a) The nervous system consists of the brain, spinal cord, nerves, and sense organs.
         (b) It receives impulses from sensory parts, interprets these impulses, and acts on them by stimulating muscles or glands to respond.
      (2) Endocrine system
         (a) The endocrine system consists of glands that secrete hormones.
         (b) Hormones help regulate metabolism.
         (c) This system includes the pituitary, thyroid, parathyroid, and adrenal glands, as well as the pancreas, ovaries, testes, pineal gland, and thymus gland.
   d. Transport
      (1) Cardiovascular system
         (a) The cardiovascular system includes the heart, which pumps blood, and the blood vessels, which carry blood to and from body parts.
         (b) Blood transports oxygen, nutrients, hormones, and wastes.
      (2) Lymphatic system
         (a) The lymphatic system is composed of lymphatic vessels, lymph fluid, lymph nodes, thymus gland, and spleen.

(b) It transports lymph fluid from tissues to the bloodstream, carries certain fatty substances away from the digestive organs, and aids in defending the body against disease-causing agents.

e. Absorption and excretion
   (1) Digestive system
      (a) The digestive system receives foods, breaks down food molecules into nutrients that can pass through cell membranes, and eliminates materials that are not absorbed.
      (b) It includes the mouth, tongue, teeth, salivary glands, pharynx, esophagus, stomach, liver, gallbladder, pancreas, small intestine, and large intestine.
      (c) Some digestive organs produce hormones.
   (2) Respiratory system
      (a) The respiratory system takes in and sends out air and exchanges gases between the air and blood.
      (b) It includes the nasal cavity, pharynx, larynx, trachea, bronchi, and lungs.
   (3) Urinary system
      (a) The urinary system includes the kidneys, ureters, urinary bladder, and urethra.
      (b) It filters wastes from the blood and helps maintain water, acid-base, and electrolyte balance.

f. Reproduction
   (1) The reproductive systems produce new organisms.
   (2) The male reproductive system includes the scrotum, testes, epididymides, vasa deferentia, seminal vesicles, prostate gland, bulbourethral glands, urethra, and penis, which produce, maintain, and transport male sex cells.
   (3) The female reproductive system includes the ovaries, uterine tubes, uterus, vagina, clitoris, and vulva, which produce, maintain, and transport female sex cells.

Anatomical Terminology (p. 14)

Terms with precise meanings help investigators to communicate effectively.

1. Relative positions

   These terms describe the location of one part with respect to another part.

2. Body sections

   Body sections are planes along which the body may be cut to observe the relative locations and organizations of internal parts.

3. Body regions

   Special terms designate various body regions.

Review Exercises

Part A

1. Briefly describe the early development of knowledge about the human body. (p. 3)
2. Distinguish between anatomy and physiology. (p. 4)
3. Explain the relationship between the form and function of body parts. (p. 4)
4. List and describe ten characteristics of life. (p. 4)
5. Define *metabolism*. (p. 5)
6. List and describe five requirements of organisms. (p. 5)
7. Describe two types of pressure that may act on the outside of an organism. (p. 5)
8. Define *homeostasis,* and explain its importance. (p. 6)
9. Explain the control of body temperature. (p. 6)
10. Describe a homeostatic mechanism that helps regulate blood pressure. (p. 7)
11. List the levels of organization within the human body. (p. 7)
12. Distinguish between the axial and appendicular portions of the body. (p. 9)
13. Distinguish between the dorsal and ventral body cavities, and name the smaller cavities within each. (p. 9)
14. Explain what is meant by *viscera*. (p. 9)
15. Describe the mediastinum and its contents. (p. 9)
16. List the cavities of the head and the contents of each cavity. (p. 11)
17. Distinguish between a parietal and a visceral membrane. (p. 11)
18. Name the major organ systems, and describe the general functions of each. (p. 12)
19. List the major organs that comprise each organ system. (p. 12)

Part B

1. Name the body cavity that houses each of the following organs:
   a. Stomach
   b. Heart
   c. Brain
   d. Liver
   e. Trachea
   f. Rectum
   g. Spinal cord
   h. Esophagus
   i. Spleen
   j. Urinary bladder
2. Write complete sentences using each of the following terms correctly:
   a. Superior
   b. Inferior
   c. Anterior
   d. Posterior
   e. Medial
   f. Lateral
   g. Proximal
   h. Distal
   i. Superficial
   j. Peripheral
   k. Deep
3. Sketch a human body, and use lines to indicate each of the following sections:
   a. Sagittal
   b. Transverse
   c. Coronal

4. Sketch the abdominal area, and indicate the location of each of the following regions:
   a. Epigastric
   b. Umbilical
   c. Hypogastric
   d. Hypochondriac
   e. Lumbar
   f. Iliac
5. Provide the common name for the region to which each of the following terms refers:
   a. Acromial
   b. Antebrachial
   c. Axillary
   d. Buccal
   e. Celiac
   f. Coxal
   g. Crural
   h. Femoral
   i. Genital
   j. Gluteal
   k. Inguinal
   l. Mental
   m. Occipital
   n. Orbital
   o. Otic
   p. Palmar
   q. Pectoral
   r. Pedal
   s. Perineal
   t. Plantar
   u. Popliteal
   v. Sacral
   w. Sternal
   x. Tarsal
   y. Umbilical
   z. Vertebral

Critical Thinking

1. Which characteristics of life does an automobile have? Why is a car not alive?
2. What environmental characteristics would be necessary for a human to survive on another planet?
3. Overweight people who lose weight often find it difficult to keep the weight off because a set point for the body's fat stores changes as the body perceives itself as starving. Explain how this protective mechanism, of great frustration to dieters, might operate.
4. Put the following in size order, from smallest to largest: organ, molecule, organelle, atom, organ system, tissue, organism, cell, macromolecule.
5. Why is lung cancer that has spread to the mediastinum very dangerous?
6. You are building an android. Choose a human organ system, and state which materials you would use to model it in the android.
7. In health, body parts interact to maintain homeostasis. Illness can threaten the maintenance of homeostasis, requiring treatment. What treatments might be used to help control a patient's (*a*) body temperature, (*b*) blood oxygen concentration, and (*c*) water content?
8. Suppose two individuals develop benign (noncancerous) tumors that produce symptoms because they occupy space and crowd adjacent organs. If one of these persons has the tumor in the ventral cavity and the other has the tumor in the dorsal cavity, which would be likely to develop symptoms first? Why?
9. If a patient complained of a "stomachache" and pointed to the umbilical region as the site of discomfort, what organs located in this region might be the source of the pain?
10. How might health-care professionals provide the basic requirements of life to an unconscious patient?

Reference Plates

The Human Organism

The series of illustrations that follows shows the major parts of the human torso. The first plate illustrates the anterior surface and reveals the superficial muscles on one side. Each subsequent plate exposes deeper organs, including those in the thoracic, abdominal, and pelvic cavities.

Chapters 6–19 of this textbook describe the organ systems of the human organism in detail. As you read them, refer to these plates to visualize the locations of various organs and the three-dimensional relationships among them.

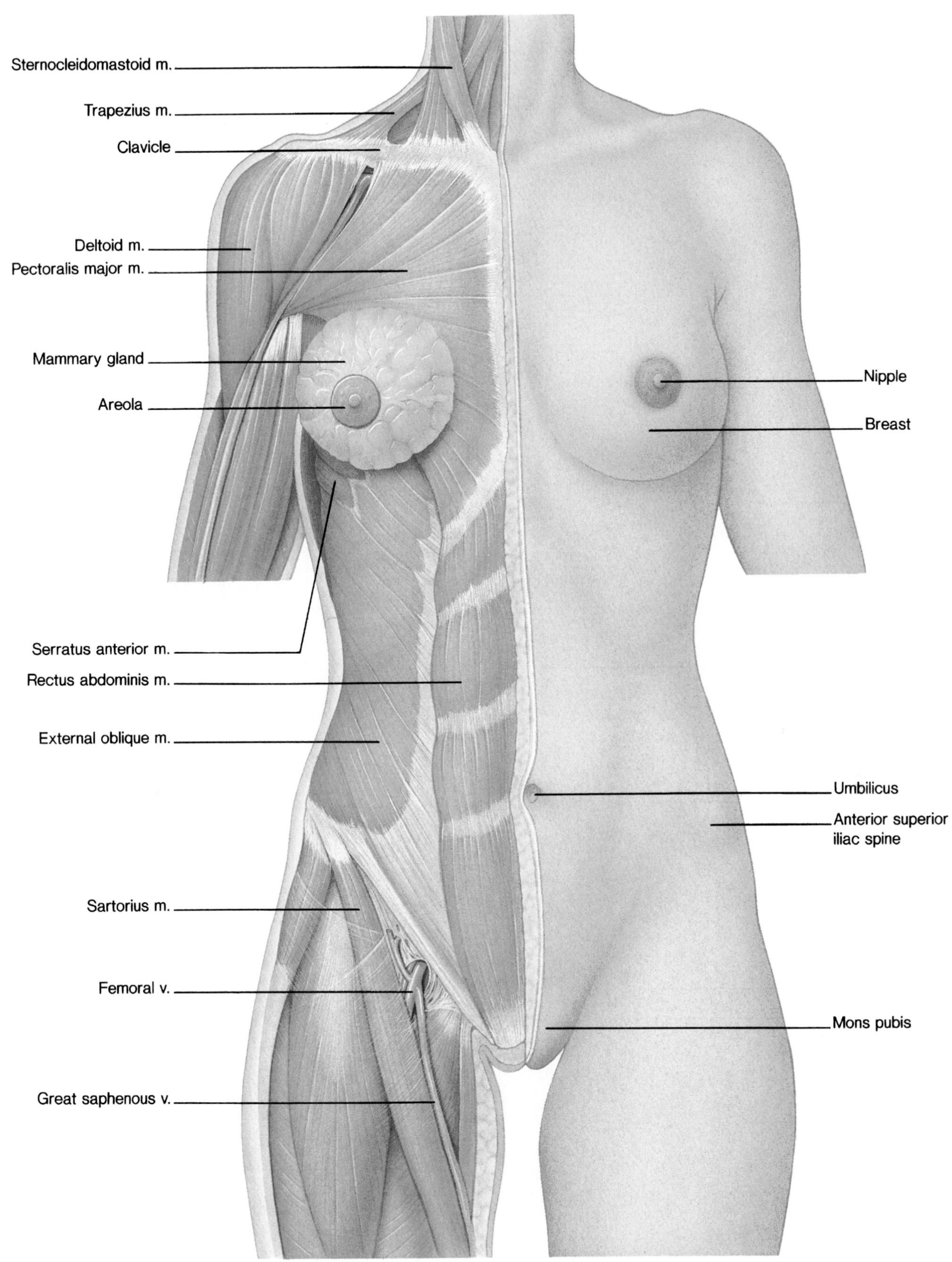

Plate 1

Human female torso, showing the anterior surface on one side and the superficial muscles exposed on the other side. (*m.* stands for *muscle,* and *v.* stands for *vein.*)

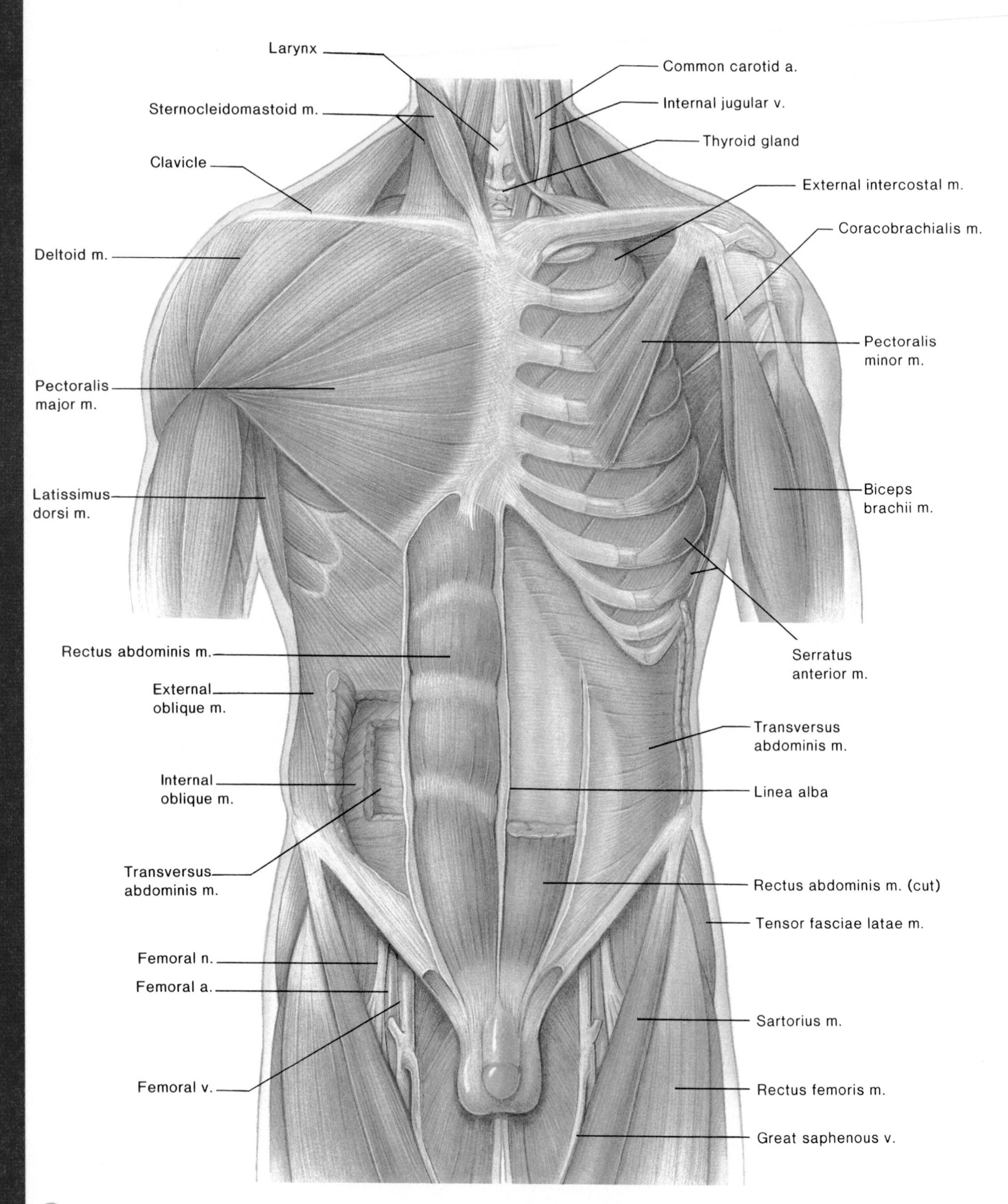

Plate 2

Human male torso with the deeper muscle layers exposed. (*n.* stands for *nerve,* and *a.* stands for *artery.*)

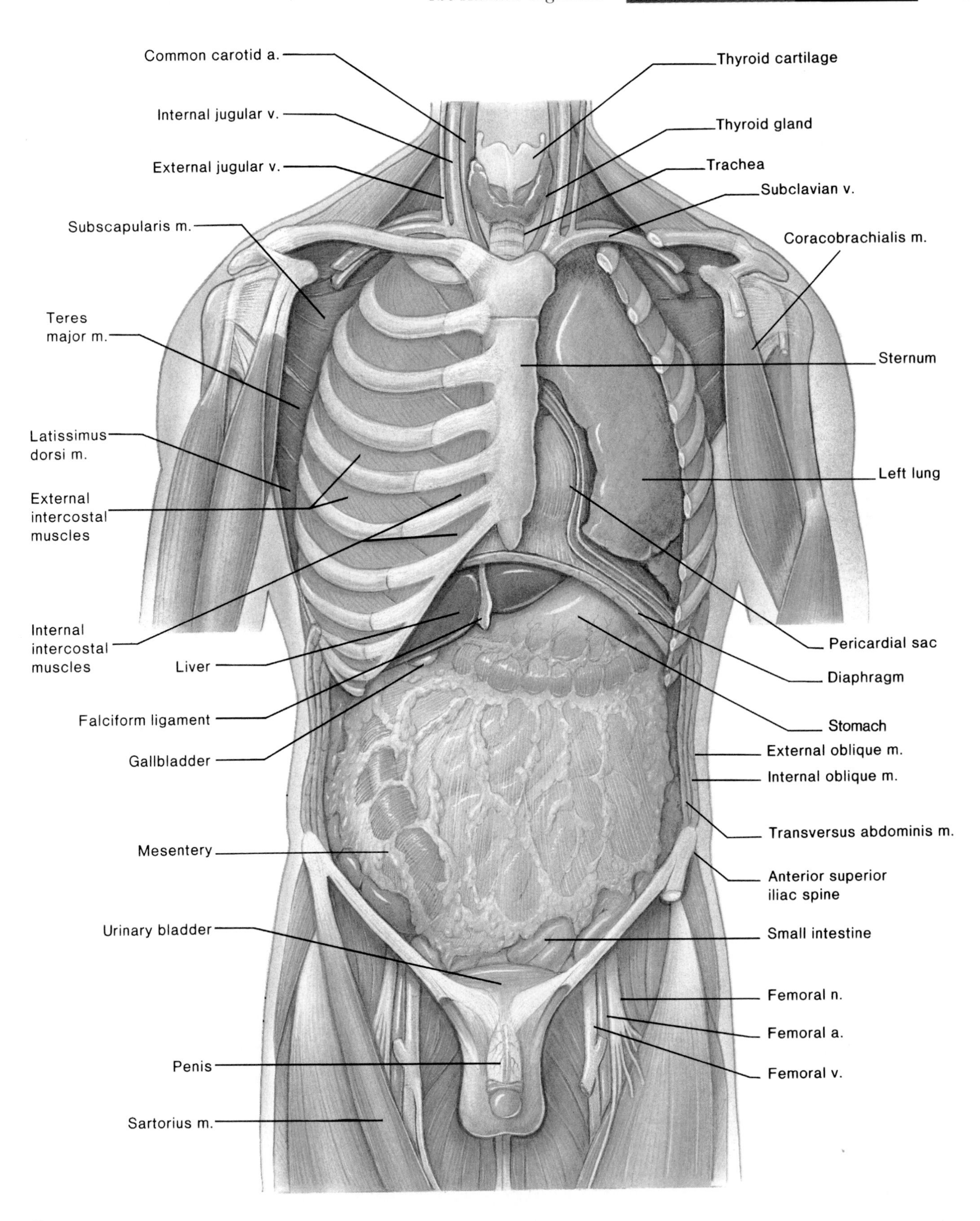

Plate 3

Human male torso with the deep muscles removed and the abdominal viscera exposed.

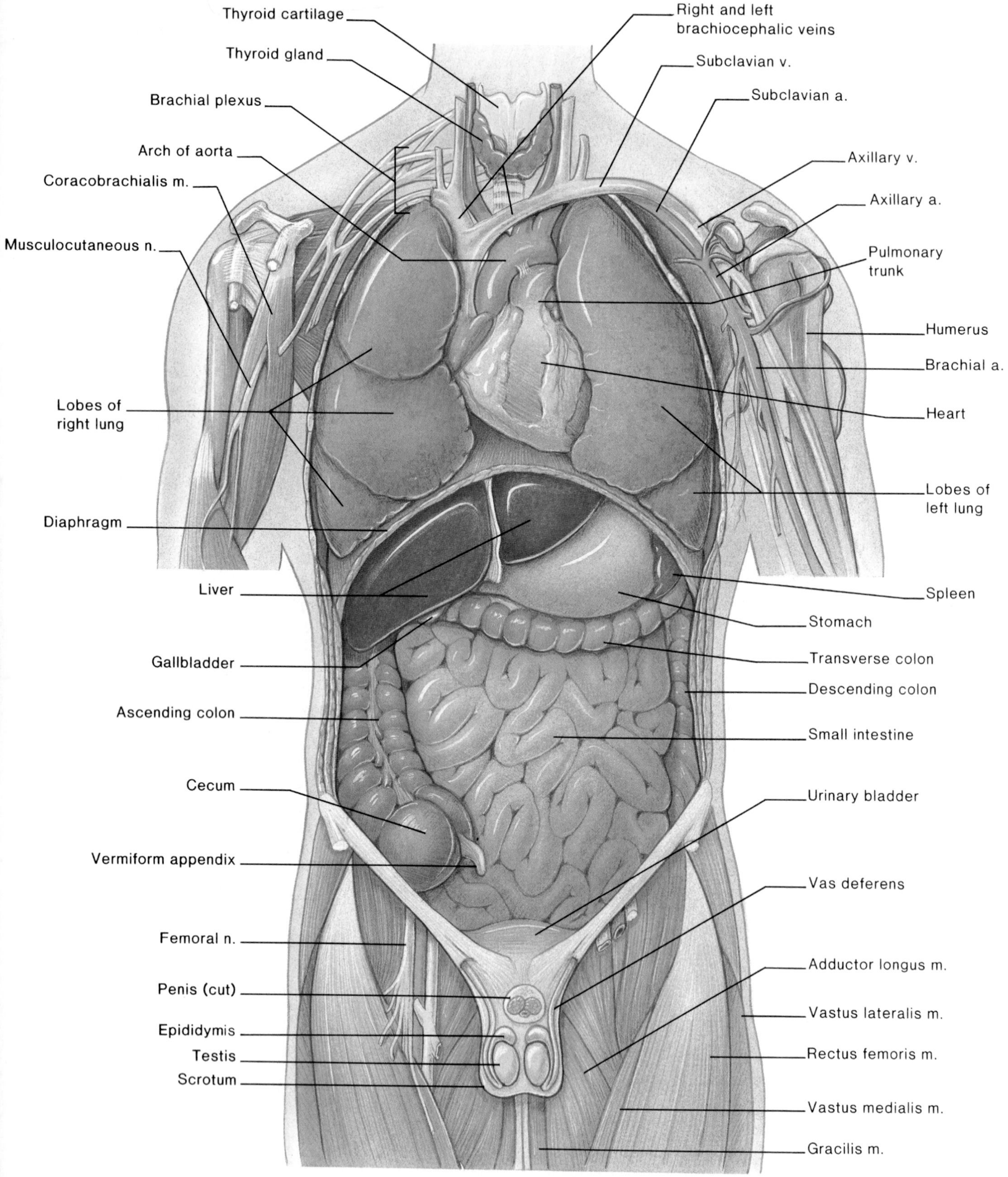

Plate 4

Human male torso with the thoracic and abdominal viscera exposed.

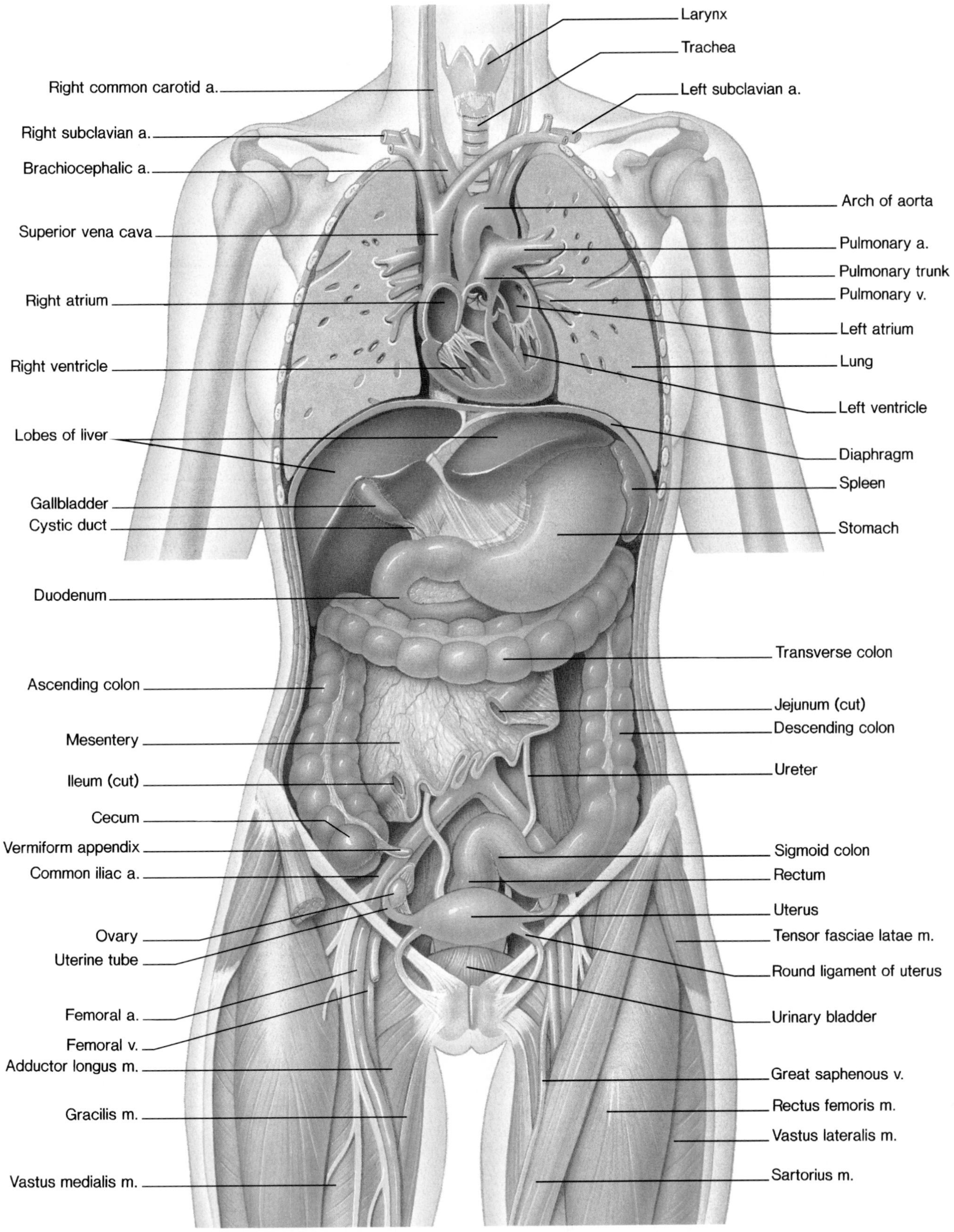

Larynx
Trachea
Right common carotid a.
Left subclavian a.
Right subclavian a.
Brachiocephalic a.
Arch of aorta
Superior vena cava
Pulmonary a.
Pulmonary trunk
Right atrium
Pulmonary v.
Left atrium
Right ventricle
Lung
Left ventricle
Lobes of liver
Diaphragm
Spleen
Gallbladder
Cystic duct
Stomach
Duodenum
Transverse colon
Ascending colon
Jejunum (cut)
Descending colon
Mesentery
Ileum (cut)
Ureter
Cecum
Vermiform appendix
Sigmoid colon
Common iliac a.
Rectum
Uterus
Ovary
Tensor fasciae latae m.
Uterine tube
Round ligament of uterus
Femoral a.
Urinary bladder
Femoral v.
Adductor longus m.
Great saphenous v.
Rectus femoris m.
Gracilis m.
Vastus lateralis m.
Vastus medialis m.
Sartorius m.

Plate 5

Human female torso with the lungs, heart, and small intestine sectioned.

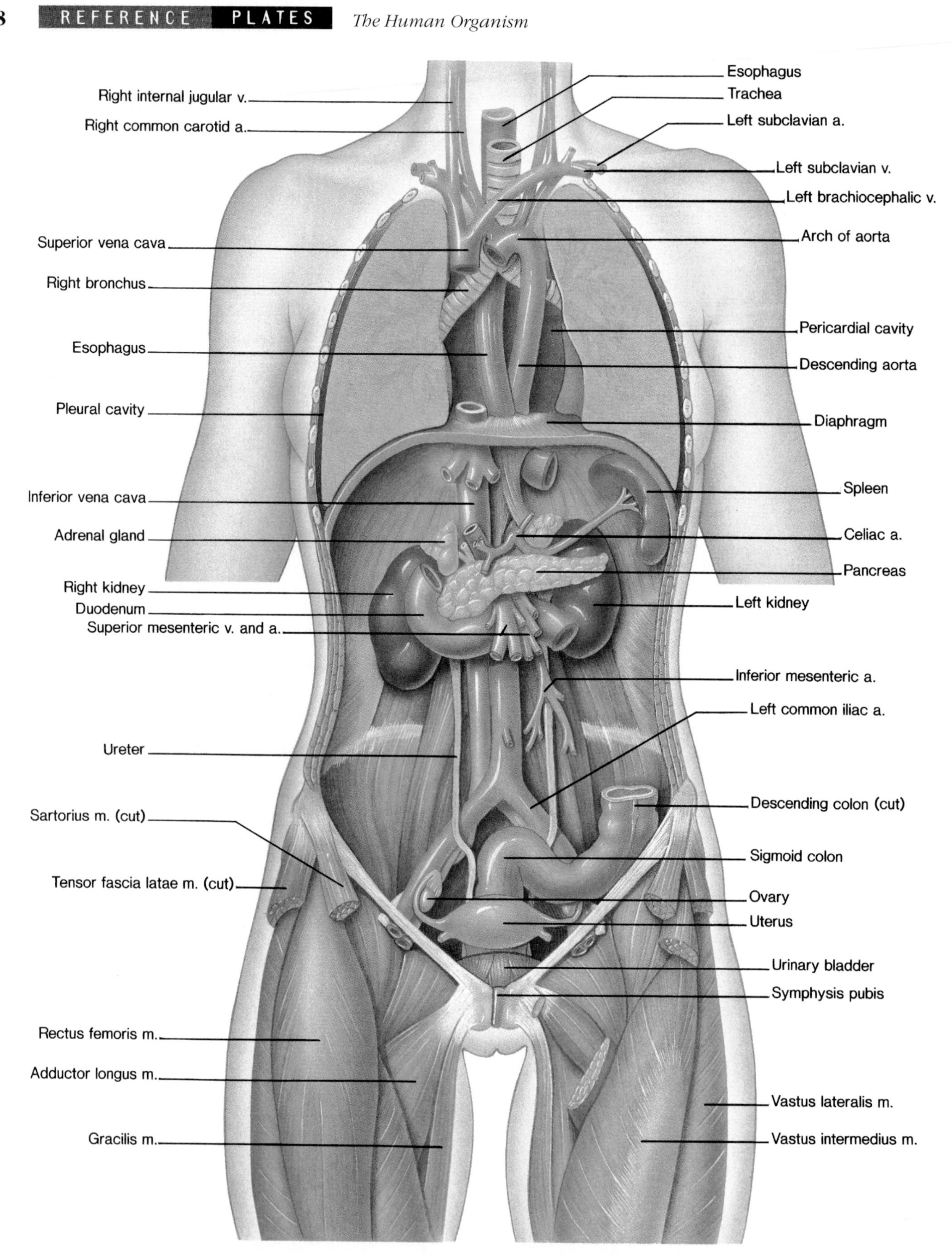

Plate 6

Human female torso with the heart, stomach, and parts of the intestine and lungs removed.

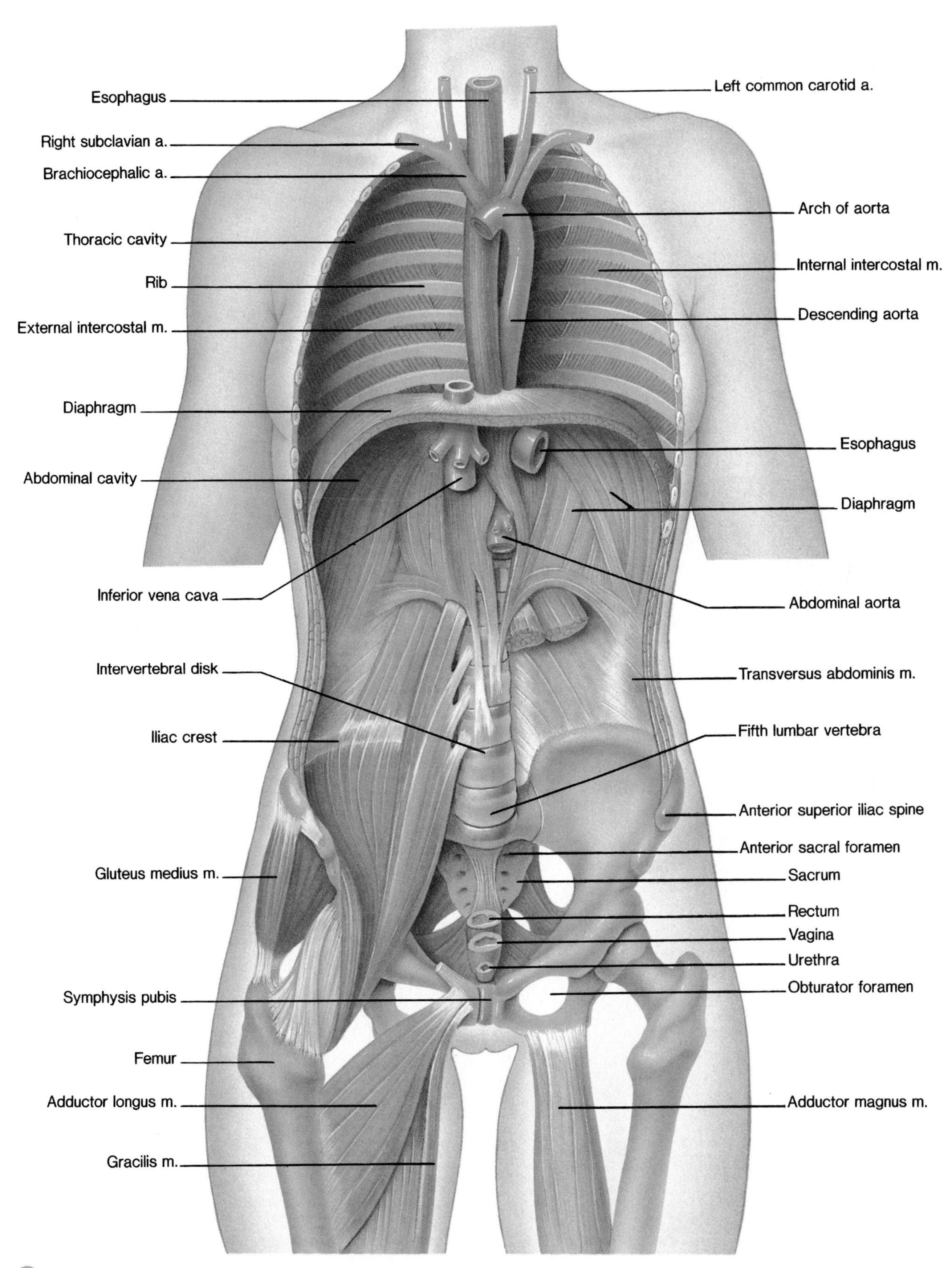

Plate 7

Human female torso with the thoracic, abdominal, and pelvic visceral organs removed.

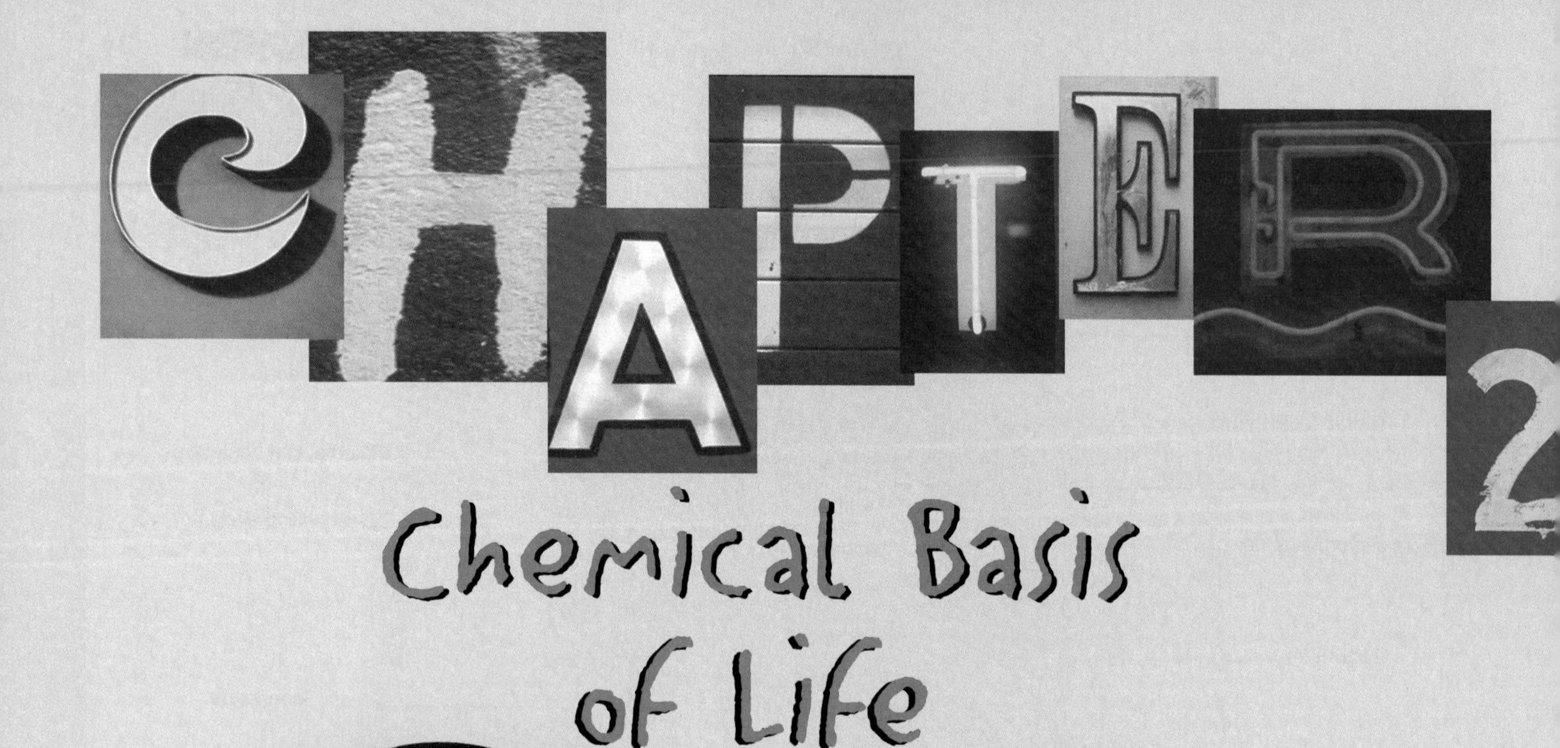

Chapter 2 Chemical Basis of Life

Lisa and Paul never expected to become intensely curious about chemistry. But when their young son was diagnosed with muscular dystrophy, they suddenly found themselves learning all about two biochemicals: proteins and nucleic acids.

Six-year-old Griffin, while not appearing ill, was clumsy and not well coordinated, walking up or down stairs haltingly. He also had an odd way of rising from a chair—pushing off with his arms, rather than using his legs. Lisa mentioned this to Griffin's physician, who grew alarmed. She asked Griffin to walk and sit in certain ways, asked questions about relatives with muscle problems, and then took a sample of Griffin's blood.

The diagnosis was devastating—Duchenne muscular dystrophy. Griffin would gradually lose muscle function, needing a wheelchair by adolescence. Despite exercises to increase his mobility and drug treatments to ease symptoms, Griffin probably would not live past age twenty.

Anxious to learn how their son's muscular system was breaking down, Lisa and Paul pored over medical journals and textbooks. They learned that Griffin's muscle cells were collapsing from lack of a protein called dystrophin.

Photo: The whole-body symptoms of muscular dystrophy stem from a missing protein, called dystrophin, in muscle cells. The pale cells here lack dystrophin. The blue cells have been given a gene for dystrophin, in an attempt at gene therapy.

Dystrophin is critical to muscle structure and function. Shaped like a rope, dystrophin attaches to the inner face of the cell membrane, which is a fatty layer enveloping the cell. One end of dystrophin dips into the inner protein filaments that fill the muscle cell. These filaments slide past each other to provide muscle action. The other end of dystrophin in the cell membrane contacts other proteins there, which, in turn, anchor the muscle cell to its surroundings.

Dystrophin's overall role is to hold the muscle cell firmly in place, much like a thumbtack holds a poster to a wall. Without the tack, the poster falls down. Likewise, without dystrophin, muscle cells collapse. When many such cells fall in on themselves, a child cannot run and play. Eventually, the heart and breathing muscles fail, too.

Dystrophin was not the only chemical causing Griffin's illness. A blood test showed that Lisa carries a gene causing the disease, which she had passed to Griffin on her X chromosome. A gene is a long chain of a chemical called deoxyribonucleic acid, or DNA. Lisa's dystrophin gene is too short, but she is healthy because she has a second, functional copy of the gene on her second X chromosome. Griffin, however, inherited only the defective dystrophin gene because, being a boy, he has only one X chromosome.

Although DNA has caused Griffin's illness, it may also hold the key to helping him, and others. Several medical centers are testing gene therapies that introduce immature muscle cells that can manufacture dystrophin into the muscles of boys with muscular dystrophy. Progress has been slow, but by understanding the chemical basis of the disorder, medical researchers know the enemy, and that knowledge will suggest new treatment strategies.

At the cellular level of organization, chemistry, in a sense, becomes biology. A cell's working parts—its organelles—are intricate assemblies of macromolecules. Because the macromolecules that build the cells that build tissues and organs are themselves composed of atoms, the study of anatomy and physiology begins with chemistry.

Chapter Objectives

After studying this chapter, you should be able to:

1. Explain how the study of living material depends on the study of chemistry. (p. 32)
2. Discuss how atomic structure determines how atoms interact. (p. 34)
3. Describe the relationships between atoms and molecules. (p. 37)
4. Explain how molecular and structural formulas symbolize the composition of compounds. (p. 38)
5. Describe three types of chemical reactions. (p. 38)
6. Discuss the concept of pH. (p. 39)
7. List the major groups of inorganic chemicals common in cells. (p. 40)
8. Describe the general roles that various types of organic chemicals play in cells. (p. 40)

Aids to Understanding Words

di- [two] *di*saccharide: A compound whose molecules are composed of two joined saccharide units.

glyc- [sweet] *glyc*ogen: A complex carbohydrate composed of many joined sugar molecules.

lip- [fat] *lip*ids: A group of organic compounds that includes fats.

-lyt [dissolvable] electro*lyte*: A substance that dissolves in water and releases ions.

mono- [one] *mono*saccharide: A compound whose molecules consist of a single saccharide unit.

poly- [many] *poly*unsaturated: A molecule with many double bonds between its carbon atoms.

sacchar- [sugar] mono*sacchar*ide: A sugar molecule composed of a single saccharide unit.

syn- [together] *syn*thesis: A process by which chemicals join to form new types of chemicals.

Key Terms

atom (at′om)
carbohydrate (kar″bo-hi′drāt)
decomposition (de″kom-po-zish′un)
electrolyte (e-lek′tro-līt)
inorganic (in″or-gan′ik)
ion (i′on)
lipid (lip′id)
molecular formula (mo-lek′u-lar for′mu-lah)
molecule (mol′ĕ-kūl)
nucleic acid (nu-kle′ik as′id)
organic (or-gan′ik)
protein (pro′tēn, pro′te-in)
structural formula (struk′cher-ol for′mu-lah)
synthesis (sin′thĕ-sis)

Introduction

Chemistry is the branch of science that considers the composition of matter and how this composition changes. While not required for understanding anatomy, chemistry is essential for understanding physiology because body functions result from chemical changes within cells.

Structure of Matter

Matter is anything that has weight and takes up space. This includes all of the solids, liquids, and gases in our surroundings, as well as inside our bodies.

Elements and Atoms

All matter is composed of basic substances called **elements.** At present, 111 elements are known, although naturally occurring matter on earth includes only 92 of them. Among these elements are such common materials as iron, copper, silver, gold, aluminum, carbon, hydrogen, and oxygen. Although some elements exist in a pure form, they occur more frequently in mixtures or chemical combinations.

Living organisms require about twenty elements. Of these, oxygen, carbon, hydrogen, and nitrogen make up more than 95% (by weight) of the human body (table 2.1). As the table shows, a one- or two-letter symbol represents each element.

Elements are composed of tiny particles called **atoms**, which are the smallest complete units of elements. Atoms of an element are similar to each other, but they differ from the atoms that make up other elements. Atoms vary in size, weight, and the ways they interact with each other. Some atoms, for instance, can combine with atoms like themselves or with other kinds of atoms, while other atoms cannot.

Table 2.1 Elements in the Human Body

Major Elements	Symbol	Approximate Percentage of the Human Body (by weight)
Oxygen	O	65.0%
Carbon	C	18.5
Hydrogen	H	9.5
Nitrogen	N	3.2
Calcium	Ca	1.5
Phosphorus	P	1.0
Potassium	K	0.4
Sulfur	S	0.3
Chlorine	Cl	0.2
Sodium	Na	0.2
Magnesium	Mg	0.1
		Total 99.9%
Trace Elements		
Chromium	Cr	
Cobalt	Co	
Copper	Cu	
Fluorine	F	Together less than 0.1%
Iodine	I	
Iron	Fe	
Manganese	Mn	
Zinc	Zn	

Atomic Structure

An atom consists of a central portion, called the **nucleus**, and one or more **electrons** that constantly move around it. The nucleus contains one or more relatively large particles called **protons.** The nucleus also usually contains one or more **neutrons**, which are similar in size to protons.

Electrons, which are extremely small, each carry a single, negative electrical charge (e^-), while protons each carry a single, positive electrical charge (p^+). Neutrons are uncharged and thus are electrically neutral (n^0) (fig. 2.1).

Because the nucleus contains the protons, it is always positively charged. However, the number of electrons outside the nucleus equals the number of protons. Therefore, a complete atom is electrically uncharged, or neutral.

The atoms of different elements contain different numbers of protons. The number of protons in the atoms of a particular element is called the element's **atomic number.** Hydrogen, for example, whose atoms contain one proton, has the atomic number 1; and carbon, whose atoms have six protons, has the atomic number 6.

The weight of an atom of an element is due primarily to the protons and neutrons in its nucleus; electrons have very little weight. For this reason, an atom of carbon with six protons and six neutrons weighs about twelve times more than an atom of hydrogen, which has only one proton and no neutrons.

The number of protons plus the number of neutrons in each of an element's atoms approximately equals the element's **atomic weight.** Thus, the atomic weight of hydrogen is 1, and the atomic weight of carbon is 12 (table 2.2).

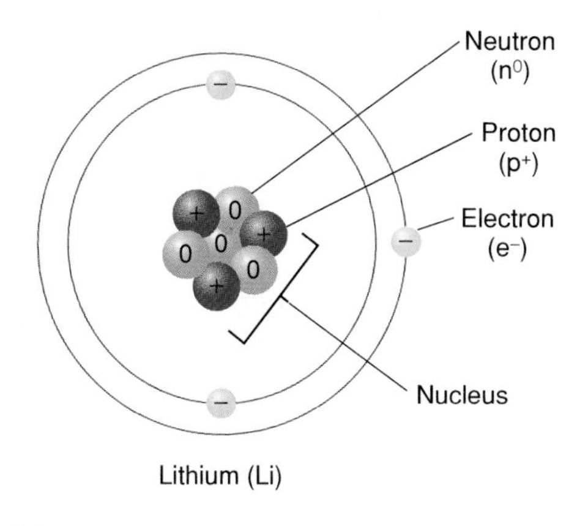

Figure 2.1

In an atom of lithium, three electrons move around a nucleus that consists of three protons and four neutrons.

1. What is the relationship between matter and the elements?
2. Which elements are most common in the human body?
3. Where are electrons, protons, and neutrons located within an atom?
4. What is the difference between atomic number and atomic weight?

Table 2.2

Atomic Structure of Elements 1 through 12

Element	Symbol	Atomic Number	Atomic Weight	Protons	Neutrons	Electrons in Shells		
						First	*Second*	*Third*
Hydrogen	H	1	1	1	0	1		
Helium	He	2	4	2	2	2 (inert)		
Lithium	Li	3	7	3	4	2	1	
Beryllium	Be	4	9	4	5	2	2	
Boron	B	5	11	5	6	2	3	
Carbon	C	6	12	6	6	2	4	
Nitrogen	N	7	14	7	7	2	5	
Oxygen	O	8	16	8	8	2	6	
Fluorine	F	9	19	9	10	2	7	
Neon	Ne	10	20	10	10	2	8 (inert)	
Sodium	Na	11	23	11	12	2	8	1
Magnesium	Mg	12	24	12	12	2	8	2

TOPIC OF INTEREST: Radioactive Isotopes

All the atoms of a particular element have the same atomic number because they have the same number of protons and electrons. However, the atoms of an element vary in the number of neutrons in their nuclei; thus, they vary in atomic weight. For example, all oxygen atoms have eight protons in their nuclei. Some, however, have eight neutrons (atomic weight 16), others have nine neutrons (atomic weight 17), and still others have ten neutrons (atomic weight 18). Atoms with the same atomic numbers but different atomic weights are called **isotopes** of an element.

How atoms interact with one another is due largely to their number of electrons. Because the number of electrons in an atom is equal to its number of protons, all the isotopes of a particular element have the same number of electrons and react chemically in the same manner. Therefore, any of the isotopes of oxygen can play the same role in an organism's metabolic reactions.

Isotopes may be stable, or they may have unstable atomic nuclei that decompose, releasing energy or pieces of themselves. Unstable isotopes are called *radioactive,* and the energy or atomic fragments they give off are called *atomic radiations.*

Atomic radiations include three common forms called alpha (α), beta (β), and gamma (γ). Alpha radiation consists of particles from atomic nuclei, each of which includes two protons and two neutrons, that travel relatively slowly and can weakly penetrate matter. Beta radiation consists of much smaller particles (electrons) that travel more rapidly and penetrate matter more deeply. Gamma radiation is similar to X-ray radiation and is the most penetrating of these forms.

Each kind of radioactive isotope produces one or more forms of radiation, and each becomes less radioactive at a particular rate. The time required for an isotope to lose one-half of its radioactivity is called its *half-life.* Thus, the isotope of iodine called iodine-131, which emits one-half of its radiation in 8.1 days, has a half-life of 8.1 days. Similarly, the half-life of phosphorus-32 is 14.3 days; that of cobalt-60 is 5.26 years; and that of radium-226 is 1,620 years.

Because atomic radiation can be detected with special equipment, such as a scintillation counter, radioactive substances are useful in studying life processes. (See fig. 2A.) A radioactive isotope, for example, can be introduced into an organism and then traced as it enters into metabolic activities. Because the human thyroid gland is unique in using the element iodine in its metabolism, radioactive iodine-131 can be used to study thyroid functions and to evaluate patients with thyroid disease (fig. 2B). Likewise, doctors commonly use thallium-201, which has a half-life of 73.5 hours, to assess heart conditions, and gallium-67, with a half-life of 78 hours, to detect and monitor the progress of certain cancers and inflammatory diseases.

Bonding of Atoms

When atoms combine with other atoms, they either gain or lose electrons, or share electrons with other atoms. The electrons of an atom occupy one or more *shells* around the nucleus (table 2.2). For the elements up to atomic number 18, the maximum number of electrons that each of the first three inner shells can hold is as follows:

First shell (closest to the nucleus)	2 electrons
Second shell	8 electrons
Third shell	8 electrons

More complex atoms may have as many as eighteen electrons in the third shell. Simplified diagrams, such as those in figure 2.2, depict electron locations within the shells of atoms.

Atoms such as helium, whose outermost electron shells are filled, have stable structures and are chemically inactive, or inert (table 2.2). Atoms such as hydrogen or lithium, whose outermost electron shells are incompletely filled, tend to gain, lose, or share electrons in ways that empty or fill their outer shells. In this way, they achieve stable structures.

An atom of sodium, for example, has eleven electrons (fig. 2.3). This atom tends to lose the single electron in its outer shell, which leaves the second shell filled and the form stable. A chlorine atom has seventeen electrons: two in the first shell, eight in the second shell, and seven in the third shell. An atom of this type

Atomic radiations also can change chemical structures, and in this way, alter vital cellular processes. For this reason, doctors sometimes use radioactive isotopes, such as cobalt-60, to treat cancers. The radiation from the cobalt preferentially kills the rapidly dividing cancer cells.

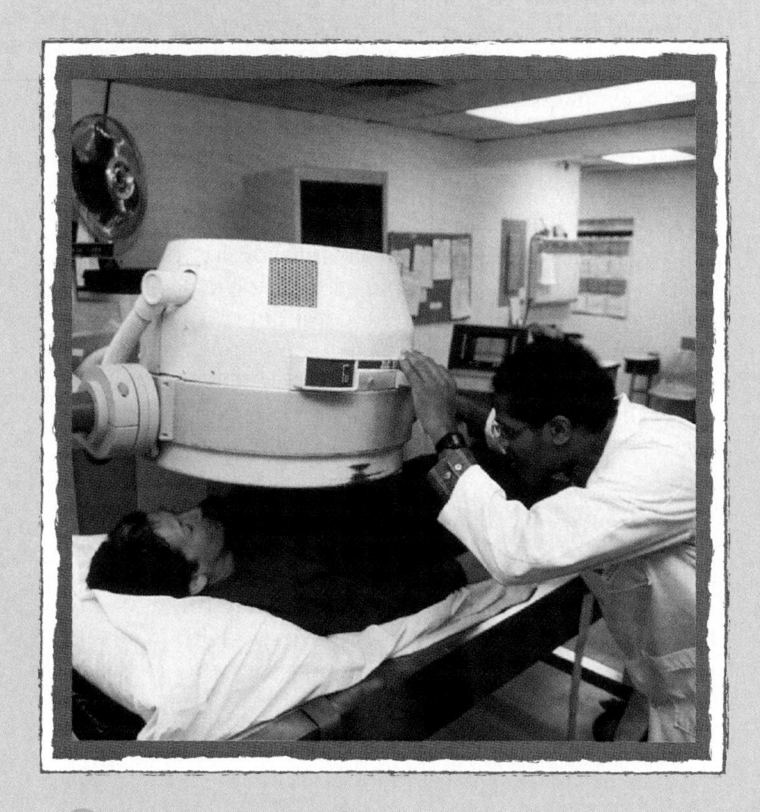

Figure 2A

Scintillation counters detect radioactive isotopes.

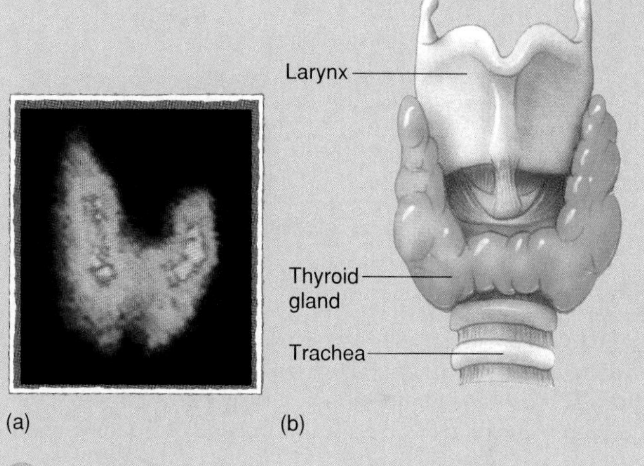

Figure 2B

(*a*) Scan of the thyroid gland 24 hours after the patient received radioactive iodine. Note how closely the scan in (*a*) resembles the shape of the thyroid gland, shown in (*b*).

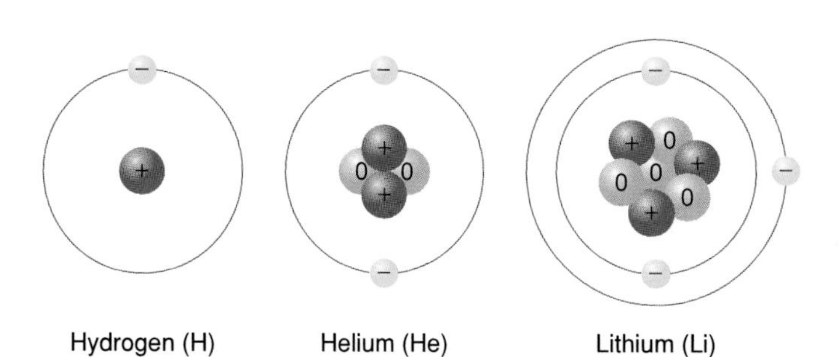

Figure 2.2

The single electron of a hydrogen atom is located in its first shell. The two electrons of a helium atom fill its first shell. Two of the three electrons of a lithium atom are in the first shell, and one is in the second shell.

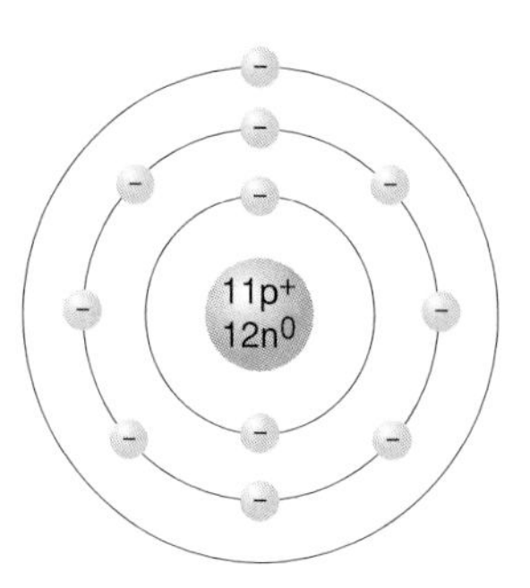

Figure 2.3

Sodium atom.

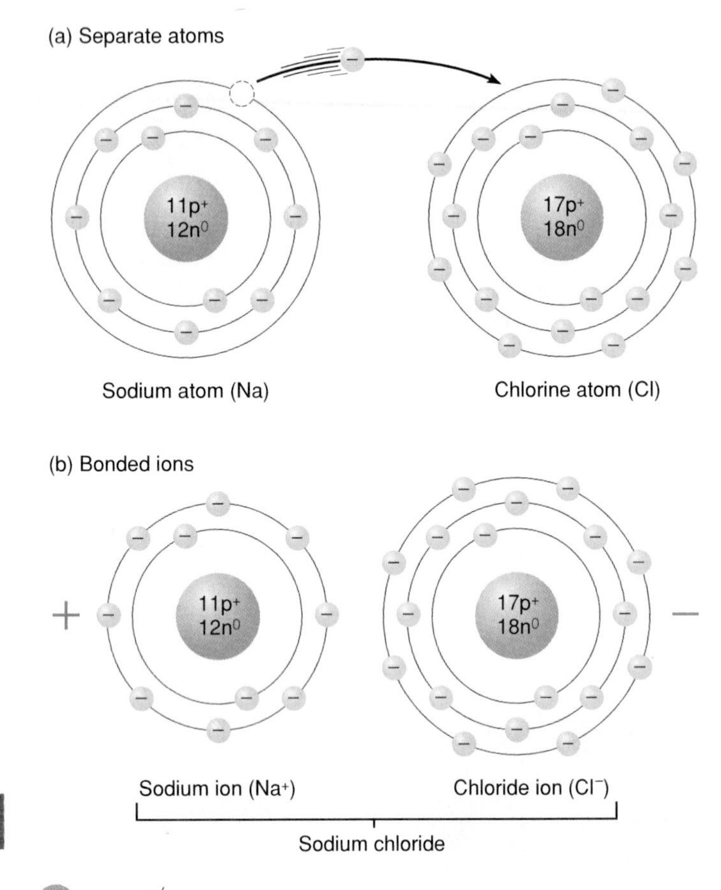

***Figure* 2.4**

(*a*) If a sodium atom loses an electron to a chlorine atom, the sodium atom becomes a sodium ion (Na^+), and the chlorine atom becomes a chloride ion (Cl^-). (*b*) These oppositely charged particles attract electrically and join by an ionic bond.

tends to accept a single electron, thus filling its outer shell and achieving a stable form (fig. 2.4).

Since each sodium atom tends to lose a single electron and each chlorine atom tends to accept a single electron, sodium and chlorine atoms will react together. During this reaction, a sodium atom loses an electron and is left with eleven protons (11^+) in its nucleus and only ten electrons (10^-). As a result, the atom develops a net electrical charge of 1^+ and is symbolized Na^+. At the same time, a chlorine atom gains an electron, which leaves it with seventeen protons (17^+) in its nucleus and eighteen electrons (18^-). Thus, it develops a net electrical charge of 1^- and is symbolized Cl^-.

Atoms that have become electrically charged by gaining or losing electrons are called **ions.** Ions with opposite electrical charges attract electrically. When this happens, a chemical bond, called an **ionic bond** (electrovalent bond), forms between the ions, as figure 2.4 shows. When sodium ions (Na^+) and chloride ions (Cl^-) unite in this manner, the compound sodium chloride (NaCl, common salt) forms. Some ions have more than one electrical charge, for example, Ca^{+2} (or Ca^{++}).

Atoms may also bond by sharing electrons, rather than by exchanging them. A hydrogen atom, for example, has one electron in its first shell but requires two electrons to achieve a stable structure (fig. 2.5). It may fill this shell by combining with another hydrogen atom in such a way that the two atoms share a pair of electrons. The two electrons then encircle the nuclei of both atoms, and each atom achieves a stable form. In this case, the chemical bond between the atoms is called a **covalent bond.**

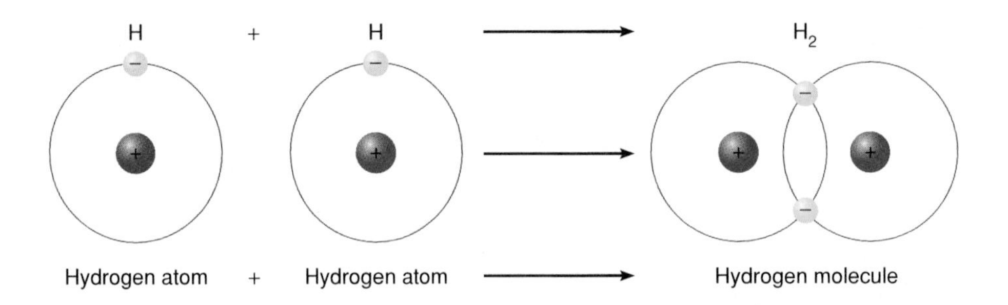

***Figure* 2.5**

A hydrogen molecule forms when two hydrogen atoms share a pair of electrons and join by a covalent bond.

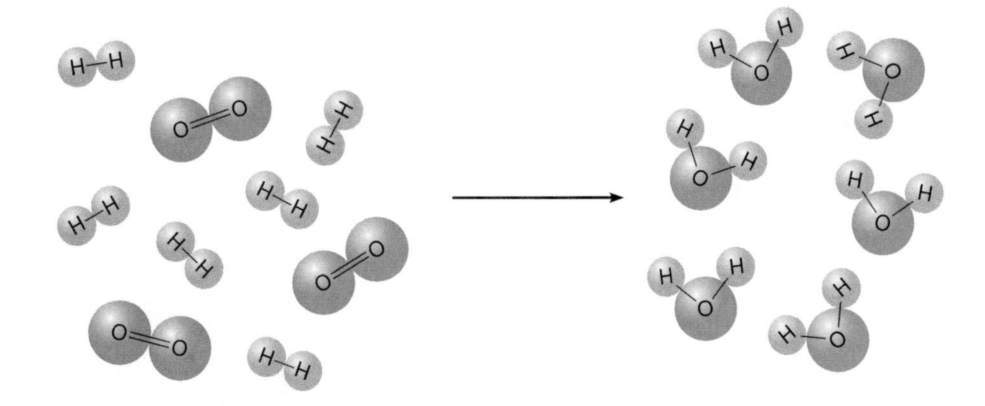

Figure 2.6

Hydrogen molecules can combine with oxygen molecules to form water molecules.

Carbon atoms, with two electrons in their first shells and four electrons in their second shells, form covalent bonds when they unite with other atoms. In fact, carbon atoms (and certain other atoms) may bond in such a way that two atoms share one or more pairs of electrons. If one pair of electrons is shared, the resulting bond is called a *single covalent bond;* if two pairs of electrons are shared, the bond is called a *double covalent bond.*

Another type of chemical bond, called a *hydrogen bond,* is described in this chapter's discussion of proteins.

Molecules and Compounds

When two or more atoms bond, they form a new kind of particle called a **molecule.** If atoms of the same element combine, they produce molecules of that element. Gases of hydrogen, oxygen, and nitrogen consist of such molecules (fig. 2.5).

When atoms of different elements combine, they form molecules called **compounds.** Two atoms of hydrogen, for example, can combine with one atom of oxygen to produce a molecule of the compound water (H_2O) (fig. 2.6). Table sugar (*sucrose*), baking soda, natural gas, beverage alcohol, and most drugs are compounds.

A molecule of a compound always contains definite kinds and numbers of atoms. A molecule of water, for instance, always contains two hydrogen atoms and one oxygen atom. If two hydrogen atoms combine with two oxygen atoms, the compound formed is not water, but hydrogen peroxide.

Table 2.3 lists some particles of matter and their characteristics.

Table 2.3 Some Particles of Matter

Particle	Characteristics
Atom	Smallest particle of an element that has the properties of that element
Electron (e^-)	Extremely small particle; carries a negative electrical charge and is in constant motion around a nucleus
Proton (p^+)	Relatively large particle; carries a positive electrical charge and is found within a nucleus
Neutron (n^0)	Relatively large particle; uncharged and thus electrically neutral; found within a nucleus
Ion	Atom that is electrically charged because it has gained or lost one or more electrons
Molecule	Particle formed by the chemical union of two or more atoms

1. What is an ion?
2. Describe two ways that atoms can combine with other atoms.
3. Distinguish between a molecule and a compound.

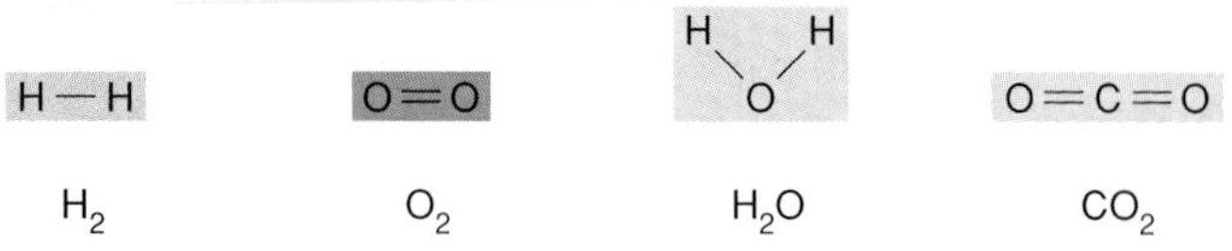

Figure 2.7

Structural and molecular formulas for molecules of hydrogen, oxygen, water, and carbon dioxide. Note the double covalent bonds.

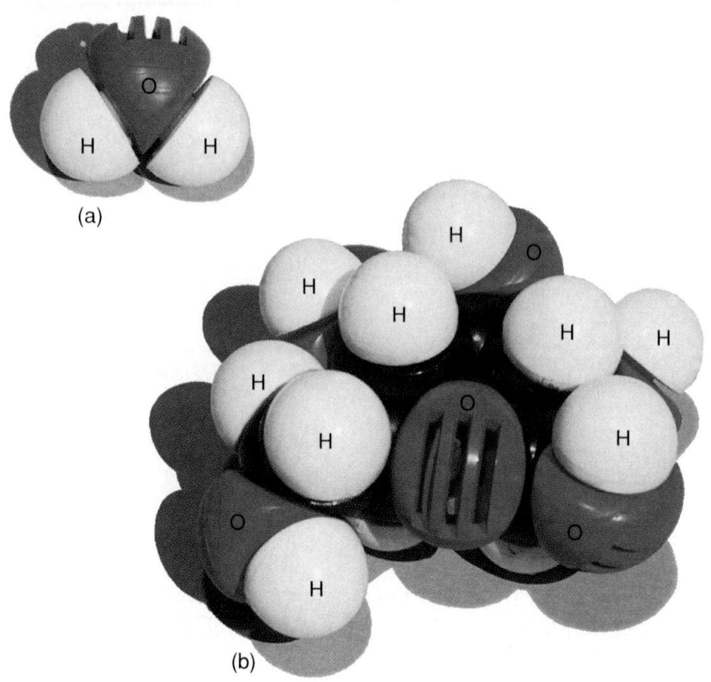

Figure 2.8

(*a*) This three-dimensional model represents a water molecule (H_2O), with the white parts depicting hydrogen atoms and the red part representing oxygen. (*b*) In this model of a glucose molecule ($C_6H_{12}O_6$), the black parts represent carbon atoms.

Formulas

A **molecular formula** represents the numbers and types of atoms in a molecule. Such a formula consists of the symbols for the elements in the molecule, together with numbers to indicate how many atoms of each element are present. For example, the molecular formula for water is H_2O, which means that each water molecule consists of two atoms of hydrogen and one atom of oxygen (fig. 2.7). The molecular formula for a sugar called glucose is $C_6H_{12}O_6$, which means that each glucose molecule consists of six atoms of carbon, twelve atoms of hydrogen, and six atoms of oxygen.

Usually, the atoms of each element will form a specific number of bonds. Hydrogen atoms form single bonds, oxygen atoms form two bonds, nitrogen atoms form three bonds, and carbon atoms form four bonds. Symbols and lines can be used as follows:

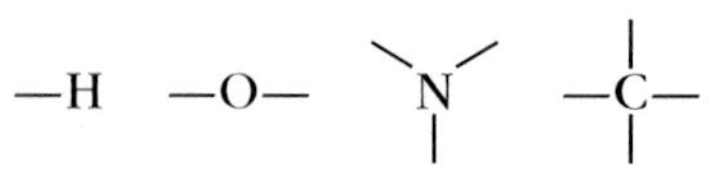

These representations depict how atoms are joined and arranged in various molecules. Single lines represent single bonds, and double lines represent double bonds. Illustrations of this type are called **structural formulas** (fig. 2.7). Three-dimensional models of structural formulas use different colors for the different kinds of atoms (fig. 2.8).

Chemical Reactions

Chemical reactions form or break bonds between atoms, ions, or molecules, generating new chemical combinations. For example, when two or more atoms (reactants) bond to form a more complex structure (end product), as when atoms of hydrogen and oxygen bond to form molecules of water, the reaction is called **synthesis.** Such a reaction is symbolized in this way:

$$A + B \rightarrow AB$$

If the bonds within a reactant molecule break so that simpler molecules, atoms, or ions form, the reaction is called **decomposition.** Thus, molecules of water can decompose to yield the products hydrogen and oxygen. Decomposition is symbolized as follows:

$$AB \rightarrow A + B$$

Synthesis, which requires energy, is particularly important in the growth of body parts and the repair of worn or damaged tissues, which require buildup of larger molecules from smaller ones. In contrast, decomposition occurs when foods are digested and energy is released from them.

A third type of chemical reaction is an **exchange reaction.** In this reaction, parts of two different types of molecules trade positions. The reaction is symbolized as follows:

$$AB + CD \rightarrow AD + CB$$

An example of an exchange reaction is when an acid reacts with a base, producing water and a salt.

Many chemical reactions are reversible. This means that the end product (or products) of the reaction can change back to the reactant (or reactants) that originally underwent the reaction. A **reversible reaction** is symbolized with a double arrow:

$$A + B \rightleftharpoons AB$$

Whether a reversible reaction proceeds in one direction or another depends on such factors as the relative proportions of the reactant (or reactants) and product (or products), as well as the amount of energy available to the reaction. The speed of the reaction may be affected by the presence or absence of **catalysts**—particular atoms or molecules that can change the rate of a reaction without themselves being consumed by the reaction.

Acids and Bases

Some compounds release ions when they dissolve in water or react with water molecules. Sodium chloride (NaCl), for example, releases sodium ions (Na^+) and chloride ions (Cl^-) when it dissolves:

$$NaCl \rightarrow Na^+ + Cl^-$$

Since the resulting solution contains electrically charged particles (ions), it will conduct an electric current. Substances that release ions in water are, therefore, called **electrolytes.** Electrolytes that release hydrogen ions (H^+) in water are called **acids.** For example, in water, the compound hydrochloric acid (HCl) releases hydrogen ions (H^+) and chloride ions (Cl^-):

$$HCl \rightarrow H^+ + Cl^-$$

Electrolytes that release ions that combine with hydrogen ions are called **bases.** The compound sodium hydroxide (NaOH), for example, releases hydroxyl ions (OH^-) when placed in water:

$$NaOH \rightarrow Na^+ + OH^-$$

The hydroxyl ions, in turn, can combine with hydrogen ions to form water; thus, sodium hydroxide is a base. (Note: Some ions, such as OH^-, contain two or more atoms. However, such a group behaves like a single atom and usually remains unchanged during a chemical reaction.)

The concentrations of hydrogen ions (H^+) and hydroxyl ions (OH^-) in body fluids greatly affect the chemical reactions that control certain physiological functions, such as blood pressure and breathing rate. Since their concentrations are inversely related (if one goes up the other goes down), we need keep track of only one of them. A value called **pH** measures hydrogen ion concentration.

The pH scale ranges from 0 to 14. A solution with a pH of 7.0, the midpoint of the scale, contains equal numbers of hydrogen and hydroxyl ions and is said to be *neutral.* A solution that contains more hydrogen than hydroxyl ions has a pH less than 7.0 and is *acidic.* A solution with fewer hydrogen than hydroxyl ions has a pH greater than 7.0 and is *basic* (alkaline). Figure 2.9 indicates the pH values of some common substances.

Each whole number on the pH scale represents a tenfold difference in the hydrogen ion concentration. For example, a solution of pH 4.0 contains 0.0001 grams of hydrogen ions per liter; a solution of pH 3.0 has 0.001 grams of hydrogen ions per liter, which is ten times as much as 0.0001 grams per liter ($0.0001 \times 10 = 0.001$).

Chapter 18 discusses the regulation of the hydrogen ion concentration in body fluids.

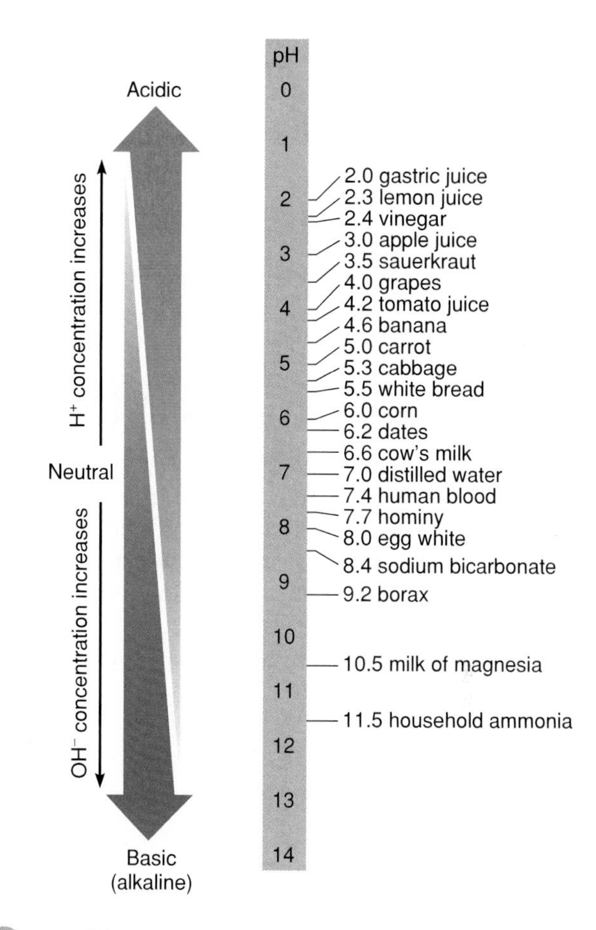

Figure 2.9

As the concentration of hydrogen ions (H^+) increases, a solution becomes more acidic, and the pH value decreases. As the concentration of ions that combine with hydrogen ions (such as hydroxyl ions) increases, a solution becomes more basic (alkaline), and the pH value increases.

Note in figure 2.9 that the pH of human blood is about 7.4 (the normal pH range is from 7.35 to 7.45). If this pH value drops below 7.35, the person is said to have *acidosis;* if it rises above 7.45, the condition is called *alkalosis.* Without medical intervention, a person usually cannot survive if blood pH drops to 6.9 or rises to 7.8 for more than a few hours. ■

1. What is a molecular formula? A structural formula?
2. Describe three kinds of chemical reactions.
3. Compare the characteristics of an acid with those of a base.
4. What does pH measure?

Chemical Constituents of Cells

Chemicals that enter into metabolic reactions or are produced by them can be divided into two large groups. Those that contain both carbon and hydrogen atoms are called **organic.** The rest are **inorganic.**

Generally, inorganic substances dissolve in water or react with water to release ions; thus, they are *electrolytes*. Some organic compounds also dissolve in water, but they are more likely to dissolve in organic liquids, such as ether or alcohol. Organic substances that dissolve in water usually do not release ions and are, therefore, called *nonelectrolytes*.

Inorganic Substances

Among the inorganic substances common in cells are water, oxygen, carbon dioxide, and a group of compounds called inorganic salts.

Water

Water is the most abundant compound in living material and accounts for about two-thirds of the weight of an adult human. It is the major component of blood and other body fluids, including those within cells.

Water is an important solvent because many substances readily dissolve in it. When a substance (*solute*) dissolves in water, relatively large pieces of the solute break into smaller ones, and eventually, molecular-sized particles result. These tiny particles, which may be ions, are much more likely to react with one another than were the original large pieces. Consequently, most metabolic reactions occur in water.

Water also plays an important role in moving chemicals within the body. The aqueous (watery) portion of blood, for example, carries many vital substances, such as oxygen, sugars, salts, and vitamins, from the organs of digestion and respiration to the body cells.

Water can absorb and transport heat. Thus, blood carries the heat released from muscle cells during exercise from deeper parts of the body to the surface, where it may be lost to the outside.

Oxygen

Molecules of oxygen (O_2) enter the body through the respiratory organs and are transported throughout the body by blood and red blood cells. Cellular organelles use oxygen to release energy from the sugar *glucose* and other nutrients. The released energy drives the cell's metabolic activities.

Carbon Dioxide

Carbon dioxide (CO_2) is a simple, carbon-containing compound of the inorganic group. It is produced as a waste product when certain metabolic processes release energy, and it is exhaled from the lungs.

Inorganic Salts

Inorganic salts are abundant in tissues and fluids. They provide many necessary ions, including sodium (Na^+), chloride (Cl^-), potassium (K^+), calcium (Ca^{+2}), magnesium (Mg^{+2}), phosphate (PO_4^{-3}), carbonate (CO_3^{-2}), bicarbonate (HCO_3^-), and sulfate (SO_4^{-2}). These ions are important in metabolic processes, including transport of substances into and out of cells, muscle contraction, and nerve impulse conduction.

Table 2.4 summarizes the functions of some of the inorganic substances in cells.

1. How do inorganic and organic molecules differ?
2. How do electrolytes and nonelectrolytes differ?
3. Name the inorganic substances common in body fluids.

Organic Substances

Important groups of organic substances in cells include carbohydrates, lipids, proteins, and nucleic acids.

Carbohydrates

Carbohydrates provide much of the energy that cells require. They supply materials to build certain cell structures and often are stored as reserve energy supplies.

Carbohydrate molecules contain atoms of carbon, hydrogen, and oxygen. These molecules usually have twice as many hydrogen as oxygen atoms—the same ratio of hydrogen to oxygen as in water molecules (H_2O). This ratio is easy to see in the molecular formulas of the carbohydrates glucose ($C_6H_{12}O_6$) and sucrose ($C_{12}H_{22}O_{11}$).

Table 2.4

Inorganic Substances Common in Cells

Substance	Symbol or Formula	Functions
I. Inorganic molecules		
Water	H_2O	Major component of body fluids (chapter 12); medium in which most biochemical reactions occur; transports chemicals (chapter 12); helps regulate body temperature (chapter 6)
Oxygen	O_2	Used in energy release from glucose molecules (chapter 4)
Carbon dioxide	CO_2	Waste product that results from metabolism (chapter 4); reacts with water to form carbonic acid (chapter 16)
II. Inorganic ions		
Bicarbonate ions	HCO_3^-	Helps maintain acid-base balance (chapter 18)
Calcium ions	Ca^{+2}	Necessary for bone development (chapter 7), muscle contraction (chapter 9), and blood clotting (chapter 12)
Carbonate ions	CO_3^{-2}	Component of bone tissue (chapter 7)
Chloride ions	Cl^-	Helps maintain water balance (chapter 18)
Magnesium ions	Mg^{+2}	Component of bone tissue (chapter 7); required for certain metabolic processes (chapter 16)
Phosphate ions	PO_4^{-3}	Required for synthesis of ATP, nucleic acids, and other vital substances (chapter 4); component of bone tissue (chapter 7); helps maintain polarization of cell membranes (chapter 9)
Potassium ions	K^+	Required for polarization of cell membranes (chapter 9)
Sodium ions	Na^+	Required for polarization of cell membranes (chapter 9); helps maintain water balance (chapter 18)
Sulfate ions	SO_4^{-2}	Helps maintain polarization of cell membranes (chapter 9) and acid-base balance (chapter 18)

The carbon atoms of carbohydrate molecules join in chains whose lengths vary with the type of carbohydrate. For example, carbohydrates with shorter chains are called **sugars.**

Sugars with six-carbon atoms (hexoses) are known as *simple sugars,* or **monosaccharides,** and they are the building blocks of more complex carbohydrate molecules. The simple sugars include glucose, fructose, and galactose. Figure 2.10 illustrates the structural formula of glucose.

In complex carbohydrates, a number of simple sugar molecules link to form molecules of varying sizes (fig. 2.11). Some complex carbohydrates, such as sucrose (table sugar) and lactose (milk sugar), are *double sugars,* or **disaccharides,** whose molecules each contain two simple-sugar building blocks. Others are made up of many simple-sugar units joined to form **polysaccharides,** such as plant starch. Animals, including humans, synthesize a polysaccharide similar to starch, called *glycogen.*

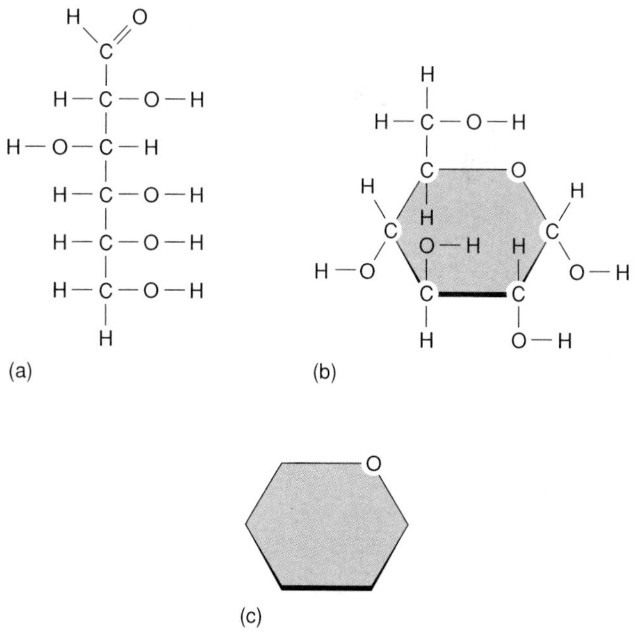

Figure 2.10

The pH values of some common substances are shown. (*a*) Some glucose molecules ($C_6H_{12}O_6$) have a straight chain of carbon atoms. (*b*) More commonly, glucose molecules form a ring structure. (*c*) This shape symbolizes the ring structure of a glucose molecule.

Lipids

Lipids are organic substances that are insoluble in water but soluble in certain organic solvents, such as ether and chloroform. Lipids include a variety of compounds—fats, phospholipids, and steroids—that have vital functions in cells. The most common lipids are fats.

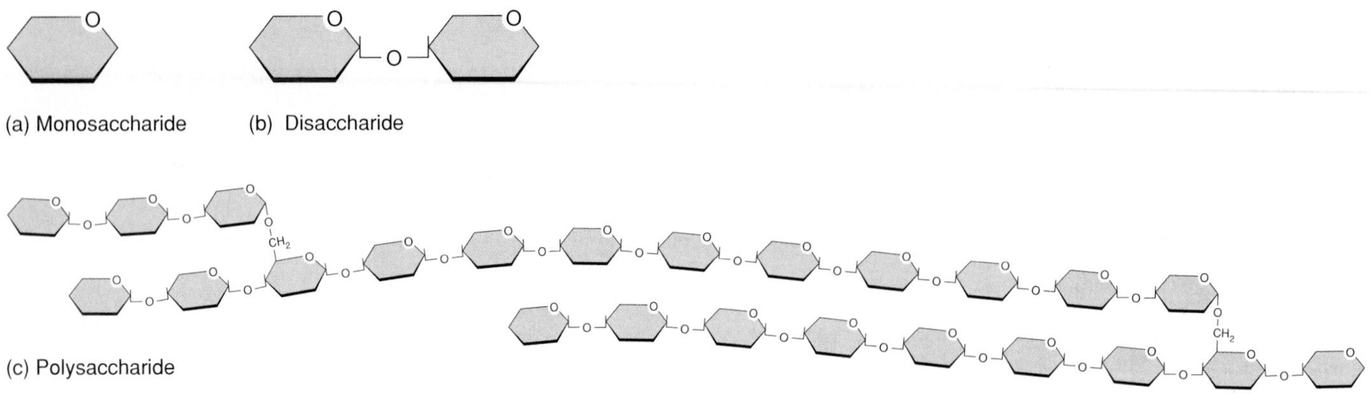

(a) Monosaccharide (b) Disaccharide

(c) Polysaccharide

Figure 2.11

(*a*) A monosaccharide molecule consists of one six-carbon atom building block. (*b*) A disaccharide molecule consists of two of these building blocks. (*c*) A polysaccharide molecule consists of many building blocks.

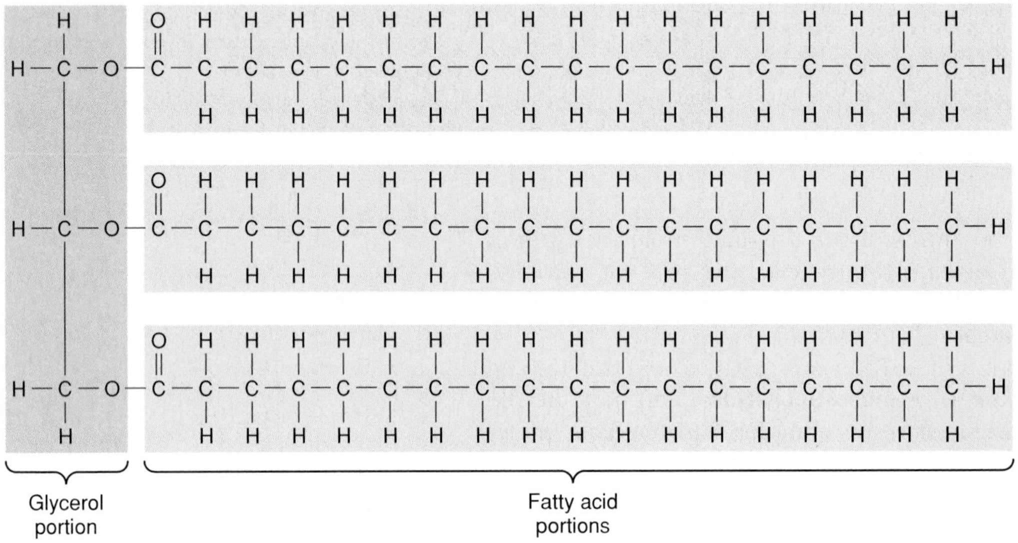

Glycerol portion

Fatty acid portions

Figure 2.12

A triglyceride molecule consists of a glycerol portion and three fatty acid portions.

Fats are used primarily to store energy for cellular activities. Fat molecules can supply more energy, gram for gram, than carbohydrate molecules.

Like carbohydrates, fat molecules are composed of carbon, hydrogen, and oxygen atoms. They contain, however, a much smaller proportion of oxygen than do carbohydrates. The formula for the fat tristearin, $C_{57}H_{110}O_6$, illustrates these characteristic proportions.

The building blocks of fat molecules are **fatty acids** and **glycerol.** Each glycerol molecule combines with three fatty acid molecules to produce a single fat, or *triglyceride,* molecule (fig. 2.12).

The glycerol portion of every fat molecule is the same, yet there are many kinds of fatty acids and, therefore, many kinds of fats. Fatty acid molecules differ in the lengths of their carbon atom chains, although such chains usually contain an even number of carbon

Table 2.5

Important Groups of Lipids

Group	Basic Molecular Structure	Characteristics
Triglycerides	Three fatty acid molecules bound to a glycerol molecule	Most common lipids in body; stored in fat tissue as an energy supply; fat tissue also provides insulation beneath the skin
Phospholipids	Two fatty acid molecules and a phosphate group bound to a glycerol molecule	Used as structural components in cell membranes; abundant in liver and parts of nervous system
Steroids	Four connected rings of carbon atoms	Widely distributed in body and have a variety of functions; include cholesterol, hormones of adrenal cortex, sex hormones, bile acids, and vitamin D

A diet high in saturated fats increases the chance of developing *atherosclerosis,* which obstructs arteries. For this reason, many nutritionists recommend that polyunsaturated fats replace some dietary saturated fats.

As a rule, saturated fats are more abundant in fatty foods that are solids at room temperature, such as butter, lard, and most other animal fats. Unsaturated fats, on the other hand, are likely to be plentiful in fatty foods that are liquids at room temperature, such as soft margarine and various seed oils, including corn, cottonseed, safflower, sesame, soybean, and sunflower oils. Exceptions include coconut and palm kernel oils, which are high in saturated fats. ■

atoms. The chains also may vary in the way the carbon atoms combine. In some cases, the carbon atoms all join by single carbon–carbon bonds. This type of fatty acid is said to be *saturated;* that is, each carbon atom is bound to as many hydrogen atoms as possible and is thus saturated with hydrogen atoms. Other fatty acid chains have not bound their maximum number of hydrogen atoms. Therefore, they have one or more double bonds between carbon atoms. Those with such double bonds are said to be *unsaturated,* and fatty acid molecules with many double-bonded carbon atoms are called *polyunsaturated.* Similarly, fat molecules that contain only saturated fatty acids are called *saturated fats,* and those that include unsaturated fatty acids are called *unsaturated fats.*

A **phospholipid** molecule is similar to a fat molecule in that it contains a glycerol portion and fatty acid chains. The phospholipid, however, has only two fatty acid chains; in place of the third is a portion containing a phosphate group. This phosphate portion is soluble in water (hydrophilic) and forms the "head" of the molecule, while the fatty acid portion is insoluble in water (hydrophobic) and forms a "tail." Phospholipids are important in cellular structures.

Steroid molecules are complex structures that include four connected rings of carbon atoms. Among the more important steroids are cholesterol, which occurs in all body cells and is used to synthesize other steroids; sex hormones, such as estrogen, progesterone, and testosterone; and several hormones from the adrenal glands. Chapters 11 and 19 discuss these steroids.

Table 2.5 lists three important groups of lipids and their characteristics.

Proteins

Some **proteins** serve as structural materials, energy sources, and hormones. Others combine with carbohydrates (glycoproteins) and function on cell surfaces as receptors that are specialized to bond to particular kinds of molecules. Other proteins (antibodies) act against foreign substances that enter the body. Still others—**enzymes**—catalyze vital metabolic processes. That is, they speed specific chemical reactions without being consumed by these reactions. (Enzymes are discussed in more detail in chapter 4.)

Amino acid	Structural formula
Alanine	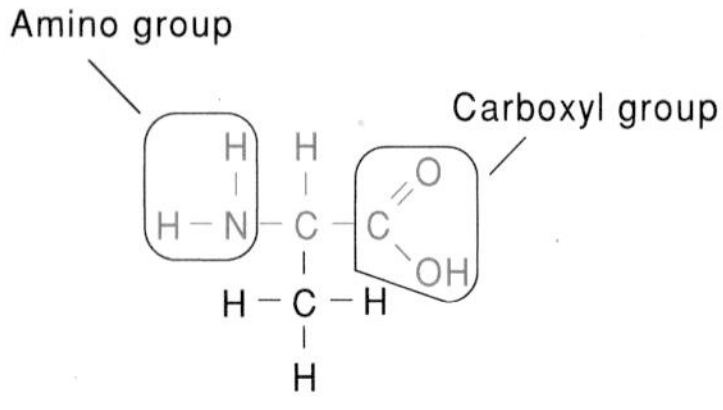 Amino group; Carboxyl group H H H–N–C–C=O OH H–C–H H
Valine	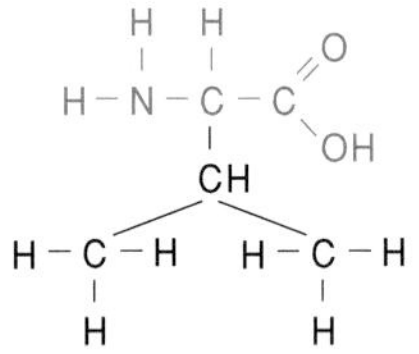 H H H–N–C–C=O OH CH H–C–H H–C–H H H
Cysteine	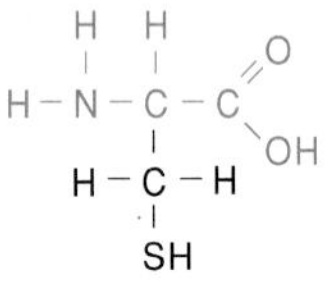 H H H–N–C–C=O OH H–C–H SH

***Figure* 2.13**

Some representative amino acids and their structural formulas. Each amino acid molecule has a particular shape. The portion shown in red is common to all amino acid molecules. (The remaining portion, in black, is unique to each amino acid type and is called an R group.)

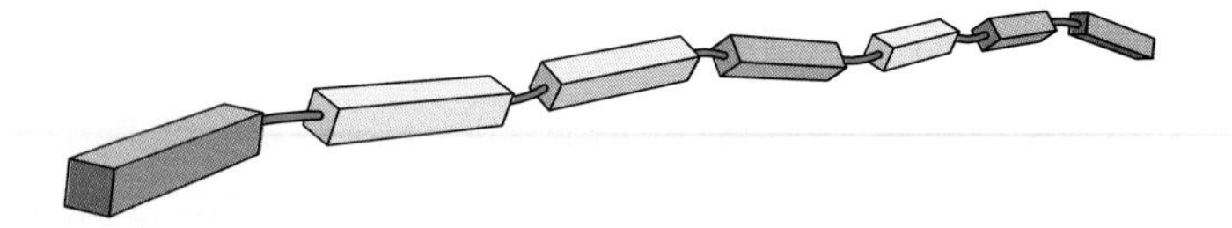

***Figure* 2.14**

Each different color in this chain represents a different kind of amino acid molecule. The whole chain represents a portion of a protein molecule.

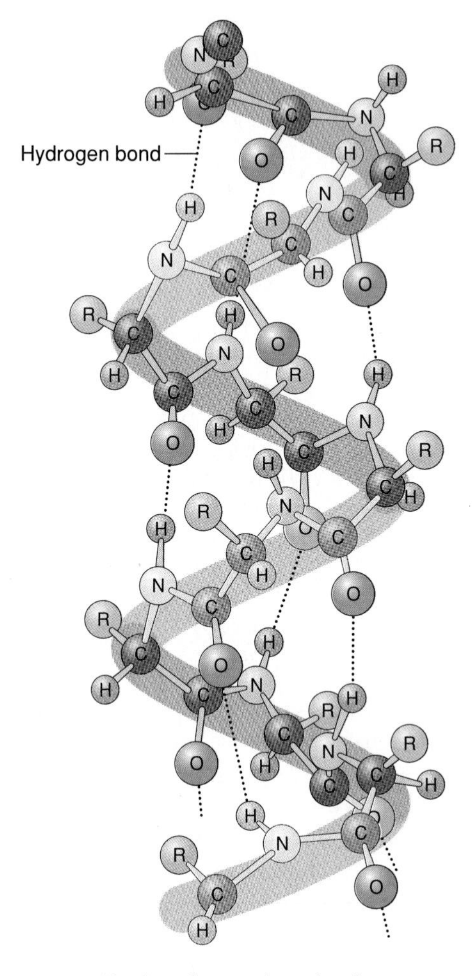

***Figure* 2.15**

The amino acid chain of a protein molecule is sometimes twisted to form a coil. The circles labeled "R" represent R groups (see fig. 2.13).

Like carbohydrates and lipids, proteins are composed of atoms of carbon, hydrogen, and oxygen. In addition, proteins always contain nitrogen atoms and, sometimes, sulfur atoms. The building blocks of proteins are smaller molecules called **amino acids**, each of which has an *amino group* (—NH_2) at one end and a *carboxyl group* (—COOH) at the other (fig. 2.13).

About twenty different kinds of amino acids occur commonly in the proteins of life. Within the protein molecules, the amino acids join in chains that vary in length from less than 100 to more than 50,000 amino acids (fig. 2.14).

Each protein molecule includes specific numbers and types of different amino acids in a particular linear sequence. This amino acid chain twists to form a coil (fig. 2.15), and it, in turn, may fold into a unique three-dimensional shape (fig. 2.16). Consequently, different kinds of protein molecules have different shapes, and their shapes make possible particular functions. Destroy a biological protein's three-dimensional form, and it can no longer function.

Hydrogen bonds maintain the special shape of a protein molecule by determining how the molecule binds within itself. A hydrogen bond is a weak electrical attraction between a hydrogen atom (covalently bound to a nitrogen or oxygen atom) and another nitrogen or oxygen atom nearby (see fig. 2.15).

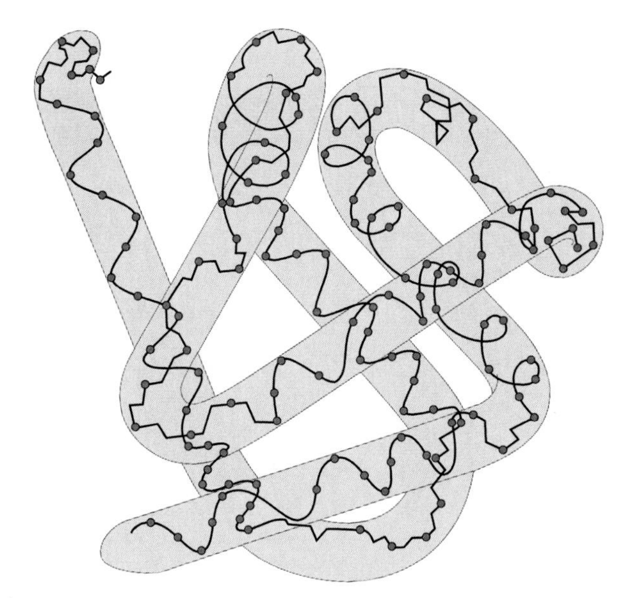

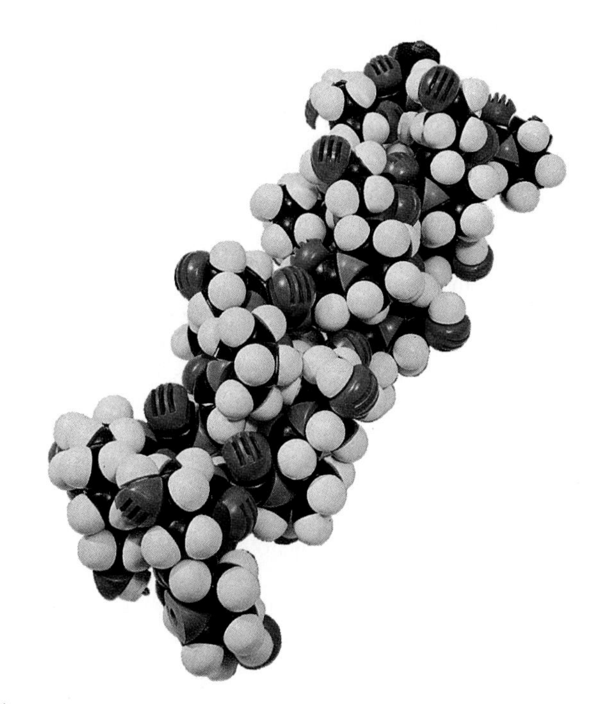

(b)

Figure 2.16

(*a*) The coiled amino acid chain of a protein molecule folds into a unique three-dimensional structure. (*b*) A model of a portion of the protein collagen.

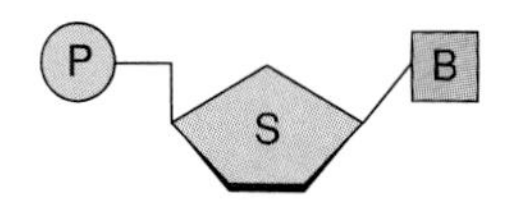

Figure 2.17

A nucleotide consists of a five-carbon sugar (S), a phosphate group (P), and an organic base (B).

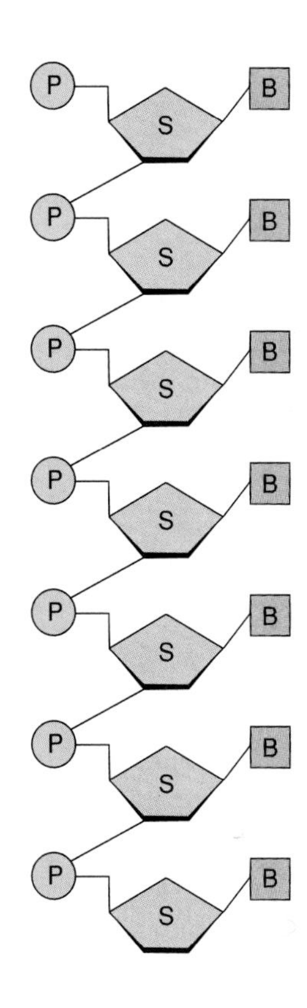

Figure 2.18

A nucleic acid molecule consists of nucleotides joined in a chain.

When hydrogen bonds break, as a result of exposure to excessive heat, radiation, electricity, pH changes, or various chemicals, this can change a protein's unique shape. Such proteins are said to be *denatured.* At the same time, the proteins lose their special properties. For example, the protein in egg white (albumin) is denatured when heated. This treatment causes the egg white to change from a liquid to a solid—an irreversible change. Similarly, cellular proteins that are denatured may be permanently changed and become nonfunctional.

Nucleic Acids

Nucleic acids form genes and take part in protein synthesis. These molecules are generally very large and complex. They contain atoms of carbon, hydrogen, oxygen, nitrogen, and phosphorus, which are bound into building blocks called **nucleotides.** Each nucleotide consists of a five-carbon *sugar* (ribose or deoxyribose), a *phosphate group,* and one of several *organic bases* (fig. 2.17). A nucleic acid molecule consists of a chain of many nucleotides (fig. 2.18).

Table 2.6

Organic Compounds in Cells

Compound	Elements Present	Building Blocks	Functions	Examples
Carbohydrates	C,H,O	Simple sugars	Provide energy, cell structure	Glucose, starch
Lipids	C,H,O (often P)	Glycerol, fatty acids, phosphate groups	Provide energy, cell structure	Triglycerides, phospholipids, steroids
Proteins	C,H,O,N (often S)	Amino acids	Provide cell structure, enzymes, energy	Albumins, hemoglobin
Nucleic acids	C,H,O,N,P	Nucleotides	Store information for protein synthesis; control cell activities	RNA, DNA

Nucleic acids are of two major types. One type—**RNA** (ribonucleic acid)—is composed of molecules whose nucleotides contain ribose. The second type—**DNA** (deoxyribonucleic acid)—contains deoxyribose.

DNA molecules store information in a type of molecular code. Cells use this information to synthesize protein molecules. RNA molecules help in protein synthesis. (Nucleic acids are discussed in more detail in chapter 4.)

Table 2.6 summarizes these four groups of organic compounds.

1. Compare the chemical composition of carbohydrates, lipids, proteins, and nucleic acids.
2. How does an enzyme affect a chemical reaction?
3. What is the chemical basis of the great diversity of proteins?
4. What are the functions of nucleic acids?

Summary Outline

Introduction (p. 32)

Chemistry describes the composition of substances and how chemicals react with each other.

Structure of Matter (p. 32)

1. Elements and atoms
   a. Naturally occurring matter on earth is composed of ninety-two elements.
   b. Some elements occur in pure form, but many are found in mixtures or are united in chemical combinations.
   c. Elements are composed of atoms, which are the smallest complete units of elements.
   d. Atoms of different elements vary in size, weight, and ways of interacting.
2. Atomic structure
   a. An atom consists of one or more electrons surrounding a nucleus, which contains one or more protons and usually contains one or more neutrons.
   b. Electrons are negatively charged, protons are positively charged, and neutrons are uncharged.
   c. A complete atom is electrically neutral.
   d. An element's atomic number is equal to the number of protons in each atom. The atomic weight is equal to the number of protons plus the number of neutrons in each atom.
3. Bonding of atoms
   a. When atoms combine, they gain, lose, or share electrons.
   b. Electrons occupy shells around a nucleus.
   c. Atoms with completely filled outer shells are inert, but atoms with incompletely filled outer shells tend to gain, lose, or share electrons and thus achieve stable structures.
   d. Atoms that lose electrons become positively charged ions. Atoms that gain electrons become negatively charged ions.
   e. Ions with opposite electrical charges attract and form ionic bonds. Atoms that share electrons form covalent bonds.
4. Molecules and compounds
   a. Two or more atoms of the same element may bond to form a molecule of that element. Atoms of different elements may bond to form a molecule of a compound.
   b. Molecules contain definite kinds and numbers of atoms.
5. Formulas
   a. A molecular formula represents the numbers and types of atoms in a molecule.
   b. A structural formula depicts the arrangement of atoms within a molecule.
6. Chemical reactions
   a. A chemical reaction breaks or forms bonds between atoms, ions, or molecules.
   b. Three types of chemical reactions are: synthesis, in which larger molecules form from smaller particles; decomposition, in which smaller particles form from larger molecules; and exchange reactions, in which the parts of two different molecules trade positions.
   c. Many reactions are reversible. The direction of a reaction depends on the proportions of reactants and end products, the energy available, and the presence of catalysts.

7. Acids and bases
   a. Compounds that release ions when they dissolve in water are electrolytes.
   b. Electrolytes that release hydrogen ions are acids, and those that release hydroxyl or other ions that react with hydrogen ions are bases.
   c. A value called pH represents a solution's concentration of hydrogen ions (H^+) and hydroxyl ions (OH^-).
   d. A solution with equal numbers of H^+ and OH^- is neutral and has a pH of 7.0. A solution with more H^+ than OH^- is acidic and has a pH less than 7.0. A solution with fewer H^+ than OH^- is basic and has a pH greater than 7.0.
   e. Each whole number on the pH scale represents a tenfold difference in the hydrogen ion concentration.

Chemical Constituents of Cells (p. 40)

Molecules containing carbon and hydrogen atoms are organic and are usually nonelectrolytes. Other molecules are inorganic and are usually electrolytes.

1. Inorganic substances
   a. Water is the most abundant compound in cells and is a solvent in which chemical reactions occur. Water transports chemicals and heat.
   b. Oxygen releases energy from glucose and other nutrients. This energy drives metabolism.
   c. Carbon dioxide is produced when metabolism releases energy.
   d. Inorganic salts provide a variety of ions that metabolic processes require.
2. Organic substances
   a. Carbohydrates provide much of the energy that cells require and also contribute to cell structure. Their basic building blocks are simple sugar molecules.
   b. Lipids, such as fats, phospholipids, and steroids, supply energy and build cell parts. The basic building blocks of fats—the most common lipid—are molecules of glycerol and fatty acids.
   c. Proteins serve as structural materials, energy sources, hormones, cell surface receptors, and enzymes.
      (1) Enzymes speed chemical reactions without themselves being consumed.
      (2) The building blocks of proteins are amino acids.
      (3) Proteins vary in the numbers and types of amino acids they contain and in the sequence of these amino acids.
      (4) The amino acid chain of a protein molecule folds into a complex shape that is maintained largely by hydrogen bonds.
      (5) Excessive heat, radiation, electricity, altered pH, or various chemicals can denature proteins.
   d. Nucleic acids are the genetic material and control cellular activities.
      (1) Nucleic acid molecules are composed of nucleotides.
      (2) Two major types of nucleic acids are RNA and DNA.
      (3) DNA molecules store information that cell parts use to construct specific protein molecules. RNA molecules help to synthesize proteins.

Review Exercises

1. Define *chemistry*. (p. 32)
2. Define *matter* (p. 32)
3. Explain the relationship between elements and atoms. (p. 32)
4. List the four most abundant elements in the human body. (p. 32)
5. Describe the major parts of an atom. (p. 32)
6. Explain why a complete atom is electrically neutral. (p. 33)
7. Define *atomic number* and *atomic weight*. (p. 33)
8. Explain how electrons are arranged within an atom. (p. 34)
9. Distinguish between an ionic bond and a covalent bond. (p. 36)
10. Explain the relationship between molecules and compounds. (p. 37)
11. Distinguish between a molecular formula and a structural formula. (p. 38)
12. Explain what the formula $C_6H_{12}O_6$ means. (p. 38)
13. Describe three major types of chemical reactions. (p. 38)
14. Explain what a reversible reaction is. (p. 38)
15. Define *catalyst*. (p. 39)
16. Define *acid* and *base*. (p. 39)
17. Explain what pH measures, and describe the pH scale. (p. 39)
18. Distinguish between electrolytes and nonelectrolytes. (p. 39)
19. Distinguish between inorganic and organic substances. (p. 40)
20. Describe the roles water and oxygen play in the human body. (p. 40)
21. List several of the ions found in body fluids. (p. 40)
22. Describe the general characteristics of carbohydrates. (p. 40)
23. Distinguish between simple and complex carbohydrates. (p. 41)
24. Describe the general characteristics of lipids. (p. 41)
25. Define *triglyceride*. (p. 42)
26. Distinguish between saturated and unsaturated fats. (p. 43)
27. Describe the general characteristics of proteins. (p. 43)
28. Define *enzyme*. (p. 43)
29. Explain how protein molecules may denature. (p. 45)
30. Describe the structure of nucleic acids. (p. 45)
31. Explain the major functions of nucleic acids. (p. 46)

Critical Thinking

1. If a shampoo is labeled "nonalkaline," would it more likely have a pH of 3, 7, or 12?
2. A topping for ice cream contains fructose, hydrogenated soybean oil, salt, and cellulose. What types of chemicals are in it?
3. An advertisement for a supposedly healthful cookie claims that it contains an "organic carbohydrate." Why is this statement silly?
4. At a restaurant, a waiter recommends a sparkling carbonated beverage, claiming that it contains no carbohydrates. The product label lists water and fructose as ingredients. Is the waiter correct?
5. A Horta is a fictional animal (from "Star Trek") whose biochemistry is based on the element silicon. Consult a periodic table. Is silicon a likely substitute for carbon in a life-form? Cite a reason for your answer.
6. A man on a very low-fat diet proclaims to his friend, "I'm going to get my cholesterol down to zero!" Why is this an impossible (and undesirable) goal?
7. What acidic and basic substances do you encounter in your everyday activities? What acidic foods do you eat regularly? What basic foods do you eat?
8. How would you explain the importance of amino acids and proteins in a diet to a person who is following a diet composed primarily of carbohydrates?
9. What clinical laboratory tests with which you are acquainted require a knowledge of chemistry to understand the significance of what they measure?

Cells

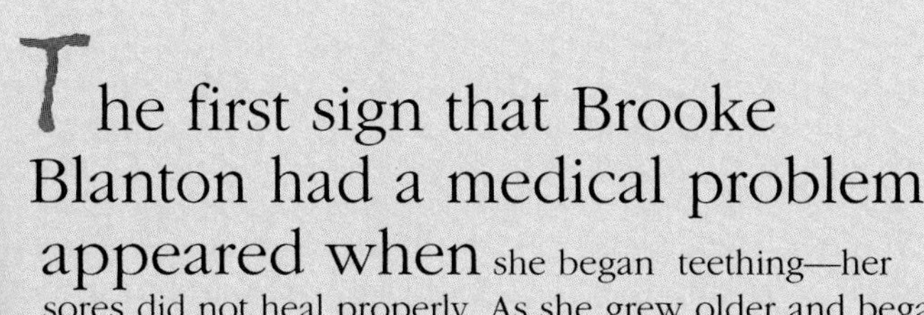

The first sign that Brooke Blanton had a medical problem appeared when she began teething—her sores did not heal properly. As she grew older and began getting the bumps and bruises that come with crawling and then walking, the problem worsened. Wounds simply remained open, instead of swelling and then gradually closing. They never accumulated pus, which consists of bacteria, cellular debris, and white blood cells, a sure sign that the body is fighting infection.

It took years before puzzled physicians determined that Brooke's problem is slippery cells. Normally, white blood cells migrate in the bloodstream to a site of injury or infection. The cells halt at the wound site, with the proteins on their surfaces sticking to proteins protruding from the blood vessel's inner wall, like Velcro to a sweater. Then, other proteins pull the white blood cells between the tilelike cells forming the blood vessel wall, so that the white blood cells can migrate to where they are needed.

Brooke's very rare defect in cell adhesion means that her white blood cells zip past injury sites. Now a young adult, Brooke must be very careful to stay as healthy and injury-free as possible because of her slippery cells.

Photo: White blood cells are chemically attracted to the site of an injury. Unless they have certain cellular adhesion molecules, however, they cannot be directed to the site.

Recipe for a human being: cells, their products, and fluids. A cell, as the unit of life, is a world unto itself. To build a human, trillions of cells connect and interact, forming dynamic tissues, organs, and organ systems.

Chapter Objectives

After studying this chapter, you should be able to:

1. Explain how cells differ from one another. (p. 51)
2. Describe the characteristics of a composite cell. (p. 51)
3. Explain how the structure of a cell membrane makes possible its functions. (p. 53)
4. Describe each type of cytoplasmic organelle, and explain its function. (p. 54)
5. Describe the cell nucleus and its parts. (p. 57)
6. Explain how substances move through cell membranes. (p. 59)
7. Describe the cell cycle. (p. 65)
8. Explain how a cell reproduces. (p. 65)

Aids to Understanding Words

cyt- [cell] *cyt*oplasm: The fluid (cytosol) and organelles that occupy the space between the cell membrane and nuclear envelope.

endo- [within] *endo*plasmic reticulum: A complex of membranous structures within the cytoplasm.

hyper- [above] *hyper*tonic: A solution that has a greater concentration of dissolved particles than another solution (such as the extracellular fluid).

hypo- [below] *hypo*tonic: A solution that has a lesser concentration of dissolved particles than another solution.

inter- [between] *inter*phase: The stage that occurs between mitotic divisions of a cell.

iso- [equal] *iso*tonic: A solution that has a concentration of dissolved particles equal to that of another solution.

mit- [thread] *mit*osis: The process of cell division when threadlike chromosomes become visible within a cell.

phag- [to eat] *phag*ocytosis: The process by which a cell takes in solid particles.

pino- [to drink] *pino*cytosis: The process by which a cell takes in tiny droplets of liquid.

-som [body] ribo*som*e: A tiny, spherical organelle consisting of protein and RNA.

Key Terms

active transport (ak′tiv trans′port)
centrosome (sent′tro-sōm)
chromosome (kro′mo-sōm)
cytoplasm (si′to-plazm)
differentiation (dif″er-en″she-a′shun)
diffusion (dĭ-fu′zhun)
endocytosis (en″do-si-to′sis)
endoplasmic reticulum (en′do-plaz′mik rĕ-tik′u-lum)
equilibrium (e″kwĭ-lib′re-um)
exocytosis (ex-o-si-to′sis)
facilitated diffusion (fah-sil″ĭ-tāt′ed dĭ-fu′zhun)
filtration (fil-tra′shun)
Golgi apparatus (gol′je ap″ah-ra′tus)
lysosome (li′so-sōm)
mitochondrion (mi″to-kon′dre-un); plural: mitochondria (mi″to-kon′dre-ah)
mitosis (mi-to′sis)
nucleolus (nu-kle′o-lus)
nucleus (nu′kle-us)
osmosis (oz-mo′sis)
phagocytosis (fag″o-si-to′sis)
pinocytosis (pi″no-si-to′sis)
ribosome (ri′bo-sōm)
selectively permeable (se-lek′tiv-le per′me-ah-bl)
vesicle (ves′ĭ-k′l)

Introduction

The estimated 75 trillion cells that make up an adult human body have much in common. Yet, cells in different tissues vary considerably in size and shape, and typically, their shapes make possible their functions. For instance, nerve cells often have long, threadlike extensions that transmit nerve impulses from one part of the body to another. Epithelial cells that line the inside of the mouth are thin, flattened, and tightly packed, somewhat like floor tiles, an arrangement that enables them to protect underlying cells. Muscle cells, which pull parts closer together, are slender and rodlike, with their ends attached to the structures they move (fig. 3.1).

Composite Cell

Because cells vary so greatly in size, shape, content, and function, describing a "typical" cell is impossible. The cell in figure 3.2 and described in this chapter is a composite cell that includes many known cell structures. In reality, cells have many, but not all, of these structures.

Commonly, a cell consists of two major parts—the **nucleus** and the **cytoplasm.** The nucleus is innermost and is surrounded by a thin nuclear envelope. The cytoplasm surrounds the nucleus and is itself encircled by an even thinner cell membrane (also called the plasma membrane).

Within the cytoplasm are specialized structures called cytoplasmic organelles, suspended in a liquid called *cytosol.* These organelles perform specific functions, in a sense dividing the labor of the cell. The nucleus, on the other hand, directs overall cell activities.

1. Give two examples to illustrate that the shape of a cell makes possible its function.
2. Name the two major parts of a cell.
3. What are the general functions of the two major cell parts?

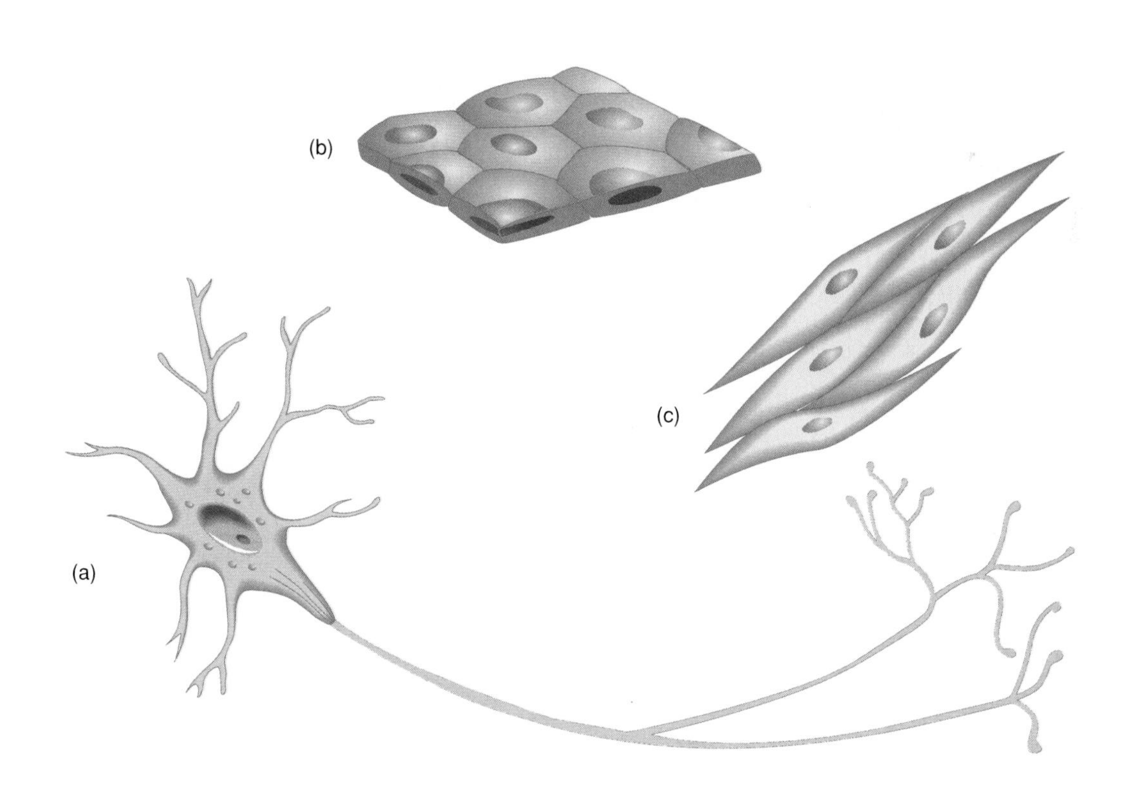

***Figure* 3.1**

Cells vary in structure and function. (*a*) A nerve cell transmits impulses from one body part to another. (*b*) Epithelial cells protect underlying cells. (*c*) Muscle cells pull parts closer together.

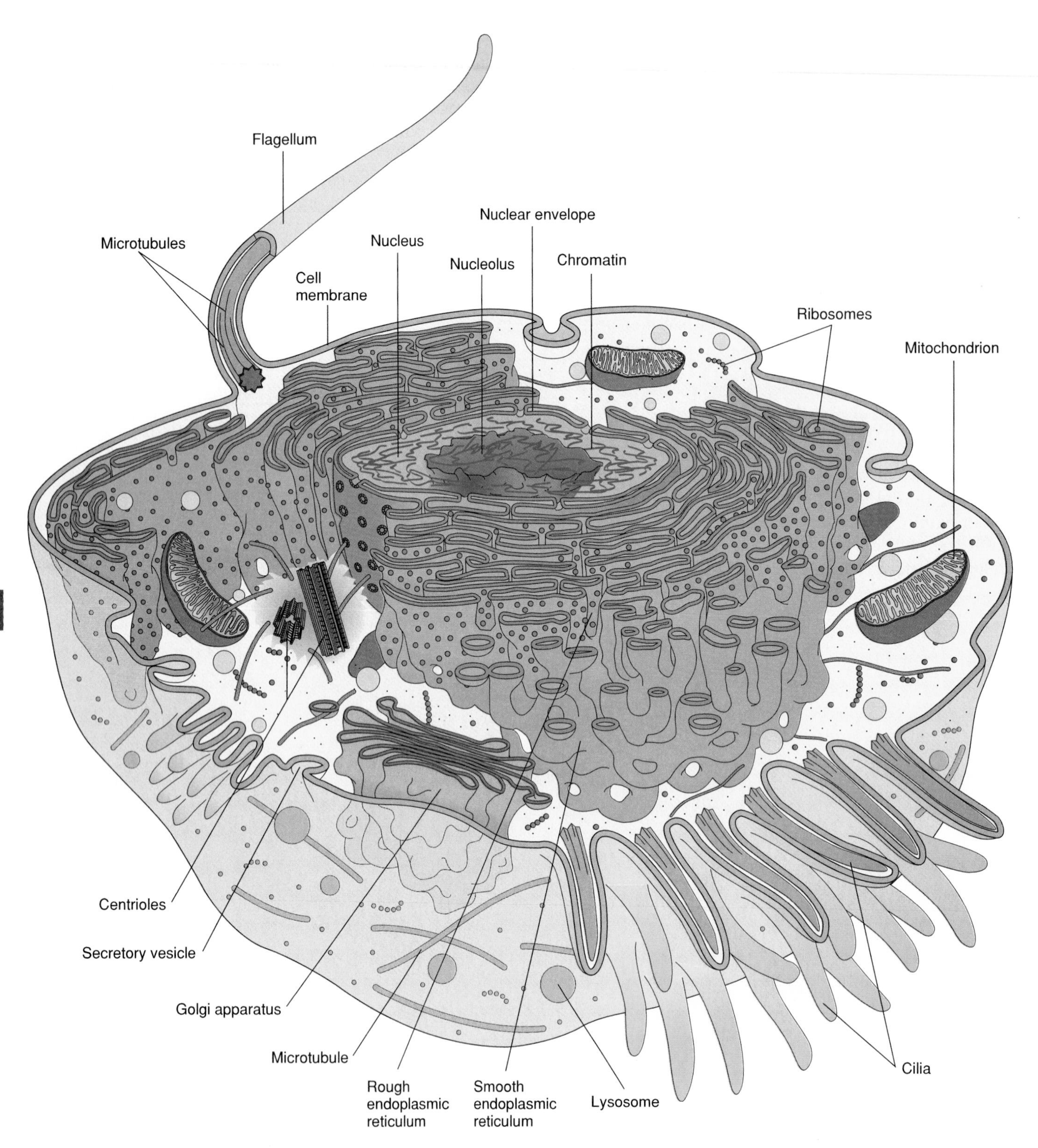

Figure 3.2

A composite cell. Organelles are not drawn to scale.

Cell Membrane

The **cell membrane** is more than a simple boundary surrounding the cellular contents; it is an actively functioning part of the living material, and many important metabolic reactions take place on its surfaces.

General Characteristics

The cell membrane is extremely thin—visible only with the aid of an electron microscope—but it is flexible and somewhat elastic. It typically has complex surface features with many outpouchings and infoldings that increase surface area (fig. 3.2).

In addition to maintaining cell integrity, the membrane controls which substances exit and enter. A membrane that functions in this way is called **selectively permeable** (also known as *semipermeable* or *differentially permeable*).

Membrane Structure

The cell membrane is composed mainly of lipids, proteins, and a small quantity of carbohydrates. Its basic framework consists of a double layer (bilayer) of phospholipid molecules. Each of these molecules includes a phosphate group and two fatty acids bound to a glycerol molecule (see chapter 2). The lipid molecules are relatively free to move sideways within the plane of the membrane, and together, they form a soft and flexible, but stable, fluid film.

Because the membrane's interior consists largely of the fatty acid portions of the phospholipid molecules (fig. 3.3), it is oily. Molecules such as oxygen and carbon dioxide, which are soluble in lipids, can pass through this layer easily. However, the layer is impermeable to water-soluble molecules, such as amino acids, sugars, proteins, nucleic acids, and various ions. Many cholesterol molecules embedded in the membrane's interior also help make the membrane less permeable to water-soluble substances. In addition, the relatively rigid structure of the cholesterol molecules helps stabilize the membrane.

The cell membrane includes only a few types of lipid molecules but many kinds of proteins. These proteins provide special functions, and they can be classified according to their shapes. One group consists of tightly coiled, rodlike (fibrous) molecules embedded in the phospholipid bilayer. These rodlike proteins may completely span the membrane; that is, they may extend outward from the surface on one side, while their opposite ends communicate with the cell's interior. Such proteins often function as *receptors* that combine with specific kinds of molecules, such as hormones (see chapter 11).

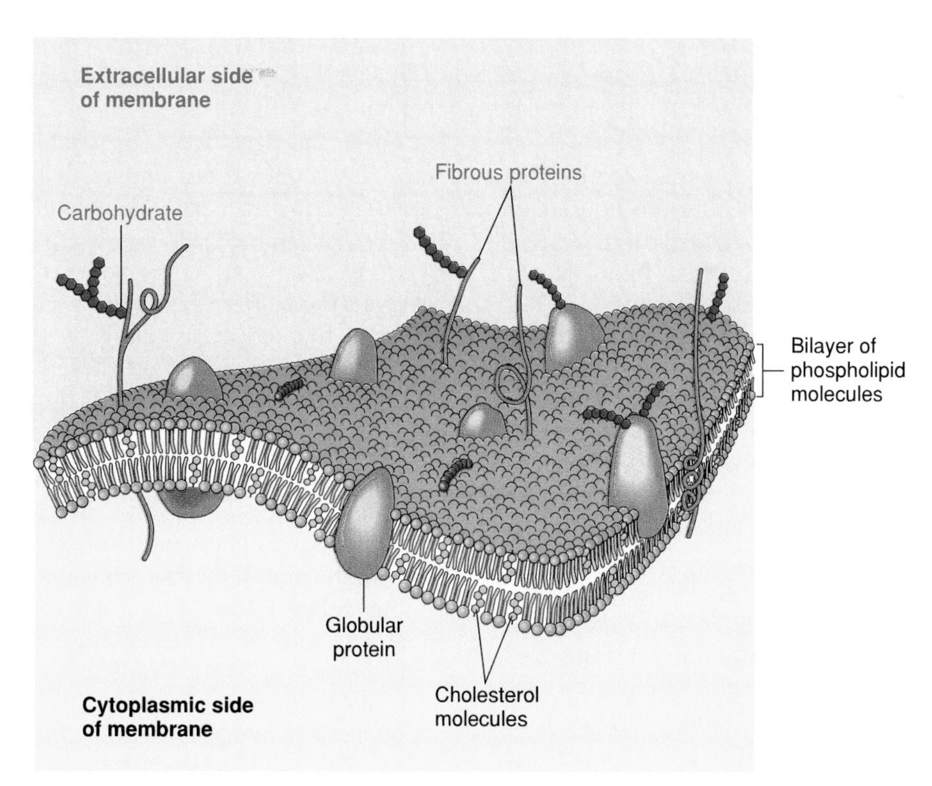

Figure 3.3

The cell membrane is composed primarily of phospholipids, with proteins scattered throughout the lipid bilayer.

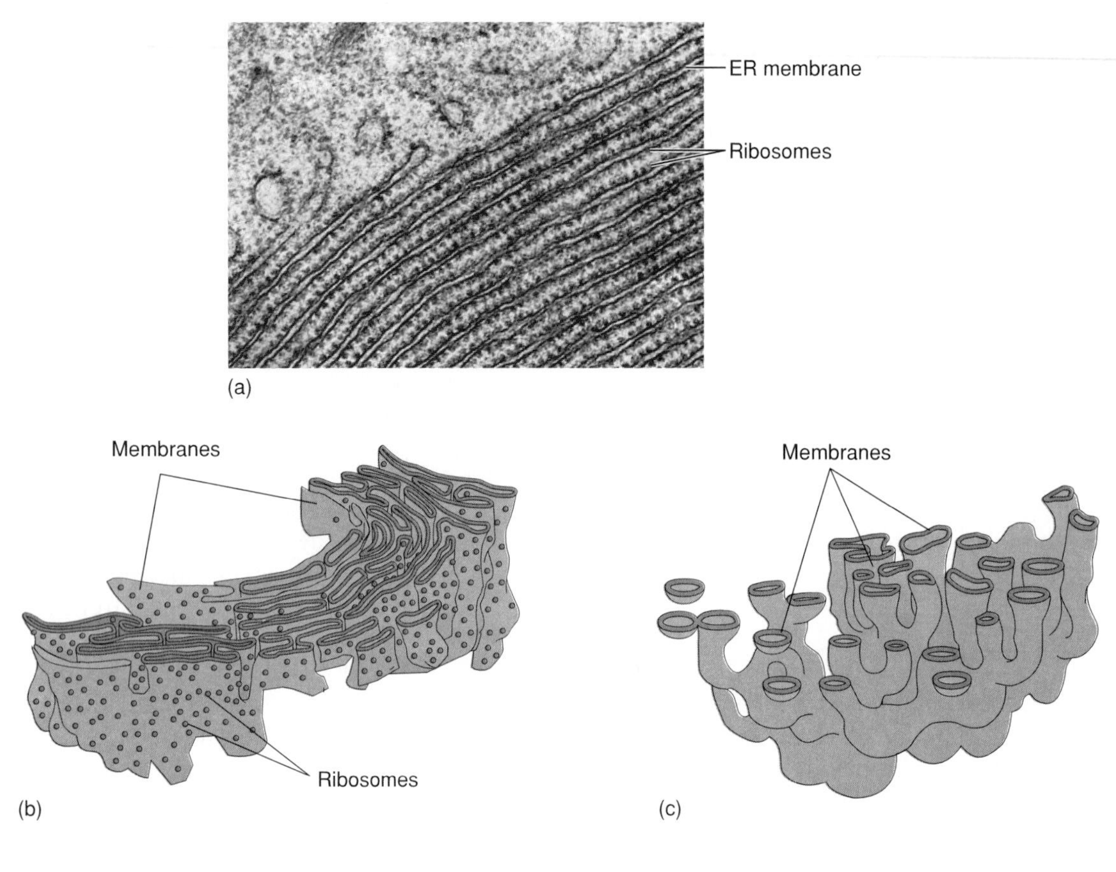

Figure 3.4

(*a*) Transmission electron micrograph of rough endoplasmic reticulum (ER) (100,000×).
(*b*) Rough ER is dotted with ribosomes, while (*c*) smooth ER lacks ribosomes.

Another group of membrane proteins are more compact and globular and are embedded in the interior of the phospholipid bilayer. Typically, they span the membrane and form narrow passageways, through which various ions and molecules can cross the bilayer. For example, some globular proteins form "pores" in the membrane that allow the movement of water molecules. Others serve as selective *channels* that only allow particular ions to pass through. In muscle and nerve cells, for example, such selective channels control the movements of sodium and potassium ions, which play important roles in muscle contraction and nerve impulse conduction (see chapters 8 and 9, respectively).

Carbohydrate molecules bound to proteins (glycoproteins) on membrane surfaces help cells to recognize and bind to one another. This is important in tissue formation and also in the immune system's ability to recognize cells of the body as "self" and the infection-causing bacteria or virus as "nonself."

Cytoplasm

When viewed through a light microscope, **cytoplasm** usually appears as a clear liquid with specks scattered throughout. However, an electron microscope, which produces much greater magnification and the ability to distinguish fine detail (resolution), reveals that cytoplasm is filled with networks of membranes and cytoplasmic organelles (see fig. 3.2).

Cell activities occur mainly in the cytoplasm, where nutrients are received, processed, and used. The following cytoplasmic organelles play specific roles:

1. **Endoplasmic reticulum** The endoplasmic reticulum (ER) is a complex organelle composed of membrane-bound, flattened sacs and elongated canals. These membranous parts interconnect and communicate with the cell membrane, nuclear envelope, and other organelles. ER provides a tubular communication system through which molecules can be transported from one cell part to another.

   ER participates in the synthesis of certain types of molecules. In some regions, the ER's outer membranous surface has many tiny, spherical organelles called ribosomes attached to it, which impart a textured appearance when the ER is viewed with an electron microscope. Such ER is termed *rough ER* because of this appearance. ER that lacks ribosomes is called *smooth ER* (figs. 3.2 and 3.4).

The ribosomes of rough ER help synthesize proteins. The proteins may then move through ER tubules to the Golgi apparatus for further processing. Smooth ER, on the other hand, contains enzymes required to manufacture certain lipid molecules, including some steroid hormones.

2. **Ribosomes** Although many ribosomes are attached to ER membranes, others are scattered throughout the cytoplasm. All ribosomes are composed of protein and RNA molecules. Ribosomes provide a structural support as well as enzymes, which link amino acids to form proteins. (Protein synthesis is described in chapter 4.)

3. **Golgi apparatus** The Golgi apparatus consists of a stack of about six flattened, membranous sacs whose membranes are continuous with those of the ER. The Golgi apparatus refines and "packages" proteins synthesized on ribosomes associated with the ER. Proteins arrive at the Golgi apparatus enclosed in tiny sacs (vesicles) composed of ER membrane. These sacs fuse with the membrane at the beginning of the Golgi apparatus, which is specialized to receive glycoproteins.

   The glycoproteins pass from layer to layer through the stacks of Golgi membrane and are modified chemically; for example, some sugar molecules may be added or removed. When they reach the outermost layer, the altered glycoproteins are packaged in bits of Golgi membrane, which bud off and form transport vesicles. Such a vesicle may then move to the cell membrane, where it fuses with the membrane and releases its contents to the outside of the cell as a secretion (figs. 3.2 and 3.5). Other vesicles may transport glycoproteins to various organelles.

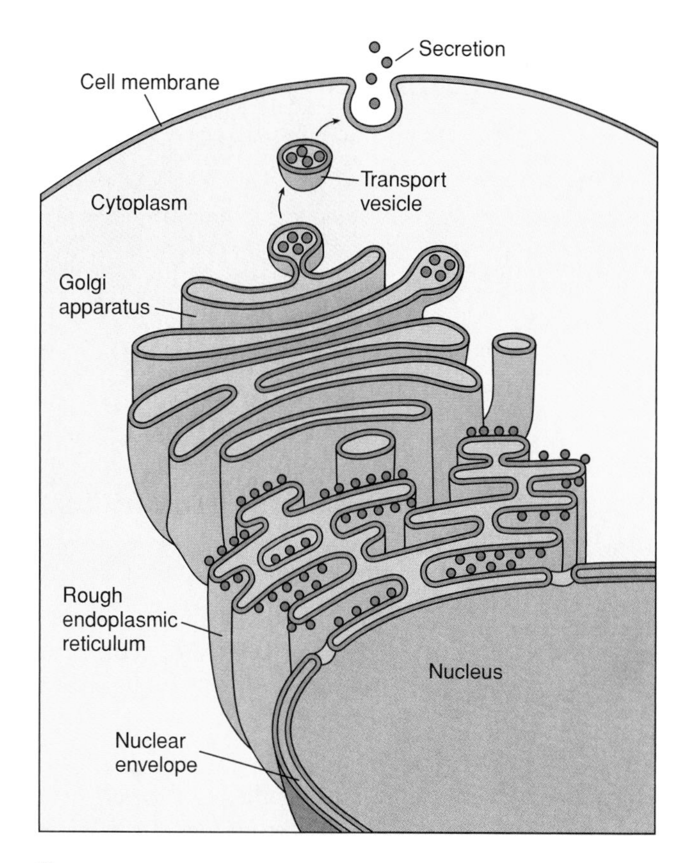

Figure 3.5

The Golgi apparatus packages glycoproteins, which then may be moved to the cell membrane and released as secretions.

1. What is a selectively permeable membrane?
2. Describe the chemical structure of a cell membrane.
3. What are the functions of the endoplasmic reticulum?
4. Describe the functions of the Golgi apparatus.

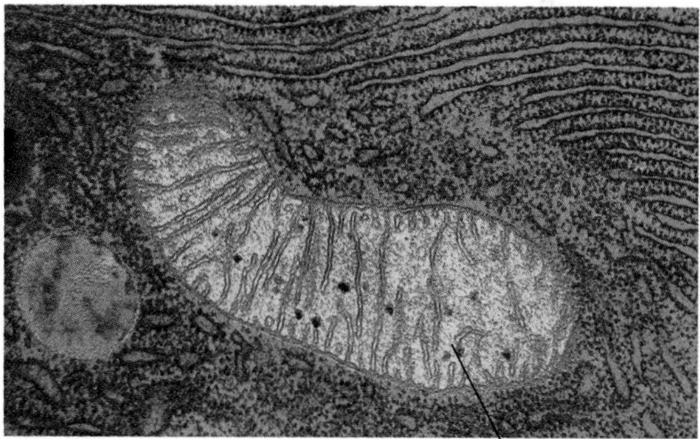

Figure 3.6

Falsely colored transmission electron micrograph of a mitochondrion.

4. **Mitochondria** Mitochondria are elongated, fluid-filled sacs that vary in size and shape. They often move about slowly in the cytoplasm and can reproduce by dividing. The membranes surrounding a mitochondrion include an outer and inner layer (figs. 3.2 and 3.6). The inner layer folds extensively to form partitions called *cristae*. Connected to the cristae are enzymes that control some of the chemical reactions that release energy from glucose and other nutrient molecules. Mitochondria are the major sites of chemical reactions that transform this energy into a chemical form the cell can use. (Chapter 4 describes this energy-releasing function in more detail.)

Centriole (cross section)

Centriole (longitudinal section)

(a) (b)

Figure 3.7

(*a*) Transmission electron micrograph of the two centrioles in a centrosome (142,000×).
(*b*) The centrioles lie at right angles to one another.

5. **Lysosomes** Lysosomes, the "garbage disposals of the cell," are tiny membranous sacs (see fig. 3.2). They contain powerful enzymes that break down nutrient molecules or foreign particles that may enter cells. Certain white blood cells, for example, can engulf bacteria, which are then digested by the lysosomal enzymes. This is one way that white blood cells fight bacterial infections. Lysosomes also destroy worn cellular parts. The lysosomes of some scavenger cells may even digest entire engulfed cells.

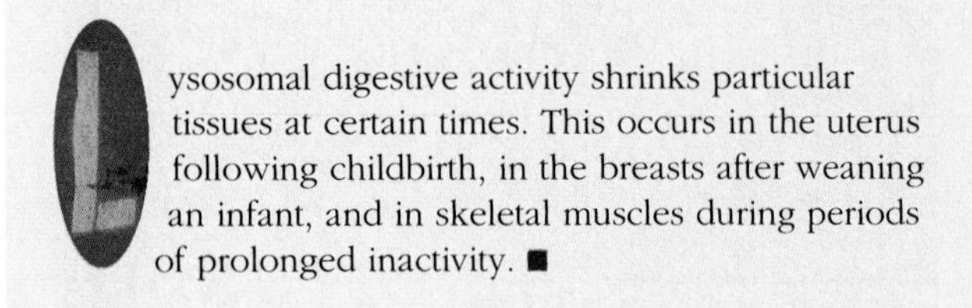

ysosomal digestive activity shrinks particular tissues at certain times. This occurs in the uterus following childbirth, in the breasts after weaning an infant, and in skeletal muscles during periods of prolonged inactivity. ■

6. **Centrosome** The centrosome is located in the cytoplasm near the Golgi apparatus and nucleus. It is nonmembranous and consists of two hollow cylinders, called *centrioles* (figs. 3.2 and 3.7). The centrioles lie at right angles to each other and function in cell reproduction by distributing chromosomes to newly forming cells.
7. **Cilia and flagella** Cilia and flagella are motile extensions from the surfaces of certain cells. They are similar structures that differ mainly in length and abundance.

   Cilia fringe the free surfaces of some epithelial (lining) cells. Each cilium is a tiny, hairlike structure that is attached beneath the cell membrane (see fig. 3.2). Cilia occur in precise patterns, and they have a "to-and-fro" type of movement, coordinated so that rows of cilia beat in succession, producing a wave of motion that sweeps over the ciliated surface. This wave moves fluids, such as mucus, over the surface of certain tissues, including those that form the inner linings of the respiratory tubes (fig. 3.8*a*).

   Flagella are considerably longer than cilia, and usually a cell has only a single flagellum. Flagella have an undulating wavelike motion, which begins at their base. The tail of a sperm cell is a flagellum that enables this motile cell to "swim" (fig. 3.8*b*).
8. **Vesicles** Vesicles (vacuoles) are membranous sacs formed by part of the cell membrane folding inward and pinching off. As a result, a tiny, bubblelike vesicle, containing some liquid or solid material formerly outside the cell, appears in the cytoplasm. The Golgi apparatus and ER also form vesicles that play a role in secretion (see fig. 3.2).
9. **Microfilaments and microtubules** Microfilaments and microtubules are two types of thin, threadlike processes within the cytoplasm. Microfilaments are tiny rods of actin protein that form meshworks or bundles. They cause various kinds of cell movement. In muscle cells, for example, microfilaments are

(a)

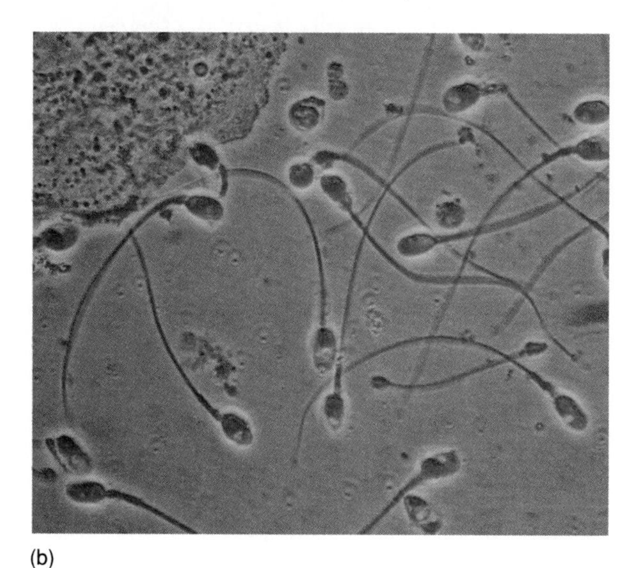

(b)

Figure 3.8

(*a*) Cilia such as these (*see arrow*) are common on the surfaces of certain cells, including those that form the inner lining of respiratory tubes (10,000×). (*b*) Flagella form the tails of these human sperm cells.

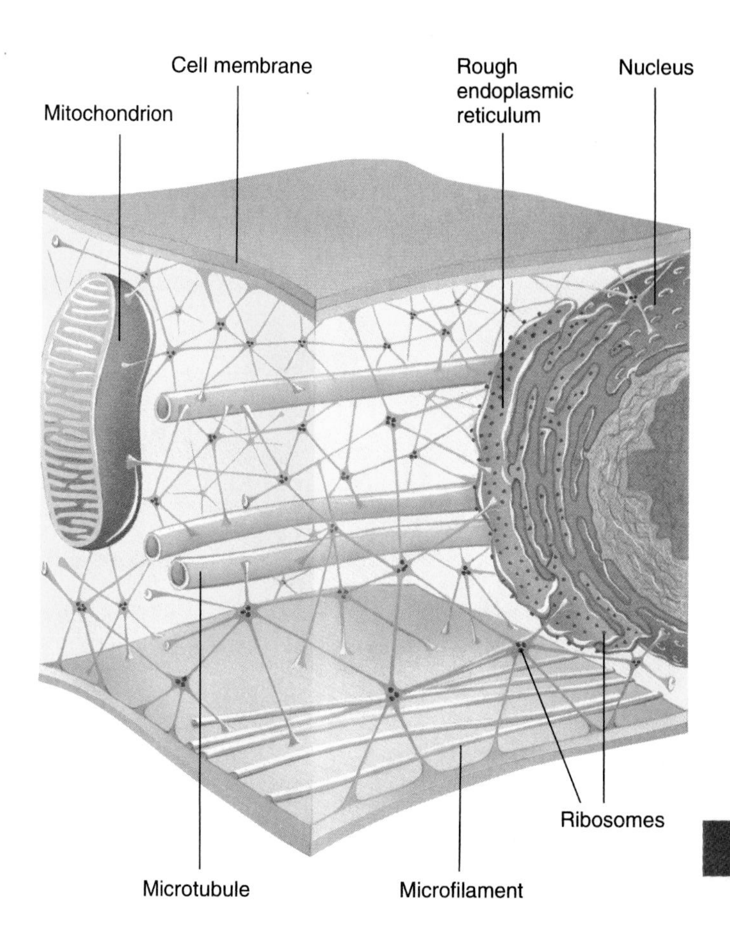

Figure 3.9

Microtubules help maintain the shape of a cell by forming an "internal skeleton" within the cytoplasm.

1. Describe a mitochondrion.
2. What is the function of a lysosome?
3. Distinguish between a centrosome and a centriole.
4. How do microfilaments and microtubules differ?

highly developed as *myofibrils,* which help these cells to shorten or contract (see chapter 8).

Microtubules are long, slender tubes with diameters two or three times that of microfilaments. Microtubules are composed of globular tubulin proteins. These tiny tubules provide support, form cilia and flagella, and pull duplicated chromosomes apart when a cell reproduces (fig. 3.9).

Cell Nucleus

The **nucleus** is an organelle usually located near the center of a cell (figs. 3.2 and 3.10). It is a relatively large, spherical structure enclosed in a double-layered **nuclear envelope,** consisting of inner and outer lipid bilayer membranes. The envelope has protein-lined channels called nuclear pores that allow various substances

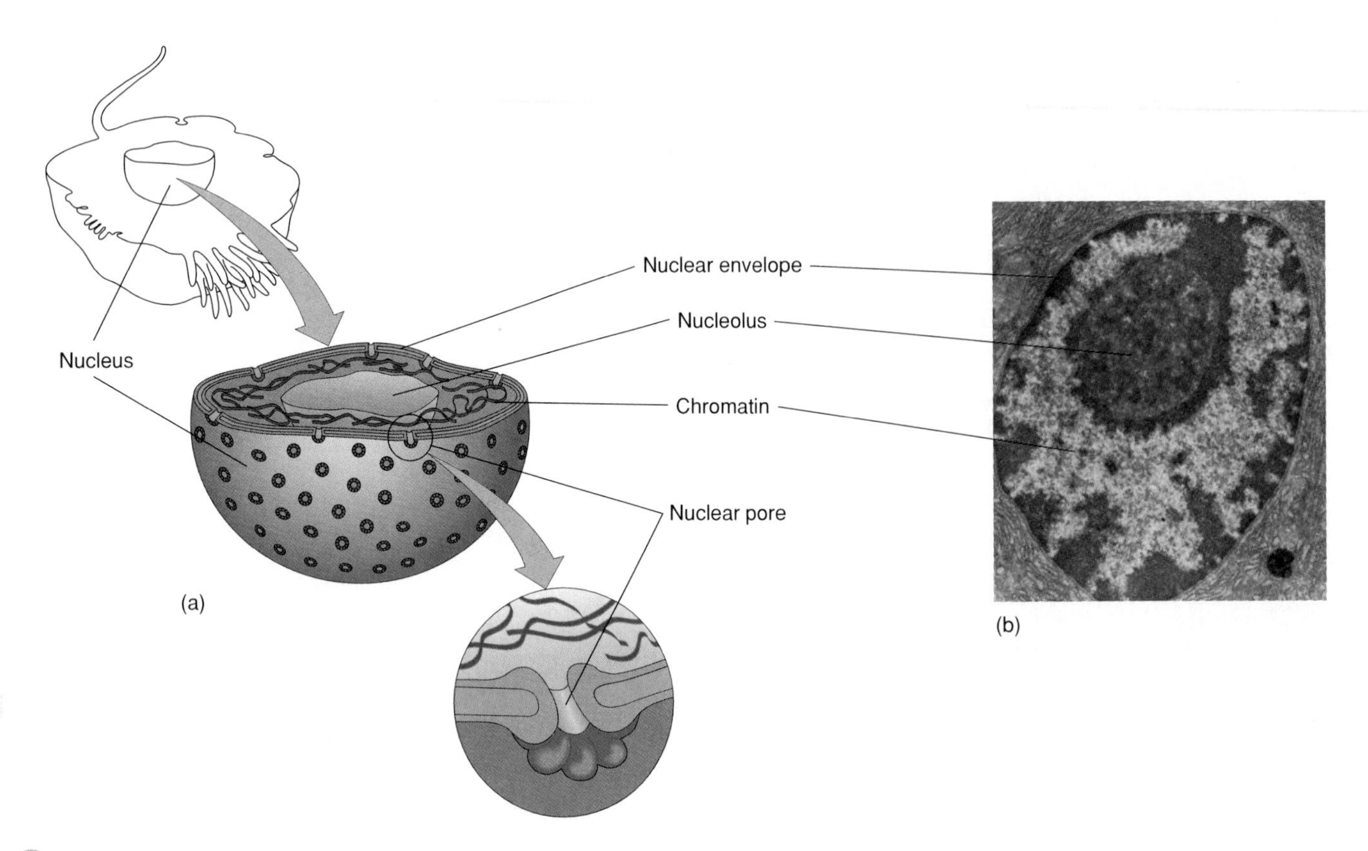

Figure 3.10

(*a*) The nuclear envelope is selectively permeable and allows certain substances to pass between the nucleus and the cytoplasm. (*b*) Transmission electron micrograph of a cell nucleus (8,000×). It contains a nucleolus and masses of chromatin.

to move between the nucleus and the cytoplasm. The nucleus contains a fluid in which the following structures float:

1. **Nucleolus** A nucleolus ("little nucleus") is a small, dense body composed largely of RNA and protein. It has no surrounding membrane and forms in specialized regions of certain chromosomes. It is the site of ribosome production. Once ribosomes form, they migrate through nuclear pores to the cytoplasm.
2. **Chromatin** Chromatin consists of loosely coiled fibers of protein and DNA. These fibers contain the information for protein synthesis. When the cell begins to reproduce, chromatin fibers coil into tiny, rodlike chromosomes.

Table 3.1 summarizes the structures and functions of cell organelles.

1. How are the nuclear contents separated from the cytoplasm?
2. What is the function of the nucleolus?
3. What is chromatin?

In a form of programmed cell death called apoptosis, organelles change in a characteristic way. Chromatin is cut into many equal-sized pieces, and the nuclear envelope breaks down. The cell membrane pinches in, and the cell and the organelles within it collapse in upon themselves. Apoptosis is a normal part of prenatal development, carving fingers and toes from webbed precursors. In contrast is necrosis, a form of cell death that is harmful and not normal. ■

Table 3.1

Structures and Functions of Cell Organelles

Organelle(s)	Structure	Function
Cell membrane	Membrane composed of protein and lipid molecules	Maintains integrity of cell and controls passage of materials into and out of cell
Endoplasmic reticulum	Complex of interconnected membrane-bound sacs and canals	Transports materials within cell, provides attachment for ribosomes, and synthesizes lipids
Ribosomes	Particles composed of protein and RNA molecules	Synthesize proteins
Golgi apparatus	Group of flattened, membranous sacs	Packages protein molecules for transport and secretion
Mitochondria	Membranous sacs with inner partitions	Release energy from nutrient molecules and transform energy into usable form
Lysosomes	Membranous sacs	Digest worn cellular parts or substances that enter cells
Centrosome	Nonmembranous structure composed of two rodlike centrioles	Helps distribute chromosomes to new cells during cell reproduction
Cilia and flagella	Motile projections attached beneath the cell membrane	Propel fluid over cellular surfaces and enable sperm cells to move
Vesicles	Membranous sacs	Contain various substances
Microfilaments and microtubules	Thin rods and tubules	Support the cytoplasm and help move substances and organelles within the cytoplasm
Nuclear envelope	Double membrane that separates the nuclear contents from the cytoplasm	Maintains integrity of nucleus and controls passage of materials between nucleus and cytoplasm
Nucleolus	Dense, nonmembranous body composed of protein and RNA molecules	Is site of ribosome synthesis
Chromatin	Fibers composed of protein and DNA molecules	Contains information for synthesizing proteins

Movements through Cell Membranes

The cell membrane is a selective barrier that controls which substances enter and leave the cell. These movements involve both passive mechanisms, not requiring cellular energy (diffusion, facilitated diffusion, osmosis, and filtration), and active mechanisms that use cellular energy (active transport, endocytosis, and exocytosis).

Passive Mechanisms

Diffusion

Diffusion is the process by which molecules or ions scatter or spread spontaneously from regions where they are in higher concentrations toward regions where they are in lower concentrations. Under natural conditions, molecules and ions constantly move at high speeds. Each particle travels in a separate path along a straight line until it collides and bounces off some other particle. Then it moves in another direction, only to collide again and change direction once more. Such random motion mixes molecules.

For example, when sugar (a solute) is first put into a glass of water (a solvent), the sugar remains highly concentrated at the bottom of the glass (fig. 3.11). Then, as it slowly disappears into solution, the moving water and sugar molecules collide, and in time, they are evenly mixed. As a result of diffusion, the sugar molecules eventually become uniformly distributed in the water, a state called **equilibrium.** Although molecules continue to move after equilibrium is achieved, their concentrations no longer change.

To better understand how diffusion accounts for the movement of molecules through a cell membrane, imagine a container of water that is separated into two compartments by a completely permeable membrane (fig. 3.12). This membrane has many pores that are large enough for water and sugar molecules to pass through. The sugar molecules are placed in one compartment (*A*) but not in the other (*B*). Although the

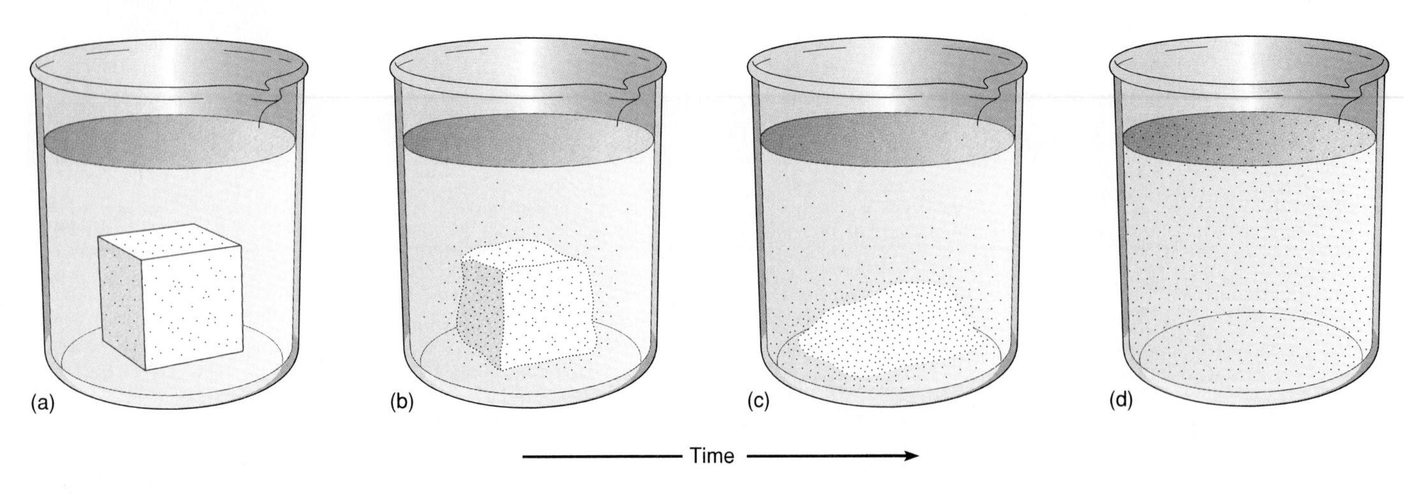

Figure 3.11

An example of diffusion. (*a, b,* and *c*) A sugar cube placed in water slowly disappears because the sugar molecules diffuse from regions where they are more concentrated toward regions where they are less concentrated. (*d*) Eventually, the sugar molecules distribute evenly throughout the water.

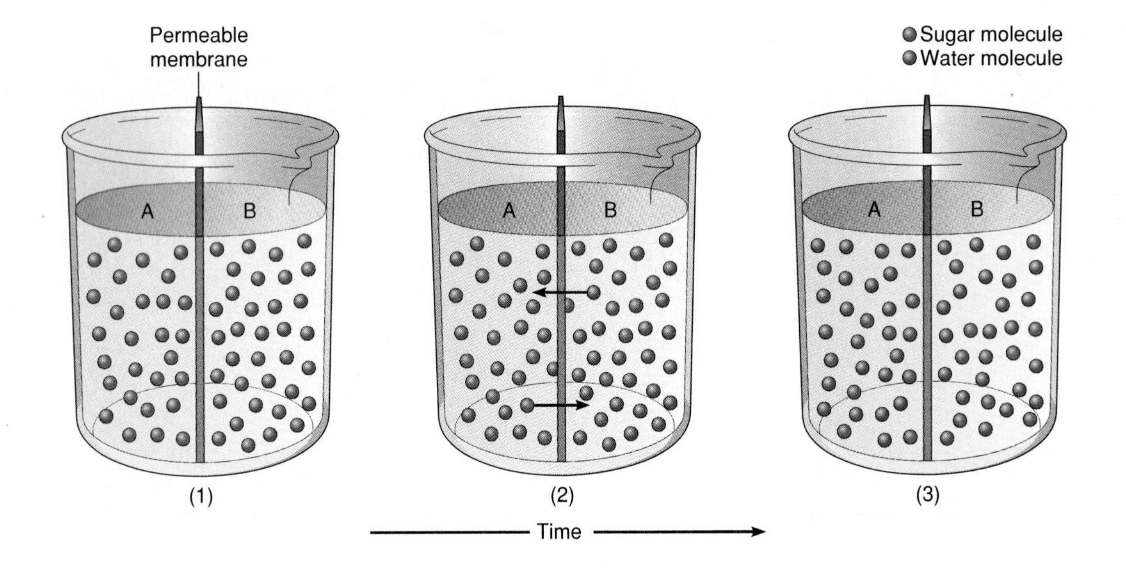

Figure 3.12

(1) A membrane permeable to water and sugar molecules separates the container into two compartments. Compartment *A* contains both types of molecules, while compartment *B* contains only water molecules. (2) As a result of molecular motions, sugar molecules tend to diffuse from compartment *A* into compartment *B*. Water molecules tend to diffuse from compartment *B* into compartment *A*. (3) Eventually, equilibrium prevails.

sugar molecules move in all directions, more diffuse from compartment *A* (where they are in greater concentration) through the pores in the membrane and into compartment *B* (where they are in lesser concentration) than move in the other direction. At the same time, the water molecules tend to diffuse from compartment *B* (where they are in greater concentration) through the pores into compartment *A* (where they are in lesser concentration). Eventually, equilibrium is achieved with equal concentrations of water and sugar in each compartment.

Similarly, oxygen molecules diffuse through cell membranes and enter cells if these molecules are more highly concentrated on the outside than on the inside. Carbon dioxide molecules, too, diffuse through cell membranes and leave cells if they are more concentrated on the

Dialysis uses diffusion to separate smaller molecules from larger ones in a liquid. The artificial kidney uses a variant of this process—*hemodialysis*—to treat patients suffering from kidney damage or failure. When an artificial kidney (dialyzer) is operating, blood from a patient is passed through a long, coiled tubing composed of porous cellophane. The size of the pores allows smaller molecules carried in the blood, such as urea, to pass out through the tubing, while larger molecules, such as those of blood proteins, remain inside the tubing. The tubing is submerged in a tank of dialyzing fluid (wash solution), which contains varying concentrations of different chemicals. The solution has low concentrations of substances that should leave the blood and higher concentrations of those that should remain in the blood.

To remove blood urea, a waste product, the dialyzing fluid must have a lower urea concentration than the blood; to maintain blood glucose concentration, glucose concentration in the dialyzing fluid must be at least equal to that of the blood. Thus, altering the concentrations of molecules in the dialyzing fluid can control which molecules diffuse out of blood and which remain in it. ■

inside than on the outside. Thus, diffusion enables oxygen and carbon dioxide molecules to be exchanged between the air and blood in the lungs, and between blood and the cells of various tissues.

Facilitated Diffusion

Most sugars are insoluble in lipids and are too large to pass through membrane pores. However, glucose may still enter through the cell membrane by a process called **facilitated diffusion** (fig. 3.13). In this process, which occurs in most cells, the glucose molecule combines with a special protein carrier molecule at the surface of the membrane. This union of the glucose and carrier molecules changes the shape of the carrier, which moves glucose to the other side of the membrane. The glucose portion is released, and the carrier molecule can return to its original shape to pick up another glucose molecule. The hormone *insulin,* discussed in chapter 11, promotes facilitated diffusion of glucose through the membranes of certain cells.

Facilitated diffusion is similar to simple diffusion in that it only moves molecules from regions of higher concentration toward regions of lower concentration. The number of carrier molecules in the cell membrane limits the rate of facilitated diffusion.

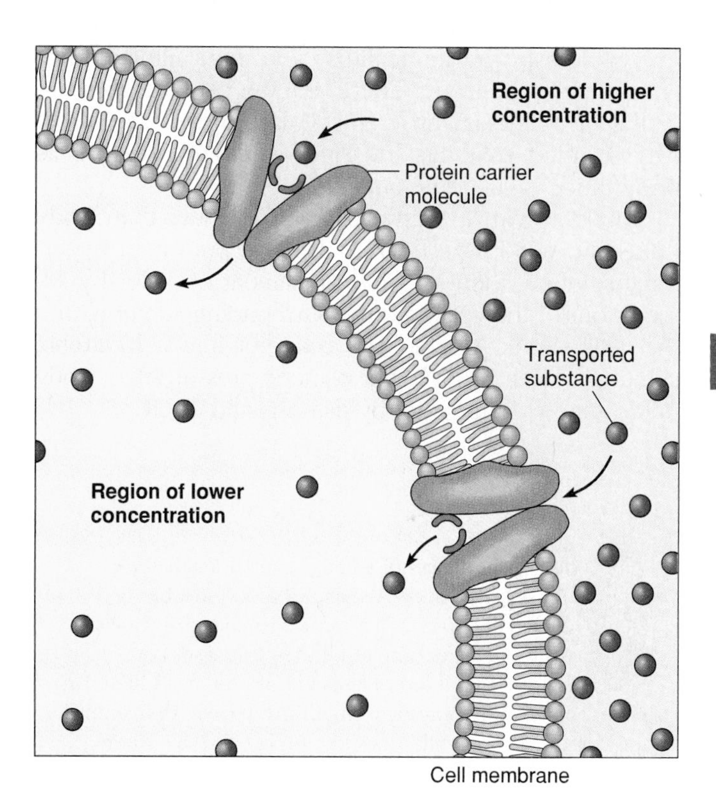

Figure 3.13

Some substances move across a cell membrane by facilitated diffusion. Carrier molecules transport them from a region of higher concentration to one of lower concentration.

Osmosis

Osmosis is a special case of diffusion. It occurs whenever water molecules diffuse from a region of higher water concentration to a region of lower water concentration across a selectively permeable membrane, such as a cell membrane. In the example that follows, assume that the selectively permeable membrane is permeable to water molecules (the solvent), but impermeable to glucose molecules (the solute).

In solutions, a higher concentration of solute (glucose in this case) means a lower concentration of water; a lower concentration of solute means a higher concentration of water. This is because solute molecules take up space that water molecules would otherwise occupy.

Just like molecules of other substances, molecules of water diffuse from areas of higher concentration to areas of lower concentration. In figure 3.14, the presence of glucose in compartment *A* means that the water concentration there is less than the concentration of pure water in compartment *B*. Therefore, water diffuses from compartment *B* across the selectively permeable membrane and into compartment *A*. In other words, water moves from compartment *B* into compartment *A* by osmosis. Glucose, on the other hand, cannot diffuse out of compartment *A* because the selectively permeable membrane is impermeable to it.

Note in figure 3.14 that as osmosis occurs, the water level on side A rises. This ability of osmosis to generate enough pressure to lift a volume of water is called *osmotic pressure.*

The greater the concentration of nonpermeable solute particles (glucose in this case) in a solution, the *lower* the water concentration of that solution and the *greater* the osmotic pressure. Water always tends to diffuse toward solutions of greater osmotic pressure.

Since cell membranes are generally permeable to water, water equilibrates by osmosis throughout the body, and the concentration of water and solutes everywhere in the intracellular and extracellular fluids is essentially the same. Therefore, the osmotic pressure of the intracellular and extracellular fluids is the same. Any solution that has the same osmotic pressure as body fluids is called **isotonic.**

Solutions with a higher osmotic pressure than body fluids are called **hypertonic.** If cells are put into a hypertonic solution, there is a net movement of water by osmosis out of the cells into the surrounding solution, and the cells shrink. Conversely, cells put into a **hypotonic** solution, which has a lower osmotic pressure than body fluids, tend to gain water by osmosis and swell.

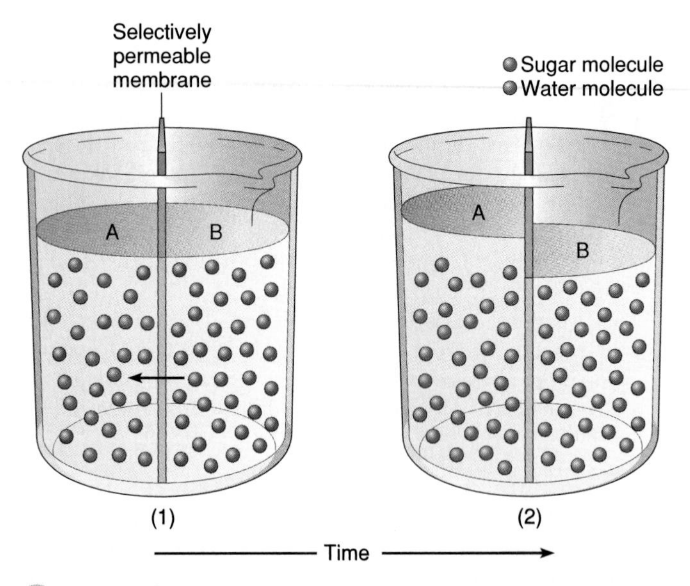

Figure 3.14

Osmosis. (1) A selectively permeable membrane separates the container into two compartments. At first, compartment *A* contains water and sugar molecules, while compartment *B* contains only water. As a result of molecular motions, water diffuses by osmosis from compartment *B* into compartment *A*. Sugar molecules remain in compartment *A* because they are too large to pass through the pores of the membrane. (2) Because more water enters compartment *A* than leaves it, water accumulates in this compartment. The level of liquid in compartment *A* rises above that of compartment *B*.

The concentration of solute in solutions that are infused into body tissues or blood must be controlled. Otherwise, osmosis may cause cells to swell or shrink, impairing their function. For instance, if red blood cells are placed in distilled water (which is hypotonic to them), water will diffuse into the cells, and they will burst (hemolyze). On the other hand, if red blood cells are exposed to 0.9% NaCl solution (normal saline), they will remain unchanged because this solution is isotonic to human cells. ■

Filtration

Molecules pass through membranes by diffusion or osmosis because of random molecular motion. In other instances, molecules are forced through membranes by the hydrostatic pressure, called *blood pressure,* that is greater on one side of the membrane than the other. **Filtration** is this process that forces molecules through membranes.

In the laboratory, filtration commonly separates solids from water. One filtration method is to pour a mixture of solids and water onto filter paper in a funnel. The paper is a porous membrane through which the small water molecules can pass, leaving behind the larger solid particles. Hydrostatic pressure, created by the weight of water on the paper due to gravity, forces the water molecules through to the other side.

In the body, tissue fluid forms when water and small dissolved substances are forced out through the thin, porous walls of blood capillaries, but larger particles, such as blood protein molecules, are left inside (fig. 3.15). The force for this movement comes from blood pressure, created largely by heart action, which is greater within the vessel than outside it.

1. What kinds of substances diffuse most readily through a cell membrane?
2. Explain the differences between diffusion and osmosis.
3. Distinguish between hypertonic, hypotonic, and isotonic solutions.
4. Explain how filtration occurs within the body.

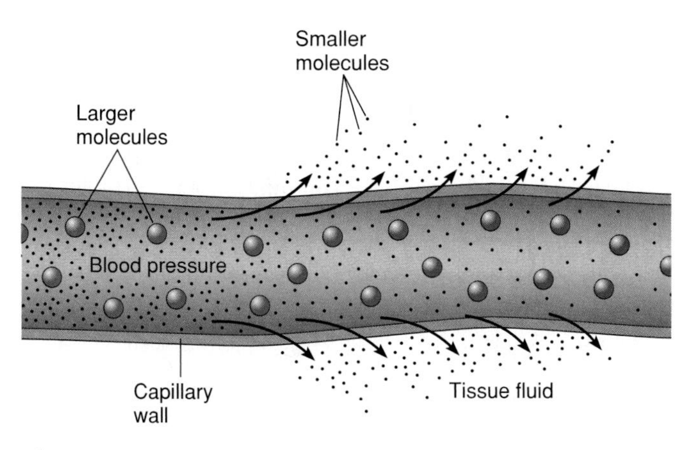

Figure 3.15

In this example of filtration, blood pressure forces smaller molecules through tiny openings in the capillary wall. Larger molecules remain inside.

Active Mechanisms

Active Transport

When molecules or ions pass through cell membranes by diffusion, facilitated diffusion, or osmosis, their net movements are from regions of higher concentration toward regions of lower concentration. Sometimes, however, particles move from a region of lower concentration to one of higher concentration.

Sodium ions, for example, can diffuse through cell membranes. Yet, the concentration of these ions typically remains many times greater on the outside of cells than on the inside of cells. This is because sodium ions are continually moved through cell membranes from regions of lower concentration (inside) to regions of higher concentration (outside) by a process called **active transport.** Energy from cellular metabolism powers active transport. Up to 40% of a cell's energy supply may be used for active transport of particles through its membranes.

Active transport is similar to facilitated diffusion in that it uses specific carrier molecules in cell membranes (fig. 3.16). These carrier molecules are proteins with binding sites that combine with the particles being transported. Such a union triggers the release of some cellular energy, and this alters the shape of the carrier protein. As a result, the "passenger" molecule moves through the membrane. Once on the other side, the transported molecule is released, and the carrier can return it to its original position to accept another passenger molecule at its binding site. Because they transport substances from regions of low concentration to regions of high concentration, these carrier proteins are sometimes called "pumps."

Particles that are actively transported across cell membranes include sugars, amino acids, and sodium, potassium, calcium, and hydrogen ions. Active transport also absorbs nutrient molecules into cells that line intestinal walls.

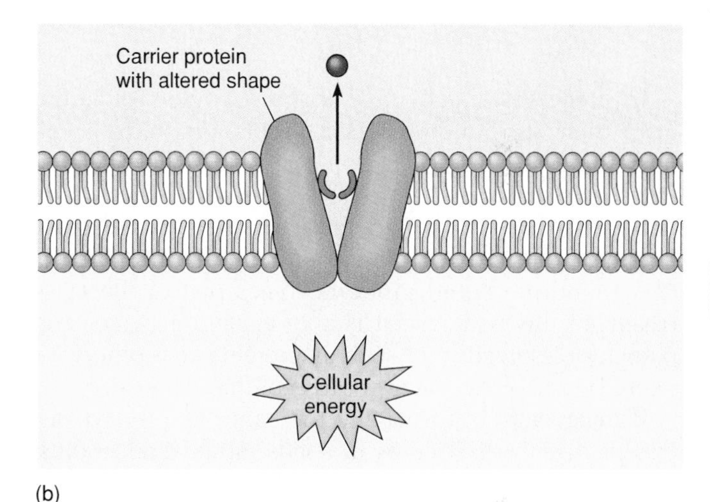

Figure 3.16

(*a*) During active transport, a molecule or an ion combines with a carrier protein, whose shape is altered as a result.
(*b*) This process, which requires energy, transports the particle through the cell membrane.

Endocytosis and Exocytosis

Two processes use cellular energy to move substances into (endocytosis) or out of (exocytosis) a cell without actually crossing the cell membrane. In **endocytosis,** molecules or other particles that are too large to enter a cell by diffusion or active transport may be conveyed within a vesicle formed from a section of the cell membrane. In **exocytosis,** the reverse process secretes from the cell a substance stored in a vesicle. Two forms of endocytosis are pinocytosis and phagocytosis.

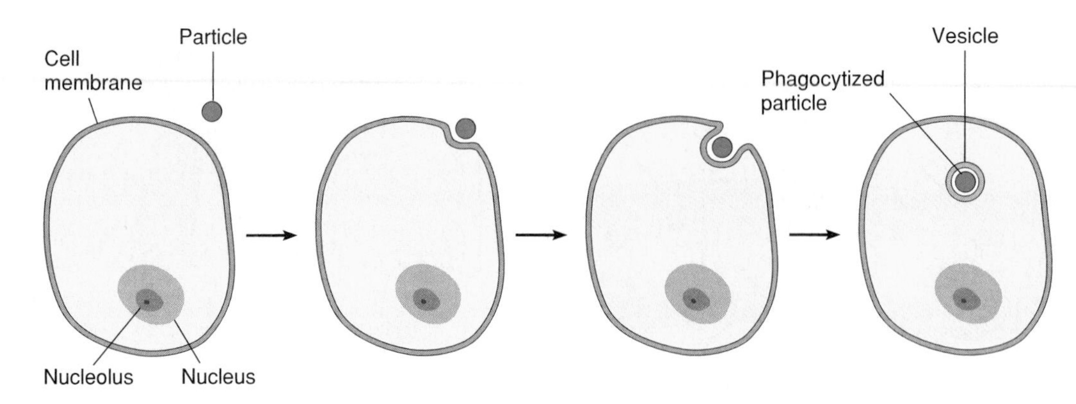

Figure 3.17

A cell may take in a solid particle from its surroundings by phagocytosis.

In **pinocytosis**, cells take in tiny droplets of liquid from their surroundings, as a small portion of a cell membrane indents. The open end of the tubelike part that forms seals off and produces a small vesicle, which detaches from the surface and moves into the cytoplasm. Eventually, the vesicular membrane breaks down, and the liquid inside becomes part of the cytoplasm. In this way, a cell is able to take in water and particles dissolved in it, such as proteins, that otherwise might be unable to enter because of their large size.

Phagocytosis is essentially the same as pinocytosis, except that the cell takes in solids rather than liquids. Certain kinds of white blood cells are called *phagocytes* because they can take in tiny solid particles, such as bacteria or cell debris, by phagocytosis. When a phagocyte first encounters a particle, the particle attaches to the phagocyte's cell membrane. This stimulates a portion of the membrane to project outward, surround the particle, and slowly draw it inside. The part of the membrane surrounding the solid detaches from the cell's surface, forming a vesicle containing the particle (fig. 3.17).

Commonly, a lysosome then combines with such a newly formed vesicle, and the lysosomal digestive enzymes decompose the vesicular contents. The products of this decomposition may diffuse out of the lysosome and into the cytoplasm. Exocytosis usually expels remaining residue from the cell.

Table 3.2 summarizes the types of movements through cell membranes.

1. What type of mechanism maintains unequal concentrations of ions on opposite sides of a cell membrane?
2. How are facilitated diffusion and active transport similar? How are they different?
3. What is the difference between endocytosis and exocytosis?

In inherited lysosomal storage diseases, one of the forty types of lysosomal enzymes is absent or abnormal, causing buildup of the starting material of the reaction that the enzyme normally catalyzes. Tay-Sachs disease is an example of such a condition. Lipids accumulate abnormally in certain nerve cells. A child with Tay-Sachs disease appears normal for the first six months of life, then gradually loses motor skills, becoming deaf and blind by age four. Microscopic signs of Tay-Sachs disease—swollen lysosomes—are present even before birth. ■

Table 3.2

Movements through Cell Membranes

Process	Characteristics	Source of Energy	Example
Passive mechanisms			
Diffusion	Molecules or ions move from regions of higher concentration toward regions of lower concentration.	Molecular motion	Exchange of oxygen and carbon dioxide in lungs
Facilitated diffusion	Carrier molecules move molecules through a membrane from a region of higher concentration to one of lower concentration.	Molecular motion	Movement of glucose through cell membrane
Osmosis	Water molecules move from regions of higher concentration toward regions of lower concentration through a selectively permeable membrane.	Molecular motion	Distilled water entering a cell
Filtration	Molecules are forced from regions of higher pressure to regions of lower pressure.	Hydrostatic pressure	Water molecules leaving blood capillaries
Active mechanisms			
Active transport	Carrier molecules move molecules or ions through membranes from regions of lower concentration toward regions of higher concentration.	Cellular energy	Movement of various ions, sugars, and amino acids through membranes
Endocytosis			
Pinocytosis	Membrane engulfs tiny droplets of liquid from surroundings.	Cellular energy	Membrane forming vesicles containing liquid and dissolved particles
Phagocytosis	Membrane engulfs solid particles from surroundings.	Cellular energy	White blood cell engulfing bacterial cell
Exocytosis	Vesicle fuses with membrane to expel substances from cell.	Cellular energy	Secretion of certain hormones

The Cell Cycle

The series of changes that a cell undergoes from the time it forms until it reproduces is called the *cell cycle.* Superficially, this cycle seems rather simple: A newly formed cell grows for a time and then divides to form two new cells, which, in turn, may grow and divide. Yet, the stages of the cycle are quite complex and include interphase, mitosis, cytoplasmic division, and differentiation.

Interphase

Before a cell actively divides, it must grow and duplicate much of its contents, so that two cells can form from one. This period of preparedness is called **interphase.**

Once thought to be a time of rest, interphase is actually a time of great synthetic activity. During interphase, the cell obtains nutrients, utilizes them to manufacture new living material, maintains routine "housekeeping" functions, and takes on the tremendous task of replicating its genetic material. The cell also duplicates membranes, ribosomes, lysosomes, and mitochondria.

Mitosis

Many kinds of body cells constantly grow and reproduce, thus increasing the number of cells. Such activity is responsible for the growth and development of an embryo and fetus into a child, and of a child into an adult. Growth replaces cells with relatively short life

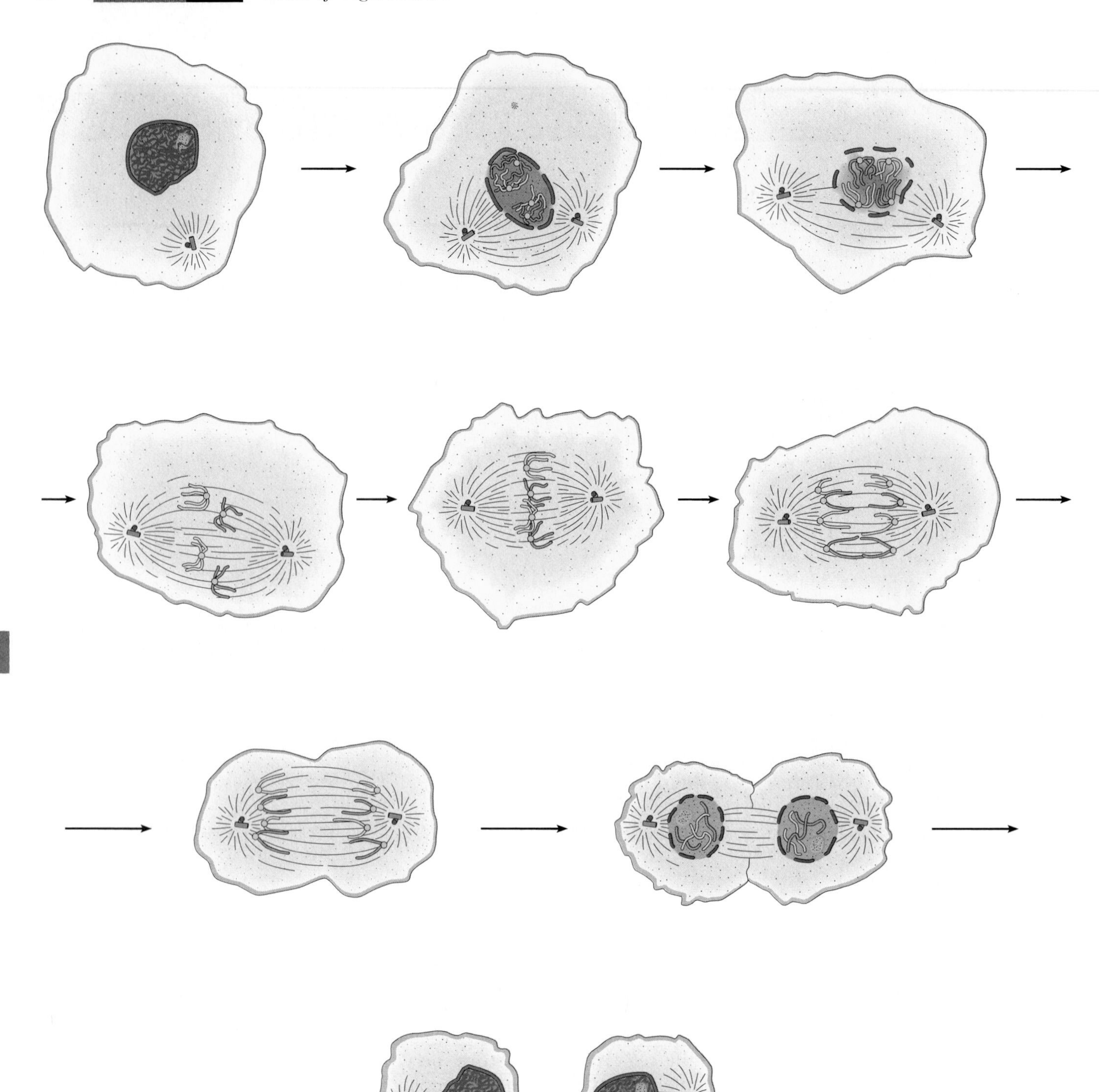

Figure 3.18

Mitosis is a continuous process during which the replicated genetic material divides into two identical sets.

spans, such as those that form the skin and stomach lining, as well as cells that are worn out or lost due to injury or disease.

In cell reproduction, a cell divides into two portions in two separate processes: (1) division of the nucleus, called **mitosis,** and (2) division of the cytoplasm. (Another type of cell division, called *meiosis,* occurs during sex cell formation and is described in chapter 19.)

Division of the nucleus must be very precise because the nucleus contains the DNA that directs life processes. Each new cell resulting from mitosis must have a complete and accurate copy of this information to survive.

Although mitosis is often described in terms of stages, the process is actually continuous, without marked changes between one step and the next (fig. 3.18). The idea of stages is useful, however, to indicate the sequence of major events. The stages of mitosis include:

1. **Prophase** One of the first indications that a cell is going to reproduce is the appearance of **chromosomes** scattered throughout the nucleus. These structures, initially fibers of chromatin in the nucleus, become visible as they condense into tightly coiled rods. Chromosomes consist of DNA and protein. Each prophase chromosome is composed of two identical portions (chromatids), which are temporarily attached at a region on each called the *centromere.*

   The centrioles of the centrosome replicate just before mitosis begins, and during prophase, the two newly formed centriole pairs move to opposite sides of the cytoplasm. Soon, the nuclear envelope and the nucleolus break up, disperse, and are no longer visible. Microtubules assemble from protein subunits in the cytoplasm and associate with the centrioles and chromosomes. A spindle-shaped group of microtubules (spindle fibers) forms between the centrioles as they move apart.
2. **Metaphase** The chromosomes line up about midway between the centrioles, as a result of microtubule activity. Spindle fibers have attached to the centromeres of each chromosome so that a fiber from one pair of centrioles attaches to one centromere, and a fiber from the other pair of centrioles attaches to the other centromere.
3. **Anaphase** Soon the centromeres are pulled apart. As the chromatids separate they become individual chromosomes. The separated chromosomes now move in opposite directions, once again guided by microtubule activity. The spindle fibers shorten and pull their attached chromosomes toward the centrioles at opposite sides of the cell.

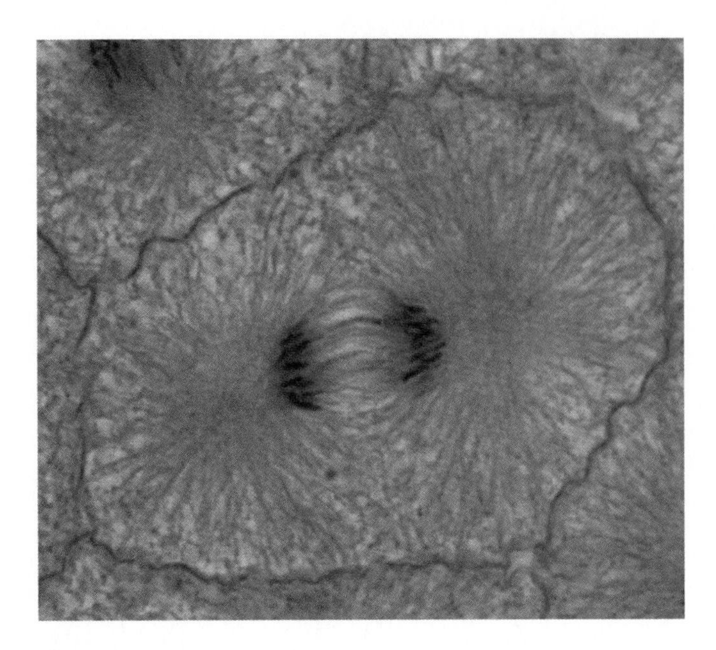

Figure 3.19

Note the two groups of darkly stained chromosomes within the cell near the center of this micrograph (280×).

4. **Telophase** The final stage of mitosis begins when the chromosomes complete their migration toward the centrioles. It is much like prophase, but in reverse. As the chromosomes approach the centrioles, the chromosomes begin to elongate and unwind from rodlike into threadlike structures. A nuclear envelope forms around each chromosome set, and nucleoli appear within the newly formed nuclei. Finally, the microtubules disassemble into their component protein building blocks (fig. 3.19).

Cytoplasmic Division

Cytoplasmic division (cytokinesis) begins during anaphase, when the cell membrane starts to constrict, an action that continues through telophase. A muscle-like contraction of a ring of microfilaments, which assemble in the cytoplasm and attach to the inner surface of the cell membrane, divides the cytoplasm. The contractile ring lies at right angles to the microtubules that pulled the chromosomes to opposite sides of the cell during mitosis. As it pinches inward, the ring separates the two newly formed nuclei and divides about half of the organelles into each of the new cells.

Although the newly formed cells may differ slightly in size and number of organelles, they have identical chromosomes and thus contain identical genetic information. Except for size, they are copies of the parent cell.

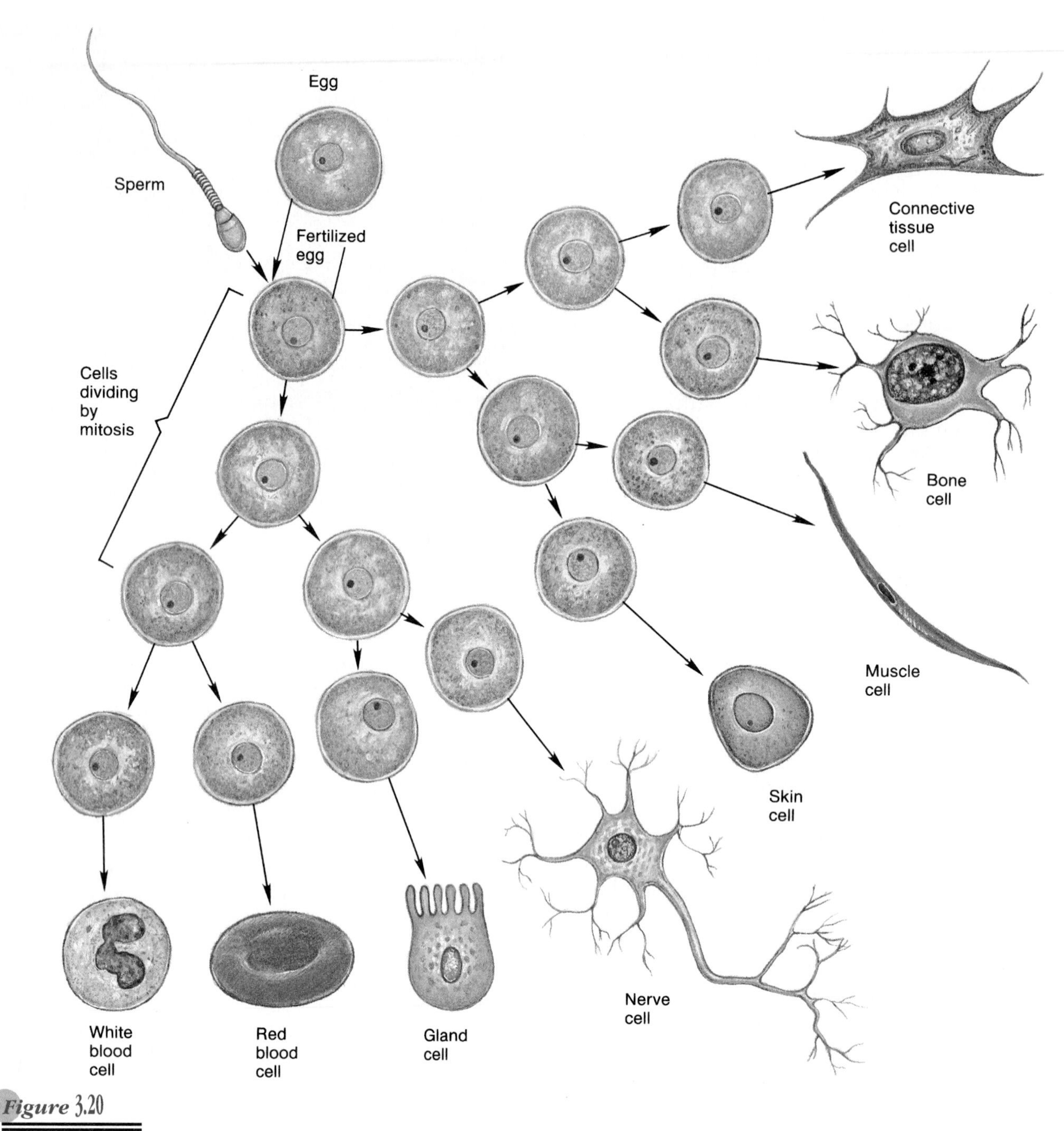

Figure 3.20

During development, many body cells are produced from a single fertilized egg cell by mitosis. As these cells differentiate, they become different kinds of cells with specialized functions. (Relative cell sizes are not to scale.)

Cell Differentiation

All body cells form by mitosis and contain the same DNA information, yet they do not look or act alike. A fertilized egg cell reproduces to form two new cells; they, in turn, divide into four cells, the four yield eight, and so forth. Then, sometime during development, the cells begin to *specialize* (fig. 3.20). That is, they develop special structures or begin to function in different ways. Some cells become muscle cells, others become bone cells, and still others become nerve cells.

The process by which cells develop different characteristics in structure and function is called **differentiation.** Cells differentiate by expressing some of the DNA information and repressing other DNA information. Thus, the DNA information required for general cell activities is activated in both nerve and bone cells, but information important to specific bone cell functions is activated in bone cells, yet repressed in nerve cells. Similarly, the information necessary for specific nerve cell functions is repressed in bone cells.

TOPIC OF INTEREST

Cancer

Cancer is a group of closely related diseases that can occur in many different tissues. These conditions result from changes in cells that inactivate some of the mechanisms that regulate cell activities. Cancers have certain common characteristics, including the following:

1. **Hyperplasia** Hyperplasia is uncontrolled cell reproduction. In normal cells, chromosome tips, called telomeres, normally wear down (shorten) with each mitosis. When the tips shorten to a certain point, this signals the cell to cease reproducing. The tips of cancer cells do not shorten, and as a result, they do not "know" to stop reproducing.
2. **Dedifferentiation** Cancer cells typically resemble undifferentiated cells, a state termed dedifferentiation. That is, they lose the specialized structures of the normal type of cell from which they descend. In addition, dedifferentiated cells lose specialized functions. Cancer cells also form disorganized masses, rather than organize in orderly groups as normal cells do.
3. **Metastasis** Metastasis is a tendency to spread into other tissues. Normal cells are usually cohesive; that is, they stick together in groups of similar kinds. Small numbers of cancer cells often detach from their original (primary) cellular mass and move from their place of origin (fig. 3A). If metastatic cells evade the immune system, they may establish new (secondary) cancerous growths in other body parts.

Mutations in certain genes cause many cancers. A cancer-causing mutation may activate a cancer-causing oncogene (a gene that normally controls mitotic rate) or inactivate a protective gene called a tumor suppressor. Environmental insults thought to cause cancer may cause these types of mutations in somatic (nonsex) cells, leading to the runaway cell reproduction that is cancer.

One in three of us will develop some form of cancer. Fortunately, doctors can treat the majority of cases.

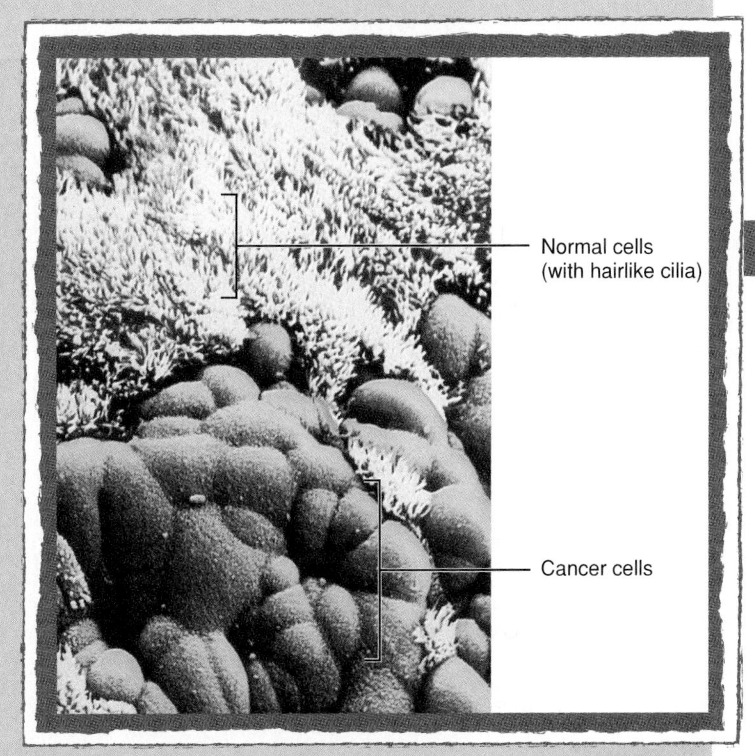

***Figure* 3A**

Cancer cells, misshapen and unspecialized, encroach upon normal cells.

1. Why is it important that the division of nuclear materials during mitosis be precise?
2. Describe the events that occur during mitosis.
3. Name the process by which some cells become muscle cells and others become nerve cells.

Summary Outline

Introduction (p. 51)

Cells vary considerably in size, shape, and function. The shapes of cells make possible their functions.

Composite Cell (p. 51)

A cell includes a nucleus, cytoplasm, and a cell membrane. Cytoplasmic organelles perform specific metabolic functions, but the nucleus controls overall cell activities.

1. Cell membrane
   a. The cell membrane forms the outermost limit of the living material.
   b. It is a selectively permeable passageway that controls the entrance and exit of substances.
   c. The cell membrane includes protein, lipid, and carbohydrate molecules.
   d. The cell membrane's framework is mainly a double layer of phospholipid molecules.
   e. Molecules that are soluble in lipids pass through the cell membrane easily, but water-soluble molecules do not.
   f. Proteins function as receptors on membrane surfaces and form channels for the passage of ions and molecules.
   g. Carbohydrates associated with membrane proteins enable certain cells to recognize one another.
2. Cytoplasm
   a. The endoplasmic reticulum is a tubular communication system in the cytoplasm that transports lipids and proteins.
   b. Ribosomes function in protein synthesis.
   c. The Golgi apparatus packages glycoproteins for secretion.
   d. Mitochondria contain enzymes that catalyze reactions that release energy from food molecules.
   e. Lysosomes contain digestive enzymes that can decompose substances that enter cells.
   f. The centrosome contains centrioles that aid in distributing chromosomes during cell reproduction.
   g. Cilia and flagella are motile processes that extend from cell surfaces.
   h. Vesicles contain substances that recently entered the cell or that are to be secreted from the cell.
   i. Microfilaments and microtubules aid cellular movements and provide support and stability to the cytoplasm and cell organelles.
3. Cell nucleus
   a. The nucleus is enclosed in a double-layered nuclear envelope.
   b. It contains a nucleolus, which is the site of ribosome production.
   c. It contains chromatin, which is composed of loosely coiled fibers of DNA and protein. As chromatin fibers condense, chromosomes become visible during cell reproduction.

Movements through Cell Membranes (p. 59)

The cell membrane is a barrier through which substances enter and leave a cell.

1. Passive mechanisms do not require cellular energy.
   a. Diffusion
      (1) Diffusion is the movement of molecules or ions from regions of higher concentration toward regions of lower concentration.
      (2) It is responsible for exchanges of oxygen and carbon dioxide within the body.
   b. Facilitated diffusion
      (1) In facilitated diffusion, special carrier molecules move substances through the cell membrane.
      (2) This process moves substances only from regions of higher concentration toward regions of lower concentration.
   c. Osmosis
      (1) Osmosis is diffusion of water molecules from regions of higher water concentration toward regions of lower water concentration through a selectively permeable membrane.
      (2) Osmotic pressure increases as the number of particles dissolved in a solution increases.
      (3) A solution is isotonic to a cell when it has the same concentration of solute particles as the cell.
      (4) Cells lose water when placed in hypertonic solutions and gain water when placed in hypotonic solutions.
   d. Filtration
      (1) Filtration is the movement of molecules from regions of higher hydrostatic pressure toward regions of lower hydrostatic pressure.
      (2) Blood pressure causes filtration through porous capillary walls.
2. Active mechanisms require cellular energy.
   a. Active transport
      (1) Active transport moves molecules or ions from regions of lower concentration toward regions of higher concentration.
      (2) It requires cellular energy and carrier molecules in the cell membrane.
   b. Endocytosis and exocytosis
      (1) Endocytosis may convey relatively large particles into a cell.
      (2) In pinocytosis, a cell membrane engulfs tiny droplets of liquid.
      (3) In phagocytosis, a cell membrane engulfs solid particles.
      (4) Exocytosis is the reverse of endocytosis.

The Cell Cycle (p. 65)

The cell cycle includes interphase, mitosis, cytoplasmic division, and differentiation.

1. Interphase
   a. Interphase is the stage when a cell grows and forms new organelles.
   b. It terminates when the cell begins mitosis.
2. Mitosis
   a. Mitosis is the division and distribution of the nucleus to new cells during cell reproduction.
   b. The stages of mitosis include prophase, metaphase, anaphase, and telophase.
3. Cytoplasmic division distributes cytoplasm into two portions following mitosis.
4. Cell differentiation is the development of specialized structures and functions.

Review Exercises

1. Describe how the shapes of nerve and muscle cells are important for their functions. (p. 51)
2. Name the two major portions of a cell, and describe their relationship to one another. (p. 51)
3. Define *selectively permeable*. (p. 53)
4. Describe the chemical structure of a cell membrane. (p. 53)
5. Explain how the structure of a cell membrane determines which substances can pass through it. (p. 53)
6. Describe the structures and functions of each of the following: (p. 54)
   a. Endoplasmic reticulum
   b. Ribosome
   c. Golgi apparatus
   d. Mitochondrion
   e. Lysosome
   f. Centrosome
   g. Cilium and flagellum
   h. Vesicle
   i. Microfilament and microtubule
7. Describe the structure of the nucleus and the functions of its parts. (p. 57)
8. Define *diffusion*. (p. 59)
9. Explain how diffusion aids in the exchange of gases within the body. (p. 60)
10. Distinguish between diffusion and facilitated diffusion. (p. 61)
11. Define *osmosis*. (p. 61)
12. Define *osmotic pressure*. (p. 62)
13. Distinguish between solutions that are hypertonic, hypotonic, and isotonic. (p. 62)
14. Define *filtration*. (p. 62)
15. Explain how filtration moves substances through capillary walls. (p. 62)
16. Explain the function of carrier molecules in active transport. (p. 63)
17. Distinguish between pinocytosis and phagocytosis. (p. 64)
18. List the phases in the cell cycle. (p. 65)
19. Explain what happens during interphase. (p. 65)
20. Name the two processes of cell reproduction. (p. 67)
21. Describe the major events of mitosis. (p. 67)
22. Explain how the cytoplasm divides during cell reproduction. (p. 67)
23. Define *differentiation*. (p. 68)

Critical Thinking

1. Organelles compartmentalize a cell, much as a department store displays related items together. What advantage does such compartmentalization offer a large cell? Cite two examples of organelles and the activities that they compartmentalize.
2. Liver cells are packed with glucose. What mechanism could be used to transport more glucose into a liver cell? Why would only this mode of transport work?
3. In an inherited condition called glycogen cardiomyopathy, teenagers develop muscle weakness, which affects the heart as well as other muscles. Samples of the affected muscle cells contain huge lysosomes, swollen with the carbohydrate glycogen. How might this condition arise?
4. Why does a muscle cell contain many mitochondria, and a white blood cell contain many lysosomes?
5. Many drugs used to treat cancer stop working because of a large protein, called P-gp, in the cell membranes of some cancer cells. This protein causes a phenomenon called "multidrug resistance" because it pumps many types of drugs out of the cell membrane before they can enter the cytoplasm. P-gp requires ATP (the biological energy molecule) to function, and it continually cycles from its site of synthesis in the cytoplasm to the cell membrane and then is recycled inside the cell. What structures most likely transport P-gp?
6. Which process—diffusion, osmosis, or filtration—is utilized in the following situations?
   a. Injection of a drug that is hypertonic to the tissues stimulates pain.
   b. The urea concentration in the dialyzing fluid of an artificial kidney is decreased.
7. What characteristic of cell membranes may explain why fat-soluble substances like chloroform and ether rapidly affect cells?
8. Exposure to tobacco smoke immobilizes cilia, and they eventually disappear. How might this effect account for smokers having an increased incidence of coughing and respiratory infections?

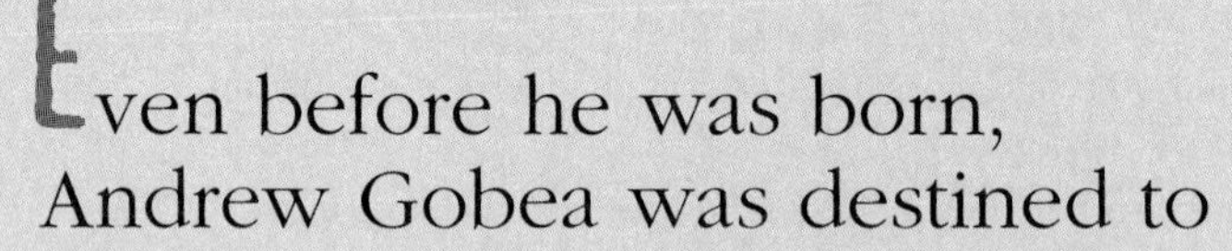

Even before he was born, Andrew Gobea was destined to die young—but instead, he lived to make history. When his parents learned from prenatal genetic tests that Andrew had inherited the same severe combined immune deficiency (SCID) that had killed his brother, they desperately sought help. Fortunately, the type of SCID in the family was the target of the first gene therapy experiments.

Andrew's body lacked an enzyme that normally controls breakdown of a particular toxin. Without the enzyme, the toxin accumulates and destroys immune system cells called T cells. T cells are vital to the immune response. When they die, protection against infection topples. (SCID is like an inherited form of AIDS.)

Doctors had already successfully treated a four-year-old and eight-year-old for the same condition by removing some of their white blood cells and bolstering them with the gene that tells the cells to manufacture the missing enzyme. With Andrew, they hoped to intervene earlier. So at his birth, physicians took certain immature blood cells from his umbilical cord, "fixed" them with the missing gene, and on Andrew's fourth day of life, infused his patched cells into him. Andrew was also given the missing enzyme directly to prevent symptoms from developing.

Over the next few years, as doctors detected healthy enzyme-producing T cells in Andrew's bloodstream, they gradually decreased the enzyme supplements. Today, Andrew and two other youngsters treated at the same time are healthy preschoolers.

Photo: Each enzyme that controls a metabolic reaction is vital to health. Newborn Andrew Gobea lacked a key enzyme, and without gene therapy, buildup of a toxic biochemical would have shut down his immune system. Gene therapy at four days of age enabled some of his white blood cells to manufacture the missing enzyme, and he survived.

Cells require energy and information to build bodies. Cells house the many chemical reactions of metabolism. These reactions break down nutrients to release energy and also build molecules to store energy. Cells carry genetic information that encodes the amino acid sequences of enzymes and other proteins. Enzymes control metabolism.

Chapter Objectives

After studying this chapter, you should be able to:

1. Define anabolism and catabolism. (p. 74)
2. Explain how enzymes control metabolic reactions. (p. 75)
3. Explain how cellular respiration releases chemical energy. (p. 77)
4. Describe how energy becomes available for cellular activities. (p. 77)
5. Describe the general metabolic pathways of carbohydrates, lipids, and proteins. (p. 78)
6. Explain how nucleic acid molecules (DNA and RNA) store and carry genetic information. (p. 82)
7. Explain how genetic information controls cellular processes. (p. 82)
8. Describe how DNA molecules replicate. (p. 89)

Aids to Understanding Words

an- [without] *an*aerobic respiration: A respiratory process that proceeds without oxygen.

ana- [up] *ana*bolism: Cellular processes that use smaller molecules to build larger ones.

cata- [down] *cata*bolism: Cellular processes that break larger molecules into smaller ones.

de- [undoing] *de*amination: A process that removes the nitrogen-containing portions of amino acid molecules.

mut- [change] *mut*ation: A change in the genetic information of a cell.

-zym [causing to ferment] en*zym*e: A protein that initiates or speeds a chemical reaction without itself being consumed.

Key Terms

aerobic respiration (a″er-o′bik res″pĭ-ra′shun)

anabolism (an″ah-bol′lizm)

anaerobic respiration (an″a-er-o′bik res″pĭ-ra′shun)

anticodon (an″tĭ-ko′don)

catabolism (kat″ah-bol′lizm)

codon (ko′don)

dehydration synthesis (de″hi-dra′shun sin′thĕ-sis)

enzyme (en′zīm)

gene (jēn)

hydrolysis (hi-drol′ĭ-sis)

oxidation (ok″sĭ-da′shun)

replication (rĕ″plĭ-ka′shun)

substrate (sub′strāt)

Introduction

A living cell is the site of many ongoing metabolic reactions that maintain life. A special type of protein called an **enzyme** controls each of the interrelated reactions of metabolism.

Metabolic Reactions

Metabolic reactions are of two major types. **Anabolism** is the buildup of larger molecules from smaller ones and requires energy. **Catabolism** is the breakdown of larger molecules into smaller ones and releases energy.

Anabolism

Anabolism provides the biochemicals required for cell growth and repair. For example, cells join many simple sugar molecules (monosaccharides) into a chain to form larger molecules of glycogen using an anabolic process called **dehydration synthesis.** As adjacent monosaccharide units join, an —OH (hydroxyl group) from one monosaccharide molecule and an —H (hydrogen atom) from an —OH group of another are removed. The —H and —OH react to produce a water molecule, and the monosaccharides unite by a shared oxygen atom (fig. 4.1). As this process repeats, the molecular chain grows.

Dehydration synthesis also links glycerol and fatty acid molecules in fat (adipose) cells to form fat molecules (triglycerides). In this case, three hydrogen atoms are removed from a glycerol molecule, and an —OH group is removed from each of the three fatty acid molecules (fig. 4.2). The result is three water molecules and a single fat molecule, whose glycerol and fatty acid portions are bound by shared oxygen atoms.

Cells also join amino acid molecules by dehydration synthesis to build protein molecules (fig. 4.3). When two amino acid molecules unite, an —OH from one and an —H from the $—NH_2$ group of another are removed. A water molecule forms, and the amino acid molecules are joined by a bond between a carbon atom and a nitrogen atom, called a *peptide bond.* Two bound amino acids form a *dipeptide,* and many joined in a chain form a *polypeptide.* Generally, a polypeptide that has a specific function and consists of perhaps 100 or more amino acid molecules is called a *protein,* although the boundary distinguishing between polypeptides and proteins is not defined precisely.

Catabolism

Physiological processes that break larger molecules into smaller ones constitute catabolism. An example of catabolism is **hydrolysis,** which decomposes carbohydrates, lipids, and proteins, and splits a water molecule in the process.

Hydrolysis of a disaccharide such as sucrose, for instance, yields two monosaccharides—glucose and fructose—as a molecule of water splits:

$$\underset{\text{(sucrose)}}{C_{12}H_{22}O_{11}} + \underset{\text{(water)}}{H_2O} \rightarrow \underset{\text{(glucose)}}{C_6H_{12}O_6} + \underset{\text{(fructose)}}{C_6H_{12}O_6}$$

In this case, the bond between the simple sugars within the sucrose molecule breaks, and the water molecule supplies a hydrogen atom to one sugar molecule and a hydroxyl group to the other. Thus, hydrolysis is the opposite of dehydration synthesis (see figs. 4.1, 4.2, and 4.3). Each of these reactions is reversible and can be summarized as follows:

$$\text{Disaccharide} + \text{Water} \underset{\text{Dehydration Synthesis}}{\overset{\text{Hydrolysis}}{\rightleftharpoons}} \text{Monosaccharide} + \text{Monosaccharide}$$

Hydrolysis, which occurs during *digestion,* breaks down carbohydrates into monosaccharides, fats into glycerol and fatty acids, proteins into amino acids, and nucleic acids into nucleotides. (Chapter 15 discusses digestion in more detail.)

1. What are the functions of anabolism? Of catabolism?
2. What is the product of anabolism of monosaccharides? Of glycerol and fatty acids? Of amino acids?
3. Distinguish between dehydration synthesis and hydrolysis.

A*nabolic steroids* are a group of lipids that stimulate anabolism and, thus, promote the growth of certain tissues. Doctors prescribe anabolic steroids to treat various diseases. Some individuals, however, use these powerful drugs illicitly to increase muscle mass, often with the hope of enhancing athletic performance. Such nonmedical use of steroids can cause dangerous side effects, including adverse changes in liver functions, increased risk of heart and blood vessel diseases, upsets in normal hormonal balances, infertility, and severe psychological disorders. ■

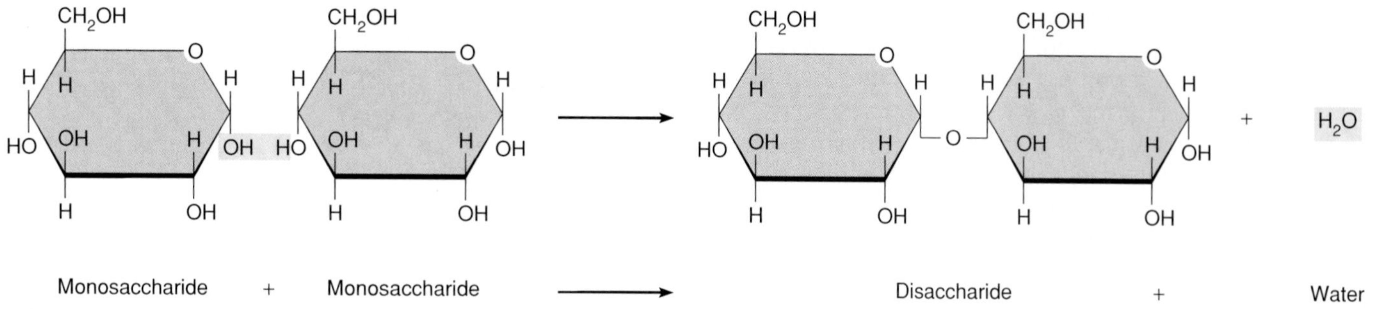

Monosaccharide + Monosaccharide → Disaccharide + Water

Figure 4.1

A synthesis reaction.

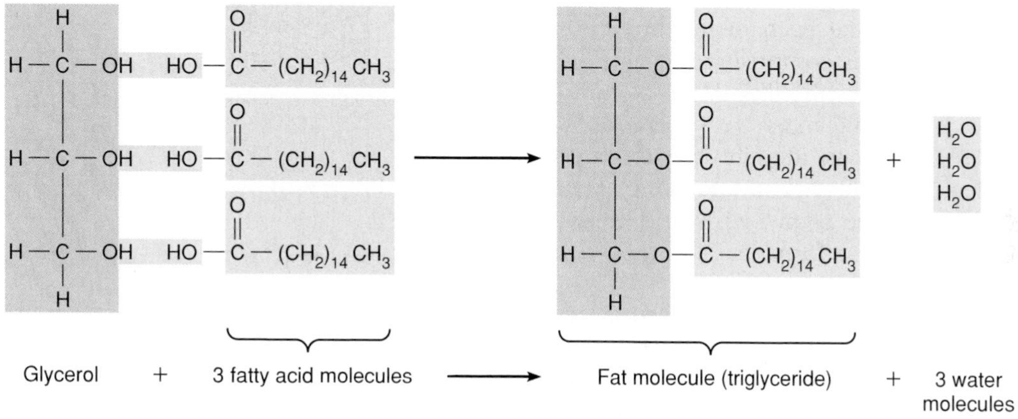

Glycerol + 3 fatty acid molecules → Fat molecule (triglyceride) + 3 water molecules

Figure 4.2

Dehydration synthesis joins a glycerol molecule and three fatty acid molecules to form a fat molecule (triglyceride).

Peptide bond

Amino acid + Amino acid → Dipeptide molecule + Water

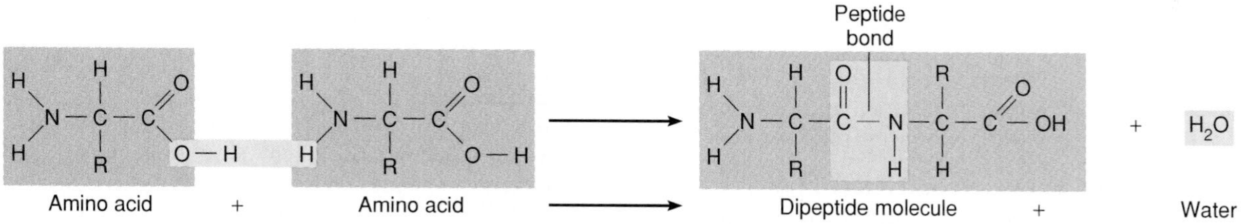

Figure 4.3

When dehydration synthesis unites two amino acid molecules, a peptide bond forms between a carbon atom and a nitrogen atom.

Control of Metabolic Reactions

Specialized cells, such as nerve, muscle, or blood cells, have distinctive chemical reactions. However, all cells perform certain basic reactions, such as buildup and breakdown of carbohydrates, lipids, proteins, and nucleic acids. These reactions include hundreds of specific chemical changes that occur rapidly—yet in a coordinated fashion—thanks to enzymes.

Enzyme Action

Like other chemical reactions, metabolic reactions require energy to proceed. The temperature conditions in cells, however, are usually too mild to adequately promote the reactions required to support life. Enzymes make these reactions possible.

Enzymes are complex molecules, almost always proteins, that promote chemical reactions within cells by lowering the amount of energy (activation energy) required to start these reactions. Thus, enzymes speed the rates of metabolic reactions. This acceleration is called catalysis, and an enzyme is a catalyst. Enzymes are required in very small quantities because, as they function, they are not consumed and can, therefore, be recycled.

The reaction a particular enzyme catalyzes is very specific. Each enzyme acts only on a particular chemical, which is called its **substrate.** For example, the substrate of the enzyme *catalase* is hydrogen peroxide, a toxic by-product of certain metabolic reactions. This enzyme's only function is to speed up the breakdown of hydrogen peroxide into water and oxygen. Thus, catalase helps prevent an accumulation of hydrogen peroxide, which might damage cells.

Specific enzymes catalyze each of the hundreds of different chemical reactions comprising cellular metabolism. Thus, every cell contains hundreds of different enzymes, and each enzyme must "recognize" its specific substrate. This ability to identify its substrate depends on the shape of the enzyme molecule. That is, each enzyme's polypeptide chain twists and coils into a unique three-dimensional form that fits the specific shape of its substrate.

During an enzyme-controlled reaction, parts of the enzyme molecule called **active sites** temporarily combine with portions of the substrate, forming an enzyme-substrate complex (fig. 4.4). This interaction between the molecules distorts or strains chemical bonds within the substrate, increasing the likelihood that the reaction will occur. When the substrate changes, the reaction takes place, its products form, and the enzyme is released in its original form.

An enzyme-catalyzed reaction can be summarized as follows:

$$\text{Substrate molecule} + \text{Enzyme molecule} \rightarrow \text{Enzyme-substrate complex} \rightarrow \text{Product (changed substrate)} + \text{Enzyme molecule}$$

The speed of an enzyme-controlled reaction depends on the number of enzyme and substrate molecules in the cell. Generally, the reaction occurs more rapidly if the concentrations of the enzyme or the substrate increase. Also, the efficiency of different kinds of enzymes varies greatly. Thus, some enzymes can process only a few substrate molecules per second, while others can process thousands or hundreds of thousands of substrate molecules per second. Many enzymatic reactions are reversible.

Factors That Alter Enzymes

Almost all enzymes are proteins, and like other proteins, they can be denatured by exposure to heat, radiation, electricity, certain chemicals, or fluids with extreme pH values. For example, many enzymes become inactive at 45° C, and nearly all of them are denatured at 55° C. Some poisons work by denaturing enzymes. Cyanides, for instance, interfere with respiratory enzymes, impairing a cell's ability to release energy from nutrient molecules.

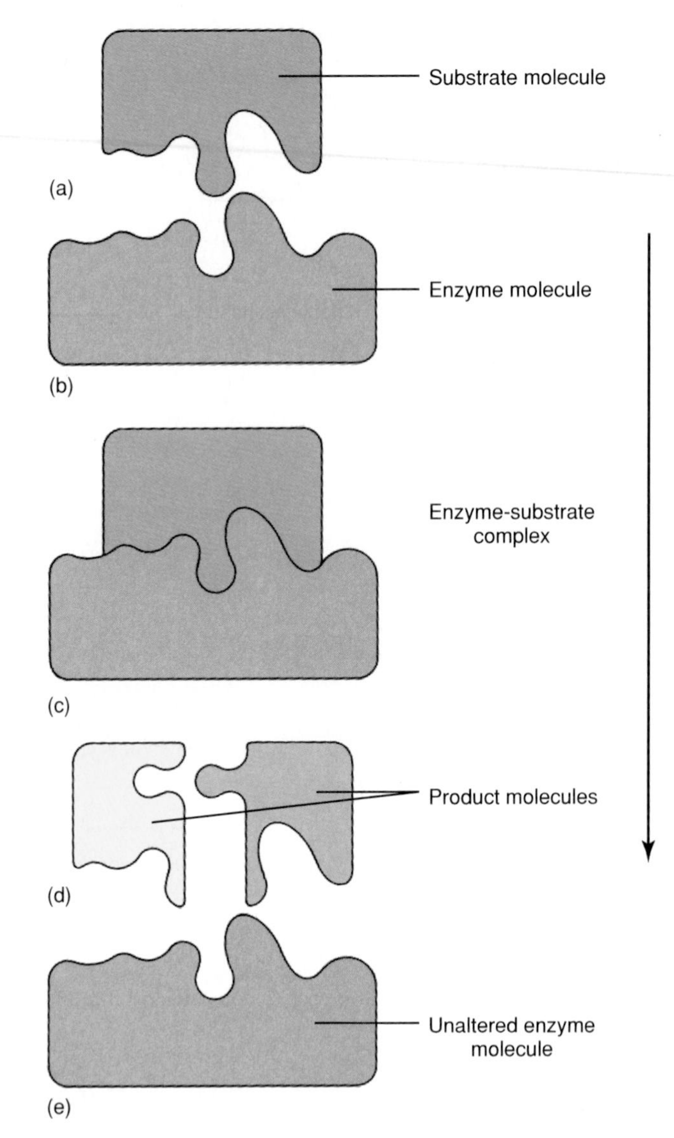

Figure 4.4

(*a*) The shape of a substrate molecule fits (*b*) the shape of the enzyme's active site. (*c*) When the substrate molecule temporarily combines with the enzyme, a chemical reaction occurs. The result is (*d*) product molecules and (*e*) an unaltered enzyme.

1. What is an enzyme?
2. How does an enzyme recognize its substrate?
3. What factors affect the speed of an enzyme-controlled reaction?
4. What factors can denature enzymes?

Often, an enzyme is inactive until it combines with another molecule. This nonprotein part completes the shape of the enzyme's active site or helps bind the enzyme to its substrate. Such a nonprotein portion is called a **cofactor.** It may be an ion of copper, iron, or zinc, or it may be a small organic molecule, in which case it is called a **coenzyme.** Most coenzymes are vitamin molecules or have vitamin molecules incorporated into their structures. Vitamins are organic molecules—other than carbohydrates, lipids, or proteins—that are required in small amounts for normal metabolism. However, body cells cannot synthesize vitamins in adequate amounts. Vitamins must thus come from foods. ■

Energy for Metabolic Reactions

Energy is the capacity to change or move matter; that is, energy is the ability to do work. Therefore, we recognize energy by what it has done, by whatever changes occur. Common forms of energy include heat, light, sound, electrical energy, mechanical energy, and chemical energy.

Release of Chemical Energy

Most metabolic processes use chemical energy. This form of energy is held in the bonds between the atoms of molecules and is released when these bonds break. For example, burning releases the chemical energy of many chemicals. Such a reaction usually starts by applying heat to activate the burning process. As the chemical burns, bonds break, and energy escapes as heat and light.

Cells "burn" glucose molecules in a process more correctly called **oxidation.** The energy released by the oxidation of glucose powers the reactions of cellular metabolism. However, the oxidation of biochemicals inside cells and the burning of substances outside cells differ in some ways.

Burning usually requires a relatively large amount of energy to begin, and most of the energy released escapes as heat or light. In cells, enzymes reduce the amount of energy (activation energy) required for oxidation. Also, by transferring energy to special energy-carrying molecules, cells can capture in the form of chemical energy about half of the energy released. The rest escapes as heat, which helps maintain body temperature.

1. What is energy?
2. How does cellular oxidation differ from burning?

Anaerobic Respiration

During the first part of *cellular respiration,* enzymes control a series of reactions that break down six-carbon glucose into two three-carbon pyruvic acid molecules (fig. 4.5). This phase of the process occurs in the cytosol

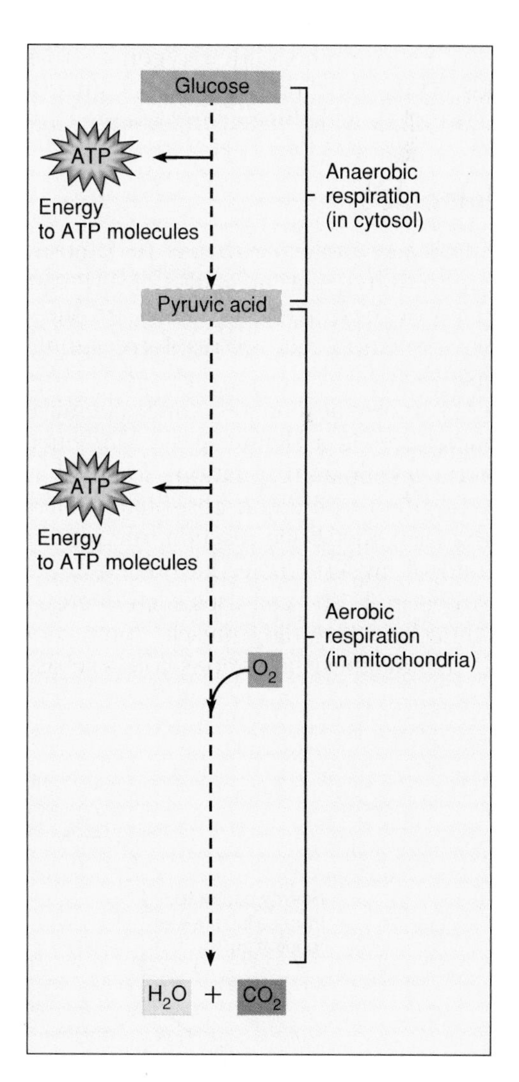

Figure 4.5

Anaerobic respiration occurs in the cytosol and does not require oxygen, while aerobic respiration occurs in the mitochondria and only in the presence of oxygen.

(the liquid portion of the cytoplasm), and because it does not require oxygen, it is called **anaerobic respiration** (glycolysis).

Although some energy is required to activate the reactions of anaerobic respiration, more energy is released than is consumed. The excess energy is used to synthesize the energy-carrying molecule **ATP** (adenosine triphosphate).

Aerobic Respiration

Following the anaerobic phase of cellular respiration, oxygen must be available for the process to continue. For this reason, the second phase is called **aerobic respiration.** It takes place within mitochondria and transfers energy to considerably more ATP molecules.

When glucose breakdown is complete, carbon dioxide molecules and hydrogen atoms remain. The carbon dioxide diffuses out of the cell as a waste, and the hydrogen atoms combine with oxygen to form water molecules. Thus, the final products of glucose oxidation are carbon dioxide, water, and energy (fig. 4.5).

ATP Molecules

For each glucose molecule that is decomposed completely, up to thirty-eight molecules of ATP can be produced. Two of these ATP molecules are the result of anaerobic respiration, and the rest form during the aerobic phase.

Each ATP molecule contains three chemical groups called phosphates in a chain (fig. 4.6). As energy is released during cellular respiration, some of it is captured in the bond of the end phosphate. When energy is required for a metabolic reaction, the terminal phosphate bond of an ATP molecule breaks, releasing the energy stored in it. The cell uses such energy for a variety of functions, including muscle contraction, active transport, and synthesis of various compounds.

An ATP molecule that has lost its terminal phosphate becomes an **ADP** (adenosine diphosphate) molecule. The ADP molecule can convert back into an ATP, however, by capturing some energy and a phosphate. Thus, as figure 4.6 shows, ATP and ADP molecules shuttle back and forth between the energy-releasing reactions of cellular respiration and the energy-utilizing reactions of the cell.

1. What is anaerobic respiration? Aerobic respiration?
2. What are the final products of cellular respiration?
3. What is the general function of ATP?

Metabolic Pathways

Anabolic and catabolic reactions usually have a number of steps that must occur in a particular sequence, such as the reactions of aerobic respiration. The enzymes that control the rates of these reactions must act in a specific sequence. Such coordination suggests that the enzymes are positioned in the exact sequence as that of the reactions they control. For example, the enzymes responsible for aerobic respiration are located in tiny, stalked particles on the membranes (cristae) within the mitochondria, in the sequence in which they function. A sequence of enzyme-controlled reactions is called a **metabolic pathway** (fig. 4.7).

Carbohydrate Pathways

The average human diet consists largely of carbohydrates, which are digested into monosaccharides such as glucose. Carbohydrates are used primarily as cellular

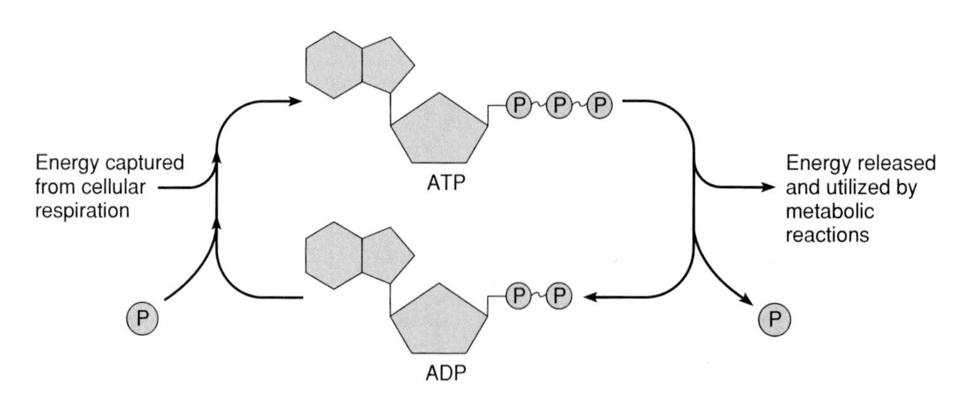

***Figure* 4.6**

ATP provides energy for cellular reactions. Cellular respiration regenerates ATP.

energy sources, which means they usually enter the catabolic pathways of cellular respiration (fig. 4.8). As discussed previously, the first phase of this process, glycolysis, occurs in the cytosol and is anaerobic. These reactions break down glucose into *pyruvic acid* (pyruvate). Each glucose molecule breaks down to yield two pyruvic acid molecules.

In the second phase of carbohydrate breakdown, the pyruvic acid reacts to form a two-carbon acetyl group, which combines with a molecule called coenzyme A to form *acetyl coenzyme A*. It, in turn, is transported into a mitochondrion, where a series of chemical reactions known as the **citric acid cycle** (Krebs cycle), changes it into a number of intermediate products. These changes

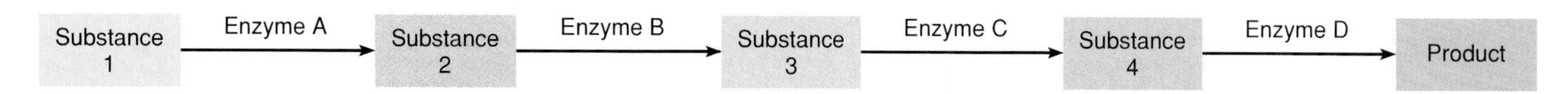

Figure 4.7

A metabolic pathway consists of a series of enzyme-controlled reactions leading to formation of a product.

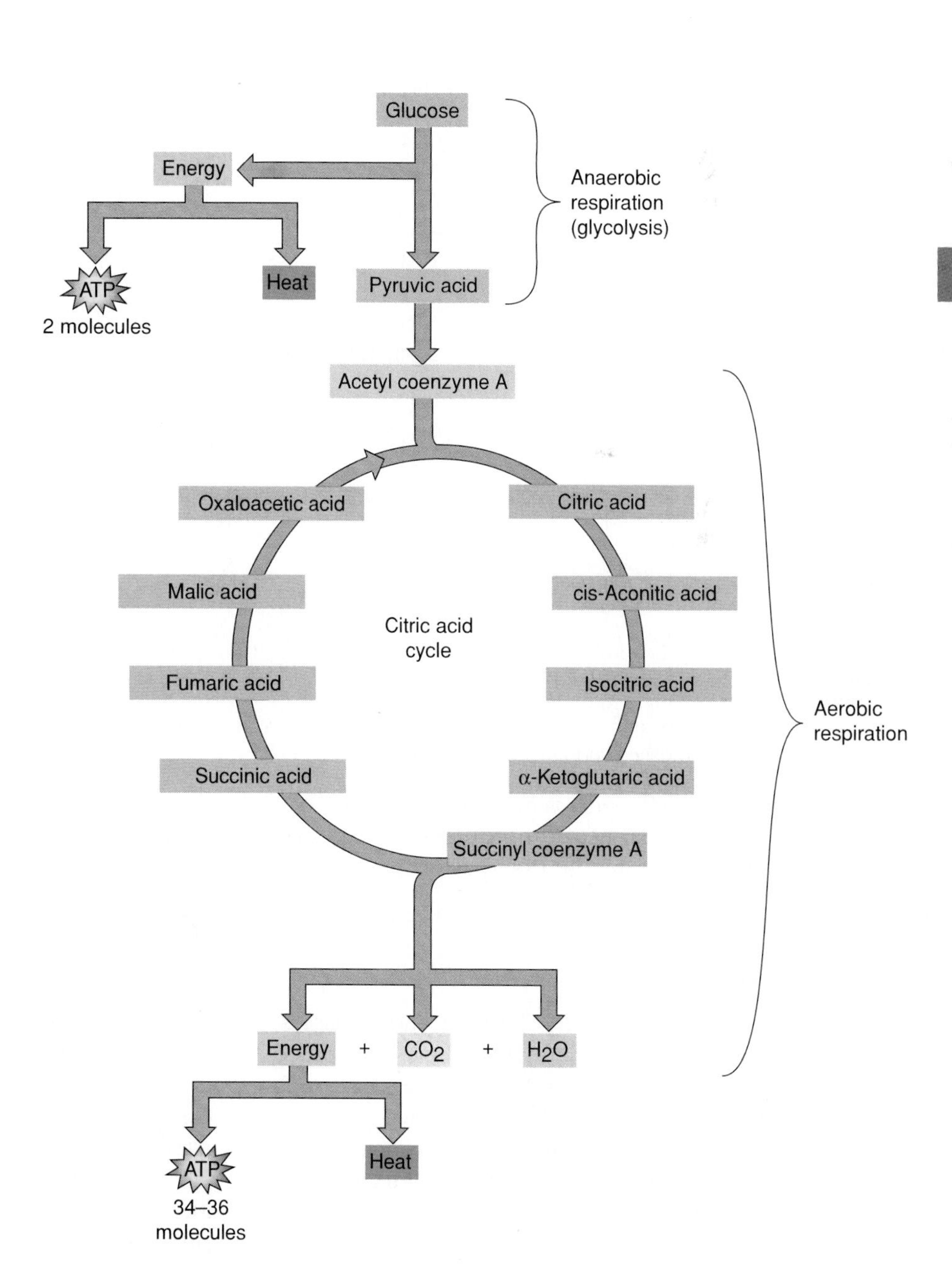

Figure 4.8

Glycolysis, the first phase of cellular respiration, occurs in the cytosol. The second phase, aerobic respiration, occurs in the mitochondria and includes a series of chemical reactions called the citric acid cycle. The total number of ATP molecules that cellular respiration produces from one glucose molecule is either thirty-six or thirty-eight, depending on the cell type.

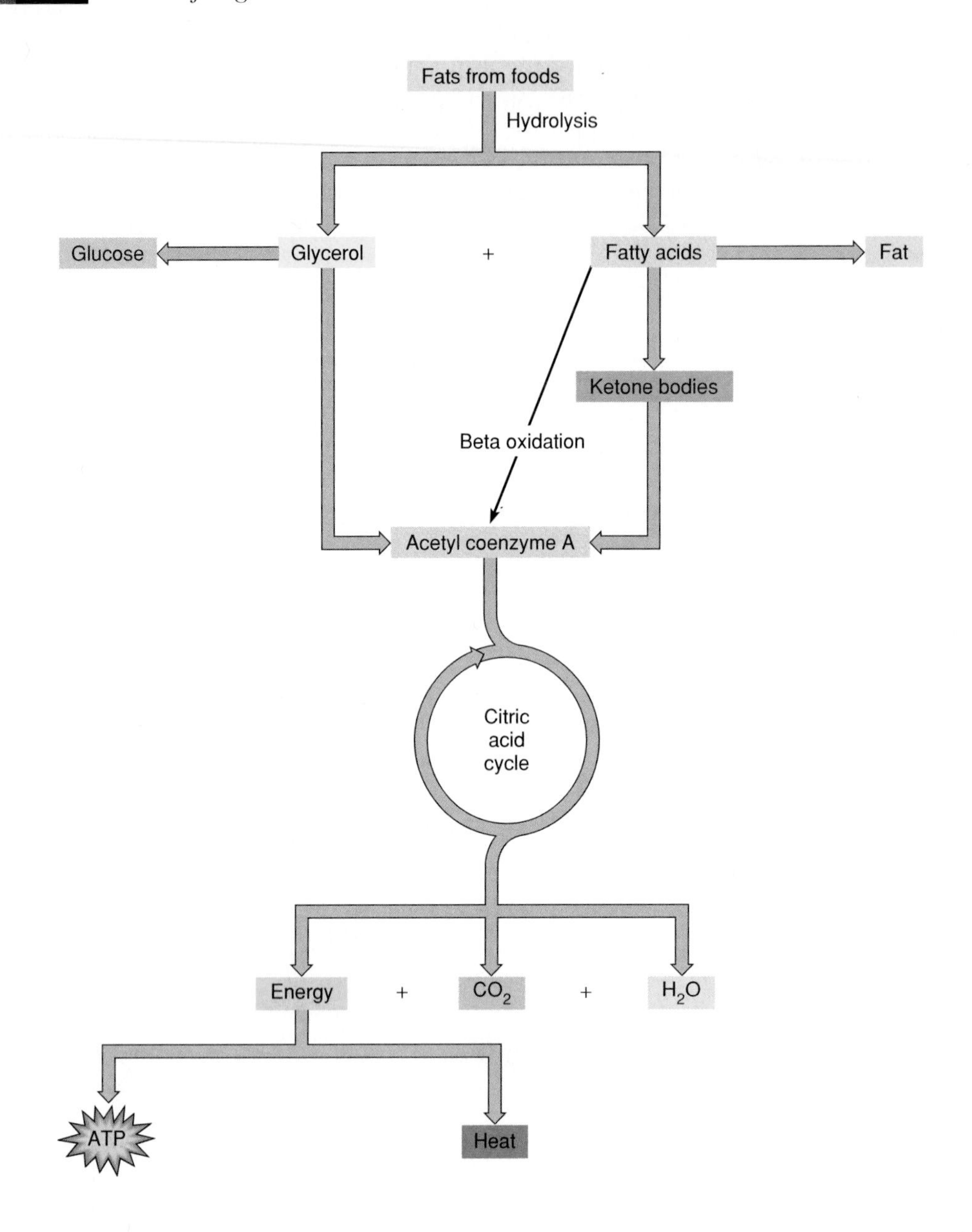

Figure 4.9

The body digests fats from foods into glycerol and fatty acids, which may enter catabolic pathways and be used as energy sources.

release energy, some of which is transferred to ATP molecules. The rest is lost as heat. Any excess glucose may enter anabolic pathways and be stored as glycogen or fat.

Lipid Pathways

Many foods contain lipids in the form of phospholipids, cholesterol, or the most common dietary lipids, fats called *triglycerides*. Recall from chapter 2 that a triglyceride molecule consists of a glycerol and three fatty acids.

Lipids serve a variety of physiological functions but mainly supply energy. Gram for gram, fats contain more than twice as much chemical energy as carbohydrates or proteins.

A fat molecule must undergo hydrolysis before it can release energy. Figure 4.9 shows that, upon hydrolysis, some of the resulting fatty acid portions can react to form molecules of acetyl coenzyme A by a series of reactions called *beta oxidation*. Other breakdown products of fat hydrolysis can react to yield compounds called *ketone bodies,* such as acetone, which later may react to form acetyl coenzyme A. In either case, the resulting acetyl coenzyme A can be oxidized in the citric acid cycle. The glycerol portions of the triglyceride molecules can also enter metabolic pathways leading to the citric acid cycle, or they can be used to synthesize glucose.

Fatty acid molecules released from fat hydrolysis can also combine to form fat molecules by anabolic processes and are stored in fat tissue.

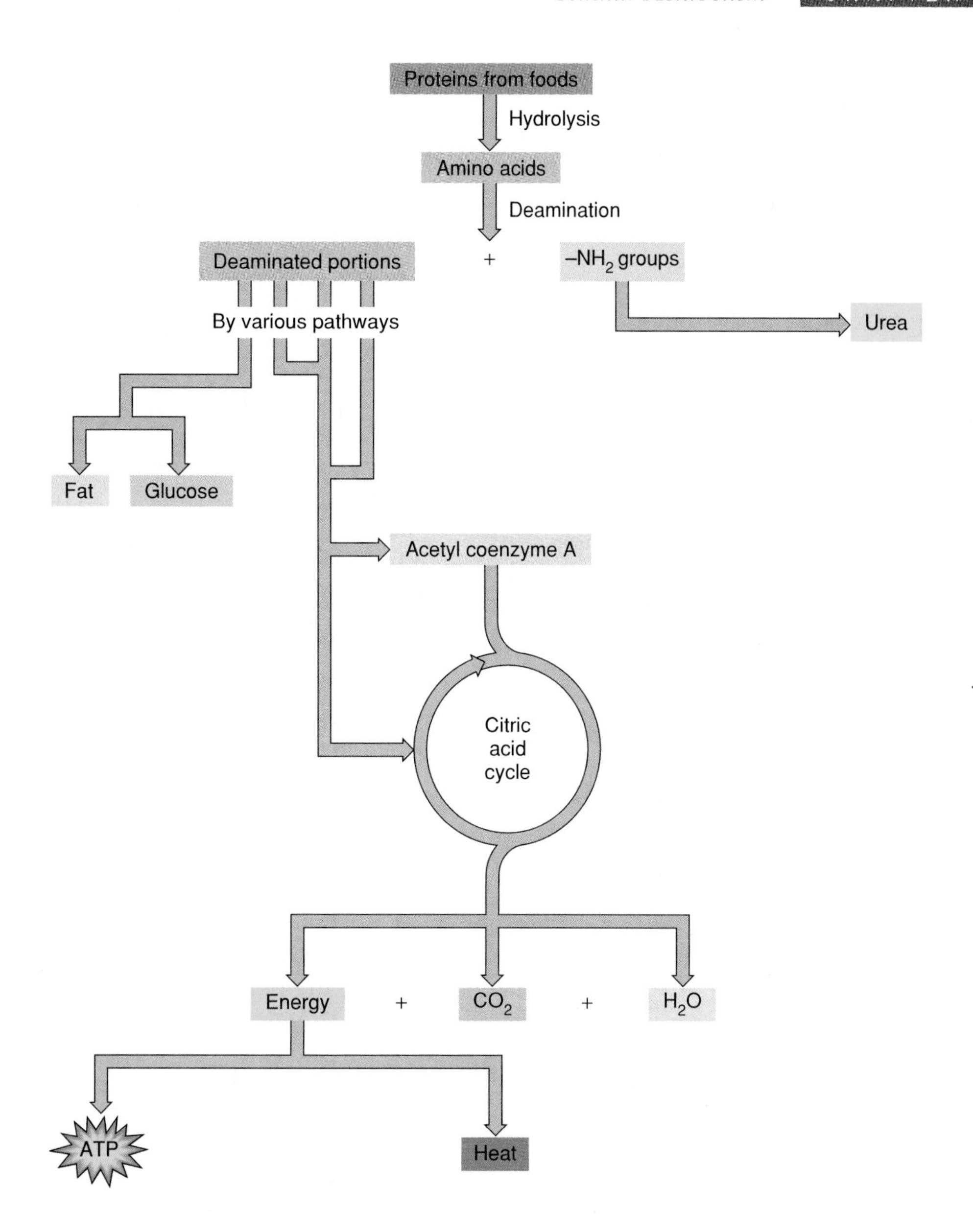

Figure 4.10

The body digests proteins from foods into amino acids but must deaminate these smaller molecules before they can be used as energy sources.

When ketone bodies form more rapidly than they can be decomposed, some of them are eliminated through the lungs and kidneys. The ketone acetone may then cause the breath and urine to develop a fruity odor. This sometimes happens when a person fasts to lose weight, thus forcing cells to metabolize body fat. Persons suffering from *diabetes mellitus* also may metabolize excess fat, producing acetone in the breath and urine. At the same time, they may develop a serious pH imbalance called *acidosis,* which is due to accumulation of still other acidic ketone bodies. ■

Protein Pathways

When dietary proteins are digested, the blood absorbs and transports the resulting amino acids to cells. Many of these amino acids reunite to form protein molecules, which may be incorporated into cell parts, be used as enzymes, or form any of the other diverse types of proteins found in the body. Some proteins are broken down, supplying energy.

When protein molecules are used as energy sources, they break down into amino acids. Then the amino acids undergo *deamination,* a process that occurs in the liver and removes their nitrogen-containing portions ($—NH_2$ groups) (see fig. 2.13). These $—NH_2$

groups then react to form the waste *urea,* which is excreted in urine.

Depending upon the particular amino acids involved, the remaining deaminated portions are decomposed in one of several pathways (fig. 4.10). Some of these pathways lead to formation of acetyl coenzyme A, and others lead more directly to steps of the citric acid cycle. As energy is released from the cycle, some of it is captured in ATP molecules. If energy is not required immediately, the deaminated portions of the amino acids may react to form glucose or fat molecules in other metabolic pathways.

Regulation of Metabolic Pathways

Recall that the rate of an enzyme-controlled reaction usually increases if either the number of substrate molecules or the number of enzyme molecules increases. However, the rate of a metabolic pathway is often determined by a regulatory enzyme responsible for one of its steps. This regulatory enzyme is present in limited quantity. Consequently, it can become saturated with substrate molecules whenever the substrate concentration increases above a certain level. Once the enzyme is saturated, increasing the number of substrate molecules will no longer affect the reaction rate. In this way, a single enzyme in a pathway can control the whole pathway.

As a rule, a *rate-limiting enzyme* is the first enzyme in a series. This position is important because some intermediate chemical in the pathway might accumulate if an enzyme occupying another location in the sequence were rate limiting.

1. What is a metabolic pathway?
2. How do cells use carbohydrates, lipids, and proteins?
3. What is a rate-limiting enzyme?

Nucleic Acids and Protein Synthesis

Because enzymes control the metabolic processes that enable cells to survive, cells must have instructions for producing these specialized proteins. **DNA** (deoxyribonucleic acid) molecules hold such information in the form of a *genetic code.* This code "instructs" cells how to synthesize enzymes and other specific protein molecules.

Genetic Information

Children resemble their parents because of inherited traits, but what actually passes from parents to child is *genetic information* in the form of DNA molecules from the parents' sex cells. As an offspring develops, mitosis passes the information from cell to cell. Genetic information "tells" cells of the developing body how to construct specific protein molecules, which, in turn, function as structural materials, enzymes, or other vital biochemicals.

The portions of DNA molecules that contain the genetic information for making particular proteins are called **genes.** Thus, inherited traits are determined by the genes contained in the parents' sex cells, which fuse to form the first cell of an offspring's body. Genes instruct cells to synthesize the enzymes that control metabolic pathways.

DNA Molecules

As described in chapter 2, the building blocks of nucleic acids are nucleotides joined so that the sugar and phosphate portions alternate. They form a long "backbone" to the polynucleotide chain (see fig. 2.18).

In a DNA molecule, the organic bases project from this backbone and bind weakly to the bases of the second strand (fig. 4.11). The resulting structure is like a ladder in which the uprights represent the sugar and phosphate backbones of the two strands, and the rungs represent the organic bases. The organic base of a DNA nucleotide can be one of four types: *adenine* (A), *thymine* (T), *cytosine* (C), or *guanine* (G).

Both strands of a DNA molecule consist of nucleotides in a particular sequence (fig. 4.12). Because of their molecular shapes, the nucleotide bases pair in specific ways. An adenine will bond only to a thymine, and a cytosine will bond only to a guanine. As a consequence of this *complementary base pairing,* a DNA strand with the base sequence A, T, G, C joins a second strand with the complementary base sequence T, A, C, G (see the upper region of DNA in fig. 4.12). The sequence of base pairs along a DNA molecule is what encodes the genetic information that specifies a particular protein's amino acid sequence.

The DNA molecule twists to form a double helix (figs. 4.13 and 4.14). A molecule of DNA may be millions of base pairs long.

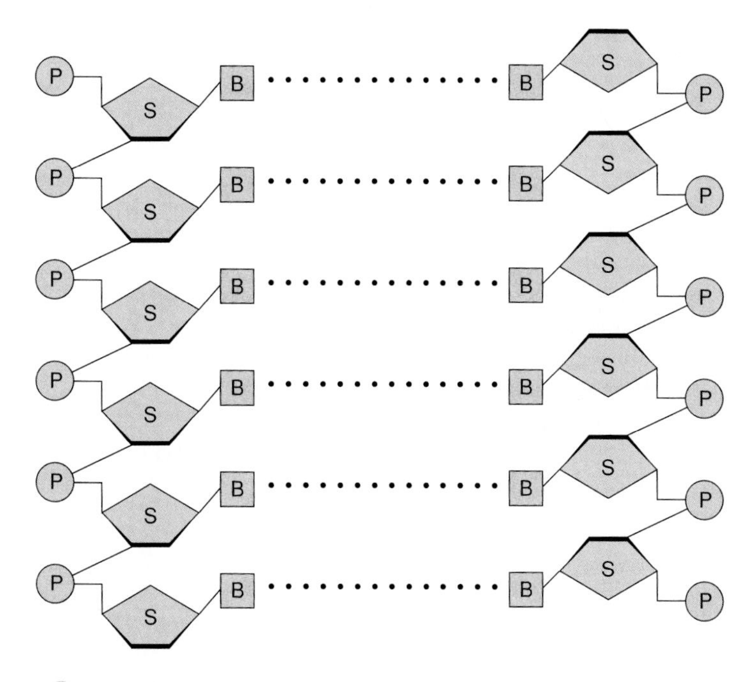

***Figure* 4.11**

The nucleotides of a DNA strand join to form a sugar-phosphate (S-P) backbone. The organic bases (B) of the nucleotides of one strand extend from this backbone and are weakly held to the bases of the second strand by hydrogen bonds (dotted lines).

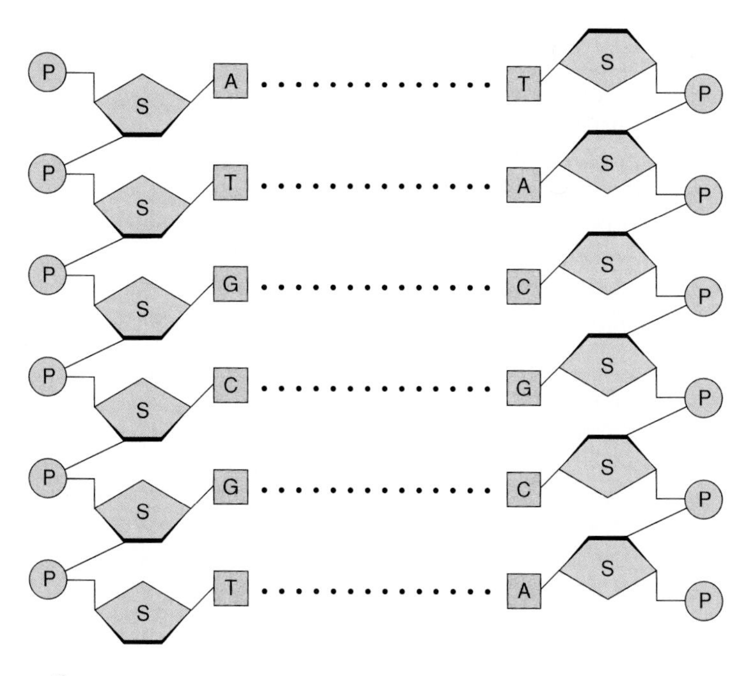

***Figure* 4.12**

DNA consists of two chains of nucleotides in a particular sequence. Within the chains are four types of nucleotides, containing the bases adenine (A), thymine (T), cytosine (C), or guanine (G).

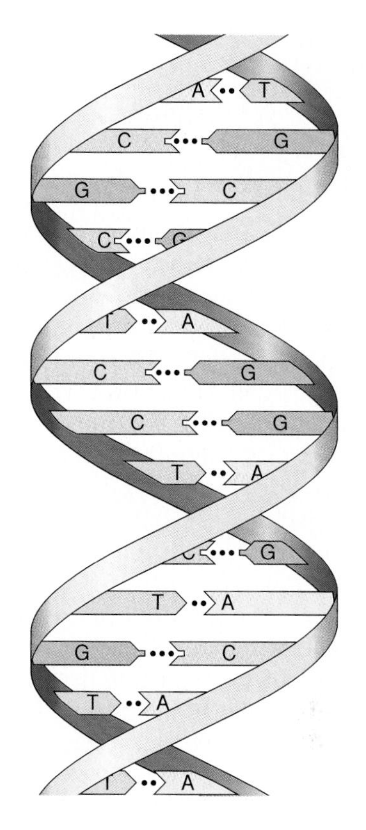

***Figure* 4.13**

The molecular ladder of a double-stranded DNA molecule twists into a double helix, held together by "rungs" consisting of complementary base pairs—A with T (or T with A) and G with C (or C with G).

***Figure* 4.14**

James Watson (left) and Francis Crick (right) used models to work out the three-dimensional structure of the DNA molecule in 1953.

TOPIC OF INTEREST The Human Genome Project

Imagine the task of deciphering the entries in a thousand 1,000-page telephone directories. Genetic researchers worldwide are rising to this challenge in a massive effort called the human genome project (HGP). The term *genome* refers to the total genetic information within a somatic cell of a particular species. For humans, the genome includes about 3 billion DNA nucleotides, which form approximately 60,000–70,000 protein-encoding genes.

The genome project is regularly revealing the genetic underpinnings of a myriad of human conditions, leading to development of new diagnostic tests and treatments. Early breakthroughs included identifying the genes that cause Duchenne muscular dystrophy and cystic fibrosis. More recent discoveries include identification of two genes that cause breast cancer.

The Project Nears Completion

Although headlines often depict the HGP as a 1990's phenomenon, it is actually a continuation of genetic research that has been ongoing throughout the twentieth century. The idea to sequence the human genome came in 1986 from noted virologist Renato Dulbecco, who thought the information would reveal how cancer arises. That summer, enthusiastic geneticists convened at Cold Spring Harbor on New York's Long Island, and soon, worldwide planning began.

The project officially started in 1990, under the auspices of the United States Department of Energy and the National Institutes of Health. The focus at first was to develop tools and technologies to divide the genome into pieces small enough to determine the nucleotide sequence. The original goal for completion was the year 2005, but progress has been so remarkable that the project may be completed by the year 2003 (fig. 4A).

Shortcuts

Just as you would not look up an entry in a phone book by reading the entire document from the beginning, but would search for key words located near the entry of interest, the HGP has proceeded in a series of steps to make the huge amount of information accessible. Today's gene searches often begin with decades-old approaches to localizing genes, such as studying people with certain inherited disorders who also have visibly altered chromosomes that tell researchers where to look in the genome for a causative gene. Then, molecular-level techniques obtain small pieces of DNA containing genes of interest. Other techniques are used to "map" genes, too, assigning specific protein-encoding DNA sequences to approximate locations on particular chromosomes. Finally, automated sequencing technologies decipher genes nucleotide by nucleotide.

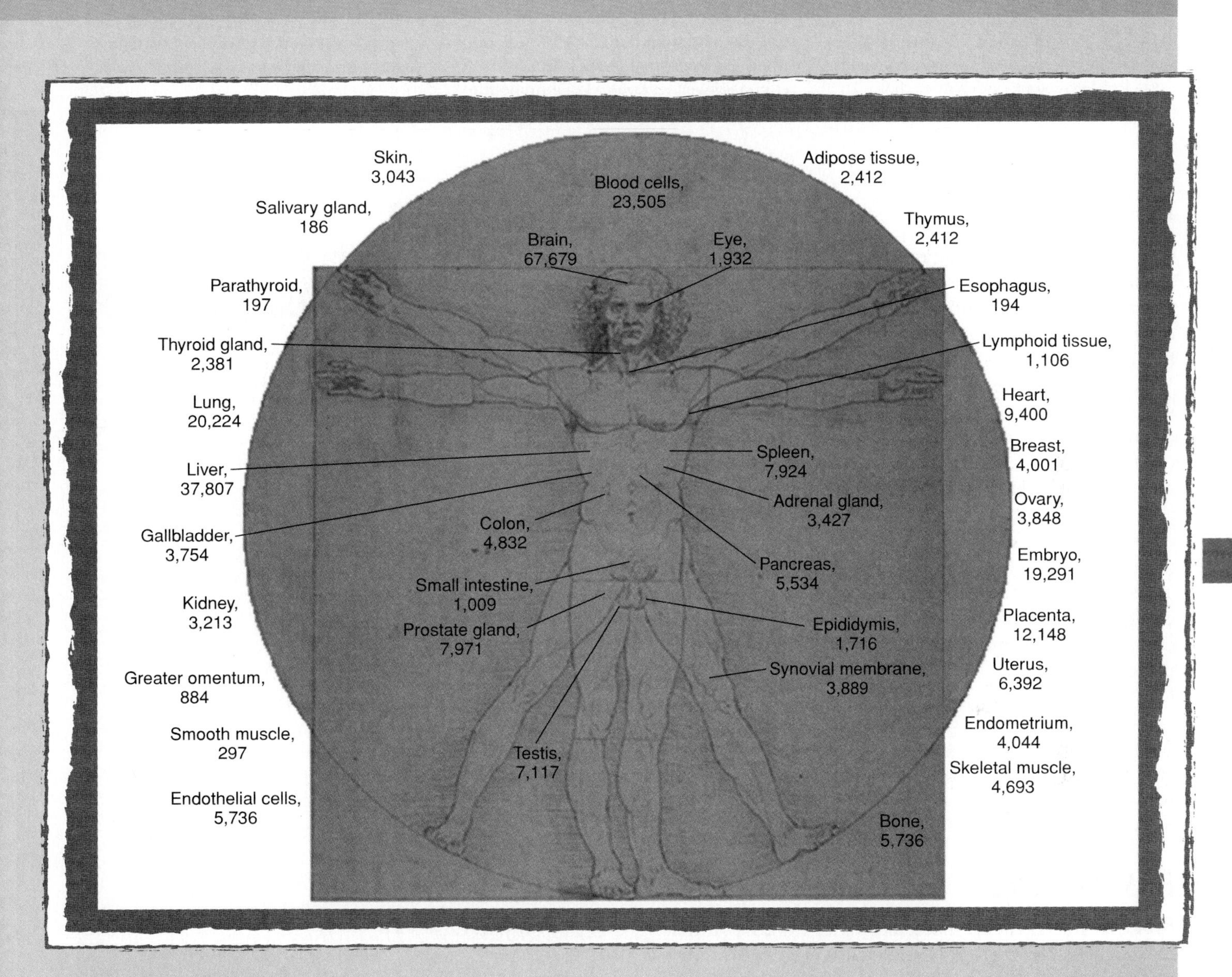

***Figure* 4A**

The human genome project spawned results faster than expected, prompting this late 1995 survey of the sites of action of known genes, which provides a whole new way of looking at human anatomy and physiology. Not surprisingly, the largest number of genes affects brain function.

The patterns of tiny ridges in a person's fingertips are so distinctive that people can be identified by their fingerprints. Similarly, because the sequence of DNA nucleotides in each person's cells is unique (with the exception of identical twins), a person can be matched to a sample of body cells by analyzing the cells' DNA. Some DNA sequences are analyzed in criminal investigations to determine the likelihood that a particular individual was the source of blood, semen, or other forensic evidence at a crime scene. This technique is called DNA fingerprinting. ■

Genetic Code

Genetic information contains the instructions for synthesizing proteins, and because proteins consist of twenty types of amino acids joined in particular sequences, the genetic information must tell how to position the amino acids correctly in a polypeptide chain.

Each of the twenty different types of amino acids is represented in a DNA molecule by a particular sequence of three nucleotides. That is, the sequence G, G, T in a DNA strand represents one type of amino acid; the sequence G, C, A represents another type, and T, T, A still another type (table 4.1). Other nucleotide sequences represent the instructions for beginning or ending the synthesis of a protein molecule. Thus, the sequence of nucleotides in a DNA molecule denotes the order of amino acids of a protein molecule, as well as indicates where to start or stop that protein's synthesis. This method of storing information for protein synthesis is the **genetic code.**

Because DNA molecules are part of the chromatin within a cell's nucleus and protein synthesis occurs in the cytoplasm, the genetic information must be transferred from the nucleus into the cytoplasm. Certain RNA molecules accomplish this transfer of information.

Table 4.1 Some Nucleotide Sequences of the Genetic Code

	DNA Sequence	RNA Sequence
Amino Acids		
Alanine	CGT	GCA
Arginine	GCA	CGU
Asparagine	TTA	AAU
Aspartic acid	CTA	GAU
Cysteine	ACA	UGU
Glutamic acid	CTT	GAA
Glutamine	GTT	CAA
Glycine	CCG	GGC
Histidine	GTA	CAU
Isoleucine	TAG	AUC
Leucine	GAA	CUU
Lysine	TTT	AAA
Methionine	TAC	AUG
Phenylalanine	AAA	UUU
Proline	GGA	CCU
Serine	AGG	UCC
Threonine	TGC	ACG
Tryptophan	ACC	UGG
Tyrosine	ATA	UAU
Valine	CAA	GUU
Instructions		
Start protein synthesis	TAC	AUG
Stop protein synthesis	ATT	UAA

RNA Molecules

RNA (ribonucleic acid) molecules differ from DNA molecules in several ways. RNA molecules are usually single-stranded, and their nucleotides contain ribose rather than deoxyribose sugar. Like DNA, each RNA nucleotide contains one of four organic bases, but while adenine, cytosine, and guanine nucleotides occur in both DNA and RNA, thymine nucleotides are found only in DNA. In place of thymine nucleotides, RNA molecules contain *uracil* (U) nucleotides.

A type of RNA called **messenger RNA** (mRNA) carries the information in a gene's nucleotide sequence from the nucleus to the cytoplasm. Synthesis of mRNA begins when the enzyme RNA polymerase associates with the DNA base sequence at the beginning of a gene—that is, the instructions for synthesizing a particular protein. At this site, other enzymes unwind and pull apart the double-stranded DNA molecule, exposing the first portion of the gene. RNA polymerase then moves along the strand, exposing other portions of the gene and stringing together (polymerizing) a molecule of mRNA from nucleotides complementary to those along the unwound DNA strand. For example, if the sequence of DNA bases is A, T, G, C, G, T, A, A, C, A, then the complementary bases in the developing mRNA molecule are U, A, C, G, C, A, U, U, G, U (fig. 4.15). RNA polymerase somehow "knows" which of the two DNA strands contains the information. It also knows the correct direction to read the DNA. Just as a sentence must be read in one direction, so must a strand of DNA.

RNA polymerase continues to move along the DNA strand, exposing the gene, until it reaches a special DNA base sequence (termination signal) that represents the end of the gene. At this point, the enzyme releases the newly formed mRNA molecule and leaves the DNA. This process of copying DNA information into the structure of mRNA is called **transcription.**

Because an amino acid is encoded by a particular sequence of three nucleotides in a DNA molecule, that amino acid will be represented in the transcribed mRNA by the complementary set of three nucleotides. Such a "triplet" of nucleotides in mRNA that specifies a particular amino acid is called a **codon.**

Once formed, mRNA molecules (each of which consists of hundreds or even thousands of nucleotides) can move out of the nucleus through pores in the nuclear envelope and enter the cytoplasm (fig. 4.16). There, mRNA molecules associate with ribosomes and act as patterns or templates for the synthesis of protein molecules—a process called **translation.** In this process, codons specify amino acids, according to the genetic code (fig. 4.17).

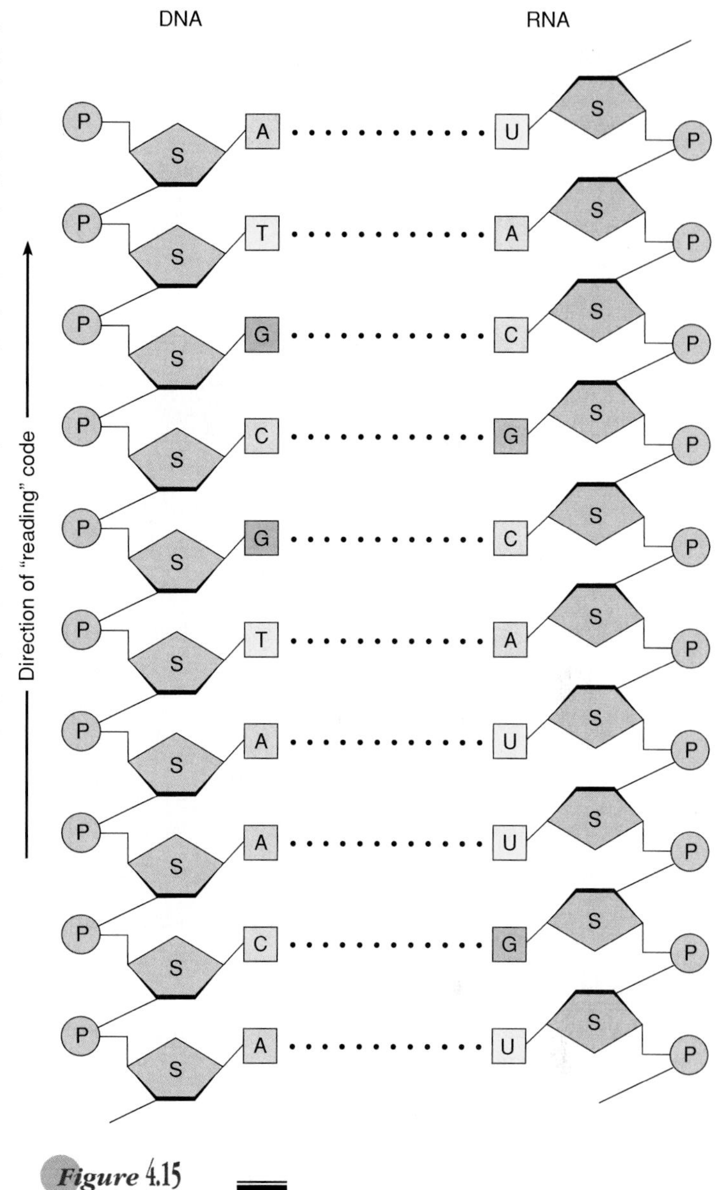

Figure 4.15

When an RNA molecule is synthesized beside a strand of DNA, complementary nucleotides bond as in a double-stranded DNA molecule, with one exception: RNA contains uracil nucleotides (U) in place of thymine nucleotides (T).

Protein Synthesis

Building a protein molecule requires that the correct amino acids are present in the cytoplasm and that they are positioned in the proper locations along a strand of mRNA. A second kind of RNA molecule, synthesized in the nucleus and called **transfer RNA** (tRNA), correctly aligns amino acids to form proteins.

Because twenty different types of amino acids form biological proteins, at least twenty different types of tRNA molecules must be available to serve as guides, one for each type of amino acid. Each type of tRNA has a region at one end that consists of three nucleotides in a particular sequence. These nucleotides bond a specific complementary set of three nucleotides of an mRNA molecule—a codon. Thus, the set of three nucleotides in the tRNA is called an **anticodon.** In this way, tRNA carries its amino acid to a correct position on an mRNA strand. This action occurs in close association with a ribosome.

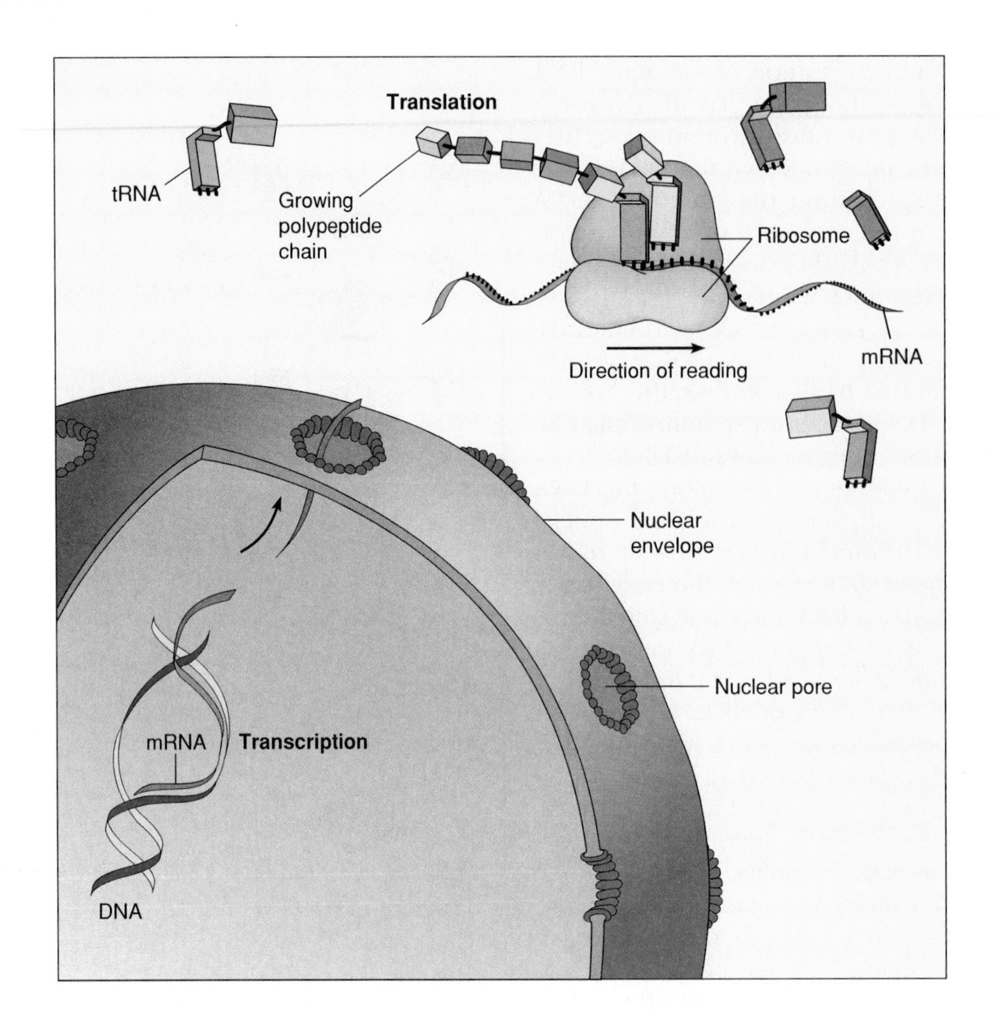

Figure 4.16

A section of DNA information is transcribed into a messenger RNA (mRNA) molecule, which moves out of the nucleus and into the cytoplasm, where it associates with a ribosome.

Shortly after protein synthesis begins, a ribosome is bound to an mRNA molecule. A tRNA molecule with the complementary anticodon holds its amino acid and the growing polypeptide chain. A second tRNA complementarily binds to the next codon bringing its amino acid to an adjacent site on the ribosome. Then a peptide bond forms, adding the new amino acid to the chain. The first tRNA molecule releases from its amino acid and returns to the cytoplasm (fig. 4.18).

This process repeats again and again as the ribosome moves along the mRNA molecule. The amino acids, which the tRNA molecules release, are added one at a time to the developing protein. Enzymes associated with the ribosome control this addition of amino acids.

As the protein molecule forms, it folds into its unique shape and is then released to become a separate functional molecule. The tRNA molecules can pick up other amino acids from the cytoplasm, and like the mRNA molecules can function repeatedly.

Table 4.2 compares RNA and DNA, and table 4.3 summarizes protein synthesis.

1. What is the function of DNA?
2. How is information carried from the nucleus to the cytoplasm?
3. How are protein molecules synthesized?

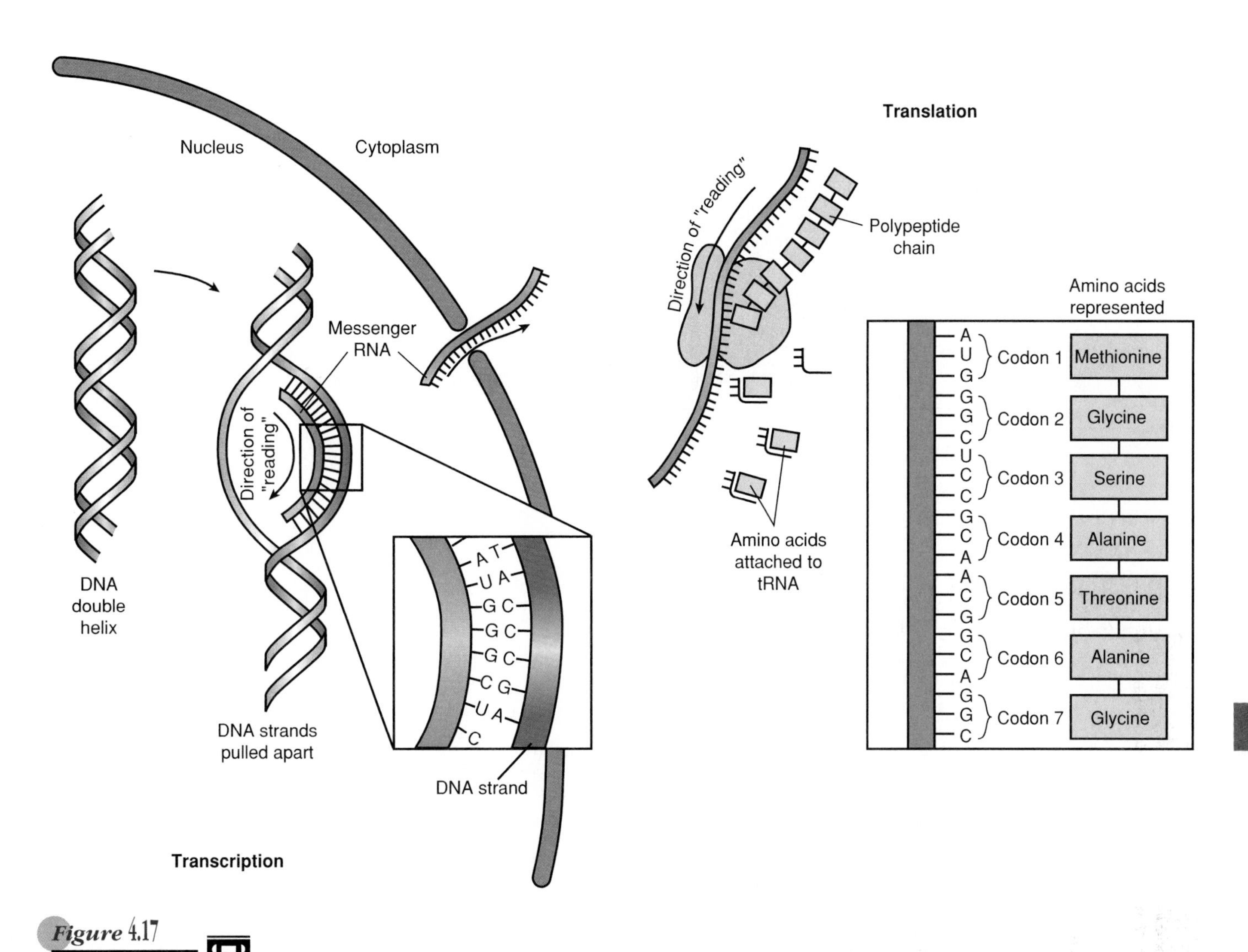

Figure 4.17

Transcription copies DNA's information into the structure of mRNA. Translation uses the information in mRNA to construct a specific protein molecule from amino acids.

DNA Replication

When a cell reproduces, each newly formed cell requires a copy of the parent cell's genetic information so that the new cell can synthesize the proteins necessary to maintain life functions, build cell parts, and metabolize. To accomplish this, DNA molecules are **replicated** (duplicated) during interphase of the cell cycle.

As replication begins, bonds between complementary base pairs of the double strands in each DNA molecule break (fig. 4.19). Then the double-stranded structure pulls apart and unwinds, exposing the organic bases of its nucleotides. New DNA nucleotides form complementary pairs with the exposed bases, and enzymes knit together the sugar-phosphate backbone. In this way, a new strand of complementary nucleotides forms along each of the old strands, producing two complete DNA molecules, each with one old strand of the original molecule and one new strand. These two DNA molecules become incorporated into replicate copies of a chromosome and separate during mitosis so that one passes to each of the newly forming cells.

1. Why must DNA molecules be replicated?
2. How is replication accomplished?

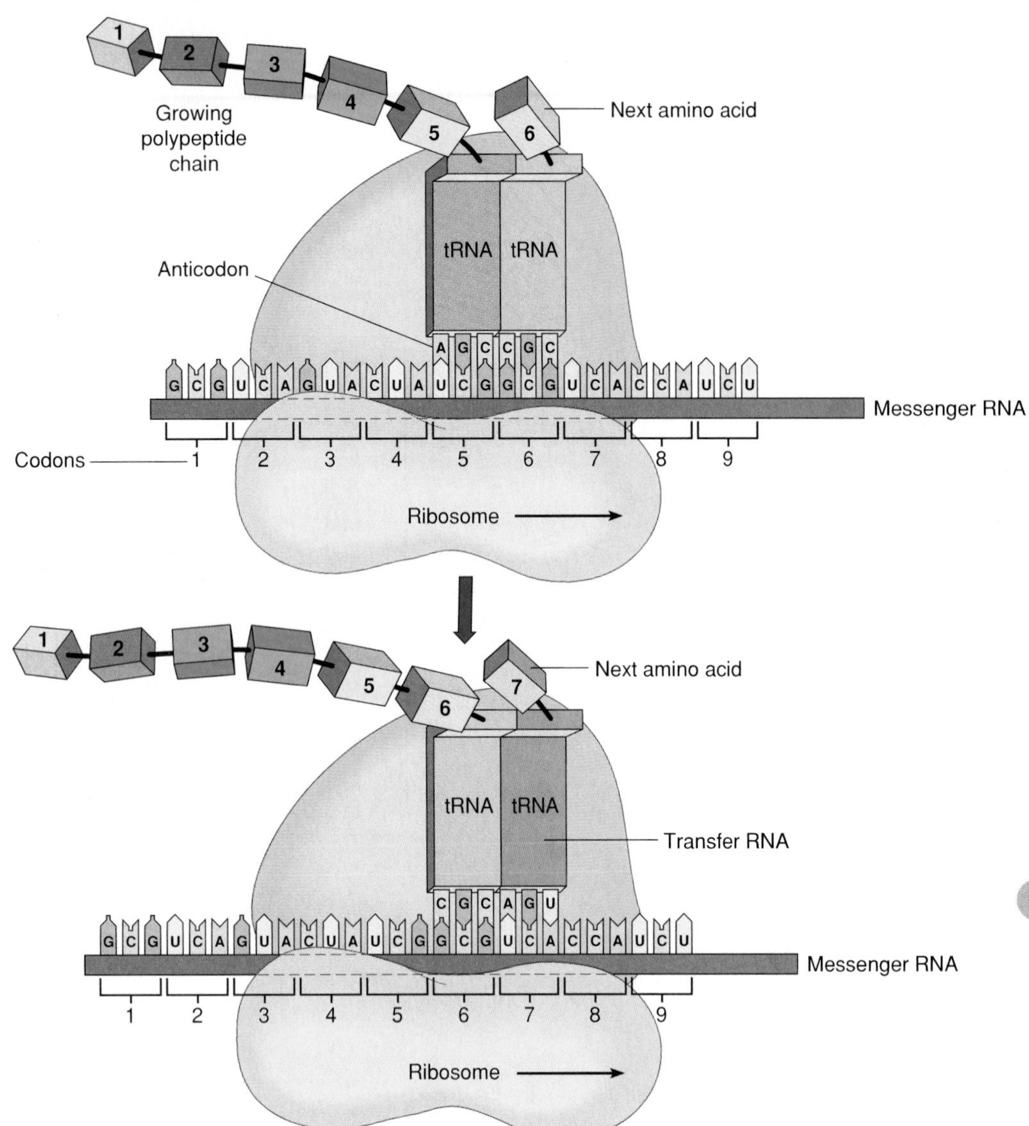

***Figure* 4.18**

Molecules of transfer RNA (tRNA) place specific amino acids in the sequence that the codons in an mRNA molecule specify. These amino acids bond and form the polypeptide chain of a protein molecule.

Table 4.2

A Comparison of DNA and RNA Molecules

	DNA	RNA
Main location	Part of chromosomes in nucleus	In the cytoplasm
Five-carbon sugar	Deoxyribose	Ribose
Basic molecular structure	Double-stranded	Single-stranded
Organic bases included	Adenine, thymine, cytosine, guanine	Adenine, uracil, cytosine, guanine
Major functions	Contains genetic code for protein synthesis; replicates prior to cell division	Messenger RNA carries transcribed DNA information to cytoplasm and acts as template for synthesis of protein molecules; transfer RNA carries amino acids to messenger RNA

Table 4.3

Protein Synthesis

Transcription (occurs within the nucleus)

1. RNA polymerase associates with the base sequence of one strand of a gene.
2. Other enzymes unwind the DNA molecule, exposing a portion of the gene.
3. RNA polymerase moves along the exposed gene and polymerizes an mRNA molecule, whose nucleotides are complementary to those of the gene strand.
4. When the RNA polymerase reaches the end of the gene, the newly formed mRNA molecule is released.
5. The mRNA molecule passes through a pore in the nuclear envelope and enters the cytoplasm.

Translation (occurs within the cytoplasm)

1. A ribosome binds to the mRNA molecule near the codon at the beginning of the messenger strand.
2. A tRNA molecule with the complementary anticodon associates with the ribosome and the amino acid it carries becomes part of the chain.
3. This process repeats for each codon in the mRNA sequence as the ribosome moves along the mRNA's length.
4. Enzymes associated with the ribosome catalyze peptide bonds, forming a chain of amino acids.
5. As the chain of amino acids grows, it folds into the unique shape of a functional protein molecule.
6. The completed protein molecule is released. The mRNA molecule, ribosome, and tRNA molecules can function repeatedly to synthesize other protein molecules.

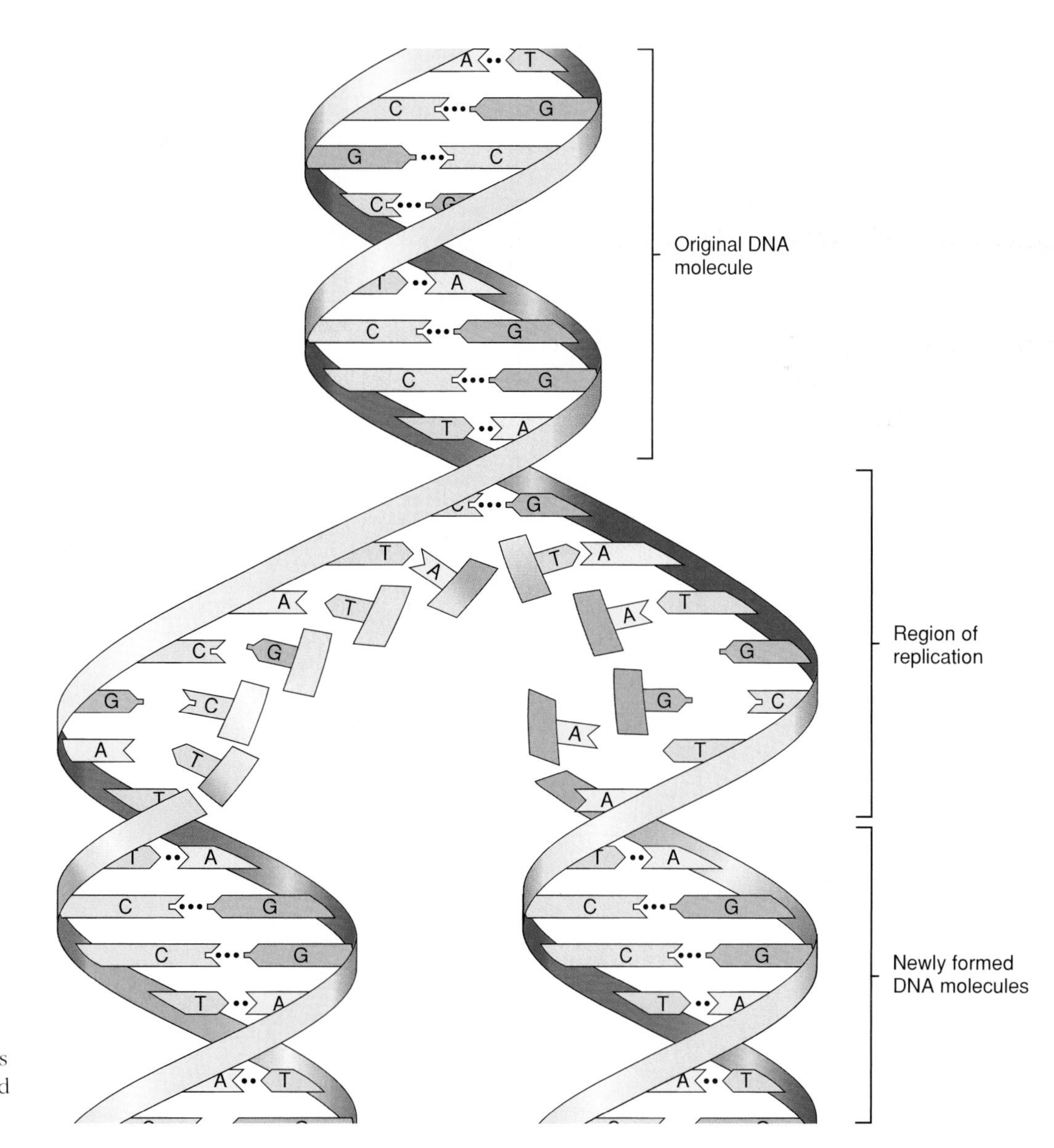

Figure 4.19

When a DNA molecule replicates, its strands pull apart, and a new strand of complementary nucleotides forms along each of the old strands.

TOPIC OF INTEREST Mutations

It is easy to make an error when typing a sentence consisting of several hundred letters. DNA replication, which is similar to copying such a sentence, is also error-prone. A newly replicated gene may have too many or too few bases, or an "A" where the parent DNA sequence indicates it should be "C." Fortunately, cells have several mechanisms to scan newly replicated DNA, detect such changes (*mutations*) in genetic information, and correct them. When this fails, health may suffer. Mutations may occur spontaneously or may be induced by agents called mutagens, such as certain toxic chemicals and ionizing radiation (fig. 4B).

Hundreds of inherited illnesses result from mutations. Genetic researchers are developing tests that detect particular mutations. Sometimes, this leads to "presymptomatic" tests because the mutated gene is present in an individual from the time of conception. For example, a genetic test may identify the neurological disorder Huntington disease in an eighteen-year-old, even though the symptoms—personality changes and uncontrollable, constant movements—probably will not appear for another two decades. Such testing is controversial, particularly when the illness is not treatable.

Another complication of testing for mutations is that a single gene may mutate in several ways. This is hampering the effort to develop tests for cystic fibrosis and inherited breast cancer. If a disease-causing gene can be abnormal in dozens or even hundreds of ways, of what use will a test be that detects only a few ways?

Because of these uncertainties, medical groups are calling for caution in using genetic tests. Stay tuned—one of these tests may be in your (or your children's) future!

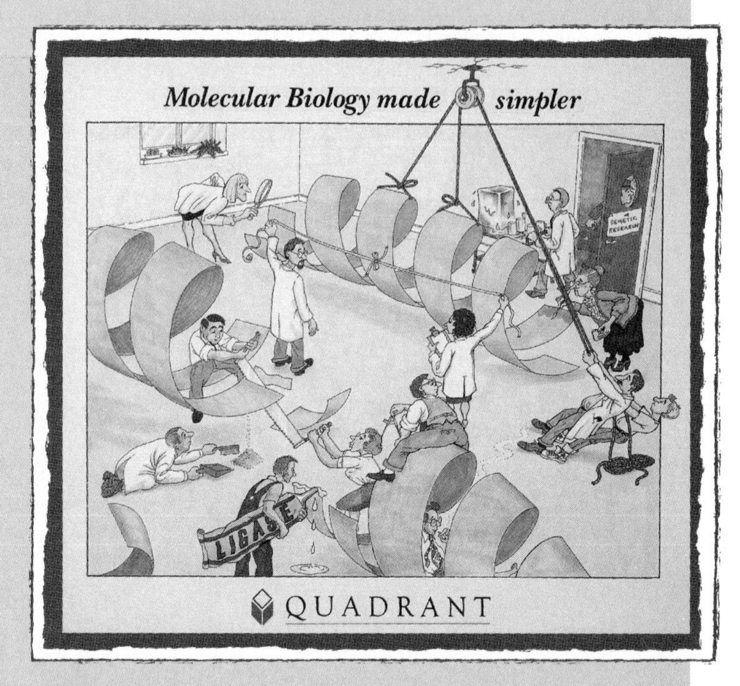

Figure 4B

Mutations occur when failure of natural DNA repair systems or exposure to toxic chemicals or ionizing radiation allows errors in the DNA sequence of a gene to persist. The errors (mutations) may affect health or alter an inherited trait. This ad whimsically depicts the arduous task of DNA replication. Notice that the artist did not understand the process described in figure 4.19. Reproduced with the permission of Quadrant Holdings Cambridge Ltd.

Summary Outline

Introduction (p. 74)

A cell continuously carries on metabolic reactions.

Metabolic Reactions (p. 74)

1. Anabolism
   a. Anabolism builds large molecules from smaller molecules.
   b. In dehydration synthesis, water forms, and smaller molecules attach by sharing atoms.
   c. Complex carbohydrates are synthesized from monosaccharides, fats are synthesized from glycerol and fatty acids, and proteins are synthesized from amino acids.
2. Catabolism
   a. Catabolism breaks down larger molecules into smaller ones.
   b. In hydrolysis, a water molecule is split as an enzyme breaks the bond between two portions of a molecule.
   c. Hydrolysis breaks down complex carbohydrates into monosaccharides, fats into glycerol and fatty acids, proteins into amino acids, and nucleic acids into nucleotides.

Control of Metabolic Reactions (p. 75)

Enzymes control metabolic reactions, which consist of many specific chemical changes.

1. Enzyme action
   a. Enzymes are molecules that promote metabolic reactions without being consumed.
   b. An enzyme acts upon a specific substrate.
   c. The shape of an enzyme molecule fits the shape of its substrate molecule.
   d. When an enzyme combines with its substrate, the substrate changes, lowering the energy necessary for a reaction to proceed. A product forms, and the enzyme is released in its original form.
   e. The speed of an enzyme-controlled reaction depends upon the number of enzyme and substrate molecules present and the enzyme's efficiency.
2. Factors that alter enzymes
   a. Almost all enzymes are proteins.
   b. Factors that may denature enzymes include heat, radiation, electricity, certain chemicals, and extreme pH values.

Energy for Metabolic Reactions (p. 77)

Energy is the capacity to do work. Common forms of energy include heat, light, sound, electrical energy, mechanical energy, and chemical energy.

1. Release of chemical energy
   a. Most metabolic processes use chemical energy released when molecular bonds break.
   b. The energy released from glucose breakdown during cellular respiration drives the reactions of cellular metabolism.
2. Anaerobic respiration
   a. The first phase of glucose decomposition does not require oxygen.
   b. Some of the energy released is transferred to ATP molecules.
3. Aerobic respiration
   a. The second phase of glucose decomposition requires oxygen.
   b. Considerably more ATP molecules form during this phase than during the anaerobic phase.
   c. The final products of glucose breakdown are carbon dioxide, water, and energy.
4. ATP molecules
   a. Energy is captured in the bond of the terminal phosphate of each ATP molecule.
   b. When a cell requires energy, the terminal phosphate bond of an ATP molecule breaks, releasing stored energy.
   c. An ATP molecule that loses its terminal phosphate becomes an ADP molecule.
   d. An ADP molecule that captures energy and a phosphate becomes ATP.

Metabolic Pathways (p. 78)

Metabolic processes consist of chemical reactions that occur in a certain sequence. A sequence of enzyme-controlled reactions constitutes a metabolic pathway.

1. Carbohydrate pathways
   a. Carbohydrates may enter catabolic pathways and be used as energy sources.
   b. Carbohydrates may enter anabolic pathways and be converted into glycogen or fat.
2. Lipid pathways
   a. Before fats can be used as an energy source, they must break down into glycerol and fatty acids.
   b. Fatty acids can react to yield acetyl coenzyme A, which, in turn, is oxidized in the citric acid cycle.
   c. Glycerol enters metabolic pathways leading to the citric acid cycle or is used to synthesize glucose.
3. Protein pathways
   a. Proteins are used as cell building materials, as enzymes, and as energy sources.
   b. Before proteins can be used as energy sources, they must break down into amino acids, and the amino acids must be deaminated.
   c. The deaminated portions of amino acids can break down into carbon dioxide and water or can be converted into glucose or fat.
4. Regulation of metabolic pathways
   a. Rate-limiting enzymes present in limited quantities determine the rates of metabolic pathways.
   b. Rate-limiting enzymes become saturated when substrate concentrations increase above a certain level.

Nucleic Acids and Protein Synthesis (p. 82)

DNA molecules contain information that instructs a cell how to synthesize enzymes and other proteins.

1. Genetic information
   a. Inherited traits result from DNA information passed from parents to offspring.
   b. A gene is a portion of a DNA molecule that contains the genetic information for making a particular protein.
2. DNA molecules
   a. A DNA molecule consists of two strands of nucleotides twisted into a double helix.
   b. The nucleotides of a DNA strand are in a particular sequence.

c. The nucleotides of each strand pair with those of the other strand in a complementary fashion.
3. Genetic code
   a. The sequence of nucleotides in a DNA molecule encodes the sequence of amino acids in a protein molecule.
   b. RNA molecules transfer genetic information from the nucleus to the cytoplasm.
4. RNA molecules
   a. RNA molecules are usually single-stranded and contain ribose instead of deoxyribose, and uracil nucleotides in place of thymine nucleotides.
   b. Messenger RNA molecules consist of nucleotide sequences that are complementary to those of exposed strands of DNA.
   c. Messenger RNA molecules associate with ribosomes and provide patterns for the synthesis of protein molecules.
5. Protein synthesis
   a. A ribosome binds to a messenger RNA molecule and allows a transfer RNA molecule to recognize its correct position on the messenger RNA.
   b. Molecules of transfer RNA position amino acids along a strand of messenger RNA.
   c. Amino acids released from the transfer RNA molecules join and form a protein molecule that folds into a unique shape.
6. DNA replication
   a. Each new cell requires a copy of the parent cell's genetic information.
   b. DNA molecules replicate during interphase of the cell cycle.
   c. Each new DNA molecule contains one old strand and one new strand.

Review Exercises

1. Distinguish between anabolism and catabolism. (p. 74)
2. Distinguish between dehydration synthesis and hydrolysis. (p. 74)
3. Define *enzyme*. (p. 75)
4. Describe how an enzyme interacts with its substrate. (p. 76)
5. Explain how an enzyme can be denatured. (p. 76)
6. Explain how oxidation of molecules inside cells differs from the burning of substances outside cells. (p. 77)
7. Distinguish between anaerobic and aerobic respiration. (p. 77)
8. Explain the importance of ATP to cell processes. (p. 78)
9. Describe the relationship between ATP and ADP molecules. (p. 78)
10. Explain what a *metabolic pathway* is. (p. 78)
11. Describe what happens to carbohydrates that enter catabolic pathways. (p. 78)
12. Explain how fats may provide energy. (p. 80)
13. Define *deamination*, and explain its importance. (p. 81)
14. Explain how one enzyme can control the rate of a metabolic pathway. (p. 82)
15. Explain how DNA encodes genetic information. (p. 82)
16. Describe the relationship between a DNA molecule and a gene. (p. 82)
17. Distinguish between messenger RNA and transfer RNA. (p. 86)
18. Define *transcription*. (p. 87)
19. Define *translation*. (p. 87)
20. Distinguish between a codon and an anticodon. (p. 87)
21. Calculate the number of amino acids that a DNA sequence of twenty-seven nucleotides encodes. (p. 87)
22. Explain the function of a ribosome in protein synthesis. (p. 88)
23. Explain why a new cell requires a copy of the parent cell's genetic information. (p. 89)
24. Describe how DNA molecules replicate. (p. 89)

Critical Thinking

1. How can the same biochemical be both a reactant (starting material) and a product in aerobic respiration?
2. After finishing a grueling marathon, a runner exclaims, "Whew, I think I've used up all my ATP!" Could this be possible?
3. DNA fingerprinting could theoretically identify a perpetrator of a crime if the entire genome was sequenced and found to match cells at the crime scene or on the victim. However, only certain highly variable DNA sequences typically are considered in forensic uses of DNA fingerprinting. Given this limited use of the technology, could a man accused of murder plead guilty to cover for his son, who actually committed the crime, and be found guilty because of the DNA evidence?
4. Mutations in proteins that participate in ATP formation in mitochondria are never seen. Why might this be?
5. A mutation that deletes one or two DNA nucleotides changes gene function more drastically than a substitution of one nucleotide for another type, or removal or addition of three DNA nucleotides. Why?
6. A portion of a protein molecule has the following amino acid sequence: serine-lysine-glycine-proline-tyrosine-alanine-glutamine-valine. Write the DNA and RNA sequences specifying this chain of amino acids.
7. What effect might changes in pH of body fluids that accompany illness have on enzymes?
8. Some weight-reduction diets drastically limit intake of carbohydrates but allow foods high in fat and protein. What changes would such a diet cause in the dieter's cellular metabolism? What changes might be noted in this person's urine?

Tissues

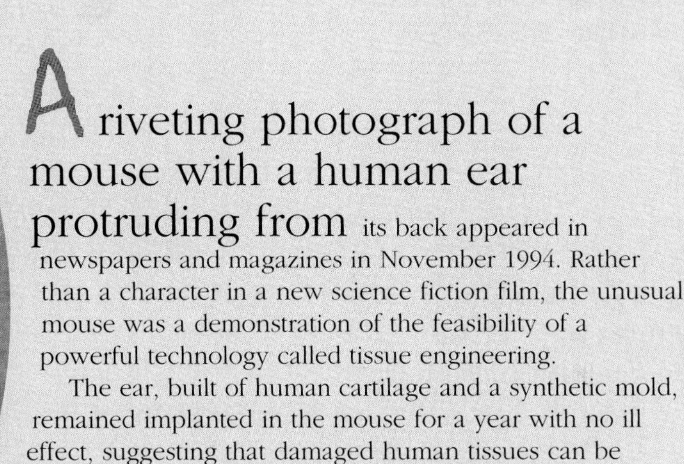

A riveting photograph of a mouse with a human ear protruding from its back appeared in newspapers and magazines in November 1994. Rather than a character in a new science fiction film, the unusual mouse was a demonstration of the feasibility of a powerful technology called tissue engineering.

The ear, built of human cartilage and a synthetic mold, remained implanted in the mouse for a year with no ill effect, suggesting that damaged human tissues can be replaced with counterparts engineered from cells plus synthetic materials. The idea behind tissue engineering is to construct a replacement part that is similar enough to its real counterpart to function in the body, but not so realistic that it triggers the immune system to reject it. The general recipe for an engineered tissue includes: human cells stripped of immune-stimulating molecules, a synthetic, biodegradable scaffolding or support, adhesion molecules to hold the tissue in place in the body, and molecules that stimulate the recipient's cells to infiltrate the engineered tissue, gradually replacing it with the real thing. Variations on the tissue-engineering theme include skin, liver tissue, cartilage, blood vessels, and brain implants that secrete needed biochemicals.

Tissue engineering is good news for the many people each year who could use a spare part: 15,000 burn victims needing skin grafts, 400,000 people with foot ulcers from diabetes, a million people requiring surgery to repair damaged cartilage. One day, maybe even the thousands of people awaiting organ transplants will benefit from this technology.

Photo: This replacement blood vessel is part biological, part synthetic. It has an inner single layer of cells stripped of molecules that might provoke an immune response, a middle layer of smooth muscle cells in collagen, and an outer layer of connective tissue. The cell-containing layers are molded around a cylindrical synthetic material.

Cells, the basic units of structure and function within the human organism, are organized into groups and layers called *tissues*. Each type of tissue is composed of similar cells specialized to carry on particular functions. For example, epithelial tissues form protective coverings and function in secretion and absorption. Connective tissues support softer body parts and bind structures together. Muscle tissues produce body movements, and nervous tissues conduct impulses that help control and coordinate body activities.

Besides cells, all tissues contain a nonliving portion called the *matrix* or intercellular substance. This material varies in composition from tissue to tissue and supports the cells within.

Chapter Objectives

After studying this chapter, you should be able to:

1. List the four major tissue types, and provide examples of where each occurs in the body. (p. 97)
2. Describe the general characteristics and functions of epithelial tissues. (p. 97)
3. Name the types of epithelium, and identify an organ in which each is found. (p. 97)
4. Explain how to classify glands. (p. 102)
5. List the types of connective tissues in the body. (p. 104)
6. Describe the major functions of each type of connective tissue. (p. 105)
7. Describe the general cellular components, structures, fibers, and matrix (where applicable) of each type of connective tissue. (p. 105)
8. Distinguish among the three types of muscle tissues. (p. 110)
9. Describe the general characteristics and functions of nervous tissues. (p. 112)

Aids to Understanding Words

adip- [fat] *adip*ose tissue: A tissue that stores fat.

chondr- [cartilage] *chondr*ocyte: A cartilage cell.

-cyt [cell] osteo*cyt*e: A bone cell.

epi- [upon] *epi*thelial tissue: A tissue that covers all the free body surfaces.

-glia [glue] neuro*glia*: Cells that bind nervous tissue together and help it to function.

inter- [between] *inter*calated disk: A band between cardiac muscle cells.

macro- [large] *macro*phage: A large phagocytic cell.

oss- [bone] *oss*eous tissue: Bone tissue.

pseudo- [false] *pseudo*stratified epithelium: A tissue whose cells appear to be in layers, but are not.

squam- [scale] *squam*ous epithelium: A tissue whose cells are flattened or scalelike.

strat- [layer] *strat*ified epithelium: A tissue whose cells are in layers.

Key Terms

adipose tissue (ad′ĭ-pōs tish′u)
cartilage (kar′tĭ-lij)
chondrocyte (kon′dro-sīt)
connective tissue (kŏ-nek′tiv tish′u)
epithelial tissue (ep″ĭ-the′le-al tish′u)
fibroblast (fi′bro-blast)
macrophage (mak′ro-fāj)
muscle tissue (mus′el tish′u)
nervous tissue (ner′vus tish′u)
neuroglial cell (nu-rog′le-ahl sel)
neuron (nu′ron)
osteocyte (os′te-o-sīt″)
osteon (os′te-on)

Introduction

Tissues are groups of cells that have specialized structural and functional roles. Although the cells of different tissues vary in size, shape, organization, and function, those within each type of tissue are quite similar in function.

The tissues of the human body are of four major types: epithelial tissues, connective tissues, muscle tissues, and nervous tissues. Table 5.1 compares these tissue types.

1. What is a tissue?
2. List the four major tissue types.

Epithelial Tissues

General Characteristics

Epithelial tissues cover all body surfaces, line most internal organs, and are the major tissues of glands. Because epithelium covers organs, forms the inner lining of body cavities, and lines hollow organs, it always has a free surface—one that is exposed to the outside or to an open space internally. The underside of this tissue is always anchored to connective tissue by a thin, noncellular layer, called the *basement membrane.*

As a rule, epithelial tissues lack blood vessels; however, substances diffuse from underlying connective tissues to nourish epithelial cells. Connective tissue is usually well supplied with blood vessels.

Epithelial cells reproduce readily. Injuries to epithelium heal rapidly as new cells replace lost or damaged ones. For example, skin cells and cells that line the stomach and intestines are continually damaged and replaced.

Epithelial cells are tightly packed, with little intercellular material between them. Consequently, these cells are effective barriers in such structures as the outer layer of the skin and the lining of the mouth. Other epithelial functions include secretion, absorption, excretion, and sensory reception. In the descriptions that follow, note that the free surfaces of various epithelial cells are modified in ways that reflect their specialized functions.

Simple Squamous Epithelium

Simple squamous epithelium consists of a single layer of thin, flattened cells. These cells fit tightly together, somewhat like floor tiles, and their nuclei are usually broad and thin (fig. 5.1).

Substances easily pass through simple squamous epithelium, which occurs commonly at sites of diffusion and filtration. For instance, simple squamous epithelium lines the air sacs of the lungs, where oxygen and carbon dioxide are exchanged. It also forms the walls of capillaries, lines the insides of blood and lymph vessels, and covers membranes that line body cavities. However, its thin and delicate nature makes simple squamous epithelium relatively easy to damage.

Simple Cuboidal Epithelium

Simple cuboidal epithelium consists of a single layer of cube-shaped cells. These cells usually have centrally located, spherical nuclei (fig. 5.2).

Table 5.1 Tissues

Type	Function	Location	Distinguishing Characteristics
Epithelial	Protects, secretes, absorbs, excretes	Covers body surfaces (inside and out), composes glands	Lacks blood vessels, reproduces readily, has tightly packed cells
Connective	Binds, supports, protects, fills spaces, stores fat, produces blood cells	Widely distributed throughout body	Mostly has good blood supply, cells are spaced farther apart with matrix in between
Muscle	Allows movement	Attached to bones, in the walls of hollow internal organs, heart	Contractile
Nervous	Transmits impulses for coordination, regulation, integration, and sensory reception	Brain, spinal cord, nerves	Cells connect to each other and other body parts

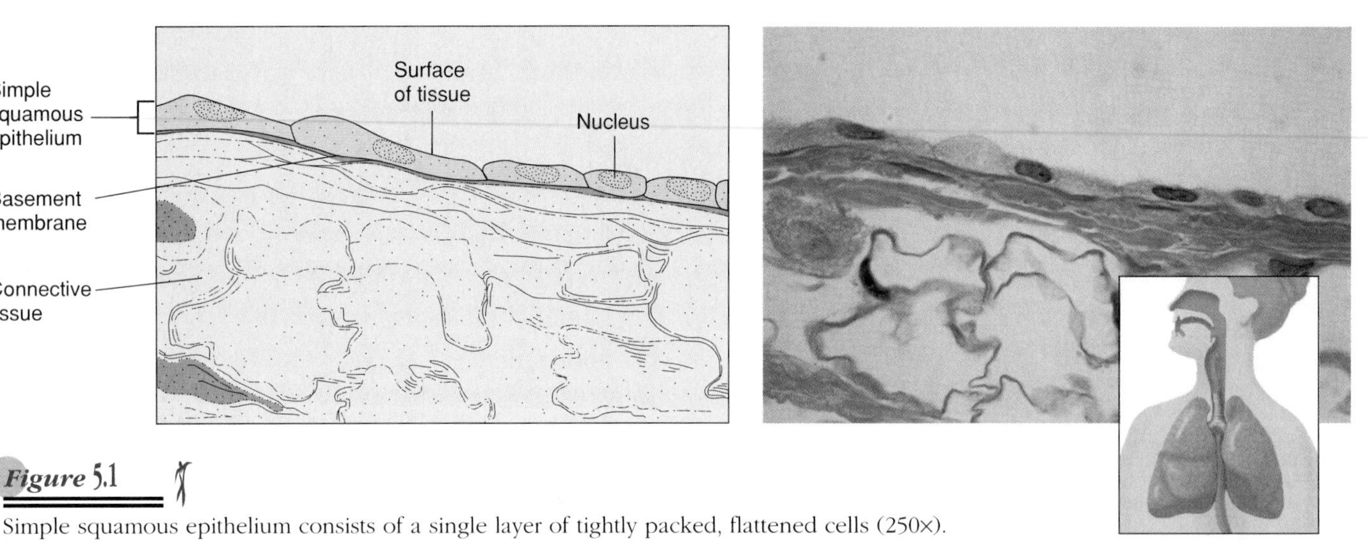

Figure 5.1

Simple squamous epithelium consists of a single layer of tightly packed, flattened cells (250×).

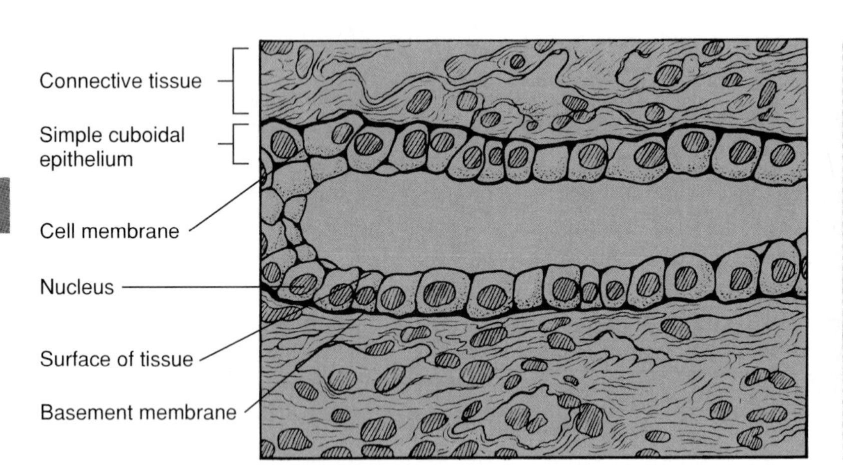

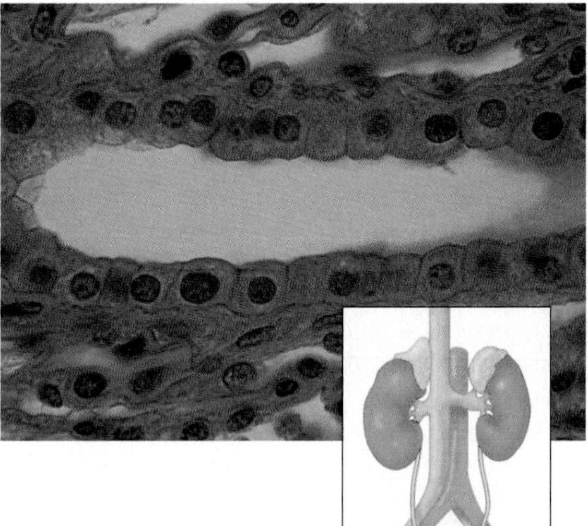

Figure 5.2

Simple cuboidal epithelium consists of a single layer of tightly packed, cube-shaped cells (250×).

Simple cuboidal epithelium covers the ovaries and lines most of the kidney tubules and the ducts of certain glands, such as the salivary glands, thyroid gland, pancreas, and liver. In the kidneys, this tissue functions in secretion and absorption; in glands, it secretes glandular products.

Simple Columnar Epithelium

The cells of **simple columnar epithelium** are elongated; that is, they are longer than they are wide. This tissue is composed of a single layer of cells whose nuclei are usually located at about the same level near the basement membrane (fig. 5.3).

Simple columnar epithelium lines the uterus and most organs of the digestive tract, including the stomach and small and large intestines. Its elongated cells make this tissue relatively thick and protective of underlying tissues. Simple columnar epithelium also secretes digestive fluids and absorbs nutrients from digested food.

Simple columnar cells, specialized for absorption, often have many minute, cylindrical processes extending outward. These processes, called *microvilli,* increase the surface area of the cell membrane where it is exposed to the substances being absorbed.

Typically, specialized, flask-shaped glandular cells are scattered among the columnar cells of this tissue. These cells, called *goblet cells,* secrete a protective fluid (*mucus*) onto the tissue's free surface (fig. 5.3).

Figure 5.3

Simple columnar epithelium consists of a single layer of elongated cells (400×).

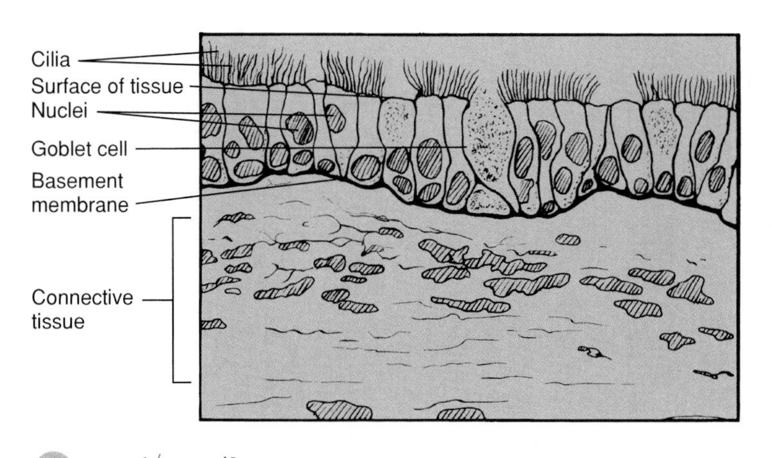

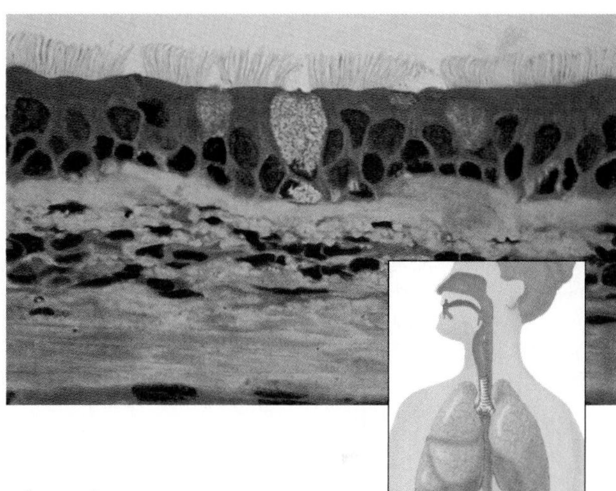

Figure 5.4

Pseudostratified columnar epithelium appears stratified because nuclei are located at different levels (500×).

Pseudostratified Columnar Epithelium

The cells of **pseudostratified columnar epithelium** appear stratified or layered, but they are not. Although each cell has only one nucleus, the layered effect occurs because the nuclei are located at two or more levels within the cells of the tissue (fig. 5.4).

Pseudostratified columnar epithelial cells commonly possess cilia, which extend from the cells' free surfaces and move constantly (see chapter 3). Goblet cells are scattered throughout this tissue, and cilia sweep the secreted mucus along.

Pseudostratified columnar epithelium lines the passages of the respiratory and reproductive systems. In the respiratory passages, the mucus-covered linings are sticky and trap particles of dust and microorganisms that enter with the air. The cilia move the mucus and its captured particles upward and out of the airways. In the reproductive tubes, the cilia aid in moving sex cells from one region to another.

Stratified Squamous Epithelium

The many cell layers of **stratified squamous epithelium** make this tissue relatively thick. Cells reproduce in the deeper layers, and newer cells push older ones farther and farther outward (fig. 5.5).

Stratified squamous epithelium forms the outer layer of the skin (*epidermis*). As skin cells age, they accumulate a protein called *keratin* and then harden and die. This "keratinization" produces a covering of dry, tough, protective material that prevents water and other substances from escaping from underlying tissues and blocks various chemicals and microorganisms from entering.

Stratified squamous epithelium also lines the mouth, throat, vagina, and anal canal. Here, the tissue is not keratinized; it stays moist, and the cells on its free surfaces remain alive.

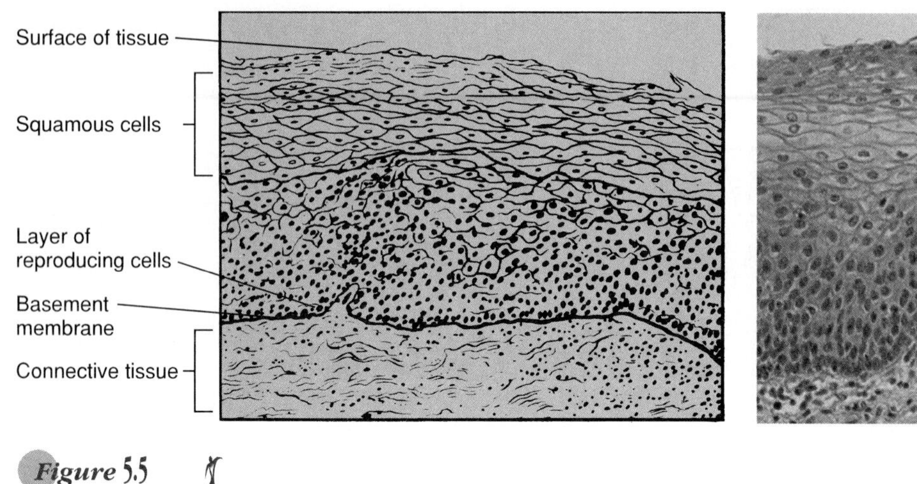

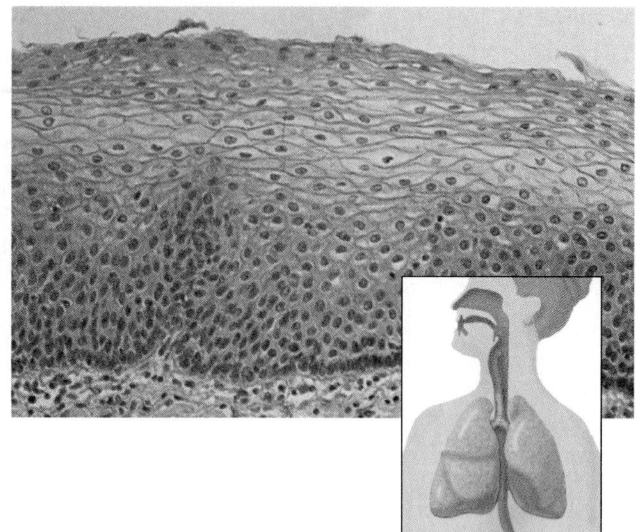

Figure 5.5

Stratified squamous epithelium consists of many layers of cells (70×).

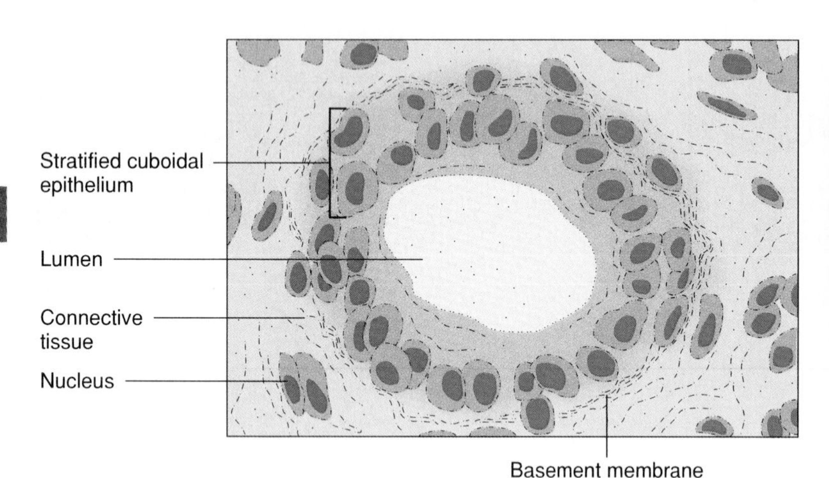

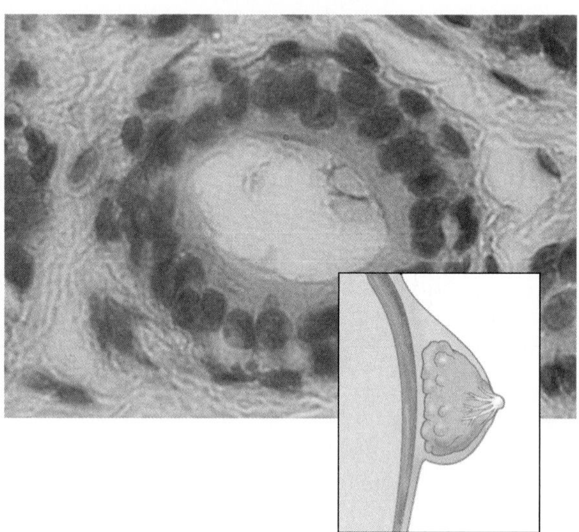

Figure 5.6

Stratified cuboidal epithelium consists of two to three layers of cube-shaped cells surrounding a lumen (250×).

Stratified Cuboidal Epithelium

Stratified cuboidal epithelium consists of two or three layers of cuboidal cells that line a lumen (fig. 5.6). Cell layering protects more than a single layer.

Stratified cuboidal epithelium lines the larger ducts of the mammary glands, sweat glands, salivary glands, and pancreas. It also lines developing ovarian follicles and seminiferous tubules, parts of the female and male reproductive systems, respectively.

Stratified Columnar Epithelium

Stratified columnar epithelium consists of layers of cells (fig. 5.7). The superficial cells are elongated, while the basal layers consist of cube-shaped cells. Stratified columnar epithelium is found in the male urethra and vas deferens, and in parts of the pharynx.

Transitional Epithelium

Transitional epithelium is specialized to change in response to increased tension. It forms the inner lining of the urinary bladder and the passageways of the urinary system. The walls of these organs consist of several layers of cuboidal cells. When distended, they flatten, and when relaxed, they assume the more characteristic cuboidal shape (fig. 5.8). In addition to providing an expandable lining, transitional epithelium forms a barrier that helps prevent the contents of the urinary tract from diffusing out of its passageways.

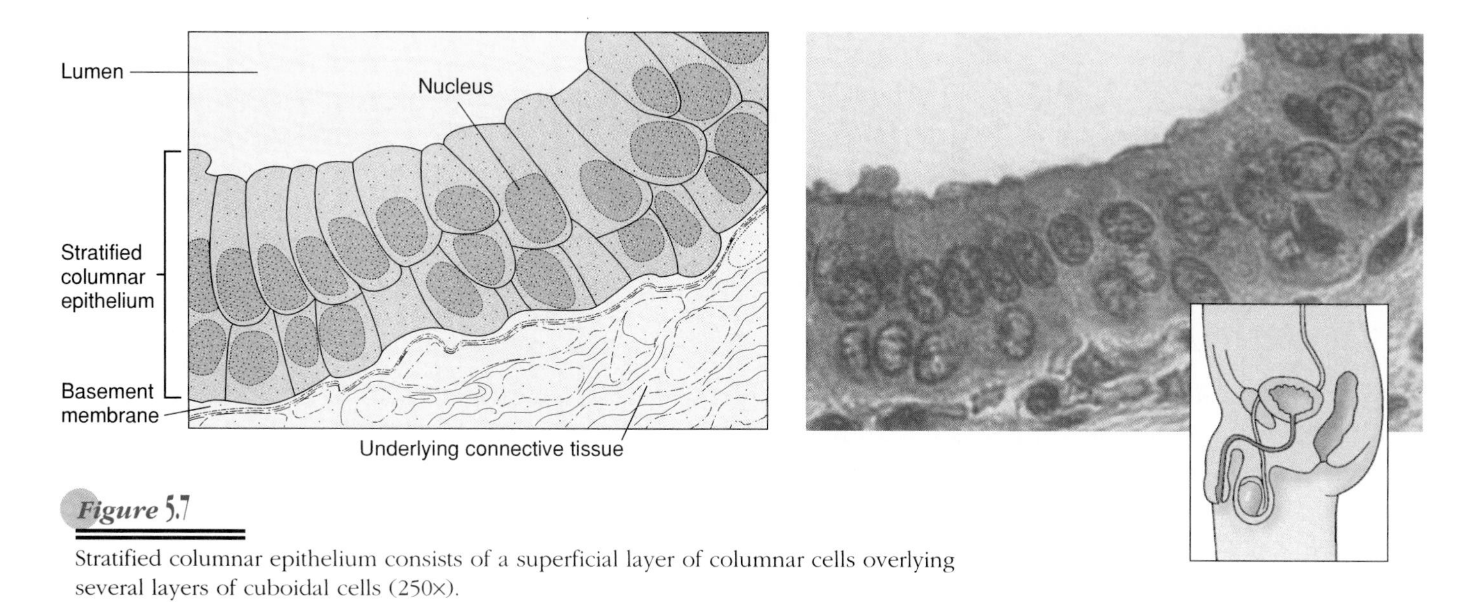

Figure 5.7

Stratified columnar epithelium consists of a superficial layer of columnar cells overlying several layers of cuboidal cells (250×).

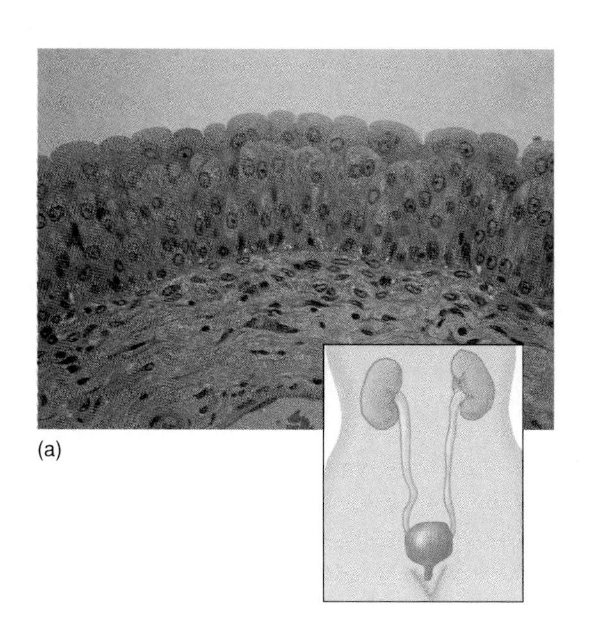

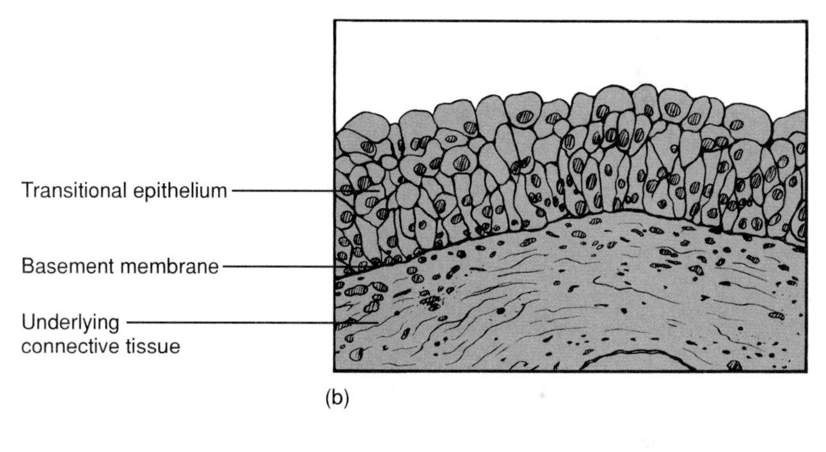

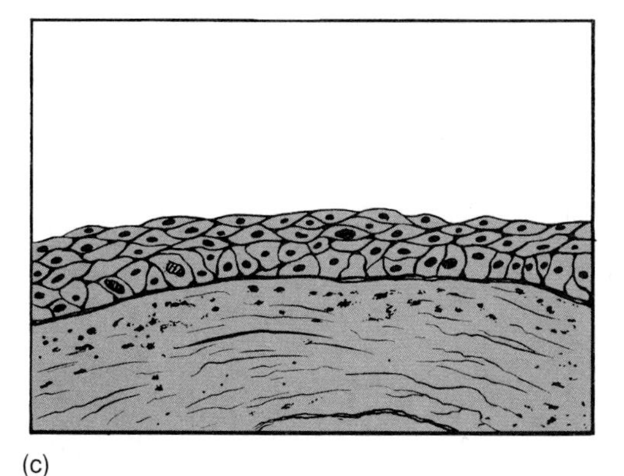

Figure 5.8

(*a*) Micrograph of transitional epithelium (100×). (*b*) This tissue consists of many layers when the organ wall is contracted. (*c*) The tissue is thinner when the wall is stretched.

A cancer originating in epithelium is called a *carcinoma,* and up to 90% of all human cancers are of this type. Most carcinomas begin on surfaces that contact the external environment, such as the skin, the linings of the airways in the respiratory tract, or the linings of the stomach or intestines in the digestive tract. This observation suggests that the more common cancer-causing agents may not penetrate tissues very deeply. Cytotechnologists identify cancerous cells under the microscope. Increasingly, computerized image analysis systems are minimizing human error in classifying cells. ■

Table 5.2

Epithelial Tissues

Type	Function	Location
Simple squamous epithelium	Filtration, diffusion	Air sacs of the lungs, walls of capillaries, linings of blood and lymph vessels
Simple cuboidal epithelium	Secretion, absorption	Surface of ovaries, linings of kidney tubules, and linings of ducts of certain glands
Simple columnar epithelium	Absorption, secretion	Linings of uterus and organs of the digestive tract
Pseudostratified columnar epithelium	Protection, secretion, movement of mucus and sex cells	Linings of respiratory passages and reproductive tract
Stratified squamous epithelium	Protection	Outer layer of skin, linings of oral cavity, throat, vagina, and anal canal
Stratified cuboidal epithelium	Protection	Linings of larger ducts of mammary glands, sweat glands, salivary glands, and pancreas
Stratified columnar epithelium	Protection, secretion	Male urethra, parts of the pharynx
Transitional epithelium	Distensibility, protection	Inner linings of urinary bladder and passageways of urinary tract

Table 5.2 summarizes the characteristics of the different types of epithelial tissues.

1. List the general characteristics of epithelial tissues.
2. Describe the structure of each type of epithelium.
3. Describe the special functions of each type of epithelium.

Glandular Epithelium

Glandular epithelium is composed of cells that are specialized to produce and secrete substances. Such cells occur most commonly within columnar and cuboidal epithelia, and one or more of these cells constitute a *gland*. Glands that secrete their products into ducts opening onto internal or external surfaces are called **exocrine glands.** Those that secrete their products into tissue fluids or blood are called **endocrine glands.** (Endocrine glands are discussed in chapter 11.)

Exocrine glands are classified according to the composition of their secretions and the method by which secretion occurs (fig. 5.9). Glands that secrete watery, protein-rich serous fluids by exocytosis are *merocrine glands.* Those gland cells that lose small portions of their cytoplasm as fluid-filled packets are *apocrine glands. Holocrine glands* are those in which the entire cell lyses during secretion. Table 5.3 summarizes these glands and their secretions.

Most exocrine secretory cells are merocrine, which can be further subdivided into either *serous cells* or *mucous cells.* The secretion of serous cells is typically watery, has a high concentration of enzymes, and is called *serous fluid.* Serous cells are common in the glands of body cavities. Mucous cells secrete a thicker *mucus,* which is rich in the glycoprotein *mucin.* Glands from the inner linings of the digestive and respiratory systems secrete abundant mucus.

1. Distinguish between exocrine glands and endocrine glands.
2. Explain how exocrine glands are classified.
3. Distinguish between a serous cell and a mucous cell.

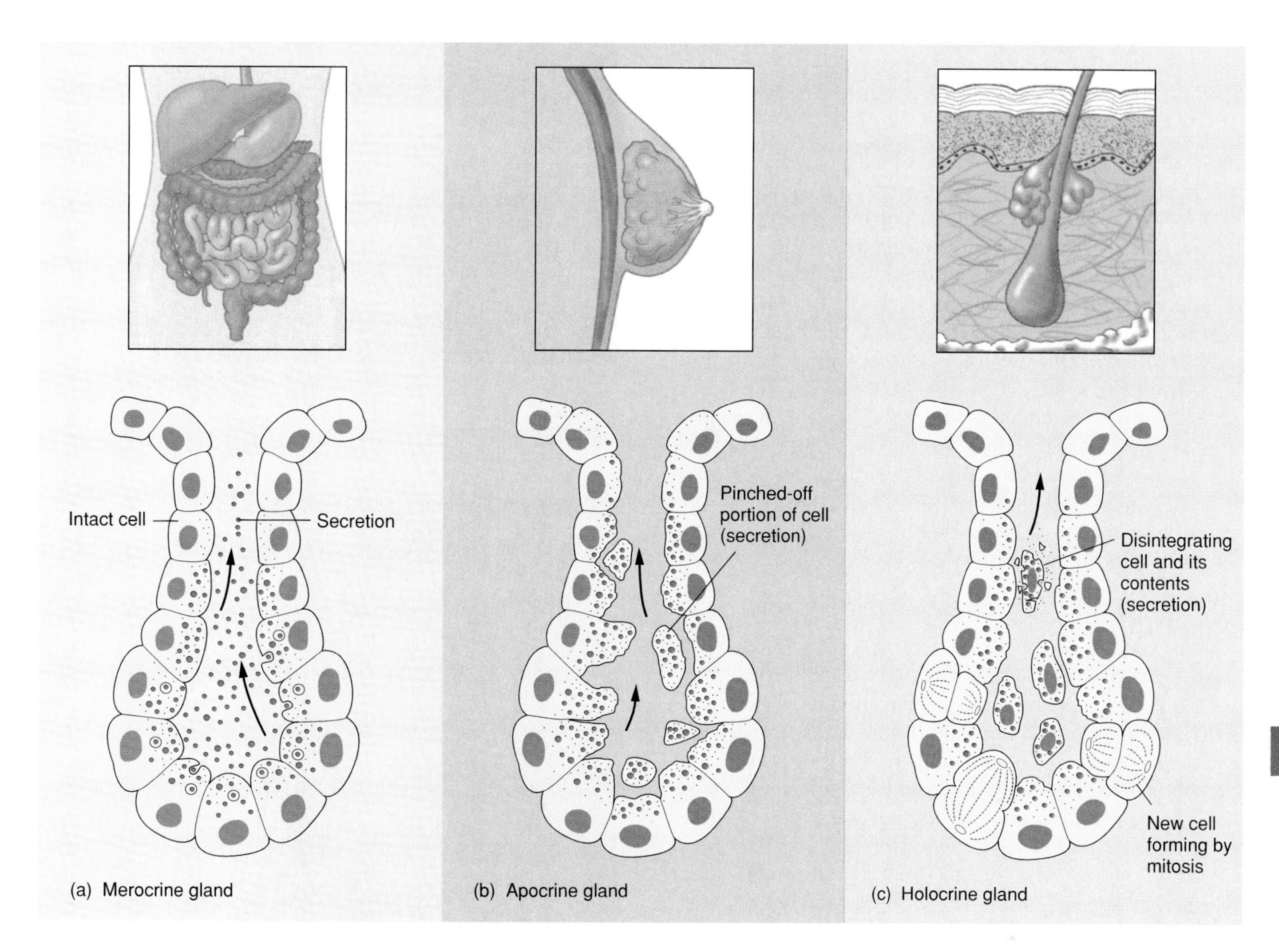

Figure 5.9

(*a*) Merocrine glands release secretions without losing cytoplasm. (*b*) Apocrine glands lose small portions of their cell bodies during secretion. (*c*) Holocrine glands release entire cells filled with secretory products.

Table 5.3

Types of Glandular Secretions

Secretory Type	Description of Secretion	Example
Merocrine glands	A fluid product released by exocytosis	Salivary glands, pancreatic glands, certain sweat glands
Apocrine glands	Cellular product and portions of free ends of glandular cells pinched off during secretion	Mammary glands, certain sweat glands
Holocrine glands	Entire cells that are filled with secretory products	Sebaceous glands

Connective Tissues

General Characteristics

Connective tissues bind structures, provide support and protection, serve as frameworks, fill spaces, store fat, produce blood cells, protect against infections, and help repair tissue damage. Connective tissue cells are usually farther apart than epithelial cells, and they have an abundance of intercellular material, or *matrix,* between them. This matrix consists of *fibers* and a *ground substance* whose consistency varies from fluid to semisolid to solid.

Connective tissue cells are usually able to reproduce. In most cases, they have good blood supplies and are well nourished. Some connective tissues, such as bone and cartilage, are quite rigid. Loose fibrous connective tissue, adipose tissue, and dense fibrous connective tissue are more flexible.

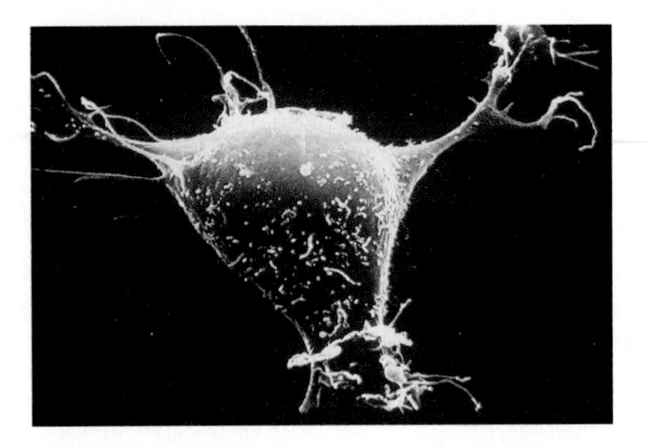

Figure 5.10

Scanning electron micrograph of a fibroblast (6,000×).

Major Cell Types

Connective tissues contain a variety of cell types. *Resident cells* are usually present in relatively stable numbers and include fibroblasts and mast cells. In contrast, *wandering cells* move temporarily into the tissues from the bloodstream, usually in response to an injury or infection. Wandering cells include several types of white blood cells.

Fibroblasts are the most common type of resident cell in connective tissue. These large and usually star-shaped cells produce fibers by secreting proteins into the matrix around them (fig. 5.10).

Macrophages are almost as numerous as fibroblasts in some connective tissues. They are specialized to carry on phagocytosis. Macrophages can move about and function as scavenger and defensive cells that clear foreign particles from tissues (fig. 5.11).

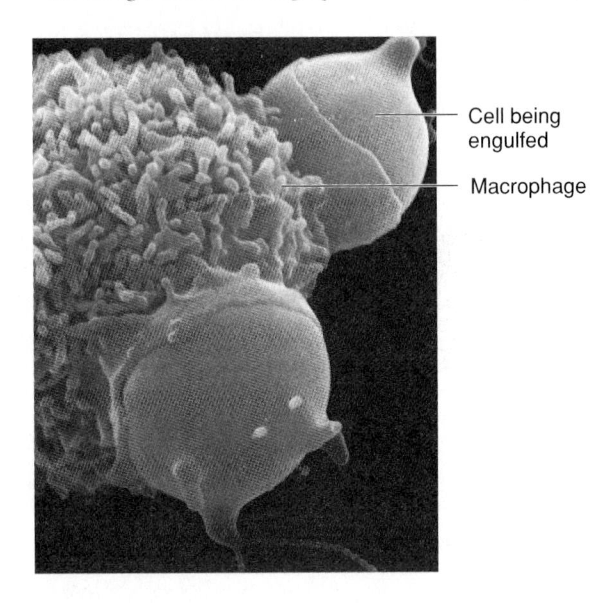

Figure 5.11

Macrophages are scavenger cells common in connective tissues. In this scanning electron micrograph, a macrophage engulfs two cells (about 5,600×).

Mast cells are large and widely distributed in connective tissues. They are usually located near blood vessels (fig. 5.12). Mast cells release heparin, which prevents blood clotting, and histamine, which promotes reactions associated with inflammation and allergies (see chapter 14).

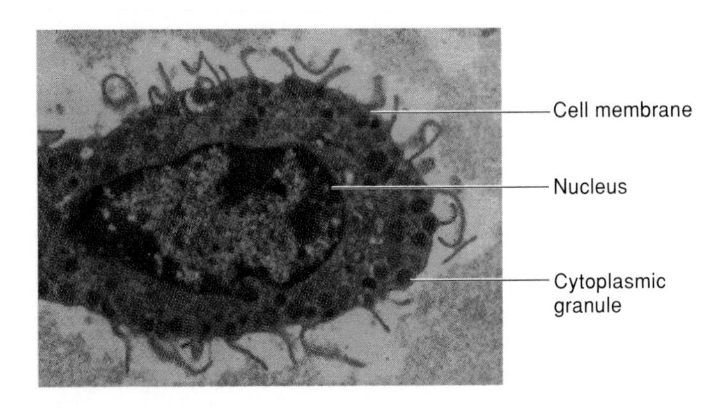

Figure 5.12

Transmission electron micrograph of a mast cell (4,000×).

Connective Tissue Fibers

Fibroblasts produce three types of connective tissue fibers: collagenous fibers, elastic fibers, and reticular fibers. Of these, collagenous and elastic fibers are the most abundant (fig. 5.13).

Collagenous fibers are thick threads made from molecules of the protein *collagen.* These fibers are grouped in long, parallel bundles, and they are flexible

Ground substance
Fibroblast
Elastic fiber
Collagenous fiber

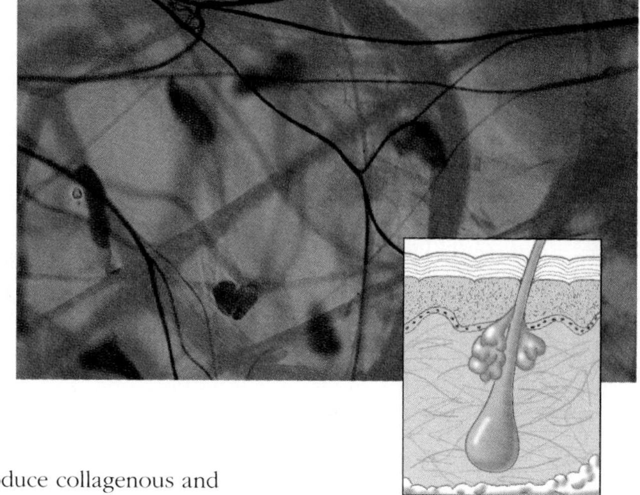

Figure 5.13

Loose fibrous connective tissue contains numerous fibroblasts that produce collagenous and elastic fibers (250×).

but only slightly elastic. More importantly, they have great tensile strength—that is, they resist considerable pulling force. Thus, collagenous fibers are important components of body parts that hold structures together, such as **tendons** that connect muscles to bones.

Tissue containing abundant collagenous fibers is called *dense connective tissue*. Such tissue appears white, and for this reason, collagenous fibers are sometimes called *white fibers*.

Elastic fibers are composed of a protein called *elastin*. These thin fibers branch, forming complex networks. Elastic fibers are weaker than collagenous fibers, but they stretch easily and can resume their original lengths and shapes. Elastic fibers are common in body parts that are frequently stretched, such as the vocal cords. They are sometimes called *yellow fibers* because tissues well supplied with them appear yellowish.

When skin is exposed to prolonged and intense sunlight, connective tissue fibers lose elasticity, and the skin stiffens and becomes leathery. In time, the skin may sag and wrinkle. Collagen injections may temporarily smooth out wrinkles. Collagen applied as a cream to the skin does not combat wrinkles because collagen molecules are too large for the skin to absorb. ■

Reticular fibers are very thin collagenous fibers. They are highly branched and form delicate supporting networks in a variety of tissues.

Loose Fibrous Connective Tissue

Loose fibrous connective tissue forms delicate, thin membranes throughout the body. The cells of this tissue, mainly fibroblasts, are located some distance apart and are separated by a gel-like matrix that contains many collagenous and elastic fibers (fig. 5.13).

Loose fibrous connective tissue binds the skin to the underlying organs and fills spaces between muscles. It lies beneath most layers of epithelium, where its many blood vessels nourish nearby epithelial cells.

Adipose Tissue

Adipose tissue, or fat, is a specialized form of loose fibrous connective tissue that develops when certain cells store fat in droplets within their cytoplasm and enlarge (fig. 5.14). When such cells are so numerous that they crowd other cell types, adipose tissue results. Adipose tissue lies beneath the skin, in spaces between muscles, around the kidneys, behind the eyeballs, in certain abdominal membranes, on the surface of the heart, and around certain joints.

Adipose tissue cushions joints and some organs, such as the kidneys. It also insulates beneath the skin, and it stores energy in fat molecules.

Overeating and lack of exercise can increase the size of adipose cells, leading to overweight or obesity. During periods of fasting, however, fat supplies energy, and adipose cells lose fat, shrink, and become more like fibroblasts. ■

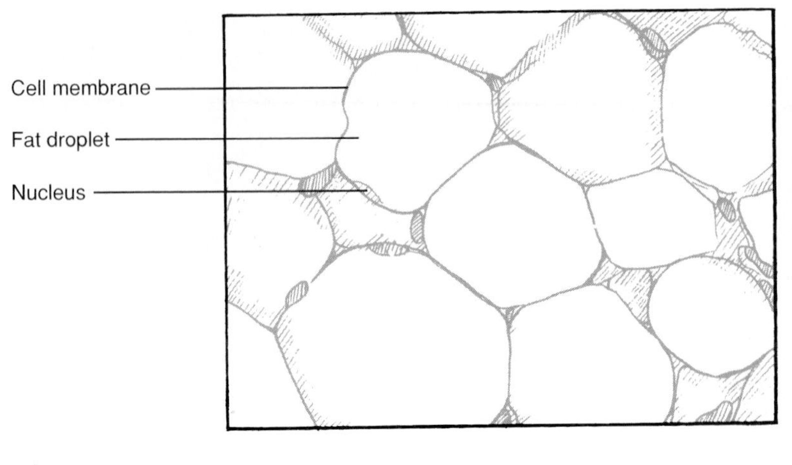

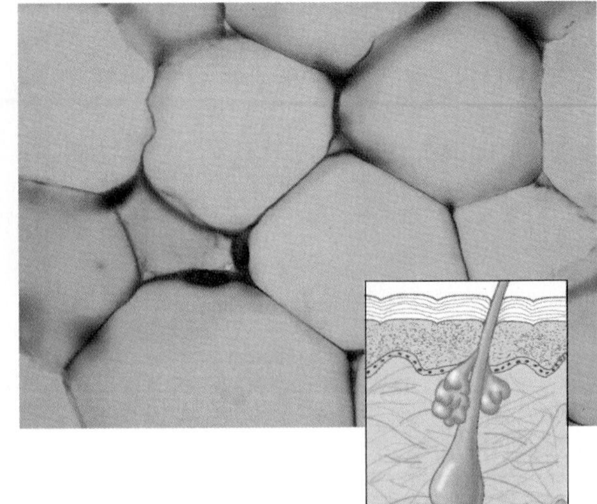

Figure 5.14

Adipose tissue cells contain large fat droplets that push the nuclei close to the cell membranes (250×).

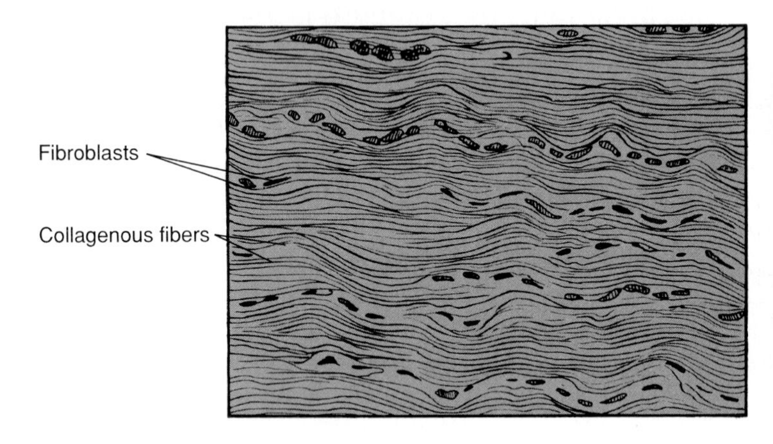

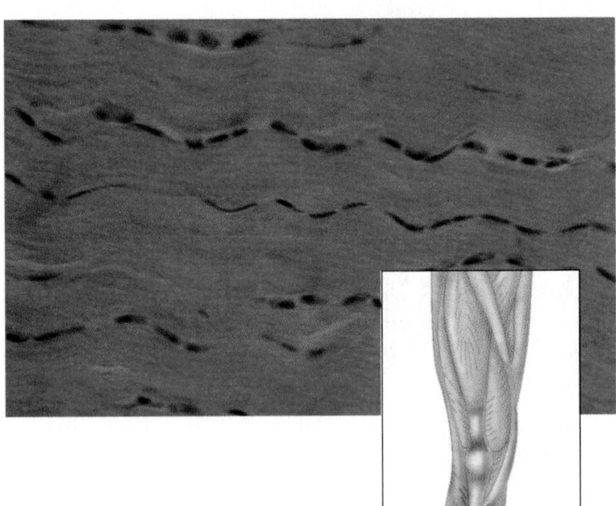

Figure 5.15

Regular dense fibrous connective tissue consists largely of tightly packed collagenous fibers (100×).

Dense Fibrous Connective Tissue

Dense fibrous connective tissue consists of many closely packed, thick, collagenous fibers and a fine network of elastic fibers. It has relatively few cells, most of which are fibroblasts (fig. 5.15).

Because collagenous fibers are very strong, dense fibrous connective tissue is able to withstand pulling forces. It often binds body parts together as a component of tendons (which connect muscles to bones) and **ligaments** (which connect bones to bones at joints). This type of tissue is also in the protective white layer of the eyeball and in the deeper skin layers. The blood supply to dense fibrous connective tissue is relatively poor, so tissue repair is slow or incomplete.

1. What are the general characteristics of connective tissues?
2. What are the characteristics of collagen and elastin?
3. How are loose fibrous connective tissue and adipose tissue related?
4. Distinguish between a tendon and a ligament.

Cartilage

Cartilage is a partly rigid, partly flexible connective tissue. It provides support, frameworks and attachments, protects underlying tissues, and forms structural models for many developing bones.

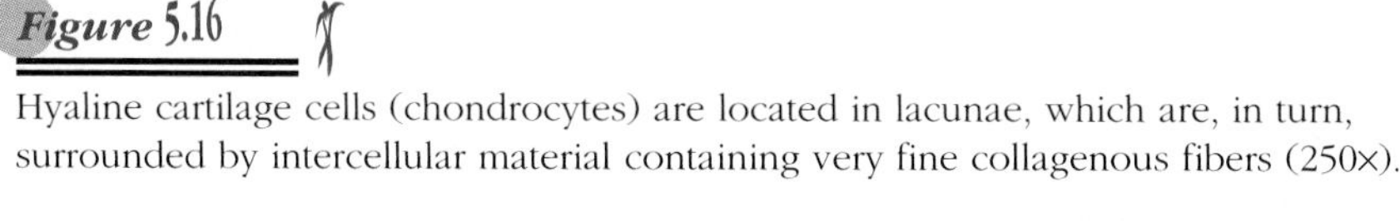

Figure 5.16

Hyaline cartilage cells (chondrocytes) are located in lacunae, which are, in turn, surrounded by intercellular material containing very fine collagenous fibers (250×).

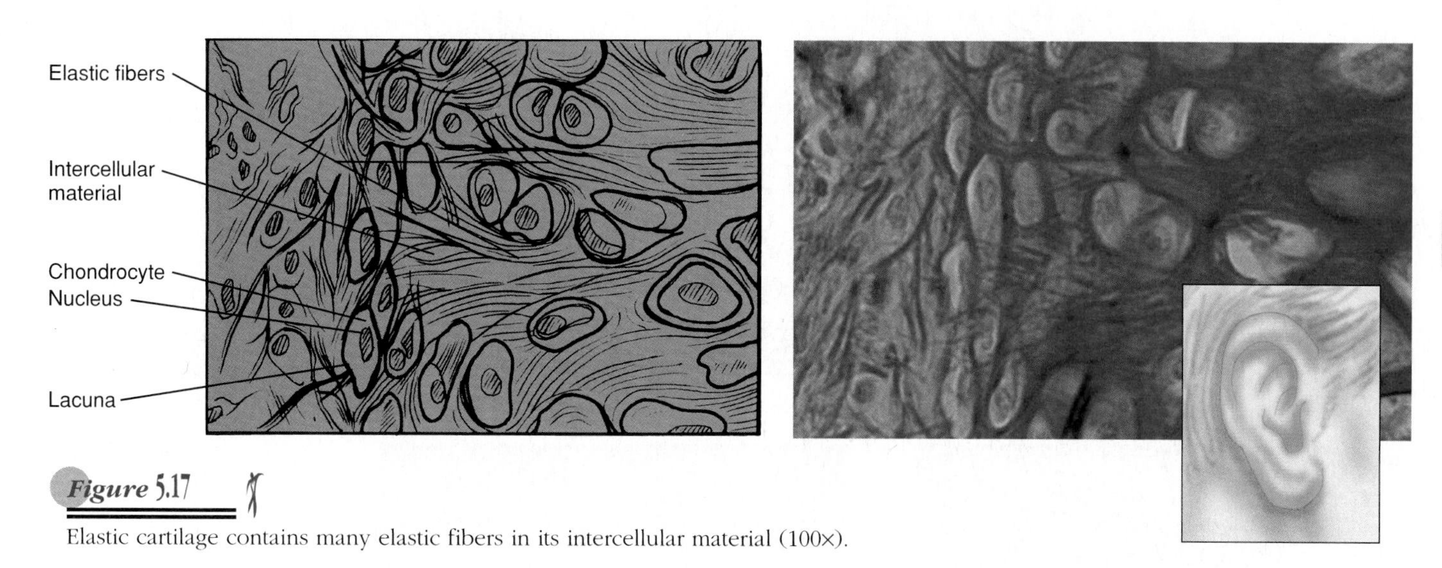

Figure 5.17

Elastic cartilage contains many elastic fibers in its intercellular material (100×).

The intercellular material of cartilage is abundant and is composed largely of collagenous fibers embedded in a gel-like ground substance. Cartilage cells, or **chondrocytes**, occupy small chambers called *lacunae* and are completely surrounded by matrix (fig. 5.16).

A cartilaginous structure is enclosed in a covering of fibrous connective tissue called the *perichondrium*. The perichondrium contains blood vessels that provide cartilage cells with nutrients by diffusion. The lack of a direct blood supply to cartilage tissue is why torn cartilage heals slowly and why chondrocytes do not divide frequently.

Different types of intercellular material (matrix) distinguish three types of cartilage. **Hyaline cartilage**, the most common type, has very fine collagenous fibers in its matrix and looks somewhat like white plastic (fig. 5.16). It is found on the ends of bones in many joints, in the soft part of the nose, and in the supporting rings of the respiratory passages. Hyaline cartilage is also important in the development of most bones (see chapter 7).

Elastic cartilage contains a dense network of elastic fibers and thus is more flexible than hyaline cartilage (fig. 5.17). It provides the framework for the external ears and for parts of the larynx.

Fibrocartilage, a very tough tissue, contains many collagenous fibers (fig. 5.18). It is often a shock absorber for structures subjected to pressure. For example, fibrocartilage forms pads (intervertebral disks) between the individual parts of the backbone. It also cushions bones in the knees and in the pelvic girdle.

Bone

Bone is the most rigid connective tissue. Its hardness is due largely to mineral salts between cells. This matrix also contains abundant collagen, whose strong but flexible fibers reinforce bone's mineral components.

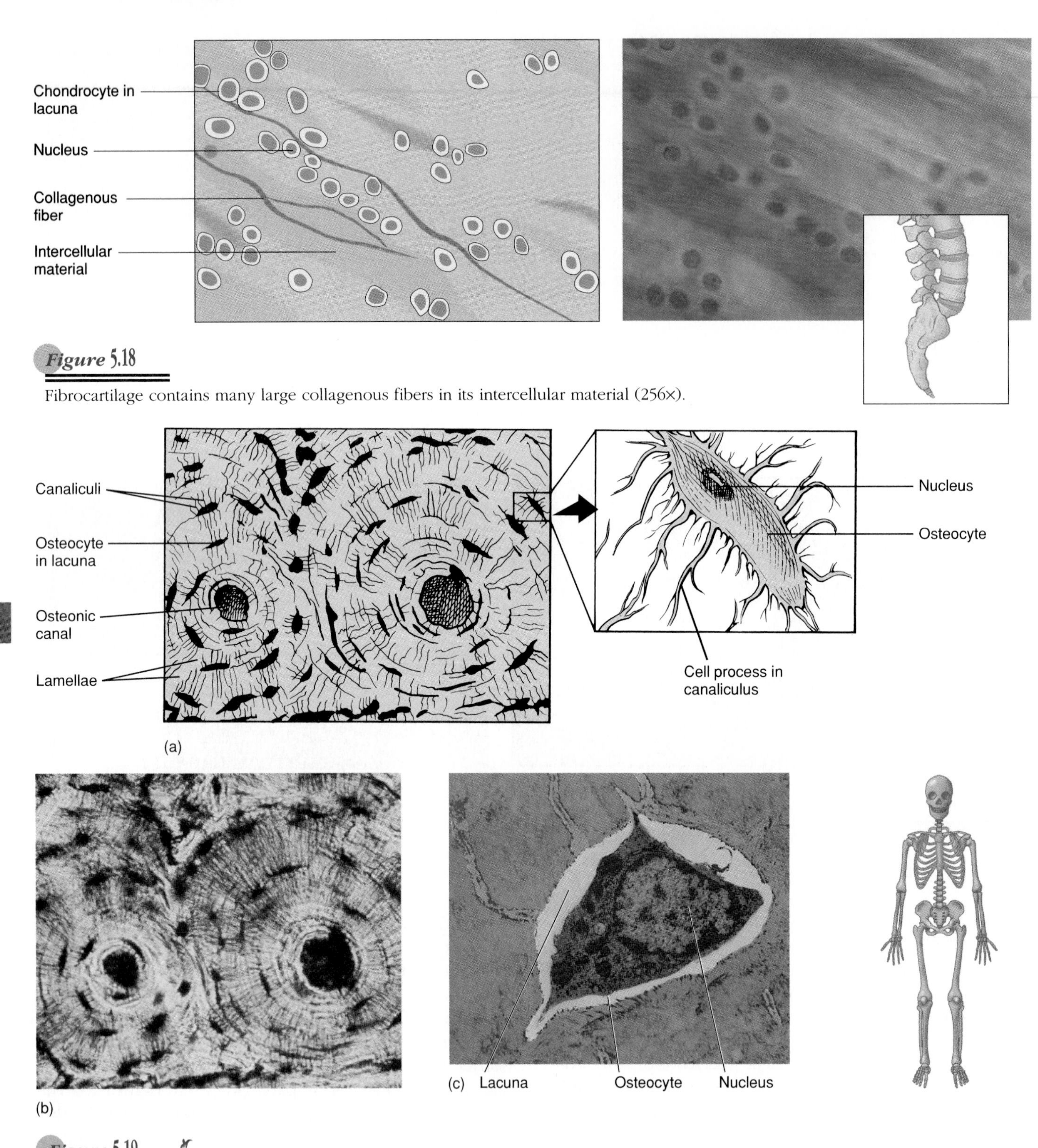

Figure 5.18

Fibrocartilage contains many large collagenous fibers in its intercellular material (256×).

Figure 5.19

(*a*) Bone matrix is deposited in concentric layers around osteonic canals. (*b*) Micrograph of bone tissue (160×). (*c*) Transmission electron micrograph of an osteocyte within a lacuna.

Bone supports body structures. It protects vital parts in various cavities and is an attachment for muscles. Bone houses marrow, where blood cells form, and stores inorganic chemicals such as calcium and phosphorus.

Bone matrix is deposited in thin layers called *lamellae,* which form concentric patterns around tiny longitudinal tubes called *osteonic canals* (Haversian canals) (fig. 5.19). Bone cells, or **osteocytes,** are located in lacunae, which are rather evenly spaced between the

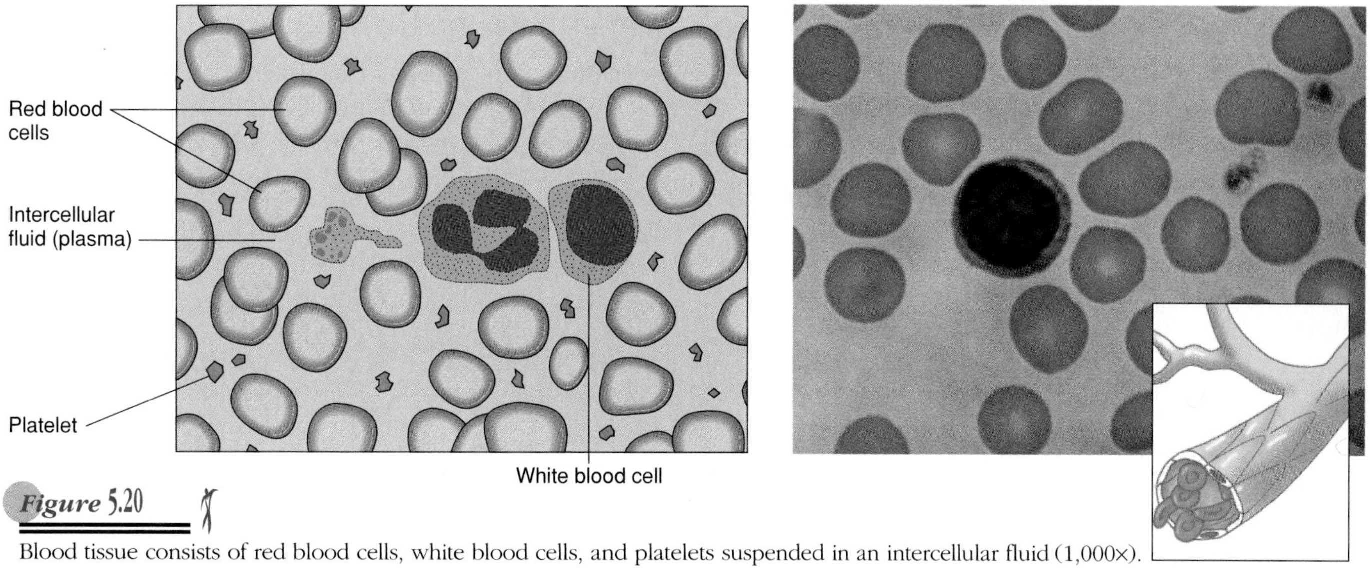

Red blood cells
Intercellular fluid (plasma)
Platelet
White blood cell

Figure 5.20

Blood tissue consists of red blood cells, white blood cells, and platelets suspended in an intercellular fluid (1,000×).

Table 5.4

Connective Tissues

Type	Function	Location
Loose fibrous connective tissue	Binds organs together, holds tissue fluids	Beneath skin, between muscles, beneath epithelial tissues
Adipose tissue	Protects, insulates, stores fat	Beneath skin, around kidneys, behind eyeballs, on surface of heart
Dense fibrous connective tissue	Binds organs together	Tendons, ligaments, deeper layers of skin
Hyaline cartilage	Supports, protects, provides framework	Nose, ends of bones, rings in the walls of respiratory passages
Elastic cartilage	Supports, protects, provides flexible framework	Framework of external ear and parts of larynx
Fibrocartilage	Supports, protects, absorbs shock	Between bony parts of backbone, knee, and pelvic girdle
Bone	Supports, protects, provides framework	Bones of skeleton
Blood	Transports substances, helps maintain stable internal environment	Throughout body within a closed system of blood vessels and heart chambers

lamellae. Consequently, osteocytes too, form patterns of concentric circles.

In a bone, osteocytes and layers of intercellular material clustered concentrically around an osteonic canal form a cylinder-shaped unit, called an **osteon** (Haversian system). Many osteons cemented together form the substance of bone.

Each osteonic canal contains a blood vessel; thus, every bone cell is fairly close to a nutrient supply. In addition, bone cells have numerous cytoplasmic processes, which extend outward and pass through very small tubes in the matrix called *canaliculi*. These cellular processes connect with processes of nearby cells. As a result, materials can move rapidly between blood vessels and bone cells and from one osteocyte to the next. Thus, in spite of its inert appearance, bone is a very active tissue that heals much more rapidly than cartilage. (The microscopic structure of bone is described in more detail in chapter 7.)

Blood

Blood transports a variety of substances between interior body cells and those that exchange with the external environment. Thus, it helps maintain stable internal environmental conditions. This tissue is composed of cells suspended in a fluid matrix called *blood plasma*. These cells include *red blood cells, white blood cells,* and cellular fragments called *platelets* (fig. 5.20). Most blood cells form in red marrow within the hollow parts of certain long bones. Blood is described in chapter 12.

Table 5.4 lists the characteristics of the connective tissues.

1. Describe the general characteristics of cartilage.
2. Explain why injured bone heals more rapidly than injured cartilage.
3. Name the types of blood cells.

Muscle Tissues

General Characteristics

Muscle tissues can contract; that is, their elongated cells, or *muscle fibers,* can shorten. As they contract, muscle fibers pull at their attached ends, which moves body parts. The three types of muscle tissue—skeletal muscle, smooth muscle, and cardiac muscle—are discussed in more detail in chapter 8.

Skeletal Muscle Tissue

Skeletal muscle tissue is found in muscles that attach to bones and that we can consciously control. For this reason, it is often called *voluntary* muscle tissue. The long, threadlike cells have alternating light and dark cross-markings called *striations.* Each muscle fiber has many nuclei located just beneath its cell membrane (fig. 5.21). When stimulated by a nerve impulse, the muscle fiber contracts and then relaxes.

The muscles containing skeletal muscle tissue move the head, trunk, and limbs. They also enable us to make facial expressions, write, talk, sing, chew, swallow, and breathe.

Smooth Muscle Tissue

Smooth muscle tissue is called smooth because its cells lack striations. Smooth muscle cells are shorter than those of skeletal muscle and are spindle-shaped, with a single, centrally located nucleus (fig. 5.22). This tissue is located in the walls of hollow internal organs, such as the stomach, intestines, urinary bladder, uterus, and blood vessels. Unlike skeletal muscle, smooth muscle usually cannot be consciously stimulated to contract. Thus, its actions are *involuntary.* For example, smooth muscle tissue moves food through the digestive tract, constricts blood vessels, and empties the urinary bladder.

Cardiac Muscle Tissue

Cardiac muscle tissue is found only in the heart. Its cells, which are striated, are joined end to end. The resulting muscle fibers are branched and interconnected in complex networks. Each cell within a cardiac muscle fiber has a single nucleus (fig. 5.23). At its end, where it touches another cell, is a specialized intercellular junction called an *intercalated disk,* discussed further in chapter 8.

Cardiac muscle, like smooth muscle, is controlled involuntarily. This tissue makes up the bulk of the heart and pumps blood through the heart chambers and into blood vessels.

1. List the general characteristics of muscle tissues.
2. Distinguish among skeletal, smooth, and cardiac muscle tissues.

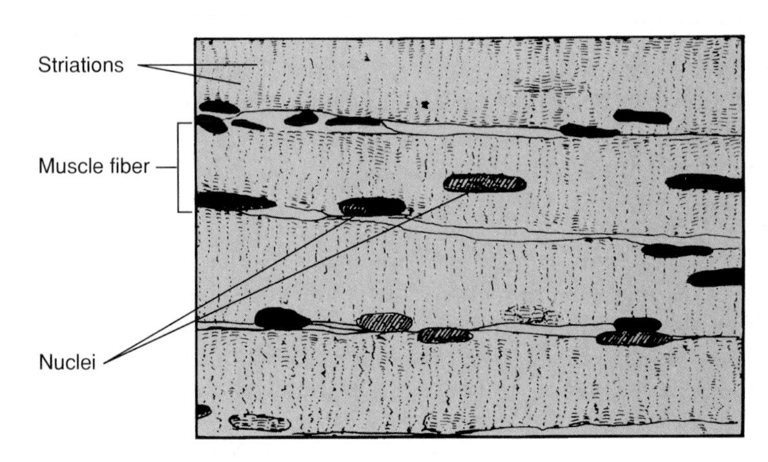

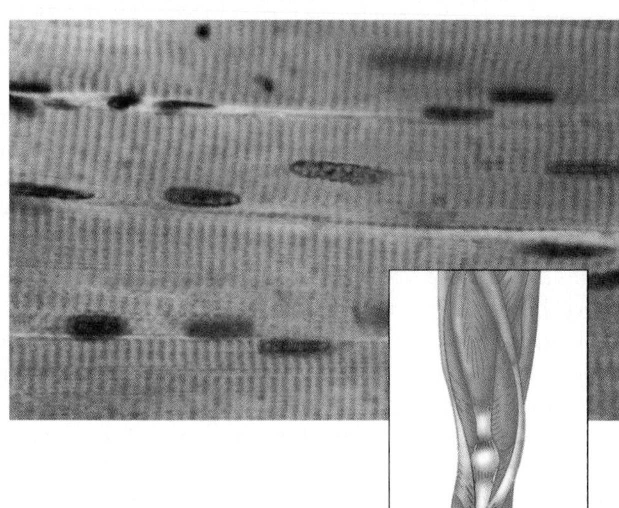

Figure 5.21

Skeletal muscle tissue is composed of striated muscle fibers with many nuclei (250×).

The cells of different tissues vary greatly in their reproductive abilities. Epithelial cells of the skin and the inner lining of the digestive tract, and the connective tissue cells that form blood cells in red bone marrow, reproduce continuously. However, striated and cardiac muscle cells and nerve cells do not reproduce at all after differentiating.

Fibroblasts respond rapidly to injuries by increasing in numbers and increasing fiber production. They are often the principal agents of repair in tissues with limited abilities to regenerate. For instance, cardiac muscle tissue typically degenerates in the regions a heart attack damages. Connective tissue formed by fibroblasts replaces the oxygen-starved cardiac muscle tissue, forming a scar. ■

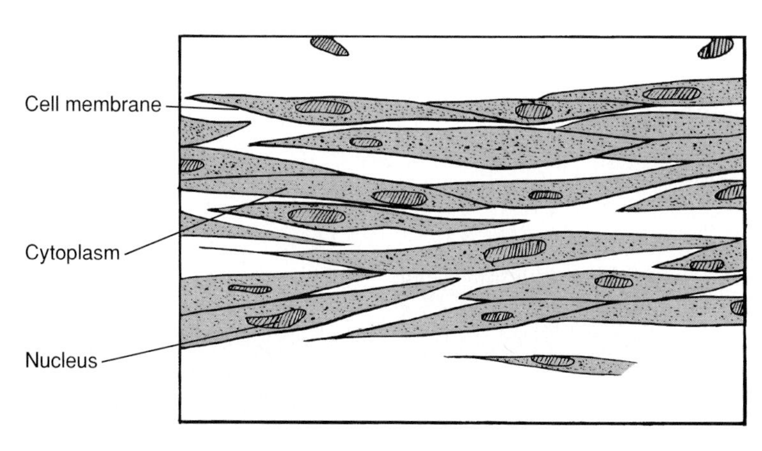

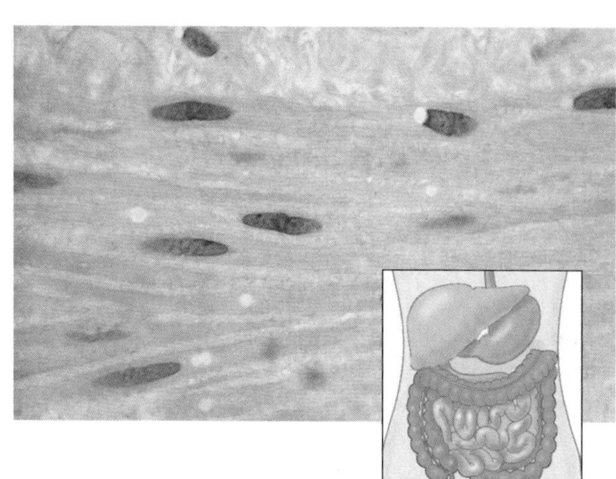

***Figure* 5.22**

Smooth muscle tissue consists of spindle-shaped cells, each with a large nucleus (250×).

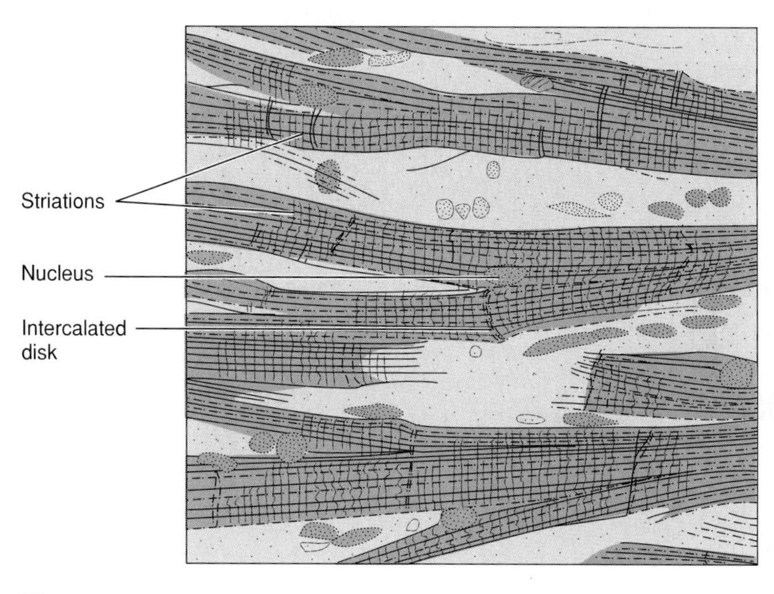

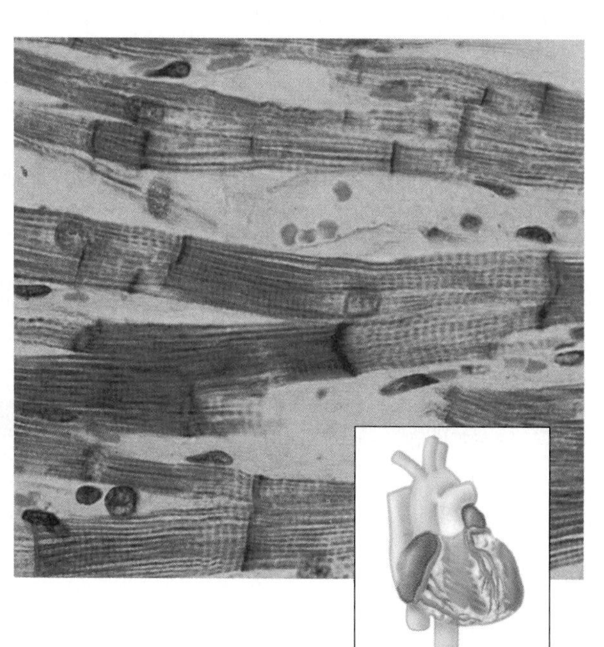

***Figure* 5.23**

Cardiac muscle fibers are branched and interconnected, with a single nucleus each (400×).

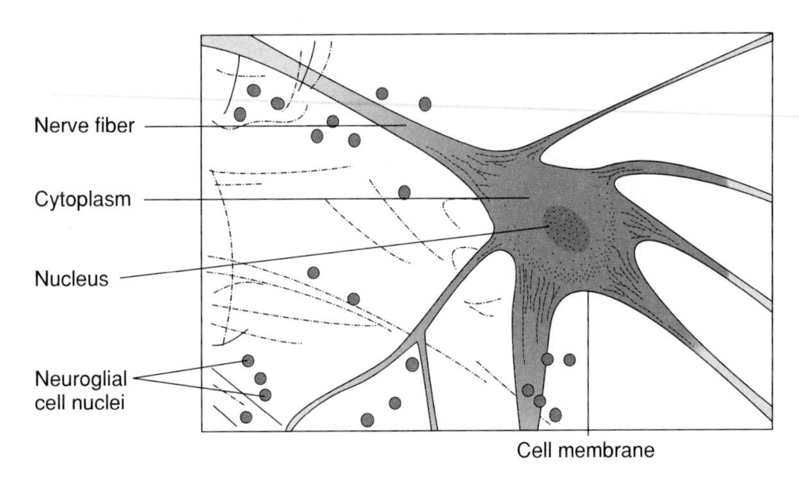

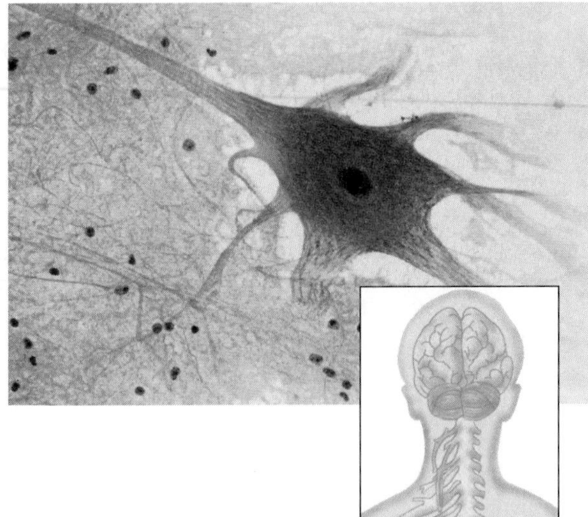

Figure 5.24

A nerve cell with nerve fibers extending into its surroundings (50×).

Nervous Tissues

Nervous tissues are found in the brain, spinal cord, and peripheral nerves. The basic cells are called **neurons** (nerve cells) (fig. 5.24). Neurons are sensitive to certain types of changes in their surroundings. They respond by transmitting nerve impulses along cellular extensions (nerve fibers) to other neurons or to muscles or glands. Because neurons communicate with each other and with various body parts, they coordinate, regulate, and integrate many body functions.

In addition to neurons, nervous tissue includes **neuroglial cells** (fig. 5.24). These cells support and bind the components of nervous tissue, carry on phagocytosis, and help supply nutrients to neurons by connecting them to blood vessels. Nervous tissue is discussed in more detail in chapter 9.

Table 5.5 summarizes the general characteristics of muscle and nervous tissues.

1. Describe the general characteristics of nervous tissues.
2. Distinguish between neurons and neuroglial cells.

Table 5.5 Muscle and Nervous Tissues

Type	Function	Location
Skeletal muscle tissue	Voluntary movements of skeletal parts	Muscles usually attached to bones
Smooth muscle tissue	Involuntary movements of internal organs	Walls of hollow internal organs
Cardiac muscle tissue	Heart movements	Heart muscle
Nervous tissue	Sensitivity and conduction of nerve impulses	Brain, spinal cord, and peripheral nerves

Summary Outline

Introduction (p. 97)

Tissues are groups of cells with specialized structural and functional roles. The four major types of human tissue are epithelial tissues, connective tissues, muscle tissues, and nervous tissues. An intercellular matrix supports cells in tissues.

Epithelial Tissues (p. 97)

1. General characteristics
   a. Epithelial tissues cover all free body surfaces and are the major tissues of glands.
   b. Epithelium is anchored to connective tissue by a basement membrane, lacks blood vessels, contains little intercellular material (matrix), and is replaced continuously.
   c. It protects, secretes, absorbs, and excretes.
2. Simple squamous epithelium
   a. This tissue consists of a single layer of thin, flattened cells.
   b. It functions in gas exchange in the lungs and lines the blood and lymph vessels and various body cavities.
3. Simple cuboidal epithelium
   a. This tissue consists of a single layer of cube-shaped cells.
   b. It carries on secretion and absorption in the kidneys and various glands.
4. Simple columnar epithelium
   a. This tissue is composed of elongated cells whose nuclei are located near the basement membrane.
   b. It lines the uterus and digestive tract.
   c. Absorbing cells often possess microvilli.
   d. This tissue contains goblet cells that secrete mucus.
5. Pseudostratified columnar epithelium
   a. This tissue appears stratified because the nuclei are located at two or more levels.
   b. Its cells may have cilia that move mucus or sex cells.
   c. It lines passageways of the respiratory and reproductive systems.
6. Stratified squamous epithelium
   a. This tissue is composed of many layers of cells.
   b. It protects underlying cells.
   c. It covers the skin and lines the mouth, throat, vagina, and anal canal.
7. Stratified cuboidal epithelium
   a. This tissue is composed of two or three layers of cube-shaped cells.
   b. It lines the larger ducts of the mammary glands, sweat glands, salivary glands, and pancreas.
   c. It protects.
8. Stratified columnar epithelium
   a. The top layer of cells in this tissue contains elongated columns. Cube-shaped cells make up the bottom layers.
   b. It is found in the male urethra and vas deferens, and parts of the pharynx.
   c. This tissue protects and secretes.
9. Transitional epithelium
   a. This tissue is specialized to change under tension.
   b. It is in the walls of various organs of the urinary system.
10. Glandular epithelium
    a. This tissue is composed of cells specialized to secrete substances.
    b. A gland consists of one or more cells.
       (1) Exocrine glands secrete into ducts.
       (2) Endocrine glands secrete into tissue fluid or blood.
    c. Exocrine glands are classified according to the method by which they release secretions.
       (1) Merocrine glands secrete fluid without loss of cytoplasm.
          (a) Serous cells secrete watery fluid with a high enzyme content.
          (b) Mucous cells secrete mucus.
       (2) Apocrine glands lose portions of their cells during secretion.
       (3) Holocrine glands release cells filled with secretory products.

Connective Tissues (p. 104)

1. General characteristics
   a. Connective tissues connect, support, protect, provide frameworks, fill spaces, store fat, produce blood cells, protect against infection, and help repair damaged tissues.
   b. Connective tissue cells are separated by considerable intercellular material (matrix).
   c. This intercellular matrix consists of fibers and a ground substance.
   d. Major cell types
      (1) Fibroblasts produce collagenous and elastic fibers.
      (2) Macrophages are phagocytes.
      (3) Mast cells may release heparin and histamine, and usually are located near blood vessels.
   e. Connective tissue fibers
      (1) Collagenous fibers are composed of collagen and have great tensile strength.
      (2) Tissue consisting largely of collagenous fibers is called dense connective tissue.
      (3) Elastic fibers are composed of microfibrils embedded in elastin and are very elastic.
      (4) Reticular fibers are very fine collagenous fibers.
2. Loose fibrous connective tissue
   a. This tissue forms thin membranes between organs.
   b. It is found beneath the skin and between muscles.
3. Adipose tissue
   a. This tissue is a specialized form of loose fibrous connective tissue that stores fat.
   b. It is found beneath the skin, in certain abdominal membranes, and around the kidneys, heart, and various joints.
4. Dense fibrous connective tissue
   a. This tissue is composed largely of strong, collagenous fibers.
   b. It is found in the tendons, ligaments, white portions of the eyes, and the deep layer of the skin.
5. Cartilage
   a. Cartilage provides a supportive framework for various structures.
   b. Its intercellular material is composed of fibers and a gel-like ground substance.

c. Cartilaginous structures are enclosed in a perichondrium.
d. Cartilage lacks a direct blood supply and is slow to heal.
e. The major types are hyaline cartilage, elastic cartilage, and fibrocartilage.

6. Bone
   a. The intercellular matrix of bone contains mineral salts and collagen.
   b. Its cells are usually organized in concentric circles around osteonic canals and are interconnected by canaliculi.
   c. Bone is an active tissue that heals rapidly.
7. Blood
   a. Blood transports substances and helps maintain a stable internal environment.
   b. Blood is composed of red cells, white cells, and platelets suspended in plasma.
   c. Blood cells develop in red marrow in the hollow parts of long bones.

Muscle Tissues (p. 110)

1. General characteristics
   a. Muscle tissues contract, moving structures that are attached to them.
   b. Three types are skeletal, smooth, and cardiac muscle tissues.
2. Skeletal muscle tissue
   a. Muscles containing this tissue usually are attached to bones and controlled by conscious effort.
   b. Cells, or muscle fibers, are long and threadlike.
   c. Muscle fibers contract when stimulated by nerve action and then relax immediately.
3. Smooth muscle tissue
   a. This tissue is in the walls of hollow internal organs.
   b. Usually, it is controlled by involuntary activity.
4. Cardiac muscle tissue
   a. This tissue is found only in the heart.
   b. Cells are joined by intercalated disks and form branched networks.

Nervous Tissues (p. 112)

1. Nervous tissues are found in the brain, spinal cord, and peripheral nerves.
2. Neurons (nerve cells)
   a. Neurons sense changes and respond with nerve impulses to other neurons or body parts.
   b. They coordinate, regulate, and integrate body activities.
3. Neuroglial cells
   a. Some forms bind and support nervous tissue.
   b. Others carry on phagocytosis.
   c. Still others connect neurons to blood vessels.

Review Exercises

1. Define *tissue*. (p. 97)
2. Name the four major types of tissue in the human body. (p. 97)
3. Describe the general characteristics of epithelial tissues. (p. 97)
4. Explain how the structure of simple squamous epithelium provides its function. (p. 97)
5. Name an organ in which each of the following tissues is found, and give the function of each tissue. (p. 97)
   a. Simple squamous epithelium
   b. Simple cuboidal epithelium
   c. Simple columnar epithelium
   d. Pseudostratified columnar epithelium
   e. Stratified squamous epithelium
   f. Stratified cuboidal epithelium
   g. Stratified columnar epithelium
   h. Transitional epithelium
6. Define *gland*. (p. 102)
7. Distinguish between exocrine and endocrine glands. (p. 102)
8. Explain how glandular secretions differ. (p. 102)
9. Define *mucus*. (p. 102)
10. Describe the general characteristics of connective tissues. (p. 104)
11. Define *matrix*. (p. 104)
12. Describe three major types of connective tissue cells. (p. 104)
13. Distinguish between collagen and elastin. (p. 104)
14. Define *dense connective tissue*. (p. 105)
15. Explain the relationship between loose fibrous connective tissue and adipose tissue. (p. 105)
16. Explain why injured dense fibrous connective tissue and cartilage are usually slow to heal. (p. 106)
17. Name the types of cartilage, and describe their differences and similarities. (p. 107)
18. Describe how bone cells are organized in bone tissue. (p. 108)
19. Describe the composition of blood. (p. 109)
20. Describe the general characteristics of muscle tissues. (p. 110)
21. Distinguish among skeletal, smooth, and cardiac muscle tissues. (p. 110)
22. Describe the general characteristics of nervous tissues. (p. 112)
23. Distinguish between neurons and neuroglial cells. (p. 112)

Critical Thinking

1. Tissue engineering combines living cells with synthetic materials to create functional substitutes for human tissues. What components would you use to engineer replacement (*a*) skin, (*b*) blood, (*c*) bone, and (*d*) muscle?
2. Bone, blood, and cartilage look different and have different functions, yet they are all connective tissues. What is the basis for their similar classification?
3. Collagen and elastin are added to many beauty products. What type of tissue are they normally part of?
4. Why do injuries to cartilage take a long time to heal?
5. In the lungs of smokers, a process called metaplasia occurs where the normal lining cells of the lung are replaced by squamous metaplastic cells (many layers of squamous epithelial cells). Functionally, why is this an undesirable body reaction to tobacco smoke?
6. Cancer-causing agents (carcinogens) usually act on cells that are dividing. Which of the four tissues would carcinogens most influence? Least influence?

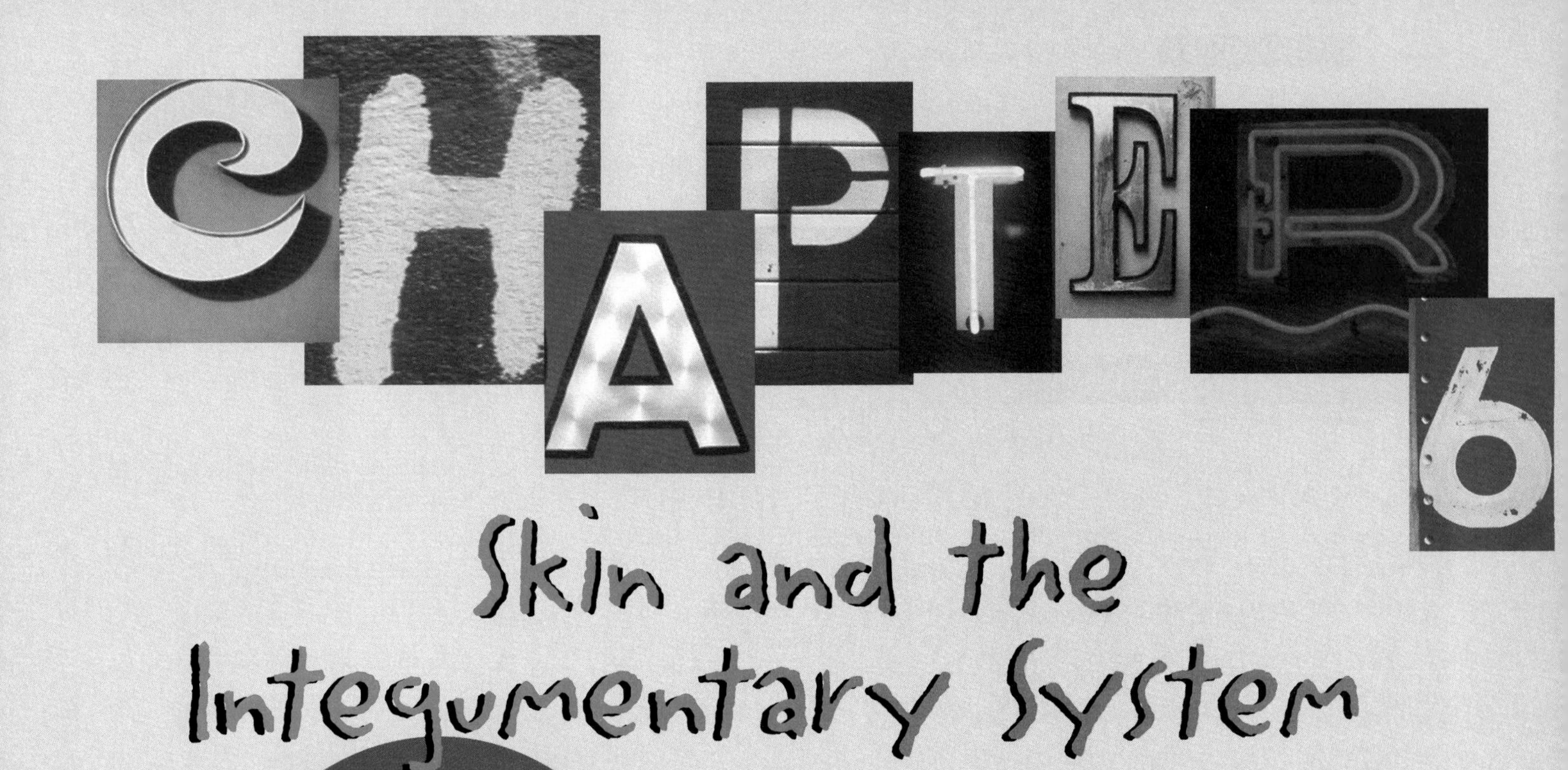

Skin and the Integumentary System

"The sun is a monster!" So says five-year-old Katie Maher. A rare disorder makes sunlight extremely dangerous to Katie. She and her family think constantly about skin.

Caren and Dan Maher first noticed that Katie had incredibly sensitive skin when she was a month old, after they had put her under a tree in the backyard to enjoy the warm spring weather. Almost immediately, Katie broke out everywhere in spots. As the little girl shrieked in pain, the spots turned to blisters, and later, scabs. Some of them might one day become skin cancers.

At first, the Mahers thought the skin reaction was an isolated incident, perhaps an allergic reaction to something in their yard. But it happened every time Katie encountered sunlight. Even a shaft of light entering through a window could cause painful blisters to form instantaneously on the little girl.

Eventually, Katie was diagnosed with *xeroderma pigmentosum,* an inherited inability of her skin to repair damage caused by ultraviolet radiation in sunlight. She is one of only about 250 people in the world with the condition.

Today, Katie lives a shaded, indoor existence to prevent skin cancer. She is liberally smeared with sunblock up to eight times a day. To go to a doctor, she bundles up and travels in a car with blocked windows, at night. The family has adapted their way of life to make Katie's easier. Katie and her three siblings play in the yard at night or on a jungle gym in the garage, where the windows are covered and low-ultraviolet incandescent lightbulbs are used. Caren and Dan run a special camp where children with the disorder can turn night into day so that they, like other children, can enjoy the outdoors.

Photo: Children with xeroderma pigmentosum have an extremely high risk of developing skin cancer. However, excess sun exposure raises the risk of developing skin cancer in all of us, even people with dark skin.

Chemicals, cells, tissues, organs, and finally, organ systems build a human body. Tissues associate and function together to form organs. The skin, along with hair follicles and glands that produce oils and sweat, constitute the integumentary system.

Chapter Objectives

After studying this chapter, you should be able to:

1. Describe the four major types of membranes. (p. 117)
2. Describe the structure of the skin's layers. (p. 117)
3. List the general functions of each layer of skin. (p. 117)
4. Summarize the factors that determine skin color. (p. 119)
5. Describe the accessory organs associated with the skin. (p. 120)
6. Explain how the skin regulates body temperature. (p. 123)
7. Describe the events that are part of wound healing. (p. 125)

Aids to Understanding Words

cut- [skin] sub*cut*aneous: Beneath the skin.

derm- [skin] *derm*is: The inner layer of the skin.

epi- [upon] *epi*dermis: The outer layer of the skin.

follic- [small bag] hair *follic*le: A tubelike depression in which a hair develops.

kerat- [horn] *kerat*in: A protein produced as epidermal cells die and harden.

melan- [black] *melan*in: A dark pigment that certain cells produce.

seb- [grease] *seb*aceous gland: A gland that secretes an oily substance.

Key Terms

cutaneous membrane (ku-ta′ne-us mem′brān)

dermis (der′mis)

epidermis (ep″ĭ-der′mis)

hair follicle (hār fol′ĭ-kl)

integumentary (in-teg-u-men′tar-e)

keratinization (ker″ah-tin″ĭ-za′shun)

melanin (mel′ah-nin)

mucous membrane (mu′kus mem′brān)

sebaceous gland (se-ba′shus gland)

serous membrane (se′rus mem′brān)

subcutaneous (sub″ku-ta′ne-us)

sweat gland (swet gland)

synovial membrane (sĭ-no′ve-al mem′brān)

Introduction

Two or more kinds of tissues grouped together and performing specialized functions constitute an organ. Thus, the thin, sheetlike membranes, composed of epithelium and connective tissue that cover body surfaces and line body cavities are organs. The cutaneous membrane, for example, together with certain accessory organs, make up the **integumentary system.**

Types of Membranes

The four major types of membranes are serous, mucous, synovial, and cutaneous. **Serous membranes** line body cavities that lack openings to the outside. They form the inner linings of the thorax (parietal pleura) and abdomen (parietal peritoneum), and they cover the organs within these cavities (visceral pleura and visceral peritoneum, respectively). A serous membrane consists of a layer of simple squamous epithelium and a thin layer of loose connective tissue. Cells of a serous membrane secrete watery *serous fluid,* which lubricates membrane surfaces.

Mucous membranes line cavities and tubes that open to the outside of the body. These include the oral and nasal cavities and the tubes of the digestive, respiratory, urinary, and reproductive systems. A mucous membrane consists of epithelium overlying a layer of loose connective tissue. Specialized cells within a mucous membrane secrete *mucus.*

Synovial membranes form the inner linings of the joint cavities between the ends of bones at freely movable joints (synovial joints). These membranes usually include fibrous connective tissue overlying loose connective tissue and adipose tissue. Cells of a synovial membrane secrete a thick, colorless *synovial fluid* into the joint cavity, which lubricates the ends of the bones within the joint.

The **cutaneous membrane** is more commonly called *skin.* This chapter describes skin in detail.

Skin and Its Tissues

The skin is one of the larger and more versatile organs of the body, and it is vital in maintaining homeostasis. For example, the skin is a protective covering, helps regulate body temperature, retards water loss from the deeper tissues, houses sensory receptors, synthesizes various biochemicals, and excretes small quantities of wastes.

The skin includes two distinct tissue layers (fig. 6.1). The outer layer, called the **epidermis,** is composed of stratified squamous epithelium. The inner layer, or **dermis,** is thicker than the epidermis, and it contains fibrous connective tissue, epithelial tissue, smooth muscle tissue, nervous tissue, and blood. A *basement membrane* that is anchored to the dermis separates the two skin layers.

Beneath the dermis are masses of loose connective and adipose tissues that bind the skin to the underlying organs. These tissues form the **subcutaneous layer** (hypodermis).

1. Name the four types of membranes, and explain how they differ.
2. List the general functions of the skin.
3. Name the tissue in the outer layer of the skin.
4. Name the tissues in the inner layer of the skin.

Epidermis

Since the epidermis is composed of stratified squamous epithelium, it lacks blood vessels. However, the deepest layer of epidermal cells, called the *stratum basale,* is close to the dermis and is nourished by dermal blood vessels (fig. 6.1). As the cells of this layer divide and grow, the older epidermal cells are pushed slowly away from the dermis toward the skin surface. The farther the cells travel, the poorer their nutrient supply becomes, and in time, they die.

Subcutaneous injections are administered through a hollow needle into the subcutaneous layer beneath the skin. *Intradermal injections* are injected into tissues within the skin. Subcutaneous injections and *intramuscular injections* (administered into muscles) are sometimes called hypodermic injections.

Some substances are administered through the skin by means of an adhesive transdermal patch with a small drug reservoir. The drug passes from the reservoir through a permeable membrane at a known rate. It then diffuses into the epidermis and enters the blood vessels of the dermis. Transdermal patches are often worn behind the ear, where the epidermis is the thinnest. Such patches are used to deliver estrogen to postmenopausal women, the motion sickness medication scopalamine, nicotine to help smokers quit, and medications to treat heart disease and elevated blood pressure. ■

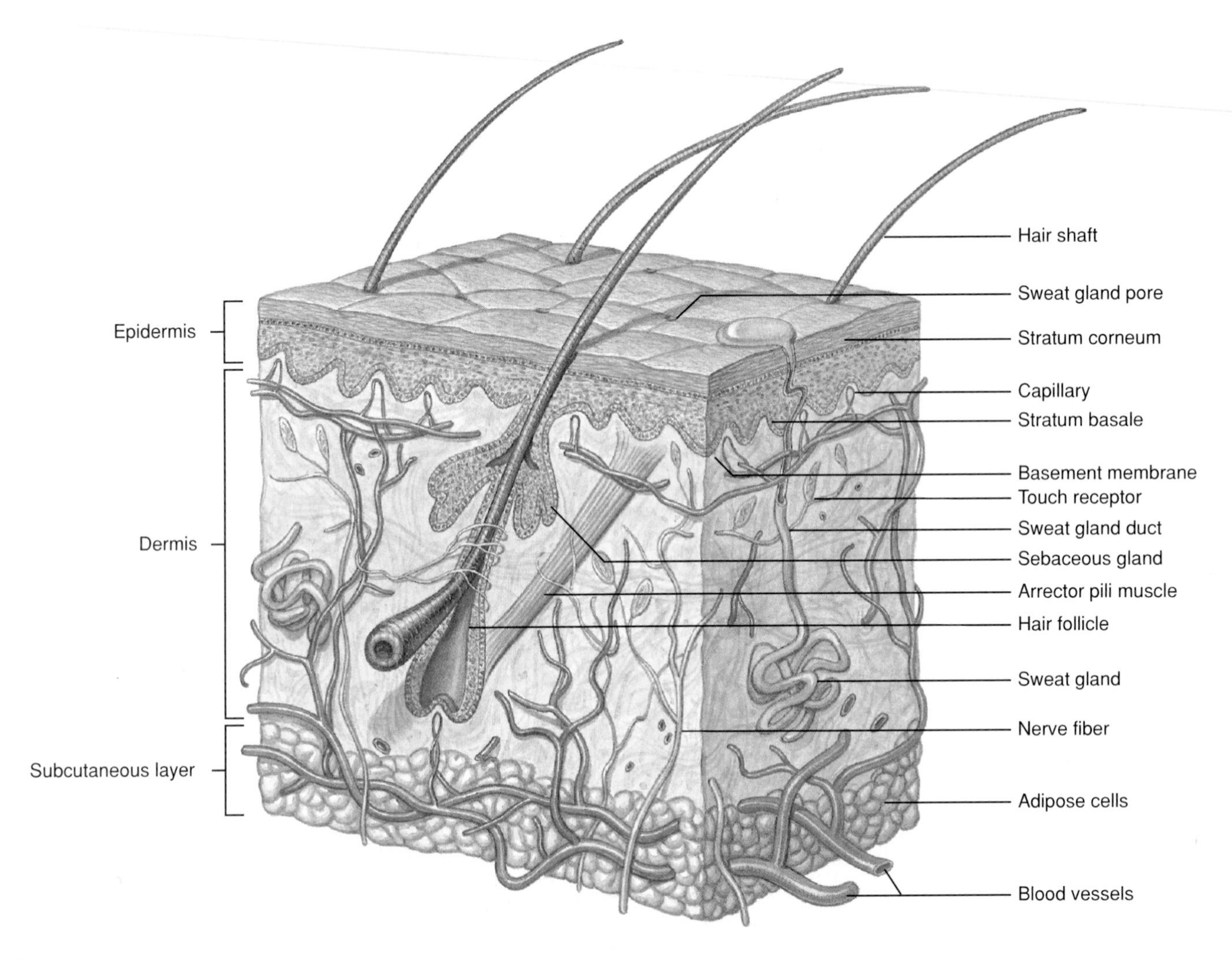

Figure 6.1

A section of skin.

The older cells (keratinocytes) harden in a process called **keratinization.** The cytoplasm fills with strands of a tough, fibrous, waterproof *keratin* protein. As a result, many layers of tough, dead cells accumulate in the outer epidermis, a layer called the *stratum corneum*. Dead cells that compose it are often rubbed away.

In healthy skin, production of epidermal cells is closely balanced with the loss of stratum corneum, so that the skin seldom wears away completely. In fact, the rate of cellular reproduction increases in regions where the skin is rubbed or pressed regularly, causing growth of thickened areas called *calluses* on the palms and soles. ■

The epidermis has important protective functions. It shields the moist underlying tissues against excessive water loss, mechanical injury, and the effects of harmful chemicals. When unbroken, the epidermis also keeps out disease-causing microorganisms.

Specialized cells known as *melanocytes* produce **melanin,** a dark pigment in the deeper layers of the epidermis (fig. 6.2). Melanin absorbs light energy and in this way helps protect still deeper cells from the damaging effects of the sun's ultraviolet light. Melanocytes lie in the deepest portion of the epidermis and in the underlying connective tissue of the dermis. Although they are the only cells that produce melanin, the pigment also may appear in nearby epidermal cells. This happens because melanocytes have long, pigment-containing cellular extensions that pass upward between epidermal cells. The extensions transfer melanin granules into these other cells by a

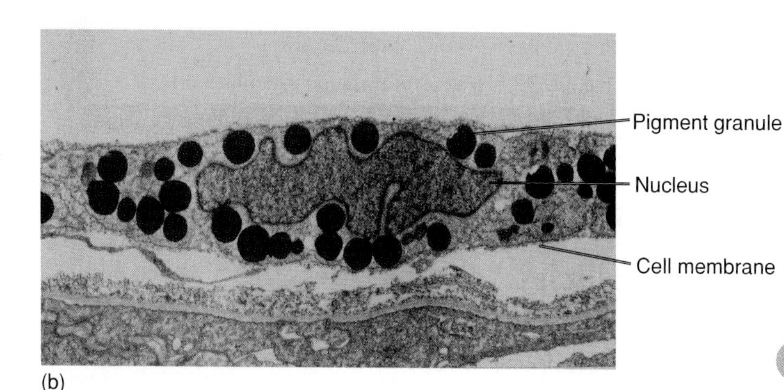

Figure 6.2

(*a*) Melanocytes (*arrow*) in the deepest layer of the epidermis produce a pigment called melanin (160×). (*b*) Transmission electron micrograph of a melanocyte with melanin-containing granules (10,000×).

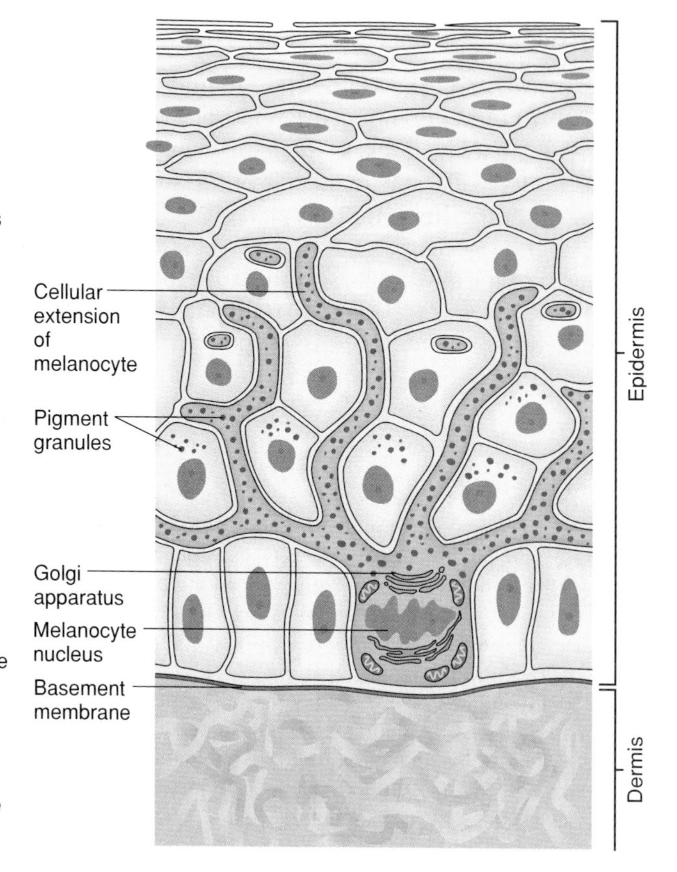

Figure 6.3

A melanocyte may have cellular extensions that transfer melanin granules into nearby epidermal cells.

process called *cytocrine secretion*. Nearby epidermal cells may actually contain more melanin than melanocytes (fig. 6.3).

Skin Color

Skin color is due largely to melanin. All people have about the same number of melanocytes in their skin. Differences in skin color result from differences in the amount of melanin that melanocytes produce and in the size of the pigment granules. Skin color is mostly genetically determined. If genes instruct melanocytes to produce abundant melanin, then the skin is dark.

Environmental and physiological factors also influence skin color. Sunlight, ultraviolet light from sunlamps, and X rays stimulate production of additional pigment. Blood in the dermal vessels may affect skin color as physiological changes occur. When blood is well oxygenated, the blood pigment (hemoglobin) is bright red, making the skin of light-complexioned persons appear pinkish. On the other hand, when blood oxygen concentration is relatively low, the blood pigment is darker, and the skin may appear bluish—a condition called *cyanosis*.

1. Explain how the epidermis forms.
2. Distinguish between the stratum basale and the stratum corneum.
3. How does melanin protect the skin?
4. What factors affect skin color?

Dermis

The dermis binds the epidermis to underlying tissues (see fig. 6.1). It is composed largely of fibrous connective tissue that includes tough collagenous fibers and elastic fibers within a gel-like ground substance. Networks of both these fiber types give the dermis its strength and elasticity.

Dermal blood vessels supply nutrients to all skin cells. These vessels also help regulate body temperature, as explained later in this chapter.

Because dermal blood vessels supply nutrients to the epidermis, any interference with blood flow may kill epidermal cells. For example, when a person lies in one position for a long time, the weight of the body pressing against the bed interferes with the skin's blood supply. If cells die, the tissues begin to break down, and a *pressure ulcer* (also called a decubitus ulcer or bedsore) may appear.

Pressure ulcers usually occur in the skin overlying bony projections, such as on the hip, heel, elbow, or shoulder. Changing the body position frequently or massaging the skin to stimulate blood flow in the regions associated with bony prominences can prevent ulcers. ■

Nerve fibers are scattered throughout the dermis. Those called motor fibers carry impulses out from the brain or spinal cord to the dermal muscles and glands. Other nerve fibers, called sensory fibers, carry impulses away from specialized sensory receptors, such as touch receptors located within the dermis, and into the brain or spinal cord. The dermis also contains hair follicles, sebaceous (oil-producing) glands, and sweat glands, which are discussed later in the chapter (see fig. 6.1).

Skin cells help produce vitamin D, which is necessary for normal bone and tooth development. This vitamin can form from a substance (dehydrocholesterol) that is synthesized by cells in the digestive system or obtained in the diet. When dehydrocholesterol reaches the skin by means of the blood and is exposed to ultraviolet light from the sun, it is converted to another chemical, which becomes vitamin D.

Certain skin cells (keratinocytes) assist the immune system by producing hormonelike substances that stimulate development of certain white blood cells (T lymphocytes) that defend against infection by disease-causing bacteria and viruses (see chapter 14). ■

Subcutaneous Layer

The subcutaneous layer (hypodermis) beneath the dermis consists of loose connective and adipose tissues (see fig. 6.1). The collagenous and elastic fibers of this layer are continuous with those of the dermis. Most of these fibers run parallel to the surface of the skin and extend in all directions. As a result, no sharp boundary separates the dermis and the subcutaneous layer.

The adipose tissue of the subcutaneous layer insulates, which conserves body heat and impedes the entrance of outside heat. The subcutaneous layer also contains the major blood vessels that supply the skin and underlying adipose tissue.

1. Describe the dermis.
2. What are the functions of the dermis?
3. What are the functions of the subcutaneous layer?

Accessory Organs of the Skin

Hair Follicles

Hair is present on all skin surfaces except the palms, soles, lips, nipples, and parts of the external reproductive organs. Each hair develops from a group of epidermal cells at the base of a tubelike depression called a **hair follicle** (figs. 6.1 and 6.4). This follicle extends from the surface into the dermis and contains the hair *root*. The epidermal cells at its base are nourished by dermal blood vessels in a projection of connective tissue at the deep end of the follicle. As these epidermal cells divide and grow, older cells are pushed toward the surface. The cells that move upward and away from their nutrient supply become keratinized and die. Their remains form a developing hair, whose *shaft* extends away from the skin surface (fig. 6.5). In other words, a hair is composed of dead epidermal cells.

A bundle of smooth muscle cells, forming the *arrector pili muscle,* is attached to each hair follicle (see figs. 6.1 and 6.4*a*). This muscle is positioned so that a short hair within the follicle stands on end when the muscle contracts. If a person is emotionally upset or very cold, nerve impulses may stimulate the arrector pili muscles to contract, causing gooseflesh or goose bumps.

Hair color, like skin color, is determined by the genes that direct the type and amount of pigment the epidermal melanocytes produce. If these cells, which lie at the deep end of a follicle, produce an abundance of melanin, the hair is dark; if an intermediate quantity of pigment is produced, the hair is blond; if no pigment appears, the hair is white. Another pigment, trichosiderin, is found only in red hair.

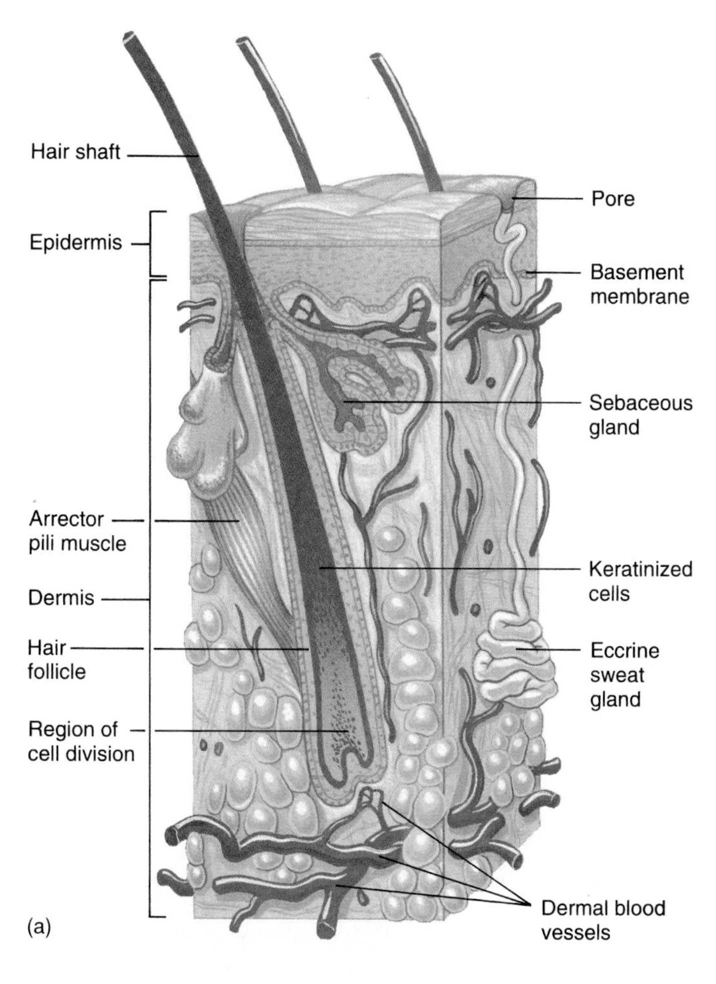

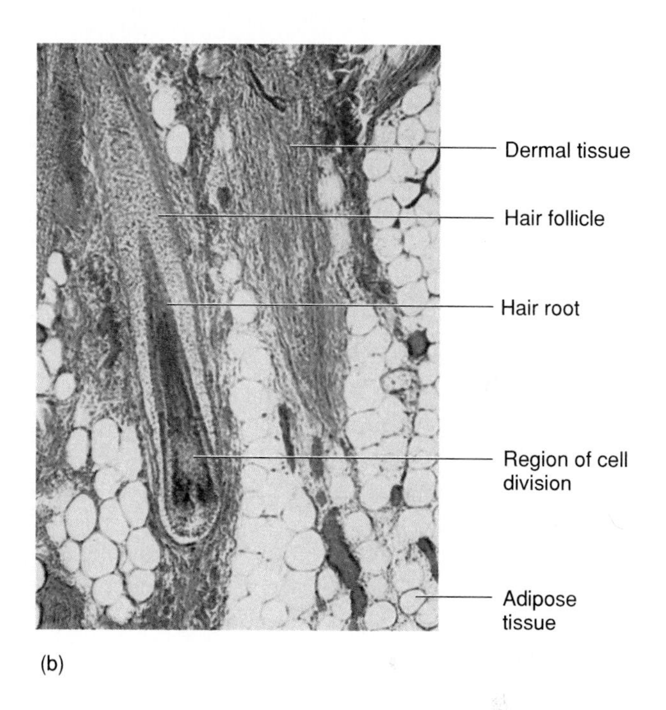

Figure 6.4

(*a*) A hair grows from the base of a hair follicle when epidermal cells divide and older cells move outward and become keratinized. (*b*) Light micrograph of a hair follicle (160×).

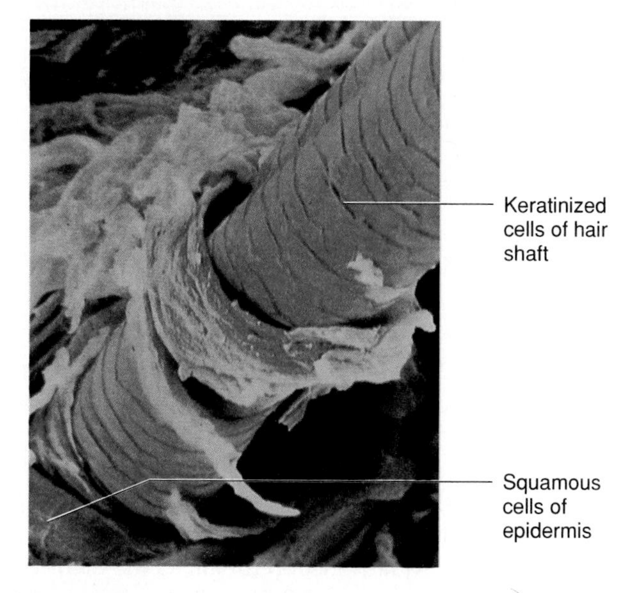

Figure 6.5

Scanning electron micrograph of a hair emerging from the epidermis (340×).

Men exhibited in circus sideshows as "werewolves" actually may have an inherited condition called *congenital generalized hypertrichosis,* which causes the skin to have many more hair follicles than normal. As a result, long, very dense hair grows on the face and upper body. Females have milder cases. ■

Sebaceous Glands

Sebaceous glands contain groups of specialized epithelial cells and are usually attached to hair follicles (figs. 6.4*a* and 6.6). They are holocrine glands (see chapter 5) that secrete an oily mixture of fatty material and cellular debris called *sebum* through small ducts into the hair follicles. Sebum helps keep the hair and skin soft, pliable, and relatively waterproof.

Topic of Interest: Skin Cancer

Skin cancer usually arises from nonpigmented epithelial cells within the deep layer of the epidermis or from melanocytes. Skin cancers originating from epithelial cells are called *cutaneous carcinomas* (basal cell carcinoma and squamous cell carcinoma); those arising from melanocytes are *cutaneous melanomas* (melanocarcinomas or malignant melanomas) (fig. 6A).

Cutaneous carcinomas are the most common type of skin cancer, occurring most frequently in light-skinned people over forty. These cancers usually occur in persons who are exposed to sunlight regularly, such as farmers, sailors, athletes, and sunbathers, and may be a result of failure of a normal protective function—peeling of sun-damaged cells.

Cutaneous carcinomas often develop from hard, dry, scaly growths (lesions) with reddish bases. Such lesions may be either flat or raised, and they adhere firmly to the skin. Fortunately, cutaneous carcinomas are typically slow growing and can usually be cured completely by surgical removal or radiation treatment.

Because melanomas develop from melanocytes, they are pigmented with melanin, often with a variety of colored areas, such as variegated brown, black, gray, or blue. They usually have irregular rather than smooth outlines.

Cutaneous melanomas may appear in young as well as in older adults and occur most often in light-skinned people who burn rather than tan. Short, intermittent exposure to high-intensity sunlight seems to initiate these growths. Thus, melanoma risk is higher for persons who stay indoors but occasionally sustain blistering sunburns.

A cutaneous melanoma may arise from normal-appearing skin or from a mole (nevus). Typically, the growth spreads through the skin horizontally, but eventually, it may thicken and invade deeper tissues. Removing a melanoma while it is in its horizontal growth phase can often stop the cancer. Once the melanoma thickens and grows downward, treatment is difficult, and the survival rate is very low.

To reduce the chances of developing skin cancer, avoid exposing the skin to high-intensity sunlight, use sunscreens and sunblocks, and examine the skin regularly. Report any unusual growths—particularly those that change in color, shape, or surface texture—to a physician.

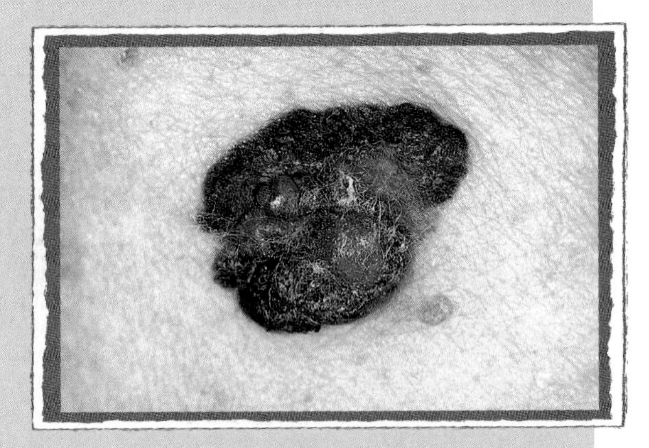

Figure 6A
A cutaneous malignant melanoma.

Many teens are all too familiar with a disorder of the sebaceous glands called *acne* (acne vulgaris). Overactive and inflamed glands in some body regions become plugged and surrounded by small red elevations containing blackheads (comedones) or pimples (pustules). ■

Nails

Nails are protective coverings on the ends of the fingers and toes (fig. 6.7). Each nail consists of keratinized stratified squamous epithelial cells with very hard keratin. These cells form by cell division in a region called the *nail root,* located in the nail's proximal end. A whitish, half-moon-shaped area at the base of the nail, called the *lunula,* is the growing region.

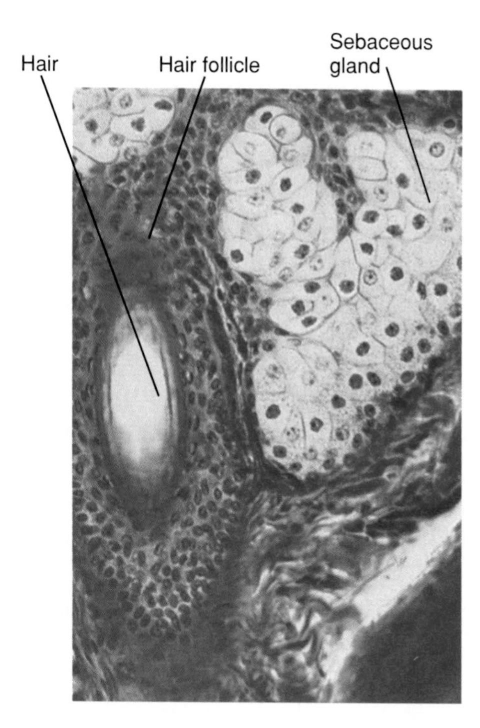

Figure 6.6

A sebaceous gland secretes sebum into a hair follicle (shown here in oblique section: 175×).

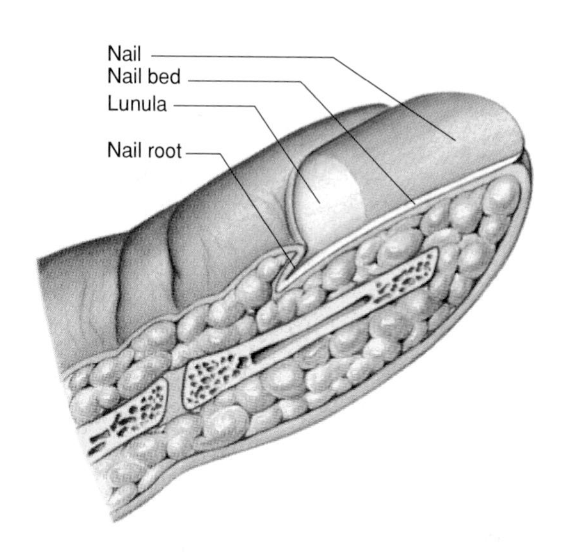

Figure 6.7

Nails grow from epithelial cells that reproduce and undergo keratinization in the root of the nail.

As the nail develops, it slides forward over a layer of epithelium called the *nail bed,* to which the nail remains attached. This epithelial layer is continuous with the epithelium of the skin.

Sweat Glands

Sweat glands (sudoriferous glands) are exocrine glands that are widespread in the skin but are most numerous in the palms and soles. Each gland consists of a tiny tube that originates as a ball-shaped coil in the deeper dermis or superficial subcutaneous layer. The coiled portion of the gland is closed at its end and lined with sweat-secreting epithelial cells.

The most numerous sweat glands, the *eccrine glands,* are not associated with hair follicles (see fig. 6.4*a*). They function throughout life and respond primarily to elevated body temperatures associated with environmental heat or physical exercise. These glands are common on the forehead, neck, and back, where they produce profuse sweating on hot days and during physical exertion.

The fluid (sweat) that eccrine glands secrete is carried away in a duct, which opens at the surface as a *pore.* Sweat is mostly water, but it also contains small quantities of salt and wastes, such as urea and uric acid. Thus, sweating is also an excretory function.

Other sweat glands, known as *apocrine glands,* become active when a person is emotionally upset, frightened, or in pain. They are most numerous in the armpits and groin, and usually connect to hair follicles. Sex hormones stimulate development of these glands, which begin to function as an individual becomes sexually mature (puberty). Certain apocrine glands are structurally and functionally modified to secrete specific fluids, such as the ceruminous glands of the external ear canal that secrete earwax.

1. Explain how a hair forms.
2. What is the function of the sebaceous glands?
3. Distinguish between the eccrine and apocrine sweat glands.

Regulation of Body Temperature

Regulation of body temperature is vitally important because even slight shifts can disrupt rates of metabolic reactions. Normally, the temperature of deeper body parts remains close to a set point of 37° C (98.6° F).

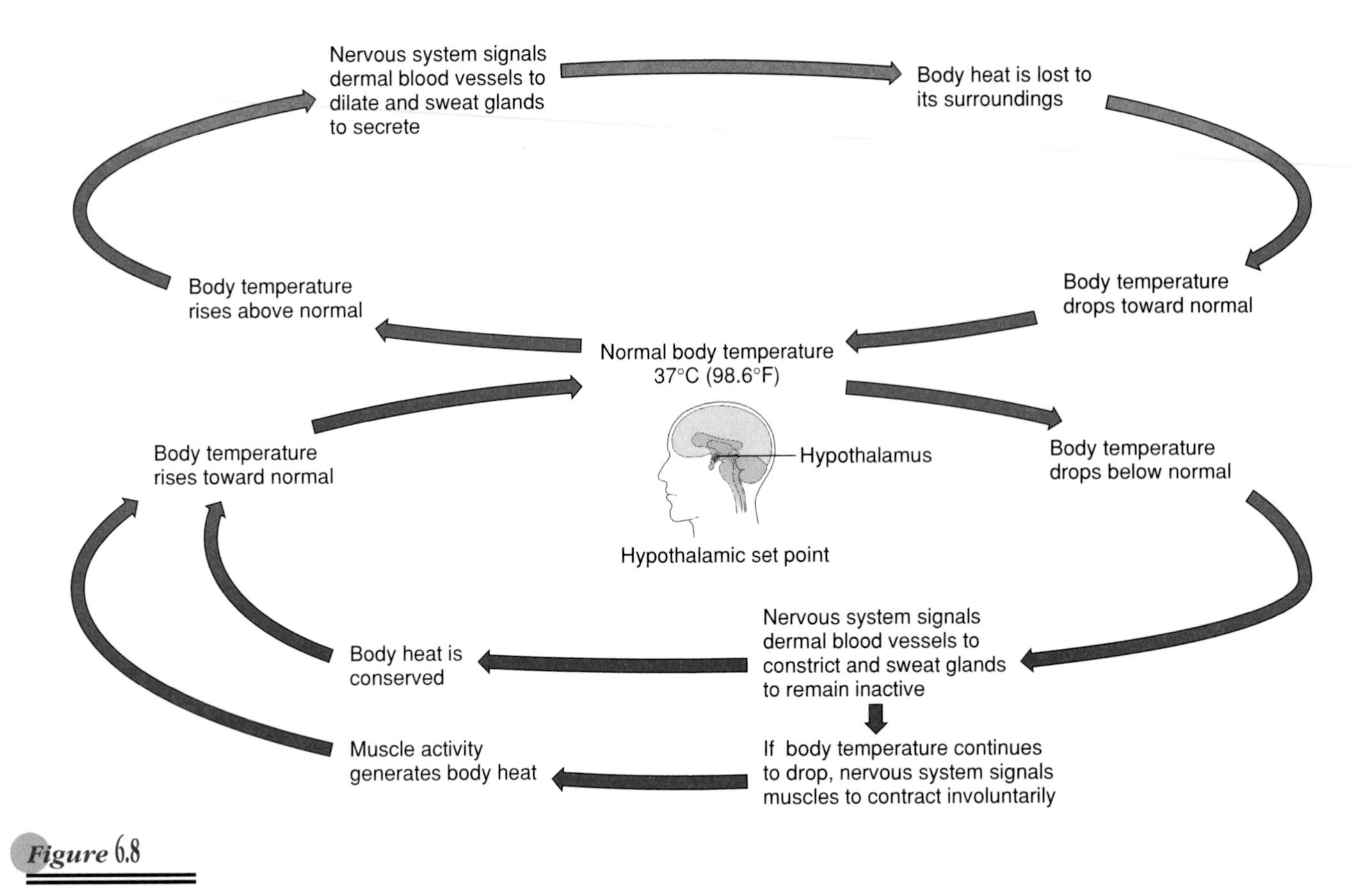

Figure 6.8

Body temperature regulation.

Maintenance of a stable temperature requires that the amount of heat the body loses be balanced by the amount it produces. The skin plays a key role in the homeostatic mechanism that regulates body temperature.

Heat is a product of cellular metabolism; thus, the more active cells of the body are the major heat producers. These cells include skeletal and cardiac muscle cells and the cells of certain glands, such as the liver.

When heat production is excessive, nerve impulses stimulate structures in the skin and other organs to react in ways that dissipate heat. For example, during physical exercise, active muscles release heat, which blood carries away. The brain's temperature control center (in the hypothalamus) senses the increasing blood temperature, and signals from the brain relax the muscles of the walls of the dermal blood vessels. As these blood vessels dilate, more heat-carrying blood enters the dermis, and more of the heat dissipates to the outside.

At the same time as the skin loses heat, the nervous system stimulates the eccrine sweat glands to become active and to release fluid onto the skin surface. As this fluid evaporates (changes from a liquid to a gas), it carries heat away from the surface, cooling the skin further.

If too much heat is lost, as may occur in a very cold environment, the brain triggers different responses in the skin structures. Muscles in the walls of dermal blood vessels contract; this decreases the flow of heat-carrying blood through the skin and helps to reduce heat loss. Also, the sweat glands remain inactive, decreasing heat loss by evaporation. If body temperature continues to drop, the nervous system may stimulate muscle fibers in the skeletal muscles throughout the body to contract slightly. This action requires an increase in the rate of cellular respiration and produces heat as a by-product. If this response does not raise body temperature to normal, small groups of muscles may contract rhythmically with still greater force, and the person begins to shiver, generating more heat. Figure 6.8 summarizes the body's temperature regulating mechanism.

Body temperature regulation depends, in large part, on sweat evaporation from the skin's surface. Because high humidity hinders evaporation, athletes should slow down their activities on hot, humid days, stay out of the sunlight whenever possible, and drink enough fluids to avoid dehydration. Such precautions can prevent the symptoms of *heat exhaustion,* which include fatigue, dizziness, headache, muscle cramps, and nausea. ■

TOPIC OF INTEREST Elevated Body Temperature

On a warm June morning, the harried and hurried father strapped his five-month-old son Bryan into the backseat of his car and headed for work. Tragically, the father forgot to drop his son off at the babysitter's. When his wife called him at work late that afternoon to inquire why the child was not at the sitter's, the shocked father realized his mistake and hurried down to his parked car. But it was too late—Bryan had died. Left for 10 hours in the car in the sun, all windows shut, the baby's temperature had quickly soared. Two hours after he was discovered, the child's temperature still exceeded 41° C (106° F).

Sarah's elevated body temperature was more typical. She awoke with a fever of 40° C (104° F) and a terribly painful sore throat. Peering down the five-year-old's throat with a flashlight, her mother spotted the whitish lesions that can indicate a *Streptococcus* infection. Sarah indeed had strep throat, and the fever was her body's attempt to fight the infection.

The true cases of young Bryan and Sarah illustrate two reasons why body temperature may rise: (1) the temperature homeostatic mechanism's inability to handle an extreme environment and (2) an immune system response to infection.

In Bryan's case, sustained exposure to very high heat overwhelmed the temperature-regulating mechanism, causing hyperthermia. Body heat built up faster than it could dissipate, and body temperature rose, even though the set point of the thermostat was normal. Bryan's blood vessels dilated so greatly in an attempt to dissipate the excess heat that after a few hours his circulatory system collapsed.

In a fever, molecules on the surfaces of the infectious agents (usually bacteria or viruses) stimulate phagocytes to release a substance called interleukin-1 (IL-1, also called endogenous pyrogen, meaning "fire maker from within"). The bloodstream carries IL-1 to the hypothalamus, where it raises the set point controlling temperature. In response, the brain signals skeletal muscles to increase heat production, blood flow to the skin to decrease, and sweat glands to decrease secretion. As a result, body temperature rises to the new set point, and a fever develops. The increased body temperature helps the immune system kill the pathogens.

Rising body temperature requires different treatments, depending on the degree of elevation. Hyperthermia in response to exposure to intense, sustained heat should be rapidly treated by administering liquids to replace lost body fluids and electrolytes, sponging the skin with water to increase cooling by evaporation, and covering the person with a refrigerated blanket. Some health professionals believe that a slight fever should not be reduced (with medication or cold baths) because it may be part of a normal immune response. A high or prolonged fever, however, requires medical attention.

1. Why is body temperature regulation so important?
2. How does the body dissipate excess heat?
3. What actions help the body to conserve heat?

Healing of Wounds

A wound and the area surrounding it usually become red and painfully swollen. This is the result of **inflammation**, which is a normal response to injury or stress. Blood vessels in affected tissues dilate and become more permeable, forcing fluids to leave the blood vessels and enter the damaged tissues. Inflamed skin may become reddened, swollen, warm, and painful to touch. However, the dilated blood vessels provide the tissues with more nutrients and oxygen, which aids healing.

The specific events in healing depend on the nature and extent of the injury. If a break in the skin is shallow, epithelial cells along the break's margin are stimulated to reproduce more rapidly than usual, and the newly formed cells simply fill the gap.

If the injury extends into the dermis or subcutaneous layer, blood vessels break, and the escaping blood forms a clot in the wound. The blood clot and dried tissue fluids form a *scab* that covers and protects underlying tissues. Before long, fibroblasts migrate into the injured region and begin forming new collagenous fibers that bind the wound edges together.

Suturing or otherwise closing a large break in the skin speeds this process.

As healing continues, blood vessels extend into the area beneath the scab. Phagocytic cells remove dead cells and other debris. Eventually, the damaged tissues are replaced, and the scab sloughs off. If the wound is extensive, the newly formed connective tissue may appear on the surface as a *scar.*

In large, open wounds, healing may be accompanied by formation of small, rounded masses called *granulations* in the exposed tissues. A granulation consists of a new branch of a blood vessel and a cluster of collagen-secreting fibroblasts that the vessel nourishes. In time, some of the blood vessels are resorbed, and the fibroblasts move away, leaving a scar.

1. Describe how inflammation helps a wound to heal.
2. Distinguish between the activities necessary to heal a wound in the epidermis from those necessary to heal a wound in the dermis.
3. Explain the role of phagocytic cells in wound healing.
4. Define granulation.

Common Skin Disorders

Acne (ak′ne) A disease of the sebaceous glands that produces blackheads and pimples.

Alopecia (al″o-pe′she-ah) Hair loss, usually sudden.

Athlete's foot (ath′-lētz foot) (tinea pedis) A fungus infection usually in the skin of the toes and soles.

Birthmark (berth′ mark) A vascular tumor of the skin and subcutaneous tissues, visible at birth or soon after.

Boil (boil) (furuncle) A bacterial infection of a hair follicle and/or sebaceous glands.

Carbuncle (kar′bung-kl) A bacterial infection, similar to a boil, that spreads into subcutaneous tissues.

Cyst (sist) A liquid-filled sac or capsule.

Dermatitis (der″mah-ti′tis) Inflammation of the skin.

Eczema (ek′zĕ-mah) A noncontagious skin rash that produces itching, blistering, and scaling.

Erythema (er″ĭ-the′mah) Reddening of the skin due to dilation of dermal blood vessels in response to injury or inflammation.

Herpes (her′pēz) An infectious disease of the skin, usually caused by a *herpes simplex* virus and characterized by recurring formations of small clusters of vesicles.

Impetigo (im″pĕ-ti′go) A contagious disease of bacterial origin, characterized by pustules that rupture and become covered with loosely held crusts.

Keloid (ke′loid) An elevated, enlarging fibrous scar usually initiated by an injury.

Mole (mōl) (nevus) A fleshy skin tumor that usually is pigmented; colors range from brown to black.

Pediculosis (pĕ-dik″u-lo′sis) A disease produced by lice infestation.

Pruritus (proo-ri′tus) Itching.

Psoriasis (so-ri′ah-sis) A chronic skin disease characterized by red patches covered with silvery scales.

Pustule (pus′tūl) Elevated, pus-filled area on the skin.

Scabies ska′bēz) A disease resulting from mite infestation.

Seborrhea (seb″o-re′ah) Hyperactivity of the sebaceous glands, causing greasy skin and dandruff.

Ulcer (ul′ser) An open sore.

Urticaria (ur″tĭ-ka′re-ah) An allergic reaction of the skin that produces reddish, elevated patches (hives).

Wart (wort) A flesh-colored, raised area caused by a viral infection.

organization

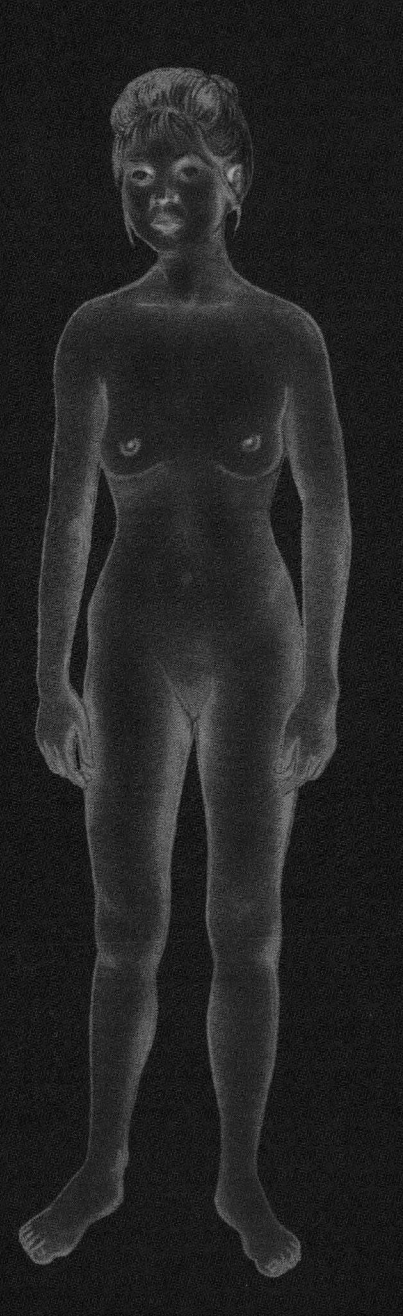

Integumentary System

The skin provides protection, contains sensory organs, and helps control body temperature.

Skeletal System

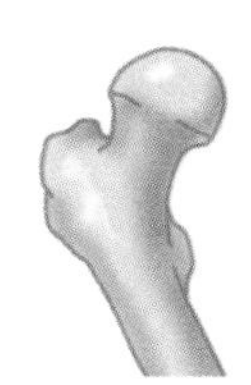

Vitamin D activated by the skin helps provide calcium for bone matrix.

Lymphatic System

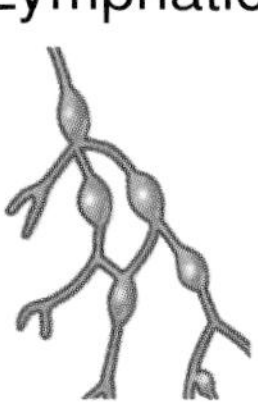

The skin provides an important first line of defense for the immune system.

Muscular System

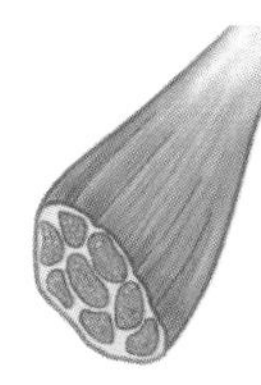

Involuntary muscle contractions (shivering) work with the skin to control body temperature. Muscles act on facial skin to create expressions.

Digestive System

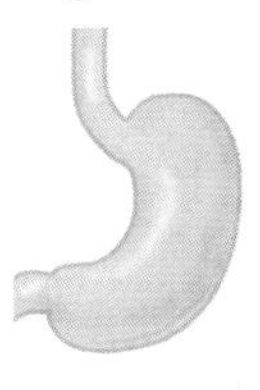

Excess calories may be stored as subcutaneous fat. Vitamin D activated by the skin stimulates dietary calcium absorption.

Nervous System

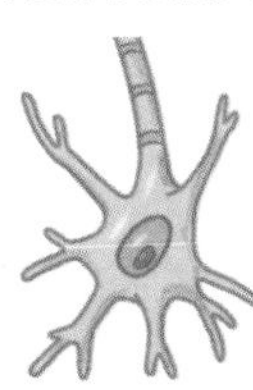

Sensory receptors provide information about the outside world to the nervous system. Nerves control the activity of sweat glands.

Respiratory System

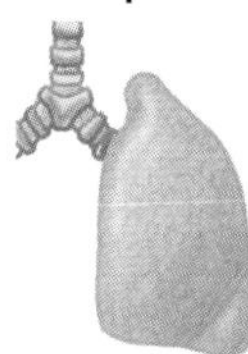

Stimulation of skin receptors may alter respiratory rate.

Endocrine System

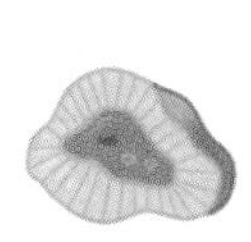

Hormones help to increase skin blood flow during exercise. Other hormones stimulate either the synthesis or the decomposition of subcutaneous fat.

Urinary System

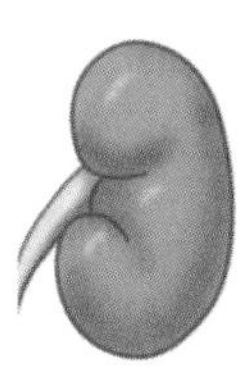

The kidneys help compensate for water and electrolytes lost in sweat.

Cardiovascular System

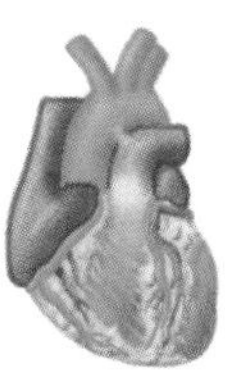

Skin blood vessels play a role in regulating body temperature.

Reproductive System

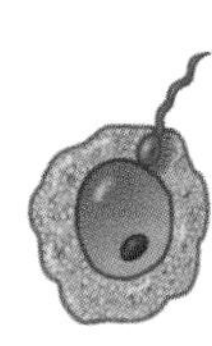

Sensory receptors play an important role in sexual activity and in the suckling reflex.

Summary Outline

Introduction (p. 117)

Organs, such as membranes, are composed of two or more kinds of tissues. The skin is an organ. Together with its accessory organs, it constitutes the integumentary system.

Types of Membranes (p. 117)

1. Serous membranes
   a. Serous membranes line body cavities that lack openings to the outside.
   b. Cells of serous membranes secrete watery serous fluid that lubricates membrane surfaces.
2. Mucous membranes
   a. Mucous membranes line cavities and tubes that open to the outside.
   b. Cells of mucous membranes secrete mucus.
3. Synovial membranes
   a. Synovial membranes line joint cavities.
   b. They secrete synovial fluid that lubricates the ends of the bones at joints.
4. The cutaneous membrane is the external body covering commonly called the skin.

Skin and Its Tissues (p. 117)

Skin is a protective covering, helps regulate body temperature, retards water loss, houses sensory receptors, synthesizes various biochemicals, and excretes wastes. It is composed of an epidermis and a dermis separated by a basement membrane.

1. Epidermis
   a. The deepest layer of the epidermis, called the stratum basale, contains cells undergoing mitosis.
   b. Epidermal cells undergo keratinization as they mature and are pushed toward the surface.
   c. The outermost layer, called the stratum corneum, is composed of dead epidermal cells.
   d. The epidermis protects underlying tissues against water loss, mechanical injury, and the effects of harmful chemicals.
   e. Melanin protects underlying cells from the effects of ultraviolet light.
   f. Melanocytes transfer melanin to nearby epidermal cells.
2. Skin color
   a. All people possess about the same concentration of melanocytes.
   b. Skin color is due largely to the amount of melanin and the size of the pigment granules in the epidermis.
   c. Skin color is influenced by environmental and physiological factors, as well as by genes.
3. Dermis
   a. The dermis binds the epidermis to underlying tissues.
   b. Dermal blood vessels supply nutrients to all skin cells and help regulate body temperature.
   c. Nerve fibers are scattered throughout the dermis.
      (1) Some dermal nerve fibers carry impulses to muscles and glands of the skin.
      (2) Other dermal nerve fibers are associated with sensory receptors in the skin, and carry impulses to the brain and spinal cord.
   d. The dermis also contains hair follicles, sebaceous glands, and sweat glands.
4. Subcutaneous layer
   a. The subcutaneous layer beneath the dermis consists of loose connective and adipose tissues.
   b. Adipose tissue helps conserve body heat.
   c. The subcutaneous layer contains blood vessels that supply the skin and underlying adipose tissue.

Accessory Organs of the Skin (p. 120)

1. Hair follicles
   a. Each hair develops from epidermal cells at the base of a tubelike hair follicle.
   b. As newly formed cells develop and grow, older cells are pushed toward the surface and undergo keratinization.
   c. A bundle of smooth muscle cells is attached to each hair follicle.
   d. Hair color is determined by genes that direct the amount of melanin produced by melanocytes associated with hair follicles.
2. Sebaceous glands
   a. Sebaceous glands are usually attached to hair follicles.
   b. Sebaceous glands secrete sebum, which helps keep the skin and hair soft and waterproof.
3. Nails
   a. Nails are protective covers on the ends of fingers and toes.
   b. Keratinized epidermal cells produce nails.
   c. The keratin of nails is harder than that produced by the skin's epidermal cells.
4. Sweat glands
   a. Each sweat gland consists of a coiled tube.
   b. Sweat is primarily water but also contains salts and waste products.
   c. Eccrine sweat glands respond to elevated body temperature, while apocrine glands respond to emotional stress.

Regulation of Body Temperature (p. 123)

Regulation of body temperature is vital because heat affects the rates of metabolic reactions. The normal temperature of deeper body parts is about 37° C (98.6° F).

1. When body temperature rises above the normal set point, dermal blood vessels dilate, and sweat glands secrete sweat.
2. If body temperature drops below the normal set point, dermal blood vessels constrict, and sweat glands become inactive.
3. Excessive heat loss stimulates skeletal muscles to contract involuntarily.
4. Fever results from an elevated temperature set point.

Healing of Wounds (p. 125)

Skin injuries trigger inflammation. The affected area becomes red, warm, swollen, and tender.

1. Dividing epithelial cells fill in shallow cuts in the epidermis.
2. Clots close deeper cuts, sometimes leaving a scar where connective tissue replaces skin.
3. Granulations form as part of the healing process.

Review Exercises

1. Explain why a membrane is an organ. (p. 117)
2. Define *integumentary system.* (p. 117)
3. Distinguish between serous and mucous membranes. (p. 117)
4. Define *synovial membrane.* (p. 117)
5. List six functions of skin. (p. 117)
6. Distinguish between the epidermis and the dermis. (p. 117)
7. Explain what happens to epidermal cells as they undergo keratinization. (p. 118)
8. Describe the function of melanocytes. (p. 118)
9. List three factors that affect skin color. (p. 119)
10. Review the functions of nervous tissue within skin. (p. 120)
11. Describe the subcutaneous layer and its functions. (p. 120)
12. Explain how blood is supplied to various skin layers. (p. 120)
13. Distinguish between a hair and a hair follicle. (p. 120)
14. Explain the function of sebaceous glands. (p. 121)
15. Describe how nails are formed. (p. 122)
16. Distinguish between eccrine and apocrine sweat glands. (p. 123)
17. Explain how body heat is produced. (p. 124)
18. Explain how sweat glands help regulate body temperature. (p. 124)
19. Describe the body's responses to decreasing body temperature. (p. 124)
20. Distinguish between the healing of shallow and deeper breaks in the skin. (p. 125)

Critical Thinking

1. Everyone's skin contains about the same number of melanocytes even though people come in many different colors. How is this possible?
2. Why would collagen and elastin added to skin creams be unlikely to penetrate the skin—as some advertisements imply they do?
3. A severe form of the inherited illness epidermolysis bullosa causes extreme blistering of the skin. The person lacks a protein called laminin, which normally attaches the dermis to the epidermis. Explain how lack of this protein disrupts the skin's structure.
4. How is skin peeling after a severe sunburn protective? How might a fever be protective?
5. What special problems would result from loss of 50% of a person's functional skin surface? How might this person's environment be modified to compensate partially for such a loss?
6. A premature infant typically lacks subcutaneous adipose tissue. Also, the surface area of an infant's small body is relatively large compared to its volume. How do you think these factors influence an infant's ability to regulate body temperature?
7. Which of the following would result in the more rapid absorption of a drug: a subcutaneous injection or an intradermal injection? Why?
8. How would you explain to an athlete the importance of keeping the body hydrated when exercising in warm weather?

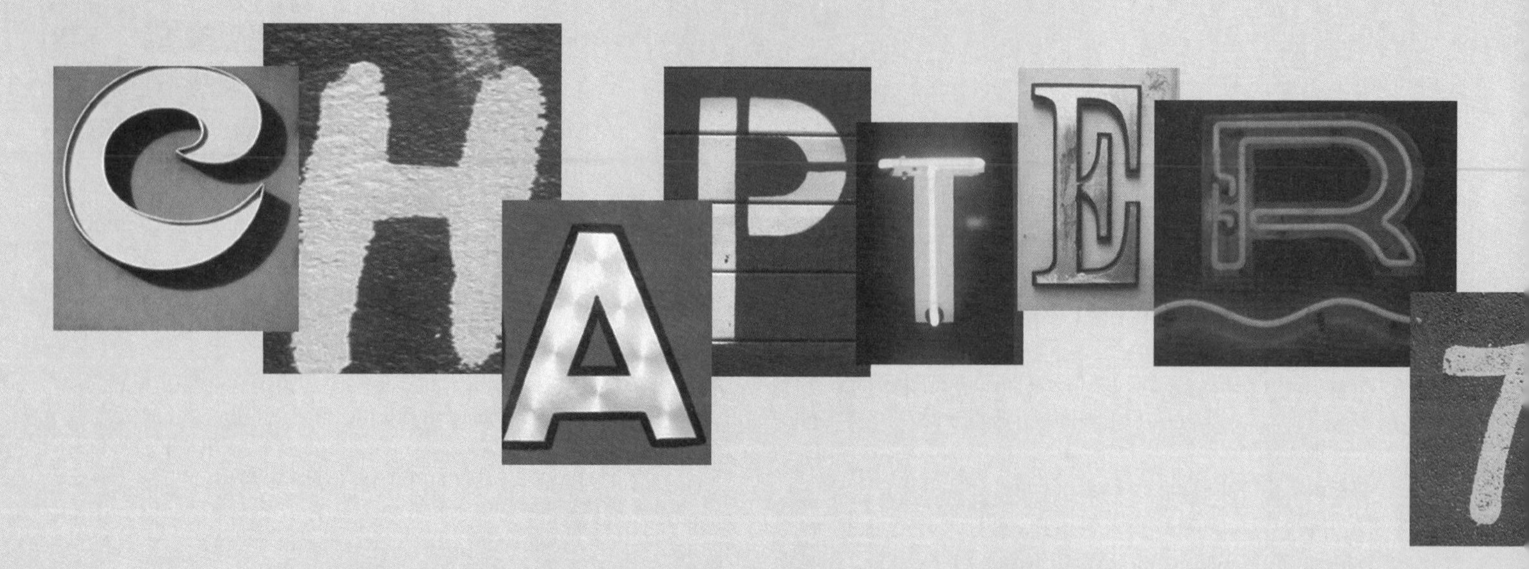

Skeletal System

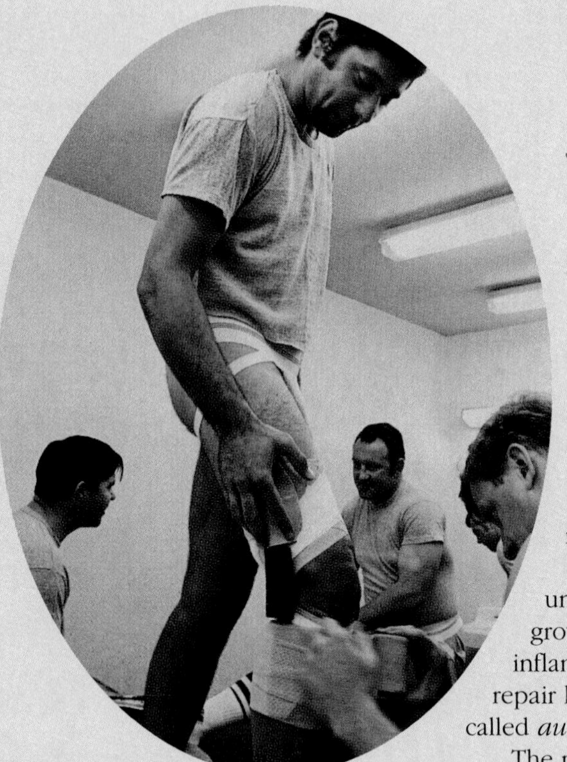

The knee is a marvel of biological engineering. Its parts meet, but with a friction that is one-fifth that of ice on ice. Its lining cleanses and lubricates the joint, while secreting complex molecules in just the right amounts. This almost-perfect piece of anatomy has one shortcoming—it cannot heal. Hippocrates noted the inability of cartilage to heal in the fifth century A.D. In the United States today, 170,000 people receive synthetic knee replacements each year.

Why knee cartilage cannot repair injuries is still unknown. Perhaps, lack of a blood supply robs the tissue of growth factors, clotting factors, and cells to carry out inflammation. But researchers now have invented a way to repair knee cartilage using a patient's healthy tissue, a procedure called *autologous chondrocyte transplantation*.

The procedure is straightforward in concept. First, a physician inserts a needle into the knee joint and withdraws healthy cartilage cells. The cells are then sent to a biotechnology company, where over a period of four to six weeks, they are cultured to yield many cells. Then the patient enters the hospital for a day, where a surgeon opens a 6-inch incision over the injured knee area, removes the damaged tissue, and inserts the cultured cells. A flap of outer bone from the patient's tibia (shinbone) is then used to cover the area. For the next four months, as the patient undergoes rehabilitative therapy, the new cells form cartilage, secreting the appropriate matrix and protecting the joint.

Photo: Football legend Joe Namath might have had a longer athletic career had the technique to replace damaged knee cartilage been available in the 1970s. The healthy cartilage comes from the patient's own body.

Halloween skeletons and the skull-and-crossbones symbol of poison and pirates may make bones seem like lifeless objects, but in actuality, bones are not only very much alive but also multifunctional. Bones, the organs of the skeletal system, provide points of attachment for muscles, protect and support softer tissues, house blood-producing cells, store inorganic salts, and contain passageways for blood vessels and nerves.

Chapter Objectives

After studying this chapter, you should be able to:

1. Describe the general structure of a bone, and list the functions of its parts. (p. 132)
2. Distinguish between intramembranous and endochondral bones, and explain how such bones develop and grow. (p. 134)
3. Discuss the major functions of bones. (p. 137)
4. Distinguish between the axial and appendicular skeletons, and name the major parts of each. (p. 140)
5. Locate and identify the bones and the major features of the bones that comprise the skull, vertebral column, thoracic cage, pectoral girdle, upper limb, pelvic girdle, and lower limb. (p. 142)
6. List three classes of joints, describe their characteristics, and name an example of each. (p. 162)
7. List six types of synovial joints, and describe the actions of each. (p. 163)
8. Explain how skeletal muscles produce movements at joints, and identify several types of such movements. (p. 166)

Aids to Understanding Words

acetabul- [vinegar cup] *acetabul*um: A depression in the coxal bone that articulates with the head of a femur.

ax- [axis] *ax*ial skeleton: The upright portion of the skeleton that supports the head, neck, and trunk.

-blast [budding] osteo*blast*: A cell that forms bone tissue.

carp- [wrist] *carp*als: Wrist bones.

-clast [broken] osteo*clast*: A cell that breaks down bone tissue.

condyl- [knob] *condyl*e: A rounded, bony process.

corac- [beaklike] *corac*oid process: A beaklike process of the scapula.

cribr- [sievelike] *cribr*iform plate: A portion of the ethmoid bone with many small openings.

crist- [ridge] *crist*a galli: A bony ridge that projects upward into the cranial cavity.

fov- [pit] *fov*ea capitis: The pit in the head of a femur.

glen- [joint socket] *glen*oid cavity: A depression in the scapula that articulates with the head of the humerus.

hema- [blood] *hema*toma: A blood clot.

inter- [between] *inter*vertebral disk: A structure located between adjacent vertebra.

intra- [inside] *intra*membranous bone: Bone that forms within sheetlike masses of connective tissue.

meat- [passage] auditory *meat*us: A canal of the temporal bone that leads inward to parts of the ear.

odont- [tooth] *odont*oid process: A toothlike process of the second cervical vertebra.

poie- [making] hemato*poie*sis: A process by which blood cells are formed.

Key Terms

articular cartilage (ar-tik′u-lar kar′tĭ-lij)
bursa (ber′sah)
cartilaginous joint (kar″tĭ-lah′jin-us joint)
compact bone (kom′pakt bōn)
diaphysis (di-af′ĭ-sis)
endochondral bone (en″do-kon′dral bōn)
epiphyseal disk (ep″ĭ-fiz′e-al disk)
epiphysis (e-pif′ĭ-sis)
fibrous joint (fi′brus joint)
hematopoiesis (hem″ah-to-poi-e′sis)
intramembranous bone (in″trah-mem′brah-nus bōn)
lever (lev′er)
marrow (mar′o)
meatus (me-a′tus)
medullary cavity (med′u-lār″e kav′ĭ-te)
meniscus (mĕ-nis′kus)
osteoblast (os′te-o-blast)
osteoclast (os′te-o-klast)
osteocyte (os′te-o-sīt)
periosteum (per″e-os′te-um)
spongy bone (spun′je bōn)
synovial joint (sĭ-no′ve-al joint)

Introduction

Nonliving material in the matrix of bone tissue makes the whole organ appear to be inactive. However, a bone contains a variety of very active tissues: bone tissue, cartilage, fibrous connective tissue, blood, and nervous tissue.

Bone Structure

The bones of the skeletal system differ greatly in size and shape, yet they are similar in structure, development, and functions.

Parts of a Long Bone

The femur, a long bone in the thigh, illustrates the structure of bone (fig. 7.1). At each end of such a bone is an expanded portion called an **epiphysis** (plural, *epiphyses*), which articulates (forms a joint) with another bone. On its outer surface, the articulating portion of the epiphysis is coated with a layer of hyaline cartilage called **articular cartilage.** The shaft of the bone, which is located between the epiphyses, is called the **diaphysis.**

A tough, vascular covering of fibrous tissue called the **periosteum** completely encloses the bone, except for the articular cartilage on the bone's ends. The periosteum is firmly attached to the bone, and periosteal fibers are continuous with ligaments and tendons that connect to the membrane. The periosteum also helps form and repair bone tissue.

A bone's shape makes possible its functions. For example, bony projections called *processes* provide sites for ligaments and tendons to attach; grooves and openings provide passageways for blood vessels and nerves; a depression of one bone might articulate with a process of another.

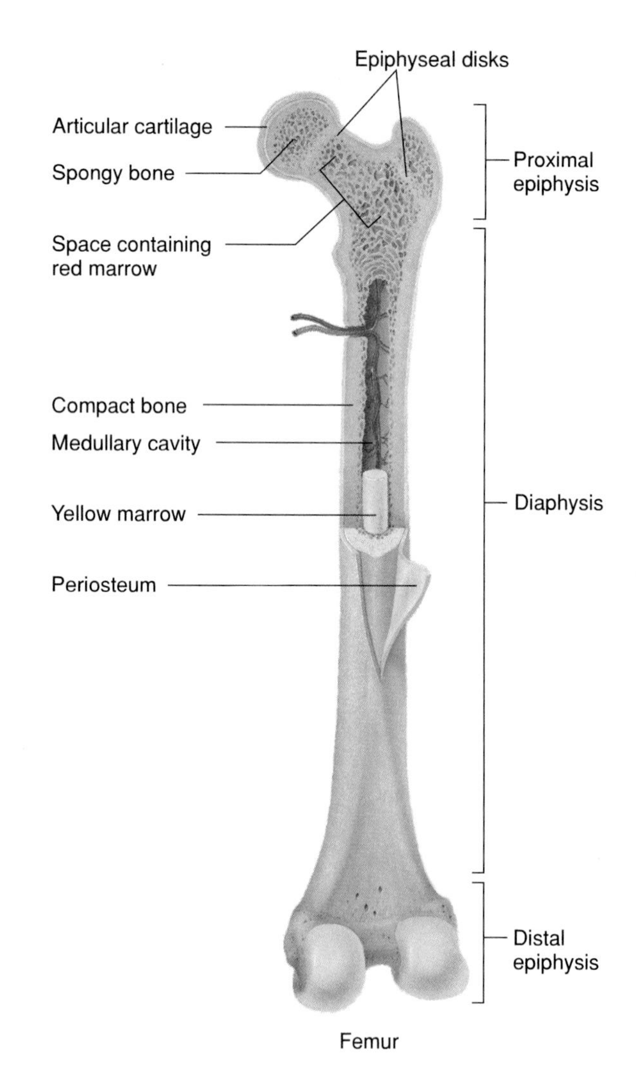

Figure 7.1

Major parts of a long bone.

The wall of the diaphysis is composed mainly of tightly packed tissue called **compact bone** (cortical bone). This type of bone has a continuous matrix without spaces. The epiphyses, in contrast, are composed largely of **spongy bone** (cancellous bone) with thin layers of compact bone on their surfaces. Spongy bone consists of many branching bony plates. Irregular interconnecting spaces between these plates help reduce the bone's weight. The bony plates are most highly developed in the regions of the epiphyses that are subjected to compressive forces. Both types of bone are strong and resist bending (fig. 7.2).

Compact bone in the diaphysis of a long bone forms a semirigid tube with a hollow chamber called the **medullary cavity** that is continuous with the spaces of the spongy bone. A thin layer of cells called **endosteum** lines these areas, and a specialized type of soft connective tissue called **marrow** fills them.

Microscopic Structure

Recall from chapter 5 that bone cells (osteocytes) are located in very small, bony chambers called *lacunae,* which form concentric circles around *osteonic canals* (Haversian canals) (figs. 5.19 and 7.3). Osteocytes

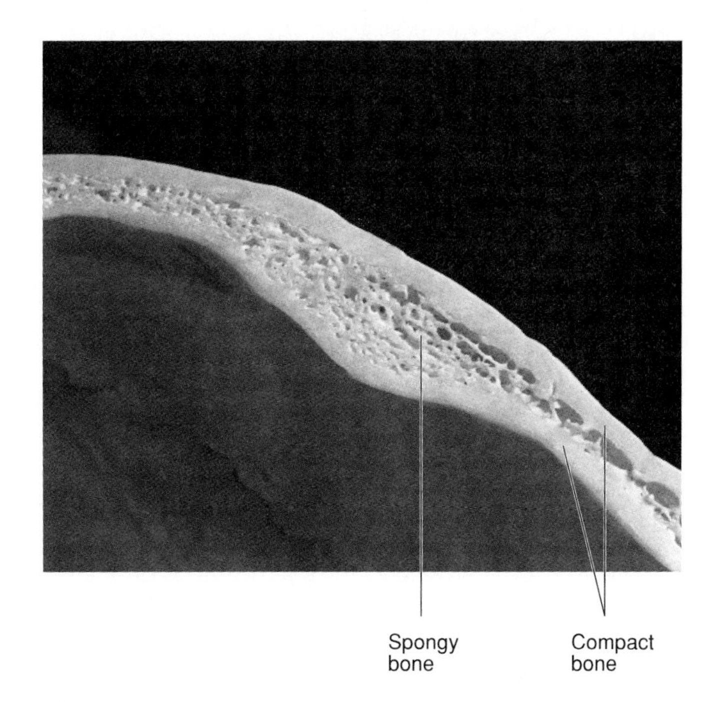

Figure 7.2

This skull bone contains a layer of spongy bone sandwiched between thin plates of compact bone.

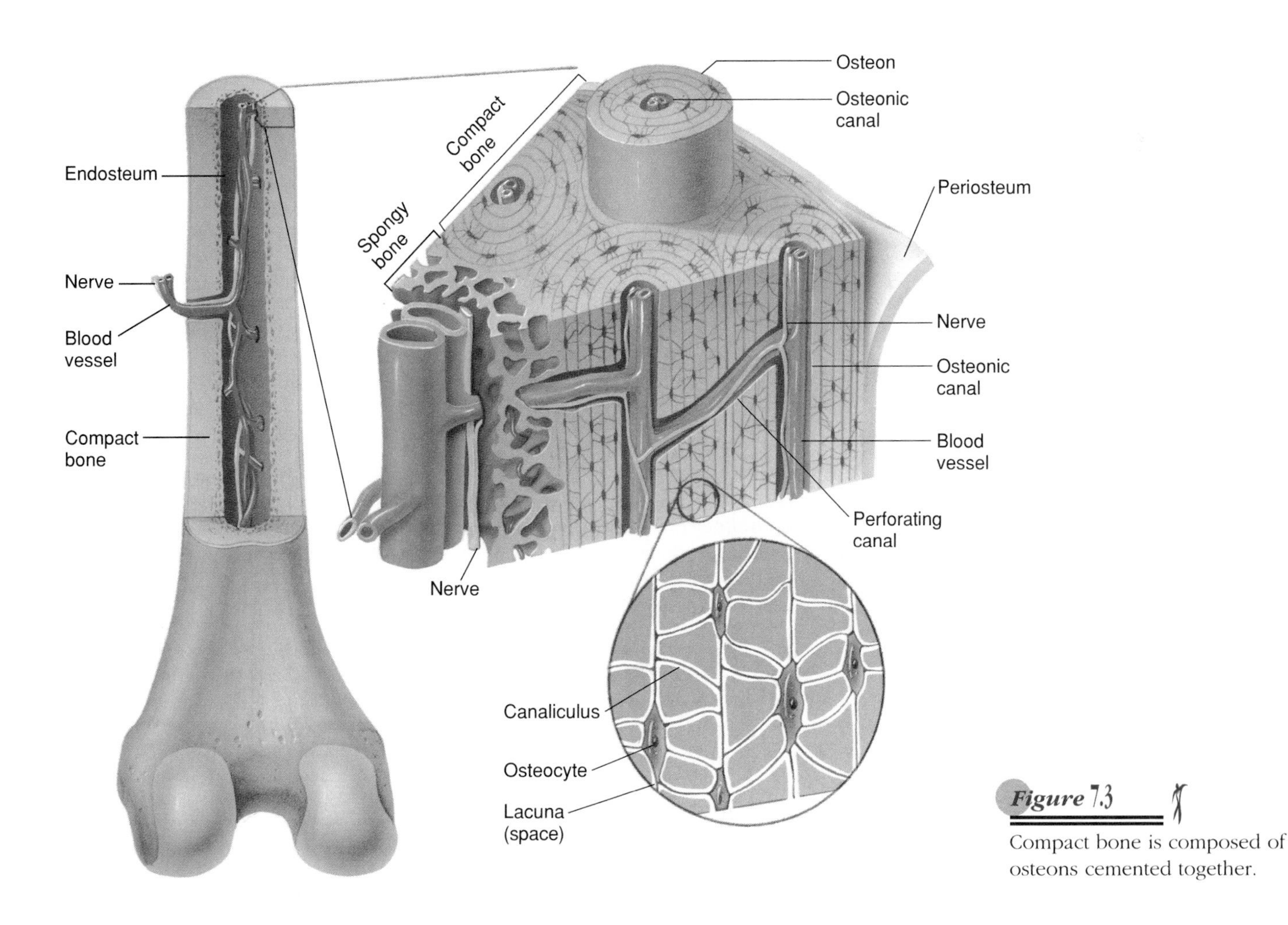

Figure 7.3

Compact bone is composed of osteons cemented together.

communicate with nearby cells by means of cellular processes passing through canaliculi. The intercellular material of bone tissue is largely collagen and inorganic salts. Collagen gives bone its strength and resilience, and inorganic salts make it hard and resistant to crushing.

In compact bone, the osteocytes and layers of intercellular material clustered concentrically around an osteonic canal form a cylinder-shaped unit called an *osteon* (Haversian system). Many of these units cemented together form the substance of compact bone.

Each osteonic canal houses one or two small blood vessels (usually capillaries) and a nerve within loose connective tissue. Blood in these vessels nourishes bone cells associated with the osteonic canal.

Osteonic canals extend longitudinally through bone tissue, and transverse *perforating canals* (Volkmann's canals) connect them. Perforating canals house larger blood vessels and nerves that the vessels and nerves in osteonic canals use to communicate with the bone surface and the medullary cavity (fig. 7.3).

Spongy bone is also composed of osteocytes and intercellular material, but the bone cells are not organized around osteonic canals. Instead, substances diffusing into canaliculi that lead to the surface of these thin, bony plates nourish the cells.

1. List five major parts of a long bone.
2. How do compact and spongy bone differ in structure?
3. Describe the microscopic structure of compact bone.

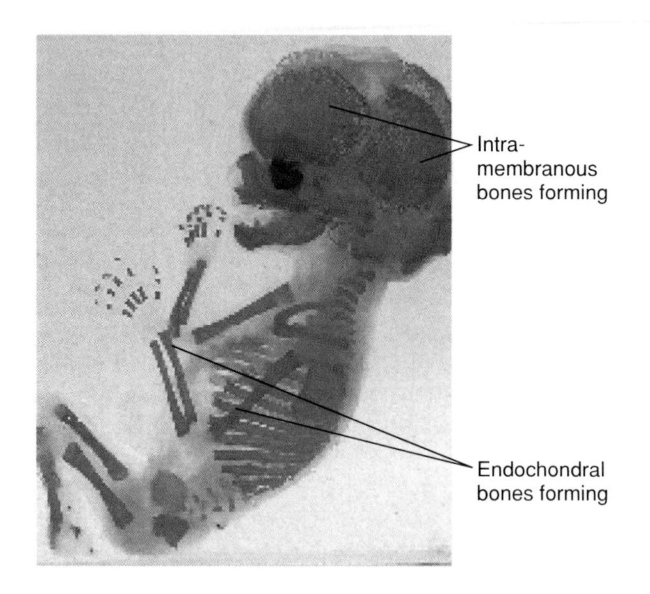

***Figure* 7.4**

Note the stained, developing bones of this fourteen-week fetus.

Bone Development and Growth

Parts of the skeletal system begin to form during the first few weeks of prenatal development, and bony structures continue to develop and grow into adulthood. Bones form by replacing existing connective tissues in either of two ways: (1) **Intramembranous bones** originate between sheetlike layers of connective tissues; (2) **Endochondral bones** begin as masses of cartilage that bone tissue later replaces.

Intramembranous Bones

The broad, flat bones of the skull are intramembranous bones (fig. 7.4). During their development, membranelike layers of connective tissues appear at the sites of the future bones. Then, some of the primitive connective tissue cells enlarge and differentiate into bone-forming cells called **osteoblasts.** The osteoblasts become active within the membranes and deposit bony matrix around themselves. As a result, spongy bone tissue forms in all directions within the layers of primitive connective tissues. Eventually, cells of the membranous tissues that persist outside the developing bone give rise to the periosteum. Osteoblasts on the inside of the periosteum form a layer of compact bone over the surface of the newly built spongy bone. When matrix completely surrounds osteoblasts, they are called **osteocytes.**

Endochondral Bones

Most of the bones of the skeleton are endochondral bones. They develop from masses of hyaline cartilage with shapes similar to future bony structures (fig. 7.4). These cartilaginous models grow rapidly for a time and then begin to extensively change. In a long bone, for example, the changes begin in the center of the diaphysis, where the cartilage slowly breaks down and disappears (fig. 7.5). At about the same time, a periosteum forms from connective tissue that encircles the developing diaphysis. Blood vessels and osteoblasts from the periosteum invade the disintegrating cartilage, and spongy bone forms in its place. This region of bone formation is called the *primary ossification center,* and bone tissue develops from it toward the ends of the cartilaginous structure.

Meanwhile, osteoblasts from the periosteum deposit a thin layer of compact bone around the primary ossi-

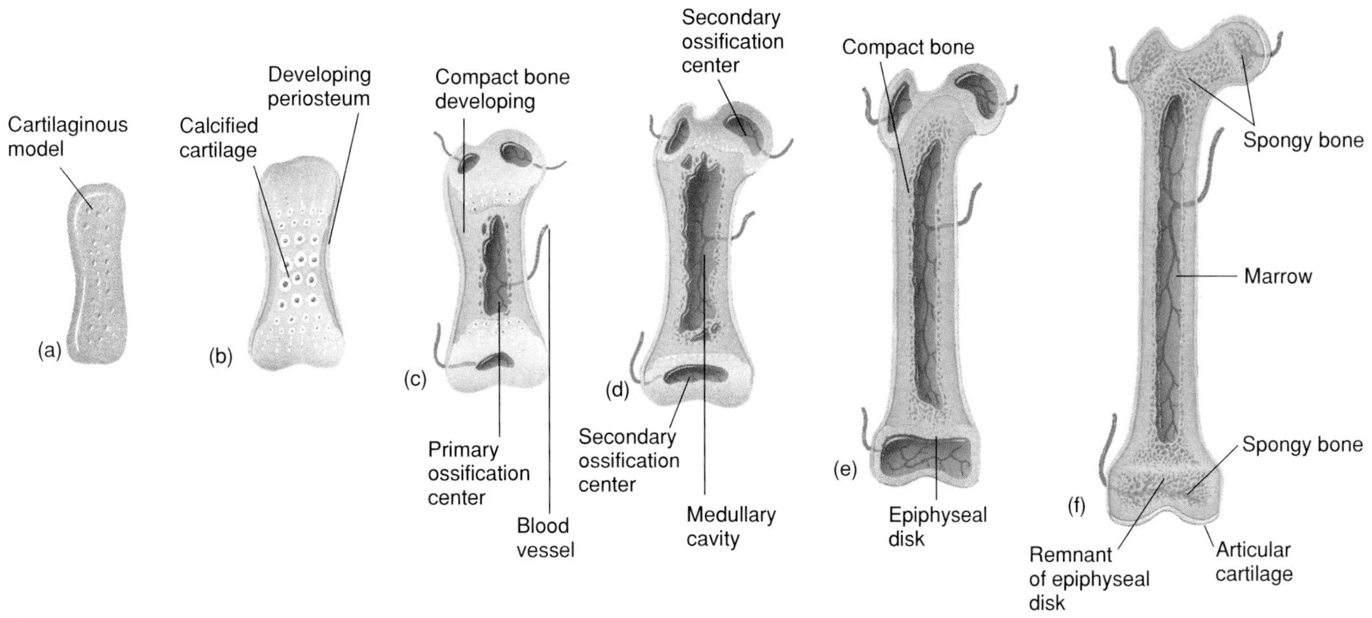

Figure 7.5

Major stages (*a–f*) in the development of an endochondral bone. (Bones are not shown to scale.)

cation center. The epiphyses of the developing bone remain cartilaginous and continue to grow. Later, *secondary ossification centers* appear in the epiphyses, and spongy bone forms in all directions from them. As bone is deposited in the diaphysis and in the epiphysis, a band of cartilage called the **epiphyseal disk** (physis) remains between these two ossification centers.

The cartilaginous cells of the epiphyseal disk include layers of young cells that are undergoing mitosis and producing new cells. As these cells enlarge and matrix forms around them, the cartilaginous disk thickens, lengthening the bone. At the same time, calcium salts accumulate in the matrix adjacent to the oldest cartilaginous cells, and as the matrix calcifies, the cells begin to die.

In time, large, multinucleated cells called **osteoclasts** break down the calcified matrix. These large cells originate in bone marrow when certain single-nucleated white blood cells (monocytes) fuse (see chapter 12).

Osteoclasts secrete an acid that dissolves the inorganic component of the calcified substance, and their lysosomal enzymes digest the organic components. After osteoclasts remove the matrix, bone-building osteoblasts invade the region and deposit bone tissue in place of the calcified cartilage.

A long bone continues to lengthen while the cartilaginous cells of the epiphyseal disks are active. However, once the ossification centers of the diaphysis and epiphyses meet and the epiphyseal disks ossify, lengthening is no longer possible.

If an epiphyseal disk is damaged before it ossifies, elongation of the long bone may cease prematurely, or growth may be uneven. For this reason, injuries to the epiphyses of a young person's bones are of special concern. Surgeons can alter an epiphysis to equalize the growth rate of bones developing at very different rates. ■

A developing long bone thickens as compact bone is deposited on the outside, just beneath the periosteum. As this compact bone forms on the surface, osteoclasts erode other bone tissue on the inside. The resulting space becomes the medullary cavity of the diaphysis, which later fills with marrow. The bone in the central regions of the epiphyses and diaphysis remains spongy, and hyaline cartilage on the ends of the epiphyses persists throughout life as articular cartilage.

Homeostasis of Bone Tissue

After the intramembranous and endochondral bones form, the actions of osteoclasts and osteoblasts continually remodel them. Thus, throughout life, osteoclasts resorb bone tissue, and osteoblasts replace the bone. These opposing processes of resorption and deposition are well regulated by hormones (see chapter 11) so that the total mass of bone tissue within an adult skeleton normally remains nearly constant, even though 3–5% of bone calcium is exchanged each year.

TOPIC OF INTEREST Repair of a Bone Fracture

A *fracture* is a break in a bone. Whenever a bone breaks, blood vessels within it and its periosteum rupture, and the periosteum may tear. Blood escaping from the broken vessels spreads through the damaged area and soon forms a blood clot, or *hematoma*. Vessels in surrounding tissues dilate, which swells and inflames tissues.

Within days or weeks, developing blood vessels and many osteoblasts from the periosteum invade the hematoma. The osteoblasts multiply rapidly in the regions close to the new blood vessels, building spongy bone nearby. Granulation tissue develops, and in regions further from the blood supply, fibroblasts produce masses of fibrocartilage. Meanwhile, phagocytic cells begin to remove the blood clot, as well as any dead or damaged cells in the affected area. Osteoclasts also appear and resorb bone fragments, aiding in "cleaning up" debris.

In time, fibrocartilage fills the gap between the ends of the broken bone. This mass, termed a *cartilaginous callus,* is later replaced by bone tissue in much the same way that the hyaline cartilage of a developing endochondral bone is replaced. That is, the cartilaginous callus breaks down, blood vessels and osteoblasts invade the area, and a *bony callus* fills the space.

Typically, more bone is produced at the site of a healing fracture than is required to replace the damaged tissues. Osteoclasts remove the excess, and the final result is a bone shaped very much like the original (fig. 7A).

Physicians help the bone-healing process. The first casts to immobilize fractured bones were introduced in Philadelphia in 1876, and soon after, doctors began using screws and plates internally to align healing bone parts. Today, orthopedic surgeons also use rods, wires, and nails. These devices have become lighter and smaller; many are built of titanium. A new approach, called a hybrid fixator, treats a broken leg using metal pins internally to align bone pieces. The pins are anchored to a metal ring device worn outside the leg.

In bone cancers, abnormally active osteoclasts destroy bone tissue. Interestingly, cancer of the prostate gland can have the opposite effect if the cancer cells reach the bone marrow (as they do in most cases of advanced prostatic cancer). These cells stimulate osteoblast activity, which promotes formation of new bone on the surfaces of the bony plates. ■

1. Describe the development of an intramembranous bone.
2. Explain how an endochondral bone develops.
3. Explain how osteoclasts and osteoblasts remodel bone.

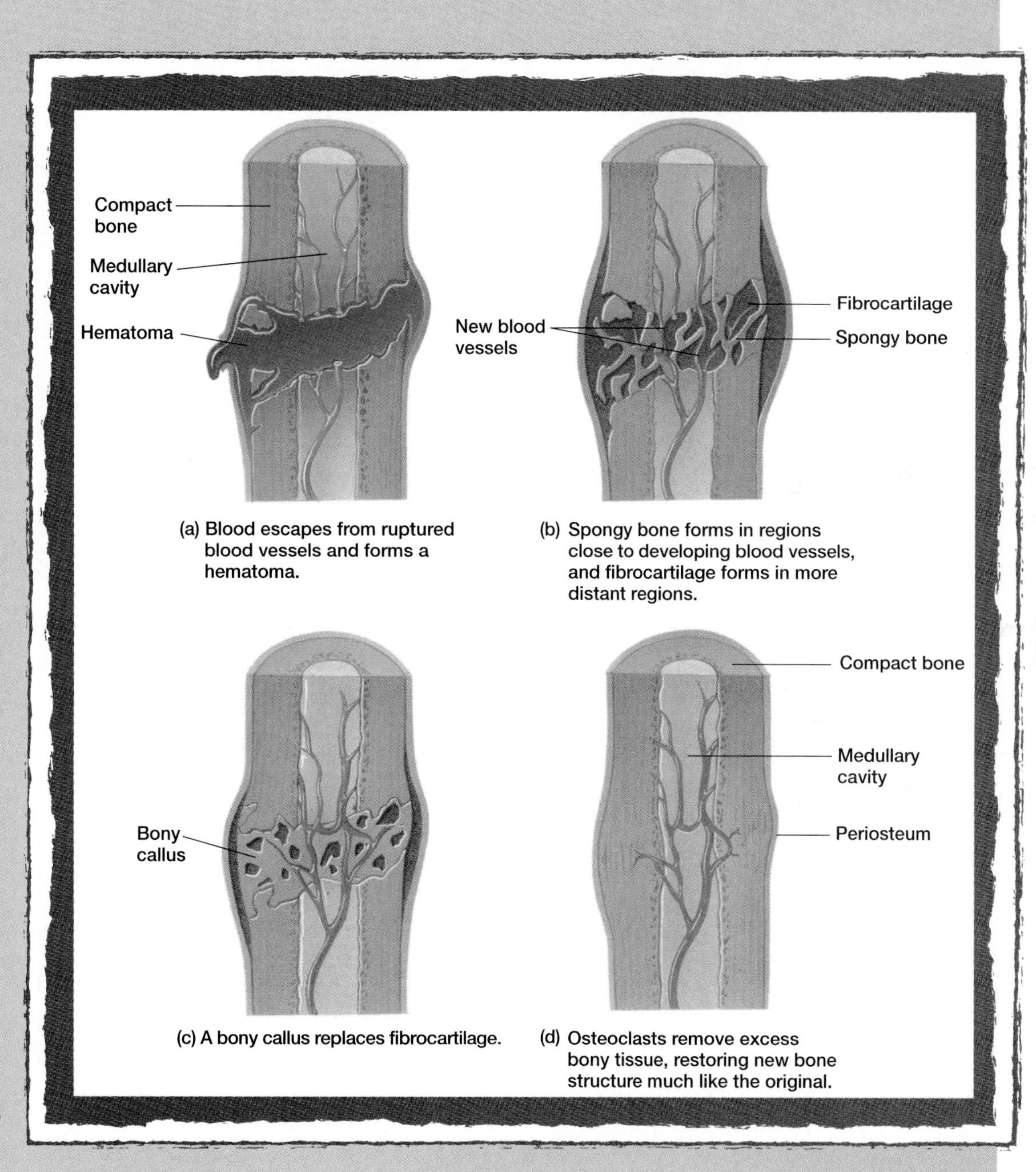

***Figure* 7A**

Major steps (*a–d*) in fracture repair.

Bone Function

Bones shape, support, and protect body structures. They also aid body movements, house tissues that produce blood cells, and store various inorganic salts.

Support and Protection

Bones give shape to such structures as the head, face, thorax, and limbs and also provide support and protection. For example, the bones of the lower limbs, pelvis, and backbone support the body's weight. The bones of the skull protect the eyes, ears, and brain. Those of the

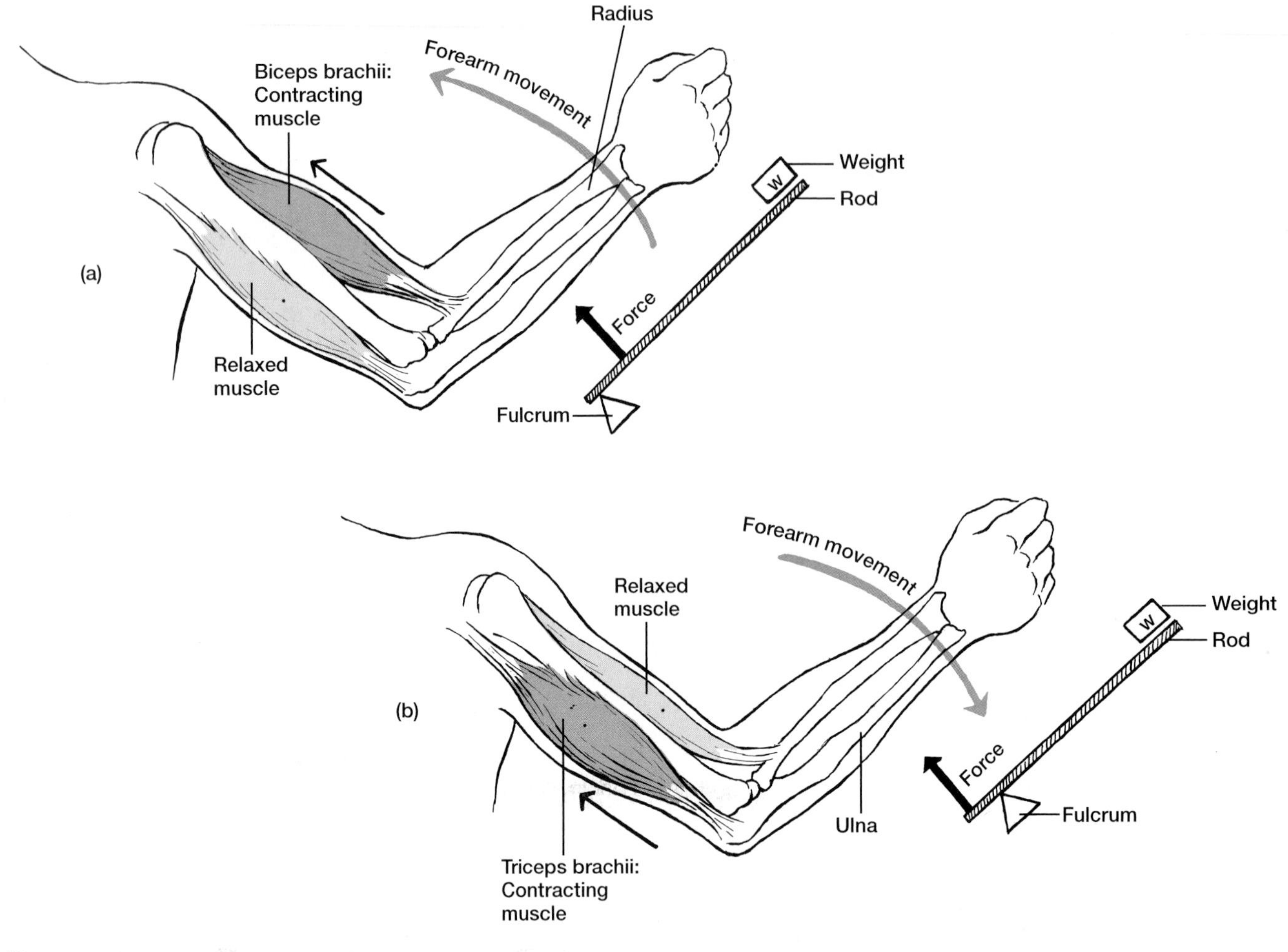

Figure 7.6

When the upper limb (*a*) bends at the elbow or (*b*) when the upper limb straightens at the elbow, bones and muscles function as a lever.

rib cage and shoulder girdle protect the heart and lungs, while bones of the pelvic girdle protect the lower abdominal and internal reproductive organs.

Body Movement

Whenever limbs or other body parts move, bones and muscles interact as simple mechanical devices called **levers.** A lever has four basic components: (1) a rigid rod or bar, (2) a fulcrum or pivot on which the bar turns, (3) an object or weight (resistance) that is moved, and (4) a force that supplies energy for the movement of the bar.

The actions of bending and straightening the upper limb at the elbow illustrate bones and muscles functioning as levers (fig. 7.6). When the upper limb bends, the forearm bones represent the rigid rod, the elbow joint is the fulcrum, the hand is the weight that is moved, and muscles on the anterior side of the arm supply the force. One of these muscles, the *biceps brachii,* attaches by a tendon to a projection on a bone (radius) in the forearm, a short distance below the elbow.

When the upper limb straightens at the elbow, the forearm bones again serve as a rigid rod and the elbow joint as the fulcrum. However, this time, the *triceps brachii,* a muscle located on the posterior side of the arm, supplies the force. A tendon of this muscle attaches to a projection on a bone (ulna) at the point of the elbow.

Blood Cell Formation

Very early in life, the process of blood cell formation, called **hematopoiesis,** occurs in the *yolk sac* that lies outside the human embryo (see chapter 20). Later in development, blood cells are manufactured in the liver and spleen, and still later, they form in bone marrow.

Marrow is a soft, netlike mass of connective tissue within the medullary cavities of long bones, in the irregular spaces of spongy bone, and in the larger os-

Osteoporosis

In *osteoporosis,* the skeletal system loses bone volume and mineral content. The affected bones develop spaces and canals that enlarge and fill with fibrous and fatty tissues. Such bones easily fracture and may break spontaneously because they are no longer able to support body weight. For example, a person with osteoporosis may suffer a spontaneous fracture of the thighbone (femur) at the hip or a collapse of sections of the backbone (vertebrae).

Osteoporosis is associated with aging and causes many fractures in persons over age forty-five. It is most common in light-skinned females past menopause (see chapter 19).

Factors that increase the risk of osteoporosis include low intake of dietary calcium, lack of physical exercise (particularly during early growing years), and in females, decrease in blood estrogen concentration. (Estrogen is a hormone the ovaries produce until menopause.) Drinking alcohol, smoking cigarettes, and inheriting certain genes may also increase a person's risk of developing osteoporosis.

Bone mass usually peaks at about age thirty-five. Thereafter, bone loss may exceed bone formation in both males and females. To reduce such loss, people in their mid-twenties and older should take in 1,000–1,500 milligrams of calcium daily. In addition, people should regularly engage in exercises, such as walking or jogging, that require the bones to support body weight. Postmenopausal women may also require estrogen replacement therapy to prevent osteoporosis.

A bone marrow transplant can correct damage from X rays, certain drugs, cancer, sickle cell disease, or other blood disorders. In this procedure, a hollow needle and syringe remove normal red marrow cells from the spongy bone of a donor, who is selected because the pattern of molecules on his or her cell surfaces closely matches that of the recipient. The donor cells are injected into the bloodstream of the recipient, whose own marrow has been intentionally destroyed. The hope is that the donor cells will lodge in the spaces that red marrow normally inhabits and replace the damaged tissue.

To avoid the pain of removing donor marrow from a bone, a new approach gives the donor a biochemical that coaxes blood-forming cells to leave marrow and enter the bloodstream. Then the blood-forming cells are harvested from the donor's bloodstream. ■

teonic canals of compact bone tissue. The two kinds of marrow are red marrow and yellow marrow. *Red marrow* forms red blood cells (erythrocytes), white blood cells (leukocytes), and blood platelets (thrombocytes). It is red because of the red, oxygen-carrying pigment **hemoglobin** that red blood cells contain.

Red marrow fills most bone cavities in an infant. With age, however, *yellow marrow* replaces much of it. Yellow marrow stores fat and does not produce blood cells. In an adult, red marrow is primarily in spongy bone of the skull, ribs, sternum, clavicles, vertebrae, and pelvis. Chapter 12 describes blood cell formation in more detail.

Storage of Inorganic Salts

The intercellular matrix of bone tissue is rich in calcium salts, mostly in the form of calcium phosphate. Vital metabolic processes require calcium. A low blood calcium ion concentration stimulates osteoclasts to break down bone tissue, which releases calcium ions into the blood. A high blood calcium ion concentration inhibits osteoclast activity, which stimulates osteoblasts to form bone tissue. As a result, the intercellular matrix of bone stores excess calcium. Chapter 11 describes the details of this mechanism.

Bone tissue contains lesser amounts of magnesium, sodium, potassium, and carbonate ions. Bones also accumulate metallic elements, such as lead, radium, or strontium, which are not normally present in the body but are sometimes ingested accidentally.

1. Name three functions of bones.
2. Distinguish between the functions of red marrow and yellow marrow.
3. List the chemicals that bone stores.

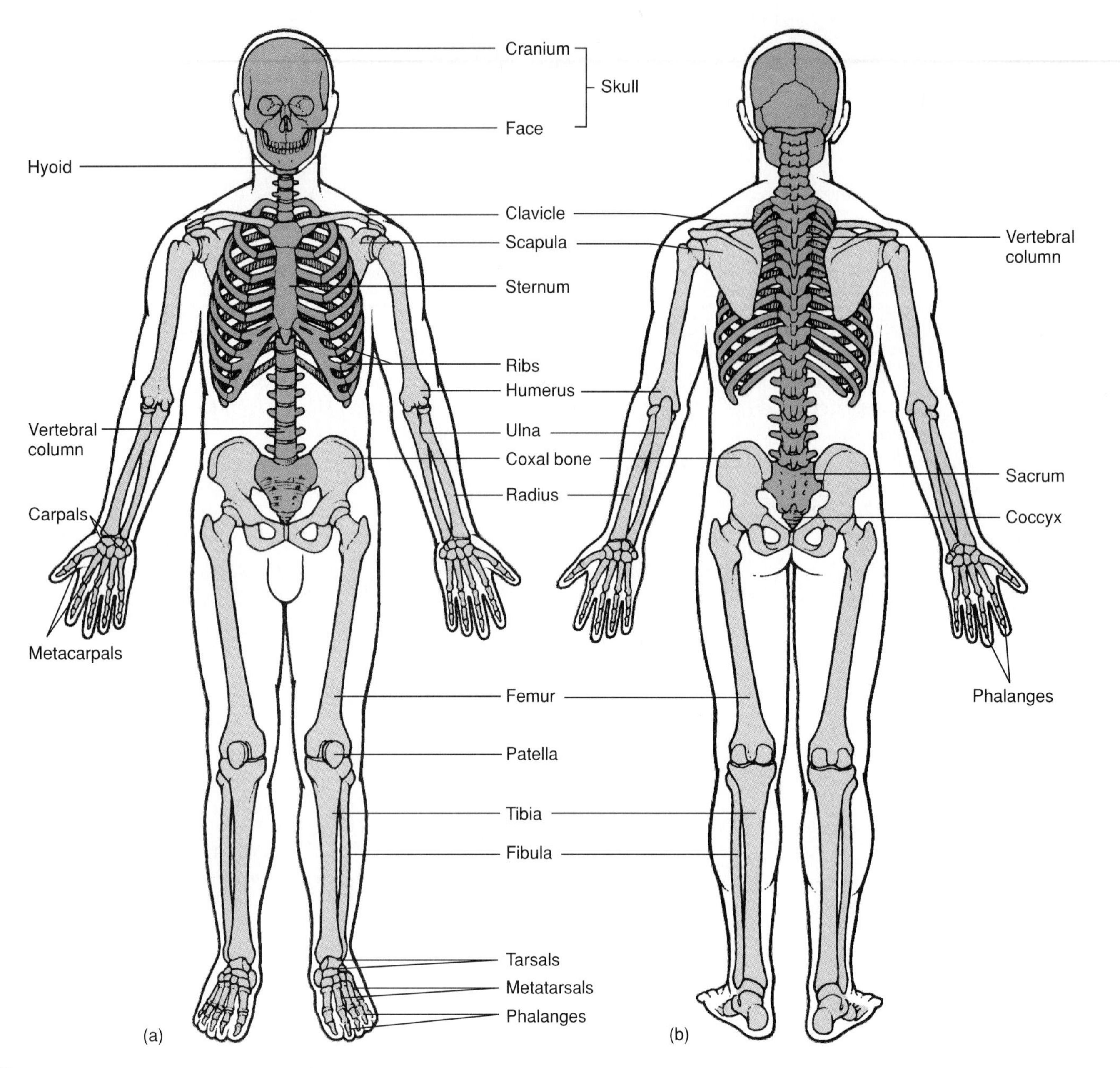

Figure 7.7

Major bones of the skeleton. (*a*) Anterior view. (*b*) Posterior view. (Note: The axial portions are shown in orange, and the appendicular portions are shown in yellow.)

Skeletal Organization

For purposes of study, dividing the skeleton into two major portions—an axial skeleton and an appendicular skeleton—is helpful (fig. 7.7). The **axial skeleton** consists of the bony and cartilaginous parts that support and protect the organs of the head, neck, and trunk. These parts include:

1. **Skull** The skull is composed of the **cranium** (kra′ne-um) (braincase) and the *facial bones.*
2. **Hyoid** (hi′oid) **bone** The hyoid bone is located in the neck between the lower jaw and the larynx. It supports the tongue and is an attachment for certain muscles that help to move the tongue during swallowing.
3. **Vertebral column** The vertebral column (backbone) consists of many vertebrae separated by cartilaginous *intervertebral disks.* Near its distal end, several vertebrae fuse to form the **sacrum** (sa′krum), which is part of the pelvis. A small,

Table 7.1 Bones of the Adult Skeleton

1. Axial Skeleton			2. Appendicular Skeleton	
a. Skull			a. Pectoral girdle	
8 cranial bones			scapula 2	
frontal 1	temporal 2		clavicle 2	
parietal 2	sphenoid 1			4 bones
occipital 1	ethmoid 1		b. Upper limbs	
13 facial bones			humerus 2	
maxilla 2	lacrimal 2		radius 2	
zygomatic 2	nasal 2		ulna 2	
palatine 2	vomer 1		carpal 16	
inferior nasal concha 2			metacarpal 10	
1 mandible			phalanx 28	
		22 bones		60 bones
b. Middle ear bones			c. Pelvic girdle	
malleus 2			coxal bone 2	
incus 2				2 bones
stapes 2			d. Lower limbs	
		6 bones	femur 2	
c. Hyoid			tibia 2	
hyoid bone 1			fibula 2	
		1 bone	patella 2	
d. Vertebral column			tarsal 14	
cervical vertebra 7			metatarsal 10	
thoracic vertebra 12			phalanx 28	
lumbar vertebra 5				60 bones
sacrum 1			Total	206 bones
coccyx 1				
		26 bones		
e. Thoracic cage				
rib 24				
sternum 1				
		25 bones		

rudimentary tailbone called the **coccyx** (kok′siks) attaches to the end of the sacrum.

4. **Thoracic cage** The thoracic cage protects the organs of the thoracic cavity and the upper abdominal cavity. It is composed of twelve pairs of **ribs**, which articulate posteriorly with thoracic vertebrae. The thoracic cage also includes the **sternum** (ster′num), to which most of the ribs attach anteriorly.

The **appendicular skeleton** consists of the bones of the upper and lower limbs and the bones that anchor the limbs to the axial skeleton. It includes:

1. **Pectoral girdle** A **scapula** (scap′u-lah) and a **clavicle** (klav′ĭ-k′l) bone form the pectoral girdle on both sides of the body. The pectoral girdle connects the bones of the upper limbs to the axial skeleton and aids in upper limb movements.
2. **Upper limbs** Each upper limb consists of a **humerus** (hu′mer-us), or arm bone, and two forearm bones—a **radius** (ra′de-us) and an **ulna** (ul′nah). These three bones articulate with each other at the elbow joint. At the distal end of the radius and ulna are eight **carpals** (kar′pals) or wrist bones. The bones of the palm are called **metacarpals** (met″ah-kar′pals), and the finger bones are called **phalanges** (fah-lan′jēz).
3. **Pelvic girdle** Two **coxal** (kok′sal) **bones** (hipbones) form the pelvic girdle and are attached to each other anteriorly and to the sacrum posteriorly. They connect the bones of the lower limbs to the axial skeleton and, with the sacrum and coccyx, form the **pelvis.**
4. **Lower limbs** Each lower limb consists of a **femur** (fe′mur), or thighbone, and two leg bones—a large **tibia** (tib′e-ah) and a slender **fibula** (fib′u-lah). The femur and tibia articulate with each other at the knee joint, where the **patella** (pah-tel′ah) covers the anterior surface. At the distal ends of the tibia and fibula are seven **tarsals** (tahr′sals), or anklebones. The bones of the instep are called **metatarsals** (met″ah-tahr′sals), and those of the toes (like the fingers) are called **phalanges.**

Table 7.1 lists the bones of the adult skeleton, and table 7.2 lists terms that describe skeletal structures.

Table 7.2

Terms Used to Describe Skeletal Structures

Term	Definition	Example
Condyle (kon′dīl)	A rounded process that usually articulates with another bone	Occipital condyle of occipital bone (fig. 7.11)
Crest (krest)	A narrow, ridgelike projection	Iliac crest of ilium (fig. 7.27)
Epicondyle (ep″ĭ-kon′dīl)	A projection situated above a condyle	Medial epicondyle of humerus (fig. 7.23)
Facet (fas′et)	A small, nearly flat surface	Rib facet of thoracic vertebra (fig. 7.15)
Fontanel (fon″tah-nel′)	A soft spot in the skull where membranes cover the space between bones	Anterior fontanel between frontal and parietal bones (fig. 7.14)
Foramen (fo-ra′men)	An opening through a bone that usually is a passageway for blood vessels, nerves, or ligaments	Foramen magnum of occipital bone (fig. 7.11)
Fossa (fos′ah)	A relatively deep pit or depression	Olecranon fossa of humerus (fig. 7.23)
Fovea (fo′ve-ah)	A tiny pit or depression	Fovea capitis of femur (fig. 7.29)
Head (hed)	An enlargement of the end of a bone	Head of humerus (fig. 7.23)
Meatus (me-a′tus)	A tubelike passageway within a bone	External auditory meatus of ear (fig. 7.10)
Process (pros′es)	A prominent projection on a bone	Mastoid process of temporal bone (fig. 7.10)
Sinus (si′nus)	A cavity within a bone	Frontal sinus of frontal bone (fig. 7.13)
Spine (spīn)	A thornlike projection	Spine of scapula (fig. 7.22)
Suture (soo′cher)	An interlocking line of union between bones	Lambdoidal suture between occipital and parietal bones (fig. 7.10)
Trochanter (tro-kan′ter)	A relatively large process	Greater trochanter of femur (fig. 7.29)
Tubercle (tu′ber-kl)	A small, knoblike process	Greater tubercle of humerus (fig. 7.23)
Tuberosity (tu″bĕ-ros′ĭ-te)	A knoblike process usually larger than a tubercle	Radial tuberosity of radius (fig. 7.24)

1. Distinguish between the axial and appendicular skeletons.
2. List the bones of the axial skeleton.
3. List the bones of the appendicular skeleton.

Skull

A human skull usually consists of twenty-two bones that, except for the lower jaw, firmly interlock along lines called *sutures* (soo′cherz) (fig. 7.8). Eight of these interlocked bones form the cranium, and thirteen form the facial skeleton. The **mandible** (man′dĭ-b′l), or lower jawbone, is a movable bone held to the cranium by ligaments. (Three other bones found in each middle ear are discussed in chapter 10.)

Reference plates 8–11 on pages 173–175 are a set of photographs of the human skull and its parts.

Cranium

The **cranium** encloses and protects the brain, and its surface provides attachments for muscles that make chewing and head movements possible. Some of the cranial bones contain air-filled cavities called *paranasal sinuses,* which are lined with mucous membranes and connect by passageways to the nasal cavity (fig. 7.9). Sinuses reduce the skull's weight and increase the intensity of the voice by serving as resonant sound chambers.

The eight bones of the cranium are (figs. 7.8 and 7.10):

1. **Frontal bone** The frontal (frun′tal) bone forms the anterior portion of the skull above the eyes. On the upper margin of each orbit (the bony socket of the eye), the frontal bone is marked by a *supraorbital foramen* (or *supraorbital notch* in some skulls), through which blood vessels and nerves pass to the tissues of the forehead. Within the frontal bone are two *frontal sinuses,* one above each eye near the midline (fig. 7.9).
2. **Parietal bones** One parietal (pah-ri′ĕ-tal) bone is located on each side of the skull just behind the frontal bone (fig. 7.10). Together, the parietal bones form the bulging sides and roof of the cranium. They fuse at the midline along the *sagittal suture,* and they meet the frontal bone along the *coronal suture.*
3. **Occipital bone** The occipital (ok-sip′ĭ-tal) bone joins the parietal bones along the *lambdoidal* (lam′doid-al) *suture* (figs. 7.10 and 7.11). It forms the back of the skull and the base of the cranium.

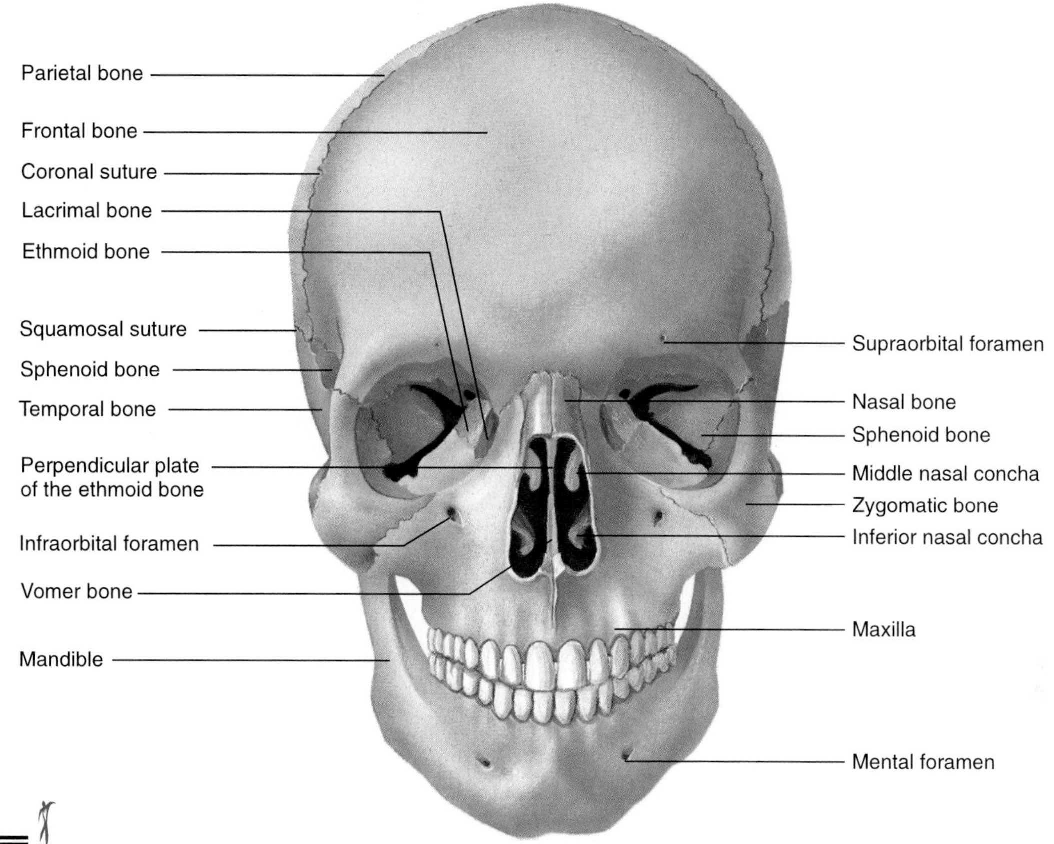

Figure 7.8

Anterior view of the skull.

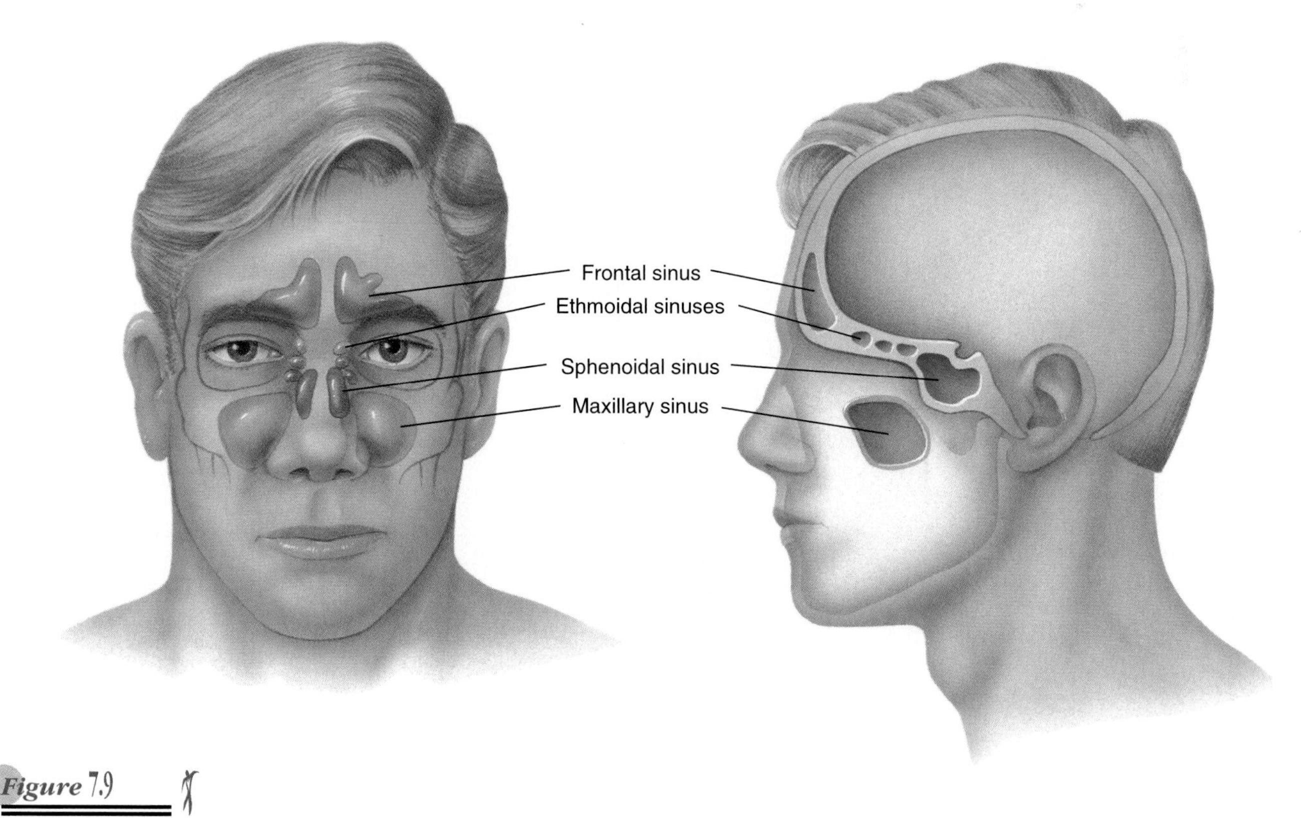

Figure 7.9

Locations of the sinuses.

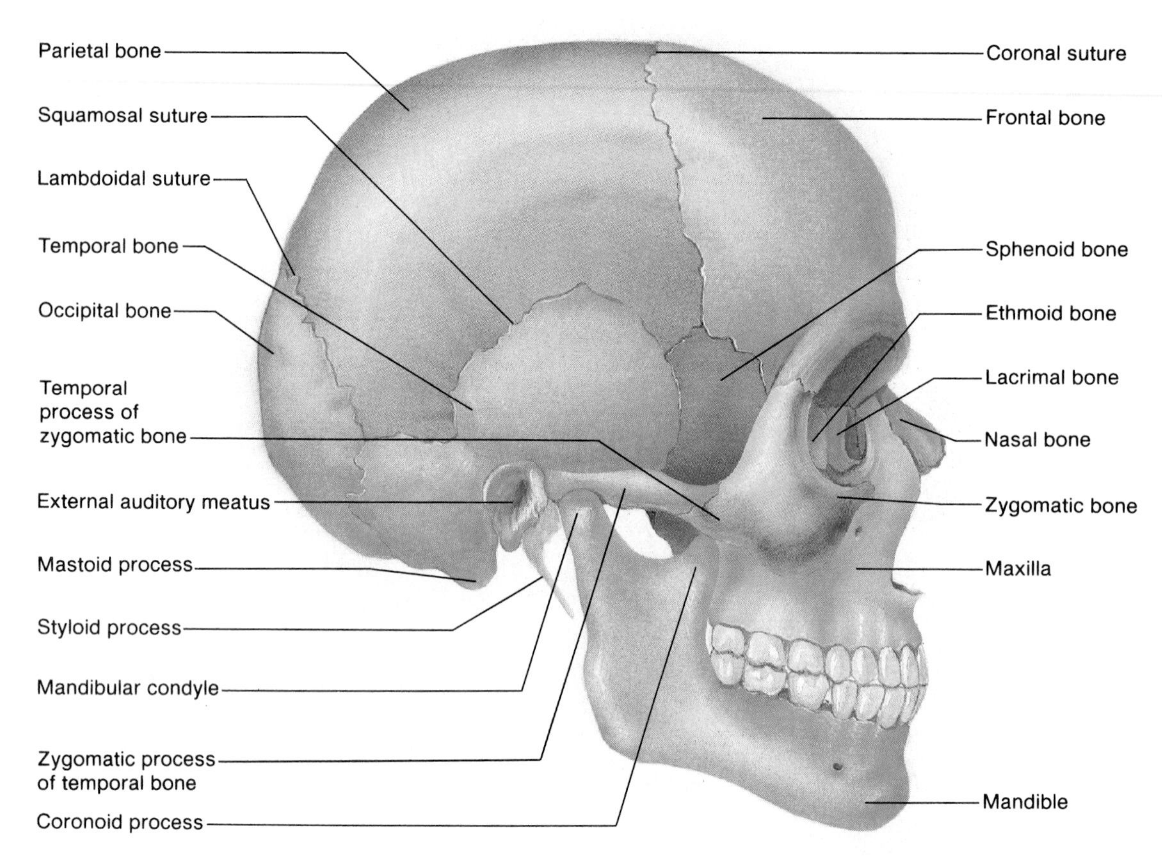

Figure 7.10

Lateral view of the skull.

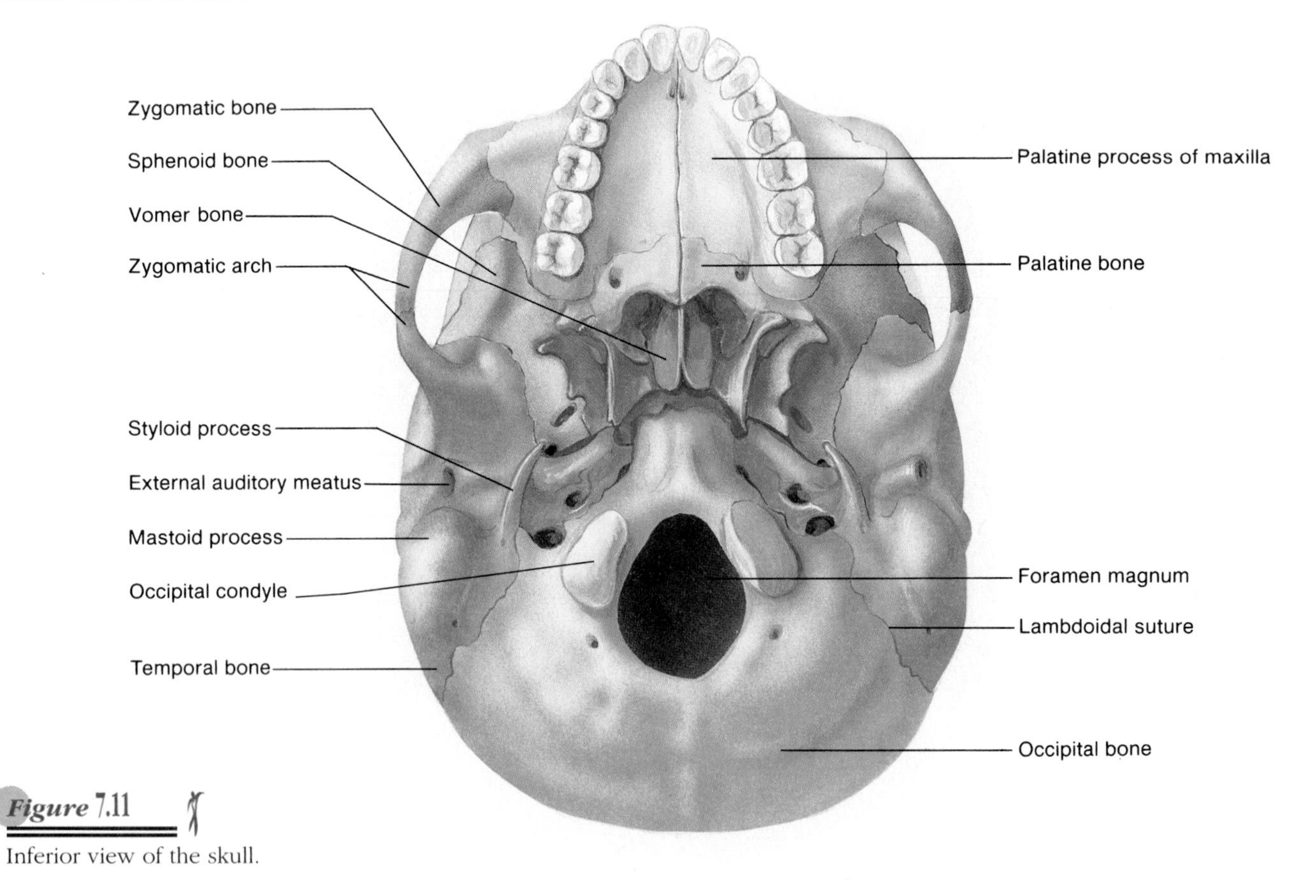

Figure 7.11

Inferior view of the skull.

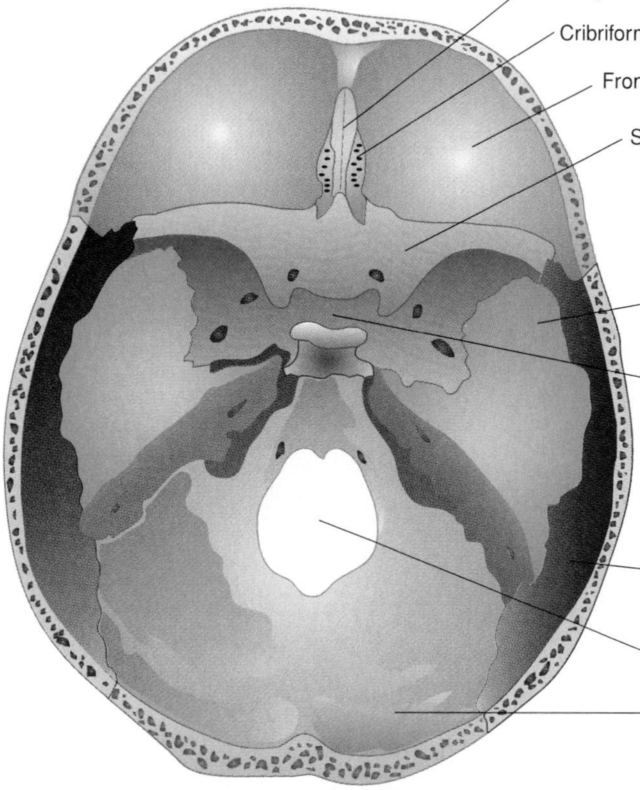

Crista galli
Cribriform plate (ethmoid bone)
Frontal bone
Sphenoid bone
Temporal bone
Sella turcica
Parietal bone
Foramen magnum
Occipital bone

Figure 7.12

Floor of the cranial cavity, viewed from above.

Through a large opening on its lower surface called the *foramen magnum* pass nerve fibers from the brain, which enter the vertebral canal to become part of the spinal cord. Rounded processes called *occipital condyles,* located on each side of the foramen magnum, articulate with the first vertebra (atlas) of the vertebral column.

4. **Temporal bones** A temporal (tem′po-ral) bone on each side of the skull joins the parietal bone along a *squamosal* (skwa-mo′sal) *suture* (fig. 7.10). The temporal bones form parts of the sides and the base of the cranium. Located near the inferior margin is an opening, the *external auditory meatus,* which leads inward to parts of the ear. The temporal bones have depressions called the *mandibular fossae* that articulate with condyles of the mandible. Below each external auditory meatus are two projections—a rounded *mastoid process* and a long, pointed *styloid process.* The mastoid process provides an attachment for certain muscles of the neck, while the styloid process anchors muscles associated with the tongue and pharynx. A *zygomatic process* projects anteriorly from the temporal bone, joins the *zygomatic bone,* and helps form the prominence of the cheek.
5. **Sphenoid bone** The sphenoid (sfe′noid) bone is wedged between several other bones in the anterior portion of the cranium (figs. 7.10 and 7.11). It consists of a central part and two winglike structures that extend laterally toward each side of the skull. This bone helps form the base of the cranium, the sides of the skull, and the floors and sides of the orbits. Along the midline within the cranial cavity, a portion of the sphenoid bone indents to form the saddle-shaped *sella turcica* (sel′ah tur′si-ka). The pituitary gland occupies this depression. The sphenoid bone also contains two *sphenoidal sinuses* (fig. 7.9).
6. **Ethmoid bone** The ethmoid (eth′moid) bone is located in front of the sphenoid bone (figs 7.10 and 7.12). It consists of two masses, one on each side of the nasal cavity, which are joined horizontally by thin *cribriform* (krib′rĭ-form) *plates.* These plates form part of the roof of the nasal cavity (fig. 7.12).

Projecting upward into the cranial cavity between the cribriform plates is a triangular process of the ethmoid bone called the *crista galli* (kris′tă gal′li) (cock's comb). Membranes that enclose the brain attach to this process (figs. 7.12 and 7.13). Portions of the ethmoid bone also form sections of the cranial floor, the orbital walls, and the nasal cavity walls. A *perpendicular plate* projects downward in the midline from the cribriform plates and forms most of the nasal septum (fig. 7.13).

Delicate scroll-shaped plates, called the *superior nasal concha* (kong′kah) and the *middle nasal concha,* project inward from the lateral portions of the ethmoid bone toward the perpendicular plate

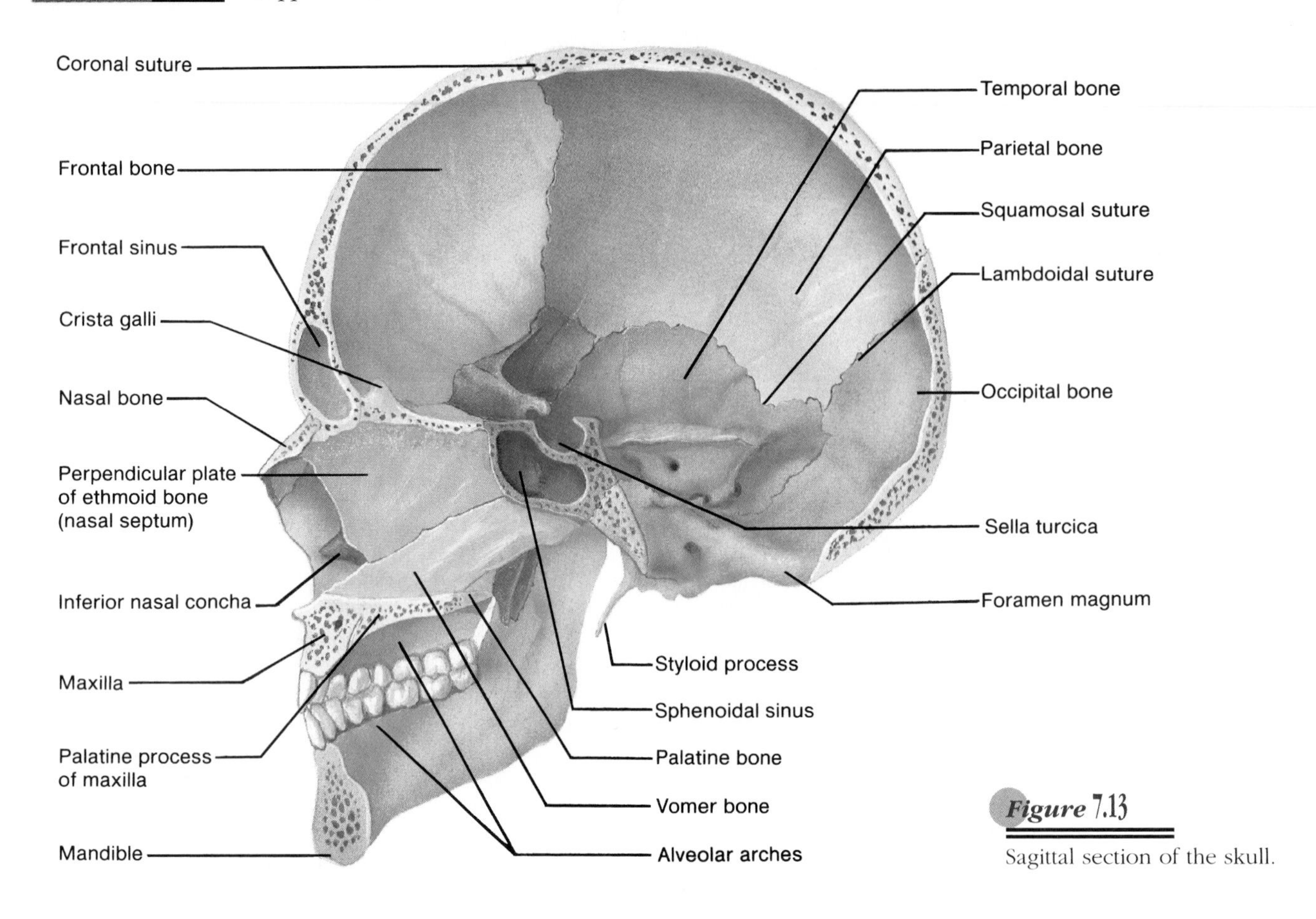

Figure 7.13

Sagittal section of the skull.

(fig. 7.13). The lateral portions of the ethmoid bone contain many small air spaces, the *ethmoidal sinuses* (fig. 7.9).

Facial Skeleton

The **facial skeleton** consists of thirteen immovable bones and a movable lower jawbone. These bones form the basic shape of the face and provide attachments for muscles that move the jaw and control facial expressions.

The bones of the facial skeleton are:

1. **Maxillae** The maxillae (mak-sil′e; singular, *maxilla,* mak-sil′ah) form the upper jaw (see figs. 7.10 and 7.11). Portions of these bones comprise the anterior roof of the mouth (*hard palate*), the floors of the orbits, and the sides and floor of the nasal cavity. They also contain the sockets of the upper teeth. Inside the maxillae, lateral to the nasal cavity, are *maxillary sinuses,* the largest paranasal sinuses (see fig. 7.9).

   During development, portions of the maxillae called *palatine processes* grow together and fuse along the midline to form the anterior section of the hard palate. The inferior border of each maxillary bone projects downward, forming an *alveolar* (al-ve′o-lar) *process.* Together, these processes form a horseshoe-shaped *alveolar arch* (dental arch) (fig. 7.13). Teeth occupy cavities in this arch (dental alveoli). Fibrous connective tissue binds teeth to the bony sockets.

Sometimes, fusion of the palatine processes of the maxillae is incomplete at birth; the result is a *cleft palate.* Infants with a cleft palate may have trouble suckling because of the opening between the oral and nasal cavities. Surgery can usually correct a cleft palate, and special bottles are available to help these children eat. ■

2. **Palatine bones** The L-shaped palatine (pal′ah-tīn) bones are located behind the maxillae (see figs. 7.11 and 7.13). The horizontal portions form the posterior section of the hard palate and the floor of the nasal cavity. The perpendicular portions help form the lateral walls of the nasal cavity.
3. **Zygomatic bones** The zygomatic (zi″go-mat′ik) bones form the prominences of the cheeks below and to the sides of the eyes (see figs. 7.10 and 7.11). These bones also help form the lateral walls and the floors of the orbits. Each bone has a *temporal process,* which extends posteriorly to join the zygomatic process of a temporal bone. Together, these processes form a *zygomatic arch.*

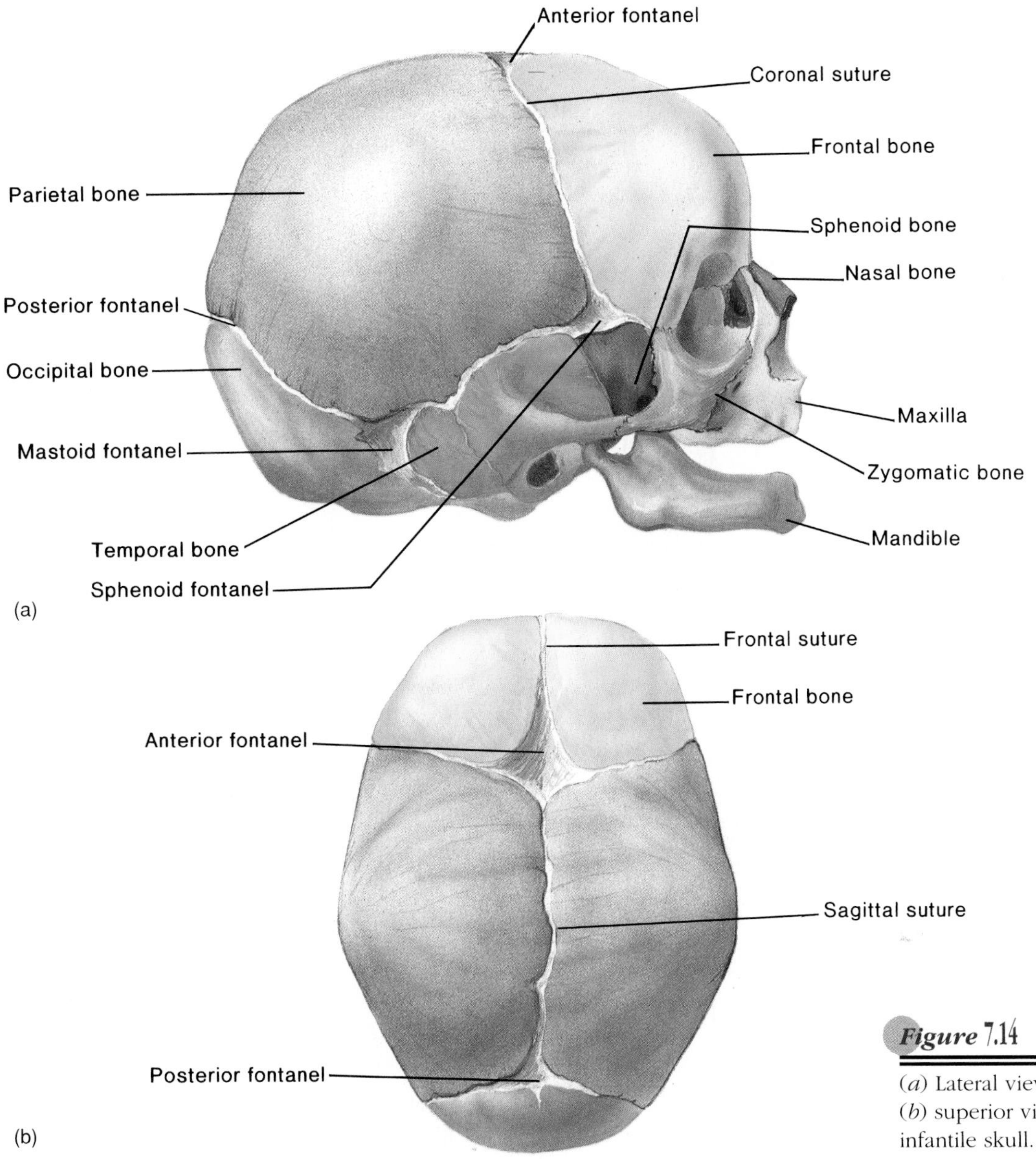

(a)

(b)

Figure 7.14

(*a*) Lateral view and (*b*) superior view of the infantile skull.

4. **Lacrimal bones** A lacrimal (lak′rĭ-mal) bone is a thin, scalelike structure located in the medial wall of each orbit between the ethmoid bone and the maxilla (see figs. 7.8 and 7.10).
5. **Nasal bones** The nasal (na′zal) bones are long, thin, and nearly rectangular (see figs. 7.8 and 7.10). They lie side by side and are fused at the midline, where they form the bridge of the nose.
6. **Vomer bone** The thin, flat vomer (vo′mer) bone is located along the midline within the nasal cavity (see figs. 7.8 and 7.13). Posteriorly, it joins the perpendicular plate of the ethmoid bone, and together, they form the nasal septum.
7. **Inferior nasal conchae** The inferior nasal conchae (kong′ke) are fragile, scroll-shaped bones attached to the lateral walls of the nasal cavity (see figs. 7.8 and 7.13). Like the ethmoidal conchae, the inferior conchae support mucous membranes within the nasal cavity.
8. **Mandible** The mandible is a horizontal, horseshoe-shaped body with a flat portion projecting upward at each end (see figs. 7.8 and 7.10). This projection is divided into two processes—a posterior *mandibular condyle* and an anterior *coronoid process.* The mandibular condyles articulate with the mandibular fossae of the temporal bones, while the coronoid processes provide attachments for muscles used in chewing. Other large chewing muscles insert on the lateral surface of the mandible. A curved bar of bone on the superior border of the mandible, the *alveolar arch,* contains the hollow sockets (dental alveoli) that bear the lower teeth (fig. 7.13).

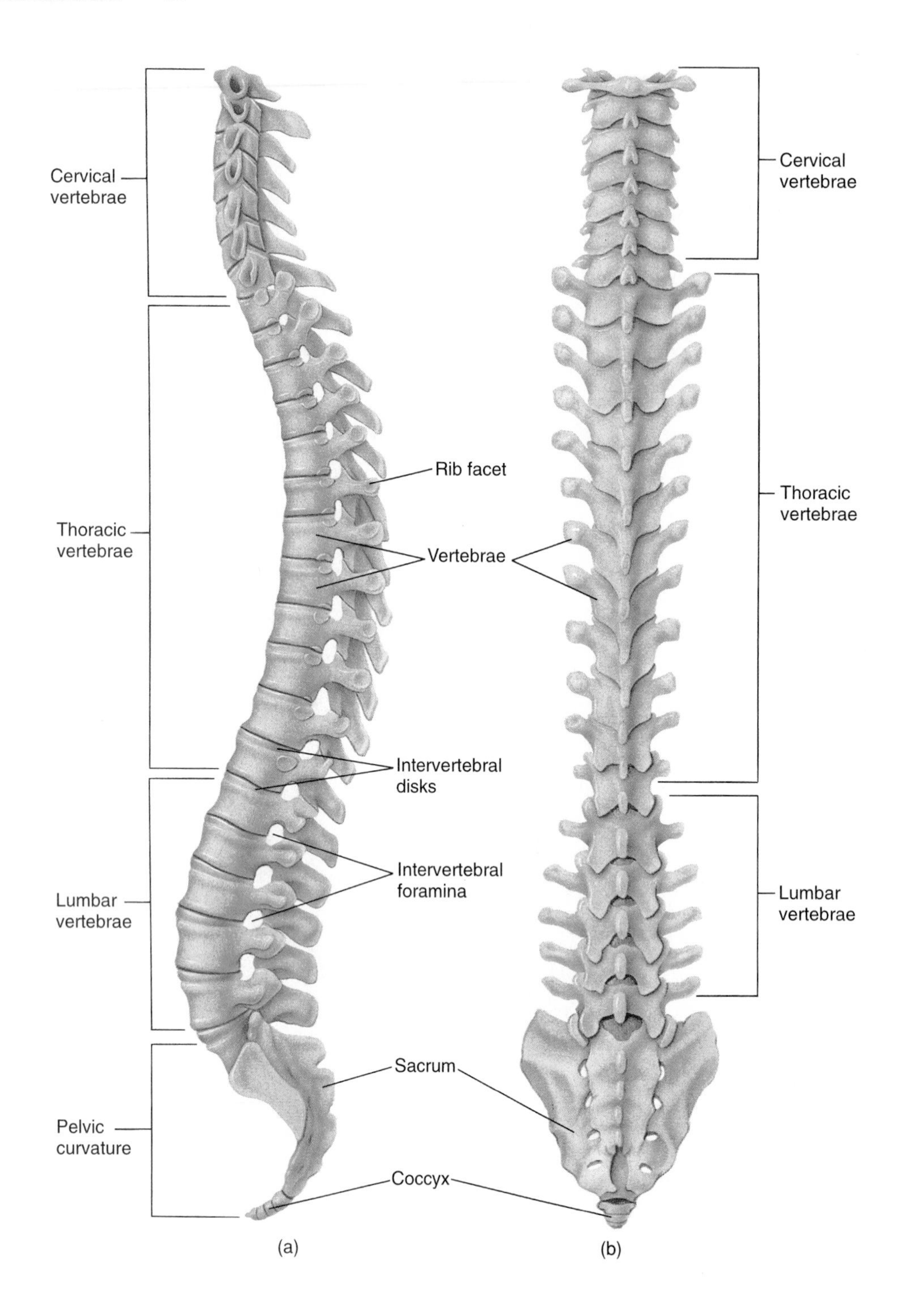

Figure 7.15

The curved vertebral column consists of many vertebrae, separated by intervertebral disks. (*a*) Left lateral view. (*b*) Posterior view.

Infantile Skull

At birth, the skull is incompletely developed, with fibrous membranes connecting the cranial bones. These membranous areas are called **fontanels** (fon″tah-nels) or, more commonly, soft spots (fig. 7.14). They permit some movement between the bones, so that the developing skull is partially compressible and can change shape slightly. This enables an infant's skull to pass more easily through the birth canal. Eventually, the fontanels close as the cranial bones grow together.

Other characteristics of an infantile skull include a relatively small face with a prominent forehead and large orbits. The jaw and nasal cavity are small, the paranasal sinuses are incompletely formed, and the frontal bone is in two parts. The skull bones are thin but also somewhat flexible and thus fracture less easily than do adult bones.

1. Locate and name each of the bones of the cranium.
2. Locate and name each of the facial bones.
3. Explain how an adult skull differs from that of an infant.

Vertebral Column

The **vertebral column** extends from the skull to the pelvis and forms the vertical axis of the skeleton. It is composed of many bony parts called **vertebrae** (ver′tĕ-brā) that are separated by masses of fibrocartilage called *intervertebral disks* and are connected to one another by ligaments (fig. 7.15). The vertebral column supports the head and trunk of the body. It also protects the spinal cord, which passes through a *vertebral canal* formed by openings in the vertebrae.

A Typical Vertebra

Vertebrae in different regions of the vertebral column have special characteristics, but they also have features in common. A typical vertebra has a drum-shaped *body,* which forms the thick, anterior portion of the bone (fig. 7.16). A longitudinal row of these vertebral bodies supports the weight of the head and trunk. The intervertebral disks, which separate vertebral bodies, cushion and soften the forces from movements such as walking and jumping.

Projecting posteriorly from each vertebral body are two short stalks called *pedicles* (ped′ĭ-k'lz). Two plates called *laminae* (lam'i-ne) arise from the pedicles and fuse in the back to become a *spinous process.* The pedicles, laminae, and spinous process together complete a bony *vertebral arch* around a *vertebral foramen,* through which the spinal cord passes.

If the laminae of the vertebrae fail to unite during development, the vertebral arch remains incomplete, resulting in a condition called *spina bifida.* Spina bifida may cause the contents of the vertebral canal to protrude outward. This problem occurs most frequently in the lumbosacral region. ■

Between the pedicles and laminae of a typical vertebra is a *transverse process,* which projects laterally and posteriorly. Ligaments and muscles attach to the dorsal spinous process and the transverse processes. Projecting upward and downward from each vertebral arch are *superior* and *inferior articulating processes.* These processes bear cartilage-covered facets that join each vertebra to the ones above and below it.

On the lower surfaces of the vertebral pedicles are notches that align to form openings called *intervertebral foramina* (in″ter-ver′tĕ-bral fo-ram′ĭ-nah). These openings provide passageways for spinal nerves that proceed between vertebrae and connect to the spinal cord (see fig. 7.15).

Cervical Vertebrae

Seven **cervical vertebrae** comprise the bony axis of the neck (see fig. 7.15). The transverse processes of these vertebrae are distinctive because they have *transverse foramina,* which are passageways for arteries leading to the brain (fig. 7.17). Also, the spinous processes of the second through the fifth cervical vertebrae are uniquely forked (bifid). These processes provide attachments for muscles.

Two cervical vertebrae are of special interest (fig. 7.17). The first vertebra, or **atlas** (at′las), supports the head. On its superior surface are two kidney-shaped *facets* that articulate with the occipital condyles.

The second cervical vertebra, or **axis** (ak′sis), bears a toothlike *dens* (odontoid process) on its body. This process projects upward and lies in the ring of the atlas. As the head turns from side to side, the atlas pivots around the dens.

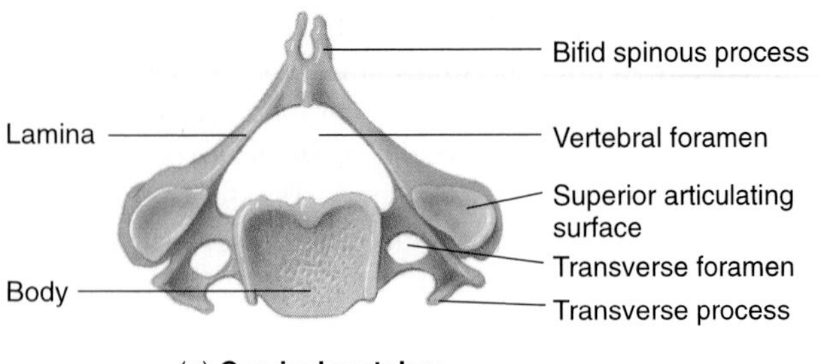

(a) **Cervical vertebra**

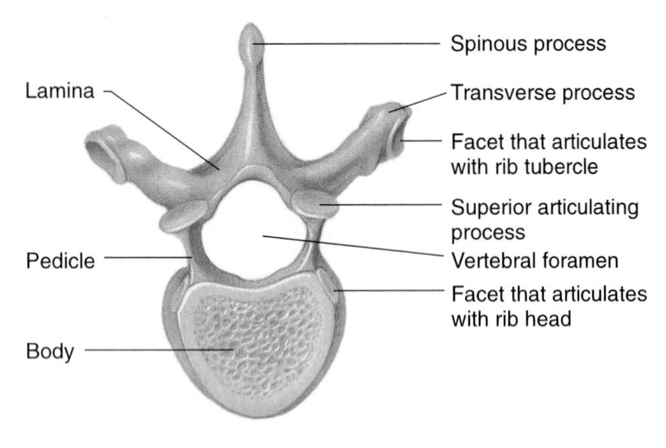

(b) **Thoracic vertebra**

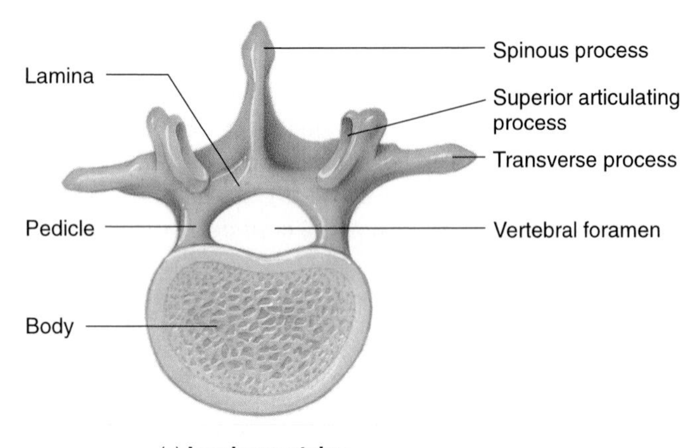

(c) **Lumbar vertebra**

Figure 7.16

Superior view of (*a*) a cervical vertebra, (*b*) a thoracic vertebra, and (*c*) a lumbar vertebra.

Thoracic Vertebrae

The twelve **thoracic vertebrae** are larger than the cervical vertebrae (see fig. 7.15). Each vertebra has a long, pointed spinous process, which slopes downward, and facets on the sides of its body, which articulate with a rib.

Beginning with the third thoracic vertebra and moving inferiorly, the bodies of these bones increase in size. Thus, they are adapted to bear increasing loads of body weight.

Lumbar Vertebrae

Five **lumbar vertebrae** are in the small of the back (loin) (see fig. 7.15). The lumbars are adapted with larger and stronger bodies to support more weight than the vertebrae above them.

Sacrum

The **sacrum** is a triangular structure, composed of five fused vertebrae, that forms the base of the vertebral column (fig. 7.18). The spinous processes of these fused bones form a ridge of *tubercles.* To the sides of the tubercles are rows of openings, the *dorsal sacral foramina,* through which nerves and blood vessels pass.

The vertebral foramina of the sacral vertebrae form the *sacral canal,* which continues through the sacrum to an opening of variable size at the tip, called the *sacral hiatus* (sa′kral hi-a′tus). On the sacrum's anterior surface, four pairs of *pelvic sacral foramina* provide passageways for nerves and blood vessels.

Coccyx

The **coccyx,** the lowest part of the vertebral column, is usually composed of four fused vertebrae (fig. 7.18). Ligaments attach it to the margins of the sacral hiatus.

Changes in the intervertebral disks can cause back problems. Each disk is composed of a tough, outer layer of fibrocartilage and an elastic central mass. As a person ages, the disks degenerate. The central masses lose their firmness, and the outer layers thin, weaken, and develop cracks. Extra pressure, such as falling or lifting, may break the outer layer of a disk and allow the central mass to squeeze out. Such a rupture may press on the spinal cord or on a spinal nerve that branches from it. This condition—a ruptured or herniated disk—may cause back pain and numbness or the loss of muscular function in areas that the affected spinal nerve innervates. ■

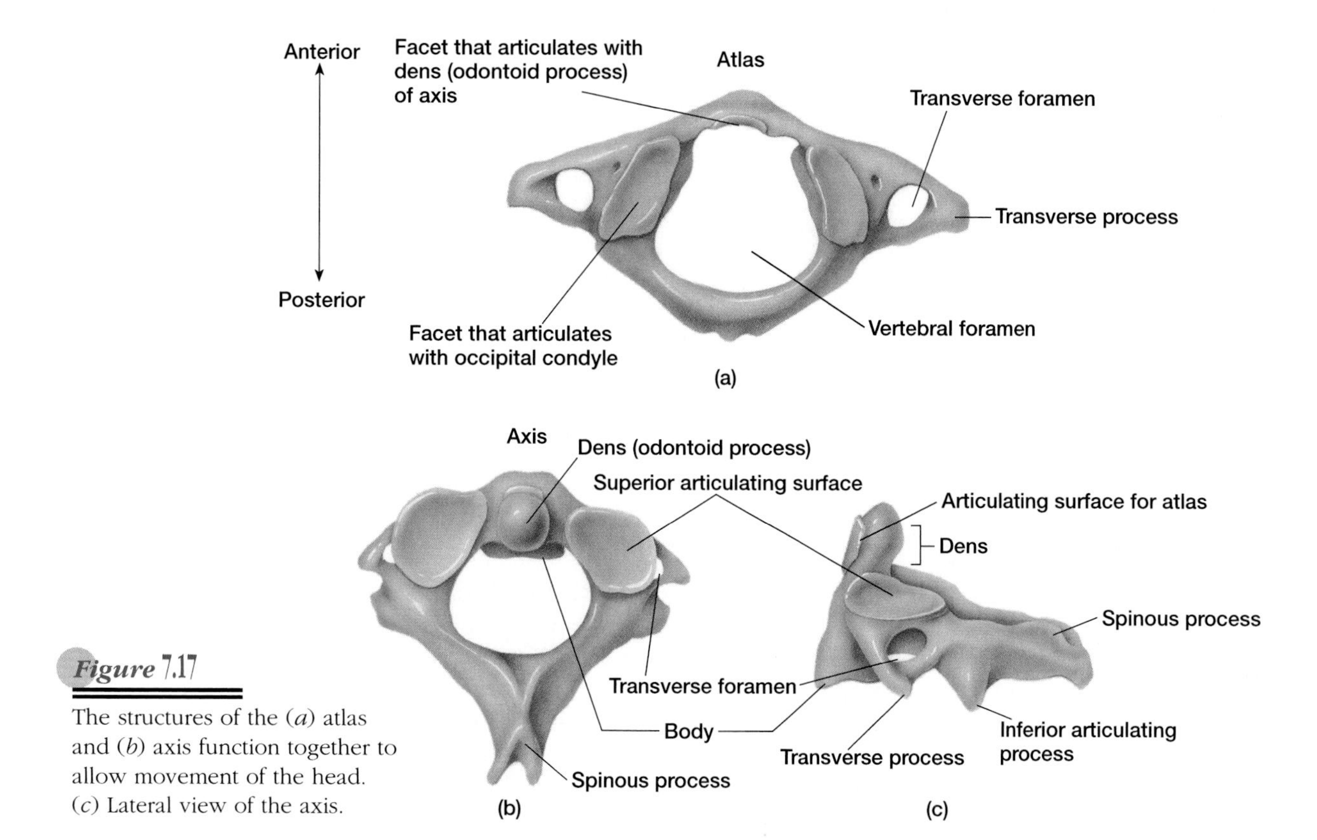

Figure 7.17

The structures of the (*a*) atlas and (*b*) axis function together to allow movement of the head. (*c*) Lateral view of the axis.

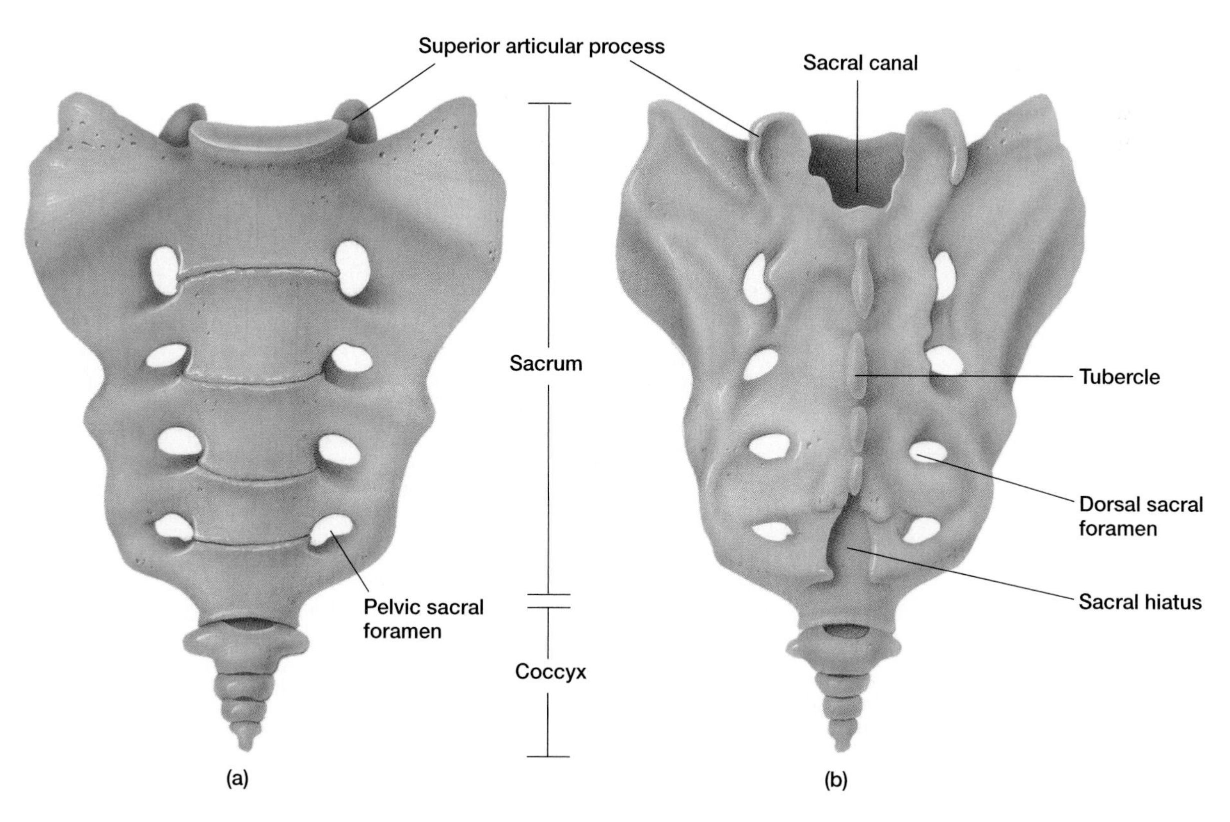

Figure 7.18

(*a*) Anterior view and (*b*) posterior view of the sacrum and coccyx.

1. Describe the structure of the vertebral column.
2. Describe a typical vertebra.
3. How do the structures of cervical, thoracic, and lumbar vertebrae differ?

Thoracic Cage

The **thoracic cage** includes the ribs, thoracic vertebrae, sternum, and costal cartilages that attach the ribs to the sternum (fig. 7.19). These parts support the shoulder girdle and upper limbs, protect the viscera in the thoracic and upper abdominal cavities, and play a role in breathing.

Ribs

Most people have twelve pairs of **ribs**—one pair attached to each of the twelve thoracic vertebrae. The first seven rib pairs, *true ribs* (vertebrosternal ribs), join the sternum directly by their costal cartilages. The remaining five pairs are called *false ribs* because their cartilages do not reach the sternum directly. Instead, the cartilages of the upper three false ribs (vertebrochondral ribs) join the cartilages of the seventh rib. The last two (or sometimes three) rib pairs are called *floating ribs* (vertebral ribs) because they have no cartilaginous attachments to the sternum.

A typical rib has a long, slender shaft, which curves around the chest and slopes downward. On the posterior end is an enlarged *head* by which the rib articulates with a *facet* on the body of its own vertebra and with the body of the next higher vertebra. A *tubercle,* close to the head of the rib, articulates with the transverse process of the vertebra.

Sternum

The **sternum** or breastbone is located along the midline in the anterior portion of the thoracic cage (fig. 7.19). This flat, elongated bone develops in three parts—an upper *manubrium* (mah-nu′bre-um), a middle *body,* and a lower *xiphoid* (zīf′oid) *process* that projects downward. The manubrium articulates with the clavicles by facets on its superior border.

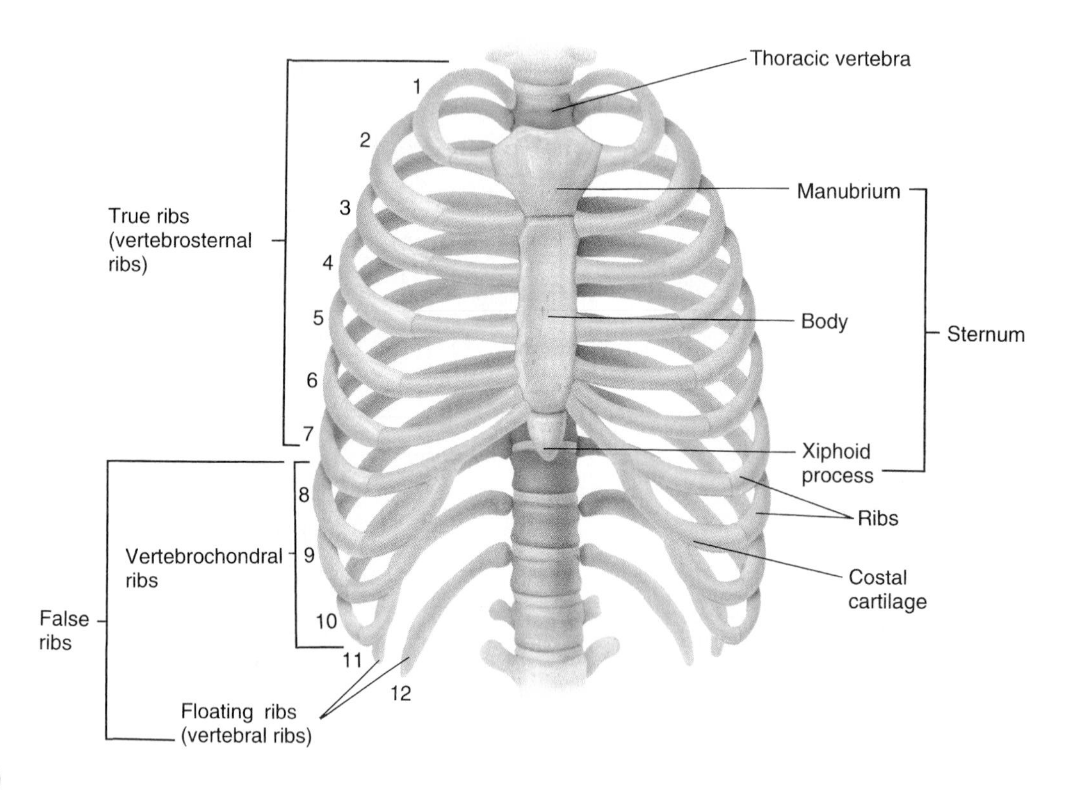

Figure 7.19

The thoracic cage includes the thoracic vertebrae, the sternum, the ribs, and the costal cartilages that attach the ribs to the sternum.

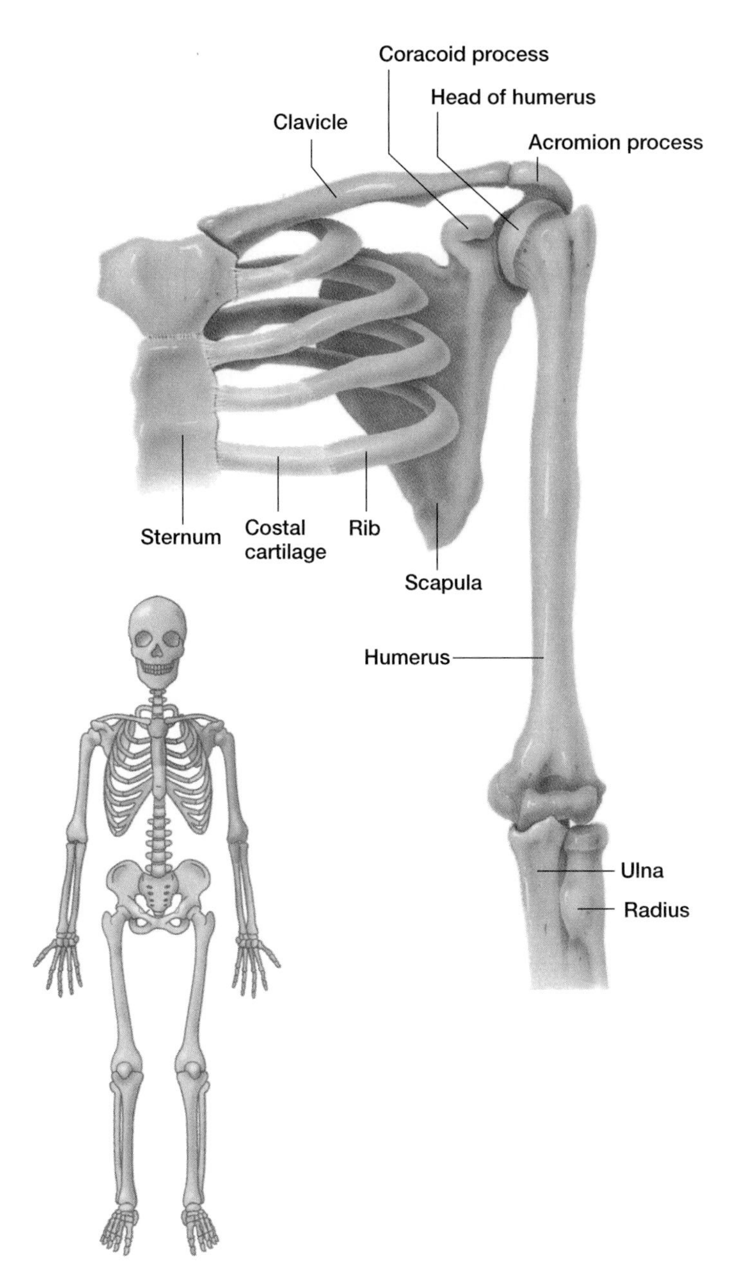

Figure 7.20

The pectoral girdle, to which the upper limbs attach, consists of a clavicle and a scapula on each side.

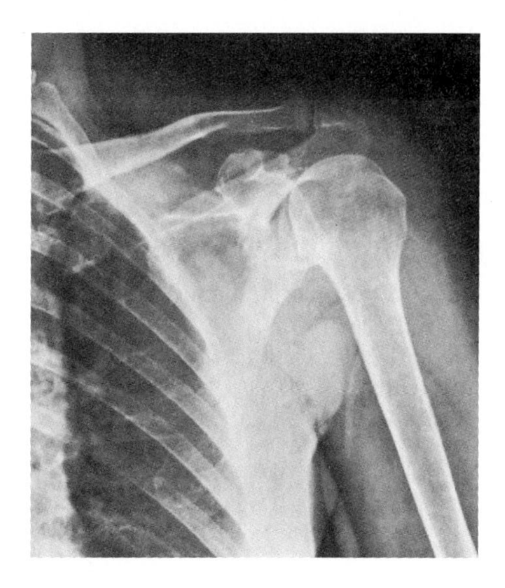

Figure 7.21

X-ray film of the left shoulder region, viewed from the front.

1. Which bones comprise the thoracic cage?
2. Describe a typical rib.
3. What are the differences among true, false, and floating ribs?

Red marrow within the spongy bone of the sternum produces blood cells into adulthood. Since the sternum has a thin covering of compact bone and is easy to reach, samples of its marrow may be removed to diagnose diseases. This procedure, a *sternal puncture,* suctions (aspirates) some marrow through a hollow needle. (Marrow may also be removed from the iliac crest of a coxal bone.) ■

Pectoral Girdle

The **pectoral** (pek′to-ral) **girdle** or shoulder girdle is composed of four parts—two clavicles and two scapulae (figs. 7.20 and 7.21). Although the word *girdle* suggests a ring-shaped structure, the pectoral girdle is an incomplete ring. It is open in the back between the scapulae, and the sternum separates its bones in front. The pectoral girdle supports the upper limbs and is an attachment for several muscles that move them.

Clavicles

The **clavicles** or collarbones are slender, rodlike bones with elongated S-shapes (see fig. 7.20). Located at the base of the neck, they run horizontally between the manubrium and scapulae.

The clavicles brace the freely movable scapulae, helping to hold the shoulders in place. They also provide attachments for muscles of the upper limbs, chest, and back.

Scapulae

The **scapulae** (skap′u-le) or shoulder blades are broad, somewhat triangular bones located on either side of the upper back (figs. 7.20 and 7.22). A *spine* divides the posterior surface of each scapula into unequal portions. This spine leads to two processes—an *acromion* (ah-kro′me-on) *process* that forms the tip of the shoulder and a *coracoid* (kor′ah-koid) *process* that curves anteriorly and inferiorly to the clavicle. The acromion process articulates with the clavicle and provides attachments for muscles of the upper limb and chest. The coracoid process also provides attachments for upper limb and chest muscles. Between the processes is a depression called the *glenoid cavity* that articulates with the head of the arm bone (humerus).

1. Which bones form the pectoral girdle?
2. What is the function of the pectoral girdle?

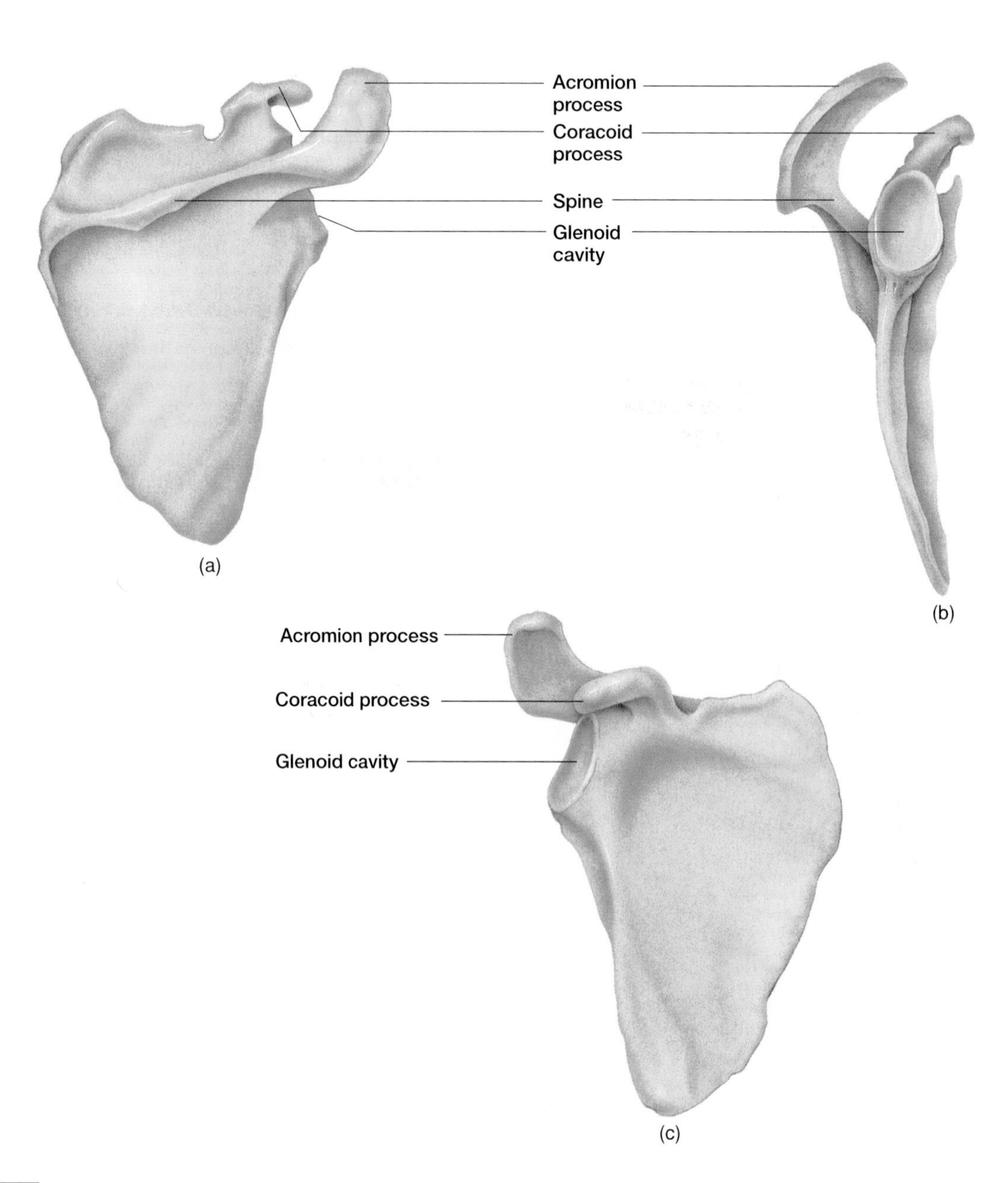

Figure 7.22

(*a*) Posterior surface of the right scapula. (*b*) Lateral view showing the glenoid cavity that articulates with the head of the humerus. (*c*) Anterior surface.

Upper Limb

The bones of the upper limb form the framework of the arm, forearm, and hand. They also provide attachments for muscles, and they function in levers that move limb parts. These bones include a humerus, a radius, an ulna, carpals, metacarpals, and phalanges (see fig. 7.7).

Humerus

The **humerus** is a heavy bone that extends from the scapula to the elbow (fig. 7.23). At its upper end is a smooth, rounded *head* that fits into the glenoid cavity of the scapula. Just below the head are two processes—a *greater tubercle* on the lateral side and a *lesser tubercle* on the anterior side. These tubercles provide attachments for muscles that move the upper limb at the shoulder. Between them is a narrow furrow, the *intertubercular groove*.

The narrow depression along the lower margin of the humerus head separates it from the tubercles and is called the *anatomical neck*. Just below the head and tubercles is a tapering region called the *surgical neck*, so named because fractures are common there. Near the middle of the bony shaft on the lateral side is a rough V-shaped area called the *deltoid tuberosity*. It provides an attachment for the muscle (deltoid) that raises the upper limb horizontally to the side.

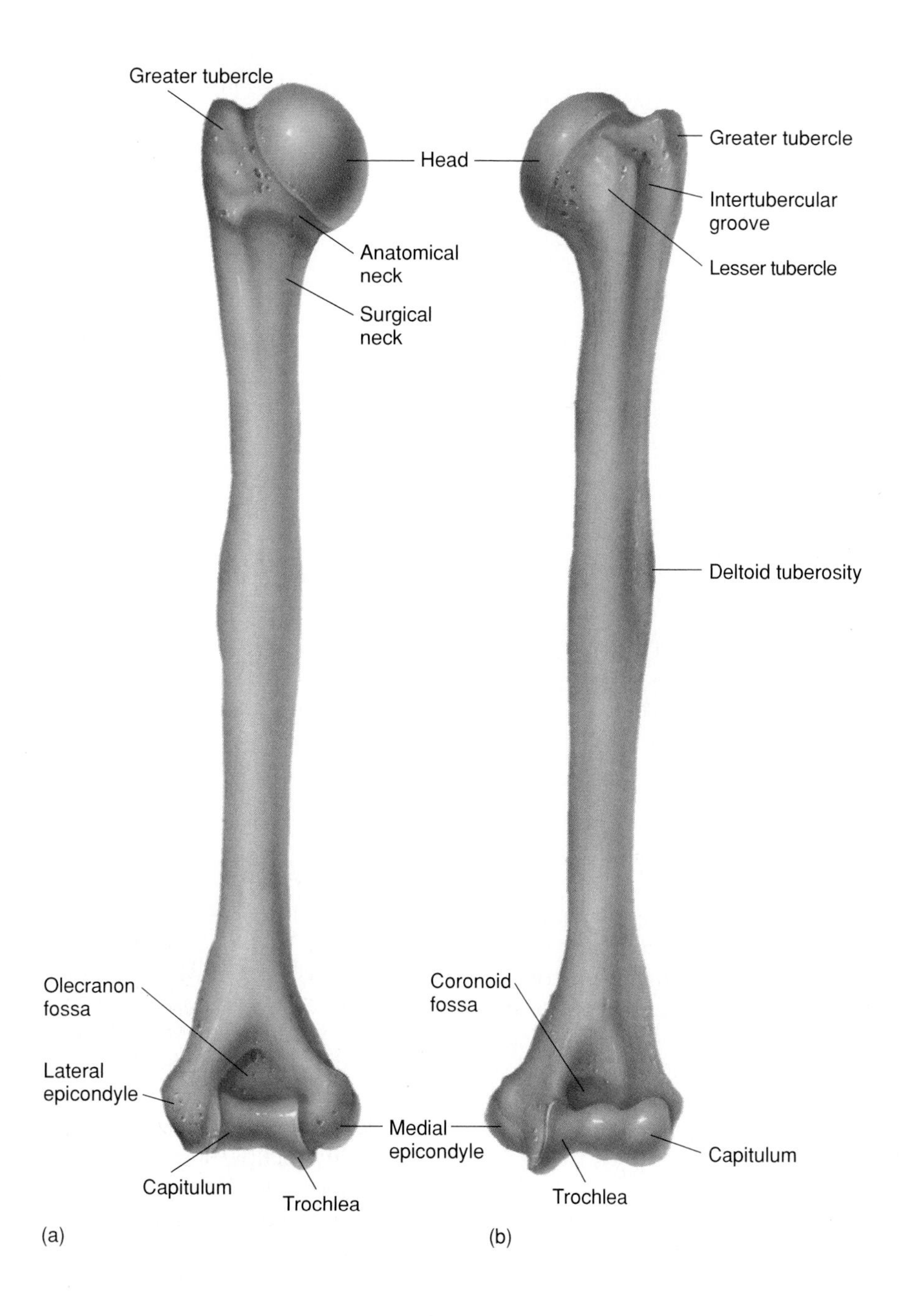

Figure 7.23

(*a*) Posterior and (*b*) anterior surfaces of the left humerus.

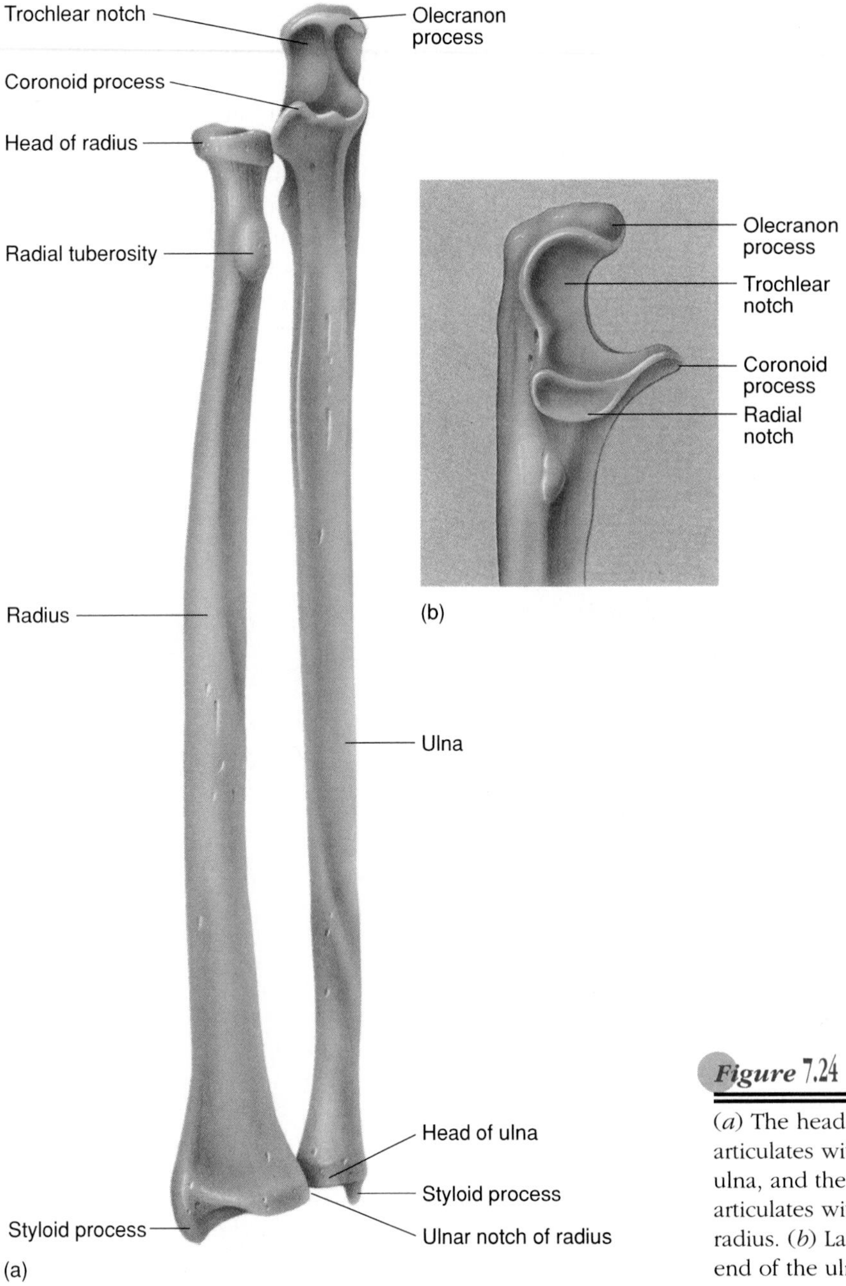

Figure 7.24

(*a*) The head of the right radius articulates with the radial notch of the ulna, and the head of the ulna articulates with the ulnar notch of the radius. (*b*) Lateral view of the proximal end of the ulna.

At the lower end of the humerus are two smooth *condyles* (a lateral *capitulum* and a medial *trochlea*) that articulate with the radius on the lateral side and the ulna on the medial side. Above the condyles on either side are *epicondyles,* which provide attachments for muscles and ligaments of the elbow. Between the epicondyles anteriorly is a depression, the *coronoid* (kor′o-noid) *fossa,* that receives a process of the ulna (coronoid process) when the elbow bends. Another depression on the posterior surface, the *olecranon* (o″lek′ra-non) *fossa,* receives an ulnar process (olecranon process) when the upper limb straightens at the elbow.

Radius

The **radius**, located on the thumb side of the forearm, extends from the elbow to the wrist and crosses over the ulna when the hand turns so that the palm faces backward (fig. 7.24). A thick, disklike *head* at the upper end of the radius articulates with the humerus and a notch of the ulna (radial notch). This organization allows the radius to rotate freely.

On the radial shaft just below the head is a process called the *radial tuberosity.* It is an attachment for a muscle (biceps brachii) that bends the upper limb at the elbow. At the distal end of the radius, a lateral *styloid* (sti′loid) *process* provides attachments for wrist ligaments.

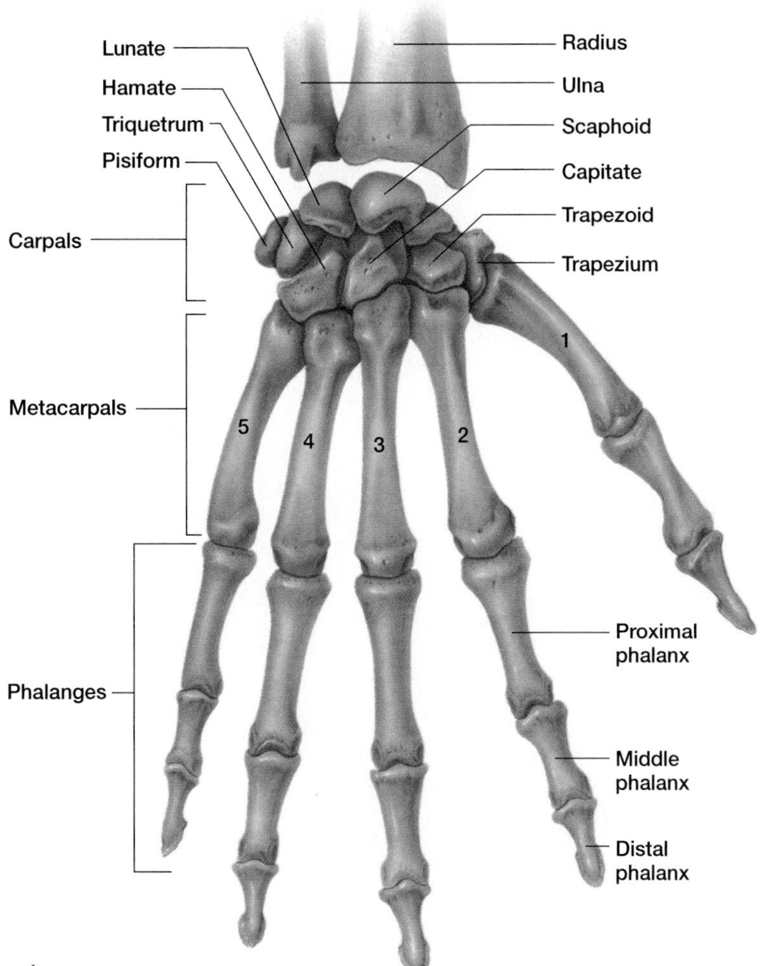

Figure 7.25

Posterior view of the right hand.

Ulna

The **ulna** is longer than the radius and overlaps the end of the humerus posteriorly (fig. 7.24). At its proximal end, the ulna has a wrenchlike opening, the *trochlear* (trok′le-ar) *notch,* that articulates with the humerus. Two processes on either side of this notch, the *olecranon process* and the *coronoid process,* provide attachments for muscles.

At the distal end of the ulna, its knoblike *head* articulates with a notch of the radius (ulnar notch) laterally and with a disk of fibrocartilage inferiorly. This disk, in turn, joins a wrist bone (triangular). A medial *styloid process* at the distal end of the ulna provides attachments for wrist ligaments.

Hand

The hand is composed of a wrist, a palm, and five fingers (fig. 7.25). The skeleton of the wrist consists of eight small **carpal bones** that are firmly bound in two rows of four bones each. The resulting compact mass is called a *carpus* (kar′pus). The carpus articulates with the radius and with the fibrocartilaginous disk on the ulnar side. Its distal surface articulates with the metacarpal bones. Figure 7.25 names the individual bones of the carpus.

Five **metacarpal bones,** one in line with each finger, form the framework of the palm. These bones are cylindrical, with rounded distal ends that form the knuckles of a clenched fist. They are numbered 1–5, beginning with the metacarpal of the thumb (fig. 7.25). The metacarpals articulate proximally with the carpals and distally with the phalanges.

The **phalanges** are the finger bones. Each finger has three phalanges—a proximal, a middle, and a distal phalanx—except the thumb, which has two (it lacks a middle phalanx).

1. Locate and name each of the bones of the upper limb.
2. Explain how the bones of the upper limb articulate with one another.

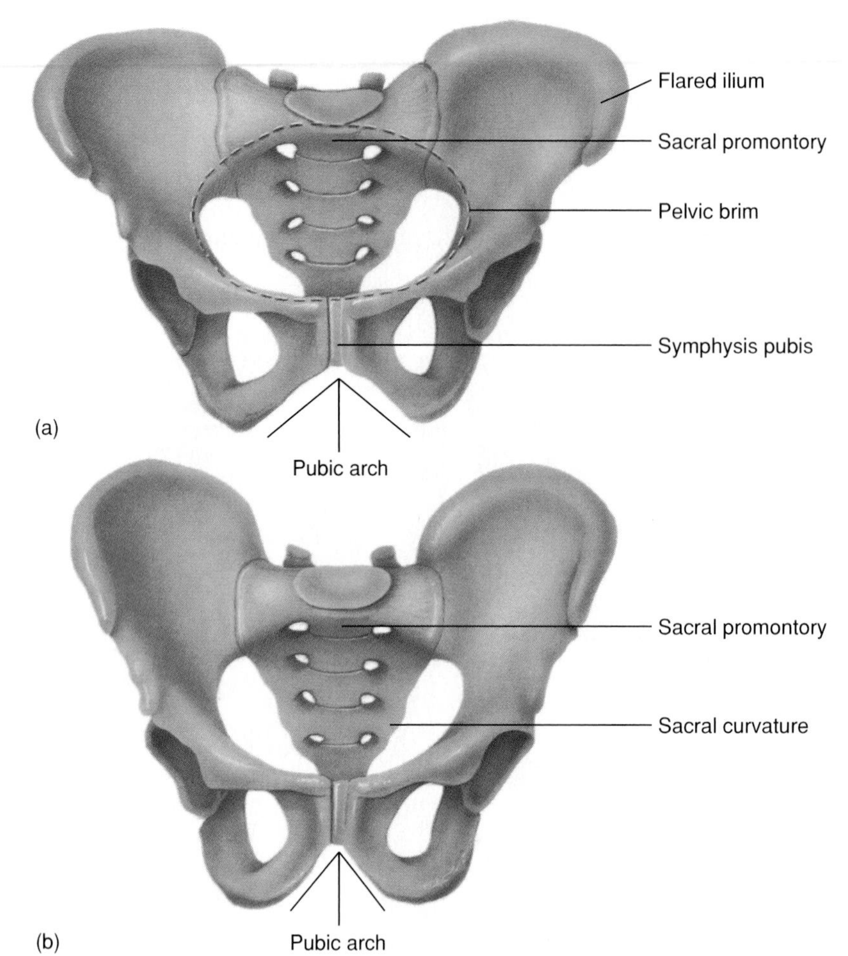

Figure 7.26

(*a*) Female pelvis. (*b*) Male pelvis. The female pelvis is usually wider in all diameters and roomier than that of the male.

Pelvic Girdle

The **pelvic girdle** consists of two coxal bones that articulate with each other anteriorly and with the sacrum posteriorly. The sacrum, coccyx, and pelvic girdle together form the bowl-shaped **pelvis** (fig. 7.26). The pelvic girdle supports the trunk of the body, provides attachments for the lower limbs, and protects the urinary bladder, the distal end of the large intestine, and the internal reproductive organs.

Each **coxal bone** (os coxa) develops from three parts—an ilium, an ischium, and a pubis (fig. 7.27). These parts fuse in the region of a cup-shaped cavity called the *acetabulum* (as″ĕ-tab′u-lum). This depression, on the lateral surface of the hipbone, receives the rounded head of the femur (thighbone).

The **ilium** (il′e-um), which is the largest and uppermost portion of the coxal bone, flares outward, forming the prominence of the hip. The margin of this prominence is called the *iliac crest.*

Posteriorly, the ilium joins the sacrum at the *sacroiliac* (sa″kro-il′e-ak) *joint.* A projection of the ilium, the *anterior superior iliac spine,* can be felt lateral to the groin and provides attachments for ligaments and muscles.

The **ischium** (is′ke-um), which forms the lowest portion of the coxal bone, is L-shaped, with its angle, the *ischial tuberosity,* pointing posteriorly and downward. This tuberosity has a rough surface that provides attachments for ligaments and lower limb muscles. It also supports the weight of the body during sitting. Above the ischial tuberosity, near the junction of the ilium and ischium, is a sharp projection called the *ischial spine.*

The **pubis** (pu′bis) constitutes the anterior portion of the coxal bone. The two pubic bones join at the midline, forming a joint called the *symphysis pubis* (sim′fi-sis pu′bis). The angle these bones form below the symphysis is the *pubic arch* (see fig. 7.26).

A portion of each pubis passes posteriorly and downward to join an ischium. Between the bodies of these bones on either side is a large opening, the *obturator foramen,* which is the largest foramen in the skeleton (figs. 7.27 and 7.28).

If a line were drawn along each side of the pelvis, from the sacral promontory downward and anteriorly to the upper margin of the symphysis pubis, it would mark the *pelvic brim* (linea terminalis) (see fig. 7.26). This margin separates the lower, or lesser (true), pelvis from the upper, or greater (false), pelvis.

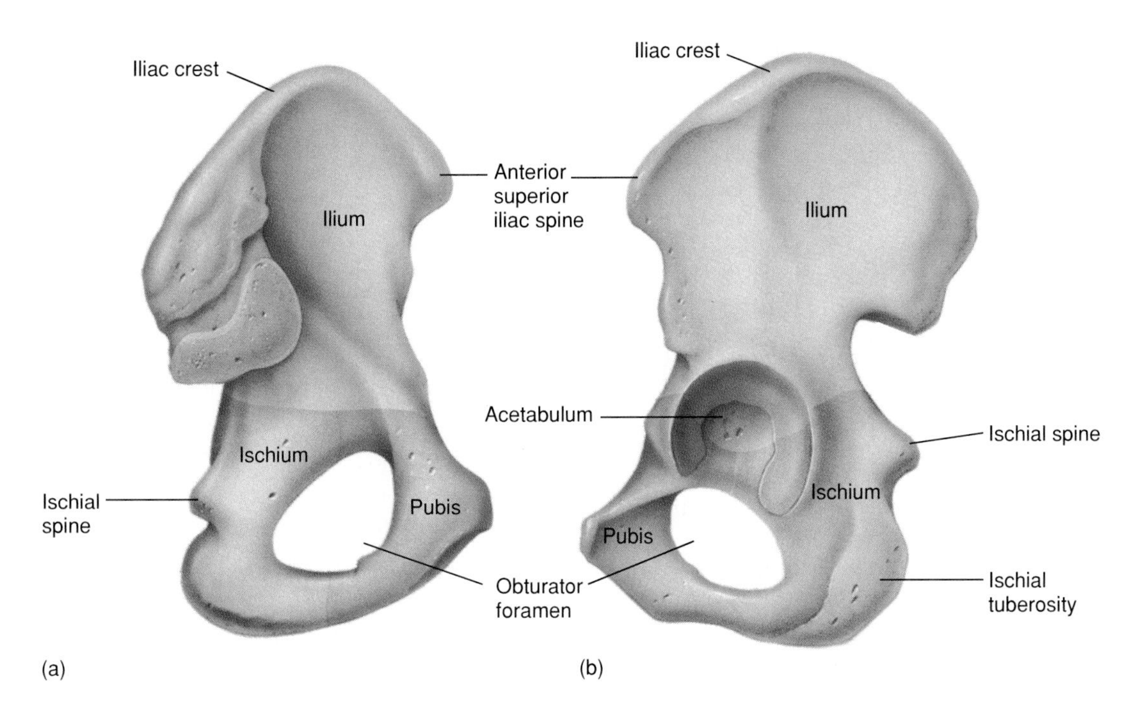

Figure 7.27

(*a*) Medial surface of the left coxal bone. (*b*) Lateral view.

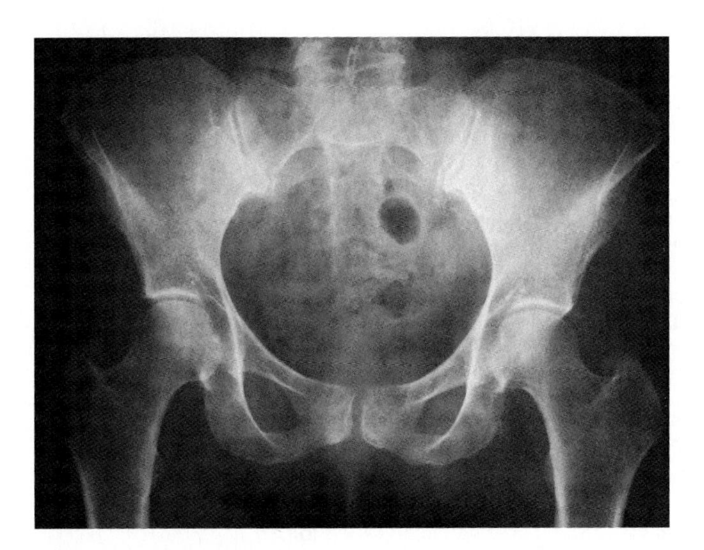

Figure 7.28

X-ray film of the pelvic girdle.

Table 7.3 summarizes some differences in the female and male pelves and other skeletal structures.

1. Locate and name each bone that forms the pelvis.
2. Name the bones that fuse to form a coxal bone.

Table 7.3 Differences between the Female and Male Skeletons

Part	Differences
Skull	Female skull is smaller and lighter, with less conspicuous muscular attachments. Female facial area is rounder, jaw is smaller, and mastoid process is less prominent than those of a male.
Pelvic girdle	Female coxal bones are lighter, thinner, and have less obvious muscular attachments. The obturator foramina and acetabula are smaller and farther apart than those of a male.
Pelvic cavity	Female pelvic cavity is wider in all diameters and is shorter, roomier, and less funnel-shaped. The distances between the ischial spines and ischial tuberosities are greater than in a male.
Sacrum	Female sacrum is wider, the first sacral vertebra projects forward to a lesser degree, and sacral curvature is bent more sharply posteriorly than in a male.
Coccyx	Female coccyx is more movable than that of a male.

Lower Limb

Bones of the lower limb form the frameworks of the thigh, leg, and foot. They include a femur, a tibia, a fibula, tarsals, metatarsals, and phalanges (see fig. 7.7).

Femur

The **femur** is the longest bone in the body and extends from the hip to the knee (fig. 7.29). A large, rounded *head* at its proximal end projects medially into the acetabulum of the coxal bone. On the head, a pit called the *fovea capitis* marks the attachment of a ligament (ligamentum capitis). Just below the head are a constriction, or *neck,* and two large processes—a superior, lateral *greater trochanter* and an inferior, medial *lesser trochanter.* These processes provide attachments for muscles of the lower limbs and buttocks.

At the distal end of the femur, two rounded processes, the *lateral* and *medial condyles,* articulate with the tibia of the leg. A **patella**, or kneecap, also articulates with the femur on its distal anterior surface (see fig. 7.7). It is located in a tendon that passes anteriorly over the knee.

Hip fracture is one of the more serious causes of hospitalization among elderly persons. The site of hip fracture is most commonly the neck of a femur or the region between the trochanters of a femur. Falls often cause hip fracture, especially in people who have osteoporosis. ■

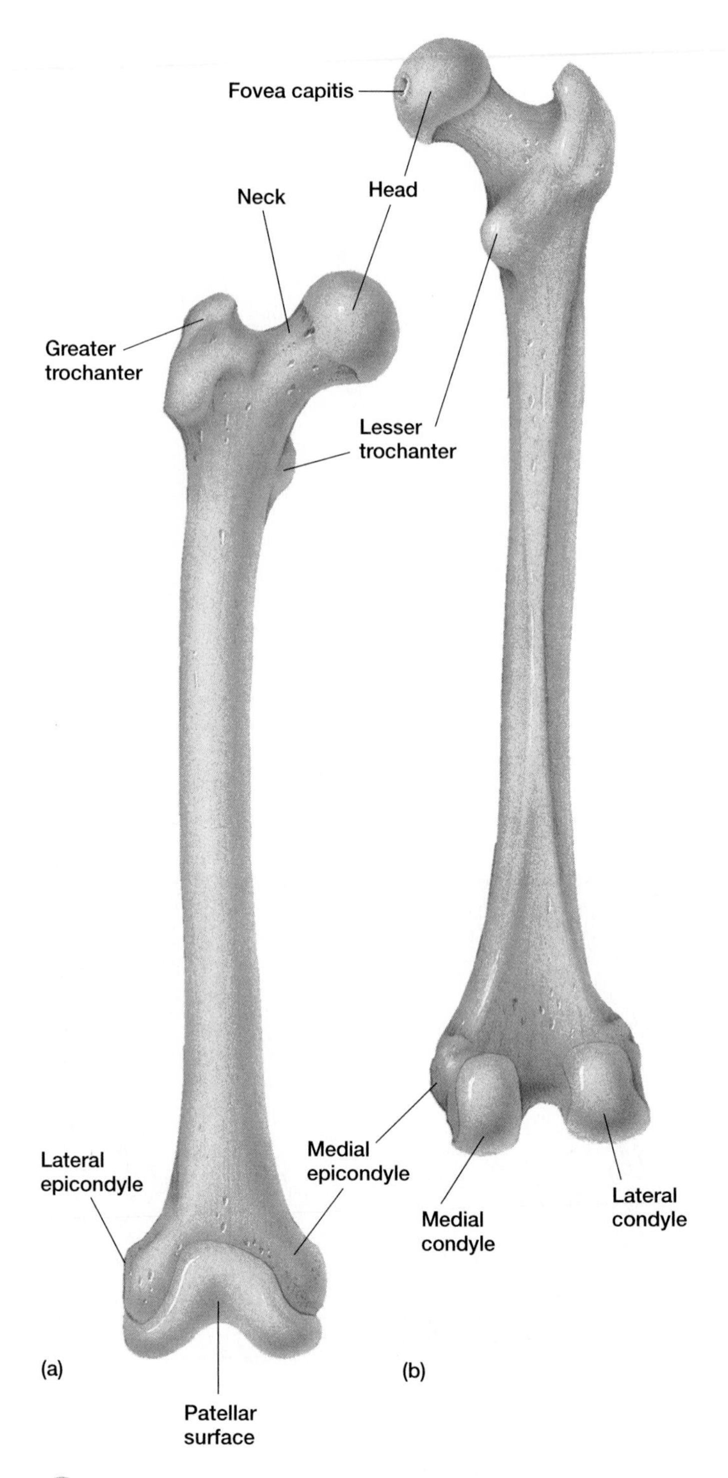

Figure 7.29

(*a*) Anterior and (*b*) posterior surfaces of the right femur.

Tibia

The **tibia** or shinbone is the larger of the two leg bones and is located on the medial side (fig. 7.30). Its proximal end expands into the *medial* and *lateral condyles,* which have concave surfaces and articulate with the condyles of the femur. Below the condyles, on the anterior surface, is a process called the *tibial tuberosity,* which provides an attachment for the *patellar ligament*—a continuation of the patella-bearing tendon.

At its distal end, the tibia expands to form a prominence on the inner ankle called the *medial malleolus* (mah-le′o-lus), which is an attachment for ligaments. On its lateral side is a depression that articulates with the fibula. The inferior surface of the tibia's distal end articulates with a large bone (the talus) in the foot.

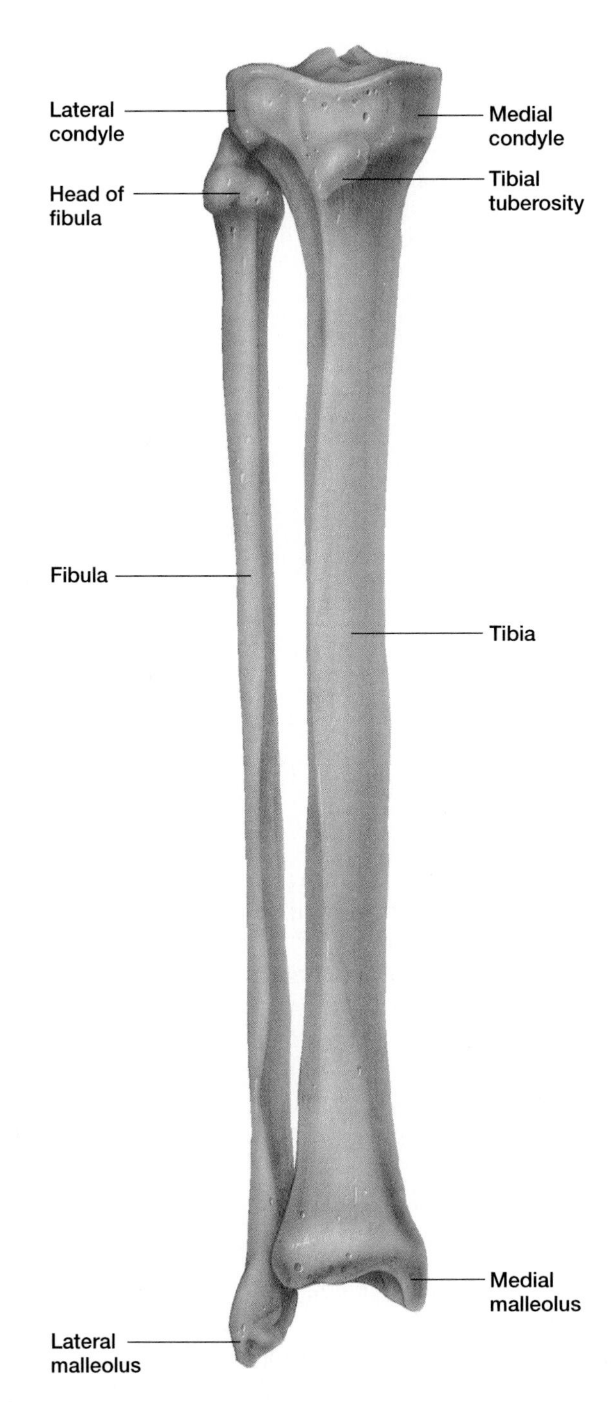

Figure 7.30

Bones of the right leg, viewed from the front.

Fibula

The **fibula** is a long, slender bone located on the lateral side of the tibia (fig. 7.30). Its ends slightly enlarge into a proximal *head* and a distal *lateral malleolus*. The head articulates with the tibia just below the lateral condyle; however, it does not enter into the knee joint and does not bear any body weight. The lateral malleolus articulates with the ankle and protrudes on the lateral side.

Foot

The foot consists of an ankle, an instep, and five toes. The ankle is composed of seven **tarsal bones**, forming a group called the *tarsus* (tahr′sus) (figs. 7.31 and 7.32). These bones are organized so that one of them, the **talus** (ta′lus), can move freely where it joins the tibia and fibula. The remaining tarsal bones are bound firmly together, forming a mass supporting the talus. Figure 7.32 names the individual bones of the tarsus.

The largest of the ankle bones, the **calcaneus** (kal-ka′ne-us), or heel bone, is located below the talus, where it projects backward to form the base of the heel. The calcaneus helps support body weight and provides an attachment for muscles that move the foot.

The instep consists of five, elongated **metatarsal bones**, which articulate with the tarsus. They are numbered 1–5, beginning on the medial side (fig. 7.32). The heads at the distal ends of these bones form the ball of the foot. The tarsals and metatarsals are organized and bound by ligaments, forming the arches of the foot. A longitudinal arch extends from the heel to the toe, and a transverse arch stretches across the foot. These arches provide a stable, springy base for the body.

The **phalanges** of the toes are similar to those of the fingers and align and articulate with the metatarsals. Each toe has three phalanges—a proximal, a middle, and a distal phalanx—except the great toe, which lacks a middle phalanx.

1. Locate and name each of the bones of the lower limb.
2. Explain how the bones of the lower limb articulate with one another.
3. Describe how the foot is adapted to support the body.

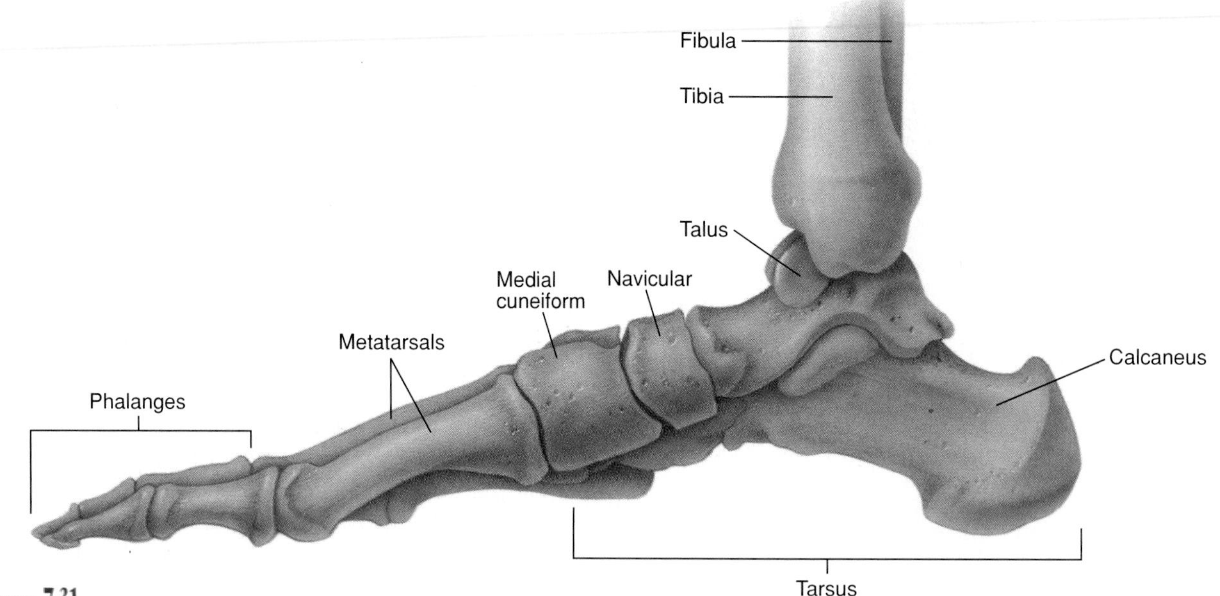

Figure 7.31

The talus moves freely where it articulates with the tibia and fibula.

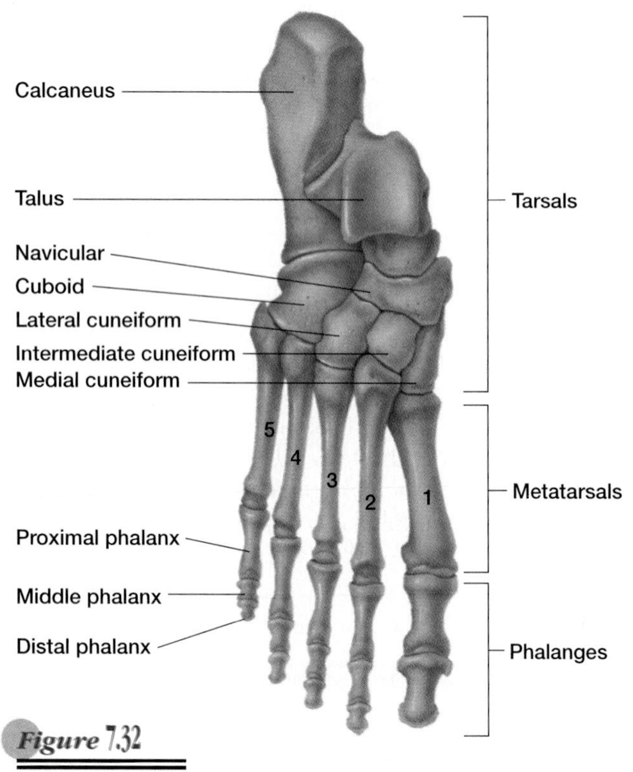

Figure 7.32

The right foot, viewed superiorly.

Joints

Joints (articulations) are functional junctions between bones. They vary considerably in structure and function. If classified according to the degree of movement they make possible, joints can be immovable, slightly movable, or freely movable. Joints also can be grouped by the type of tissue (fibrous, cartilaginous, or synovial) that binds the bones together at each junction. Currently, structural classification by tissue type is more commonly used.

Fibrous Joints

Fibrous joints lie between bones that closely contact one another. A thin layer of fibrous connective tissue joins the bones at such joints, as in the case of a *suture* between a pair of flat bones of the skull (fig. 7.33). No appreciable movement takes place at a fibrous joint. Some fibrous joints, such as the joint in the leg between the distal ends of the tibia and fibula, have limited movement.

Cartilaginous Joints

Disks of fibrocartilage or hyaline cartilage connect the bones of **cartilaginous joints.** For example, joints of this type separate the vertebrae of the vertebral column. Each intervertebral disk is composed of a band of fibrous fibrocartilage (annulus fibrosus) surrounding a pulpy or gelatinous core (nucleus pulposus). The disk absorbs shocks and helps equalize pressures between adjacent vertebrae when the body moves (see fig. 7.15).

Due to the slight flexibility of the disks, cartilaginous joints allow limited movement, as when the back is bent forward or to the side or is twisted. Other examples of cartilaginous joints include the symphysis pubis and the first rib with the sternum.

Synovial Joints

Most joints within the skeletal system are **synovial joints** and allow free movement. They are more complex structurally than fibrous or cartilaginous joints.

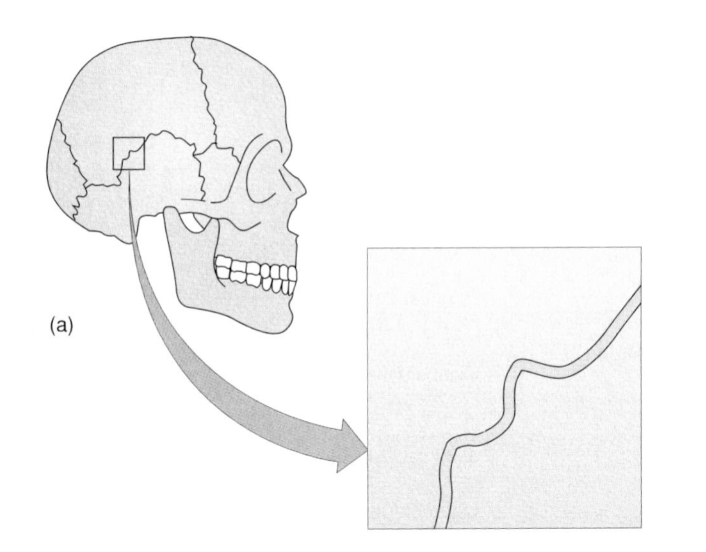

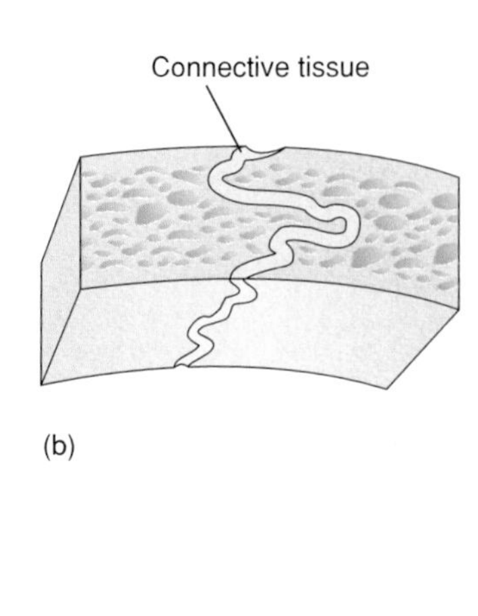

Figure 7.33

(*a*) The fibrous joints between the bones of the skull are immovable and are called sutures.
(*b*) A thin layer of connective tissue connects the bones at the suture.

The articular ends of the bones in a synovial joint are covered with hyaline cartilage (articular cartilage), and a surrounding, tubular capsule of dense fibrous tissue holds them together (fig. 7.34). This *joint capsule* is composed of an outer layer of ligaments and an inner lining of *synovial membrane,* which secretes synovial fluid. Synovial fluid, with a consistency similar to uncooked egg white, lubricates joints.

Some synovial joints have flattened, shock-absorbing pads of fibrocartilage called **menisci** (singular, *meniscus*) between the articulating surfaces of the bones (fig. 7.35). Such joints may also have fluid-filled sacs called **bursae** associated with them. Each bursa is lined with synovial membrane, which may be continuous with the synovial membrane of a nearby joint cavity.

Bursae are commonly located between the skin and underlying bony prominences, as in the case of the patella of the knee or the olecranon process of the elbow. They aid movement of tendons that pass over these bony parts or over other tendons. Figures 7.35 and 7.36 show and name some of the bursae associated with the knee and shoulder.

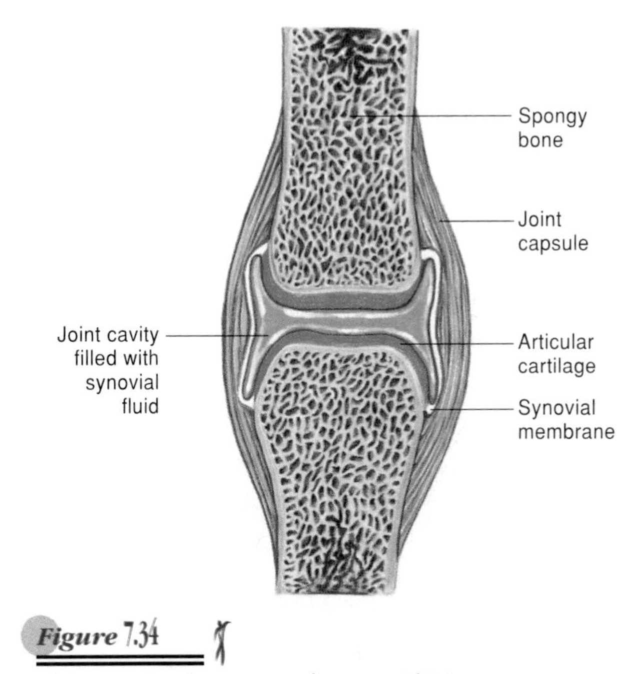

Figure 7.34

The generalized structure of a synovial joint.

Based on the shapes of their parts and the movements they permit, synovial joints are classified as follows:

1. A **ball-and-socket joint** consists of a bone with a ball-shaped head that articulates with a cup-shaped cavity of another bone. Such a joint allows a wider range of motion than does any other kind, permitting movements in all planes, as well as rotational movement around a central axis. The shoulder and hip contain joints of this type (figs. 7.36 and 7.37).
2. In a **condyloid joint,** an oval-shaped condyle of one bone fits into an elliptical cavity of another bone, as in the joints between the metacarpals and phalanges (see fig. 7.25). This type of joint permits a variety of movements in different planes; rotational movement, however, is not possible.
3. The articulating surfaces of **gliding joints** are nearly flat or slightly curved. Most of the joints within the wrist (see fig. 7.25) and ankle, as well as those between the articular processes of adjacent vertebrae, belong to this group. They allow sliding and twisting movements. The sacroiliac joint and the joints formed by ribs 2–7 connecting with the sternum are also gliding joints.

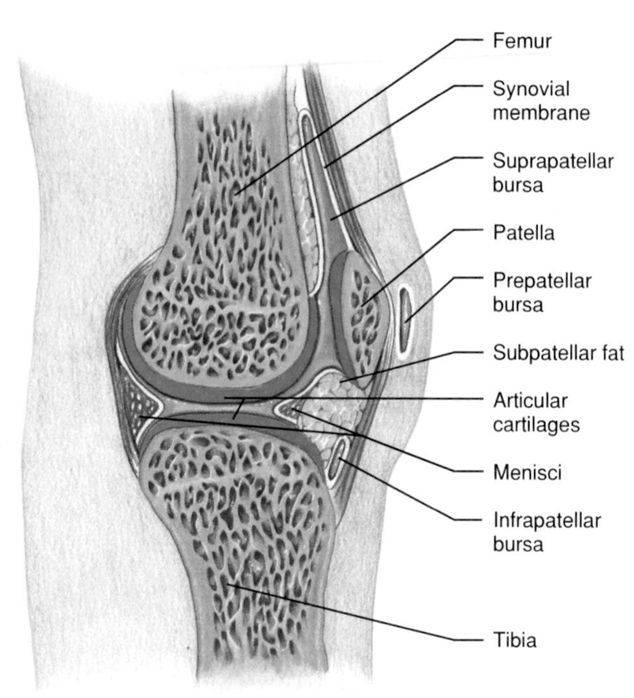

Figure 7.35

Menisci separate the articulating surfaces of the femur and tibia. Several bursae are associated with the knee joint.

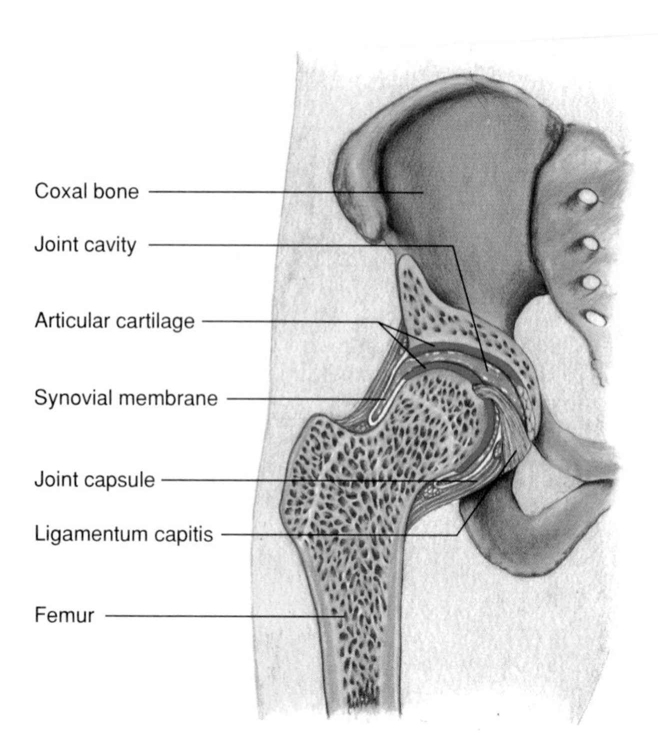

Figure 7.37

The hip is a ball-and-socket joint.

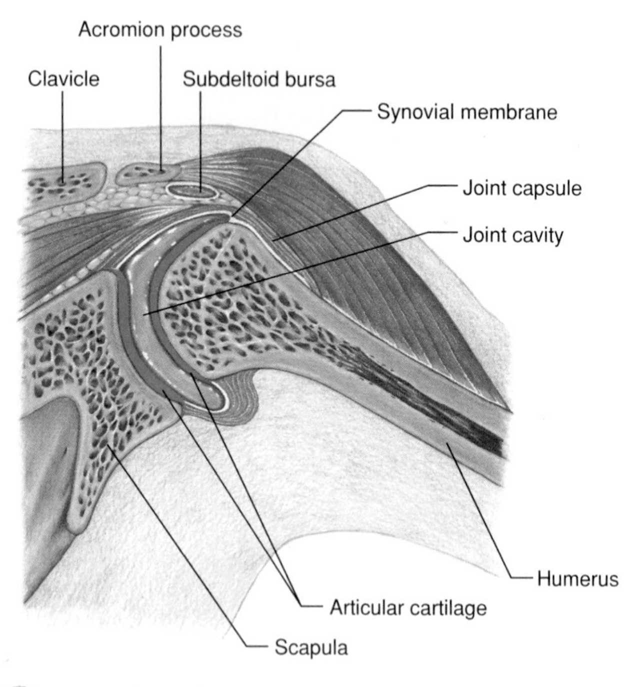

Figure 7.36

The shoulder joint allows movements in all directions. Several bursae are associated with the shoulder joint (not all are shown).

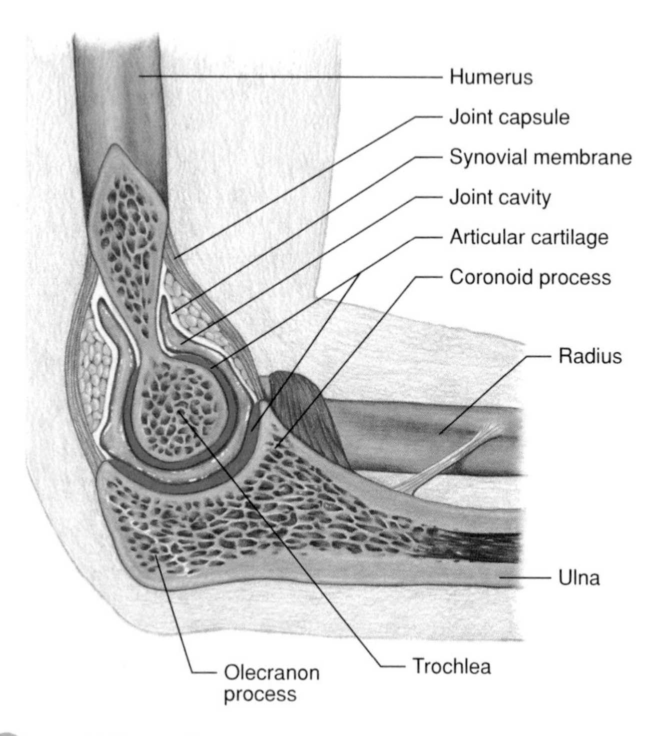

Figure 7.38

The elbow is a hinge joint.

Table 7.4

Types of Joints

Type	Description	Possible Movements	Example
Fibrous	Articulating bones are fastened together by thin layer of fibrous connective tissue.	None	Suture between bones of skull, joint between the distal ends of tibia and fibula
Cartilaginous	Articulating bones are connected by hyaline cartilage or fibrocartilage.	Limited movements, as when back is bent or twisted	Joints between the vertebrae, symphysis pubis
Synovial	Articulating bones are surrounded by a joint capsule of ligaments and synovial membranes; ends of articulating bones are covered by hyaline cartilage and separated by synovial fluid.		
1. Ball-and-socket	Ball-shaped head of one bone articulates with cup-shaped cavity of another.	Movements in all planes and rotation	Shoulder, hip
2. Condyloid	Oval-shaped condyle of one bone articulates with elliptical cavity of another.	Variety of movements in different planes, but no rotation	Joints between the metacarpals and phalanges
3. Gliding	Articulating surfaces are nearly flat or slightly curved.	Sliding or twisting	Joints between various bones of wrist and ankle, sacroiliac joint, joints between ribs 2–7 and sternum
4. Hinge	Convex surface of one bone articulates with concave surface of another.	Flexion and extension	Elbow, joints of phalanges
5. Pivot	Cylindrical surface of one bone articulates with ring of bone and fibrous tissue.	Rotation around a central axis	Joint between the proximal ends of radius and ulna
6. Saddle	Articulating surfaces have both concave and convex regions; the surface of one bone fits the complementary surface of another.	Variety of movements	Joint between the carpal and metacarpal of thumb

Arthritis is a disease that causes inflamed, swollen, and painful joints. The more than a hundred different types of arthritis affect 50 million people in the United States. The most common forms are *rheumatoid arthritis* and *osteoarthritis*.

In rheumatoid arthritis, which is the most painful and debilitating of the arthritic diseases, the synovial membrane of a freely movable joint becomes inflamed and thickens. Then the articular cartilage is damaged, and fibrous tissue infiltrates, interfering with joint movements. In time, the joint may ossify, fusing the articulating bones. Rheumatoid arthritis is an autoimmune disorder in which the immune system attacks the body's healthy tissues.

Osteoarthritis is a degenerative disorder that occurs as a result of aging, but an inherited form may appear as early as one's thirties. In osteoarthritis, articular cartilage softens and disintegrates gradually, roughening the articular surfaces. Joints become painful with restricted movement. Osteoarthritis usually affects the most-used joints, such as those of the fingers, hips, knees, and lower parts of the vertebral column. ■

4. In a **hinge joint,** the convex surface of one bone fits into the concave surface of another, as in the elbow (fig. 7.38) and the joints of the phalanges. Such a joint resembles the hinge of a door in that it permits movement in one plane only.
5. In a **pivot joint,** the cylindrical surface of one bone rotates within a ring formed of bone and fibrous tissue. Movement is limited to the rotation around a central axis. The joint between the proximal ends of the radius and the ulna is of this type (see fig. 7.24).
6. A **saddle joint** forms between bones whose articulating surfaces have both concave and convex regions. The surface of one bone fits the complementary surface of the other. This organization permits a variety of movements, as in the case of the joint between the carpal (trapezium) and the metacarpal of the thumb (see fig. 7.25).

Table 7.4 summarizes the characteristics of different types of joints.

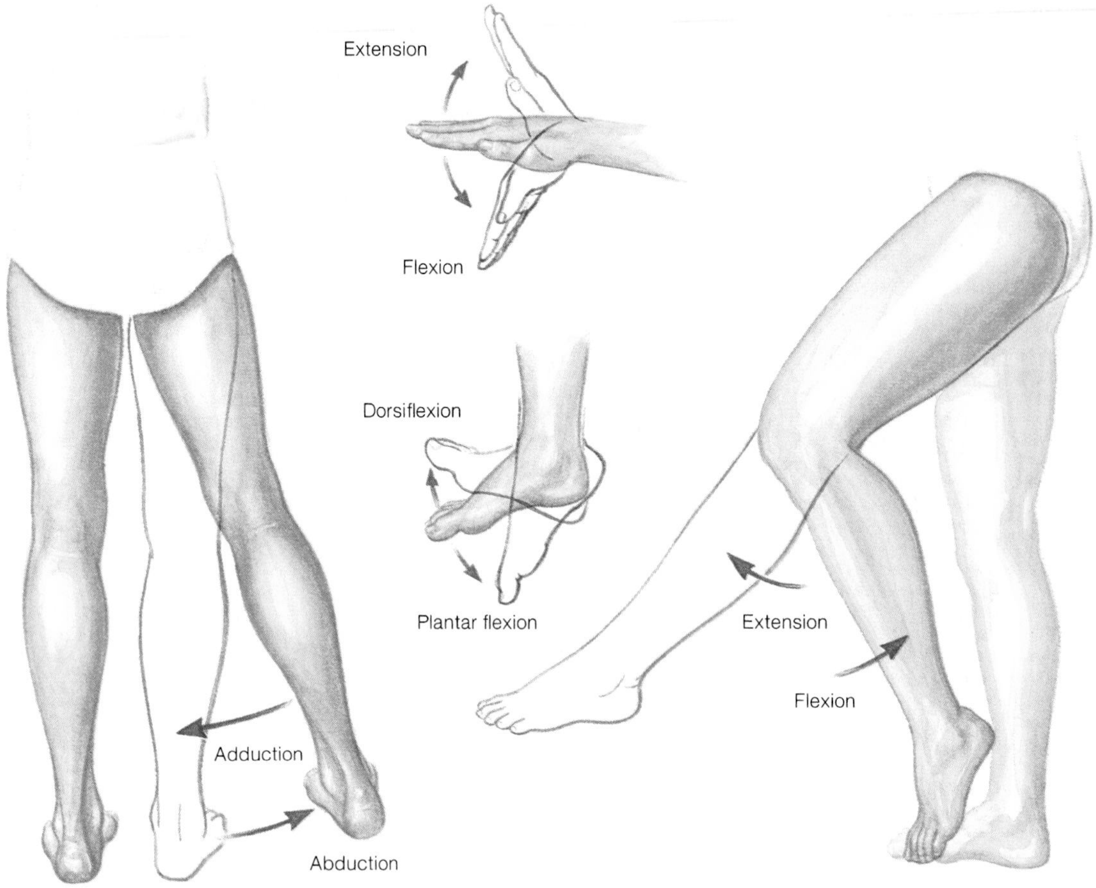

Figure 7.39

Joint movements illustrating flexion, extension, dorsiflexion, plantar flexion, abduction, and adduction.

Types of Joint Movements

Skeletal muscle action produces movements at synovial joints. Typically, one end of a muscle is attached to a relatively immovable or fixed part on one side of a joint, and the other end of the muscle is fastened to a movable part on the other side. When the muscle contracts, its fibers pull its movable end **(insertion)** toward its fixed end **(origin)**, and a movement occurs at the joint.

The following terms describe movements at joints (figs. 7.39, 7.40, and 7.41):

Flexion (flek′shun) Bending parts at a joint so that the angle between them decreases and the parts come closer together (bending the lower limb at the knee).

Extension (ek-sten′shun) Straightening parts at a joint so that the angle between them increases and the parts move farther apart (straightening the lower limb at the knee).

Dorsiflexion (dor″sĭ-flek′shun) Flexing the foot at the ankle toward the shin (bending the foot upward).

Plantar flexion (plan′tar flek′shun) Flexing the foot at the ankle toward the sole (bending the foot downward).

Hyperextension (hi″per-ek-sten′shun) Excessive extension of the parts at a joint, beyond the normal range of motion (overextending the elbow).

Abduction (ab-duk′shun) Moving a part away from the midline (lifting the upper limb horizontally to form a right angle with the side of the body).

Adduction (ah-duk′shun) Moving a part toward the midline (returning the upper limb from the horizontal position to the side of the body).

Rotation (ro-ta′shun) Moving a part around an axis (twisting the head from side to side).

Circumduction (ser″kum-duk′shun) Moving a part so that its end follows a circular path (moving the finger in a circular motion without moving the hand).

Pronation (pro-na′shun) Turning the hand so that the palm is downward or turning the foot so that the medial margin is lowered.

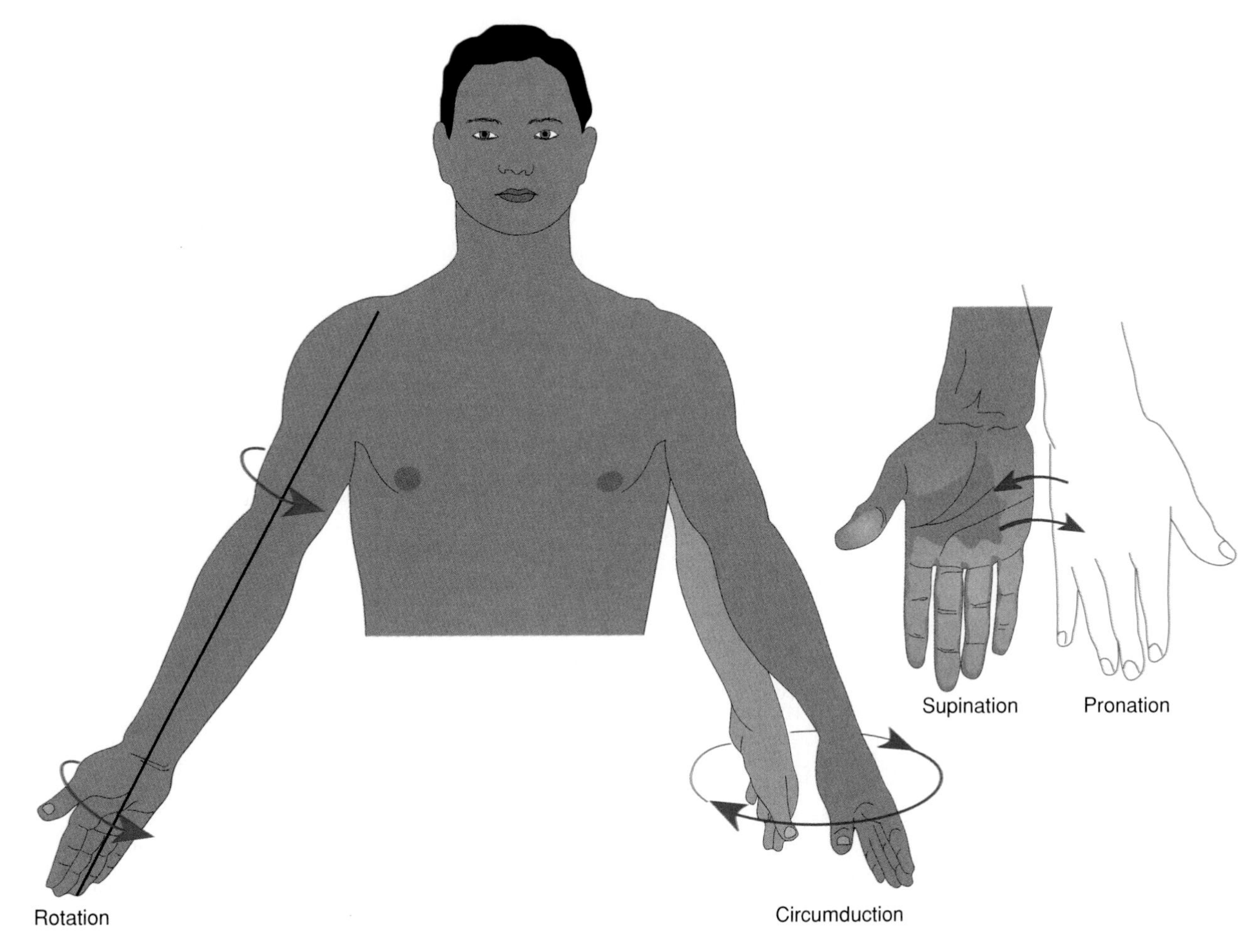

Figure 7.40

Joint movements illustrating rotation, circumduction, pronation, and supination.

Supination (soo″pĭ-na′shun) Turning the hand so that the palm is upward or turning the foot so that the medial margin is raised.

Eversion (e-ver′zhun) Turning the foot so that the sole is outward.

Inversion (in-ver′zhun) Turning the foot so that the sole is inward.

Retraction (re-trak′shun) Moving a part backward (pulling the chin backward).

Protraction (pro-trak′shun) Moving a part forward (thrusting the chin forward).

Elevation (el″ĕ-va′shun) Raising a part (shrugging the shoulders).

Depression (de-presh′un) Lowering a part (drooping the shoulders).

1. Describe the characteristics of the three major types of joints.
2. Name six types of synovial joints.
3. What terms describe movements possible at synovial joints?

Injuries to the elbow, shoulder, and knee are commonly diagnosed and treated using a procedure called *arthroscopy*. Arthroscopy enables a surgeon to visualize a joint's interior and even perform diagnostic or therapeutic procedures, guided by the image on a video screen. An arthroscope is a thin, tubular instrument about 25 centimeters long containing optical fibers that transmit an image. The surgeon inserts the device through a small incision in the joint capsule. Arthroscopy is far less invasive than conventional surgery. Many runners have undergone arthroscopy and raced just weeks later. ■

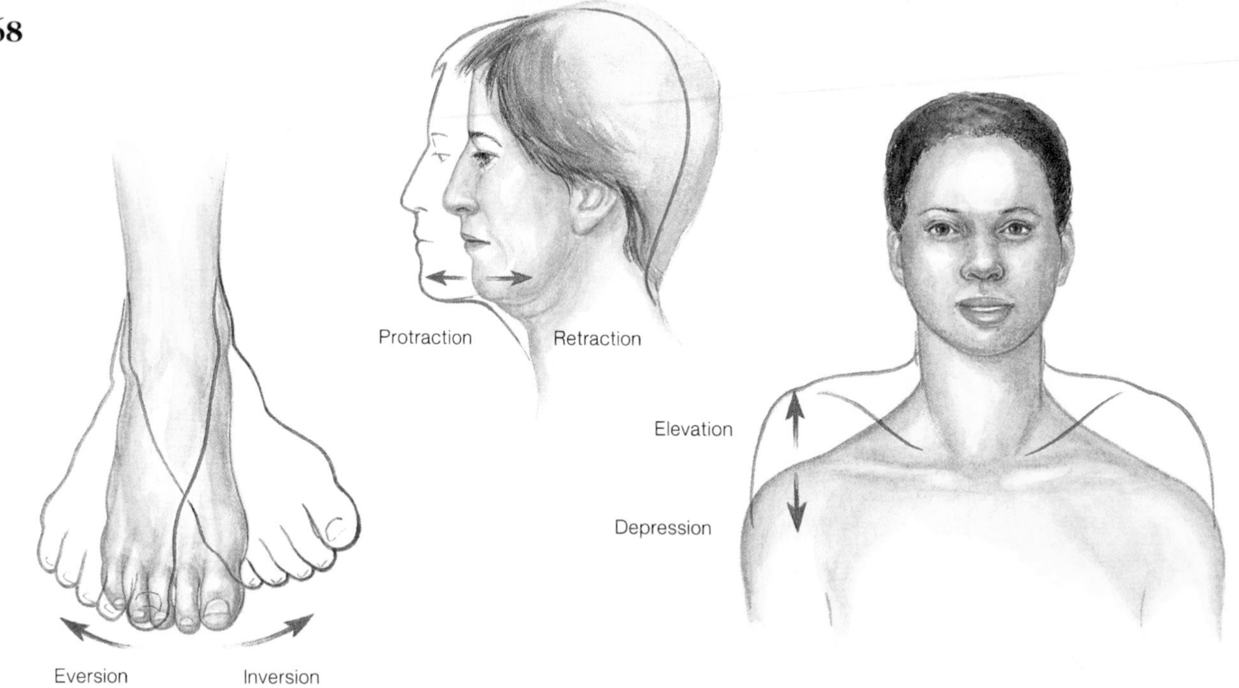

Figure 7.41

Joint movements illustrating eversion, inversion, retraction, protraction, elevation, and depression.

Clinical Terms Related to the Skeletal System

Acromegaly (ak″ro-meg′ah-le) Abnormal enlargement of facial features, hands, and feet in adults as a result of overproduction of growth hormone.

Ankylosis (ang″kĭ-lo′sis) Abnormal stiffness of a joint or fusion of bones at a joint, often due to damage of the joint membranes from chronic rheumatoid arthritis.

Arthralgia (ar-thral′je-ah) Pain in a joint.

Arthrocentesis (ar″thro-sen-te′sis) Puncture of and removal of fluid from a joint cavity.

Arthrodesis (ar″thro-de′sis) Surgery to fuse the bones at a joint.

Arthroplasty (ar′thro-plas″te) Surgery to make a joint more movable.

Colles' fracture (kol′ēz frak′ture) A fracture at the distal end of the radius that displaces the smaller fragment posteriorly.

Epiphysiolysis (ep″ĭ-fiz″e-ol′ĭ-sis) A separation or loosening of the epiphysis from the diaphysis of a bone.

Hemarthrosis (hem″ar-thro′sis) Blood in a joint cavity.

Laminectomy (lam″ĭ-nek′to-me) Surgical removal of the posterior arch of a vertebra, usually to relieve symptoms of a ruptured intervertebral disk.

Lumbago (lum-ba′go) A dull ache in the lumbar region of the back.

Orthopedics (or″tho-pe′diks) Medical specialty that prevents, diagnoses, and treats diseases and abnormalities of the skeletal and muscular systems.

Ostealgia (os″te-al′je-ah) Pain in a bone.

Ostectomy (os-tek′to-me) Surgical removal of a bone.

Osteitis (os″te-i′tis) Inflammation of bone tissue.

Osteochondritis (os″te-o-kon-dri′tis) Inflammation of bone and cartilage tissues.

Osteogenesis (os″te-o-jen′ĕ-sis) Bone development.

Osteogenesis imperfecta (os″te-o-jen′ĕ-sis im-per-fek′ta) An inherited condition of deformed and abnormally brittle bones.

Osteoma (os″te-o′mah) A tumor composed of bone tissue.

Osteomalacia (os″te-o-mah-la′she-ah) A softening of adult bone due to a disorder in calcium and phosphorus metabolism, usually caused by vitamin D deficiency.

Osteomyelitis (os″te-o-mi″ĕ-li′tis) Bone inflammation caused by the body's reaction to bacterial or fungal infection.

Osteonecrosis (os″te-o-ne-kro′sis) Death of bone tissue. This condition occurs most commonly in the femur head in elderly persons and may be due to obstructed arteries supplying the bone.

Osteopathology (os″te-o-pah-thol′o-je) The study of bone diseases.

Osteotomy (os″te-ot′o-me) Cutting a bone.

Synovectomy (sin″o-vek′to-me) Surgical removal of the synovial membrane of a joint.

organization

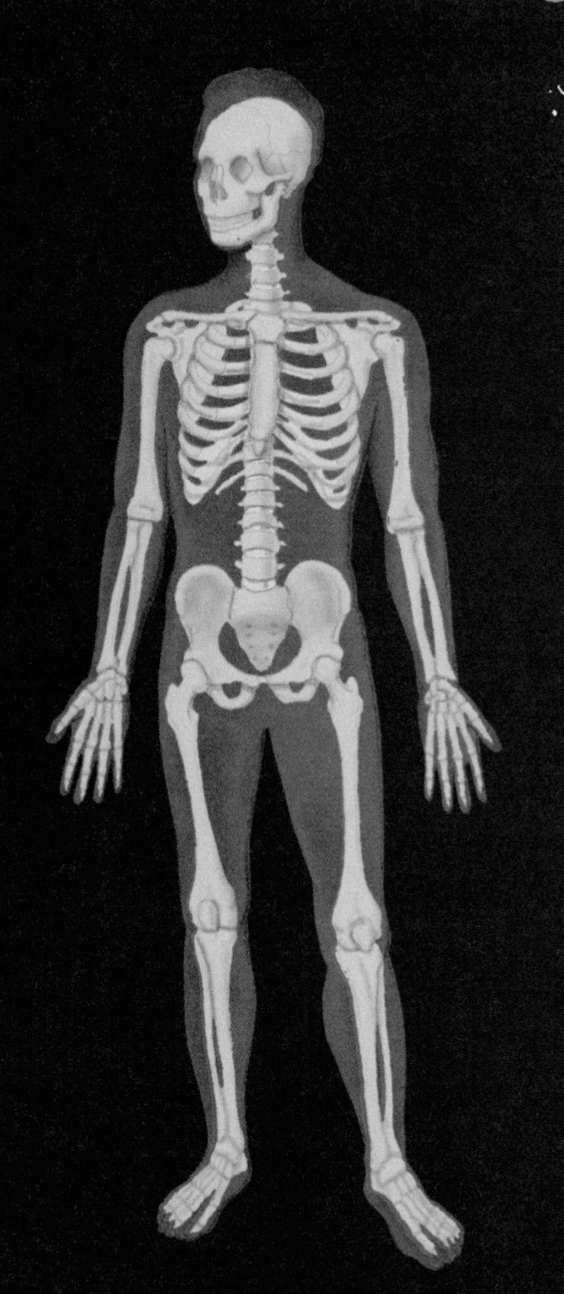

Skeletal System

Bones provide support, protection, and movement and also play a role in calcium balance.

Integumentary System

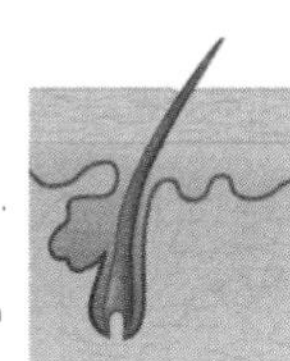

Vitamin D, activated in the skin, plays a role in calcium availability for bone matrix.

Lymphatic System

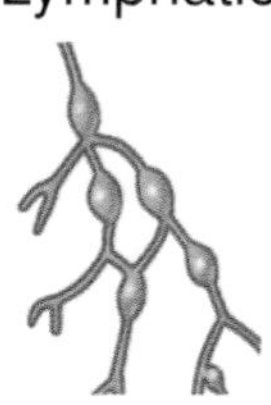

Cells of the immune system originate in the bone marrow.

Muscular System

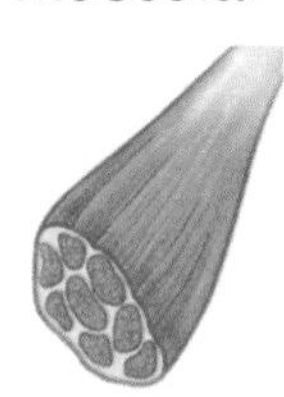

Muscles pull on bones to cause movement.

Digestive System

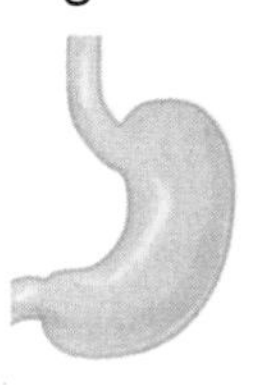

Absorption of dietary calcium provides material for bone matrix.

Nervous System

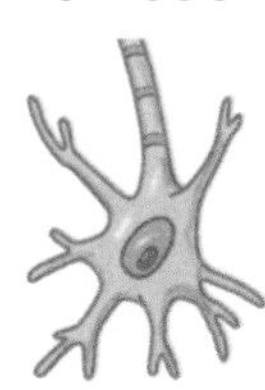

Proprioreceptors sense the position of body parts. Pain receptors warn of trauma to bone. Bones protect the brain and spinal cord.

Respiratory System

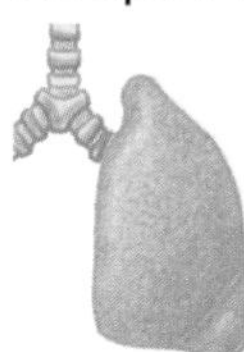

Ribs and muscles work together in breathing.

Endocrine System

Some hormones act on bone to help regulate blood calcium levels.

Urinary System

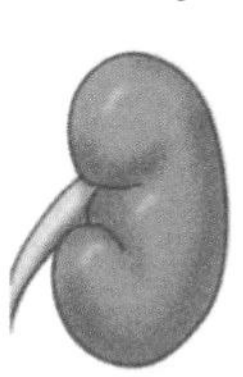

The kidneys and bones work together to help regulate blood calcium levels.

Cardiovascular System

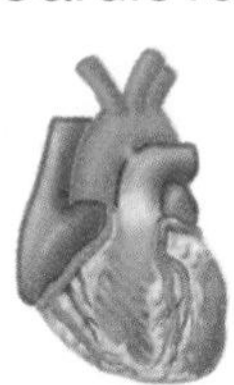

Blood transports nutrients to bone cells. Bone helps regulate plasma calcium levels, important to heart function.

Reproductive System

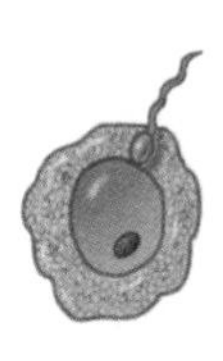

The pelvis helps support the uterus during pregnancy. Bone may provide a source of calcium during lactation.

Summary Outline

Introduction (p. 132)

Individual bones are the organs of the skeletal system. A bone contains very active tissues.

Bone Structure (p. 132)

Bone structure reflects its function.

1. Parts of a long bone
   a. Epiphyses at each end are covered with articular cartilage and articulate with other bones.
   b. The shaft of a bone is called the diaphysis.
   c. Except for the articular cartilage, a bone is covered by a periosteum.
   d. Compact bone has a continuous matrix with no gaps.
   e. Spongy bone has irregular interconnecting spaces between bony plates that reduce the weight of bone.
   f. Both compact and spongy bone are strong and resist bending.
   g. The diaphysis contains a medullary cavity filled with marrow.
2. Microscopic structure
   a. Compact bone contains osteons cemented together.
   b. Osteonic canals contain blood vessels that nourish the cells of osteons.
   c. Diffusion from the surface of the thin, bony plates nourishes the cells of spongy bone.

Bone Development and Growth (p. 134)

1. Intramembranous bones
   a. Intramembranous bones develop from layers of connective tissues.
   b. Osteoblasts within the membranous layers form bone tissue.
   c. Mature bone cells are called osteocytes.
2. Endochondral bones
   a. Endochondral bones develop first as hyaline cartilage, which later is replaced by bone tissue.
   b. The primary ossification center appears in the diaphysis, while secondary ossification centers appear in the epiphyses.
   c. An epiphyseal disk remains between the primary and secondary ossification centers.
   d. The epiphyseal disks are responsible for lengthening.
   e. Long bones continue to lengthen until the epiphyseal disks ossify.
   f. Growth in thickness is due to intramembranous ossification beneath the periosteum.
3. Homeostasis of bone tissue
   a. Osteoclasts and osteoblasts continually remodel bone.
   b. The total mass of bone remains nearly constant.

Bone Function (p. 137)

1. Support and protection
   a. Bones shape and form body structures.
   b. Bones support and protect softer, underlying tissues.
2. Body movement
   a. Bones and muscles function together as levers.
   b. A lever consists of a rod, a pivot (fulcrum), a movable weight, and a force that supplies energy.
3. Blood cell formation
   a. At different ages, hematopoiesis occurs in the yolk sac, liver and spleen, and red bone marrow.
   b. Red marrow produces red blood cells, white blood cells, and blood platelets. Yellow marrow stores fat.
4. Storage of inorganic salts
   a. The intercellular material of bone tissue contains large quantities of calcium phosphate.
   b. When blood calcium is low, osteoclasts break down bone. When blood calcium is high, osteoblasts build bone.
   c. Bone stores small amounts of magnesium, sodium, potassium, and carbonate ions.

Skeletal Organization (p. 140)

1. The skeleton can be divided into axial and appendicular portions.
2. The axial skeleton consists of the skull, hyoid bone, vertebral column, and thoracic cage.
3. The appendicular skeleton consists of the pectoral girdle, upper limbs, pelvic girdle, and lower limbs.

Skull (p. 142)

The skull consists of twenty-two bones: eight cranial bones, thirteen facial bones, and one mandible.

1. Cranium
   a. The cranium encloses and protects the brain.
   b. Some cranial bones contain air-filled paranasal sinuses.
   c. Cranial bones include the frontal bone, parietal bones, occipital bone, temporal bones, sphenoid bone, and ethmoid bone.
2. Facial skeleton
   a. Facial bones form the basic shape of the face and provide attachments for muscles.
   b. Facial bones include the maxillae, palatine bones, zygomatic bones, lacrimal bones, nasal bones, vomer bone, inferior nasal conchae, and mandible.
3. Infantile skull
   a. Fontanels connect incompletely developed bones.
   b. Proportions of the infantile skull are different from those of an adult skull.

Vertebral Column (p. 149)

The vertebral column extends from the skull to the pelvis and protects the spinal cord. It is composed of vertebrae, separated by intervertebral disks.

1. A typical vertebra
   a. A typical vertebra consists of a body and a bony vertebral arch, which surrounds the spinal cord.
   b. Notches on the upper and lower surfaces provide intervertebral foramina through which spinal nerves pass.
2. Cervical vertebrae
   a. Transverse processes bear transverse foramina.
   b. The atlas (first vertebra) supports and balances the head.
   c. The dens of the axis (second vertebra) provides a pivot for the atlas when the head is turned from side to side.
3. Thoracic vertebrae
   a. Thoracic vertebrae are larger than cervical vertebrae.
   b. Facets on the sides articulate with the ribs.
4. Lumbar vertebrae
   a. Vertebral bodies are large and strong.
   b. They support more body weight than other vertebrae.

5. Sacrum
   a. The sacrum is a triangular structure formed of five fused vertebrae.
   b. Vertebral foramina form the sacral canal.
6. Coccyx
   a. The coccyx, composed of four fused vertebrae, forms the lowest part of the vertebral column.
   b. It acts as a shock absorber when a person sits.

Thoracic Cage (p. 152)

The thoracic cage includes the ribs, thoracic vertebrae, sternum, and costal cartilages. It supports the pectoral girdle and upper limbs, protects viscera, and functions in breathing.

1. Ribs
   a. Twelve pairs of ribs attach to the twelve thoracic vertebrae.
   b. Costal cartilages of the true ribs join the sternum directly. Those of the false ribs join it indirectly or not at all.
   c. A typical rib has a shaft, a head, and tubercles that articulate with the vertebrae.
2. Sternum
   a. The sternum consists of a manubrium, body, and xiphoid process.
   b. It articulates with the clavicles.

Pectoral Girdle (p. 153)

The pectoral girdle is composed of two clavicles and two scapulae. It forms an incomplete ring that supports the upper limbs and provides attachments for muscles.

1. Clavicles
   a. Clavicles are rodlike bones located between the manubrium and scapulae.
   b. They hold the shoulders in place and provide attachments for muscles.
2. Scapulae
   a. The scapulae are broad, triangular bones.
   b. They articulate with the humerus of each upper limb and provide attachments for muscles.

Upper Limb (p. 155)

Bones of the upper limb provide the frameworks and attachments of muscles, and function in levers that move the limb and its parts.

1. Humerus
   a. The humerus extends from the scapula to the elbow.
   b. It articulates with the radius and ulna at the elbow.
2. Radius
   a. The radius is located on the thumb side of the forearm between the elbow and the wrist.
   b. It articulates with the humerus, ulna, and wrist.
3. Ulna
   a. The ulna is longer than the radius and overlaps the humerus posteriorly.
   b. It articulates with the radius laterally and with a disk of fibrocartilage inferiorly.
4. Hand
   a. The hand is composed of a wrist, a palm, and five fingers.
   b. It includes eight carpal bones that form a carpus, five metacarpal bones, and fourteen phalanges.

Pelvic Girdle (p. 158)

The pelvic girdle consists of two coxal bones that articulate with each other anteriorly and with the sacrum posteriorly.

1. The sacrum, coccyx, and pelvic girdle form the bowl-shaped pelvis.
2. Each coxal bone consists of an ilium, ischium, and pubis, which are fused in the region of the acetabulum.
   a. The ilium
      (1) The ilium is the largest portion of the coxal bone.
      (2) It joins the sacrum at the sacroiliac joint.
   b. The ischium
      (1) The ischium is the lowest portion of the coxal bone.
      (2) It supports body weight when sitting.
   c. The pubis
      (1) The pubis is the anterior portion of the coxal bone.
      (2) The pubic bones are fused anteriorly at the symphysis pubis.

Lower Limb (p. 160)

Bones of the lower limb provide frameworks of the thigh, leg, and foot.

1. Femur
   a. The femur extends from the hip to the knee.
   b. The patella articulates with the femur's anterior surface.
2. Tibia
   a. The tibia is located on the medial side of the leg.
   b. It articulates with the talus of the ankle.
3. Fibula
   a. The fibula is located on the lateral side of the tibia.
   b. It articulates with the ankle but does not bear body weight.
4. Foot
   a. The foot consists of an ankle, an instep, and five toes.
   b. It includes seven tarsal bones that form the tarsus, five metatarsal bones, and fourteen phalanges.

Joints (p. 162)

Joints can be classified according to the type of tissue that binds the bones together.

1. Fibrous joints
   a. Bones at fibrous joints are tightly joined by a layer of fibrous connective tissue.
   b. Little or no movement occurs at a fibrous joint.
2. Cartilaginous joints
   a. A layer of cartilage joins bones of cartilaginous joints.
   b. Such joints allow limited movement.
3. Synovial joints
   a. The bones of a synovial joint are covered with hyaline cartilage and held together by a fibrous joint capsule.
   b. The joint capsule consists of an outer layer of ligaments and an inner lining of synovial membrane.
   c. Bursae are located between the skin and underlying bony prominences.
   d. Types of synovial joints include: ball-and-socket, condyloid, gliding, hinge, pivot, and saddle.
4. Types of joint movements
   a. Muscles fastened on either side of a joint produce movements of synovial joints.
   b. The movements include flexion, extension, dorsiflexion, plantar flexion, hyperextension, abduction, adduction, rotation, circumduction, pronation, supination, eversion, inversion, retraction, protraction, elevation, and depression.

Review Exercises

Part A

1. Sketch a typical long bone, and label its epiphyses, diaphysis, medullary cavity, periosteum, and articular cartilages. (p. 132)
2. Distinguish between spongy and compact bone. (p. 133)
3. Explain how osteonic canals and perforating canals are related. (p. 134)
4. Explain how the development of intramembranous bone differs from that of endochondral bone. (p. 134)
5. Distinguish between osteoblasts and osteocytes. (p. 134)
6. Explain the function of an epiphyseal disk. (p. 135)
7. Explain how a bone grows in thickness. (p. 135)
8. Provide several examples to illustrate how bones support and protect body parts. (p. 137)
9. Describe a lever. (p. 137)
10. Explain how upper limb movements involve levers. (p. 138)
11. Describe the functions of red and yellow bone marrow. (p. 138)
12. Explain the mechanism that regulates the concentration of blood calcium ions. (p. 139)
13. Distinguish between the axial and appendicular skeletons. (p. 140)
14. Name the bones of the cranium and facial skeleton. (p. 142)
15. Explain the importance of fontanels. (p. 147)
16. Describe a typical vertebra. (p. 149)
17. Explain the differences among the cervical, thoracic, and lumbar vertebrae. (p. 149)
18. Name the bones that comprise the thoracic cage. (p. 152)
19. List the bones that form the pectoral and pelvic girdles. (p. 153)
20. Name the bones of the upper limb. (p. 155)
21. Define *coxal bone*. (p. 158)
22. List the bones of the lower limb. (p. 160)
23. Define *joint*. (p. 162)
24. Describe a fibrous joint, a cartilaginous joint, and a synovial joint. (p. 162)
25. Define *bursa*. (p. 163)
26. List six types of synovial joints, and name an example of each type. (p. 163)

Part B

Match the terms in column I with the movements in column II.

I	II
1. Rotation	a. Turning palm upward
2. Supination	b. Decreasing angle between parts
3. Extension	c. Moving part forward
4. Eversion	d. Moving part around an axis
5. Protraction	e. Turning sole of foot outward
6. Flexion	f. Increasing angle between parts
7. Pronation	g. Lowering a part
8. Abduction	h. Turning palm downward
9. Depression	i. Moving part away from midline

Part C

Match the parts in column I with the bones in column II.

I	II
1. Coronoid process	a. Ethmoid bone
2. Cribriform plate	b. Frontal bone
3. Foramen magnum	c. Mandible
4. Mastoid process	d. Maxillary bone
5. Palatine process	e. Occipital bone
6. Sella turcica	f. Temporal bone
7. Supraorbital foramen	g. Sphenoid bone
8. Temporal process	h. Zygomatic bone
9. Acromion process	i. Femur
10. Deltoid tuberosity	j. Fibula
11. Greater trochanter	k. Humerus
12. Lateral malleolus	l. Radius
13. Medial malleolus	m. Scapula
14. Olecranon process	n. Sternum
15. Radial tuberosity	o. Tibia
16. Xiphoid process	p. Ulna

Critical Thinking

1. How does the structure of a bone make it strong yet lightweight?
2. A fifty-five-year-old woman sees a television program on osteoporosis and decides to take action to lower her risk of developing this bone-weakening condition. She starts to take daily calcium supplements. Why might her precaution not be very helpful?
3. Babies who are put to sleep only on their backs for several months sometimes develop flattened heads, and their fontanels do not close properly. Why might this happen in an infant, but not in an adult who likes to sleep in this position?
4. Archaeologists discover skeletal remains of humanlike animals in Ethiopia. Examination of the bones suggests that the remains represent four types of individuals. Two of the skeletons have bone densities that are 30% less than those of the other two skeletons. The skeletons with the lower bone mass also have broader front pelvic bones. Within the two groups defined by bone mass, smaller skeletons have bones with evidence of epiphyseal plates, but larger bones have only a thin line where the epiphyseal plates should be. Give the age group and gender of the individuals in this find.
5. What steps do you think should be taken to reduce the chances of people accumulating foreign metallic elements, such as lead or radium, in their bones?
6. When a child's bone is fractured, growth may be stimulated at the epiphyseal disk of that bone. What problems might this extra growth cause in an upper or lower limb before the growth of the other limb compensates for the difference in length?
7. Why are women more likely to develop osteoporosis than men? What steps might reduce the risk of developing this condition?

Reference Plates

Human Skull

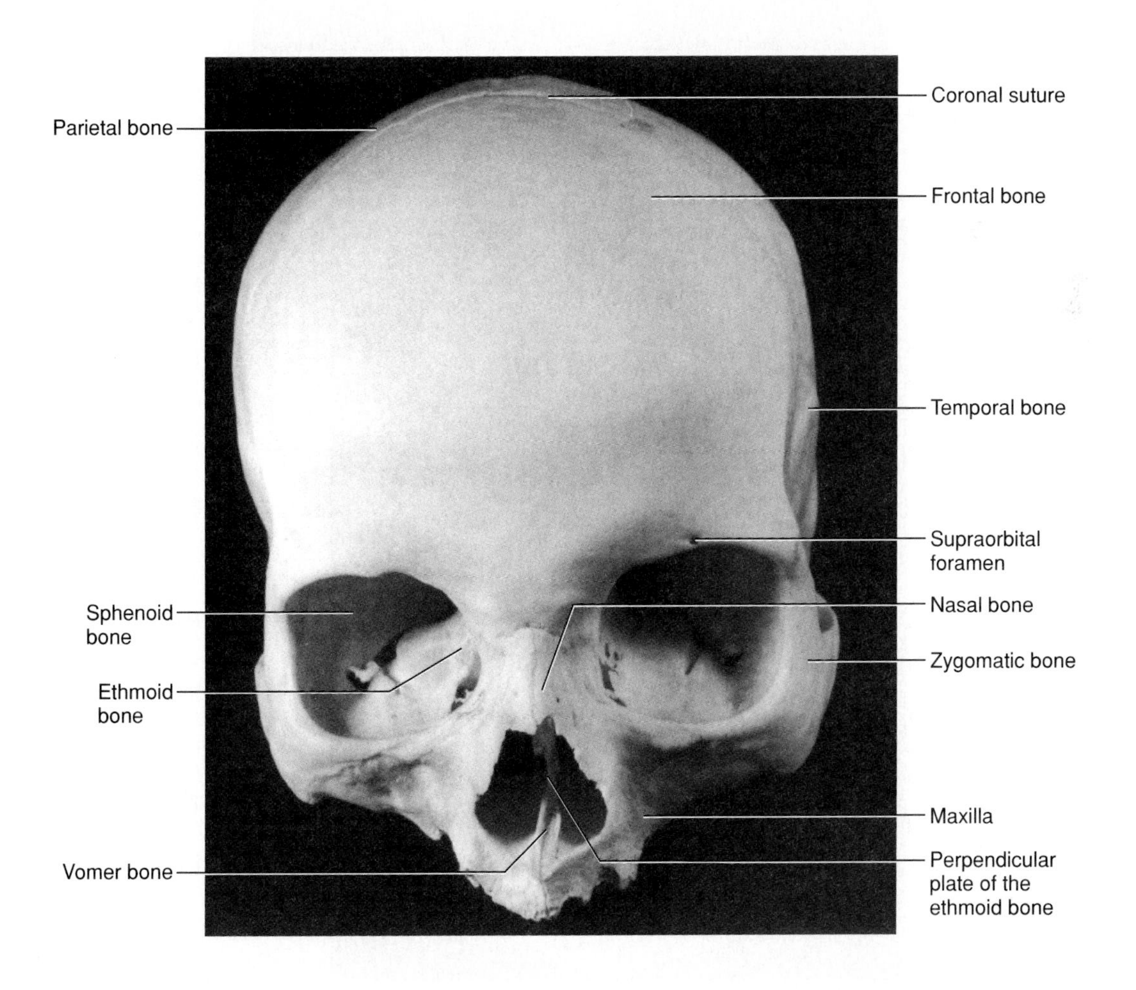

Plate 8

The skull (minus the mandible and teeth), frontal view.

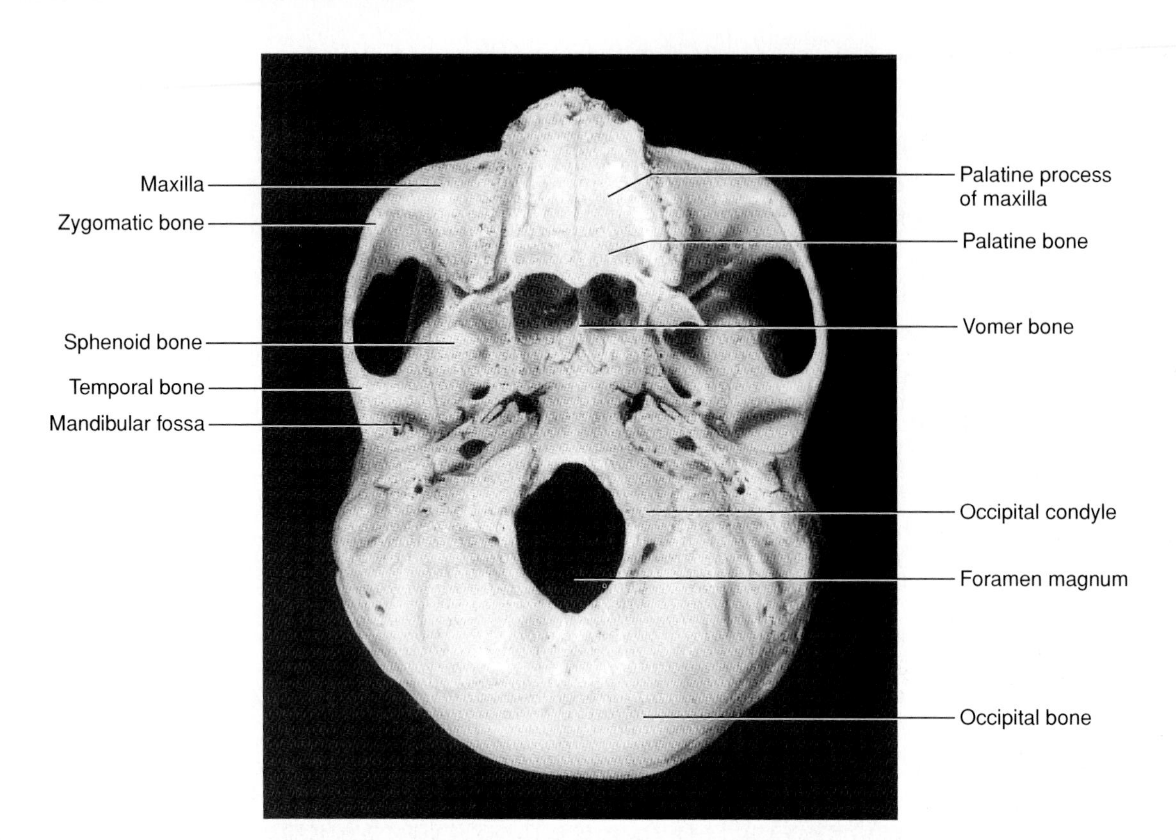

***Plate* 9**

The skull, inferior view.

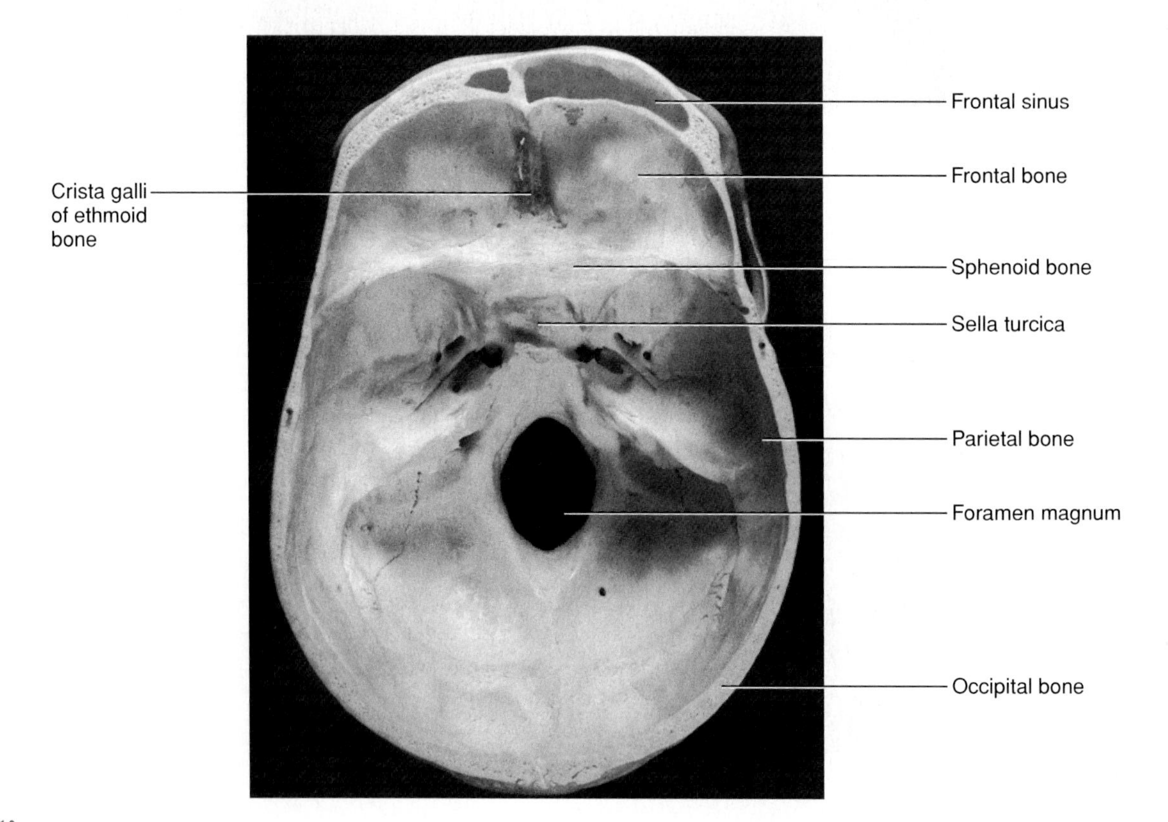

***Plate* 10**

The skull, floor of the cranial cavity.

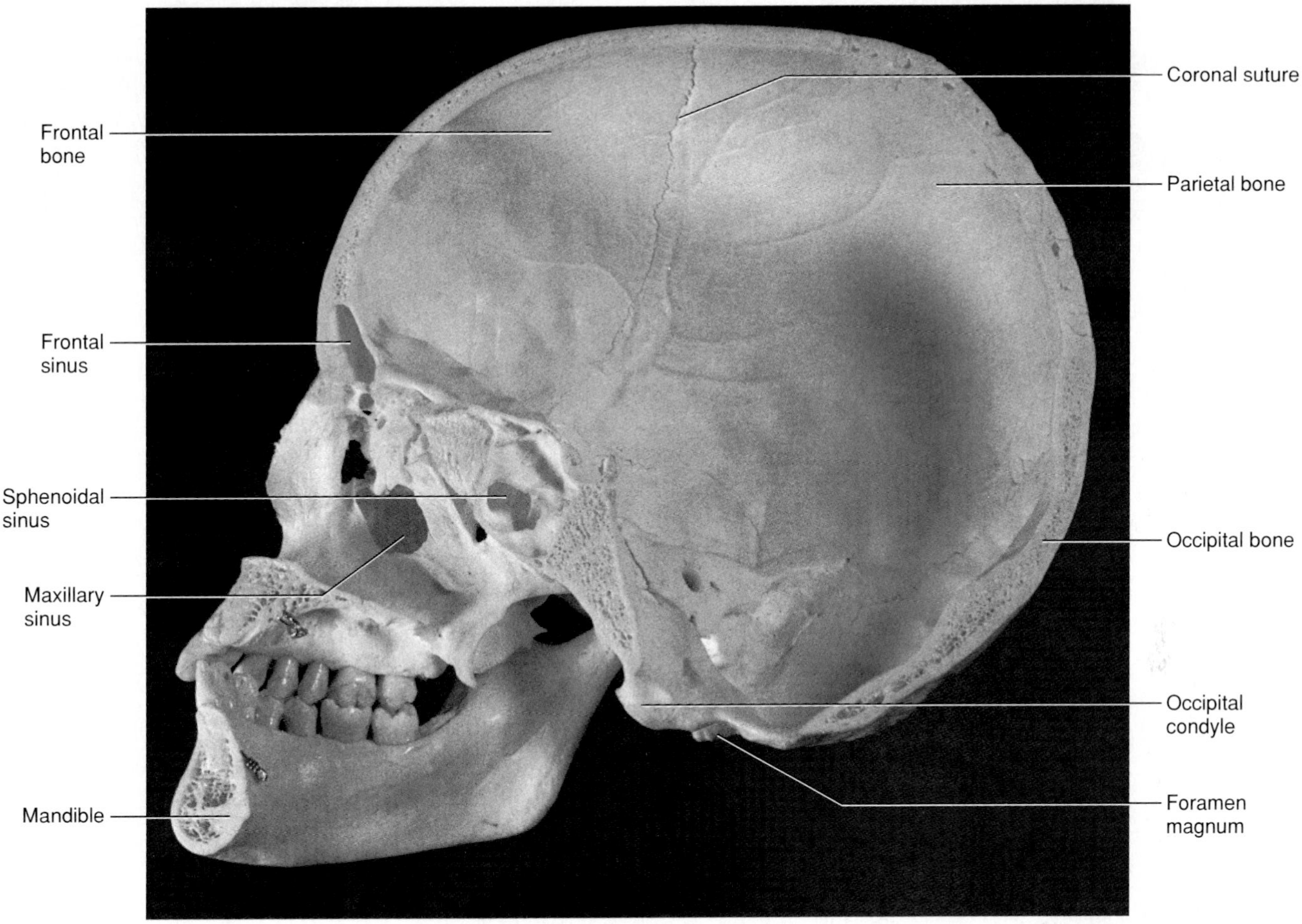

***Plate* 11**

The skull, sagittal section.

Muscular System

Smiling is an important part of childhood. Chelsey Thomas, however, could not smile, nor could she pout, frown, or otherwise use her face to convey her feelings. Chelsey is one of only a thousand or so people in the United States who have a condition called Moebius syndrome. She lacks the nerves that carry impulses from the brain to the muscles of her face. But Chelsey is beginning to smile, thanks to surgery.

In 1995 and 1996, when she was seven years old, Chelsey underwent two operations to transplant nerve and muscle tissue from her legs to either side of her mouth. The transplant gave her the missing "smile apparatus." Gradually, Chelsey is acquiring the subtle, and not-so-subtle, muscular movements of the mouth that make the human face so expressive. Like many things in life, individual muscles are not appreciated until we see what happens when they do not work!

Photo: Just after her first surgery, seven-year-old Chelsey Thomas tried out her smile. It was a few months, however, before the muscle and nerve transplant from her leg to her mouth gave her this ability.

Talking and walking, breathing and sneezing—in fact, all movements—require muscles. Muscles are organs composed of specialized cells that use the chemical energy stored in nutrients to contract. Muscular actions also provide muscle tone, propel body fluids and food, provide the heartbeat, and distribute heat.

Chapter Objectives

After studying this chapter, you should be able to:

1. Describe how connective tissue is part of a skeletal muscle. (p. 178)
2. Name the major parts of a skeletal muscle fiber, and describe the function of each. (p. 179)
3. Explain the major events of skeletal muscle fiber contraction. (p. 182)
4. Explain how the muscle fiber contraction mechanism obtains energy. (p. 184)
5. Describe how oxygen debt develops and how a muscle may become fatigued. (p. 185)
6. Distinguish between a twitch and a sustained contraction. (p. 187)
7. Explain how the types of muscular contractions produce body movements and help maintain posture. (p. 188)
8. Distinguish between the structures and functions of a multiunit smooth muscle and a visceral smooth muscle. (p. 189)
9. Compare the fiber contraction mechanisms of skeletal, smooth, and cardiac muscles. (p. 189)
10. Explain how the locations and interactions of skeletal muscles make possible certain movements. (p. 190)
11. Describe the locations and actions of the major skeletal muscles of each body region. (p. 191)

Aids to Understanding Words

calat- [something inserted] inter*calat*ed disk: A membranous band that connects cardiac muscle cells.

erg- [work] syn*erg*ist: A muscle that works with a prime mover to produce a movement.

hyper- [over, more] muscular *hyper*trophy: The enlargement of muscle fibers.

inter- [between] *inter*calated disk: A membranous band that connects cardiac muscle cells.

laten- [hidden] *laten*t period: The time between when a stimulus is applied and the beginning of a muscle contraction.

myo- [muscle] *myo*fibril: A contractile structure within a muscle cell.

sarco- [flesh] *sarco*plasm: The material (cytoplasm) within a muscle fiber.

syn- [together] *syn*ergist: A muscle that works with a prime mover to produce a movement.

tetan- [stiff] *tetan*ic contraction: A sustained muscular contraction.

-troph [well fed] muscular hyper*troph*y: The enlargement of muscle fibers.

Key Terms

actin (ak′tin)
antagonist (an-tag′o-nist)
aponeurosis (ap″o-nu-ro′sēz)
fascia (fash′e-ah)
insertion (in-ser′shun)
motor neuron (mo′tor nu′ron)
motor unit (mo′tor u′nit)
muscle impulse (mus′el im′puls)
myofibril (mi″o-fi′bril)
myosin (mi′o-sin)
neurotransmitter (nu″ro-trans′mit-er)
origin (or′ĭ-jin)
oxygen debt (ok′sĭ-jen det)
prime mover (prīm mo͞ov′er)
recruitment (re-kro͞ot′ment)
synergist (sin′er-jist)
threshold stimulus (thresh′old stim′u-lus)

Introduction

Muscles are of three types—skeletal muscle, smooth muscle, and cardiac muscle. This chapter focuses on skeletal muscle, which attaches to bones and is under conscious control.

Structure of a Skeletal Muscle

A skeletal muscle is an organ of the muscular system. It is composed of skeletal muscle tissue, nervous tissue, blood, and connective tissues.

Connective Tissue Coverings

Layers of fibrous connective tissue called **fascia** separate an individual skeletal muscle from adjacent muscles and hold it in position (fig. 8.1). This connective tissue surrounds each muscle and may project beyond the end of its muscle fibers to form a cordlike tendon. Fibers in a tendon may intertwine with those in a bone's periosteum, attaching the muscle to the bone. In other cases, the connective tissue forms broad fibrous sheets called **aponeuroses**, which may attach to the coverings of adjacent muscles (see figs. 8.16 and 8.18).

The layer of connective tissue that closely surrounds a skeletal muscle is called *epimysium* (fig. 8.1). Other layers of connective tissue, called *perimysium,* extend inward from the epimysium and separate the muscle tissue into small compartments. These compartments contain bundles of skeletal muscle fibers called *fascicles* (fasciculi). Each muscle fiber within a fascicle (fasciculus) lies within a layer of connective tissue in the form of a thin covering called *endomysium*. Thus, all parts of a skeletal muscle are enclosed in layers of connective tissue, which form a network extending throughout the muscular system.

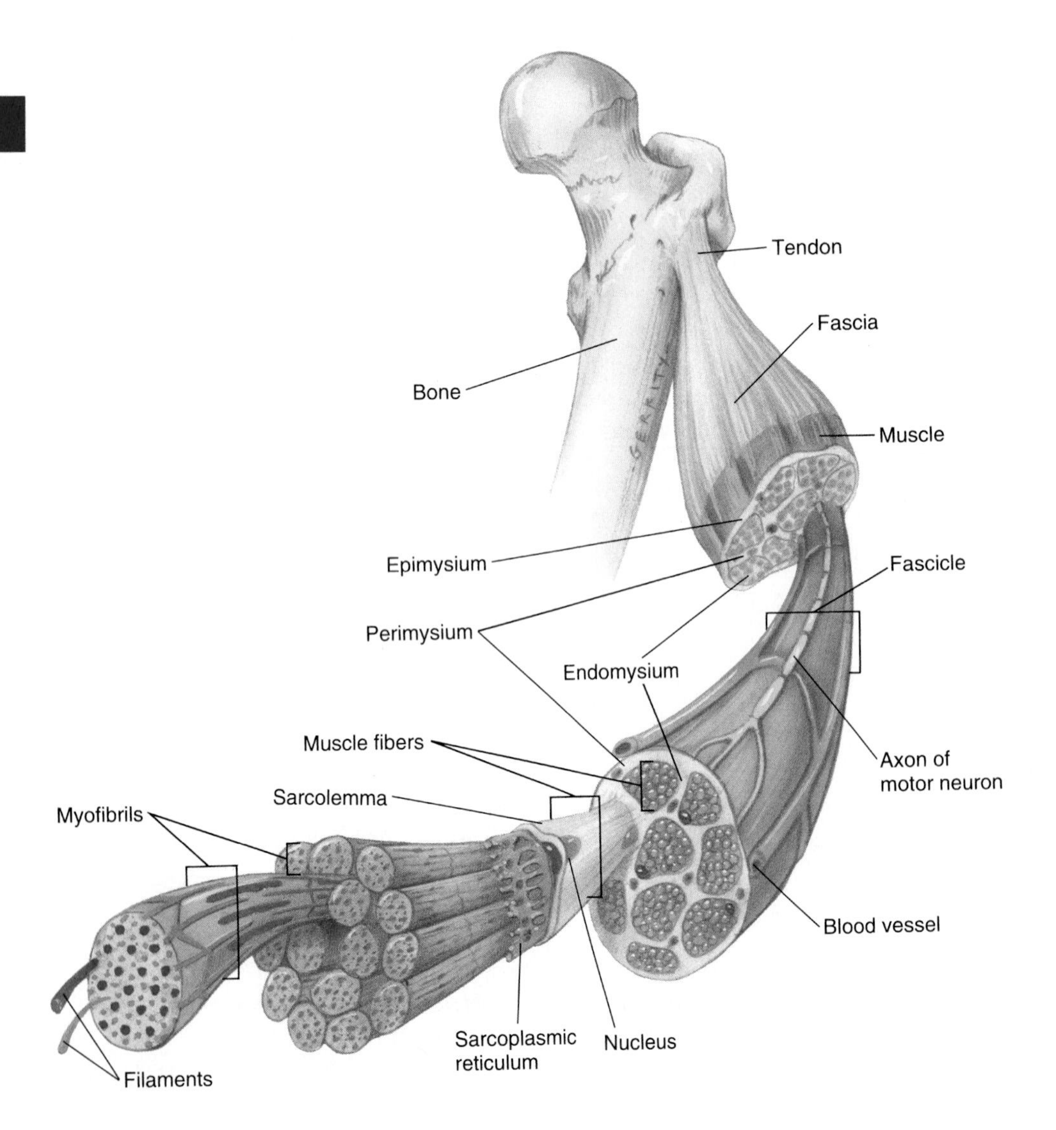

Figure 8.1

A skeletal muscle is composed of a variety of tissues, including layers of connective tissue. Fascia covers the surface of the muscle, epimysium lies beneath the fascia, and perimysium extends into the structure of the muscle, where it separates muscle cells into fascicles. Endomysium separates individual muscle fibers.

A tendon, or the connective tissue sheath of a tendon (tenosynovium), may become painfully inflamed and swollen following injury or the repeated stress of athletic activity. These conditions are called *tendinitis* and *tenosynovitis,* respectively. The tendons most commonly affected are those associated with the joint capsules of the shoulder, elbow, and hip, and those that move the wrist, hand, thigh, and foot. ■

Skeletal Muscle Fibers

A skeletal muscle fiber is a single cell that contracts in response to stimulation and then relaxes when the stimulation ends. Each skeletal muscle fiber is a thin, elongated cylinder with rounded ends, and it may extend the full length of the muscle. Just beneath its cell membrane (or *sarcolemma*), the cytoplasm (or *sarcoplasm*) of the fiber contains many small, oval nuclei and mitochondria (fig. 8.1). The sarcoplasm also contains many threadlike **myofibrils** that lie parallel to one another.

Myofibrils play a fundamental role in muscle contraction. They contain two kinds of protein filaments—thick ones composed of the protein **myosin** and thin ones composed of the protein **actin** (figs. 8.2 and 8.3). The arrangement of these filaments produces the characteristic alternating light and dark *striations,* or bands, of a skeletal muscle fiber.

The striation pattern of skeletal muscle fibers has two main parts. The first, the *I bands* (the light bands), are composed of thin actin filaments directly attached to structures called *Z lines.*

The second part consists of the *A bands* (the dark bands), which are composed of thick myosin filaments overlapping with thin actin filaments. The A band consists not only of a region where the thick and thin filaments overlap, but also a central region (*H zone*) consisting [illegible] of thick filaments, plus a thickening known as the M [illegible] fig. 8.2). The segment of a myofibril that extends from [illegible] line to the next Z line is called a **sarcomere** (figs. [illegible] 8.3).

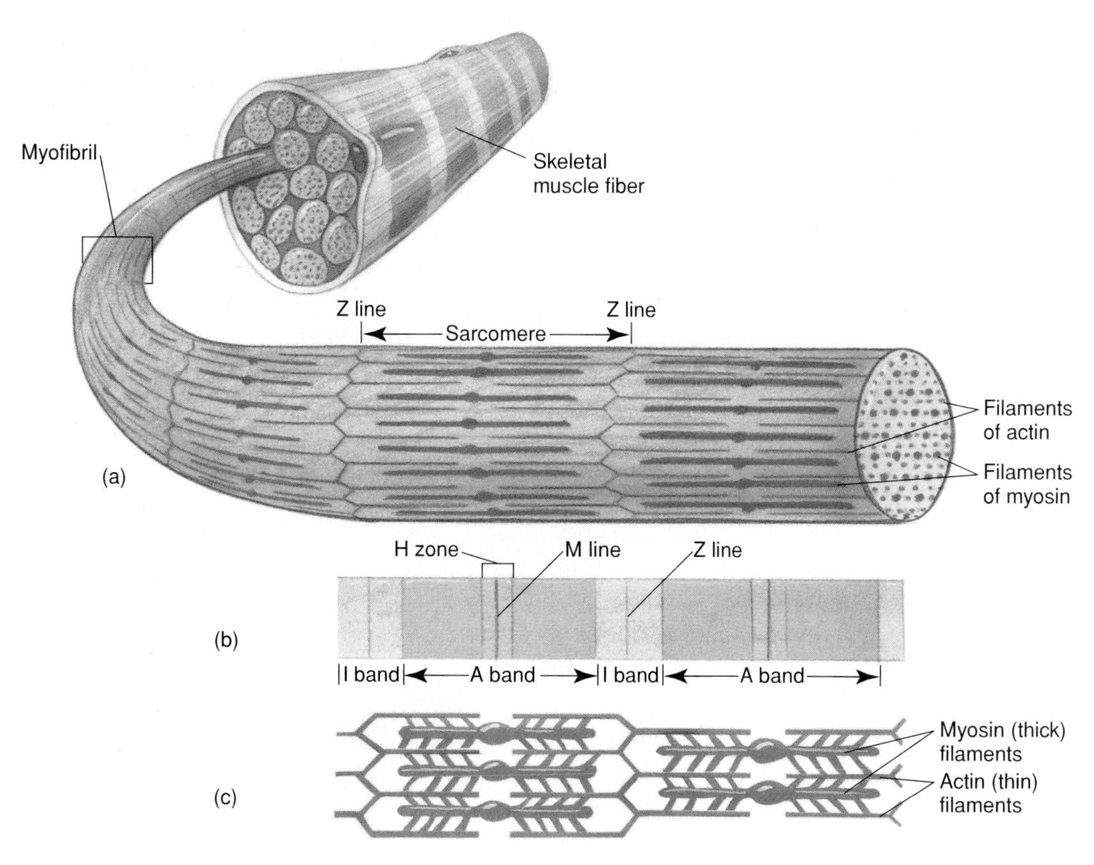

Figure 8.2

(*a*) A skeletal muscle fiber contains numerous myofibrils, each consisting of (*b*) units called sarcomeres. (*c*) The characteristic striations of a sarcomere are due to the spatial arrangement of actin and myosin filaments.

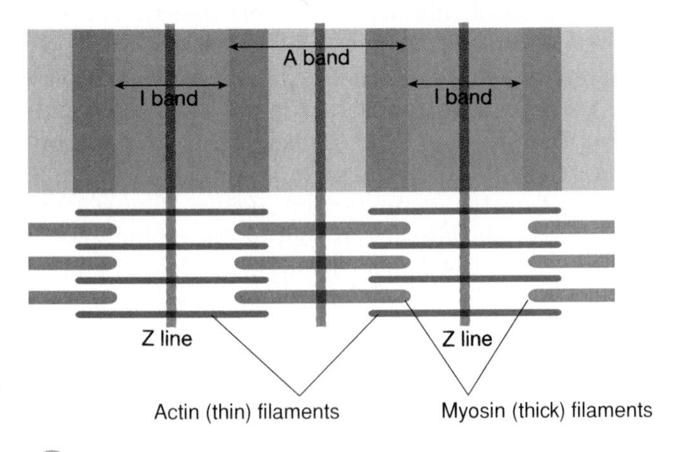

Figure 8.3

A sarcomere (80,000×).

Within the sarcoplasm of a muscle fiber is a network of membranous channels that surrounds each myofibril and runs parallel to it (fig. 8.4). These membranes form the **sarcoplasmic reticulum**, which corresponds to the endoplasmic reticulum of other cells. Another set of membranous channels, called **transverse tubules** (T-tubules), extends inward as invaginations from the fiber's membrane and passes all the way through the fiber. Thus, each tubule opens to the outside of the muscle fiber and contains extracellular fluid. Furthermore, each transverse tubule lies between two enlarged portions of the sarcoplasmic reticulum called *cisternae,* near the region where the actin and myosin filaments overlap. The sarcoplasmic reticulum and transverse tubules activate the muscle contraction mechanism when the fiber is stimulated.

1. Describe how connective tissue is part of a skeletal muscle.
2. Describe the general structure of a skeletal muscle fiber.
3. Explain why skeletal muscle fibers appear striated.
4. Explain the relationship between the sarcoplasmic reticulum and the transverse tubules.

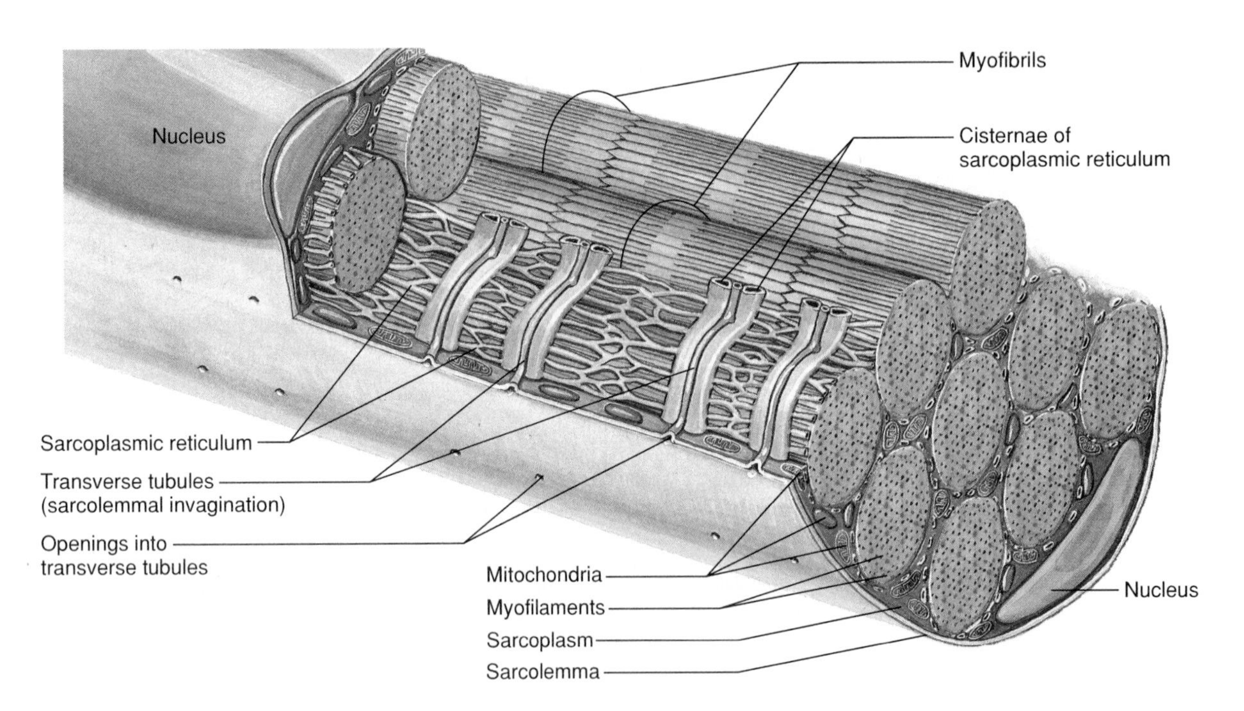

Figure 8.4

Within the sarcoplasm of a skeletal muscle fiber are a network of sarcoplasmic reticulum and a system of transverse tubules.

Muscle fibers and their associated connective tissues are flexible but can tear if overstretched. This type of injury, common in athletes, is called *muscle strain*. The seriousness of the injury depends on the degree of damage the tissues sustain. A mild strain injures only a few muscle fibers, the fascia remains intact, and loss of function is minimal. In a severe strain, however, many muscle fibers as well as the fascia tear, and muscle function may be completely lost. Such a severe strain is painful and produces discoloration and swelling. ■

Neuromuscular Junction

Each skeletal muscle fiber connects to a fiber from a nerve cell called a **motor neuron.** This nerve fiber extends outward from the brain or spinal cord, and a muscle fiber contracts only when a motor neuron stimulates it.

The connection between the motor neuron and muscle fiber is called a **neuromuscular junction.** Here, the muscle fiber membrane is specialized to form a **motor end plate.** In this region of the muscle fiber, nuclei and mitochondria are abundant, and the cell membrane (sarcolemma) is extensively folded (fig. 8.5).

The end of the motor neuron branches and projects into recesses of the muscle fiber membrane. The cytoplasm at the distal ends of these motor neuron fibers is rich in mitochondria and contains many tiny vesicles (synaptic vesicles) that store chemicals called **neurotransmitters.**

When a nerve impulse traveling from the brain or spinal cord reaches the end of a motor neuron fiber, some of the vesicles release a neurotransmitter into the gap (synaptic cleft) between the neuron and the motor end plate of the muscle fiber. This action stimulates the muscle fiber to contract.

Motor Units

A muscle fiber usually has a single motor end plate. The nerve fibers of motor neurons, however, are densely branched. By means of these branches, one motor neuron may connect to many muscle fibers. When a motor neuron transmits an impulse, all of the muscle fibers it links to are stimulated to contract simultaneously. Together, a motor neuron and the muscle fibers that it controls constitute a **motor unit** (fig. 8.6).

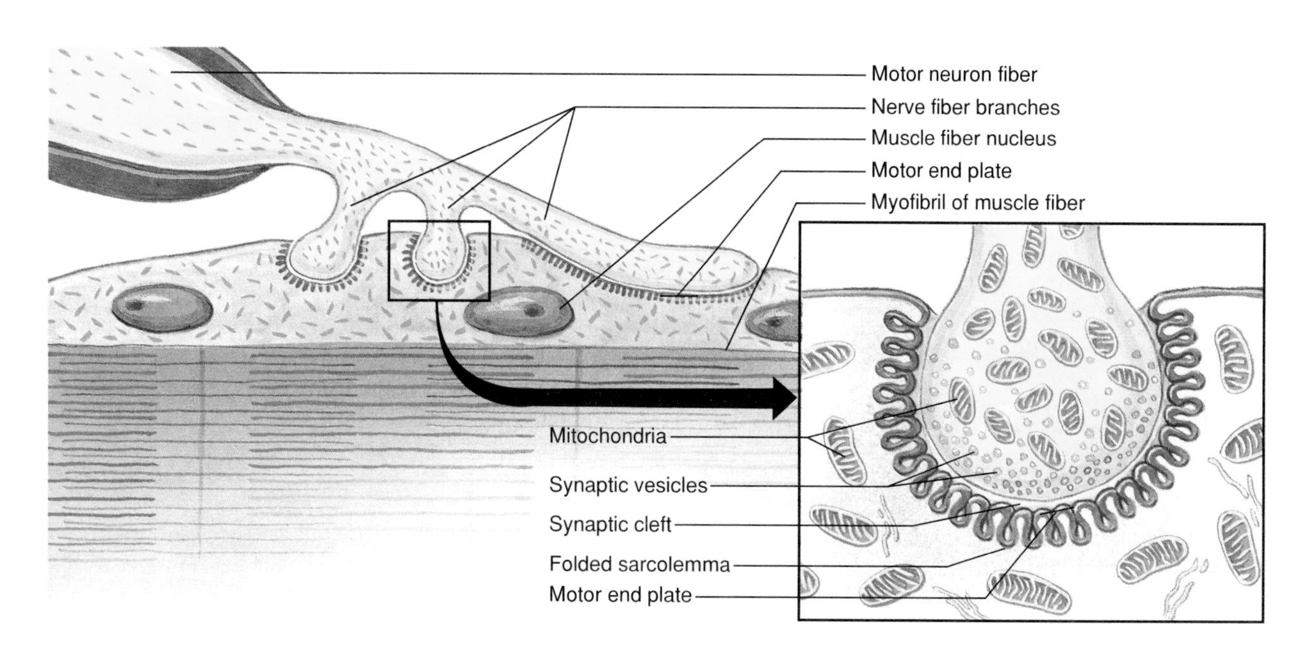

Figure 8.5

A neuromuscular junction includes the end of a motor neuron fiber and the motor end plate of a muscle fiber.

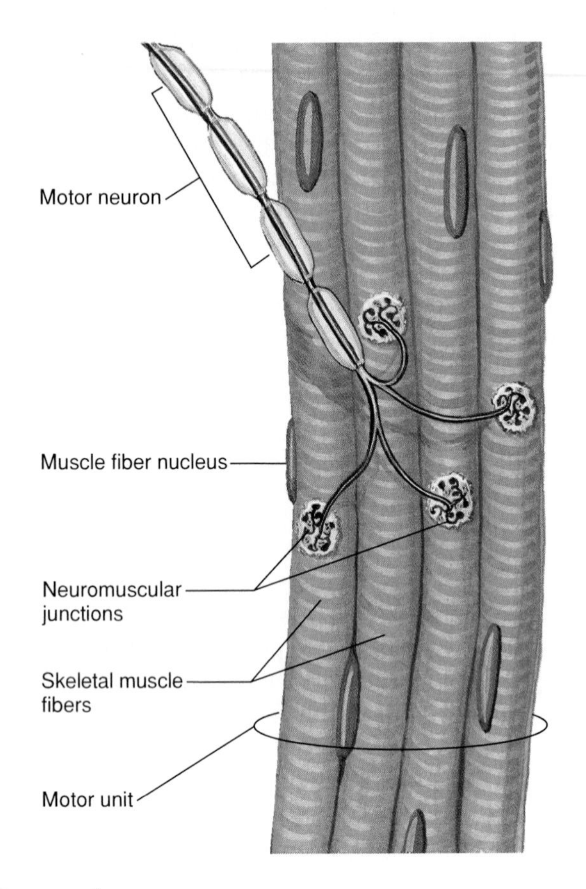

Figure 8.6

A motor unit consists of one motor neuron and all the muscle fibers with which it communicates.

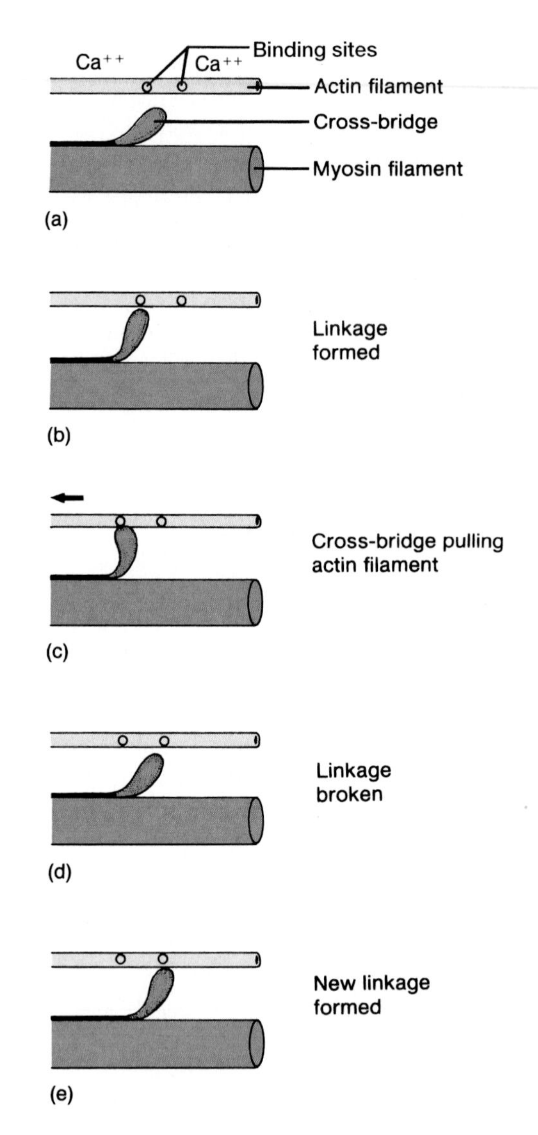

Figure 8.7

According to the sliding filament theory, (*a*) when calcium ions are present, binding sites on an actin filament are exposed. (*b*) Cross-bridges on a myosin filament form linkages at the binding sites. (*c*) A myosin cross-bridge bends slightly, pulling an actin filament, using energy from ATP. (*d*) The linkage breaks, and (*e*) the myosin cross-bridge forms a linkage with the next binding site.

Skeletal Muscle Contraction

A muscle fiber contraction is a complex interaction of cell components and biochemicals. The final result is a movement within the myofibrils in which the filaments of actin and myosin slide past one another, shortening the muscle fiber and pulling on its attachments.

Role of Myosin and Actin

A myosin molecule is composed of two twisted protein strands with globular parts called cross-bridges projecting outward along their lengths. Many of these molecules comprise a myosin filament. An actin molecule is a globular structure with a binding site to which the myosin cross-bridges can attach. Many of these actin molecules twist into a double strand (helix), forming an actin filament.

The **sliding filament theory** of muscle contraction states that the head of a myosin cross-bridge can attach to an actin binding site and bend slightly, pulling the actin filament with it. Then the head releases, straightens, and combines with another binding site farther down the actin filament (fig. 8.7).

The globular portions of the myosin filaments contain an enzyme called **ATPase.** ATPase catalyzes the breakdown of ATP to ADP and phosphate, releasing energy (see chapter 4). This energy provides the force with which the cross-bridges pull. In addition, ATP binding to the cross-bridge causes the cross-bridge to release from actin even before the ATP splits.

Presumably, this cycle repeats again and again, as long as ATP is present as an energy source and as long as the muscle fiber is stimulated to contract. When the actin filament moves toward the center of the sarcomere, the sarcomere shortens (figs. 8.7 and 8.8).

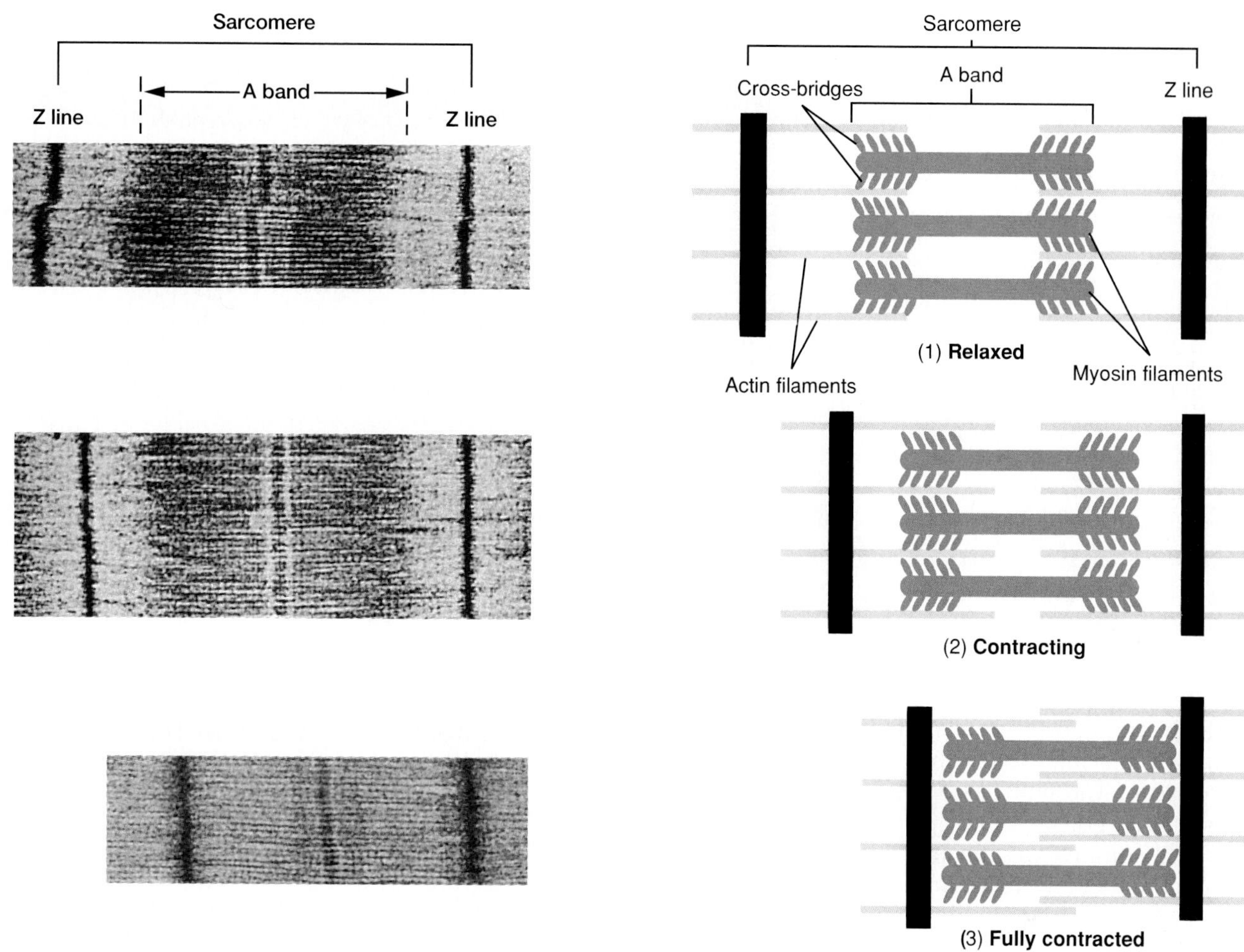

Figure 8.8

When a skeletal muscle contracts, individual sarcomeres shorten as thick and thin filaments slide past one another (about 60,000×).

Stimulus for Contraction

A skeletal muscle fiber normally does not contract until a neurotransmitter stimulates it. In skeletal muscle, the neurotransmitter is **acetylcholine.** Acetylcholine is synthesized in the cytoplasm of the motor neuron and is stored in vesicles at the distal end of the motor neuron fibers. When a nerve impulse (described in chapter 9) reaches the end of a motor neuron fiber, some of the vesicles release their acetylcholine into the synaptic cleft between the motor neuron fiber and motor end plate (see fig. 8.5).

Acetylcholine diffuses rapidly across the synaptic cleft and combines with certain protein molecules (receptors) in the muscle fiber membrane, stimulating a **muscle impulse**, which is very much like a nerve impulse. The impulse passes in all directions over the surface of the muscle fiber membrane and travels through the transverse tubules, deep into the fiber, and reaches the sarcoplasmic reticulum (see fig. 8.4).

The sarcoplasmic reticulum contains a high concentration of calcium ions. In response to a muscle impulse, the membranes of the cisternae become more permeable to these ions, and the calcium ions diffuse into the sarcoplasm of the muscle fiber.

When a relatively high concentration of calcium ions is present in the sarcoplasm, linkages form between the actin and myosin filaments, and a muscle contracts (figs. 8.7 and 8.8). The contraction, which also requires ATP, continues as long as nerve impulses cause acetylcholine release. When the nerve impulses cease, two events lead to muscle relaxation.

First, the acetylcholine that stimulated the muscle fiber is rapidly decomposed by the enzyme **acetylcholinesterase.** This enzyme is present at the neuromuscular junction on the membranes of the motor end plate. Acetylcholinesterase prevents a single nerve impulse from continuously stimulating the muscle fiber.

Second, when acetylcholine breaks down, the stimulus to the muscle fiber ceases. As a result, calcium ions

Table 8.1

Major Events of Muscle Contraction and Relaxation

Muscle Fiber Contraction	Muscle Fiber Relaxation
1. The distal end of a motor neuron fiber releases acetylcholine.	1. Acetylcholinesterase decomposes acetylcholine, and the muscle fiber membrane is no longer stimulated.
2. Acetylcholine diffuses across the gap at the neuromuscular junction.	2. Calcium ions are actively transported into the sarcoplasmic reticulum.
3. The muscle fiber membrane is stimulated, and a muscle impulse travels deep into the fiber through the transverse tubules and reaches the sarcoplasmic reticulum.	3. Linkages between actin and myosin filaments break.
4. Calcium ions diffuse from the sarcoplasmic reticulum into the sarcoplasm.	4. Actin and myosin filaments slide apart.
5. Linkages form between actin and myosin filaments.	5. Muscle fiber relaxes.
6. Myosin cross-bridges pull actin filaments inward.	
7. Muscle fiber shortens as a contraction occurs.	

are actively transported back into the sarcoplasmic reticulum, decreasing the calcium ion concentration of the sarcoplasm. The linkages break, and consequently, the muscle fiber relaxes.

Table 8.1 summarizes the major events leading to muscle contraction and relaxation.

The bacterium *Clostridium botulinum* produces a poison, called botulinum toxin, that can prevent the release of acetylcholine from motor nerve fibers at neuromuscular junctions, causing a very serious form of food poisoning called *botulism*. This condition is most likely to result from eating home-processed food that has not been heated enough to kill the bacteria in it or to inactivate the toxin.

Botulinum toxin blocks stimulation of muscle fibers, paralyzing muscles, including those responsible for breathing. Without prompt medical treatment, the fatality rate for botulism is high. ■

1. Describe a neuromuscular junction.
2. Define *motor unit*.
3. Explain how the filaments of a myofibril interact during muscle contraction.
4. Explain how a motor nerve impulse can trigger a muscle contraction.

Energy Sources for Contraction

ATP molecules supply the energy for muscle fiber contraction. However, a muscle fiber has only enough ATP to contract for a very short time, so when a fiber is active, ATP must be regenerated.

The initial source of energy available to regenerate ATP from ADP and phosphate is **creatine phosphate.** Like ATP, creatine phosphate contains high-energy phosphate bonds, and it is four to six times more abundant in muscle fibers than ATP. Creatine phosphate, however, cannot directly supply energy to a cell's energy-utilizing reactions. Instead, it stores excess energy released from the mitochondria. Thus, whenever ATP supply is sufficient, an enzyme in the mitochondria (creatine phosphokinase) promotes the synthesis of creatine phosphate, which stores excess energy in its phosphate bonds (fig. 8.9).

As ATP decomposes, the energy from creatine phosphate can be transferred to ADP molecules, converting them back into ATP. Active muscle, however, rapidly exhausts the supply of creatine phosphate. When this happens, the muscle fibers depend on cellular respiration of glucose as an energy source for synthesizing ATP.

Oxygen Supply and Cellular Respiration

As described in chapter 4, the early phase of cellular respiration can take place in the absence of oxygen. The more complete breakdown of glucose, however, occurs in the mitochondria and requires oxygen. The blood carries the oxygen required to support this

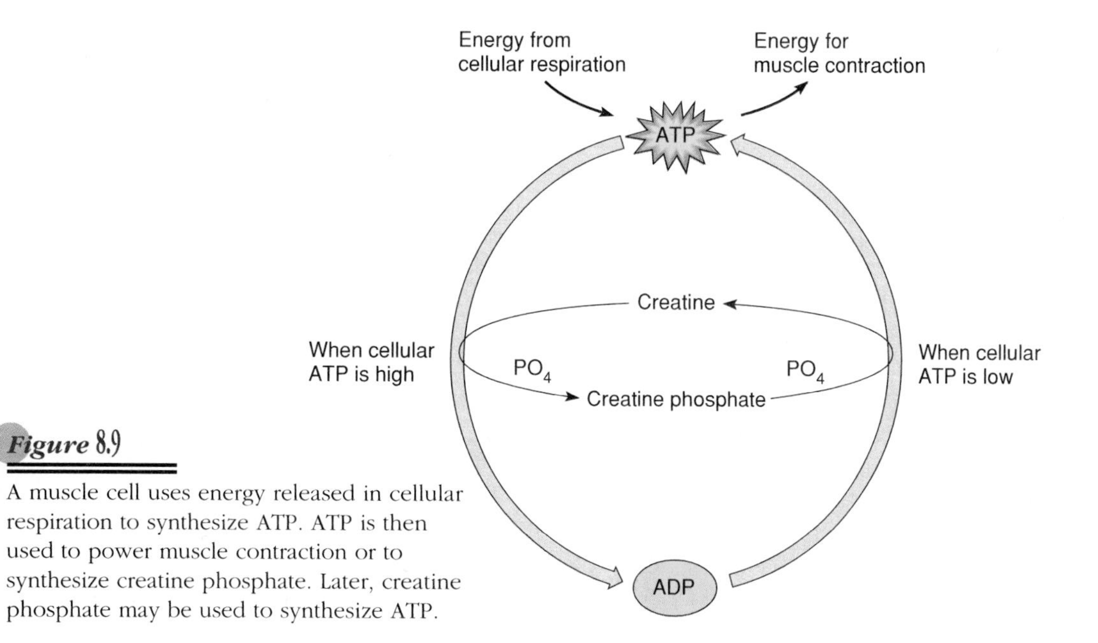

Figure 8.9

A muscle cell uses energy released in cellular respiration to synthesize ATP. ATP is then used to power muscle contraction or to synthesize creatine phosphate. Later, creatine phosphate may be used to synthesize ATP.

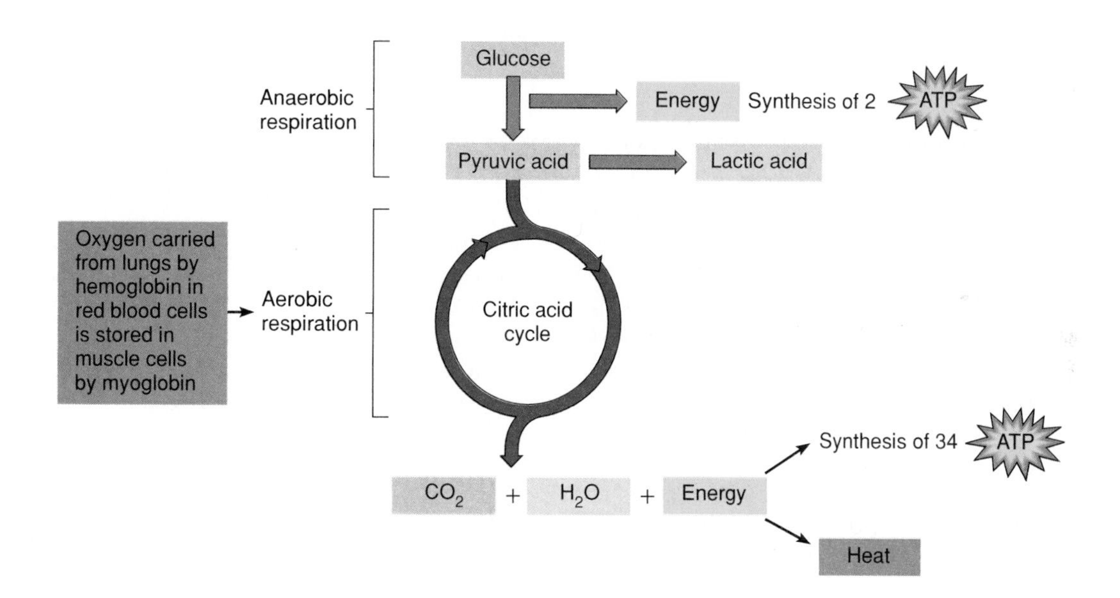

Figure 8.10

The oxygen required to support aerobic respiration is carried in the blood and stored in myoglobin. In the absence of sufficient oxygen, pyruvic acid is converted to lactic acid. The maximum number of ATPs generated per glucose molecule varies with cell type and is thirty-six (2 + 34) in skeletal muscle.

aerobic respiration from the lungs to body cells. Red blood cells carry the oxygen, loosely bonded to molecules of **hemoglobin**, the pigment responsible for the red color of blood.

Another pigment, **myoglobin**, is synthesized in muscle cells and causes the reddish brown color of skeletal muscle tissue. Like hemoglobin, myoglobin can combine loosely with oxygen. This ability to store oxygen temporarily reduces a muscle's requirement for a continuous blood supply during muscular contraction (fig. 8.10).

Oxygen Debt

When a person is resting or moderately active, the respiratory and circulatory systems can usually supply sufficient oxygen to skeletal muscles to support aerobic respiration. However, this is not the case when skeletal muscles are used strenuously for even a minute or two. Consequently, muscle fibers must depend increasingly on the anaerobic phase of respiration for energy.

In anaerobic respiration, glucose molecules are broken down to yield *pyruvic acid* (see chapter 4). If the oxygen supply is low, however, the pyruvic acid reacts to produce *lactic acid* that may accumulate in the muscles. The blood transports lactic acid, which diffuses out of muscle fibers, and goes to the liver (fig. 8.10). In liver cells, reactions requiring ATP synthesize glucose from lactic acid.

During strenuous exercise, available oxygen is used primarily to synthesize the ATP the muscle fiber requires to contract, rather than to make ATP for synthesizing glucose from lactic acid. Consequently, as lactic acid accumulates, a person develops an **oxygen debt** that must be repaid. Oxygen debt equals the amount of oxygen liver cells require to convert the accumulated lactic acid into glucose, plus the amount muscle cells require to restore ATP and creatine phosphate to their original concentrations.

The conversion of lactic acid back into glucose is slow. Thus, repaying an oxygen debt following vigorous exercise may take several hours.

Muscle Fatigue

A muscle exercised strenuously for a prolonged period may lose its ability to contract, a condition called *fatigue*. Interruption in the muscle's blood supply or, rarely, lack of acetylcholine in motor nerve fibers may cause fatigue. Fatigue, however, is most likely to arise from accumulation of lactic acid in the muscle as a result of anaerobic respiration. The lactic acid buildup lowers pH, and as a result, muscle fibers no longer respond to stimulation.

Occasionally, a muscle becomes fatigued and cramps at the same time. A cramp is a painful condition in which the muscle contracts spasmodically but does not relax completely in between contractions. Cramping is due to a lack of ATP, which is required to return calcium ions to the sarcoplasmic reticulum and to break the linkages between the actin and myosin filaments before the muscle fibers can relax.

Several hours after death, the skeletal muscles undergo a partial contraction that fixes the joints. This condition, *rigor mortis,* may continue for 72 hours or more. It results from an increase in membrane permeability to calcium ions and a decrease in ATP in muscle fibers, which prevents relaxation. Thus, the actin and myosin filaments of the muscle fibers remain linked until the muscles begin to decompose. ■

Heat Production

Only about 25% of the energy released in cellular respiration is available for use in metabolic processes; the rest becomes heat. Although all active cells generate heat, muscle tissue is a major heat source because muscle is such a large proportion of the total body mass. Thus, active muscles release large amounts of heat. Blood transports this heat to other tissues, helping to maintain body temperature.

1. Which biochemicals provide the energy to regenerate ATP?
2. What are the sources of oxygen for aerobic respiration?
3. How are lactic acid, oxygen debt, and muscle fatigue related?
4. What is the relationship between cellular respiration and heat production?

Muscular Responses

One way to observe muscle contraction is to remove a single muscle fiber from a skeletal muscle and connect it to a device that records changes in the fiber's length. Such experiments usually require the use of an electrical device that can produce stimuli of varying strengths and frequencies.

Threshold Stimulus

When an isolated muscle fiber is exposed to a series of stimuli of increasing strength, the fiber remains unresponsive until a certain strength of stimulation is applied. This minimal strength required to cause a contraction is called the **threshold stimulus.**

All-or-None Response

A muscle fiber exposed to a stimulus of threshold strength (or above) responds to its fullest extent. Increasing the strength of the stimulus does not affect the fiber's degree of contraction. In other words, a muscle fiber cannot partially contract; if it contracts at all, it contracts completely, even though in some instances, it may not shorten completely. This phenomenon is called the **all-or-none response.**

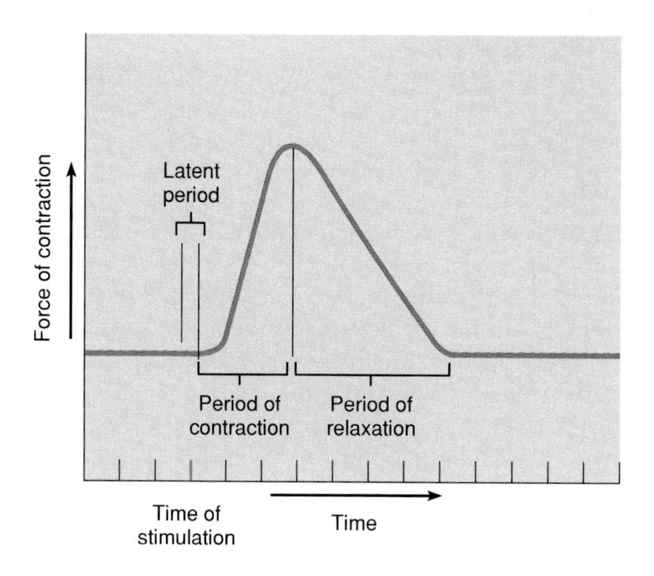

Figure 8.11

Myogram of a single muscle twitch.

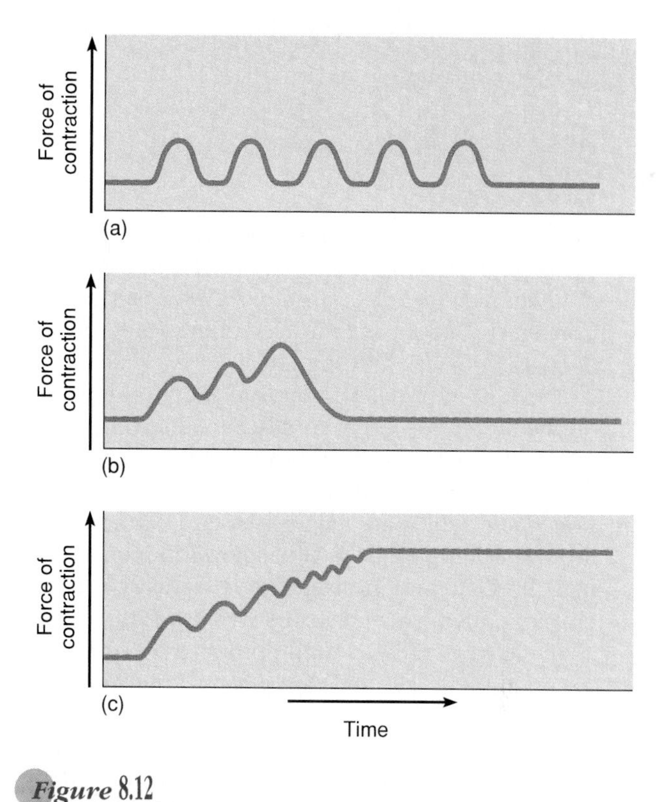

Figure 8.12

Myograms of (*a*) a series of twitches, (*b*) summation, and (*c*) a tetanic contraction. Note that stimulation frequency increases from one myogram to the next.

Recording a Muscle Contraction

A skeletal muscle removed from a frog or other small animal can show how a whole muscle responds to stimulation. The muscle is stimulated electrically, and when it contracts, its movement is recorded. The resulting pattern is called a **myogram.**

If a muscle is exposed to a single stimulus of sufficient strength to activate some of its motor units, the muscle will contract and then relax. This action—a single contraction that lasts only a fraction of a second—is called a **twitch.** A twitch produces a myogram like that in figure 8.11. The delay between the time the stimulus was applied and the time the muscle responded is the **latent period.** In a frog muscle, the latent period lasts about 0.01 second; in a human muscle, it is even shorter. The latent period is followed by a *period of contraction* when the muscle pulls at its attachments, and a *period of relaxation* when it returns to its former length.

Summation

The force that a muscle fiber can generate is not limited to the maximum force of a single twitch. A muscle fiber exposed to a series of stimuli of increasing frequency reaches a point when it is unable to completely relax before the next stimulus in the series arrives. When this happens, the force of individual twitches combines by the process of **summation.** When the resulting forceful, sustained contraction lacks even partial relaxation, it is called a **tetanic** (tĕ-tan′ik) **contraction** (tetanus) (fig. 8.12).

Recruitment of Motor Units

Since the muscle fibers within a muscle are organized into motor units, and a single motor neuron controls each motor unit, all the muscle fibers in a motor unit are stimulated at the same time. Therefore, a motor unit also responds in an all-or-none manner. A whole muscle, however, does not respond like this because it is composed of many motor units controlled by different motor neurons, which respond to different thresholds of stimulation. Thus, if only the motor neurons with low thresholds are stimulated, few motor units contract. At higher intensities of stimulation, other motor neurons respond, and more motor units are activated. Such an increase in the number of motor units being activated is called **recruitment.** As the intensity of stimulation increases, recruitment of motor units continues until, finally, all possible motor units are activated and the muscle contracts with maximal tension.

Sustained Contractions

At the same time that twitches combine, the strength of the contractions may increase due to the recruitment of motor units. The smaller motor units, which have finer fibers, are most easily stimulated and tend to respond earlier in the series of stimuli. The larger motor units,

TOPIC OF INTEREST

Use and Disuse of Skeletal Muscles

Skeletal muscles are very responsive to use and disuse. Forcefully exercised muscles enlarge, a phenomenon called *muscular hypertrophy.* Conversely, an unused muscle undergoes *atrophy,* decreasing in size and strength.

The way a muscle responds to use also depends on the type of exercise. A muscle contracting relatively weakly, as during swimming and running, activates a specialized group of muscle fibers called *slow fibers,* which are fatigue-resistant. With use, these specialized muscle fibers develop more mitochondria, and more extensive capillary networks envelope them. Such changes increase slow fibers' ability to resist fatigue during prolonged exercise, although their sizes and strengths may remain unchanged.

Forceful exercise, such as weight lifting, in which a muscle exerts more than 75% of its maximum tension, utilizes another group of specialized muscle fibers called *fast fibers,* which are fatigable. In response to strenuous exercise, these fibers produce new filaments of actin and myosin, the diameter of the muscle fibers increases, and the entire muscle enlarges. However, the muscular hypertrophy does not produce new muscle fibers.

The strength of a muscular contraction is directly proportional to the diameter of the activated muscle fibers. Consequently, an enlarged muscle can produce stronger contractions than before. Such a change, however, does not increase the muscle's ability to resist fatigue during activities like swimming or running.

If regular exercise stops, the capillary networks shrink, and the number of mitochondria within the muscle fibers drops. The number of actin and myosin filaments decreases, and the entire muscle atrophies. Such atrophy commonly occurs when accidents or diseases interfere with motor nerve impulses and prevent them from reaching muscle fibers. An unused muscle may decrease to less than half its usual size within a few months.

The fibers of muscles whose motor neurons are severed not only shrink, but also may fragment and, in time, be replaced by fat or fibrous connective tissue. However, reinnervation within the first few months following an injury may restore muscle function.

which contain thicker fibers, respond later and more forcefully. Summation and recruitment together can produce a *sustained contraction* of increasing strength.

Although twitches may occur occasionally in human skeletal muscles, as when an eyelid twitches, such contractions are of limited use. More commonly, muscular contractions are sustained. Lifting a weight or walking, for example, maintains sustained contractions in the upper limb or lower limb muscles for varying lengths of time. These contractions are responses to a rapid series of stimuli transmitted from the brain and spinal cord on motor neuron fibers.

Even when a muscle appears to be at rest, a certain amount of sustained contraction is occurring in its fibers. This is called **muscle tone** (tonus). Muscle tone is a response to nerve impulses originating repeatedly from the spinal cord and traveling to small numbers of muscle fibers.

Muscle tone is particularly important in maintaining posture. If tone is suddenly lost, as happens when a person loses consciousness, the body collapses.

When skeletal muscles contract very forcefully, they may generate up to 50 pounds of pull for each square inch of muscle cross section. Consequently, large muscles such as those in the thigh can pull with several hundred pounds of force. Occasionally, this force is so great that the tendons of muscles tear away from their attachments to the bones (*muscle pull*). ■

1. Define *threshold stimulus.*
2. What is an all-or-none response?
3. Distinguish between a twitch and a sustained contraction.
4. How is muscle tone maintained?

Smooth Muscles

The contractile mechanisms of smooth and cardiac muscles are essentially the same as those of skeletal muscles. The cells of these tissues, however, have some important structural and functional differences.

Smooth Muscle Fibers

Recall from chapter 5 that smooth muscle cells are elongated with tapering ends. They contain filaments of actin and myosin in myofibrils that extend the lengths of the cells. However, these filaments are organized differently and occur more randomly than those in skeletal muscle. Consequently, smooth muscle cells lack striations.

The two major types of smooth muscles are multiunit and visceral. In **multiunit smooth muscle**, the muscle fibers occur as separate fibers rather than in organized sheets. Smooth muscle of this type is in the irises of the eyes and in the walls of blood vessels. Typically, multiunit smooth muscle tissue contracts only after stimulation by motor nerve impulses or certain hormones.

Visceral smooth muscle is composed of sheets of spindle-shaped cells in close contact with one another (see fig. 5.22). This type, which is more common, is found in the walls of hollow organs, such as the stomach, intestines, urinary bladder, and uterus.

The fibers of visceral smooth muscles can stimulate each other. When one fiber is stimulated, the impulse moving over its surface may excite adjacent fibers, which, in turn, stimulate still others. Visceral smooth muscles also display *rhythmicity*—a pattern of repeated contractions. This phenomenon is due to self-exciting fibers that deliver spontaneous impulses periodically into surrounding muscle tissue. These two features—transmission of impulses from cell to cell and rhythmicity—are largely responsible for the wavelike motion, called **peristalsis**, that occurs in certain tubular organs, such as the intestines, and helps force the contents of these organs along their lengths.

Smooth Muscle Contraction

Smooth muscle contraction resembles skeletal muscle contraction in a number of ways. Both mechanisms include reactions of actin and myosin, both are triggered by membrane impulses and an increase in intracellular calcium ions, and both use energy from ATP. These two types of muscle tissue, however, also have significant differences.

Recall that acetylcholine is the neurotransmitter in skeletal muscle. Two neurotransmitters affect smooth muscle—acetylcholine and norepinephrine. Each of these neurotransmitters stimulates contractions in some smooth muscles and inhibits contractions in others (see chapter 9). Also, a number of hormones affect smooth muscle, stimulating contractions in some cases and altering the degree of response to neurotransmitters in others.

Smooth muscle is slower to contract and to relax than skeletal muscle. On the other hand, smooth muscle can maintain a forceful contraction longer with a given amount of ATP. Also, unlike skeletal muscle, smooth muscle fibers can change length without changing tautness; therefore, smooth muscles in the stomach and intestinal walls can stretch as these organs fill, maintaining the pressure inside these organs.

1. Describe two major types of smooth muscle.
2. What special characteristics of visceral smooth muscle make peristalsis possible?
3. How does smooth muscle contraction differ from that of skeletal muscle?

Cardiac Muscle

Cardiac muscle occurs only in the heart. It is composed of striated cells joined end to end, forming fibers (see fig. 5.23). These fibers interconnect in branching, three-dimensional networks. Each cell contains many filaments of actin and myosin, similar to those in skeletal muscle. A cardiac muscle cell also has a well-developed sarcoplasmic reticulum, many mitochondria, and a system of transverse tubules. The cisternae of cardiac muscle fibers, however, are less well developed and store less calcium than those of skeletal muscle. On the other hand, the transverse tubules of cardiac muscle are larger, and they release large numbers of calcium ions into the sarcoplasm in response to muscle impulses. This extra calcium from the transverse tubules comes from fluid outside the muscle fibers and causes cardiac muscle twitches to be longer than skeletal muscle twitches.

The opposing ends of cardiac muscle cells connect by cross-bands called *intercalated disks*. These bands are the result of elaborate junctions between cell membranes. The disks help to join cells and to transmit the force of contraction from cell to cell. Intercalated disks also allow muscle impulses to pass freely so that they travel rapidly from cell to cell.

When one portion of the cardiac muscle network is stimulated, the resulting impulse passes to the other fibers of the network, and the whole structure contracts as a unit; that is, the network responds to stimulation in an all-or-none manner. Cardiac muscle is also self-exciting and rhythmic. Consequently, a pattern of contraction and relaxation repeats again and again and causes the rhythmic contractions of the heart.

Table 8.2

Types of Muscle Tissue

	Skeletal	Smooth	Cardiac
Major location	Skeletal muscles	Walls of hollow viscera	Wall of the heart
Major function	Movement of bones at joints, maintenance of posture	Movement of viscera, peristalsis	Pumping action of the heart
Cellular characteristics			
Striations	Present	Absent	Present
Nucleus	Many nuclei	Single nucleus	Single nucleus
Special features	Well-developed transverse tubule system	Lacks transverse tubules	Well-developed transverse tubule system; intercalated disks separating adjacent cells
Mode of control	Voluntary	Involuntary	Involuntary
Contraction characteristics	Contracts and relaxes rapidly	Contracts and relaxes slowly; self-exciting; rhythmic	Network of fibers contracts as a unit; self-exciting; rhythmic

Table 8.2 summarizes the characteristics of the three types of muscle tissues.

1. How is cardiac muscle similar to smooth muscle?
2. How is cardiac muscle similar to skeletal muscle?
3. What is the function of intercalated disks?
4. What characteristic of cardiac muscle causes contraction of the heart as a unit?

Skeletal Muscle Actions

Skeletal muscles provide a variety of body movements. Each muscle's movement depends largely on the kind of joint it is associated with and the way the muscle attaches on either side of that joint.

Origin and Insertion

One end of a skeletal muscle usually fastens to a relatively immovable or fixed part, and the other end connects to a movable part on the other side of a joint. The immovable end of the muscle is called its **origin,** and the movable one is its **insertion.** When a muscle contracts, its insertion is pulled toward its origin.

Some muscles have more than one origin or insertion. The *biceps brachii* in the arm, for example, has two origins. This is reflected in the name *biceps,* which means *two heads.* (Note: The head of a muscle is the part nearest its origin.) One head of the muscle attaches to the coracoid process of the scapula, and the other head arises from a tubercle above the glenoid cavity of the scapula. The muscle extends along the front surface of the humerus and is inserted by means of a tendon on the radial tuberosity of the radius. When the biceps brachii contracts, its insertion is pulled toward its origin, and the upper limb bends at the elbow (fig. 8.13).

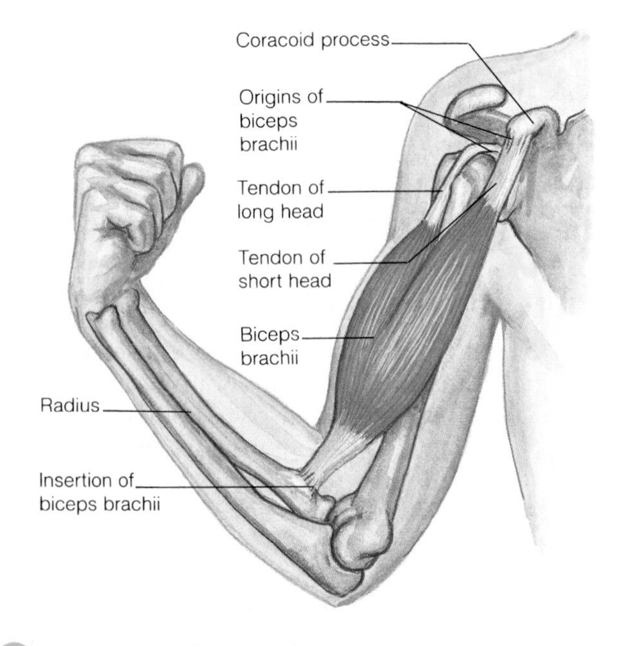

Figure 8.13

The biceps brachii has two heads that originate on the scapula. A tendon inserts this muscle on the radius.

Interaction of Skeletal Muscles

Skeletal muscles almost always function in groups. Consequently, for a particular body movement to occur, a person must do more than contract a single muscle; instead, after learning to make a particular movement, the person wills the movement to occur, and the nervous system stimulates the appropriate group of muscles.

Careful observation of body movements indicates the special roles of muscles. For instance, when the upper limb is lifted horizontally away from the side, a contracting *deltoid* muscle provides most of the movement and is said to be the **prime mover.** However, while a prime mover is acting, certain nearby muscles are also contracting. In the case of the contracting deltoid muscle, nearby muscles help hold the shoulder steady and in this way make the prime mover's action more effective. Muscles that contract and assist the prime mover are called **synergists.**

Still other muscles act as **antagonists** to prime movers. These muscles can resist a prime mover's action and cause movement in the opposite direction. For example, the antagonist of the prime mover that raises the upper limb can lower the upper limb, or the antagonist of the prime mover that bends the upper limb can straighten it (see fig. 7.6). If both a prime mover and its antagonist contract simultaneously, the part they act upon remains rigid. Consequently, smooth body movements depend on antagonists relaxing and, thus, giving way to the prime movers whenever the prime movers contract. Once again, the nervous system controls these complex actions, as chapter 9 describes.

1. Distinguish between the origin and insertion of a muscle.
2. Define *prime mover*.
3. What is the function of a synergist? An antagonist?

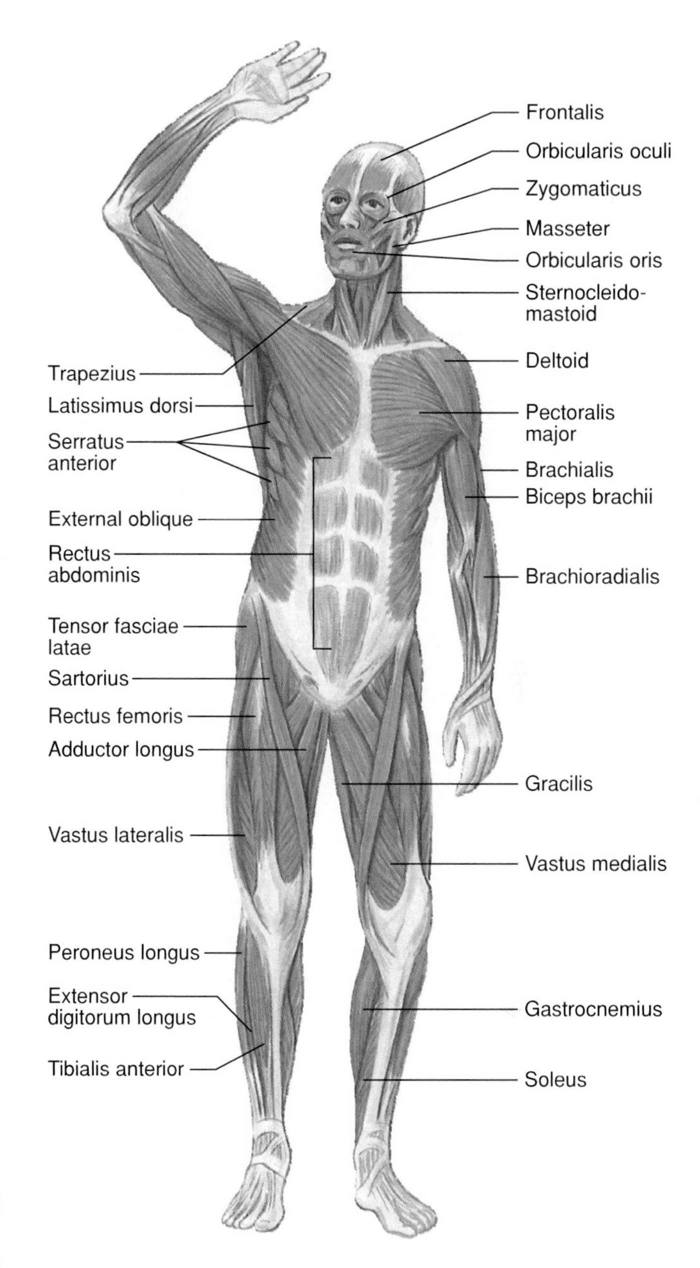

Figure 8.14

Anterior view of superficial skeletal muscles.

Major Skeletal Muscles

The section that follows discusses the locations, actions, and attachments of some of the major skeletal muscles. (Figures 8.14 and 8.15 and reference plates 1 and 2 show the locations of the superficial skeletal muscles—those near the surface.)

Note that the names of these muscles often describe them. A name may indicate a muscle's relative size, shape, location, action, number of attachments, or the direction of its fibers, as in the following examples:

Pectoralis major Of large size (major) located in the pectoral region (chest).

Deltoid Shaped like a delta or triangle.

Extensor digitorum Extends the digits (fingers or toes).

Biceps brachii With two heads (biceps) or points of origin and located in the brachium (arm).

Sternocleidomastoid Attached to the sternum, clavicle, and mastoid process.

External oblique Located near the outside with fibers that run obliquely (in a slanting direction).

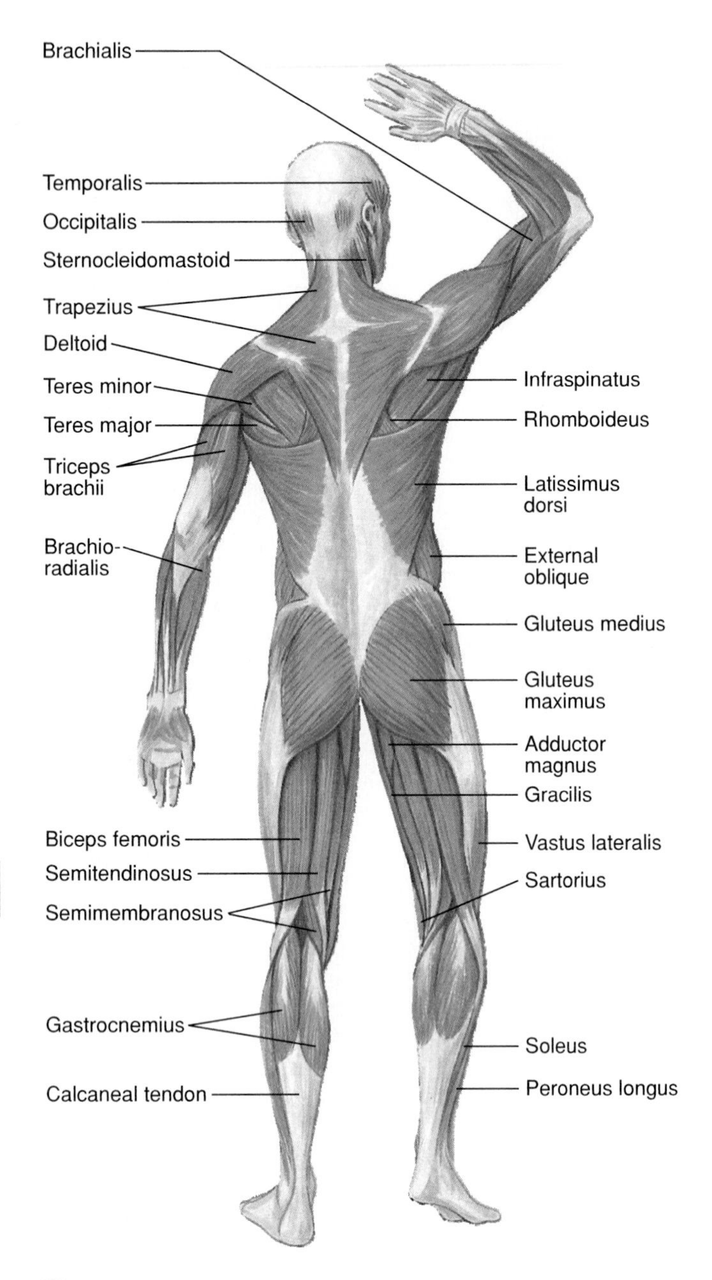

Figure 8.15

Posterior view of superficial skeletal muscles.

Muscles of Facial Expression

A number of small muscles that lie beneath the skin of the face and scalp enable us to communicate feelings through facial expression (fig. 8.16). Many of these muscles, located around the eyes and mouth, are responsible for such expressions as surprise, sadness, anger, fear, disgust, and pain. As a group, the muscles of facial expression join the bones of the skull to connective tissue in various regions of the overlying skin. They include:

Epicranius (ep″ĭ-kra′ne-us) (composed of two parts, the *frontalis* [frun-ta′lis] and the *occipitalis* [ok-sip″ĭ-ta′lis])
Orbicularis oculi (or-bik′u-la-rus ok′u-li)
Orbicularis oris (or-bik′u-la-rus o′ris)
Buccinator (buk′sĭ-na″tor)
Zygomaticus (zi″go-mat′ik-us)
Platysma (plah-tiz′mah)

Table 8.3 lists the origins, insertions, and actions of muscles of facial expression. (The muscles that move the eyes are listed in chapter 10.)

Muscles of Mastication

Muscles attached to the mandible produce chewing movements. Two pairs of these muscles close the lower jaw, a motion used in biting. These muscles are the *masseter* (mas-se′ter) and the *temporalis* (tem-po-ra′lis) (fig. 8.16).

Table 8.4 lists the origins, insertions, and actions of muscles of mastication.

Grinding the teeth, a common response to stress, may strain the temporomandibular joint—the articulation between the mandibular condyle of the mandible and the mandibular fossa of the temporal bone. This condition, called temporomandibular joint syndrome (TMJ syndrome), may produce headache, earache, and pain in the jaw, neck, or shoulder. ■

Muscles That Move the Head

Head movements result from the actions of paired muscles in the neck and upper back. These muscles flex, extend, and rotate the head. They include (fig. 8.16):

Sternocleidomastoid (ster″no-kli″do-mas′toid)
Splenius capitis (sple′ne-us kap′ĭ-tis)
Semispinalis capitis (sem″e-spi-na′lis kap′ĭ-tis)

Table 8.5 lists the origins, insertions, and actions of muscles that move the head.

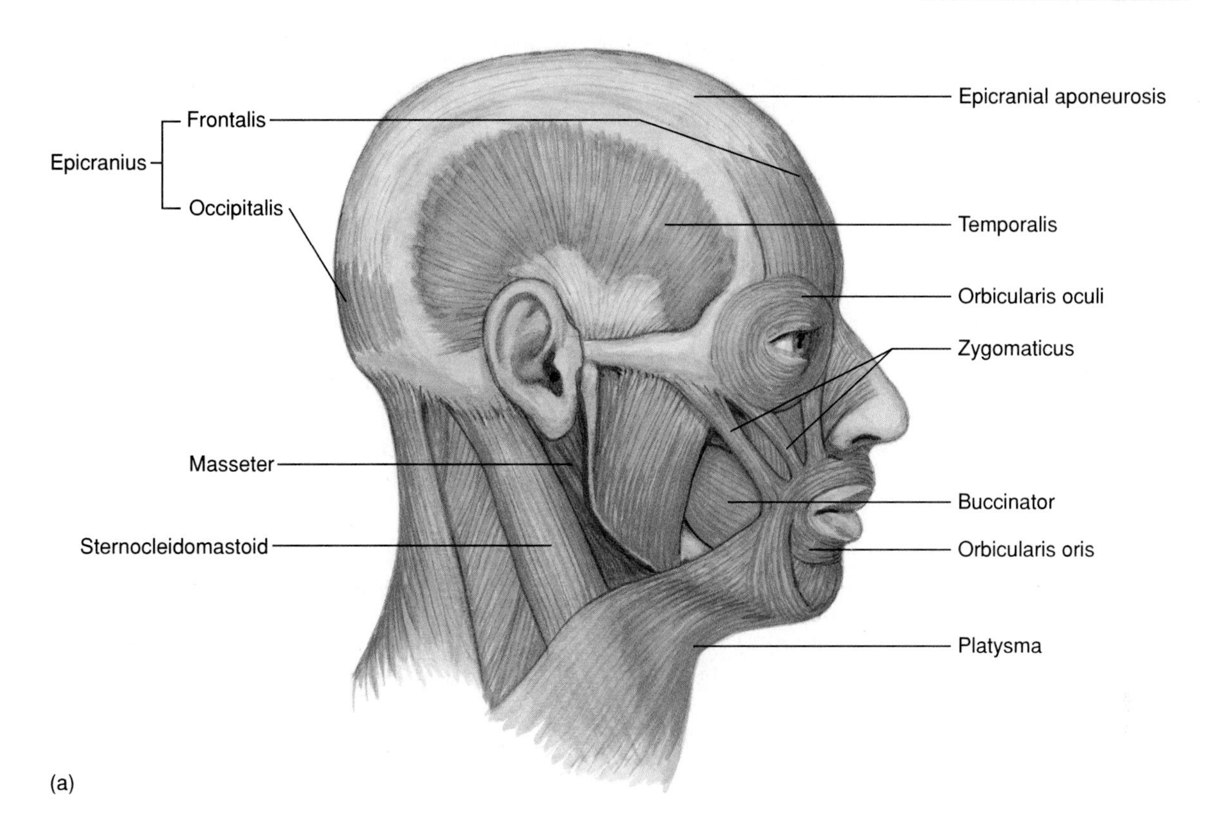

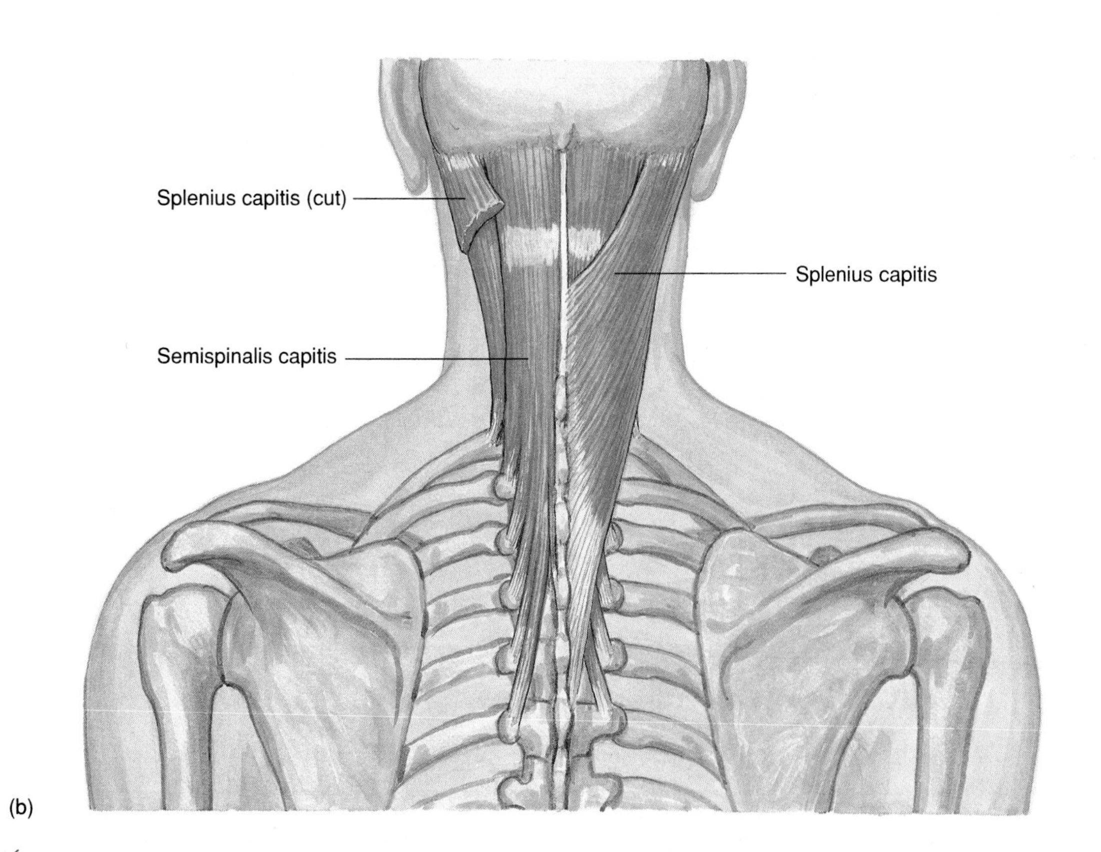

***Figure* 8.16**

(*a*) Muscles of facial expression and mastication. (*b*) Posterior view of muscles that move the head.

TOPIC OF INTEREST A New Muscle Discovered?

An unusual view of a cadaver provided a new perspective that may have revealed a previously undiscovered muscle. When in 1995 two dentists were examining a cadaver's skull whose eyes had been dissected out, they discovered what they believe is a new muscle in the head. The muscle, named the *sphenomandibularis,* extends about an inch and a half from behind the eyes to the inside of the jawbone and may help provide the movements of chewing. The muscle has a unique combination of the five characteristics used to describe muscles: an origin, insertion, innervation, a blood vessel supply, and specific function.

In traditional dissection from the side, the new muscle's origin and insertion are not visible, so it may have appeared to be part of the larger and overlying temporalis muscle. Although the sphenomandibularis inserts on the inner side of the jawbone, as does the temporalis, it originates differently, on the sphenoid bone.

Following their discovery of the sphenomandibularis in the eyeless cadaver head, the dentists quickly identified it in twenty-five other cadavers. Other researchers soon found it in live patients undergoing magnetic resonance imaging scans. If yet other researchers confirm that the muscle is actually newly found, it will certainly change the commonly held view that anatomy is a "dead" science.

Table 8.3 Muscles of Facial Expression

Muscle	Origin	Insertion	Action
Epicranius	Occipital bone	Skin and muscles around eye	Raises eyebrow
Orbicularis oculi	Maxillary and frontal bones	Skin around eye	Closes eye
Orbicularis oris	Muscles near the mouth	Skin of lips	Closes and protrudes lips
Buccinator	Outer surfaces of maxilla and mandible	Orbicularis oris	Compresses cheeks inward
Zygomaticus	Zygomatic bone	Orbicularis oris	Raises corner of mouth
Platysma	Fascia in upper chest	Lower border of mandible	Draws angle of mouth downward

Table 8.4 Muscles of Mastication

Muscle	Origin	Insertion	Action
Masseter	Lower border of zygomatic arch	Lateral surface of mandible	Closes jaw
Temporalis	Temporal bone	Coronoid process and lateral surface of mandible	Closes jaw

Table 8.5 Muscles That Move the Head

Muscle	Origin	Insertion	Action
Sternocleidomastoid	Anterior surface of sternum and upper surface of clavicle	Mastoid process of temporal bone	Pulls head to one side, pulls head toward chest, or raises sternum
Splenius capitis	Spinous processes of lower cervical and upper thoracic vertebrae	Mastoid process of temporal bone	Rotates head, bends head to one side, or brings head into an upright position
Semispinalis capitis	Processes of lower cervical and upper thoracic vertebrae	Occipital bone	Extends head, bends head to one side, or rotates head

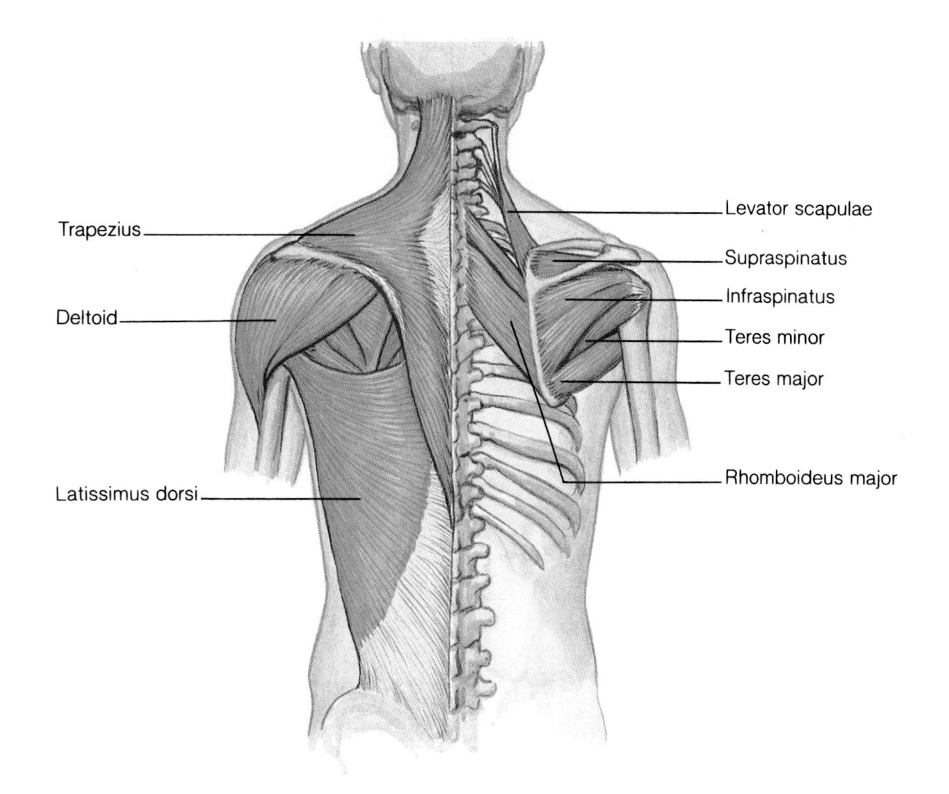

Figure 8.17

Muscles of the posterior shoulder. The right trapezius is removed to show underlying muscles.

Muscles That Move the Pectoral Girdle

Muscles that move the pectoral girdle are closely associated with those that move the arm. A number of these chest and shoulder muscles connect the scapula to nearby bones and move the scapula upward, downward, forward, and backward. They include (figs. 8.17 and 8.18):

Trapezius (trah-pe′ze-us)
Rhomboideus major (rom-boid′e-us)
Levator scapulae (le-va′tor scap′u-lē)
Serratus anterior (ser-ra′tus an-te′re-or)
Pectoralis minor (pek″to-ra′lis)

Table 8.6 lists the origins, insertions, and actions of muscles that move the pectoral girdle.

Muscles That Move the Arm

The arm is one of the more freely movable parts of the body. Muscles that connect the humerus to various regions of the pectoral girdle, ribs, and vertebral column

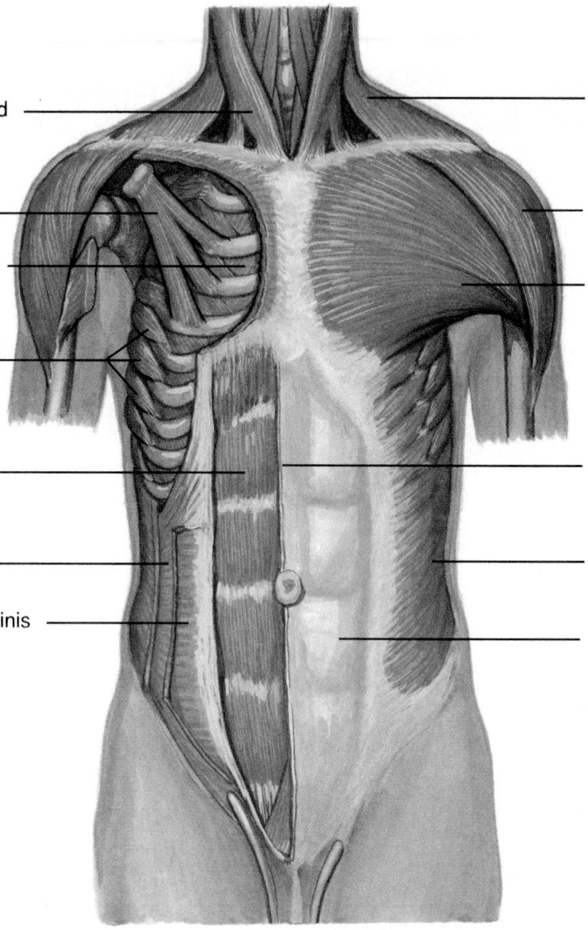

Figure 8.18

Muscles of the anterior chest and abdominal wall. The right pectoralis major is removed to show the pectoralis minor.

Table 8.6 Muscles That Move the Pectoral Girdle

Muscle	Origin	Insertion	Action
Trapezius	Occipital bone and spines of cervical and thoracic vertebrae	Clavicle; spine and acromion process of scapula	Rotates scapula and raises arm; raises scapula; pulls scapula medially; or pulls scapula and shoulder downward
Rhomboideus major	Spines of upper thoracic vertebrae	Medial border of scapula	Raises and adducts scapula
Levator scapulae	Transverse processes of cervical vertebrae	Medial margin of scapula	Elevates scapula
Serratus anterior	Outer surfaces of upper ribs	Ventral surface of scapula	Pulls scapula anteriorly and downward
Pectoralis minor	Sternal ends of upper ribs	Coracoid process of scapula	Pulls scapula anteriorly and downward or raises ribs

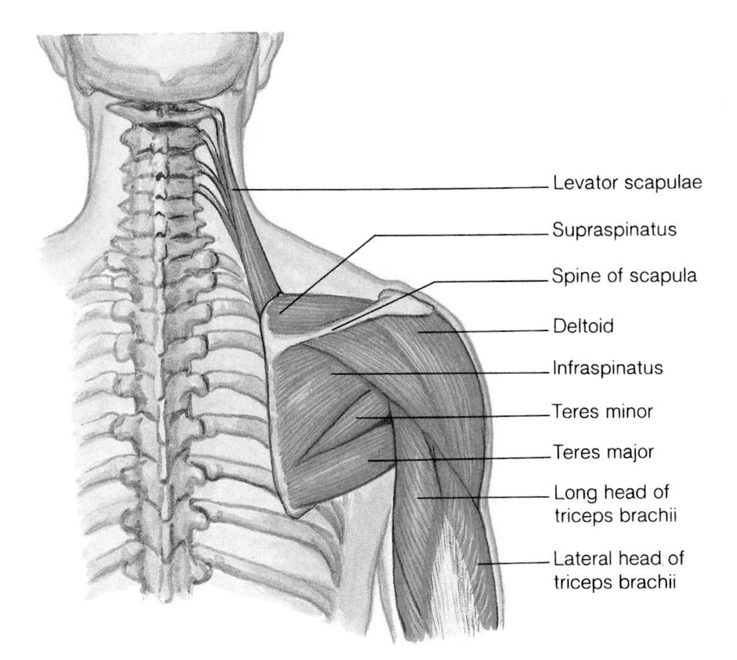

Figure 8.19

Muscles of the posterior surface of the scapula and arm.

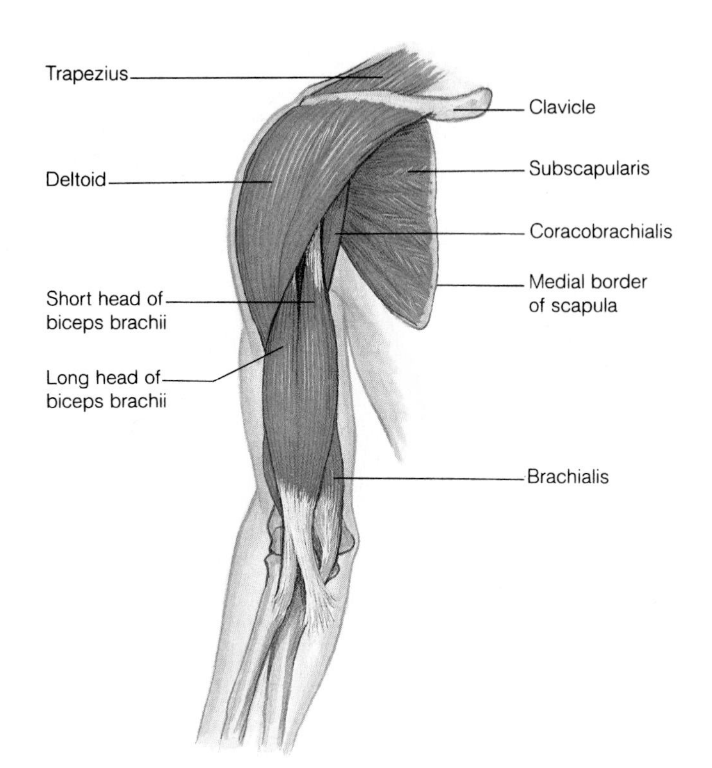

Figure 8.20

Muscles of the anterior shoulder and arm, with the rib cage removed.

make these movements possible (figs. 8.17, 8.18, 8.19, and 8.20). These muscles can be grouped according to their primary actions—flexion, extension, abduction, and rotation—as follows:

Flexors

Coracobrachialis (kor″ah-ko-bra′ke-al-is)

Pectoralis major (pek″to-ra′lis)

Extensors

Teres major (te′rēz)

Latissimus dorsi (lah-tis′ĭ-mus dor′si)

Abductors

Supraspinatus (su″prah-spi′na-tus)

Deltoid (del′toid)

Rotators

Subscapularis (sub-scap′u-lar-is)

Infraspinatus (in″frah-spi′na-tus)

Teres minor (te′rēz)

Table 8.7 lists the origins, insertions, and actions of muscles that move the arm.

Muscles That Move the Forearm

Muscles that connect the radius or ulna to the humerus or pectoral girdle produce most forearm movements. A group of muscles located along the anterior surface of the humerus flexes the elbow, and a single posterior muscle extends this joint. Other muscles move the radioulnar joint and rotate the forearm.

Muscles that move the forearm include (figs. 8.19, 8.20, and 8.21):

Flexors

Biceps brachii (bi′seps bra′ke-i)

Brachialis (bra′ke-al-is)

Brachioradialis (bra″ke-o-ra″de-a′lis)

Extensor

Triceps brachii (tri′seps bra′ke-i)

Rotators

Supinator (su′pĭ-na-tor)

Pronator teres (pro-na′tor te′rēz)

Pronator quadratus (pro-na′tor kwod-ra′tus)

Table 8.7

Muscles That Move the Arm

Muscle	Origin	Insertion	Action
Coracobrachialis	Coracoid process of scapula	Shaft of humerus	Flexes and adducts arm
Pectoralis major	Clavicle, sternum, and costal cartilages of upper ribs	Intertubercular groove of humerus	Pulls arm anteriorly and across chest, rotates humerus, or adducts arm
Teres major	Lateral border of scapula	Intertubercular groove of humerus	Extends humerus, or adducts and rotates arm medially
Latissimus dorsi	Spines of sacral, lumbar, and lower thoracic vertebrae, iliac crest, and lower ribs	Intertubercular groove of humerus	Extends and adducts arm and rotates humerus inwardly, or pulls shoulder downward and posteriorly
Supraspinatus	Posterior surface of scapula	Greater tubercle of humerus	Abducts arm
Deltoid	Acromion process, spine of scapula, and clavicle	Deltoid tuberosity of humerus	Abducts arm, extends or flexes humerus
Subscapularis	Anterior surface of scapula	Lesser tubercle of humerus	Rotates arm medially
Infraspinatus	Posterior surface of scapula	Greater tubercle of humerus	Rotates arm laterally
Teres minor	Lateral border of scapula	Greater tubercle of humerus	Rotates arm laterally

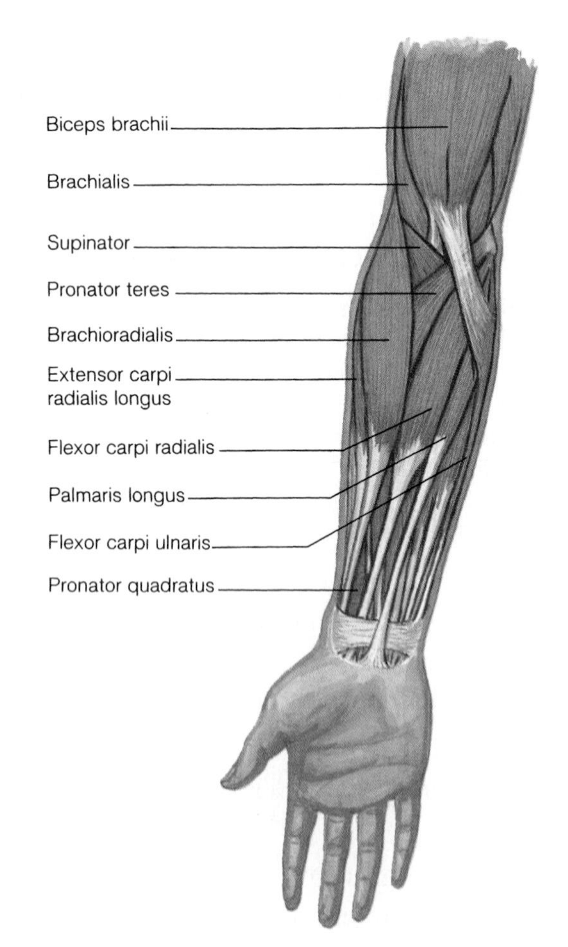

***Figure* 8.21**

Muscles of the anterior forearm.

Table 8.8 lists the origins, insertions, and actions of muscles that move the forearm.

Muscles That Move the Wrist, Hand, and Fingers

Many muscles move the wrist, hand, and fingers. They originate from the distal end of the humerus and from the radius and ulna. The two major groups of these muscles are flexors on the anterior side of the forearm and extensors on the posterior side. These muscles include (figs. 8.21 and 8.22):

Flexors

- *Flexor carpi radialis* (flex′sor kar-pi′ ra″de-a′lis)
- *Flexor carpi ulnaris* (flex′sor kar-pi′ ul-na′ris)
- *Palmaris longus* (pal-ma′ris long′gus)
- *Flexor digitorum profundus* (flex′sor dij″ĭ-to′rum pro-fun′dus)

Extensors

- *Extensor carpi radialis longus* (eks-ten′sor kar-pi′ ra″de-a′lis long′gus)
- *Extensor carpi radialis brevis* (eks-ten′sor kar-pi′ ra″de-a′lis brev′ĭs)
- *Extensor carpi ulnaris* (eks-ten′sor kar-pi′ ul-na′ris)
- *Extensor digitorum* (eks-ten′sor dij″ĭ-to′rum)

Table 8.9 lists the origins, insertions, and actions of muscles that move the wrist, hand, and fingers.

Table 8.8

Muscles That Move the Forearm

Muscle	Origin	Insertion	Action
Biceps brachii	Coracoid process and tubercle above glenoid cavity of scapula	Radial tuberosity of radius	Flexes forearm at elbow and rotates hand laterally
Brachialis	Anterior shaft of humerus	Coronoid process of ulna	Flexes forearm at elbow
Brachioradialis	Distal lateral end of humerus	Lateral surface of radius above styloid process	Flexes forearm at elbow
Triceps brachii	Tubercle below glenoid cavity and lateral and medial surfaces of humerus	Olecranon process of ulna	Extends forearm at elbow
Supinator	Lateral epicondyle of humerus and crest of ulna	Lateral surface of radius	Rotates forearm laterally
Pronator teres	Medial epicondyle of humerus and coronoid process of ulna	Lateral surface of radius	Rotates forearm medially
Pronator quadratus	Anterior distal end of ulna	Anterior distal end of radius	Rotates forearm medially

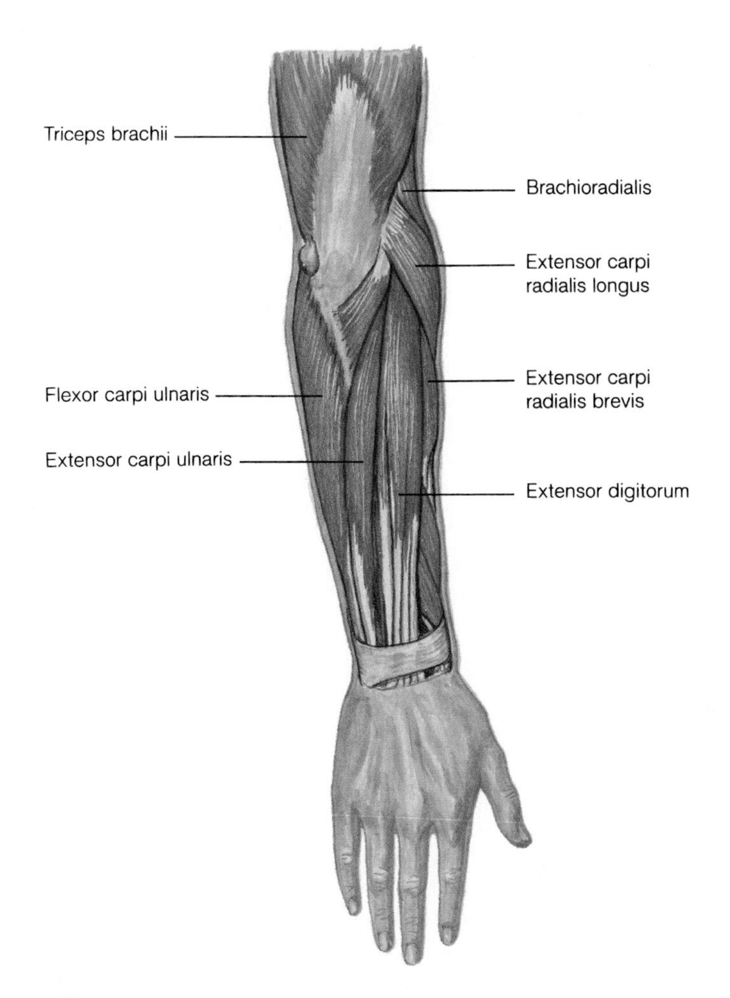

Figure 8.22

Muscles of the posterior forearm.

Muscles of the Abdominal Wall

Bone supports the walls of the chest and pelvic regions, but not those of the abdomen. Instead, the anterior and lateral walls of the abdomen are composed of layers of broad, flattened muscles. These muscles connect the rib cage and vertebral column to the pelvic girdle. A band of tough connective tissue called the **linea alba** extends from the xiphoid process of the sternum to the symphysis pubis (see fig. 8.18). It is an attachment for some of the abdominal wall muscles.

Contraction of these muscles decreases the size of the abdominal cavity and increases the pressure inside. These actions help press air out of the lungs during forceful exhalation and aid in the movements of defecation, urination, vomiting, and childbirth.

The abdominal wall muscles include (see fig. 8.18):

External oblique (eks-ter′nal o-blēk)
Internal oblique (in-ter′nal o-blēk)
Transversus abdominis (trans-ver′sus ab-dom′ĭ-nis)
Rectus abdominis (rek′tus ab-dom′ĭ-nis)

Table 8.10 lists the origins, insertions, and actions of muscles of the abdominal wall.

Muscles of the Pelvic Outlet

Two muscular sheets—a deeper **pelvic diaphragm** and a more superficial **urogenital diaphragm**—span the outlet of the pelvis. The pelvic diaphragm forms the floor of the pelvic cavity, and the urogenital diaphragm fills

Table 8.9 Muscles That Move the Wrist, Hand, and Fingers

Muscle	Origin	Insertion	Action
Flexor carpi radialis	Medial epicondyle of humerus	Base of second and third metacarpals	Flexes and abducts wrist
Flexor carpi ulnaris	Medial epicondyle of humerus and olecranon process	Carpal and metacarpal bones	Flexes and adducts wrist
Palmaris longus	Medial epicondyle of humerus	Fascia of palm	Flexes wrist
Flexor digitorum profundus	Anterior surface of ulna	Bases of distal phalanges in fingers 2–5	Flexes distal joints of fingers
Extensor carpi radialis longus	Distal end of humerus	Base of second metacarpal	Extends wrist and abducts hand
Extensor carpi radialis brevis	Lateral epicondyle of humerus	Base of second and third metacarpals	Extends wrist and abducts hand
Extensor carpi ulnaris	Lateral epicondyle of humerus	Base of fifth metacarpal	Extends and adducts wrist
Extensor digitorum	Lateral epicondyle of humerus	Posterior surface of phalanges in fingers 2–5	Extends fingers

Table 8.10 Muscles of the Abdominal Wall

Muscle	Origin	Insertion	Action
External oblique	Outer surfaces of lower ribs	Outer lip of iliac crest and linea alba	Tenses abdominal wall and compresses abdominal contents
Internal oblique	Crest of ilium and inguinal ligament	Cartilages of lower ribs, linea alba, and crest of pubis	Tenses abdominal wall and compresses abdominal contents
Transversus abdominis	Costal cartilages of lower ribs, processes of lumbar vertebrae, lip of iliac crest, and inguinal ligament	Linea alba and crest of pubis	Tenses abdominal wall and compresses abdominal contents
Rectus abdominis	Crest of pubis and symphysis pubis	Xiphoid process of sternum and costal cartilages	Tenses abdominal wall and compresses abdominal contents; also flexes vertebral column

the space within the pubic arch (see fig. 7.26). The muscles of the male and female pelvic outlets include (fig. 8.23):

Pelvic diaphragm

Levator ani (le-va′tor ah-ni′)

Urogenital diaphragm

Superficial transversus perinei (su″per-fish′al trans-ver′sus per″ĭ-ne′i)

Bulbospongiosus (bul″bo-spon″je-o′sus)

Ischiocavernosus (is″ke-o-kav″er-no′sus)

Table 8.11 lists the origins, insertions, and actions of pelvic outlet muscles.

Muscles That Move the Thigh

Muscles that move the thigh are attached to the femur and to some part of the pelvic girdle. These muscles occur in anterior and posterior groups. Muscles of the anterior group primarily flex the thigh; those of the posterior group extend, abduct, or rotate the thigh. The muscles in these groups include (figs. 8.24, 8.25, and 8.26):

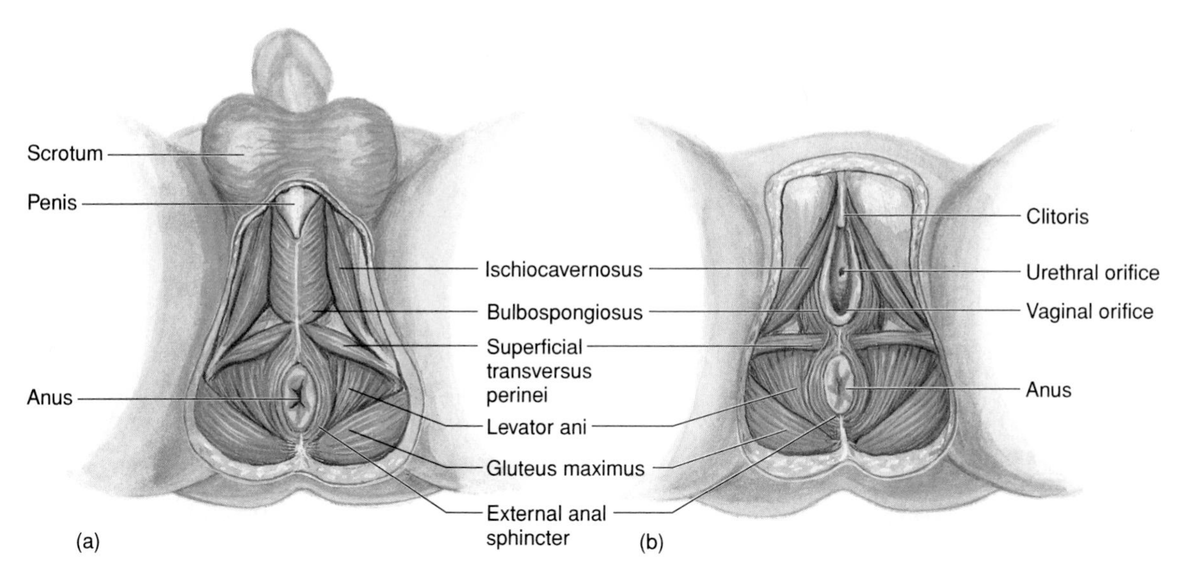

Figure 8.23

External view of muscles of (*a*) the male pelvic outlet and (*b*) the female pelvic outlet.

Table 8.11 Muscles of the Pelvic Outlet

Muscle	Origin	Insertion	Action
Levator ani	Pubic bone and ischial spine	Coccyx	Supports pelvic viscera and provides sphincterlike action in anal canal and vagina
Superficial transversus perinei	Ischial tuberosity	Central tendon	Supports pelvic viscera
Bulbospongiosus	Central tendon	Males: Urogenital diaphragm and fascia of the penis Females: Pubic arch and root of clitoris	Males: Assists emptying of urethra Females: Constricts vagina
Ischiocavernosus	Ischial tuberosity	Pubic arch	Assists function of bulbospongiosus

Anterior group
- *Psoas major* (so′as)
- *Iliacus* (il′e-ak-us)

Posterior group
- *Gluteus maximus* (gloo′te-us mak′si-mus)
- *Gluteus medius* (gloo′te-us me′de-us)
- *Gluteus minimus* (gloo′te-us min′ĭ-mus)
- *Tensor fasciae latae* (ten′sor fash′e-e lah-tē)

Still another group of muscles attached to the femur and pelvic girdle adduct the thigh. They include (figs. 8.24 and 8.26):

- *Adductor longus* (ah-duk′tor long′gus)
- *Adductor magnus* (ah-duk′tor mag′nus)
- *Gracilis* (gras′il-is)

Table 8.12 lists the origins, insertions, and actions of muscles that move the thigh.

Muscles That Move the Leg

Muscles that move the leg connect the tibia or fibula to the femur or to the pelvic girdle. They can be separated into two major groups—those that flex the knee and those that extend the knee. Muscles of these

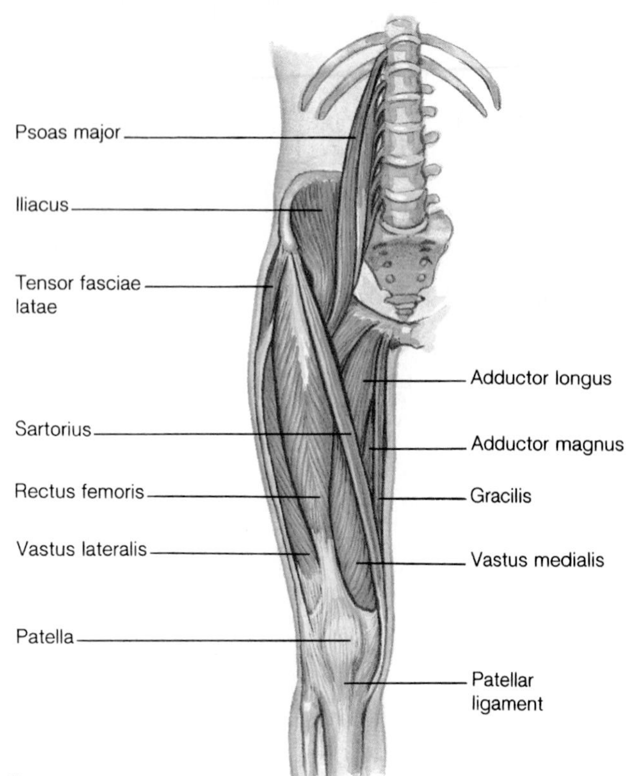

Figure 8.24

Muscles of the anterior right thigh. (Note that vastus intermedius is a deep muscle not visible in this view.)

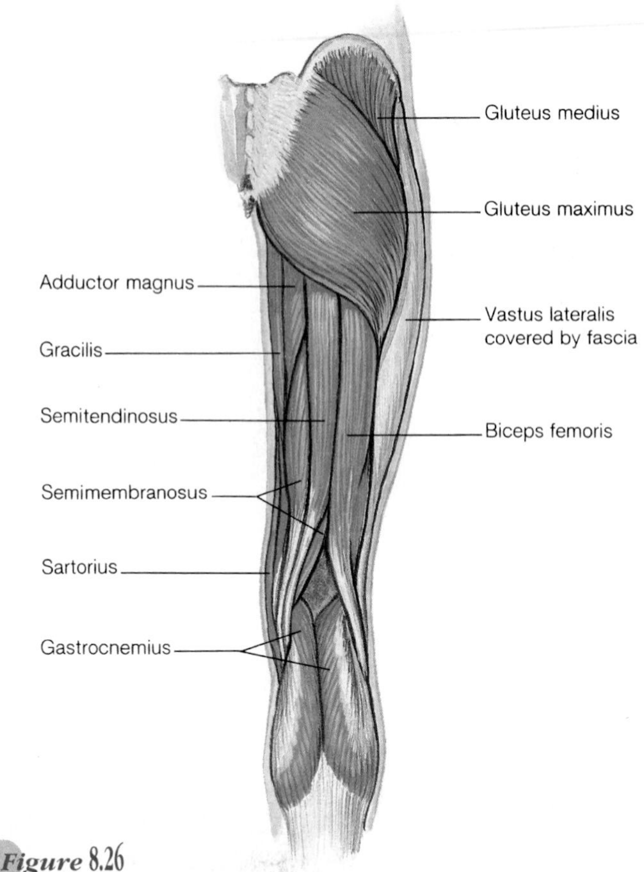

Figure 8.26

Muscles of the posterior right thigh.

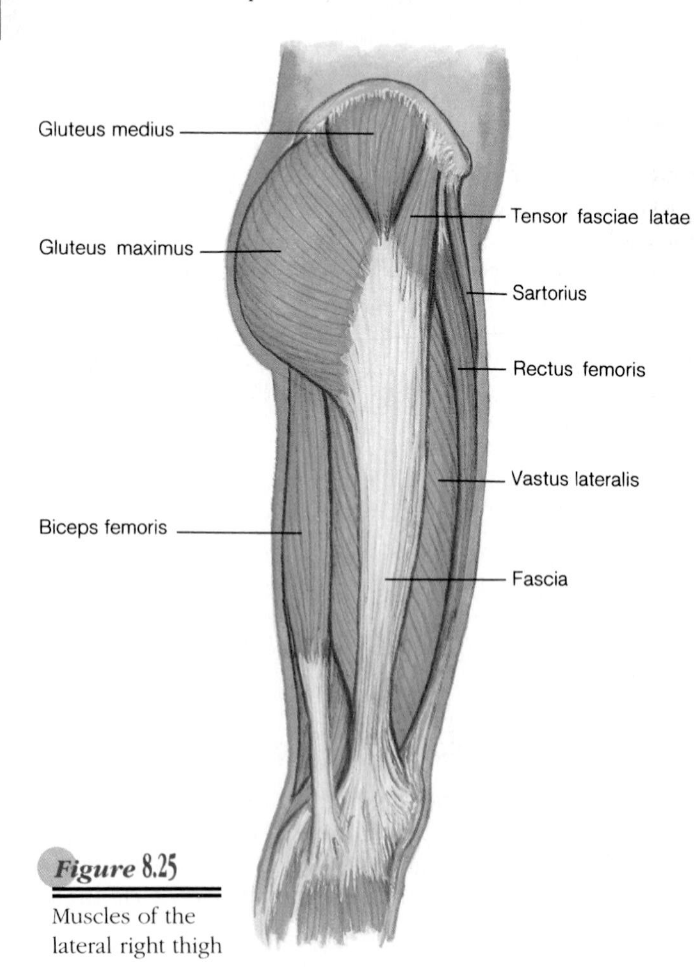

Figure 8.25

Muscles of the lateral right thigh

groups include the hamstring group and the quadriceps femoris group (figs. 8.24, 8.25, and 8.26):

Flexors

Biceps femoris (bi′seps fem′or-is)

Semitendinosus (sem″e-ten′dĭ-no-sus)

Semimembranosus (sem″e-mem′brah-no-sus)

Sartorius (sar-to′re-us)

Extensor

Quadriceps femoris group (kwod′rĭ-seps fem′or-is) (composed of four parts—the rectus femoris, vastus lateralis, vastus medialis, and vastus intermedius)

Table 8.13 lists the origins, insertions, and actions of muscles that move the leg.

Table 8.12 Muscles That Move the Thigh

Muscle	Origin	Insertion	Action
Psoas major	Lumbar intervertebral disks, bodies and transverse processes of lumbar vertebrae	Lesser trochanter of femur	Flexes thigh
Iliacus	Iliac fossa of ilium	Lesser trochanter of femur	Flexes thigh
Gluteus maximus	Sacrum, coccyx, and posterior surface of ilium	Posterior surface of femur and fascia of thigh	Extends thigh
Gluteus medius	Lateral surface of ilium	Greater trochanter of femur	Abducts and rotates thigh medially
Gluteus minimus	Lateral surface of ilium	Greater trochanter of femur	Abducts and rotates thigh medially
Tensor fasciae latae	Anterior iliac crest	Fascia of thigh	Abducts, flexes, and rotates thigh medially
Adductor longus	Pubic bone near symphysis pubis	Posterior surface of femur	Adducts, flexes, and rotates thigh laterally
Adductor magnus	Ischial tuberosity	Posterior surface of femur	Adducts, extends, and rotates thigh laterally
Gracilis	Lower edge of symphysis pubis	Medial surface of tibia	Adducts thigh, flexes and rotates lower limb medially

Table 8.13 Muscles That Move the Leg

Muscle	Origin	Insertion	Action
Hamstring group			
Biceps femoris	Ischial tuberosity and posterior surface of femur	Head of fibula and lateral condyle of tibia	Flexes leg, extends thigh
Semitendinosus	Ischial tuberosity	Medial surface of tibia	Flexes leg, extends thigh
Semimembranosus	Ischial tuberosity	Medial condyle of tibia	Flexes leg, extends thigh
Sartorius	Anterior superior iliac spine	Medial surface of tibia	Flexes leg and thigh, abducts thigh, rotates thigh laterally, and rotates leg medially
Quadriceps femoris group			
Rectus femoris	Spine of ilium and margin of acetabulum	Patella by the tendon, which continues as patellar ligament to tibial tuberosity	Extends leg at knee
Vastus lateralis	Greater trochanter and posterior surface of femur	Patella by the tendon, which continues as patellar ligament to tibial tuberosity	Extends leg at knee
Vastus medialis	Medial surface of femur	Patella by the tendon, which continues as patellar ligament to tibial tuberosity	Extends leg at knee
Vastus intermedius	Anterior and lateral surfaces of femur	Patella by the tendon, which continues as patellar ligament to tibial tuberosity	Extends leg at knee

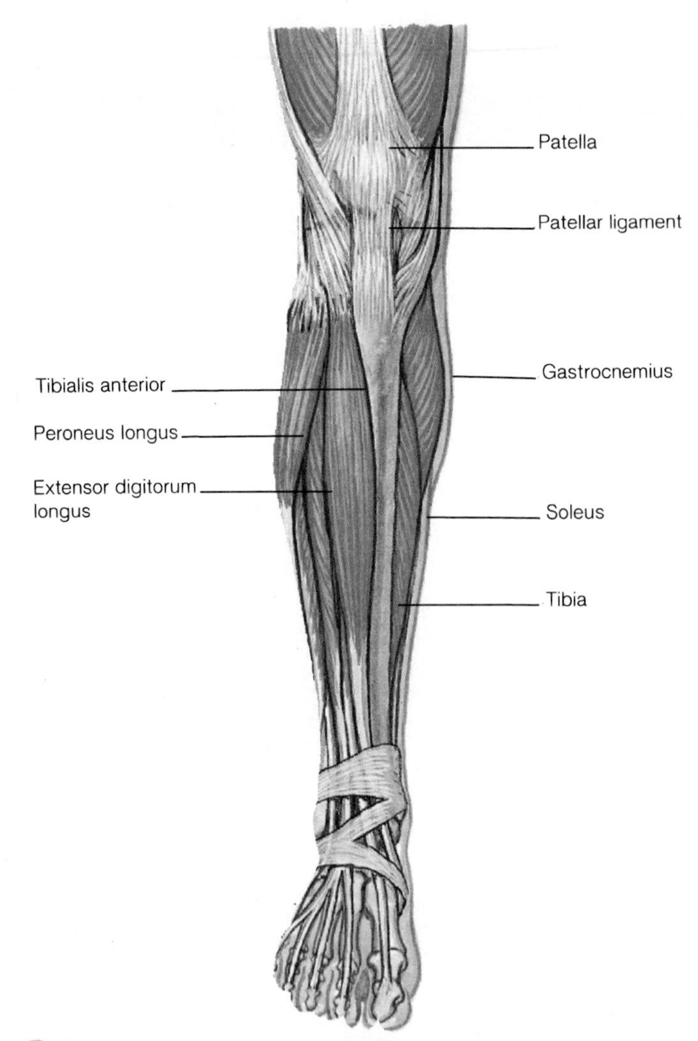

Figure 8.27

Muscles of the anterior right leg.

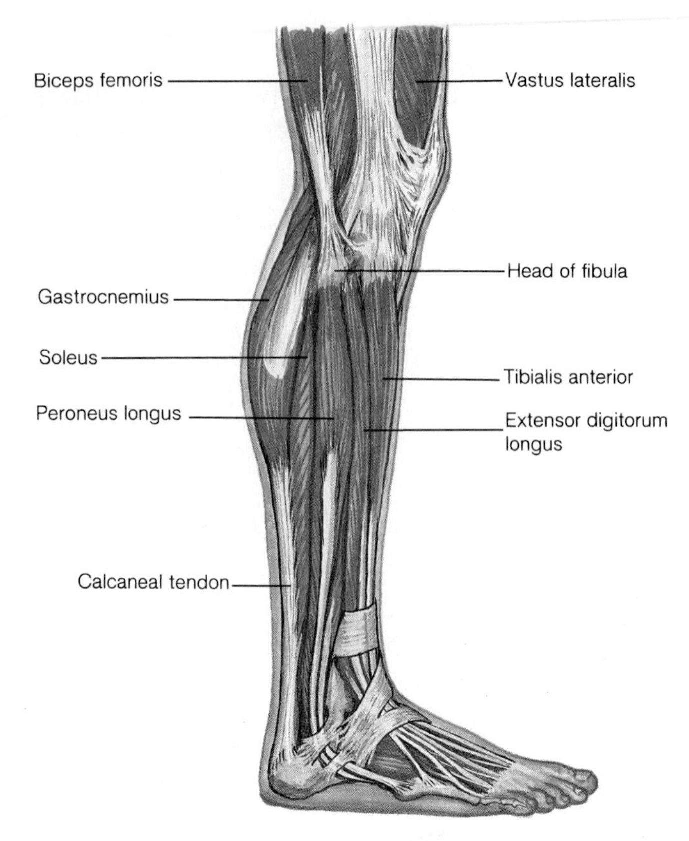

Figure 8.28

Muscles of the lateral right leg. (Note that tibialis posterior is a deep muscle not visible in this view.)

Muscles That Move the Ankle, Foot, and Toes

A number of muscles that move the ankle, foot, and toes are located in the leg. They attach the femur, tibia, and fibula to bones of the foot and are responsible for moving the foot upward (dorsiflexion) or downward (plantar flexion), and turning the sole of the foot inward (inversion) or outward (eversion). These muscles include (figs. 8.27, 8.28, and 8.29):

Dorsal flexors

Tibialis anterior (tib″e-a′lis an-te′re-or)

Peroneus tertius (per″o-ne′us ter′shus)

Extensor digitorum longus (eks-ten′sor dij″ĭ-to′rum long′gus)

Plantar flexors

Gastrocnemius (gas″trok-ne′me-us)

Soleus (so′le-us)

Flexor digitorum longus (flek′sor dij″ĭ-to′rum) long′gus)

Invertor

Tibialis posterior (tib″e-a′lis pos-tēr′e-or)

Evertor

Peroneus longus (per″o-ne′us long′gus)

Table 8.14 lists the origins, insertions, and actions of muscles that move the ankle, foot, and toes.

Clinical Terms Related to the Muscular System

Contracture (kon-trak′tur) A condition of great resistance to the stretch of a muscle.

Convulsion (kun-vul′shun) A series of involuntary contractions of various voluntary muscles.

Electromyography (e-lek″tro-mi-og′rah-fe) A technique for recording electrical changes in muscle tissues.

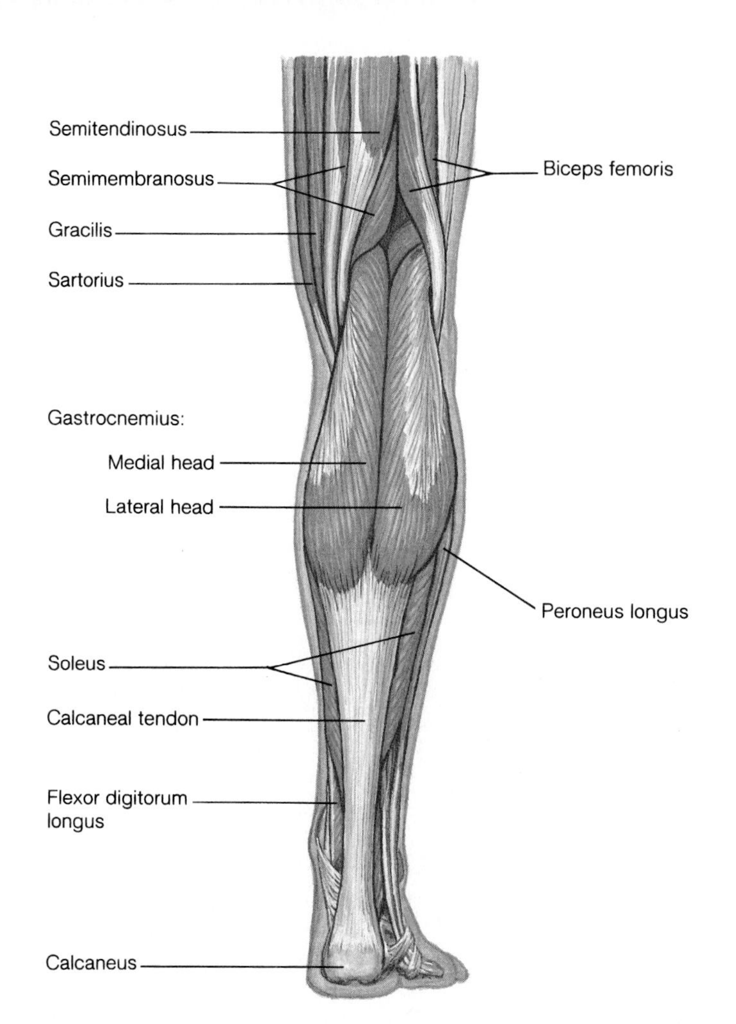

Figure 8.29

Muscles of the posterior right leg.

Fibrillation (fi″brĭ-la′shun) Spontaneous contractions of individual muscle fibers, producing rapid and uncoordinated activity within a muscle.

Fibrosis (fi-bro′sis) A degenerative disease in which fibrous connective tissue replaces skeletal muscle tissue.

Fibrositis (fi″bro-si′tis) Inflammation of fibrous connective tissues, especially in the muscle fascia. This disease is also called *muscular rheumatism*.

Muscular dystrophies (mus′ku-lar dis′tro-fes) A group of inherited disorders in which deficiency of cytoskeletal protein (or glycoprotein) collapses muscle cells, leading to progressive loss of function.

Myalgia (mi-al′je-ah) Pain from any muscular disease or disorder.

Myasthenia gravis (mi″as-the′ne-ah gra′vis) A chronic disease in which muscles are weak and easily fatigued because of malfunctioning neuromuscular junctions.

Myokymia (mi″o-ki′me-ah) Persistent quivering of a muscle.

Myology (mi-ol′o-je) The study of muscles.

Myoma (mi-o′mah) A tumor composed of muscle tissue.

Myopathy (mi-op′ah-the) Any muscular disease.

Myositis (mi″o-si′tis) Inflammation of skeletal muscle tissue.

Table 8.14 Muscles That Move the Ankle, Foot, and Toes

Muscle	Origin	Insertion	Action
Tibialis anterior	Lateral condyle and lateral surface of tibia	Tarsal bone (cuneiform) and first metatarsal	Dorsiflexion and inversion of foot
Peroneus tertius	Anterior surface of fibula	Dorsal surface of fifth metatarsal	Dorsiflexion and eversion of foot
Extensor digitorum longus	Lateral condyle of tibia and anterior surface of fibula	Dorsal surfaces of second and third phalanges of the four lateral toes	Dorsiflexion and eversion of foot and extension of toes
Gastrocnemius	Lateral and medial condyles of femur	Posterior surface of calcaneus	Plantar flexion of foot and flexion of leg at knee
Soleus	Head and shaft of fibula and posterior surface of tibia	Posterior surface of calcaneus	Plantar flexion of foot
Flexor digitorum longus	Posterior surface of tibia	Distal phalanges of the four lateral toes	Plantar flexion and inversion of foot, and flexion of the four lateral toes
Tibialis posterior	Lateral condyle and posterior surface of tibia, and posterior surface of fibula	Tarsal and metatarsal bones	Plantar flexion and inversion of foot
Peroneus longus	Lateral condyle of tibia and head and shaft of fibula	Tarsal and metatarsal bones	Plantar flexion and eversion of foot; also supports arch

TOPIC OF INTEREST: Steroids and Athletes—An Unhealthy Combination

Canadian track star Ben Johnson flew past his competitors in the 100-meter run at the 1988 Summer Olympics in Seoul, Korea. But 72 hours later, officials rescinded the gold medal awarded for his record-breaking time of 9.79 seconds after a urine test revealed traces of the drug stanozolol (fig. 8A).

Stanozolol is one of several synthetic versions of the steroid hormone testosterone. Like testosterone, these drugs promote signs of masculinity (their androgenic effect) and increased synthesis of muscle proteins (their anabolic effect). Used in the past to treat a handful of medical conditions—anemia and breast cancer among them—steroids are used by some professional and amateur athletes to build muscle tissue easily.

Steroid users may improve their performances and physiques in the short term, but in the long run, they may suffer for it. Steroids hasten adulthood, stunting height and causing early baldness. In males, they lead to breast development and in females to a deepened voice, hairiness, and a male physique. Steroid use may damage the kidneys, liver, and heart, and atherosclerosis may develop because steroids raise LDL and lower HDL—the opposite of a healthy cholesterol profile. In males, the body mistakes the synthetic steroids for the natural hormone and lowers its own testosterone production. The price of athletic prowess today may be infertility later.

Steroid use began in Nazi Germany, where Hitler used the drugs to fashion his "super race." Ironically, steroids were used shortly after to build up the emaciated bodies of concentration camp survivors. In the 1950s, Soviet athletes began using steroids in the Olympics, and a decade later, U.S. athletes did the same. In 1976, the International Olympic Committee banned steroid use and required urine tests.

Ben Johnson was caught in his tracks by such a test, refined by 1988 so that part-per-billion traces of synthetic steroids can be detected even weeks after they are taken. Even though Johnson at first claimed the stanozolol in his urine was the result of a spiked drink of an approved anti-inflammatory drug used on his ankle, a test of his natural testosterone level showed it to be only 15% of normal—a sure sign that this athlete had been taking steroids for a long time.

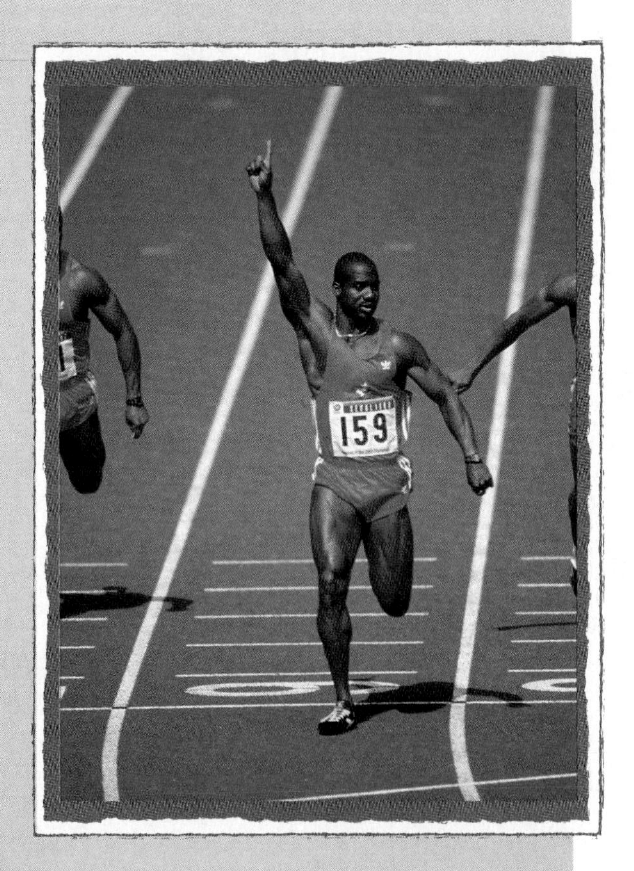

***Figure* 8A**

Canadian track star Ben Johnson ran away with the gold medal in the 100-meter race at the 1988 Summer Olympics—but had to return it after a urine test revealed traces of a steroid drug.

Myotomy (mi-ot′o-me) Cutting of muscle tissue.

Myotonia (mi″o-to′ne-ah) A prolonged muscular spasm.

Paralysis (pah-ral′ĭ-sis) Loss of ability to move a body part.

Paresis (pah-re′sis) Partial or slight paralysis of muscles.

Shin splints (shin splints) Soreness on the front of the leg due to straining the flexor digitorum longus, often as a result of walking up and down hills.

Torticollis (tor″tĭ-kol′is) A condition in which the neck muscles, such as the sternocleidomastoids, contract involuntarily. It is more commonly called *wryneck*.

organization

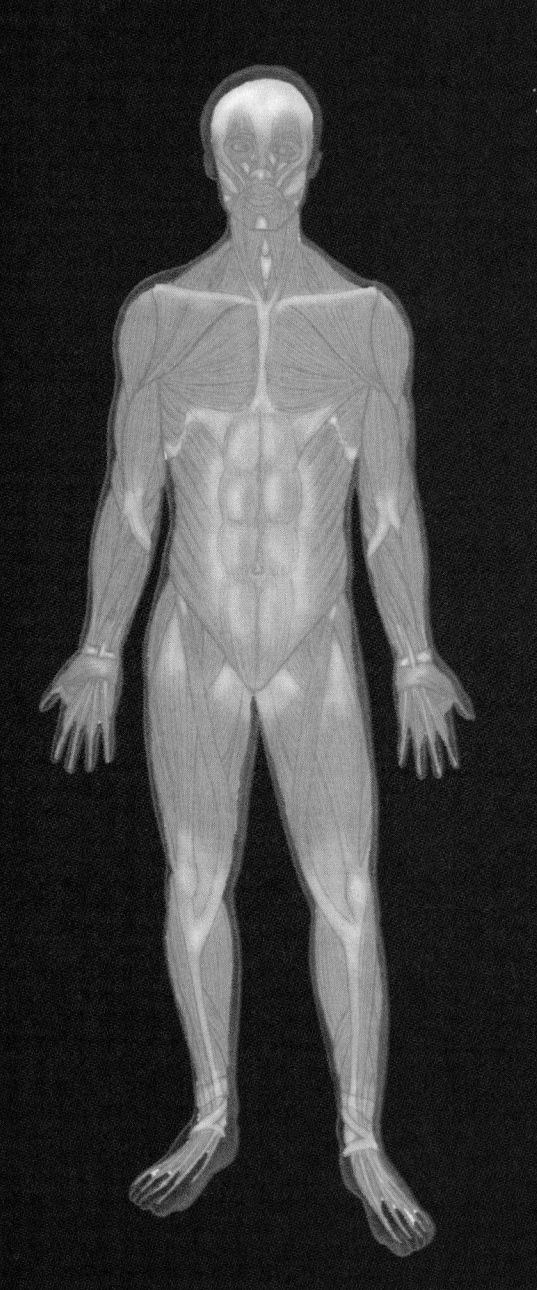

Muscular System

Muscles provide the force for moving body parts.

Integumentary System

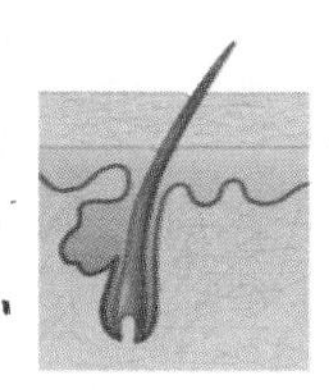

The skin increases heat loss during skeletal muscle activity. Sensory receptors function in the reflex control of skeletal muscles.

Lymphatic System

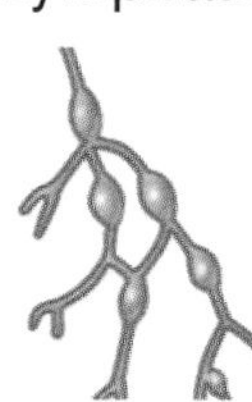

Muscle action pumps lymph through lymphatic vessels.

Skeletal System

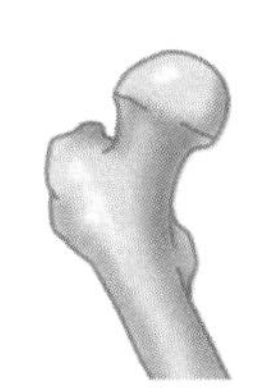

Bones provide attachments that allow skeletal muscles to cause movement.

Digestive System

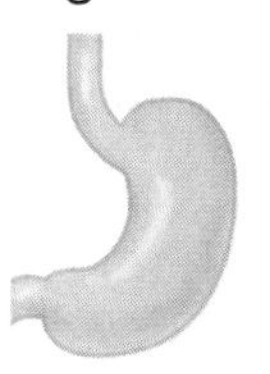

Skeletal muscles are important in swallowing. The digestive system absorbs needed nutrients.

Nervous System

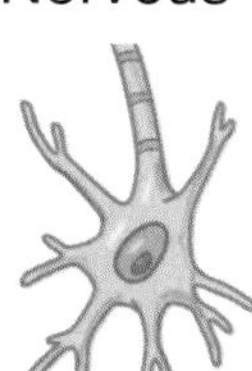

Neurons control muscle contractions.

Respiratory System

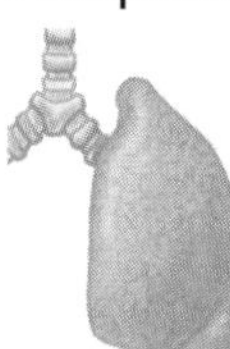

Breathing depends on skeletal muscles. The lungs provide oxygen for body cells and eliminate carbon dioxide.

Endocrine System

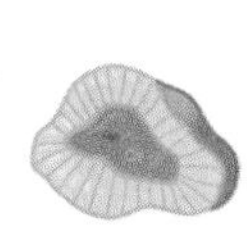

Hormones help increase blood flow to exercising skeletal muscles.

Urinary System

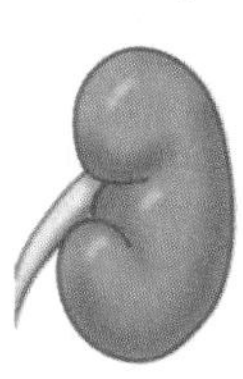

Skeletal muscles help control urine elimination.

Cardiovascular System

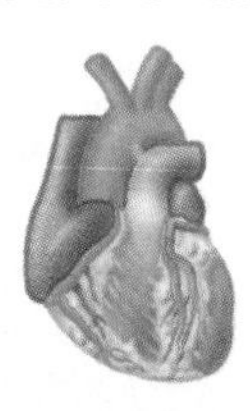

Blood flow delivers oxygen and nutrients and removes wastes.

Reproductive System

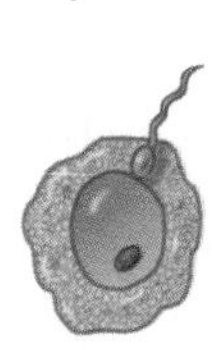

Skeletal muscles are important in sexual activity.

Summary Outline

Introduction (p. 178)

The three types of muscle tissue are skeletal, smooth, and cardiac.

Structure of a Skeletal Muscle (p. 178)

Individual muscles are the organs of the muscular system. They contain skeletal muscle tissue, nervous tissue, blood, and connective tissues.

1. Connective tissue coverings
   a. Fascia cover skeletal muscles.
   b. Other connective tissues attach muscles to bones or to other muscles.
   c. A network of connective tissue extends throughout the muscular system.
2. Skeletal muscle fibers
   a. Each skeletal muscle fiber is a single muscle cell, which is the unit of contraction.
   b. The cytoplasm contains mitochondria, sarcoplasmic reticulum, and myofibrils of actin and myosin.
   c. The organization of actin and myosin filaments produces striations.
   d. Transverse tubules extend inward from the cell membrane and associate with the sarcoplasmic reticulum.
3. Neuromuscular junction
   a. Motor neurons stimulate muscle fibers to contract.
   b. In response to a nerve impulse, the end of a motor neuron fiber secretes a neurotransmitter, which stimulates the muscle fiber to contract.
4. Motor units
   a. One motor neuron and the muscle fibers associated with it constitute a motor unit.
   b. All the muscle fibers of a motor unit contract together.

Skeletal Muscle Contraction (p. 182)

Muscle fiber contraction results from a sliding movement of actin and myosin filaments.

1. Role of myosin and actin
   a. Cross-bridges of myosin filaments form linkages with actin filaments.
   b. The reaction between actin and myosin filaments generates the force of contraction.
2. Stimulus for contraction
   a. Acetylcholine released from the distal end of a motor neuron fiber stimulates a skeletal muscle fiber.
   b. Acetylcholine causes the muscle fiber to conduct an impulse over the surface of the fiber that reaches deep within the fiber through the transverse tubules.
   c. A muscle impulse signals the sarcoplasmic reticulum to release calcium ions.
   d. Linkages form between actin and myosin, and the actin filaments move inward, shortening the fiber.
   e. The muscle fiber relaxes when calcium ions are actively transported (requiring ATP) back into the sarcoplasmic reticulum.
   f. Acetylcholinesterase breaks down acetylcholine.
3. Energy sources for contraction
   a. ATP supplies the energy for muscle fiber contraction.
   b. Creatine phosphate stores energy that can be used to synthesize ATP.
   c. ATP is needed for muscle relaxation.
4. Oxygen supply and cellular respiration
   a. Aerobic respiration requires oxygen.
   b. Red blood cells carry oxygen to body cells.
   c. Myoglobin in muscle cells temporarily stores oxygen.
5. Oxygen debt
   a. During rest or moderate exercise, muscles receive enough oxygen to respire aerobically.
   b. During strenuous exercise, oxygen deficiency may cause lactic acid to accumulate.
   c. Oxygen debt is the amount of oxygen required to convert accumulated lactic acid to glucose and to restore supplies of ATP and creatine phosphate.
6. Muscle fatigue
   a. A fatigued muscle loses its ability to contract.
   b. Muscle fatigue is usually due to accumulation of lactic acid.
7. Heat production
   a. Most of the energy released in cellular respiration is lost as heat.
   b. Muscle action is an important source of body heat.

Muscular Responses (p. 186)

1. Threshold stimulus
   This is the minimal stimulus required to elicit a muscular contraction.
2. All-or-none response
   a. If a muscle fiber contracts at all, it will contract completely.
   b. Motor units respond in an all-or-none manner.
3. Recording a muscle contraction
   a. A myogram is a recording of an electrically stimulated isolated muscle.
   b. A twitch is a single, short contraction reflecting stimulation of some motor units in a muscle.
   c. The latent period, the time between stimulus and responding muscle contraction, is followed by a period of contraction and a period of relaxation.
4. Summation
   a. A rapid series of stimuli may produce summation of twitches.
   b. Forceful, sustained contraction without relaxation is a tetanic contraction.
5. Recruitment of motor units
   a. At a low intensity of stimulation, small numbers of motor units contract.
   b. At increasing intensities of stimulation, other motor units are recruited until the muscle contracts with maximal tension.
6. Sustained contractions
   a. Summation and recruitment together can produce a sustained contraction of increasing strength.
   b. Even when a muscle is at rest, its fibers usually remain partially contracted.

Smooth Muscles (p. 189)

The contractile mechanisms of smooth and cardiac muscles are similar to those of skeletal muscle.

1. Smooth muscle fibers
   a. Smooth muscle cells contain filaments of actin and myosin.
   b. Types include multiunit smooth muscle and visceral smooth muscle.

c. Visceral smooth muscle displays rhythmicity and is self-exciting.
2. Smooth muscle contraction
   a. Two neurotransmitters—acetylcholine and norepinephrine—and hormones affect smooth muscle function.
   b. Smooth muscle can maintain a contraction longer with a given amount of energy than can skeletal muscle.
   c. Smooth muscles can change length without changing tension.

Cardiac Muscle (p. 189)

1. Cardiac muscle twitches last longer than skeletal muscle twitches.
2. Intercalated disks connect cardiac muscle cells.
3. A network of fibers contracts as a unit and responds to stimulation in an all-or-none manner.
4. Cardiac muscle is self-exciting and rhythmic.

Skeletal Muscle Actions (p. 190)

The type of movement a skeletal muscle produces depends on the way the muscle attaches on either side of a joint.

1. Origin and insertion
   a. The movable end of a skeletal muscle is its insertion, and the immovable end is its origin.
   b. Some muscles have more than one origin.
2. Interaction of skeletal muscles
   a. Skeletal muscles function in groups.
   b. A prime mover is responsible for most of a movement. Synergists aid prime movers. Antagonists can resist the action of a prime mover.
   c. Smooth movements depend on antagonists giving way to the actions of prime movers.

Major Skeletal Muscles (p. 191)

1. Muscles of facial expression
   a. These muscles lie beneath the skin of the face and scalp and are used to communicate feelings through facial expression.
   b. They include the epicranius, orbicularis oculi, orbicularis oris, buccinator, zygomaticus, and platysma.
2. Muscles of mastication
   a. These muscles attach to the mandible and are used in chewing.
   b. They include the masseter and temporalis.
3. Muscles that move the head
   a. Muscles in the neck and upper back move the head.
   b. They include the sternocleidomastoid, splenius capitis, and semispinalis capitis.
4. Muscles that move the pectoral girdle
   a. Most of these muscles connect the scapula to nearby bones and closely associate with muscles that move the arm.
   b. They include the trapezius, rhomboideus major, levator scapulae, serratus anterior, and pectoralis minor.
5. Muscles that move the arm
   a. These muscles connect the humerus to various regions of the pectoral girdle, ribs, and vertebral column.
   b. They include the coracobrachialis, pectoralis major, teres major, latissimus dorsi, supraspinatus, deltoid, subscapularis, infraspinatus, and teres minor.
6. Muscles that move the forearm
   a. These muscles connect the radius and ulna to the humerus or pectoral girdle.
   b. They include the biceps brachii, brachialis, brachioradialis, triceps brachii, supinator, pronator teres, and pronator quadratus.
7. Muscles that move the wrist, hand, and fingers
   a. These muscles arise from the distal end of the humerus and from the radius and ulna.
   b. They include the flexor carpi radialis, flexor carpi ulnaris, palmaris longus, flexor digitorum profundus, extensor carpi radialis longus, extensor carpi radialis brevis, extensor carpi ulnaris, and extensor digitorum.
8. Muscles of the abdominal wall
   a. These muscles connect the rib cage and vertebral column to the pelvic girdle.
   b. They include the external oblique, internal oblique, transversus abdominis, and rectus abdominis.
9. Muscles of the pelvic outlet
   a. These muscles form the floor of the pelvic cavity and fill the space within the pubic arch.
   b. They include the levator ani, superficial transversus perinei, bulbospongiosus, and ischiocavernosus.
10. Muscles that move the thigh
    a. These muscles attach to the femur and to some part of the pelvic girdle.
    b. They include the psoas major, iliacus, gluteus maximus, gluteus medius, gluteus minimus, tensor fasciae latae, adductor longus, adductor magnus, and gracilis.
11. Muscles that move the leg
    a. These muscles connect the tibia or fibula to the femur or pelvic girdle.
    b. They include the biceps femoris, semitendinosus, semimembranosus, sartorius, and the quadriceps femoris group.
12. Muscles that move the ankle, foot, and toes
    a. These muscles attach the femur, tibia, and fibula to bones of the foot.
    b. They include the tibialis anterior, peroneus tertius, extensor digitorum longus, gastrocnemius, soleus, flexor digitorum longus, tibialis posterior, and peroneus longus.

Review Exercises

Part A

1. List the three types of muscle tissue. (p. 178)
2. Distinguish between a tendon and an aponeurosis. (p. 178)
3. Describe how connective tissue associates with skeletal muscle. (p. 178)
4. List the major parts of a skeletal muscle fiber, and describe the function of each part. (p. 179)
5. Describe a neuromuscular junction. (p. 181)
6. Explain the function of a neurotransmitter. (p. 181)
7. Define *motor unit*. (p. 181)
8. Describe the major events of muscle fiber contraction. (p. 182)
9. Explain how ATP and creatine phosphate interact. (p. 184)
10. Describe how muscles obtain oxygen. (p. 184)

11. Describe how an oxygen debt may develop. (p. 185)
12. Explain how muscles may become fatigued. (p. 186)
13. Explain how skeletal muscle function affects the maintenance of body temperature. (p. 186)
14. Define *threshold stimulus.* (p. 186)
15. Explain an *all-or-none response.* (p. 186)
16. Sketch a myogram of a single muscular twitch, and identify the latent period, period of contraction, and period of relaxation. (p. 187)
17. Explain *motor unit recruitment.* (p. 187)
18. Explain how skeletal muscle stimulation produces a sustained contraction. (p. 187)
19. Distinguish between tetanic contraction and muscle tone. (p. 189)
20. Distinguish between multiunit and visceral smooth muscle fibers. (p. 189)
21. Compare smooth and skeletal muscle contractions. (p. 189)
22. Compare the structure of cardiac and skeletal muscle fibers. (p. 189)
23. Distinguish between a muscle's origin and its insertion. (p. 190)
24. Define *prime mover, synergist,* and *antagonist.* (p. 191)

Part B

Match the muscles in column I with the descriptions and functions in column II.

I	II
1. Buccinator	a. Inserted on coronoid process of mandible
2. Epicranius	b. Draws corner of mouth upward
3. Orbicularis oris	c. Can raise and adduct scapula
4. Platysma	d. Can pull head into an upright position
5. Rhomboideus major	e. Raises eyebrow
6. Splenius capitis	f. Compresses cheeks
7. Temporalis	g. Extends over neck from chest to face
8. Zygomaticus	h. Closes lips
9. Biceps brachii	i. Extends forearm at elbow
10. Brachialis	j. Pulls shoulder back and downward
11. Deltoid	k. Abducts arm
12. Latissimus dorsi	l. Inserted on radial tuberosity
13. Pectoralis major	m. Pulls arm forward and across chest
14. Pronator teres	n. Rotates forearm medially
15. Teres minor	o. Inserted on coronoid process of ulna
16. Triceps brachii	p. Rotates arm laterally
17. Biceps femoris	q. Inverts foot
18. External oblique	r. Member of quadriceps femoris group
19. Gastrocnemius	s. Plantar flexor of foot
20. Gluteus maximus	t. Compresses contents of abdominal cavity
21. Gluteus medius	u. Extends thigh
22. Gracilis	v. Hamstring muscle
23. Rectus femoris	w. Adducts thigh
24. Tibialis anterior	x. Abducts thigh

Part C

Which muscles can you identify in the bodies of these models?

(a)

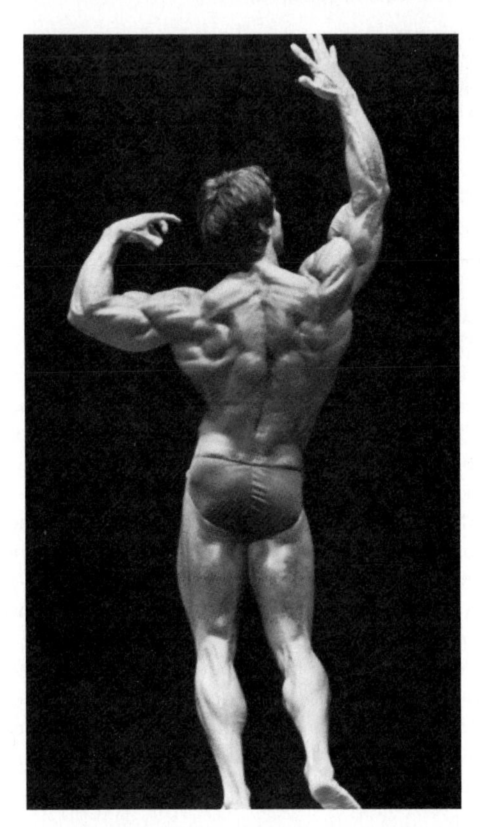
(b)

Critical Thinking

1. A person with severe, lifelong constipation finally receives an accurate diagnosis: He is missing some nerves in his lower digestive tract, resulting in sluggish and intermittent peristalsis (rhythmic contractions of the digestive system wall). How can a problem with muscles really be a problem with nerves?
2. A man exercises extensively, building up his muscles. He believes he will pass this hypertrophy on to his future children. Why is he mistaken?
3. A woman takes her daughter to a sports medicine specialist and requests that she determine the percent of fast and slow fibers in the girl's leg muscles. The parent wants to know if the healthy girl should try out for soccer or cross-country running. Do you think that this is a valid reason to test muscle tissue? Why or why not?
4. Why do you think athletes generally perform better if they warm up by exercising before a competitive event?
5. What steps might minimize atrophy of the skeletal muscles in patients confined to bed for prolonged times?
6. Lactic acid and other biochemicals accumulating in an active muscle stimulate pain receptors, and the muscle may feel sore. How might the application of heat or substances that dilate blood vessels relieve such soreness?
7. A nerve injury may paralyze the muscle it supplies. How would you explain to a patient the importance of moving the disabled muscles passively or contracting them using electrical stimulation?

Chapter 9

Nervous System

Christopher Reeve's life changed forever in a split second on May 27, 1995. Reeve, best known for his portrayal of Superman in four films, was one of 300 equestrians competing on that bright Saturday in Culpeper County, Virginia. He and his horse Eastern Express were poised to clear the third of fifteen hurdles in a 2-mile event. The horse's front legs went over the hurdle, Reeve's back arched as he propelled himself forward, and then the horse stopped. Reeve hurled forward over the horse's head, striking his head on the fence and then landing on the grass—unconscious, not moving or breathing.

Reeve had broken the first and second cervical vertebrae in his neck. He was quickly resuscitated, stabilized, and flown to a major medical center. A week later, he could move a few muscles in his chest and back, indicating that his spinal cord had not been severed. On June 5, he underwent surgery to implant U-shaped wires in his neck to limit further damage. By mid-June, he could sit and had some sensation in his upper body. But it was not until January 1996 that he could breathe unaided for longer than a few minutes.

Each year, tens of thousands of people sustain spinal cord injuries, most in automobile accidents, sports, or violent actions. Many more people live with partial paralysis due to damage in the lower back or near-total paralysis due to damage to the upper spine. Treatment includes surgery to remove bone chips and to insert rings or plates to stabilize the spine; and drugs to limit inflammation, which causes further damage. Love and hope are also important. When Reeve regained consciousness, he asked to die. His wife, touching his head, said, "This is who you are, not your body."

Christopher Reeve has made slow but steady progress, and has inspired many. He is, and will always be, a super man.

Photo: A horseback riding accident severely damaged the spinal cord of actor Christopher Reeve, shown here in happier times portraying Superman.

Feeling, thinking, remembering, moving, and being aware of the world require nervous system activity. This vast collection of cells also helps coordinate all other body functions to maintain homeostasis and to enable the body to respond to changing conditions. Sensory receptors bring information from within and outside the body to the brain and spinal cord, which send nerve fibers to stimulate responses from muscles and glands.

Chapter Objectives

After studying this chapter, you should be able to:

1. Explain the general functions of the nervous system. (p. 214)
2. Describe the general structure of a neuron. (p. 215)
3. Explain how differences in structure and function are used to classify neurons. (p. 217)
4. Name four types of neuroglial cells, and describe the functions of each. (p. 218)
5. Explain how a membrane becomes polarized. (p. 219)
6. Describe the events that lead to the conduction of a nerve impulse. (p. 220)
7. Explain how information passes from one neuron to another. (p. 223)
8. Describe how nerve fibers in peripheral nerves are classified. (p. 227)
9. Name the parts of a reflex arc, and describe the function of each part. (p. 228)
10. Describe the coverings of the brain and spinal cord. (p. 230)
11. Describe the structure of the spinal cord and its major functions. (p. 231)
12. Name the major parts and functions of the brain. (p. 234)
13. Distinguish among motor, sensory, and association areas of the cerebral cortex. (p. 235)
14. Describe the formation and function of cerebrospinal fluid. (p. 238)
15. List the major parts of the peripheral nervous system. (p. 243)
16. Name the cranial nerves, and list their major functions. (p. 244)
17. Describe the structure of a spinal nerve. (p. 247)
18. Describe the functions of the autonomic nervous system. (p. 248)
19. Distinguish between the sympathetic and parasympathetic divisions of the autonomic nervous system. (p. 249)
20. Describe a sympathetic and a parasympathetic nerve pathway. (p. 251)

Aids to Understanding Words

ax- [axle] *ax*on: A cylindrical nerve fiber that carries impulses away from a neuron cell body.

dendr- [tree] *dendr*ite: A branched nerve fiber that serves as a receptor surface of a neuron.

funi- [small cord or fiber] *funi*culus: A major nerve tract or bundle of myelinated nerve fibers within the spinal cord.

gangli- [a swelling] *gangli*on: A mass of neuron cell bodies.

-lemm [rind or peel] neuri*lemm*a: A sheath that surrounds the myelin of a nerve fiber.

mening- [membrane] *mening*es: Membranous coverings of the brain and spinal cord.

moto- [moving] *moto*r neuron: A neuron that stimulates a muscle to contract or a gland to secrete.

peri- [all around] *peri*pheral nervous system: The portion of the nervous system that consists of nerves branching from the brain and spinal cord.

plex- [interweaving] choroid *plex*us: A mass of specialized capillaries associated with spaces in the brain.

sens- [feeling] *sens*ory neuron: A sensory receptor stimulates a neuron to conduct impulses into the brain or spinal cord.

syn- [together] *syn*apse: The junction between two neurons.

ventr- [belly or stomach] *ventr*icle: A fluid-filled space within the brain.

Key Terms

action potential (ak′shun po-ten′shal)
autonomic nervous system (aw″to-nom′ik ner′vus sis′tem)
axon (ak′son)
central nervous system (CNS) (sen′tral ner′vus sis′tem)
convergence (kon-ver′jens)
dendrite (den′drīt)
divergence (di-ver′jens)
effector (e-fek′tor)
facilitation (fah-sil″ĭ-ta′shun)
ganglion (gang′gle-on)
meninges (mĕ-nin′jēz)
myelin (mi′ĕ-lin)
neurilemma (nu″rĭ-lem′mah)
neurotransmitter (nu″ro-trans-mit′er)
Nissl body (nis′l bod′e)
parasympathetic nervous system (par″ah-sim″pah-thet′ik ner′vus sis′tem)
peripheral nervous system (PNS) (pĕ-rif′er-al ner′vus sis′tem)
plexus (plek′sus)
reflex (re′fleks)
sensory receptor (sen′sor-e re-sep′tor)
somatic nervous system (so-mat′ik ner′vus sis′tem)
sympathetic nervous system (sim″pah-thet′ik ner′vus sis′tem)
synapse (sin′aps)

Introduction

Recall from chapter 5 that nervous tissue consists of masses of nerve cells, or **neurons.** These cells are the structural and functional units of the nervous system and are specialized to react to physical and chemical changes in their surroundings. Neurons transmit information in the form of electrochemical changes, called **nerve impulses**, along **nerve fibers** to other neurons and to cells outside the nervous system. **Nerves** are bundles of nerve fibers. Nervous tissue also includes **neuroglial cells** (glial cells or neuroglia), which provide neurons with certain physiological requirements and function like the connective tissue cells in other systems (fig. 9.1).

The organs of the nervous system can be divided into two groups. One group, consisting of the brain and spinal cord, forms the **central nervous system (CNS)**, and the other, composed of the nerves (peripheral nerves) that connect the central nervous system to other body parts, is called the **peripheral nervous system (PNS)**. Together, these systems provide three general functions: sensory, integrative, and motor.

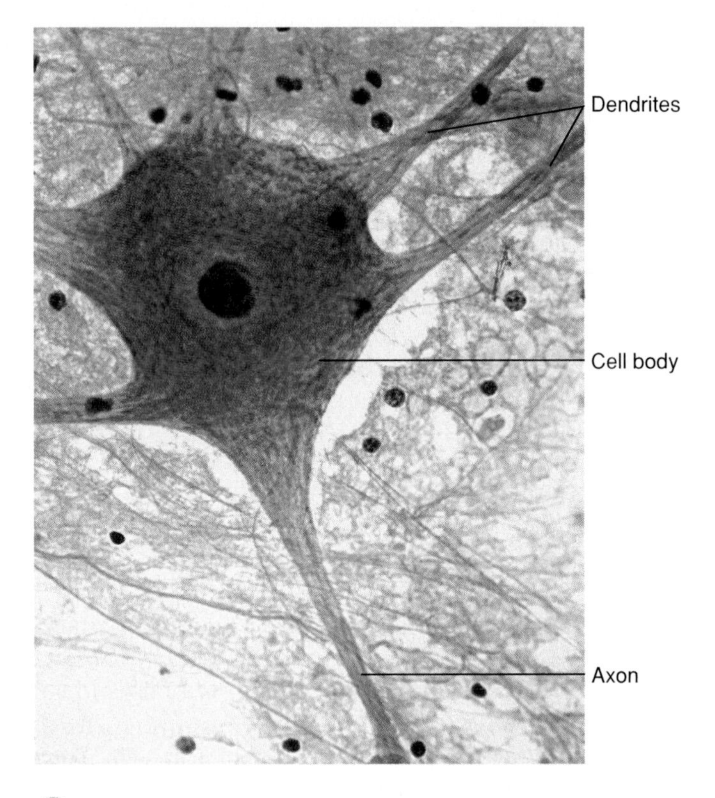

Figure 9.1

Neurons are the structural and functional units of the nervous system (50×). The dark spots in the area surrounding the neuron are nuclei of neuroglial cells. Note the location of nerve fibers (dendrites and a single axon).

General Functions of the Nervous System

The **sensory function** of the nervous system derives from **sensory receptors** at the ends of peripheral neurons (see chapter 10). These receptors gather information by detecting changes inside and outside the body.

They monitor external environmental factors, such as light and sound intensities, and conditions of the body's internal environment, such as temperature and oxygen concentration.

Sensory receptors convert information into nerve impulses, which are then transmitted over peripheral nerves to the central nervous system. There, the signals are integrated; that is, they are brought together, creating sensations, adding to memory, or helping to produce thoughts that translate sensations into perceptions. As a result of this **integrative function,** we make conscious or subconscious decisions and then act on them, using **motor functions.**

The motor functions of the nervous system employ peripheral neurons, which carry impulses from the central nervous system to responsive structures called **effectors.** Effectors are outside the nervous system and include muscles that contract and glands that secrete when stimulated by nerve impulses.

The nervous system can detect changes outside and within the body, make decisions based on the information received, and stimulate muscles or glands to respond. Typically, these responses counteract the effects of the changes detected, and in this way, the nervous system helps to maintain homeostasis.

Neuron Structure

Neurons vary considerably in size and shape, but they have certain features in common. Every neuron has a **cell body** and tubular, cytoplasm-filled nerve fibers, which conduct nerve impulses to or from the cell body (fig. 9.2).

The neuron cell body consists of granular cytoplasm, a cell membrane, and organelles such as mitochondria, lysosomes, a Golgi apparatus, and a network of fine threads called **neurofibrils,** which extend into the nerve fibers. Scattered throughout the cytoplasm are many membranous sacs called **Nissl bodies,** which are similar to rough endoplasmic reticulum in other cells. Ribosomes attached to Nissl bodies function in protein synthesis, as they do elsewhere. Near the center of the cell body is a large spherical nucleus with a conspicuous nucleolus. Mature neurons do not divide.

Two kinds of nerve fibers, **dendrites** and **axons,** extend from the cell bodies of most neurons. A neuron may have many dendrites but only one axon.

Dendrites are usually short and highly branched. These fibers, together with the membrane of the cell body, are the neuron's main receptive surfaces with which fibers from other neurons communicate.

The axon usually arises from a slight elevation of the cell body called the *axonal hillock*. It conducts nerve impulses away from the cell body. Many mitochondria, microtubules, and neurofibrils are in the axon cytoplasm. An axon begins as a single fiber but may give off side branches. Near its end, it may have many fine extensions that contact the receptive surfaces of other cells.

Larger axons of peripheral neurons are enclosed in sheaths composed of many neuroglial cells, called **Schwann cells** (fig. 9.3). These cells wind tightly around axons, somewhat like a bandage wrapped around a finger. As a result, such axons are coated with many layers of cell membrane that have little or no cytoplasm between them. These membrane layers are composed largely of a lipid-protein (lipoprotein) called **myelin.** These layers, which have a higher proportion of lipid than other cell membranes, form a *myelin sheath* around an axon. In addition, the portions of the Schwann cells that contain most of the cytoplasm and the nuclei remain outside the myelin sheath and comprise a **neurilemma** (neurilemmal sheath), which surrounds the myelin sheath. Narrow gaps in the myelin sheath between Schwann cells are called **nodes of Ranvier.**

Schwann cells also enclose the smaller axons of peripheral neurons but may not wind around these axons. Consequently, such axons lack myelin sheaths. Instead, the axon or a small group of axons may lie in a longitudinal groove of a Schwann cell.

Fibers with myelin sheaths are called *myelinated* nerve fibers, and those that lack sheaths are *unmyelinated*. Myelin is also found in the CNS. There groups of myelinated fibers appear white, and masses of such fibers form the *white matter* in the central nervous system. Unmyelinated nerve fibers and neuron cell bodies form *gray matter* within the central nervous system.

1. List the general functions of the nervous system.
2. Describe a neuron.
3. Distinguish between an axon and a dendrite.
4. Describe how a myelin sheath forms.

Types of Neurons and Neuroglial Cells

Neurons differ in the structure, size, and shape of their cell bodies. They also vary in length and size of their axons and dendrites and in the number of connections they make with other neurons.

Neurons vary in function. Some neurons carry impulses into the brain or spinal cord, others transmit impulses out of the brain or spinal cord, and still others conduct impulses from neuron to neuron within the brain or spinal cord.

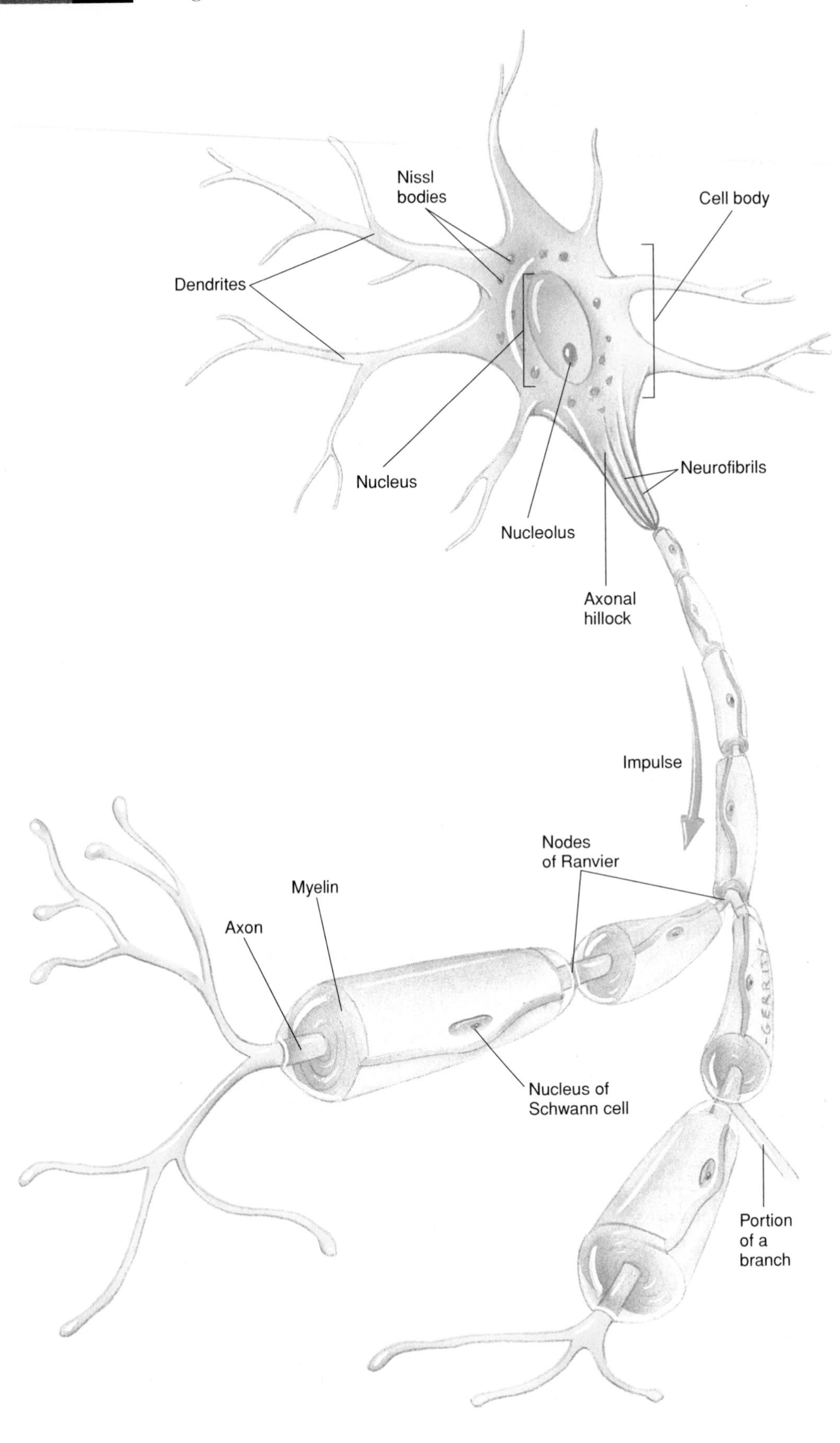

Figure 9.2

A common neuron.

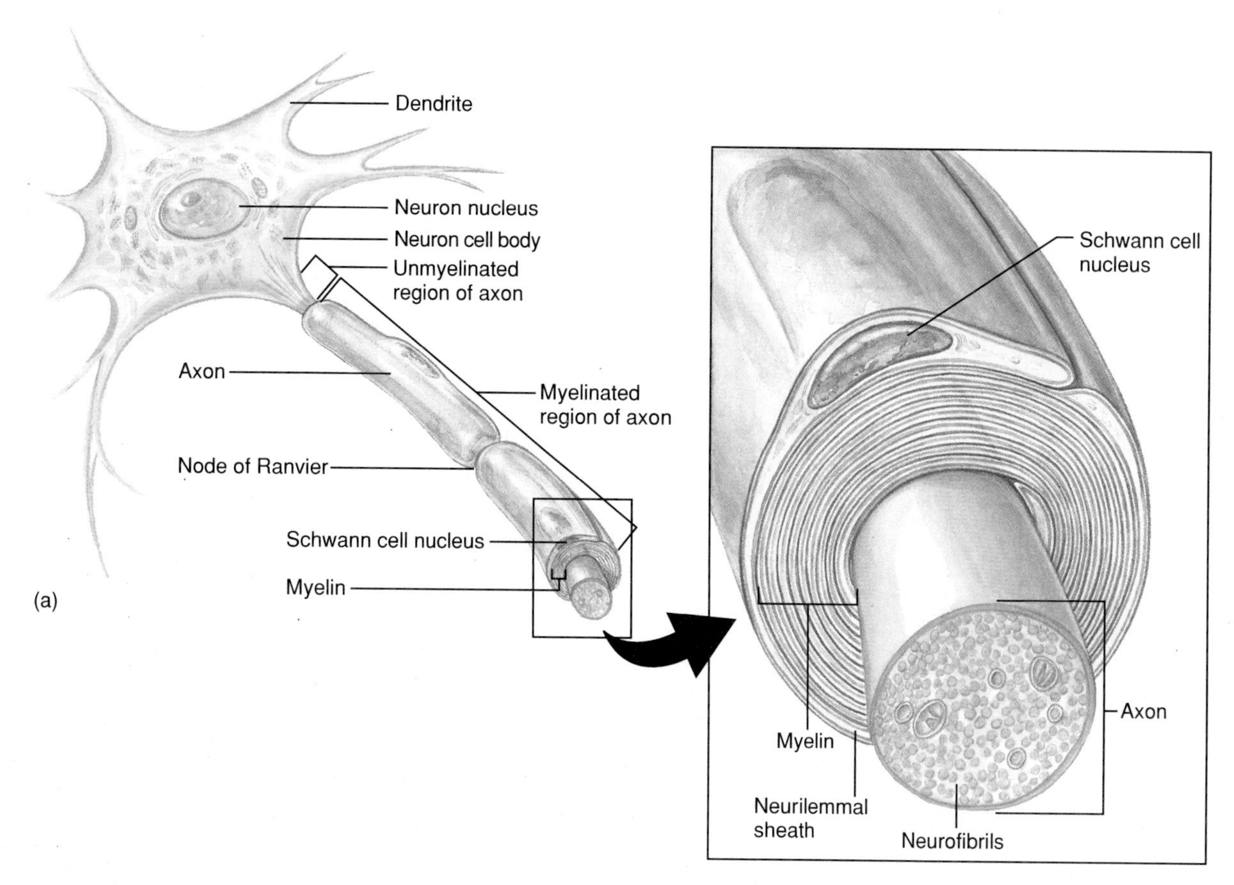

Figure 9.3

The portion of a Schwann cell that winds tightly around an axon forms a myelin sheath, and the cytoplasm and nucleus of the Schwann cell, remaining outside, form a neurilemmal sheath.

Classification of Neurons

On the basis of structural differences, neurons are classified into the following three major groups (fig. 9.4):

1. **Bipolar neurons** The cell body of a bipolar neuron has only two nerve fibers, one arising from each end. Although these fibers are structurally similar, one is an axon and the other is a dendrite. Neurons within specialized parts of the eyes, nose, and ears are bipolar.
2. **Unipolar neurons** A unipolar neuron has a single nerve fiber that extends from the cell body and then divides into two branches, one connected to a peripheral body part and functioning as a dendrite, and the other entering the brain or spinal cord and functioning as an axon. The cell bodies of some unipolar neurons aggregate in specialized masses of nervous tissue called **ganglia** (singular, *ganglion*), located outside the brain and spinal cord.
3. **Multipolar neurons** Multipolar neurons have many nerve fibers arising from their cell bodies. Only one fiber of each neuron is an axon; the rest are dendrites. Most neurons whose cell bodies lie within the brain or spinal cord are multipolar.

On the basis of functional differences, neurons are grouped as follows:

1. **Sensory neurons** (afferent neurons) These carry nerve impulses from peripheral body parts into the brain or spinal cord. These neurons either have specialized *receptor ends* at the tips of their dendrites, or they have dendrites that are closely associated with *receptor cells* in the skin or in sensory organs.

   Changes that occur inside or outside the body stimulate receptor ends or receptor cells, triggering sensory nerve impulses. The impulses travel along the sensory neuron fibers, which lead to the brain or spinal cord, where other neurons process the impulses. Most sensory neurons are unipolar, although some are bipolar.
2. **Interneurons** (also called *internuncial* or *association neurons*) These neurons lie within the brain or spinal cord. They are multipolar and link other neurons. Interneurons transmit impulses from one part of the brain or spinal cord to another. That is, they may direct incoming sensory impulses to appropriate parts for processing and interpreting. Other incoming impulses are transferred to motor neurons.

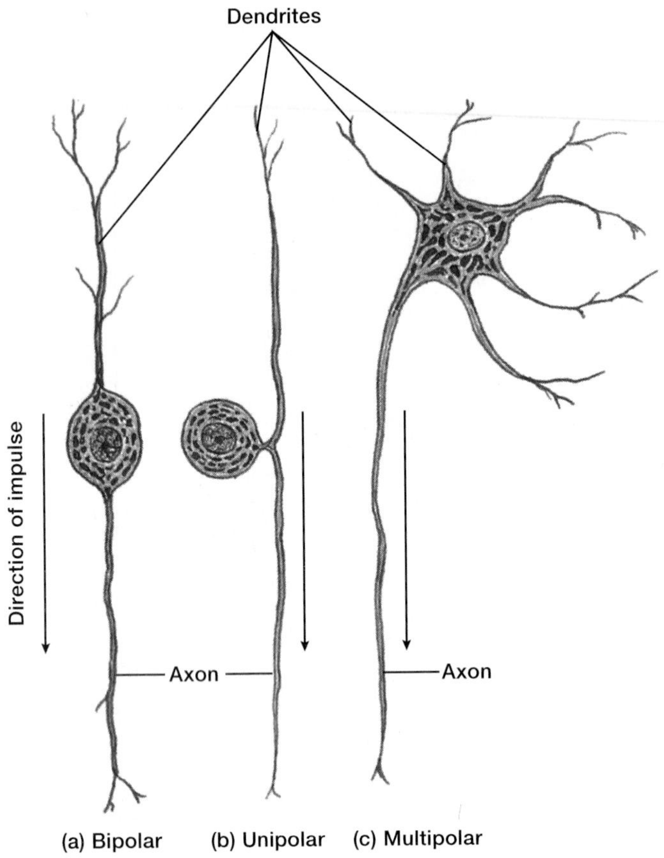

Figure 9.4

Structural types of neurons include (*a*) the bipolar neuron, (*b*) the unipolar neuron, and (*c*) the multipolar neuron.

3. **Motor neurons** (efferent neurons) Motor neurons are multipolar and carry nerve impulses out of the brain or spinal cord to effectors. Motor impulses stimulate muscles to contract and glands to release secretions.

Neurons deprived of oxygen irreversibly change, with altered shapes and shrunken nuclei. This *ischemic cell change* eventually disintegrates the cells. Oxygen deficiency can result from lack of blood flow through nerve tissue (ischemia), an abnormally low blood oxygen concentration (hypoxemia), or toxins that block aerobic respiration. ■

Classification of Neuroglial Cells

Neuroglial cells are part of the nervous system. They fill spaces, support neurons, provide structural frameworks, produce myelin, and carry on phagocytosis.

Within the peripheral nervous system, neuroglial cells include the Schwann cells, previously described. In the central nervous system, neuroglial cells greatly outnumber neurons and are of the following types (fig. 9.5):

1. **Microglial cells** These cells are scattered throughout the central nervous system. They support neurons and phagocytize bacterial cells and cellular debris.
2. **Oligodendrocytes** These cells occur in rows along nerve fibers, and they form myelin within the brain and spinal cord. Unlike Schwann cells in the peripheral nervous system, oligodendrocytes do not form neurilemmal sheaths.
3. **Astrocytes** These cells are commonly found between neurons and blood vessels. They provide structural support, join parts by numerous cellular processes, and help regulate the concentrations of nutrients and ions within the tissue. Astrocytes also form scar tissue that fills spaces following injury to the CNS.
4. **Ependymal cells** These cells form an epithelial-like membrane that covers specialized brain parts (choroid plexuses) and forms the inner linings that enclose spaces within the brain (ventricles) and spinal cord (central canal).

Myelin begins to form on nerve fibers during the fourteenth week of prenatal development. Yet, many of the nerve fibers in newborns are not completely myelinated. Consequently, the infants' nervous systems are unable to function as effectively as those of older children or adults. Their responses to stimuli are coarse and undifferentiated, and may involve the whole body. All myelinated fibers have begun to develop sheaths by the time a child starts to walk, and myelination continues into adolescence. Interference with the supply of essential nutrients during the developmental years may limit myelin formation, which may impair nervous system function later in life. ■

1. Distinguish between a neuron and a neuroglial cell.
2. Explain how neurons are classified according to structure or function.
3. Name and describe four types of neuroglial cells.
4. Describe some functions of neuroglial cells.

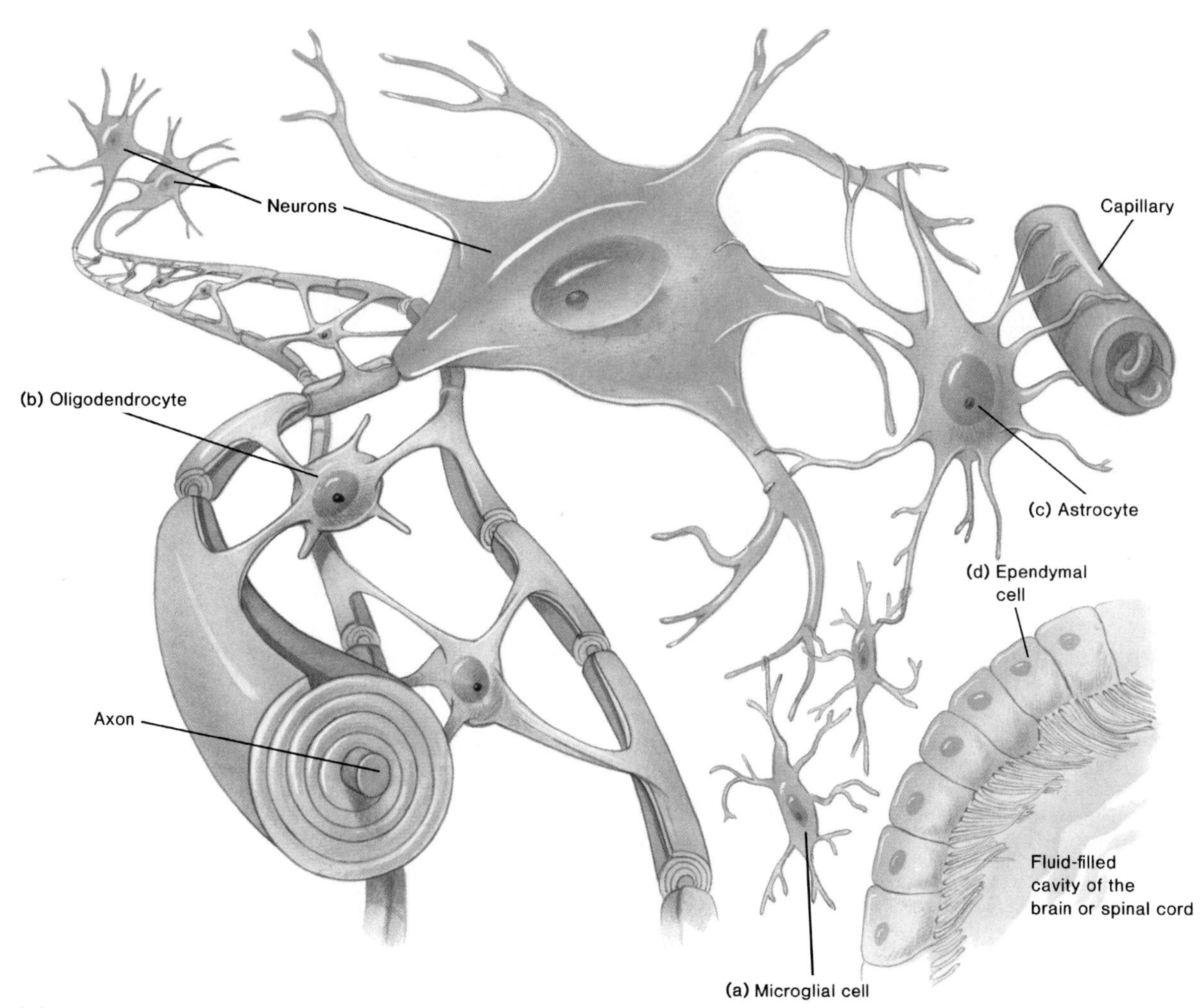

Figure 9.5

Types of neuroglial cells in the central nervous system include (*a*) microglial cell, (*b*) oligodendrocyte, (*c*) astrocyte, and (*d*) ependymal cell.

Cell Membrane Potential

The surface of a cell membrane (including a nonstimulated or *resting* neuron) is usually electrically charged, or *polarized,* with respect to the inside. This polarization is due to an unequal distribution of positive and negative ions between sides of the membrane, and it is particularly important in the conduction of muscle and nerve impulses.

Distribution of Ions

The distribution of ions inside and outside cell membranes is determined in part by pores or channels in those membranes (see chapter 3). Some channels are always open, and others can be opened or closed. Furthermore, channels can be selective; that is, a channel may allow one kind of ion to pass through and exclude other kinds (fig. 9.6).

Potassium ions tend to pass through cell membranes much more easily than sodium ions. This makes potassium ions a major contributor to membrane polarization. Calcium ions are less able to cross the resting cell membrane than either sodium ions or potassium ions, and have a special role in nerve function, described later.

Resting Potential

Because of the active transport of sodium and potassium ions, cells throughout the body have a relatively greater concentration of sodium ions (Na^+) outside and a relatively greater concentration of potassium ions (K^+) inside. The cytoplasm of these cells has many large negatively charged particles, including phosphate ions (PO_4^{-2}), sulfate ions (SO_4^{-2}), and proteins, that cannot diffuse across the cell membranes (fig. 9.7*a*).

Sodium and potassium ions follow the laws of diffusion stated earlier and show a net movement from

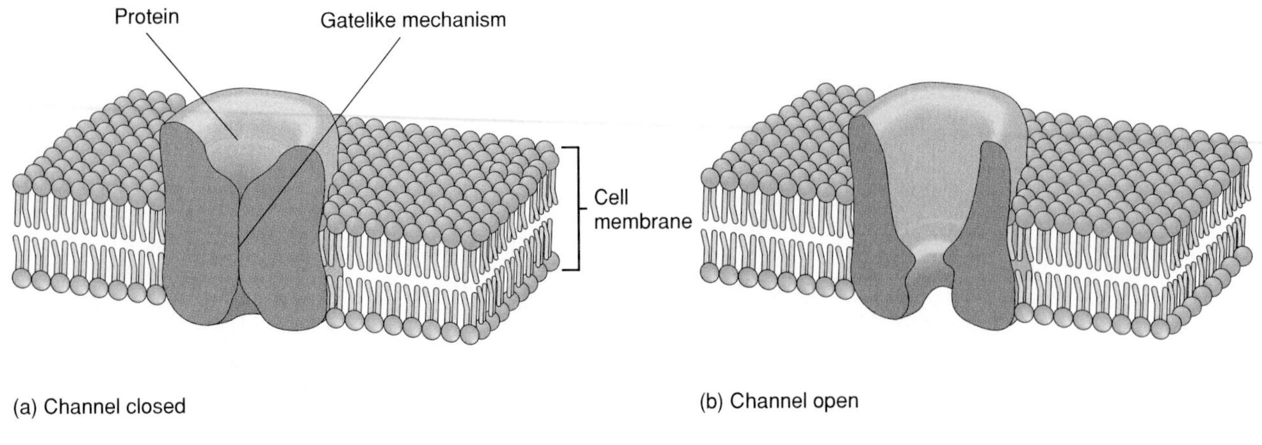

Figure 9.6

A gatelike mechanism can (*a*) close or (*b*) open some of the channels in cell membranes through which ions pass.

high concentration to low concentration as permeabilities permit. Because a resting cell membrane is more permeable to potassium ions than to sodium ions, potassium ions tend to diffuse out of the cell more rapidly than sodium ions can diffuse in (fig. 9.7*a*). This means that, every millisecond, more positive charges leave the cell by diffusion than enter it. As a result, the outside of the cell membrane gains a slight surplus of positive charges, and the inside is left with a slight surplus of impermeable negative charges (fig. 9.7*b*). At the same time, the cell continues to expend energy to drive the Na^+/K^+ "pumps" that actively transport sodium and potassium ions in the opposite directions, thus maintaining the concentration gradients for these ions responsible for their diffusion in the first place.

The difference in electrical charge between two regions is called a *potential difference.* In a resting nerve cell, the potential difference between the region inside and the region outside the membrane is called a **resting potential.** As long as a nerve cell membrane is undisturbed, the membrane remains in this polarized state (fig. 9.8*a*).

Potential Changes

Nerve cells are excitable; that is, they can respond to changes in their surroundings. Some nerve cells, for example, are specialized to detect changes in temperature, light, or pressure occurring outside the body. Many neurons respond to signals from other neurons. Such changes (or stimuli) usually affect the resting potential in a particular region of a nerve cell membrane. If the membrane's resting potential decreases (as the inside of the membrane becomes less negative when compared to the outside), the membrane is said to be *depolarizing.*

Changes in the resting potential of a membrane are graded. This means that the amount of change in potential is directly proportional to the intensity of stimulation. If additional stimulation arrives before the effect of previous stimulation subsides, the change in potential is still greater. This additive phenomenon is called *summation,* and as a result of summated potentials, a level called **threshold potential** may be reached. Many subthreshold potential changes must combine to reach threshold, and once threshold is achieved, an event called an **action potential** occurs.

Action Potential

At the threshold potential, permeability suddenly changes in the region of cell membrane being stimulated. Channels highly selective for sodium ions open and allow sodium to diffuse freely inward (fig. 9.8*b*). This movement is aided by the negative electrical condition on the inside of the membrane, which attracts the positively charged sodium ions.

As sodium ions diffuse inward, the membrane loses its negative electrical charge and becomes depolarized. At almost the same time, however, membrane channels open that allow potassium ions to pass through, and as these positive ions diffuse outward, the inside of the membrane becomes negatively charged once more (fig. 9.8*c*). Thus, the membrane becomes *repolarized,* and it remains in this state until stimulated again.

This rapid sequence of depolarization and repolarization takes about one-thousandth of a second and is an action potential. Because only a small proportion of the sodium and potassium ions move through a membrane during an action potential, many action potentials can occur before the original concentrations of these ions change significantly. Also, the active transport mechanism within the membrane works to maintain the original concentrations of sodium and potassium ions on either side; thus, the resting potential quickly returns.

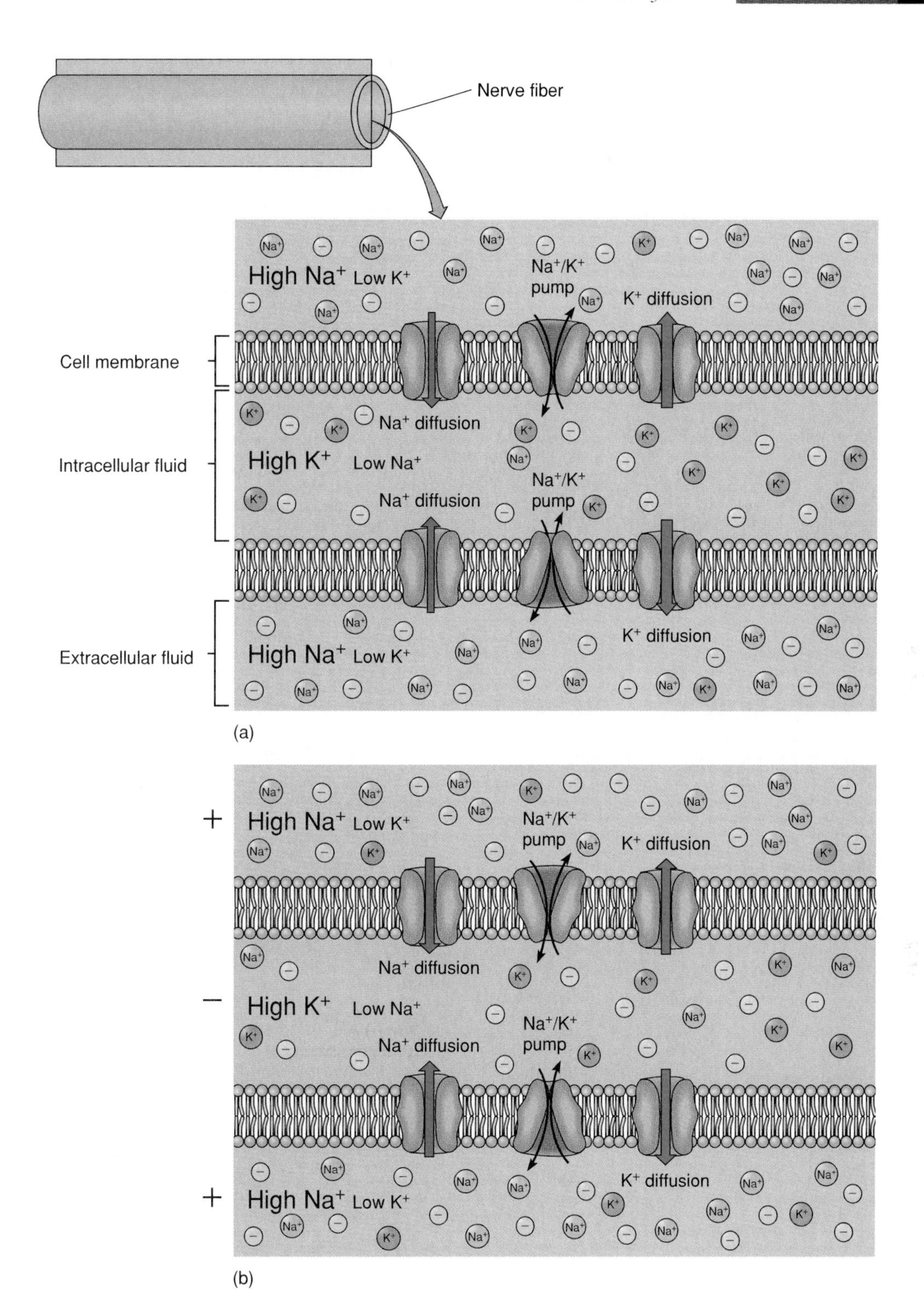

Figure 9.7

Development of the resting membrane potential. (*a*) Active transport creates a concentration difference across the cell membrane for sodium ions (Na^+) and potassium ions (K^+); K^+ diffuses out of the cell rather slowly but nonetheless faster than Na^+ can diffuse in. (*b*) This unequal diffusion results in a net loss of positive charge and a resultant excess of negative charge inside the membrane.

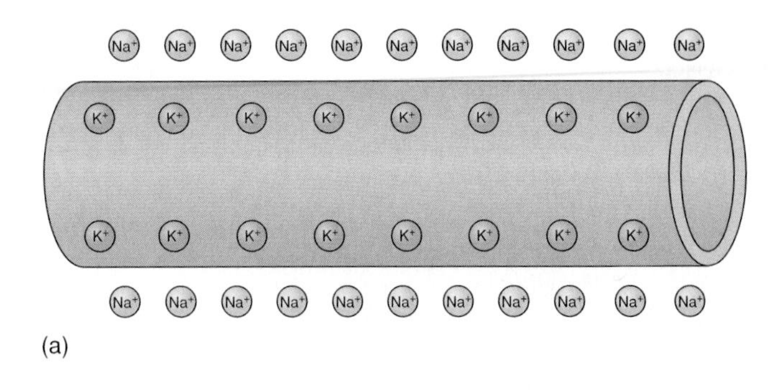

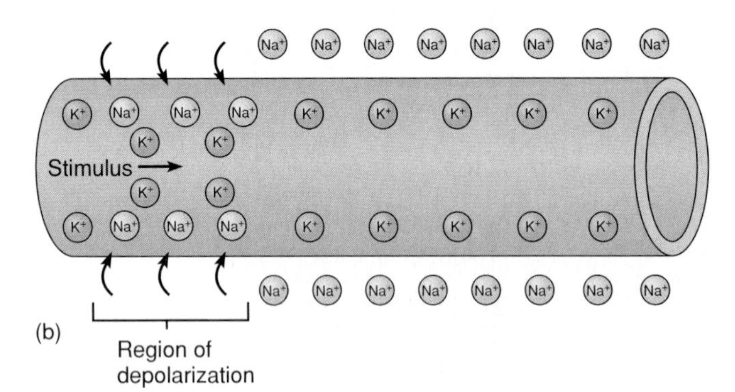

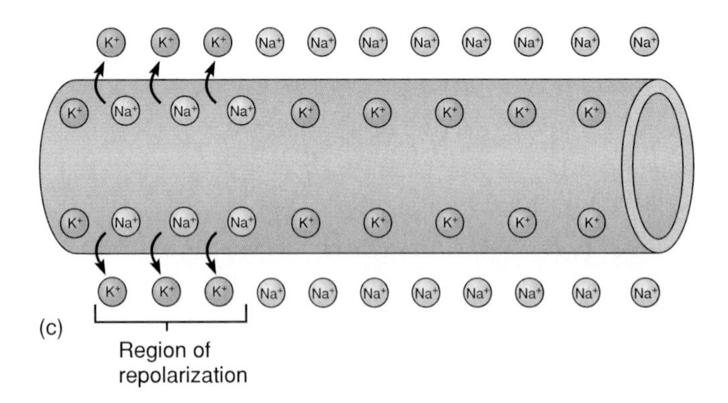

Figure 9.8

(*a*) At rest, the membrane potential is negative. (For simplicity, negative ions are not shown.) (*b*) When the membrane reaches threshold, sodium channels open, some sodium (Na^+) diffuses in, and the membrane is depolarized. (*c*) Soon afterward, potassium channels open, potassium (K^+) diffuses out, and the membrane is repolarized.

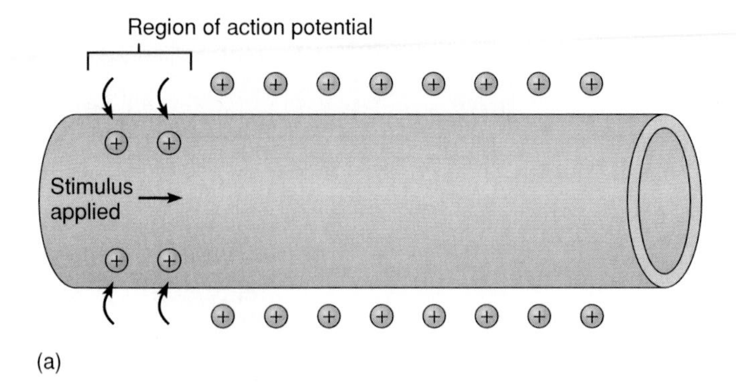

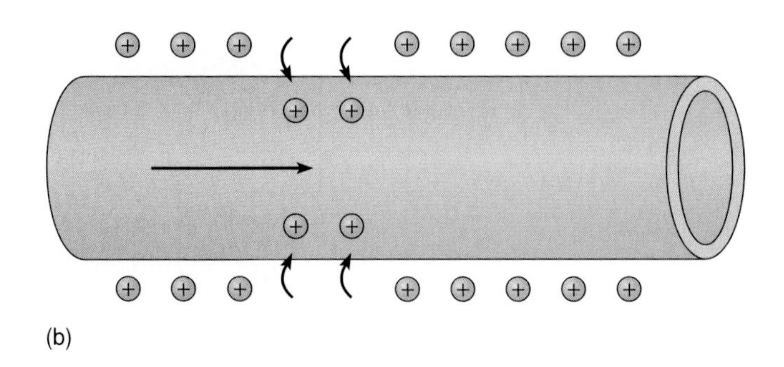

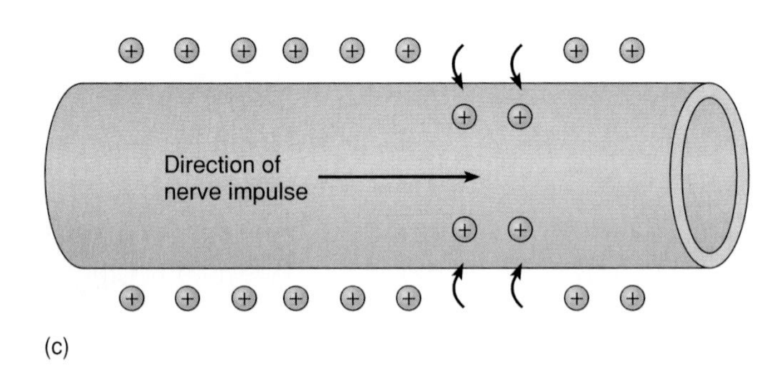

Figure 9.9

(*a*) An action potential in one region stimulates the adjacent region, and (*b* and *c*) a wave of action potentials (a nerve impulse) moves along the fiber.

Nerve Impulse

When an action potential occurs a in one region of a nerve fiber membrane, it causes a bioelectric current to flow to adjacent portions of the membrane. This *local current* stimulates the adjacent membrane to its threshold level and triggers another action potential. This, in turn, stimulates the next adjacent region. A wave of action potentials moves down the fiber to the end. This propagation of action potentials along a nerve fiber constitutes a **nerve impulse** (fig. 9.9).

Table 9.1 summarizes the events leading to the conduction of a nerve impulse.

Impulse Conduction

An unmyelinated nerve fiber conducts an impulse over its entire surface. A myelinated fiber functions differently because myelin is an insulator and prevents almost all ion flow through the membrane it encloses. The myelin sheath would prevent the conduction of a nerve impulse altogether if the sheath were continu-

Table 9.1

Events Leading to the Conduction of a Nerve Impulse

1. Nerve fiber membrane maintains resting potential.
2. Threshold stimulus is received.
3. Sodium channels in a local region of the membrane open.
4. Sodium ions diffuse inward, depolarizing the membrane.
5. Potassium channels in the membrane open.
6. Potassium ions diffuse outward, repolarizing the membrane.
7. The resulting action potential causes a local bioelectric current that stimulates adjacent portions of the membrane.
8. Wave of action potentials travels the length of the nerve fiber as a nerve impulse.

Certain local anesthetic drugs decrease membrane permeability to sodium ions. Such a drug in the fluids surrounding a nerve fiber prevents impulses from passing through the affected region. This keeps impulses from reaching the brain, preventing sensations of touch and pain. ■

1. Summarize how a nerve fiber becomes polarized.
2. List the major events that occur during an action potential.
3. Explain how impulse conduction differs in myelinated and unmyelinated nerve fibers.
4. Define *all-or-none response.*

ous. However, nodes of Ranvier between adjacent Schwann cells interrupt the sheath (see fig. 9.2). Action potentials occur at these nodes, where the exposed nerve fiber membrane contains sodium and potassium channels. Thus, a nerve impulse traveling along a myelinated fiber appears to jump from node to node. This type of impulse conduction (saltatory) is many times faster than conduction on an unmyelinated fiber.

The speed of nerve impulse conduction is proportional to the diameter of the fiber—the greater the diameter, the faster the impulse. For example, an impulse on a relatively thick, myelinated nerve fiber, such as a motor fiber associated with a skeletal muscle, might travel 120 meters per second. An impulse on a thin, unmyelinated nerve fiber, such as a sensory fiber associated with the skin, might move only 0.5 meter per second.

All-or-None Response

Like muscle fiber contraction, nerve impulse conduction is an *all-or-none response.* That is, if a nerve fiber responds at all, it responds completely. Thus, a nerve impulse is conducted whenever a stimulus of threshold intensity or above is applied to a nerve fiber, and all impulses carried on that fiber are of the same strength. A greater intensity of stimulation does not produce a stronger impulse, but rather, more impulses per second.

The Synapse

Within the nervous system, nerve impulses travel from neuron to neuron along complex nerve pathways. The junction between two communicating neurons is called a **synapse.** Two such neurons are not in direct contact at a synapse. A gap called a *synaptic cleft* separates them. An impulse continuing along a nerve pathway must cross this gap (fig. 9.10).

Synaptic Transmission

Within a neuron (*presynaptic neuron*), an impulse travels from a dendrite to the cell body and then moves along the axon to the end. There, the impulse encounters a synapse separating it from a dendrite or cell body of another neuron (*postsynaptic neuron*). The process of crossing the synapse is called *synaptic transmission.*

Transmission from an axon of one neuron to a dendrite or cell body of another neuron is one-way because axons usually have several rounded *synaptic knobs* at their distal ends, which dendrites lack. These knobs contain many membranous sacs, called *synaptic vesicles.* When a nerve impulse reaches a synaptic knob, some of the synaptic vesicles release a biochemical called a **neurotransmitter** (figs. 9.11 and 9.12). The neurotransmitter diffuses across the synaptic cleft and reacts with specific receptors on the postsynaptic neuron membrane.

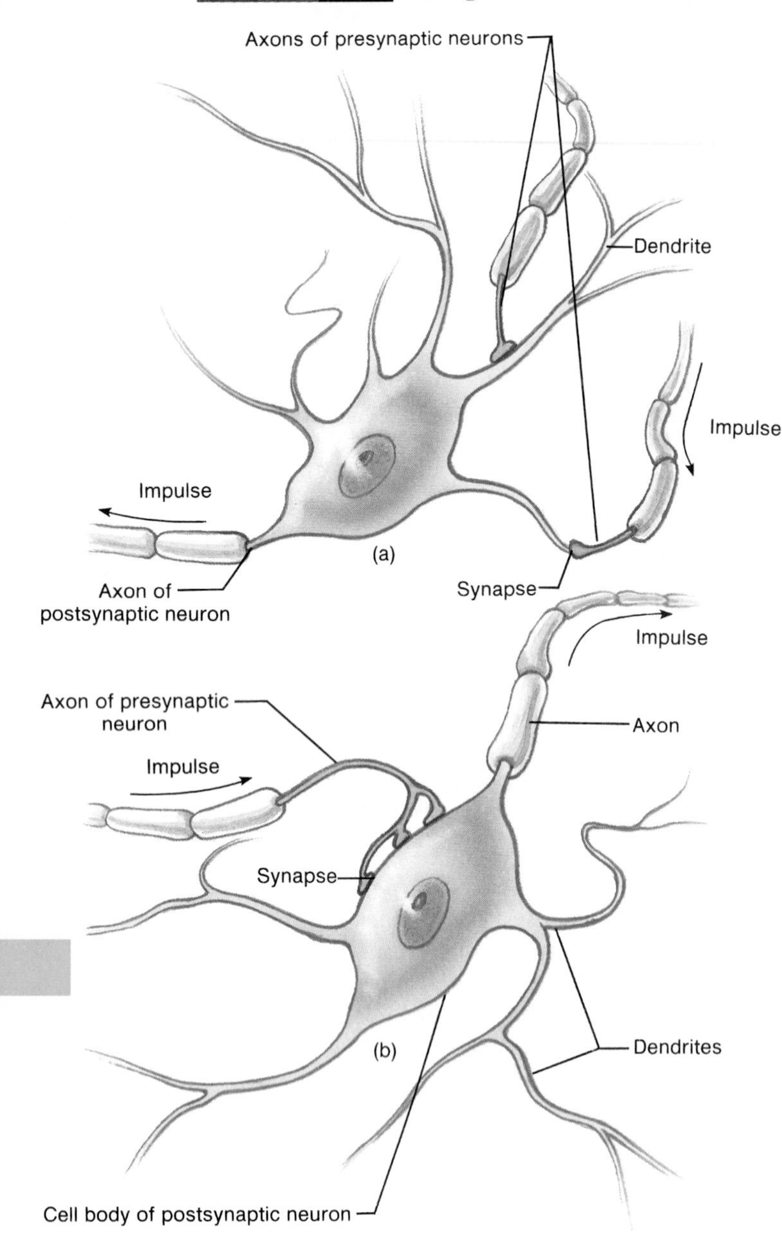

Figure 9.10

For an impulse to continue from one neuron to another, it must cross the synaptic cleft at a synapse. A synapse usually occurs (*a*) between an axon and a dendrite or (*b*) between an axon and a cell body.

Excitatory and Inhibitory Actions

Neurotransmitters that increase postsynaptic membrane permeability to sodium ions may trigger nerve impulses and thus are said to be **excitatory.** Other neurotransmitters decrease membrane permeability to sodium ions, thus making it less likely that threshold will be reached. This action is called **inhibitory** because it lessens the chance that a nerve impulse will occur.

The synaptic knobs of a thousand or more neurons may communicate with the dendrites and cell body of a single postsynaptic neuron. Neurotransmitters released by some of these knobs have an excitatory action, but those from other knobs have an inhibitory action. The effect on the postsynaptic neuron depends on which presynaptic knobs are activated from moment to moment. In other words, if more excitatory than inhibitory neurotransmitters are released, the postsynaptic neuron's threshold may be reached, and a nerve impulse will be triggered. Conversely, if most of the neurotransmitters released are inhibitory, no impulse will be initiated.

Neurotransmitters

About fifty types of neurotransmitters have been identified in the nervous system. Some neurons release only one type, while others produce two or three kinds. The neurotransmitters include *acetylcholine,* which stimulates skeletal muscle contractions (see chapter 8); a group of compounds called *monoamines* (such as epinephrine, norepinephrine, dopamine, and serotonin), which form from modified amino acids; several *amino acids* (such as glycine, glutamic acid, aspartic acid, and gamma-aminobutyric acid—GABA); and a large group of *peptides,* which are short chains of amino acids. Acetylcholine and norepinephrine are excitatory. Dopamine, GABA, and glycine are inhibitory. Neurotransmitters are usually synthesized in the cytoplasm of the synaptic knobs and stored in the synaptic vesicles.

When an action potential reaches the membrane of a synaptic knob, it increases the membrane's permeability to calcium ions by opening the membrane's calcium ion channels. Consequently, calcium ions diffuse inward, and in response, some synaptic vesicles fuse with the membrane and release their contents into the synaptic cleft. A vesicle that has released its neurotransmitter eventually breaks away from the membrane and reenters the cytoplasm, where it quickly picks up more neurotransmitter.

After being released, some neurotransmitters are decomposed by enzymes in the synaptic cleft. Other neurotransmitters are transported back into the synaptic knob that released them (reuptake) or into nearby neurons or neuroglial cells. The enzyme *acetylcholinesterase,* for example, decomposes acetylcholine and is present at synapses that separate neurons using this neurotransmitter. Similarly, the enzyme *monoamine oxidase* inactivates monoamines. Decomposition or removal of neurotransmitters prevents continuous stimulation of postsynaptic neurons.

Table 9.2 summarizes the events leading to the release of a neurotransmitter.

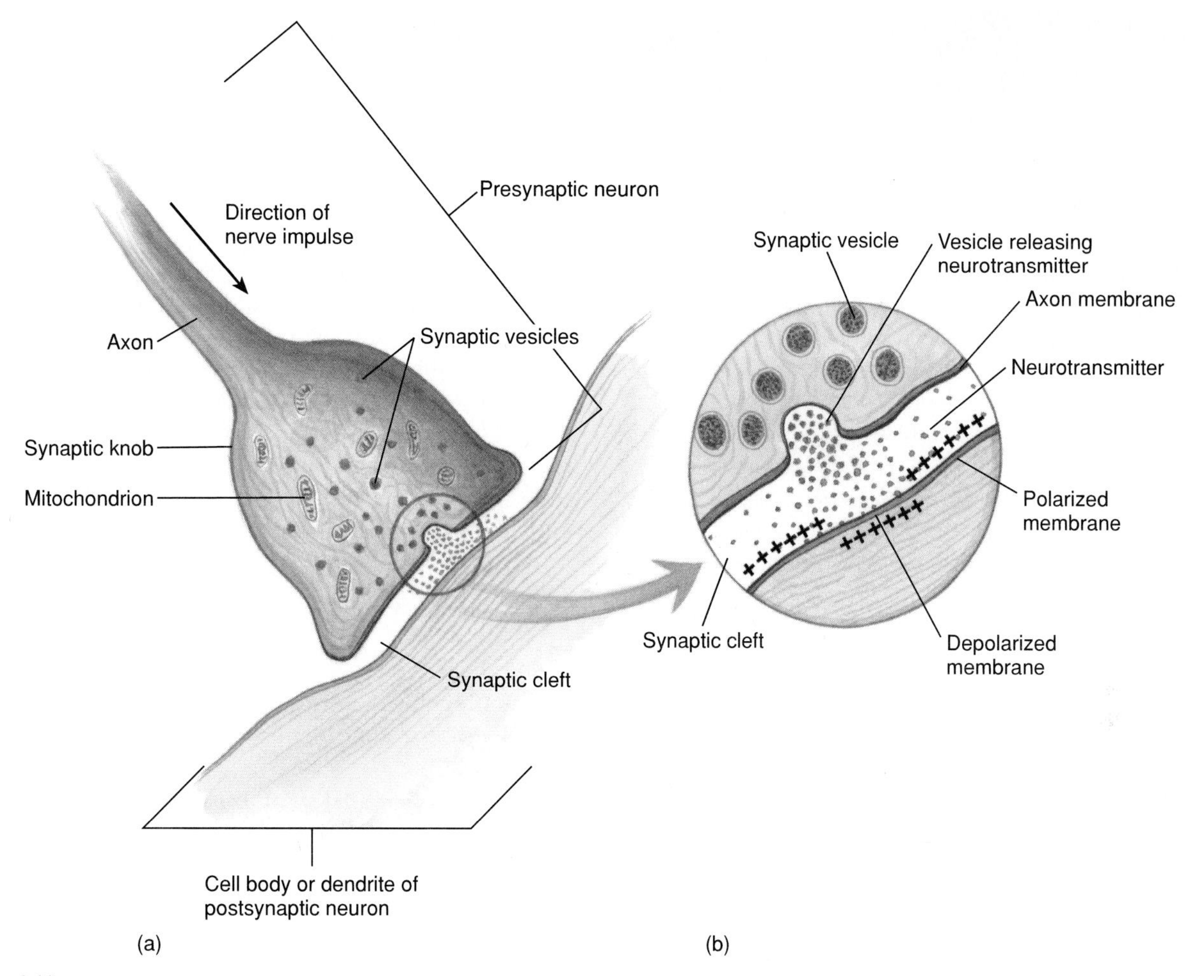

Figure 9.11

(*a*) When a nerve impulse reaches the synaptic knob at the end of an axon, (*b*) synaptic vesicles release a neurotransmitter that diffuses across the synaptic cleft.

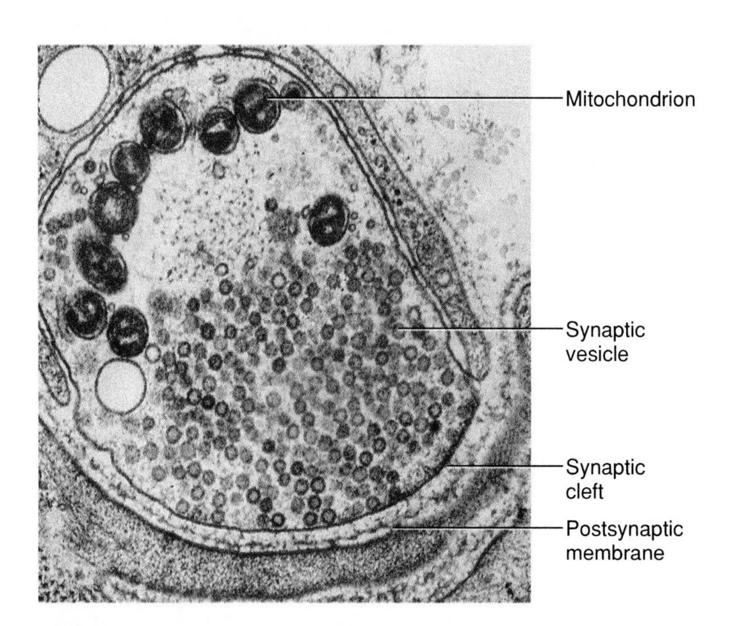

Figure 9.12

Transmission electron micrograph of a synaptic knob filled with synaptic vesicles.

Table 9.2

Events Leading to the Release of a Neurotransmitter

1. Action potential passes along a nerve fiber and over the surface of its synaptic knob.
2. Synaptic knob membrane becomes more permeable to calcium ions, and they diffuse inward.
3. In the presence of calcium ions, synaptic vesicles fuse to synaptic knob membrane.
4. Synaptic vesicles release their neurotransmitter into synaptic cleft.
5. Synaptic vesicles reenter cytoplasm of nerve fiber and pick up more neurotransmitter.

Topic of Interest: Factors Affecting Synaptic Transmission

Nerve impulses reaching synaptic knobs at too rapid a rate can exhaust neurotransmitter supplies, and impulse conduction ceases until more neurotransmitters are synthesized. This happens during an epileptic seizure. Abnormal and too rapid impulses originate from certain brain cells and reach skeletal muscle fibers, stimulating violent contractions. In time, the synaptic knobs run out of neurotransmitters, and the seizure subsides.

A drug called Dilantin (diphenylhydantoin) treats seizure disorders by increasing the effectiveness of the sodium active transport mechanism. Sodium ions transported from inside the neurons stabilize membrane thresholds against too intense stimulation.

Many other drugs affect synaptic transmission. For example, caffeine in coffee, tea, and cola drinks stimulates nervous system activity by lowering the thresholds at synapses so that neurons are more easily excited. Cocaine exerts its effects by blocking reuptake of dopamine and serotonin at synapses, enhancing the effect of these substances on postsynaptic cells.

Impulse Processing

The way the nervous system processes and responds to nerve impulses reflects, in part, the organization of neurons and their nerve fibers within the brain and spinal cord.

Neuronal Pools

Neurons within the central nervous system are organized into groups called *neuronal pools* that have varying numbers of cells. Each neuronal pool receives impulses from input nerve fibers. These impulses are processed according to the special characteristics of the pool, and any resulting impulses are conducted away on output fibers.

Each input fiber divides many times as it enters, and its branches spread over a certain region of the neuronal pool. The branches give off smaller branches, and their terminals form hundreds of synapses with the dendrites and cell bodies of the neurons in the pool.

Facilitation

As a result of incoming impulses, and neurotransmitter release, a particular neuron of a neuronal pool may receive excitatory and inhibitory input. If the net effect of the input is excitatory, threshold may be reached, and an outgoing impulse triggered. If the net effect is excitatory but subthreshold, an impulse is not triggered, but the neuron is more excitable to incoming stimulation than before, a state called **facilitation.**

Convergence

Any single neuron in a neuronal pool may receive impulses from two or more incoming fibers. Fibers originating from different parts of the nervous system and leading to the same neuron exhibit **convergence** (fig. 9.13*a*).

Convergence makes it possible for impulses arriving from different sources to have an additive effect on a neuron. For example, if a neuron is facilitated by receiving subthreshold stimulation from one input fiber, it may reach threshold if it receives additional stimulation from a second input fiber. As a result, a nerve impulse may travel to a particular effector and evoke a response.

Incoming impulses often bring information from several sensory receptors that detect changes. Convergence allows the nervous system to collect a variety of kinds of information, to process it, and to respond to it in a special way.

Divergence

Impulses leaving a neuron of a neuronal pool often exhibit **divergence** by passing into several other output fibers (fig. 9.13*b*). For example, an impulse from one neuron may stimulate two others; each of these, in turn, may stimulate several others, and so forth. Such a pattern of diverging nerve fibers can amplify an impulse—that is, spread it to more neurons within the pool. As a result of divergence, an impulse originating from a single neuron in the central nervous system may be amplified so that impulses reach enough motor units within a skeletal muscle to cause forceful contraction (see chapter 8). Similarly, an impulse originating from a sensory receptor may diverge and reach several different regions of the central nervous system, where the resulting impulses are processed and acted upon.

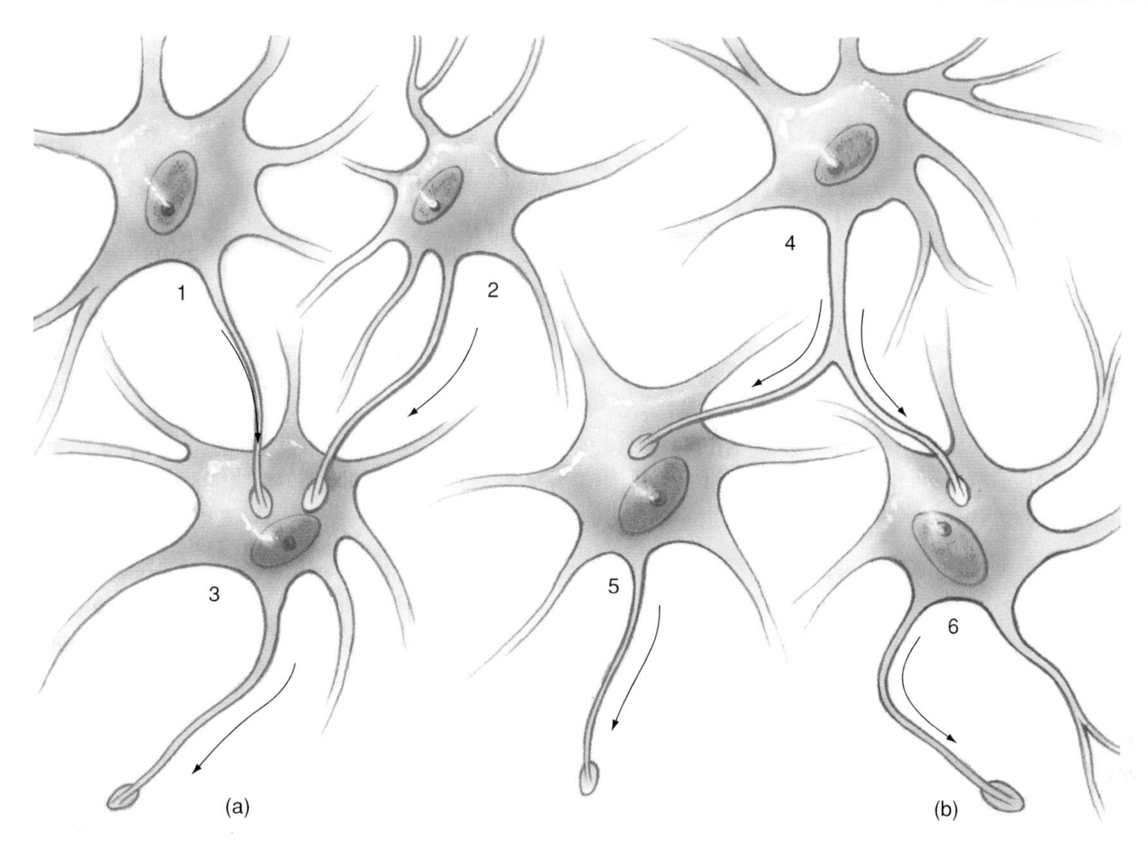

Figure 9.13 

(*a*) Nerve fibers of neurons 1 and 2 converge to the cell body of neuron 3. (*b*) The nerve fiber of neuron 4 diverges to the cell bodies of neurons 5 and 6.

1. Describe the function of a neurotransmitter.
2. Distinguish between excitatory and inhibitory actions of neurotransmitters.
3. Define *neuronal pool.*
4. Distinguish between convergence and divergence.

Types of Nerves

A nerve fiber is an extension of a neuron. A **nerve** is a cordlike bundle (or group of bundles) of nerve fibers held together by layers of connective tissue (fig. 9.14).

Like nerve fibers, nerves that conduct impulses into the brain or spinal cord are called **sensory nerves,** and those that carry impulses to muscles or glands are termed **motor nerves.** Most nerves include both sensory and motor fibers and are called **mixed nerves.**

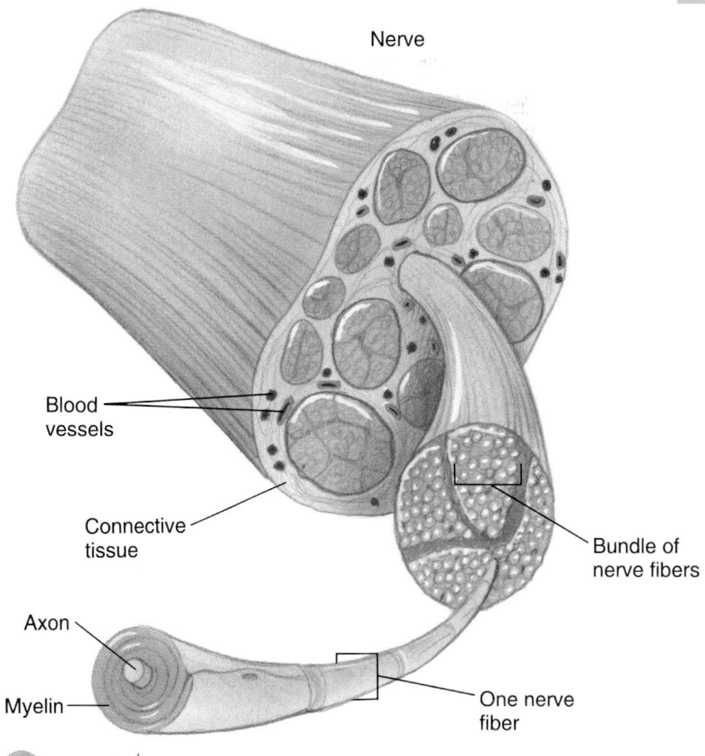

Figure 9.14

Connective tissue holds together a bundle of nerve fibers, forming a nerve.

1. What is a nerve?
2. How does a mixed nerve differ from a sensory nerve? From a motor nerve?

Nerve Pathways

The routes nerve impulses follow as they travel through the nervous system are called *nerve pathways*. The simplest of these pathways includes only a few neurons and is called a **reflex arc.** It constitutes the structural and functional basis for involuntary actions called reflexes.

Reflex Arcs

A reflex arc begins with a receptor at the end of a sensory nerve fiber. This fiber usually leads to several interneurons within the CNS, which serve as a processing center, or *reflex center*. Fibers from these interneurons may connect with interneurons in other parts of the nervous system. They also communicate with motor neurons, whose fibers pass outward from the CNS to effectors.

Reflex Behavior

Reflexes are automatic subconscious responses to changes (stimuli) within or outside the body. They help maintain homeostasis by controlling many involuntary processes, such as heart rate, breathing rate, blood pressure, and digestion. Reflexes also carry out the automatic actions of swallowing, sneezing, coughing, and vomiting.

The *knee-jerk reflex* (patellar tendon reflex) is an example of a simple reflex that employs only two neurons—a sensory neuron communicating directly to a motor neuron. Striking the patellar ligament just below the patella initiates this reflex. As a result, the quadriceps femoris muscle group, which a tendon attaches to the patella, is pulled slightly, stimulating stretch receptors within the muscles. These receptors, in turn, trigger impulses that pass along the fibers of a sensory neuron into the spinal cord. Within the spinal cord, the sensory axon forms a synapse with a motor neuron. The impulse then continues along the axon of the motor neuron and travels back to the quadriceps femoris. The

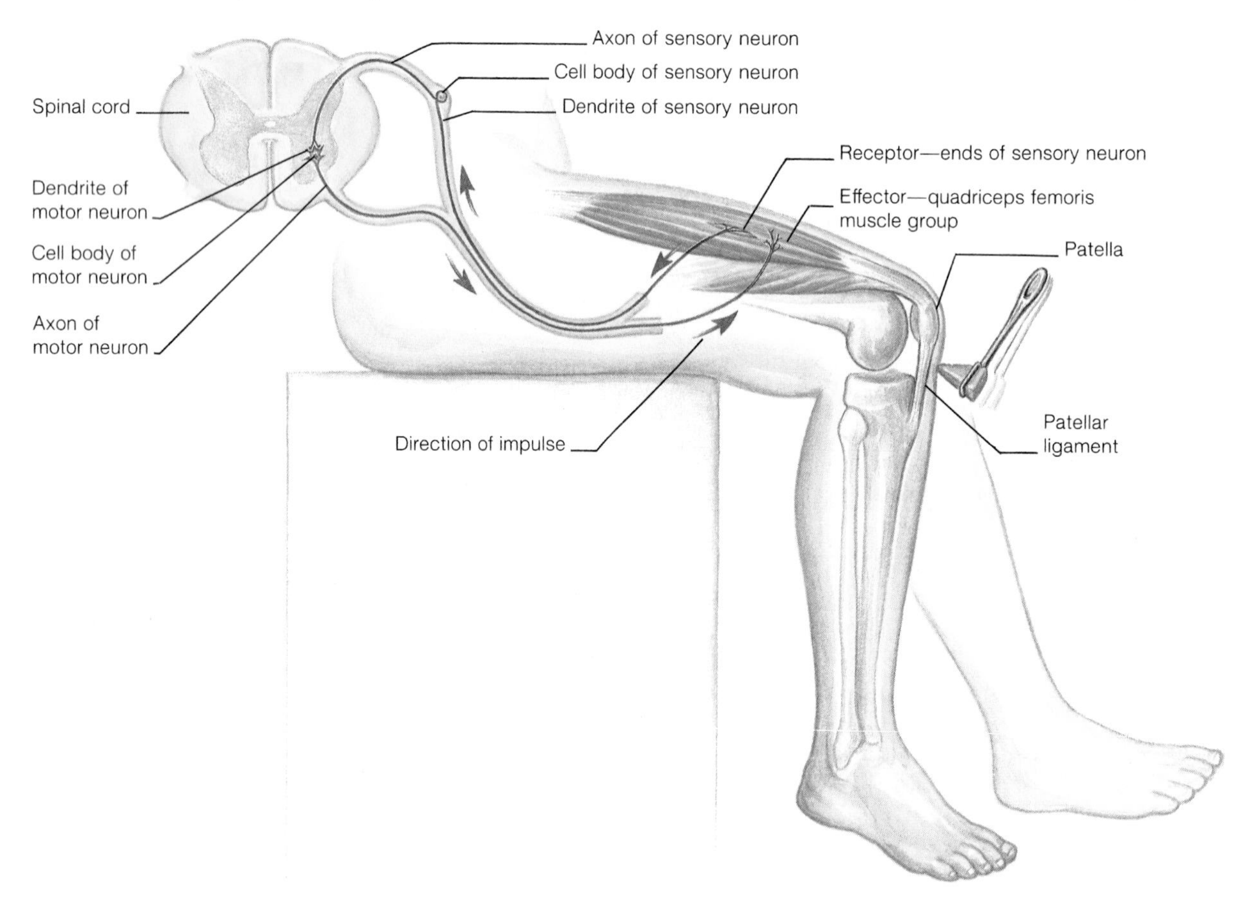

Figure 9.15

The knee-jerk reflex involves two neurons—a sensory neuron and a motor neuron.

muscle group contracts in response, and the reflex completes as the leg extends (fig. 9.15).

The knee-jerk reflex helps maintain upright posture. If the knee begins to bend from the force of gravity when a person is standing still, the quadriceps femoris is stretched, the reflex is triggered, and the leg straightens again.

Another type of reflex, called a *withdrawal reflex,* occurs when a person unexpectedly touches a finger to something painful. This activates skin receptors and sends sensory impulses to the spinal cord. There, the impulses pass to interneurons of a reflex center and are directed to motor neurons. The motor neurons transmit signals to flexor muscles in the arm, and the muscles contract in response. At the same time, the antagonistic extensor muscles are inhibited, and the hand is rapidly and unconsciously withdrawn from the painful stimulation. Concurrent with the withdrawal reflex, other interneurons carry sensory impulses to the brain, and the person becomes aware of the experience and may feel pain (fig. 9.16). A withdrawal reflex is protective because it may limit tissue damage from touching something potentially harmful.

Table 9.3 summarizes the parts of a reflex arc.

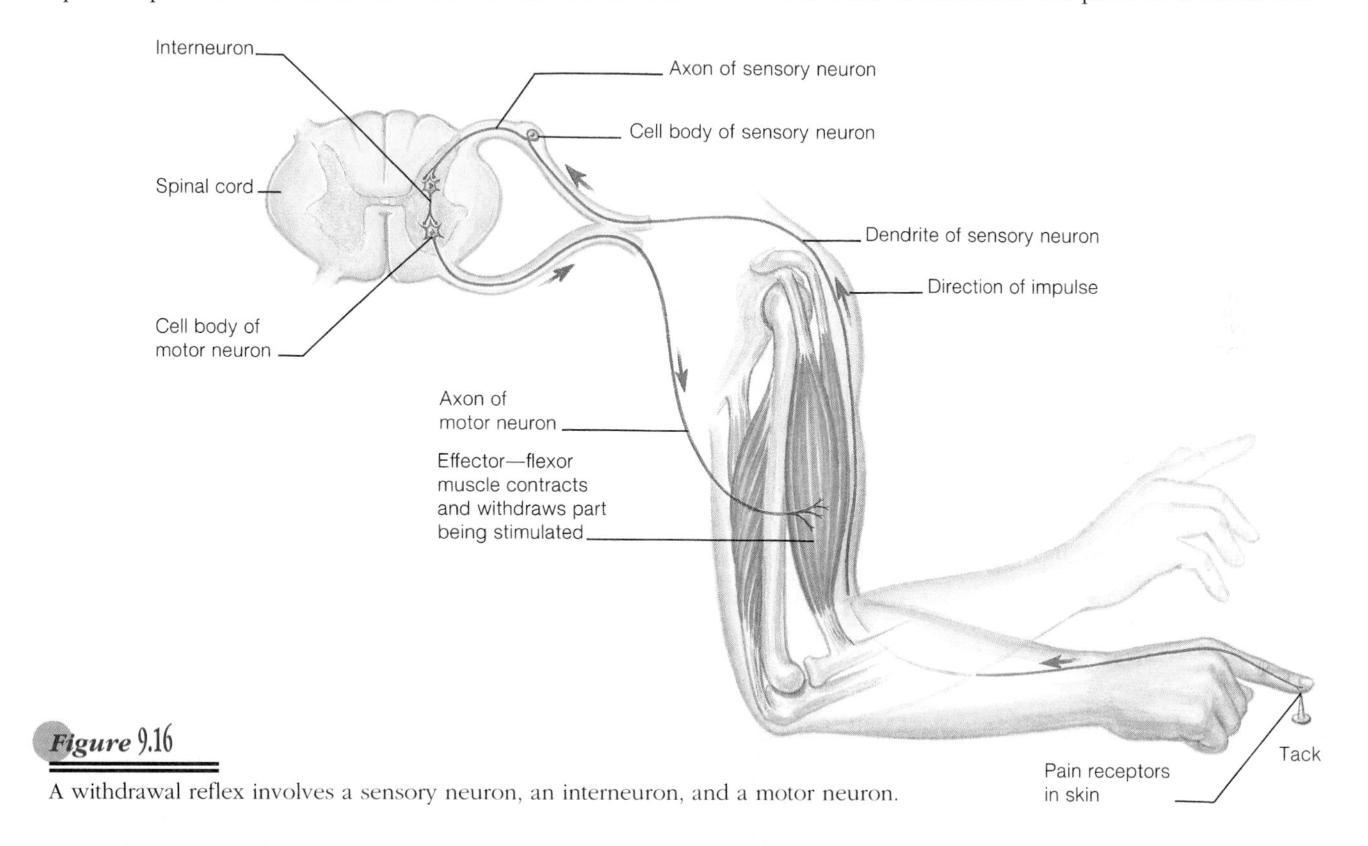

Figure 9.16

A withdrawal reflex involves a sensory neuron, an interneuron, and a motor neuron.

Table 9.3 Parts of a Reflex Arc

Part	Description	Function
Receptor	Receptor end of a dendrite or a specialized receptor cell in a sensory organ	Sensitive to a specific type of internal or external change
Sensory neuron	Dendrite, cell body, and axon of a sensory neuron	Transmits nerve impulse from receptor into brain or spinal cord
Interneuron	Dendrite, cell body, and axon of a neuron within the brain or spinal cord	Conducts nerve impulse from sensory neuron to motor neuron
Motor neuron	Dendrite, cell body, and axon of a motor neuron	Transmits nerve impulse from brain or spinal cord out to effector
Effector	Muscle or gland	Responds to stimulation by motor neuron and produces reflex or behavioral action

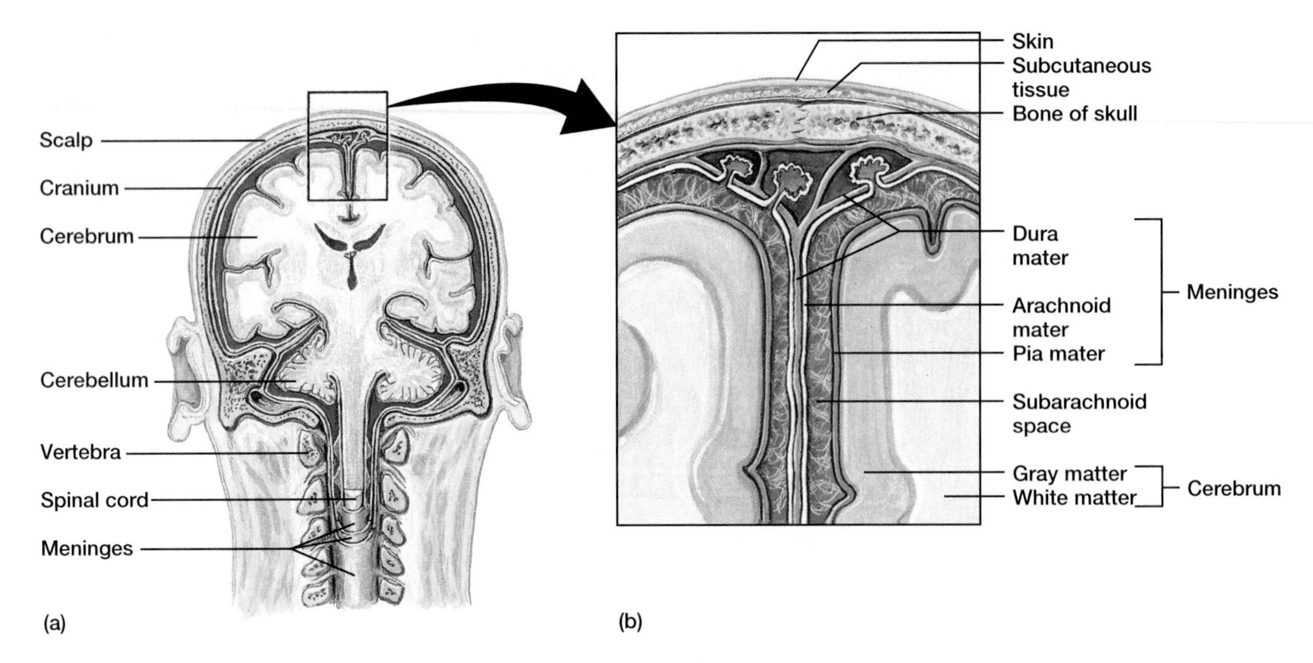

Figure 9.17

(*a*) Membranes called meninges enclose the brain and spinal cord. (*b*) The meninges include three layers: dura mater, arachnoid mater, and pia mater.

Because normal reflexes depend on normal neuron functions, reflexes provide information on the condition of the nervous system. An anesthesiologist, for instance, may try to initiate a reflex in a patient being anesthetized to determine how well the anesthetic drug is affecting nerve functions. Also, in the case of injury to some part of the nervous system, reflexes may be tested to discover the location and extent of damage. ■

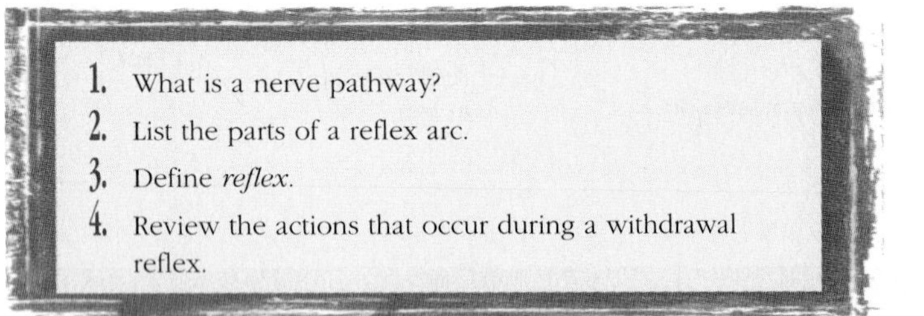

1. What is a nerve pathway?
2. List the parts of a reflex arc.
3. Define *reflex*.
4. Review the actions that occur during a withdrawal reflex.

Meninges

Bones, membranes, and fluid surround the organs of the CNS. The brain lies within the cranial cavity of the skull, and the spinal cord occupies the vertebral canal within the vertebral column. Beneath these bony coverings, membranes called **meninges** (singular, *meninx*), located between the bone and soft tissues of the nervous system, protect the brain and spinal cord (fig. 9.17*a*).

The meninges have three layers—dura mater, arachnoid mater, and pia mater (fig. 9.17*b*). The **dura mater** is the outermost layer of the meninges. It is composed primarily of tough, white fibrous connective tissue and contains many blood vessels and nerves. It attaches to the inside of the cranial cavity and forms the internal periosteum of the surrounding skull bones. In some regions, the dura mater extends inward between lobes of the brain and forms partitions that support and protect these parts.

The dura mater continues into the vertebral canal as a strong, tubular sheath that surrounds the spinal cord. It terminates as a blind sac below the end of the cord. The membrane around the spinal cord is not attached directly to the vertebrae but is separated by an *epidural space,* which lies between the dural sheath and the bony walls (fig. 9.18). This space contains loose connective and adipose tissues, which provide a protective pad around the spinal cord.

The **arachnoid mater** is a thin, weblike membrane that lacks blood vessels and is located between the dura and pia maters. It spreads over the brain and spinal cord but generally does not dip into the grooves and depressions on their surfaces.

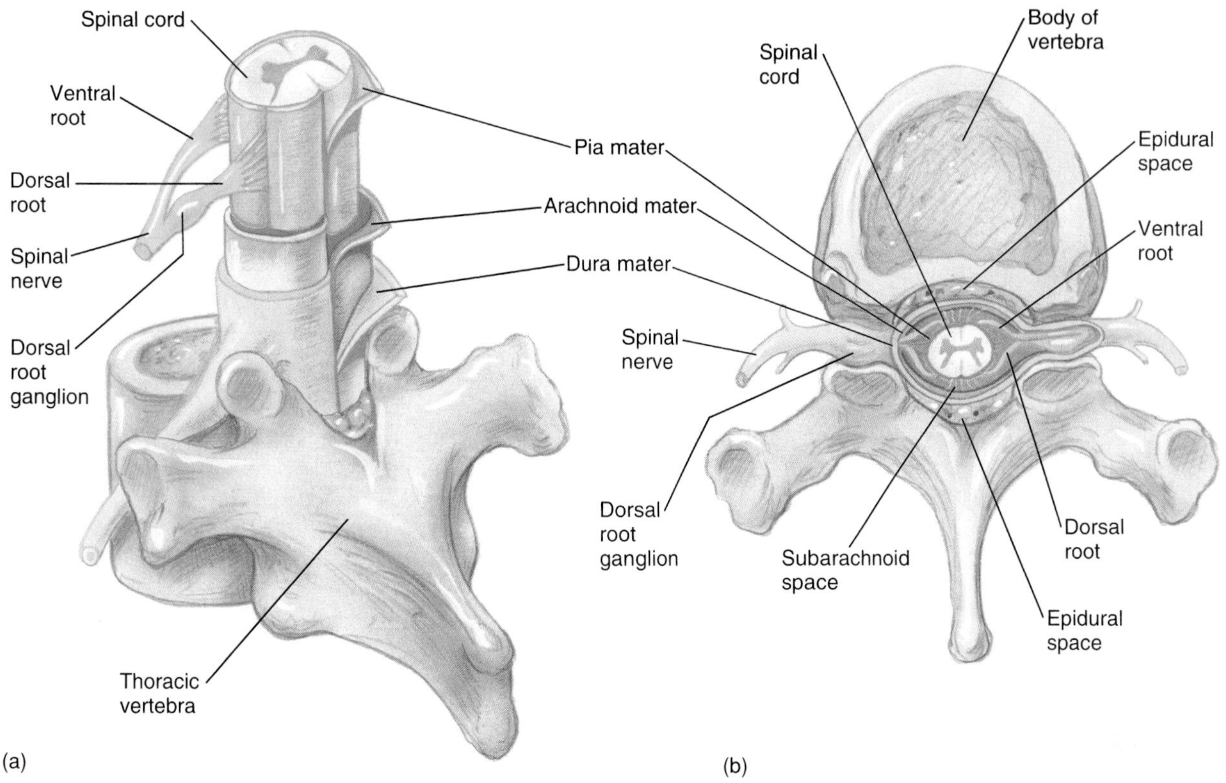

Figure 9.18

(*a*) The dura mater ensheaths the spinal cord. (*b*) Tissues forming a protective pad around the cord fill the epidural space between the dural sheath and the bone of the vertebra.

Between the arachnoid and pia maters is a *subarachnoid space* that contains the clear, watery **cerebrospinal fluid** (CSF). The **pia mater** is very thin and contains many nerves and blood vessels that nourish underlying cells of the brain and spinal cord. This layer hugs the surfaces of these organs and follows their irregular contours, passing over high areas and dipping into depressions.

A blow to the head may break some blood vessels associated with the brain, and escaping blood may collect in the space beneath the dura mater. Such a *subdural hematoma* increases pressure between the rigid bones of the skull and the soft tissues of the brain. Unless the accumulating blood is evacuated, compression of the brain may lead to functional losses or even death. ■

1. Describe the meninges.
2. Name the layers of the meninges.
3. State the location of cerebrospinal fluid.

Spinal Cord

The **spinal cord** is a slender nerve column that passes downward from the brain into the vertebral canal. Although continuous with the brain, the spinal cord is said to begin where nervous tissue leaves the cranial cavity at the level of the foramen magnum. The cord tapers to a point and terminates near the intervertebral disk that separates the first and second lumbar vertebrae (fig. 9.19).

Structure of the Spinal Cord

The spinal cord consists of thirty-one segments, each of which gives rise to a pair of **spinal nerves.** These nerves branch to various body parts and connect them with the central nervous system.

In the neck region, a thickening in the spinal cord, called the *cervical enlargement,* supplies nerves to the upper limbs. A similar thickening in the lower back, the *lumbar enlargement,* gives off nerves to the lower limbs (fig. 9.19).

Two grooves, a deep *anterior median fissure* and a shallow *posterior median sulcus,* extend the length of the spinal cord, dividing it into right and left halves (fig. 9.20). A cross section of the cord reveals that it consists

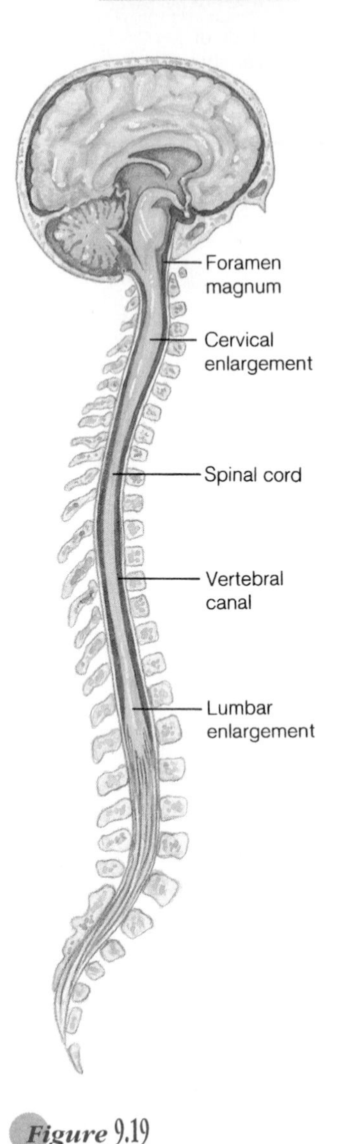

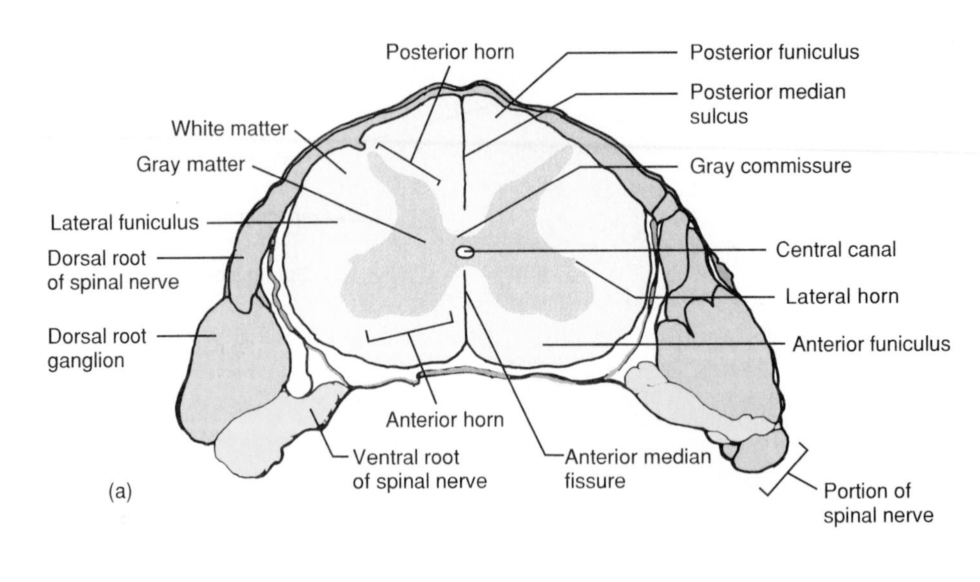

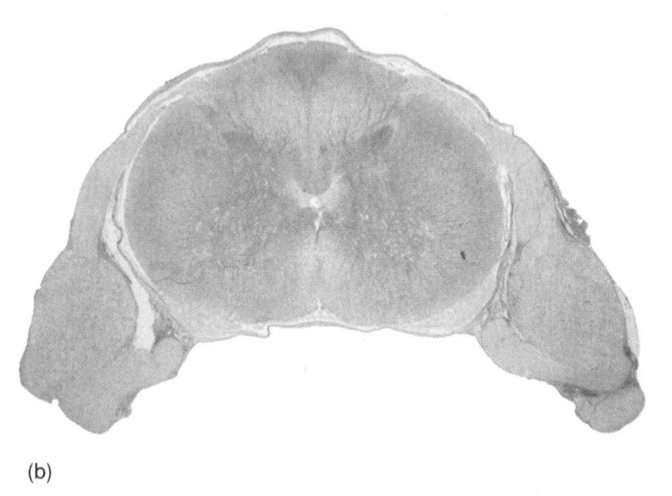

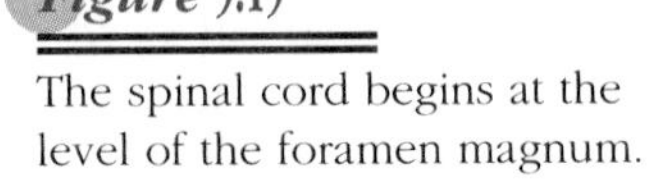

Figure 9.19

The spinal cord begins at the level of the foramen magnum.

Figure 9.20

(*a*) Cross section of the spinal cord. (*b*) Artificially stained micrograph of the cord (7.5×).

of a core of gray matter within white matter. The pattern the gray matter produces roughly resembles a butterfly with its wings spread. The upper and lower wings of gray matter are called the *posterior horns* and *anterior horns,* respectively. Between them on either side in the thoracic and upper lumbar segments is a protrusion of gray matter called the *lateral horn.*

Neurons with large cell bodies located in the anterior horns give rise to motor fibers that pass out through spinal nerves to skeletal muscles. The majority of neurons in the gray matter of the spinal cord, however, are interneurons.

Gray matter divides the white matter of the spinal cord into three regions on each side—the *anterior, lateral,* and *posterior funiculi* (fig. 9.20*a*). Each funiculus consists of longitudinal bundles of myelinated nerve fibers that comprise major nerve pathways called **nerve tracts.**

A horizontal bar of gray matter in the middle of the spinal cord, the *gray commissure,* connects the wings of the gray matter on the right and left sides. This bar surrounds the **central canal**, which contains cerebrospinal fluid.

Functions of the Spinal Cord

The spinal cord has two major functions—conducting nerve impulses and serving as a center for spinal reflexes. The nerve tracts of the spinal cord provide a two-way communication system between the brain and body parts outside the nervous system. The tracts that carry sensory information to the brain are called **ascending tracts** (fig. 9.21); those that conduct motor impulses from the brain to muscles and glands are called **descending tracts** (fig. 9.22).

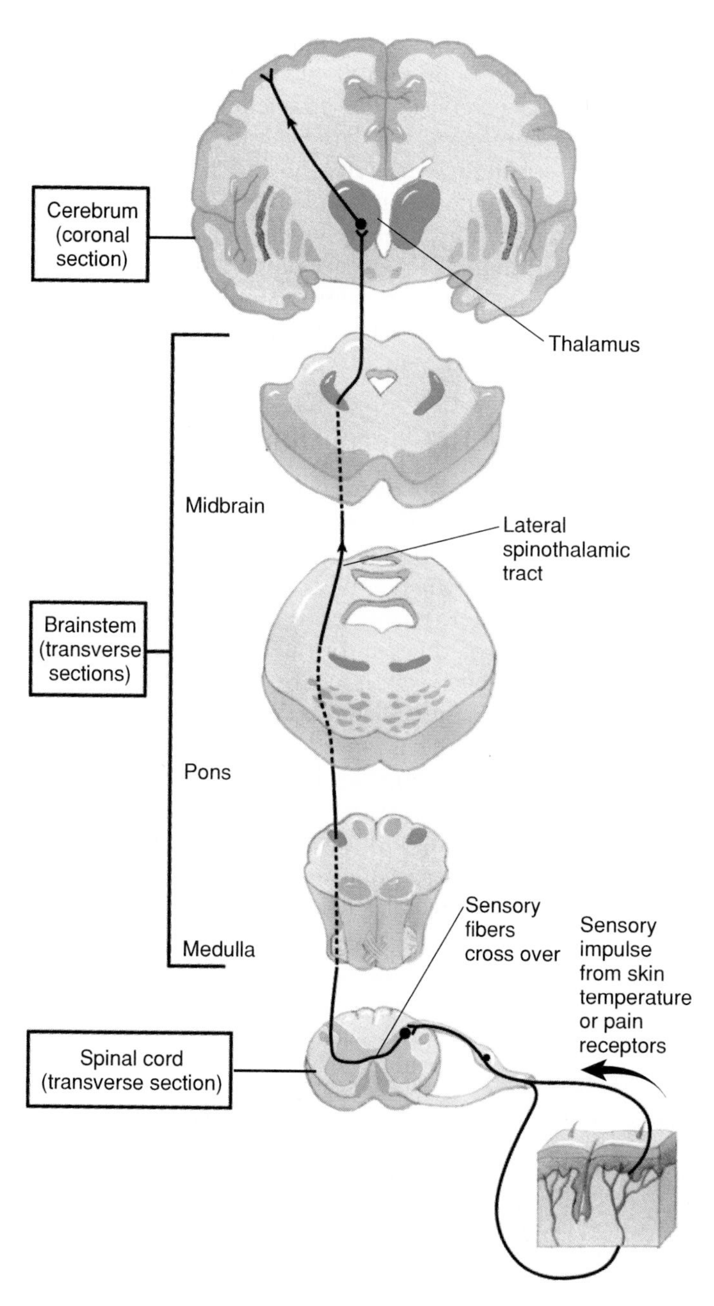

Figure 9.21

Sensory impulses originating in skin receptors cross over in the spinal cord and ascend to the brain.

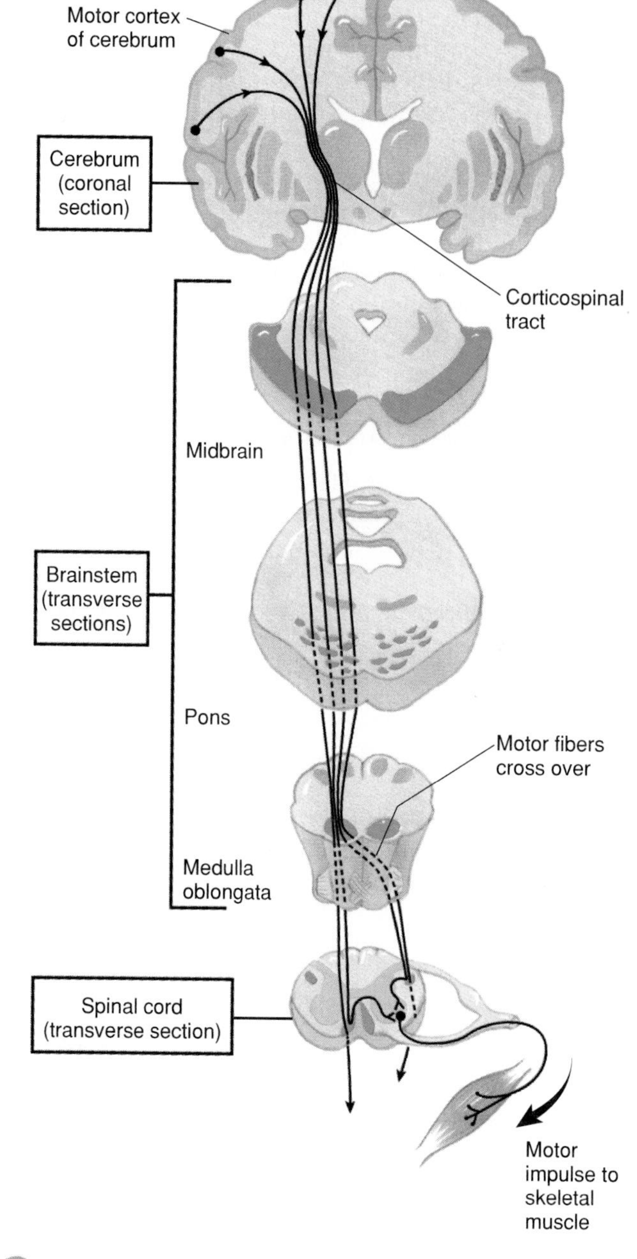

Figure 9.22

Motor fibers of the corticospinal tract begin in the cerebral cortex, cross over in the medulla oblongata, and descend in the spinal cord. There, they synapse with neurons whose fibers lead to the spinal nerves that supply skeletal muscles.

The nerve fibers within ascending and descending tracts are axons. Typically, all the axons within a given tract originate from neuron cell bodies located in the same part of the nervous system and terminate together in some other part. The names that identify nerve tracts often reflect these common origins and terminations. For example, a *spinothalamic tract* begins in the spinal cord and carries sensory impulses associated with the sensations of pain and touch to the thalamus of the brain. A *corticospinal tract* originates in the cortex of the brain and carries motor impulses downward through the spinal cord and spinal nerves. These impulses control skeletal muscle movements.

Corticospinal tracts are also called *pyramidal tracts* after the pyramid-shaped areas in the medulla oblongata of the brain through which they pass. Other descending tracts, called *extrapyramidal tracts,* control motor activities associated with maintaining balance and posture.

In addition to providing a pathway for nerve tracts, the spinal cord functions in many reflexes, like the knee-jerk and withdrawal reflexes described previously. These are called **spinal reflexes** because their reflex arcs pass through the spinal cord.

1. Describe the structure of the spinal cord.
2. Describe the general functions of the spinal cord.
3. Distinguish between an ascending and a descending tract.

Brain

The **brain** is composed of about 100 billion (10^{11}) multipolar neurons and innumerable nerve fibers by which these neurons communicate with one another and with neurons in other parts of the nervous system. As figure 9.23 shows, the brain can be divided into three major portions—the cerebrum, the cerebellum, and the brain stem. The **cerebrum,** the largest part, contains nerve centers associated with sensory and motor functions and provides higher mental functions, including memory and reasoning. The **diencephalon** also processes sensory information. The **cerebellum** includes centers that coordinate voluntary muscular movements. Nerve pathways in the **brain stem** connect various parts of the nervous system and regulate certain visceral activities.

Structure of the Cerebrum

The cerebrum consists of two large masses called **cerebral hemispheres,** which are essentially mirror images of each other. A deep bridge of nerve fibers called the **corpus callosum** connects the cerebral hemispheres. A layer of dura mater (falx cerebri) separates them.

The surface of the cerebrum has many ridges, or **convolutions** (gyri), separated by grooves. A shallow groove is called a **sulcus,** and a deep groove is called a **fissure.** Although the arrangement of these elevations and depressions is complex, they form distinct patterns in all normal brains. For example, a *longitudinal fissure* separates the right and left cerebral hemispheres, a *transverse fissure* separates the cerebrum from the cerebellum, and several sulci divide each hemisphere into lobes.

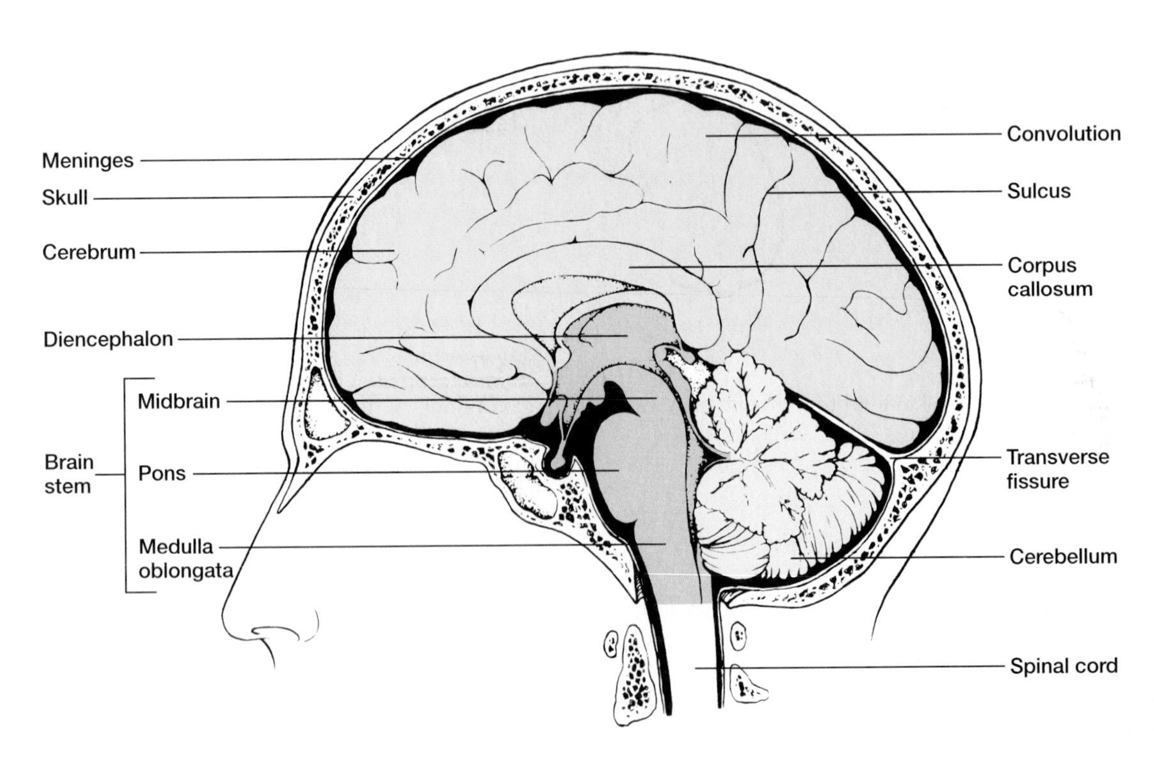

Figure 9.23

The major portions of the brain include the cerebrum, the cerebellum, and the brain stem.

The lobes of the cerebral hemispheres are named after the skull bones they underlie (fig. 9.24). They include:

1. **Frontal lobe** The frontal lobe forms the anterior portion of each cerebral hemisphere. It is bordered posteriorly by a *central sulcus,* which extends from the longitudinal fissure at a right angle, and inferiorly by a *lateral sulcus,* which extends from the undersurface of the brain along its sides.
2. **Parietal lobe** The parietal lobe is posterior to the frontal lobe and separated from it by the central sulcus.
3. **Temporal lobe** The temporal lobe lies below the frontal and parietal lobes and is separated from them by the lateral sulcus.
4. **Occipital lobe** The occipital lobe forms the posterior portion of each cerebral hemisphere and is separated from the cerebellum by a shelflike extension of dura mater (tentorium cerebelli). The boundary between the occipital lobe and the parietal and temporal lobes is not distinct.
5. **Insula** The insula is located deep within the lateral sulcus and is covered by parts of the frontal, parietal, and temporal lobes. A *circular sulcus* separates the insula from the lobes.

A thin layer of gray matter called the **cerebral cortex** is the outermost portion of the cerebrum. This layer covers the convolutions and dips into the sulci and fissures. It contains nearly 75% of all the neuron cell bodies in the nervous system.

Just beneath the cerebral cortex is a mass of white matter that makes up the bulk of the cerebrum. This mass contains bundles of myelinated nerve fibers that connect neuron cell bodies of the cortex with other parts of the nervous system. Some of these fibers pass from one cerebral hemisphere to the other by way of the corpus callosum, and others carry sensory or motor impulses from portions of the cortex to nerve centers in the brain or spinal cord.

Functions of the Cerebrum

The cerebrum provides higher brain functions. It contains centers for interpreting sensory impulses arriving from sense organs and centers for initiating voluntary muscular movements. The cerebrum stores the information of memory and utilizes it to reason. Intelligence and personality also stem from cerebral activity.

Functional Regions of the Cerebral Cortex

Specific regions of the cerebral cortex perform specific functions. Although functions overlap among regions, the cortex can be divided into motor, sensory, and association areas.

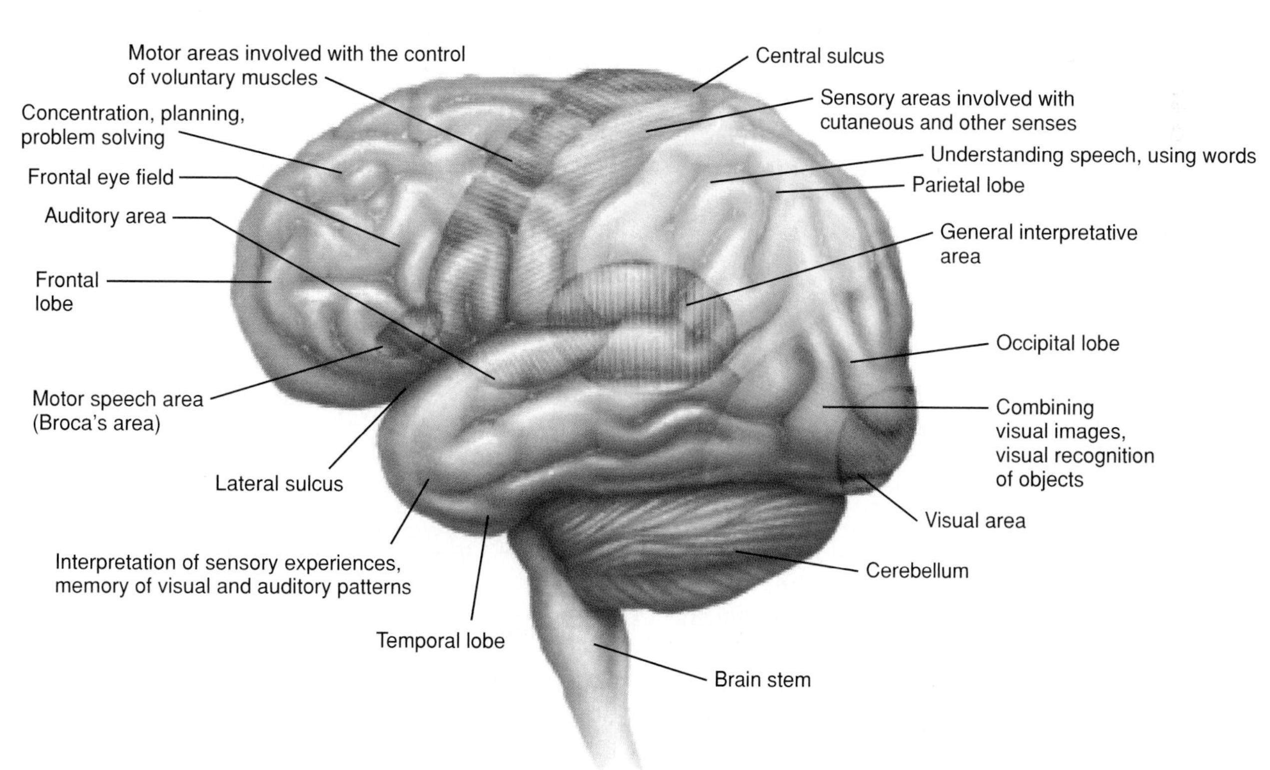

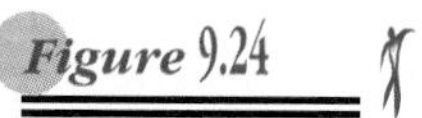

Figure 9.24

Some motor, sensory, and association areas of the left cerebral cortex.

The effects of injuries to the cerebral cortex depend on the locations and extent of the damage. The abilities that become impaired can indicate the site of damage. For example, injury to the motor areas of one frontal lobe causes partial or complete paralysis on the opposite side of the body.

A person with damage to the association areas of the frontal lobe may have difficulty concentrating on complex mental tasks and may appear disorganized and easily distracted. A person who suffers damage to association areas of the temporal lobes may have difficulty recognizing printed words or arranging words into meaningful thoughts. ■

The primary **motor areas** of the cerebral cortex lie in the frontal lobes, just in front of the central sulcus (fig. 9.24). The nervous tissue in these regions contains numerous, large *pyramidal cells,* named for their pyramid-shaped cell bodies.

Impulses from the pyramidal cells travel downward through the brain stem and into the spinal cord on the corticospinal tracts (see fig. 9.22). Most of the nerve fibers in these tracts cross over from one side of the brain to the other within the brain stem. As a result, the motor area of the right cerebral hemisphere generally controls skeletal muscles on the left side of the body, and vice versa.

In addition to the primary motor areas, certain other regions of the frontal lobe affect motor functions. For example, a region called the *motor speech area* (*Broca's area*) is just anterior to the primary motor cortex and superior to the lateral sulcus (fig. 9.24). It coordinates the complex muscular actions of the mouth, tongue, and larynx, which make speech possible. Above Broca's area is a region called the *frontal eye field.* The motor cortex in this area controls voluntary movements of the eyes and eyelids. Another region just in front of the primary motor area controls the muscular movements of the hands and fingers that make skills such as writing possible.

Sensory areas located in several lobes of the cerebrum interpret impulses that arrive from sensory receptors, producing feelings or sensations. For example, sensations from all parts of the skin (cutaneous senses) arise in the anterior portions of the parietal lobes along the central sulcus (fig. 9.24). The posterior parts of the occipital lobes affect vision (visual area), and the temporal lobes contain the centers for hearing (auditory area). The sensory areas for taste are located near the bases of the central sulci along the lateral sulci, and the sense of smell arises from centers deep within the cerebrum.

Like motor fibers, sensory fibers cross over either in the spinal cord or the brain stem (see fig. 9.21). Thus, the centers in the right cerebral hemisphere interpret impulses originating from the left side of the body, and vice versa.

Association areas are neither primarily sensory nor motor. They connect with one another and with other brain structures. These areas analyze and interpret sensory experiences and oversee memory, reasoning, verbalizing, judgment, and emotion. Association areas occupy the anterior portions of the frontal lobes and are widespread in the lateral portions of the parietal, temporal, and occipital lobes (fig. 9.24).

Association areas of the frontal lobes control a number of higher intellectual processes. These include concentrating, planning, complex problem solving, and judging the possible consequences of behavior. Association areas of the parietal lobes help in understanding speech and choosing words to express thoughts and feelings. Association areas of the temporal lobes and regions at the posterior ends of the lateral fissures interpret complex sensory experiences, such as those needed to understand speech and to read. These regions also provide memory of visual scenes, music, and other complex sensory patterns. Association areas of the occipital lobes that are adjacent to the visual centers are important in analyzing visual patterns and combining visual images with other sensory experiences, as when one recognizes another person or an object.

The parietal, temporal, and occipital association areas meet near the posterior end of the lateral sulcus. This important region is called the *general interpretative area,* and it plays the primary role in complex thought processing (fig. 9.24).

1. List the major divisions of the brain.
2. Describe the cerebral cortex.
3. What are the major functions of the cerebrum?
4. Locate the major functional regions of the cerebral cortex.

Hemisphere Dominance

Both cerebral hemispheres participate in basic functions, such as receiving and analyzing sensory impulses, controlling skeletal muscles, and storing memory. However, in most persons, one side of the cerebrum is the **dominant hemisphere,** controlling certain other functions.

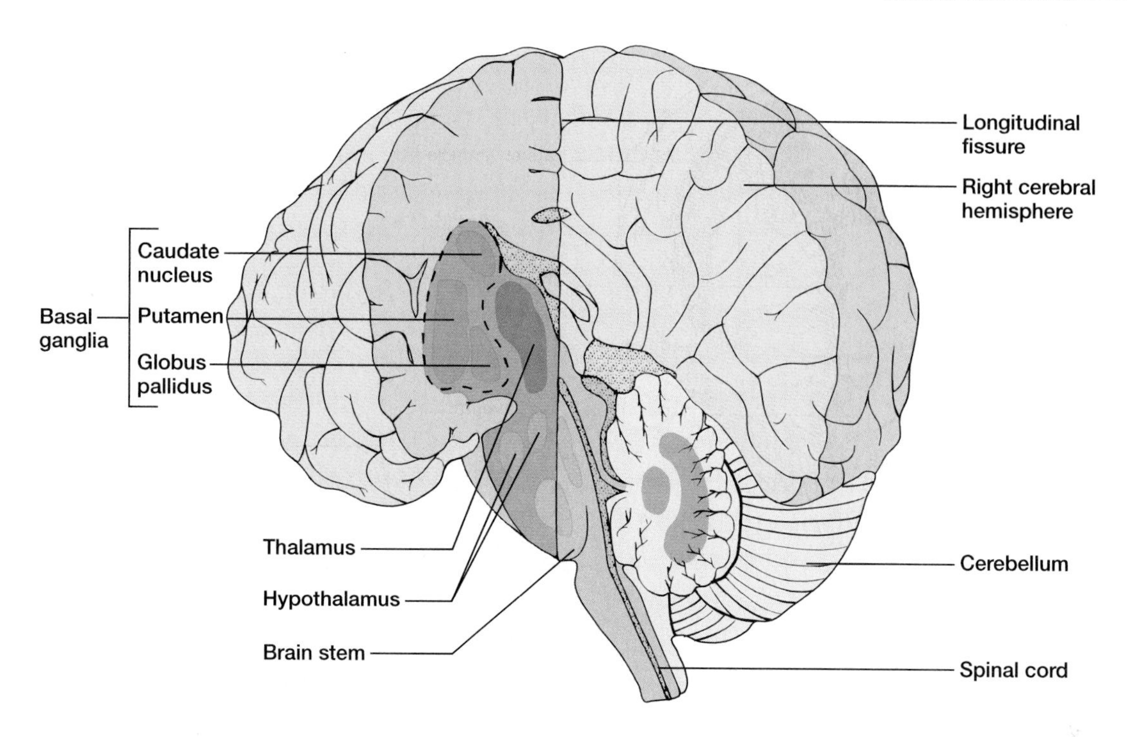

Figure 9.25

A coronal section of the left cerebral hemisphere reveals some of the basal ganglia.

The left hemisphere is dominant in 90% of right-handed adults and in 64% of left-handed ones. The right hemisphere is dominant in 10% of right-handed adults and in 20% of left-handed ones. The hemispheres are equally dominant in the remaining 16% of left-handed persons. Because of hemisphere dominance, Broca's area on one side almost completely controls the motor activities associated with speech. For this reason, over 90% of patients with language impairment involving the cerebrum have disorders in the left hemisphere. ■

In over 90% of the population, the left hemisphere is dominant for the language-related activities of speech, writing, and reading, and for complex intellectual functions requiring verbal, analytical, and computational skills. In other persons, the right hemisphere is dominant, or the hemispheres are equally dominant.

In addition to carrying on basic functions, the nondominant hemisphere specializes in nonverbal functions, such as motor tasks that require orientation of the body in space, understanding and interpreting musical patterns, and nonverbal visual experiences. The nondominant hemisphere also controls emotional and intuitive thinking.

Nerve fibers of the corpus callosum, which connect the cerebral hemispheres, allow the dominant hemisphere to control the motor cortex of the nondominant hemisphere. These fibers also transfer sensory information reaching the nondominant hemisphere to the dominant one, where the information can be used in decision making.

Deep within each cerebral hemisphere are several masses of gray matter called **basal ganglia** (basal nuclei) (fig. 9.25). They are the *caudate nucleus,* the *putamen,* and the *globus pallidus.* Their neuron cell bodies serve as relay stations for motor impulses originating in the cerebral cortex and passing into the brain stem and spinal cord. The basal ganglia produce most of the inhibitory neurotransmitter dopamine. Impulses from the basal ganglia normally inhibit motor functions and thus control various skeletal muscle activities.

The uncontrollable movements of Parkinson disease and Huntington disease result from lesions in the basal ganglia. The lack of inhibiting impulses causes the excessive movement. ■

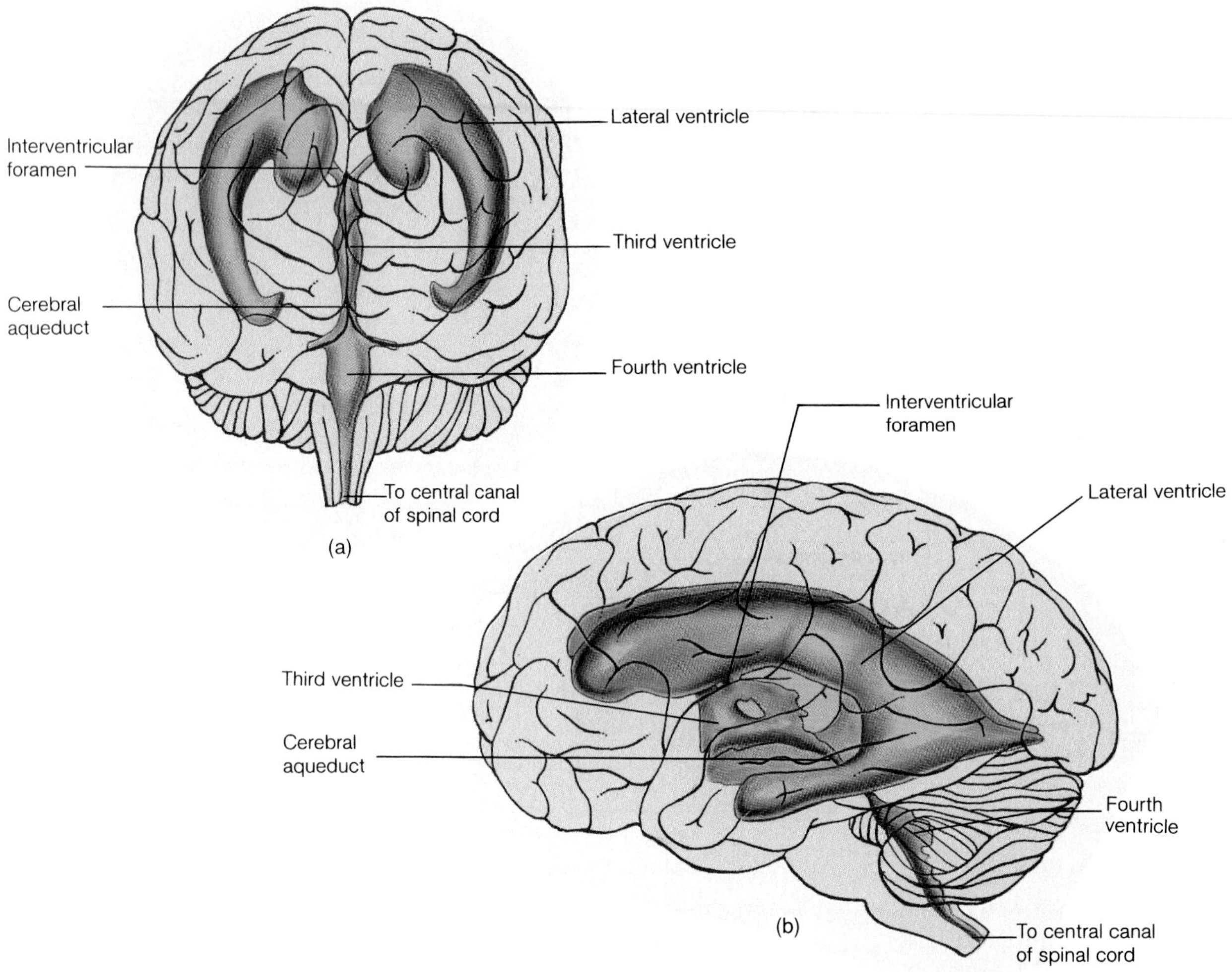

Figure 9.26

(*a*) Anterior view of the ventricles within the cerebral hemispheres and brain stem.
(*b*) Lateral view.

Ventricles and Cerebrospinal Fluid

Within the cerebral hemispheres and brain stem is a series of interconnected cavities called **ventricles** (fig. 9.26). These spaces are continuous with the central canal of the spinal cord, and like it, they contain cerebrospinal fluid.

The largest ventricles are the *lateral ventricles* (first and second ventricles), which extend into the cerebral hemispheres and occupy portions of the frontal, temporal, and occipital lobes. A narrow space that constitutes the *third ventricle* is in the midline of the brain, beneath the corpus callosum. This ventricle communicates with the lateral ventricles through openings (interventricular foramina) in its anterior end. The *fourth ventricle* is in the brain stem just anterior to the cerebellum. A narrow canal, the *cerebral aqueduct,* connects it to the third ventricle and passes lengthwise through the brain stem. This ventricle is continuous with the central canal of the spinal cord and has openings in its roof that lead into the subarachnoid space of the meninges.

Tiny, reddish cauliflower-like masses of specialized capillaries from the pia mater, called **choroid plexuses,** secrete cerebrospinal fluid (fig. 9.27). These structures project into the ventricles. Most of the cerebrospinal fluid arises in the lateral ventricles. From there, it circulates slowly into the third and fourth ventricles and into the central canal of the spinal cord. Cerebrospinal fluid also enters the subarachnoid space of the meninges through the wall of the fourth ventricle near the cerebellum and completes its circuit by being reabsorbed into the blood.

Because it occupies the subarachnoid space of the meninges, cerebrospinal fluid completely surrounds the brain and spinal cord. In effect, these organs float in the fluid, which supports and protects them by absorbing forces that might otherwise jar and damage them. Cerebrospinal fluid also maintains a stable ionic concentration in the central nervous system and provides a pathway to the blood for wastes.

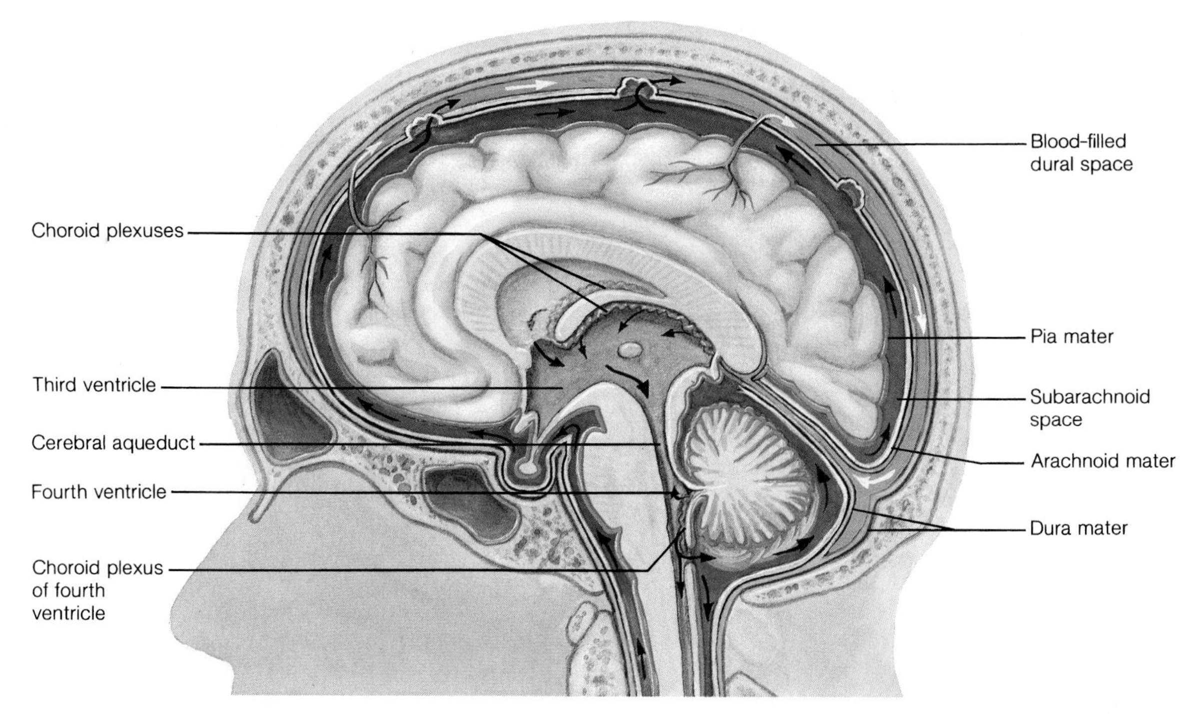

Figure 9.27

The choroid plexuses in the walls of the ventricles secrete cerebrospinal fluid. The fluid circulates through the ventricles and central canal, enters the subarachnoid space, and is reabsorbed into the blood.

Because cerebrospinal fluid is secreted and reabsorbed continuously, the fluid pressure in the ventricles normally remains relatively constant. An infection, a tumor, or a blood clot can interfere with fluid circulation, increasing pressure within the ventricles and thus in the cranial cavity (intracranial pressure). This can injure the brain by forcing it against the rigid skull.

A *lumbar puncture* (spinal tap) is used to measure the pressure of cerebrospinal fluid. In this procedure, a fine, hollow needle is inserted into the subarachnoid space between the third and fourth or between the fourth and fifth lumbar vertebrae. An instrument called a *manometer* measures the pressure. ■

1. What is hemisphere dominance?
2. What are the major functions of the dominant hemisphere? The nondominant one?
3. Where are the ventricles of the brain?
4. Describe the circulation of cerebrospinal fluid.

Diencephalon

The **diencephalon** is located between the cerebral hemispheres and above the midbrain. It surrounds the third ventricle and is composed largely of gray matter. Within the diencephalon, a dense mass called the **thalamus** bulges into the third ventricle from each side (figs. 9.25 and 9.28). Another region of the diencephalon that includes many nuclei is the hypothalamus. It lies below the thalamus and forms the lower walls and floor of the third ventricle.

Other parts of the diencephalon include: (1) the **optic tracts** and the **optic chiasma** that is formed by optic nerve fibers crossing over each other; (2) the **infundibulum**, a conical process behind the optic chiasma to which the pituitary gland attaches; (3) the **posterior pituitary gland**, which hangs from the floor of the hypothalamus; (4) the **mammillary bodies**, which appear as two rounded structures behind the infundibulum; and (5) the **pineal gland**, a cone-shaped structure attached to the upper portion of the diencephalon (see chapter 11).

The thalamus is a central relay station for sensory impulses ascending from other parts of the nervous system to the cerebral cortex. It receives all sensory impulses (except those associated with the sense of

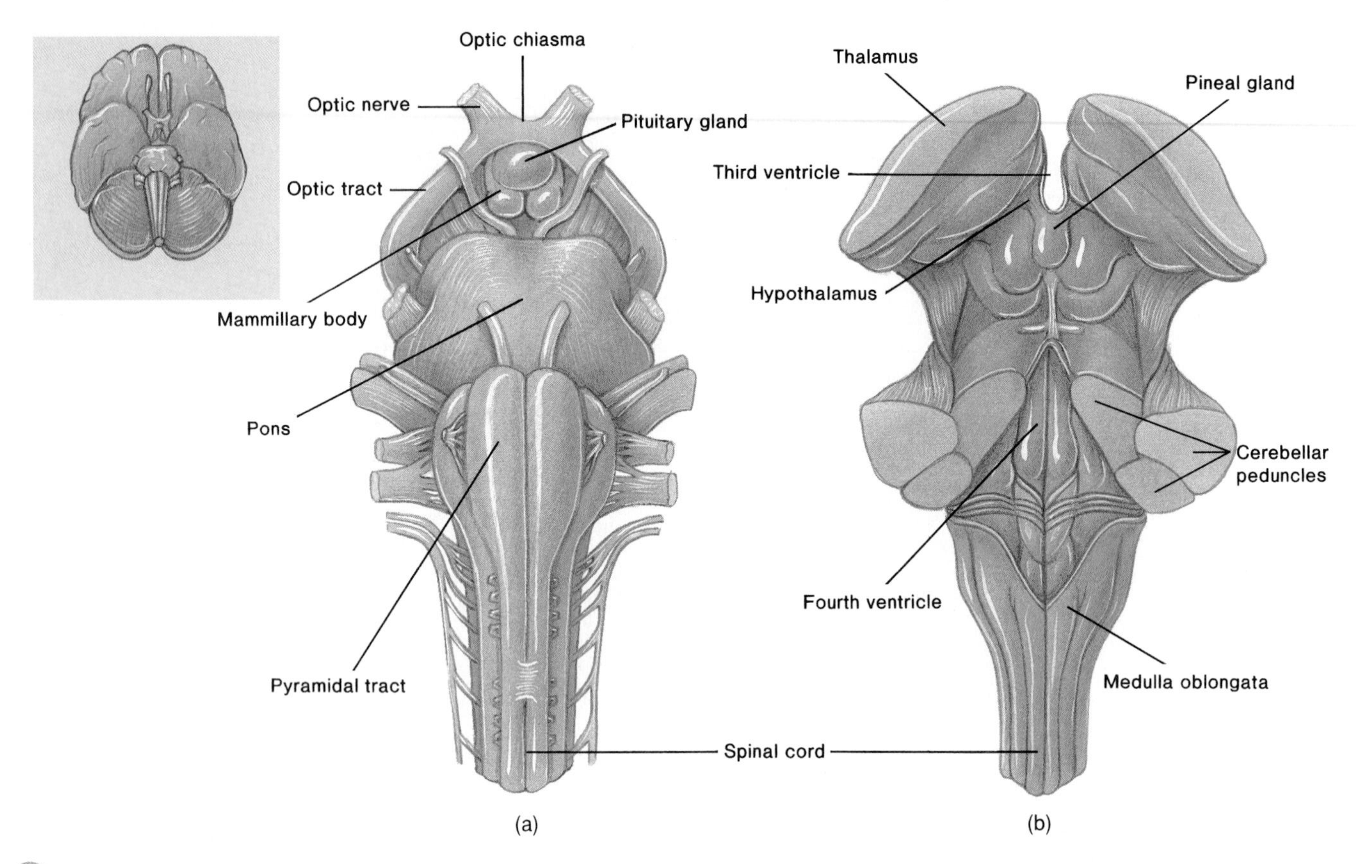

Figure 9.28

(*a*) Ventral view of the brain stem. (*b*) Dorsal view of the brain stem with the cerebellum removed, exposing the fourth ventricle.

smell) and channels them to the appropriate regions of the cortex for interpretation. In addition, all regions of the cerebral cortex can communicate with the thalamus by means of descending fibers.

The cerebral cortex pinpoints the origin of sensory stimulation. The thalamus produces a general awareness of certain sensations, such as pain, touch, and temperature.

Nerve fibers connect the **hypothalamus** to the cerebral cortex, thalamus, and other parts of the brain stem. The hypothalamus maintains homeostasis by regulating a variety of visceral activities and by linking the nervous and endocrine systems.

The hypothalamus regulates:

1. Heart rate and arterial blood pressure
2. Body temperature
3. Water and electrolyte balance
4. Control of hunger and body weight
5. Control of movements and glandular secretions of the stomach and intestines
6. Production of neurosecretory substances that stimulate the pituitary gland to secrete hormones
7. Sleep and wakefulness

Structures in the general region of the diencephalon also control emotional responses. For example, portions of the cerebral cortex in the medial parts of the frontal and temporal lobes interconnect with the hypothalamus, thalamus, basal ganglia, and other deep masses of gray matter called *nuclei.* Together, these structures comprise a complex called the **limbic system.**

The limbic system controls emotional experience and expression. It can modify the way a person acts by producing such feelings as fear, anger, pleasure, and sorrow. The limbic system recognizes upsets in a person's physical or psychological condition that might threaten life. By causing pleasant or unpleasant feelings about experiences, the limbic system guides a person into behavior that is likely to increase the chance of survival.

Brain Stem

The **brain stem** is a bundle of nervous tissue that connects the cerebrum to the spinal cord. It consists of numerous tracts of nerve fibers and several nuclei. The parts of the brain stem include the midbrain, pons, and medulla oblongata (figs. 9.23 and 9.28).

Midbrain

The **midbrain** is a short section of the brain stem between the diencephalon and the pons (see fig. 9.23). It contains bundles of myelinated nerve fibers that join lower parts of the brain stem and spinal cord with higher parts of the brain. Two prominent bundles of nerve fibers on the underside of the midbrain include the corticospinal tracts and are the main motor pathways between the cerebrum and lower parts of the nervous system.

The midbrain includes several masses of gray matter that serve as reflex centers. For example, the midbrain contains the centers for certain visual reflexes, such as those responsible for moving the eyes to view something as the head turns. It also contains the auditory reflex centers that enable a person to move the head to hear sounds more distinctly.

Pons

The **pons** is a rounded bulge on the underside of the brain stem, where it separates the midbrain from the medulla oblongata (see fig. 9.23). The dorsal portion of the pons consists largely of longitudinal nerve fibers, which relay impulses to and from the medulla oblongata and the cerebrum. The ventral portion of the pons contains large bundles of transverse nerve fibers, which transmit impulses from the cerebrum to centers within the cerebellum.

Several nuclei of the pons relay sensory impulses from peripheral nerves to higher brain centers. Other nuclei function with centers of the medulla oblongata to regulate the rate and depth of breathing (see chapter 16).

Medulla Oblongata

The **medulla oblongata** extends from the pons to the foramen magnum of the skull (see fig. 9.23). Its dorsal surface flattens to form the floor of the fourth ventricle, and its ventral surface is marked by the corticospinal tracts, most of whose fibers cross over at this level (see fig. 9.22).

All the ascending and descending nerve fibers connecting the brain and spinal cord must pass through the medulla oblongata because of its location. As in the spinal cord, the white matter of the medulla oblongata surrounds a central mass of gray matter. Here, however, the gray matter breaks into nuclei separated by nerve fibers. Some of these nuclei relay ascending impulses to the other side of the brain stem and then on to higher brain centers.

Other nuclei within the medulla oblongata control vital visceral activities. These centers include:

1. **Cardiac center** Impulses originating in the cardiac center are transmitted to the heart on peripheral nerves, altering heart rate.
2. **Vasomotor center** Certain cells of the vasomotor center initiate impulses that travel to smooth muscles in the walls of certain blood vessels and stimulate them to contract. This constricts the blood vessels (vasoconstriction), elevating blood pressure. Other cells of the vasomotor center produce the opposite effect—dilating blood vessels (vasodilation), and consequently dropping blood pressure.
3. **Respiratory center** The respiratory center acts with centers in the pons to regulate the rate, rhythm, and depth of breathing.

Still other nuclei within the medulla oblongata are centers for the reflexes associated with coughing, sneezing, swallowing, and vomiting.

Reticular Formation

Scattered throughout the medulla oblongata, pons, and midbrain is a complex network of nerve fibers associated with tiny islands of gray matter. This network, the **reticular formation** (reticular activating system), extends from the upper portion of the spinal cord into the diencephalon. Its nerve fibers join centers of the hypothalamus, basal nuclei, cerebellum, and cerebrum with fibers in all the major ascending and descending tracts.

When sensory impulses reach the reticular formation, it responds by activating the cerebral cortex into a state of wakefulness. Without this arousal, the cortex remains unaware of stimulation and cannot interpret sensory information or carry on thought processes. Thus, decreased activity in the reticular formation results in sleep. If the reticular formation is injured so that it cannot function, the person remains unconscious and cannot be aroused, even with strong stimulation. This is called a comatose state.

1. What are the major functions of the thalamus? The hypothalamus?
2. How may the limbic system influence behavior?
3. List the structures of the brain stem.
4. What vital reflex centers are located in the brain stem?
5. What is the function of the reticular formation?

Cerebellum

The **cerebellum** is a large mass of tissue located below the occipital lobes of the cerebrum and posterior to the

TOPIC OF INTEREST Drug Abuse

Drug abuse is the chronic self-administration of a drug in doses high enough to cause *addiction*—a physical or psychological dependence on the drug where the user is preoccupied with locating and taking the drug. Stopping drug use causes intense, unpleasant withdrawal symptoms. Prolonged and repeated abuse of a drug may also result in *drug tolerance,* in which the physiological response to a particular dose of the drug becomes less intense over time. Drug tolerance results as the drug increases synthesis of certain liver enzymes, which metabolize the drug more rapidly, so that the addict needs the next dose sooner. Drug tolerance also arises from physiological changes that lessen the drug's effect on its target cells.

The most commonly abused drugs are central nervous system (CNS) depressants ("downers"), CNS stimulants ("uppers"), hallucinogens, and anabolic steroids. *CNS depressants* include barbiturates, benzodiazepines, opiates, and cannabinoids.

Barbiturates, such as amytal, nembutal, and seconal, act uniformly throughout the brain; however, the reticular formation is particularly sensitive to their effects. CNS depression occurs due to inhibited secretion of certain excitatory and inhibitory neurotransmitters. Effects range from mild calming of the nervous system (sedation) to sleep, loss of sensory sensations (anesthesia), respiratory distress, cardiovascular collapse, and death.

The *benzodiazepines,* such as diazepam (Valium), depress activity in the limbic system and the reticular formation. Low doses are commonly prescribed to relieve anxiety. Higher doses cause sedation, sleep, or anesthesia. These drugs increase either the activity or release of the inhibitory neurotransmitter GABA. When benzodiazepines are metabolized, they may form other biochemicals that have depressing effects.

The *opiates* include heroin (which has no legal use in the United States), codeine, hydromorphone (Dilaudid), meperidine (Demerol), and methadone. These drugs stimulate certain receptors (opioid receptors) in the CNS, and in prescribed dosages, sedate and relieve pain (analgesia). Opiates cause both physical and psychological dependence. Effects include a feeling of well-being (euphoria), respiratory distress, convulsions, coma, and possible death.

The *cannabinoids* include marijuana and hashish, both derived from the hemp plant. Hashish is several times more potent than marijuana. These drugs depress higher brain centers and release lower brain centers from the normal inhibitory influence of the higher centers. This induces an anxiety-free state, characterized by euphoria and a distorted perception of time and space. *Hallucinations* (sensory perceptions that have no external stimuli), respiratory distress, and vasomotor depression may occur with higher doses.

CNS stimulants include amphetamines and cocaine (including "crack"). These drugs have great abuse potential and may quickly produce psychological dependence. Cocaine, especially when smoked or inhaled, produces euphoria but may also change personality, cause seizures, and constrict certain blood vessels, leading to sudden death from stroke or cardiac arrhythmias. Cocaine's very rapid effect, and perhaps its addictiveness, reflect its rapid entry and metabolism in the brain. Cocaine arrives at the basal ganglia in 4 to 6 minutes and is cleared mostly within 30 minutes. The drug inhibits transporter molecules that remove dopamine from synapses after it is released.

Hallucinogens alter perceptions. They cause *illusions,* which are distortions of vision, hearing, taste, touch, and smell; *synesthesia,* such as "hearing" colors or "feeling" sounds; and hallucinations.

The most commonly abused and most potent hallucinogen is lysergic acid diethylamide (LSD). LSD may act as an excitatory neurotransmitter. A person under the influence of LSD may greatly overestimate physical capabilities, such as believing he or she can fly off the top of a high building.

Phencyclidine (PCP) is another commonly abused hallucinogen. Its use may lead to prolonged psychosis that may provoke assault, murder, and suicide.

Anabolic steroids are lipids that have malelike hormonal effects. They stimulate anabolic metabolism and thus promote the growth of certain tissues, including skeletal muscle tissue. These drugs are sometimes used to treat disease conditions, such as osteoporosis. However, their ability to increase skeletal muscle mass has led to their abuse, especially among athletes hoping to improve performance. Such abuse may change liver functions, increase risk of heart and blood vessel disease and certain cancers, atrophy the testes, and cause severe personality disorders.

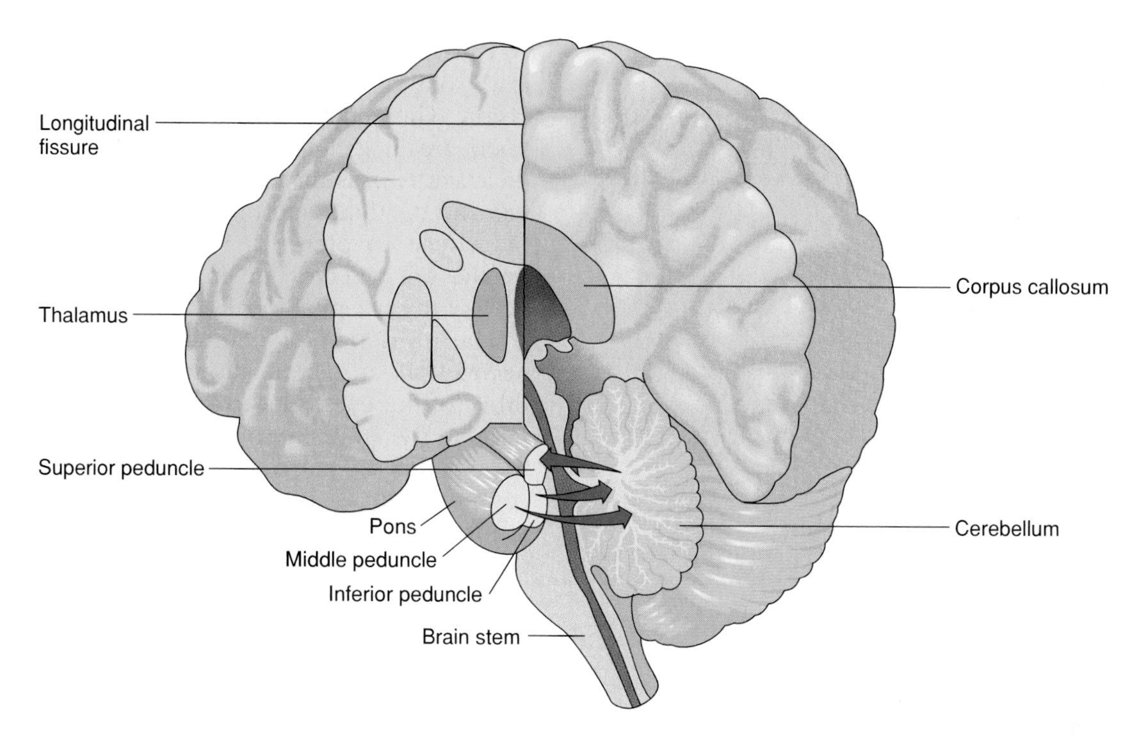

Figure 9.29

The cerebellum, which is located below the occipital lobes of the cerebrum, communicates with other parts of the nervous system by means of the cerebellar peduncles.

pons and medulla oblongata (see fig. 9.23). It consists of two lateral hemispheres partially separated by a layer of dura mater (falx cerebelli) and connected in the midline by a structure called the *vermis*. Like the cerebrum, the cerebellum is composed primarily of white matter with a thin layer of gray matter, the **cerebellar cortex**, on its surface.

The cerebellum communicates with other parts of the central nervous system by means of three pairs of nerve tracts called *cerebellar peduncles* (figs. 9.28 and 9.29). One pair (the inferior peduncles) brings sensory information concerning the position of the limbs, joints, and other body parts to the cerebellum. Another pair (the middle peduncles) transmits signals from the cerebral cortex to the cerebellum concerning the desired positions of these parts. After integrating and analyzing this information, the cerebellum sends correcting impulses via a third pair (the superior peduncles) to the midbrain. These corrections are incorporated into motor impulses that travel downward through the pons, medulla oblongata, and spinal cord in the appropriate patterns to move the body in the desired way.

Thus, the cerebellum is a reflex center for integrating sensory information concerning the position of body parts and for coordinating complex skeletal muscle movements. It also helps to maintain posture. Damage to the cerebellum is likely to result in tremors, inaccurate movements of voluntary muscles, loss of muscle tone, a reeling walk, and loss of equilibrium.

1. Where is the cerebellum located?
2. What are the major functions of the cerebellum?

Peripheral Nervous System

The **peripheral nervous system (PNS)** consists of nerves that branch out from the CNS and connect it to other body parts. The PNS includes the cranial nerves, which arise from the brain, and the spinal nerves, which arise from the spinal cord.

The PNS can also be subdivided into the somatic and autonomic nervous systems. Generally, the **somatic nervous system** consists of the cranial and spinal nerve fibers that connect the CNS to the skin and skeletal muscles; it oversees conscious activities. The **autonomic nervous system** includes fibers that connect the CNS to viscera, such as the heart, stomach, intestines, and glands; it controls unconscious activities. Table 9.4 outlines the subdivisions of the nervous system.

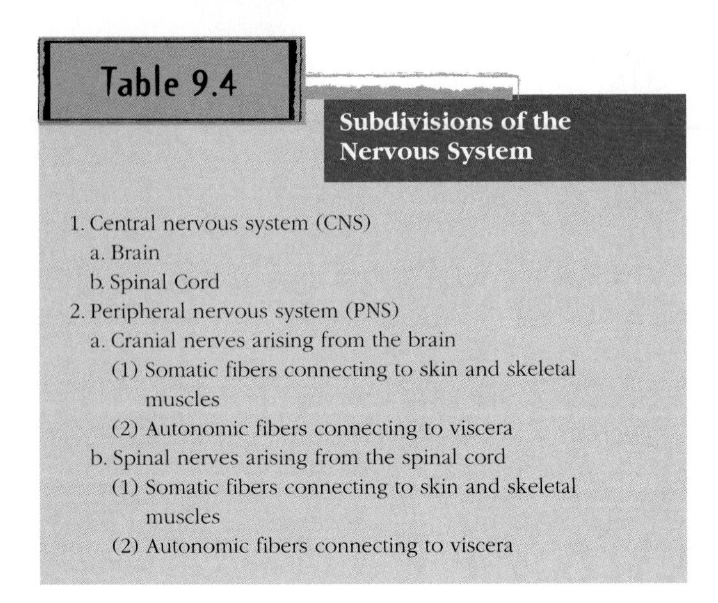

Table 9.4 Subdivisions of the Nervous System

1. Central nervous system (CNS)
   a. Brain
   b. Spinal Cord
2. Peripheral nervous system (PNS)
   a. Cranial nerves arising from the brain
      (1) Somatic fibers connecting to skin and skeletal muscles
      (2) Autonomic fibers connecting to viscera
   b. Spinal nerves arising from the spinal cord
      (1) Somatic fibers connecting to skin and skeletal muscles
      (2) Autonomic fibers connecting to viscera

Cranial Nerves

Twelve pairs of **cranial nerves** arise from the underside of the brain (fig. 9.30). Except for the first pair, which begins within the cerebrum, these nerves originate from the brain stem. They pass from their sites of origin through foramina of the skull and lead to parts of the head, neck, and trunk.

Most of the cranial nerves are mixed nerves, but some of those associated with special senses, such as smell and vision, contain only sensory fibers. Other cranial nerves that affect muscles and glands are composed primarily of motor fibers.

Sensory fibers present in the cranial nerves have neuron cell bodies that are outside the brain, usually in groups called *ganglia*. On the other hand, motor neuron cell bodies are typically located within the gray matter of the brain.

Numbers or names designate the cranial nerves. The numbers indicate the order in which the nerves arise from the front to the back of the brain, and the names describe their primary functions or the general distribution of their fibers (fig. 9.30).

The first pair of cranial nerves, the **olfactory nerves (I)**, are associated with the sense of smell and contain only sensory neurons. These neurons process impulses received from *olfactory receptor cells* in the lining of the upper nasal cavity. Axons from these receptors pass upward through the cribriform plates of the ethmoid bone. Their impulses reach the olfactory neurons in the *olfactory bulbs,* which are extensions of the cerebral cortex located just beneath the frontal lobes. Sensory impulses travel from the olfactory bulbs along *olfactory tracts* to cerebral centers where they are interpreted. The result of this interpretation is the sensation of smell.

The second pair of cranial nerves, the **optic nerves (II)**, lead from the eyes to the brain and are associated with vision. The sensory nerve cell bodies of these nerve fibers are in ganglion cell layers within the eyes, and their axons pass through the *optic foramina* of the orbits and continue into the visual nerve pathways of the brain (see chapter 10). Sensory impulses transmitted on the optic nerves are interpreted in the visual cortices of the occipital lobes.

The third pair of cranial nerves, the **oculomotor nerves (III)**, arise from the midbrain and pass into the orbits of the eyes. One component of each nerve connects to the voluntary muscles that raise the eyelid and four of the six muscles that move the eye. A second component of each oculomotor nerve is part of the autonomic nervous system and supplies involuntary muscles within the eyes. These muscles adjust the amount of light entering the eyes and focus the lenses.

The fourth pair, the **trochlear nerves (IV)**, are the smallest cranial nerves. They arise from the midbrain and each nerve carries motor impulses to a fifth voluntary muscle that moves the eye that the oculomotor nerve does not supply.

The fifth pair, the **trigeminal nerves (V)**, are the largest cranial nerves and arise from the pons. They are mixed nerves, with the sensory portions more extensive than the motor portions. Each sensory component includes three large branches, called the ophthalmic, maxillary, and mandibular divisions.

The *ophthalmic division* of the trigeminal nerves consists of sensory fibers that bring impulses to the brain from the surface of the eyes, the tear glands, and the skin of the anterior scalp, forehead, and upper eyelids. The fibers of the *maxillary division* carry sensory impulses from the upper teeth, upper gum, and upper lip, as well as from the mucous lining of the palate and the skin of the face. The *mandibular division* includes both motor and sensory fibers. The sensory branches transmit impulses from the scalp behind the ears, the skin of the jaw, the lower teeth, the lower gum, and the lower lip. The motor branches supply the muscles of mastication and certain muscles in the floor of the mouth.

The sixth pair of cranial nerves, the **abducens nerves (VI)**, are quite small and originate from the pons near the medulla oblongata. Each nerve enters the orbit of the eye and supplies motor impulses to the remaining muscle that moves the eye.

The seventh pair of cranial nerves, the **facial nerves (VII)**, arise from the lower part of the pons and emerge on the sides of the face. Their sensory branches are associated with taste receptors on the anterior two-thirds of the tongue, and some of their motor fibers transmit impulses to muscles of facial expression. Still other motor fibers of these nerves function in the autonomic nervous system and stimulate secretions from tear glands and salivary glands.

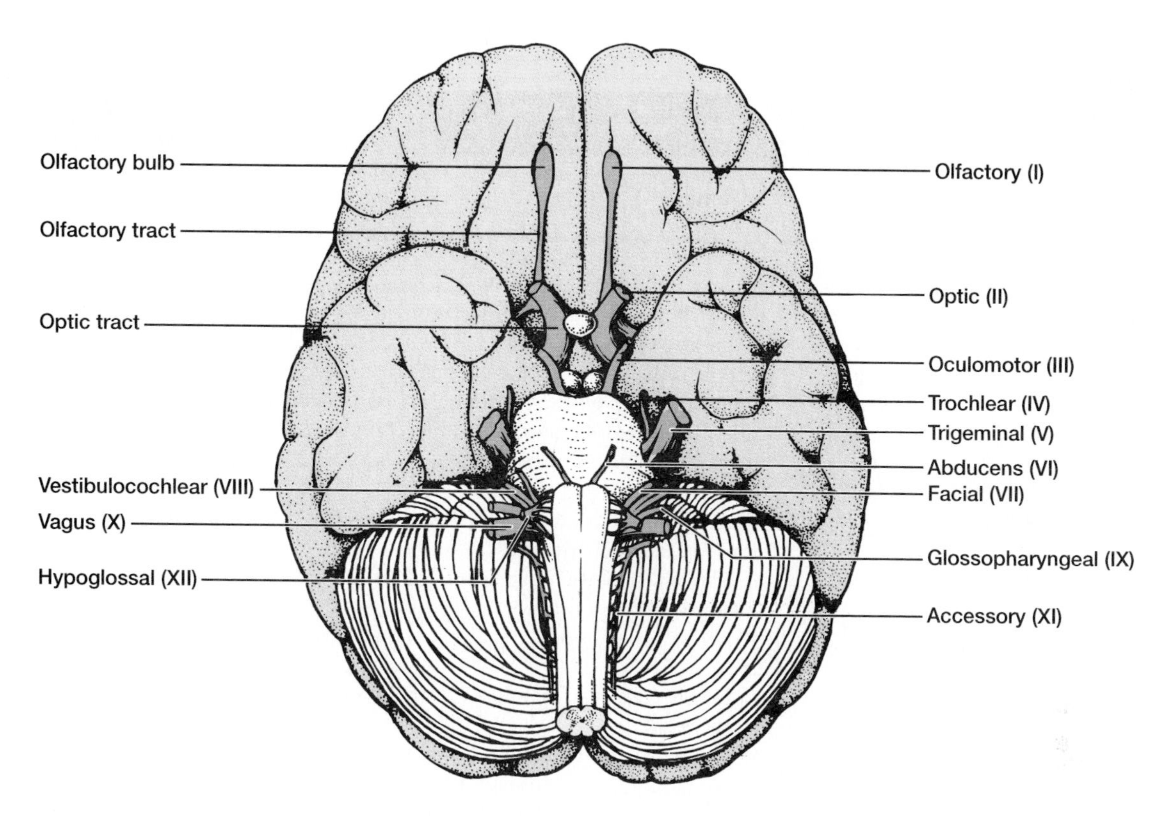

Figure 9.30

The cranial nerves, except for the first pair, arise from the brain stem. They are identified either by numbers indicating their order or by names describing their function or the general distribution of their fibers.

The eighth pair of cranial nerves, the **vestibulocochlear nerves (VIII)**, are sensory nerves that arise from the medulla oblongata. Each of these nerves has two distinct parts—a vestibular branch and a cochlear branch.

The neuron cell bodies of the *vestibular branch* fibers are located in ganglia associated with parts of the inner ear. These parts contain the receptors involved with reflexes that help to maintain equilibrium. The neuron cell bodies of the *cochlear branch* fibers are located in the parts of the inner ear that house the hearing receptors. Impulses from these branches pass through the pons and medulla oblongata on their way to the temporal lobes, where they are interpreted.

The ninth pair of cranial nerves, the **glossopharyngeal nerves (IX)**, are associated with the tongue and pharynx. These mixed nerves arise from the medulla oblongata, with predominantly sensory fibers. These sensory fibers carry impulses from the linings of the pharynx, tonsils, and posterior third of the tongue to the brain. Fibers in the motor component innervate muscles of the pharynx that function in swallowing.

The tenth pair of cranial nerves, the **vagus nerves (X)**, originate in the medulla oblongata and extend downward through the neck into the chest and abdomen. These nerves are mixed, containing both somatic and autonomic branches, with autonomic fibers predominant. Certain somatic motor fibers carry impulses to muscles of the larynx that are associated with speech and swallowing. Autonomic motor fibers of the vagus nerves supply the heart and many smooth muscles and glands in the thorax and abdomen.

The eleventh pair of cranial nerves, the **accessory nerves (XI)**, originate in the medulla oblongata and the spinal cord; thus, they have both cranial and spinal branches. Each *cranial branch* joins a vagus nerve and carries impulses to muscles of the soft palate, pharynx, and larynx. The *spinal branch* descends into the neck and supplies motor fibers to the trapezius and sternocleidomastoid muscles.

The twelfth pair of cranial nerves, the **hypoglossal nerves (XII)**, arise from the medulla oblongata and pass into the tongue. They include motor fibers that carry impulses to muscles that move the tongue in speaking, chewing, and swallowing.

Table 9.5 summarizes the functions of the cranial nerves.

Table 9.5

Functions of Cranial Nerves

Nerve	Type	Function
I Olfactory	Sensory	Sensory fibers transmit impulses associated with the sense of smell.
II Optic	Sensory	Sensory fibers transmit impulses associated with the sense of vision.
III Oculomotor	Primarily motor	Motor fibers transmit impulses to muscles that raise eyelids, move eyes, adjust the amount of light entering the eyes, and focus lenses. Some sensory fibers transmit impulses associated with the condition of muscles.
IV Trochlear	Primarily motor	Motor fibers transmit impulses to muscles that move the eyes. Some sensory fibers transmit impulses associated with the condition of muscles.
V Trigeminal	Mixed	
Opthalmic division		Sensory fibers transmit impulses from the surface of the eyes, tear glands, scalp, forehead, and upper eyelids.
Maxillary division		Sensory fibers transmit impulses from the upper teeth, upper gum, upper lip, lining of the palate, and skin of the face.
Mandibular division		Sensory fibers transmit impulses from the skin of the jaw, lower teeth, lower gum, and lower lip. Motor fibers transmit impulses to muscles of mastication and to muscles in the floor of the mouth.
VI Abducens	Primarily motor	Motor fibers transmit impulses to muscles that move the eyes. Some sensory fibers transmit impulses associated with the condition of muscles.
VII Facial	Mixed	Sensory fibers transmit impulses associated with taste receptors of the anterior tongue. Motor fibers transmit impulses to muscles of facial expression, tear glands, and salivary glands.
VIII Vestibulocochlear	Sensory	
Vestibular branch		Sensory fibers transmit impulses associated with the sense of equilibrium.
Cochlear branch		Sensory fibers transmit impulses associated with the sense of hearing.
IX Glossopharyngeal	Mixed	Sensory fibers transmit impulses from the pharynx, tonsils, posterior tongue, and carotid arteries. Motor fibers transmit impulses to muscles of the pharynx used in swallowing and to salivary glands.
X Vagus	Mixed	Somatic motor fibers transmit impulses to muscles associated with speech and swallowing; autonomic motor fibers transmit impulses to the heart, smooth muscles, and glands in the thorax and abdomen. Sensory fibers transmit impulses from the pharynx, larynx, esophagus, and viscera of the thorax and abdomen.
XI Accessory	Primarily motor	
Cranial branch		Motor fibers transmit impulses to muscles of the soft palate, pharynx, and larynx.
Spinal branch		Motor fibers transmit impulses to muscles of the neck and back.
XII Hypoglossal	Primarily motor	Motor fibers transmit impulses to muscles that move the tongue.

The consequences of a cranial nerve injury depend on the injury's location and extent. For example, damage to one member of a nerve pair limits loss of function to the affected side, but injury to both nerves affects both sides. Also, if a nerve is severed completely, the functional loss is total; if the cut is incomplete, the loss may be partial. ■

1. Define *peripheral nervous system*.
2. Distinguish between somatic and autonomic nerve fibers.
3. Name the cranial nerves, and list the major functions of each.

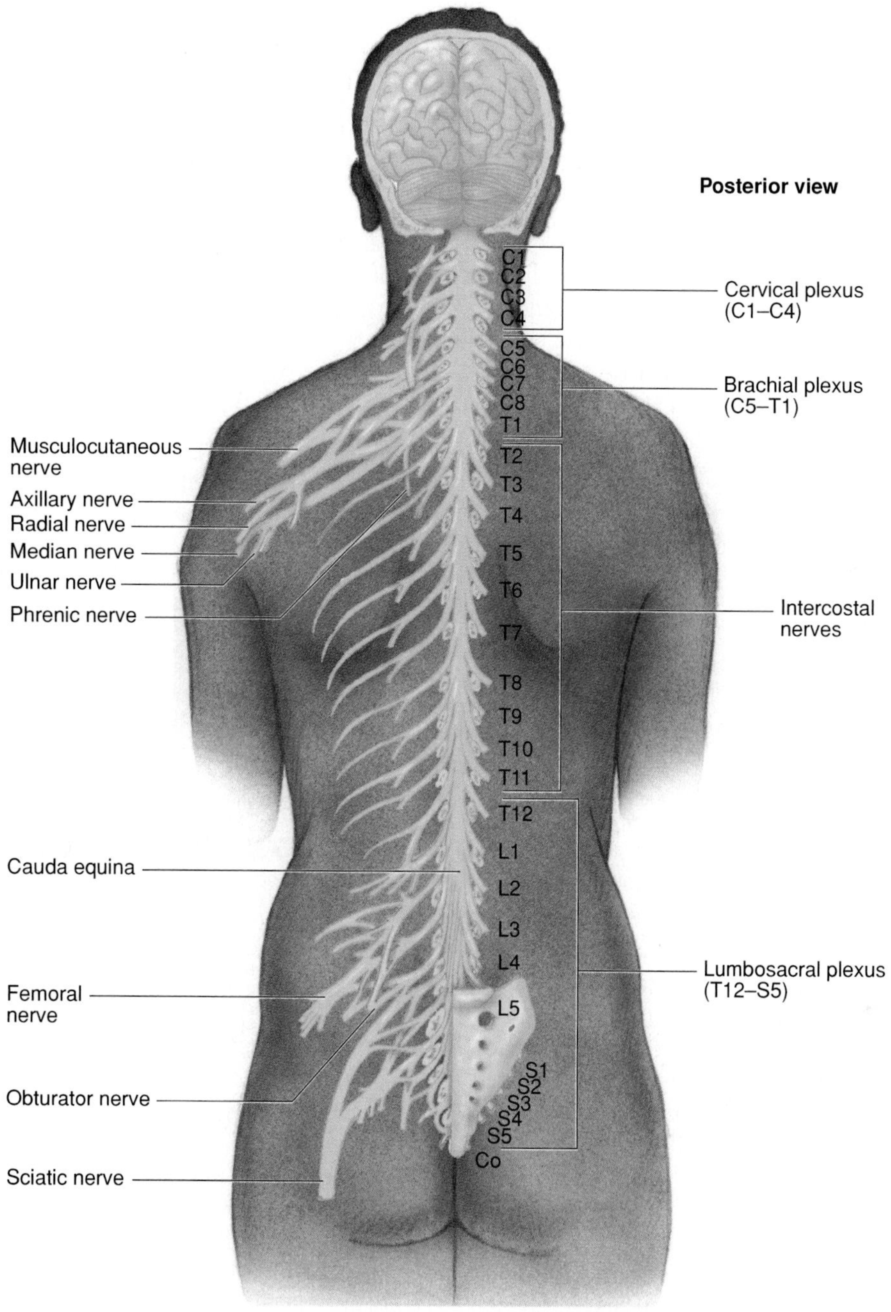

***Figure* 9.31**

The anterior branches of the spinal nerves in the thoracic region give rise to intercostal nerves. Those in other regions combine to form complex networks called plexuses.

Spinal Nerves

Thirty-one pairs of **spinal nerves** originate from the spinal cord (fig. 9.31). They are mixed nerves that provide two-way communication between the spinal cord and parts of the upper and lower limbs, neck, and trunk.

Spinal nerves are not named individually, but are grouped according to the level from which they arise. Each nerve is numbered in sequence. Thus, there are eight pairs of *cervical nerves* (numbered C1 to C8), twelve pairs of *thoracic nerves* (numbered T1 to T12), five pairs of *lumbar nerves* (numbered L1 to L5), five pairs of *sacral nerves* (numbered S1 to S5), and one pair of *coccygeal nerves* (Co).

The adult spinal cord ends at the level between the first and second lumbar vertebrae, so the lumbar, sacral, and coccygeal nerves descend beyond the end of the cord, forming a structure called the *cauda equina* (horse's tail).

Each spinal nerve emerges from the cord by two short branches, or *roots,* which lie within the vertebral column. The **dorsal root** (posterior or sensory root) can be identified by an enlargement called the *dorsal root ganglion* (see fig. 9.20). This ganglion contains the cell bodies of the sensory neurons whose dendrites conduct impulses inward from the peripheral body parts. The axons of these neurons extend through the dorsal root and into the spinal cord, where they form synapses with dendrites of other neurons. The **ventral root** (anterior or motor root) of each spinal nerve consists of axons from the motor neurons whose cell bodies are located within the gray matter of the cord.

A ventral root and a dorsal root unite to form a spinal nerve, which extends outward from the vertebral canal through an *intervertebral foramen* (see fig. 7.15). Just beyond its foramen, each spinal nerve divides into several parts.

Except in the thoracic region, the main portions of the spinal nerves combine to form complex networks called **plexuses** instead of continuing directly to peripheral body parts (fig. 9.31). In a plexus, spinal nerve fibers are sorted and recombined so that fibers that innervate a particular peripheral body part reach it in the same nerve, even though the fibers originate from different spinal nerves.

Cervical Plexuses

The **cervical plexuses** lie deep in the neck on either side and form from the branches of the first four cervical nerves. Fibers from these plexuses supply the muscles and skin of the neck. In addition, fibers from the third, fourth, and fifth cervical nerves pass into the right and left **phrenic nerves,** which conduct motor impulses to the muscle fibers of the diaphragm.

Brachial Plexuses

Branches of the lower four cervical nerves and the first thoracic nerve give rise to the **brachial plexuses.** These networks of nerve fibers are located deep within the shoulders between the neck and axillae (armpits). The major branches emerging from the brachial plexuses supply the muscles and skin of the arm, forearm, and hand, and include the **musculocutaneous, ulnar, median, radial,** and **axillary nerves.**

Lumbosacral Plexuses

The **lumbosacral plexuses** are formed on either side by the last thoracic nerve and the lumbar, sacral, and coccygeal nerves. These networks of nerve fibers extend from the lumbar region of the back into the pelvic cavity, giving rise to a number of motor and sensory fibers associated with the muscles and skin of the lower abdominal wall, external genitalia, buttocks, thighs, legs, and feet. The major branches of these plexuses include the **obturator, femoral,** and **sciatic nerves.**

The anterior branches of the thoracic spinal nerves do not enter a plexus. Instead, they enter spaces between the ribs and become **intercostal nerves.** These nerves supply motor impulses to the intercostal muscles and the upper abdominal wall muscles. They also receive sensory impulses from the skin of the thorax and abdomen.

Spinal nerves may be injured in a variety of ways, including stabs, gunshot wounds, birth injuries, dislocations and fractures of the vertebrae, and pressure from tumors in surrounding tissues. The nerves of the cervical plexuses, for example, are sometimes compressed by a sudden bending of the neck, called *whiplash,* which may occur during rear-end automobile collisions. Whiplash may cause continuing headaches and pain in the neck and skin, which the cervical nerves supply. ■

1. How are spinal nerves grouped?
2. Describe how a spinal nerve joins the spinal cord.
3. Name and locate the major nerve plexuses.

Autonomic Nervous System

The **autonomic nervous system** is the portion of the PNS that functions independently (autonomously) and continuously without conscious effort. This system controls visceral functions by regulating the actions of smooth muscles, cardiac muscles, and glands. It regulates heart rate, blood pressure, breathing rate, body temperature, and other visceral activities that maintain homeostasis. Portions of the autonomic nervous system respond to emotional stress and prepare the body to meet the demands of strenuous physical activity.

General Characteristics

Reflexes in which sensory signals originate from receptors within the viscera and the skin regulate autonomic activities. Nerve fibers transmit these signals to nerve centers within the brain or spinal cord. In response, motor impulses travel out from these centers on peripheral nerve fibers within cranial and spinal nerves.

Typically peripheral nerve fibers lead to ganglia outside the CNS. The impulses they carry are integrated within these ganglia and relayed to viscera (muscles and glands) that respond by contracting, releasing secretions, or being inhibited. The integrative function of the ganglia provides the autonomic system with a degree of independence from the brain and spinal cord.

The autonomic nervous system includes two divisions—the **sympathetic** and **parasympathetic divisions.** Some viscera have nerve fibers from each division. In such cases, impulses on one set of fibers may activate an organ, while impulses on the other set inhibit it. Thus, the divisions may act antagonistically, alternately activating or inhibiting the actions of some viscera.

The functions of the autonomic divisions are mixed; that is, each activates some organs and inhibits others. However, the divisions have important functional differences. The sympathetic division prepares the body for energy-expending, stressful, or emergency situations. Conversely, the parasympathetic division is most active under ordinary, restful conditions. It also counterbalances the effects of the sympathetic division and restores the body to a resting state following a stressful experience. For example, during an emergency, the sympathetic division increases heart and breathing rates; following the emergency, the parasympathetic division decreases these activities.

Autonomic Nerve Fibers

The nerve fibers of the autonomic nervous system are motor fibers. Unlike the motor pathways of the somatic nervous system, however, which usually include a single neuron between the brain or spinal cord and a skeletal muscle, those of the autonomic system include two neurons (fig. 9.32). The cell body of one neuron is

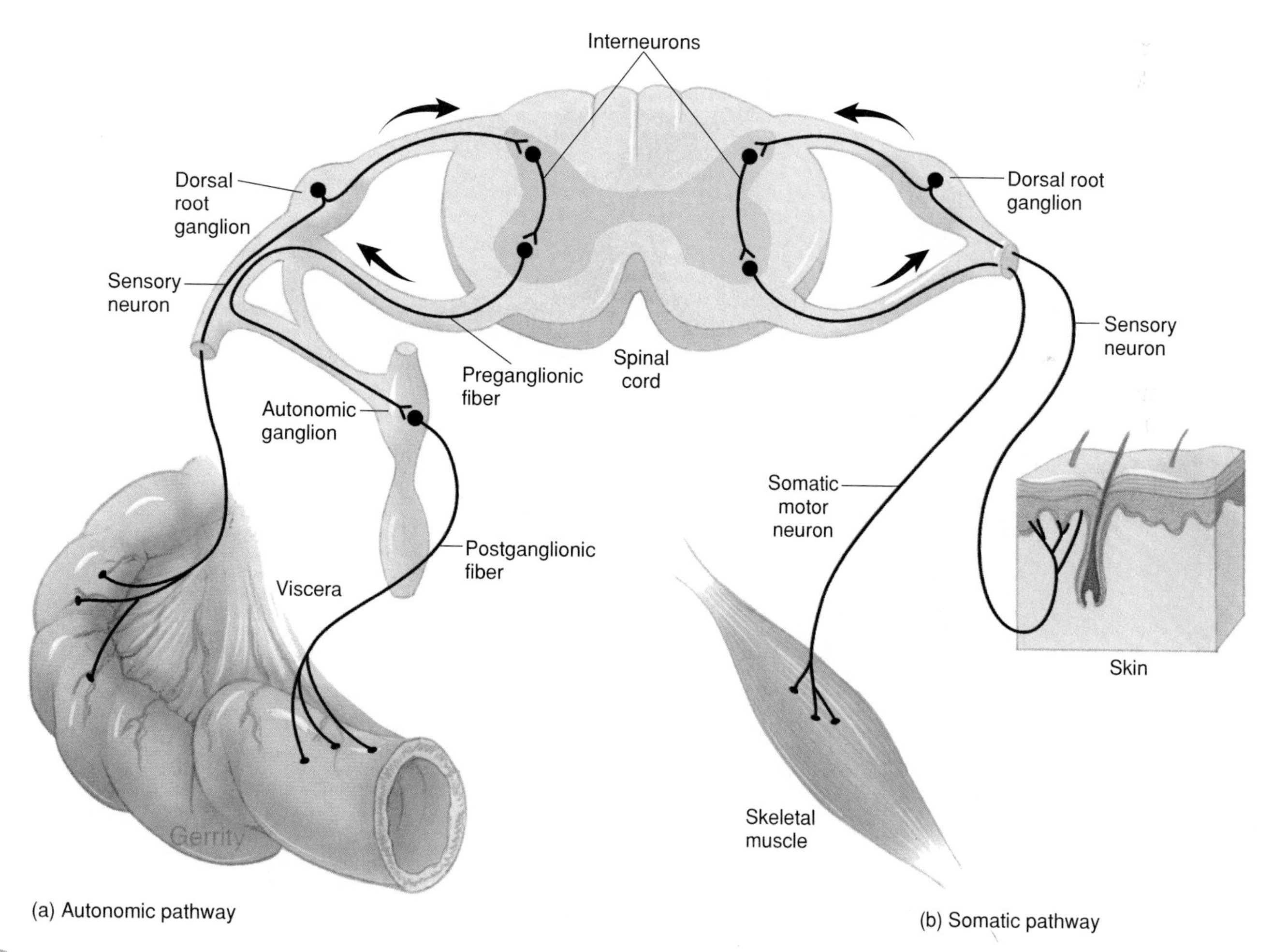

Figure 9.32

(*a*) Autonomic pathways include two neurons between the central nervous system and an effector. (*b*) Somatic pathways usually have a single neuron between the central nervous system and an effector.

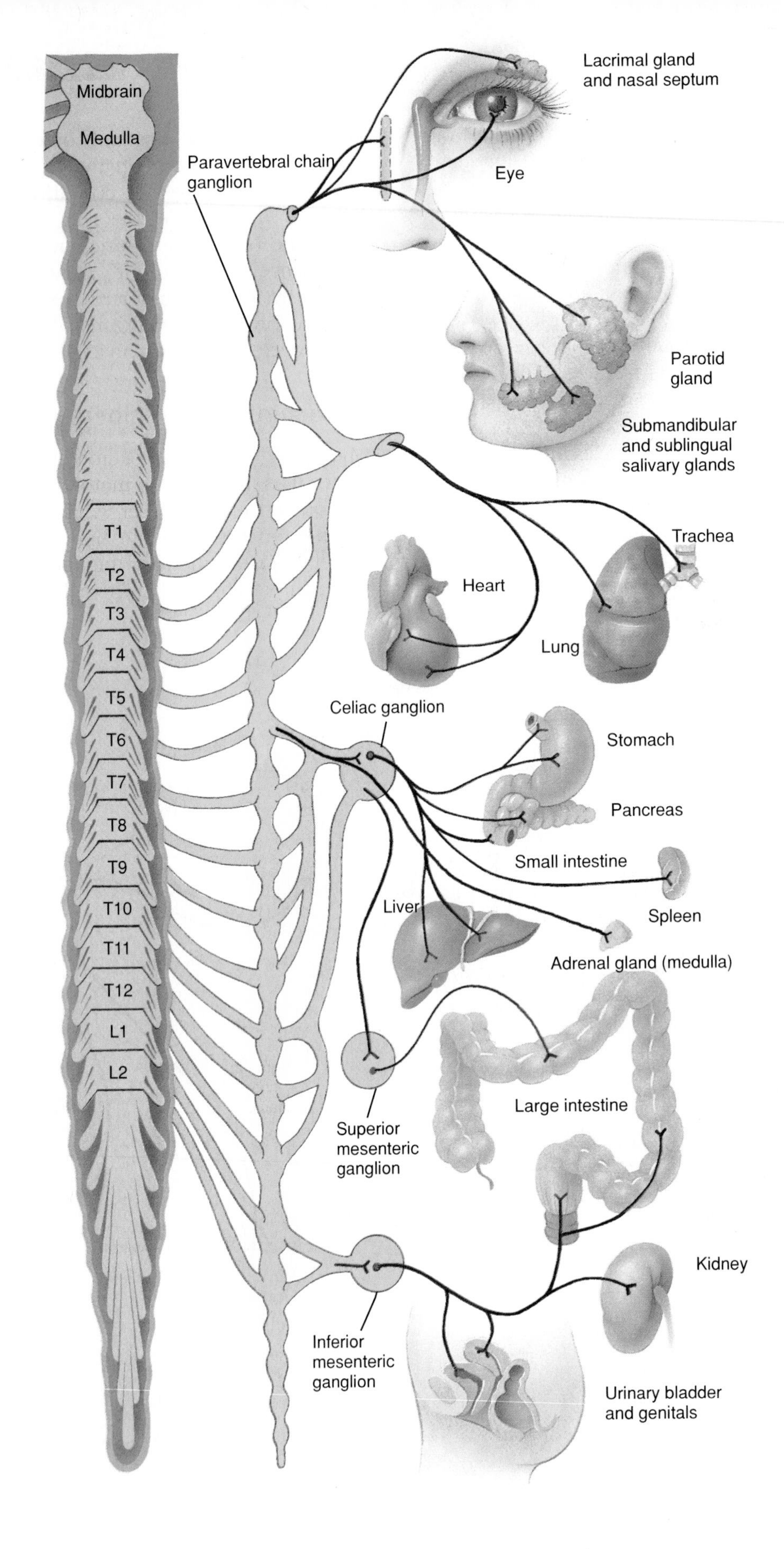

Figure 9.33

The preganglionic fibers of the sympathetic division of the autonomic nervous system arise from the thoracic and lumbar regions of the spinal cord. Note that the adrenal medulla is innervated directly by a preganglionic fiber.

located in the brain or spinal cord. Its axon, the **preganglionic fiber**, leaves the CNS and synapses with one or more neurons whose cell bodies are housed within an autonomic ganglion. The axon of such a second neuron is called a **postganglionic fiber**, and it extends to a visceral effector.

Sympathetic Division

Within the sympathetic division, the preganglionic fibers originate from neurons in the gray matter of the spinal cord (fig. 9.33). Their axons leave the cord through the ventral roots of spinal nerves in the first thoracic through the second lumbar segments. After traveling a short distance, these fibers leave the spinal nerves, and each enters a member of a chain of *paravertebral ganglia*. One of these chains extends longitudinally along each side of the vertebral column.

Within a paravertebral ganglion, a preganglionic fiber forms a synapse with a second neuron. The axon of this neuron, the postganglionic fiber, typically returns to a spinal nerve and extends with it to a visceral effector.

Parasympathetic Division

The preganglionic fibers of the parasympathetic division arise from the brain stem and sacral region of the spinal cord (fig. 9.34). From there, they lead outward in cranial or sacral nerves to ganglia located near or within various viscera. The relatively short postganglionic fibers continue from the ganglia to specific muscles or glands within these viscera.

1. What parts of the nervous system are included in the autonomic nervous system?
2. How are the divisions of the autonomic system distinguished?
3. Describe a sympathetic nerve pathway and a parasympathetic nerve pathway.

Autonomic Neurotransmitters

The preganglionic fibers of the sympathetic and parasympathetic divisions secrete *acetylcholine* and are therefore called **cholinergic fibers.** The parasympathetic postganglionic fibers are also cholinergic fibers (fig. 9.35). Most sympathetic postganglionic fibers, however, secrete *norepinephrine* (noradrenalin) and are called **adrenergic fibers.** The different postganglionic neurotransmitters cause the different effects that the sympathetic and parasympathetic divisions have on their effector organs.

Although each division of the autonomic nervous system can activate some effectors and inhibit others, most viscera are controlled primarily by one division or the other. That is, the divisions are not always actively antagonistic. For example, the sympathetic division regulates the diameter of most blood vessels, which lack parasympathetic innervation. Smooth muscles in the walls of these vessels are continuously stimulated and thus are in a state of partial contraction (sympathetic tone). Decreasing sympathetic stimulation increases (dilates) the diameter of vessels, which relaxes their muscular walls. Conversely, increasing sympathetic stimulation constricts vessels.

Similarly, the parasympathetic division dominates in controlling movements in the digestive system. Parasympathetic impulses stimulate stomach and intestinal motility. When the impulses decrease, motility lessens.

Table 9.6 summarizes the effects of adrenergic and cholinergic fibers on some visceral effectors.

Control of Autonomic Activity

The brain and spinal cord largely control the autonomic nervous system, despite the system's independence resulting from the integrative function of its ganglia. Consider control centers in the medulla oblongata for cardiac, vasomotor, and respiratory activities. These reflex centers receive sensory impulses from viscera on vagus nerve fibers, and they employ autonomic nerve pathways to stimulate motor responses in muscles and glands. Similarly, the hypothalamus helps regulate body temperature, hunger, thirst, and water and electrolyte balance by influencing autonomic pathways.

Still higher levels within the brain, including the limbic system and cerebral cortex, control the autonomic nervous system during emotional stress. These structures utilize autonomic pathways to regulate emotional expression and behavior.

1. Which neurotransmitters operate in the autonomic nervous system?
2. How do the divisions of the autonomic system regulate visceral activities?
3. How are autonomic activities controlled?

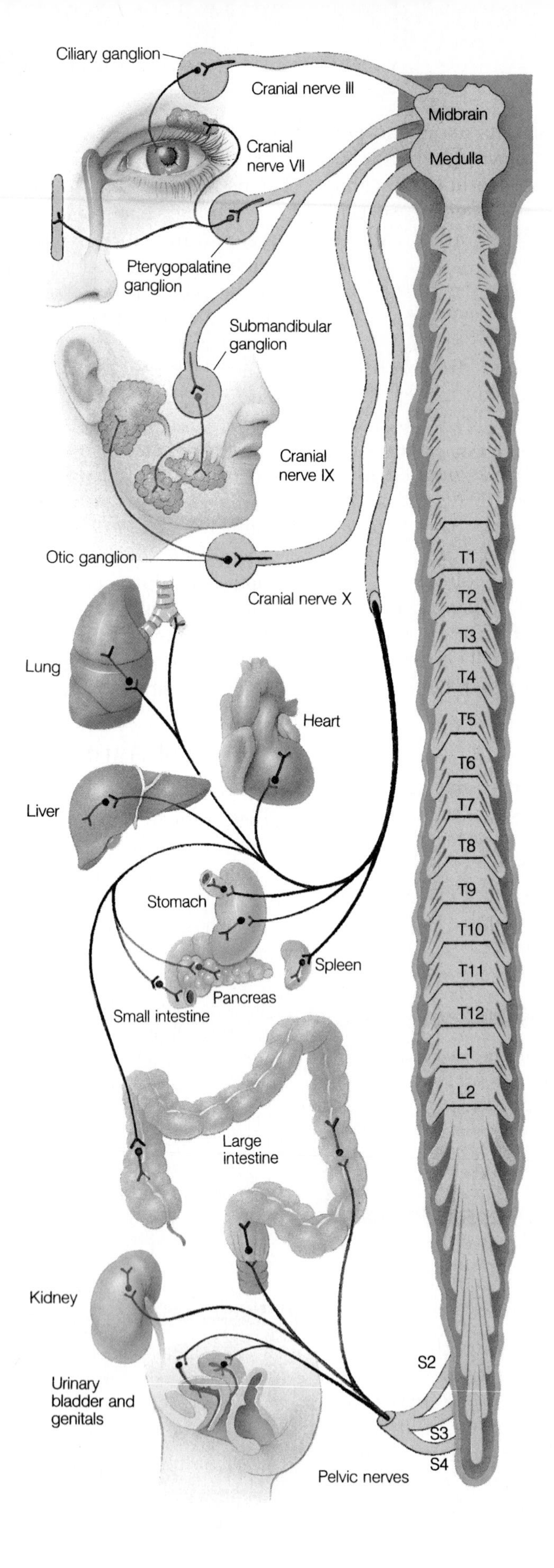

Figure 9.34

The preganglionic fibers of the parasympathetic division of the autonomic nervous system arise from the brain and sacral region of the spinal cord.

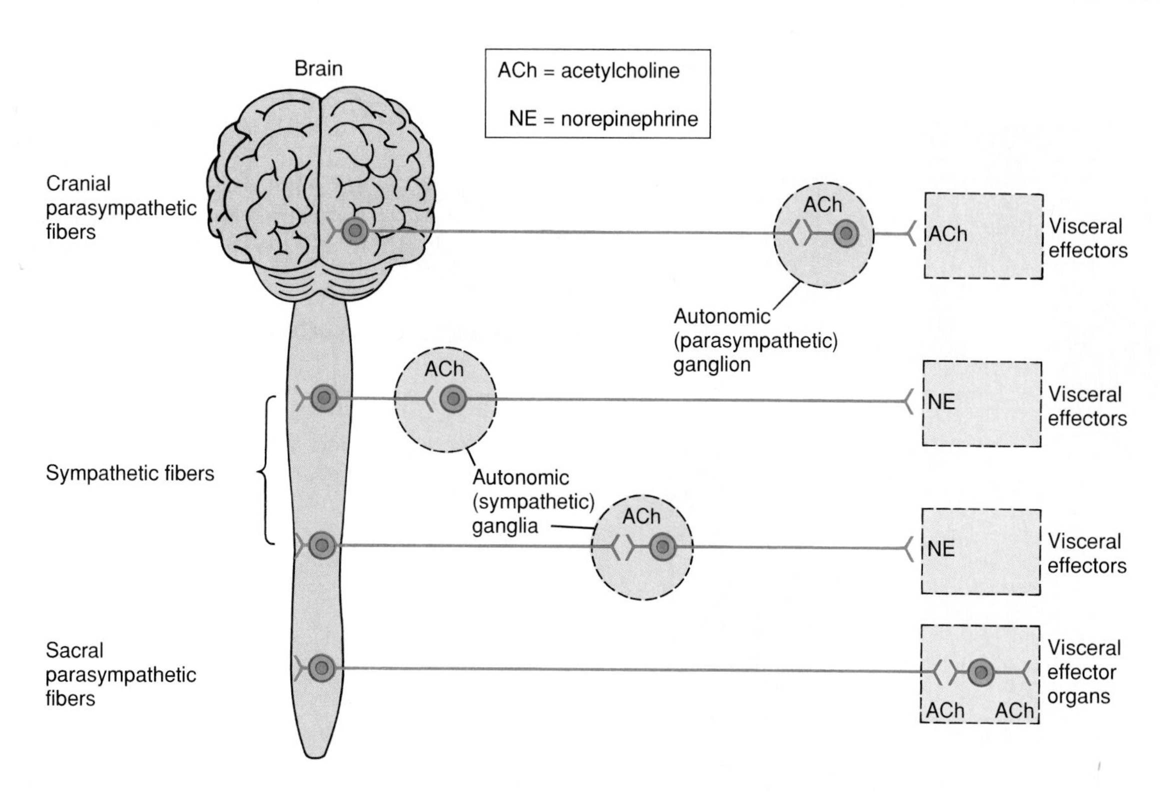

Figure 9.35

Parasympathetic fibers are cholinergic and secrete acetylcholine at the ends of postganglionic fibers. Most sympathetic fibers are adrenergic and secrete norepinephrine at the ends of postganglionic fibers.

Table 9.6

Effects of Neurotransmitter Substances on Visceral Effectors or Actions

Visceral Effector or Action	Response to Adrenergic Stimulation (sympathetic)	Response to Cholinergic Stimulation (parasympathetic)
Pupil of the eye	Dilation	Constriction
Heart rate	Increases	Decreases
Bronchioles of lungs	Dilation	Constriction
Muscles of intestinal wall	Slows peristaltic action	Speeds peristaltic action
Intestinal glands	Secretion decreases	Secretion increases
Blood distribution	More blood to skeletal muscles; less blood to digestive organs	More blood to digestive organs; less blood to skeletal muscles
Blood glucose concentration	Increases	Decreases
Salivary glands	Secretion decreases	Secretion increases
Tear glands	No action	Secretion
Muscles of gallbladder	Relaxation	Contraction
Muscles of urinary bladder	Relaxation	Contraction

organization

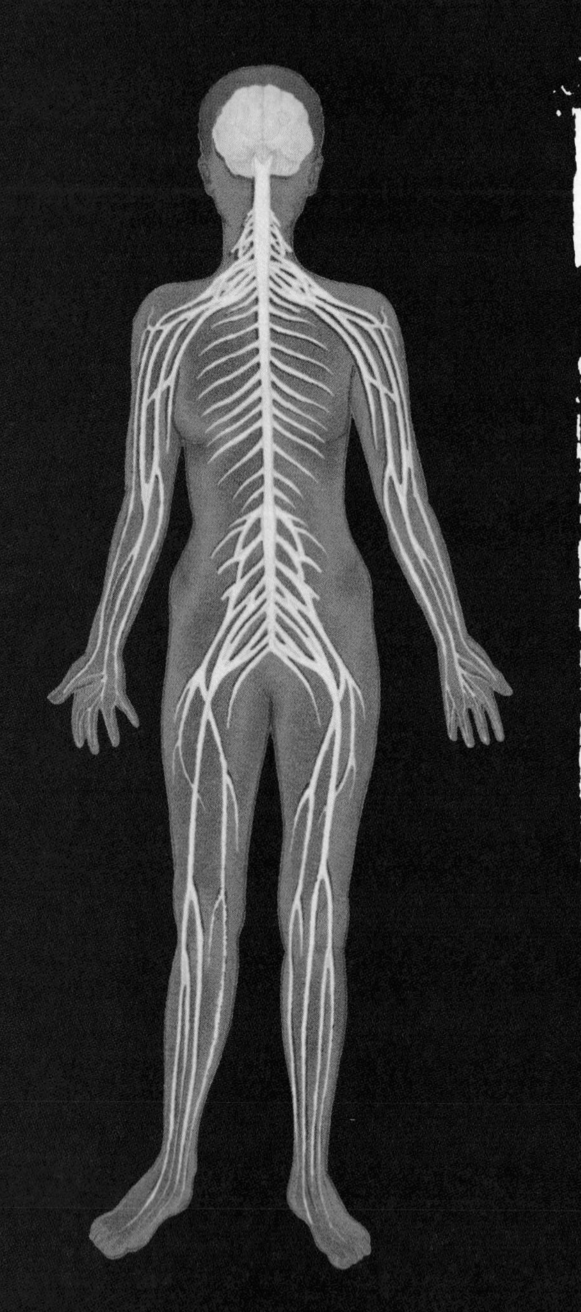

Nervous System

Neurons carry impulses that allow body systems to communicate.

Integumentary System

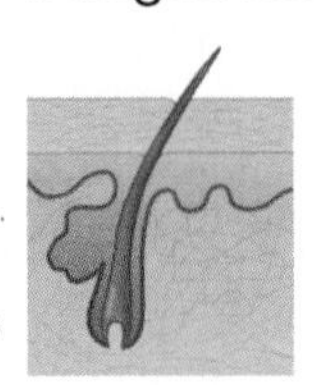

Sensory receptors provide the nervous system with information about the outside world.

Lymphatic System

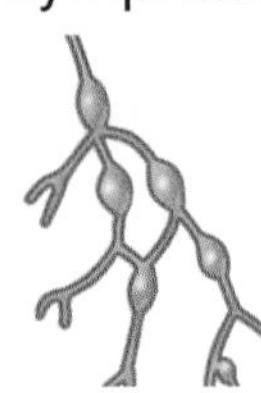

Stress may impair the immune response.

Skeletal System

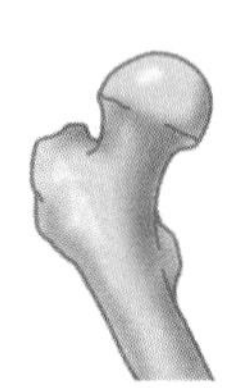

Bones protect the brain and spinal cord and help maintain plasma calcium, which is important to neuron function.

Digestive System

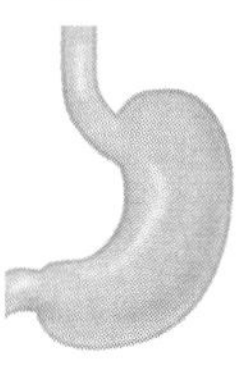

The nervous system can influence digestive function.

Muscular System

Nerve impulses control movement and carry information about the position of body parts.

Respiratory System

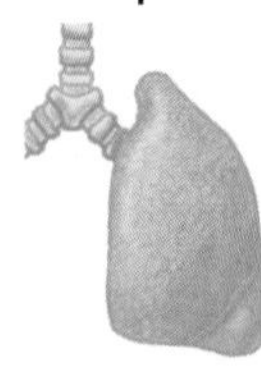

The nervous system alters respiratory activity to control oxygen levels and blood pH.

Endocrine System

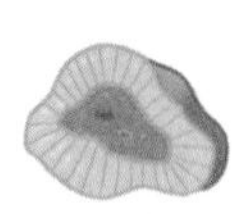

The hypothalamus controls secretion of many hormones.

Urinary System

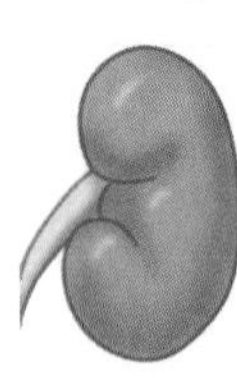

Nerve impulses affect urine production and elimination.

Cardiovascular System

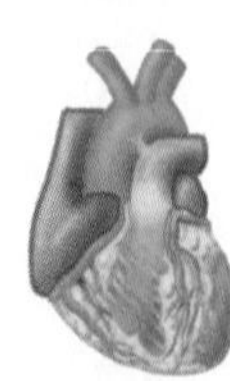

Nerve impulses help control blood flow and blood pressure.

Reproductive System

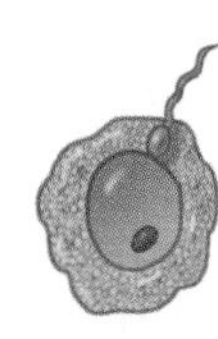

The nervous system plays a role in egg and sperm formation, sexual pleasure, childbirth, and nursing.

Clinical Terms Related to the Nervous System

Analgesia (an″al-je′ze-ah) Loss or reduction in the ability to sense pain, but without loss of consciousness.

Analgesic (an″al-je′sik) A pain-relieving drug.

Anesthesia (an″es-the′ze-ah) A loss of feeling.

Aphasia (ah-fa′ze-ah) Disturbance or loss in the ability to use words or to understand them, usually due to damage to cerebral association areas.

Apraxia (ah-prak′se-ah) Impairment in ability to use objects.

Ataxia (ah-tak′se-ah) Partial or complete inability to coordinate voluntary movements.

Cerebral palsy (ser′ĕ-bral pawl′ze) Partial paralysis and lack of muscular coordination caused by damage to the cerebrum.

Coma (ko′mah) An unconscious condition in which a person does not respond to stimulation.

Cordotomy (kor-dot′o-me) Surgical procedure that severs a nerve tract within the spinal cord to relieve intractable pain.

Craniotomy (kra″ne-ot′o-me) Surgical procedure that opens part of the skull.

Electroencephalogram (EEG) (e-lek″tro-en-sef′ah-lo-gram″) A recording of the brain's electrical activity.

Encephalitis (en″sef-ah-li′tis) Inflammation of the brain and meninges, producing drowsiness and apathy.

Epilepsy (ep′ĭ-lep″se) A disorder of the central nervous system that temporarily disturbs brain impulses, producing convulsive seizures and loss of consciousness.

Hemiplegia (hem″ĭ-ple′je-ah) Paralysis on one side of the body and of the limbs on that side.

Huntington disease (hunt′ing-tun diz-ēz′) An inherited disorder of the brain causing involuntary dancelike movements and personality changes.

Laminectomy (lam″ĭ-nek′to-me) Surgical removal of the posterior arch of a vertebra, usually to relieve symptoms of a ruptured intervertebral disk pressing on a spinal nerve.

Monoplegia (mon″o-ple′je-ah) Paralysis of a single limb.

Multiple sclerosis (mul′tĭ-pl skle-ro′sis) Loss of myelin and the appearance of scarlike patches throughout the brain or spinal cord or both.

Neuralgia (nu-ral′je-ah) A sharp, recurring pain associated with a nerve; usually caused by inflammation or injury.

Neuritis (nu-ri′tis) Inflammation of a nerve.

Paraplegia (par″ah-ple′je-ah) Paralysis of both lower limbs.

Quadriplegia (kwod″rĭ-ple′je-ah) Paralysis of all four limbs.

Vagotomy (va-got′o-me) Surgical severing of a vagus nerve.

Summary Outline

Introduction (p. 214)

Organs of the nervous system are divided into the central and peripheral nervous systems. These structures provide sensory, integrative, and motor functions. Nervous tissue includes neurons, which are the structural and functional units of the nervous system, and neuroglial cells.

General Functions of the Nervous System (p. 214)

1. Sensory functions employ receptors that detect internal and external changes.
2. Integrative functions collect sensory information and make decisions that motor functions carry out.
3. Motor functions stimulate effectors to respond.

Neuron Structure (p. 215)

1. A neuron includes a cell body, nerve fibers, and organelles.
2. Dendrites and the cell body provide receptive surfaces.
3. A single axon arises from the cell body and may be enclosed in a myelin sheath and a neurilemma.

Types of Neurons and Neuroglial Cells (p. 215)

1. Classification of neurons
   a. On the basis of structure, neurons can be classified as multipolar, bipolar, or unipolar.
   b. On the basis of function, neurons can be classified as sensory neurons, interneurons, or motor neurons.
2. Classification of neuroglial cells
   a. Neuroglial cells fill spaces, support neurons, hold nervous tissue together, produce myelin, help regulate the concentrations of nutrients and ions, and carry on phagocytosis.
   b. They include Schwann cells, astrocytes, oligodendrocytes, microglial cells, and ependymal cells.

Cell Membrane Potential (p. 219)

A cell membrane is usually polarized as a result of unequal ion distribution.

1. Distribution of ions
   a. Ion distribution is due to pores and channels in the membranes that allow passage of some ions but not others.
   b. Potassium ions pass more easily through cell membranes than do sodium ions.
2. Resting potential
   a. A high concentration of sodium ions is on the outside of a membrane, and a high concentration of potassium ions is on the inside.
   b. Many negatively charged ions are inside a cell.

c. In a resting cell, more positive ions leave than enter, so the outside of the cell membrane develops a positive charge, while the inside develops a negative charge.
3. Potential changes
   a. Stimulation of a membrane affects the membrane's resting potential.
   b. When its resting potential becomes more positive, a membrane becomes depolarized.
   c. Potential changes are subject to summation.
   d. Achieving threshold potential triggers an action potential.
4. Action potential
   a. At threshold, sodium channels open, and sodium ions diffuse inward, depolarizing the membrane.
   b. About the same time, potassium channels open, and potassium ions diffuse outward, repolarizing the membrane.
   c. This rapid change in potential is an action potential.
   d. Many action potentials can occur before active transport reestablishes the resting potential.

Nerve Impulse (p. 222)

A wave of action potentials is a nerve impulse.

1. Impulse conduction
   a. Unmyelinated fibers conduct impulses over their entire surfaces.
   b. Myelinated fibers conduct impulses more rapidly.
   c. Nerves with larger diameters conduct impulses faster than those with smaller diameters.
2. All-or-none response
   a. A nerve impulse is conducted in an all-or-none manner whenever a stimulus of threshold intensity is applied to a fiber.
   b. All the impulses conducted on a fiber are of the same strength.

The Synapse (p. 223)

A synapse is a junction between two neurons.

1. Synaptic transmission
   a. Impulses usually travel from a dendrite to a cell body, then along the axon to a synapse.
   b. Axons have synaptic knobs at their distal ends, which secrete neurotransmitters.
   c. A neurotransmitter is released when a nerve impulse reaches the end of an axon.
   d. A neurotransmitter reaching the nerve fiber on the distal side of the synaptic cleft triggers a nerve impulse.
2. Excitatory and inhibitory actions
   a. Neurotransmitters that trigger nerve impulses are excitatory. Those that inhibit impulses are inhibitory.
   b. The net effect of synaptic knobs communicating with a neuron depends on which knobs are activated from moment to moment.
3. Neurotransmitters
   a. The nervous system produces many different neurotransmitters.
   b. Neurotransmitters include acetylcholine, monoamines, amino acids, and peptides.
   c. A synaptic knob releases neurotransmitters when an action potential increases membrane permeability to calcium ions.
   d. After being released, neurotransmitters are decomposed or removed from synaptic clefts.

Impulse Processing (p. 226)

How the nervous system processes and responds to nerve impulses reflects the organization of neurons in the brain and spinal cord.

1. Neuronal pools
   a. Neurons form pools within the central nervous system.
   b. Each pool receives impulses, processes them, and conducts impulses away.
2. Facilitation
   a. Each neuron in a pool may receive excitatory and inhibitory stimuli.
   b. A neuron is facilitated when it receives subthreshold stimuli and becomes more excitable.
3. Convergence
   a. Impulses from two or more incoming fibers may converge on a single neuron.
   b. Convergence enables impulses from different sources to have an additive effect on a neuron.
4. Divergence
   a. Impulses leaving a pool may diverge by passing into several output fibers.
   b. Divergence amplifies impulses.

Types of Nerves (p. 227)

1. Nerves are cordlike bundles of nerve fibers.
2. Nerves can be classified as sensory nerves, motor nerves, or mixed nerves, depending on which type of fibers they contain.

Nerve Pathways (p. 228)

A nerve pathway is a route an impulse follows through the nervous system.

1. Reflex arcs
   A reflex arc usually includes a sensory neuron, a reflex center composed of interneurons, and a motor neuron.
2. Reflex behavior
   a. Reflexes are automatic, subconscious responses to changes.
   b. They help maintain homeostasis.
   c. The knee-jerk reflex employs only two neurons.
   d. Withdrawal reflexes are protective.

Meninges (p. 230)

1. Bone and meninges surround the brain and spinal cord.
2. The meninges consist of the dura mater, arachnoid mater, and pia mater.
3. Cerebrospinal fluid occupies the space between the arachnoid and pia maters.

Spinal Cord (p. 231)

The spinal cord is a nerve column that extends from the brain into the vertebral canal.

1. Structure of the spinal cord
   a. The spinal cord is composed of thirty-one segments, each of which gives rise to a pair of spinal nerves.
   b. The spinal cord has a cervical enlargement and a lumbar enlargement.
   c. It has a central core of gray matter within white matter.
   d. The white matter is composed of bundles of myelinated nerve fibers.

2. Functions of the spinal cord
   a. The spinal cord provides a two-way communication system between the brain and other body parts.
   b. Ascending tracts carry sensory impulses to the brain. Descending tracts carry motor impulses to muscles and glands.

Brain (p. 234)

The brain is subdivided into the cerebrum, cerebellum, and brain stem.

1. Structure of the cerebrum
   a. The cerebrum consists of two cerebral hemispheres connected by the corpus callosum.
   b. The cerebral cortex is a thin layer of gray matter near the surface.
   c. White matter consists of myelinated nerve fibers that connect neurons within the nervous system and communicate with other body parts.
2. Functions of the cerebrum
   a. The cerebrum provides higher brain functions.
   b. The cerebral cortex consists of sensory, motor, and association areas.
   c. One cerebral hemisphere usually dominates for certain intellectual functions.
3. Ventricles and cerebrospinal fluid
   a. Ventricles are interconnected cavities within the cerebral hemispheres and brain stem.
   b. Cerebrospinal fluid fills the ventricles.
   c. The choroid plexuses in the walls of the ventricles secrete cerebrospinal fluid.
4. Diencephalon
   a. The diencephalon contains the thalamus, which is a central relay station for incoming sensory impulses, and the hypothalamus, which maintains homeostasis.
   b. The limbic system produces emotions and modifies behavior.
5. Brain stem
   a. The brain stem consists of the midbrain, pons, and medulla oblongata.
   b. The midbrain contains reflex centers associated with eye and head movements.
   c. The pons transmits impulses between the cerebrum and other parts of the nervous system and contains centers that help regulate the rate and depth of breathing.
   d. The medulla oblongata transmits all ascending and descending impulses and contains several vital and nonvital reflex centers.
   e. The reticular formation filters incoming sensory impulses, arousing the cerebral cortex into wakefulness when significant impulses arrive.
6. Cerebellum
   a. The cerebellum consists of two hemispheres.
   b. It functions primarily as a reflex center for integrating sensory information required in the coordination of skeletal muscle movements and maintenance of equilibrium.

Peripheral Nervous System (p. 243)

The peripheral nervous system consists of cranial and spinal nerves that branch from the brain and spinal cord to all body parts. It is subdivided into the somatic and autonomic systems.

1. Cranial nerves
   a. Twelve pairs of cranial nerves connect the brain to parts in the head, neck, and trunk.
   b. Most cranial nerves are mixed, but some are purely sensory, and others are primarily motor.
   c. The names of the cranial nerves indicate their primary functions or the general distributions of their fibers.
   d. Some cranial nerves are somatic, and others are autonomic.
2. Spinal nerves
   a. Thirty-one pairs of spinal nerves originate from the spinal cord.
   b. These mixed nerves provide a two-way communication system between the spinal cord and parts of the upper and lower limbs, neck, and trunk.
   c. Spinal nerves are grouped according to the levels from which they arise, and they are numbered in sequence.
   d. Each spinal nerve emerges by a dorsal and a ventral root.
   e. Each spinal nerve divides into several branches just beyond its foramen.
   f. Most spinal nerves combine to form plexuses in which nerve fibers are sorted and recombined so that those fibers associated with a particular part reach it together.

Autonomic Nervous System (p. 248)

The autonomic nervous system functions without conscious effort. It regulates the visceral activities that maintain homeostasis.

1. General characteristics
   a. Autonomic functions are reflexes controlled from nerve centers in the brain and spinal cord.
   b. The autonomic nervous system consists of two divisions—sympathetic and parasympathetic.
   c. The sympathetic division responds to stressful and emergency conditions.
   d. The parasympathetic division is most active under ordinary conditions.
2. Autonomic nerve fibers
   a. Autonomic nerve fibers are motor fibers.
   b. Sympathetic fibers leave the spinal cord and synapse in paravertebral ganglia.
   c. Parasympathetic fibers begin in the brain stem and sacral region of the spinal cord and synapse in ganglia near viscera.
3. Autonomic neurotransmitters
   a. Sympathetic and parasympathetic preganglionic fibers secrete acetylcholine.
   b. Parasympathetic postganglionic fibers secrete acetylcholine. Sympathetic postganglionic fibers secrete norepinephrine.
   c. The different effects of the autonomic divisions are due to the different neurotransmitters the postganglionic fibers release.
   d. One division predominantly controls most viscera.
4. Control of autonomic activity
   a. The autonomic nervous system is somewhat independent.
   b. Control centers in the medulla oblongata and hypothalamus utilize autonomic nerve pathways.
   c. The limbic system and cerebral cortex control the autonomic system during emotional stress.

Review Exercises

1. Distinguish between neurons and neuroglial cells. (p. 214)
2. Explain the relationship between the central nervous system and the peripheral nervous system. (p. 214)
3. List three general functions of the nervous system. (p. 214)
4. Describe the generalized structure of a neuron, and explain the functions of its parts. (p. 215)
5. Distinguish between myelinated and unmyelinated nerve fibers. (p. 215)
6. Explain how neurons can be classified on the basis of their structure. (p. 217)
7. Discuss the functions of each type of neuroglial cell. (p. 218)
8. Explain how a membrane becomes polarized. (p. 219)
9. Describe how ions associated with nerve cell membranes are distributed. (p. 219)
10. Define *resting potential.* (p. 219)
11. Explain how threshold potential may be achieved. (p. 220)
12. List the events during an action potential. (p. 220)
13. Explain how nerve impulses are related to action potentials. (p. 222)
14. Explain how impulses are conducted on unmyelinated and myelinated nerve fibers. (p. 222)
15. Define *synapse.* (p. 223)
16. Explain how information passes from one neuron to another. (p. 223)
17. Distinguish between excitatory and inhibitory actions of neurotransmitters. (p. 224)
18. List four types of neurotransmitters. (p. 224)
19. Explain what happens to neurotransmitters after they are released. (p. 224)
20. Describe a neuronal pool. (p. 226)
21. Distinguish between convergence and divergence. (p. 226)
22. Distinguish among sensory, motor, and mixed nerves. (p. 227)
23. Define *reflex.* (p. 228)
24. Describe a reflex arc that consists of two neurons. (p. 228)
25. Name the layers of the meninges, and explain their functions. (p. 230)
26. Describe the structure of the spinal cord. (p. 231)
27. Distinguish between the ascending and descending tracts of the spinal cord. (p. 232)
28. Name the three major portions of the brain, and describe the general functions of each. (p. 234)
29. Describe the general structure of the cerebrum. (p. 234)
30. Describe the location of the motor, sensory, and association areas of the cerebral cortex, and describe the general functions of each. (p. 235)
31. Define *hemisphere dominance.* (p. 236)
32. Explain the function of the corpus callosum. (p. 237)
33. Distinguish between the cerebral cortex and basal nuclei. (p. 237)
34. Describe the location of the ventricles of the brain. (p. 238)
35. Explain how cerebrospinal fluid is produced and how it functions. (p. 238)
36. Define *limbic system,* and explain its functions. (p. 240)
37. Name the parts of the brain stem, and describe the general functions of each part. (p. 240)
38. Name the parts of the midbrain, and describe the general functions of each part. (p. 241)
39. Describe the pons and its functions. (p. 241)
40. Describe the medulla oblongata and its functions. (p. 241)
41. Describe the functions of the cerebellum. (p. 241)
42. Name, locate, and describe the major functions of each pair of cranial nerves. (p. 244)
43. Explain how the spinal nerves are grouped and numbered. (p. 247)
44. Describe the structure of a spinal nerve. (p. 248)
45. Define *plexus,* and locate the major plexuses of the spinal nerves. (p. 248)
46. Describe the general functions of the autonomic nervous system. (p. 248)
47. Distinguish between the sympathetic and parasympathetic divisions of the autonomic nervous system. (p. 249)
48. Distinguish between preganglionic and postganglionic nerve fibers. (p. 249)
49. Explain why the effects of the sympathetic and parasympathetic autonomic divisions differ. (p. 251)
50. Describe how portions of the central nervous system control autonomic activities. (p. 251)

Critical Thinking

1. List four skills encountered in everyday life that depend on nervous system function, and the part of the nervous system responsible for them.
2. What is the role of the cerebellum in athletics? The cerebrum?
3. A fetus or newborn with anencephaly lacks higher brain structures, possessing only a brain stem. What functions would such an individual have? What functions would he or she lack?

4. Narcolepsy is a condition in which a person suddenly falls fast asleep, even in the midst of an activity or conversation. What part of the brain is probably responsible for narcolepsy?
5. What nervous system functions contribute to thinking?
6. What functional losses would you expect in a patient who has suffered injury to the right occipital lobe of the cerebral cortex? The right temporal lobe?
7. A reflex called the *biceps-jerk reflex* employs motor neurons that exit from the spinal cord in the fifth spinal nerve (C5)—that is, fifth from the top of the cord. Another reflex called the *triceps-jerk reflex* utilizes motor neurons in the seventh spinal nerve (C7). How might these reflexes help locate the site of damage in a patient with a neck injury?
8. In multiple sclerosis, nerve fibers in the central nervous system lose their myelin. Why would this loss affect control of skeletal muscle function?
9. Why are rapidly growing cancers that originate in nervous tissue most likely composed of neuroglial cells rather than neurons?
10. Intravenous drug abusers sometimes dissolve and then inject tablets that contain fillers, such as talc or cornstarch, in addition to the drug. These fillers may obstruct tiny blood vessels in the cerebrum. What problems might such obstructions create?

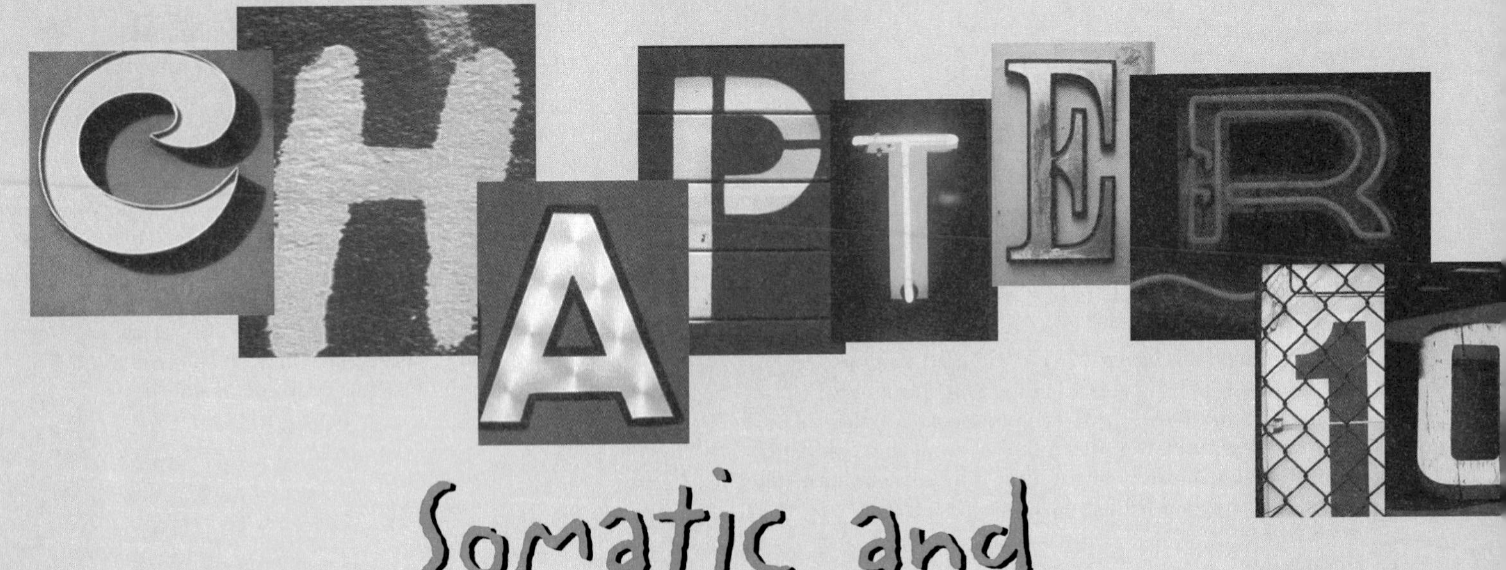

Somatic and Special Senses

"His name was lavender."

"The song was full of glittering orange diamonds."

"The paint smelled blue."

"The sunset was salty."

In people with synesthesia, sensation and perception become mixed up, so that a stimulus to one sense is perceived as another. Most commonly, visions take on characteristic smells or sounds, or are linked to specific colors. These associations are involuntary and persist over a lifetime.

Causes of synesthesia are unknown, although it seems to be inherited. People have sheepishly reported the condition to psychologists and physicians for at least 200 years. Various theories, all unproven, attribute synesthesia to an immature nervous system that cannot sort out sensory stimuli, altered brain circuitry that routes stimuli to the wrong part of the cerebral cortex, or simply an exaggerated use of metaphor, taking such descriptions as "a sharp flavor" too literally.

PET (positron emission tomography) scanning reveals a physical basis to synesthesia. A brain scan of a woman who tasted in geometric shapes discovered that blood flow in her left temporal lobe plummeted 18% during a tasting experience. The left hemisphere is the site of the language center. In addition, people with temporal lobe epilepsy are often synesthetic. PET studies indicate that synesthesia reflects a breakdown in the translation of a perception into language—but much remains unknown about this fascinating mixing of the senses.

Photo: Our senses enable us to hear, see, smell, taste, and feel the world around us.

How dull life would be without sight and sound, smell and taste, touch and balance. Our senses are not only necessary to enjoy life, but to survive. Sensory receptors detect environmental changes and transmit them to the nervous system so that the body can react appropriately.

Sensory receptors vary greatly but fall into two major categories. Receptors associated with the *somatic senses* of touch, pressure, temperature, and pain form one group. These receptors are widely distributed throughout the skin and deeper tissues, and are structurally simple. Receptors of the second type are parts of complex, specialized sensory organs that provide the *special senses* of smell, taste, hearing, equilibrium, and vision.

Chapter Objectives

After studying this chapter, you should be able to:

1. Name five kinds of receptors and explain their functions. (p. 262)
2. Explain how a sensation arises. (p. 262)
3. Describe the somatic senses. (p. 262)
4. Describe the receptors associated with the senses of touch, pressure, temperature, and pain. (p. 263)
5. Describe how the sense of pain is produced. (p. 264)
6. Explain the relationship between the senses of smell and taste. (p. 266)
7. Name the parts of the ear, and explain the function of each part. (p. 270)
8. Distinguish between static and dynamic equilibrium. (p. 274)
9. Name the parts of the eye, and explain the function of each part. (p. 280)
10. Explain how the eye refracts light. (p. 285)
11. Describe the visual nerve pathway. (p. 286)

Aids to Understanding Words

choroid [skinlike] *choroid* coat: The middle, vascular layer of the eye.

cochlea [snail] *cochlea*: The coiled tube in the inner ear.

iris [rainbow] *iris*: The colored, muscular part of the eye.

labyrinth [maze] *labyrinth*: A complex system of connecting chambers and tubes of the inner ear.

lacri- [tears] *lacri*mal gland: A tear gland.

macula [spot] *macula* lutea: A yellowish spot on the retina.

olfact- [to smell] *olfact*ory: Pertaining to the sense of smell.

scler- [hard] *scler*a: The tough, outer protective layer of the eye.

tympan- [drum] *tympan*ic membrane: The eardrum.

vitre- [glass] *vitre*ous humor: A clear, jellylike substance within the eye.

Key Terms

accommodation (ah-kom″o-da′shun)
ampulla (am-pul′lah)
chemoreceptor (ke″mo-re-sep′tor)
cochlea (kok′le-ah)
cornea (kor′ne-ah)
dynamic equilibrium (di-nam′ik e″kwĭ-lib′re-um)
labyrinth (lab′i-rinth)
macula (mak′u-lah)
mechanoreceptor (mek″ah-no-re-sep′tor)
pain receptor (pān re-sep′tor)
photoreceptor (fo″to-re-sep′tor)
projection (pro-jek′shun)
referred pain (re-furd′pān)
refraction (re-frak′shun)
retina (ret′ĭ-nah)
rhodopsin (ro-dop′sin)
sclera (skle′rah)
sensory adaptation (sen′so-re ad″ap-ta′shun)
static equilibrium (stat′ik e″kwĭ-lib′re-um)
thermoreceptor (therm′o-re-sep″tor)

Introduction

Changes occurring within the body and in its surroundings stimulate *sensory receptors,* and they, in turn, trigger nerve impulses. These impulses travel on sensory pathways into the central nervous system for processing and interpretation. As a result, a person experiences or perceives a particular feeling or sensation.

Receptors and Sensations

The many kinds of sensory receptors share common features. For example, each type of receptor is particularly sensitive to a distinct type of environmental change and is much less sensitive to other forms of stimulation.

Types of Receptors

Sensory receptors are categorized into five types according to their sensitivities: **Chemoreceptors** are stimulated by changes in the chemical concentration of substances, **pain receptors** by tissue damage, **thermoreceptors** by changes in temperature, **mechanoreceptors** by changes in pressure or movement, and **photoreceptors** by light energy.

Sensations

A **sensation** (perception) is a feeling that occurs when the brain interprets sensory impulses. Because all the nerve impulses that travel away from sensory receptors into the central nervous system are alike, the resulting sensation depends on which region of the brain receives the impulse. For example, impulses reaching one region are always interpreted as sounds, and those reaching another are always sensed as touch.

At the same time that a sensation forms, the cerebral cortex causes the feeling to seem to come from the stimulated receptors. This process is called **projection** because the brain projects the sensation back to its apparent source. Projection allows a person to pinpoint the region of stimulation; thus, the eyes seem to see and the ears seem to hear.

Sensory Adaptation

When sensory receptors are continuously stimulated, many of them undergo an adjustment called **sensory adaptation.** As receptors adapt, impulses leave them at decreasing rates, until finally, these receptors may stop sending signals. Once receptors have adapted, impulses can be triggered only if the stimulus strength changes. A person entering a room with a strong odor experiences sensory adaptation. At first, the scent seems intense, but it becomes less and less noticeable as the smell (olfactory) receptors adapt.

1. List five general types of sensory receptors.
2. Explain how a sensation occurs.
3. What is sensory adaptation?

Somatic Senses

Somatic senses are associated with receptors in the skin, muscles, joints, and viscera.

Touch and Pressure Senses

The senses of touch and pressure derive from three kinds of receptors (fig. 10.1). These receptors sense

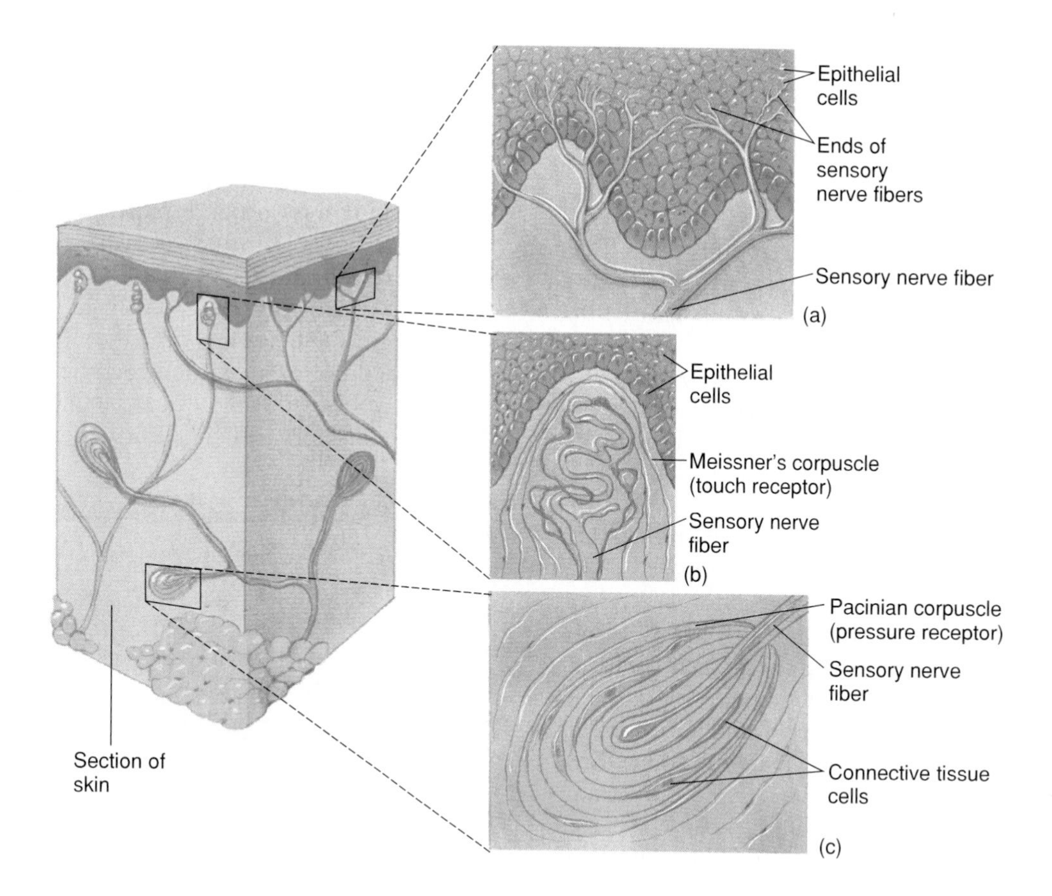

Figure 10.1

Touch and pressure receptors include (*a*) free ends of sensory nerve fibers, (*b*) Meissner's corpuscles, and (*c*) Pacinian corpuscles.

mechanical forces that deform or displace tissues. Touch and pressure receptors include:

1. **Sensory nerve fibers** These receptors are common in epithelial tissues, where their free ends are between epithelial cells. They are associated with the sensations of touch and pressure.
2. **Meissner's corpuscles** These are small, oval masses of flattened connective tissue cells within connective tissue sheaths. Two or more sensory nerve fibers branch into each corpuscle and end within it as tiny knobs.

   Meissner's corpuscles are abundant in the hairless portions of the skin, such as the lips, fingertips, palms, soles, nipples, and external genital organs. They respond to the motion of objects that barely contact the skin, interpreting impulses from them as the sensation of light touch.
3. **Pacinian corpuscles** These sensory bodies are relatively large structures composed of connective tissue fibers and cells. They are common in the deeper subcutaneous tissues and in muscle tendons and joint ligaments. Pacinian corpuscles respond to heavy pressure and are associated with the sensation of deep pressure.

Temperature Senses

Temperature sensation depends on two types of free nerve endings in the skin. Those that respond to warmer temperatures are called *heat receptors,* and those that respond to colder temperatures are called *cold receptors.*

Heat receptors are most sensitive to temperatures above 25° C (77° F) and become unresponsive at temperatures above 45° C (113° F). Temperatures near and above 45° C stimulate pain receptors, producing a burning sensation.

Cold receptors are most sensitive to temperatures between 10° C (50° F) and 20° C (68° F). Temperatures below 10° C stimulate pain receptors, producing a freezing sensation.

Both heat and cold receptors rapidly adapt. Within about a minute of continuous stimulation, the sensation of heat or cold begins to fade.

Sense of Pain

Other receptors that consist of free nerve endings sense pain. These receptors are widely distributed throughout the skin and internal tissues, except in the nervous tissue of the brain, which lacks pain receptors.

Pain receptors protect in that tissue damage stimulates them. Pain sensation is usually perceived as unpleasant, and it signals a person to take action to remove the stimulation. Pain receptors adapt poorly, if at all. Thus, once a pain receptor is activated, even by a single stimulus, it may send impulses into the central nervous system for some time. Pain may persist.

The way in which tissue damage stimulates pain receptors is poorly understood. Injuries are believed to promote release of certain chemicals that build up and stimulate pain receptors. Deficiency of oxygen-rich blood (ischemia) in a tissue or stimulation of certain mechanoreceptors also triggers pain sensations. The pain elicited during a muscle cramp, for example, stems from sustained contraction that squeezes capillaries and interrupts blood flow. Stimulation of mechanical-sensitive pain receptors also contributes to the sensation.

Injuries to bones, tendons, or ligaments stimulate pain receptors that may also contract nearby skeletal muscles. The contracting muscles may become ischemic, which may trigger still other pain receptors within the muscle tissue, further increasing muscular contraction in a "vicious circle." ■

Visceral Pain

As a rule, pain receptors are the only receptors in viscera whose stimulation produces sensations. Pain receptors in these organs respond differently to stimulation than those associated with surface tissues. For example, localized damage to intestinal tissue during surgical procedures may not elicit pain sensations, even in a conscious person. More widespread stimulation of visceral tissues, however, as when intestinal tissues are stretched or when smooth muscles in intestinal walls undergo spasms, may produce a strong pain sensation. Once again, the resulting pain seems to stem from stimulation of mechanoreceptors and to decreased blood flow, producing lower tissue oxygen concentration and accumulation of pain-stimulating chemicals.

Visceral pain may feel as if it is coming from some part of the body other than the part being stimulated—a phenomenon called **referred pain.** For example, pain originating in the heart may be referred to the left shoulder or left upper limb (fig. 10.2). Referred pain arises from common nerve pathways that carry sensory impulses from skin areas as well as viscera. In other words, pain impulses from the heart travel over the same nerve pathways as those from the skin of the left shoulder and left upper limb (fig. 10.3). Consequently, the cerebral cortex may incorrectly interpret the source of pain impulses during a heart attack as the left shoulder or upper limb, rather than the heart.

1. Describe the three types of touch and pressure receptors.
2. Describe the receptors that sense temperature.
3. What types of stimuli excite pain receptors?
4. What is referred pain?

Pain Nerve Fibers

Nerve fibers that conduct impulses away from pain receptors are of two main types: acute pain fibers and chronic pain fibers. *Acute pain fibers* are relatively thin, myelinated nerve fibers. They conduct nerve impulses rapidly and are associated with the sensation of sharp pain, which typically originates from a restricted area of the skin and seldom continues after the pain-producing stimulus stops. *Chronic pain fibers* are thin, unmyelinated nerve fibers. They conduct impulses more slowly and produce a dull, aching pain sensation that may be diffuse and difficult to pinpoint. Such pain may continue for some time after the original stimulus ceases. Acute pain is usually sensed as coming only from the skin; chronic pain is felt in deeper tissues as well.

An event that stimulates pain receptors usually triggers impulses on both acute and chronic pain fibers. This causes a dual sensation—a sharp, pricking pain, followed shortly by a dull, aching one. The aching pain is usually more intense and may worsen with time. Chronic pain can cause prolonged suffering.

Pain impulses that originate from the head reach the brain on sensory fibers of cranial nerves. All other pain impulses travel on the sensory fibers of spinal nerves, and they pass into the spinal cord by way of the dorsal roots of these spinal nerves. Within the spinal cord, neurons process pain impulses in the gray matter of the dorsal horn, and the impulses are transmitted to the brain. Here, most pain fibers terminate in the reticular formation (see chapter 9). From there, other neurons conduct impulses to the thalamus, hypothalamus, and cerebral cortex.

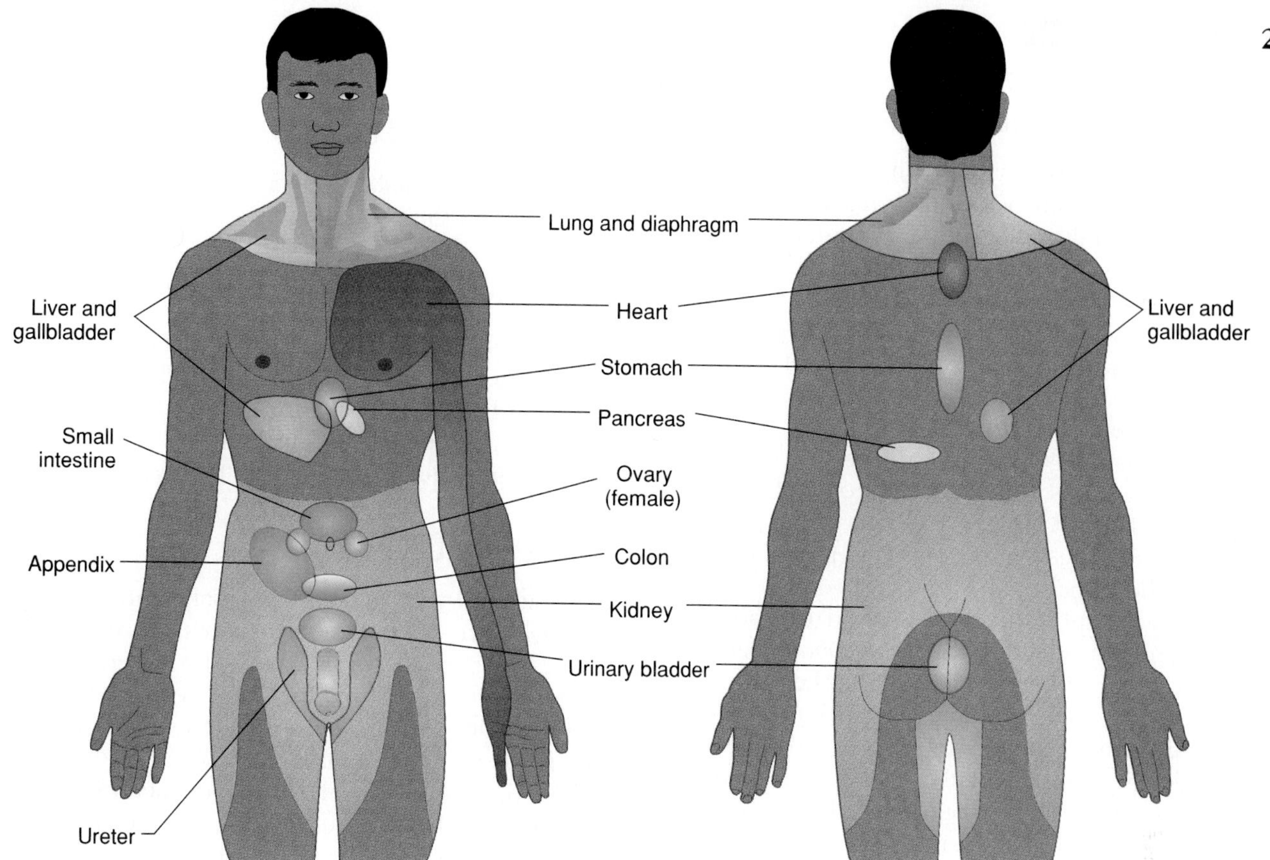

Figure 10.2

Surface regions to which visceral pain originating from various internal organs may be referred.

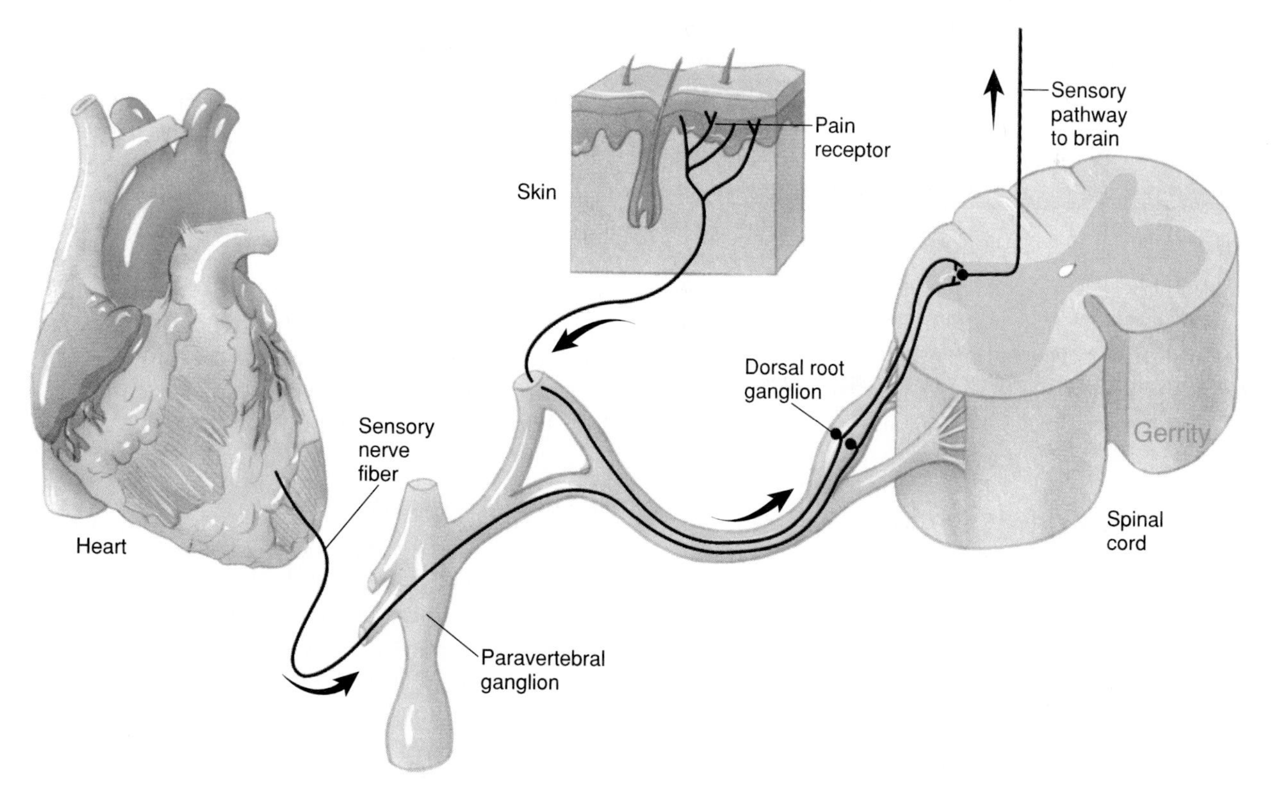

Figure 10.3

Pain originating in the heart may feel as if it is coming from the skin because sensory impulses from those two regions follow common nerve pathways to the brain.

Regulation of Pain Impulses

Awareness of pain arises when pain impulses reach the thalamus—that is, even before they reach the cerebral cortex. The cerebral cortex, however, determines pain intensity, locates the pain source, and mediates emotional and motor responses to the pain.

Areas of gray matter in the midbrain, pons, and medulla oblongata regulate movement of pain impulses from the spinal cord. Impulses from special neurons in these brain areas descend in the lateral funiculus (see chapter 9) to various levels of the spinal cord. These impulses stimulate ends of certain nerve fibers to release biochemicals that can block pain signals by inhibiting presynaptic nerve fibers in the posterior horn of the spinal cord.

The inhibiting substances released in the posterior horn include neuropeptides called *enkephalins* and the monoamine *serotonin* (see chapter 9). Enkephalins can suppress acute and chronic pain impulses and thus can relieve severe pain, much as morphine and other opiate drugs do. In fact, enkephalins bind to the same receptor sites on neuron membranes as does morphine. Serotonin stimulates other neurons to release enkephalins.

Endorphins are another group of neuropeptides with pain-suppressing, morphinelike actions. Endorphins are in the pituitary gland and the hypothalamus. Enkephalins and endorphins are released in response to extreme pain and provide natural pain control.

1. Describe two types of pain fibers.
2. How do acute pain and chronic pain differ?
3. What parts of the brain interpret pain impulses?
4. How do neuropeptides help control pain?

Special Senses

Special senses are those whose sensory receptors are within relatively large, complex sensory organs of the head. These organs and their respective senses include:

- Olfactory organs ——> Smell
- Taste buds ——> Taste
- Ears ——> Hearing
- Organs of equilibrium ——> Static equilibrium, Dynamic equilibrium
- Eyes ——> Sight

Sense of Smell

The sense of smell is associated with complex sensory structures in the upper region of the nasal cavity.

Olfactory Receptors

Smell (olfactory) receptors and taste receptors are chemoreceptors, which means that chemicals dissolved in liquids stimulate them. Smell and taste function closely together and aid in food selection because we usually smell food at the same time we taste it.

Olfactory Organs

The **olfactory organs**, which contain the olfactory receptors, are yellowish brown masses that cover the upper parts of the nasal cavity, the superior nasal conchae, and a portion of the nasal septum. **Olfactory receptor cells** are neurons surrounded by columnar epithelial cells (fig. 10.4). Hairlike cilia cover tiny knobs at the distal ends of these neurons' dendrites. The cilia project into the nasal cavity and are the sensitive parts of the receptors. Chemicals that stimulate olfactory receptors enter the nasal cavity as gases, but they must dissolve at least partially in the watery fluids that surround the cilia before receptors can detect them.

Olfactory Nerve Pathways

Stimulated olfactory receptors send nerve impulses along the axons of the receptor cells to neurons located in enlargements called **olfactory bulbs.** These structures lie on either side of the crista galli of the ethmoid bone (see fig. 7.12). Within the olfactory bulbs, the impulses are analyzed, and as a result, additional impulses travel along the **olfactory tracts** to the limbic system (see chapter 9). The major interpreting areas (olfactory cortex) for these impulses are located within the temporal lobes and at the bases of the frontal lobes, anterior to the hypothalamus.

Olfactory Stimulation

Biologists are uncertain how substances that trigger the sense of smell—called odorant molecules—stimulate olfactory receptors. One hypothesis suggests that the shapes of gaseous molecules fit complementary shapes of membrane receptor sites. According to this idea, a molecule binding to its particular receptor site triggers a nerve impulse. (Recall that membrane receptors are molecules such as glycoproteins on cell membranes. Sensory receptors, on the other hand, may be as small as cells or as large as organs, such as the eye.)

A limited number of receptors can detect a much greater variety of odors if each odor stimulates a distinct

TOPIC OF INTEREST: Headache

Headaches are a common type of pain. Although the nervous tissue of the brain lacks pain receptors, nearly all the other tissues of the head, including the meninges and blood vessels, are richly innervated.

Many headaches are associated with stressful life situations that cause fatigue, emotional tension, anxiety, or frustration. These conditions can trigger various physiological changes, such as prolonged contraction of the skeletal muscles in the forehead, sides of the head, or back of the neck. Such contractions may stimulate pain receptors and produce a *tension headache*. More severe *vascular headaches* accompany constriction or dilation of the cranial blood vessels. The throbbing headache of a "hangover" from drinking too much alcohol, for example, may be due to blood pulsating through dilated cranial vessels.

Migraine is another form of vascular headache. In this disorder, certain cranial blood vessels constrict, producing a localized cerebral blood deficiency. This causes a variety of symptoms, such as seeing patterns of bright light that obstruct vision or numbness in the limbs or face. Typically, vasoconstriction subsequently leads to vasodilation of the affected vessels, causing a severe headache, which usually affects one side of the head and may last for several hours. Effective drug treatment is now available for migraines.

Other causes of headaches include sensitivity to food additives, high blood pressure, increased intracranial pressure due to a tumor or to blood escaping from a ruptured vessel, decreased cerebrospinal fluid pressure following a lumbar puncture, or sensitivity to or withdrawal from certain drugs.

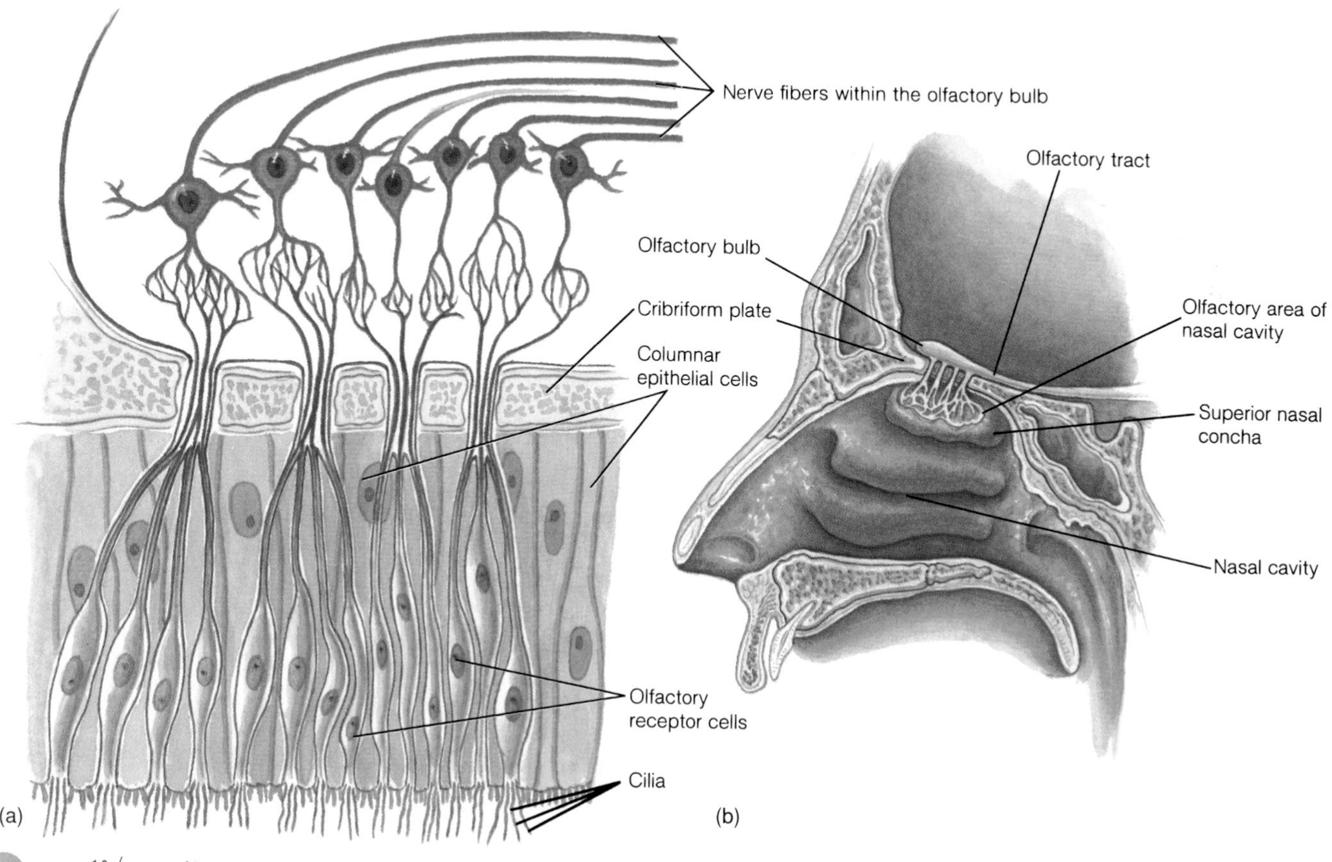

Figure 10.4

(*a*) Columnar epithelial cells support olfactory receptor cells, which have cilia at their distal ends. (*b*) The olfactory area is associated with the superior nasal concha.

set of receptor subtypes. The brain then interprets different receptor combinations as an *olfactory code*. If there are ten odor receptors, for example, parsley might stimulate receptors 3, 4, and 8; chocolate might stimulate receptors 1, 5, and 10.

Because the olfactory organs are located high in the nasal cavity above the usual pathway of inhaled air, a person may have to sniff and force air up to the receptor areas to smell a faint odor. Also, olfactory receptors undergo sensory adaptation rather rapidly, but even if they have adapted to one scent, their sensitivity to other odors persists.

Partial or complete loss of smell is called *anosmia*. It may result from a variety of factors, including inflammation of the nasal cavity lining from a respiratory infection, tobacco smoking, or use of certain drugs, such as cocaine. ■

1. Where are olfactory receptors located?
2. Trace the pathway of an olfactory impulse from a receptor to the cerebrum.

Sense of Taste

Taste buds are the special organs of taste (fig. 10.5). They occur primarily on the surface of the tongue and are associated with tiny elevations called *papillae*. They are also found in smaller numbers in the roof of the mouth and walls of the pharynx.

Taste Receptors

Each taste bud includes a group of modified epithelial cells, the **taste cells** (gustatory cells), which function as receptors. The taste bud also includes epithelial supporting cells. The entire structure is somewhat spherical with an opening, the **taste pore**, on its free surface. Tiny projections called **taste hairs** protrude from the outer ends of the taste cells and extend from the taste pore. These taste hairs are believed to be the sensitive parts of the receptor cells.

Interwoven among the taste cells and wrapped around them is a network of nerve fibers. Stimulation of a receptor cell triggers an impulse on a nearby nerve fiber, and the impulse then travels into the brain.

Before a particular chemical can be tasted, it must dissolve in the watery fluid surrounding the taste buds. The salivary glands provide this fluid. As is the case for smell, researchers hypothesize that food molecules combine with specific receptor sites on taste hair surfaces, stimulating the sense of taste. Such a combination could then generate sensory impulses on nearby nerve fibers.

The taste cells in all taste buds appear alike microscopically but are of at least four types. Each type is most sensitive to a particular kind of chemical stimulus, producing at least four primary taste (gustatory) sensations.

Taste Sensations

The four primary taste sensations are:

1. *Sweet,* such as table sugar
2. *Sour,* such as a lemon
3. *Salty,* such as table salt
4. *Bitter,* such as caffeine or quinine

Some investigators recognize two other taste sensations—*alkaline* and *metallic*. Each of the four major types of taste receptors is most highly concentrated in certain regions of the tongue's surface (fig. 10.6).

A flavor results from one of the primary sensations or from a combination of two or more of them. Experiencing flavors involves taste (concentrations of stimulating chemicals), as well as the sensations of odor, texture (touch), and temperature. Furthermore, the chemicals in some foods—chili peppers and ginger, for instance—may stimulate pain receptors, which cause a burning sensation.

Taste receptors, like olfactory receptors, undergo sensory adaptation relatively rapidly. Moving bits of food over the surface of the tongue to stimulate different receptors at different moments avoids the resulting loss of taste.

Taste Nerve Pathways

Sensory impulses from taste receptors in the tongue travel on fibers of the facial, glossopharyngeal, and vagus nerves into the medulla oblongata. From there, the impulses ascend to the thalamus and are directed to the gustatory cortex, which is located in the parietal lobe of the cerebrum, along a deep portion of the lateral sulcus (see fig. 9.24).

1. Why is saliva necessary for the sense of taste?
2. Name the four primary taste sensations.
3. Trace a sensory impulse from a taste receptor to the cerebral cortex.

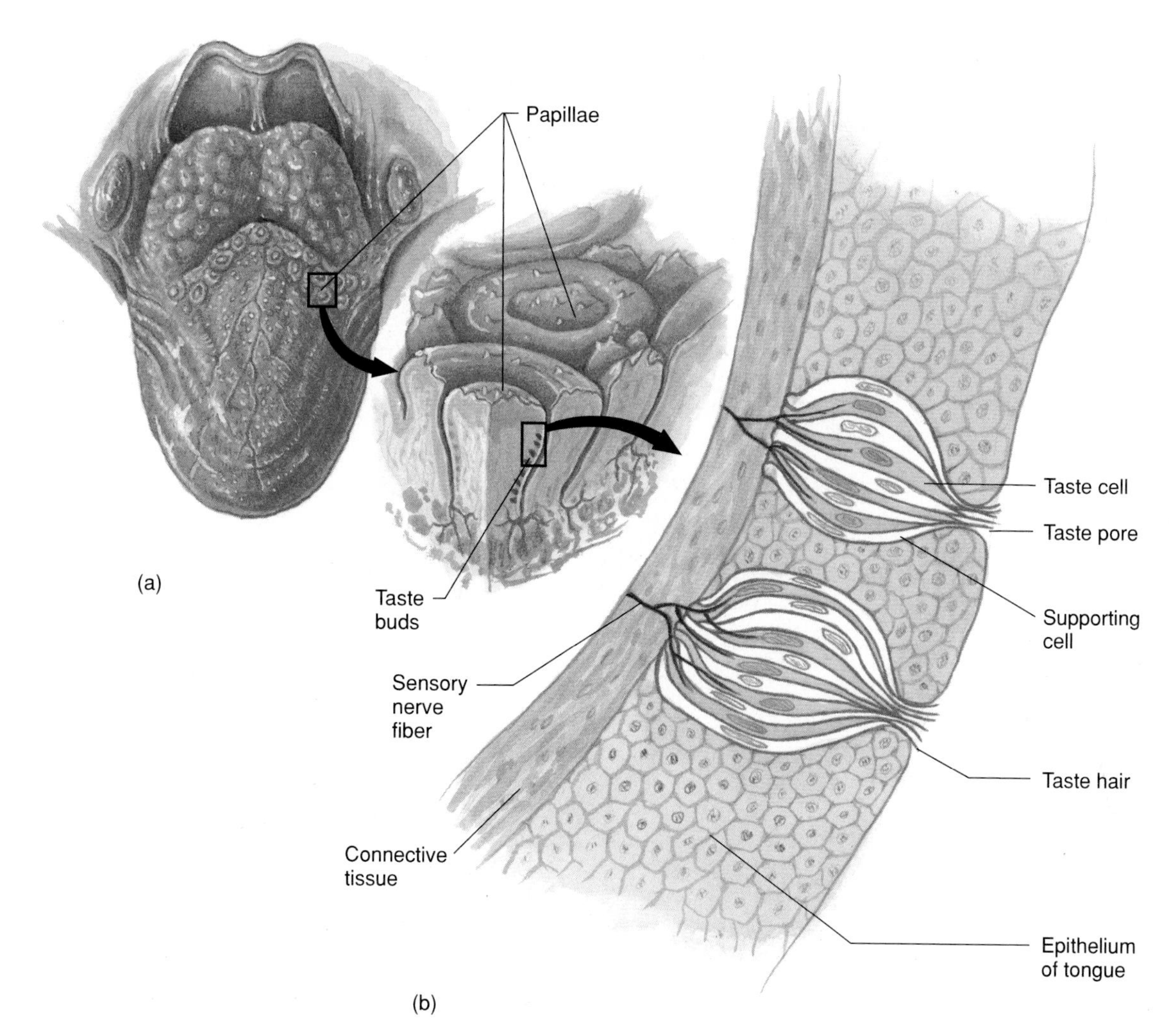

Figure 10.5

(*a*) Taste buds on the surface of the tongue are associated with nipplelike elevations called papillae. (*b*) A taste bud contains taste cells and has an opening, the taste pore, at its free surface.

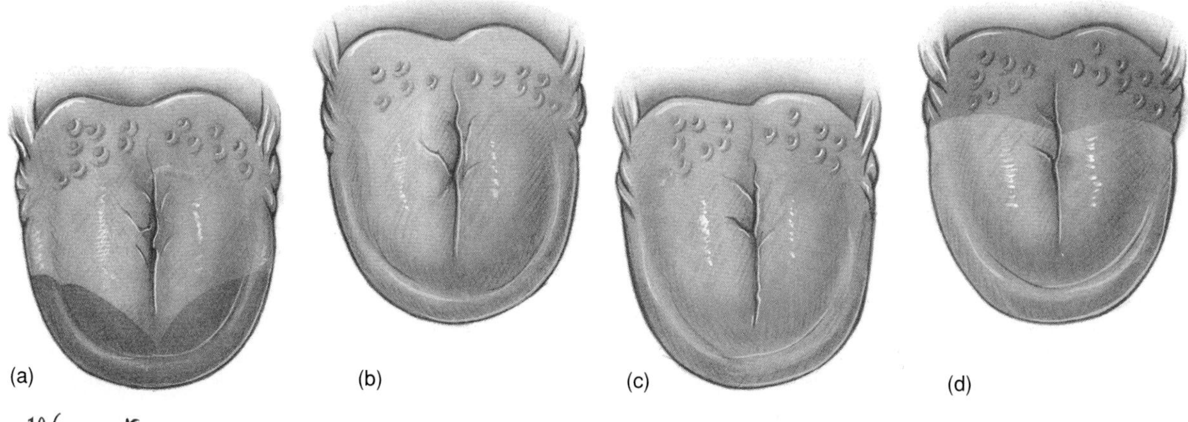

Figure 10.6

Colored sections indicate common patterns of taste receptors. (*a*) Sweet receptors. (*b*) Sour receptors. (*c*) Salt receptors. (*d*) Bitter receptors.

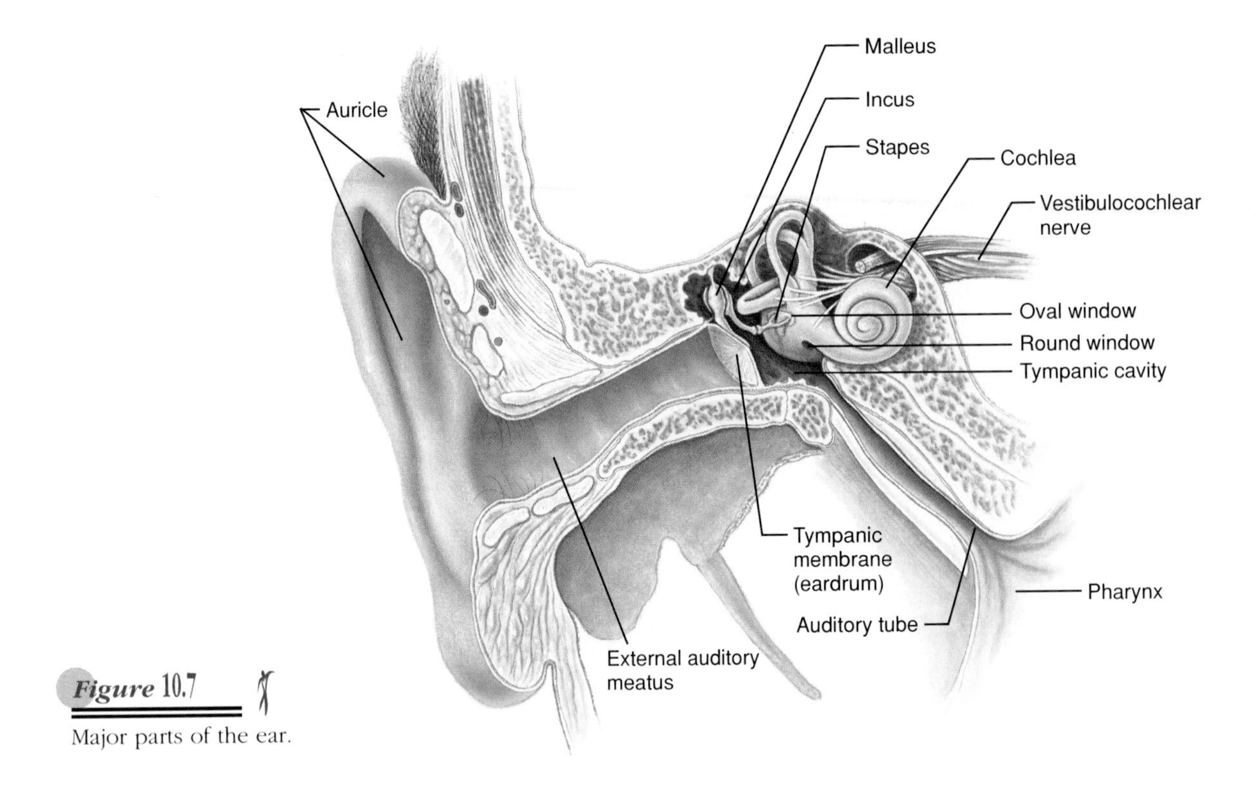

Figure 10.7

Major parts of the ear.

Sense of Hearing

The organ of hearing, the ear, has external, middle, and inner parts. The ear also functions in the sense of equilibrium.

External Ear

The external ear consists of two parts. The first is an outer, funnel-like structure called the **auricle** (pinna). The second is an S-shaped tube called the **external auditory meatus** (external auditory canal) that leads inward through the temporal bone for about 2.5 centimeters (fig. 10.7).

Vibrating objects produce sounds, and the vibrations are transmitted through matter in the form of sound waves. For example, vibrating strings or reeds produce the sounds of some musical instruments, and vibrating vocal folds in the larynx produce the voice. The auricle of the ear helps collect sound waves traveling through air and directs them into the external auditory meatus.

Middle Ear

The middle ear includes an air-filled space in the temporal bone called the *tympanic cavity,* an eardrum (tympanic membrane), and three small bones called auditory ossicles. The **eardrum** is a semitransparent membrane covered by a thin layer of skin on its outer surface and by mucous membrane on the inside. It has an oval margin and is cone-shaped, with the apex of the cone directed inward. The attachment of one of the auditory ossicles (malleus) maintains the eardrum's cone shape. Sound waves that enter the external auditory meatus change the pressure on the eardrum, which moves back and forth in response and thus reproduces the vibrations of the sound wave source.

The three **auditory ossicles** are the *malleus, incus,* and *stapes* (fig. 10.8). Tiny ligaments attach them to the wall of the tympanic cavity, and they are covered by mucous membrane. These bones bridge the eardrum and the inner ear, transmitting vibrations between these parts. Specifically, the malleus attaches to the eardrum, and when the eardrum vibrates, the malleus vibrates in unison. The malleus causes the incus to vibrate, and the incus passes the movement onto the stapes. Ligaments hold the stapes to an opening in the wall of the tympanic cavity called the **oval window,** which leads into the inner ear. Vibration of the stapes at the oval window moves a fluid within the inner ear, which stimulates the hearing receptors.

In addition to transmitting vibrations, the auditory ossicles help to increase (amplify) the force of vibrations as they pass from the eardrum to the oval window. Because the ossicles transmit vibrations from the relatively large surface of the eardrum to a much smaller area at the oval window, the vibrational force concentrates as it moves from the external to the inner ear. As a result, the pressure (per square millimeter) that the stapes applies on the oval window is many times greater than the pressure that sound waves exert on the eardrum.

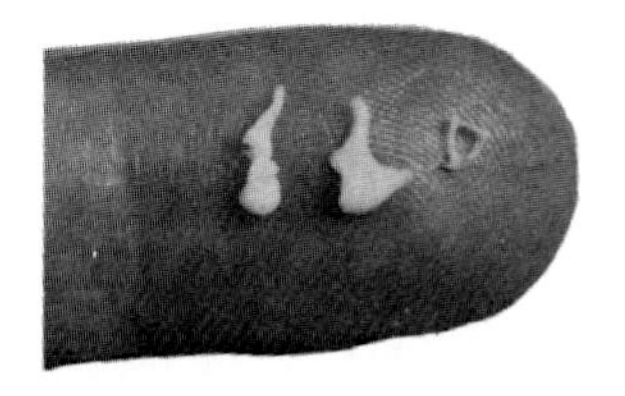

Figure 10.8

Auditory ossicles—malleus, incus, and stapes. These bones bridge the tympanic membrane and the inner ear (see figure 10.7).

Auditory Tube

An **auditory tube** (eustachian tube) connects each middle ear to the throat. This tube conducts air between the tympanic cavity and the outside of the body by way of the throat (nasopharynx) and mouth. The auditory tube helps maintain equal air pressure on both sides of the eardrum, which is necessary for normal hearing.

The function of the auditory tube is noticeable during rapid changes in altitude. As a person moves from a high altitude to a lower one, air pressure on the outside of the eardrum increases. This may push the eardrum inward, impairing hearing.

When the air pressure difference is great enough, air movement through the auditory tube equalizes the pressure on both sides of the eardrum, and the membrane moves back into its regular position. This produces a popping sound, which restores normal hearing.

Because auditory tube mucous membranes connect directly with middle ear linings, mucous membrane infections of the throat may spread through these tubes and cause a middle ear infection. For this reason, pinching a nostril when blowing the nose is poor practice because the pressure in the nasal cavity may force material from the throat up the auditory tube and into the middle ear. ■

Inner Ear

The inner ear is a complex system of communicating chambers and tubes called a **labyrinth.** Each ear has two such structures—the *osseous labyrinth* and the *membranous labyrinth* (fig. 10.9). The osseous labyrinth is a bony canal in the temporal bone. The membranous labyrinth is a tube that lies within the osseous labyrinth and has a similar shape. Between the osseous and membranous labyrinths is a fluid called *perilymph* that cells in the wall of the bony canal secrete. The membranous labyrinth contains another fluid called *endolymph*.

The parts of the labyrinths include three **semicircular canals,** which provide a sense of equilibrium (discussed in a subsequent section), and a **cochlea,** which functions in hearing. The cochlea contains a bony core and a thin, bony shelf that winds around the core like the threads of a screw. The shelf divides the osseous labyrinth of the cochlea into upper and lower compartments. The upper compartment, called the *scala vestibuli,* leads from the oval window to the apex of the spiral. The lower compartment, the *scala tympani,* extends from the apex of the cochlea to a membrane-covered opening in the wall of the inner ear called the **round window** (fig. 10.9).

The portion of the membranous labyrinth within the cochlea is called the *cochlear duct*. It lies between the two bony compartments and ends as a closed sac at the apex of the cochlea. The cochlear duct is separated from the scala vestibuli by a *vestibular membrane* (Reissner's membrane) and from the scala tympani by a *basilar membrane* (fig. 10.10).

The basilar membrane contains many thousands of stiff, elastic fibers, whose lengths progressively increase from the base of the cochlea to its apex. Sound vibrations entering the perilymph at the oval window travel along the scala vestibuli and pass through the vestibular membrane and into the endolymph of the cochlear duct, where they move the basilar membrane.

After passing through the basilar membrane, the vibrations enter the perilymph of the scala tympani. Movements of the membrane covering the round window dissipate the vibrations into the tympanic cavity.

The **organ of Corti,** which contains the hearing receptors, is located on the upper surface of the basilar membrane and stretches from the apex to the base of the cochlea (fig. 10.10). Its receptor cells (*hair cells*) are organized in rows and have many hairlike processes that project into the endolymph of the cochlear duct. Above these hair cells is a *tectorial membrane* attached to the bony shelf of the cochlea, passing over the receptor cells and contacting the tips of their hairs.

As sound vibrations pass through the inner ear, the hairs shear back and forth against the tectorial membrane, and the resulting mechanical deformation of the hairs stimulates the receptor cells (figs. 10.10 and 10.11). Various receptor cells, however, have slightly different sensitivities to deformation. Thus, a particular sound frequency excites certain receptor cells, and a sound of another frequency stimulates a different set of hair cells.

Hearing receptor cells are epithelial, but function somewhat like neurons. For example, when a hearing receptor cell is at rest, its membrane is polarized. When it is stimulated, selective ion channels open,

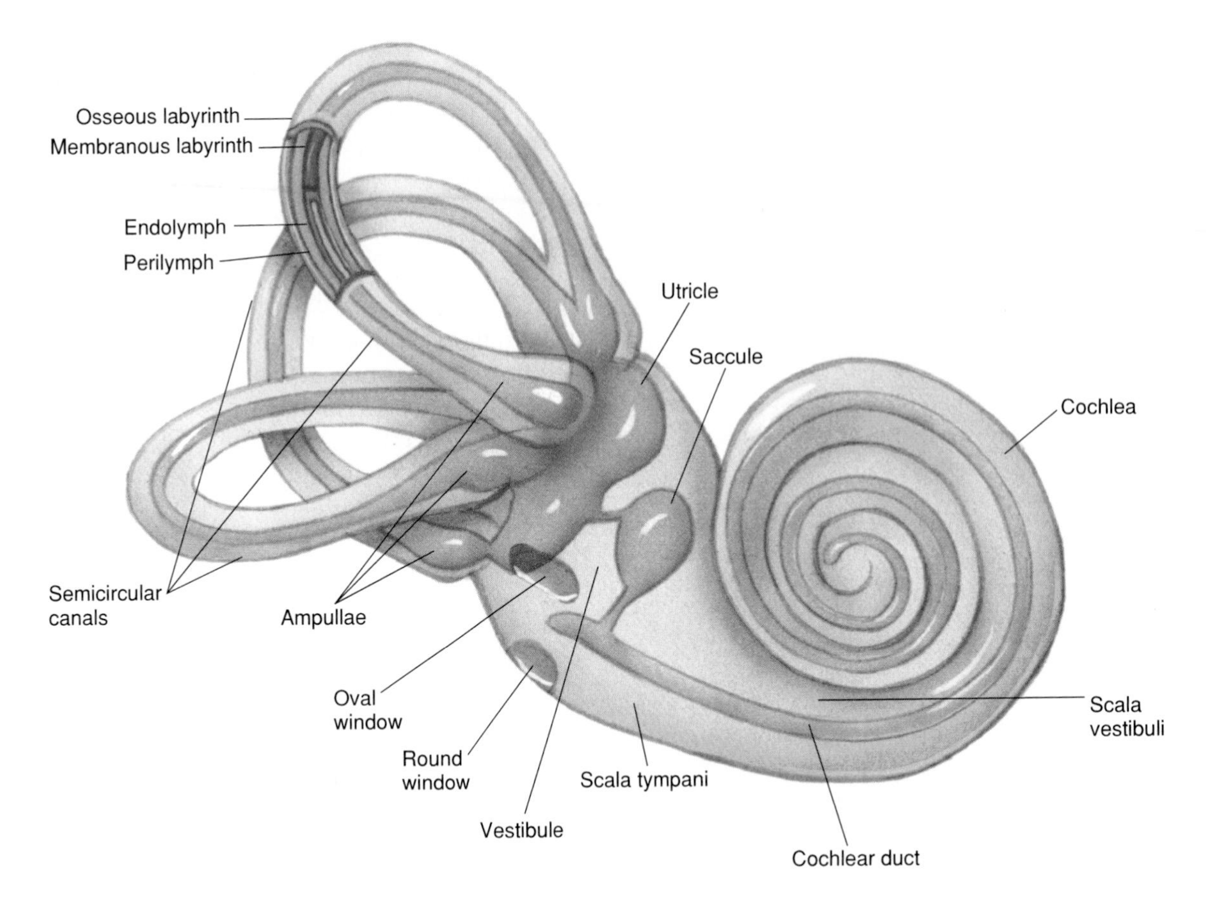

Figure 10.9

Perilymph separates the osseous labyrinth of the inner ear from the membranous labyrinth, which contains endolymph.

depolarizing the membrane and making it more permeable to calcium ions. The receptor cell has no axon or dendrites; however, it has neurotransmitter-containing vesicles near its base. As calcium ions diffuse into the cell, some of these vesicles fuse with the cell membrane and release neurotransmitter. The neurotransmitter stimulates the ends of nearby sensory nerve fibers, and in response, they transmit impulses along the cochlear branch of the vestibulocochlear nerve to the auditory cortex of the temporal lobe of the brain.

The ear of a young person with normal hearing can detect sound waves with frequencies from 20–20,000 or more vibrations per second. The range of greatest sensitivity is 2,000–3,000 vibrations per second.

Table 10.1 summarizes the steps of hearing.

Units called *decibels* (dB) measure sound intensity. The decibel scale begins at 0 dB, which is the intensity of the sound that is least perceptible by a normal human ear. The decibel scale is logarithmic. Thus, a sound of 10 dB is 10 times as intense as the least perceptible sound; a sound of 20 dB is 100 times as intense; and a sound of 30 dB is 1,000 times as intense.

On this scale, a whisper has an intensity of about 40 dB, normal conversation measures 60–70 dB, and heavy traffic produces about 80 dB. A sound of 120 dB, common at a rock concert, produces discomfort, and a sound of 140 dB, such as that emitted by a jet plane at takeoff, causes pain. Frequent or prolonged exposure to sounds with intensities above 90 dB can damage hearing receptors and cause permanent hearing loss. Pete Townshend of the 1960's rock band The Who has suffered hearing loss from years of exposure to loud sounds. ■

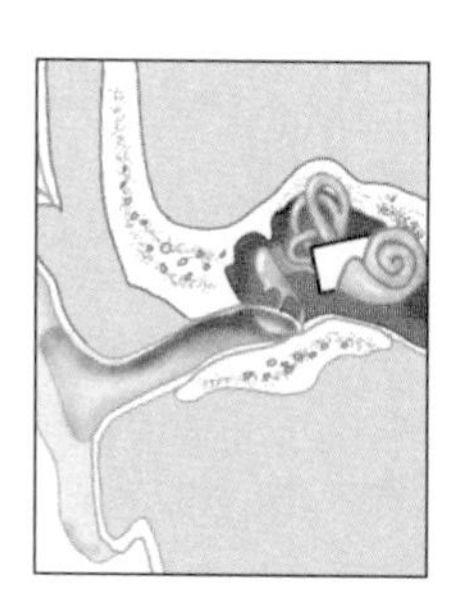

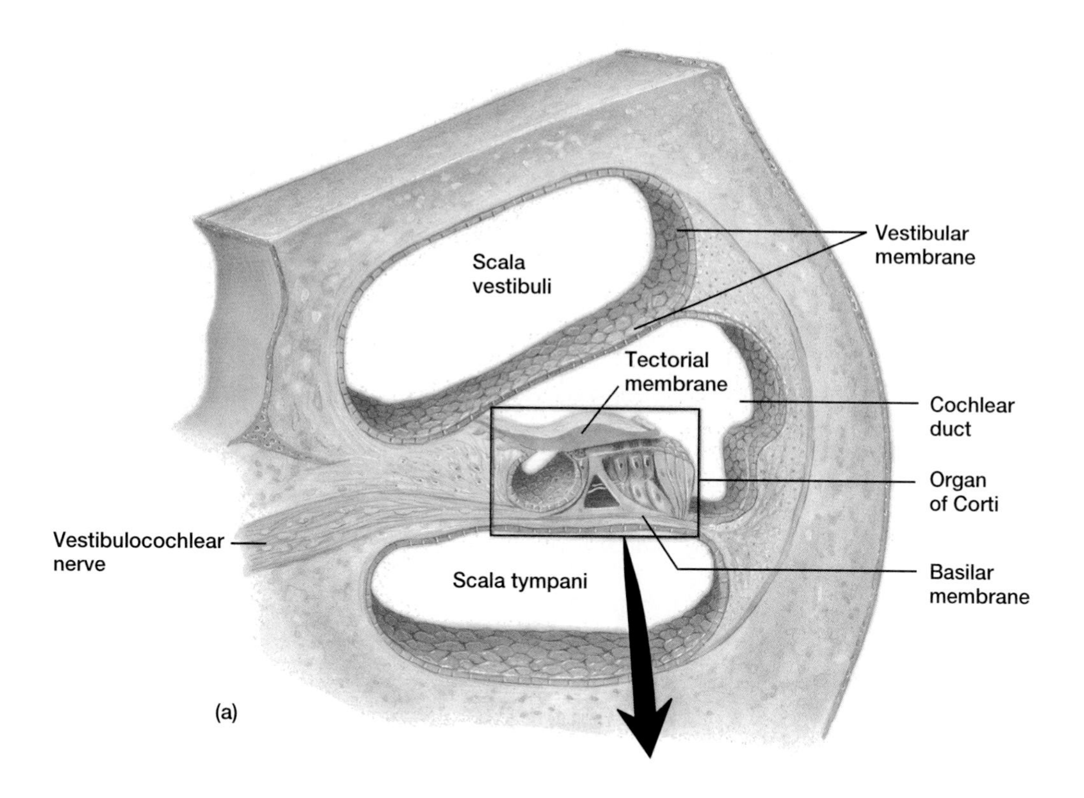

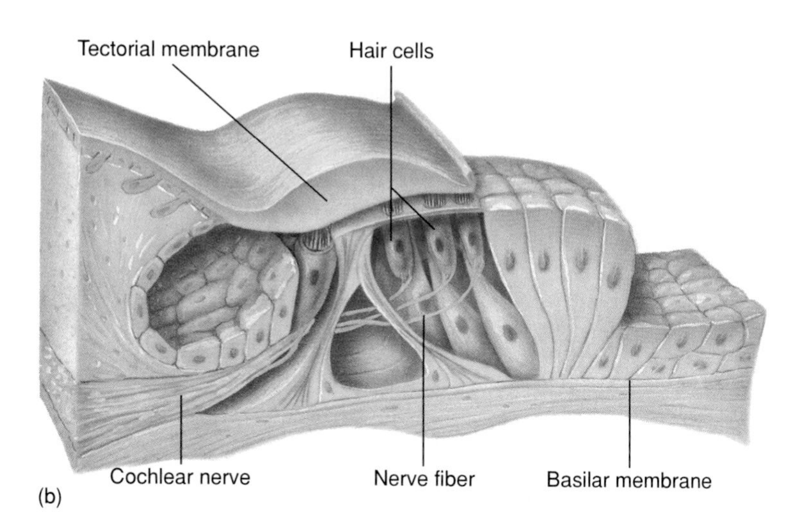

Figure 10.10

(*a*) Cross section of the cochlea. (*b*) Organ of Corti and the tectorial membrane.

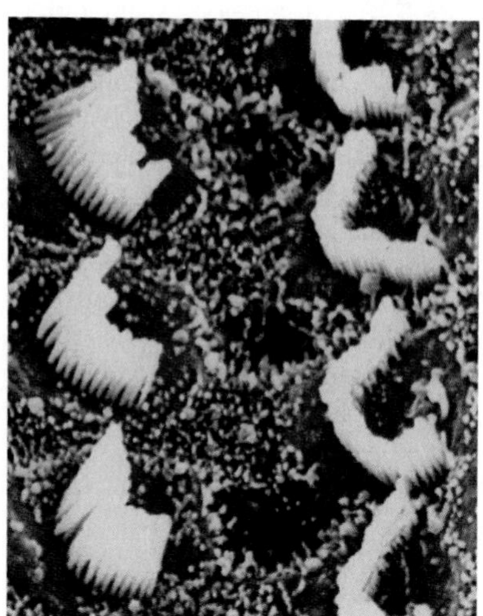

Figure 10.11

Scanning electron micrograph of hair cells in the organ of Corti (13,000×).

Table 10.1

Steps in the Generation of Sensory Impulses from the Ear

1. Sound waves enter external auditory meatus.
2. Waves of changing pressures cause eardrum to reproduce vibrations coming from sound wave source.
3. Auditory ossicles amplify and transmit vibrations to end of stapes.
4. Movement of stapes at oval window transmits vibrations to perilymph in scala vestibuli.
5. Vibrations pass through vestibular membrane and enter endolymph of cochlear duct.
6. Different frequencies of vibration in endolymph stimulate different sets of receptor cells.
7. As a receptor cell depolarizes, its membrane becomes more permeable to calcium ions.
8. Inward diffusion of calcium ions causes vesicles at the base of the receptor cell to release neurotransmitter.
9. Neurotransmitter stimulates ends of nearby sensory neurons.
10. Sensory impulses are triggered on fibers of cochlear branch of vestibulocochlear nerve.
11. Auditory cortex of temporal lobe interprets sensory impulses.

Auditory Nerve Pathways

The nerve fibers associated with hearing enter the auditory nerve pathways, which pass into the auditory cortices of the temporal lobes of the cerebrum, where they are interpreted. On the way, some of these fibers cross over, so that impulses arising from each ear are interpreted on both sides of the brain. Consequently, damage to a temporal lobe on one side of the brain does not necessarily cause complete hearing loss in the ear on that side.

A variety of factors can cause partial or complete hearing loss, including interference with the transmission of vibrations to the inner ear (*conductive deafness*) or damage to the cochlea, auditory nerve, or auditory nerve pathways (*sensorineural deafness*). Conductive deafness may be due to plugging of the external auditory meatus or to changes in the eardrum or auditory ossicles. The eardrum, for example, may harden as a result of disease and thus be less responsive to sound waves, or disease or injury may tear or perforate the eardrum.

Sensorineural deafness can be caused by loud sounds, tumors in the central nervous system, brain damage as a result of vascular accidents, or use of certain drugs. ■

1. How are sound waves transmitted through the external, middle, and inner ears?
2. Distinguish between the osseous and membranous labyrinths.
3. Describe the organ of Corti.

Sense of Equilibrium

The sense of equilibrium is really two senses—static equilibrium and dynamic equilibrium—that come from different sensory organs. The organs of **static equilibrium** sense the position of the head, maintaining stability and posture when the head and body are still. When the head and body suddenly move or rotate, the organs of **dynamic equilibrium** detect such motion and aid in maintaining balance.

Static Equilibrium

The organs of static equilibrium are located within the **vestibule**, a bony chamber between the semicircular canals and cochlea. The membranous labyrinth inside the vestibule consists of two expanded chambers—a **utricle** and a **saccule** (see fig. 10.9).

Each of these chambers has a tiny structure called a **macula** on its anterior wall. Maculae contain numerous

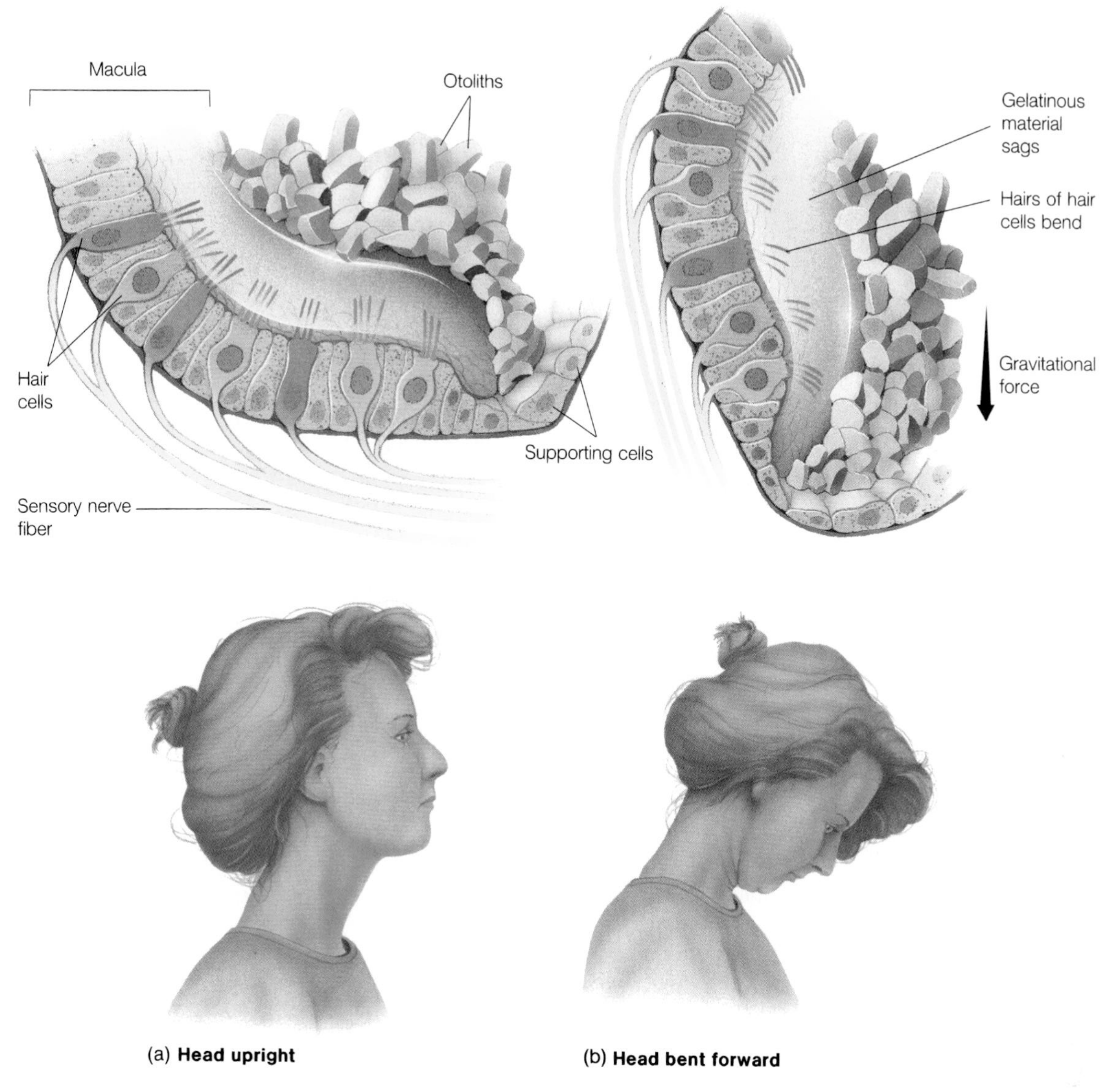

Figure 10.12

The macula responds to changes in head position. (*a*) Macula with the head in an upright position. (*b*) Macula with the head bent forward.

hair cells, which serve as sensory receptors. When the head is upright, the hairs of the hair cells project upward into a mass of gelatinous material, which has grains of calcium carbonate (otoliths) embedded in it. These particles add weight to the gelatinous structure.

The head bending forward, backward, or to one side stimulates hair cells. Such movements tilt the gelatinous masses of the maculae, and as they sag in response to gravity, the hairs projecting into them bend. This action stimulates the hair cells, and they signal the nerve fibers associated with them in a manner similar to that of hearing receptors. The resulting nerve impulses travel into the central nervous system on the vestibular branch of the vestibulocochlear nerve. These impulses inform the brain of the head's new position. The brain responds by sending motor impulses to skeletal muscles, which contract or relax to maintain balance (fig. 10.12).

Dynamic Equilibrium

The three **semicircular canals** detect motion of the head, and they aid in balancing the head and body during sudden movement. These canals lie at right angles to each other, and each corresponds to a different anatomical plane (see fig. 10.9).

Suspended in the perilymph of the bony portion of each semicircular canal is a membranous canal that

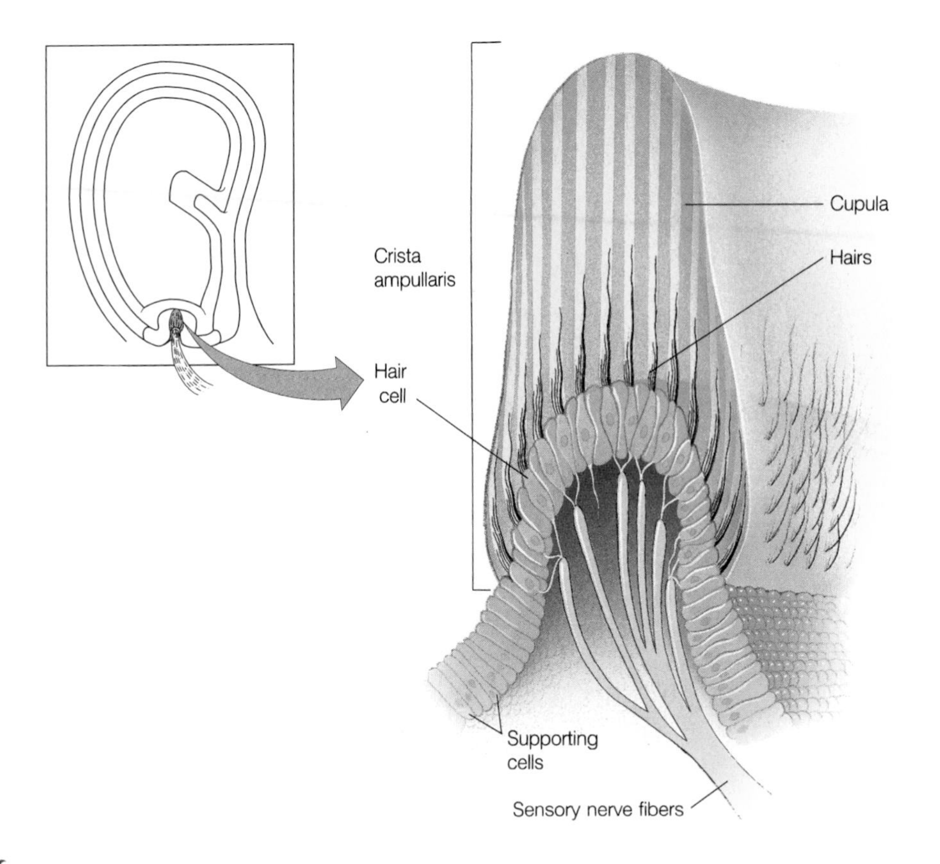

Figure 10.13

A crista ampullaris is located within the ampulla of each semicircular canal.

ends in a swelling called an **ampulla** that contains the sensory organs of the semicircular canals. Each of these organs, called a **crista ampullaris**, contains a number of sensory hair cells and supporting cells (fig. 10.13). As in the hairs of the maculae, the hair cells extend upward into a dome-shaped gelatinous mass called the *cupula*.

Rapid turns of the head or body stimulate the hair cells of the crista ampullaris. At such times, the semicircular canals move with the head or body, but the fluid inside the membranous canals remains stationary. This bends the cupula in one or more of the canals in a direction opposite that of the head or body movement, and the hairs embedded in it also bend. This stimulates the hair cells to signal their associated nerve fibers, and as a result, impulses travel to the brain (fig. 10.14).

Parts of the cerebellum are particularly important in interpreting impulses from the semicircular canals. Analysis of such information allows the brain to predict the consequences of rapid body movements, and by modifying signals to appropriate skeletal muscles, the cerebellum can maintain balance.

Other sensory structures aid in maintaining equilibrium. For example, certain mechanoreceptors (proprioceptors), particularly those associated with the joints of the neck, inform the brain about the position of body parts. In addition, the eyes detect changes in posture that result from body movements. Such visual information is so important that even if the organs of equilibrium are damaged, a person may be able to maintain normal balance by keeping the eyes open and moving slowly.

Some people experience *motion sickness* in a moving boat, airplane, or automobile. The cause is abnormal and irregular body motions that disturb the organs of equilibrium. Symptoms of motion sickness include nausea, vomiting, dizziness, headache, and prostration. ■

1. Distinguish between static and dynamic equilibrium.
2. What structures provide the sense of static equilibrium? Of dynamic equilibrium?
3. How does sensory information from other receptors help maintain equilibrium?

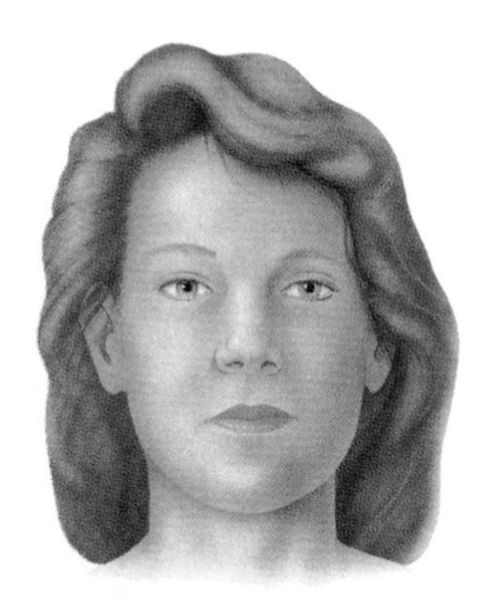

(a) Head in still position

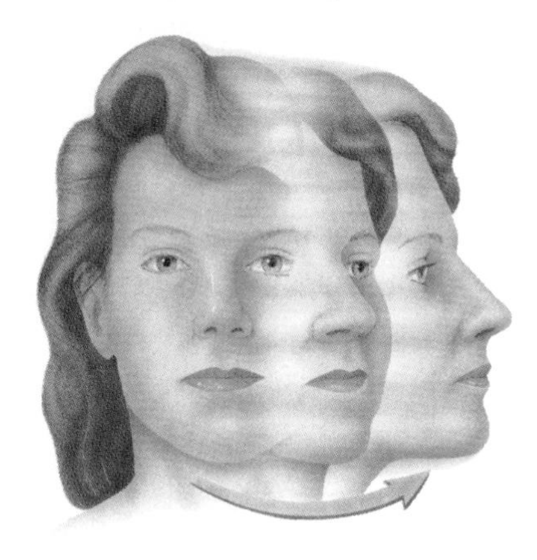

(b) Head rotating

Figure 10.14

(*a*) When the head is stationary, the cupula of the crista ampullaris remains upright.
(*b*) When the head is moving rapidly, the cupula bends opposite the motion of the head, stimulating sensory receptors.

Sense of Sight

Accessory organs assist the eye, the organ containing visual receptors, in providing vision. These accessory organs include the eyelids and lacrimal apparatus, which protect the eye, and a set of extrinsic muscles, which move the eye.

Visual Accessory Organs

The eye, lacrimal gland, and associated extrinsic muscles are housed within the pear-shaped orbital cavity of the skull. This orbit, which is lined with the periosteum of various bones, also contains fat, blood vessels, nerves, and connective tissues.

Each **eyelid** has four layers—skin, muscle, connective tissue, and conjunctiva (fig. 10.15). The skin of the eyelid, which is the thinnest skin of the body, covers the lid's outer surface and fuses with its inner lining near the margin of the lid. The eyelids are moved by the *orbicularis oculi* muscle (see fig. 8.16), which acts as a sphincter and closes the lids when it contracts, and by the *levator palpebrae superioris* muscle, which raises the upper lid and thus helps open the eye (fig. 10.15). The **conjunctiva** is a mucous membrane that lines the inner surfaces of the eyelids and folds back to cover the anterior surface of the eyeball, except for its central portion (cornea).

The *lacrimal apparatus* consists of the **lacrimal gland**, which secretes tears, and a series of ducts that carry tears into the nasal cavity (fig. 10.16). The gland is located in the orbit and secretes tears continuously. The tears exit through tiny tubules and flow downward and medially across the eye.

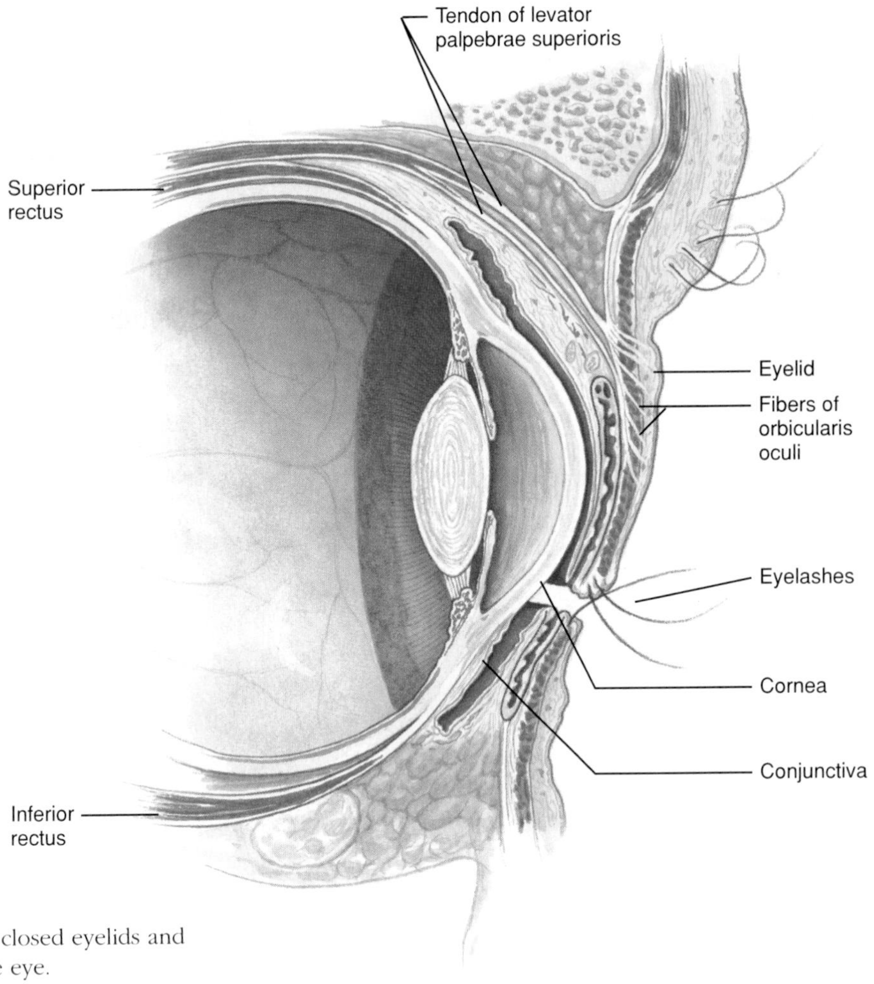

***Figure* 10.15**

Sagittal section of the closed eyelids and anterior portion of the eye.

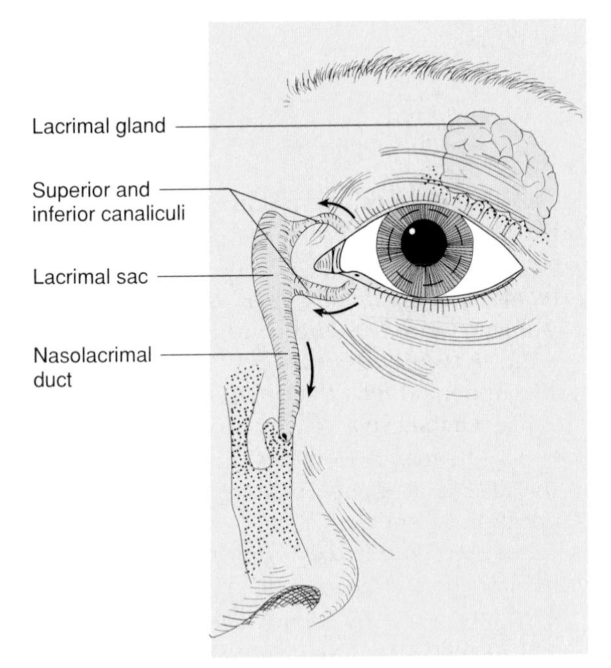

***Figure* 10.16**

The lacrimal apparatus consists of a tear-secreting gland and a series of ducts.

Two small ducts (the superior and inferior canaliculi) collect tears, which flow into the *lacrimal sac,* which lies in a deep groove of the lacrimal bone, and then into the *nasolacrimal duct,* which empties into the nasal cavity. Secretion of the lacrimal gland moistens and lubricates the surface of the eye and the lining of the lids. Tears also contain an enzyme (*lysozyme*) that is an antibacterial agent, reducing the risk of eye infections.

The **extrinsic muscles** arise from the bones of the orbit and insert by broad tendons on the eye's tough outer surface. Six extrinsic muscles move the eye in various directions. Any given eye movement may utilize more than one extrinsic muscle, but each muscle is associated with one primary action. Figure 10.17 illustrates the locations of these extrinsic muscles, and table 10.2 lists their functions.

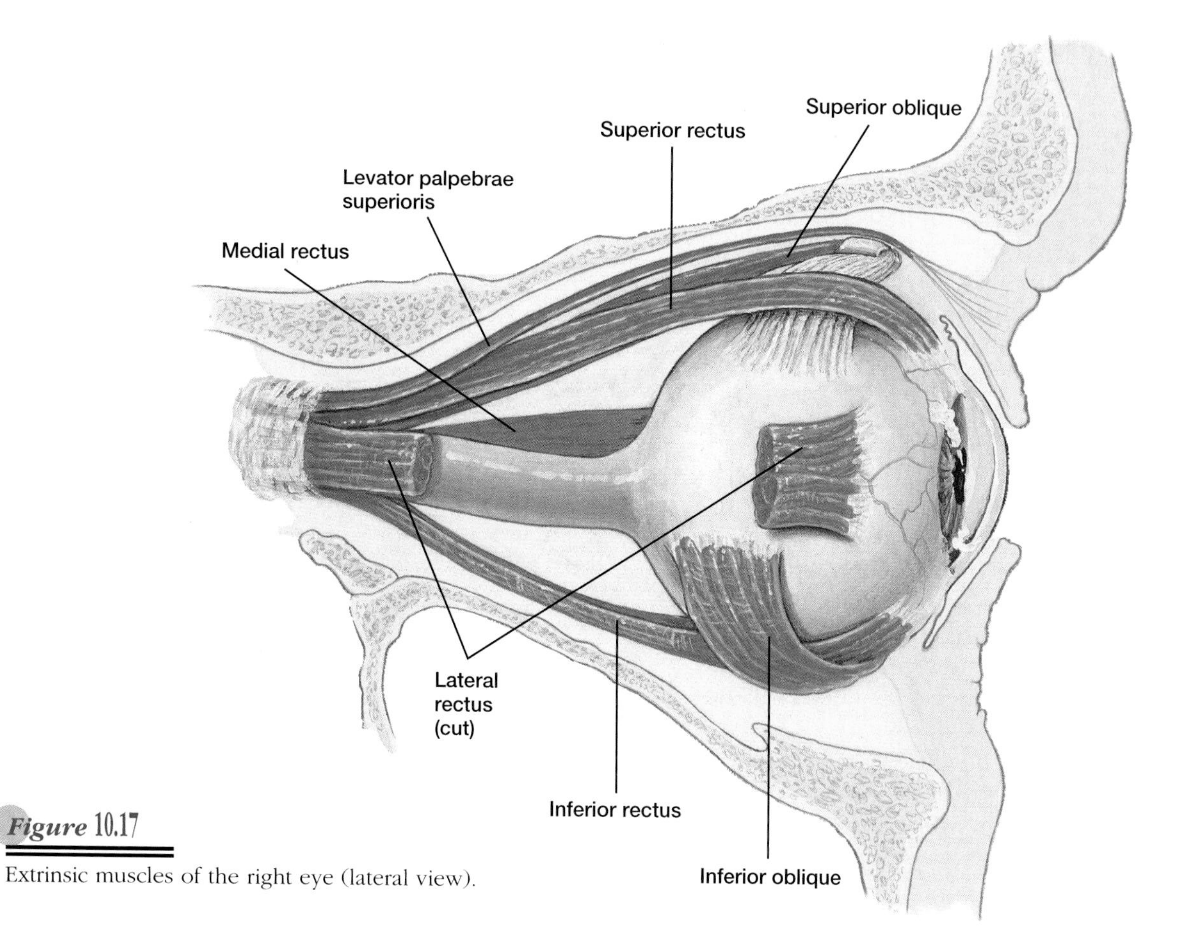

Figure 10.17

Extrinsic muscles of the right eye (lateral view).

Table 10.2

Muscles Associated with the Eyelids and Eyes

Name	Innervation	Function
Muscles of the eyelids		
Orbicularis oculi	Facial nerve (VII)	Closes eye
Levator palpebrae superioris	Oculomotor nerve (III)	Opens eye
Extrinsic muscles of the eyes		
Superior rectus	Oculomotor nerve (III)	Rotates eye upward and toward midline
Inferior rectus	Oculomotor nerve (III)	Rotates eye downward and toward midline
Medial rectus	Oculomotor nerve (III)	Rotates eye toward midline
Lateral rectus	Abducens nerve (VI)	Rotates eye away from midline
Superior oblique	Trochlear nerve (IV)	Rotates eye downward and away from midline
Inferior oblique	Oculomotor nerve (III)	Rotates eye upward and away from midline

One eye deviating from the line of vision may result in double vision (diplopia). If this condition persists, the brain may suppress the image from the deviated eye. As a result, the turning eye may become blind (suppression amblyopia). Treating eye deviation early in life with exercises, eyeglasses, and surgery can prevent such monocular (one eye) blindness. ■

1. Explain how the eyelid moves.
2. Describe the conjunctiva.
3. What is the function of the lacrimal apparatus?

Structure of the Eye

The eye is a hollow, spherical structure about 2.5 centimeters in diameter. Its wall has three distinct layers—a fibrous *outer tunic,* a vascular *middle tunic,* and a nervous *inner tunic*. The spaces within the eye are filled with fluids that support its wall and internal parts, and help maintain its shape. Figure 10.18 shows the major parts of the eye.

Outer Tunic

The anterior sixth of the outer tunic (fibrous tunic) bulges forward as the transparent **cornea,** which is the window of the eye and helps focus entering light rays. The cornea is composed largely of connective tissue with a thin layer of epithelium on its surface. It is transparent because it contains few cells and no blood vessels, and its cells and collagenous fibers form unusually regular patterns.

Along its circumference, the cornea is continuous with the **sclera,** the white portion of the eye. The sclera makes up the posterior five-sixths of the outer tunic and is opaque due to many large, disorganized, collagenous and elastic fibers. The sclera protects the eye and is an attachment for the extrinsic muscles. In the back of the eye, the **optic nerve** and certain blood vessels pierce the sclera.

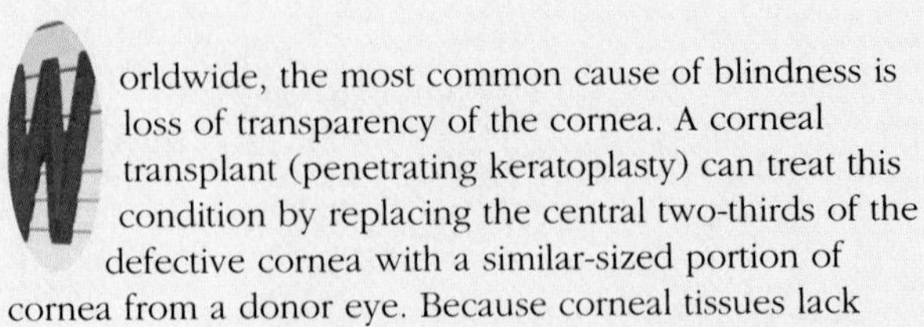

Worldwide, the most common cause of blindness is loss of transparency of the cornea. A corneal transplant (penetrating keratoplasty) can treat this condition by replacing the central two-thirds of the defective cornea with a similar-sized portion of cornea from a donor eye. Because corneal tissues lack blood vessels, transplanted tissue is usually not rejected. The success rate of the procedure is very high. ■

Middle Tunic

The middle tunic (vascular tunic) includes the choroid coat, ciliary body, and iris (fig. 10.18). The **choroid coat,** in the posterior five-sixths of the globe of the

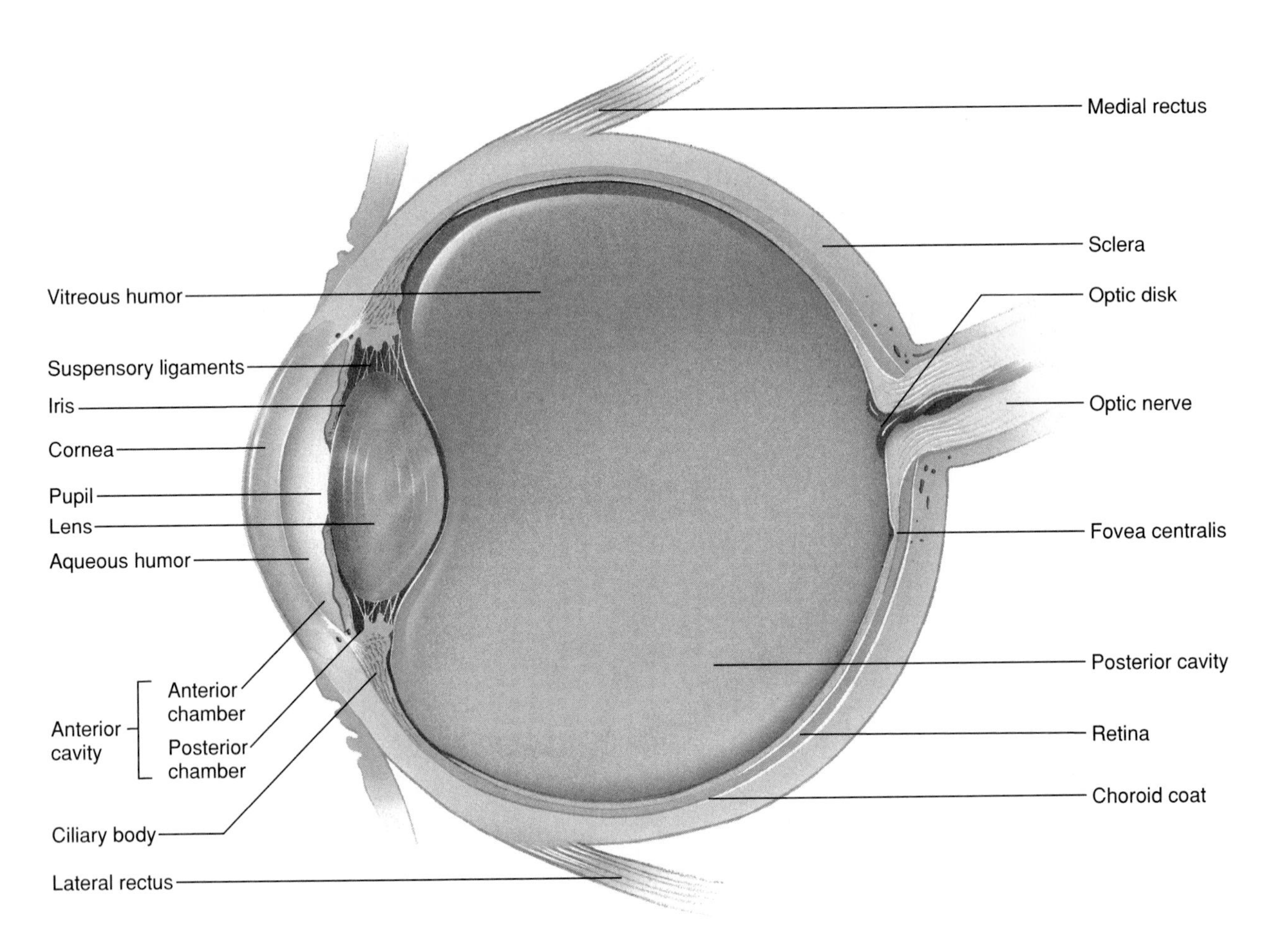

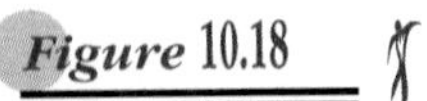
Figure 10.18

Transverse section of the left eye (superior view).

eye, is loosely joined to the sclera and is honeycombed with blood vessels, which nourish surrounding tissues. The choroid coat also contains many pigment-producing melanocytes. The melanin that these cells produce absorbs excess light and thus helps to keep the inside of the eye dark.

The **ciliary body**, which is the thickest part of the middle tunic, extends forward from the choroid coat and forms an internal ring around the front of the eye. Within the ciliary body are many radiating folds called *ciliary processes* and groups of muscle fibers that constitute the *ciliary muscles*.

Many strong but delicate fibers, called *suspensory ligaments,* extend inward from the ciliary processes and hold the transparent **lens** in position (fig. 10.19). The distal ends of these fibers attach along the margin of a thin capsule that surrounds the lens. The body of the lens lies directly behind the iris and pupil and is composed of differentiated epithelial cells called *lens fibers*. The cytoplasm of these cells is the transparent substance of the lens.

The lens capsule is a clear, membranelike structure composed largely of intercellular material. Its elastic nature keeps it under constant tension. As a result, the lens can assume a globular shape. The suspensory ligaments attached to the margin of the capsule are also under tension. They pull outward, flattening the capsule and the lens inside (fig. 10.20).

If the tension on the suspensory ligaments relaxes, the elastic capsule rebounds, and the lens surface becomes more convex. The ciliary muscles accomplish this. For example, one set of these muscle fibers extends back from fixed points in the sclera to the choroid coat. When the fibers contract, the choroid coat is pulled forward and the ciliary body shortens. This relaxes the suspensory ligaments; the lens thickens in response and is now focused for viewing closer objects than before. To allow focus on more distant objects, the ciliary muscles relax, tension on the suspensory ligaments increases, and the lens becomes thinner and less convex again. This ability of the lens to adjust shape to facilitate focusing is called **accommodation** (fig. 10.20).

A common eye disorder, particularly in older people, is *cataract*. The lens or its capsule slowly becomes cloudy and opaque. Without treatment, cataracts eventually cause blindness. In the past, cataracts were treated surgically, with a two-week recovery period. Today, cataracts are treated on an outpatient basis with a laser. ■

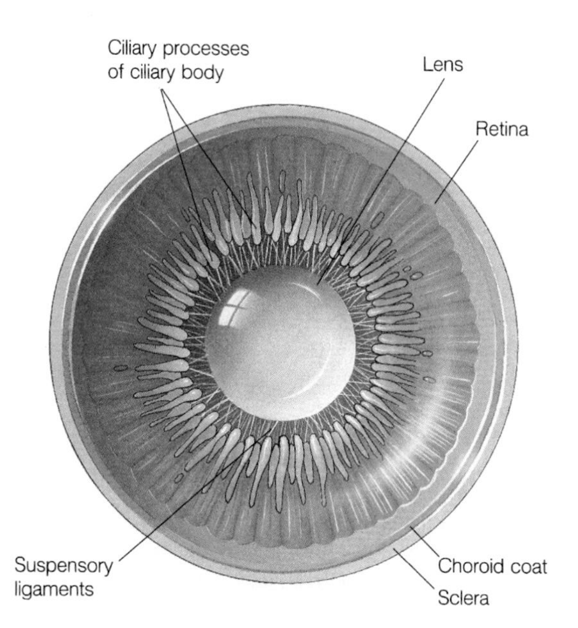

Figure 10.19

Lens and ciliary body viewed from behind.

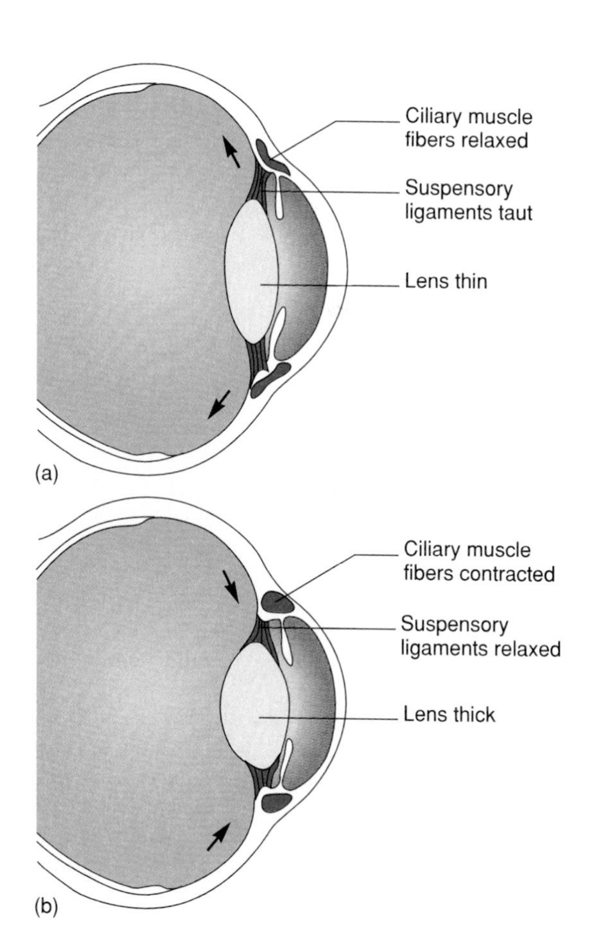

Figure 10.20

(*a*) The lens thins as ciliary muscle fibers relax. (*b*) The lens thickens as ciliary muscle fibers contract.

1. Describe the outer and middle tunics of the eye.
2. What factors contribute to the transparency of the cornea?
3. How does the shape of the lens change during accommodation?

The **iris** is a thin diaphragm composed mostly of connective tissue and smooth muscle fibers. From the outside, the iris is the colored portion of the eye. The iris extends forward from the periphery of the ciliary body and lies between the cornea and lens (see fig. 10.18). The iris divides the space (anterior cavity) separating these parts into an *anterior chamber* (between the cornea and iris) and a *posterior chamber* (between the iris and vitreous body, and containing the lens).

The epithelium on the inner surface of the ciliary body secretes a watery fluid called **aqueous humor** into the posterior chamber. The fluid circulates from this chamber through the **pupil**, a circular opening in the center of the iris, and into the anterior chamber. Aqueous humor fills the space between the cornea and lens, helps nourish these parts, and aids in maintaining the shape of the front of the eye. It subsequently leaves the anterior chamber through veins and a special drainage canal, the scleral venous sinus (canal of Schlemm) located in its wall.

An eye disorder called *glaucoma* develops when the rate of aqueous humor formation exceeds the rate of its removal. As fluid accumulates in the anterior chamber of the eye, fluid pressure rises and is transmitted to all parts of the eye. In time, the building pressure squeezes shut blood vessels that supply the receptor cells of the retina, and cells that are robbed of nutrients and oxygen may die. Permanent blindness may result.

When diagnosed early, glaucoma can usually be treated successfully with drugs, laser therapy, or surgery, all of which promote the outflow of aqueous humor. Since glaucoma in its early stages typically produces no symptoms, discovery of the condition usually depends on measuring intraocular pressure, using an instrument called a *tonometer*. ■

The smooth muscle fibers of the iris are organized into two groups, a *circular set* and a *radial set*. These muscles control the size of the pupil, which is the opening that light passes through as it enters the eye. The circular set of muscle fibers acts as a sphincter. When it contracts, the pupil gets smaller, and the amount of light entering decreases. When the radial muscle fibers contract, the pupil's diameter increases, and the amount of light entering increases (fig. 10.21).

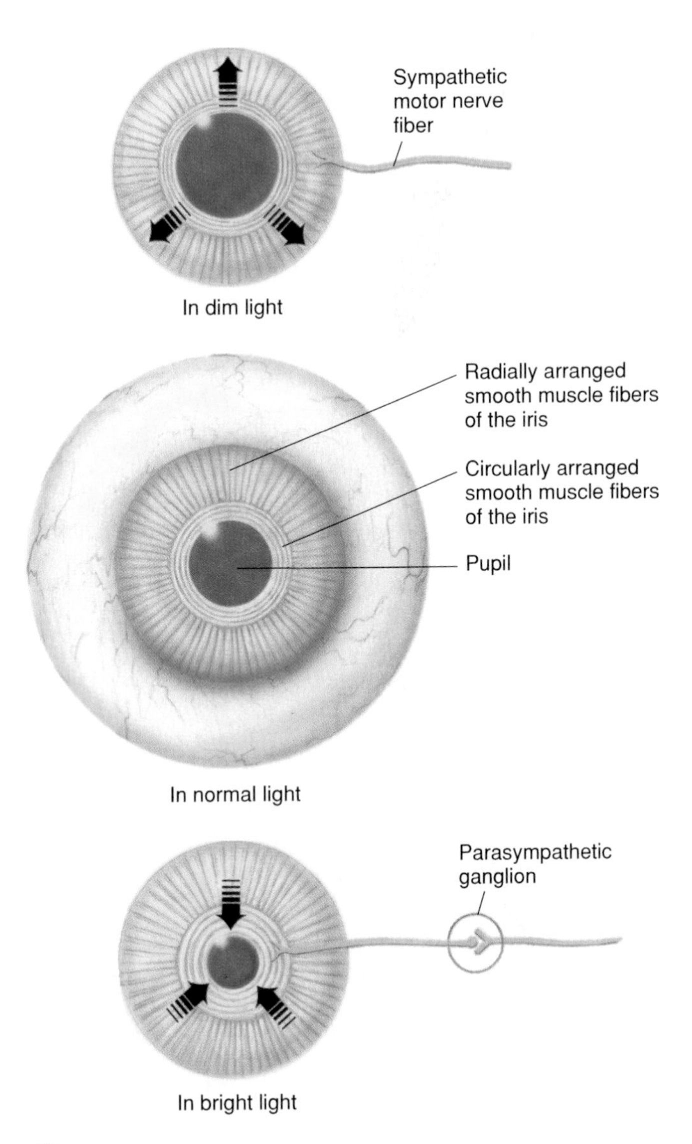

Figure 10.21

Dim light stimulates the radial muscles of the iris to contract, and the pupil dilates. Bright light stimulates the circular muscles of the iris to contract, and the pupil constricts.

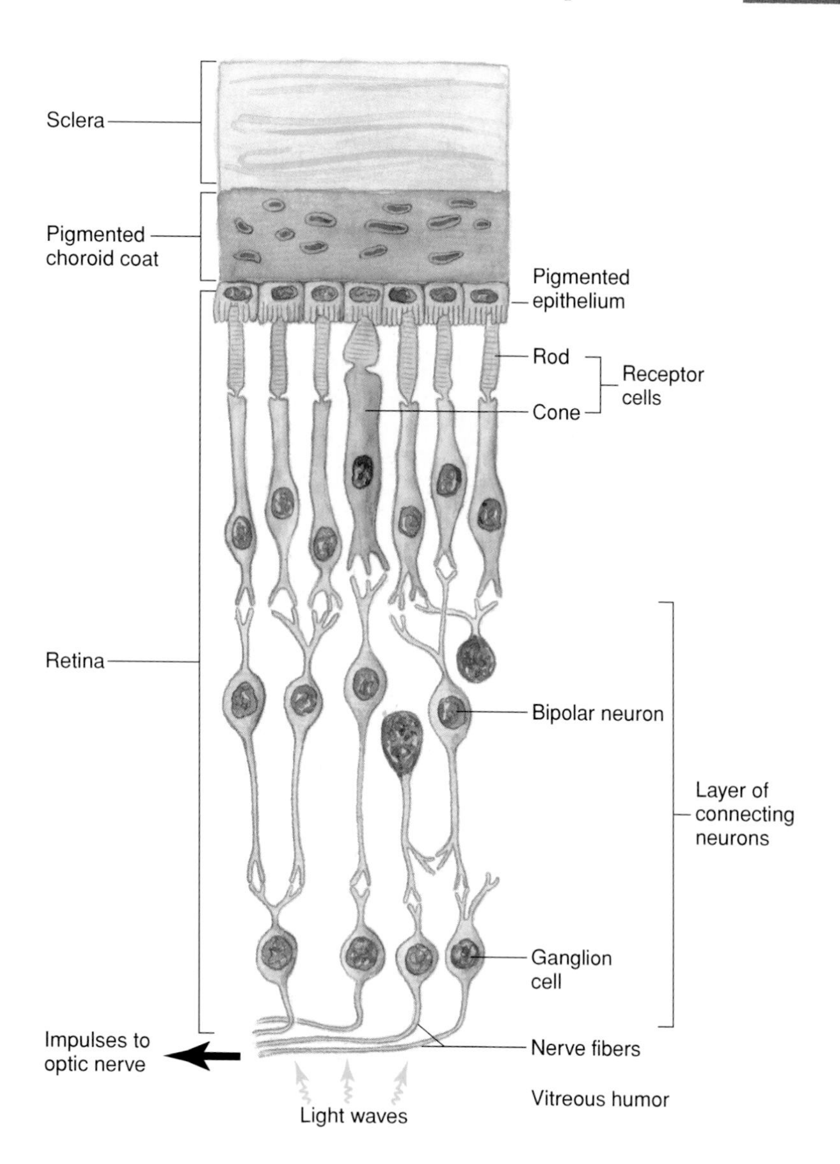

Figure 10.22

The retina consists of several cell layers.

Inner Tunic

The inner tunic consists of the **retina**, which contains the visual receptor cells (photoreceptors). This nearly transparent sheet of tissue is continuous with the optic nerve in the back of the eye and extends forward as the inner lining of the eyeball. It ends just behind the margin of the ciliary body.

The retina is thin and delicate, but its structure is quite complex. It has a number of distinct layers, as figures 10.22 and 10.23 illustrate.

In the central region of the retina is a yellowish spot called the *macula lutea*. A depression in its center, called the **fovea centralis**, is in the region of the retina that produces the sharpest vision (see fig. 10.18).

Just medial to the fovea centralis is an area called the **optic disk** (fig. 10.24). Here, nerve fibers from the retina leave the eye and join the optic nerve. A central artery and vein also pass through the optic disk. These vessels are continuous with the capillary networks of the retina, and with vessels in the underlying choroid coat, they supply blood to the cells of the inner tunic. Because the optic disk region has no receptor cells, it is commonly known as the *blind spot* of the eye.

The space bounded by the lens, ciliary body, and retina is the largest compartment of the eye and is called the *posterior cavity* (see fig. 10.18). The posterior cavity is filled with a transparent, jellylike fluid called **vitreous humor**, which with collagenous fibers comprise the *vitreous body*. The vitreous body supports the internal parts of the eye and helps maintain its shape.

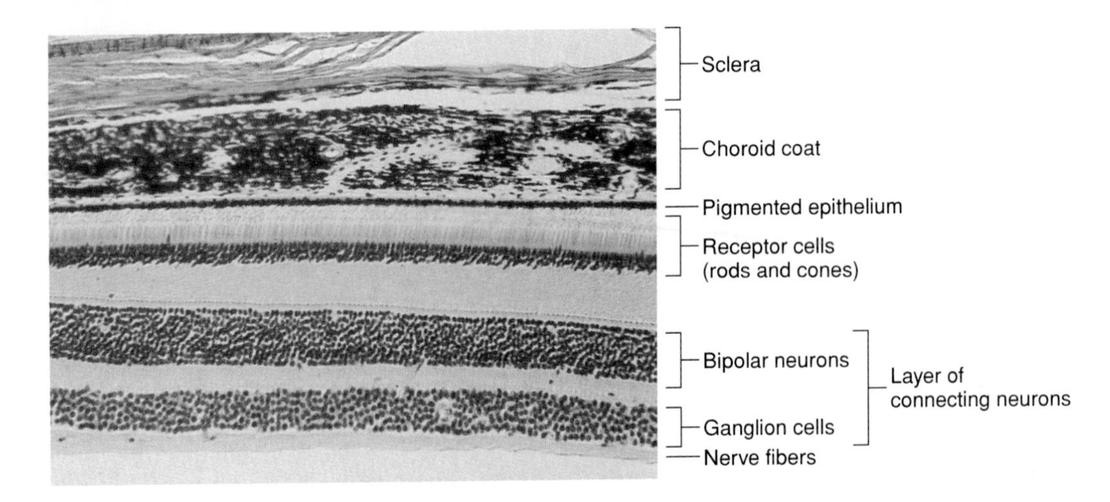

Figure 10.23

Note the layers of cells and nerve fibers in this light micrograph of the retina (200×).

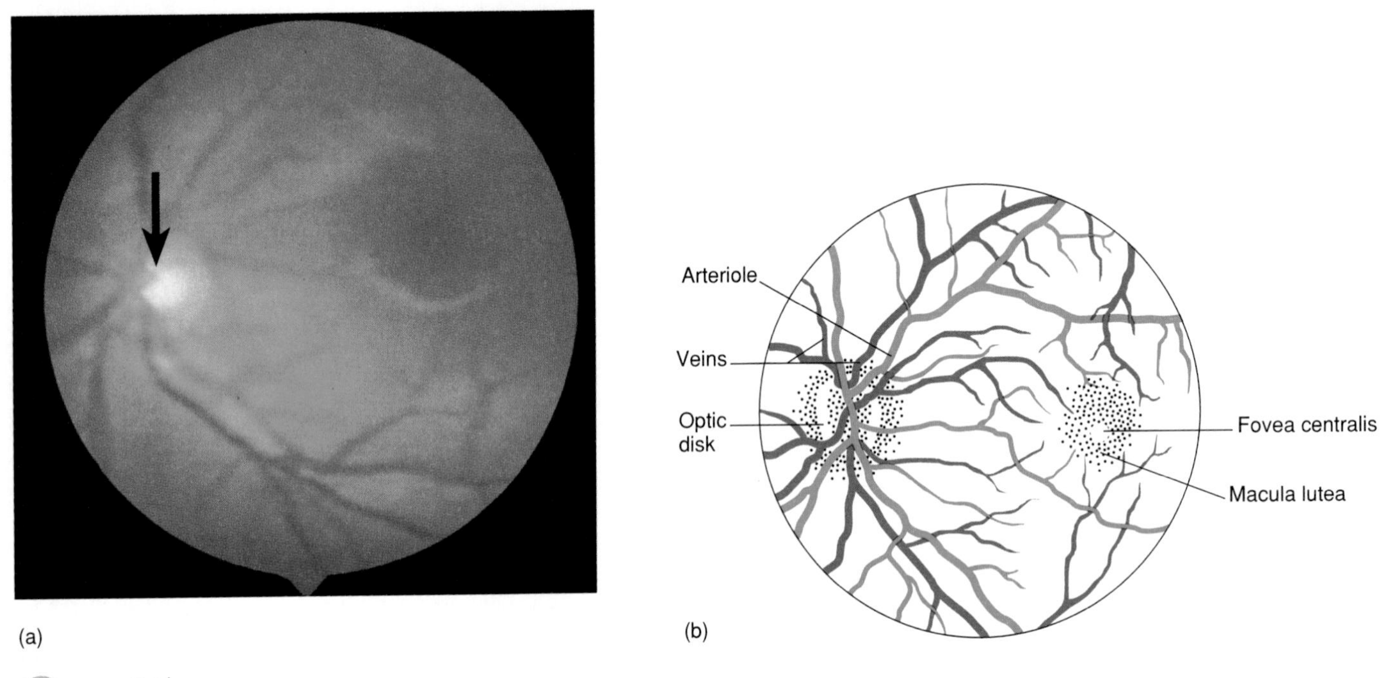

Figure 10.24

(*a*) Nerve fibers leave the eye in the area of the optic disk (arrow) to form the optic nerve (53×). (*b*) Major features of the retina.

As a person ages, tiny, dense clumps of gel or deposits of crystal-like substances form in the vitreous humor. When these clumps cast shadows on the retina, the person sees small, moving specks in the field of vision. Such specks, known as *floaters,* are most apparent when looking at a plain background, such as the sky or a wall. ■

1. Explain the origin of aqueous humor, and trace its path through the eye.
2. How is the size of the pupil regulated?
3. Describe the structure of the retina.

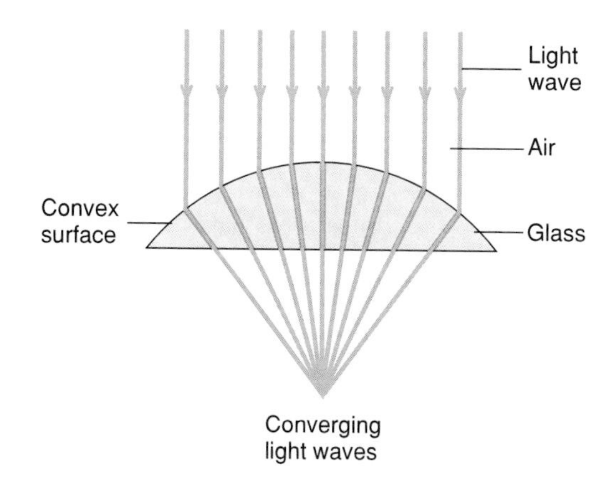

Figure 10.25

A lens with a convex surface causes light waves to converge.

Light Refraction

When a person sees something, either the object is giving off light, or light waves are reflected from it. These light waves enter the eye, and an image of the object is focused on the retina. Focusing bends the light waves—a phenomenon called **refraction.**

Refraction occurs when light waves pass at an oblique angle from a medium of one optical density into a medium of a different optical density. This occurs at the curved surface between the air and the cornea and at the curved surface of the lens itself. A lens with a *convex* surface (such as in the eye) causes light waves to converge (fig. 10.25).

The convex surface of the cornea refracts light waves from outside objects. The convex surface of the lens and, to a lesser extent, the surfaces of the fluids within the chambers of the eye then refract the light again.

If eye shape is normal, light waves focus sharply on the retina, much as a motion picture image is focused on a screen for viewing. Unlike the motion picture image, however, an image that forms on the retina is upside down and reversed from left to right. The visual cortex interprets the image in its proper position.

1. What is refraction?
2. What parts of the eye provide refracting surfaces?

Visual Receptors

Visual receptor cells are modified neurons of two distinct kinds, as figure 10.22 illustrates. One group, called *rods,* have long, thin projections at their ends, while the other group, *cones,* have short, blunt projections.

Rods and cones are in a deep portion of the retina, closely associated with a layer of pigmented epithelium (see fig. 10.23). The epithelial pigment absorbs light waves not absorbed by the receptor cells, and together with the pigment of the choroid coat, keeps light from reflecting off surfaces inside the eye. Projections from receptors, which are loaded with light-sensitive visual pigments, extend into this pigmented layer.

Visual receptors are stimulated only when light reaches them. A light image focused on an area of the retina stimulates some receptors, and impulses travel from them to the brain. The impulse leaving each activated receptor, however, provides only a fragment of the information required for the brain to interpret a total scene.

Rods and cones function differently. Rods are hundreds of times more sensitive to light than cones and therefore can provide vision in dim light. Rods produce colorless vision, while cones detect color.

Rods and cones also differ in visual acuity—the sharpness of the perceived images. Cones provide sharp images, and rods provide more general outlines of objects. Rods give less precise images because nerve fibers from many rods converge, their impulses transmitted to the brain on the same nerve fiber. Thus, if a point of light stimulates a rod, the brain cannot tell which one of many receptors has been stimulated. Convergence of impulses is less common among cones. When a cone is stimulated, the brain can pinpoint the stimulation more accurately (fig. 10.26).

The fovea centralis, the area of sharpest vision, lacks rods but contains densely packed cones with few or no converging fibers (see fig. 10.18). Also in the fovea centralis, the overlying layers of the retina and the retinal blood vessels are displaced to the sides, more fully exposing receptors to incoming light. Consequently, to view something in detail, a person moves the eyes so that the important part of an image falls on the fovea centralis.

Visual Pigments

Both rods and cones contain light-sensitive pigments that decompose when they absorb light energy. The light-sensitive biochemical in rods is called **rhodopsin**

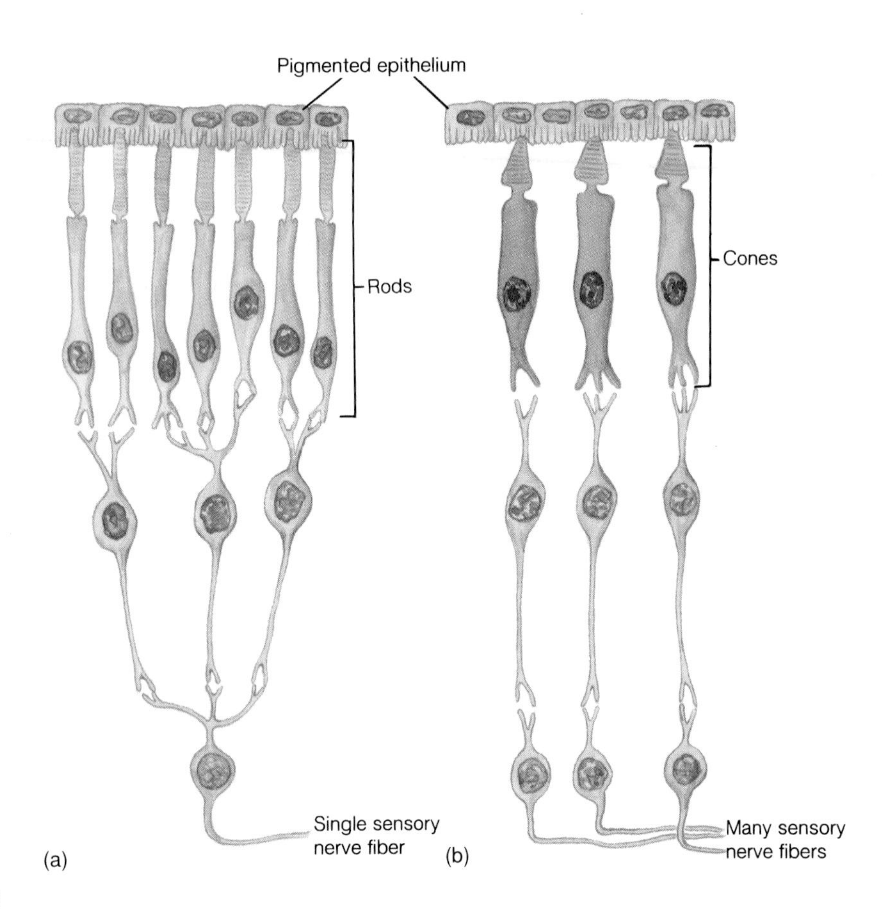

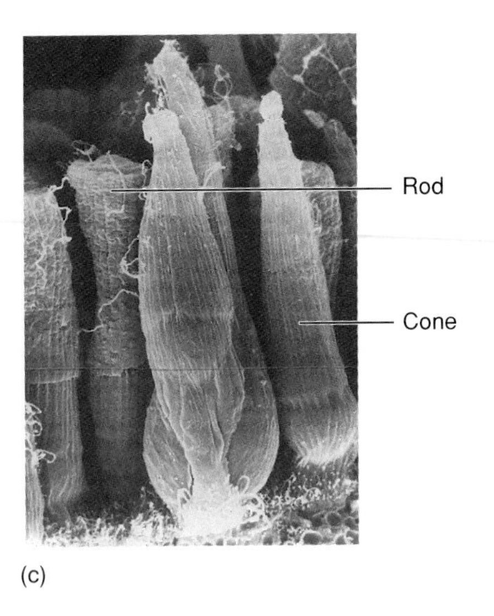

Figure 10.26

(*a*) A single sensory nerve fiber transmits impulses from several rods to the brain. (*b*) Separate sensory nerve fibers transmit impulses from cones to the brain. (*c*) Scanning electron micrograph of rods and cones.

(visual purple). In the presence of light, rhodopsin molecules break down into molecules of a colorless protein called *opsin* and a yellowish substance called *retinal* (retinene) that is synthesized from vitamin A.

Decomposition of rhodopsin molecules activates an enzyme that initiates a series of reactions that alters the permeability of the rod cell membrane. As a result, a complex pattern of nerve impulses originates in the retina. The impulses travel away from the retina along the optic nerve into the brain, where they are interpreted as vision.

In bright light, nearly all of the rhodopsin in the rods of the retina decomposes, greatly reducing rod sensitivity. In dim light, however, regeneration of rhodopsin from opsin and retinal is faster than rhodopsin breakdown. This regeneration requires cellular energy, provided by ATP (see chapter 4).

Poor vision in dim light, called night blindness, results from vitamin A deficiency. Lack of the vitamin reduces the supply of retinal, rhodopsin production falls, and rod sensitivity is low. Vitamin A treats the condition. ■

The light-sensitive pigments of the cones are similar to rhodopsin in that they are composed of retinal combined with a protein; the protein, however, differs from the protein in the rods. Three different sets of cones each contain an abundance of one of three different visual pigments.

The wavelength of light determines the color that the brain perceives from it. For example, the shortest wavelengths of visible light are perceived as violet, and the longest are perceived as red. One type of cone pigment (erythrolabe) is most sensitive to red light waves, another (chlorolabe) to green light waves, and a third (cyanolabe) to blue light waves. The color a person perceives depends on which set of cones or combination of sets the light in a given image stimulates. If all three sets of cones are stimulated, the person senses the light as white, and if none are stimulated, the person senses black. Different forms of color blindness result from lack of different types of cone pigments.

Visual Nerve Pathways

The axons of the retinal neurons leave the eyes to form the *optic nerves* (fig. 10.27). Just anterior to the pituitary gland, these nerves give rise to the X-shaped *optic chiasma,* and within the chiasma, some of the

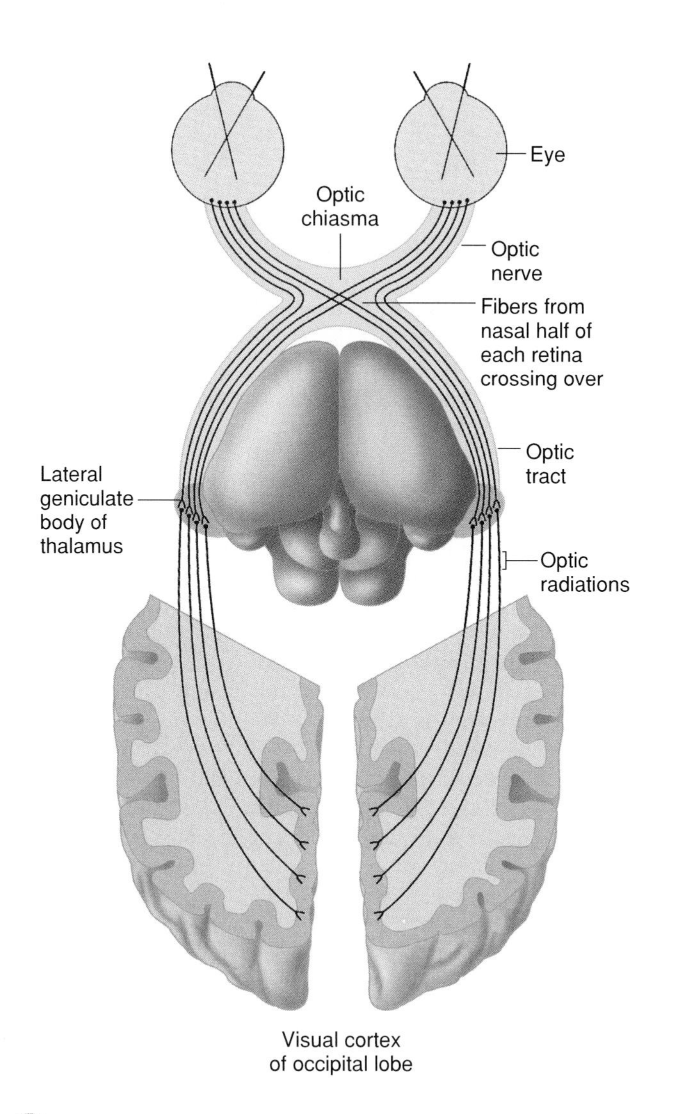

Figure 10.27

The visual pathway includes the optic nerve, optic chiasma, optic tract, and optic radiations.

fibers cross over. More specifically, the fibers from the nasal (medial) half of each retina cross over, but those from the temporal (lateral) sides do not. Thus, fibers from the nasal half of the left eye and the temporal half of the right eye form the *right optic tract,* and fibers from the nasal half of the right eye and the temporal half of the left eye form the *left optic tract.*

Just before the nerve fibers reach the thalamus, a few of them enter nuclei that function in various visual reflexes. Most of the fibers, however, enter the thalamus and synapse in its posterior portion (lateral geniculate body). From this region, the visual impulses enter nerve pathways called *optic radiations,* which lead to the visual cortex of the occipital lobes.

1. Distinguish between the rods and cones of the retina.
2. Explain the roles of visual pigments.
3. Trace a nerve impulse from the retina to the visual cortex.

Clinical Terms Related to the Senses

Amblyopia (am″ble-o′pe-ah) Dim vision due to a cause other than a refractive disorder or lesion.

Anopia (an-o′pe-ah) Absence of an eye.

Audiometry (aw″de-om′ĕ-tre) Measurement of auditory acuity for various frequencies of sound waves.

Blepharitis (blef″ah-ri′tis) Inflammation of the eyelid margins.

Causalgia (kaw-zal′je-ah) A persistent, burning pain usually associated with injury to a limb.

Conjunctivitis (kon-junk″tĭ-vi′tis) Inflammation of the conjunctiva.

Diplopia (dĭ-plo′pe-ah) Double vision.

Emmetropia (em″ĕ-tro′pe-ah) Normal condition of the eyes; eyes with no refractive defects.

Enucleation (e-nu″kle-a′shun) Removal of the eyeball.

Exophthalmos (ek″sof-thal′mos) Condition in which the eyes protrude abnormally.

Hemianopsia (hem″e-an-op′se-ah) Defective vision affecting half of the visual field.

Hyperalgesia (hi″per-al-je′ze-ah) Heightened sensitivity to pain.

Iridectomy (ir″ĭ-dek′to-me) Surgical removal of part of the iris.

Iritis (i-ri′tis) Inflammation of the iris.

Keratitis (ker″ah-ti′tis) Inflammation of the cornea.

Labyrinthectomy (lab″ĭ-rin-thek′to-me) Surgical removal of the labyrinth.

Labyrinthitis (lab″ĭ-rin-thi′tis) Inflammation of the labyrinth.

Ménière's disease (men″e-ārz′ dĭ-zēz) Inner ear disorder that causes ringing in the ears, increased sensitivity to sounds, dizziness, and hearing loss.

Neuralgia (nu-ral′je-ah) Pain resulting from inflammation of a nerve or a group of nerves.

Neuritis (nu-ri′tis) Inflammation of a nerve.

Nystagmus (nis-tag′mus) Involuntary oscillation of the eyes.

Otitis media (o-ti′tis me′de-ah) Inflammation of the middle ear.

Otosclerosis (o″to-skle-ro′sis) Formation of spongy bone in the inner ear, which often causes deafness by fixing the stapes to the oval window.

Pterygium (tĕ-rij′e-um) Abnormally thickened patch of conjunctiva that extends over part of the cornea.

Retinitis pigmentosa (ret″ĭ-ni′tis pig″men-to′sa) Inherited, progressive retinal sclerosis characterized by pigment deposits in the retina and by retinal atrophy.

Retinoblastoma (ret″ĭ-no-blas-to′mah) Inherited, highly malignant tumor arising from immature retinal cells.

Tinnitus (tĭ-ni′tus) Ringing or buzzing noise in the ears.

Trachoma (trah-ko′mah) A bacterial disease of the eye that causes conjunctivitis, which may lead to blindness.

Tympanoplasty (tim″pah-no-plas′te) Surgical reconstruction of the middle ear bones and the establishment of continuity from the eardrum to the oval window.

Uveitis (u″ve-i′tis) Inflammation of the uvea, the region of the eye that includes the iris, ciliary body, and choroid coat.

Vertigo (ver′tĭ-go) A sensation of dizziness.

Summary Outline

Introduction (p. 262)

Sensory receptors sense changes in their surroundings.

Receptors and Sensations (p. 262)

1. Types of receptors
   a. Each type of receptor is most sensitive to a distinct type of stimulus.
   b. The major types of receptors are chemoreceptors, pain receptors, thermoreceptors, mechanoreceptors, and photoreceptors.
2. Sensations
   a. Sensations are feelings resulting from sensory stimulation.
   b. A particular part of the sensory cortex interprets every impulse reaching it in the same way.
   c. The cerebral cortex projects a sensation back to the region of stimulation.
3. Sensory adaptations
   Sensory adaptations are adjustments of sensory receptors to continuous stimulation. Impulses are triggered at decreasing rates.

Somatic Senses (p. 262)

Somatic senses are associated with receptors in the skin, muscles, joints, and viscera.

1. Touch and pressure senses
   a. Free ends of sensory nerve fibers are receptors for the sensations of touch and pressure.
   b. Meissner's corpuscles are receptors for the sensation of light touch.
   c. Pacinian corpuscles are receptors for the sensation of heavy pressure.
2. Temperature senses
   Temperature receptors include two sets of free nerve endings that are heat and cold receptors.
3. Sense of pain
   a. Pain receptors are free nerve endings that tissue damage stimulates.
   b. Visceral pain
      (1) Pain receptors are the only receptors in viscera that provide sensations.
      (2) Sensations produced from visceral receptors feel as if they are coming from some other body part.
      (3) Visceral pain may be referred because sensory impulses from the skin and viscera travel on common nerve pathways.
   c. Pain nerve fibers
      (1) The two main types of pain fibers are acute pain fibers and chronic pain fibers.
      (2) Acute pain fibers are fast conducting. Chronic pain fibers are slower conducting.
      (3) Pain impulses are processed in the gray matter of the spinal cord and ascend to the brain.
      (4) Within the brain, pain impulses pass through the reticular formation before conduction to the cerebral cortex.
   d. Regulation of pain impulses
      (1) Awareness of pain occurs when pain impulses reach the thalamus.
      (2) The cerebral cortex determines pain intensity and locates its source.
      (3) Impulses descending from the brain stimulate neurons to release pain-relieving neuropeptides, such as enkephalins.

Special Senses (p. 266)

Special senses are those whose receptors are within relatively large, complex sensory organs of the head.

Sense of Smell (p. 266)

1. Olfactory receptors
   a. Olfactory receptors are chemoreceptors that chemicals dissolved in liquid stimulate.
   b. Olfactory receptors function together with taste receptors and aid in food selection.
2. Olfactory organs
   a. Olfactory organs consist of receptors and supporting cells in the nasal cavity.
   b. Olfactory receptors are neurons with cilia.
3. Olfactory nerve pathways
   Nerve impulses travel from the olfactory receptors through the olfactory nerves, olfactory bulbs, and olfactory tracts to interpreting centers in the temporal and frontal lobes of the cerebrum.
4. Olfactory stimulation
   a. Olfactory impulses may result when gaseous molecules combine with specific sites on cilia of receptor cells.
   b. Olfactory receptors adapt rapidly.

Sense of Taste (p. 268)

1. Taste receptors
   a. Taste buds consist of taste (receptor) cells and supporting cells.
   b. Taste cells have taste hairs.
   c. Taste hair surfaces have receptor sites to which chemicals combine.
2. Taste sensations
   a. The four primary taste sensations are sweet, sour, salty, and bitter.
   b. Various taste sensations result from the stimulation of one or more sets of taste receptors.
3. Taste nerve pathways
   a. Sensory impulses from taste receptors travel on fibers of the facial, glossopharyngeal, and vagus nerves.
   b. These impulses are carried to the medulla oblongata and then ascend to the thalamus, from which they travel to the gustatory cortex in the parietal lobes.

Sense of Hearing (p. 270)

1. External ear
   The external ear collects sound waves of vibrating objects.
2. Middle ear
   Auditory ossicles of the middle ear conduct sound waves from the eardrum to the oval window of the inner ear.
3. Auditory tube
   Auditory tubes connect the middle ears to the throat and help maintain equal air pressure on both sides of the eardrums.
4. Inner ear
   a. The inner ear is a complex system of connected tubes and chambers—the osseous and membranous labyrinths.
   b. The organ of Corti contains the hearing receptors that vibrations in the fluids of the inner ear stimulate.
   c. Different frequencies of vibrations stimulate different sets of receptor cells.
5. Auditory nerve pathways
   a. Auditory nerves carry impulses to the auditory cortices of the temporal lobes.
   b. Some auditory nerve fibers cross over, so that impulses arising from each ear are interpreted on both sides of the brain.

Sense of Equilibrium (p. 274)

1. Static equilibrium
   Static equilibrium maintains the stability of the head and body when they are motionless.
2. Dynamic equilibrium
   a. Dynamic equilibrium balances the head and body when they are moved or rotated suddenly.
   b. Other structures that help maintain equilibrium include the eyes and mechanoreceptors associated with certain joints.

Sense of Sight (p. 277)

1. Visual accessory organs
   Visual accessory organs include the eyelids, lacrimal apparatus, and extrinsic muscles of the eyes.
2. Structure of the eye
   a. The wall of the eye has an outer, a middle, and an inner layer (tunic).
      (1) The outer tunic (sclera) is protective, and its transparent anterior portion (cornea) refracts light entering the eye.
      (2) The middle tunic (choroid coat) is vascular and contains pigments that keep the inside of the eye dark.
      (3) The inner tunic (retina) contains the visual receptor cells.
   b. The lens is a transparent, elastic structure. Ciliary muscles control its shape.
   c. The lens must thicken to focus on close objects.
   d. The iris is a muscular diaphragm that controls the amount of light entering the eye.
   e. Spaces within the eye are filled with fluids that help maintain its shape.
3. Light refraction
   The cornea and lens refract light waves to focus an image on the retina.
4. Visual receptors
   a. Visual receptors are rods and cones.
   b. Rods are responsible for colorless vision in dim light, and cones provide color vision.
5. Visual pigments
   a. A light-sensitive pigment in rods decomposes in the presence of light and triggers a complex series of reactions that initiate nerve impulses.
   b. Color vision comes from three sets of cones containing different light-sensitive pigments.
6. Visual nerve pathways
   a. Nerve fibers from the retina form the optic nerves.
   b. Some fibers cross over in the optic chiasma.
   c. Most of the fibers enter the thalamus and synapse with others that continue to the visual cortex in the occipital lobes.

Review Exercises

1. List five groups of sensory receptors, and name the kind of change to which each is sensitive. (p. 262)
2. Define *sensation*. (p. 262)
3. Explain projection of a sensation. (p. 262)
4. Define *sensory adaptation,* and provide an example. (p. 262)
5. Describe the functions of sensory nerve fibers, Meissner's corpuscles, and Pacinian corpuscles. (p. 263)
6. Define *referred pain,* and provide an example. (p. 264)
7. Explain why pain may be referred. (p. 264)
8. Describe the olfactory organs and their functions. (p. 266)
9. Trace a nerve impulse from an olfactory receptor to the interpreting center of the cerebrum. (p. 266)
10. Explain how salivary glands aid the function of taste receptors. (p. 268)

11. Name the four primary taste sensations. (p. 268)
12. Trace the pathway of a taste impulse from a taste receptor to the cerebral cortex. (p. 268)
13. Distinguish among the external, middle, and inner ears. (p. 270)
14. Trace the path of a sound wave from the eardrum to the hearing receptors. (p. 270)
15. Describe the functions of the auditory ossicles. (p. 270)
16. Explain the function of the auditory tube. (p. 271)
17. Distinguish between the osseous and membranous labyrinths. (p. 271)
18. Describe the cochlea and its function. (p. 271)
19. Describe a hearing receptor. (p. 271)
20. Explain how a hearing receptor stimulates a sensory neuron. (p. 271)
21. Trace a nerve impulse from the organ of Corti to the interpreting centers of the cerebrum. (p. 271)
22. Describe the organs of static and dynamic equilibrium and their functions. (p. 274)
23. List the visual accessory organs, and describe the functions of each organ. (p. 277)
24. Name the three layers of the eye wall, and describe the functions of each layer. (p. 280)
25. Describe how accommodation is accomplished. (p. 281)
26. Explain how the iris functions. (p. 282)
27. Distinguish between the aqueous humor and vitreous humor. (p. 282)
28. Distinguish between the fovea centralis and optic disk. (p. 283)
29. Explain how light waves are focused on the retina. (p. 285)
30. Distinguish between rods and cones. (p. 285)
31. Explain why cone vision is generally more acute than rod vision. (p. 285)
32. Describe the function of rhodopsin. (p. 285)
33. Describe the relationship between light wavelengths and color vision. (p. 286)
34. Trace a nerve impulse from the retina to the visual cortex. (p. 286)

Critical Thinking

1. Loss of the sense of smell often precedes the major symptoms of Alzheimer disease and Parkinson disease. What additional information is needed to use this association to prevent or treat these diseases?
2. Why is dietary vitamin A good for eyesight?
3. PET (positron emission tomography) scans of the brains of people blind since birth reveal high neural activity in the visual centers of the cerebral cortex when these people read Braille. However, when sighted individuals run their fingers over the raised letters of Braille, the visual centers do not show increased activity. Explain these experimental results.
4. People who are deaf due to cochlea damage do not suffer from motion sickness. Why not?
5. We have relatively few sensory systems. How, then, do we experience such a huge and diverse number of sensory perceptions?
6. We humans love sucrose (table sugar), but armadillos, hedgehogs, lions, and seagulls do not respond to it. Opossums love lactose (milk sugar), but rats avoid it, and chickens hate the sugar xylose, while cattle love it, and we are indifferent. In what way might these diverse tastes in the animal kingdom help an organism to survive?
7. Why are astronauts unable to taste their food while eating in zero-gravity conditions?
8. Why does a fish market at first seem to have a strong odor that in time becomes less offensive?
9. Why are some serious injuries, such as a bullet entering the abdomen, relatively painless, but others, such as crushing the skin, considerably more painful?
10. Labyrinthitis is an inflammation of the inner ear. What symptoms would you expect in a patient with this disorder?
11. A patient with heart disease experiences pain at the base of the neck and in the left shoulder and upper limb during exercise. How would you explain the probable origin of this pain to the patient?

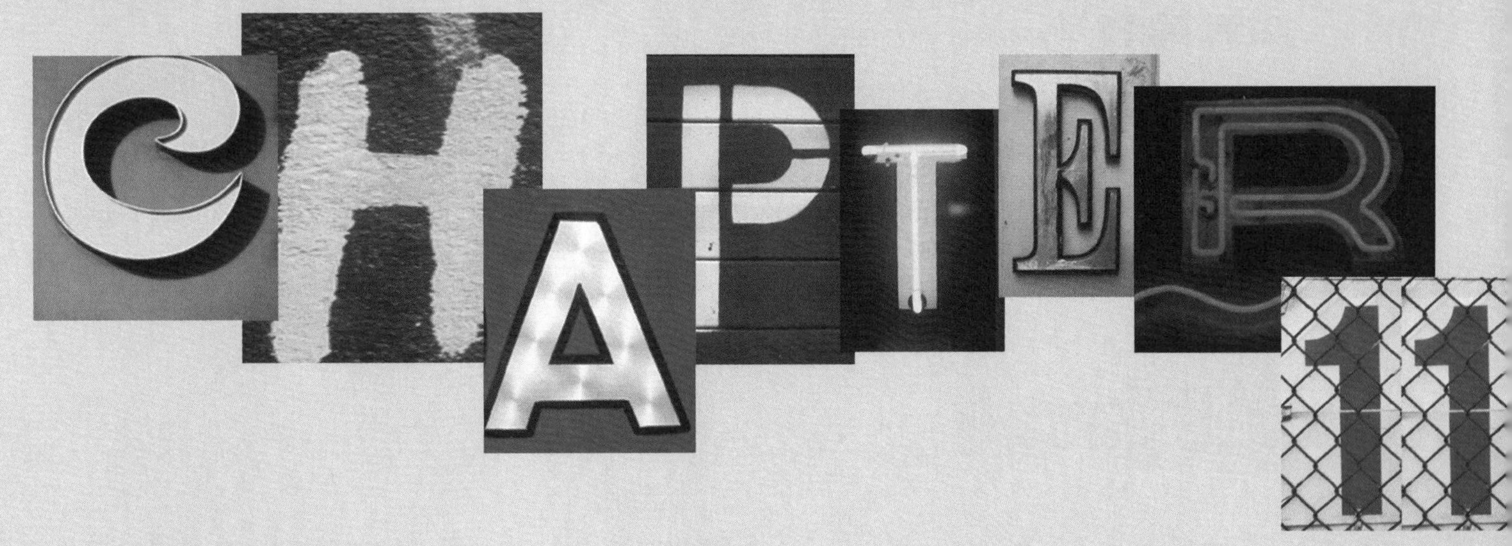

Chapter 11

Endocrine System

In the Shetland Islands, north of Scotland and just south of the Arctic Circle, daylight is fleeting in the winter, making the moonlight seem even brighter. An age-old belief among the people here is that moonlight shining on the face of a sleeping person will make the person feel ill the next day.

The superstition of the Shetland Islanders is more fact than fancy. A flash of light during a normally dark night can upset natural biochemical fluctuations in the body. A hormone called melatonin controls the body's response to light-dark cycles. It is evolution's way of dealing with existence in a world where aspects such as light and darkness of the environment cycle. The study of circadian (24-hour) and other rhythms is called chronobiology.

A very successful Massachusetts company has built a business around recognizing the effects of light on the human body. The company installs computer-controlled lighting in places where people work at night so that employees remain alert and feel well. Chronobiologists claim that it is no coincidence that the nuclear explosions at Three Mile Island and Chernobyl, the oil spill from the Exxon *Valdez,* and the majority of truck and train accidents occur in the early morning hours.

Melatonin is just one of several hormones that regulate physiology. As anyone who has ever been hungry or thirsty or who has endured the bodywide changes of puberty can attest, hormones exert powerful effects on the human body.

Photo: The endocrine system controls physiological responses to cycles of sunlight and darkness.

Regulating the functions of the human body to maintain homeostasis is an enormous job. Two organ systems function coordinately to enable body parts to communicate with each other and to adjust constantly to changing incoming signals. The nervous system is one biological communication system, utilizing nerve impulses carried on nerve fibers. The other is the endocrine system, which uses biochemicals called hormones that travel in body fluids and act as chemical messengers to their target cells.

Chapter Objectives

After studying this chapter, you should be able to:

1. Distinguish between endocrine and exocrine glands. (p. 293)
2. Explain how steroid and nonsteroid hormones affect target cells. (p. 294)
3. Discuss how negative feedback mechanisms regulate hormonal secretions. (p. 296)
4. Explain how the nervous system controls secretion. (p. 296)
5. Name and describe the location of the major endocrine glands, and list the hormones they secrete. (p. 298)
6. Describe the general functions of the hormones that endocrine glands secrete. (p. 298)
7. Explain how the secretion of each hormone is regulated. (p. 298)
8. Define stress, and describe how the body responds to it. (p. 312)

Aids to Understanding Words

-crin [to secrete] endo*crine*: Pertaining to internal secretions.

diuret- [to pass urine] *diuret*ic: A substance that promotes urine production.

endo- [within] *endo*crine gland: A gland that releases its secretion internally into a body fluid.

exo- [outside] *exo*crine gland: A gland that releases its secretion to the outside through a duct.

hyper- [above] *hyper*thyroidism: A condition resulting from an above-normal secretion of thyroid hormone.

hypo- [below] *hypo*thyroidism: A condition resulting from a below-normal secretion of thyroid hormone.

para- [beside] *para*thyroid glands: A set of glands located on the surface of the thyroid gland.

toc- [birth] oxy*toc*in: A hormone that stimulates the uterine muscles to contract during childbirth.

-tropic [influencing] *tropic* hormone: A hormone that stimulates secretion of another hormone.

Key Terms

adrenal cortex (ah-dre′nal kor′teks)

adrenal medulla (ah-dre′nal me-dul′ah)

anterior pituitary (an-ter′e-or pĭ-tu′ĭ-tar″e)

cAMP (si′ klik ay em pee)

hormone (hor′mōn)

negative feedback system (neg′ah-tiv fēd′bak sis′tem)

pancreas (pan′kre-as)

parathyroid gland (par″ah-thi′roid gland)

pineal gland (pin′e-al gland)

posterior pituitary (pos-ter′e-or pĭ-tu′ĭ-tar″e)

prostaglandin (pros″tah-glan′din)

target cell (tar′get sel)

thymus gland (thi′mus gland)

thyroid gland (thi′roid gland)

Introduction

The **endocrine system** includes cells, tissues, and organs that secrete hormones directly into body fluids. In contrast, exocrine secretions reach some internal or external body surface through ducts. The thyroid and parathyroid glands secrete hormones into the blood and are therefore endocrine glands, while sweat glands and salivary glands are exocrine (see chapter 5).

General Characteristics of the Endocrine System

Endocrine glands and their hormones help regulate metabolic processes. They control the rates of certain chemical reactions, aid in transport of substances across membranes, and help regulate water and electrolyte balances. They also play vital roles in reproduction, development, and growth.

Endocrine glands secrete the hormones described in this chapter. These glands include the pituitary gland, thyroid gland, parathyroid glands, adrenal glands, and pancreas (fig. 11.1). Subsequent chapters discuss several other hormone-secreting glands and tissues, such as those that participate in digestion and reproduction. Certain small groups of specialized cells also secrete hormones.

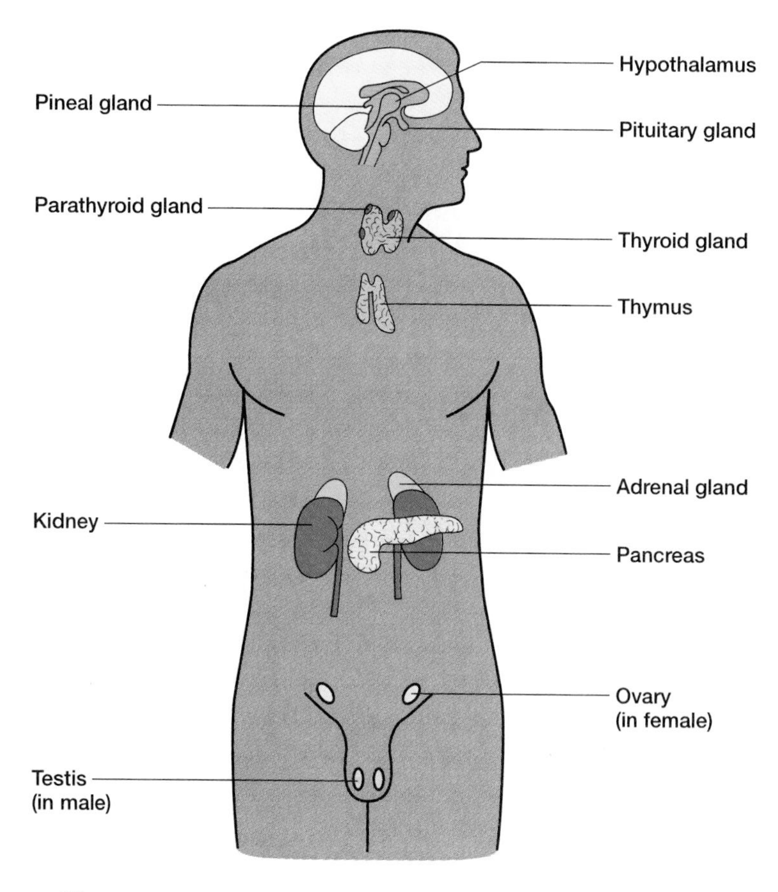

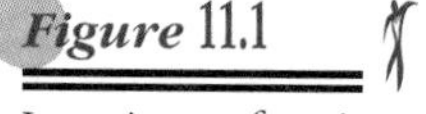

Figure 11.1

Locations of major endocrine glands.

Hormone Action

A **hormone** is a biochemical that a cell secretes to affect the functions of another cell. Secretory cells release hormones into surrounding extracellular spaces. Some hormones travel only short distances and affect nearby cells (paracrine secretion). Blood transports other hormones to all parts of the body, where they may produce general effects. In either case, a particular hormone's physiological action is restricted to the hormone's **target cells**—those cells with specific receptors for the hormone molecules. A hormone's target cells have receptors that other cells lack. These receptors are proteins or glycoproteins with binding sites for a specific hormone.

Even in extremely low concentrations, hormones can stimulate changes in target cells. Chemically, most hormones are either steroids or steroidlike substances synthesized from cholesterol, or they are amines, peptides, proteins, or glycoproteins synthesized from amino acids (table 11.1).

Table 11.1 Types of Hormones

Type of Compound	Formed from	Examples
Amines	Amino acids	Norepinephrine, epinephrine
Peptides	Amino acids	Antidiuretic hormone, oxytocin, thyrotropin-releasing hormone
Proteins	Amino acids	Parathyroid hormone, growth hormone, prolactin
Glycoproteins	Protein and carbohydrate	Follicle-stimulating hormone, luteinizing hormone, thyroid-stimulating hormone
Steroids	Cholesterol	Estrogen, testosterone, aldosterone, cortisol

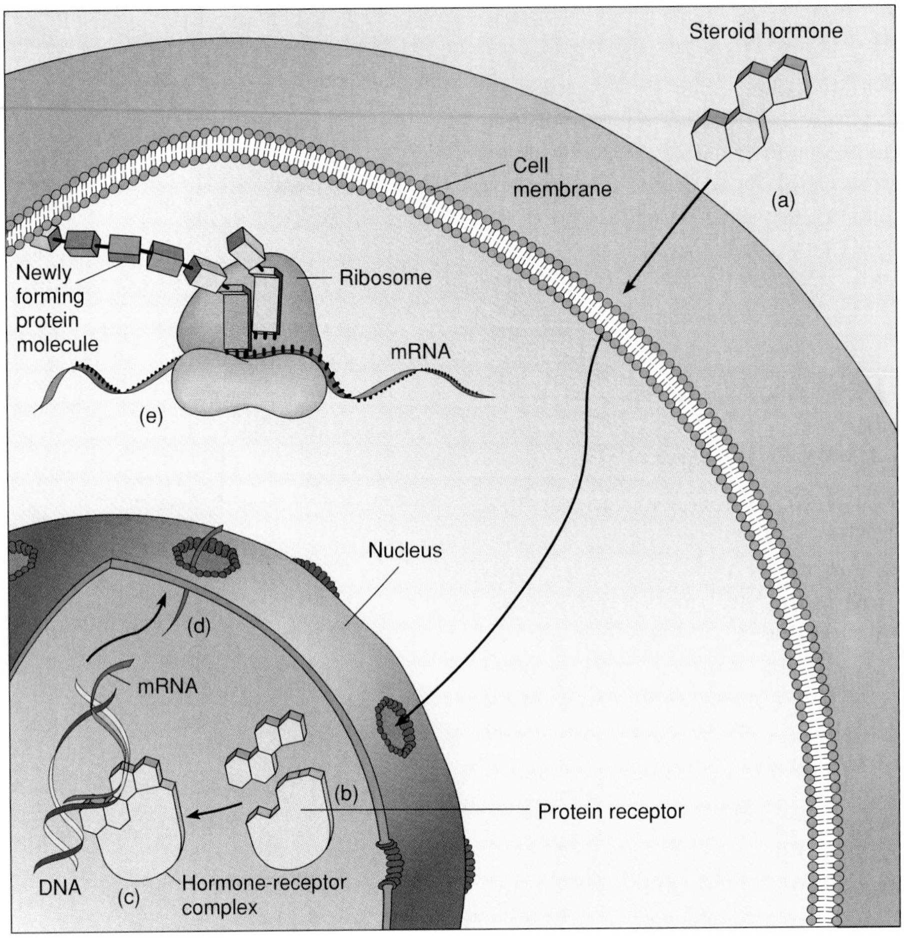

***Figure* 11.2**

(*a*) A steroid hormone crosses a cell membrane and (*b*) combines with a protein receptor, usually in the nucleus. (*c*) The hormone-receptor complex activates messenger RNA (mRNA) synthesis. (*d*) The mRNA leaves the nucleus and (*e*) guides protein synthesis.

1. Distinguish between an endocrine and an exocrine gland.
2. Describe the general function of the endocrine system.
3. What is a hormone?

Steroid Hormones

Steroids are compounds whose molecules contain complex rings of carbon and hydrogen atoms. Steroids differ according to the kinds and numbers of atoms attached to these rings and the ways they are joined.

Steroid hormones, unlike amines, peptides, and proteins, are soluble in the lipids that make up the bulk of cell membranes. For this reason, steroid molecules can diffuse into cells relatively easily. Once inside a target cell, steroid hormones may combine with specific protein molecules—the receptors. This hormone-receptor complex binds within the nucleus to particular regions of the target cell's DNA and activates specific genes. The activated genes, in turn, are transcribed into messenger RNAs (mRNA), which enter the cytoplasm and instruct the cell to synthesize specific proteins. These proteins—which may be enzymes affecting metabolism—exert the hormone's characteristic effects (fig. 11.2).

Nonsteroid Hormones

Nonsteroid hormones, such as amines, peptides, and proteins, usually combine with receptors in target cell membranes. Each of these receptor molecules is a protein with a *binding site* and an *activity site*. A hormone molecule delivers its message to its target cell by uniting with the binding site of its receptor. This combination stimulates the receptor's activity site to interact with other membrane proteins. Receptor binding may alter the function of enzymes or membrane transport

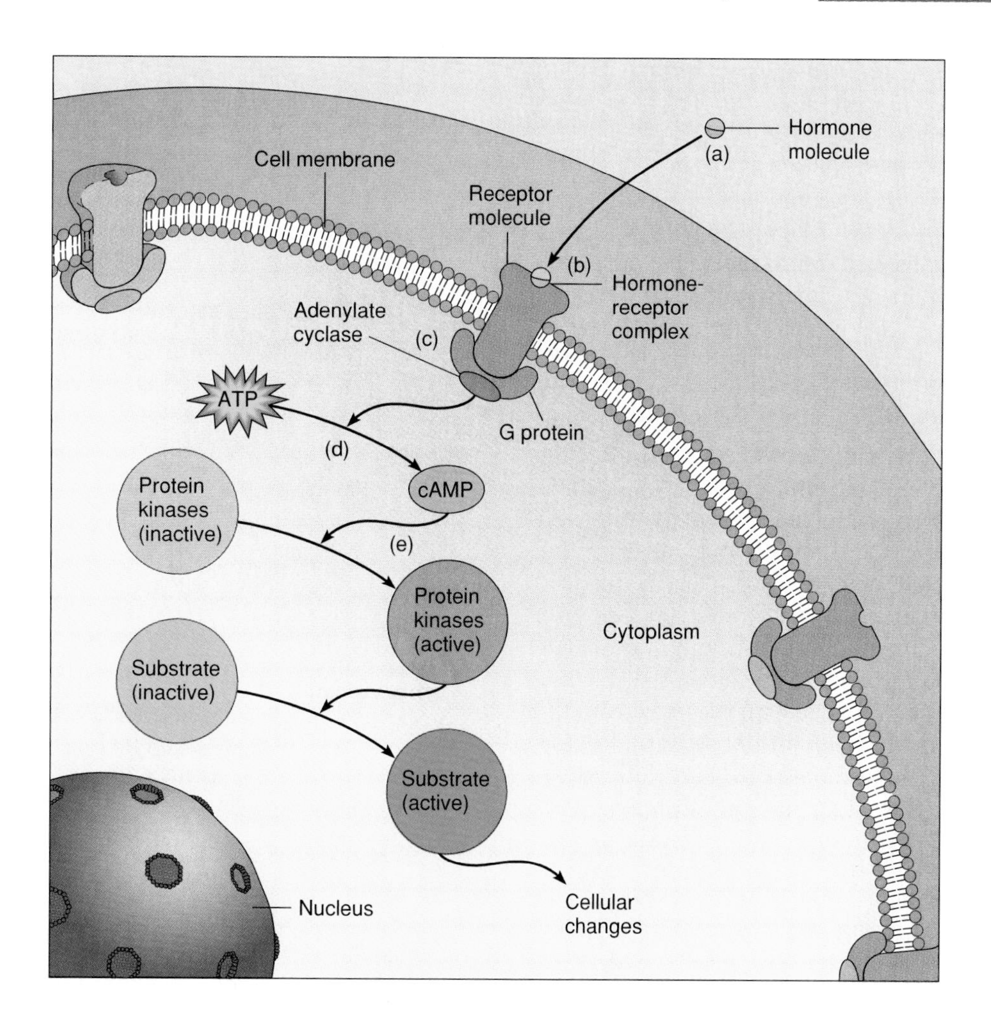

Figure 11.3

(*a*) Body fluids carry nonsteroid hormone molecules to the target cell, where they combine (*b*) with receptor sites on the cell membrane. (*c*) This activates molecules of adenylate cyclase, which (*d*) catalyze conversion of ATP into cyclic adenosine monophosphate (cAMP). (*e*) The cAMP promotes a series of reactions leading to the cellular changes associated with the hormone's action.

mechanisms, changing the concentrations of still other cellular components. The hormone that triggers this cascade of biochemical activity is called a *first messenger.* The biochemicals in the cell that induce changes in response to the hormone's binding are called *second messengers.*

The second messenger associated with one group of hormones is *cyclic adenosine monophosphate* or cyclic AMP (cAMP). In this mechanism, a hormone binds to its receptor, and the resulting hormone-receptor complex activates a protein called a *G protein*. The G protein, in turn, activates an enzyme called *adenylate cyclase,* which is bound to the inside of the cell membrane. Activated adenylate cyclase causes ATP molecules within the cytoplasm to form a circle, making cAMP. The cAMP, in turn, activates another set of enzymes called *protein kinases,* which transfer phosphate groups from ATP molecules to various protein substrate molecules. This action (phosphorylation) alters the shapes of the substrate molecules and activates some of them. The activated protein molecules then alter various cellular processes (fig. 11.3). The type of membrane receptors present and also the kinds of protein substrate molecules that a cell contains determine the cell's response to a hormone. Cellular responses to second messenger activation include altering membrane permeabilities, activating enzymes, promoting synthesis of certain proteins, stimulating or inhibiting specific metabolic pathways, moving the cell, and initiating secretion of hormones or other substances.

Another enzyme (phosphodiesterase) quickly inactivates cAMP, so that its action is short-lived. For this reason, a continuing response within a target cell depends on a continuing signal produced by hormone molecules combining with the target cell's membrane receptors.

A number of other second messengers work in much the same way. These include diacylglycerol (DAG) and inositol triphosphate (IP_3).

Prostaglandins

A group of biochemicals called **prostaglandins** also regulate cells. Prostaglandins are lipids synthesized from a fatty acid (arachidonic acid) in cell membranes. A great variety of cells produce prostaglandins, including those of the liver, kidneys, heart, lungs, thymus gland, pancreas, brain, and reproductive organs. Prostaglandins usually act more locally than hormones, often affecting only the organ where they are produced.

Prostaglandins are potent and are present in very small quantities. They are not stored in cells but rather synthesized just before release. They are rapidly inactivated.

Prostaglandins produce diverse and even opposite effects. Some, for example, relax smooth muscles in the airways of the lungs and in blood vessels, while others contract smooth muscles in the walls of the uterus and intestines. Prostaglandins stimulate hormone secretion from the adrenal cortex and inhibit secretion of hydrochloric acid from the stomach wall. They also influence the movements of sodium ions and water molecules in the kidneys, help regulate blood pressure, and have powerful effects on male and female reproductive physiology.

1. How does a steroid hormone promote cellular changes? A nonsteroid hormone?
2. What is a second messenger?
3. What are prostaglandins?
4. What kinds of effects do prostaglandins produce?

Control of Hormonal Secretions

Since hormones are very potent and maintain homeostasis, regulation of their release by endocrine cells must be precise.

Negative Feedback Systems

Negative feedback systems control many hormonal secretions (see chapter 1). In such a system, an endocrine gland or the system controlling it is sensitive to the concentration of a substance the gland secretes or to the product of a process it controls. Whenever this concentration reaches a certain level, the endocrine gland is inhibited (a negative effect), and its secretory activity decreases (fig. 11.4). Then, as the concentration of the gland's hormone decreases, the concentration of the regulated product decreases, too, and the inhibition of the gland ceases. When the gland is no longer inhibited, it begins to secrete its hormone again. Such negative feedback systems stabilize the concentrations of some hormones (fig. 11.5).

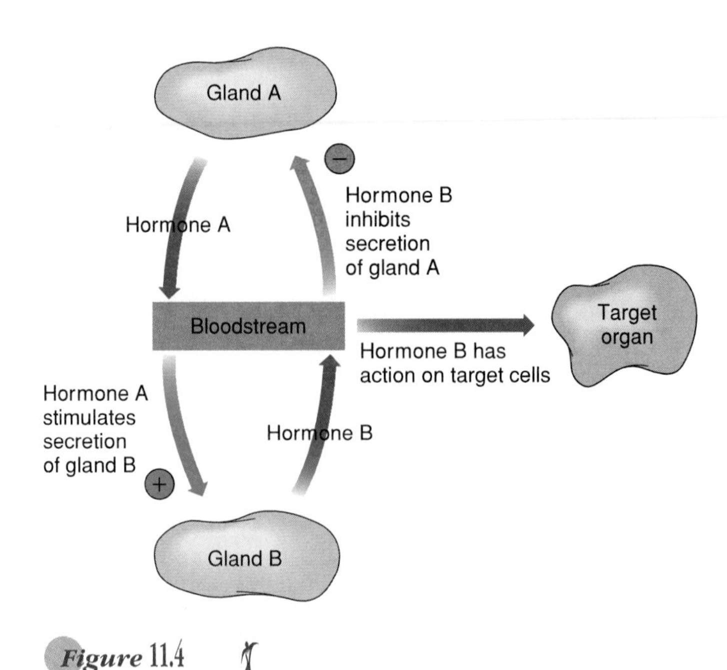

Figure 11.4

An example of a negative feedback system. Gland A secretes a hormone that stimulates gland B to increase secretion of another hormone. The hormone from gland B alters its target cells and inhibits activity of gland A. (Note: ⊕ = stimulation; ⊖ = inhibition.)

Control Sources

Hormone control occurs in three ways:

1. The hypothalamus controls the anterior pituitary gland's release of hormones that stimulate other endocrine glands to release hormones. Its location near the thalamus and the third ventricle allows the hypothalamus to constantly receive information about the internal environment from neural connections and cerebrospinal fluid (fig. 11.6).
2. The nervous system stimulates some glands directly. The adrenal medulla, for example, secretes its hormones in response to sympathetic nerve impulses.

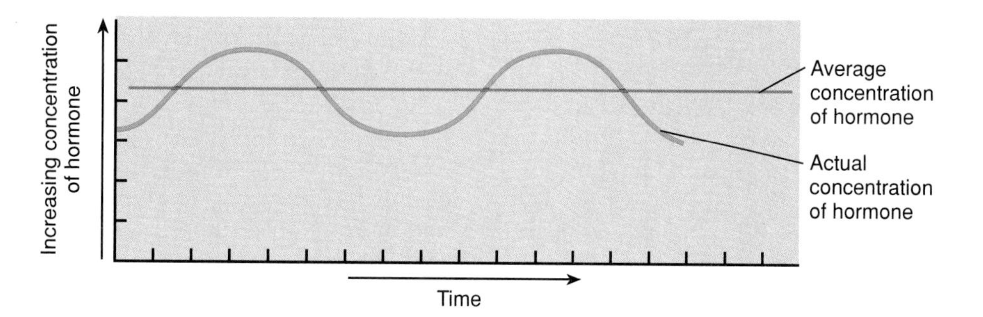

Figure 11.5

As a result of negative feedback, some hormone concentrations remain relatively stable, although they may fluctuate slightly above and below average concentrations.

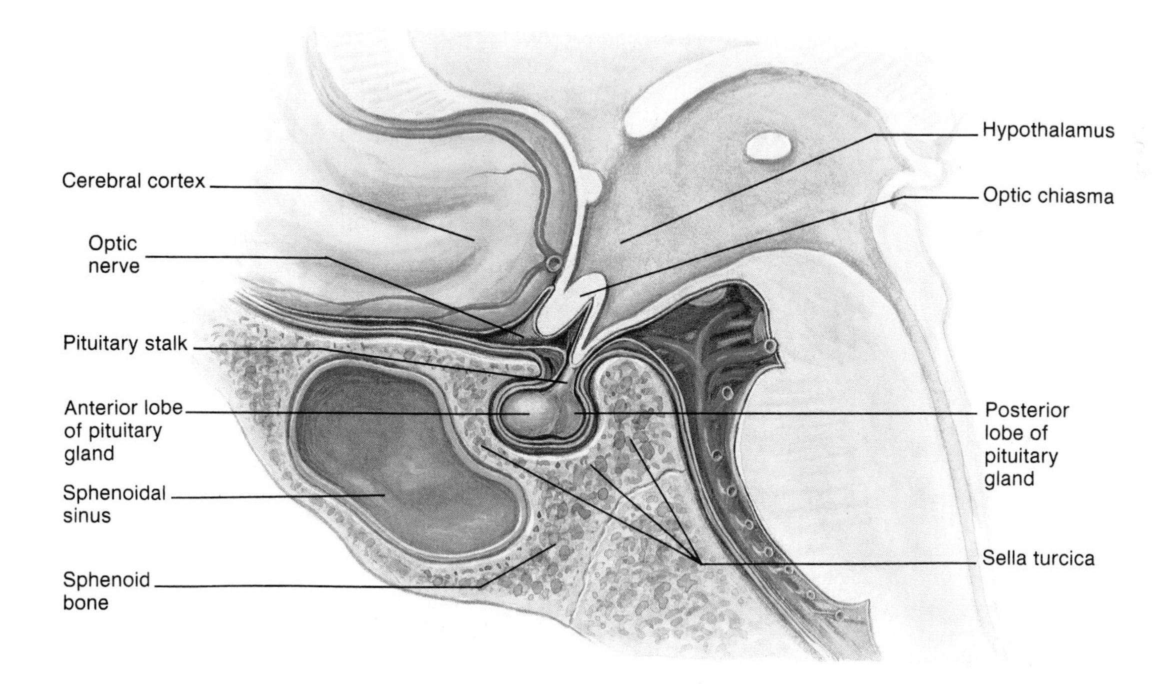

Figure 11.6

The pituitary gland is attached to the hypothalamus and lies in the sella turcica of the sphenoid bone.

3. Another group of glands responds directly to changes in the composition of the internal environment. For example, when the blood glucose level rises, the pancreas secretes insulin, and when the blood glucose level falls, it secretes glucagon, as discussed later in the chapter.

1. Describe a negative feedback system.
2. Explain three mechanisms that help control hormonal secretions.

Pituitary Gland

The **pituitary gland** (hypophysis) is located at the base of the brain, where a pituitary stalk (infundibulum) attaches it to the hypothalamus. The gland is about 1 centimeter in diameter and consists of an **anterior pituitary** (anterior lobe) and a **posterior pituitary** (posterior lobe) (fig. 11.7).

During fetal development, a narrow region develops between the anterior and posterior lobes of the pituitary gland. Called the *intermediate lobe* (pars intermedia), it produces melanocyte-stimulating hormone (MSH), which regulates the synthesis of melanin—the pigment in skin and in portions of the eyes and brain. The intermediate lobe atrophies during prenatal development and appears only as a vestige in adults. ■

The brain controls most of the pituitary gland's activities. For example, the posterior pituitary releases hormones when nerve impulses from the hypothalamus signal the axon ends of neurosecretory cells in the posterior pituitary (fig. 11.7*b*). On the other hand, releasing hormones from the hypothalamus control secretion from the anterior pituitary (fig. 11.7*a*). These releasing hormones travel in a capillary network associated with the hypothalamus. The capillaries merge to form the **hypophyseal portal veins**, which pass downward along the pituitary stalk and give rise to a capillary net in the anterior pituitary. Thus, the hypothalamus releases substances that the blood carries directly to the anterior pituitary.

Upon reaching the anterior pituitary, each of the hypothalamic releasing hormones acts on a specific population of cells there. Some of the resulting actions are inhibitory, but most stimulate the anterior pituitary to release hormones that stimulate secretions from peripheral endocrine glands. In many of these cases, important negative feedback relationships regulate hormone levels in the bloodstream.

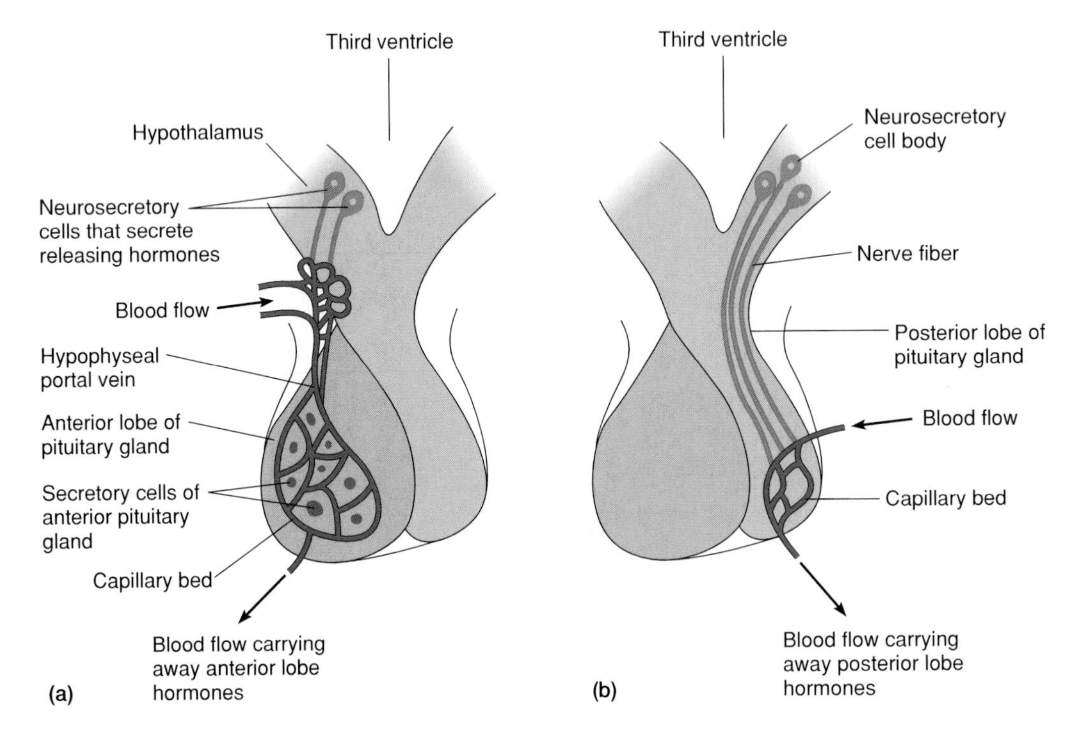

Figure 11.7

(*a*) Hypothalamic releasing hormones stimulate cells of the anterior lobe to secrete hormones. (*b*) Nerve impulses originating in the hypothalamus stimulate nerve endings in the posterior lobe of the pituitary gland to release hormones.

1. Where is the pituitary gland located?
2. Explain how the hypothalamus controls the actions of the posterior and anterior lobes of the pituitary gland.

Anterior Pituitary Hormones

The anterior pituitary is enclosed in a capsule of dense collagenous connective tissue and consists largely of epithelial tissue organized in blocks around many thin-walled blood vessels. So far, researchers have identified five types of secretory cells within the epithelium. Four of these each secrete a different hormone—growth hormone (GH), prolactin (PRL), thyroid-stimulating hormone (TSH), and adrenocorticotropic hormone (ACTH). The fifth type of cell secretes follicle-stimulating hormone (FSH) and luteinizing hormone (LH). (Note: In males, luteinizing hormone is known as interstitial cell stimulating hormone or ICSH.)

Growth hormone (GH) stimulates cells to increase in size and more rapidly divide. It also enhances the movement of amino acids across cell membranes and speeds the rate that cells utilize carbohydrates and fats. The hormone's effect on amino acids, however, seems to be more important.

Control of GH secretion involves two secretions from the hypothalamus (GH releasing hormone and GH release-inhibiting hormone). Nutritional state also influences control of GH. For example, more GH is released during periods of protein deficiency and abnormally low blood glucose concentration. Conversely, when blood protein and glucose concentrations increase, GH secretion decreases.

Prolactin (PRL) stimulates and sustains a woman's milk production following the birth of an infant (see chapter 20). The effect of PRL in males is less well understood, although excess PRL secretion may cause a deficiency of male sex hormones.

Thyroid-stimulating hormone (TSH) controls thyroid gland secretions, described later in this chapter. The hypothalamus partially regulates TSH secretion by producing *thyrotropin-releasing hormone* (TRH) (fig. 11.8). Circulating thyroid hormones inhibit release of TRH and TSH. As blood concentration of thyroid hormones increases, secretions of TRH and TSH decrease.

1. How does growth hormone affect protein synthesis?
2. What is the function of prolactin?
3. How is secretion of thyroid-stimulating hormone regulated?

Insufficient secretion of growth hormone (GH) during childhood limits growth, causing a type of *dwarfism* (hypopituitary dwarfism). In this condition, body parts are usually correctly proportioned, and mental development is normal.

However, abnormally low GH secretion is usually accompanied by deficient secretion of other anterior pituitary hormones, leading to additional hormone deficiency symptoms. For example, a person with hypopituitary dwarfism often does not develop adult sexual features. Hormone therapy can help.

Oversecretion of GH during childhood causes gigantism—height may exceed 8 feet. Gigantism, which is relatively rare, usually is a result of a pituitary gland tumor, which may also cause oversecretion of other pituitary hormones. As a result, a person with gigantism often has several metabolic disturbances. ■

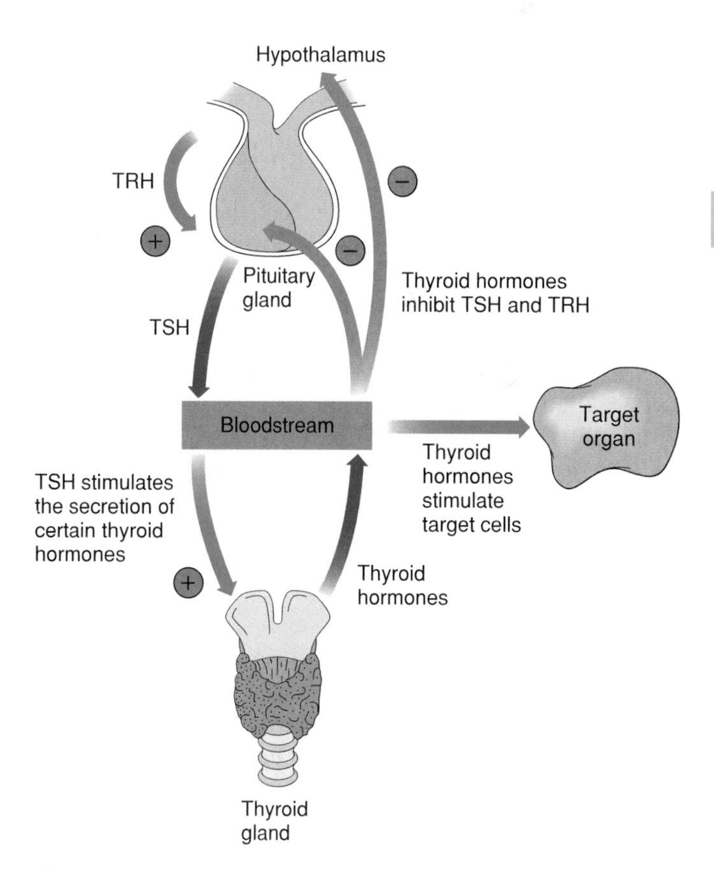

Figure 11.8

Thyrotropin-releasing hormone (TRH) from the hypothalamus stimulates the anterior pituitary gland to release thyroid-stimulating hormone (TSH), which stimulates the thyroid gland to release hormones. These thyroid hormones reduce the secretion of TSH and TRH. (Note: ⊕ = stimulation; ⊖ = inhibition.)

Adrenocorticotropic hormone (ACTH) controls the manufacture and secretion of certain hormones from the outer layer, or *cortex,* of the adrenal gland. These adrenal cortical hormones are discussed later in the chapter.

ACTH secretion is regulated in part by *corticotropin-releasing hormone* (CRH), which the hypothalamus releases in response to decreased concentrations of adrenal cortical hormones. Also, stress may increase ACTH secretion by stimulating the release of CRH.

Follicle-stimulating hormone (FSH) and **luteinizing hormone** (LH) are *gonadotropins,* which means they exert their actions on the gonads or reproductive organs. Chapter 19 discusses the functions of these gonadotropins and the ways they interact with each other.

1. What is the function of adrenocorticotropic hormone?
2. What is a gonadotropin?

Posterior Pituitary Hormones

Unlike the anterior pituitary, which is composed primarily of glandular epithelial cells, the posterior pituitary consists largely of nerve fibers and neuroglial cells. The neuroglial cells support the nerve fibers, which originate in the hypothalamus.

Specialized neurons in the hypothalamus produce the two hormones associated with the posterior pituitary—**antidiuretic hormone** (ADH) and **oxytocin** (OT) (see fig. 11.7). These hormones travel down axons through the pituitary stalk to the posterior lobe, and vesicles (secretory granules) near the ends of the axons store them. Nerve impulses from the hypothalamus release the hormones into the blood.

A *diuretic* is a chemical that increases urine production, whereas an *antidiuretic* decreases urine formation. ADH produces an antidiuretic effect by reducing the amount of water the kidneys excrete. In this way, ADH regulates the water concentration of body fluids.

The hypothalamus regulates ADH secretion. Certain neurons in this part of the brain, called *osmoreceptors,* sense changes in the osmotic pressure of body fluids. Dehydration due to lack of water intake increasingly concentrates blood solutes. Osmoreceptors, sensing the resulting increase in osmotic pressure, signal the posterior pituitary to release ADH, which travels in the blood to the kidneys. As a result, the kidneys produce less urine, conserving water. On the other hand, drinking too much water dilutes body fluids, inhibiting ADH release. The kidneys excrete more dilute urine until the water concentration of body fluids returns to normal.

If an injury or tumor damages any parts of the ADH-regulating mechanism, too little ADH may be synthesized or released, producing *diabetes insipidus.* An affected individual may produce as much as 25–30 liters of very dilute urine per day, and solute concentrations in body fluids rise. ■

OT contracts smooth muscles in the uterine wall and stimulates uterine contractions in the later stages of childbirth. Stretching of uterine and vaginal tissues late in pregnancy triggers OT release during childbirth. In the breast, OT contracts certain cells associated with the milk-producing glands and their ducts. In lactating breasts, this action forces liquid from the milk glands into the milk ducts and ejects the milk from the breasts for breast-feeding. In addition, OT is an antidiuretic, but it is much weaker than ADH.

Table 11.2 reviews the hormones of the pituitary gland.

If the uterus is not contracting sufficiently to expel a fully developed fetus, commercial preparations of oxytocin are sometimes used to stimulate uterine contractions, inducing labor. Such preparations are also often administered to the mother following childbirth to ensure that uterine muscles contract enough to squeeze broken blood vessels closed, minimizing the risk of hemorrhage. ■

1. What is the function of antidiuretic hormone?
2. How is secretion of antidiuretic hormone controlled?
3. What effects does oxytocin produce in females?

Table 11.2

Hormones of the Pituitary Gland

Hormone	Action	Source of Control
Anterior lobe		
Growth hormone (GH)	Stimulates an increase in the size and division rate of body cells; enhances movement of amino acids across membranes	Growth hormone-releasing hormone and growth hormone release-inhibiting hormone from hypothalamus
Prolactin (PRL)	Sustains milk production after birth	Secretion restrained by prolactin release inhibiting factor and stimulated by prolactin-releasing factor from hypothalamus
Thyroid-stimulating hormone (TSH)	Controls secretion of hormones from thyroid gland	Thyrotropin-releasing hormone (TRH) from hypothalamus
Adrenocorticotropic hormone (ACTH)	Controls secretion of certain hormones from adrenal cortex	Corticotropin-releasing hormone (CRH) from hypothalamus
Follicle-stimulating hormone (FSH)	In females, responsible for the development of egg-containing follicles in ovaries and stimulates follicular cells to secrete estrogen; in males, stimulates production of sperm cells	Gonadotropin-releasing hormone from hypothalamus
Luteinizing hormone (LH)	Promotes secretion of sex hormones; plays a role in releasing an egg cell in females	Gonadotropin-releasing hormone from hypothalamus
Posterior lobe		
Antidiuretic hormone (ADH)	Causes kidneys to conserve water; in high concentration, increases blood pressure	Hypothalamus in response to changes in blood water concentration
Oxytocin (OT)	Contracts muscles in the uterine wall; contracts muscles associated with milk-secreting glands	Hypothalamus in response to stretch in the uterine and vaginal walls and stimulation of breasts

Thyroid Gland

The **thyroid gland** is a very vascular structure that consists of two large lobes connected by a broad isthmus (fig. 11.9 and reference plate 4). It is just below the larynx on either side and in front of the trachea.

Structure of the Gland

A capsule of connective tissue covers the thyroid gland, which is made up of many secretory parts called *follicles.* The cavities within these follicles are lined with a single layer of cuboidal epithelial cells and filled with a clear, viscous substance called *colloid.* The follicular cells produce and secrete hormones that may either be stored in the colloid or released into the blood in nearby capillaries.

Thyroid Hormones

The follicular cells of the thyroid gland synthesize two hormones—**thyroxine** (tetraiodothyronine), also known as T_4 because it contains four atoms of iodine, and **triiodothyronine**, known as T_3 because it includes three atoms of iodine. Thyroxine and triiodothyronine have similar actions, although triiodothyronine is five times more potent. These hormones help regulate the metabolism of carbohydrates, lipids, and proteins: They increase the rate at which cells release energy from carbohydrates; they increase the rate of protein synthesis; and they stimulate breakdown and mobilization of lipids. Thyroid hormones are required for normal growth and development, and are essential to nervous system maturation.

Follicular cells require iodine salts (iodides) to produce thyroxine and triiodothyronine. Foods normally provide iodides, and after the iodides have been absorbed from the intestine, blood transports them to the thyroid gland. An efficient active transport mechanism moves the iodides into the follicular cells, where they are used to synthesize the hormones. The hypothalamus and pituitary gland control release of thyroid hormones. Once in the blood, thyroxine and triiodothyronine combine with proteins in the blood (plasma proteins) and are transported to body cells.

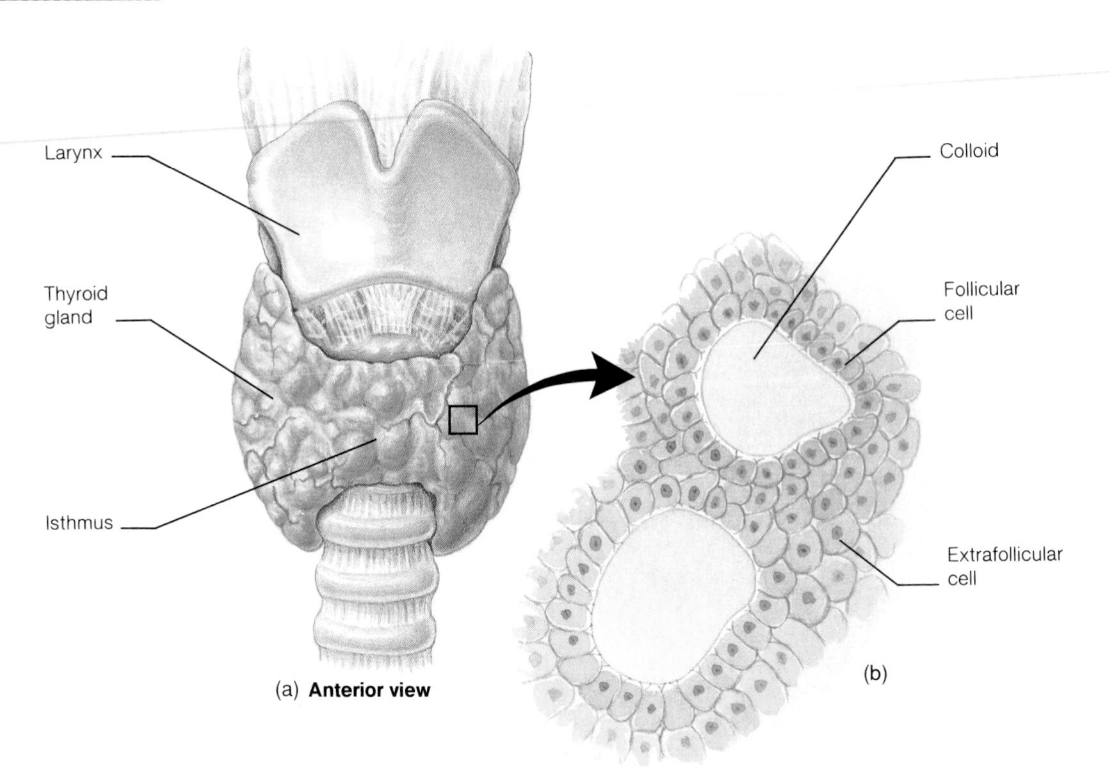

Figure 11.9

(*a*) The thyroid gland consists of two lobes connected anteriorly by an isthmus.
(*b*) Follicular cells secrete thyroid hormones.

Table 11.3

Hormones of the Thyroid Gland

Hormone	Action	Source of Control
Thyroxine (T_4)	Increases rate of energy release from carbohydrates; increases rate of protein synthesis; accelerates growth; stimulates activity in nervous system	Thyroid-stimulating hormone from the anterior pituitary gland
Triiodothyronine (T_3)	Same as above, but five times more potent than thyroxine	Thyroid-stimulating hormone from the anterior pituitary gland
Calcitonin	Lowers blood calcium and phosphate ion concentrations by inhibiting release of calcium and phosphate ions from bones and by increasing excretion of these ions by kidneys	Blood calcium concentration

A third hormone, **calcitonin**, is often not considered a thyroid hormone because the gland's extrafollicular (other than follicle) cells produce it. Along with parathyroid hormone (PTH) from the parathyroid glands, calcitonin regulates the concentrations of blood calcium and phosphate ions.

Blood concentration of calcium ions regulates calcitonin release. As this concentration increases, so does calcitonin secretion. Calcitonin inhibits the bone-resorbing activity of osteoclasts (see chapter 7) and increases the kidneys' excretion of calcium and phosphate ions—actions that lower the blood calcium and phosphate ion concentrations.

Table 11.3 reviews the actions and controls of the thyroid hormones.

Many thyroid disorders produce overactivity (*hyperthyroidism*) or underactivity (*hypothyroidism*) of the glandular cells. One form of hypothyroidism appears in infants whose thyroid glands do not function normally. An affected child may appear normal at birth because the mother provided an adequate supply of thyroid hormones for the child in utero. When the infant's own thyroid gland does not produce sufficient quantities of these hormones, a condition called cretinism develops. Symptoms include stunted growth, abnormal bone formation, retarded mental development, low body temperature, and sluggishness. Without treatment within a month or so following birth, the child may suffer permanent mental retardation.

Hyperthyroidism produces an elevated metabolic rate, restlessness, and overeating. The eyes protrude (exophthalmos) because of swelling in the tissues behind them, and the thyroid gland enlarges, producing a bulge in the neck called a *goiter*. ■

1. Where is the thyroid gland located?
2. What hormones of the thyroid gland affect carbohydrate metabolism and protein synthesis?
3. How does the thyroid gland influence the concentrations of blood calcium and phosphate ions?

Parathyroid Glands

The **parathyroid glands** are on the posterior surface of the thyroid gland, as figure 11.10 shows. Usually, there are four parathyroid glands—a superior and an inferior gland associated with each of the thyroid's lateral lobes.

Structure of the Glands

A thin capsule of connective tissue covers each small, yellowish brown parathyroid gland. The body of the gland consists of many tightly packed secretory cells closely associated with capillary networks.

Parathyroid Hormone

The parathyroid glands secrete **parathyroid hormone** (PTH), which increases blood calcium concentration and decreases blood phosphate ion concentration. PTH affects the bones, kidneys, and intestine.

The intercellular matrix of bone tissue is rich in mineral salts, including calcium phosphate (see chapter 7). PTH inhibits the activity of osteoblasts and stimulates osteocytes and osteoclasts to resorb bone and release calcium and phosphate ions into the blood. At the same time, PTH causes the kidneys to conserve blood calcium and to excrete more phosphate ions in the urine. It also stimulates calcium absorption from food in the intestine, further increasing blood calcium concentration.

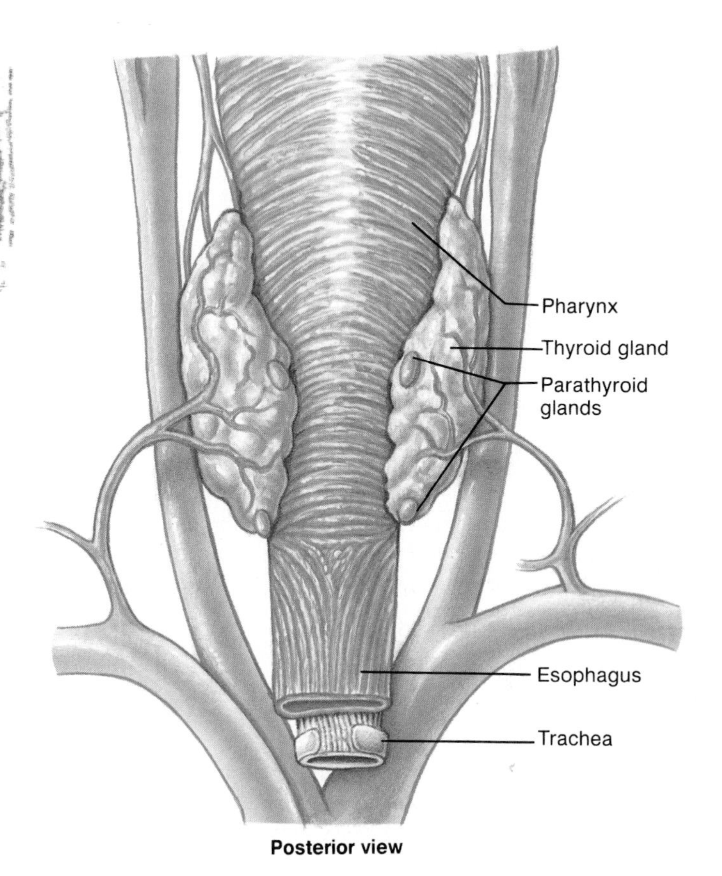

Figure 11.10

The parathyroid glands are embedded in the posterior surface of the thyroid gland.

Negative feedback between the parathyroid glands and the blood calcium concentration regulates PTH secretion. As blood calcium concentration drops, more PTH is secreted; as blood calcium concentration rises, less PTH is released (fig. 11.11).

To summarize, calcitonin and PTH activities maintain stable blood calcium concentration. Calcitonin decreases an above-normal blood calcium concentration, while PTH increases a below-normal blood calcium concentration.

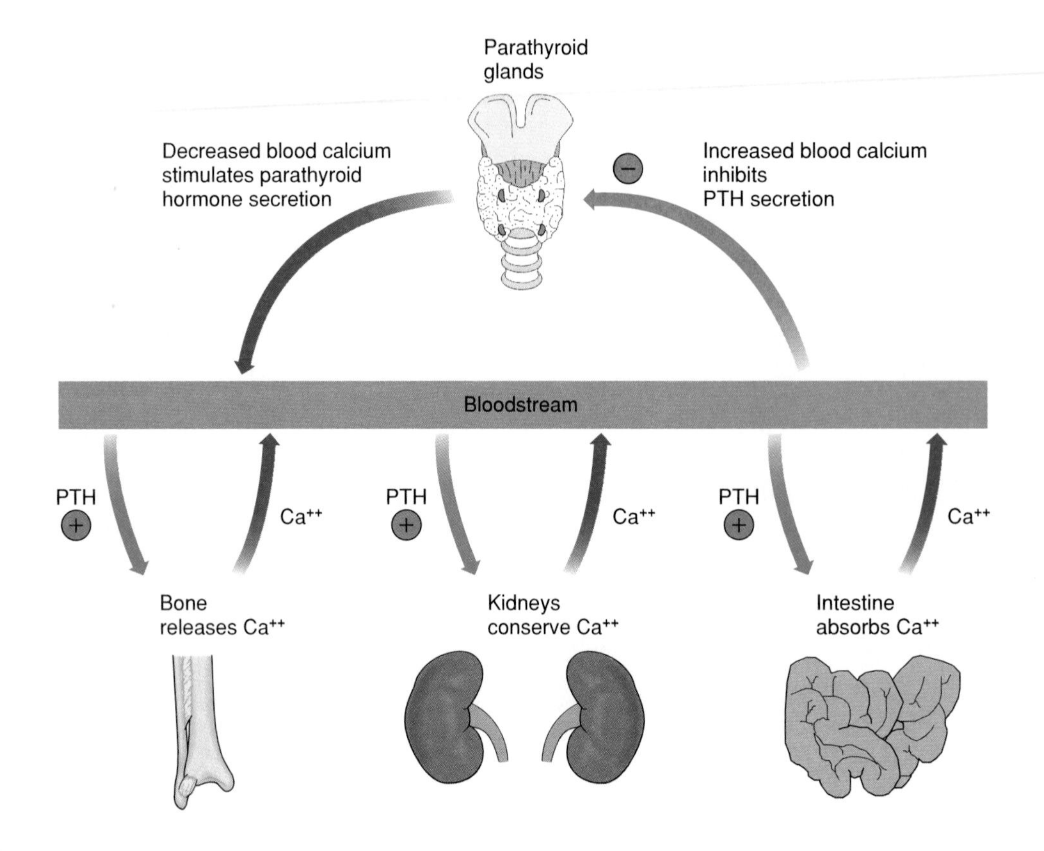

Figure 11.11

Parathyroid hormone (PTH) stimulates bone to release calcium (Ca^{++}) and the kidneys to conserve calcium. It indirectly stimulates the intestine to absorb calcium. The resulting increase in blood calcium concentration inhibits secretion of PTH. (Note: ⊕ = stimulation; ⊖ = inhibition.)

A tumor in a parathyroid gland may cause *hyperparathyroidism,* which increases PTH secretion. This stimulates osteoclast activity, and as bone tissue is resorbed, the bones soften, deform, and more easily fracture spontaneously. In addition, excess calcium and phosphate released into body fluids may deposit in abnormal places, causing new problems, such as kidney stones.

Injury to the parathyroids or their surgical removal can cause *hypoparathyroidism,* in which decreased PTH secretion reduces osteoclast activity. Although bones remain strong, blood calcium concentration decreases. The nervous system may become abnormally excitable, triggering spontaneous impulses. As a result, muscles may undergo tetanic contractions, possibly leading to respiratory failure and death. ■

1. Where are the parathyroid glands?
2. How does parathyroid hormone help regulate concentrations of blood calcium and phosphate ions?

Adrenal Glands

The **adrenal glands** are closely associated with the kidneys (fig. 11.12 and reference plate 6). A gland sits atop each kidney like a cap and is embedded in the mass of adipose tissue that encloses the kidney.

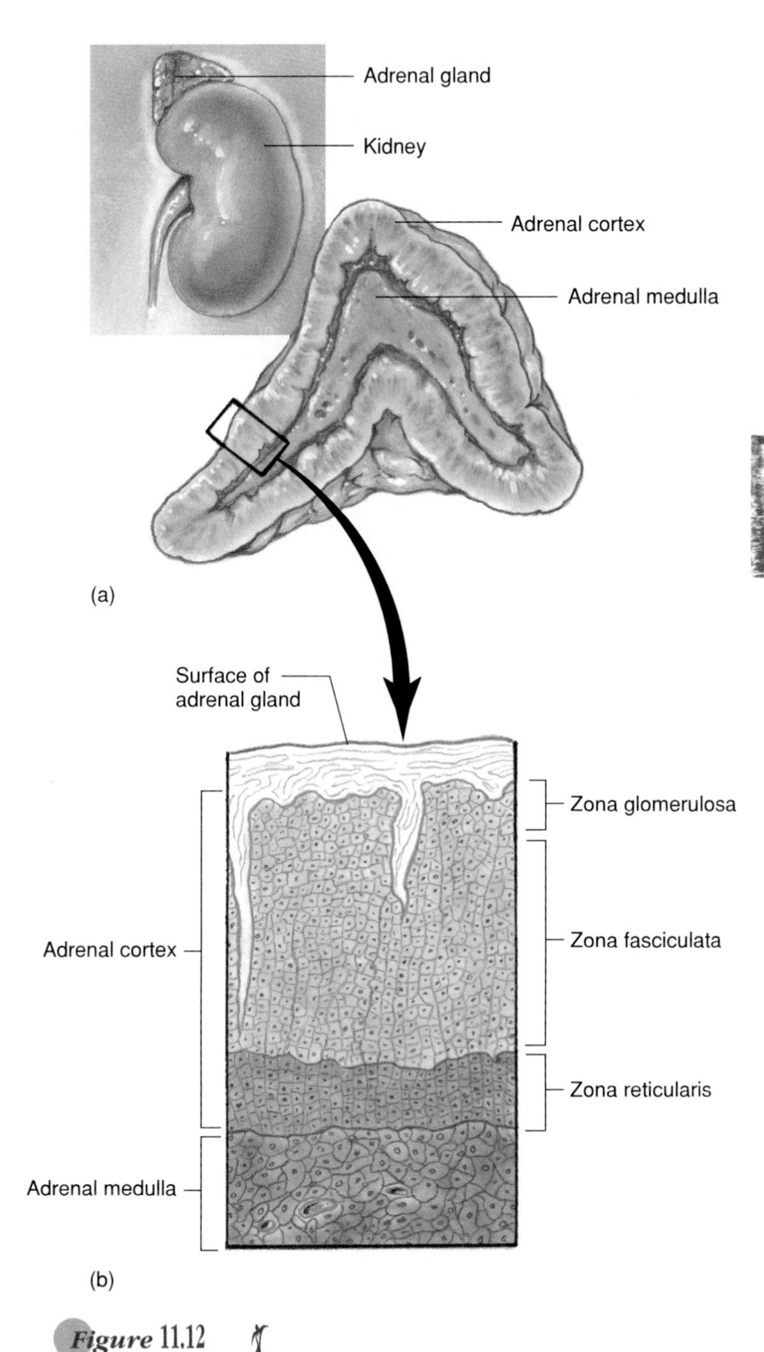

Figure 11.12

(*a*) An adrenal gland consists of an outer cortex and an inner medulla. (*b*) The cortex consists of the three layers, or zones, of cells shown.

Structure of the Glands

Each adrenal gland is very vascular and consists of two parts: The central portion is the **adrenal medulla**, and the outer part is the **adrenal cortex.** These regions are not sharply divided, but they are functionally distinct glands that secrete different hormones.

The adrenal medulla consists of irregularly shaped cells organized in groups around blood vessels. These cells are intimately connected with the sympathetic division of the autonomic nervous system. Adrenal medullary cells are actually modified postganglionic neurons. Preganglionic autonomic nerve fibers lead to them from the central nervous system (see chapter 9).

The adrenal cortex, which makes up the bulk of the adrenal gland, is composed of closely packed masses of epithelial cells, organized in layers. These layers form an outer, middle, and inner zone of the cortex (fig. 11.12*b*). As in the adrenal medulla, the cells of the adrenal cortex are well supplied with blood vessels.

1. Where are the adrenal glands?
2. Describe the two portions of an adrenal gland.

Hormones of the Adrenal Medulla

The cells of the adrenal medulla secrete two closely related hormones—**epinephrine** (adrenaline) and **norepinephrine** (noradrenaline). These hormones have similar molecular structures and physiological functions. In fact, epinephrine, which makes up 80% of the adrenal medullary secretion, is synthesized from norepinephrine.

Effects of the adrenal medullary hormones resemble those of sympathetic nerve fibers stimulating their effectors. Hormonal effects, however, last up to ten times longer because hormones are broken down less rapidly than neurotransmitters. The effects of epinephrine and norepinephrine include: increased heart rate; increased force of cardiac muscle contraction; increased breathing rate; elevated blood pressure; and decreased digestive activity.

Impulses arriving on sympathetic nerve fibers stimulate the adrenal medulla to release its hormones at the same time sympathetic impulses are stimulating other effectors. These sympathetic impulses originate in the hypothalamus in response to stress. Thus, adrenal medullary secretions function with the sympathetic division of the autonomic nervous system in preparing the body for energy-expending action—"fight or flight."

Table 11.4 compares some of the effects of the adrenal medullary hormones.

Tumors in the adrenal medulla can increase hormonal secretion. Release of norepinephrine usually predominates, prolonging sympathetic responses—high blood pressure, increased heart rate, elevated blood sugar, and so forth. Surgical removal of the tumor corrects the condition. ■

Table 11.4

Comparative Effects of Epinephrine and Norepinephrine

Part or Function Affected	Epinephrine	Norepinephrine
Heart	Rate increases; force of contraction increases	Rate increases; force of contraction increases
Blood vessels	Vessels in skeletal muscle dilate, decreasing resistance to blood flow	Blood flow to skeletal muscles increases, resulting from constriction of blood vessels in skin and viscera
Systemic blood pressure	Some increase due to increased cardiac output	Great increase due to vasoconstriction
Airways	Dilation	Some dilation
Reticular formation of brain	Activated	Little effect
Liver	Promotes breakdown of glycogen to glucose, increasing blood sugar	Little effect on blood sugar
Metabolic rate	Increases	Increases

1. Name the hormones the adrenal medulla secretes.
2. What effects do hormones from the adrenal medulla produce?
3. What stimulates release of hormones from the adrenal medulla?

Hormones of the Adrenal Cortex

The cells of the adrenal cortex produce more than thirty different steroids, including several hormones. Unlike the adrenal medullary hormones, without which a person can survive, some adrenal cortical hormones are vital. Without adrenal cortical secretions, a person usually dies within a week unless extensive electrolyte therapy is provided. The most important adrenal cortical hormones are aldosterone, cortisol, and certain sex hormones.

Aldosterone

Cells in the outer zone of the adrenal cortex synthesize **aldosterone.** Aldosterone is a *mineralocorticoid* because it helps regulate the concentration of mineral electrolytes. More specifically, aldosterone causes the kidney to conserve sodium ions and excrete potassium ions. By conserving sodium ions, aldosterone stimulates water retention indirectly by osmosis. This helps maintain blood volume and blood pressure.

A decrease in the blood concentration of sodium ions or an increase in the blood concentration of potassium ions stimulates the cells that secrete aldosterone. The kidneys also indirectly stimulate aldosterone secretion if blood pressure falls (see chapter 17).

Cortisol

Cortisol (hydrocortisone) is a *glucocorticoid,* which means it affects glucose metabolism. It is produced in the middle zone of the adrenal cortex and, like aldosterone, is a steroid. Cortisol also influences protein and fat metabolism.

Among the more important actions of cortisol are:

1. It inhibits protein synthesis in tissues, increasing blood concentration of amino acids.
2. It promotes fatty acid release from adipose tissue, increasing utilization of fatty acids as an energy source and decreasing use of glucose.
3. It stimulates liver cells to synthesize glucose from noncarbohydrates, such as circulating amino acids and glycerol, increasing blood glucose concentration.

These actions of cortisol help keep blood glucose concentration within the normal range between meals because a few hours without food can exhaust the supply of liver glycogen, a major source of glucose.

Negative feedback controls cortisol release. This is much like control of thyroid hormones, involving the hypothalamus, anterior pituitary gland, and adrenal cortex. The hypothalamus secretes corticotropin-releasing hormone (CRH) into the hypophyseal portal veins, which carry CRH to the anterior pituitary, stimulating it to secrete ACTH. In turn, ACTH stimulates the adrenal

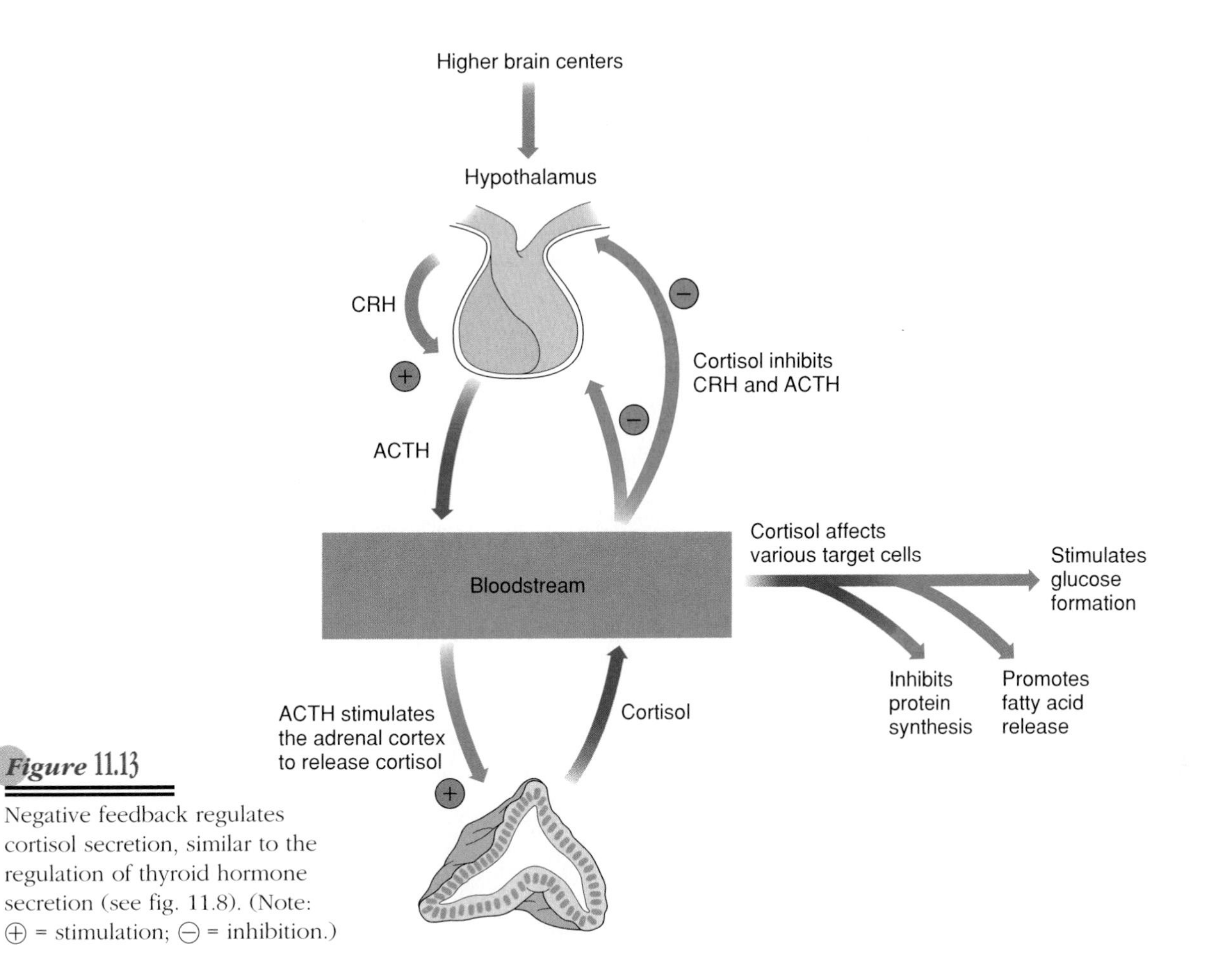

Figure 11.13

Negative feedback regulates cortisol secretion, similar to the regulation of thyroid hormone secretion (see fig. 11.8). (Note: ⊕ = stimulation; ⊖ = inhibition.)

Hyposecretion of adrenal cortical hormones leads to *Addison's disease,* a condition characterized by decreased blood sodium, increased blood potassium, low blood glucose concentration (hypoglycemia), dehydration, low blood pressure, and increased skin pigmentation. Without treatment with mineralocorticoids and glucocorticoids, Addison's disease can be lethal in days because of severe disturbances in electrolyte balance.

Hypersecretion of adrenal cortical hormones, which may be associated with an adrenal tumor or with the anterior pituitary oversecreting ACTH, causes *Cushing's syndrome.* This condition alters carbohydrate and protein metabolism and electrolyte balance. For example, when mineralocorticoids and glucocorticoids are overproduced, blood glucose concentration remains high, depleting tissue protein. Also, too much sodium is retained, increasing tissue fluids, and the skin becomes puffy. At the same time, increase in adrenal sex hormone production may cause masculinizing effects in a female, such as beard growth and deepening of the voice. ■

cortex to release cortisol. Cortisol inhibits release of CRH and ACTH, and as concentrations of these fall, cortisol production drops (fig. 11.13).

The set point of the feedback mechanism controlling cortisol secretion changes from time to time, altering hormone output to meet the demands of changing conditions. For example, under stress—injury, disease, extreme temperature, or emotional upset—nerve impulses send the brain information concerning the stressful condition. In response, brain centers signal the hypothalamus to release more CRH, leading to a higher cortisol concentration until the stress subsides (fig. 11.13).

Adrenal Sex Hormones

Cells in the inner zone of the adrenal cortex produce sex hormones. These hormones are male types (adrenal androgens), but some are converted to female hormones (estrogens) in the skin, liver, and adipose tissue. Adrenal sex hormones may supplement the supply of sex hormones from the gonads and stimulate early development of reproductive organs.

Table 11.5 summarizes the characteristics of the adrenal cortical hormones.

Table 11.5 Hormones of the Adrenal Cortex

Hormone	Action	Factor Regulating Secretion
Aldosterone	Helps regulate concentration of extracellular electrolytes by conserving sodium ions and excreting potassium ions	Electrolyte concentrations in body fluids
Cortisol	Decreases protein synthesis, increases fatty acid release, and stimulates glucose synthesis from noncarbohydrates	Corticotropin-releasing hormone from hypothalamus and adrenocorticotropic hormone from anterior pituitary
Adrenal androgens	Supplement sex hormones from the gonads; may be converted to estrogens in females	

1. Name the most important hormones of the adrenal cortex.
2. What is the function of aldosterone?
3. What actions does cortisol produce?
4. How are the blood concentrations of aldosterone and cortisol regulated?

Pancreas

The **pancreas** consists of two major types of secretory tissues. This reflects its dual function as an exocrine gland that secretes digestive juice and as an endocrine gland that releases hormones (fig. 11.14 and reference plate 6).

Structure of the Gland

The pancreas is an elongated, somewhat flattened organ posterior to the stomach and behind the parietal peritoneum. A duct joins the pancreas to the duodenum (the first section of the small intestine) and transports pancreatic digestive juice to the intestine.

The endocrine portion of the pancreas consists of groups of cells that are closely associated with blood vessels. These groups, called *islets of Langerhans,* include two distinct types of cells—alpha cells, which secrete the hormone glucagon, and beta cells, which secrete the hormone insulin (fig. 11.15).

Chapter 15 discusses the digestive functions of the pancreas.

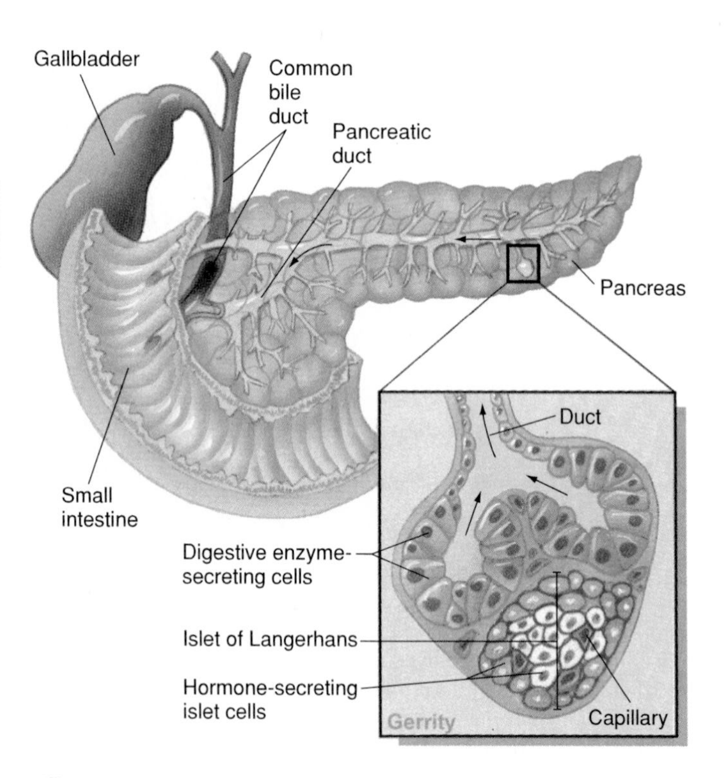

Figure 11.14

The hormone-secreting cells of the pancreas are grouped in clusters, or islets, that are closely associated with blood vessels. Other pancreatic cells secrete digestive enzymes into ducts.

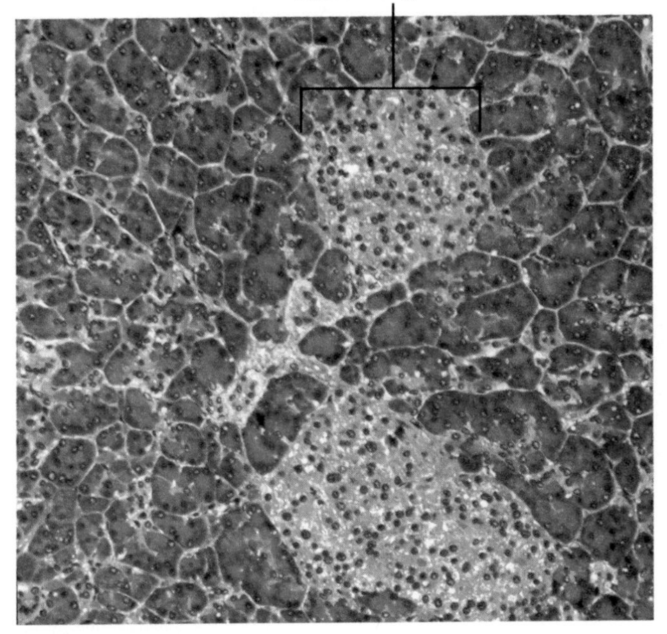

Figure 11.15

Light micrograph of an islet of Langerhans within the pancreas (50×).

Hormones of the Islets of Langerhans

Glucagon stimulates the liver to break down glycogen and certain noncarbohydrates, such as amino acids, into glucose, raising blood glucose concentration. Glucagon much more effectively elevates blood sugar than does epinephrine.

A negative feedback system regulates glucagon secretion. Low blood sugar concentration stimulates alpha cells to release glucagon. When blood sugar concentration rises, glucagon secretion falls. This control prevents hypoglycemia when glucose concentration is relatively low, such as between meals, or when glucose is used rapidly, such as during periods of exercise.

The main effect of **insulin** is exactly opposite that of glucagon. Insulin stimulates the liver to form glycogen from glucose and inhibits conversion of noncarbohydrates into glucose. Insulin also has the special effect of promoting facilitated diffusion (see chapter 3) of glucose across cell membranes with insulin receptors. These cells include those of cardiac muscle, adipose tissue, and resting skeletal muscle (glucose uptake by exercising skeletal muscle does not depend on insulin). These actions of insulin decrease blood glucose concentration. In addition, insulin secretion promotes transport of amino acids into cells, increases protein synthesis, and stimulates adipose cells to synthesize and store fat.

A negative feedback system sensitive to blood glucose concentration regulates insulin secretion. When blood glucose concentration is relatively high, as after a meal, beta cells release insulin. Insulin helps prevent too high a blood glucose concentration by promoting glycogen formation in the liver and entrance of glucose into adipose and muscle cells. When glucose concentration falls, between meals or during the night, insulin secretion decreases.

As insulin output decreases, less and less glucose enters adipose and muscle cells. Cells that lack insulin receptors, such as nerve cells, can then use glucose remaining in the blood. At the same time that insulin is decreasing, glucagon secretion is increasing. Therefore, insulin and glucagon function coordinately to maintain a relatively stable blood glucose concentration, despite great variation in the amount of carbohydrates a person eats (fig. 11.16).

Nerve cells, including those of the brain, obtain glucose by a facilitated-diffusion mechanism that is not stimulated by insulin. For this reason, nerve cells are particularly sensitive to changes in blood glucose concentration, and conditions that cause such changes—oversecretion of insulin, for example—are likely to alter brain functions.

Cancer cells that develop from nonendocrine tissues sometimes inappropriately synthesize and secrete great amounts of peptide hormones or peptide hormonelike substances. Consequently, a cancer patient may develop an endocrine disorder that seems unrelated to the cancer (endocrine paraneoplastic syndrome). Most commonly, such disorders overproduce ADH, ACTH, a PTH-like substance, or an insulin-like substance. ■

1. What is the endocrine portion of the pancreas called?
2. What is the function of glucagon?
3. What is the function of insulin?
4. How are glucagon and insulin secretions controlled?
5. Why are nerve cells particularly sensitive to changes in blood glucose concentration?

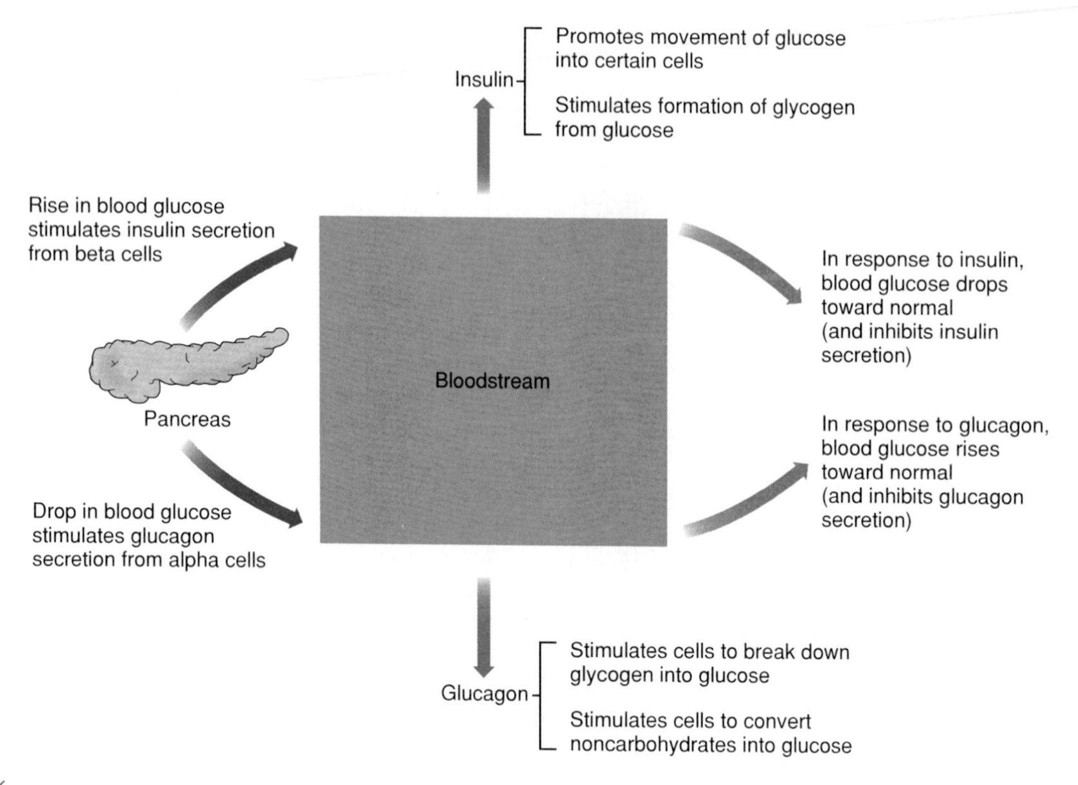

Figure 11.16

Insulin and glucagon function together to help maintain a relatively stable blood glucose concentration. Negative feedback responding to blood glucose concentration controls the levels of both hormones.

Other Endocrine Glands

Other glands that produce hormones and thus are parts of the endocrine system include the pineal gland, thymus gland, reproductive glands, and certain glands of the digestive tract, heart, and kidneys.

Pineal Gland

The **pineal gland** is a small structure located deep between the cerebral hemispheres, where it attaches to the upper portion of the thalamus near the roof of the third ventricle. Its body consists largely of specialized *pineal cells* and supportive *neuroglial cells* (see fig. 11.1).

The pineal gland secretes the hormone **melatonin** in response to light conditions outside the body. Nerve impulses originating in the retinas of the eyes send this information to the pineal gland. In the dark, nerve impulses from the eyes decrease, and melatonin secretion increases.

Melatonin helps regulate **circadian rhythms**, which are patterns of repeated activity associated with the environmental cycles of day and night. Examples include the sleep-wake rhythm and seasonal cycles of fertility in many mammals. The study of circadian and other rhythms is called chronobiology.

Although the mechanism of melatonin action is poorly understood, the hormone inhibits the secretion of gonadotropins from the anterior pituitary and helps regulate the female reproductive cycle (menstrual cycle). It may also control the onset of puberty.

Thymus Gland

The **thymus gland**, which lies in the mediastinum posterior to the sternum and between the lungs, is relatively large in young children but shrinks with age (see fig. 11.1). This gland secretes a group of hormones called **thymosins** that affect the production and differentiation of certain white blood cells (lymphocytes). In this way, the thymus plays an important role in immunity, discussed in chapter 14.

Reproductive Glands

The reproductive organs that secrete important hormones include the **ovaries**, which produce estrogens and progesterone; the **placenta**, which produces estrogens, progesterone, and gonadotropin; and the **testes**, which produce testosterone. These glands and their secretions are discussed in chapter 19.

TOPIC OF INTEREST Diabetes Mellitus

Diabetes mellitus is a condition resulting from insulin deficiency. It disturbs carbohydrate, protein, and fat metabolism. More specifically, since insulin helps glucose cross some cell membranes, movement of glucose into adipose and resting skeletal muscle cells decreases in diabetes. At the same time, glycogen formation decreases. As a result, blood sugar concentration rises (hyperglycemia). At a certain high concentration, the kidneys begin to excrete the excess. Glucose appearing in urine (glycosuria) raises the urine's osmotic pressure, and more water and electrolytes than usual are excreted. Excess urine output causes the affected person to become dehydrated and extremely thirsty (polydipsia).

Diabetes mellitus also hampers protein and fat synthesis. Glucose-starved cells increase their use of proteins as an energy source. Tissues waste away, and the person loses weight, is very hungry, fatigues easily, and has a decreasing ability to grow and repair tissues. Changes in fat metabolism cause fatty acids and ketone bodies to accumulate in the blood, which lowers pH (acidosis). The dehydration and acidosis may harm brain cells, and the person may become disoriented, slip into a coma, and die.

The two common forms of diabetes mellitus are *insulin-dependent diabetes mellitus* (also called type I or juvenile-onset diabetes mellitus) and *noninsulin-dependent diabetes mellitus* (type II or maturity-onset diabetes mellitus). Insulin-dependent diabetes mellitus (IDDM) usually appears before age twenty and is an autoimmune disease. This means that the immune system destroys the beta cells of the pancreas (see chapter 14). Treatment is injections of insulin, usually several times daily, or implanting an insulin pump to deliver the hormone. Figure 11A shows one of the first recipients of insulin treatment—a three-year-old boy who weighed only 15 pounds. Just two months of treatment doubled his weight. Pure insulin is available today from bacteria given human insulin genes. Prior to 1982, it came from pigs. In the near future, implants of insulin-producing cells may treat people with diabetes.

Noninsulin-dependent diabetes mellitus (NIDDM) accounts for 70–80% of people with diabetes. It usually develops gradually after age forty and produces milder symptoms than IDDM. Most affected individuals are overweight when they first experience symptoms. In NIDDM, the beta cells of the pancreas function, but body cells lose sensitivity to insulin. Treatment includes controlling the diet, exercising, and maintaining a desirable body weight. Drugs are available, too.

People with diabetes must monitor their blood glucose levels several times daily. Failure to control blood glucose concentration increases the chance of developing complications of diabetes mellitus, including coronary artery disease and retinal and nerve damage.

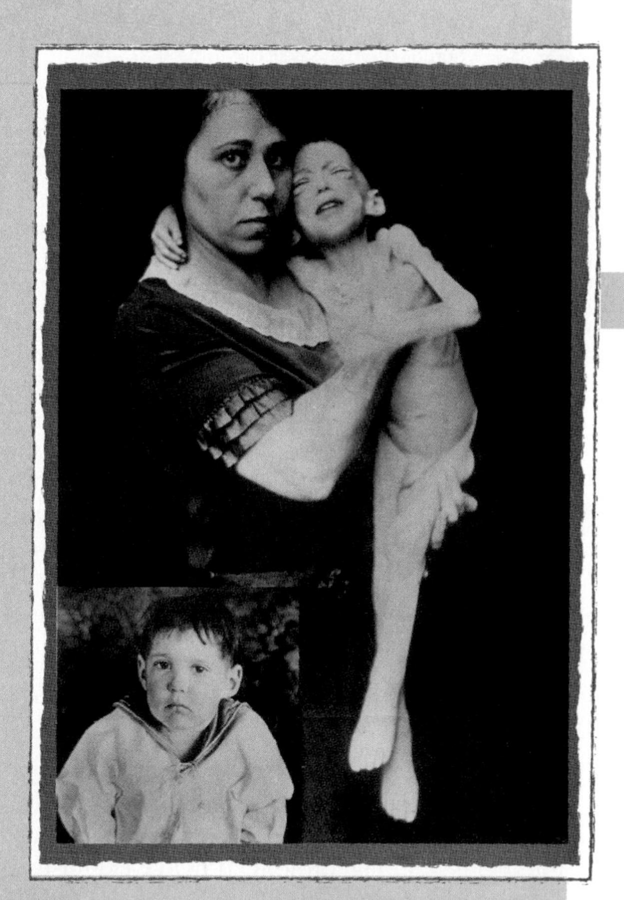

Figure 11A

Before and after insulin treatment. The boy in his mother's arms is three years old but weighs only 15 pounds because of diabetes mellitus. The inset shows the same child after just two months of receiving insulin.

Digestive Glands

The digestive glands that secrete hormones are associated with the linings of the stomach and small intestine. Chapter 15 describes these structures and their secretions.

Other Hormone-Producing Organs

Other organs that produce hormones include the heart, which secretes *atrial natriuretic peptide,* a hormone that stimulates urinary sodium excretion (see chapter 17), and the kidneys, which secrete a red blood cell growth hormone called *erythropoietin* (see chapter 12).

1. Where is the pineal gland located?
2. What is the function of the pineal gland?
3. Where is the thymus gland located?
4. Which reproductive organs secrete hormones?
5. Which other organs secrete hormones?

Stress and Health

Survival depends on the maintenance of homeostasis. Thus, factors that change the body's internal environment can threaten life. When the body senses danger, nerve impulses to the hypothalamus trigger physiological responses that preserve homeostasis. These responses include increased activity in the sympathetic division of the autonomic nervous system and increased secretion of adrenal and other hormones. A factor that can stimulate such a response is called a *stressor,* and the condition it produces in the body is called *stress.*

Types of Stress

Stressors include physical factors, such as exposure to extreme heat or cold, decreased oxygen concentration, infections, injuries, prolonged heavy exercise, and loud sounds. Stressors also include psychological factors, such as thoughts about real or imagined dangers, personal losses, and unpleasant social interactions. Feelings of anger, fear, grief, anxiety, depression, and guilt can also produce psychological stress. Sometimes, pleasant stimuli, such as friendly social contact, feelings of joy and happiness, or sexual arousal, may be stressful.

Responses to Stress

Physiological responses to stress consist of reactions called the *general stress syndrome* (general adaptation syndrome), which is under hypothalamic control. Typically, the hypothalamus activates mechanisms that prepare the body for "fight or flight." These responses include raising blood concentrations of glucose, glycerol, and fatty acids; increasing heart rate, blood pressure and breathing rate; dilating air passages; shunting blood from the skin and digestive organs to the skeletal muscles; and increasing epinephrine secretion from the adrenal medulla (fig. 11.17).

The hypothalamus also releases CRH, which, in turn, stimulates the anterior pituitary to secrete ACTH. ACTH causes the adrenal cortex to increase cortisol secretion. Cortisol increases blood amino acid concentration, fatty acid release, and glucose formation from noncarbohydrates. Thus, while the body prepares for physical activity, the actions of cortisol supply cells with biochemicals required during stress (fig. 11.18).

Other hormones whose secretions increase with stress include glucagon, GH, and ADH. Glucagon and GH mobilize energy sources, such as glucose, glycerol, fatty acids, and amino acids. ADH stimulates the kidneys to retain water, which increases blood volume—particularly important if a person is bleeding or sweating heavily.

Increased cortisol secretion may be accompanied by a decrease in the number of certain white blood cells (lymphocytes), which lowers resistance to infectious diseases and some cancers. Also, excess cortisol production may raise the risk of developing high blood pressure, atherosclerosis, and gastrointestinal ulcers.

1. What is stress?
2. Distinguish between physical stress and psychological stress.
3. Describe the general stress syndrome.

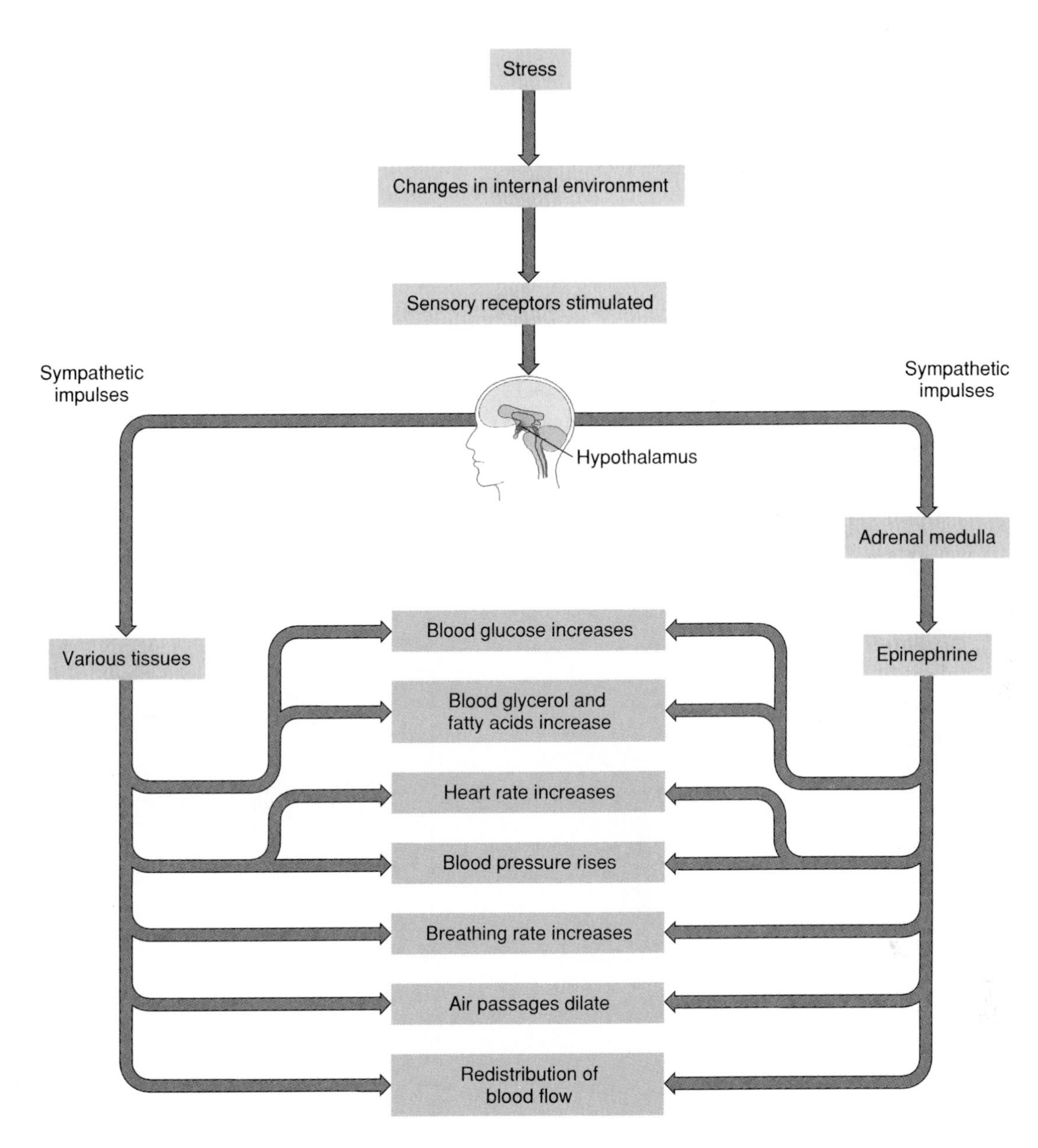

Figure 11.17

During stress, the hypothalamus helps prepare the body for "fight or flight" by triggering sympathetic impulses to various organs. It also stimulates epinephrine release, intensifying the sympathetic responses.

Clinical Terms Related to the Endocrine System

Adrenalectomy (ah-dre″nah-lek′to-me) Surgical removal of the adrenal glands.

Adrenogenital syndrome (ah-dre″no-jen′ĭ-tal sin′drōm) A group of symptoms associated with changes in sexual characteristics as a result of increased secretion of adrenal androgens.

Diabetes insipidus (di″ah-be′tēz in-sip′ĭdus) A metabolic disorder, not involving blood sugar, characterized by a large output of dilute urine and caused by the posterior pituitary's decreased secretion of antidiuretic hormone.

Diabetes mellitus (di″ah-be′tēz mel′i-tus) A condition due to insulin deficiency or the inability to respond to insulin. This condition disturbs carbohydrate, protein, and lipid metabolism.

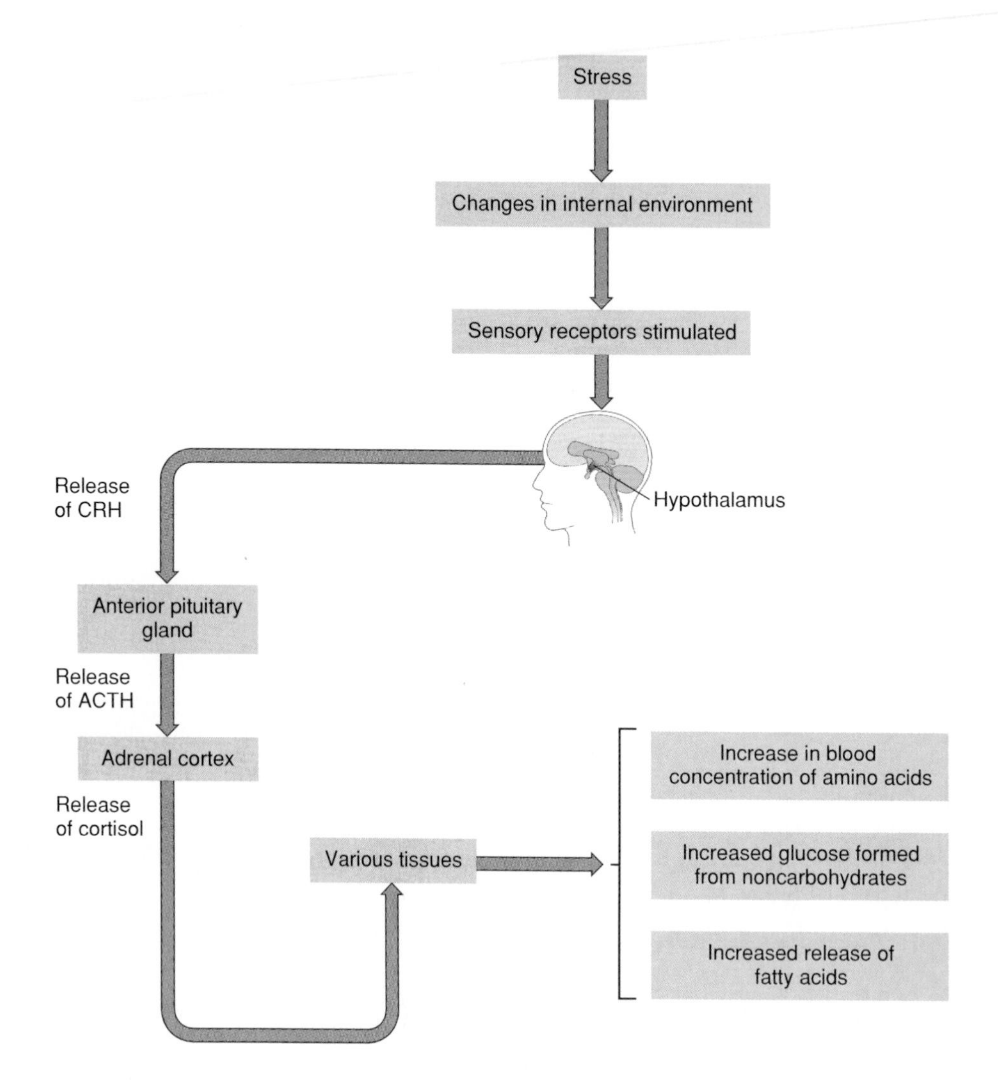

Figure 11.18

As a result of stress, the hypothalamus stimulates the adrenal cortex to release cortisol, which promotes responses that resist the effects of stress. Compare this to figure 11.13.

Exophthalmos (ek″sof-thal′mos) Abnormal protrusion of the eyes.

Goiter (goi′ter) A bulge in the neck resulting from an enlarged thyroid gland.

Hirsutism (her′sŭt-izm) Excess hair growth, especially in women.

Hypercalcemia (hi″per-kal-se′me-ah) Excess blood calcium.

Hyperglycemia (hi″per-gli-se′me-ah) Excess blood glucose.

Hypocalcemia (hi″po-kal-se′me-ah) Deficiency of blood calcium.

Hypoglycemia (hi″po-gli-se′me-ah) Deficiency of blood glucose.

Hypophysectomy (hi-pof″ĭ-sek′to-me) Surgical removal of the pituitary gland.

Parathyroidectomy (par″ah-thi″roi-dek′to-me) Surgical removal of the parathyroid glands.

Pheochromocytoma (fe-o-kro″mo-si-to′mah) A type of tumor in the adrenal medulla usually associated with high blood pressure.

Polyphagia (pol″e-fa′je-ah) Excessive eating.

Thymectomy (thi-mek′to-me) Surgical removal of the thymus gland.

Thyroidectomy (thi″roi-dek′to-me) Surgical removal of the thyroid gland.

Thyroiditis (thi″roi-di′tis) Inflammation of the thyroid gland.

Virilism (vir′ĭ-lizm) Masculinization of a female.

TOPIC OF INTEREST

Biological Rhythms

Biological rhythms are changes that systematically recur in living systems. They include the daily ebb and flow of biochemical levels in blood, reproductive cycles, and migration schedules. The period of any rhythm is the duration of one complete cycle. The frequency of a rhythm is the number of cycles per time unit.

Three common types of rhythms in humans are ultradian, infradian, and circadian rhythms. *Ultradian rhythms* have periods shorter than 24 hours and include the cardiac cycle and the breathing cycle. Periods of *infradian rhythms,* such as the menstrual cycle, are longer than 24 hours. Periods of circadian rhythms, such as the sleep-wake cycle, are approximately 24 hours.

Both external (exogenous) and internal (endogenous) factors regulate human biological rhythms. Exogenous factors are environmental components, such as daily temperature changes and the light-dark cycle. Endogenous factors are genetically programmed internal "clocks" found in virtually all organisms. Researchers have identified such "clock" genes in a range of species, including fruit flies, mice, bread mold, and plants. Although melatonin is thought to regulate human biological clocks because it can relieve insomnia, researchers caution that much remains to be learned about this hormone.

The sleep-wake cycle is the most obvious circadian rhythm in humans. Most individuals sleep 6–8 hours each night, but culture largely determines this pattern. Under conditions that remove environmental cues, the body adjusts to a 25-hour or longer day. As we age, the sleep-wake cycle changes, and many people require less sleep.

Other circadian rhythms in humans affect body temperature, cardiovascular functioning, and hormone secretion. Body temperature is, for the most part, endogenously regulated, but light exposure and physical activity also help keep this rhythm on a 24- rather than 25-hour cycle. Body temperature is usually lowest between 4 and 6 A.M., then increases and peaks between 5 and 11 P.M. It drops during the late evening hours and into the night.

Cardiovascular functioning is least efficient between 6 and 9 A.M. Platelet cohesion, blood pressure, and pulse rate are typically highest 2 hours after awakening, which may explain why heart attacks and strokes are more likely between 8 and 10 A.M. than at other times.

Hormones may have ultradian and infradian as well as circadian rhythms. Plasma cortisol, for example, surges and peaks at about 6 A.M., then gradually declines to its minimum level in late evening before increasing again in the early morning. Growth hormone secretion peaks during the night. Antidiuretic hormone is greater at night, decreasing urine formation then.

Flying across time zones can disrupt melatonin secretion, causing jet lag because body rhythms may require several days to adjust to local time. Jet lag symptoms include insomnia, fatigue, indigestion, difficulty concentrating, errors in judgment, and irritability.

Melatonin is also implicated in seasonal affective disorder (SAD), a type of depression that occurs during winter months, when daylight is shorter and less intense. Exposure to bright light for several hours each day can relieve SAD symptoms.

organization

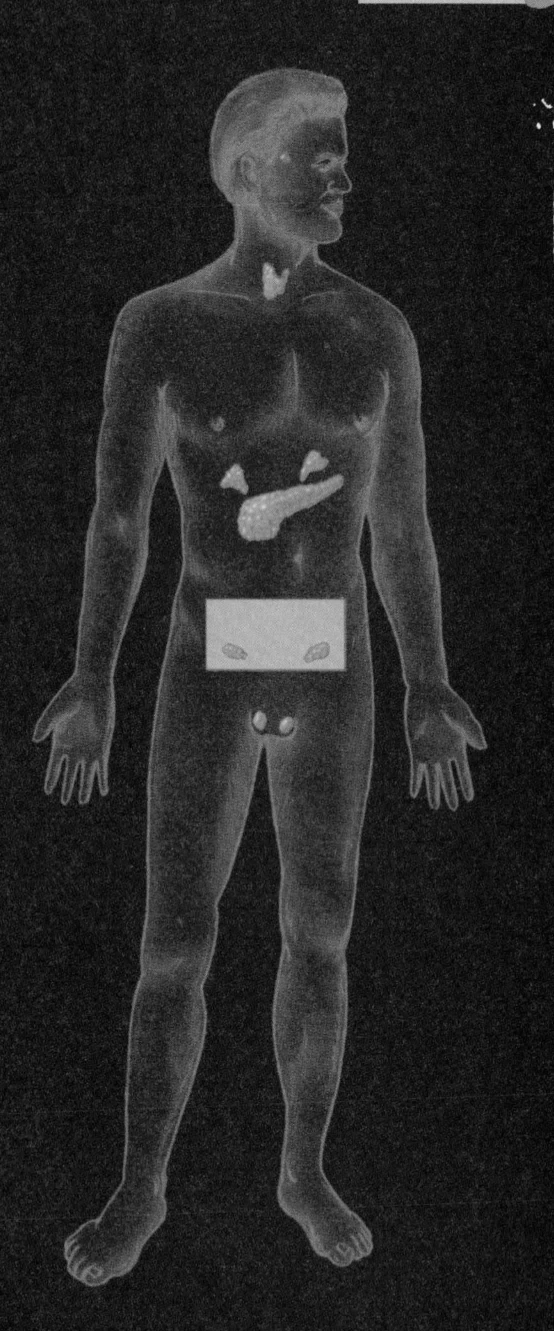

Endocrine System

Glands secrete hormones that have a variety of effects on cells, tissues, organs, and organ systems.

Integumentary System

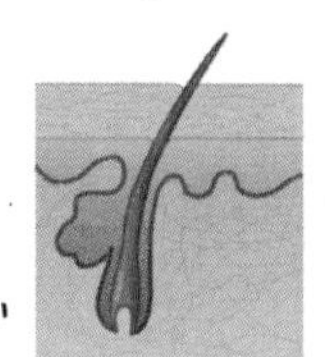

Melanocytes produce skin pigment in response to hormonal stimulation.

Lymphatic System

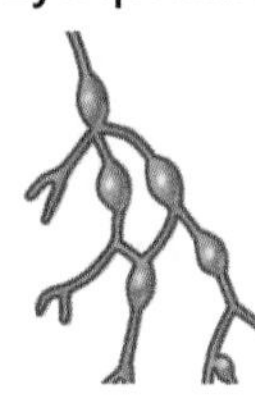

Hormones stimulate lymphocyte production.

Skeletal System

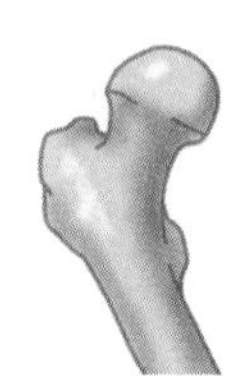

Hormones act on bones to control calcium balance.

Digestive System

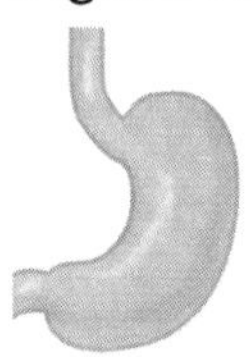

Hormones help control digestive system activity.

Muscular System

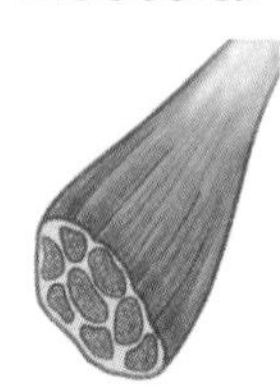

Hormones help increase blood flow to exercising muscles.

Respiratory System

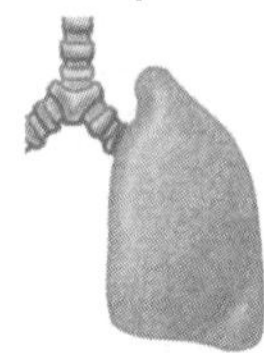

Decreased oxygen causes hormonal stimulation of red blood cell production; red blood cells transport oxygen and carbon dioxide.

Nervous System

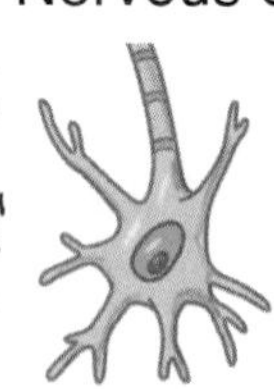

Neurons control the secretions of the anterior and posterior pituitary glands and the adrenal medulla.

Urinary System

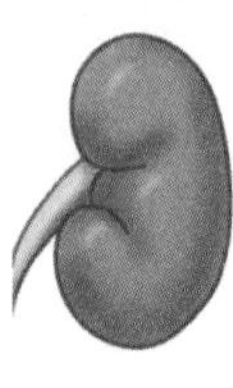

Hormones act on the kidneys to help control water and electrolyte balance.

Cardiovascular System

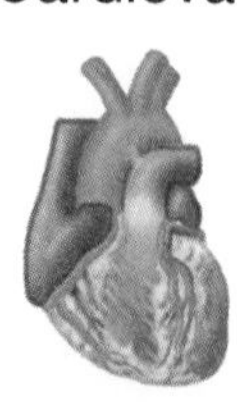

Hormones are carried in the bloodstream; some have direct actions on the heart and blood vessels.

Reproductive System

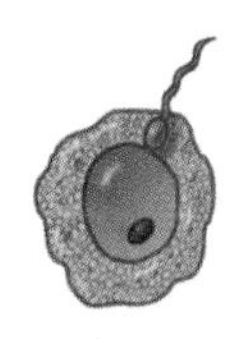

Sex hormones play a major role in development of secondary sex characteristics, egg, and sperm.

Summary Outline

Introduction (p. 293)

Endocrine glands secrete their products into body fluids. Exocrine glands secrete their products into ducts that lead to the outside of the body.

General Characteristics of the Endocrine System (p. 293)

As a group, endocrine glands secrete hormones that regulate metabolism.

Hormone Action (p. 293)

Endocrine glands secrete hormones that affect target cells with specific receptors. Hormones are very potent.

1. Chemically, hormones are steroids, amines, peptides, proteins, or glycoproteins.
2. Steroid hormones
   a. Steroid hormones enter a target cell and combine with receptors to form complexes within the nucleus.
   b. These complexes activate specific genes, which cause protein synthesis.
3. Nonsteroid hormones
   a. Nonsteroid hormones combine with receptors in the target-cell membrane.
   b. The hormone-receptor complex signals a G protein to stimulate a membrane protein, such as adenylate cyclase, to induce formation of second messenger molecules.
   c. A second messenger, such as cyclic adenosine monophosphate (cAMP), diacylglycerol (DAG), or inositol triphosphate (IP_3), activates protein kinases.
   d. Protein kinases activate protein substrate molecules, which, in turn, change a cellular process.
4. Prostaglandins
   a. Prostaglandins act on the cells of the organs that produce them.
   b. Prostaglandins are present in small quantities and have powerful hormonelike effects.

Control of Hormonal Secretions (p. 296)

The concentration of each hormone in body fluids is regulated.

1. Negative feedback systems
   a. In a negative feedback system, a gland is sensitive to the concentration of a substance it regulates.
   b. When the concentration of the regulated substance reaches a certain point, it inhibits the gland.
   c. As the gland secretes less hormone, the controlled substance also decreases.
   d. Negative feedback systems maintain relatively stable hormone concentrations.
2. Control sources
   a. Some endocrine glands secrete hormones in response to releasing hormones the hypothalamus secretes.
   b. Other glands secrete their hormones in response to nerve impulses.
   c. Some glands respond to levels of a substance in the bloodstream.

Pituitary Gland (p. 298)

The pituitary gland has an anterior lobe and a posterior lobe. The hypothalamus controls most pituitary secretions.

1. Anterior pituitary hormones
   a. The anterior pituitary secretes growth hormone (GH), prolactin (PRL), thyroid-stimulating hormone (TSH), adrenocorticotropic hormone (ACTH), follicle-stimulating hormone (FSH), and luteinizing hormone (LH).
   b. Growth hormone
      (1) GH stimulates cells to increase in size and divide more frequently.
      (2) GH releasing hormone and GH release-inhibiting hormone from the hypothalamus control GH secretion.
   c. PRL stimulates and sustains a woman's milk production.
   d. Thyroid-stimulating hormone
      (1) TSH controls secretion of hormones from the thyroid gland.
      (2) The hypothalamus secretes thyrotropin-releasing hormone (TRH), which regulates TSH secretion.
   e. Adrenocorticotropic hormone
      (1) ACTH controls secretion of hormones from the adrenal cortex.
      (2) The hypothalamus secretes corticotropin-releasing hormone (CRH), which regulates ACTH secretion.
   f. FSH and LH are gonadotropins.
2. Posterior pituitary hormones
   a. The posterior lobe of the pituitary gland consists largely of neuroglial cells and nerve fibers.
   b. The hypothalamus produces the hormones of the posterior pituitary.
   c. Antidiuretic hormone (ADH)
      (1) ADH reduces the amount of water the kidneys excrete.
      (2) The hypothalamus regulates ADH secretion.
   d. Oxytocin (OT)
      (1) OT contracts muscles in the uterine wall.
      (2) OT also contracts cells associated with producing and ejecting milk.

Thyroid Gland (p. 301)

The thyroid gland is located in the neck and consists of two lobes.

1. Structure of the gland
   a. The thyroid gland consists of many follicles.
   b. The follicles are fluid-filled and store hormones.
2. Thyroid hormones
   a. Thyroxine and triiodothyronine increase the metabolic rate of cells, enhance protein synthesis, and stimulate lipid utilization.
   b. Calcitonin helps regulate concentrations of blood calcium and phosphate ions.

Parathyroid Glands (p. 303)

The parathyroid glands are on the posterior surface of the thyroid gland.

1. Structure of the glands
   Each parathyroid gland consists of secretory cells that are well supplied with capillaries.

2. Parathyroid hormone (PTH)
   a. PTH increases blood calcium level and decreases blood phosphate ion concentration.
   b. A negative feedback mechanism operates between the parathyroid glands and the blood.

Adrenal Glands (p. 304)

The adrenal glands are located atop the kidneys.

1. Structure of the glands
   a. Each gland consists of an adrenal medulla and an adrenal cortex.
   b. The adrenal medulla and adrenal cortex are functionally distinct glands that secrete different hormones.
2. Hormones of the adrenal medulla
   a. The adrenal medulla secretes epinephrine and norepinephrine, which have similar effects.
   b. Sympathetic impulses stimulate secretion of these hormones.
3. Hormones of the adrenal cortex
   a. The adrenal cortex produces several steroid hormones.
   b. Aldosterone is a mineralocorticoid that causes the kidneys to conserve sodium ions and water, and to excrete potassium ions.
   c. Cortisol is a glucocorticoid, which affects carbohydrate, protein, and fat metabolism.
   d. Adrenal sex hormones
      (1) These hormones are of the male type but may be converted to female hormones.
      (2) They may supplement the sex hormones the gonads produce.

Pancreas (p. 308)

The pancreas secretes digestive juices as well as hormones.

1. Structure of the gland
   a. The pancreas is attached to the small intestine.
   b. The islets of Langerhans secrete glucagon and insulin.
2. Hormones of the islets of Langerhans
   a. Glucagon stimulates the liver to produce glucose from glycogen and noncarbohydrates.
   b. Insulin moves glucose across some cell membranes, stimulates glucose and fat storage, and promotes protein synthesis.
   c. Nerve cells are not dependent on insulin for a glucose supply.

Other Endocrine Glands (p. 310)

1. Pineal gland
   a. The pineal gland attaches to the thalamus.
   b. It secretes melatonin in response to varying light conditions.
   c. Melatonin may help regulate the female reproductive cycle by inhibiting gonadotropin secretion from the anterior pituitary.
2. Thymus gland
   a. The thymus gland lies behind the sternum and between the lungs.
   b. It secretes thymosins, which affect the production of certain lymphocytes that function in immunity.
3. Reproductive glands
   a. The ovaries secrete estrogens and progesterone.
   b. The placenta secretes estrogens, progesterone, and gonadotropin.
   c. The testes secrete testosterone.
4. Digestive glands
   Certain glands of the stomach and small intestine secrete hormones.
5. Other hormone-producing organs
   Other organs, such as the heart and the kidneys, also produce hormones.

Stress and Health (p. 312)

Stress occurs when the body responds to stressors that threaten the maintenance of homeostasis. Stress responses include increased activity of the sympathetic nervous system and increased secretion of adrenal hormones.

1. Types of stress
   a. Physical stress results from environmental factors that are harmful or potentially harmful to tissues.
   b. Psychological stress results from thoughts about real or imagined dangers.
2. Responses to stress
   a. Responses to stress maintain homeostasis.
   b. The hypothalamus controls a general stress syndrome.

Review Exercises

1. Define *endocrine gland.* (p. 293)
2. Define *hormone* and *target cell.* (p. 293)
3. Explain how steroid hormones produce their effects. (p. 294)
4. Explain how nonsteroid hormones employ second-messenger molecules. (p. 294)
5. Describe how prostaglandins are similar to hormones. (p. 296)
6. Describe a negative feedback system. (p. 296)
7. Describe the location and structure of the pituitary gland. (p. 298)
8. Explain how the brain controls pituitary gland activity. (p. 298)
9. Define *releasing hormone,* and give an example of one. (p. 298)
10. List the hormones that the anterior pituitary secretes. (p. 299)
11. Explain how growth hormone produces its effects. (p. 299)
12. List the major factors that affect secretion of growth hormone. (p. 299)
13. Summarize the function of prolactin. (p. 299)
14. Describe the mechanism that regulates the concentrations of circulating thyroid hormones. (p. 299)
15. Explain the control of adrenocorticotropic hormone secretion. (p. 300)

16. Compare the cellular structures of the anterior and posterior lobes of the pituitary gland. (p. 300)
17. Describe the functions of the posterior pituitary hormones. (p. 300)
18. Explain the regulation of antidiuretic hormone release. (p. 300)
19. Describe the location and structure of the thyroid gland. (p. 301)
20. Name the hormones the thyroid gland secretes, and list the general functions of each. (p. 301)
21. Describe the location and structure of the parathyroid glands. (p. 303)
22. Explain the general functions of parathyroid hormone. (p. 303)
23. Describe the mechanism that regulates parathyroid hormone secretion. (p. 303)
24. Distinguish between the adrenal medulla and the adrenal cortex. (p. 305)
25. List the hormones the adrenal medulla produces, and describe their general functions. (p. 305)
26. Name the most important hormones of the adrenal cortex, and describe the general functions of each. (p. 306)
27. Describe how the pituitary gland controls the secretion of adrenal cortical hormones. (p. 306)
28. Describe the location and structure of the pancreas. (p. 308)
29. List the hormones the islets of Langerhans secrete, and describe their functions. (p. 309)
30. Summarize the regulation of hormone secretion from the pancreas. (p. 309)
31. Describe the location and general function of the pineal gland. (p. 310)
32. Describe the location and general function of the thymus gland. (p. 310)
33. Name nine additional hormone-secreting organs. (p. 310)

Critical Thinking

1. When reactor 4 at the Chernobyl Nuclear Power Station in Ukraine exploded at 1:23 P.M. on April 26, 1986, a great plume of radioactive isotopes erupted into the air and spread for thousands of miles. Most of the isotopes immediately following the blast were of the element iodine. Which of the glands of the endocrine system would be most seriously, and immediately, affected by the blast, and how do you think this would become evident in the nearby population?
2. Human growth hormone and human insulin today are manufactured using recombinant DNA technology, which means that bacteria use human genes to produce the human proteins. Why is this safer than the former method of extracting medically useful hormones from pigs, cows, or human cadavers?
3. A young woman feels shaky, distracted, and generally ill. She lives with her mother, who is dying. A friend tells the young woman, "It's just stress, it's all in your head." Is it?
4. A pheromone is a biochemical an individual emits that affects the behavior of another individual of the same species. Insects, for example, emit pheromones that are part of mating. Pheromones have not been identified in humans. How does a pheromone differ from a hormone?
5. Growth hormone is administered to people who have pituitary dwarfism. Parents wanting their normal children to be taller have requested the treatment for them. Do you think this is a wise request? Why or why not?
6. A sixty-two-year-old woman recently gave birth after receiving hormone treatments that mimicked the hormonal environment of a woman of childbearing age. Do you think such treatments should become part of medical practice, enabling women past menopause to have children?
7. Which hormones would an adult whose anterior pituitary has been removed require to be supplemented?
8. How might the environment of a patient with hyperthyroidism be modified to minimize the drain on body energy resources?
9. The adrenal cortex of a patient who has lost a large volume of blood will increase secretion of aldosterone. What effect will this increased secretion have on the patient's blood concentrations of sodium and potassium ions?
10. Why might oversecretion of insulin actually reduce glucose uptake by nerve cells?

Blood

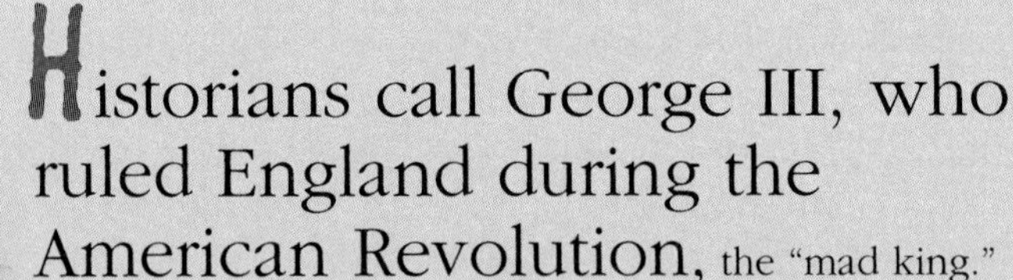

Historians call George III, who ruled England during the American Revolution, the "mad king." In the twentieth century, physicians traced the king's odd behavior to a blood disorder called porphyria.

King George was fifty when he first experienced the strange sequence of symptoms—abdominal pain and constipation, then passage of dark red urine, limb weakness, fever, hoarseness, and a rapid pulse. Next came insomnia, headache, visual problems, restlessness, delirium, convulsions, and stupor. George's reactions to his discomfort made him appear deranged. He would rip off his wig and shed his clothes in the throes of a raging fever, and had confused and racing thoughts. While Parliament and the king's baffled doctors spent four months debating the possible permanence of their ruler's apparent dementia, King George mysteriously recovered. But he relapsed thirteen years later, then again three years after that. Always, the symptoms appeared in the same order. Finally, an attack in 1811 placed George in a permanent stupor, and the Prince of Wales dethroned him. King George lived for several more years, experiencing further episodes of his bizarre illness.

In George's time, physicians were permitted to do very little to the royal body, basing diagnoses on what the patient told them. The king's red urine eventually clued medical researchers to the cause of the problem. A missing enzyme results in failure to metabolize part of the blood pigment hemoglobin, called a porphyrin ring. Although the kidneys work to excrete it, porphyrin builds up in the blood and damages the nervous system. Medical records of King George III's descendants reveal several royal relatives who also had porphyria.

Photo: An inherited blood disorder, porphyria, caused King George III's erratic behavior and physical symptoms at the time of the American Revolution.

Blood signifies life, and for good reason—it has many vital functions. This complex mixture of cells and dissolved biochemicals transports nutrients, oxygen, wastes, and hormones; helps maintain stability of the interstitial fluid; and distributes heat. The blood, heart, and blood vessels form the cardiovascular system and link the body's internal and external environments.

Chapter Objectives

After studying this chapter, you should be able to:

1. Describe the general characteristics of blood, and discuss its major functions. (p. 322)
2. Distinguish among the types of blood cells. (p. 322)
3. Explain the control of red blood cell production. (p. 324)
4. Distinguish among the five types of white blood cells and give the function(s) of each type. (p. 326)
5. List the major components of blood plasma, and describe the functions of each. (p. 331)
6. Define *hemostasis,* and explain the mechanisms that help to achieve it. (p. 333)
7. Review the major steps in blood coagulation. (p. 333)
8. Explain blood typing and how it is used to avoid adverse reactions following blood transfusions. (p. 335)
9. Describe how blood reactions may occur between fetal and maternal tissues. (p. 338)

Aids to Understanding Words

agglutin- [to glue together] *agglutin*ation: The clumping together of red blood cells.

bil- [bile] *bil*irubin: A pigment excreted in the bile.

embol- [stopper] *embol*ism: An obstruction of a blood vessel.

erythr- [red] *erythr*ocyte: A red blood cell.

hemo- [blood] *hemo*globin: The red pigment responsible for the color of blood.

leuko- [white] *leuko*cyte: A white blood cell.

-osis [a condition or process] leukocyt*osis:* A condition characterized by too many white blood cells.

-poiet [making] erythro*poiet*in: A hormone that stimulates red blood cell production.

-stas [halting] hemo*stas*is: The stoppage of bleeding.

thromb- [clot] *thromb*ocyte: A blood platelet that helps form a blood clot.

Key Terms

albumin (al-bu′min)
antibody (an′ti-bod″e)
antigen (an′ti-jen)
basophil (ba′so-fil)
coagulation (ko-ag″u-la′shun)
eosinophil (e″o-sin′o-fil)
erythrocyte (ĕ-rith′ro-sīt)
erythropoietin (ĕ-rith″ro-poi′ĕ-tin)
fibrinogen (fi-brin′o-jen)
globulin (glob′u-lin)
hemostasis (he″mo-sta′sis)
leukocyte (lu′ko-sīt)
lymphocyte (lim′fo-sīt)
monocyte (mon′o-sīt)
neutrophil (nu′tro-fil)
plasma (plaz′mah)
platelet (plāt′let)

Introduction

Blood is often considered a type of connective tissue whose cells are suspended in a liquid intercellular material. Blood is vital in transporting substances between body cells and the external environment, thereby helping to maintain a stable internal environment.

Blood and Blood Cells

Whole blood is slightly heavier and three to four times more viscous than water. Its cells, which form mostly in red bone marrow, include red blood cells and white blood cells. Blood also contains cell fragments called platelets (fig. 12.1).

Blood Volume and Composition

Blood volume varies with body size, changes in fluid and electrolyte concentrations, and the amount of adipose tissue. An average-sized adult has a blood volume of about 5 liters.

A blood sample is usually about 45% cells by volume. This percentage is called the **hematocrit** (HCT). Most blood cells are red cells, with much smaller numbers of white cells and cell fragments (platelets). The remaining 55% of a blood sample is a clear liquid called **plasma** (fig. 12.1). Plasma is a complex mixture of water, amino acids, proteins, carbohydrates, lipids, vitamins, hormones, electrolytes, and cellular wastes.

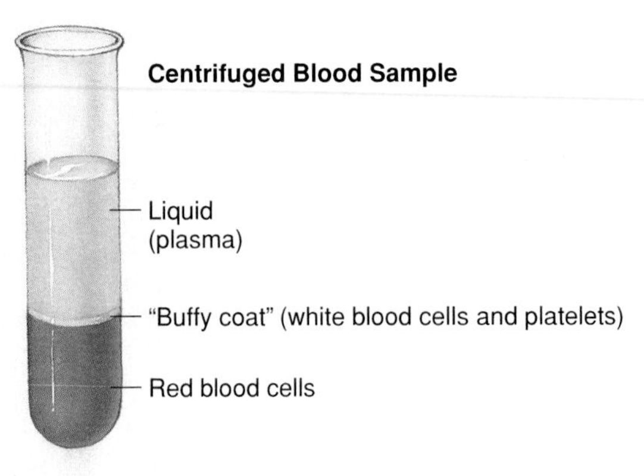

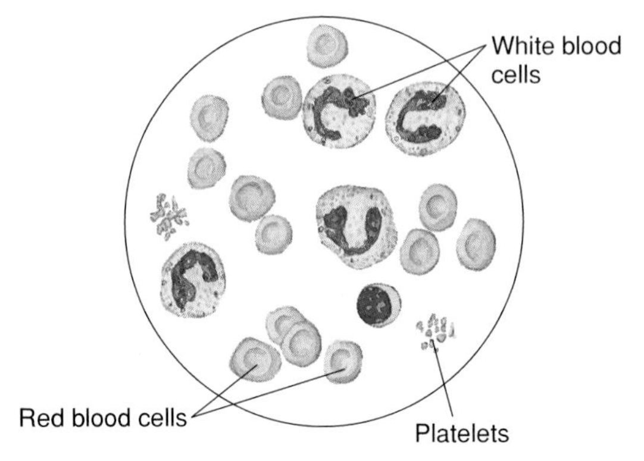

Figure 12.1

Blood consists of a liquid portion, called plasma, and a solid portion that includes red blood cells, white blood cells, and platelets. (Note: When blood components are separated, the white blood cells and platelets form a thin layer, called the "buffy coat," between the plasma and the red blood cells.)

1. What factors affect blood volume?
2. What are the major components of blood?

Characteristics of Red Blood Cells

Red blood cells, or **erythrocytes**, are biconcave disks that thin near their centers. This shape, which is an adaptation for transporting gases, provides an increased surface area through which gases can diffuse. The red blood cell's shape also places the cell membrane closer to oxygen-carrying *hemoglobin* within the cell (fig. 12.2).

Each red blood cell is about one-third hemoglobin by volume. This protein is responsible for the color of blood. When hemoglobin combines with oxygen, the resulting *oxyhemoglobin* is bright red, and when oxygen is released, the resulting *deoxyhemoglobin* is darker.

Red blood cells have nuclei during their early stages of development but extrude them as the cells mature, providing more space for hemoglobin. Because they lack nuclei, red blood cells cannot synthesize proteins or reproduce.

A person experiencing prolonged oxygen deficiency (hypoxia) may become *cyanotic*. The skin and mucous membranes appear bluish due to an abnormally high blood concentration of deoxyhemoglobin. Exposure to low temperature may also result in cyanosis. Such exposure constricts superficial blood vessels, which slows blood flow and removes more oxygen than usual from blood flowing through the vessels. ■

1. Describe a red blood cell.
2. What is the function of hemoglobin?
3. What changes occur in a red blood cell as it matures?

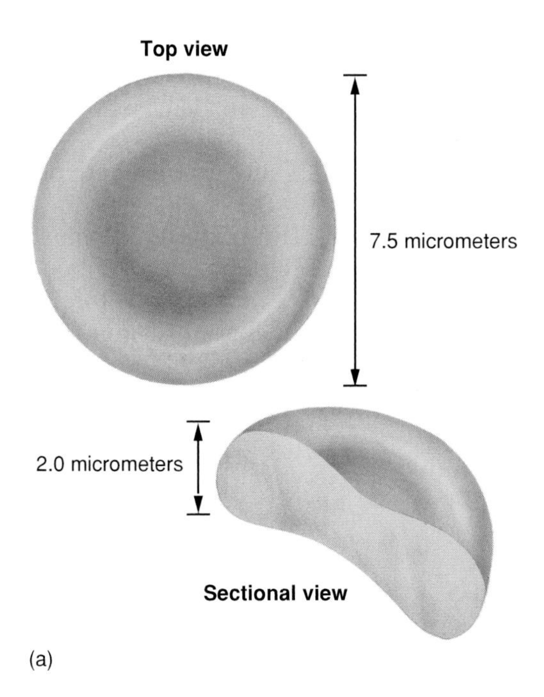

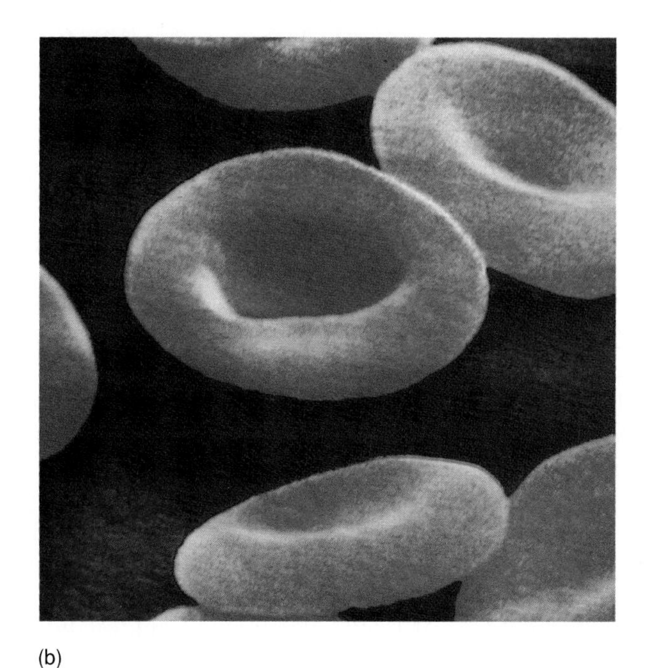

Figure 12.2

(*a*) The biconcave shape of a red blood cell is related to its function. (*b*) Scanning electron micrograph of human red blood cells (falsely colored) (5,600×).

Red Blood Cell Counts

The number of red blood cells in a cubic millimeter (mm^3) of blood is called the *red blood cell count* (RBCC or RCC). The typical range for adult males is 4,600,000–6,200,000 cells per mm^3, and that for adult females is 4,200,000–5,400,000 cells per mm^3.

The number of circulating red blood cells determines the blood's *oxygen-carrying capacity*. Therefore, any changes in this number are important. For this reason, physicians routinely consult red blood cell counts to help diagnose and evaluate the courses of certain diseases.

Destruction of Red Blood Cells

Red blood cells are quite elastic and flexible, and they readily bend as they pass through small blood vessels. With age, however, these cells become more fragile, and they are frequently damaged simply by passing through capillaries, particularly those in active muscles. **Macrophages** phagocytize and destroy damaged red blood cells, primarily in the liver and spleen (see chapter 5).

Hemoglobin molecules from the red blood cells being destroyed break down into subunits of *heme,* an iron-containing portion, and *globin,* a protein. The heme further decomposes into iron and a greenish pigment called **biliverdin.** The blood may transport the iron, combined with a protein, to the blood-cell-forming (hematopoietic) tissue in red bone marrow to be used in synthesizing new hemoglobin. Otherwise, the liver stores iron in the form of an iron-protein complex. Biliverdin eventually breaks down to an orange pigment called **bilirubin.** Biliverdin and bilirubin are excreted in the bile as bile pigments (see chapter 15).

Newborns can develop *physiologic jaundice* a few days after birth. In this condition and other forms of jaundice (icterus), accumulation of bilirubin turns the skin and eyes yellowish.

Physiologic jaundice may be the result of immature liver cells that ineffectively excrete bilirubin into the bile. Treatment includes exposure to fluorescent light, which breaks down bilirubin in the tissues, and feedings that promote bowel movements. In hospital nurseries, babies being treated for physiologic jaundice lie under "bili lights," clad only in diapers and protective goggles. The healing effect of fluorescent light was discovered in the 1950s, when an astute nurse noted that jaundiced babies improved after sun exposure, except for the areas their diapers covered. ■

1. What is the typical red blood cell count for an adult male? For an adult female?
2. What happens to damaged red blood cells?
3. What are the products of hemoglobin breakdown?

Red Blood Cell Production and Its Control

Recall from chapter 7 that red blood cell formation (hematopoiesis) occurs initially in the yolk sac, liver, and spleen. After birth, red blood cells form almost exclusively in tissue lining the spaces in bones filled with red bone marrow.

The average life span of a red blood cell is about 120 days. Many of these cells are removed from the circulation each day, yet the number of cells in the circulating blood remains relatively stable. This observation suggests a *homeostatic* control of the rate of red blood cell production.

A *negative feedback mechanism* utilizing the hormone **erythropoietin** controls the rate of red blood cell formation. The kidneys, and to a lesser extent the liver, release erythropoietin in response to prolonged oxygen deficiency (fig. 12.3). At high altitudes, for example, where the percentage of oxygen in the air is reduced, the amount of oxygen delivered to tissues decreases. This drop in oxygen concentration triggers release of erythropoietin, which travels via the blood to the red bone marrow and stimulates red blood cell production.

After a few days, many newly formed red blood cells appear in the circulating blood. The increased rate of production continues as long as the kidney and liver tissues have deficient oxygen. When the number of red blood cells in the circulation supplies sufficient oxygen to the kidneys and liver, or when the oxygen level in the air returns to normal, erythropoietin release decreases, and the rate of red blood cell production falls.

Figure 12.4 illustrates the stages in the development and differentiation of red blood cells, white blood cells, and platelets that originate from a *hemocytoblast* (stem cell).

Dietary Factors Affecting Red Blood Cell Production

B-complex vitamins—*vitamin* B_{12} and *folic acid*—significantly influence red blood cell production. These vitamins are necessary for DNA synthesis, so all cells require them to grow and reproduce. Since cell division occurs frequently in hematopoietic tissue, this tissue is especially vulnerable to deficiency of either of these vitamins.

Hemoglobin synthesis and normal red blood cell production require iron. The small intestine absorbs iron slowly from food. The body reuses much of the iron released during decomposition of hemoglobin from damaged red blood cells. Therefore, the diet need only supply small quantities of iron. Figure 12.5 summarizes the life cycle of a red blood cell.

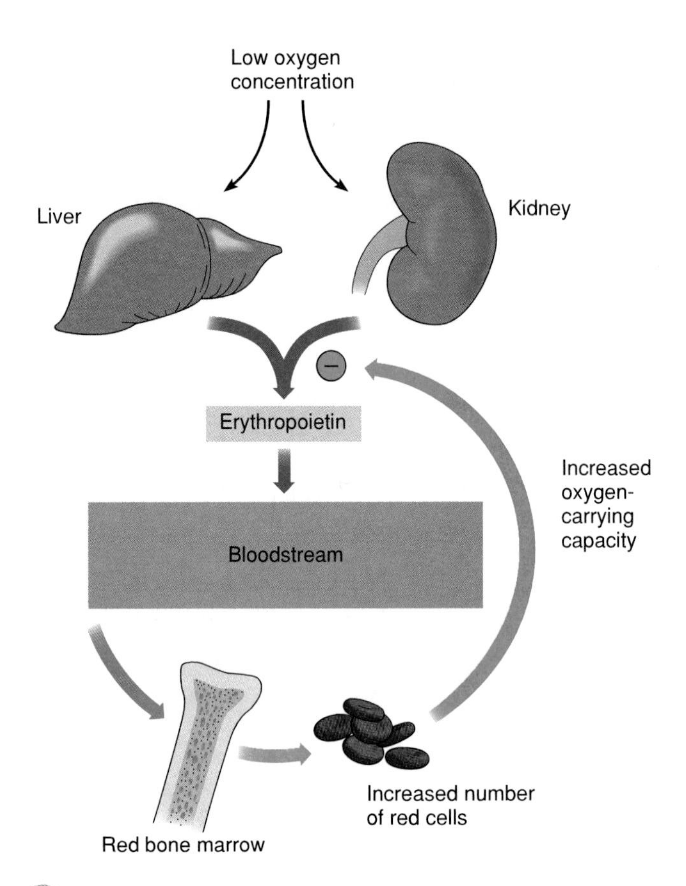

Figure 12.3

Low oxygen pressure causes the kidneys and liver to release erythropoietin, which stimulates production of red blood cells that carry oxygen to tissues.

In *sickle cell disease,* a single DNA base change causes hemoglobin to crystallize in a low-oxygen environment. This bends the red blood cells containing the hemoglobin into a sickle shape, which blocks circulation in small vessels, causing excruciating joint pain and damaging many organs. As the spleen works overtime to recycle the short-lived red blood cells, infection becomes likely.

Most children with sickle cell disease are diagnosed at birth and are given antibiotics to prevent infection. Hospitalization for blood transfusions may be necessary for painful sickling "crises" of blocked circulation. An experimental approach to treating the disorder uses drugs to activate genes that normally produce a slightly different form of hemoglobin in the fetus. Unlike the mature, sickle hemoglobin, the fetal hemoglobin functions. A bone marrow transplant can cure sickle cell disease but has a 15% risk of causing death. ■

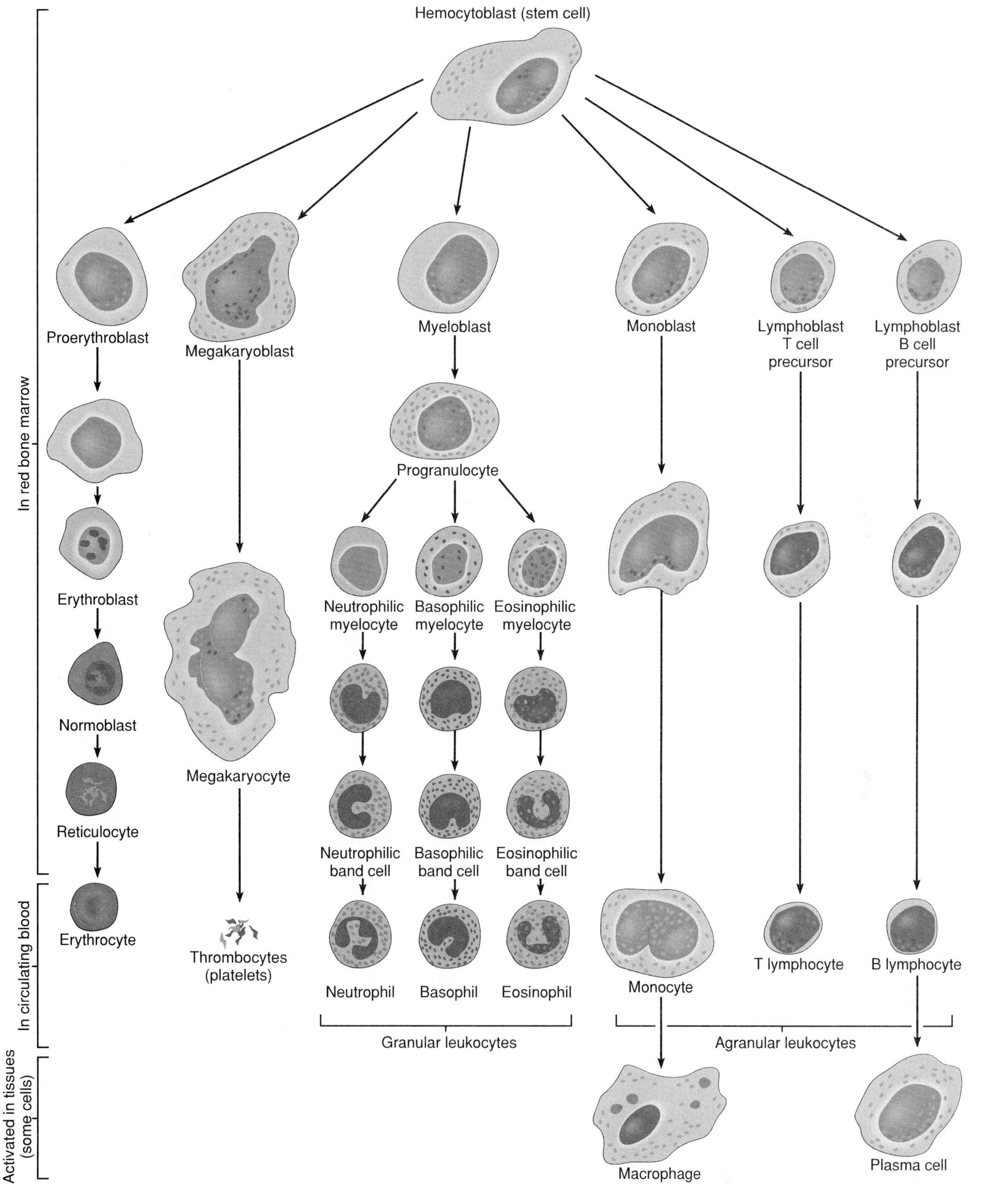

Hemocytoblast (stem cell)

In red bone marrow

Proerythroblast
Megakaryoblast
Myeloblast
Monoblast
Lymphoblast T cell precursor
Lymphoblast B cell precursor

Progranulocyte

Erythroblast
Neutrophilic myelocyte
Basophilic myelocyte
Eosinophilic myelocyte

Normoblast
Megakaryocyte
Reticulocyte

Neutrophilic band cell
Basophilic band cell
Eosinophilic band cell

In circulating blood

Erythrocyte
Thrombocytes (platelets)
Neutrophil
Basophil
Eosinophil
Monocyte
T lymphocyte
B lymphocyte

Granular leukocytes
Agranular leukocytes

Activated in tissues (some cells)

Macrophage
Plasma cell

Figure 12.4

Origin and development of blood cells from a hemocytoblast (stem cell) in bone marrow.

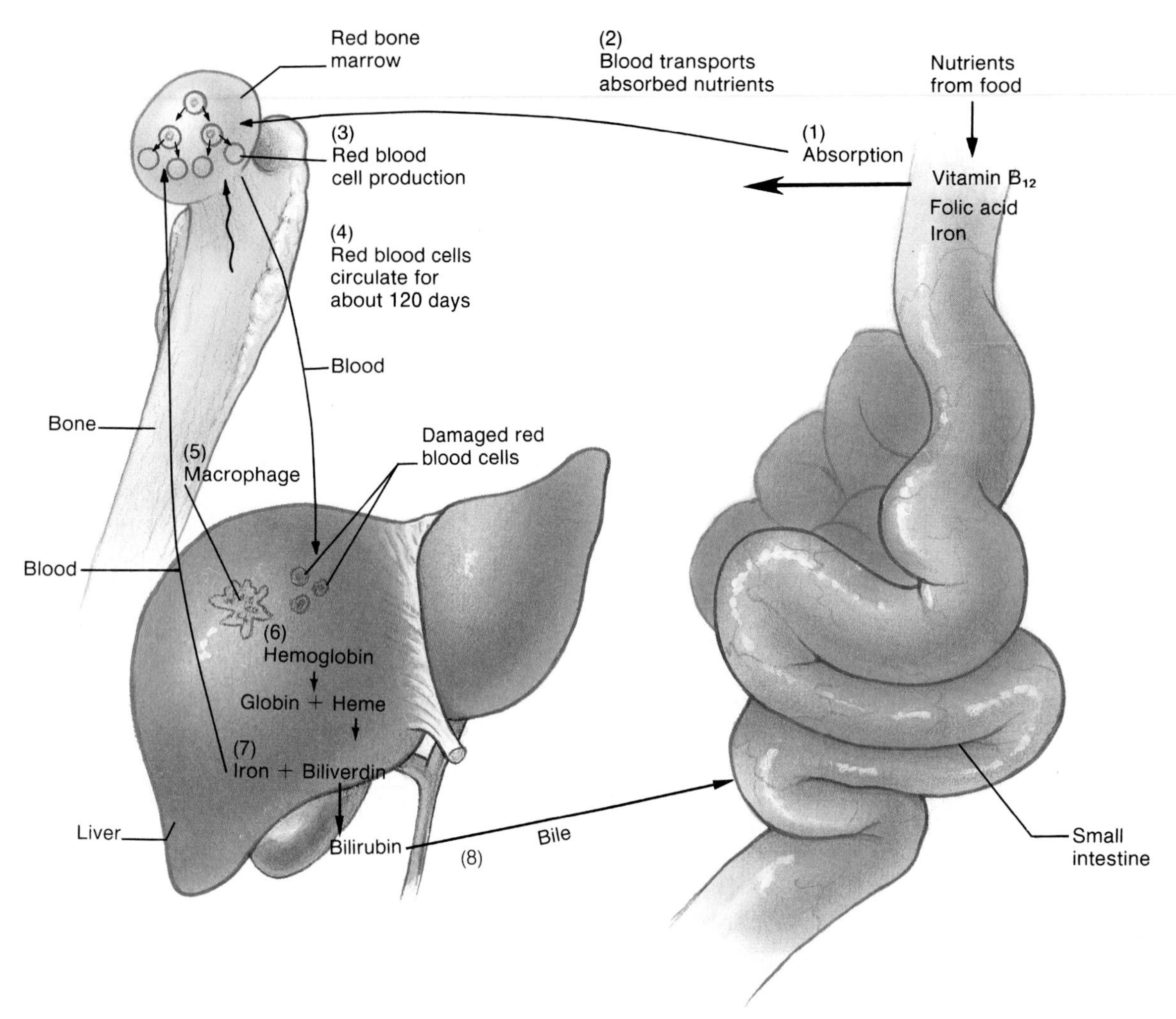

Figure 12.5

Life cycle of a red blood cell. (1) The small intestine absorbs essential nutrients. (2) Blood transports nutrients to red bone marrow. (3) In the marrow, red blood cells arise from division of less specialized cells. (4) Mature red blood cells are released into the bloodstream, where they circulate for about 120 days. (5) Macrophages destroy damaged red blood cells in the spleen and liver. (6) Hemoglobin liberated from red blood cells breaks down into heme and globin. (7) Iron from heme returns to red bone marrow and is reused. (8) Biliverdin and bilirubin are excreted in the bile.

Too few red blood cells or too little hemoglobin causes *anemia.* This reduces the blood's oxygen-carrying capacity, and the person may appear pale and lack energy. A pregnant woman may become anemic if she does not eat iron-rich foods. Furthermore, because her plasma volume increases due to fluid retention to support the fetus, her hematocrit may decrease.

1. Where are red blood cells produced?
2. How is red blood cell production controlled?
3. What vitamins are necessary for red blood cell production?
4. Why is iron required for development of red blood cells?

Types of White Blood Cells

White blood cells, or **leukocytes**, protect against disease. Blood transports white blood cells to sites of infection. White blood cells may then leave the bloodstream as described later in this chapter.

Five types of white blood cells are normally in circulating blood. They differ in size, the nature of their cytoplasm, nuclei shape, and staining characteristics. For example, leukocytes with granular cytoplasm are called **granulocytes**, while those without cytoplasmic granules are called **agranulocytes** (see fig. 12.4).

A typical granulocyte is about twice the size of a red blood cell. Members of this group include neutrophils, eosinophils, and basophils. Granulocytes develop in red bone marrow as do red blood cells, but have short life spans, averaging about 12 hours.

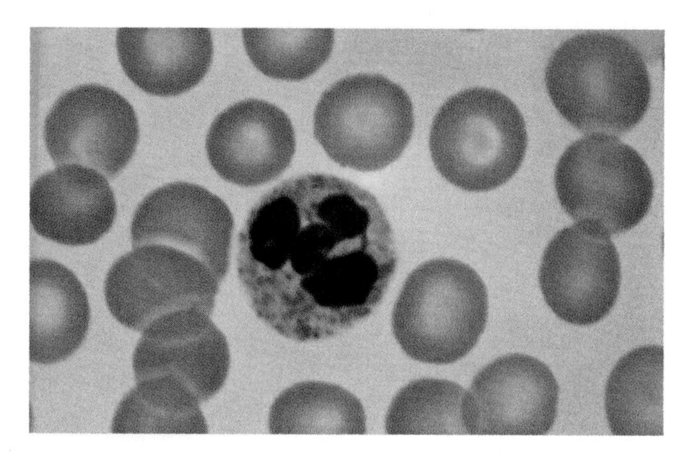

Figure 12.6

The neutrophil has a lobed nucleus with two to five sections (640×).

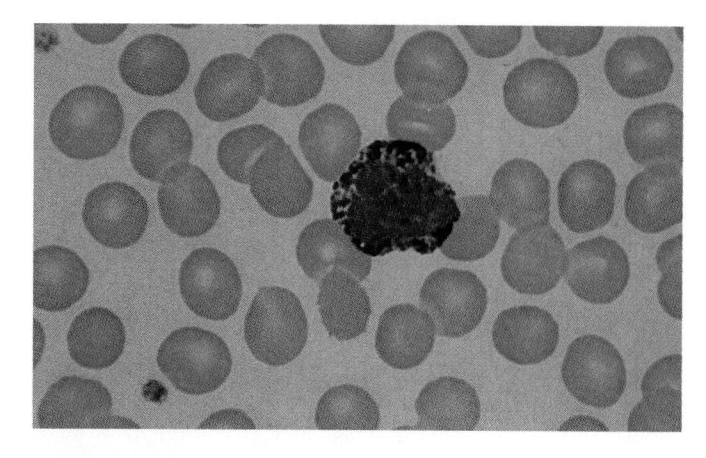

Figure 12.8

The basophil has cytoplasmic granules that stain deep blue (400×).

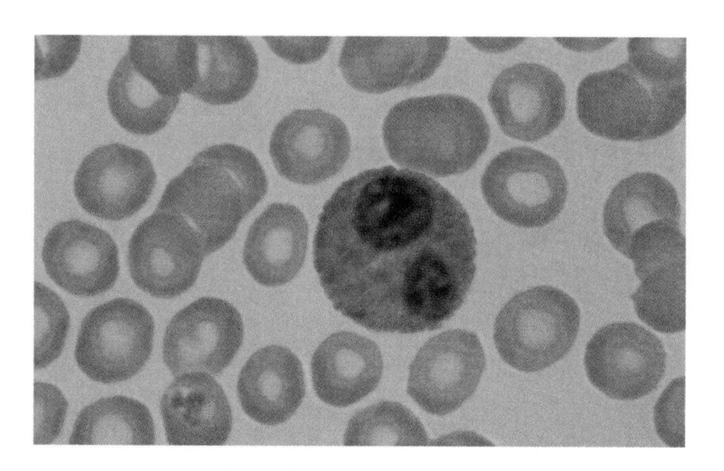

Figure 12.7

The eosinophil has red-staining cytoplasmic granules (500×).

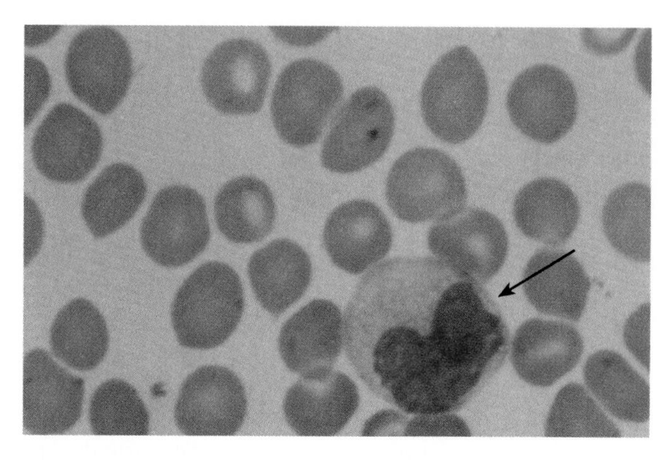

Figure 12.9

The monocyte is the largest type of blood cell (640×).

Neutrophils have fine cytoplasmic granules that stain pinkish in neutral stain. The nucleus of a neutrophil is lobed and consists of two to five sections connected by thin strands of chromatin (fig. 12.6). Neutrophils account for 54–62% of the leukocytes in a typical blood sample from an adult.

Eosinophils contain coarse, uniformly sized cytoplasmic granules that stain deep red in acid stain (fig. 12.7). The nucleus usually has only two lobes (bilobed). These cells make up 1–3% of the total number of circulating leukocytes.

Basophils are similar to eosinophils in size and in the shape of their nuclei, but they have fewer, more irregularly shaped cytoplasmic granules that stain deep blue in basic stain (fig. 12.8). Basophils usually account for less than 1% of the leukocytes.

The leukocytes of the agranulocyte group include monocytes and lymphocytes. Monocytes generally arise from red bone marrow. Lymphocytes form in the organs of the lymphatic system, as well as in the red bone marrow (see chapter 14).

Monocytes, the largest blood cells, are two to three times greater in diameter than red blood cells (fig. 12.9). Their nuclei vary in shape and can be round, kidney-shaped, oval, or lobed. They usually make up 3–9% of the leukocytes in a blood sample and live for several weeks or even months.

Lymphocytes are usually only slightly larger than red blood cells. A typical lymphocyte contains a large, round nucleus within a thin rim of cytoplasm (fig. 12.10). These cells account for 25–33% of circulating leukocytes. They may live for years.

Figure 12.10

The lymphocyte contains a large, round nucleus (500×).

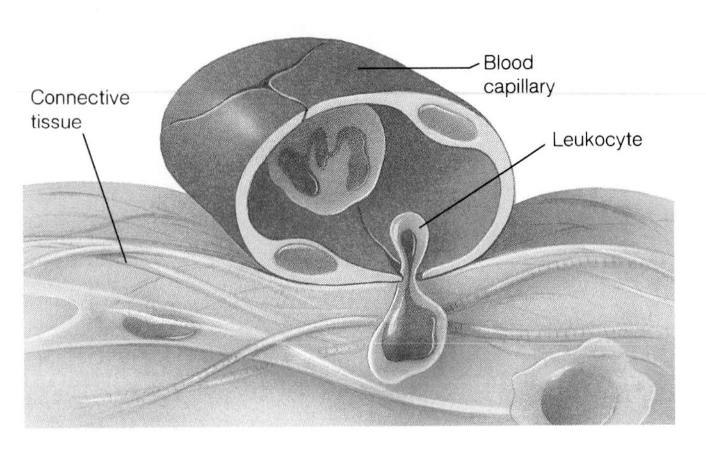

Figure 12.11

Leukocytes can squeeze between the cells of a capillary wall and enter the tissue space outside the blood vessel.

1. Distinguish between granulocytes and agranulocytes.
2. List the five types of white blood cells and their distinguishing characteristics.

White Blood Cell Counts

The number of white blood cells in a cubic millimeter of human blood, called the *white blood cell count* (WBCC), normally includes 5,000–10,000 cells. Because this number may change in response to abnormal conditions, white blood cell counts are of clinical interest. For example, a rise in the number of circulating white blood cells accompanies some infectious diseases. A total number of white blood cells exceeding 10,000 per mm^3 of blood constitutes **leukocytosis,** indicating acute infection, such as appendicitis.

A total white blood cell count below 5,000 per mm^3 of blood is called **leukopenia.** Such a deficiency may accompany typhoid fever, influenza, measles, mumps, chicken pox, AIDS, or poliomyelitis.

A *differential white blood cell count* (DIFF) lists percentages of the types of leukocytes in a blood sample. This is useful because the relative proportions of white blood cells may change in particular diseases. Neutrophils, for instance, usually increase during bacterial infections, and eosinophils may increase during certain parasitic infections and allergic reactions. In AIDS, numbers of a certain type of lymphocyte drop sharply.

1. What is the normal human white blood cell count?
2. Distinguish between leukocytosis and leukopenia.
3. What is a differential white blood cell count?

Functions of White Blood Cells

White blood cells protect against infection in various ways. Some leukocytes phagocytize bacterial cells in the body, and others produce proteins (*antibodies*) that destroy or disable foreign particles.

Leukocytes can squeeze between the cells that form blood vessel walls. This movement, called *diapedesis,* allows white blood cells to leave the circulation (fig. 12.11). Once outside the blood, they move through interstitial spaces using a form of self-propulsion called *amoeboid motion.*

The most mobile and active phagocytic leukocytes are neutrophils and monocytes. Neutrophils cannot ingest particles much larger than bacterial cells, but monocytes can engulf large objects. Both of these phagocytes contain many *lysosomes,* which are organelles filled with digestive enzymes that break down organic molecules in captured bacteria. Neutrophils and monocytes often become so engorged with digestive products and bacterial toxins that they also die.

Eosinophils are only weakly phagocytic, but they are attracted to and can kill certain parasites. Eosinophils also help control inflammation and allergic reactions by removing biochemicals associated with these reactions.

Table 12.1

Cellular Components of Blood

Component	Description	Number Present	Function
Red blood cell (erythrocyte)	Biconcave disk without a nucleus, about one-third hemoglobin	4,200,000–6,200,000 per mm^3	Transports oxygen and carbon dioxide
White blood cell (leukocyte)		5,000–10,000 per mm^3	Destroys pathogenic microorganisms and parasites and removes worn cells
Granulocytes	About twice the size of red blood cells; cytoplasmic granules are present		
1. Neutrophil	Nucleus with two to five lobes; cytoplasmic granules stain pink in neutral stain	54–62% of white blood cells present	Phagocytizes small particles
2. Eosinophil	Bilobed nucleus, cytoplasmic granules stain red in acid stain	1–3% of white blood cells present	Kills parasites and helps control inflammation and allergic reactions
3. Basophil	Bilobed nucleus, cytoplasmic granules stain blue in basic stain	Less than 1% of white blood cells present	Releases heparin and histamine
Agranulocytes	Cytoplasmic granules are absent		
1. Monocyte	Two to three times larger than a red blood cell; nuclear shape varies from round to lobed	3–9% of white blood cells present	Phagocytizes large particles
2. Lymphocyte	Only slightly larger than a red blood cell; nucleus nearly fills cell	25–33% of white blood cells present	Provides immunity
Platelet (thrombocyte)	Cytoplasmic fragment	130,000–360,000 per mm^3	Helps control blood loss from broken vessels

Some of the cytoplasmic granules of basophils contain a blood-clot-inhibiting substance called *heparin,* and other granules contain *histamine*. Basophils release heparin to help prevent intravascular blood clot formation and release histamine to increase blood flow to injured tissues. Basophils also play major roles in certain allergic reactions.

Lymphocytes are important in *immunity*. Some, for example, produce antibodies that attack specific foreign substances that enter the body. Chapter 14 discusses immunity.

1. What are the primary functions of white blood cells?
2. Which white blood cells are the most active phagocytes?
3. What are the functions of eosinophils and basophils?

Blood Platelets

Platelets, or **thrombocytes**, are not complete cells. They arise from very large cells in red bone marrow, called **megakaryocytes**, that fragment like a shattered plate, releasing small sections of cytoplasm—the platelets—into the circulation. The larger fragments of the megakaryocytes shrink and also become platelets as they pass through blood vessels in the lungs.

Each platelet lacks a nucleus and is less than half the size of a red blood cell. It is capable of amoeboid movement and may live for about ten days. In normal blood, the *platelet count* varies from 130,000–360,000 per mm^3. Platelets help close breaks in damaged blood vessels and initiate formation of blood clots, as a subsequent section of this chapter explains.

Table 12.1 summarizes the characteristics of blood cells and platelets.

1. What is the normal blood platelet count?
2. What is the function of blood platelets?

TOPIC OF INTEREST

Leukemia

The young woman at first noticed fatigue and headaches, which she attributed to studying for final exams. She had frequent colds and bouts of fever, chills, and sweats that she thought were just minor infections. When she developed several bruises and bone pain and noticed that her blood did not clot very quickly after cuts and scrapes, she consulted her physician, who examined her and took a blood sample. One glance at a blood smear under a microscope alarmed the doctor—there were far too few red blood cells and platelets, and too many white blood cells. She sent the sample to a laboratory to diagnose the type of *leukemia,* or cancer of the white blood cells, that was causing her patient's symptoms.

The young woman had *myeloid leukemia.* Her red bone marrow was producing too many granulocytes, but they were immature cells, unable to fight infection. This explained the frequent illnesses. The leukemic cells were crowding out red blood cells and their precursors in the red marrow, causing her anemia and resulting fatigue. Platelet deficiency (thrombocytopenia) led to increased tendency to bleed. Finally, spread of the cancer cells outside the marrow painfully weakened the surrounding bone. Eventually, if she was not treated, the cancer cells would spread outside the circulatory system, causing other tissues that would normally not produce white blood cells to do so.

A second type of leukemia, distinguished by the source of the cancer cells, is *lymphoid leukemia.* These cancer cells are lymphocytes, produced in lymph nodes. Many of the symptoms are similar to those of myeloid leukemia. Sometimes, a person has no leukemia symptoms at all, and a routine blood test detects the condition (fig. 12A).

Leukemia is also classified as acute or chronic. An acute condition appears suddenly, symptoms progress rapidly, and death occurs in a few months without treatment. Chronic forms begin more slowly and may remain undetected for months or even years or, in rare cases, decades. Without treatment, life expectancy after symptoms develop is about three years. With treatment, 50–80% of patients enter remission, a period of stability that may become a cure. Chemotherapy may be necessary for a year or two to increase the chances of long remission.

Leukemia treatment includes correcting symptoms with blood transfusions, treating infections, and using drugs that kill cancer cells. Several drugs in use for many years have led to spectacular increases in cure rates, particularly for acute lymphoid leukemia in children. Some newer treatments offer hope for other types of leukemia, too. If other treatments fail, a bone marrow transplant can cure leukemia, but the procedure is very risky.

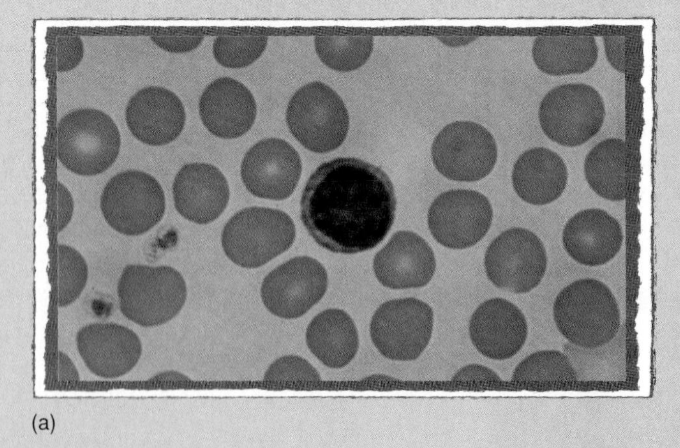

(a)

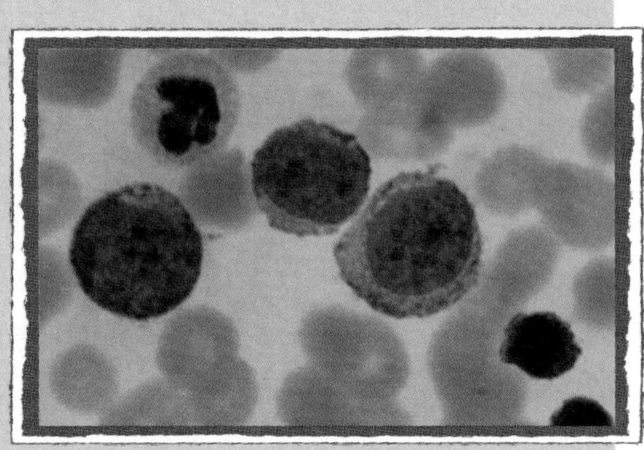

(b)

Figure 12A

(*a*) Normal blood cells (500×). (*b*) Blood cells from a person with lymphoid leukemia (500×). Note the increased number of leukocytes.

Blood Plasma

Plasma is the clear, straw-colored, liquid portion of the blood in which cells and platelets are suspended. It is approximately 92% water and contains a complex mixture of organic and inorganic biochemicals. Functions of plasma constituents include transporting nutrients, gases, and vitamins; helping to regulate fluid and electrolyte balance; and maintaining a favorable pH.

Plasma Proteins

Plasma proteins are the most abundant of the dissolved substances (solutes) in plasma. These proteins remain in the blood and interstitial fluids, and ordinarily are not used as energy sources. The three main plasma protein groups—albumins, globulins, and fibrinogen—differ in chemical composition and physiological function.

Albumins are the smallest of the plasma proteins, yet account for about 60% of these proteins by weight. They are synthesized in the liver, and because they are so plentiful, albumins are an important determinant of blood *osmotic pressure.*

Recall from chapter 3 that whenever solute concentration changes on either side of a cell membrane, water moves across the membrane toward the region where solutes are in higher concentration. Thus, the concentration of plasma solutes must remain relatively stable, or water tends to leave the blood and enter the tissues, or leave the tissues and enter the blood, by *osmosis.* Because the presence of albumins (and other plasma proteins) adds to blood osmotic pressure, albumins help regulate water movement between blood and the tissues. In doing so, they help control blood volume, which, in turn, is directly related to blood pressure (see chapter 13).

If the concentration of plasma proteins falls, tissues swell—a condition called *edema.* This may result from starvation or a protein-deficient diet, either of which requires the body to use protein for energy, or from an impaired liver that cannot synthesize plasma proteins. As blood concentration of plasma proteins drops, so does blood osmotic pressure, sending fluids into intercellular spaces. ■

Globulins, which make up about 36% of the plasma proteins, can be further subdivided into *alpha, beta,* and *gamma globulins.* The liver synthesizes alpha and beta globulins. They have a variety of functions, including transport of lipids and fat-soluble vitamins. Lymphatic tissues produce the gamma globulins, which are a type of antibody (see chapter 14).

Table 12.2 Plasma Proteins

Protein	Percentage of Total	Origin	Function
Albumin	60%	Liver	Helps maintain blood osmotic pressure
Globulin	36%		
Alpha globulins		Liver	Transport lipids and fat-soluble vitamins
Beta globulins		Liver	Transport lipids and fat-soluble vitamins
Gamma globulins		Lymphatic tissues	Constitute a type of antibody
Fibrinogen	4%	Liver	Blood coagulation

Fibrinogen, which constitutes about 4% of the plasma proteins, functions in blood coagulation, as discussed later in the chapter. Synthesized in the liver, fibrinogen is the largest of the plasma proteins.

Table 12.2 summarizes the characteristics of the plasma proteins.

1. List three types of plasma proteins.
2. How do albumins help maintain water balance between blood and tissues?
3. What are the functions of globulins?

Nutrients and Gases

The *plasma nutrients* include amino acids, simple sugars, nucleotides, and lipids absorbed from the digestive tract. For example, plasma transports glucose from the small intestine to the liver, where glucose can be stored as glycogen or react to form fat. If blood glucose concentration drops below the normal range, glycogen may be broken down into glucose, as chapter 11 describes. Plasma also carries recently absorbed amino acids to the liver, where they can be used to manufacture proteins, or deaminated and used as an energy source (see chapter 4).

Plasma lipids include fats (triglycerides), phospholipids, and cholesterol. Because lipids are not water soluble and plasma is almost 92% water, these lipids combine with proteins in **lipoprotein** complexes. Lipoprotein molecules are large and consist of a core of triglyceride surrounded by a surface layer of phospholipid, cholesterol, and protein. *Apoproteins* in the outer layer can combine with receptors on the membranes of specific target cells. Lipoprotein molecules vary in the proportions of lipids they contain.

Lipids are less dense than proteins. Therefore, as the proportion of lipids in a lipoprotein increases, the density of the particle decreases. Conversely, as the proportion of lipids decreases, the density increases. Lipoproteins are classified on the basis of their densities, which reflect their composition. *Chylomicrons* consist mainly of triglycerides absorbed from the small intestine (see chapter 15). *Very low-density lipoproteins* (*VLDL*) have a high concentration of triglycerides. *Low-density lipoproteins* (*LDL*) have a high concentration of cholesterol and are the major cholesterol-carrying lipoproteins. *High-density lipoproteins* (*HDL*) have a high concentration of protein and a lower concentration of lipids.

Chylomicrons transport dietary fats to muscle and adipose cells. Similarly, VLDL molecules, produced in the liver, transport triglycerides synthesized from excess dietary carbohydrates. After VLDL molecules deliver their loads of triglycerides to adipose cells, the remnants of the VLDL molecules break down to LDL. An enzyme called *lipoprotein lipase* catalyzes this reaction. Because most of the triglycerides have been removed, LDL molecules have a higher cholesterol content than do the original VLDL molecules. Various cells, including liver cells, have surface receptors that combine with apoproteins associated with LDL molecules. These cells slowly remove LDL from plasma by receptor-mediated endocytosis, supplying cells with cholesterol (see chapter 3).

After chylomicrons deliver their triglycerides to cells, their remnants are transferred to HDL molecules. HDL molecules, which form in the liver and small intestine, transport chylomicron remnants to the liver, where they enter cells rapidly by receptor-mediated endocytosis. The liver disposes of cholesterol it obtains in this manner by secreting it into bile or by using it to synthesize bile salts.

Table 12.3 summarizes the characteristics and functions of lipoproteins.

The intestine reabsorbs much of the cholesterol and bile salts in bile, which are then transported back to the liver, and the secretion-reabsorption cycle repeats. During each cycle, some of the cholesterol and bile salts escape reabsorption, reach the large intestine, and are eliminated with the feces.

The most important *blood gases* are oxygen and carbon dioxide. Plasma also contains a considerable amount of dissolved nitrogen, which ordinarily has no physiological function. Chapter 16 discusses the blood gases and their transport.

Table 12.3 Plasma Lipoproteins

Lipoprotein	Characteristics	Function
Chylomicron	High concentration of triglycerides	Transports dietary fats to muscle and adipose cells
Very low-density lipoprotein (VLDL)	Relatively high concentration of triglycerides; produced in the liver	Transports triglycerides synthesized in the liver from carbohydrates to adipose cells
Low-density lipoprotein (LDL)	Relatively high concentration of cholesterol; formed from remnants of VLDL molecules that have given up their triglycerides	Delivers cholesterol to various cells, including liver cells
High-density lipoprotein (HDL)	Relatively high concentration of protein and low concentration of lipids	Transports to the liver remnants of chylomicrons that have given up their triglycerides

Nonprotein Nitrogenous Substances

Molecules that contain nitrogen atoms but are not proteins comprise a group called **nonprotein nitrogenous substances.** In plasma, this group includes amino acids, urea, and uric acid. Amino acids come from protein digestion and amino acid absorption. Urea and uric acid are products of protein and nucleic acid catabolism, respectively, and are excreted in the urine.

Plasma Electrolytes

Blood plasma contains a variety of *electrolytes* that are absorbed from the intestine or released as by-products of cellular metabolism. They include sodium, potassium, calcium, magnesium, chloride, bicarbonate, phosphate, and sulfate ions. Of these, sodium and chloride ions are the most abundant. Such ions are important in maintaining the osmotic pressure and pH of plasma, and like other plasma constituents, they are regulated so that their blood concentrations remain relatively stable. Chapter 18 discusses these electrolytes in connection with water and electrolyte balance.

1. What nutrients are in blood plasma?
2. What gases are in plasma?
3. What is a nonprotein nitrogenous substance?
4. What are the sources of plasma electrolytes?

Hemostasis

Hemostasis is the stoppage of bleeding, which is vitally important when blood vessels are damaged. Following an injury to blood vessels, several actions may help prevent blood loss, including blood vessel spasm, platelet plug formation, and blood coagulation.

Blood Vessel Spasm

Cutting or breaking a blood vessel stimulates the smooth muscles in the vessel wall to contract, lessening blood loss almost immediately. In fact, such a **vasospasm** may completely close the ends of a severed vessel.

Vasospasm may last only a few minutes, but the effect of the direct stimulation usually continues for about 30 minutes. By then, a platelet plug has formed, and blood is clotting. Also, platelets release serotonin, which contracts smooth muscles in the blood vessel walls. This vasoconstriction further helps to reduce blood loss.

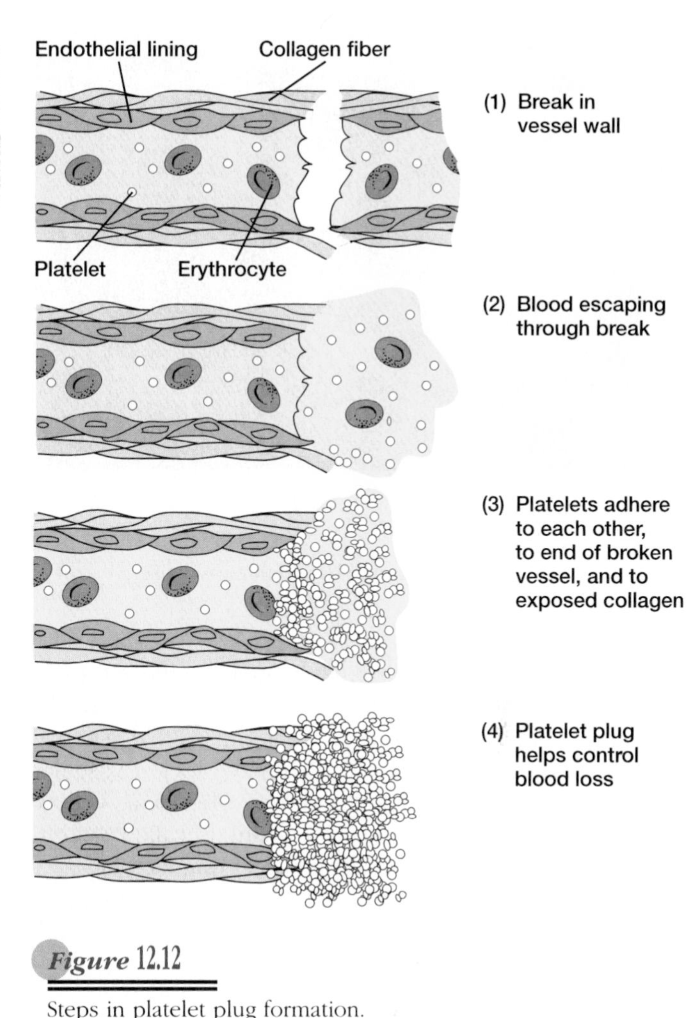

Figure 12.12
Steps in platelet plug formation.

Platelet Plug Formation

Platelets stick to any rough surface and to the collagen in connective tissue. Consequently, when a blood vessel breaks, platelets adhere to the collagen underlying the endothelial lining of blood vessels. Platelets also stick to each other, forming a platelet plug in the vascular break. A plug may control blood loss from a small break, but a larger break may require a blood clot to halt bleeding. Figure 12.12 shows the steps in platelet plug formation.

1. What is hemostasis?
2. How does a blood vessel spasm help control bleeding?
3. Describe the formation of a platelet plug.

Blood Coagulation

Coagulation, the most effective hemostatic mechanism, forms a *blood clot*. Blood coagulation is very complex and utilizes many biochemicals called *clotting factors*. Some of these factors promote coagulation, and others inhibit it. Whether or not blood coagulates depends on the balance between these two groups of factors. Normally, anticoagulants prevail, and the blood does not clot. However, as a result of injury (trauma), biochemicals that favor coagulation may increase in concentration, and the blood may coagulate.

The major event in blood clot formation is changing the soluble plasma protein fibrinogen into relatively insoluble threads of the protein **fibrin.** Damaged tissues release *tissue thromboplastin,* initiating a series of reactions resulting in the production of *prothrombin activator.* This series of changes depends on the presence of calcium ions as well as certain proteins and phospholipids for completion.

Prothrombin is an alpha globulin that the liver continually produces and thus is a normal constituent of plasma. In the presence of calcium ions, prothrombin activator converts prothrombin into **thrombin.** Thrombin, in turn, catalyzes a reaction that fragments fibrinogen. The fibrinogen fragments join, forming long threads of fibrin. Certain other proteins also enhance fibrin formation.

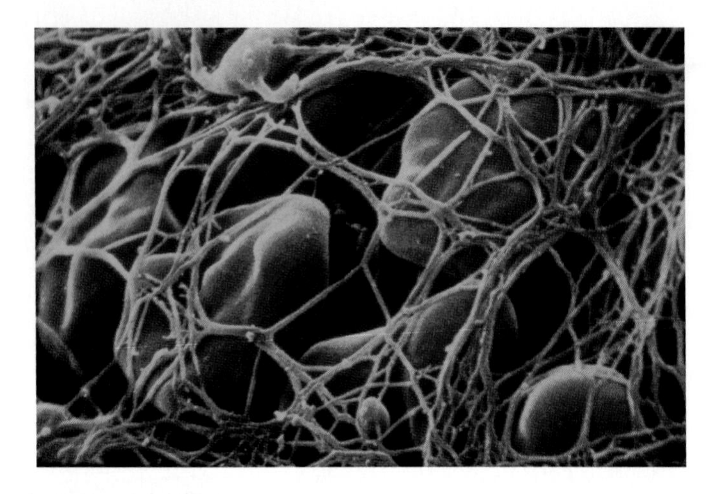

Figure 12.13

Scanning electron micrograph of fibrin threads (falsely colored) (5000×).

Once fibrin threads form, they stick to the exposed surfaces of damaged blood vessels, creating a meshwork that entraps blood cells and platelets (fig. 12.13). The resulting mass is a blood clot, which may block a vascular break and prevent further blood loss. The clear, yellow liquid that remains after the clot forms is called *serum.* Serum is the same as plasma, minus clotting factors.

The amount of prothrombin activator that appears in the blood is directly proportional to the degree of tissue damage. Once a blood clot begins to form, it promotes still more clotting because thrombin also acts directly on blood clotting factors other than fibrinogen, causing prothrombin to form still more thrombin. This is a **positive feedback system,** in which the original action stimulates more of the same type of action. Such a positive feedback mechanism produces unstable conditions and can operate for only a short time in a living system because life depends on the maintenance of a stable internal environment (see chapter 1).

Laboratory tests commonly used to evaluate the blood coagulation mechanisms include *prothrombin time (PT)* and *partial thromboplastin time (PTT).* Both of these tests measure the time it takes for fibrin threads to form in a sample of blood plasma. ■

Normally, blood flow throughout the body prevents formation of a massive clot within the cardiovascular system by rapidly carrying excess thrombin away and keeping its concentration too low to enhance further clotting. Consequently, blood coagulation is usually limited to blood that is standing still (or moving relatively slowly), and clotting ceases where a clot contacts circulating blood.

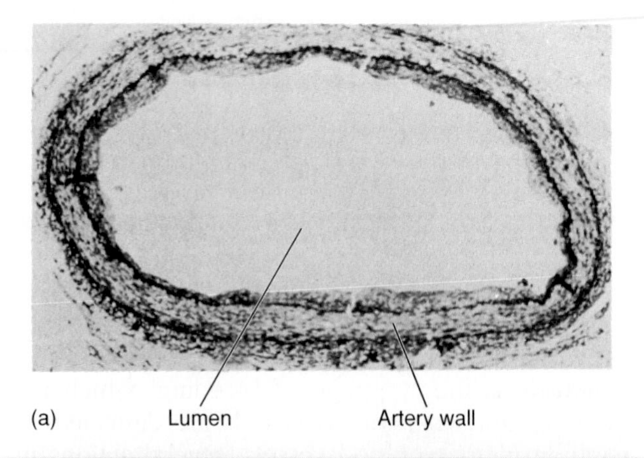

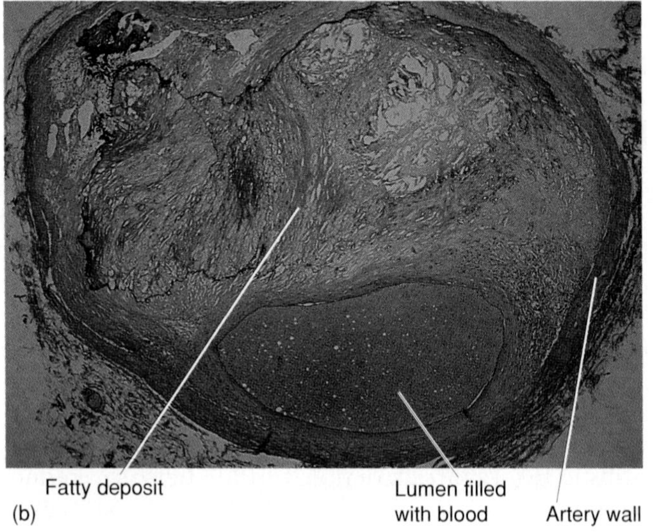

Figure 12.14

(*a*) Light micrograph of a normal artery (60×).
(*b*) Atherosclerotic plaque occludes the inner wall of this artery (60×).

Fibroblasts (see chapter 5) soon invade blood clots that form in ruptured vessels, producing fibrous connective tissue throughout the clots, which helps strengthen and seal vascular breaks. Many clots, including those that form in tissues as a result of blood leakage (hematomas), disappear in time. This dissolution depends on activation of a plasma protein that can digest fibrin threads and other proteins associated with clots. Clots that fill large blood vessels, however, are seldom removed naturally.

A blood clot forming in a vessel abnormally is a **thrombus.** If the clot dislodges or if a fragment of it breaks loose and is carried away by the blood flow, it is called an **embolus.** Generally, emboli continue to move until they reach narrow places in vessels, where they lodge and may interfere with blood flow.

Abnormal clot formations are often associated with conditions that change endothelial linings of vessels. For example, in *atherosclerosis,* fatty deposits change arterial linings, sometimes initiating inappropriate clotting (fig. 12.14).

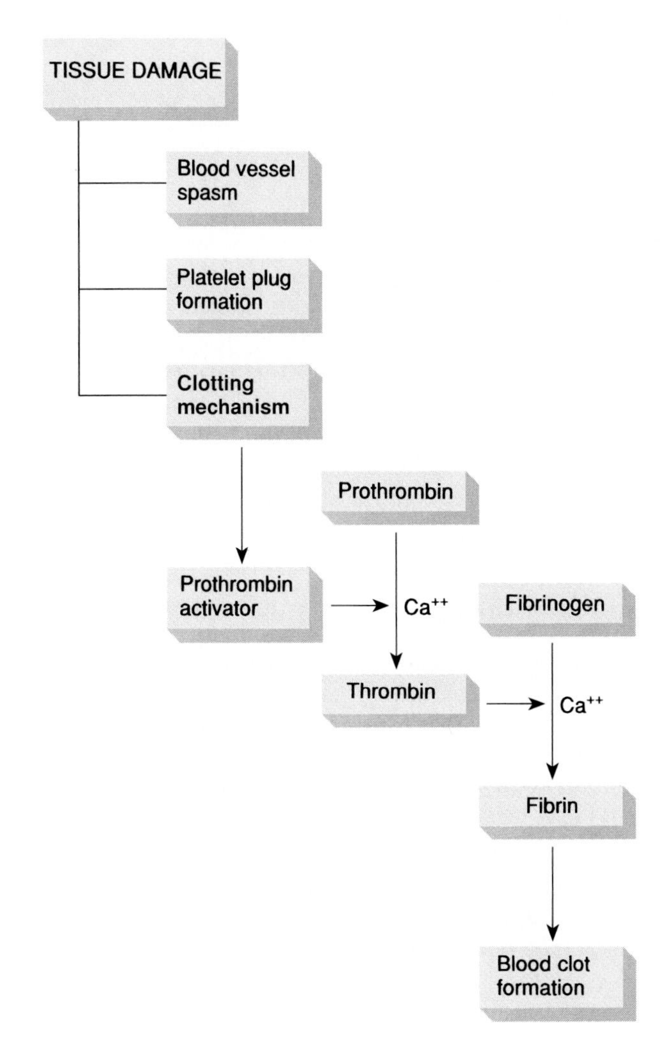

Figure 12.15

Blood vessel spasm, platelet plug formation, and blood coagulation provide hemostasis following tissue damage.

Figure 12.15 summarizes the three primary hemostatic mechanisms.

> A blood clot forming in a vessel that supplies a vital organ, such as the heart (coronary thrombosis) or the brain (cerebral thrombosis), kills tissues the vessel serves (*infarction*) and may be fatal. A blood clot that travels and then blocks a vessel that supplies a vital organ, such as the lungs (pulmonary embolism), affects the portion of the organ the blocked blood vessel supplies. Plasminogen activators break up abnormal blood clots and are used to treat heart attacks and strokes. ■

1. Review the major steps in blood clot formation.
2. What prevents the formation of massive clots throughout the cardiovascular system?
3. Distinguish between a thrombus and an embolus.

Blood Groups and Transfusions

Early attempts to transfer blood from one person to another produced varied results. Sometimes, the recipient improved. Other times, the recipient suffered a blood reaction in which the red blood cells clumped, obstructing vessels and producing other serious consequences.

Eventually, scientists determined that blood is of differing types and that only certain combinations of blood types are compatible. These discoveries led to the development of procedures for typing blood. Today, safe transfusions of whole blood depend on properly matching the blood types of donors and recipients.

Antigens and Antibodies

Agglutination is the clumping of red blood cells following a transfusion reaction. This phenomenon is due to a reaction between red blood cell surface molecule **antigens** (formerly called *agglutinogens*) and protein **antibodies** (formerly called *agglutinins*) carried in plasma.

Only a few of the many antigens on red blood cell membranes can produce serious transfusion reactions. These include the antigens of the ABO group and those of the Rh group. Avoiding the mixture of certain kinds of antigens and antibodies prevents adverse transfusion reactions.

> A person who has received mismatched blood quickly feels the effects of agglutination—anxiety, breathing difficulty, facial flushing, headache, and severe pain in the neck, chest, and lumbar area. Red blood cells burst, releasing free hemoglobin. Macrophages phagocytize the hemoglobin, converting it to bilirubin, which may accumulate sufficiently to cause the yellow skin of jaundice. Free hemoglobin in the kidneys may ultimately cause them to fail. ■

TOPIC OF INTEREST: Coagulation Disorders

Hemophilia

In 1962, five-year-old Bob Massie developed uncontrollable bleeding in his left knee, a symptom of his *hemophilia A,* an inherited clotting disorder. It took thirty transfusions of plasma over the next three months to stop the bleeding. Because the knee joint had swelled and locked into place during that time, Bob was unable to walk for the next seven years. Today, Bob still suffers from painful joint bleeds, but he injects himself with factor VIII, the coagulation protein that his body cannot make. The factor VIII soon controls the bleed.

Hemophilia has left its mark on history. One of the earliest descriptions is in the Talmud, a second century B.C. Jewish document, which reads, "If she circumcised her first child and he died, and a second one also died, she must not circumcise her third child." Queen Victoria (1819–1901) passed the hemophilia gene to several of her children, eventually spreading the condition to the royal families of England, Russia, Germany, and Spain. Hemophilia achieved notoriety when factor VIII pooled from blood donations was discovered to transmit HIV in 1985. Ninety percent of people with severe hemophilia who used such pooled factor VIII prior to then have developed AIDS.

Abnormalities of different clotting factors cause different forms of hemophilia, but hemophilia A is the most common. Symptoms of the hemophilias include tendency to hemorrhage severely following minor injuries, frequent nosebleeds, large intramuscular hematomas, and blood in the urine.

von Willebrand Disease

Easy bruising and a tendency to bleed easily are signs of *von Willebrand disease,* another inherited clotting disorder that is usually far less severe than hemophilia. Affected persons lack a plasma protein, von Willebrand factor, that endothelial cells lining blood vessels secrete and that enables platelets to adhere to damaged blood vessel walls, a key step preceding actual clotting. Sometimes, the condition can cause spontaneous bleeding from the mucous membranes of the gastrointestinal and urinary tracts.

ABO Blood Group

The *ABO blood group* is based on the presence (or absence) of two major antigens on red blood cell membranes—antigen A and antigen B. A person's erythrocytes contain one of four antigen combinations as a result of inheritance: only A, only B, both A and B, or neither A nor B.

A person with only antigen A has *type A blood.* A person with only antigen B has *type B blood.* An individual with both antigen A and B has *type AB blood.* A person with neither antigen A nor B has *type O blood.* Thus, all humans have one of four possible ABO blood types—A, B, AB, or O.

Certain antibodies are synthesized in the plasma about two to eight months following birth. Specifically, whenever antigen A is absent in red blood cells, an antibody called *anti-A* develops, and whenever antigen B is absent, an antibody called *anti-B* develops. Therefore, persons with type A blood have antibody anti-B in their plasma; those with type B blood have antibody anti-A; those with type AB blood have neither antibody; and those with type O blood have both antibody anti-A and antibody anti-B (fig. 12.16). Table 12.4 summarizes the antigens and antibodies of the ABO blood group.

An antibody of one type will react with an antigen of the same type and clump red blood cells; therefore, such combinations must be avoided. The major concern in blood transfusion procedures is that the cells in the donated blood not clump due to antibodies in the recipient's plasma. For this reason, a person with type A (anti-B) blood must not receive blood of type B or AB, either of which would clump in the presence of anti-B in the recipient's type A blood. Likewise, a person with type B (anti-A) blood must not receive type A or AB blood, and a person with type O (anti-A and anti-B) blood must not receive type A, B, or AB blood (fig. 12.17).

Because type AB blood lacks both anti-A and anti-B antibodies, an AB person can receive a transfusion of blood of any type. For this reason, type AB persons are sometimes called *universal recipients.* However, type A (anti-B) blood, type B (anti-A) blood, and type O (anti-A and anti-B) blood still contain antibodies (either anti-A and/or anti-B) that could agglutinate type AB cells if transfused rapidly. Consequently, even for AB individuals, using donor blood of the same type as the recipient is best (see table 12.5).

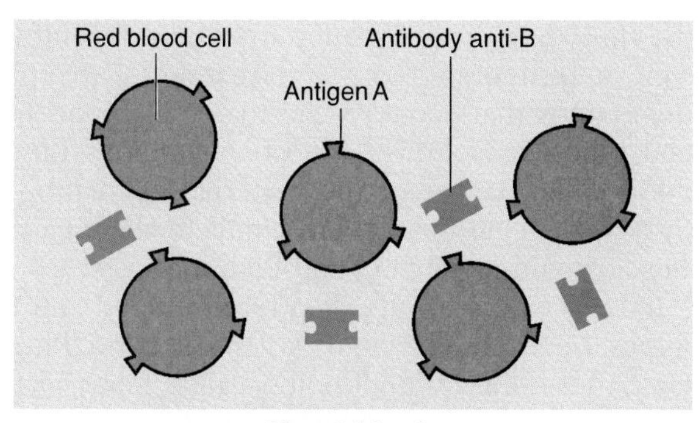

Type A blood

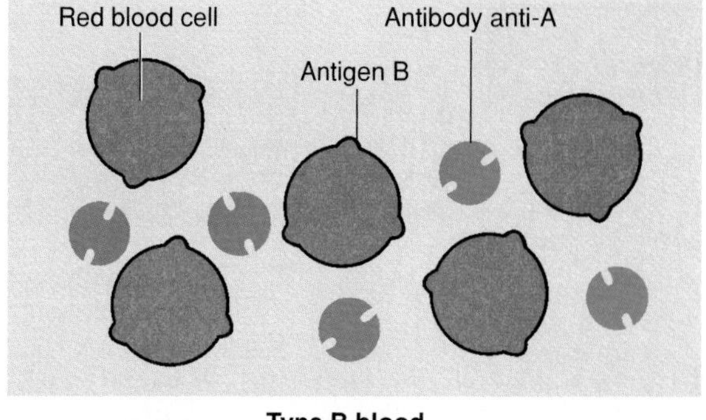

Type B blood

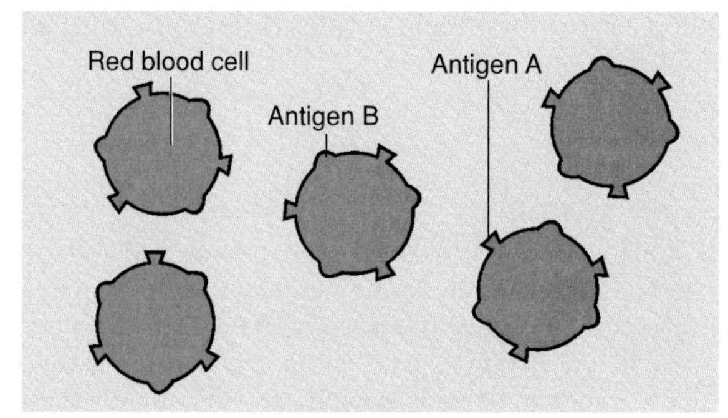

Type AB blood

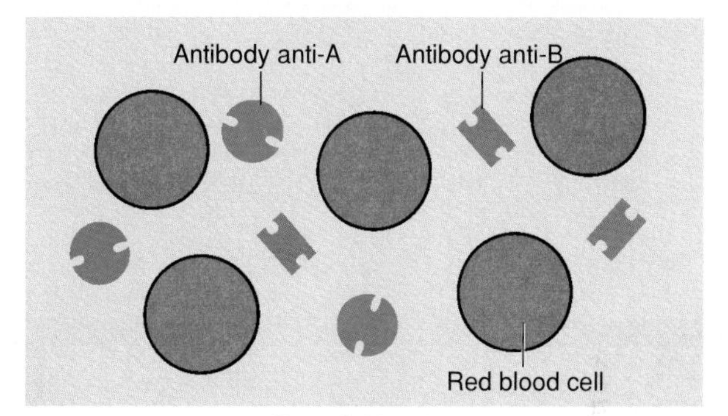

Type O blood

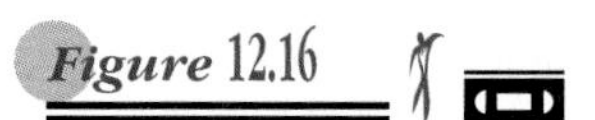

Figure 12.16

Different combinations of antigens and antibodies distinguish blood types.

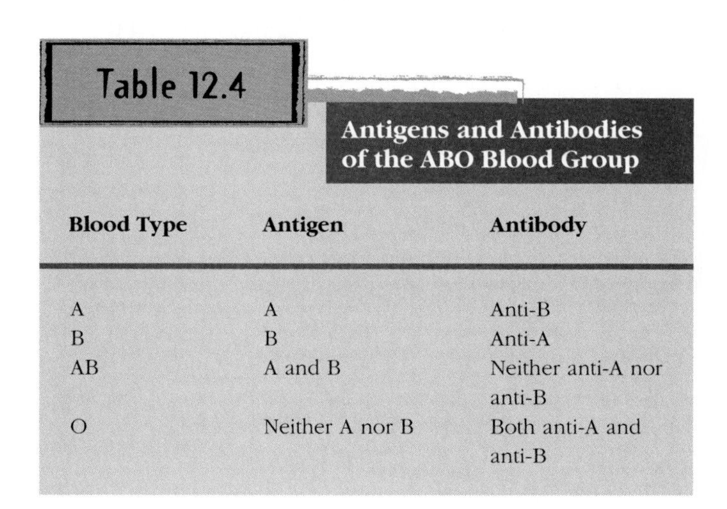

Table 12.4

Antigens and Antibodies of the ABO Blood Group

Blood Type	Antigen	Antibody
A	A	Anti-B
B	B	Anti-A
AB	A and B	Neither anti-A nor anti-B
O	Neither A nor B	Both anti-A and anti-B

Similarly, because type O blood lacks antigens A and B, this type could theoretically be transfused into persons with blood of any other type. Therefore, persons with type O blood are sometimes called *universal donors*. Type O blood, however, does contain both anti-A and anti-B antibodies. If type O blood is given to a person with blood type A, B, or AB, it should be transfused slowly so that the recipient's larger blood volume will dilute it, minimizing the chance of an adverse reaction.

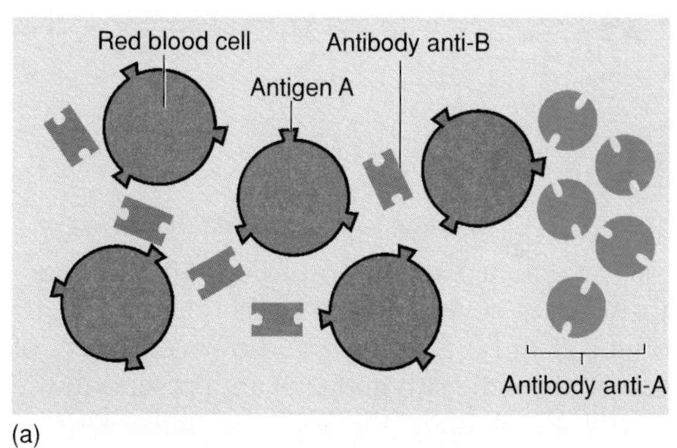

(a)

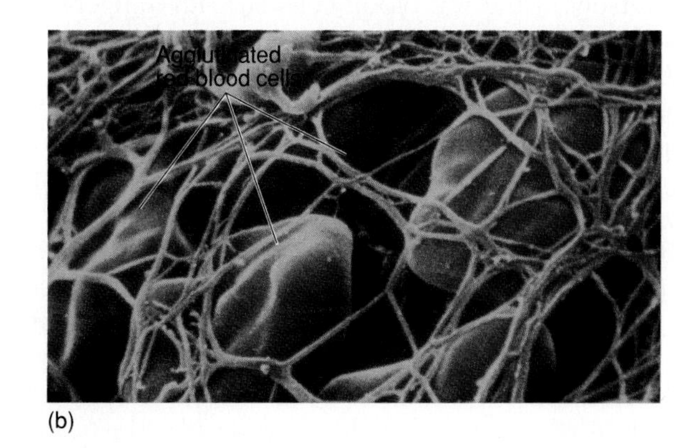

(b)

Figure 12.17

(*a*) If red blood cells with antigen A are added to blood containing antibody anti-A, (*b*) the antibodies react with the antigens, causing clumping (agglutination).

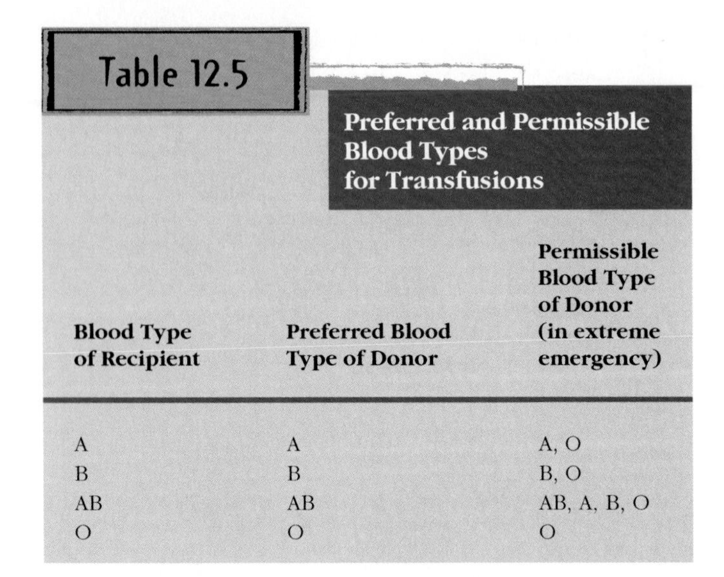

Table 12.5

Preferred and Permissible Blood Types for Transfusions

Blood Type of Recipient	Preferred Blood Type of Donor	Permissible Blood Type of Donor (in extreme emergency)
A	A	A, O
B	B	B, O
AB	AB	AB, A, B, O
O	O	O

1. Distinguish between antigens and antibodies.
2. What is the main concern when blood is transfused from one individual to another?
3. Why is a type AB person called a universal recipient?
4. Why is a type O person called a universal donor?

Rh Blood Group

The *Rh blood group* was named after the rhesus monkey, in which it was first studied. In humans, this group includes several Rh antigens (factors). The most important of these is antigen D; however, if any of the antigen D and other Rh antigens are present on the red blood cell membranes, the blood is called *Rh positive.* Conversely, if the red blood cells lack Rh antigens, the blood is called *Rh negative.*

As in the case of antigens A and B, the presence (or absence) of Rh antigens is an inherited trait. Unlike anti-A and anti-B, antibodies for Rh (*anti-Rh*) do not appear spontaneously. Instead, they form only in Rh-negative persons in response to special stimulation.

If an Rh-negative person receives a transfusion of Rh-positive blood, the Rh antigens stimulate the recipient's antibody-producing cells to begin producing anti-Rh antibodies. Generally, this initial transfusion has no serious consequences, but if the Rh-negative person—who is now sensitized to Rh-positive blood—receives another transfusion of Rh-positive blood some months later, the donated red cells are likely to agglutinate.

A related condition may occur when an Rh-negative woman is pregnant with an Rh-positive fetus for the first time. Such a pregnancy may be uneventful; however, at birth (or during a miscarriage), the placental membranes that separate the maternal blood from the fetal blood tear, and some of the infant's Rh-positive blood cells may enter the maternal circulation. These Rh-positive cells may then stimulate the maternal tissues to begin producing anti-Rh antibodies.

If a woman who has already developed anti-Rh antibodies becomes pregnant with a second Rh-positive fetus, these anti-Rh antibodies, called hemolysins, pass through the placental membrane and destroy the fetal red blood cells. The fetus then develops a condition called **erythroblastosis fetalis** (hemolytic disease of the newborn) (fig. 12.18).

Erythroblastosis fetalis is extremely rare today because physicians carefully note Rh status. An Rh-negative woman who might carry an Rh-positive fetus receives an injection of a drug called Rhogam. This is actually anti-Rh antibodies, which bind to and shield any Rh-positive fetal cells that might contact the woman's cells, sensitizing her immune system. Rhogam must be given within 72 hours of possible contact with Rh-positive cells—such situations include giving birth, terminating a pregnancy, miscarrying, or undergoing amniocentesis (a prenatal test in which a needle is inserted into the uterus). ■

1. What is the Rh blood group?
2. Under what conditions might a person with Rh-negative blood develop anti-Rh antibodies?

Clinical Terms Related to the Blood

Anisocytosis (an-i″so-si-to′sis) Abnormal variation in the size of erythrocytes.

Antihemophilic plasma (an″ti-he″mo-fil′ik plaz′mah) Normal blood plasma that has been processed to preserve an antihemophilic factor.

Citrated whole blood (sit′rāt-ed hōl blud) Normal blood to which a solution of acid citrate has been added to prevent coagulation.

Dried plasma (drīd plaz′mah) Normal blood plasma that has been vacuum dried to prevent the growth of microorganisms.

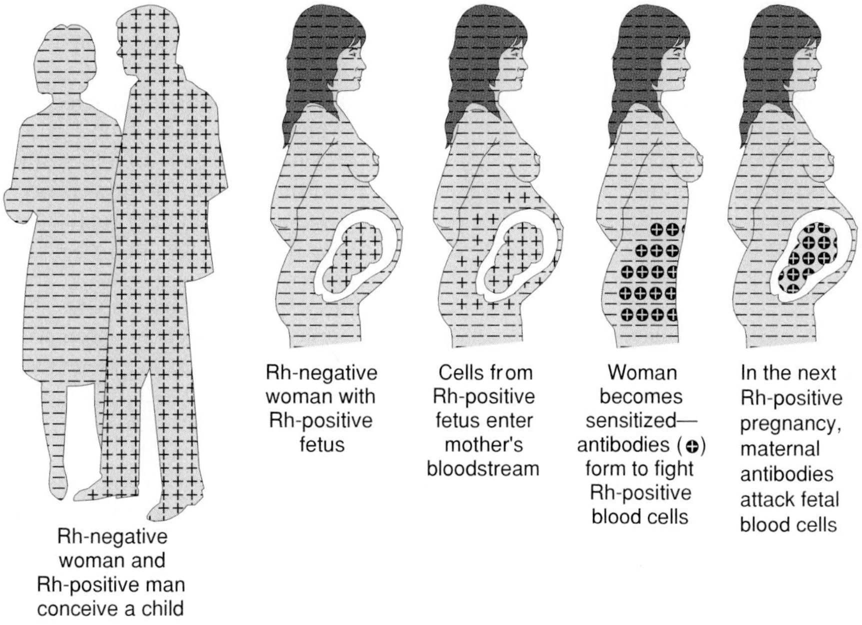

Rh-negative woman and Rh-positive man conceive a child

Rh-negative woman with Rh-positive fetus

Cells from Rh-positive fetus enter mother's bloodstream

Woman becomes sensitized—antibodies (⊕) form to fight Rh-positive blood cells

In the next Rh-positive pregnancy, maternal antibodies attack fetal blood cells

***Figure* 12.18**

If a man who is Rh positive and a woman who is Rh negative conceive a child who is Rh positive, the woman's body may manufacture antibodies that attack future Rh-positive offspring.

Hemorrhagic telangiectasia (hem″o-raj′ik tel-an″je-ek-ta′ze-ah) An inherited tendency to bleed from localized lesions of the capillaries.

Heparinized whole blood (hep′er-ĭ-nīzd″ hōl blud) Normal blood to which a solution of heparin has been added to prevent coagulation.

Macrocytosis (mak″ro-si-to′sis) Abnormally large erythrocytes.

Microcytosis (mi″kro-si-to′sis) Abnormally small erythrocytes.

Neutrophilia (nu″tro-fil′e-ah) An increase in the number of circulating neutrophils.

Packed red cells A concentrated suspension of red blood cells from which the plasma has been removed.

Pancytopenia (pan″si-to-pe′ne-ah) Abnormal depression of all the cellular components of blood.

Poikilocytosis (poi″kĭ-lo-si-to′sis) Irregularly shaped erythrocytes.

Purpura (per′pu-rah) Spontaneous bleeding into the tissues and through the mucous membranes.

Septicemia (sep″ti-se′me-ah) Presence of disease-causing microorganisms or their toxins in the blood.

Spherocytosis (sfēr″o-si-to′sis) Hemolytic anemia caused by defective proteins supporting the cell membranes of red blood cells. The cells are abnormally spherical.

Thalassemia (thal″ah-se′me-ah) A group of hereditary hemolytic anemias resulting from very thin, fragile erythrocytes.

Summary Outline

Introduction (p. 322)

Blood is a type of connective tissue whose cells are suspended in a liquid intercellular material. It transports substances between body cells and the external environment, and helps maintain a stable cellular environment.

Blood and Blood Cells (p. 322)

Blood contains red blood cells, white blood cells, and platelets.

1. Blood volume and composition
   a. Blood volume varies with body size, fluid and electrolyte balance, and adipose tissue content.
   b. Blood can be separated into formed elements and liquid portions.
      (1) The formed elements portion is mostly red blood cells.
      (2) The liquid plasma includes water, nutrients, hormones, electrolytes, and cellular wastes.
2. Characteristics of red blood cells
   a. Red blood cells are biconcave disks with shapes that increase surface area.
   b. Red blood cells contain hemoglobin, which combines with oxygen.

3. Red blood cell counts
   a. The red blood cell count equals the number of cells per cubic millimeter of blood.
   b. The average count ranges from 4–6 million cells per mm^3 of blood.
   c. Red blood cell count is related to the oxygen-carrying capacity of blood, which is used to diagnose and evaluate the courses of diseases.
4. Destruction of red blood cells
   a. Macrophages in the liver and spleen phagocytize damaged red blood cells.
   b. Hemoglobin molecules decompose, and the iron they contain is recycled.
   c. Hemoglobin releases biliverdin and bilirubin pigments.
5. Red blood cell production and its control
   a. Red bone marrow produces red blood cells.
   b. The number of red blood cells remains relatively stable.
   c. A negative feedback mechanism utilizing erythropoietin controls the rate of red blood cell production.
6. Dietary factors affecting red blood cell production
   a. Availability of vitamin B_{12} and folic acid influences red blood cell production.
   b. Hemoglobin synthesis requires iron.
7. Types of white blood cells
   a. White blood cells control infection.
   b. Granulocytes include neutrophils, eosinophils, and basophils.
   c. Agranulocytes include monocytes and lymphocytes.
8. White blood cell counts
   a. Normal total white blood cell counts vary from 5,000–10,000 cells per mm^3 of blood.
   b. The number of white blood cells may change in response to abnormal conditions, such as infections, emotional disturbances, or excessive loss of body fluids.
   c. A differential white blood cell count indicates the percentages of the types of leukocytes present.
9. Functions of white blood cells
   a. Neutrophils and monocytes phagocytize foreign particles.
   b. Eosinophils kill parasites and help control inflammation and allergic reactions.
   c. Basophils release heparin, which inhibits blood clotting, and histamine to increase blood flow to injured tissues.
   d. Lymphocytes produce antibodies that attack specific foreign substances.
10. Blood platelets
   a. Blood platelets are fragments of giant cells.
   b. The normal platelet count varies from 130,000–360,000 platelets per mm^3 of blood.
   c. Platelets help close breaks in blood vessels.

Blood Plasma (p. 331)

Plasma transports nutrients and gases, helps regulate fluid and electrolyte balance, and helps maintain stable pH.

1. Plasma proteins
   a. Plasma proteins remain in blood and interstitial fluids, and are not normally used as energy sources.
   b. Three major groups exist.
      (1) Albumins help maintain blood osmotic pressure.
      (2) Globulins include antibodies. They provide immunity and transport lipids and fat-soluble vitamins.
      (3) Fibrinogen functions in blood clotting.
2. Nutrients and gases
   a. Plasma nutrients include simple sugars, amino acids, and lipids.
      (1) The liver stores glucose as glycogen and releases glucose whenever blood glucose concentration falls.
      (2) Amino acids are used to synthesize proteins and are deaminated to provide energy.
      (3) Lipoproteins transport lipids.
   b. Gases in plasma include oxygen, carbon dioxide, and nitrogen.
3. Nonprotein nitrogenous substances
   a. Nonprotein nitrogenous substances are composed of molecules that contain nitrogen atoms but are not proteins.
   b. They include amino acids, urea, and uric acid.
4. Plasma electrolytes
   a. Plasma electrolytes include ions of sodium, potassium, calcium, magnesium, chloride, bicarbonate, phosphate, and sulfate.
   b. They are important in maintaining osmotic pressure and pH of plasma.

Hemostasis (p. 333)

Hemostasis is the stoppage of bleeding.

1. Blood vessel spasm
   a. Smooth muscles in blood vessel walls contract reflexly following injury.
   b. Platelets release serotonin, which stimulates vasoconstriction and helps maintain vessel spasm.
2. Platelet plug formation
   a. Platelets adhere to rough surfaces and exposed collagen.
   b. Platelets stick together at injury sites and form platelet plugs in broken vessels.
3. Blood coagulation
   a. Blood clotting is the most effective means of hemostasis.
   b. Clot formation depends on the balance between factors that promote clotting and those that inhibit clotting.
   c. The basic event of coagulation is changing soluble fibrinogen into insoluble fibrin.
   d. Biochemicals that promote clotting include prothrombin activator, prothrombin, and calcium ions.
   e. A thrombus is an abnormal blood clot in a vessel. An embolus is a clot or fragment of a clot that moves in a vessel.

Blood Groups and Transfusions (p. 335)

Blood can be typed on the basis of cell surface antigens.

1. Antigens and antibodies
   a. Agglutination is the clumping of red blood cells following a transfusion reaction.
   b. Red blood cell membranes may contain antigens, and blood plasma may contain antibodies.
2. ABO blood group
   a. Blood is grouped according to the presence or absence of antigens A and B.
   b. Mixing red blood cells that contain an antigen with plasma that contains the corresponding antibody, will result in an adverse transfusion reaction.

3. Rh blood group
   a. Rh antigens are present on the red blood cell membranes of Rh-positive blood. They are absent in Rh-negative blood.
   b. Mixing Rh-positive red blood cells with plasma that contains anti-Rh antibodies agglutinates the positive cells.
   c. Anti-Rh antibodies in maternal blood may pass through the placental tissues and react with the red blood cells of an Rh-positive fetus.

Review Exercises

1. List the major components of blood. (p. 322)
2. Describe a red blood cell. (p. 322)
3. Distinguish between oxyhemoglobin and deoxyhemoglobin. (p. 322)
4. Describe the life cycle of a red blood cell. (p. 322)
5. Distinguish between biliverdin and bilirubin. (p. 323)
6. Define *erythropoietin,* and explain its function. (p. 324)
7. Explain how vitamin B_{12} and folic acid deficiencies affect red blood cell production. (p. 324)
8. Distinguish between granulocytes and agranulocytes. (p. 326)
9. Name five types of leukocytes, and list the major functions of each type. (p. 327)
10. Explain the significance of white blood cell counts as aids to diagnosing diseases. (p. 328)
11. Describe a blood platelet, and explain its functions. (p. 329)
12. Name three types of plasma proteins, and list the major functions of each type. (p. 331)
13. Define *lipoprotein.* (p. 332)
14. Define *apoprotein.* (p. 332)
15. Distinguish between low-density lipoprotein and high-density lipoprotein. (p. 332)
16. Name the sources of very low-density lipoproteins, low-density lipoproteins, high-density lipoproteins, and chylomicrons. (p. 332)
17. Describe how lipoproteins are removed from plasma. (p. 332)
18. Explain how cholesterol is eliminated from plasma and from the body. (p. 332)
19. Define *nonprotein nitrogenous substances,* and name those commonly present in plasma. (p. 332)
20. Name several plasma electrolytes. (p. 332)
21. Define *hemostasis.* (p. 333)
22. Explain how an injury can stimulate blood vessel spasms. (p. 333)
23. Explain how a platelet plug forms. (p. 333)
24. List the major steps leading to the formation of a blood clot. (p. 333)
25. Distinguish between fibrinogen and fibrin. (p. 333)
26. Explain how positive feedback operates during blood clotting. (p. 334)
27. Distinguish between a thrombus and an embolus. (p. 334)
28. Distinguish between an antigen and an antibody. (p. 335)
29. Explain the basis of ABO blood types. (p. 336)
30. Explain why an exact match between donor and recipient blood is best. (p. 336)
31. Distinguish between Rh-positive and Rh-negative blood. (p. 338)
32. Describe how a person may become sensitized to Rh-positive blood. (p. 338)
33. Define *erythroblastosis fetalis,* and explain how this condition may develop. (p. 338)

Critical Thinking

1. Erythropoietin is available as a drug. Why would athletes abuse it?
2. How might a technique to remove A and B antigens from red blood cells be used to increase the supply of donated blood?
3. Why can a person receive platelets donated by anyone, but must receive a particular type of whole blood?
4. Researchers are developing several types of chemicals to be used as temporary red blood cell substitutes. What characteristics should a red blood cell substitute have?
5. How can excess dietary cholesterol lead to atherosclerosis if cells have low-density lipoprotein receptors that take up cholesterol?
6. If a patient with an inoperable cancer takes a drug that reduces the rate of cell division, how might the patient's white blood cell count change? How could the patient's environment be modified to compensate for the effects of these changes?
7. Hypochromic (iron-deficiency) anemia is relatively common among senior citizens admitted to hospitals for other conditions. What environmental and sociological factors might promote this form of anemia?
8. Why do patients with liver diseases commonly develop blood-clotting disorders?
9. How would you explain to a patient with leukemia, who has an elevated white blood count, the importance of avoiding bacterial infections?

Cardiovascular System

The rock band R.E.M. was about halfway through a show on their European tour when drummer Bill Berry suddenly developed an excruciating headache in the middle of a song. Holding his head, he stopped playing. Berry's bandmates quickly ended the show and rushed him to a hospital. Berry had an aneurysm, the sudden ballooning of a weakened area in an artery, in his head. The weakened area had probably been present since birth. Thanks to quick surgery, Berry recovered, and the band was touring the United States two months later.

David Cone was one of the all-time great baseball pitchers when, in the spring of his eleventh season, he began to feel a nagging numbness in his right hand, the one he uses to pitch. He and his doctors were relieved when dye injected into his circulatory system showed no blood clot, but his symptoms persisted, even though he was taking blood-thinning drugs. Further investigation revealed an aneurysm in the subclavian artery in his right shoulder. Years of pitching had built up the muscle in the area, which began to press on the artery, eventually injuring it. Repeated trauma caused this aneurysm. Cone was lucky—the aneurysm did not burst and was surgically repaired.

Photo: A famous pitcher developed an injury in his pitching arm from repetitive motions, which pressed muscle against an artery, eventually causing an aneurysm.

The heart pumps 7,000 liters of blood through the body each day, contracting some 2.5 billion times in an average lifetime. This muscular pump forces blood through arteries, which connect to smaller diameter vessels. The tiniest tubes, the capillaries, are the sites of nutrient, ion, gas, and waste exchange. Capillaries converge into venules and then veins that return blood to the heart, completing the closed system of blood circulation.

Chapter Objectives

After studying this chapter, you should be able to:

1. Name the organs of the cardiovascular system, and discuss their functions. (p. 344)
2. Name and describe the locations and functions of the major parts of the heart. (p. 344)
3. Trace the pathway of blood through the heart chambers. (p. 347)
4. Trace the pathway of blood through the vessels of coronary circulation. (p. 349)
5. Discuss the cardiac cycle, and explain how it is controlled. (p. 350)
6. Identify the parts of a normal ECG pattern, and discuss the significance of this pattern. (p. 355)
7. Compare the structures and functions of the major types of blood vessels. (p. 357)
8. Describe how substances are exchanged between blood in capillaries and the tissue fluid surrounding body cells. (p. 361)
9. Describe the mechanisms that return venous blood to the heart. (p. 362)
10. Explain how blood pressure arises and is controlled. (p. 363)
11. Compare the pulmonary and systemic circuits of the cardiovascular system. (p. 367)
12. Identify and locate the major arteries and veins of the pulmonary and systemic circuits. (p. 367)

Aids to Understanding Words

brady- [slow] *brady*cardia: An abnormally slow heartbeat.

diastol- [between contractions] *diastol*ic pressure: The blood pressure between contractions.

-gram [something written] electrocardio*gram:* A recording of the electrical changes in the heart muscle during a cardiac cycle.

papill- [nipple] *papill*ary muscle: A small mound of muscle within a ventricle.

syn- [together] *syn*cytium: A mass of merging cells that act together.

systol- [contraction] *systol*ic pressure: The blood pressure during a ventricular contraction.

tachy- [rapid] *tachy*cardia: An abnormally fast heartbeat.

Key Terms

arteriole (ar-te′re-ōl)
atrium (a′tre-um)
capillary (kap′i-lar″e)
cardiac conduction system (kar′de-ak kon-duk′shun sis′tem)
cardiac cycle (kar′de-ak si′kl)
cardiac output (kar′de-ak owt′poot)
diastole (di-as′to-le)
electrocardiogram (ECG) (e-lek″tro-kar′de-o-gram″)
endocardium (en″do-kar′de-um)
epicardium (ep″ĭ-kar′de-um)
functional syncytium (funk′shun-al sin-sish′e-um)
myocardium (mi″o-kar′de-um)
pericardium (per″ĭ-kar′de-um)
peripheral resistance (pĕ-rif′er-al re-zis′tans)
pulmonary circuit (pul′mo-ner″e sur′kit)
systemic circuit (sis-tem′ik sur′kit)
systole (sis′to-le)
vasoconstriction (vas″o-kon-strik′shun)
vasodilation (vas″o-di-la′shun)
ventricle (ven′tri-kl)
venule (ven′ūl)
viscosity (vis-kos′ĭ-te)

Introduction

The cardiovascular system—a powerful pump (the heart) connected to an extensive system of tubes (blood vessels)—brings oxygen and nutrients to all body cells and removes wastes. A functional cardiovascular system is vital for survival because, without circulation, tissues lack a supply of oxygen and nutrients, and wastes accumulate. Under such conditions, cells soon change irreversibly, which quickly leads to death. Figure 13.1 shows the general pattern of the cardiovascular system.

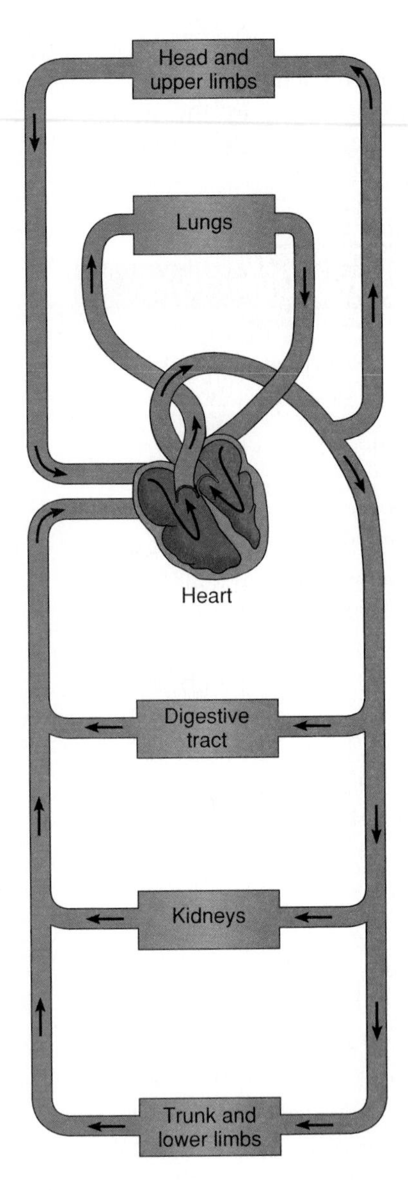

Figure 13.1 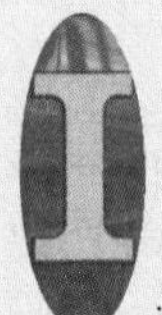

The cardiovascular system transports blood between the body cells and organs such as the lungs, intestines, and kidneys that communicate with the external environment.

Structure of the Heart

The heart is a hollow, cone-shaped, muscular pump, located within the thoracic cavity and resting on the diaphragm (fig. 13.2).

Size and Location of the Heart

Heart size varies with body size. An average adult heart is about 14 centimeters long and 9 centimeters wide.

The heart is within the mediastinum, which is bordered laterally by the lungs, posteriorly by the vertebral column, and anteriorly by the sternum. The *base* of the heart, which attaches to several large blood vessels, lies beneath the second rib. The heart's distal end extends downward and to the left, terminating as a bluntly pointed *apex* at the level of the fifth intercostal space.

Coverings of the Heart

The **pericardium** encloses the heart and the proximal ends of the large blood vessels to which it attaches. The pericardium consists of an outer fibrous bag, the *fibrous pericardium,* which surrounds a more delicate, double-layered sac. The inner layer of this sac, the *visceral pericardium* (epicardium), covers the heart. At the base of the heart, the visceral layer turns back on itself to become the *parietal pericardium* (figs. 13.2 and 13.3 and reference plate 3). The parietal pericardium, in turn, forms the inner lining of the fibrous pericardium.

The fibrous pericardium is a tough, protective sac composed largely of white fibrous connective tissue. It attaches to the central portion of the diaphragm, the posterior of the sternum, the vertebral column, and the large blood vessels that emerge from the heart. Between the parietal and visceral layers of the pericardium is a space, the *pericardial cavity,* that contains a small amount of serous fluid (fig. 13.3). This fluid reduces friction between the pericardial membranes as the heart moves within them.

In *pericarditis,* inflammation of the pericardium due to viral or bacterial infection produces adhesions that attach the layers of the pericardium. This condition is very painful and interferes with heart movements. ■

1. Where is the heart located?
2. Distinguish between the visceral pericardium and the parietal pericardium.

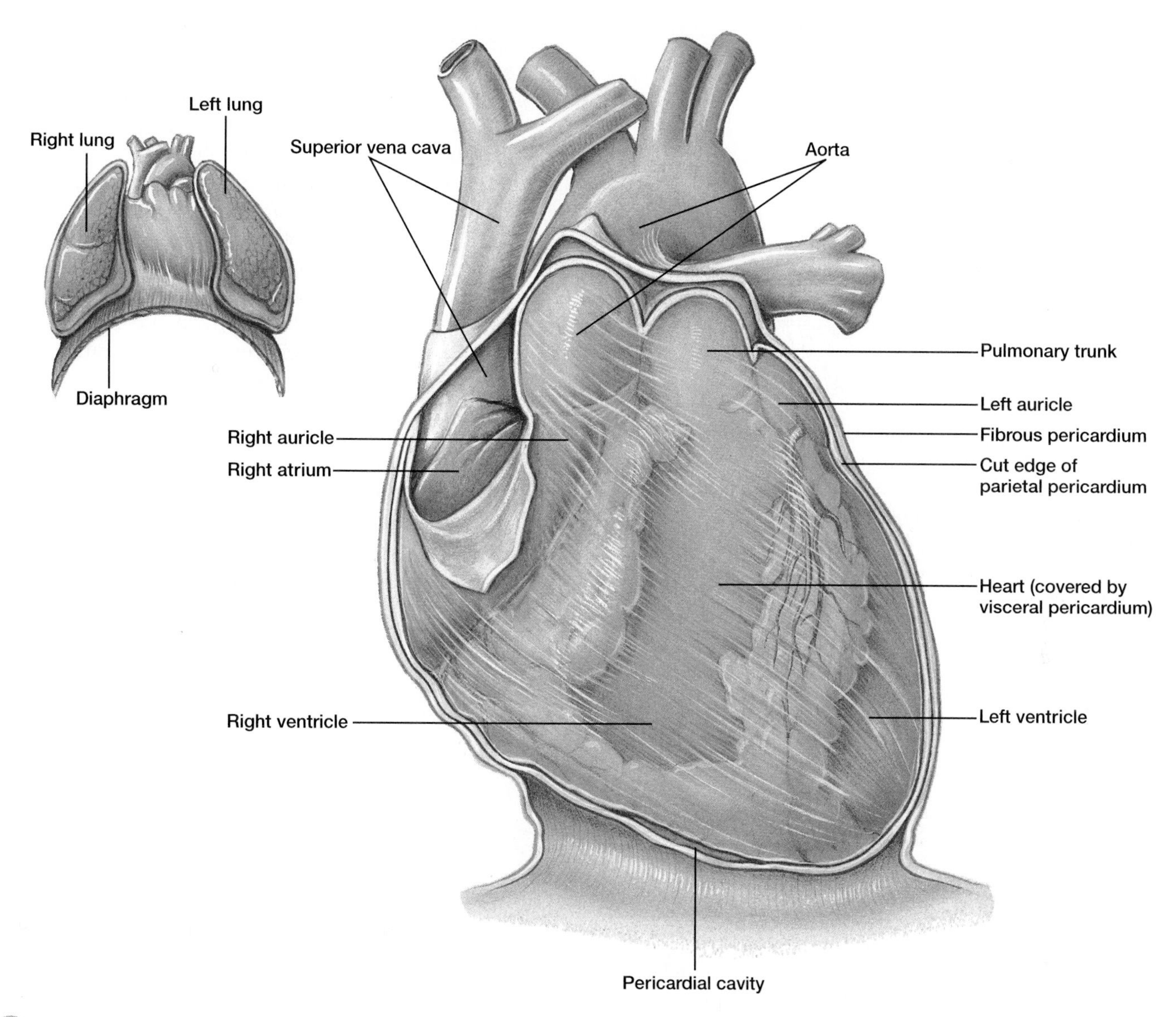

Figure 13.2

A layered pericardium encloses the heart within the mediastinum.

Wall of the Heart

The wall of the heart has three distinct layers—an outer epicardium, a middle myocardium, and an inner endocardium (fig. 13.3). The **epicardium**, which corresponds to the visceral pericardium, protects the heart by reducing friction. It is a serous membrane that consists of connective tissue beneath epithelium. Its deeper portion often contains adipose tissue, particularly along the paths of coronary arteries and cardiac veins that carry blood through the myocardium.

The thick middle layer, or **myocardium**, consists mostly of cardiac muscle that forces blood out of the heart chambers. The muscle fibers are organized in planes, separated by connective tissue richly supplied with blood capillaries, lymph capillaries, and nerve fibers.

The inner layer, or **endocardium**, consists of epithelium and connective tissue with many elastic and collagenous fibers. The endocardium also contains blood vessels and some specialized cardiac muscle fibers, called *Purkinje fibers,* described later in this chapter. The endocardium is continuous with the inner linings of blood vessels attached to the heart.

Heart Chambers and Valves

Internally, the heart is divided into four chambers—two on the left and two on the right (fig. 13.4). The upper chambers, called **atria** (singular, *atrium*), have thin

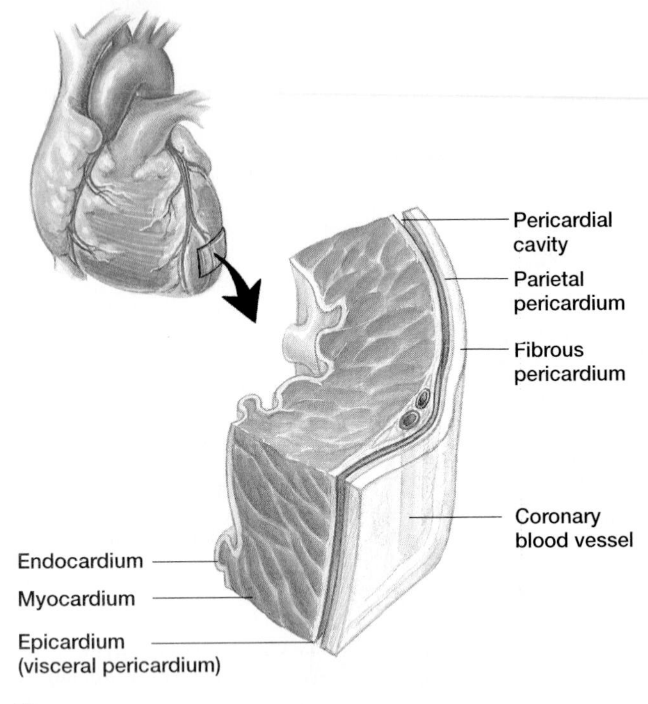

Figure 13.3

The heart wall has three layers: an endocardium, a myocardium, and an epicardium.

walls and receive blood returning to the heart. Small, earlike projections called *auricles* extend outward from the atria. The lower chambers, the **ventricles**, receive blood from the atria and contract, which forces blood out of the heart into the arteries.

A solid, wall-like **septum** separates the atrium and ventricle on the right side from those on the left. As a result, blood from one side of the heart never mixes with blood from the other side (except in the fetus, see chapter 20). An *atrioventricular valve* (A-V valve), the tricuspid on the right and the bicuspid on the left, ensures one-way blood flow between the atria and ventricles.

The right atrium receives blood from two large veins—the *superior vena cava* and the *inferior vena cava*. A smaller vein, the *coronary sinus*, also drains blood into the right atrium from the heart wall.

A large **tricuspid valve**, with three *cusps* as its name implies, lies between the right atrium and the right ventricle (fig. 13.4). The valve permits blood to move from the right atrium into the right ventricle and prevents backflow.

Strong, fibrous strings called *chordae tendineae* attach to the cusps of the tricuspid valve on the ventricular side. These strings originate from small mounds of cardiac muscle tissue, the **papillary muscles**, which project inward

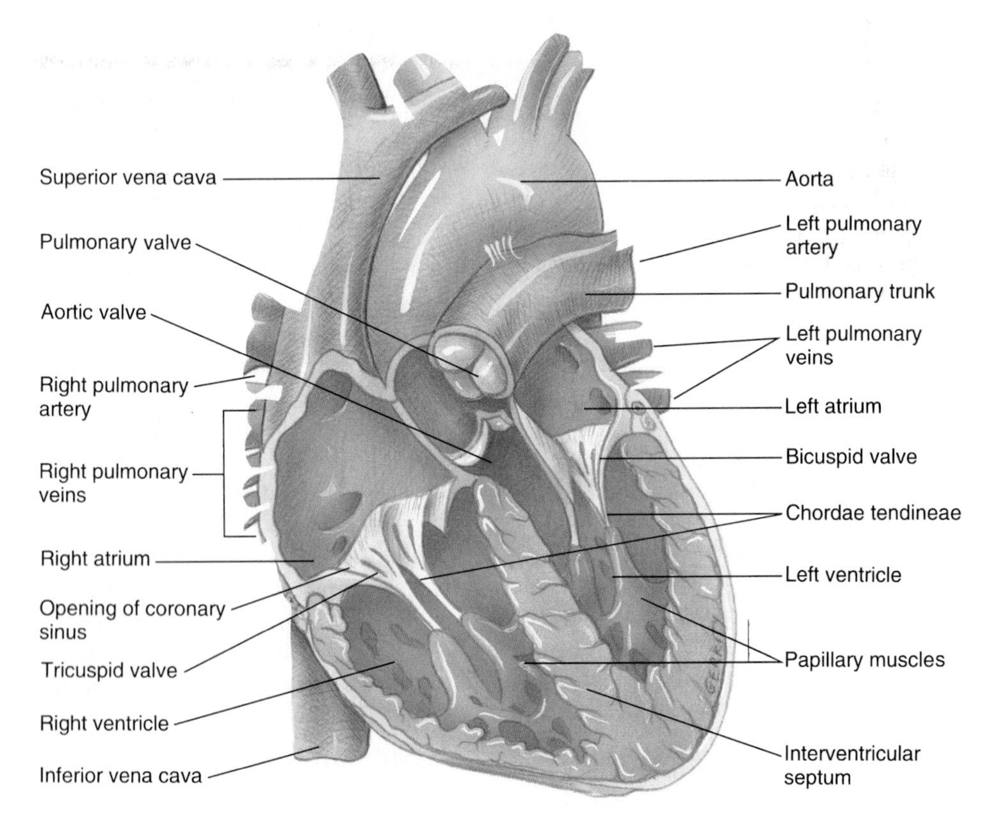

Figure 13.4

Coronal sections of the heart showing the connection between the left ventricle and the aorta.

from the walls of the ventricle. The papillary muscles contract when the ventricle contracts. As the tricuspid valve closes, they pull on the chordae tendineae and prevent the cusps from swinging back into the atrium.

The muscular wall of the right ventricle is thinner than that of the left ventricle (fig. 13.4). This right chamber pumps blood a short distance to the lungs against a low resistance to blood flow. The left ventricle, on the other hand, must force blood to all the other parts of the body against a much greater resistance to flow.

When the muscular wall of the right ventricle contracts, the blood inside is under increasing pressure, and the tricuspid valve closes as a result of the pressure difference on either side of it. The only exit for the blood is through the *pulmonary trunk*, which divides to form the left and right *pulmonary arteries* that lead to the lungs. At the base of this trunk is a **pulmonary valve** with three cusps. This valve allows blood to leave the right ventricle and prevents backflow into the ventricular chamber (fig. 13.5).

The left atrium receives blood from the lungs through four *pulmonary veins*—two from the right lung and two from the left lung. Blood passes from the left atrium into the left ventricle through the **bicuspid (mitral) valve,** which prevents blood from flowing back into the left atrium from the ventricle. As with the tricuspid valve, the papillary muscles and the chordae tendineae prevent the cusps of the bicuspid valve from swinging back into the left atrium during ventricular contraction.

When the left ventricle contracts, the bicuspid valve closes as a result of pressure differences. The only exit is through a large artery, the **aorta.** At the base of the aorta is the **aortic valve** with three cusps. As the left ventricle contracts, the valve is forced open allowing blood to leave. When the ventricular muscles relax, this valve closes and prevents blood from backing up into the ventricle (fig. 13.5).

The bicuspid and tricuspid valves are called atrioventricular valves because they are between atria and ventricles. The pulmonary and aortic valves are called "semilunar" because of the half-moon shapes of their cusps.

Table 13.1 summarizes the locations and functions of the heart valves.

1. Describe the layers of the heart wall.
2. Name and locate the four chambers and valves of the heart.
3. Describe the function of each heart valve.

Skeleton of the Heart

Rings of dense fibrous connective tissue surround the pulmonary trunk and aorta at their proximal ends (fig. 13.6). The rings provide firm attachments for the heart valves and for muscle fibers, and prevent the outlets of the atria and ventricles from dilating during contraction. The fibrous rings, together with other masses of dense fibrous tissue in the portion of the septum between the ventricles (interventricular septum), constitute the *skeleton of the heart*.

Path of Blood through the Heart

Blood that is low in oxygen concentration and high in carbon dioxide concentration enters the right atrium through the venae cavae and coronary sinus. As the right atrial wall contracts, blood passes through the tricuspid valve and enters the chamber of the right

Mitral valve prolapse (MVP) affects up to 6% of the U.S. population. In this condition, one or both of the cusps of the bicuspid valve stretch and bulge into the left atrium during ventricular contraction. The valve usually continues to function adequately, but sometimes, blood regurgitates into the left atrium. In a stethoscope, a regurgitating MVP sounds like a click at the end of ventricular contraction, then a murmur as blood goes back through the valve into the left atrium. Symptoms of MVP include chest pain, palpitations, fatigue, and anxiety. The mitral valve can be damaged by an inflammation of the endocardium due to infection by certain strains of *Streptococcus* bacteria. Endocarditis appears as a plantlike growth on the valve. Bacterial infection can lead to endocarditis because *Streptococcus* cell surfaces resemble bicuspid valve cells. When the immune system responds to the infection by producing antibodies, the antibodies also attack the valves. Prompt treatment of such an infection with antibiotics can lessen the antibody response, thereby minimizing risk of endocarditis. People with MVP are particularly susceptible to endocarditis. ■

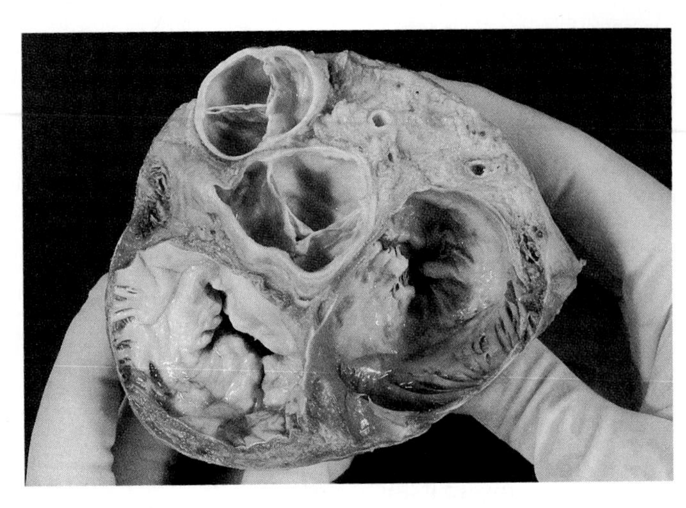

Figure 13.5

The pulmonary and aortic valves of the heart (superior view).

Table 13.1

Heart Valves

Valve	Location	Function
Tricuspid valve	Opening between right atrium and right ventricle	Prevents blood from moving from right ventricle into right atrium during ventricular contraction
Pulmonary valve	Entrance to pulmonary trunk	Prevents blood from moving from pulmonary trunk into right ventricle during ventricular relaxation
Bicuspid (mitral) valve	Opening between left atrium and left ventricle	Prevents blood from moving from left ventricle into left atrium during ventricular contraction
Aortic valve	Entrance to aorta	Prevents blood from moving from aorta into left ventricle during ventricular relaxation

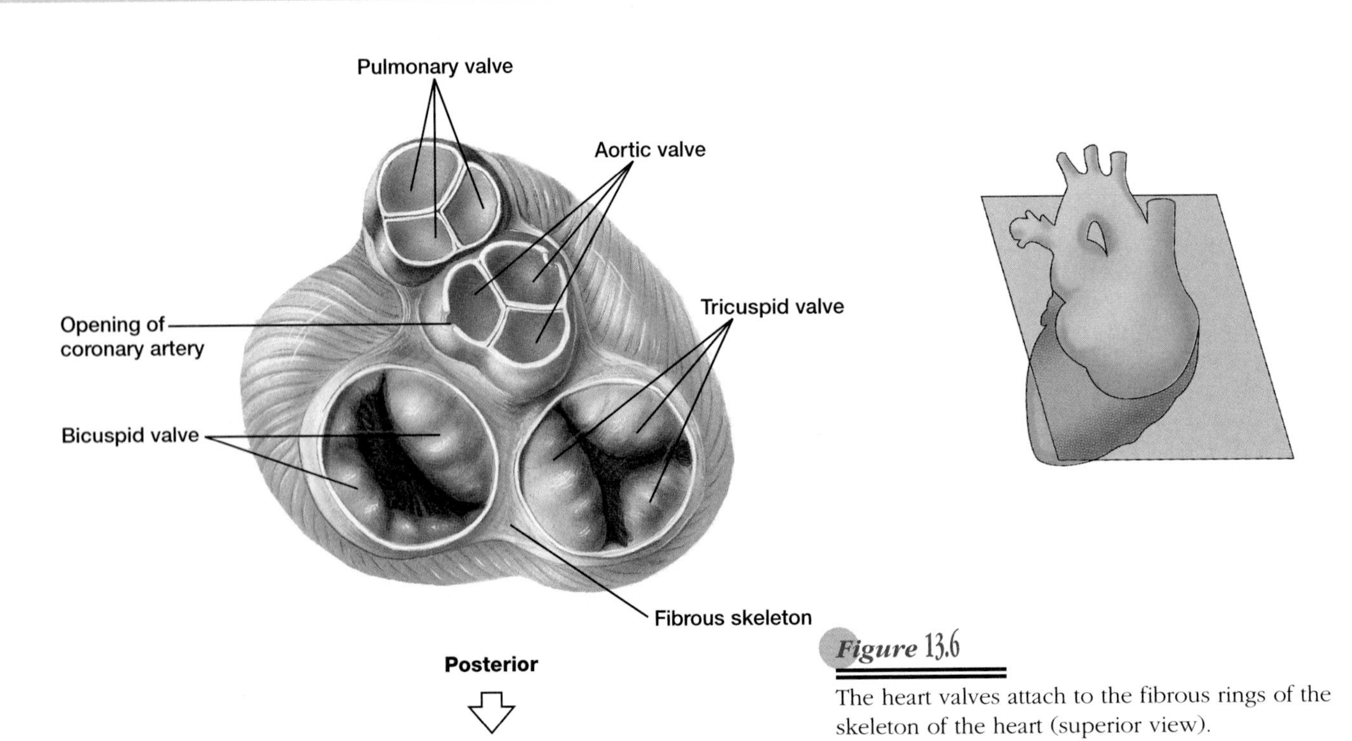

Figure 13.6

The heart valves attach to the fibrous rings of the skeleton of the heart (superior view).

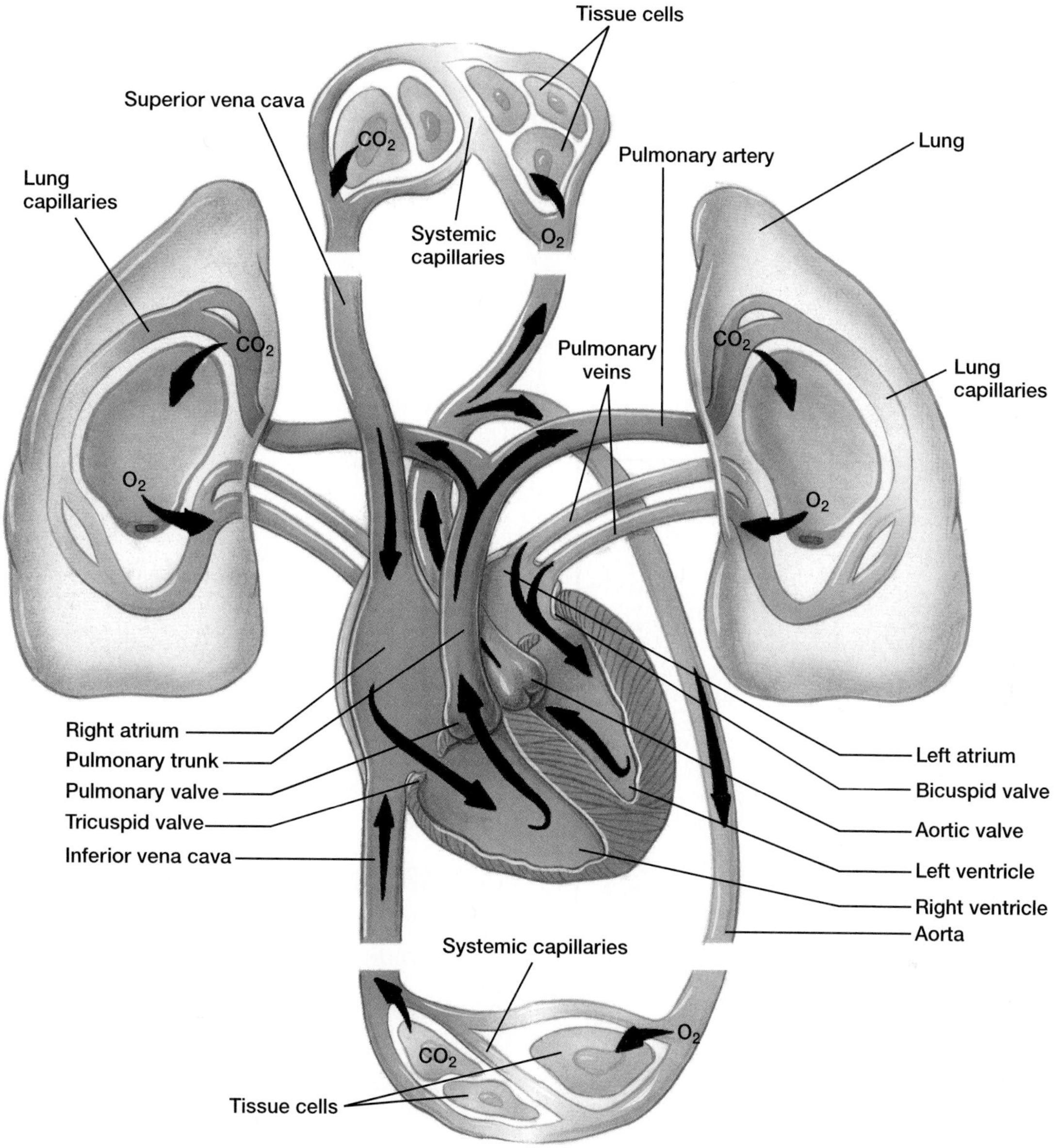

***Figure* 13.7**

The right ventricle forces blood to the lungs, while the left ventricle forces blood to all other body parts. (Structures are not drawn to scale.)

ventricle (fig. 13.7). When the right ventricular wall contracts, the tricuspid valve closes, and blood moves through the pulmonary valve and into the pulmonary trunk and its branches (pulmonary arteries).

From the pulmonary arteries, blood enters the capillaries associated with the alveoli of the lungs. Gas exchanges occur between blood in the capillaries and air in the alveoli. The freshly oxygenated blood, which is now low in carbon dioxide concentration, returns to the heart through the pulmonary veins, which lead to the left atrium.

The left atrial wall contracts, and blood moves through the bicuspid valve and into the chamber of the left ventricle. When the left ventricular wall contracts, the bicuspid valve closes, and blood moves through the aortic valve and into the aorta and its branches.

Blood Supply to the Heart

The first two branches of the aorta, called the right and left **coronary arteries**, supply blood to the heart's tissues. The openings of the coronary arteries lie just beyond the aortic valve (fig. 13.8).

If a thrombus or embolus blocks or narrows a coronary artery branch, myocardial cells are deprived of oxygen, producing ischemia and a painful condition called *angina pectoris*. The pain usually occurs during physical activity, when myocardial cells' oxygen requirements exceed their oxygen supply. Pain lessens with rest. Emotional disturbance may also trigger angina pectoris.

Angina pectoris may cause a sensation of heavy pressure, tightening, or squeezing in the chest. The pain is usually felt in the region behind the sternum or in the anterior portion of the upper thoracic cavity, but may radiate to the neck, jaw, throat, upper limb, shoulder, elbow, back, or upper abdomen. Other symptoms include profuse perspiration (diaphoresis), difficulty breathing (dyspnea), nausea, or vomiting.

A blood clot completely obstructing a coronary artery or one of its branches (coronary thrombosis) kills part of the heart. This is a *myocardial infarction* (MI), more commonly known as a heart attack. ■

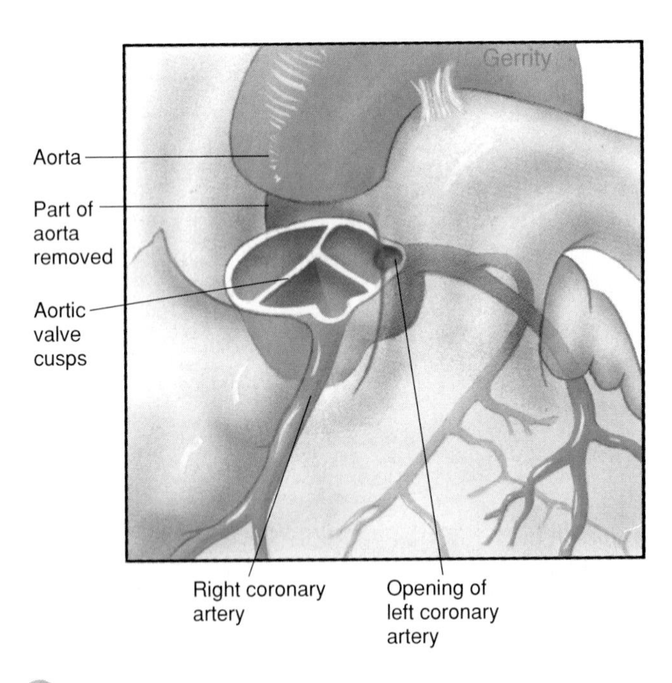

***Figure* 13.8**

The openings of the coronary arteries lie just beyond the aortic valve.

In *heart transplantation,* the recipient's failing heart is removed, except for the posterior walls of the right and left atria and their connections to the venae cavae and pulmonary veins. The donor heart is prepared similarly and is attached to the atrial cuffs remaining in the recipient's thoracic cavity. Finally, the recipient's aorta and pulmonary arteries are connected to those of the donor heart. ■

1. Which structures make up the skeleton of the heart?
2. Review the path of blood through the heart.
3. Which vessels supply blood to the myocardium?
4. How does blood return from the cardiac tissues to the right atrium?

The heart must beat continually to supply blood to body tissues. Therefore, myocardial cells require a constant supply of freshly oxygenated blood. Branches of the coronary arteries feed the many capillaries of the myocardium (fig. 13.9). The smaller branches of these arteries usually have connections (anastomoses) between vessels that provide alternate pathways for blood.

Branches of the **cardiac veins,** whose paths roughly parallel those of the coronary arteries, drain blood that has passed through myocardial capillaries. As figure 13.9*b* shows, these veins join an enlarged vein on the heart's posterior surface—the **coronary sinus**—which empties into the right atrium (see fig. 13.4).

Heart Actions

The heart chambers function in coordinated fashion. Their actions are regulated so that atrial walls contract while ventricular walls relax, then ventricular walls contract while atrial walls relax. This series of events constitutes a complete heartbeat, or **cardiac cycle.**

Cardiac Cycle

During a cardiac cycle, pressure within heart chambers rises and falls. When atria relax, blood flows into them from the large, attached veins. As these chambers fill, the pressure inside gradually increases, forcing the A-V valves open. About 70–80% of the entering blood flows directly into the ventricles before atrial walls

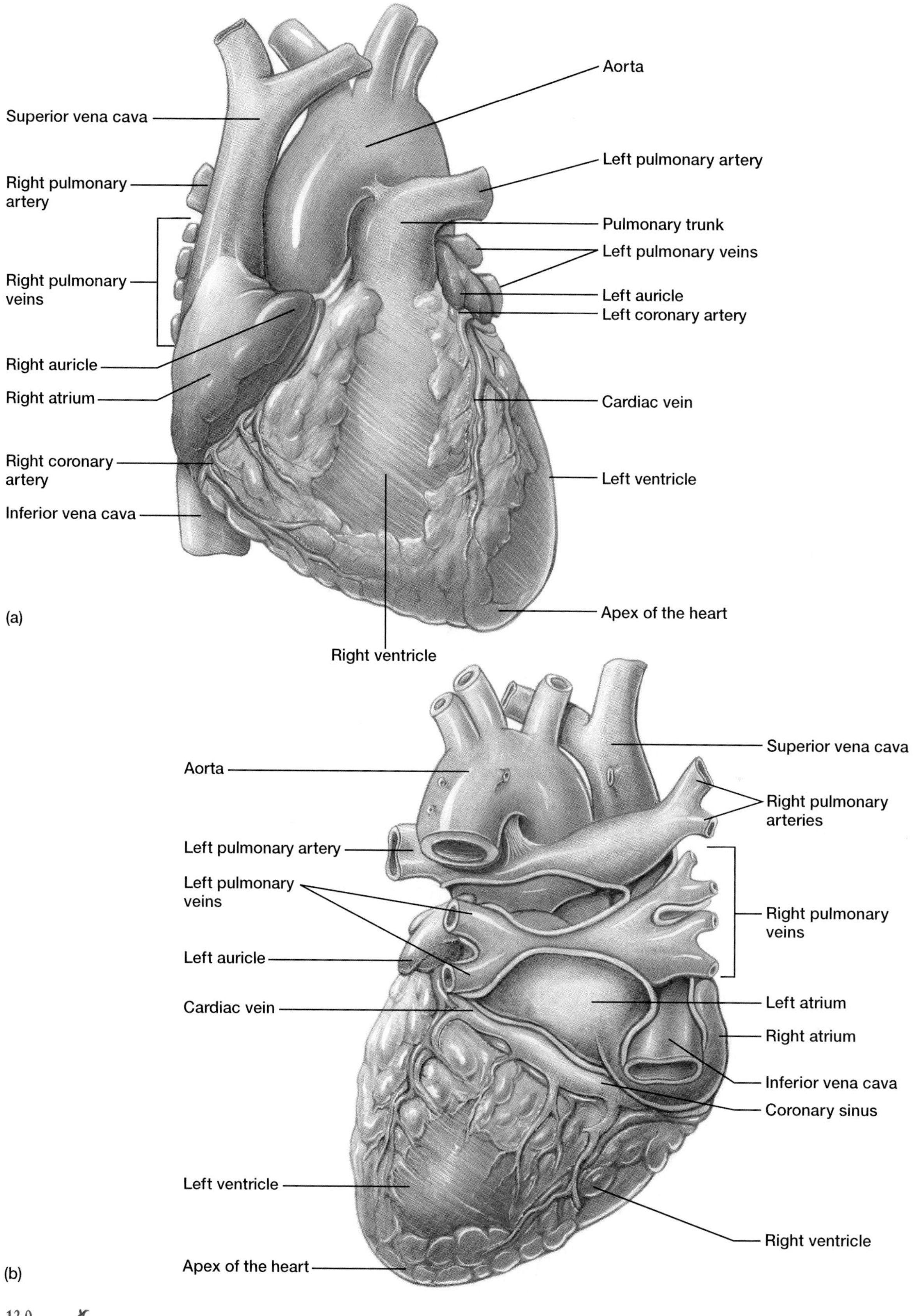

Figure 13.9

Blood vessels associated with the surface of the heart. (*a*) Anterior view. (*b*) Posterior view.

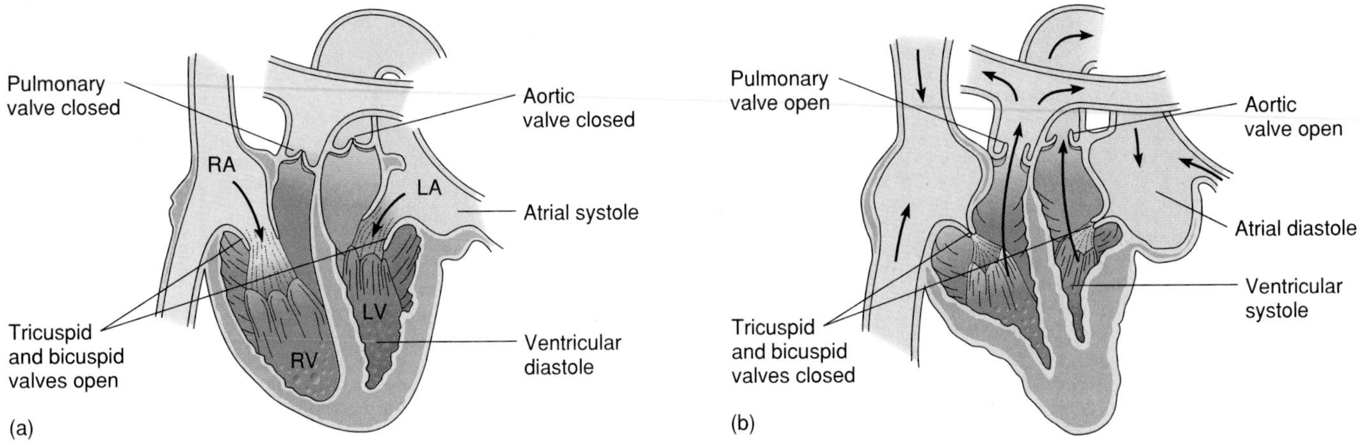

Figure 13.10

The atria (*a*) empty during atrial systole and (*b*) fill with blood during atrial diastole.

contract. Then, during atrial contraction (atrial systole), atrial pressure rises suddenly, forcing the remaining 20–30% of the atrial contents into the ventricles. Atrial relaxation (atrial diastole) follows (fig. 13.10).

As the ventricles contract (ventricular systole), the A-V valves close due to pressure differences between the ventricles and the atria and begin to bulge back into the atria, sharply increasing atrial pressure. At the same time, the papillary muscles contract, and by pulling on the chordae tendineae, they prevent the cusps of the A-V valves from bulging too far into the atria, allowing atrial pressure to return to its previous level. During ventricular contraction, the A-V valves remain closed, and atrial pressure increases again as the atria fill with blood. When the ventricles relax (ventricular diastole), the A-V valves open, blood flows through them into the ventricles, and atrial pressure drops to a low point.

Pressure in the ventricles is low while they fill, but when the atria contract, ventricular pressure increases slightly. Then, as the ventricles contract, ventricular pressure rises sharply, and as soon as the pressure exceeds that in the atria, the A-V valves close. Ventricular pressure continues to increase until it exceeds the pressure in the pulmonary trunk and aorta. Then, the pulmonary and aortic valves open, and blood is ejected from each valve's respective ventricle into these arteries. When the ventricles are nearly empty, ventricular pressure begins to drop, and it continues to drop as the ventricles relax. When ventricular pressure is less than that in the arteries, arterial blood flowing back toward the ventricles closes the pulmonary and aortic valves. As soon as ventricular pressure falls below that of the atria, the A-V valves open, and the ventricles begin to fill once more. The graph in figure 13.11 summarizes some of the changes that occur in the left ventricle during a cardiac cycle.

Heart Sounds

A heartbeat heard with a stethoscope sounds like *lub-dup*. These sounds are due primarily to vibrations in the heart tissues produced by the opening and closing of the valves.

The first part of a heart sound (*lub*) occurs during ventricular contraction, when the A-V valves are closing. The second part (*dup*) occurs during ventricular relaxation, when the pulmonary and aortic valves snap shut (fig. 13.11).

Heart sounds provide information concerning the condition of the heart valves. For example, inflammation of the endocardium (endocarditis) may erode the edges of the valvular cusps. As a result, the cusps may not close completely, and blood can leak back through the valve, producing an abnormal sound called a *murmur*. The seriousness of a murmur depends on the amount of valvular damage. Open heart surgery may repair or replace seriously damaged valves. ■

Cardiac Muscle Fibers

Cardiac muscle fibers function much like those of skeletal muscles, but the fibers connect in branching networks. Stimulation to any part of the network sends impulses throughout the heart, which contracts as a unit.

A mass of merging cells that act as a unit is called a **functional syncytium.** Two such structures are in the heart—in the atrial and ventricular walls. Portions of the heart's fibrous skeleton separate these masses of cardiac muscle fibers from each other, except for a

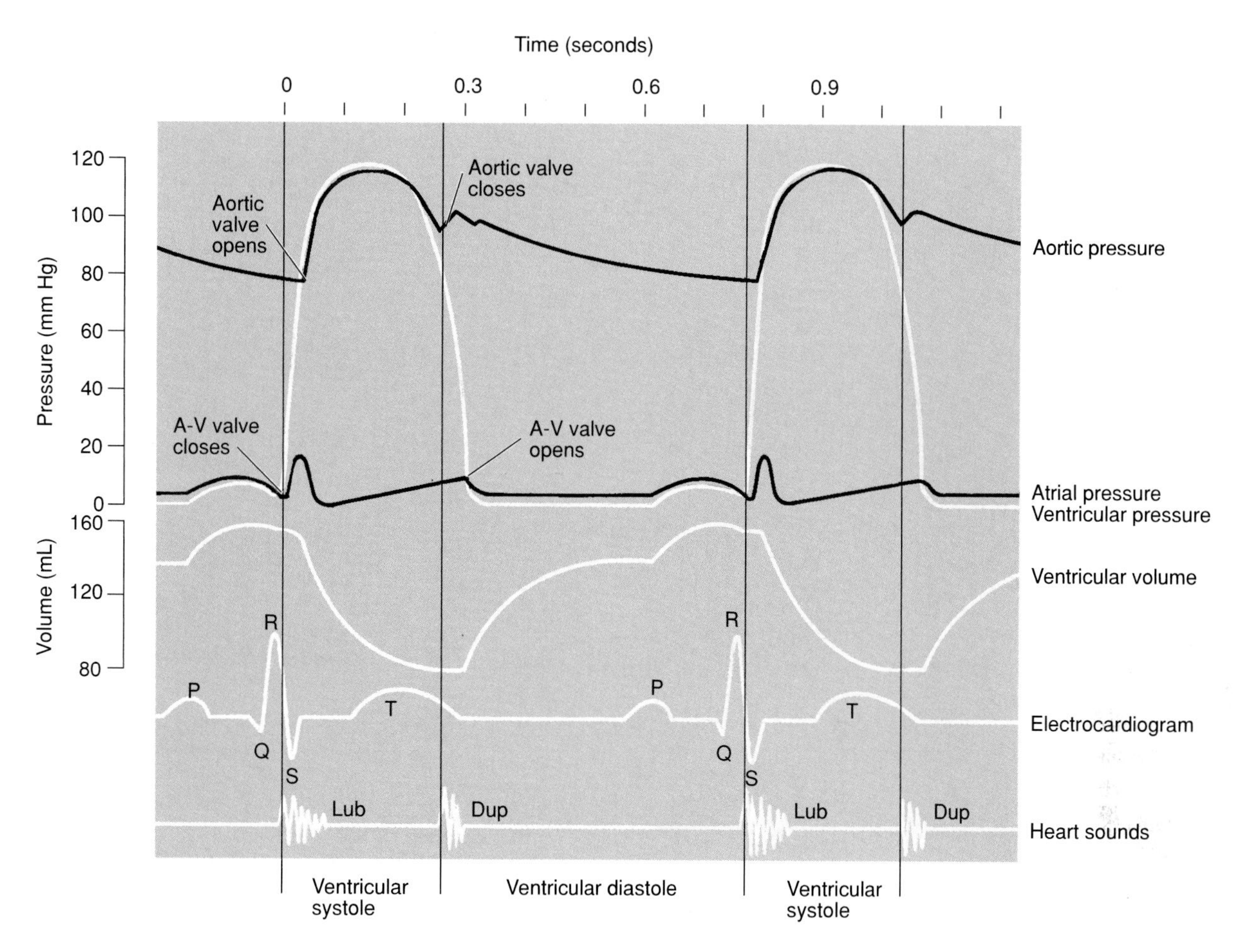

Figure 13.11

Graph of some of the changes in the left ventricle during a cardiac cycle.

small area in the right atrial floor. In this region, fibers of the cardiac conduction system connect the *atrial syncytium* and the *ventricular syncytium*.

1. Describe the pressure changes in the atria and ventricles during a cardiac cycle.
2. What causes heart sounds?
3. What is a functional syncytium?

Cardiac Conduction System

Throughout the heart are clumps and strands of specialized cardiac muscle tissue whose fibers contain only a few myofibrils. Instead of contracting, these areas initiate and distribute impulses throughout the myocardium. They comprise the **cardiac conduction system**, which coordinates events of the cardiac cycle (fig. 13.12).

A key portion of the cardiac conduction system is the **sinoatrial node** (S-A node), a small, elongated mass of specialized cardiac muscle tissue just beneath the epicardium. It is located in the right atrium near the opening of the superior vena cava, and its fibers are continuous with those of the atrial syncytium.

The cells of the S-A node can reach threshold on their own and their membranes contact one another. Without stimulation from nerve fibers or any other outside agents, the nodal cells initiate impulses that spread into the surrounding myocardium and stimulate cardiac muscle fibers to contract.

S-A node activity is rhythmic. The S-A node initiates one impulse after another, seventy to eighty times a minute in an adult. Because it generates the heart's rhythmic contractions it is often called the **pacemaker.**

As a cardiac impulse travels from the S-A node into the atrial syncytium, the right and left atria contract almost simultaneously. Instead of passing directly into

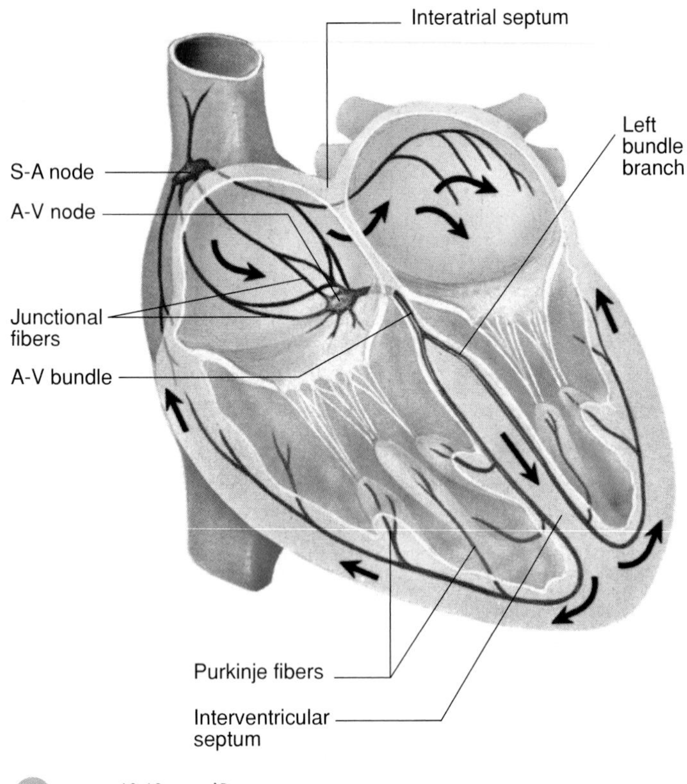

Figure 13.12
The cardiac conduction system.

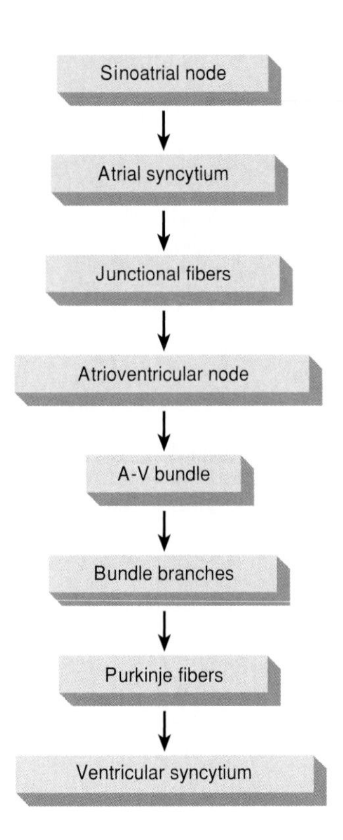

Figure 13.13
Components of the cardiac conduction system.

the ventricular syncytium, which the fibrous skeleton of the heart separates from the atrial syncytium, the cardiac impulse passes along fibers of the conduction system that are continuous with atrial muscle fibers. These conducting fibers lead to a mass of specialized cardiac muscle tissue called the **atrioventricular node** (A-V node). This node is located in the inferior portion of the septum that separates the atria (interatrial septum) and just beneath the endocardium. It is the only normal conduction pathway between the atrial and ventricular syncytia.

The fibers that conduct the cardiac impulse into the A-V node (junctional fibers) have very small diameters, and because small fibers conduct impulses slowly, they delay impulse transmission. The impulse is slowed further as it passes the A-V node, which allows time for the atria to completely empty and the ventricles to fill with blood prior to ventricular contraction.

Once the cardiac impulse reaches the distal side of the A-V node, it moves rapidly through a group of large fibers that make up the **A-V bundle** (bundle of His). The A-V bundle enters the upper part of the interventricular septum and divides into right and left bundle branches that lie just beneath the endocardium. About halfway down the septum, the branches give rise to enlarged **Purkinje fibers.**

Purkinje fibers spread from the interventricular septum into the papillary muscles, which project inward from ventricular walls, and then continue downward to the apex of the heart. There, they curve around the tips of the ventricles and pass upward over the lateral walls of these chambers. Along the way, Purkinje fibers give off many small branches, which become continuous with cardiac muscle fibers (fig. 13.13).

The muscle fibers in ventricular walls occur in irregular whorls, so that when impulses from Purkinje fibers stimulate them, ventricular walls contract with a twisting motion (fig. 13.14). This action squeezes blood out of ventricular chambers and into the aorta and pulmonary trunk.

1. What kinds of tissues make up the cardiac conduction system?
2. How is a cardiac impulse initiated?
3. How is a cardiac impulse transmitted from the right atrium to the other heart chambers?

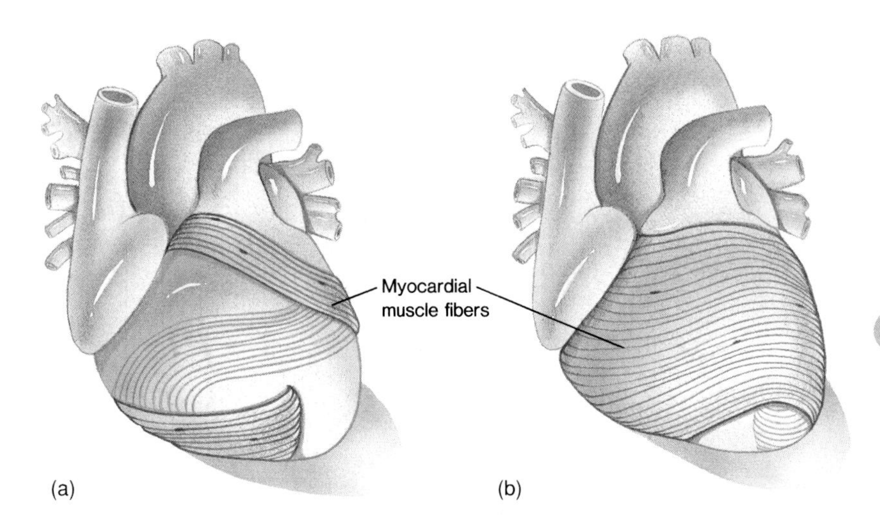

Figure 13.14

The muscle fibers within ventricular walls are organized in patterns of whorls. The fibers of groups (*a*) and (*b*) surround both ventricles in these anterior views of the heart.

Electrocardiogram

An **electrocardiogram** (ECG) is a recording of the electrical changes in the myocardium during a cardiac cycle. (This pattern occurs as action potentials stimulate cardiac muscle fibers to contract, but it is not the same as individual action potentials.) Because body fluids can conduct electrical currents, such changes can be detected on the surface of the body.

To record an ECG, a technician places electrodes on the skin that are connected by wires to an instrument that responds to very weak electrical changes by moving a pen or stylus on a moving strip of paper. Up-and-down movements of the pen correspond to electrical changes in the myocardium. Because the paper moves past the pen at a known rate, the distance between pen deflections indicates time between phases of the cardiac cycle.

As figure 13.15*a* illustrates, a normal ECG pattern includes several deflections, or *waves,* during each cardiac cycle. Between cycles, the muscle fibers remain polarized, with no detectable electrical changes, and the pen does not move and simply marks along the baseline. When the S-A node triggers a cardiac impulse, atrial fibers depolarize, producing an electrical change. The pen moves, and at the end of the electrical change, returns to the base position. This first pen movement produces a *P wave,* which corresponds to depolarization of the atrial fibers that will lead to contraction of the atria (fig. 13.15*b*).

When the cardiac impulse reaches ventricular fibers, they rapidly depolarize. Because ventricular walls are more massive than those of the atria, the electrical change is greater, and the pen deflects more than before, then returns to baseline, leaving a mark called the *QRS complex*. This mark consists of a *Q wave,* an *R wave,* and an *S wave,* and corresponds to depolarization of ventricular fibers that will contract the ventricles.

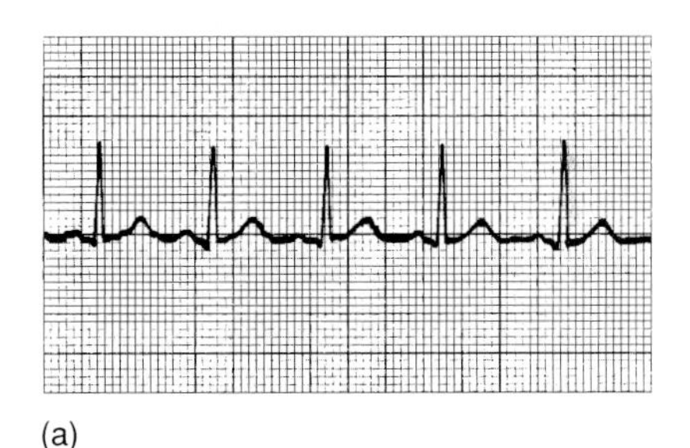

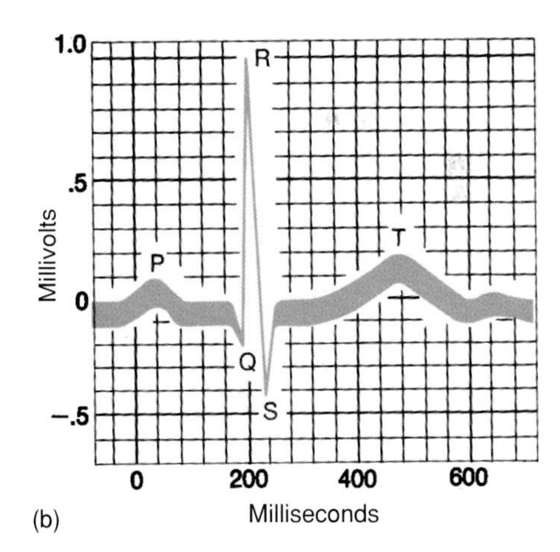

Figure 13.15

(*a*) A normal ECG. (*b*) In an ECG pattern, the P wave results from a depolarization of the atria, the QRS complex results from a depolarization of the ventricles, and the T wave results from a repolarization of the ventricles.

The electrical changes occurring as ventricular muscle fibers slowly repolarize produce a *T wave* as the pen deflects again, ending the ECG pattern. The record of atrial repolarization seems to be missing from the pattern because atrial fibers repolarize at the same time that ventricular fibers depolarize. Thus, the QRS complex obscures the recording of atrial repolarization.

Physicians use ECG patterns to assess the heart's ability to conduct impulses. For example, the time period between the beginning of a P wave and the beginning of a QRS complex (*P-Q interval,* or if the initial portion of the QRS complex is upright the P-R interval) indicates the time for the cardiac impulse to travel from the S-A node through the A-V node. Ischemia or other problems affecting fibers of the A-V conduction pathways can increase this P-Q interval. Similarly, injury to the A-V bundle can increase duration of the QRS complex because an impulse may take longer to spread throughout the ventricular walls (fig. 13.16). ■

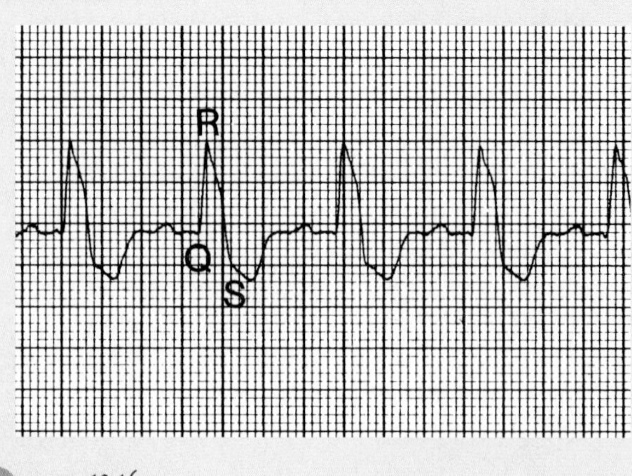

Figure 13.16

A prolonged QRS complex may result from damage to the A-V bundle fibers.

1. What is an electrocardiogram?
2. What cardiac events do the P wave, QRS complex, and T wave represent?

Regulation of the Cardiac Cycle

The quantity of blood pumped changes to accommodate cellular requirements. For example, during strenuous exercise, the amount of blood the skeletal muscles require increases greatly, and heart rate increases in response. The S-A node normally controls heart rate. Thus, changes in heart rate often reflect factors that affect the S-A node, such as the motor impulses carried on parasympathetic and sympathetic nerve fibers (see chapter 9).

Parasympathetic fibers that innervate the heart arise from neurons in the medulla oblongata (fig. 13.17). Most of these fibers branch to the S-A and A-V nodes. When nerve impulses reach their endings, these fibers secrete acetylcholine, which decreases S-A and A-V nodal activity. As a result, heart rate decreases.

Parasympathetic fibers carry impulses continually to the S-A and A-V nodes, "braking" heart action. Consequently, parasympathetic activity can change heart rate in either direction. An increase in the impulses slows the heart rate, and a decrease in the impulses releases the parasympathetic "brake" and increases heart rate.

Sympathetic fibers reach the heart and join the S-A and A-V nodes as well as other areas of the atrial and ventricular myocardium. The endings of these fibers secrete norepinephrine in response to nerve impulses, which increases the rate and force of myocardial contractions.

The *cardiac center* of the medulla oblongata maintains balance between the inhibitory effects of parasympathetic fibers and the excitatory effects of sympathetic fibers. This center receives sensory impulses from throughout the circulatory system and relays motor impulses to the heart in response. For example, receptors sensitive to stretch are located in certain regions of the aorta (aortic arch) and in the carotid arteries (carotid sinuses) (fig. 13.17). These receptors, called *baroreceptors* (pressoreceptors), can detect changes in blood pressure. Rising pressure stretches the receptors, and they signal the cardiac center in the medulla oblongata. In response, the medulla oblongata sends parasympathetic impulses to the heart via the vagus nerve, decreasing heart rate. This action helps lower blood pressure toward normal.

Impulses from the cerebrum or hypothalamus also influence the cardiac center. Such impulses may decrease heart rate, as occurs when a person faints following an emotional upset, or they may increase heart rate during a period of anxiety.

Two other factors that influence heart rate are temperature change and certain ions. Rising body temperature increases heart action, which is why heart rate usually increases during fever. Low body temperature decreases heart action. A patient's body temperature is sometimes deliberately lowered (hypothermia) to slow the heart during surgery.

Of the ions that influence heart action, the most important are potassium (K^+) and calcium (Ca^{++}) ions. Excess potassium ions (hyperkalemia) decrease the

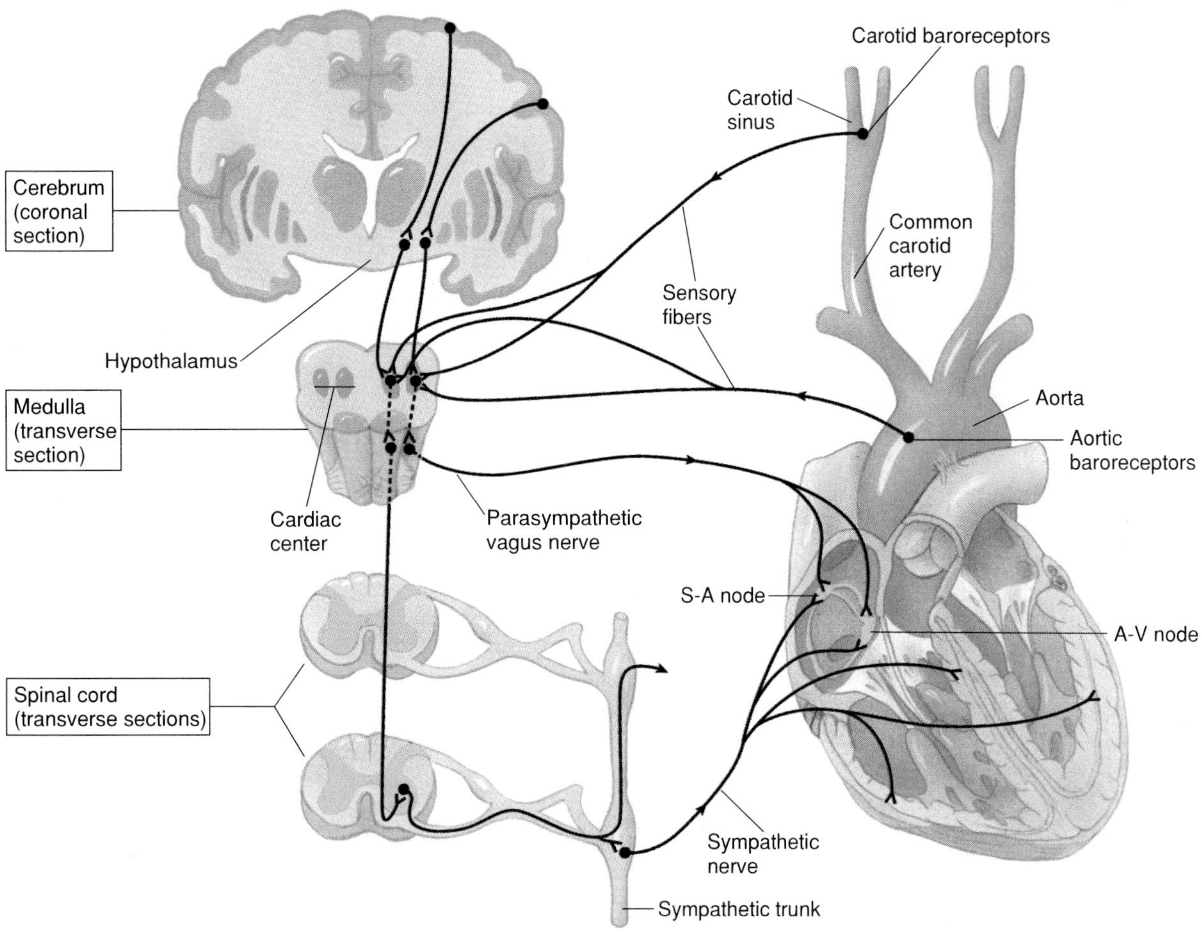

Figure 13.17

Autonomic nerve impulses alter the activities of the S-A and A-V nodes.

rate and force of contractions. If potassium concentration drops below normal (hypokalemia), the heart may develop a potentially life-threatening abnormal rhythm (arrhythmia).

Excess extracellular calcium ions (hypercalcemia) increase heart actions, posing the danger that the heart will contract for a prolonged time. Conversely, low calcium concentration (hypocalcemia) depresses heart action.

1. How do parasympathetic and sympathetic impulses help control heart rate?
2. How do changes in body temperature affect heart rate?
3. Describe the effects of abnormal concentrations of potassium and calcium ions on the heart.

Blood Vessels

The blood vessels form a closed circuit that carries blood from the heart to cells and back again. These vessels include arteries, arterioles, capillaries, venules, and veins.

Arteries and Arterioles

Arteries are strong, elastic vessels adapted for carrying blood away from the heart under high pressure. These vessels branch into progressively thinner tubes and eventually give rise to fine branches called **arterioles.**

An arterial wall consists of three distinct layers (fig. 13.18*a*). The innermost layer (*tunica interna*) is composed of simple squamous epithelium, called *endothelium,* resting on a connective tissue membrane that is rich in elastic and collagenous fibers. Endothelium helps prevent blood clotting by providing a smooth

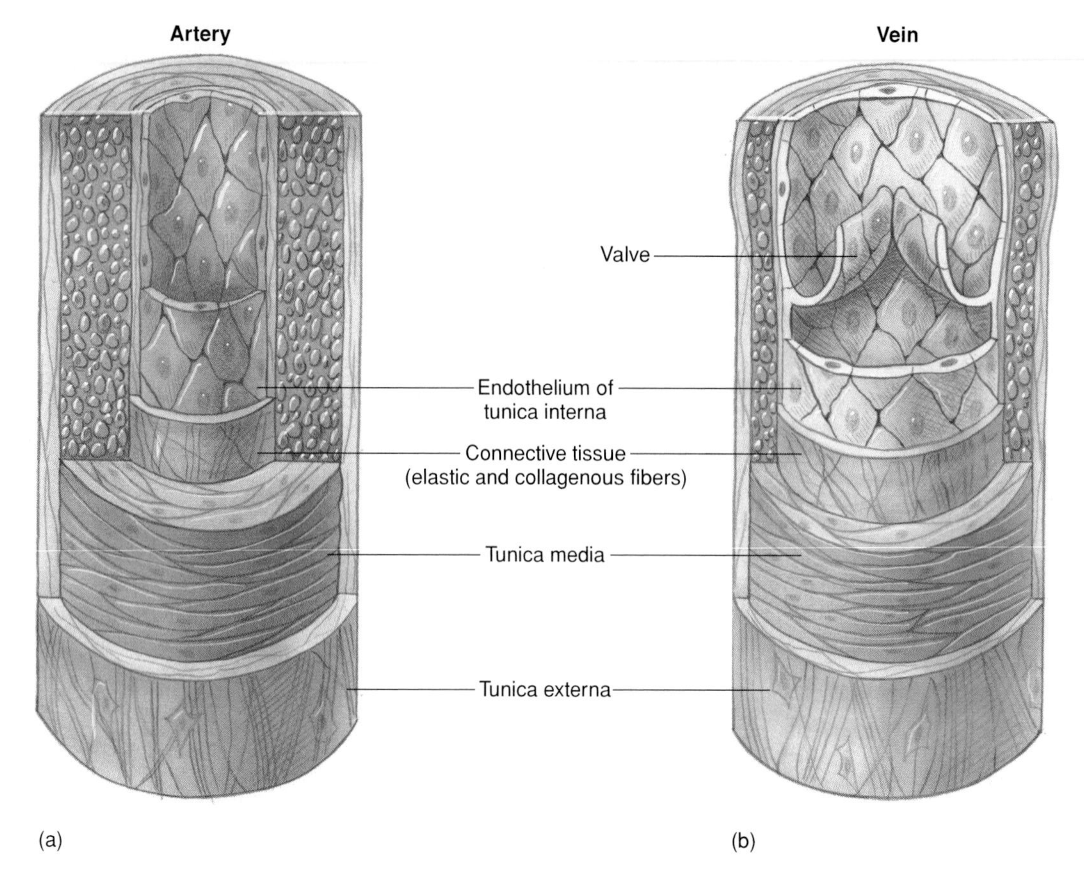

Figure 13.18

(*a*) The wall of an artery. (*b*) The wall of a vein.

surface and secreting biochemicals that inhibit platelet aggregation. Endothelium also may help regulate local blood flow by secreting substances that dilate or constrict blood vessels.

The middle layer (*tunica media*) makes up the bulk of the arterial wall. It includes smooth muscle fibers, which encircle the tube, and a thick layer of elastic connective tissue.

The outer layer (*tunica externa*) is relatively thin and consists chiefly of connective tissue with irregularly organized elastic and collagenous fibers. This layer attaches the artery to surrounding tissues.

Sympathetic branches of the autonomic nervous system innervate smooth muscle in arterial and arteriolar walls. Impulses on these *vasomotor fibers* stimulate the smooth muscles to contract, which reduces vessel diameter. This is called **vasoconstriction.** If vasomotor impulses are inhibited, the muscle fibers relax, and vessel diameter increases. This is called **vasodilation.** Changes in the diameters of arteries and arterioles greatly influence blood flow and pressure.

The walls of the larger arterioles have three layers similar to those of arteries. These walls thin as arterioles approach capillaries. The wall of a very small arteriole consists only of an endothelial lining and some smooth muscle fibers, within a thin layer of connective tissue (fig. 13.19).

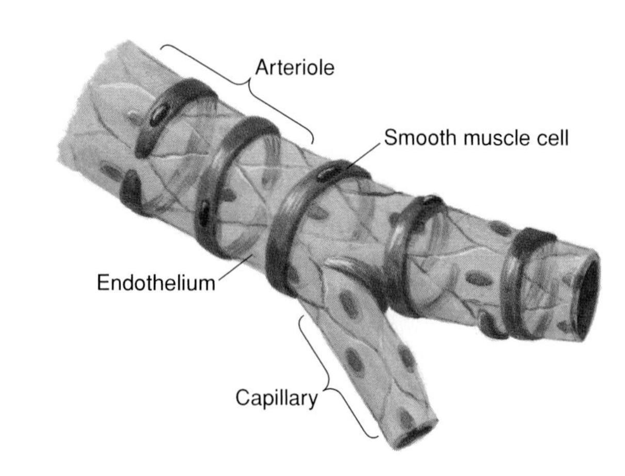

Figure 13.19

Small arterioles have smooth muscle fibers in their walls. Capillaries lack these fibers.

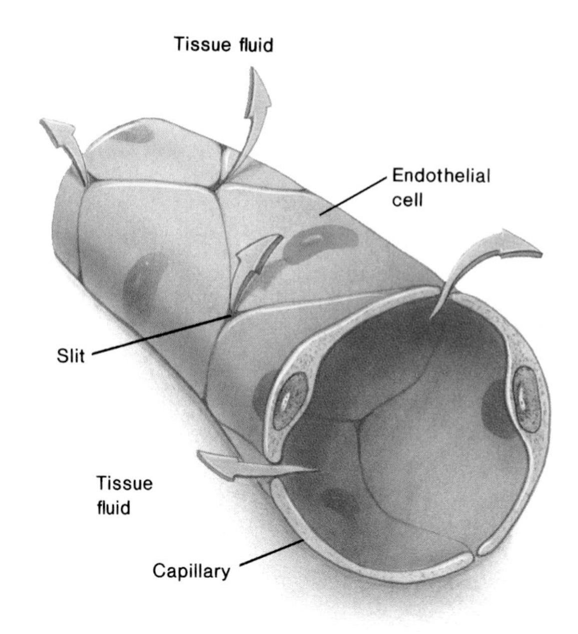

Figure 13.20

Blood and tissue fluid exchange substances through openings (slits) separating endothelial cells.

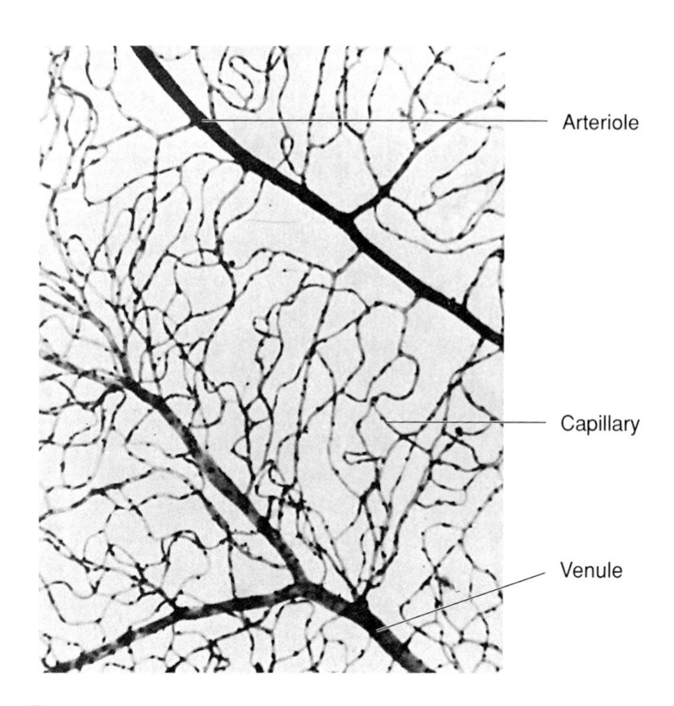

Figure 13.21

Light micrograph of a capillary network (100×).

1. Describe the wall of an artery.
2. What is the function of smooth muscle in the arterial wall?
3. How is the structure of an arteriole different from that of an artery?

Capillaries

Capillaries, the smallest diameter blood vessels, connect the smallest arterioles and the smallest venules. Capillaries are extensions of the inner linings of arterioles in that their walls are endothelium (fig. 13.19). These thin walls form the semipermeable membranes through which substances are exchanged between the blood and the tissue fluid that surrounds body cells.

The openings in capillary walls are thin slits where adjacent endothelial cells overlap. The sizes of these openings and, consequently, the permeability of the capillary wall, vary from tissue to tissue. For example, the openings are small in capillaries of smooth, skeletal, and cardiac muscle, whereas those in capillaries associated with endocrine glands, kidneys, and the lining of the small intestine are larger (fig. 13.20).

Capillary density within tissues varies directly with tissues' rates of metabolism. Muscle and nerve tissues, which use large quantities of oxygen and nutrients, are packed with capillaries. Tissues with slow metabolic rates, such as cartilaginous tissues, the epidermis, and the cornea, lack capillaries.

Patterns of capillary organization differ in various body parts. For example, some capillaries pass directly from arterioles to venules, but others lead to highly branched networks (fig. 13.21).

Smooth muscles that encircle capillary entrances regulate blood distribution in capillary pathways. These muscles form *precapillary sphincters,* which may close a capillary by contracting or open it by relaxing. Precapillary sphincters seem to respond to cells' requirements: When cells have low oxygen and nutrient concentrations, sphincters relax; when cellular requirements have been met, sphincters may contract again. Thus, blood can follow different pathways through a tissue to meet varying requirements of cells.

Routing of blood flow to different parts of the body is due to vasoconstriction and vasodilation of arterioles and precapillary sphincters. During exercise, for example, blood enters the capillary networks of the skeletal muscles, where the cells have increased oxygen and nutrient requirements. At the same time, blood can bypass some of the capillary networks in digestive tract tissues, where demand for blood is less immediate.

1. Describe a capillary wall.
2. What is the function of a capillary?
3. What structures control blood flow into capillaries?

TOPIC OF INTEREST

Atherosclerosis

Nearly half of all deaths in the United States are due to the arterial disease *atherosclerosis,* in which soft masses of fatty materials, particularly cholesterol, accumulate on the inner surface of arterial walls. Such deposits, called *plaque,* protrude into the lumens of vessels and interfere with blood flow (fig. 13A). Furthermore, plaque often forms a surface that can initiate blood clot formation, increasing the risk of developing thrombi or emboli that cause blood deficiency (ischemia) or tissue death (necrosis) beyond the obstruction.

The walls of affected arteries may degenerate, losing their elasticity and becoming hardened or *sclerotic*. During this stage of the disease, called *arteriosclerosis,* a sclerotic vessel can rupture under the force of blood pressure.

Risk factors for developing atherosclerosis include a fatty diet, elevated blood pressure, smoking, obesity, and lack of physical exercise. Emotional and genetic factors may also increase susceptibility to atherosclerosis.

Several treatments attempt to clear clogged arteries. In *percutaneous transluminal angioplasty,* a thin, plastic catheter is passed through a tiny incision in the skin and into the lumen of the affected blood vessel. The catheter, with a tiny deflated balloon at its tip, is pushed along the vessel and into the blocked region. Once in position, the balloon is inflated with high pressure for several minutes, which compresses the atherosclerotic plaque against the arterial wall, widening the arterial lumen and restoring blood flow. However, blockage can return if the underlying cause is not addressed.

Laser energy is also used to destroy atherosclerotic plaque and to channel through arterial obstructions to increase blood flow. In *laser angioplasty,* the light energy of a laser is transmitted through a bundle of optical fibers passed through a small incision in the skin and into the lumen of an obstructed artery.

Another procedure for treating arterial obstruction is *bypass graft surgery*. A surgeon uses a portion of a vein from the patient's lower limb to connect a healthy artery to the affected artery at a point beyond the obstruction. This allows blood from the healthy artery to bypass the narrowed region of the affected artery and supply the tissues downstream. The vein is connected backward, so that its valves do not impede blood flow.

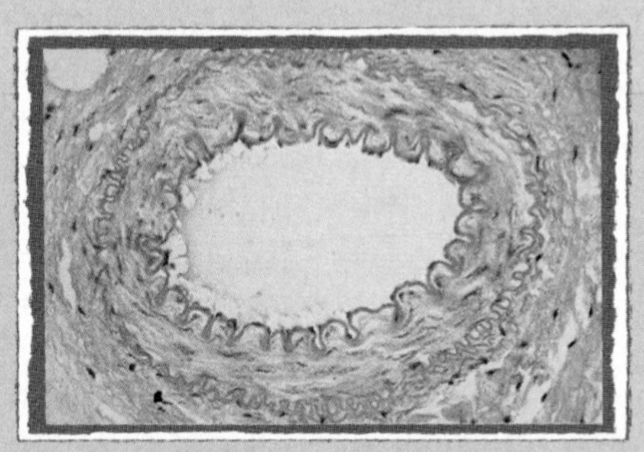

(a)

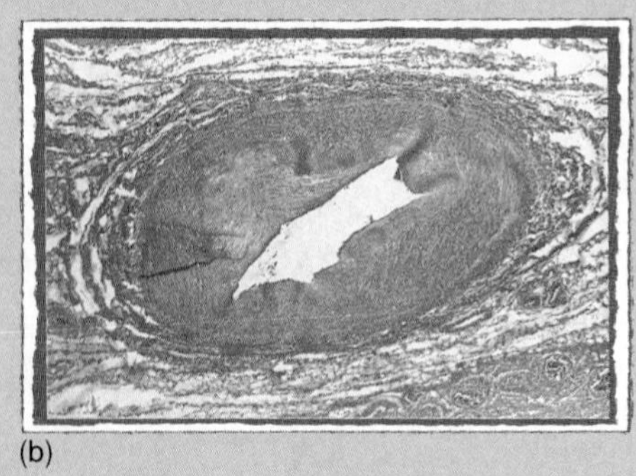

(b)

(c)

Figure 13A

(*a*) Normal arteriole. (*b* and *c*) Accumulation of plaque on the inner wall of an arteriole.

Exchanges in Capillaries

Gases, nutrients, and metabolic by-products are exchanged between blood in capillaries and the tissue fluid surrounding body cells. The substances exchanged move through capillary walls by diffusion, filtration, and osmosis (see chapter 3).

Because blood entering the systemic capillaries carries high concentrations of oxygen and nutrients, these substances diffuse through capillary walls and enter the tissue fluid. Conversely, concentrations of carbon dioxide and other wastes are greater in such tissues, and wastes diffuse into the capillary blood.

In the brain, the endothelial cells of capillary walls are more tightly fused than in other body regions. This organization forms a blood-brain barrier that protects the brain by keeping toxins out and preventing great biochemical fluctuations. Glia also help form the blood-brain barrier. Unfortunately, the barrier also keeps out many useful drugs. Researchers are developing ways to attach certain drugs to molecules that can cross the barrier. ■

Plasma proteins generally remain in blood because they are too large to diffuse through membrane pores or slitlike openings between the endothelial cells of most capillaries. Also, these bulky proteins are not soluble in the lipid parts of capillary cell membranes.

Filtration forces molecules through a membrane with hydrostatic pressure. In capillaries, blood pressure generated when ventricle walls contract provides the force for filtration.

Blood pressure also moves blood through the arteries and arterioles. This pressure decreases with distance from the heart because friction (peripheral resistance) between blood and the vessel walls slows the flow. For this reason, blood pressure is greater in the arteries than in the arterioles and greater in the arterioles than in the capillaries. Blood pressure is similarly greater at the arteriolar end of a capillary than at the venular end. Therefore, the filtration effect occurs primarily at the arteriolar ends of capillaries.

Plasma proteins, which remain in the capillaries, help make the *osmotic pressure* of blood greater (hypertonic) than that of tissue fluid. Although capillary blood has a greater osmotic attraction for water than does tissue fluid, the greater force of blood pressure overcomes this attraction at the arteriolar end of the capillary. As a result, the net movement of water and dissolved substances is outward at the arteriolar end of the capillary by filtration (fig. 13.22).

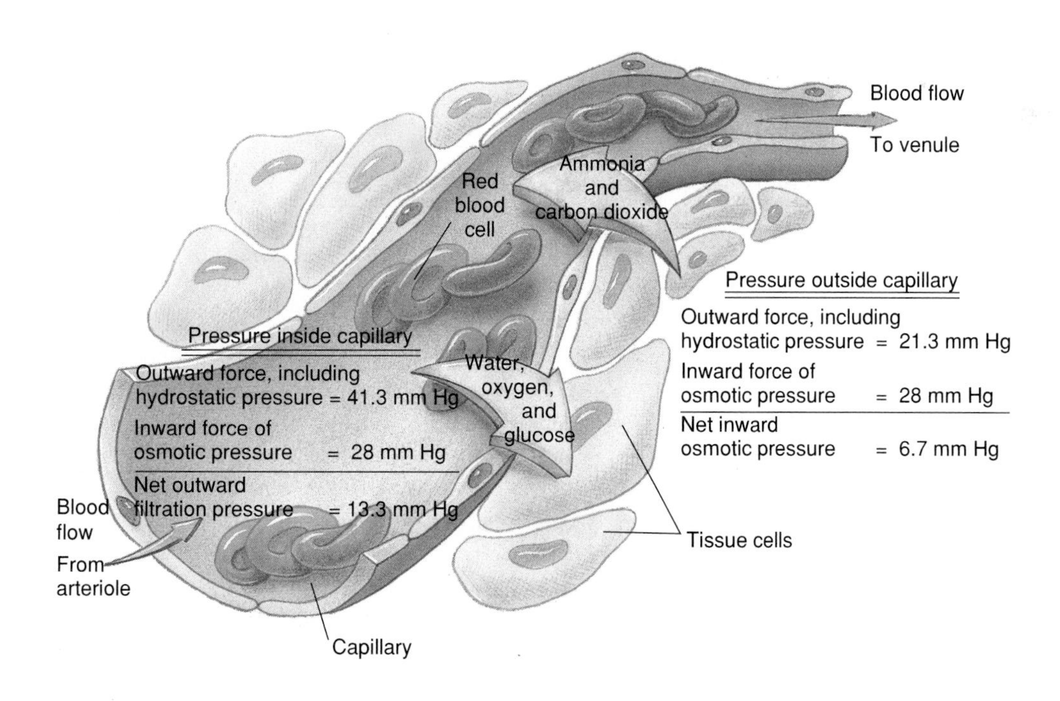

Figure 13.22

Water and other substances leave capillaries because of a net outward filtration pressure at capillaries' arteriolar ends. Water enters at capillaries' venule ends because of a net inward force of osmotic pressure. Substances move in and out along the length of capillaries according to their respective concentration gradients.

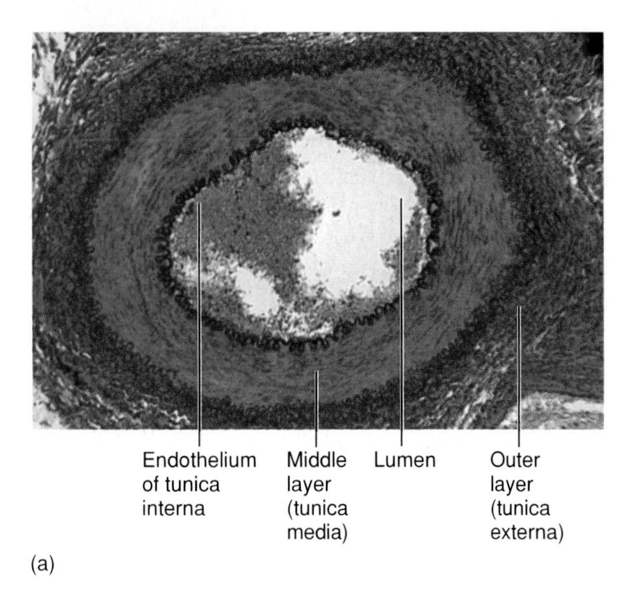

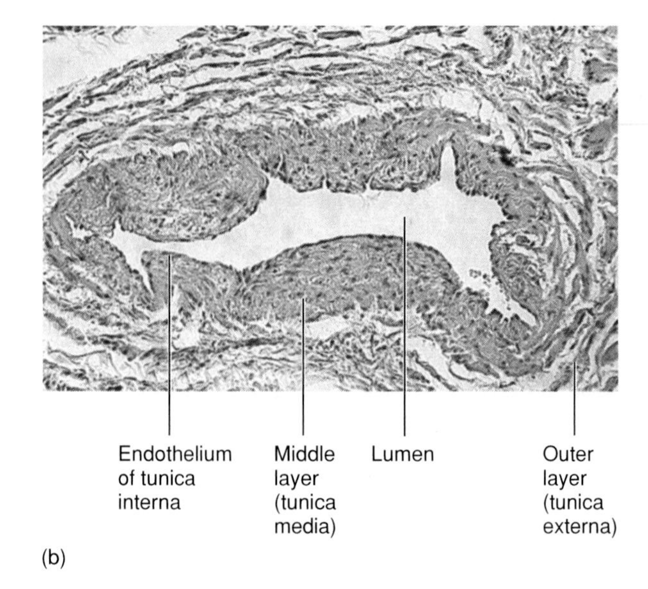

Figure 13.23

Note the structural differences in these cross sections of (*a*) an artery (50×) and (*b*) a vein (16×).

Since blood pressure decreases as blood moves through the capillary, the outward filtration force is less than the osmotic pressure of blood at the venular end. Consequently, the net movement of water and dissolved materials is inward at the venular end of the capillary by osmosis.

Normally, more fluid leaves capillaries than returns to them. Lymphatic vessels collect the excess fluid and return it to the venous circulation. Chapter 14 discusses this mechanism.

Unusual events may increase blood flow to capillaries, and excess fluid enters spaces between tissue cells. This may occur in response to certain chemicals, such as histamine, that vasodilate the arterioles near capillaries and increase capillary permeability. Enough fluid may leak out of the capillaries to overwhelm lymphatic drainage. Affected tissues become swollen (edematous) and painful.

1. What forces cause the exchange of substances between blood and tissue fluid?
2. Why is the fluid movement out of a capillary greater at the capillary's arteriolar end than at its venule end?

Venules and Veins

Venules are the microscopic vessels that continue from the capillaries and merge to form **veins.** The veins, which carry blood back to the atria, roughly parallel the pathways of the arteries.

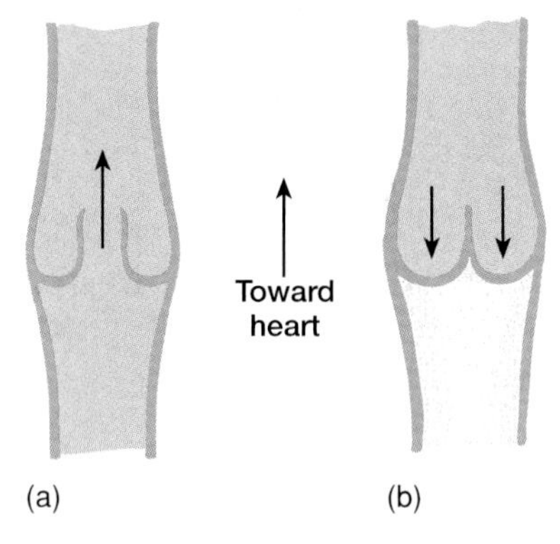

Figure 13.24

(*a*) Venous valves allow blood to move toward the heart, but (*b*) prevent blood from moving backward away from the heart.

The walls of veins are similar to those of arteries in that they have three distinct layers (see fig. 13.18*b*). The middle layer of the venous wall, however, is poorly developed. Consequently, veins have thinner walls with less smooth muscle and less elastic tissue than those of comparable arteries (fig. 13.23).

Many veins, particularly those in the upper and lower limbs, contain flaplike *valves,* which project inward from their linings. Valves usually are composed of two leaflets that close if blood begins to back up in a vein (fig. 13.24). These valves aid in returning blood to the heart because they open if blood flow is toward the heart, but close if it is in the opposite direction.

Veins also function as blood reservoirs. If a hemorrhage causes a drop in arterial blood pressure, sympathetic nerve impulses stimulate the muscular walls of

Table 13.2

Characteristics of Blood Vessels

Vessel	Type of Wall	Function
Artery	Thick, strong wall with three layers—an endothelial lining, a middle layer of smooth muscle and elastic tissue, and an outer layer of connective tissue	Carries blood under relatively high pressure from heart to arterioles
Arteriole	Thinner wall than an artery but with three layers; smaller arterioles have an endothelial lining, some smooth muscle tissue, and a small amount of connective tissue	Connects an artery to a capillary; helps control blood flow into a capillary by vasoconstricting or vasodilating
Capillary	Single layer of squamous epithelium	Provides a membrane through which nutrients, gases, and wastes are exchanged between blood and tissue fluid; connects an arteriole to a venule
Venule	Thinner wall, less smooth muscle and elastic tissue than in an arteriole	Connects a capillary to a vein
Vein	Thinner wall than an artery but with similar layers; the middle layer is more poorly developed; some have flaplike valves	Carries blood under relatively low pressure from a venule to the heart; serves as blood reservoir; valves prevent blood backflow

the veins reflexly. The resulting venous constrictions help maintain blood pressure by returning more blood to the heart. This mechanism ensures a nearly normal blood flow even when as much as 25% of blood volume is lost.

Table 13.2 summarizes the characteristics of blood vessels.

1. How does the structure of a vein differ from that of an artery?
2. How does venous circulation help to maintain blood pressure when hemorrhaging causes blood loss?

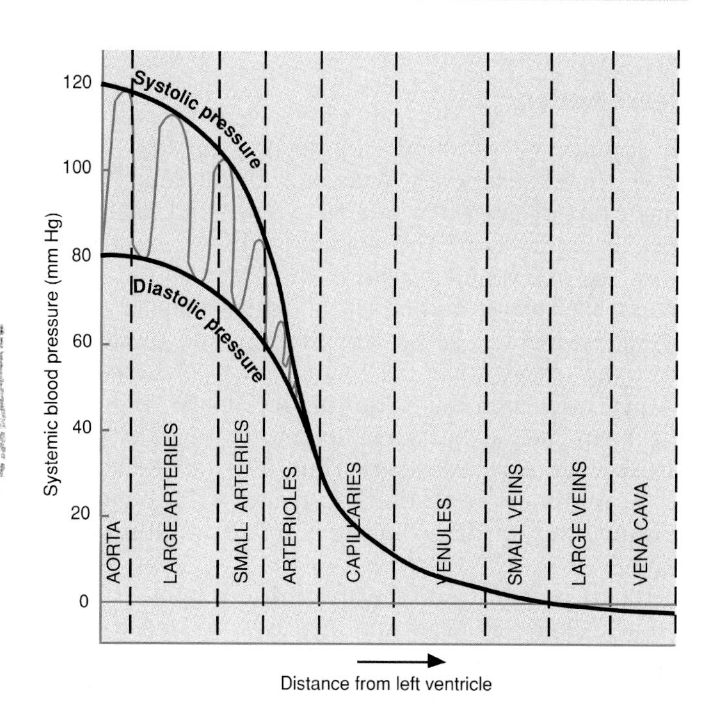

Figure 13.25

Blood pressure decreases as distance from the left ventricle increases.

Blood Pressure

Blood pressure is the force blood exerts against the inner walls of blood vessels. Although this force occurs throughout the vascular system, the term *blood pressure* most commonly refers to pressure in arteries supplied by branches of the aorta.

Arterial Blood Pressure

Arterial blood pressure rises and falls in a pattern corresponding to the phases of the cardiac cycle. That is, contracting ventricles (ventricular systole) squeeze blood out and into the pulmonary trunk and aorta, which sharply increases the pressures in these arteries. The maximum pressure during ventricular contraction is called the **systolic pressure.** When the ventricles relax (ventricular diastole), arterial pressure drops, and the lowest pressure that remains in the arteries before the next ventricular contraction is the **diastolic pressure** (fig. 13.25).

The surge of blood that enters the arterial system during a ventricular contraction distends the elastic arterial walls, but the pressure drops almost immediately as the contraction ends, and the walls recoil. This alternate expanding and recoiling generates a *pulse* in an artery that runs close to the surface.

1. What is *blood pressure?*
2. Distinguish between systolic and diastolic blood pressure.
3. What causes a pulse in an artery?

Factors That Influence Arterial Blood Pressure

Arterial blood pressure depends on a variety of factors, including heart action, blood volume, resistance to flow, and blood viscosity.

Heart Action

In addition to producing blood pressure by forcing blood into the arteries, heart action determines how much blood enters the arterial system with each ventricular contraction. The volume of blood discharged from the left ventricle with each contraction is called the **stroke volume** and equals about 70 milliliters in an average-weight male at rest. The volume discharged from the left ventricle per minute is called the **cardiac output**, calculated by multiplying the stroke volume by the heart rate in beats per minute (cardiac output = stroke volume × heart rate). Thus, if the stroke volume is 70 milliliters, and the heart rate is 72 beats per minute, the cardiac output is 5,040 milliliters per minute.

Blood pressure varies with cardiac output. If either stroke volume or heart rate increases, so does cardiac output, and as a result, blood pressure rises. Conversely, if stroke volume or heart rate decreases, cardiac output and blood pressure also decrease.

Blood Volume

Blood volume equals the sum of the formed elements and plasma volumes in the vascular system. Blood volume varies somewhat with age and sex, and varies directly with body size. Average blood volume is 5 liters, or about 8% of body weight.

Blood pressure is normally directly proportional to blood volume within the cardiovascular system. Thus, changes in blood volume alter blood pressure. For example, if a hemorrhage reduces blood volume, blood pressure drops. A transfusion that restores normal blood volume reestablishes normal blood pressure. Blood volume can also fall if the fluid balance is upset, as happens in dehydration. Fluid replacement can restore normal blood volume and pressure.

Peripheral Resistance

Friction between the blood and the walls of blood vessels creates a force called **peripheral resistance**, which hinders blood flow. Blood pressure must overcome this force if blood is to continue flowing. Therefore, factors that alter peripheral resistance change blood pressure. For example, contracting smooth muscles in arteriolar walls increase the peripheral resistance of these constricted vessels. Blood backs up into the arteries supplying the arterioles, raising arterial pressure. Dilation of arterioles has the opposite effect—peripheral resistance lessens, and arterial blood pressure drops in response.

Blood Viscosity

Viscosity is the ease with which a fluid's molecules flow past one another. The greater the viscosity, the greater the resistance to flow.

Blood cells and plasma proteins increase blood viscosity. Since the greater the blood's resistance to flow, the greater the force required to move it through the vascular system, it is not surprising blood pressure rises as blood viscosity increases, and drops as viscosity decreases.

1. What is the relationship between cardiac output and blood pressure?
2. How does blood volume affect blood pressure?
3. What is the relationship between peripheral resistance and blood pressure? Between blood viscosity and blood pressure?

Control of Blood Pressure

Two important mechanisms for maintaining normal arterial pressure are regulation of cardiac output and regulation of peripheral resistance. Cardiac output depends on blood volume discharged from the left ventricle with

*ypertension,* or high blood pressure, is persistently elevated arterial pressure. It is one of the more common diseases of the cardiovascular system.

High blood pressure with unknown cause is called *essential* (also primary or idiopathic) *hypertension*. Elevated pressure can be caused by another problem, such as kidney disease, high sodium intake, obesity, psychological stress, and arteriosclerosis. The decreased elasticity of arterial walls and narrowed arterial lumens of arteriosclerosis increase blood pressure.

The consequences of prolonged, uncontrolled hypertension can be very serious. As the left ventricle works overtime to pump sufficient blood, the myocardium thickens, enlarging the heart. If coronary blood vessels cannot support this overgrowth, parts of the heart muscle die and are replaced with fibrous tissue. Eventually, the enlarged and weakened heart dies.

Hypertension also contributes to the development of atherosclerosis. Plaque accumulation in arteries may cause a *coronary thrombosis* or *coronary embolism*. Similar changes in brain arteries increase the chances of a *cerebral vascular accident* (CVA), or *stroke,* due to a cerebral thrombosis, embolism, or hemorrhage.

Treatment of hypertension varies among patients and may include exercising regularly, controlling body weight, reducing stress, and limiting sodium in the diet. Drug treatment includes diuretics and/or inhibitors of sympathetic nerve activity. ■

each contraction (stroke volume) and heart rate. For example, the blood volume entering the ventricle affects the stroke volume. Entering blood mechanically stretches myocardial fibers in the ventricular wall. Within limits, the greater the length of these fibers, the greater the force with which they contract. This relationship between fiber length (due to stretching of the cardiac muscle cell just before contraction) and force of contraction is called *Starling's law of the heart.* Because of it, the heart can respond to the immediate demands that the varying quantities of blood that return from the venous system place on it. In other words, the more blood that enters the heart from the veins, the greater the ventricular distension, the stronger the ventricular contraction, the greater the stroke volume, and the greater the cardiac output. The less blood that returns from the veins, the lesser the ventricular distension, the weaker the ventricular contraction, the lesser the stroke volume, and the lesser the cardiac output. This mechanism ensures that blood volume discharged from the heart is equal to the volume entering its chambers.

Baroreceptors in the walls of the aorta and carotid arteries sense changes in blood pressure. If arterial pressure increases, nerve impulses travel from the baroreceptors to the cardiac center of the medulla oblongata. This center relays parasympathetic impulses to the S-A node in the heart, and heart rate decreases in response. As a result of this *cardioinhibitor reflex,* cardiac output falls, and blood pressure decreases toward normal. Conversely, decreasing arterial blood pressure initiates the *cardioaccelerator reflex,* which sends sympathetic impulses to the S-A node. As a result, the heart beats faster, increasing cardiac output and arterial pressure. Other factors that increase heart rate and blood pressure include emotional responses, such as fear and anger; physical exercise; and a rise in body temperature.

Peripheral resistance is another mechanism for blood pressure control. Changes in arteriolar diameters regulate peripheral resistance. Because blood vessels with smaller diameters offer a greater resistance to blood flow, factors that cause arteriolar vasoconstriction increase peripheral resistance, and factors causing vasodilation decrease resistance.

The *vasomotor center* of the medulla oblongata continually sends sympathetic impulses to smooth muscles in arteriolar walls, keeping them in a state of tonic contraction, which helps maintain the peripheral resistance associated with normal blood pressure. Because the vasomotor center responds to changes in blood pressure, it can increase peripheral resistance by increasing its outflow of sympathetic impulses, or it can decrease such resistance by decreasing its sympathetic outflow. In the latter case, the vessels vasodilate as sympathetic stimulation falls.

Whenever arterial blood pressure suddenly increases, baroreceptors in the aorta and carotid arteries signal the vasomotor center, and the sympathetic outflow to the arteriolar walls falls. The resulting vasodilation decreases peripheral resistance, and blood pressure decreases toward normal.

Certain chemicals, including carbon dioxide, oxygen, and hydrogen ions, also influence peripheral resistance by affecting precapillary sphincters and smooth muscles in arteriolar walls. For example, increasing concentrations of carbon dioxide, decreasing concentrations of oxygen, and lowering pH relaxes smooth muscles in the systemic circulation. This increases local blood flow to tissues with high metabolic rates, such as exercising skeletal muscles. In addition, epinephrine and norepinephrine vasoconstrict many systemic vessels, increasing peripheral resistance.

Topic of Interest: Exercise and the Cardiovascular System

The cardiovascular system adapts to exercise as a way of life. The conditioned athlete experiences increases in heart pumping efficiency, blood volume, blood hemoglobin concentration, and the number of mitochondria in muscle fibers. All of these adaptations improve oxygen delivery to, and utilization by, muscle tissue.

An athlete's heart typically changes in response to these increased demands, and may enlarge 40% or more. Myocardial mass increases, ventricular cavities expand, and ventricle walls thicken. Stroke volume increases, and heart rate decreases, as does blood pressure. To a physician unfamiliar with a conditioned cardiovascular system, a trained athlete may appear abnormal!

The cardiovascular system responds beautifully to a slow, steady buildup in exercise frequency and intensity. It does not react well to sudden demands—such as a person who never exercises suddenly shoveling snow or running 3 miles.

A recent study confirmed age-old anecdotal reports of unaccustomed exercise causing heart failure. Researchers in the United States and Germany asked about 1,000 patients hospitalized for heart attacks about their exercise habits and what they were doing in the hour before the attack. They also questioned the same number of people who had not had heart attacks about their activities during the same hours as the ill people. The people with heart attacks were much more likely to have been engaging in unaccustomed strenuous exercise. But the study also turned up good news for those who exercise regularly: Although sedentary people have a two- to sixfold increased risk of cardiac arrest while exercising than when not, people in shape have little or no excess risk while exercising.

For exercise to benefit the circulatory system, the heart rate must be elevated to 70–85% of its "theoretical maximum" for at least half an hour three times a week. You can calculate your theoretical maximum by subtracting your age from 220. If you are eighteen years old, your theoretical maximum is 202 beats per minute. Then, 70–85% of this value is 141–172 beats per minute. Good activities for raising the heart rate are tennis, skating, skiing, handball, vigorous dancing, hockey, basketball, biking, and fast walking.

Consult a physician before starting an exercise program. People over age thirty should have a stress test, which is an electrocardiogram taken during exercise. (The standard electrocardiogram is taken at rest.) An arrhythmia that appears only during exercise may indicate heart disease that has not yet produced symptoms.

1. What factors affect cardiac output?
2. What is the function of baroreceptors in the walls of the aorta and carotid arteries?
3. How does the vasomotor center control diameters of arterioles?

Venous Blood Flow

Blood pressure decreases as blood moves through the arterial system and into the capillary networks, so that little pressure remains at the venular ends of capillaries (fig. 13.25). Thus, blood flow through the venous system is only partly the direct result of heart action and depends on other factors, such as skeletal muscle contraction, breathing movements, and vasoconstriction of veins (*venoconstriction*).

Contracting skeletal muscles thicken and press on nearby vessels, squeezing the blood inside. As skeletal muscles press on veins with valves, some blood moves from one valve section to another (see fig. 13.24). This massaging action of contracting skeletal muscles helps push blood through the venous system toward the heart.

Respiratory movements also move venous blood. During inspiration, pressure within the thoracic cavity falls, as the diaphragm contracts and the rib cage moves upward and outward. At the same time, pressure within the abdominal cavity rises as the diaphragm presses downward on the abdominal viscera. Consequently, blood is squeezed out of abdominal veins and forced into thoracic veins. During exercise, these respiratory movements function with skeletal muscle contractions to increase return of venous blood to the heart.

Venoconstriction also returns venous blood to the heart. When venous pressure is low, sympathetic reflexes stimulate smooth muscles in the walls of veins to contract. The veins also provide a blood reservoir that can adapt its capacity to changes in blood volume. If some blood is lost and blood pressure falls, venoconstriction can force blood out of this reservoir. In both of these examples, venoconstriction helps maintain blood pressure by forcing more blood toward the heart.

1. What is the function of venous valves?
2. How do skeletal muscles and respiratory movements affect blood flow?
3. What factors stimulate venoconstriction?

Paths of Circulation

Blood vessels can be divided into two major pathways. The **pulmonary circuit** consists of vessels that carry blood from the heart to the lungs and back to the heart. The **systemic circuit** carries blood from the heart to all other parts of the body and back again (see fig. 13.1).

The circulatory pathways described in the sections that follow are those of an adult. Chapter 20 describes the fetal pathways, which are somewhat different.

Pulmonary Circuit

Blood enters the pulmonary circuit as it leaves the right ventricle through the pulmonary trunk. The pulmonary trunk extends upward and posteriorly from the heart. About 5 centimeters above its origin, it divides into the right and left pulmonary arteries (see fig. 13.4), which penetrate the right and left lungs, respectively. After repeated divisions, the pulmonary arteries give rise to arterioles that continue into the capillary networks associated with the walls of the alveoli, where gas is exchanged between the blood and the air (see chapter 16).

From the pulmonary capillaries, blood enters the venules, which merge to form small veins, and they, in turn, converge to form still larger veins. Four pulmonary veins, two from each lung, return blood to the left atrium. This completes the vascular loop of the pulmonary circuit.

Systemic Circuit

Freshly oxygenated blood moves from the left atrium into the left ventricle. Contraction of the left ventricle forces this blood into the systemic circuit. This circuit includes the aorta and its branches that lead to all the body tissues, as well as the companion system of veins that returns blood to the right atrium.

1. Distinguish between the pulmonary and systemic circuits of the cardiovascular system.
2. Trace the path of blood through the pulmonary circuit from the right ventricle.

Arterial System

The aorta is the largest diameter artery in the body. It extends upward from the left ventricle, arches over the heart to the left, and descends just anterior and to the left of the vertebral column. Figure 13.26 shows the aorta and its main branches.

Principal Branches of the Aorta

The first portion of the aorta is called the *ascending aorta*. At its base are the three cusps of the aortic valve, and opposite each cusp is a swelling in the aortic wall called an **aortic sinus**. The right and left coronary arteries spring from two of these sinuses (see fig. 13.8).

Three major arteries originate from the *arch of the aorta* (aortic arch). They are the **brachiocephalic** (brăk″e-o-sĕ-fal′ik) **artery**, the left **common carotid** (kah-rot′id) **artery**, and the left **subclavian** (sub-kla′ve-an) **artery**.

The upper part of the *descending aorta* lies to the left of the midline. It gradually moves medially and finally lies directly in front of the vertebral column at the level of the twelfth thoracic vertebra. The portion of the descending aorta above the diaphragm is the **thoracic aorta**. It gives off many small branches to the thoracic wall and thoracic visceral organs.

Below the diaphragm, the descending aorta becomes the **abdominal aorta**, and it gives off branches to the abdominal wall and various abdominal organs. Branches to abdominal organs include: **celiac artery** which gives rise to the **gastric**, **splenic**, and **hepatic** arteries; **superior** (supplies small intestine and superior portion of large intestine) and **inferior** (supplies inferior portion of large intestine) **mesenteric arteries**; and the **suprarenal arteries**, **renal arteries**, and **gonadal arteries** supplying blood to the adrenal glands, kidneys, and ovaries or testes, respectively. The abdominal aorta ends near the brim of the pelvis, where it divides into right and left **common iliac** (il′e-ak) **arteries**. These vessels supply blood to lower regions of the abdominal wall, the pelvic organs, and the lower extremities.

Table 13.3 summarizes the main branches of the aorta.

Arteries to the Neck, Head, and Brain

Branches of the subclavian and common carotid arteries supply blood to structures within the neck, head,

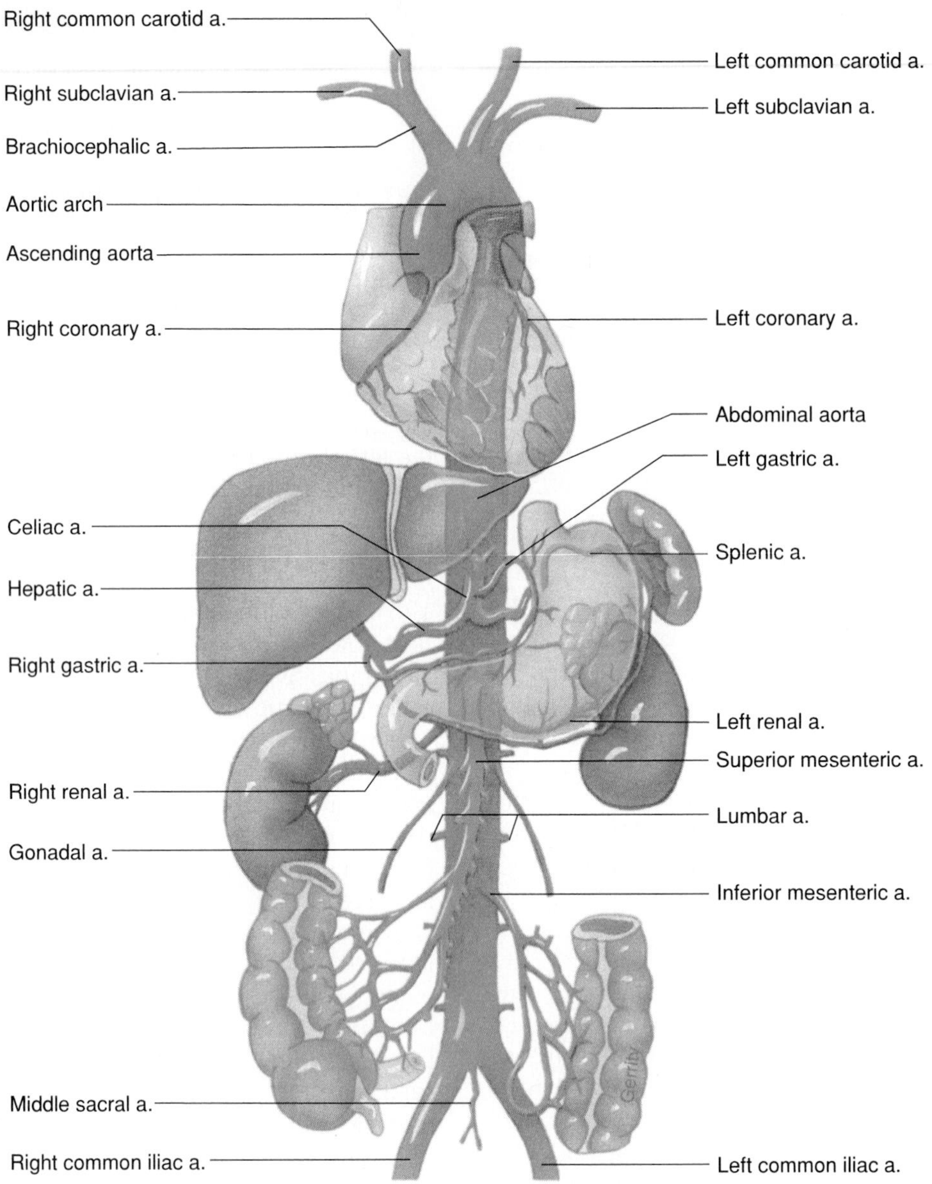

Figure 13.26

Principal branches of the aorta. (*a.* stands for *artery.*)

and brain (fig. 13.27). The main divisions of the subclavian artery to these regions include the vertebral and thyrocervical arteries. The common carotid artery communicates with these regions by means of the internal and external carotid arteries.

The **vertebral arteries** pass upward through the foramina of the transverse processes of the cervical vertebrae and enter the skull through the foramen magnum. These vessels supply blood to the vertebrae and to their associated ligaments and muscles.

Within the cranial cavity, the vertebral arteries unite to form a single *basilar artery.* This vessel passes along the ventral brain stem and branches to the pons, midbrain, and cerebellum. The basilar artery ends by dividing into two *posterior cerebral arteries* that supply portions of the occipital and temporal lobes of the cerebrum. The posterior cerebral arteries also help form at the base of the brain an arterial circle, the **circle of Willis,** which connects the vertebral artery and internal carotid artery systems (fig. 13.28). The union of these systems provides alternate pathways through which blood can reach brain tissues in the event of an arterial occlusion. It also equalizes blood pressure in the brain's blood supply.

The **thyrocervical** (thi″ro-ser′vĭ-kal) **arteries** are short vessels that at the thyrocervical axis give off branches to the thyroid gland, parathyroid glands, larynx, trachea, esophagus, and pharynx, as well as to muscles in the neck, shoulder, and back.

Table 13.3 Principal Branches of the Aorta

Portion of Aorta	Major Branch	General Regions or Organs Supplied
Ascending aorta	Right and left coronary arteries	Heart
Arch of the aorta	Brachiocephalic artery	Right upper limb, right side of head
	Left common carotid artery	Left side of head
	Left subclavian artery	Left upper limb
Descending aorta		
Thoracic aorta	Bronchial artery	Bronchi
	Pericardial artery	Pericardium
	Esophageal artery	Esophagus
	Mediastinal artery	Mediastinum
	Posterior intercostal artery	Thoracic wall
Abdominal aorta	Celiac artery	Organs of upper digestive system
	Phrenic artery	Diaphragm
	Superior mesenteric artery	Portions of small and large intestines
	Suprarenal artery	Adrenal gland
	Renal artery	Kidney
	Gonadal artery	Ovary or testis
	Inferior mesenteric artery	Lower portions of large intestine
	Lumbar artery	Posterior abdominal wall
	Middle sacral artery	Sacrum and coccyx
	Common iliac artery	Lower abdominal wall, pelvic organs, and lower limb

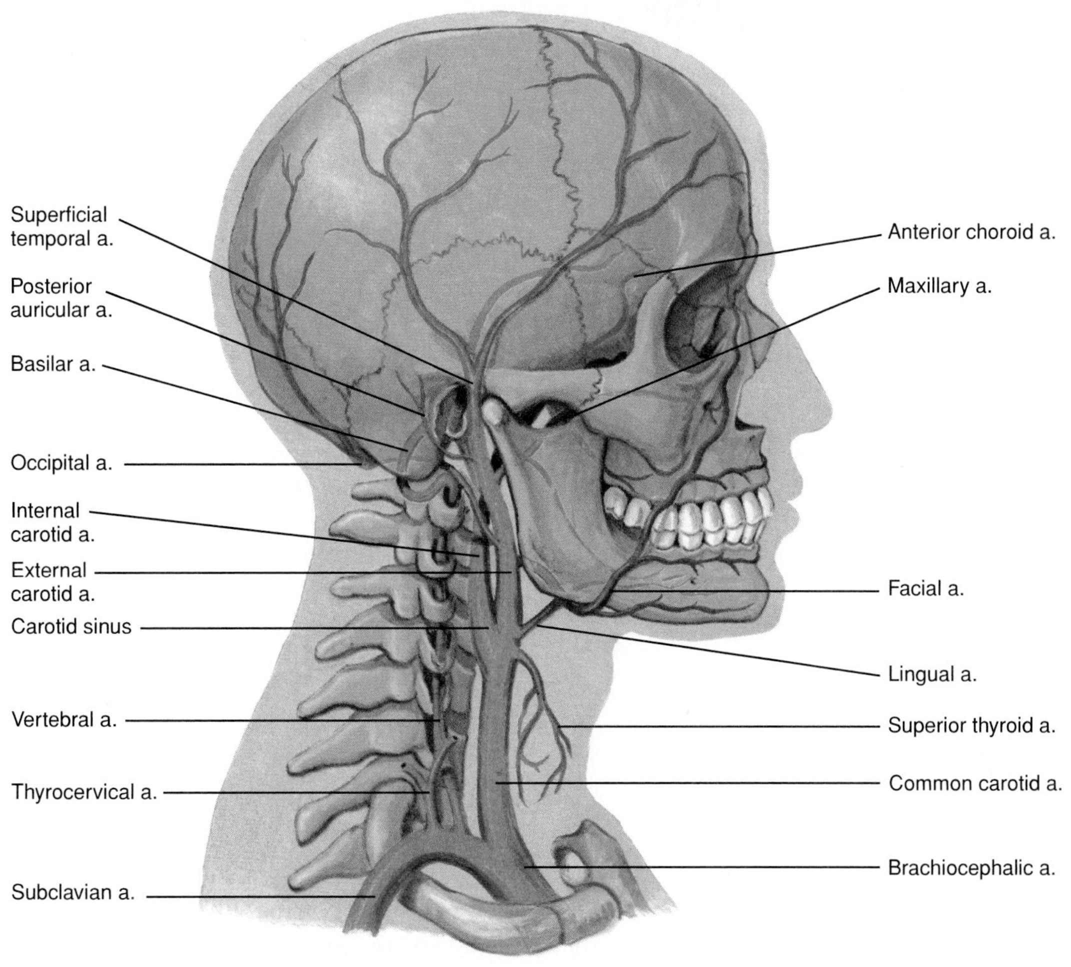

Figure 13.27

Main arteries of the head and neck. Note that the clavicle has been removed. (*a.* stands for *artery.*)

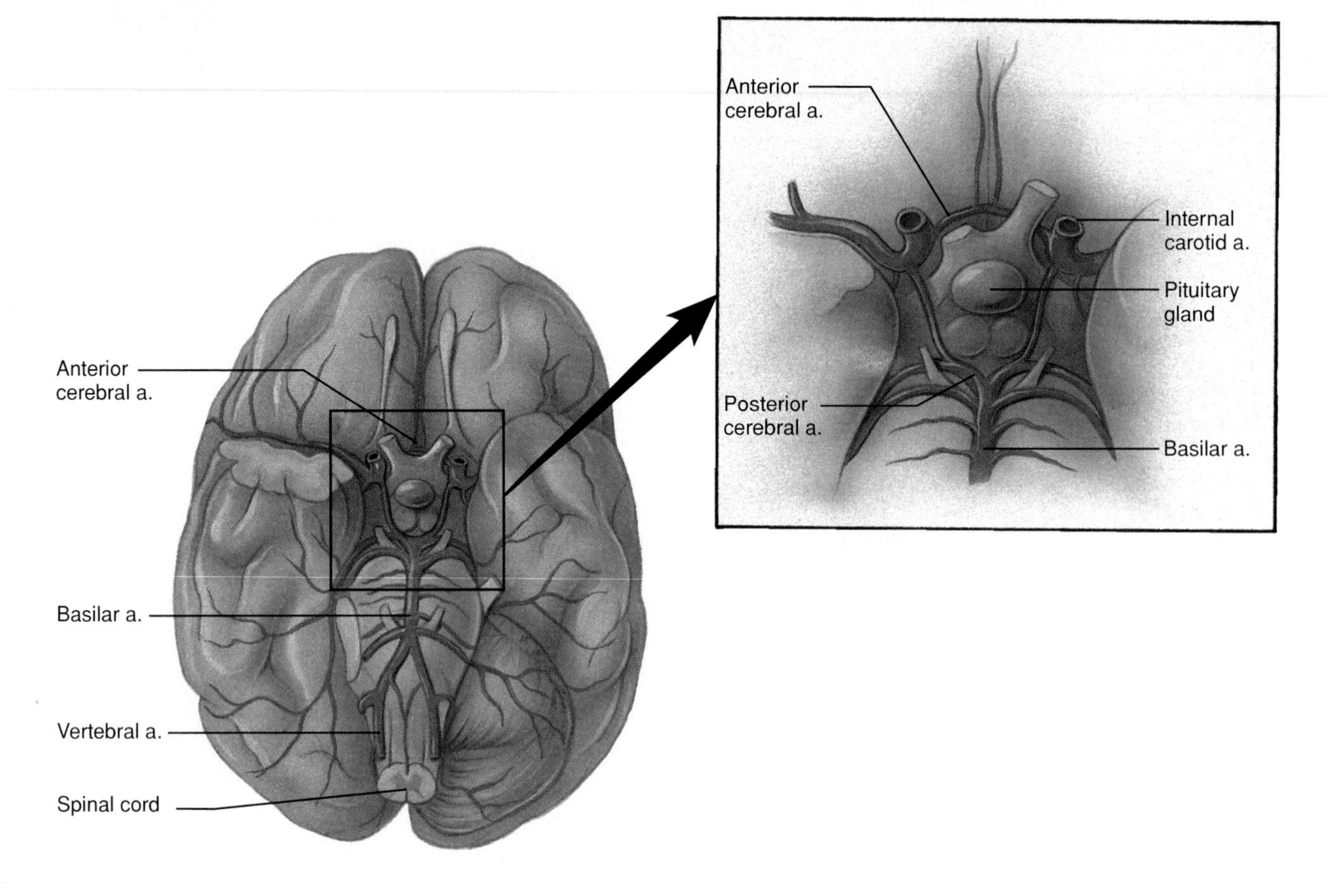

Figure 13.28

The circle of Willis forms from the anterior and posterior cerebral arteries, which join the internal carotid arteries. (*a.* stands for *artery*.)

The left and right common carotid arteries diverge into the internal and external carotid arteries. The **external carotid artery** courses upward on the side of the head, branching to structures in the neck, face, jaw, scalp, and base of the skull. The **internal carotid artery** follows a deep course upward along the pharynx to the base of the skull. Entering the cranial cavity, it provides the major blood supply to the brain. Near the base of the internal carotid arteries are enlargements called **carotid sinuses** that, like aortic sinuses, contain baroreceptors that control blood pressure.

Table 13.4 summarizes the major branches of the external and internal carotid arteries.

Arteries to the Shoulder and Upper Limb

The subclavian artery, after giving off branches to the neck, continues into the arm (fig. 13.29). It passes between the clavicle and first rib, and becomes the axillary artery. The **axillary artery** supplies branches to structures in the axilla and chest wall and becomes the **brachial artery**, which follows the humerus to the elbow and gives rise to a *deep brachial artery*. Within the elbow, the brachial artery divides into an ulnar artery and a radial artery.

The **ulnar artery** leads downward on the ulnar side of the forearm to the wrist. Some of its branches supply the elbow joint, while others supply blood to muscles in the forearm.

The **radial artery** travels along the radial side of the forearm to the wrist, supplying the lateral muscles of the forearm. As the radial artery nears the wrist, it approaches the surface and provides a convenient vessel for taking the pulse (radial pulse).

At the wrist, the branches of the ulnar and radial arteries join to form a network of vessels. Arteries arising from this network supply blood to the wrist, hand, and fingers.

Arteries to the Thoracic and Abdominal Walls

Blood reaches the thoracic wall through several vessels (fig. 13.30). The **internal thoracic artery**, a branch of the subclavian artery, gives off two *anterior intercostal* (in″ter-kos′tal) *arteries* that supply the intercostal muscles and mammary glands. The *posterior intercostal arteries*

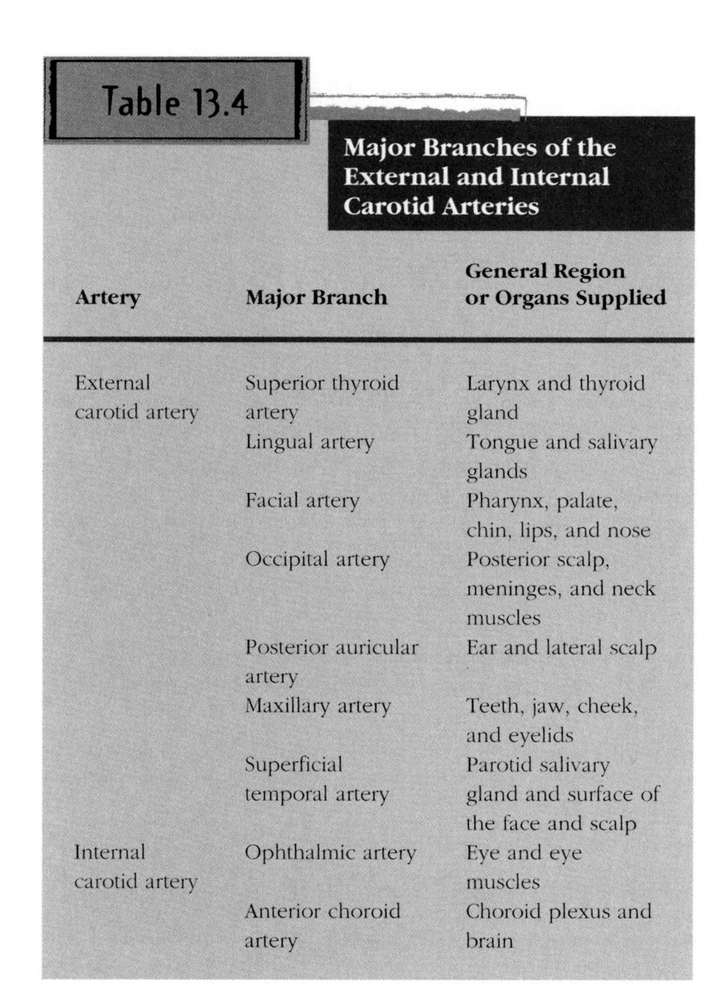

Table 13.4 Major Branches of the External and Internal Carotid Arteries

Artery	Major Branch	General Region or Organs Supplied
External carotid artery	Superior thyroid artery	Larynx and thyroid gland
	Lingual artery	Tongue and salivary glands
	Facial artery	Pharynx, palate, chin, lips, and nose
	Occipital artery	Posterior scalp, meninges, and neck muscles
	Posterior auricular artery	Ear and lateral scalp
	Maxillary artery	Teeth, jaw, cheek, and eyelids
	Superficial temporal artery	Parotid salivary gland and surface of the face and scalp
Internal carotid artery	Ophthalmic artery	Eye and eye muscles
	Anterior choroid artery	Choroid plexus and brain

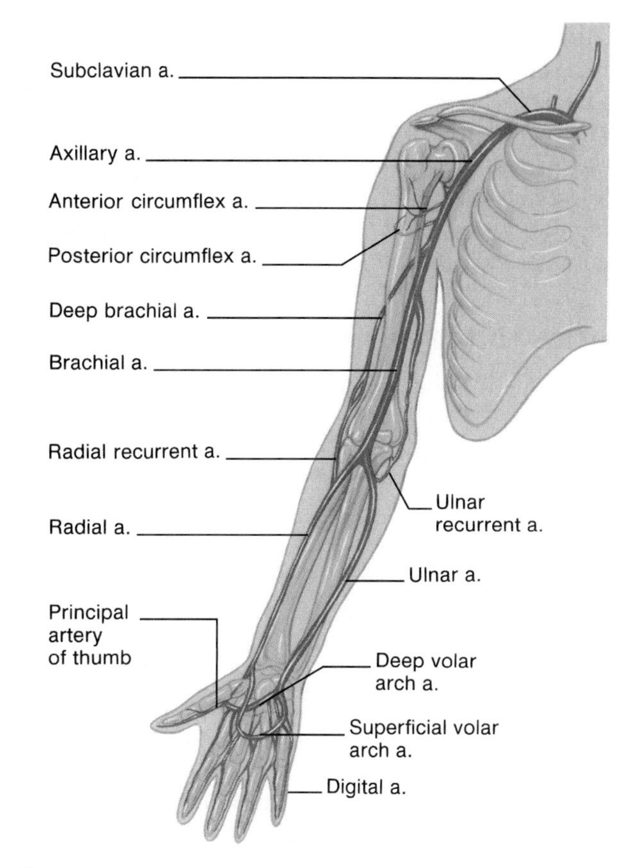

Figure 13.29

Main arteries to the shoulder and upper limb. (*a.* stands for *artery.*)

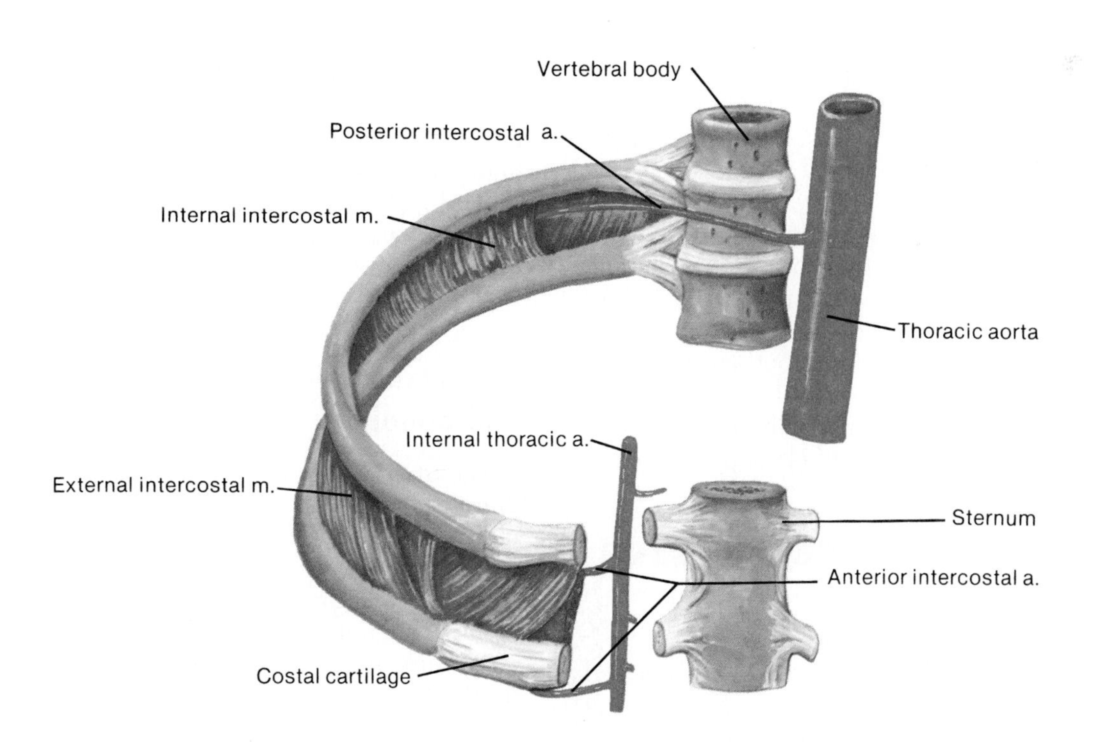

Figure 13.30

Arteries that supply the thoracic wall. (*a.* stands for *artery; m.* stands for *muscle.*)

arise from the thoracic aorta and enter the intercostal spaces. They supply the intercostal muscles, the vertebrae, the spinal cord, and deep muscles of the back.

Branches of the internal thoracic and *external iliac arteries* provide blood to the anterior abdominal wall. Paired vessels originating from the abdominal aorta, including the *phrenic* and *lumbar arteries,* supply blood to structures in the posterior and lateral abdominal wall.

Arteries to the Pelvis and Lower Limb

The abdominal aorta divides to form the common iliac arteries at the level of the pelvic brim, and these vessels provide blood to the pelvic organs, gluteal region, and lower limbs (fig. 13.31). Each common iliac artery divides into an internal and an external branch. The **internal iliac artery** gives off many branches to pelvic muscles and visceral structures, as well as to the gluteal muscles and the external reproductive organs. The **external iliac artery** provides the main blood supply to the lower limbs. It passes downward along the brim of the pelvis and branches to supply the muscles and skin in the lower abdominal wall. Midway between the symphysis pubis and the anterior superior iliac spine of the ilium, the external iliac artery becomes the femoral artery.

The **femoral** (fem′or-al) **artery,** which approaches the anterior surface of the upper thigh, branches to muscles and superficial tissues of the thigh. These branches also supply the skin of the groin and lower abdominal wall.

As the femoral artery reaches the proximal border of the space behind the knee, it becomes the **popliteal** (pop-lit′e-al) **artery.** Branches of this artery supply blood to the knee joint and to certain muscles in the thigh and calf. The popliteal artery diverges into the anterior and posterior tibial arteries.

The **anterior tibial artery** passes downward between the tibia and fibula, giving off branches to the skin and muscles in anterior and lateral regions of the leg. This vessel continues into the foot as the *dorsal pedis artery,* which supplies blood to the foot and toes. The **posterior tibial artery,** the larger of the two popliteal branches, descends beneath the calf muscles, and branches to the skin, muscles, and other tissues of the leg along the way.

1. Name the portions of the aorta.
2. List the major branches of the aortic arch.
3. Name the branches of the thoracic and abdominal aorta.
4. Which vessels supply blood to the head? To the upper limb? To the abdominal wall? To the lower limb?

Venous System

Venous circulation returns blood to the heart after blood and body cells exchange gases, nutrients, and wastes.

Characteristics of Venous Pathways

Venous vessels begin as capillaries merge into venules, venules merge into small veins, and small veins meet to form larger ones. Unlike the arterial pathways, the vessels of the venous system are difficult to follow because they connect in irregular networks. Many unnamed tributaries may join to form a relatively large vein.

Larger veins, however, typically parallel the courses of named arteries, and these veins often have the same names as their arterial counterparts. For example, the renal vein parallels the renal artery, and the common iliac vein accompanies the common iliac artery.

The veins that carry blood from the lungs and myocardium back to the heart have already been described. The veins from all the other parts of the body converge into two major pathways, the superior and inferior venae cavae, which lead to the right atrium.

Veins from the Brain, Head, and Neck

The **external jugular** (jug′u-lar) **veins** drain blood from the face, scalp, and superficial regions of the neck. These vessels descend on either side of the neck and empty into the *subclavian veins* (fig. 13.32).

The **internal jugular veins,** which are larger than the external jugular veins, arise from many veins and venous sinuses of the brain and from deep veins in parts of the face and neck. They descend through the neck and join the subclavian veins. These unions of the internal jugular and subclavian veins form large **brachiocephalic veins** on each side. The vessels then merge and form the superior vena cava, which enters the right atrium.

Veins from the Upper Limb and Shoulder

A set of deep veins and a set of superficial ones drain the upper limb. The deep veins generally parallel the arteries in each region and have similar names, such as the *radial vein, ulnar vein, brachial vein,* and *axillary vein.* The superficial veins connect in complex networks just beneath the skin. They also communicate with the deep vessels of the upper limb, providing many alternate pathways through which blood can leave the tissues (fig. 13.33). The main vessels of the superficial network are the basilic and cephalic veins.

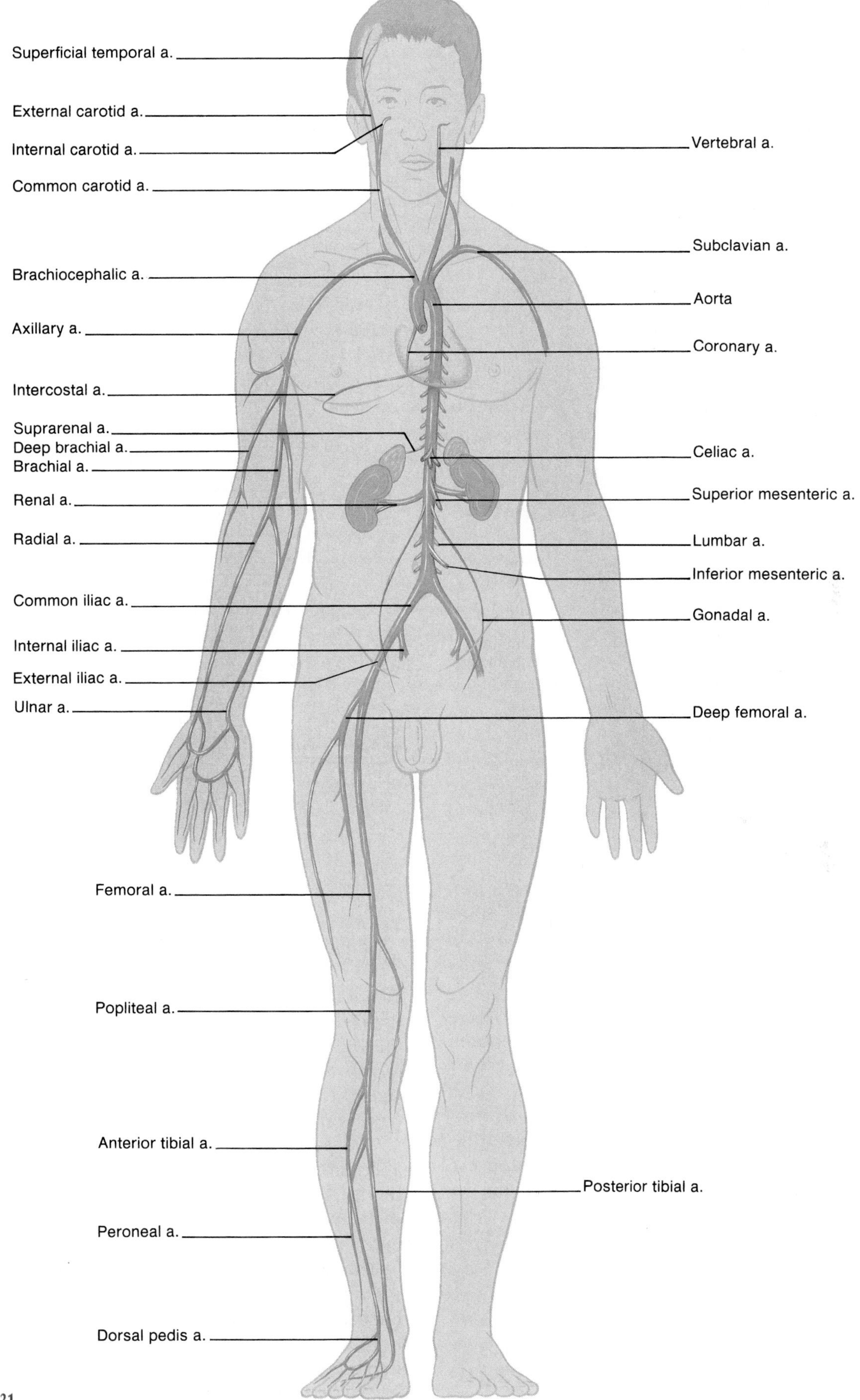

***Figure* 13.31**

Major vessels of the arterial system. (*a.* stands for *artery.*)

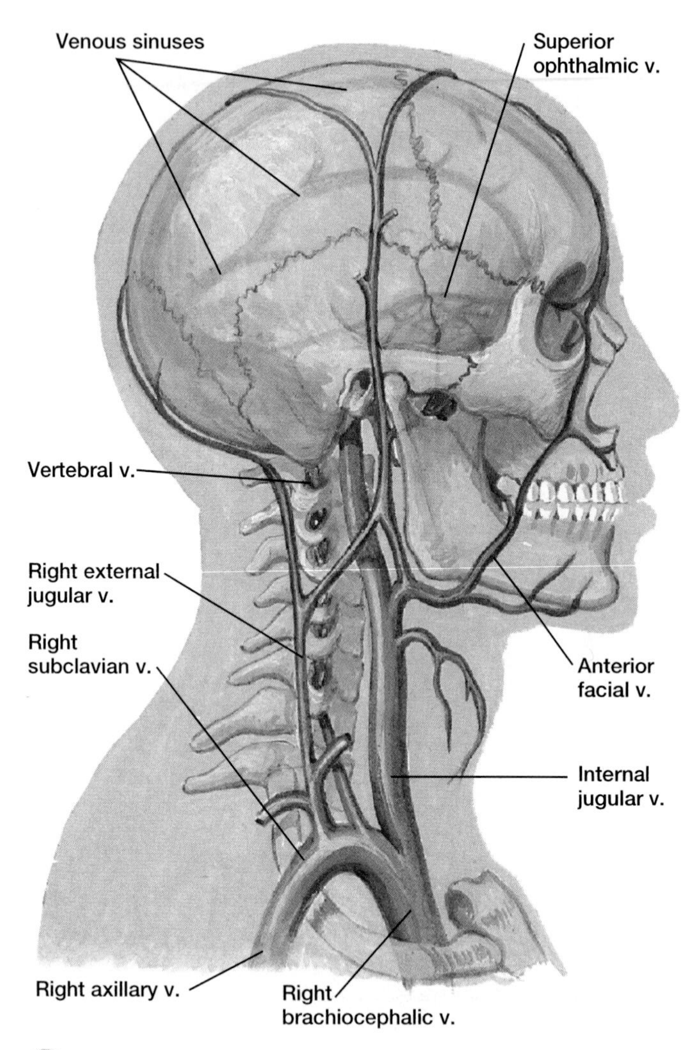

Figure 13.32

Major veins of the brain, head, and neck. Note that the clavicle has been removed. (*v.* stands for *vein.*)

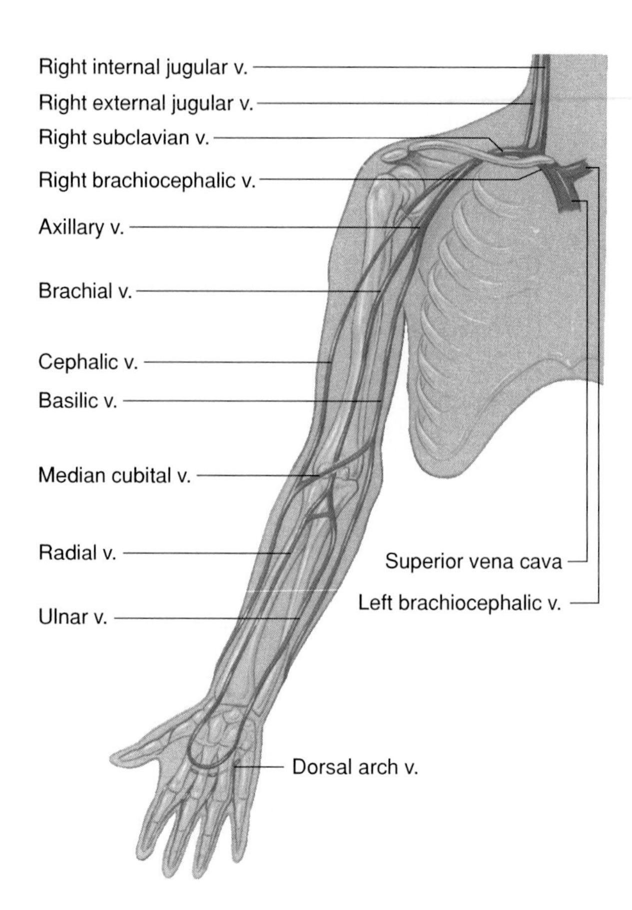

Figure 13.33

Main veins of the upper limb and shoulder. (*v.* stands for *vein.*)

The **basilic** (bah-sil′ik) **vein** ascends from the forearm to the middle of the arm, where it penetrates deeply and joins the *brachial vein.* The basilic and brachial veins merge, forming the *axillary vein.*

The **cephalic** (sĕ-fal′ik) **vein** courses upward from the hand to the shoulder. In the shoulder, it pierces the tissues and empties into the axillary vein. Beyond the axilla, the axillary vein becomes the subclavian vein.

In the bend of the elbow, a *median cubital* vein ascends from the cephalic vein on the lateral side of the arm to the basilic vein on the medial side. This vein is usually prominent. It is often a site for *venipuncture* to remove a blood sample for examination or to add fluids to blood. ■

Veins from the Abdominal and Thoracic Walls

Tributaries of the brachiocephalic and azygos veins drain the abdominal and thoracic walls. For example, the brachiocephalic vein receives blood from the *internal thoracic vein,* which generally drains the tissues the internal thoracic artery supplies. Some *intercostal veins* also empty into the brachiocephalic vein.

The **azygos** (az′ĭ-gos) **vein** originates in the dorsal abdominal wall and ascends through the mediastinum on the right side of the vertebral column to join the superior vena cava. It drains most of the muscular tissue in the abdominal and thoracic walls.

Tributaries of the azygos vein include the *posterior intercostal veins* on the right side, which drain the intercostal spaces, and the *superior* and *inferior hemiazygos veins,* which receive blood from the posterior intercostal veins on the left (fig. 13.34). The right and left *ascending lumbar veins,* with tributaries that include vessels from the lumbar and sacral regions, also connect to the azygos system.

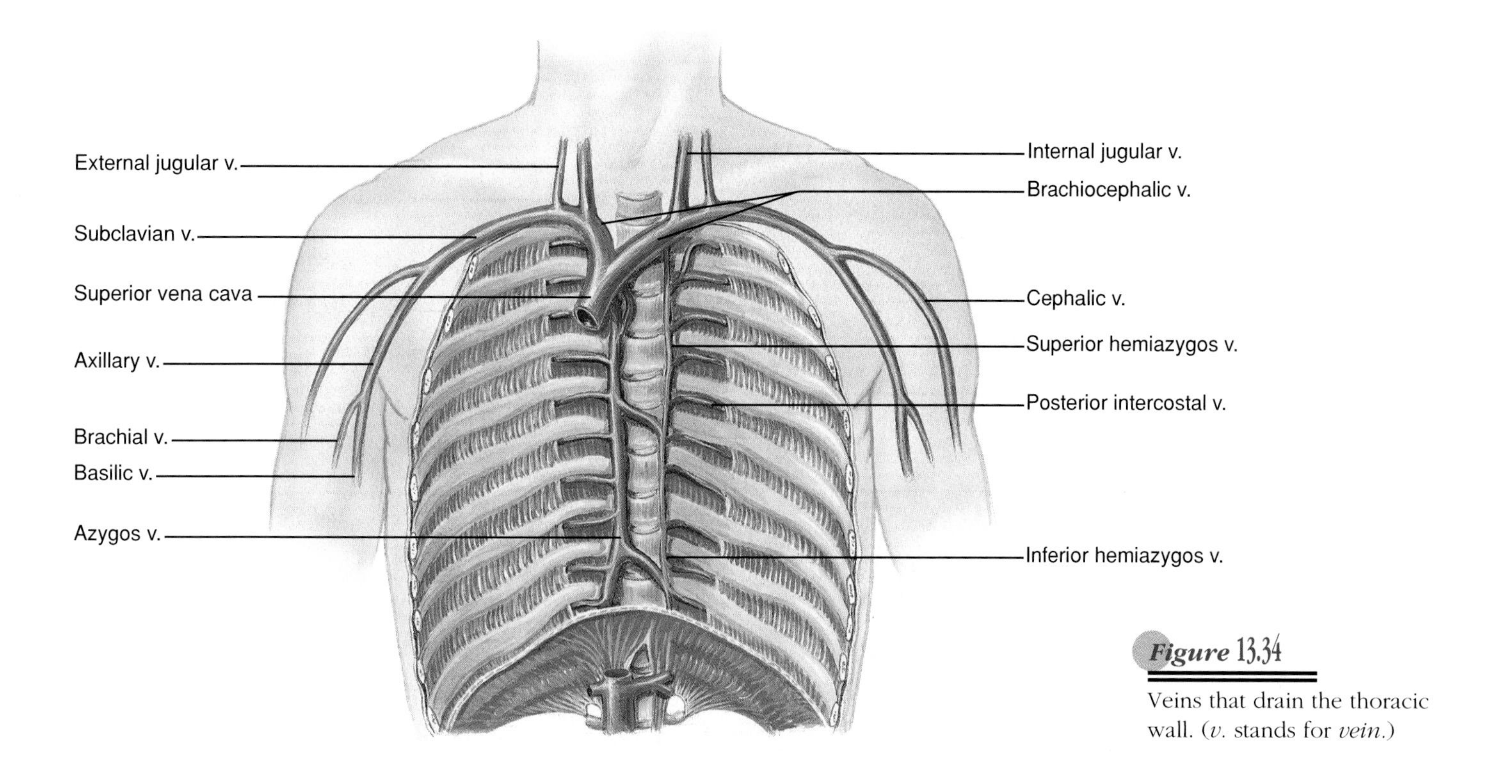

Figure 13.34

Veins that drain the thoracic wall. (*v.* stands for *vein.*)

Veins from the Abdominal Viscera

Veins usually carry blood directly to the atria of the heart. Those that drain the abdominal viscera, however, are exceptions (fig. 13.35). They originate in the capillary networks of the stomach, intestines, pancreas, and spleen and carry blood from these organs through a **portal** (por′tal) **vein** to the liver. This unusual venous pathway is called the **hepatic portal system.**

Tributaries of the portal vein include:

1. Right and left *gastric veins* from the stomach.
2. *Superior mesenteric vein* from the small intestine, ascending colon, and transverse colon.
3. *Splenic vein* from a convergence of several veins draining the spleen, the pancreas, and a portion of the stomach. Its largest tributary, the *inferior mesenteric vein,* brings blood upward from the descending colon, sigmoid colon, and rectum.

About 80% of blood flowing to the liver in the hepatic portal system comes from capillaries in the stomach and intestines, and is oxygen-poor but rich in nutrients. As discussed in chapter 15, the liver handles these nutrients in a variety of ways. It regulates blood glucose concentration by polymerizing excess glucose into glycogen for storage or by breaking down glycogen into glucose when blood glucose concentration drops below normal. The liver helps regulate blood concentrations of recently absorbed amino acids and lipids by modifying them into forms cells can use, by oxidizing them, or by changing them into storage forms. The liver also stores certain vitamins and detoxifies harmful substances. Blood in the portal vein nearly always contains bacteria that have entered through intestinal capillaries. Large *Kupffer cells* lining small vessels in the liver called hepatic sinusoids phagocytize these microorganisms, removing them from portal blood before it leaves the liver.

After passing through the hepatic sinusoids of the liver, blood in the hepatic portal system travels through a series of merging vessels into **hepatic veins.** These veins empty into the inferior vena cava, returning the blood to the general circulation.

Veins from the Lower Limb and Pelvis

As in the upper limb, veins that drain blood from the lower limb are divided into deep and superficial groups (fig. 13.36). The deep veins of the leg, such as the *anterior* and *posterior tibial veins,* are named for the arteries they accompany. At the level of the knee, these vessels form a single trunk, the **popliteal vein.** This vein continues upward through the thigh as the **femoral vein,** which, in turn, becomes the **external iliac vein.**

The superficial veins of the foot, leg, and thigh connect to form a complex network beneath the skin. These vessels drain into two major trunks—the small and great saphenous veins. The **small saphenous** (sah-fe′nus) **vein** ascends along the back of the calf, enters the popliteal fossa, and joins the popliteal vein. The

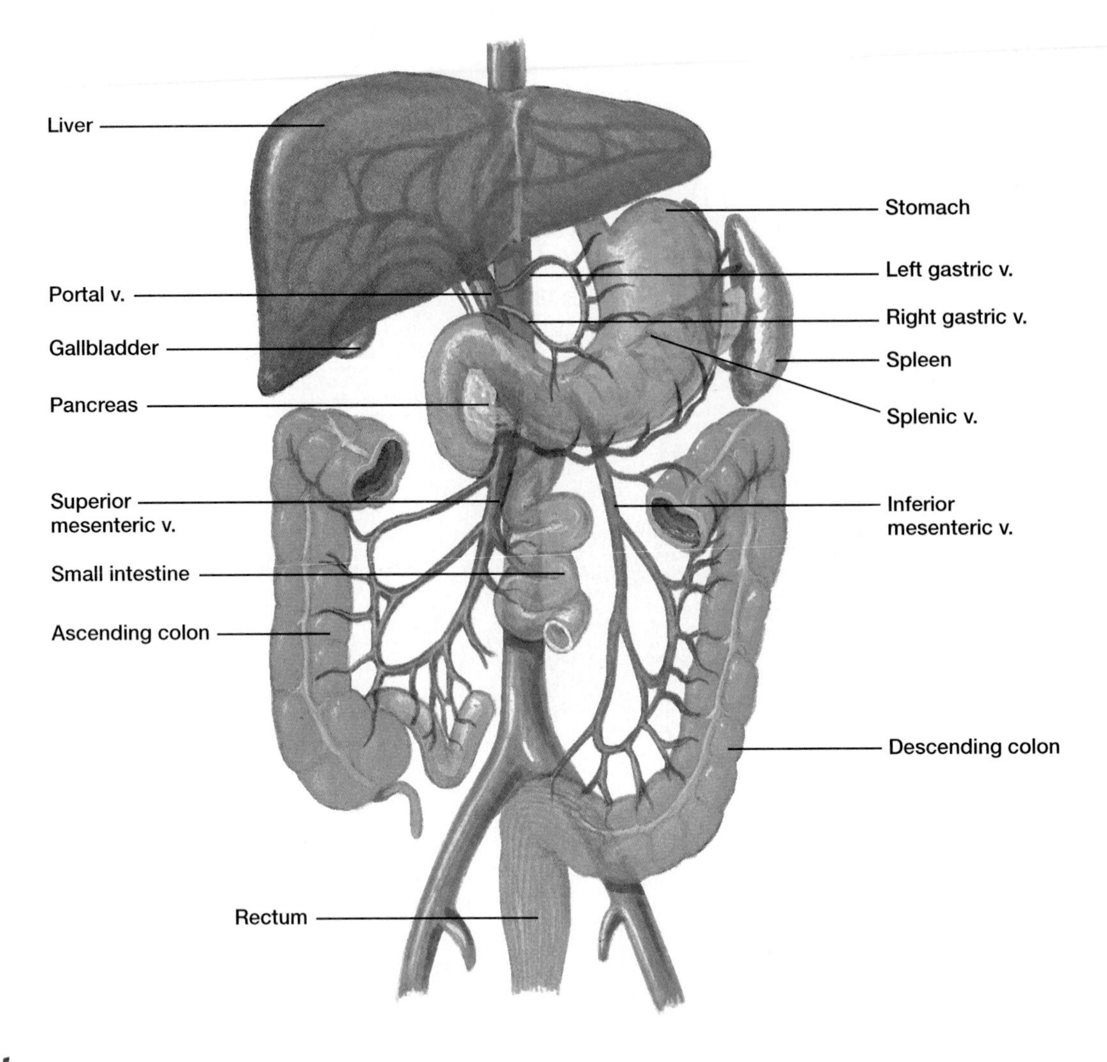

Figure 13.35

Veins that drain the abdominal viscera. (*v.* stands for *vein.*)

great saphenous vein, which is the longest vein in the body, ascends in front of the medial malleolus and extends upward along the medial side of the leg and thigh. In the thigh, it penetrates deeply and joins the femoral vein. Near its termination, the great saphenous vein receives tributaries from a number of vessels that drain the upper thigh, groin, and lower abdominal wall.

In addition to communicating freely with each other, the saphenous veins communicate extensively with the deep veins of the leg and thigh. Blood can thus take several routes to return to the heart from the lower extremities.

In the pelvic region, vessels leading to the **internal iliac vein** carry blood away from organs of the reproductive, urinary, and digestive systems. The internal iliac veins unite with the right and left external iliac veins to form the **common iliac veins.** These vessels, in turn, merge into the inferior vena cava.

aricose veins have abnormal dilations. They result from increased blood pressure in the saphenous veins due to gravity, as occurs when a person stands for a prolonged period. ■

1. Name the veins that return blood to the right atrium.
2. Which major veins drain blood from the head? From the upper limb? From the abdominal viscera? From the lower limb?

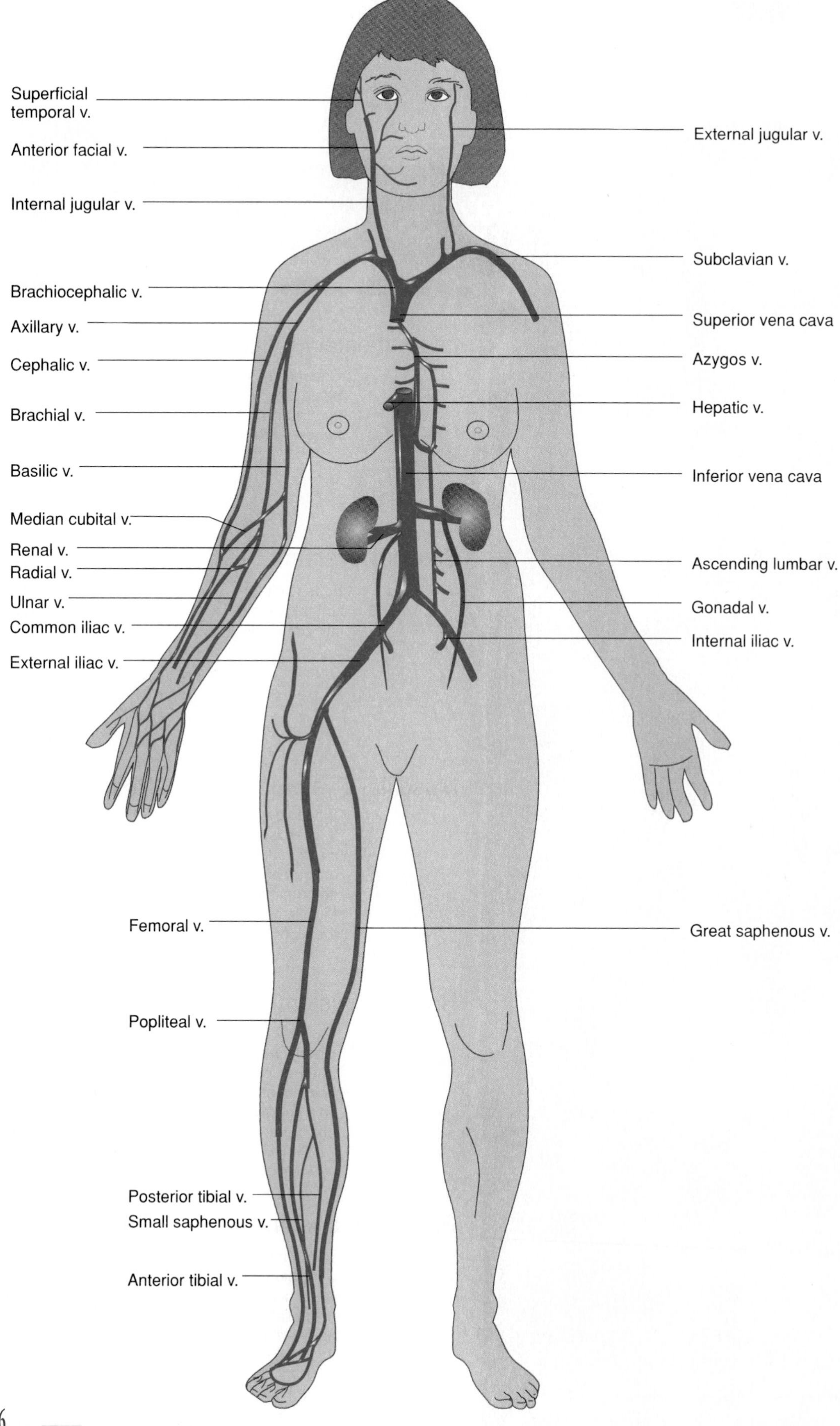

***Figure* 13.36**

Major vessels of the venous system. (*v.* stands for *vein*.)

organization

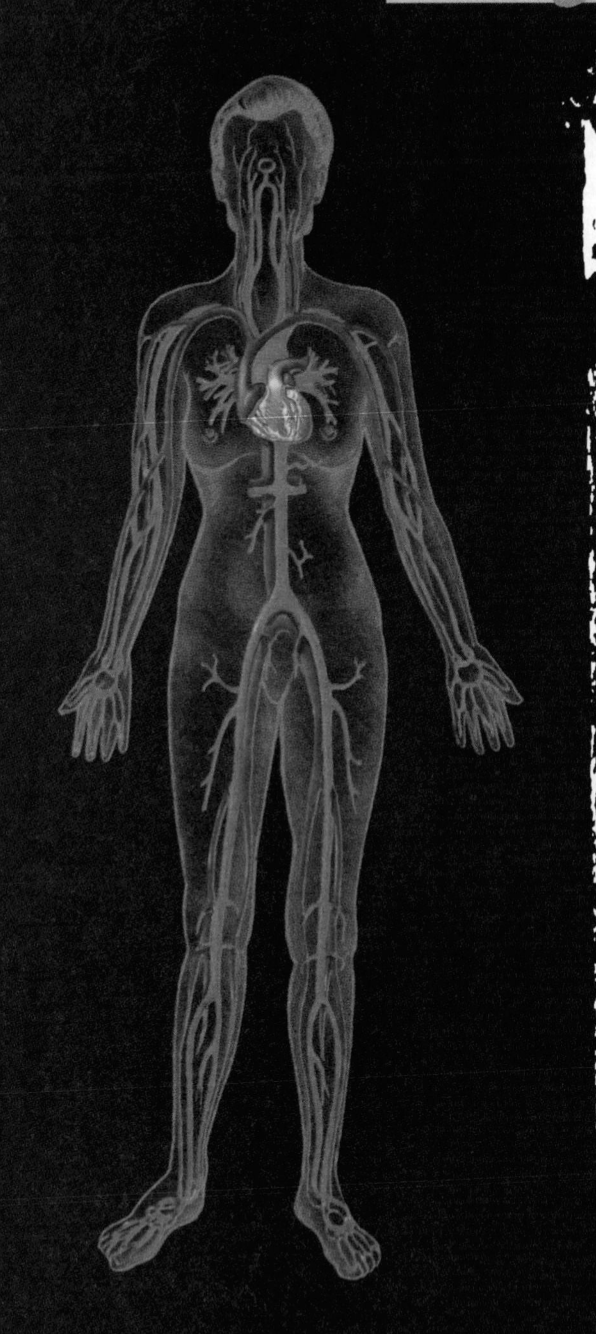

Cardiovascular System

The heart pumps blood through as much as 60,000 miles of blood vessels so that the blood reaches all body cells to deliver nutrients and remove wastes.

Integumentary System

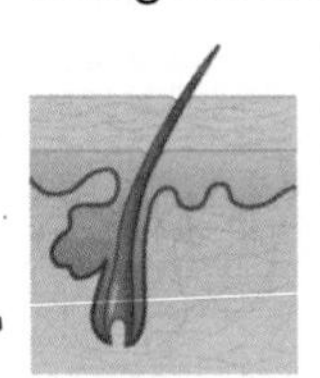

Changes in skin blood flow are important in temperature control.

Lymphatic System

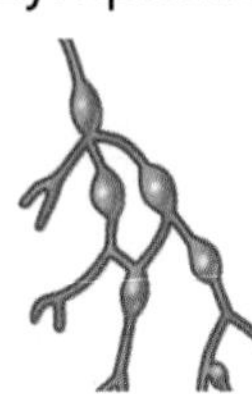

The lymphatic system returns tissue fluids to the bloodstream.

Skeletal System

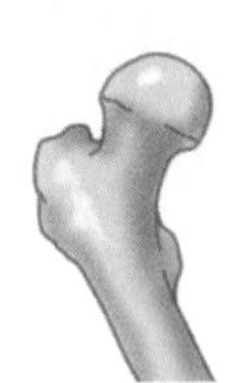

Bones help control plasma calcium levels.

Digestive System

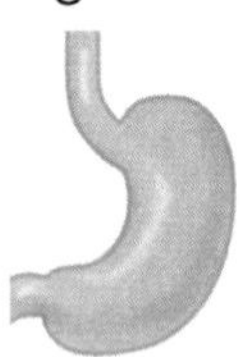

The digestive system breaks down nutrients into forms readily absorbed by the bloodstream.

Muscular System

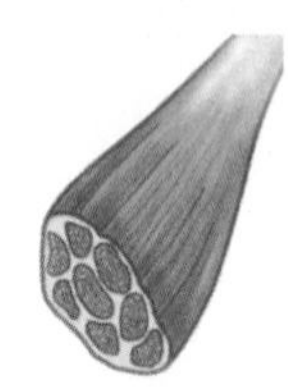

Blood flow increases to exercising skeletal muscle, delivering oxygen and nutrients and removing wastes. Muscle actions help the blood circulate.

Respiratory System

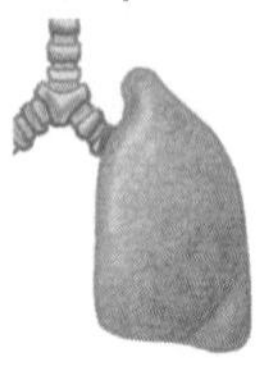

The respiratory system oxygenates the blood and removes carbon dioxide. Respiratory movements help the blood circulate.

Nervous System

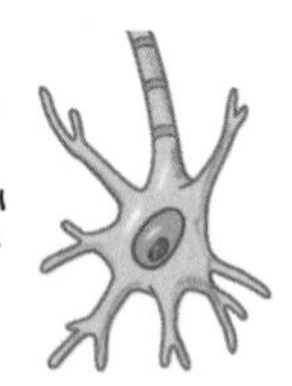

The brain depends on blood flow for survival. The nervous system helps control blood flow and blood pressure.

Urinary System

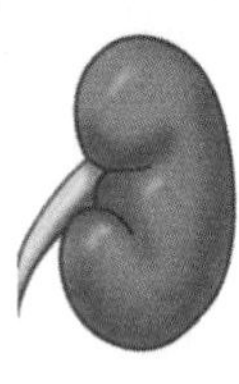

The kidneys clear the blood of wastes and excess substances. The kidneys help control blood pressure and blood volume.

Endocrine System

Hormones are carried in the bloodstream. Some hormones directly affect the heart and blood vessels.

Reproductive System

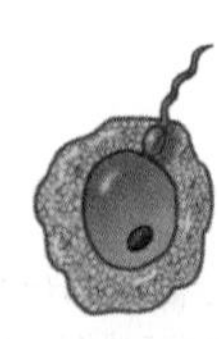

Blood pressure is important in normal function of the sex organs.

Clinical Terms Related to the Cardiovascular System

Anastomosis (ah-nas″to-mo′sis) A link between two blood vessels, sometimes produced surgically.

Angiospasm (an′je-o-spazm″) Muscular spasm in the wall of a blood vessel.

Arteriography (ar″te-re-og′rah-fe) Injection of radiopaque solution into the vascular system for X-ray examination of arteries.

Asystole (a-sis′to-le) Condition in which the myocardium fails to contract.

Cardiac tamponade (kar′de-ak tam″po-nād′) Accumulation of fluid within the pericardial cavity that compresses the heart.

Congestive heart failure (kon-jes′tiv hart fāl′yer) Inability of the heart to pump enough blood to cells.

Cor pulmonale (kor pul-mo-na′le) Heart-lung disorder of pulmonary hypertension and hypertrophy of the right ventricle.

Embolectomy (em″bo-lek′to-me) Removal of an embolus through an incision in a blood vessel.

Endarterectomy (en″dar-ter-ek′to-me) Removal of the inner wall of an artery to reduce an arterial occlusion.

Palpitation (pal″pĭ-ta′shun) Awareness of an unusually rapid, strong, or irregular heartbeat.

Pericardiectomy (per″ĭ-kar″de-ek′to-me) Excision of the pericardium.

Phlebitis (flĕ-bi′tis) Inflammation of a vein, usually in the lower limbs.

Phlebotomy (flĕ-bot′o-me) Incision of a vein to withdraw blood.

Sinus rhythm (si′nus rithm) The normal cardiac rhythm that the S-A node regulates.

Thrombophlebitis (throm″bo-flĕ-bi′tis) Formation of a blood clot in a vein in response to inflammation of the venous wall.

Valvotomy (val-vot′o-me) Incision of a valve.

Venography (ve-nog′rah-fe) Injection of radiopaque solution into the vascular system for X-ray examination of veins.

Summary Outline

Introduction (p. 344)

The cardiovascular system provides oxygen and nutrients to tissues and removes wastes.

Structure of the Heart (p. 344)

1. Size and location of the heart
   a. The heart is about 14 centimeters long and 9 centimeters wide.
   b. It is within the mediastinum and rests on the diaphragm.
2. Coverings of the heart
   a. A layered pericardium encloses the heart.
   b. The pericardial cavity is a space between the parietal and visceral layers of the pericardium.
3. Wall of the heart
   The wall of the heart has three layers—an epicardium, a myocardium, and an endocardium.
4. Heart chambers and valves
   a. The heart is divided into two atria and two ventricles.
   b. Right chambers and valves
      (1) The right atrium receives blood from the venae cavae and coronary sinus.
      (2) The tricuspid valve separates the right chambers.
      (3) A pulmonary valve guards the base of the pulmonary trunk.
   c. Left chambers and valves
      (1) The left atrium receives blood from the pulmonary veins.
      (2) The bicuspid valve separates the left chambers.
      (3) An aortic valve guards the base of the aorta.
5. Skeleton of the heart
   The skeleton of the heart consists of fibrous rings that enclose the bases of the pulmonary artery and aorta.
6. Path of blood through the heart
   a. Blood low in oxygen concentration and high in carbon dioxide concentration enters the right side of the heart and is pumped into the pulmonary circulation.
   b. After blood is oxygenated in the lungs and some carbon dioxide removed, it returns to the left side of the heart.
7. Blood supply to the heart
   a. The coronary arteries supply blood to the myocardium.
   b. Blood returns to the right atrium through the cardiac veins and coronary sinus.

Heart Actions (p. 350)

1. Cardiac cycle
   a. The atria contract while the ventricles relax. The ventricles contract while the atria relax.
   b. Pressure in the chambers rises and falls in repeated cycles.
2. Heart sounds
   Heart sounds are due to the vibrations that the valve movements produce.
3. Cardiac muscle fibers
   a. Cardiac muscle fibers connect to form a functional syncytium.
   b. If any part of the syncytium is stimulated, the whole structure contracts as a unit.
4. Cardiac conduction system
   a. This system initiates and conducts impulses throughout the myocardium.
   b. Impulses from the S-A node pass slowly to the A-V node. Impulses travel rapidly along the A-V bundle and Purkinje fibers.

5. Electrocardiogram (ECG)
   a. An ECG records electrical changes in the myocardium during a cardiac cycle.
   b. The pattern contains several waves
      (1) The P wave represents atrial depolarization.
      (2) The QRS complex represents ventricular depolarization.
      (3) The T wave represents ventricular repolarization.
6. Regulation of the cardiac cycle
   a. Physical exercise, body temperature, and concentrations of various ions affect heartbeat.
   b. Branches of sympathetic and parasympathetic nerve fibers innervate the S-A and A-V nodes.
   c. The cardiac center in the medulla oblongata regulates autonomic impulses to the heart.

Blood Vessels (p. 357)

Blood vessels form a closed circuit of tubes that carry blood from the heart to cells and back again.

1. Arteries and arterioles
   a. Arteries are adapted to carry blood under high pressure away from the heart.
   b. The walls of arteries and arterioles consist of layers of endothelium, smooth muscle, and connective tissue.
   c. Autonomic fibers that can stimulate vasoconstriction or vasodilation innervate smooth muscle in vessel walls.
2. Capillaries
   a. Capillaries connect arterioles and venules.
   b. The capillary wall is a single layer of cells that forms a semipermeable membrane.
   c. Openings in capillary walls, where endothelial cells overlap, vary in size from tissue to tissue.
   d. Precapillary sphincters regulate capillary blood flow.
3. Exchanges in capillaries
   a. Capillary blood and tissue fluid exchange gases, nutrients, and metabolic by-products.
   b. Diffusion is the most important means of transport.
   c. Filtration causes a net outward movement of fluid at the arteriolar end of a capillary.
   d. Osmosis causes a net inward movement of fluid at the venular end of a capillary.
4. Venules and veins
   a. Venules continue from capillaries and merge to form veins.
   b. Veins carry blood to the heart.
   c. Venous walls are similar to arterial walls, but are thinner and contain less smooth muscle and elastic tissue.

Blood Pressure (p. 363)

Blood pressure is the force blood exerts against the insides of blood vessels.

1. Arterial blood pressure
   a. Arterial blood pressure rises and falls with phases of the cardiac cycle.
   b. Systolic pressure is produced when the ventricle contracts. Diastolic pressure is the pressure in the arteries when the ventricle relaxes.
2. Factors that influence arterial blood pressure
   Arterial blood pressure increases as cardiac output, blood volume, peripheral resistance, or blood viscosity increases.
3. Control of blood pressure
   a. Mechanisms that regulate cardiac output and peripheral resistance partially control blood pressure.
   b. The more blood that enters the heart, the stronger the ventricular contraction, the greater the stroke volume, and the greater the cardiac output.
   c. The cardiac center of the medulla oblongata regulates heart rate.
4. Venous blood flow
   a. Venous blood flow depends on skeletal muscle contraction, breathing movements, and venoconstriction.
   b. Many veins contain flaplike valves that prevent blood from backing up.
   c. Venoconstriction can increase venous pressure and blood flow.

Paths of Circulation (p. 367)

1. Pulmonary circuit
   The pulmonary circuit consists of vessels that carry blood from the right ventricle to the lungs and back to the left atrium.
2. Systemic circuit
   a. The systemic circuit consists of vessels that lead from the heart to the body cells and back to the heart.
   b. It includes the aorta and its branches.

Arterial System (p. 367)

1. Principal branches of the aorta
   a. The aorta is the largest artery with respect to diameter.
   b. Its major branches include the coronary, brachiocephalic, left common carotid, and left subclavian arteries.
   c. The branches of the descending aorta include the thoracic and abdominal groups.
   d. The abdominal aorta diverges into the right and left common iliac arteries.
2. Arteries to the neck, head, and brain
   These include branches of the subclavian and common carotid arteries.
3. Arteries to the shoulder and upper limb
   a. The subclavian artery passes into the upper limb, and in various regions is called the axillary and brachial artery.
   b. Branches of the brachial artery include the ulnar and radial arteries.
4. Arteries to the thoracic and abdominal walls
   a. Branches of the subclavian artery and thoracic aorta supply the thoracic wall.
   b. Branches of the abdominal aorta and other arteries supply the abdominal wall.
5. Arteries to the pelvis and lower limb
   The common iliac arteries supply the pelvic organs, gluteal region, and lower limbs.

Venous System (p. 372)

1. Characteristics of venous pathways
   a. Veins return blood to the heart.
   b. Larger veins usually parallel the paths of major arteries.
2. Veins from the brain, head, and neck
   a. Jugular veins drain these regions.
   b. Jugular veins unite with subclavian veins to form the brachiocephalic veins.

3. Veins from the upper limb and shoulder
   a. Sets of superficial and deep veins drain these regions.
   b. Deep veins parallel arteries with similar names.
4. Veins from the abdominal and thoracic walls
   Tributaries of the brachiocephalic and azygos veins drain these walls.
5. Veins from the abdominal viscera
   a. Blood from abdominal viscera enters the hepatic portal system and is carried to the liver.
   b. From the liver, hepatic veins carry blood to the inferior vena cava.
6. Veins from the lower limb and pelvis
   a. Sets of deep and superficial veins drain these regions.
   b. The deep veins include the tibial veins, and the superficial veins include the saphenous veins.

Review Exercises

1. Describe the general structure, function, and location of the heart. (p. 344)
2. Describe the pericardium. (p. 344)
3. Compare the layers of the cardiac wall. (p. 345)
4. Identify and describe the locations of the chambers and the valves of the heart. (p. 345)
5. Describe the skeleton of the heart, and explain its function. (p. 347)
6. Trace the path of blood through the heart. (p. 347)
7. Trace the path of blood through the coronary circulation. (p. 349)
8. Describe a cardiac cycle. (p. 350)
9. Describe the pressure changes in the atria and ventricles during a cardiac cycle. (p. 350)
10. Explain the origin of heart sounds. (p. 352)
11. Distinguish between the roles of the S-A node and the A-V node. (p. 353)
12. Explain how the cardiac conduction system controls the cardiac cycle. (p. 353)
13. Describe and explain the normal ECG pattern. (p. 355)
14. Discuss how the nervous system regulates the cardiac cycle. (p. 356)
15. Distinguish between an artery and an arteriole. (p. 357)
16. Explain control of vasoconstriction and vasodilation. (p. 358)
17. Describe the structure and function of a capillary. (p. 359)
18. Explain control of blood flow through a capillary. (p. 359)
19. Explain how diffusion exchanges substances between blood plasma and tissue fluid. (p. 361)
20. Explain why water and dissolved substances leave the arteriolar end of a capillary and enter the venular end. (p. 361)
21. Distinguish between a venule and a vein. (p. 362)
22. Explain how veins function as blood reservoirs. (p. 362)
23. Distinguish between systolic and diastolic blood pressures. (p. 363)
24. Name several factors that influence blood pressure, and explain how each produces its effect. (p. 364)
25. Describe the control of blood pressure. (p. 364)
26. List the major factors that promote venous blood flow. (p. 366)
27. Distinguish between the pulmonary and systemic circuits of the cardiovascular system. (p. 367)
28. Trace the path of blood through the pulmonary circuit. (p. 367)
29. Describe the aorta, and name its principal branches. (p. 367)
30. Describe the relationship between the major venous pathways and the major arterial pathways. (p. 372)

Critical Thinking

1. How might the results of a cardiovascular exam differ for an athlete in top condition and a sedentary, overweight individual?
2. If you were asked to invent a blood vessel substitute, what materials might you use to build it? Include synthetic as well as natural chemicals.
3. What structures and properties should an artificial heart have?
4. Athletes tend to be slim and strong, have low blood pressure, do not smoke, and alleviate stress through exercise. How might these characteristics complicate a study to assess the effects of exercise on the cardiovascular system?
5. A man feels "flutters" in his chest, so he has a physical exam. The doctor says, "Your diastolic pressure is elevated. You seem to have angina of exertion, but the stress test does not show an arrhythmia or a blocked coronary artery. Your leukocyte count and clotting time are normal, but your red count is a bit low." What is—and is not—wrong with the man's cardiovascular system?
6. Cigarette smoke contains thousands of chemicals, including nicotine and carbon monoxide. Nicotine constricts blood vessels. Carbon monoxide prevents oxygen from binding to hemoglobin. How do these two components of smoke affect the cardiovascular system?
7. Given the way capillary blood flow is regulated, do you think it is wiser to rest or to exercise following a heavy meal? Explain.
8. If a patient develops a blood clot in the femoral vein of the left lower limb and a portion of the clot breaks loose, where is the embolus likely to go? What symptoms are likely?
9. Cirrhosis of the liver, a condition sometimes associated with alcoholism, obstructs blood flow through hepatic blood vessels. As a result, blood backs up, and capillary pressure greatly increases in organs the hepatic portal system drains. What effects might this increasing capillary pressure produce, and which organs would it affect?
10. If a cardiologist inserts a catheter into a patient's right femoral artery, which arteries will the tube pass through to reach the entrance to the left coronary artery?

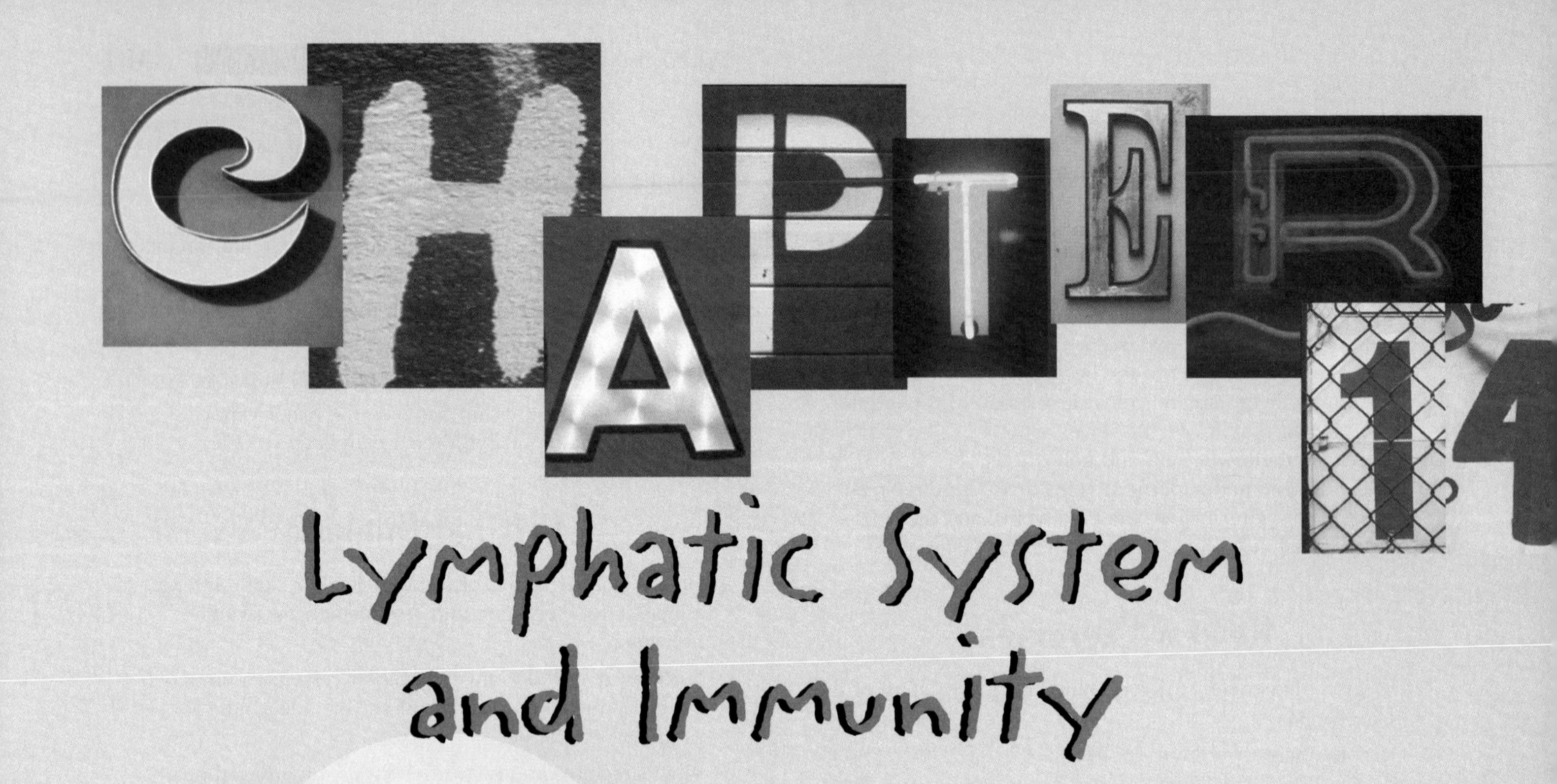

Chapter 14 Lymphatic System and Immunity

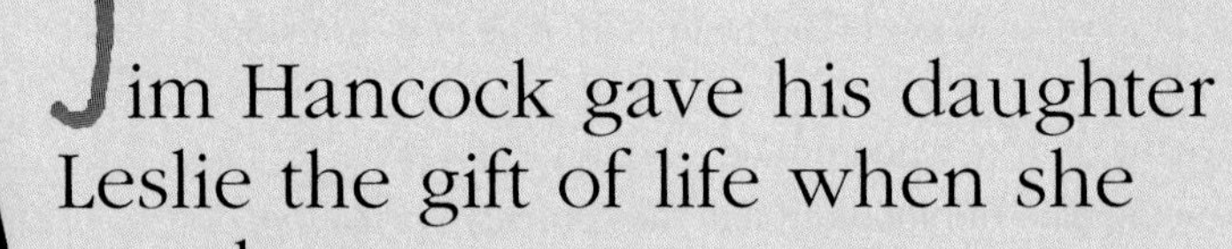

Jim Hancock gave his daughter Leslie the gift of life when she was born. Six years later, he did it again. When Leslie's liver began to fail, Jim donated part of his own, saving her life.

Leslie was born with cystic fibrosis, a hereditary disorder that usually impairs the lungs and pancreas. When in the summer of 1995 doctors discovered that Leslie's digestive difficulties were not directly due to cystic fibrosis but to a hardening liver, they suggested that a liver transplant might be necessary in a few years. The drugs they prescribed to help Leslie's liver, however, were not effective, and Leslie's condition rapidly worsened. She needed a new liver soon. The statistics were terrifying: A shortage of donor organs results in a third of the children awaiting donor livers dying before a transplant can be performed.

The Hancocks, however, were offered an alternative: Part of Jim's liver could be removed and transplanted in his daughter. "When the subject of a living-relative transplant came up, I never had to think about it. It was just something I wanted to do," said Jim. Transplants involving blood relatives have about a 90% success rate, compared to 80% for transplants involving nonrelatives. Leslie's immune system was not likely to reject her father's liver because the cell surfaces of father and daughter are very similar. Leslie also took drugs to suppress her immune system, decreasing the chances of rejection even more.

The surgery was a success. Two months later, Leslie was back in her kindergarten classroom in Dubuque, Iowa.

Photo: Jim Hancock, of Dubuque, Iowa, gave six-year-old daughter Leslie part of his liver and a new chance at life. The transplant worked because the cell surfaces of father and daughter are very similar. The little girl's immune system accepted her new liver as part of her body.

The lymphatic system transports fluid leaving the capillaries to the veins, which return the fluid to the circulation, thereby preventing it from accumulating in tissue spaces. The lymphatic system also enables us to live in a world filled with different types of organisms, some of which can take up residence in the human body and cause infections. The lymphatic system—a vast collection of cells and biochemicals that travel in lymphatic vessels—recognizes and destroys "foreign" particles, including infectious microorganisms and viruses, as well as toxins and cancer cells.

Chapter Objectives

After studying this chapter, you should be able to:

1. Describe the general functions of the lymphatic system. (p. 384)
2. Identify the locations of the major lymphatic pathways. (p. 384)
3. Describe how tissue fluid and lymph form, and explain the function of lymph. (p. 386)
4. Explain the maintenance of lymphatic circulation. (p. 386)
5. Describe a lymph node and its major functions. (p. 387)
6. Discuss the functions of the thymus and spleen. (p. 388)
7. Distinguish between specific and nonspecific immunity and provide examples of each. (p. 390)
8. Explain how two major types of lymphocytes form and how they provide immunity. (p. 391)
9. Name the major types of immunoglobulins, and discuss their origins and actions. (p. 393)
10. Distinguish between primary and secondary immune responses. (p. 397)
11. Distinguish between active and passive immunity. (p. 398)
12. Explain how allergic reactions, tissue rejection reactions, and autoimmunity involve immune mechanisms. (p. 400)

Aids to Understanding Words

gen- [to be produced] aller*gen*: A substance that stimulates an allergic response.

humor- [fluid] *humor*al immunity: Immunity resulting from antibodies in body fluids.

immun- [free] *immun*ity: Resistance to (freedom from) a specific disease.

inflamm- [setting on fire] *inflamm*ation: Localized redness, heat, swelling, and pain in tissues.

nod- [knot] *nod*ule: A small mass of lymphocytes surrounded by connective tissue.

patho- [disease] *patho*gen: A disease-causing agent.

Key Terms

allergen (al′-er-jen)
antibody (an′tĭ-bod″e)
antigen (an′tĭ-jen)
clone (klōn)
complement (kom′plĕ-ment)
hapten (hap′ten)
immunity (ĭmu′nĭ-te)
immunoglobulin (im″u-no-glob′u-lin)
lymph (limf)
lymphatic pathway (lim-fat′ik path′wa)
lymph node (limf nōd)
lymphocyte (lim′fo-sīt)
macrophage (mak′ro-fāj)
pathogen (path′o-jen)
reticuloendothelial tissue (rĕ-tik″u-lo-en″do-the′le-al tish′u)
spleen (splēn)
thymus (thī′mus)

Introduction

Like the cardiovascular system, the lymphatic system includes a network of vessels that circulates fluids. Lymphatic vessels transport excess fluid from interstitial spaces in most tissues and return it to the bloodstream (fig. 14.1). Special lymphatic capillaries (lacteals), located in the lining of the small intestine, absorb digested fats then transport the fats to venous circulation. The lymphatic system also helps defend the body against disease-causing agents.

Lymphatic Pathways

The **lymphatic pathways** begin as lymphatic capillaries. These tiny tubes merge to form larger lymphatic vessels, which, in turn, lead to larger vessels that unite with veins in the thorax.

Lymphatic Capillaries

Lymphatic capillaries are microscopic, closed-ended tubes (fig. 14.2). They extend into interstitial spaces, forming complex networks that parallel those of blood capillaries. The walls of lymphatic capillaries, like those of blood capillaries, are a single layer of squamous epithelial cells. This thin wall enables tissue fluid to enter lymphatic capillaries. Fluid inside lymphatic capillaries is called **lymph.**

Lymphatic Vessels

The walls of **lymphatic vessels** are similar to those of veins, but thinner. Also like veins, lymphatic vessels have flaplike *valves* that help prevent lymph backflow (fig. 14.3).

The larger lymphatic vessels lead to specialized organs called **lymph nodes.** After leaving the nodes, the vessels merge to form still larger **lymphatic trunks.**

Lymphatic Trunks and Collecting Ducts

Lymphatic trunks, which drain lymph, are named for the regions they serve. They join one of two **collecting ducts**—the thoracic duct or the right lymphatic duct (fig. 14.4*a*).

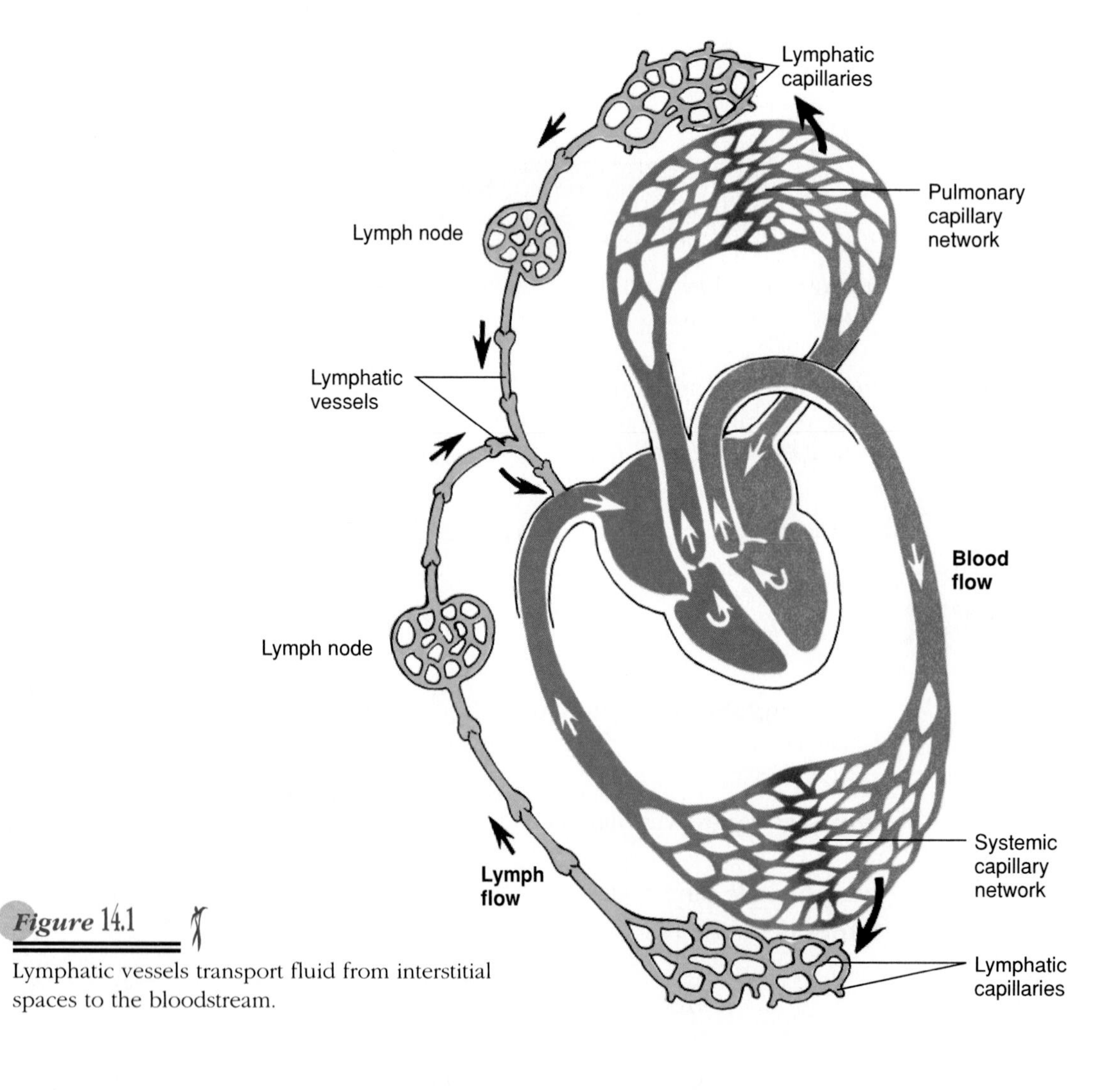

Figure 14.1

Lymphatic vessels transport fluid from interstitial spaces to the bloodstream.

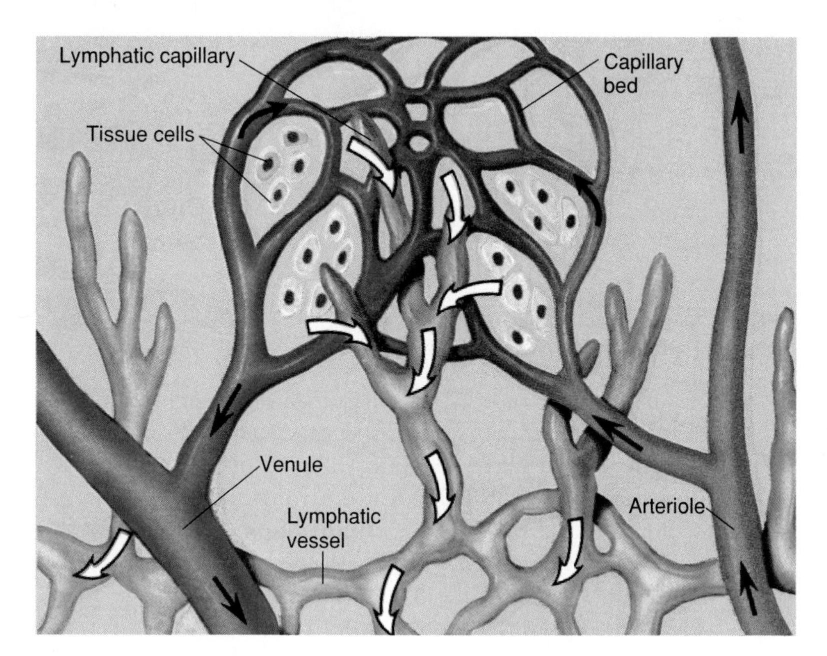

Figure 14.2

Lymphatic capillaries are microscopic, closed-ended tubes that begin in the interstitial spaces of most tissues.

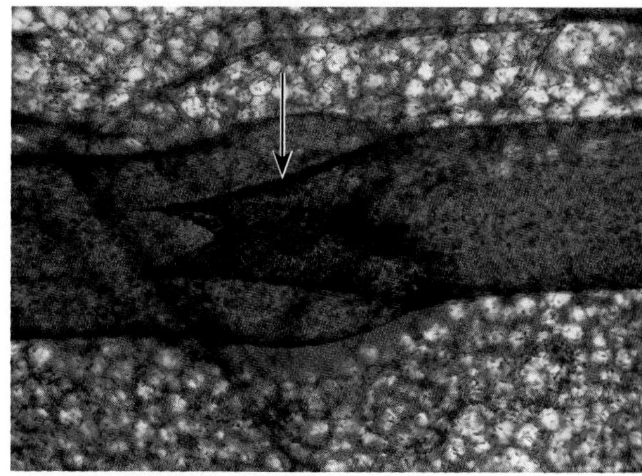

Figure 14.3

Light micrograph of the flaplike valve (arrow) within a lymphatic vessel (25×).

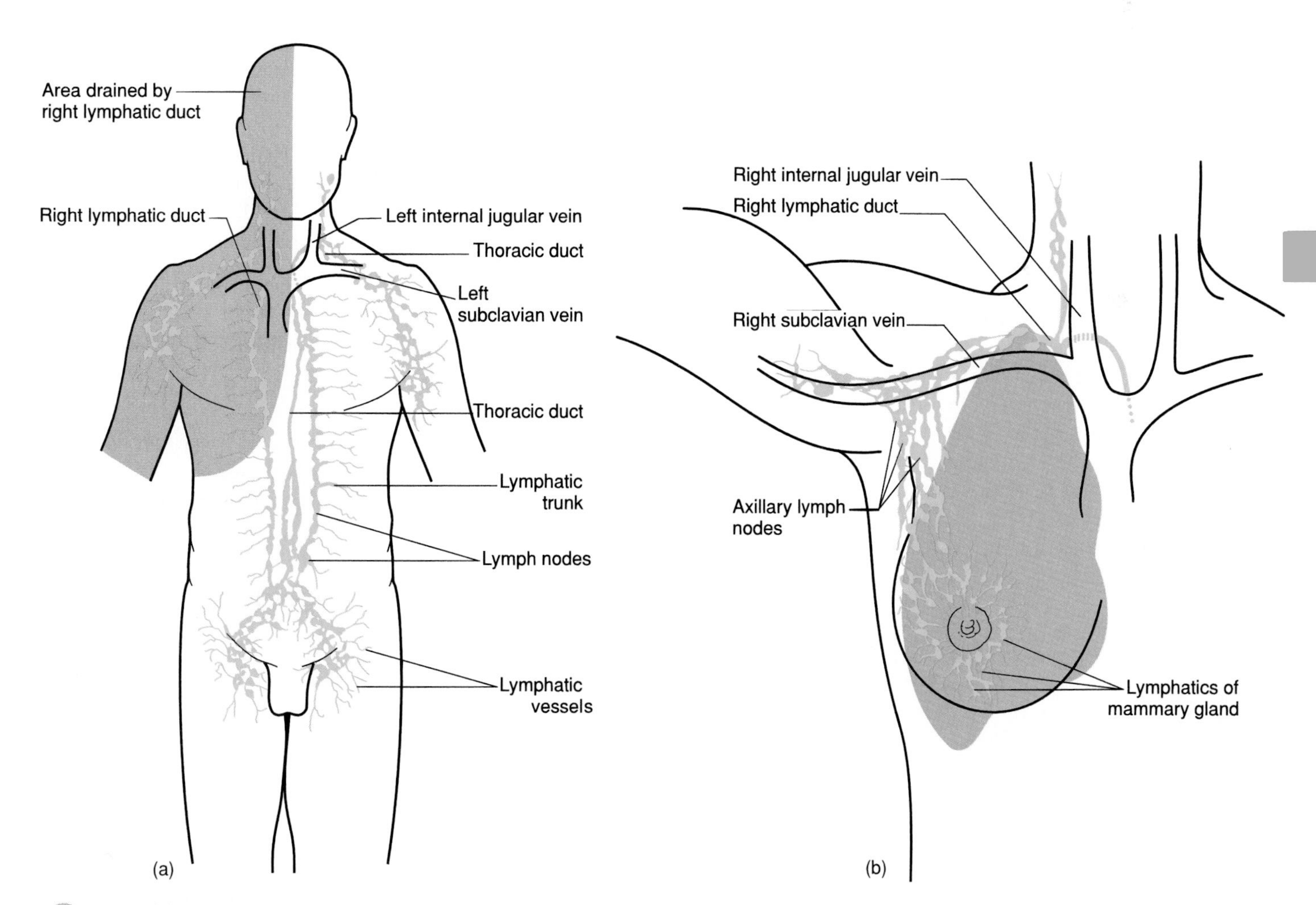

Figure 14.4

(*a*) The right lymphatic duct drains lymph from the upper right side of the body, and the thoracic duct drains lymph from the rest of the body. (*b*) Lymph drainage of the right breast.

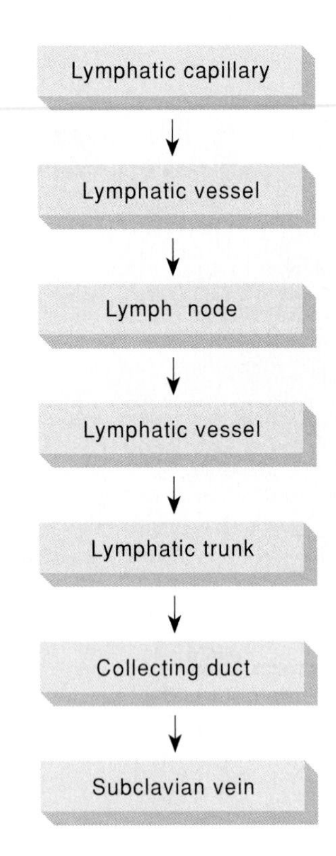

Figure 14.5

Typical lymphatic pathway.

The **thoracic duct** is the larger and longer collecting duct. It receives lymph from the lower body regions, left upper limb, and left side of the head and neck, and empties into the left subclavian vein near the junction of the left jugular vein. The **right lymphatic duct** receives lymph from the right side of the head and neck, right upper limb, and right thorax, and empties into the right subclavian vein near the junction of the right jugular vein.

After leaving the collecting ducts, lymph enters the venous system and becomes part of the plasma just before blood returns to the right atrium. Figure 14.5 summarizes a typical lymphatic pathway.

The skin is rich with lymphatic capillaries. Consequently, if the skin breaks or if something is injected into it (such as venom from a stinging insect), foreign substances enter the lymphatic system rapidly. ■

1. What are the general functions of the lymphatic system?
2. Distinguish between the thoracic duct and the right lymphatic duct.

Tissue Fluid and Lymph

Lymph is essentially tissue fluid that has entered a lymphatic capillary. Thus, lymph formation is closely associated with tissue fluid formation.

Tissue Fluid Formation

Recall from chapter 12 that tissue fluid originates from blood plasma and is composed of water and dissolved substances that leave blood capillaries by diffusion and filtration. Tissue fluid contains nutrients and gases found in plasma, but lacks large proteins. Smaller proteins, however, can leak out of blood capillaries and enter interstitial spaces. These proteins usually are not reabsorbed when water and dissolved substances move back into capillaries' venular ends by diffusion and osmosis. This raises the protein concentration of tissue fluid, which elevates the *osmotic pressure.*

Lymph Formation

Tissue fluid's increasing osmotic pressure impedes blood capillaries' osmotic reabsorption of water. This increases the volume and pressure of fluid in interstitial spaces, forcing some tissue fluid into lymphatic capillaries, where it becomes lymph (see fig. 14.2). Lymph returns to the bloodstream most of the small proteins that leak out of blood capillaries. At the same time, lymph transports foreign particles, such as bacteria and viruses, to lymph nodes.

1. What is the relationship between tissue fluid and lymph?
2. How do proteins in tissue fluid affect lymph formation?
3. What are the major functions of lymph?

Lymph Movement

Tissue fluid's hydrostatic pressure influences the entry of lymph into lymphatic capillaries. However, muscular activity largely controls lymph movement through lymphatic vessels.

Lymph, like venous blood, is under low hydrostatic pressure and may not flow readily through lymphatic vessels without outside help. This help includes contraction of skeletal muscles and of smooth muscles in the walls of the larger lymphatic trunks, and pressure changes due to the action of breathing muscles.

Contracting skeletal muscles compress lymphatic vessels and move the lymph inside lymphatic vessels. These vessels have valves that prevent backflow, so lymph can only move toward a collecting duct. Additionally, smooth muscles in the walls of larger lymphatic trunks can contract and compress the lymph inside, forcing the fluid onward.

Breathing muscles aid lymph circulation by lowering pressure in the thoracic cavity during inhalation. At the same time, the contracting diaphragm increases the pressure in the abdominal cavity. Together these actions squeeze lymph out of abdominal vessels and force it into thoracic vessels. Once again, valves within lymphatic vessels prevent lymph backflow.

The continuous movement of fluid from interstitial spaces into blood and lymphatic capillaries stabilizes the fluid volume in these spaces. Conditions that interfere with lymph movement cause tissue fluids to accumulate in interstitial spaces, producing *edema*. This may happen when surgery removes lymphatic tissue, obstructing certain lymphatic vessels. For example, surgery to remove a cancerous breast also usually removes nearby axillary lymph nodes because associated lymphatic vessels can transport cancer cells to other sites. Removing the lymphatic tissue can obstruct drainage from the upper limb, causing edema.

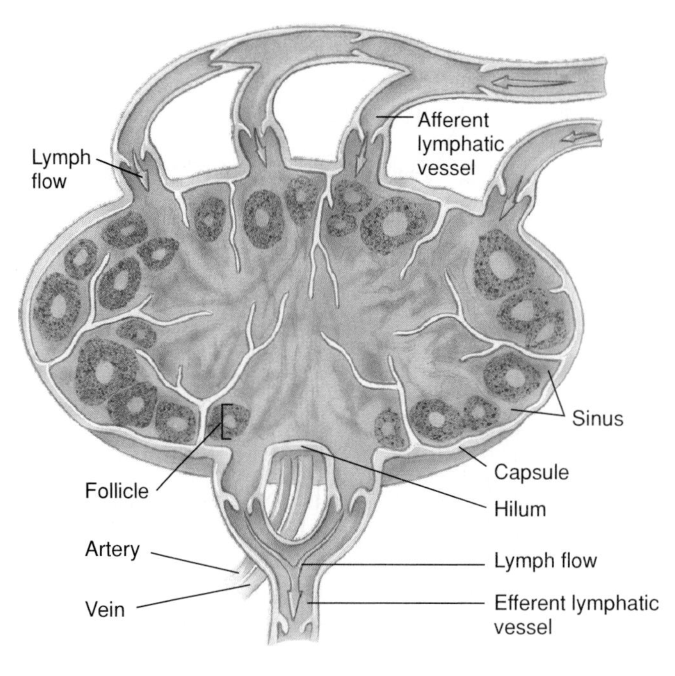

***Figure* 14.6**

Section of a lymph node.

1. What factors promote lymph flow?
2. What is the consequence of lymphatic obstruction?

Lymph Nodes

Lymph nodes (lymph glands) are located along the lymphatic pathways. They contain large numbers of **lymphocytes** and **macrophages** that fight invading microorganisms.

Structure of a Lymph Node

Lymph nodes vary in size and shape, but are usually less than 2.5 centimeters long and are bean shaped (figs. 14.6 and 14.7). Blood vessels and nerves join a lymph node through the indented region of the node, called the **hilum.** Lymphatic vessels leading to a node (afferent vessels) enter separately at various points on its convex surface, but lymphatic vessels leaving the node (efferent vessels) exit from the hilum.

A *capsule* of fibrous connective tissue encloses each lymph node and subdivides it into compartments that contain dense masses of lymphocytes and macrophages. These masses, or **nodules,** are the structural units of the lymph node. The spaces within a node, called **lymph sinuses,** form a complex network of chambers and channels through which lymph circulates.

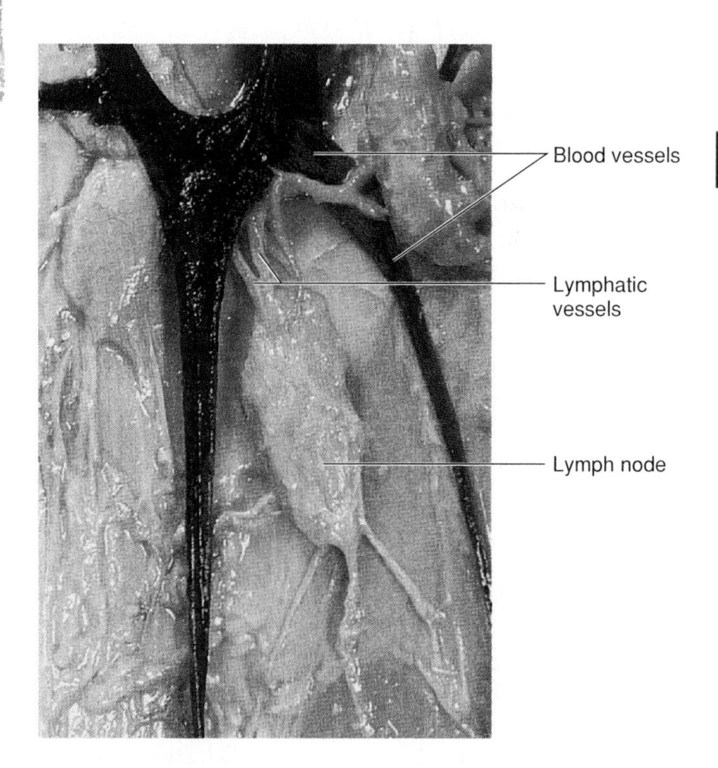

***Figure* 14.7**

Lymph enters and leaves a lymph node through lymphatic vessels.

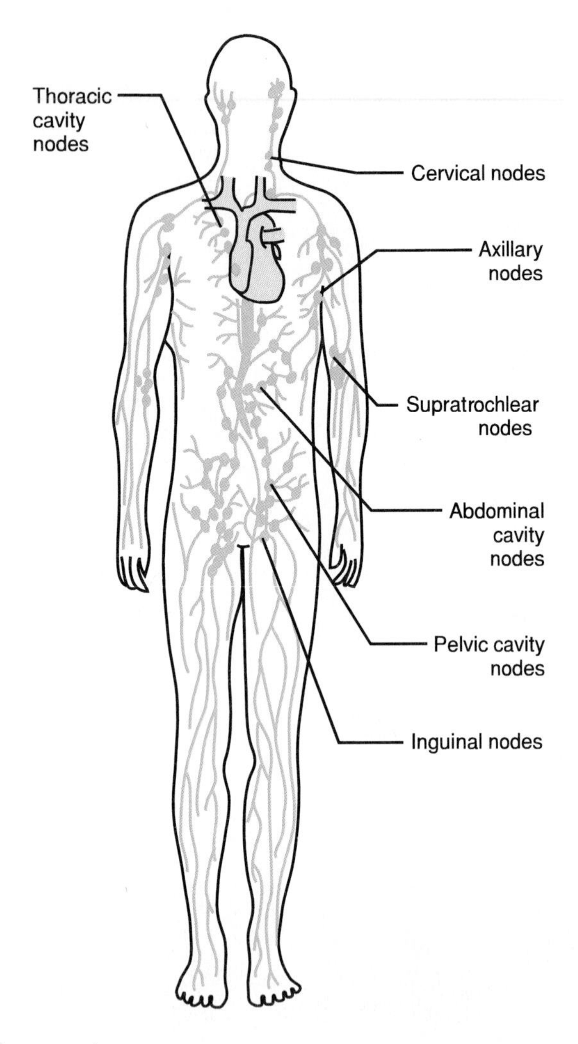

Figure 14.8

Major locations of lymph nodes.

Nodules occur singly or in groups associated with the mucous membranes of the respiratory and digestive tracts. The *tonsils,* described in chapter 15, are partially encapsulated lymph nodules. Also, aggregations of nodules called *Peyer's patches* are scattered throughout the mucosal lining of the ileum of the small intestine.

Locations of Lymph Nodes

Lymph nodes generally occur in groups or chains along the paths of the larger lymphatic vessels throughout the body, but are absent in the central nervous system. Figure 14.8 shows the locations of the major lymph nodes.

Functions of Lymph Nodes

Lymph nodes have two primary functions: (1) filtering potentially harmful particles from lymph before returning it to the bloodstream, and (2) immune surveillance, which lymphocytes and macrophages provide. Along with red bone marrow, lymph nodes are centers for lymphocyte production. Lymphocytes attack infecting viruses, bacteria, and other microorganisms brought to the nodes in lymphatic vessels. Macrophages in the nodes engulf and destroy foreign substances, damaged cells, and cellular debris.

uperficial lymphatic vessels inflamed by bacterial infection appear as red streaks beneath the skin, a condition called *lymphangitis.* Inflammation of the lymph nodes, called *lymphadenitis,* often follows. Affected nodes enlarge and may be quite painful. ■

1. Distinguish between a lymph node and a lymph nodule.
2. What are the major functions of lymph nodes?

Thymus and Spleen

The thymus and spleen are lymphatic organs whose functions closely relate to those of the lymph nodes.

Thymus

The **thymus** is a soft, bilobed gland enclosed in a connective tissue capsule and located in front of the aorta and behind the upper part of the sternum (fig. 14.9*a*). The thymus is relatively large during infancy and early childhood, but shrinks after puberty and may be quite small in an adult. In elderly persons, adipose and connective tissues replace lymphatic tissue.

Connective tissues extend inward from the thymus surface, subdividing the thymus into *lobules* (fig. 14.9*b*). The lobules contain abundant lymphocytes. Most of these cells (thymocytes) stay inactive, but some mature into **T cells** (T lymphocytes), which leave the thymus and provide immunity. Epithelial cells in the thymus secrete a hormone called *thymosin,* which stimulates maturation of T cells after they leave the thymus and migrate to other lymphatic tissues.

Spleen

The **spleen**, the largest lymphatic organ, is in the upper left portion of the abdominal cavity, just beneath the diaphragm and behind the stomach (fig. 14.9*a*). The

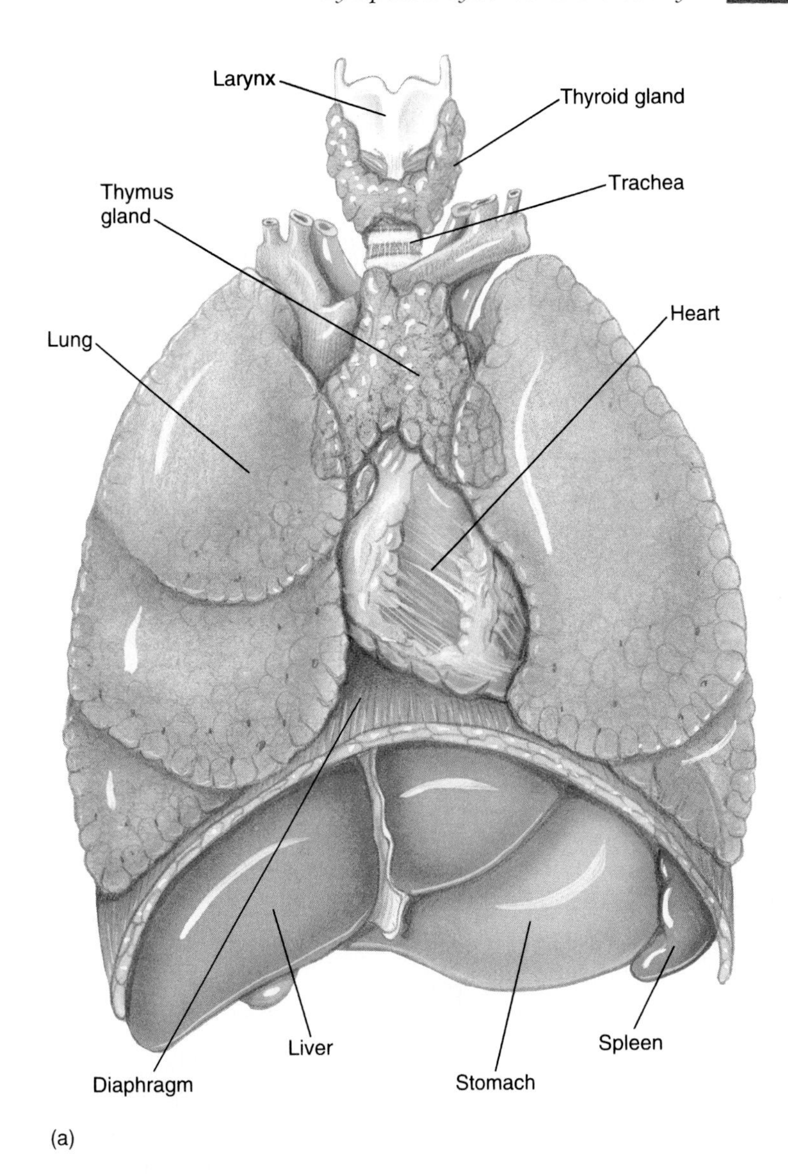

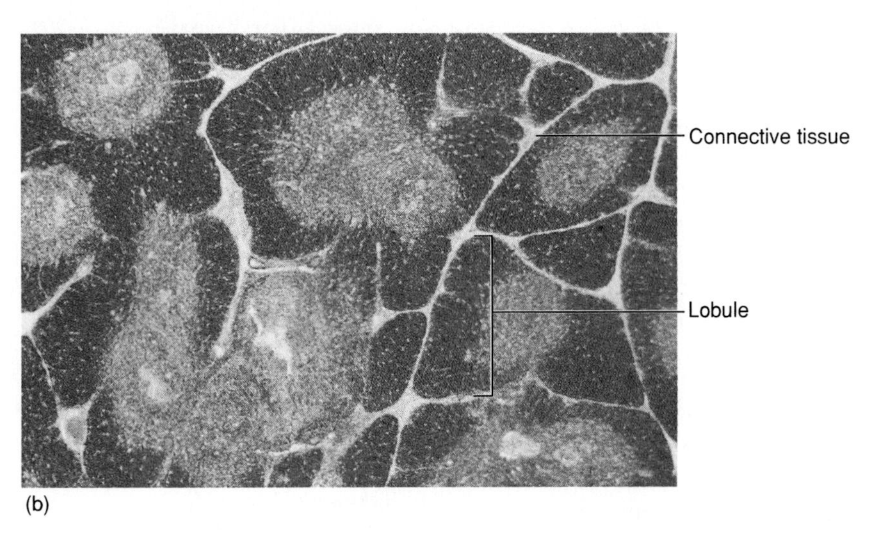

Figure 14.9

(*a*) The thymus is bilobed and between the lungs and above the heart. The spleen is beneath the diaphragm and behind the stomach. (*b*) Cross section of the thymus (10×). Note how the thymus is subdivided into lobules.

spleen resembles a large lymph node and is subdivided into lobules. Unlike the sinuses of a lymph node, however, the spaces (venous sinuses) of the spleen contain blood (fig. 14.10).

The tissues within splenic nodules are of two types. *White pulp* is distributed throughout the spleen in tiny islands. This tissue is composed of splenic nodules, which are similar to those in lymph nodes and contain many lymphocytes. The *red pulp,* which fills the remaining spaces of the lobules, surrounds the venous sinuses. This pulp contains many red blood cells, which impart its color, plus many lymphocytes and macrophages.

Blood capillaries within the red pulp are quite permeable. Red blood cells can squeeze through the pores in these capillary walls and enter the venous sinuses. The older, more fragile red blood cells may rupture as they move, and phagocytic macrophages within the splenic sinuses remove the resulting cellular debris.

Macrophages also engulf and destroy foreign particles, such as bacteria, that may be carried in blood as it flows through the sinuses. Thus, the spleen filters blood much as the lymph nodes filter lymph.

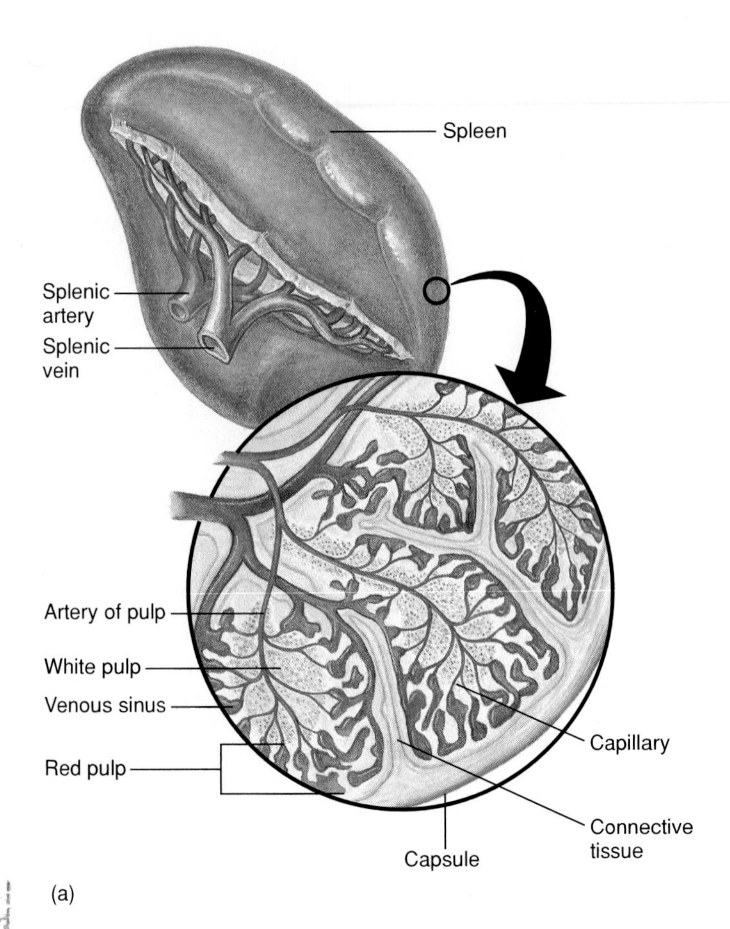

Figure 14.10

(*a*) The spleen resembles a large lymph node. (*b*) Light micrograph of the spleen (40×).

1. Why are the thymus and spleen considered organs of the lymphatic system?
2. What are the major functions of the thymus and the spleen?

Body Defenses against Infection

The presence and multiplication of a disease-causing agent, or **pathogen**, causes an infection. Pathogens include viruses and microorganisms such as bacteria, fungi, and protozoans.

The human body can prevent entry of pathogens or destroy them if they gain entrance. Some mechanisms are quite general and protect against many types of pathogens, and provide **nonspecific defense.** These mechanisms include species resistance, mechanical barriers, enzyme action, interferon, inflammation, and phagocytosis. Other defense mechanisms are very precise, targeting specific pathogens, and provide **specific defense**, or **immunity.** Specialized lymphocytes that recognize foreign molecules (nonself antigens) in the body and respond to them execute specific defense mechanisms.

Nonspecific Defenses

Species Resistance

Species resistance refers to a given kind of organism, or *species* (such as the human species, *Homo sapiens*), developing diseases that are unique to it. At the same time, a species may be resistant to diseases that affect other species because its tissues somehow fail to provide the temperature or chemical environment that a particular pathogen requires. For example, the infectious agents that cause measles, mumps, gonorrhea, and syphilis infect humans, but not other animal species.

Mechanical Barriers

The skin and mucous membranes lining the passageways of the respiratory, digestive, urinary, and reproductive systems create **mechanical barriers** that prevent the entrance of some infectious agents. As long as these barriers remain intact, many pathogens are unable to penetrate them.

Chemical Barriers

Enzymes in body fluids provide a **chemical barrier** to pathogens. Gastric juice, for example, contains the protein-splitting enzyme pepsin and has a low pH due to the presence of hydrochloric acid (HCl) (see chapter 15). The combined effect of pepsin and HCl is lethal to many pathogens that enter the stomach. Similarly, tears contain the enzyme lysozyme, which has an antibacterial action against certain pathogens that may get onto eye surfaces. Salt accumulation from perspiration kills some bacteria on the skin.

Certain cells, including lymphocytes and fibroblasts, produce hormonelike peptides called **interferons** in response to viruses or tumor cells. Interferon's effect is nonspecific. It blocks viral replication, stimulates phagocytosis, and enhances the activity of other cells that help resist infections and tumor growth.

Inflammation

Inflammation is a tissue response to injury or infection, producing localized redness, swelling, heat, and pain. The redness is a result of blood vessel dilation and the consequent increase in blood volume within the affected tissues. This effect, coupled with an increase in the permeability of nearby capillaries, causes tissue swelling (edema). The heat comes from blood from deeper body parts, which is warmer than that near the surface. Pain results from stimulation of nearby pain receptors.

Infected cells release chemicals that attract white blood cells to inflammation sites, where they engulf (phagocytize) pathogens. Body fluids also collect in inflamed tissues. These fluids contain fibrinogen and other blood clotting factors. Clotting forms a network of fibrin threads in the affected region. Later, fibroblasts may arrive and secrete fibers until the area is enclosed in a sac of fibrous connective tissue. This action inhibits the spread of pathogens and toxic substances to adjacent tissues.

Phagocytosis

Recall from chapter 12 that blood's most active phagocytic cells are *neutrophils* and *monocytes*. These cells can leave the bloodstream by squeezing between the cells of blood vessel walls (diapedesis). Chemicals released from injured tissues attract these cells (chemotaxis).

Monocytes specialize into macrophages (histiocytes), which become fixed in various tissues and attach to the inner walls of blood and lymphatic vessels. These relatively nonmotile phagocytic cells, which can divide and produce new macrophages, are found in the lymph nodes, spleen, liver, and lungs. This diffuse group of phagocytic cells constitutes the **mononuclear phagocytic (reticuloendothelial) system.**

Phagocytic activities remove foreign particles from lymph as it moves from interstitial spaces to the bloodstream. Phagocytes in the blood vessels and in the tissues of the spleen, liver, or bone marrow remove particles reaching blood.

1. What is an infection?
2. Explain six nonspecific defense mechanisms.

Specific Defenses (Immunity)

Immunity is resistance to particular pathogens or to their toxins or metabolic by-products. Lymphocytes and macrophages that recognize specific foreign molecules execute several types of immune responses.

Lymphocyte Origins

During fetal development (before birth), red bone marrow releases undifferentiated lymphocytes into the circulation. About half of these cells reach the thymus, where they specialize into T cells. Later, some of these T cells comprise 70–80% of the circulating lymphocytes in blood. Others of these T cells reside in lymphatic organs and are particularly abundant in the lymph nodes, thoracic duct, and spleen.

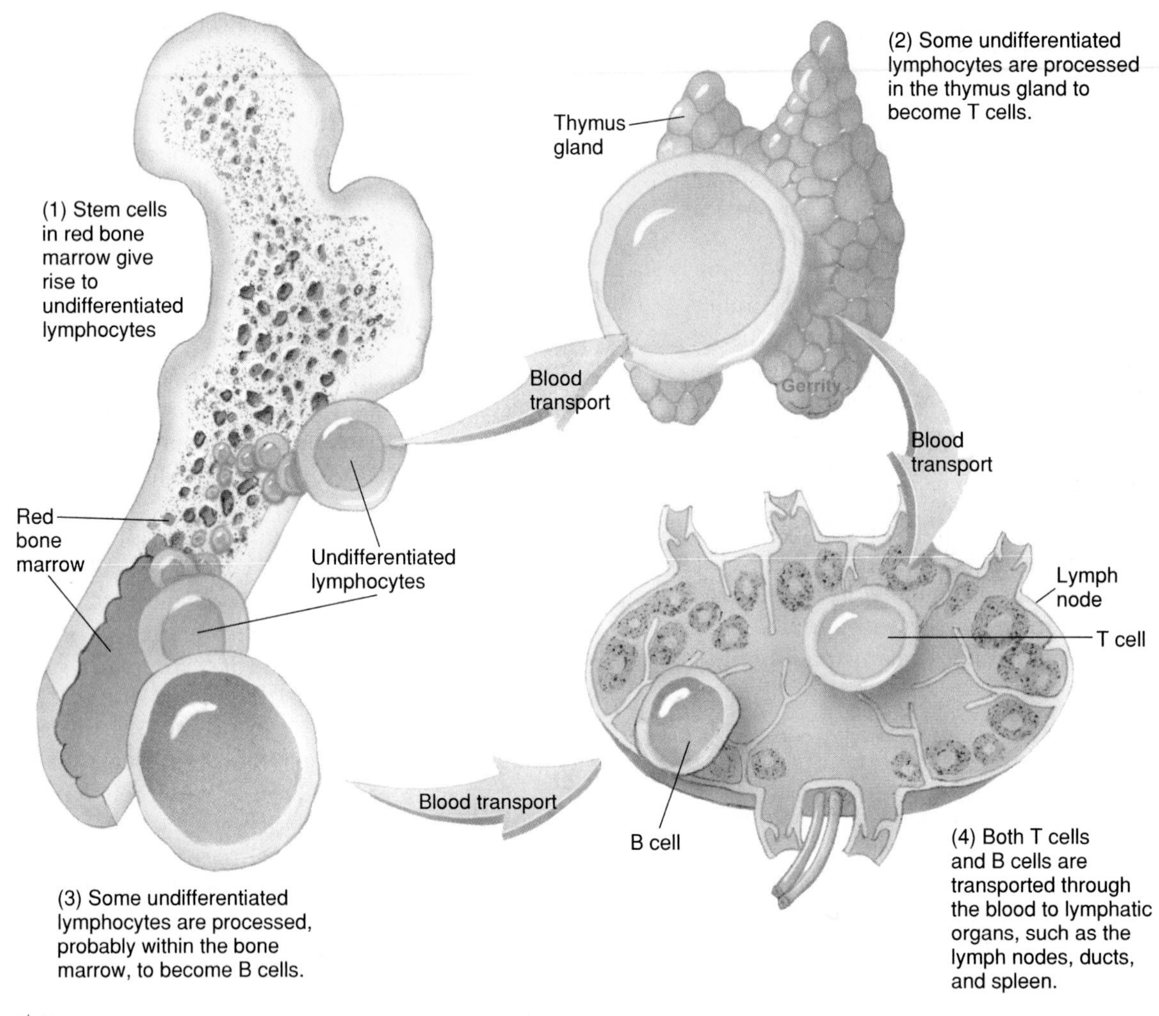

Figure 14.11

Bone marrow releases undifferentiated lymphocytes, which after processing become T cells (T lymphocytes) or B cells (B lymphocytes). Note that the medullary cavity contains red marrow in the fetus.

Other lymphocytes are thought to remain in the red bone marrow until they differentiate into **B cells** (B lymphocytes). Blood distributes B cells, which constitute 20–30% of circulating lymphocytes. B cells settle in lymphatic organs along with T cells and are abundant in the lymph nodes, spleen, bone marrow, and intestinal lining (figs. 14.11 and 14.12).

Antigens

Antigens may be proteins, polysaccharides, or glycolipids usually located on a cell's surface. Before birth, body cells inventory the proteins and other large molecules in the body, learning to distinguish these as "self." The lymphatic system responds to nonself, or foreign, antigens, but not normally to self antigens. Receptors on T and B cell surfaces enable these cells to recognize foreign antigens.

The antigens that are most effective in eliciting an immune response are large and complex, with few repeating parts. Sometimes, a smaller molecule that cannot by itself stimulate an immune response combines with a larger one, which makes it able to do so (antigenic). Such a small molecule is called a **hapten.** Stimulated lymphocytes react either to the hapten or to the larger molecule of the combination. Haptens are in certain drugs, such as penicillin; in household and industrial chemicals; in dust particles; and in products of animal skins (dander).

1. What is immunity?
2. How do T cells and B cells originate?
3. How are antigens and haptens different?

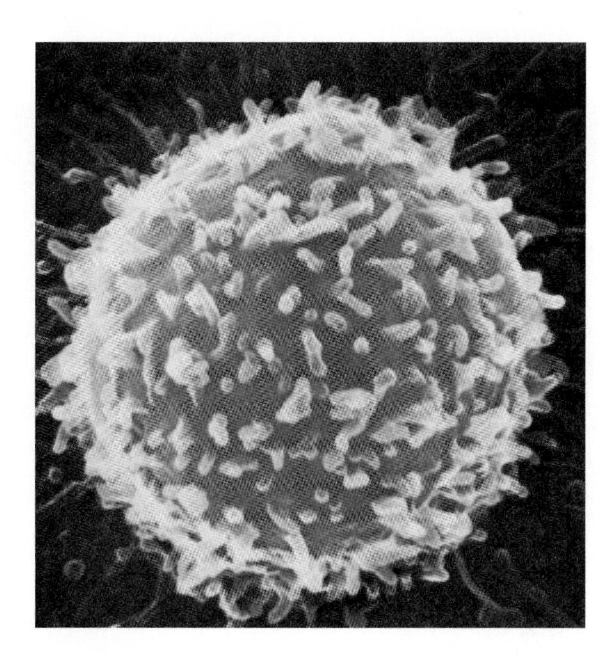

Figure 14.12

Scanning electron micrograph of a human circulating lymphocyte (36,000×).

Lymphocyte Functions

T cells and B cells respond to antigens they recognize in different ways. T cells attach to foreign, antigen-bearing cells, such as bacterial cells, and interact directly—that is, by cell-to-cell contact. This type of response is called **cell-mediated immunity** (CMI).

T cells (and some macrophages) also synthesize and secrete polypeptides called *cytokines* (or more specifically, lymphokines) that enhance certain cellular responses to antigens. For example, *interleukin-1* and *interleukin-2* stimulate synthesis of several cytokines from other T cells. In addition, interleukin-1 helps activate T cells, while interleukin-2 causes T cells to proliferate and activates another type of T cell (cytotoxic T cells). Other cytokines called *colony-stimulating factors* (CSFs) stimulate leukocyte production in red bone marrow, cause growth and maturation of B cells, and activate macrophages.

T cells may also secrete toxins that are lethal to their antigen-bearing target cells. These toxins include growth-inhibiting factors that prevent target-cell growth. They also include interferon, which inhibits the proliferation of viruses and tumor cells.

B cells attack foreign antigens in a different way. They differentiate into **plasma cells**, which produce and secrete enormous numbers of large globular proteins called **antibodies**, or **immunoglobulins.** Body fluids carry antibodies, which destroy specific antigens or antigen-bearing particles in various ways. This type of response is called **antibody-mediated immunity** (AMI) or humoral immunity ("humoral" refers to fluid).

Each person has millions of varieties of T and B cells. Because the members of each variety originate from a single early cell, they are all alike, forming a **clone** of cells. The members of each variety have a particular type of antigen receptor on their cell membranes that can respond only to a specific antigen.

Table 14.1 compares the characteristics of T cells and B cells.

B Cell Activation

A lymphocyte must be activated before it can respond to an antigen. A B cell, for example, may become activated when it encounters an antigen whose molecular shape fits the shape of the B cell's antigen receptors. In response to the receptor-antigen combination and usually stimulation by a similarly activated T cell (described later), the B cell divides repeatedly, expanding its clone (fig. 14.13).

Some members of the activated B cell's clone differentiate further into plasma cells, and they mass-produce antibody molecules. These antibodies are similar in structure to the antigen-receptor molecules on the original B cell's surface. Thus, antibodies can combine with the antigen-bearing agent in the body and react against it.

An individual's B cells can produce an estimated 10 million to 1 billion different varieties of antibodies, each reacting against a specific antigen. The enormity and diversity of the antibody response defends against many pathogens.

Types of Antibodies

Antibodies (immunoglobulins) are soluble, globular proteins that constitute the *gamma globulin* fraction of plasma proteins (see chapter 12). Of the five major types of immunoglobulins, the most abundant are immunoglobulin G, immunoglobulin A, and immunoglobulin M.

Immunoglobulin G (IgG) is in plasma and tissue fluids and is effective against bacteria, viruses, and toxins. It also activates a set of enzymes called **complement** that attacks the foreign antigens, as described in the section that follows.

Immunoglobulin A (IgA) is common in exocrine gland secretions. It is in breast milk, tears, nasal fluid, gastric juice, intestinal juice, bile, and urine.

Immunoglobulin M (IgM) develops in the blood plasma in response to contact with certain antigens in foods or bacteria. The antibodies anti-A and anti-B, described in chapter 12, are examples of IgM. IgM also activates complement.

Immunoglobulin D (IgD) is found on the surfaces of most B cells, especially those of infants. IgD is important in B cell activation.

Table 14.1 A Comparison of T Cells and B Cells

Characteristic	T Cells	B Cells
Origin of undifferentiated cell	Red bone marrow	Red bone marrow
Site of differentiation	Thymus	Probably the red bone marrow
Primary locations	Lymphatic tissues, 70–80% of the circulating lymphocytes	Lymphatic tissues, 20–30% of the circulating lymphocytes
Primary functions	Provides cell-mediated immunity in which T cells interact directly with antigen-bearing agents	Provides antibody-mediated immunity in which B cells interact indirectly with antigen-bearing agents to produce antibodies

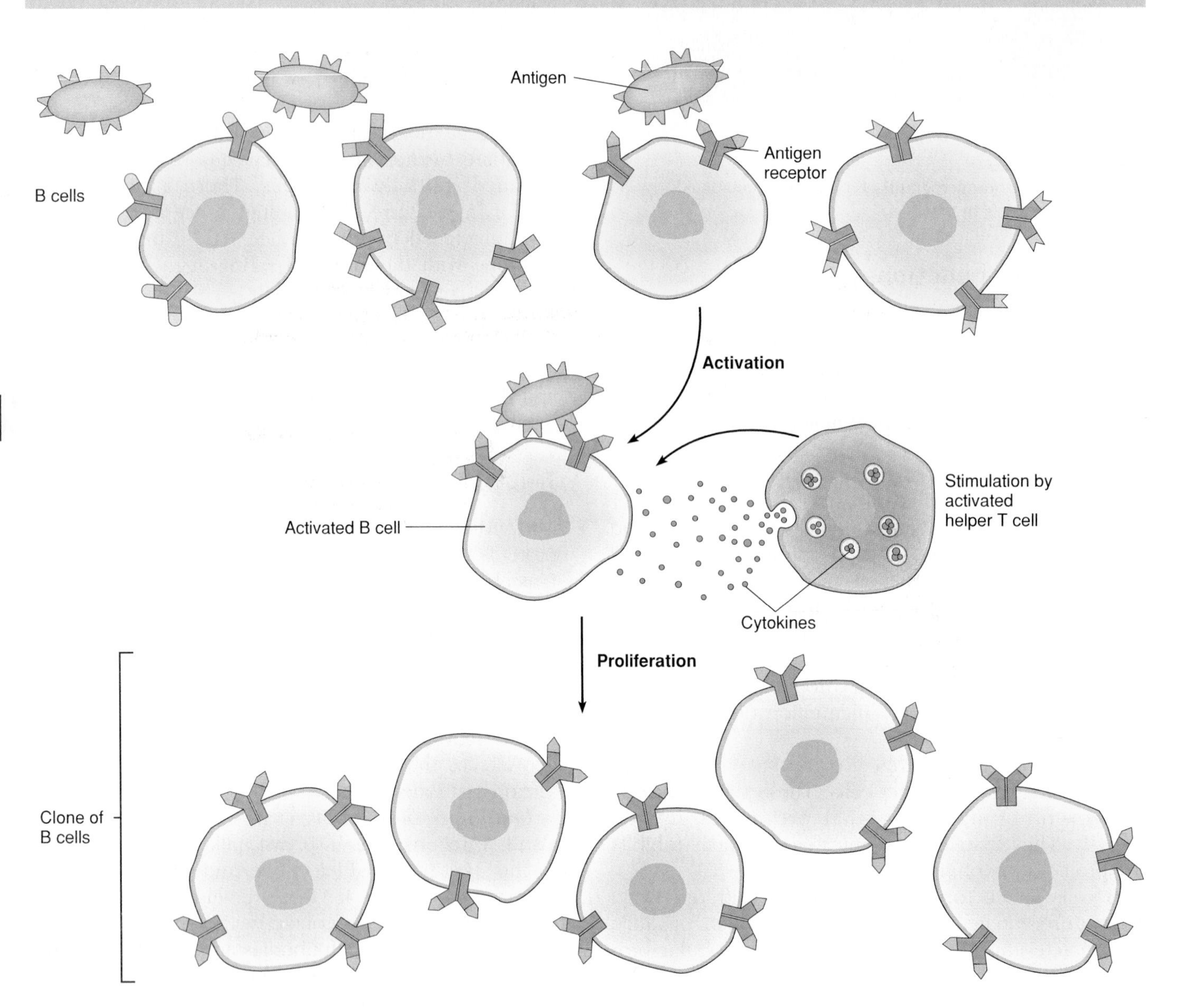

Figure 14.13

When a B cell encounters an antigen that fits its antigen receptor, it becomes activated and proliferates, enlarging its clone.

Immunoglobulin E (IgE) appears in exocrine secretions along with IgA. It is associated with allergic reactions described later in this chapter.

A newborn does not yet have its own antibodies but does retain IgG that passed through the placenta from the mother. These maternal antibodies protect the infant against some illnesses to which the mother is immune. Just as the maternal antibody supply falls, the infant begins to manufacture its own. A breastfed newborn receives IgA from colostrum, a substance in milk for the first few days after birth. Antibodies in colostrum protect against certain digestive and respiratory infections. ■

1. What are the functions of T cells and B cells?
2. How are B cells activated?
3. How does the antibody response protect against diverse infections?
4. Which immunoglobulins are most abundant, and how do they differ?

Antibody Actions

In general, antibodies attack foreign antigens directly, activate complement enzymes that attack the antigens, or stimulate changes in local areas that help prevent spread of the pathogens.

In a direct attack, antibodies combine with antigens and clump (agglutinate) them to form insoluble material (precipitation). This clumping makes it easier for phagocytic cells to engulf the antigen-bearing agents and eliminate them. In other instances, antibodies cover the toxic parts of antigen molecules and neutralize their effects (neutralization). Under normal conditions, however, complement activation is more important in protecting against infection than is direct antibody attack.

Complement is a group of proteins in plasma and other body fluids. Certain IgG or IgM antibodies combining with antigens expose reactive sites on antibody molecules. This triggers a series of reactions, leading to activation of the complement proteins, which, in turn, produce a variety of effects, including: coating the antigen-antibody complexes (opsonization), making them more susceptible to phagocytosis; attracting macrophages and neutrophils into the region (chemotaxis); clumping antigen-bearing agents; rupturing membranes of foreign cells (lysis); and rendering viruses harmless. Other proteins promote inflammation, which helps prevent the spread of infectious agents (fig. 14.14).

1. In what general ways do antibodies function?
2. What is the function of complement?
3. How is complement activated?

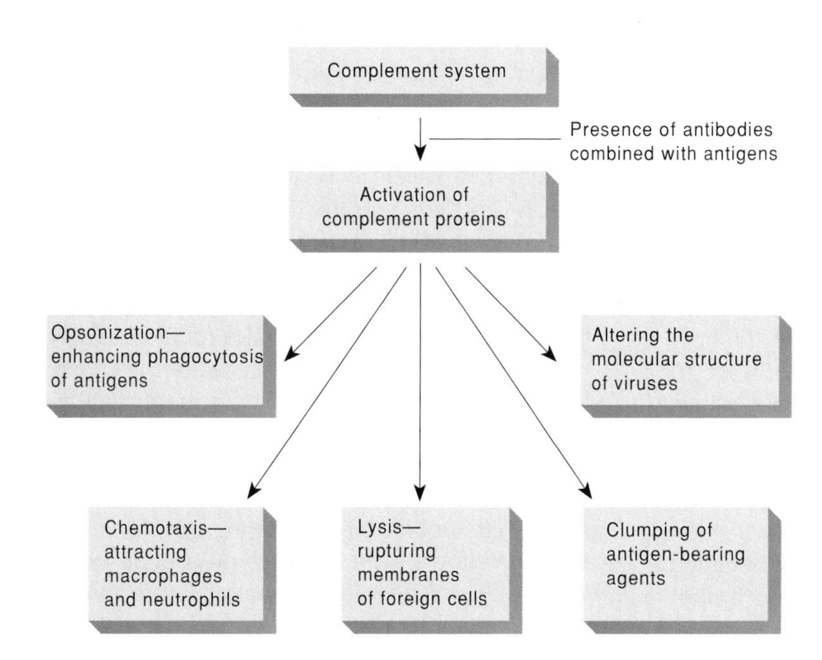

Figure 14.14

Actions of the complement system.

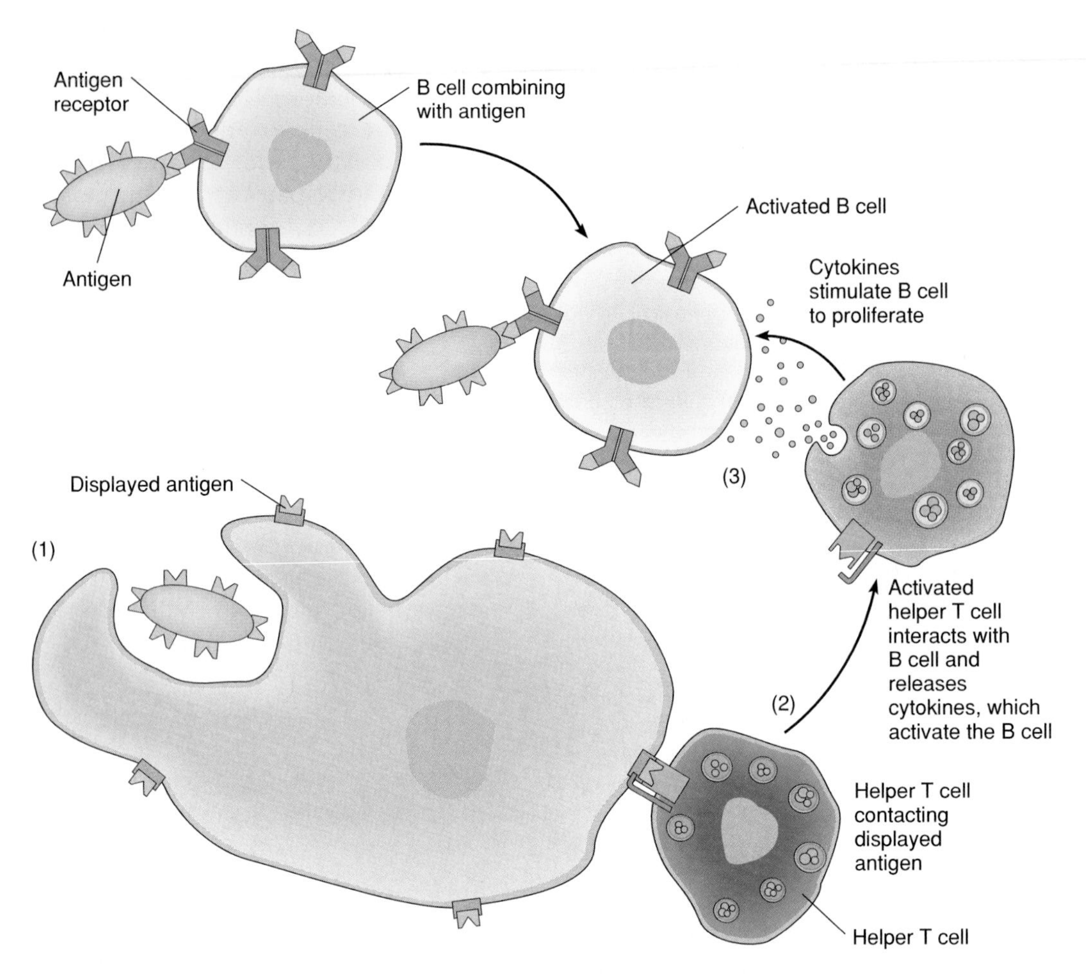

Figure 14.15

(1) After digesting antigen-bearing agents, a macrophage displays antigens on its surface. (2) Helper T cells become activated when they contact displayed antigens that fit their antigen receptors. (3) An activated helper T cell interacts with a B cell that has combined with an identical antigen and causes the B cell to proliferate.

T Cell Activation

As mentioned earlier, B cells are usually activated by activated T cells. T cell activation requires the presence of processed fragments of antigen attached to the surface of another kind of cell, called an **antigen-presenting cell** (or accessory cell). Macrophages, B cells, and several other cell types can be antigen-presenting cells.

T cell activation begins when a macrophage phagocytizes a bacterium. Lysosomal digestion of the bacterium releases some bacterial antigens, which move to the macrophage's surface. Here, they are displayed on the outside of the cell membrane near certain molecules that are part of a group of proteins called the *major histocompatibility complex* (MHC). A specialized type of T cell, called a *helper T cell,* contacts a foreign antigen displayed on a macrophage. If the displayed antigen binds to a helper T cell's receptor, the helper cell becomes activated.

When an activated helper T cell encounters a B cell that has already bound an identical foreign antigen, the helper cell releases certain cytokines. These cytokines stimulate the B cell to proliferate, thereby enlarging its clone of antibody-producing cells (fig. 14.15). The cytokines also attract macrophages and leukocytes into inflamed tissues and help keep them there.

Table 14.2 summarizes the steps leading to antibody production as a result of B cell and T cell actions.

A second type of T cell, a *cytotoxic T cell,* recognizes nonself antigens that cancerous or virally infected cells display on their surfaces near certain MHC proteins. A cytotoxic T cell becomes activated when an antigen binds its receptors. Then the activated cytotoxic T cell proliferates, enlarging its clone. Cytotoxic T cells then bind to the surfaces of antigen-bearing cells, where they release a protein that cuts porelike openings, destroying these cells. In this way, cytotoxic

Table 14.2

Steps in Antibody Production

B Cell Activity
1. Antigen-bearing agents enter tissues.
2. B cell becomes activated when it encounters an antigen that fits its antigen receptors, either alone or more often in conjunction with helper T cells.
3. Activated B cell proliferates, enlarging its clone.
4. Some of the newly formed B cells differentiate further to become plasma cells.
5. Plasma cells synthesize and secrete antibodies whose molecular structure is similar to the activated B cell's antigen receptors.
6. Antibodies combine with antigen-bearing agents, helping to destroy them.
T Cell Activity
1. Antigen-bearing agents enter tissues.
2. Accessory cell, such as a macrophage, phagocytizes antigen-bearing agent, and the macrophage's lysosomes digest the agent.
3. Antigens from the digested antigen-bearing agents are displayed on the accessory cell's membrane.
4. Helper T cell becomes activated when it encounters a displayed antigen that fits its antigen receptors.
5. Activated helper T cell releases cytokines when it encounters a B cell that has previously combined with an identical antigen-bearing agent.
6. Cytokines stimulate the B cell to proliferate.
7. Some of the newly formed B cells differentiate into antibody-secreting plasma cells.
8. Antibodies bind antigen-bearing agents, helping to destroy them.

T cells continually monitor body cells, recognizing and eliminating tumor cells and cells infected with viruses.

Some of the T cells do not respond to the antigen on first exposure. Rather, they remain as *memory cells* that immediately differentiate into cytotoxic T cells upon subsequent exposure to the same antigen.

Immune Responses

Activation of B cells or T cells after an encounter with the antigens for which they are specialized to react constitutes a **primary immune response.** During such a response, plasma cells release antibodies (IgM, followed by IgG) into lymph. The antibodies are transported to blood and then throughout the body, where they help destroy antigen-bearing agents. Production and release of antibodies continues for several weeks.

After a primary immune response, some of the B cells produced during proliferation of the clone remain dormant as *memory cells* (fig. 14.16). If the identical antigen is encountered in the future, the clones of these memory cells enlarge, and they can respond rapidly with IgG to the antigen to which they were previously sensitized. These memory B cells along with the aforementioned memory T cells produce a **secondary immune response.**

A primary immune response usually produces detectable concentrations of antibodies in body fluids within five to ten days following antigen exposure. A secondary immune response may produce additional antibodies within a day or two of exposure (fig. 14.17). New antibodies may persist in the body for only a few months or years, but memory cells live much longer. Consequently, the ability to produce a secondary immune response may be long-lasting.

Superantigens are foreign antigens that elicit unusually vigorous lymphocyte responses. The bacterium *Staphylococcus aureus* produces two such superantigens. One type causes food poisoning until digestive enzymes destroy it. The second type causes toxic shock syndrome, a potentially fatal condition producing high fever, diarrhea, vomiting, confusion, and plummeting blood pressure. ■

1. How do T cells become activated?
2. What is the function of cytokines?
3. How do cytotoxic T cells destroy antigen-bearing cells?
4. How do primary and secondary immune responses differ?

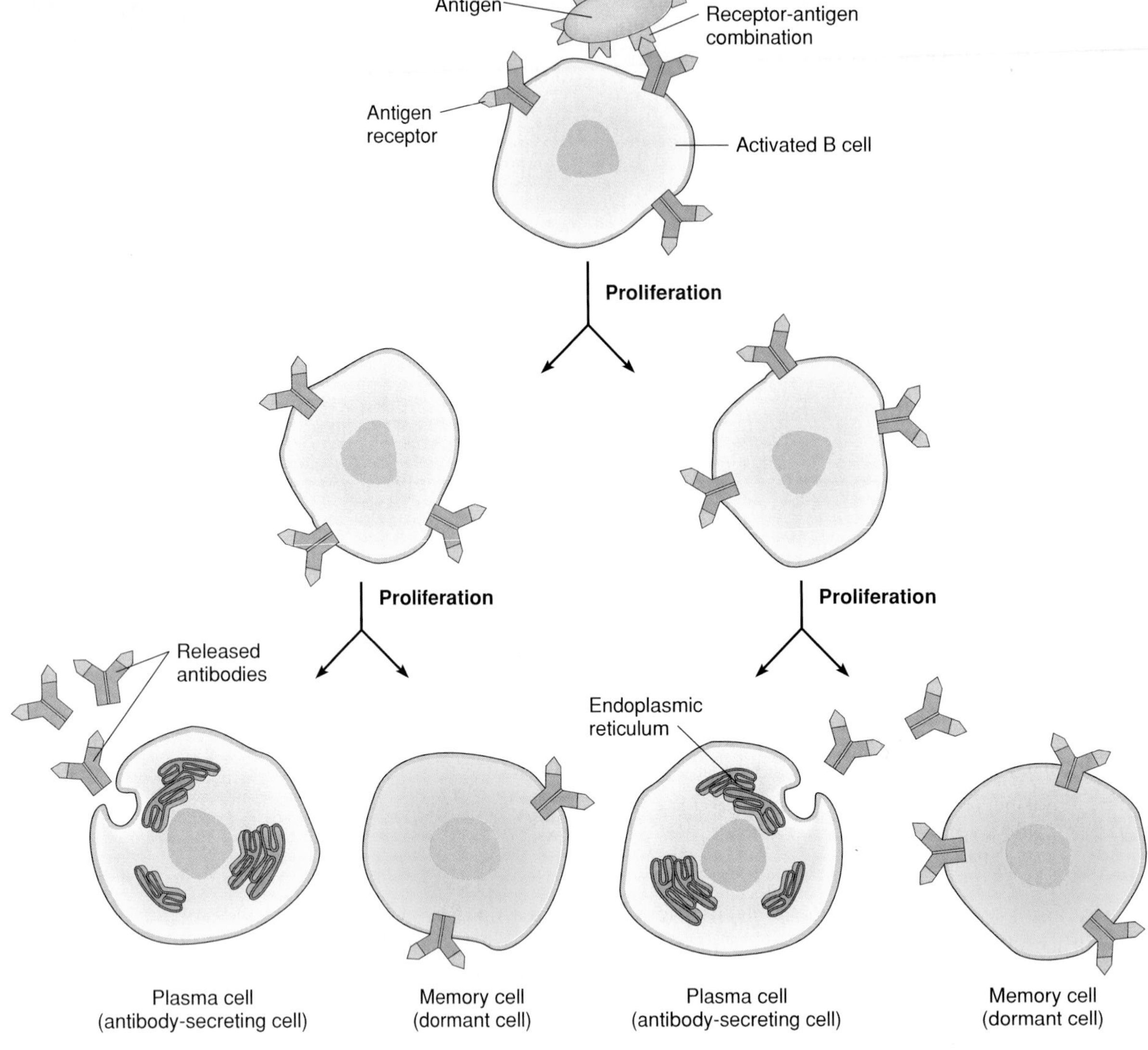

***Figure* 14.16**

Activated B cells proliferate, giving rise to antibody-secreting plasma cells and dormant memory cells.

Types of Immunity

One type of immunity, called *naturally acquired active immunity,* occurs when a person exposed to a pathogen develops a disease and becomes resistant to that pathogen as a result of a primary immune response. A **vaccine** produces another type of active immunity. A vaccine contains an antigen that can stimulate a primary immune response against a particular disease-causing agent, but does not produce the severe symptoms of that disease.

A vaccine might consist of bacteria or viruses that have been killed or weakened so that they cannot cause a serious infection, or a toxoid, which is a toxin from an infectious organism that has been chemically altered to destroy its toxic effects. Whatever its composition, a vaccine includes the antigens that stimulate a primary immune response. A vaccine causes a person to develop *artificially acquired active immunity.*

Vaccines stimulate active immunity against a variety of diseases, including typhoid fever, cholera, whooping cough, diphtheria, tetanus, polio, chicken pox, measles (rubeola), German measles (rubella), mumps, influenza, hepatitis B, and bacterial pneumonia. Vaccines have virtually eliminated smallpox from the world and greatly diminished the number of cases of poliomyelitis and infection by *Haemophilus influenzae* type B, which causes meningitis in pre-school children. Unfortunately, vaccine distribution is not equitable worldwide. Many thousands of people in underdeveloped countries die of infectious diseases for which vaccines are widely available in other nations. ■

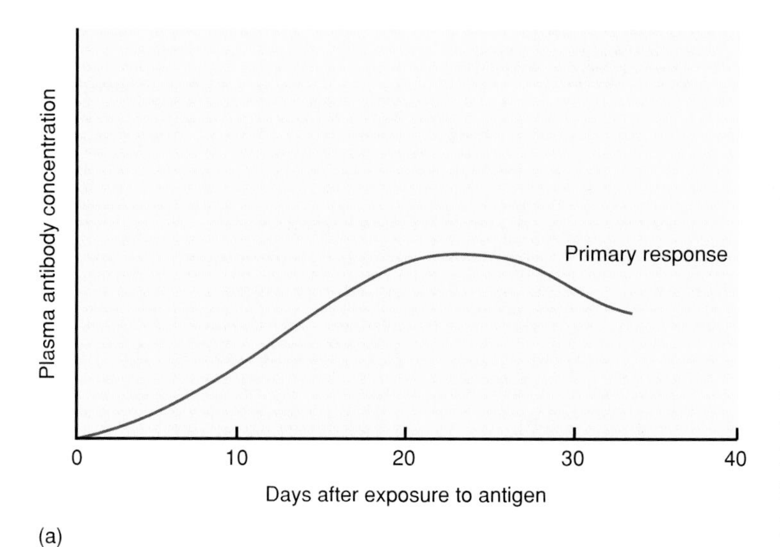

(a)

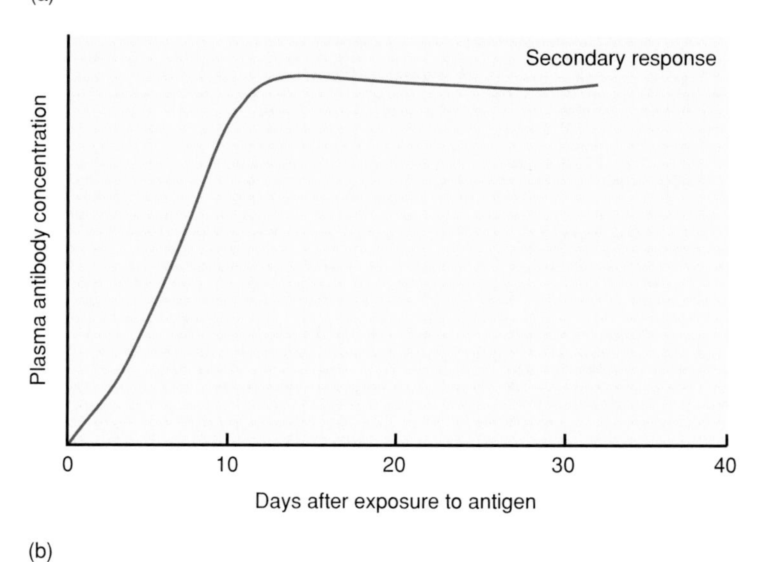

(b)

Figure 14.17

(*a*) A primary immune response produces a lesser concentration of antibodies than does (*b*) a secondary immune response.

Sometimes, a person who has been exposed to infection needs protection against a disease-causing microorganism but lacks the time to develop active immunity. An injection of antiserum (ready-made antibodies) may help. These antibodies may be obtained from gamma globulin (see chapter 12) separated from the blood plasma of persons who have already developed immunity against the particular disease.

A gamma globulin injection provides *artificially acquired passive immunity*. This type of immunity is called passive because the recipient's cells do not produce the antibodies. Passive immunity is short-term, lasting usually a few weeks. Furthermore, because the recipient's lymphocytes did not initially fight the pathogens, susceptibility may persist.

During pregnancy, certain antibodies (IgG) pass from the maternal blood into the fetal bloodstream. As a result, the fetus acquires limited immunity against pathogens that the pregnant woman has developed active immunities against. The fetus thus has *naturally acquired passive immunity*, which may last for six months to a year after birth.

Table 14.3 summarizes types of immunity.

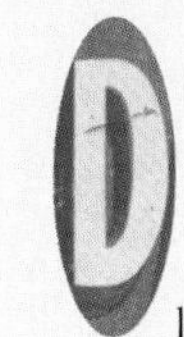eclining competence of the immune system with age is why the elderly are more likely to develop cancers and contract infections than younger people. This decline in immunity is primarily due to loss of T cells. B cell activity changes little with age. ■

Table 14.3

Types of Immunity

Type	Mechanism	Result
Naturally acquired active immunity	Exposure to pathogens	Symptoms of a disease and stimulation of an immune response
Artificially acquired active immunity	Exposure to a vaccine containing weakened or dead pathogens or their components	Stimulation of an immune response without the severe symptoms of a disease
Artificially acquired passive immunity	Injection of gamma globulin containing antibodies	Immunity for a short time without stimulating an immune response
Naturally acquired passive immunity	Antibodies passed to fetus from pregnant woman with active immunity	Short-term immunity for infant without stimulating an immune response

TOPIC OF INTEREST Immunity Breakdown: AIDS

Infection by the *human immunodeficiency virus* (HIV) causes *acquired immune deficiency syndrome* (AIDS), a progressive breakdown of the immune system. The virus typically attacks T cells first, then infects other cells. T cells are the linchpins of the immune system; when they fail, immunity collapses. Specific symptoms in different individuals reflect the infectious agents to which they are exposed. Common AIDS-related conditions include:

- Persistent lymphadenopathy (swollen lymph glands)
- Constant low-grade fever
- Nausea and vomiting
- Fatigue
- Night sweats
- Headaches
- Wasting syndrome (persistent diarrhea, severe weight loss, weakness, fever)
- Dementia (confusion, apathy, inability to concentrate, memory loss, insomnia, disorientation, sudden, strong emotions)
- Cancers (Kaposi's sarcoma, cervical cancer, lymphoma, others)
- Opportunistic infections (pneumonia, brain infection, diarrhea, spinal meningitis, tuberculosis, fungal infections, many others)

HIV infection has three stages: initial symptoms, a latency period, and AIDS. The initial, acute stage may include weakness, recurrent fever, night sweats, swollen neck glands, and weight loss. This stage varies in duration and severity. Often, it lasts only a few days, and the person thinks it is the flu. Then comes a latency period, typically lasting five to ten years, when the person feels well. This well-being is deceptive. In the lymph nodes and then the bloodstream, the immune system is struggling to contain the growing HIV population. The third stage, AIDS, brings opportunistic infections, so-called because they appear when the immune system is compromised.

Modes of Transmission

AIDS transmission requires contact with a body fluid containing abundant HIV, such as blood and semen. Although the virus has been detected in sweat, tears, and saliva, levels are so low that transmission is highly unlikely. Whether or not a person becomes infected appears to depend on the amount of infected fluid contacted, the site of exposure in the body, and the individual's health. Table 14A lists ways that HIV infection can and cannot spread.

Allergic Reactions

An allergic response is an immune response against an agent that is not pathogenic, such as chocolate. Allergic reactions are similar to immune responses because they sensitize lymphocytes, and antibodies may bind antigens. Allergic reactions, unlike normal immune responses, can damage tissues. Antigens that trigger allergic responses are called **allergens.** There are two types of allergic reactions.

A *delayed-reaction allergy* can occur in almost anyone. It results from repeated exposure of the skin to certain chemicals—commonly, household or industrial chemicals or some cosmetics. After repeated contacts, the foreign substance activates T cells, which collect in large numbers in the skin. The T cells and the macrophages they attract release chemicals that erupt and inflame the skin (dermatitis). This reaction is called *delayed* because it usually takes about 48 hours to occur.

An *immediate-reaction allergy* occurs within minutes after contact with a nonself substance. Affected persons have an inherited tendency to overproduce IgE antibodies in response to certain antigens. This type of allergic reaction activates B cells.

In an immediate-reaction allergy, B cells become sensitized when the allergen is first encountered, and subsequent exposures trigger allergic reactions. In the

Progress

Three groups of people are providing the clues that may lead to conquering HIV infection:

1. Infected individuals who never develop symptoms ("long-term nonprogressors"). These people have a weakened strain of HIV that lacks part of a gene that HIV normally uses to replicate. Enough of the virus remains to alert the immune system to protect against other strains. A vaccine might be based on this weakened strain.
2. People exposed repeatedly who never become infected. About 1% of the population has a gene variant that protects them from becoming infected with HIV.
3. Infected individuals who become apparently uninfected. Several infants infected at birth have apparently lost the virus as their immune systems matured.

Researchers are identifying how these people survive, evade, or vanquish HIV infection, and will use this knowledge to develop prevention and treatment strategies.

Several new drugs target HIV infection at various stages. Drugs called protease inhibitors prevent HIV from processing its proteins to a functional size, crippling the virus. Older drugs, such as AZT, ddI, ddC, and 3TC, block viral replication. Combining drugs can keep viral load low and delay symptom onset and progression. The goal is to enable infected people to live normal life spans in relatively good health. Some 200 drugs are also available to treat AIDS-associated opportunistic infections and cancers.

Better understanding of the biology of HIV infection, plus new drug weapons and clues from survivors, are providing what has long been lacking in the global fight against AIDS—hope.

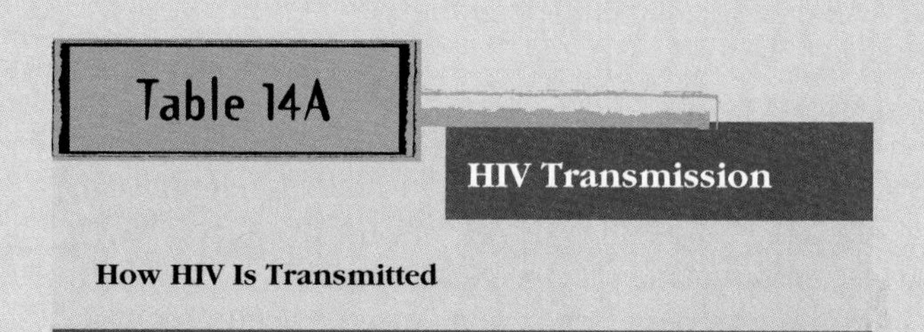

Table 14A HIV Transmission

How HIV Is Transmitted
Sexual contact, particularly anal intercourse
Contaminated needles (intravenous drug use, injection of anabolic steroids, accidental needle stick in medical setting)
During birth from infected mother
Receiving infected blood or other tissue (precautions usually prevent this)

How HIV Is NOT Transmitted
Casual contact (social kissing, hugging, handshakes)
Objects—toilet seats, deodorant sticks, doorknobs
Mosquitoes
Sneezing and coughing
Sharing food
Swimming in the same water

initial exposure, IgE attaches to the membranes of widely distributed mast cells and basophils. When a subsequent allergen-antibody reaction occurs, these cells release allergy mediators such as *histamine* and *leukotrienes*. These substances cause a variety of physiological effects, including dilation of blood vessels, increased vascular permeability that swells tissues, contraction of bronchial and intestinal smooth muscles, and increased mucus production. The result is a severe inflammation reaction that causes allergy symptoms, such as hives, hay fever, asthma, eczema, or gastric disturbances.

Transplantation and Tissue Rejection

Transplantation of tissues or an organ, such as the skin, kidney, heart, or liver, from one person to another can replace a nonfunctional, damaged, or lost body part. The danger the immune system poses to transplanted tissue is that the recipient's cells may recognize the donor's tissues as foreign and attempt to destroy the transplant. Such a response is called a **tissue rejection reaction.**

Tissue rejection resembles the cell-mediated response against a nonself antigen. The greater the antigenic difference between cell surface molecules of the recipient and donor tissues, the more rapid and severe the rejection reaction. Matching donor and recipient tissues can minimize the rejection reaction. This means locating a donor whose tissues are antigenically similar to those of the person needing a transplant—a procedure much like matching the blood of a donor with that of a recipient before a blood transfusion.

Immunosuppressive drugs reduce rejection of transplanted tissues. These drugs interfere with the recipient's immune response by suppressing antibody formation or T cell production, thereby reducing the humoral and cellular responses. Unfortunately, immunosuppressive drugs leave the recipient vulnerable to infections. A patient may survive a transplant, only to die of infection because of a weakened immune system.

Autoimmunity

Sometimes, the immune system backfires, manufacturing **autoantibodies** that attack the body's own cells, causing **autoimmune** disorders. The specific nature of an autoimmune disorder depends on the cell types the misguided immune system attacks. Juvenile diabetes, rheumatoid arthritis, and systemic lupus erythematosus are some autoimmune disorders.

Why might the immune system attack body tissues? Perhaps a virus, while replicating within a human cell, "borrows" proteins from the host cell's surface and incorporates them onto its own surface. When the immune system "learns" the surface of the virus to destroy it, it also learns to attack the human cells that normally bear the particular protein. Another explanation of autoimmunity is that somehow T cells escape their "education" in the thymus, never learning to distinguish self from nonself. A third possible route of autoimmunity is when a nonself antigen coincidentally resembles a self antigen. This is what happens when an infection by *Streptococcus* bacteria triggers inflammation of heart valves, as mentioned in chapter 13.

1. Explain the difference between active and passive immunities.
2. How is an allergic reaction similar to yet different from an immune reaction?
3. How is a tissue rejection reaction an immune response?
4. How is autoimmunity an abnormal functioning of the immune response?

Clinical Terms Related to the Lymphatic System and Immunity

Allograft (al′o-graft) Transplantation of tissue from an individual of one species to another individual of that species.

Asplenia (ah-sple′ne-ah) The absence of a spleen.

Autograft (aw′to-graft) Transplantation of tissue from one part of the body to another part of the same body.

Immunocompetence (im″u-no-kom′pe-tens) The immune system's ability to respond to foreign (nonself) antigens.

Immunodeficiency (im″u-no-de-fish′en-se) Inability to produce an immune response.

Lymphadenectomy (lim-fad″ĕ-nek′to-me) Surgical removal of lymph nodes.

Lymphadenopathy (lim-fad″ĕ-nop′ah-the) Enlargement of lymph nodes.

Lymphadenotomy (lim-fad″ĕ-not′o-me) An incision of a lymph node.

Lymphocytopenia (lim″fo-si″to-pe′ne-ah) Too few lymphocytes in blood.

Lymphocytosis (lim″fo-si″to′sis) Too many lymphocytes in blood.

Lymphoma (lim-fo′mah) A tumor composed of lymphatic tissue.

Lymphosarcoma (lim″fo-sar-ko′mah) A cancer within lymphatic tissue.

Splenectomy (sple-nek′to-me) Surgical removal of the spleen.

Splenitis (sple-ni′tis) Inflammation of the spleen.

Splenomegaly (sple″no-meg′ah-le) Abnormal enlargement of the spleen.

Splenotomy (sple-not′o-me) An incision of the spleen.

Thymectomy (thi-mek′to-me) Surgical removal of the thymus.

Thymitis (thi-mi′tis) Inflammation of the thymus.

organization

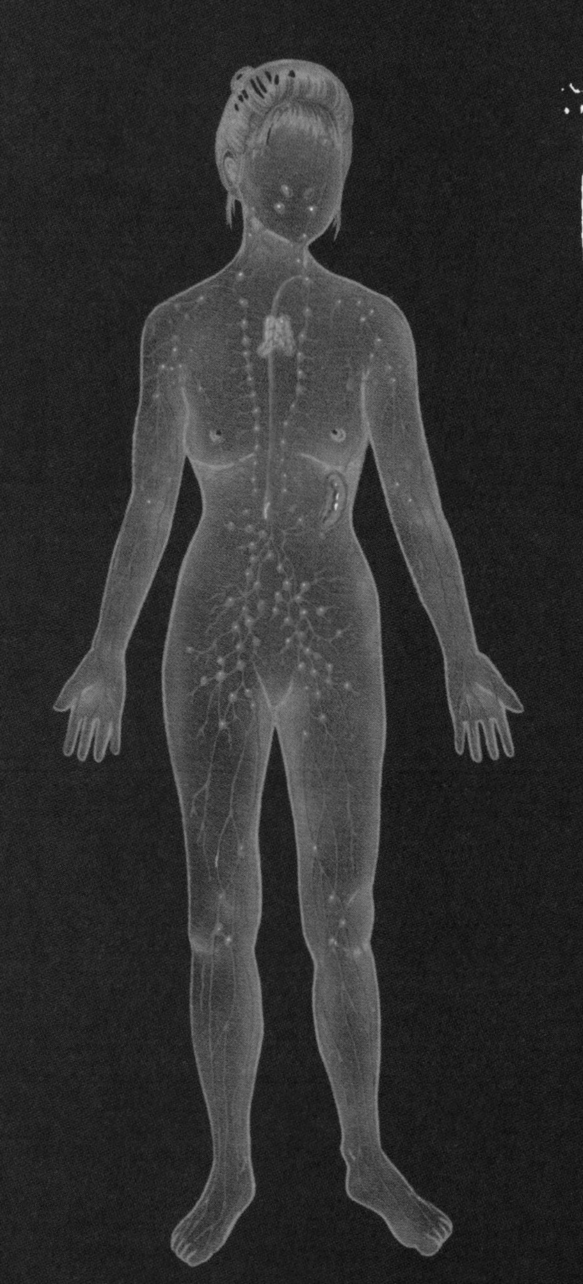

Lymphatic System

The lymphatic system is an important link between the interstitial fluid and the plasma; it also plays a major role in the response to infection.

Integumentary System

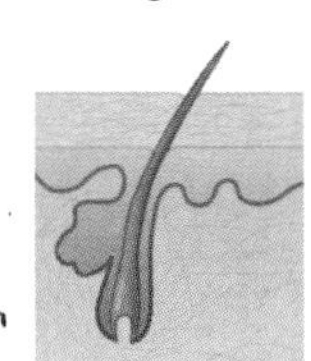

The skin is a first line of defense against infection.

Cardiovascular System

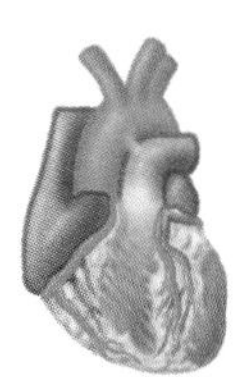

The lymphatic system returns tissue fluid to the bloodstream. Lymph originates as interstitial fluid, formed by the action of blood pressure.

Skeletal System

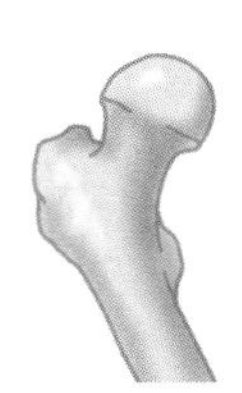

Cells of the immune system originate in the bone marrow.

Digestive System

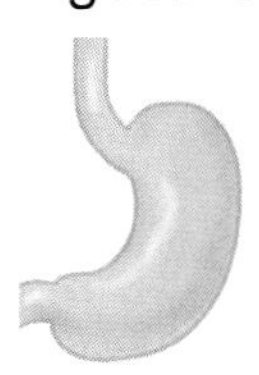

Lymph plays a major role in the absorption of fats.

Muscular System

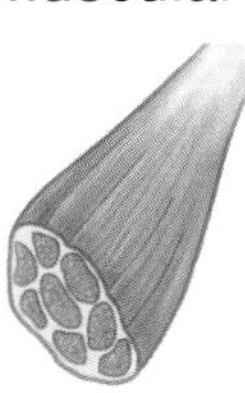

Muscle action helps pump lymph through the lymphatic vessels.

Respiratory System

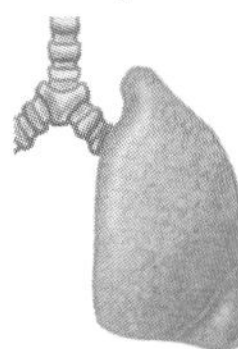

Cells of the immune system patrol the respiratory system to defend against infection.

Nervous System

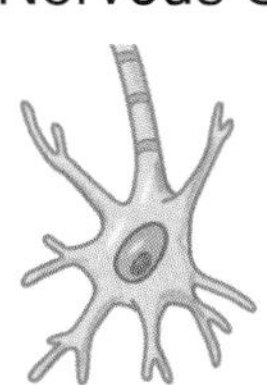

Stress may impair the immune response.

Urinary System

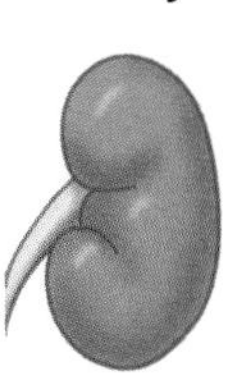

The kidneys control the volume of extracellular fluid, including lymph.

Endocrine System

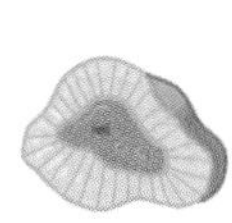

Hormones stimulate lymphocyte production.

Reproductive System

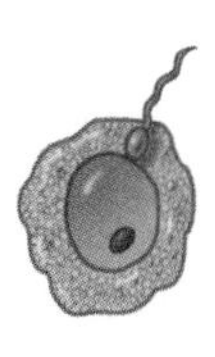

Special mechanisms inhibit the female immune system in its attack of sperm as foreign invaders.

Summary Outline

Introduction (p. 384)

The lymphatic system is closely associated with the cardiovascular system. It transports excess fluid to the bloodstream, absorbs fats, and protects against disease-causing agents.

Lymphatic Pathways (p. 384)

1. Lymphatic capillaries
   a. Lymphatic capillaries are microscopic, closed-ended tubes.
   b. They receive lymph through their thin walls.
2. Lymphatic vessels
   a. Lymphatic vessels have walls similar to veins and possess valves.
   b. Larger lymphatic vessels lead to lymph nodes and then merge to form lymphatic trunks.
3. Lymphatic trunks and collecting ducts
   a. Lymphatic trunks lead to two collecting ducts—the thoracic duct and the right lymphatic duct.
   b. Collecting ducts join the subclavian veins.

Tissue Fluid and Lymph (p. 386)

1. Tissue fluid formation
   a. Tissue fluid originates from blood plasma.
   b. It generally lacks large proteins, but some smaller proteins leak into interstitial spaces.
   c. As the protein concentration of tissue fluid increases, osmotic pressure increases.
2. Lymph formation
   a. Tissue fluid's rising osmotic pressure interferes with return of water to blood.
   b. Increasing pressure within interstitial spaces forces some tissue fluid into lymphatic capillaries, where it becomes lymph.
   c. Lymph returns protein molecules to the bloodstream and transports foreign particles to lymph nodes.

Lymph Movement (p. 386)

1. Lymph is under low pressure and may not flow readily without aid.
2. Contraction of skeletal muscles and of smooth muscles in the walls of lymphatic trunks and the low pressure in the thorax created by breathing movements move lymph.

Lymph Nodes (p. 387)

1. Structure of a lymph node
   a. Lymph nodes are subdivided into nodules.
   b. Nodules contain masses of lymphocytes and macrophages.
2. Locations of lymph nodes
   Lymph nodes are in groups along the paths of larger lymphatic vessels.
3. Functions of lymph nodes
   a. Lymph nodes filter potentially harmful particles from lymph.
   b. Lymph nodes are centers for lymphocyte production and also contain phagocytic cells.

Thymus and Spleen (p. 388)

1. Thymus
   a. The thymus is composed of lymphatic tissue subdivided into lobules.
   b. It slowly shrinks after puberty.
   c. Some lymphocytes leave the thymus and provide immunity.
2. Spleen
   a. The spleen resembles a large lymph node subdivided into lobules.
   b. Spaces within splenic lobules are filled with blood.
   c. The spleen contains many macrophages, which filter foreign particles and damaged red blood cells from blood.

Body Defenses against Infection (p. 390)

The body has nonspecific and specific defenses against infection.

Nonspecific Defenses (p. 391)

1. Species resistance
   Each species is resistant to certain diseases that may affect other species.
2. Mechanical barriers
   Mechanical barriers include the skin and mucous membranes, which block entrance of some pathogens.
3. Chemical barriers
   a. Enzymes in gastric juice and tears kill some pathogens.
   b. Interferons block viral replication, stimulate phagocytosis, and enhance the activities of cells that help resist infections and tumor growth.
4. Inflammation
   a. Inflammation is a tissue response to injury or infection, and includes localized redness, swelling, heat, and pain.
   b. Infected cells release chemicals that attract white blood cells to the site.
   c. Fibrous connective tissue may form a sac around injured tissue and thus block the spread of pathogens.
5. Phagocytosis
   a. The most active phagocytes in blood are neutrophils and monocytes. Monocytes give rise to macrophages, which remain fixed in tissues.
   b. Phagocytic cells are associated with the linings of blood vessels in the bone marrow, liver, spleen, lungs, and lymph nodes.

Specific Defenses (Immunity) (p. 391)

1. Lymphocyte origins
   a. Lymphocytes originate in red bone marrow, which releases undifferentiated lymphocytes into blood.
   b. Some lymphocytes reach the thymus, where they mature into T cells.
   c. Others, the B cells, mature in the red bone marrow.
   d. Both T cells and B cells reside in lymphatic tissues and organs.
2. Antigens
   a. Before birth, body cells inventory "self" proteins and other large molecules.
   b. After inventory, lymphocytes develop receptors that allow them to differentiate between nonself (foreign) and self antigens.

c. Nonself antigens combine with T cell and B cell surface receptors and stimulate these cells to cause an immune reaction.
d. Haptens are small molecules that can combine with larger ones to become antigenic.

3. Lymphocyte functions
   a. Some T cells interact with antigen-bearing agents directly, providing cell-mediated immunity.
   b. T cells secrete cytokines, such as interleukins, that enhance cellular responses to antigens.
   c. T cells may also secrete substances that are toxic to their target cells.
   d. B cells interact with antigen-bearing agents indirectly, providing antibody-mediated immunity (or humoral immunity).
   e. Varieties of T and B cells number in the millions.
   f. The members of each variety respond only to a specific antigen.
   g. As a group, the members of each variety form a clone.
4. B cell activation
   a. A B cell is activated when it encounters an antigen that fits its antigen receptors.
   b. An activated B cell proliferates (usually when stimulated by a T cell), enlarging its clone.
   c. Some activated B cells specialize into antibody-producing plasma cells.
   d. Antibodies react against the antigen-bearing agent that stimulated their production.
   e. An individual's diverse B cells defend against a very large number of pathogens.
5. Types of antibodies
   a. Antibodies are soluble proteins called immunoglobulins.
   b. The five major types of immunoglobulins are IgG, IgA, IgM, IgD, and IgE.
6. Antibody actions
   a. Antibodies attack antigens directly, activate complement, or stimulate local tissue changes that are unfavorable to antigen-bearing agents.
   b. Direct attacks occur by means of agglutination and precipitation, or by neutralization.
   c. Activated proteins of complement attract phagocytes, alter cells so they become more susceptible to phagocytosis, and rupture foreign cell membranes.
7. T cell activation
   a. T cells are activated when an antigen-presenting cell displays a foreign antigen.
   b. When a macrophage acts as an accessory cell, it phagocytizes an antigen-bearing agent, digests the agent, and displays the antigens on its surface membrane in association with certain MHC proteins.
   c. A helper T cell becomes activated when it encounters displayed antigens for which it is specialized to react.
   d. An activated helper T cell contacts a B cell that carries the foreign antigen that the T cell previously encountered on an antigen-presenting cell.
   e. In response, the T cell secretes cytokines, stimulates B cell proliferation, and attracts macrophages.
   f. Cytotoxic T cells recognize foreign antigens on tumor cells and virally infected cells.
   g. Memory T cells form but do not respond to the first exposure to the antigen.
8. Immune responses
   a. The first reaction to an antigen is called a primary immune response.
      (1) During this response, antibody production and release continues for several weeks.
      (2) Some B cells remain dormant as memory cells.
   b. A secondary immune response occurs rapidly as a result of memory cell response if the same antigen is encountered later.
9. Types of immunity
   a. A person who encounters a pathogen and has a primary immune response develops naturally acquired active immunity.
   b. A person who receives a vaccine containing a dead or weakened pathogen, or part of it, develops artificially acquired active immunity.
   c. A person who receives an injection of antibodies has artificially acquired passive immunity.
   d. When antibodies pass through a placental membrane from a pregnant woman to her fetus, the fetus develops naturally acquired passive immunity.
   e. Active immunity lasts much longer than passive immunity.
10. Allergic reactions
    a. Allergic reactions are misdirected immune responses that may damage tissue.
    b. Delayed-reaction allergy, which can occur in anyone and inflame the skin, results from repeated exposure to antigens.
    c. Immediate-reaction allergy is an inborn ability to overproduce IgE antibodies.
    d. Allergic reactions result from mast cells that release allergy mediators such as histamine.
    e. The released chemicals cause allergy symptoms: hives, hay fever, asthma, eczema, or gastric disturbances.
11. Transplantation and tissue rejection
    a. A transplant recipient's immune system may react against the donated tissue in a tissue rejection reaction.
    b. Matching donor and recipient tissues and using immunosuppressive drugs can minimize tissue rejection.
    c. Immunosuppressive drugs may increase susceptibility to infection.
12. Autoimmunity
    a. In autoimmune disorders, the immune system manufactures autoantibodies that attack one's own body tissues.
    b. Autoimmune disorders may result from a previous viral infection, faulty T cell development, or reaction to a nonself antigen that resembles a self antigen.

Review Exercises

1. Explain how the lymphatic system is related to the cardiovascular system. (p. 384)
2. Trace the general pathway of lymph from interstitial spaces to the bloodstream. (p. 384)
3. Describe the primary functions of lymph. (p. 386)
4. Explain why exercise promotes lymphatic circulation. (p. 386)

5. Describe the structure and functions of a lymph node. (p. 387)
6. Describe the structure and functions of the thymus. (p. 388)
7. Describe the structure and functions of the spleen. (p. 388)
8. Distinguish between specific and nonspecific body defenses against infection. (p. 390)
9. Explain *species resistance.* (p. 391)
10. Describe how enzymatic actions function as defense mechanisms. (p. 391)
11. Define *interferon,* and explain its action. (p. 391)
12. List and explain the major symptoms of inflammation. (p. 391)
13. Identify the major phagocytic cells in blood and other tissues. (p. 391)
14. Review the origin of T cells and B cells. (p. 391)
15. Distinguish between an antigen and an antibody. (p. 392)
16. Define *hapten.* (p. 392)
17. Define *cell-mediated immunity.* (p. 393)
18. Explain the function of *plasma cells.* (p. 393)
19. Define *antibody-mediated immunity.* (p. 393)
20. Define a *clone* of lymphocytes. (p. 393)
21. Explain how a B cell is activated. (p. 393)
22. List the major types of immunoglobulins, and describe where each occurs. (p. 393)
23. Explain two mechanisms by which antibodies attack antigens directly. (p. 395)
24. Explain the function of complement. (p. 395)
25. Describe how T cells become activated. (p. 396)
26. Explain the function of *memory cells.* (p. 397)
27. Distinguish between primary and secondary immune responses. (p. 397)
28. Distinguish between active and passive immunity. (p. 398)
29. Define *vaccine.* (p. 398)
30. Explain how a vaccine produces its effect. (p. 398)
31. Explain the relationship between an allergic reaction and an immune response. (p. 400)
32. Distinguish between an antigen and an allergen. (p. 400)
33. List the major events leading to a delayed-reaction allergic response. (p. 400)
34. Describe how an immediate-reaction allergic response may occur. (p. 400)
35. Explain the relationship between tissue rejection and an immune response. (p. 401)
36. Describe a method for reducing the severity of a tissue rejection reaction. (p. 402)
37. Explain the relationship between autoimmunity and an immune response. (p. 402)

Critical Thinking

1. The immune response is specific, diverse, and remembers. Give examples of each of these characteristics.
2. How should experimental AIDS vaccines be evaluated?
3. More people need transplants than organs are available. Discuss the pros and cons of the following proposed rationing systems for determining who should receive transplants: (a) first come, first served; (b) those with the best tissue and blood type match; (c) those whose need for an organ is caused by infection or disease, as opposed to people whose need for an organ was preventable, such as a lung destroyed by smoking; (d) the youngest; (e) the wealthiest; (f) the most important people.
4. Why is a transplant consisting of fetal tissue less likely to provoke an immune rejection response than tissue from an adult?
5. T cells "learn" to distinguish self from nonself during prenatal development. How could this learning process be altered to prevent allergies? To enable a person to accept a transplant?
6. Some parents keep their preschoolers away from other children to prevent them from catching illnesses. How might these well-meaning parents actually be harming their children?
7. One out of every 310,000 children who receives the vaccine for pertussis (whooping cough) develops permanent brain damage. The risk of suffering such damage from pertussis is about 1 in 30,000. Some parents refuse to vaccinate their children because of the few reported cases of adverse reaction to the vaccine. What are the dangers, both to the individual and to the population, when parents refuse to allow their children to be vaccinated against pertussis?
8. Xenotransplantation is the use of tissue from nonhuman animals to replace body parts in humans. For example, pigs are being developed to provide cardiovascular spare parts because their hearts and blood vessels are similar to ours. To increase the likelihood of such a xenotransplant working, researchers genetically engineer pigs to produce human antigens on their cell surfaces. How can this improve the chances of a human body not rejecting such a transplant?
9. How can removal of enlarged lymph nodes for microscopic examination aid in diagnosing certain diseases?
10. Why is an injection into the skin like an injection into the lymphatic system?
11. Why does vaccination provide long-lasting protection against a disease, while gamma globulin provides only short-term protection?
12. How would being born without a thymus affect susceptibility to infection?

Digestion and Nutrition

Peter Alindato does not fit the stereotype of a person with bulimia, the eating disorder that compels a person to gorge, and then vomit, take laxatives, or fanatically exercise. Peter did all three. About 95% of individuals with bulimia are young women, often teens, from middle- or upper-class homes. Peter was a boxer.

Bulimia abuses nutrition and the digestive system, yet its roots are psychological, attributed to a person's need to be in control. On the surface, that control is the maintenance of body weight; on a deeper level, bulimic behavior often expresses rage against past abuse. Psychologists estimate that up to half of all people with bulimia were sexually abused as children. Two men brutally raped Peter Alindato when he was eight, and he learned to deal with his anger at age eleven in the boxing ring.

In 1979, Peter competed at the Pan American games. One night he ate too much, so he stuck a finger down his throat and promptly returned his huge dinner. The next day, he passed the weigh-in easily. Laxatives and diuretics made it even easier to keep his weight down. Peter Alindato had bulimia.

Bulimia is much more prevalent than another eating disorder, anorexia nervosa (self-starvation). A person with bulimia is often of normal weight, but telltale signs include dental enamel loss, throat rawness, and scrapes on the fingers from inducing vomiting. Bulimia is fatal in 10% of cases, usually due to electrolyte imbalances caused by fluid depletion.

Peter Alindato suffered for ten years, sometimes vomiting more than ten times a day. Then he got help and recovered. Today, he lectures about his illness and has written a book about it. Bulimia is not a bad habit or a way to overeat while maintaining weight—it is an illness that can kill.

Photo: Boxer Peter Alindato overcame the eating disorder bulimia.

Many of us are fortunate enough to have available a vast and diverse supply of foods that can supply our bodies with the appropriate nutrients. Nutrients are food substances required to maintain health, and they include carbohydrates, lipids, proteins, vitamins, and minerals.

Because many food molecules are too large to enter cells, the organs of the digestive system mechanically and chemically break them down to a size that can cross cell membranes. Specifically, the digestive system ingests foods, breaks large particles into smaller ones, secretes enzymes that decompose food molecules, absorbs the products, and eliminates unused residues. Nutrition includes the processes that ingest, assimilate, and utilize nutrients.

Chapter Objectives

After studying this chapter, you should be able to:

1. Name and describe the locations and major parts of the organs of the digestive system. (p. 409)
2. Describe the general functions of each digestive organ. (p. 409)
3. Describe the structure of the wall of the alimentary canal. (p. 409)
4. Explain how the contents of the alimentary canal are mixed and moved. (p. 411)
5. List the enzymes the digestive organs and glands secrete, and describe the function of each. (p. 415)
6. Describe the mechanism of swallowing. (p. 416)
7. Describe the regulation of digestive secretions. (p. 419)
8. Explain how the products of digestion are absorbed. (p. 419)
9. Describe the defecation reflex. (p. 435)
10. List the major sources of carbohydrates, lipids, and proteins. (p. 435)
11. Describe how cells use carbohydrates, lipids, and proteins. (p. 435)
12. List the fat-soluble and water-soluble vitamins, and summarize the general functions of each vitamin. (p. 438)
13. List the major minerals and trace elements, and summarize the general functions of each. (p. 439)
14. Describe an adequate diet. (p. 443)

Aids to Understanding Words

aliment- [food] *aliment*ary canal: Tubelike portion of the digestive system.

chym- [juice] *chym*e: Semifluid paste of food particles and gastric juice formed in the stomach.

decidu- [falling off] *decidu*ous teeth: Teeth replaced during childhood.

gastr- [stomach] *gastr*ic gland: Portion of the stomach that secretes gastric juice.

hepat- [liver] *hepat*ic duct: Duct that carries bile from the liver to the common bile duct.

lingu- [tongue] *lingu*al tonsil: Mass of lymphatic tissue at the root of the tongue.

nutri- [nourishing] *nutri*ent: Chemical substance required to nourish body cells.

peri- [around] *peri*stalsis: Wavelike ring of contraction that moves material along the alimentary canal.

pylor- [gatekeeper] *pylor*ic sphincter: Muscle that serves as a valve between the stomach and small intestine.

vill- [hairy] *vill*i: Tiny projections of mucous membrane in the small intestine.

Key Terms

alimentary canal (al″ĭ-men′tar-e kah-nal′)
bile (bīl)
calorie (kal′o-re)
chyme (kīm)
feces (fe′sēz)
gastric juice (gas′trik jo͞os)
intestinal juice (in-tes′tĭ-nal jo͞os)
intestinal villus (in-tes′tĭ-nal vil′us); plural: villi (vil′i)
intrinsic factor (in-trin′sik fak′tor)
malnutrition (mal″nu-trish′un)
mesentery (mes′en-ter″e)
mineral (min′er-al)
nutrient (nu′tre-ent)
pancreatic juice (pan″kre-at′ik jo͞os)
peristalsis (per″ĭ-stal′sis)
vitamin (vi′tah-min)

Introduction

Digestion is the mechanical and chemical breakdown of foods into nutrients that cell membranes can absorb. The organs of the **digestive system** carry out these processes. The digestive system consists of the **alimentary canal**, which extends from the mouth to the anus, and several accessory organs, which release secretions into the canal. The alimentary canal includes the mouth, pharynx, esophagus, stomach, small intestine, large intestine, and anal canal; the accessory organs include the salivary glands, liver, gallbladder, and pancreas (fig. 15.1 and reference plates 4, 5, and 6). The digestive system makes nutrition possible.

General Characteristics of the Alimentary Canal

The alimentary canal is a 9-meter long muscular tube that passes through the body's ventral cavity. It is specialized in certain regions to carry on particular functions, but the structure of its wall, how it moves food, and its innervation are similar throughout its length (fig. 15.2).

Structure of the Wall

The wall of the alimentary canal consists of four distinct layers that are developed to different degrees from region to region. Beginning with the innermost tissues, these layers are (fig. 15.3):

1. **Mucosa (mucous membrane)** Surface epithelium, underlying connective tissue, and a small amount of smooth muscle form this layer. In some regions, the mucosa develops folds and tiny projections that extend into the passageway, or **lumen**, of the digestive tube and increase the mucosa's absorptive surface area. The mucosa also may contain glands that are tubular invaginations into which lining cells

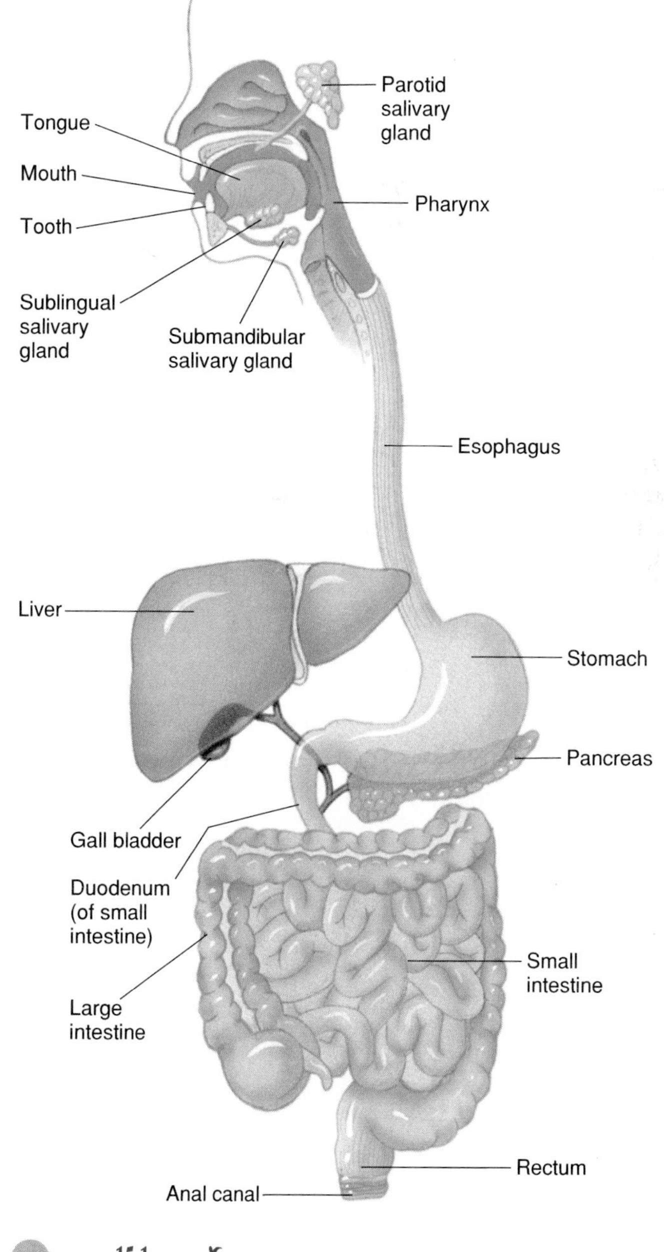

Figure 15.1

Major organs of the digestive system.

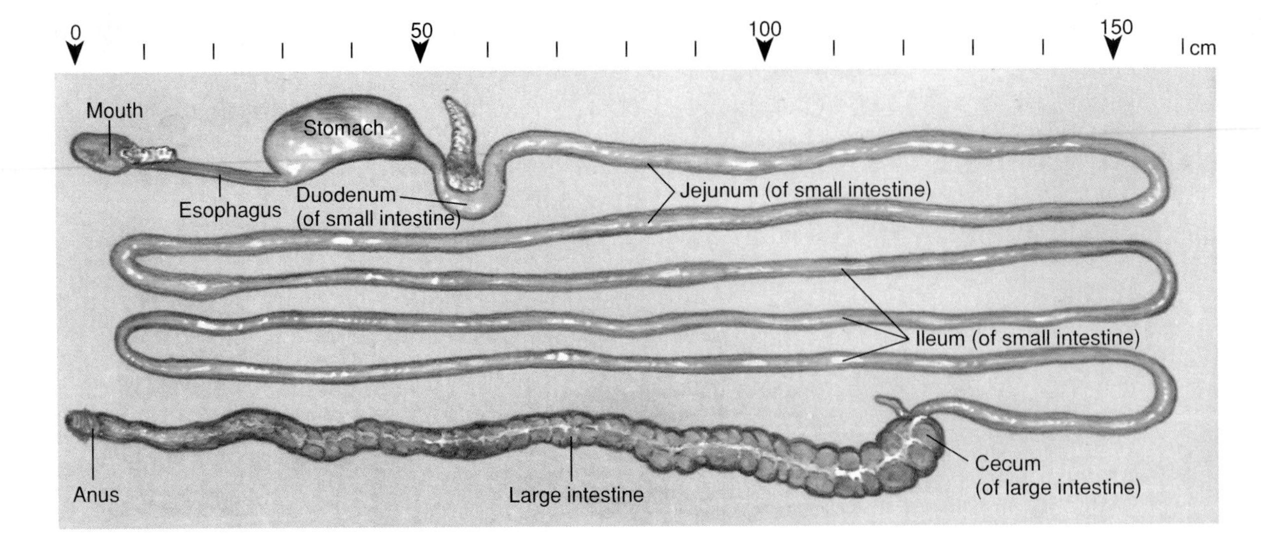

***Figure* 15.2**

The alimentary canal is a muscular tube about 9 meters long.

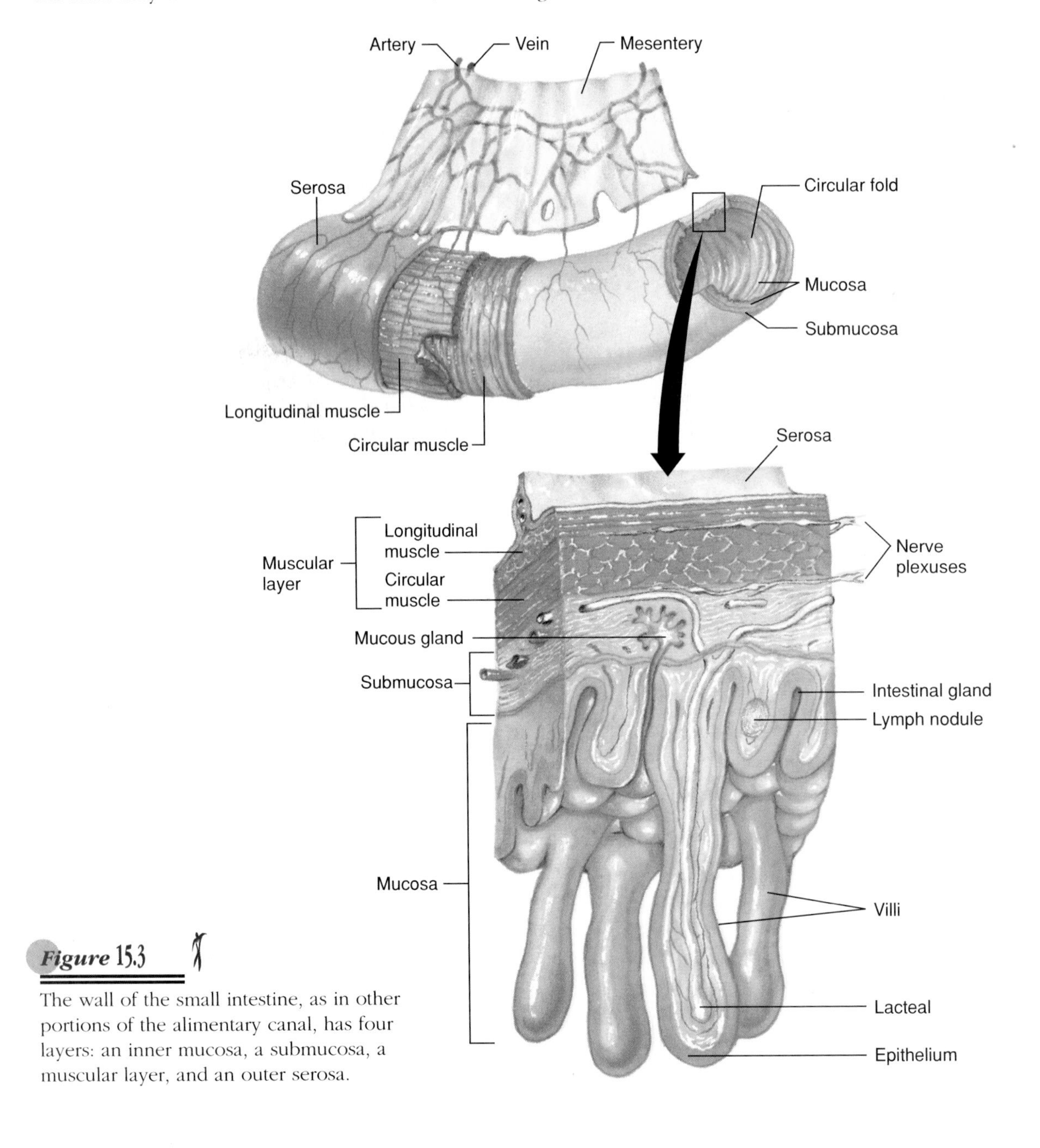

***Figure* 15.3**

The wall of the small intestine, as in other portions of the alimentary canal, has four layers: an inner mucosa, a submucosa, a muscular layer, and an outer serosa.

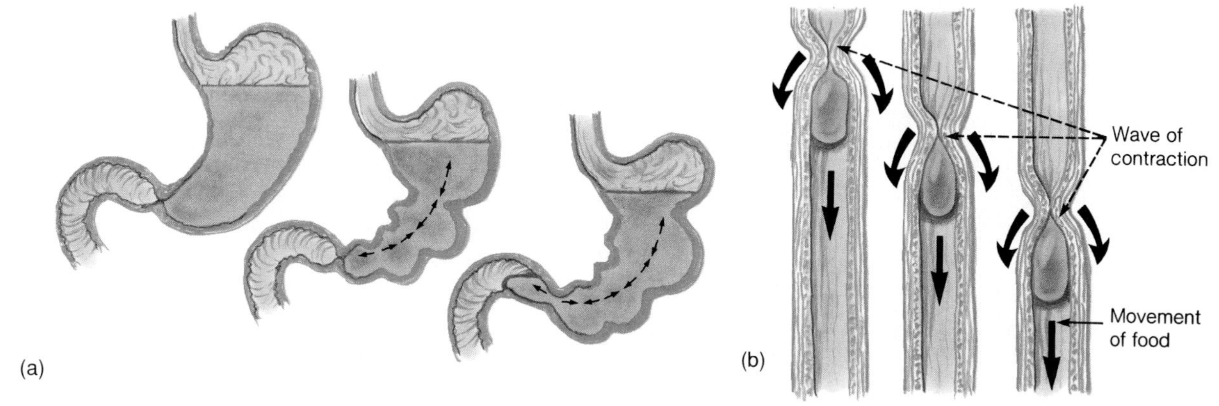

Figure 15.4

(*a*) Mixing movements occur when small segments of the muscular wall of the alimentary canal contract rhythmically. (*b*) Peristaltic waves move the contents along the canal.

secrete mucus and digestive enzymes. The mucosa secretes and absorbs, and also protects the tissues beneath it.

2. **Submucosa** The submucosa contains considerable loose connective tissue as well as glands, blood vessels, lymphatic vessels, and nerves organized into a network called a plexus. Its vessels nourish surrounding tissues and carry away absorbed materials.
3. **Muscular layer** This layer, which moves the tube, consists of two coats of smooth muscle tissue and some nerves organized into a plexus. The fibers of the inner coat encircle the tube. When these *circular fibers* contract, the tube's diameter decreases. The fibers of the outer muscular coat run lengthwise. When these *longitudinal fibers* contract, the tube shortens.
4. **Serosa (serous layer)** The *visceral peritoneum* comprises the serous layer, or outer covering of the tube. The cells of the serosa protect underlying tissues and secrete serous fluid, which moistens and lubricates the tube's outer surface so that organs within the abdominal cavity slide freely against one another.

Movements of the Tube

The motor functions of the alimentary canal are of two basic types—*mixing movements* and *propelling movements.* Mixing occurs when smooth muscles in small segments of the tube contract rhythmically. For example, when the stomach is full, waves of muscular contractions move along its walls from one end to the other. These waves mix food with digestive juices that the mucosa secretes (fig. 15.4*a*).

Propelling movements include a wavelike motion called **peristalsis.** When peristalsis occurs, a ring of contraction appears in the wall of the tube. At the same time, the muscular wall just ahead of the ring relaxes. As the peristaltic wave moves along, it pushes the tubular contents ahead of it (fig. 15.4*b*).

1. Which organs constitute the digestive system?
2. Describe the wall of the alimentary canal.
3. Describe the propelling movement in the alimentary canal.

Mouth

The **mouth** receives food and begins digestion by mechanically reducing the size of solid particles and mixing them with saliva. The lips, cheeks, tongue, and palate surround the mouth, which includes a chamber between the palate and tongue called the *oral cavity,* as well as a narrow space between the teeth, cheeks, and lips called the *vestibule* (fig. 15.5).

Cheeks and Lips

The **cheeks** consist of outer layers of skin, pads of subcutaneous fat, muscles associated with expression and chewing, and inner linings of moist stratified squamous epithelium. The **lips** are highly mobile structures that surround the mouth opening. They contain skeletal muscles and sensory receptors useful in judging food's temperature and texture. The lips' normal reddish color is due to the many blood vessels near their surfaces.

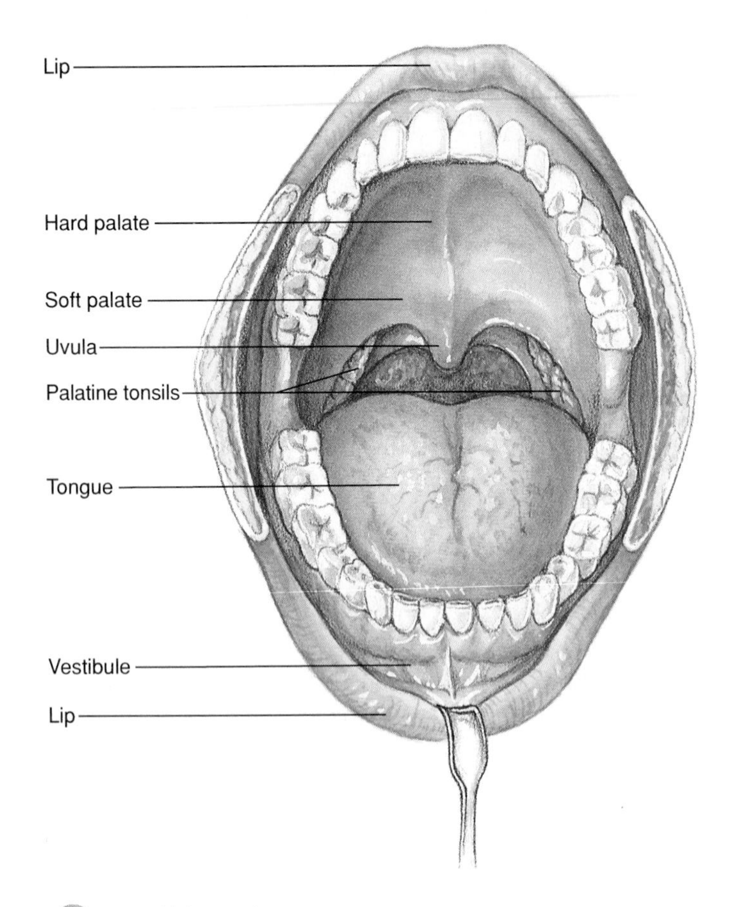

Figure 15.5

The mouth is adapted for ingesting food and preparing it for digestion.

During swallowing, muscles draw the soft palate and the uvula upward. This action closes the opening between the nasal cavity and the pharynx, preventing food from entering the nasal cavity.

In the back of the mouth, on either side of the tongue and closely associated with the palate, are masses of lymphatic tissue called **palatine tonsils** (see fig. 15.5). These structures lie beneath the epithelial lining of the mouth and, like other lymphatic tissues, help protect the body against infections.

The palatine tonsils are common sites of infections, and when inflamed, produce *tonsillitis*. Infected tonsils may swell so greatly that they block the passageways of the pharynx and interfere with breathing and swallowing. Because the mucous membranes of the pharynx, auditory tubes, and middle ears are continuous, such an infection can travel from the throat into the middle ears (*otitis media*).

When tonsillitis occurs repeatedly and does not respond to antibiotic treatment, the tonsils are sometimes removed surgically (*tonsillectomy*). Tonsillectomies are far less common today than they were a generation ago because tonsils' role in immunity is now known. ■

Other masses of lymphatic tissue, called **pharyngeal tonsils**, or *adenoids*, are on the posterior wall of the pharynx, above the border of the soft palate (fig. 15.6). If the adenoids enlarge and block the passage between the pharynx and the nasal cavity, they may be removed surgically.

1. How does the tongue function as part of the digestive system?
2. What is the soft palate's role in swallowing?
3. Where are the tonsils located?

Tongue

The **tongue** nearly fills the oral cavity when the mouth is closed. Mucous membrane covers the tongue, and a membranous fold called the **frenulum** connects the midline of the tongue to the floor of the mouth.

The *body* of the tongue is composed largely of skeletal muscle. These muscles mix food particles with saliva during chewing and move food toward the pharynx during swallowing. Rough projections called **papillae** on the tongue surface provide friction, which is useful in handling food. These papillae also contain taste buds (see chapter 10).

The posterior region, or *root,* of the tongue is anchored to the hyoid bone. It is covered with rounded masses of lymphatic tissue called **lingual tonsils** (fig. 15.6).

Palate

The **palate** forms the roof of the oral cavity and consists of a hard anterior part (*hard palate*) and a soft posterior part (*soft palate*). The soft palate forms a muscular arch, which extends posteriorly and downward as a cone-shaped projection called the **uvula.**

Teeth

Two different sets of **teeth** form during development. The members of the first set, the *primary teeth* (deciduous teeth) usually erupt through the gums at regular intervals between the ages of six months and two to four years (fig. 15.7). There are twenty deciduous teeth—ten in each jaw.

The primary teeth are usually shed in the same order they appeared. Before this happens, though, their roots are resorbed. Pressure from the developing *secondary teeth* (permanent teeth) then pushes the primary teeth out of their sockets. This secondary

Figure 15.6

Sagittal section of the mouth, nasal cavity, and pharynx.

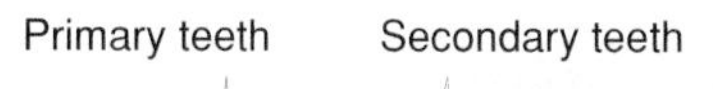

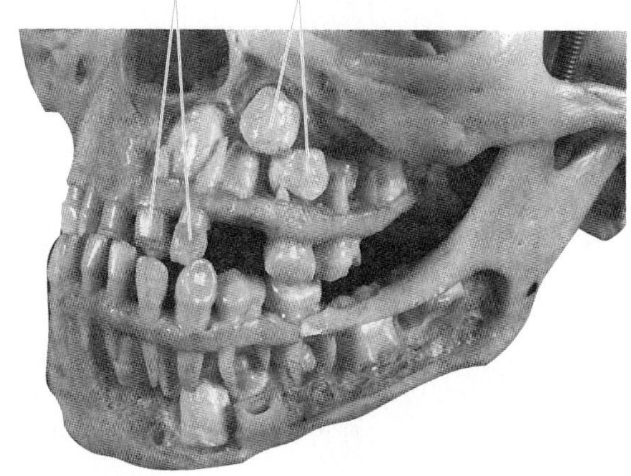

Figure 15.7

This child's skull reveals primary and secondary teeth developing in the maxilla and mandible.

set consists of thirty-two teeth—sixteen in each jaw (fig. 15.8). The secondary teeth usually begin to appear at six years, but the set may not be complete until the third molars appear between seventeen and twenty-five years.

Teeth break pieces of food into smaller pieces. This action increases the surface area of food particles and thus enables digestive enzymes to react more effectively with food molecules.

Different teeth are adapted to handle food in different ways. The *incisors* are chisel-shaped, and their sharp edges bite off large pieces of food. The *cuspids* are cone-shaped, and they grasp and tear food. The *bicuspids* and *molars* have somewhat flattened surfaces and grind food particles (fig. 15.8).

Table 15.1 summarizes the number and kinds of teeth that appear during development.

Each tooth consists of two main portions—the *crown,* which projects beyond the gum, and the *root,* which is anchored to the alveolar process of the jaw. Where these portions meet is called the *neck* of the tooth.

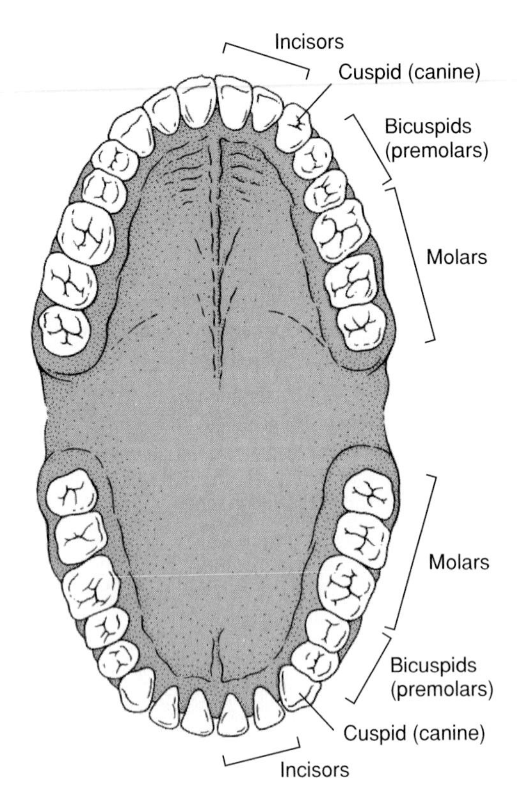

Figure 15.8

Secondary teeth of the upper and lower jaws.

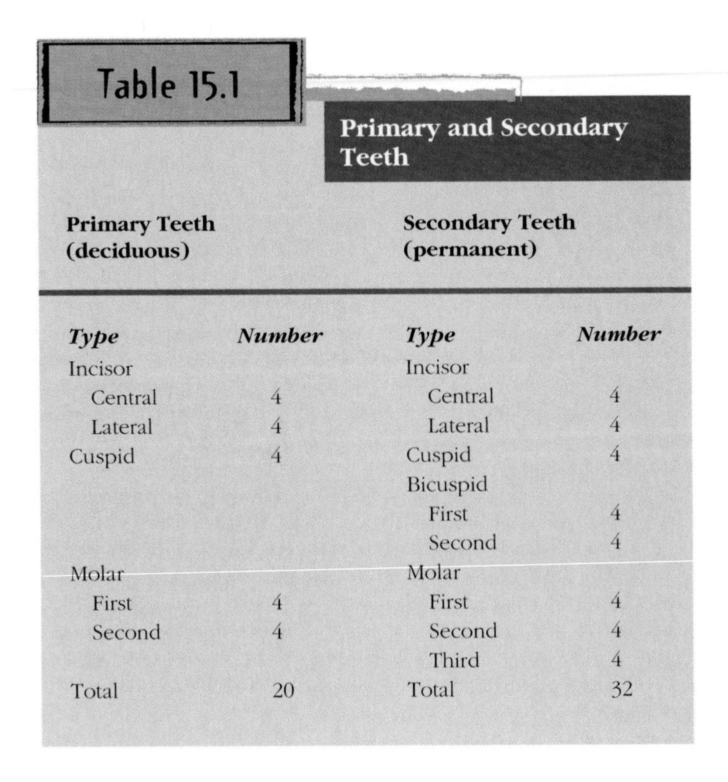

Table 15.1 Primary and Secondary Teeth

Primary Teeth (deciduous) Type	Number	Secondary Teeth (permanent) Type	Number
Incisor		Incisor	
Central	4	Central	4
Lateral	4	Lateral	4
Cuspid	4	Cuspid	4
		Bicuspid	
		First	4
		Second	4
Molar		Molar	
First	4	First	4
Second	4	Second	4
		Third	4
Total	20	Total	32

Glossy, white *enamel* covers the crown. Enamel consists mainly of calcium salts and is the hardest substance in the body. Unfortunately, if damaged by abrasive action or injury, enamel is not replaced.

The bulk of a tooth beneath the enamel is composed of *dentin,* a substance much like bone but somewhat harder. Dentin, in turn, surrounds the tooth's central cavity (pulp cavity), which contains a combination of blood vessels, nerves, and connective tissue called pulp. Blood vessels and nerves reach this cavity through tubular *root canals,* which extend into the root.

A thin layer of bonelike material called *cementum* encloses the root. A *periodontal ligament* surrounds the cementum. This ligament contains bundles of thick collagenous fibers, which pass between the cementum and the bone of the alveolar process, firmly attaching the tooth to the jaw. It also contains blood vessels and nerves (fig. 15.9).

1. How do primary teeth differ from secondary teeth?
2. Describe the structure of a tooth.
3. Explain how a tooth attaches to the bone of the jaw.

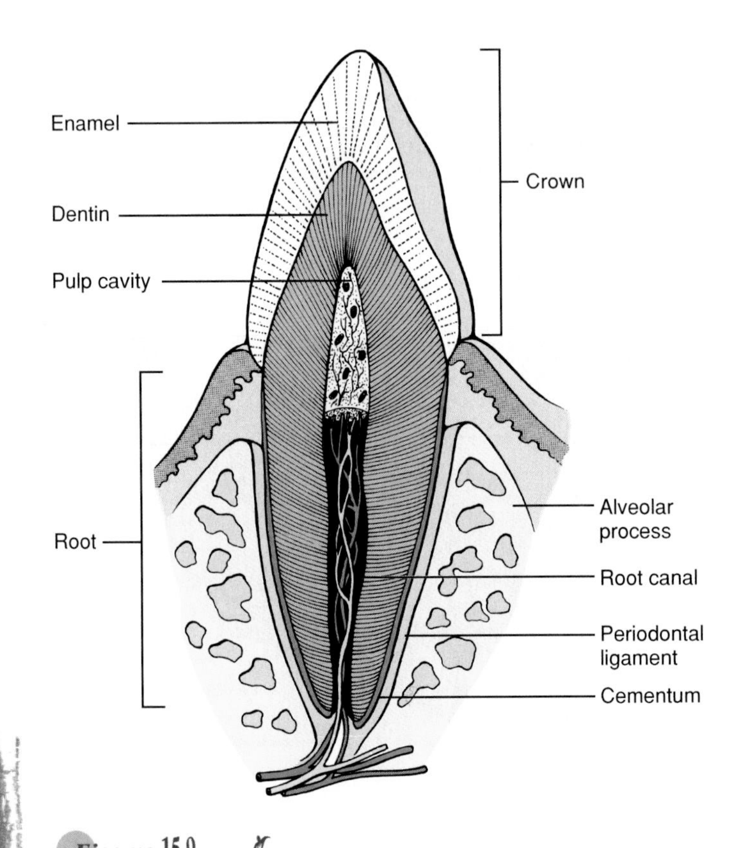

Figure 15.9

Section of a cuspid tooth.

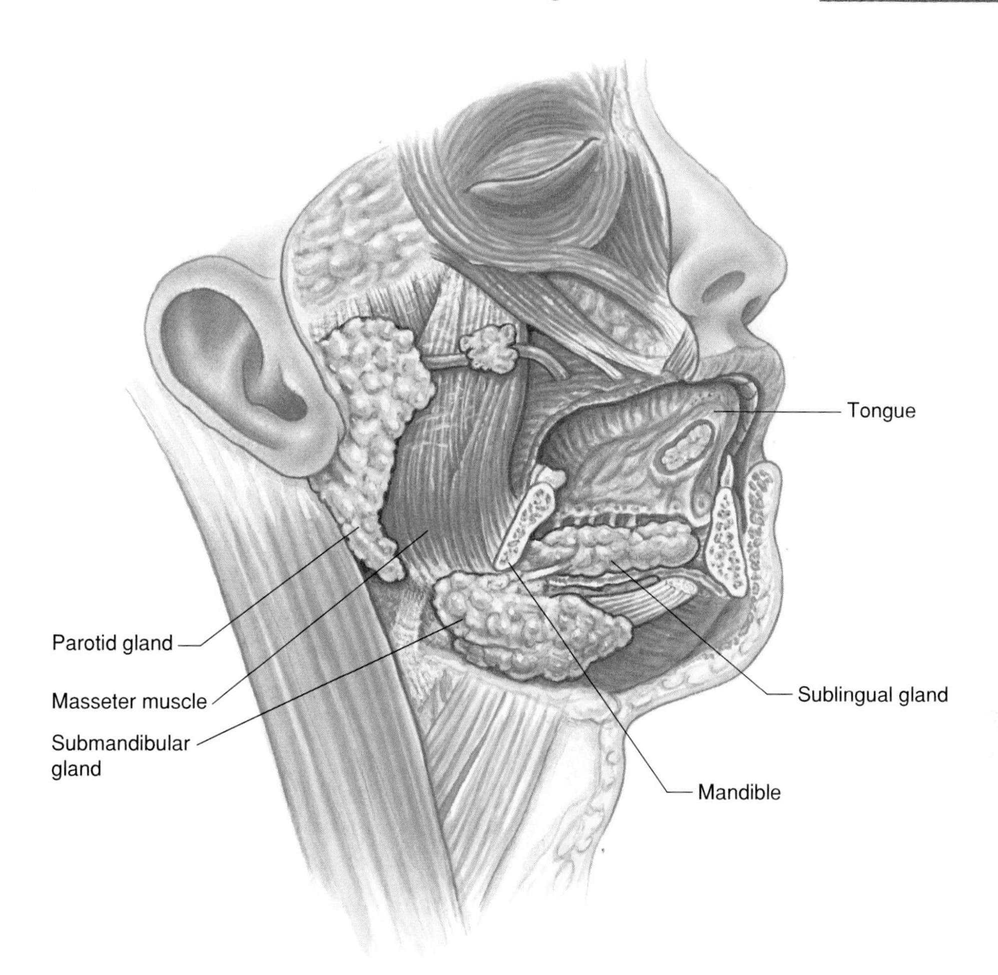

Figure 15.10
Locations of the major salivary glands.

Salivary Glands

The **salivary glands** secrete saliva. This fluid moistens food particles, helps bind them, and begins the chemical digestion of carbohydrates. Saliva is also a solvent, dissolving foods so that they can be tasted, and it helps cleanse the mouth and teeth.

Salivary Secretions

Within a salivary gland are two types of secretory cells—*serous cells* and *mucous cells*. These cells are present in varying proportions within different glands. Serous cells produce a watery fluid that contains the digestive enzyme **amylase.** This enzyme splits starch and glycogen molecules into disaccharides—the first step in the chemical digestion of carbohydrates. Mucous cells secrete a thick liquid called **mucus,** which binds food particles and lubricates during swallowing.

When a person sees, smells, tastes, or even thinks about pleasant food, parasympathetic nerve impulses elicit the secretion of a large volume of watery saliva. Conversely, food that looks, smells, or tastes unpleasant inhibits parasympathetic activity so that less saliva is produced, and swallowing may become difficult (see chapter 10).

Major Salivary Glands

Three pairs of major salivary glands—the parotid, submandibular, and sublingual glands—and many minor ones are associated with the mucous membranes of the tongue, palate, and cheeks (fig. 15.10). The **parotid glands** are the largest of the major salivary glands. Each gland lies in front of and somewhat below each ear, between the skin of the cheek and the masseter muscle. The parotid glands secrete a clear, watery fluid that is rich in amylase.

The **submandibular glands** are in the floor of the mouth on the inside surface of the lower jaw. The secretory cells of these glands are predominantly serous, with a few mucous cells. Consequently, the submandibular glands secrete a more viscous fluid than the parotid glands.

TOPIC OF INTEREST Dental Caries

At the movies, you eat a sticky candy bar. Inevitably, some of the caramel lodges between your teeth or in the crevices of your molars. The sticky snack feeds not only you, but also bacteria in your mouth. *Actinomyces, Streptococcus mutans,* and *Lactobacillus* metabolize carbohydrates in food, producing acid by-products that destroy tooth enamel and dentin. Sticky substances that the bacteria produce enable them to adhere to teeth.

If you eat candy bars but do not brush your teeth soon afterward, the actions of the acid-forming bacteria will produce *dental caries*. Unless a dentist cleans and fills the resulting cavity that forms where enamel is destroyed, the damage spreads to the underlying dentin. The tooth becomes very sensitive.

You can prevent dental caries in several ways:

1. Brush and floss teeth regularly.
2. See the dentist regularly.
3. Drink fluoridated water or receive a fluoride treatment. Fluoride incorporates into enamel's chemical structure, strengthening it.
4. Apply a sealant to children's and adolescents' teeth that have crevices that might entrap decay-causing bacteria. The sealant is a coating that keeps acids from eating away at tooth enamel.

The **sublingual glands**, the smallest of the major salivary glands, are on the floor of the mouth under the tongue. Their secretory cells are primarily the mucous type, making their secretions thick and stringy.

1. What is the function of saliva?
2. What stimulates salivary glands to secrete saliva?
3. Where are the major salivary glands located?

Pharynx and Esophagus

The **pharynx** is a cavity behind the mouth from which the tubular **esophagus** leads to the stomach (see fig. 15.1). The pharynx and esophagus do not digest food, but both are important passageways whose muscular walls function in swallowing.

Structure of the Pharynx

The pharynx connects the nasal and oral cavities with the larynx and esophagus. It has three parts (see fig. 15.6):

1. The **nasopharynx** communicates with the nasal cavity and provides a passageway for air during breathing.
2. The **oropharynx** opens behind the soft palate into the nasopharynx. It is a passageway for food moving downward from the mouth, and for air moving to and from the nasal cavity.
3. The **laryngopharynx**, just below the oropharynx, is a passageway to the esophagus.

Swallowing Mechanism

Swallowing reflexes have three stages. In the first stage, initiated voluntarily, food is chewed and mixed with saliva. Then the tongue rolls this mixture into a mass (bolus) and forces it into the pharynx.

The second stage begins as food stimulates sensory receptors around the pharyngeal opening. This triggers the swallowing reflex, which includes the following actions:

1. The soft palate raises, preventing food from entering the nasal cavity.
2. The hyoid bone and the larynx are elevated. A flaplike structure attached to the larynx, called the *epiglottis,* closes off the top of the trachea so that food is less likely to enter.
3. The tongue presses against the soft palate, sealing off the oral cavity from the pharynx.
4. The longitudinal muscles in the pharyngeal wall contract, pulling the pharynx upward toward the food.
5. Muscles in the lower portion of the pharynx relax, opening the esophagus.
6. A peristaltic wave begins in the pharyngeal muscles and forces food into the esophagus.

The swallowing reflex briefly inhibits breathing. Then, during the third stage of swallowing, peristalsis transports the food in the esophagus to the stomach.

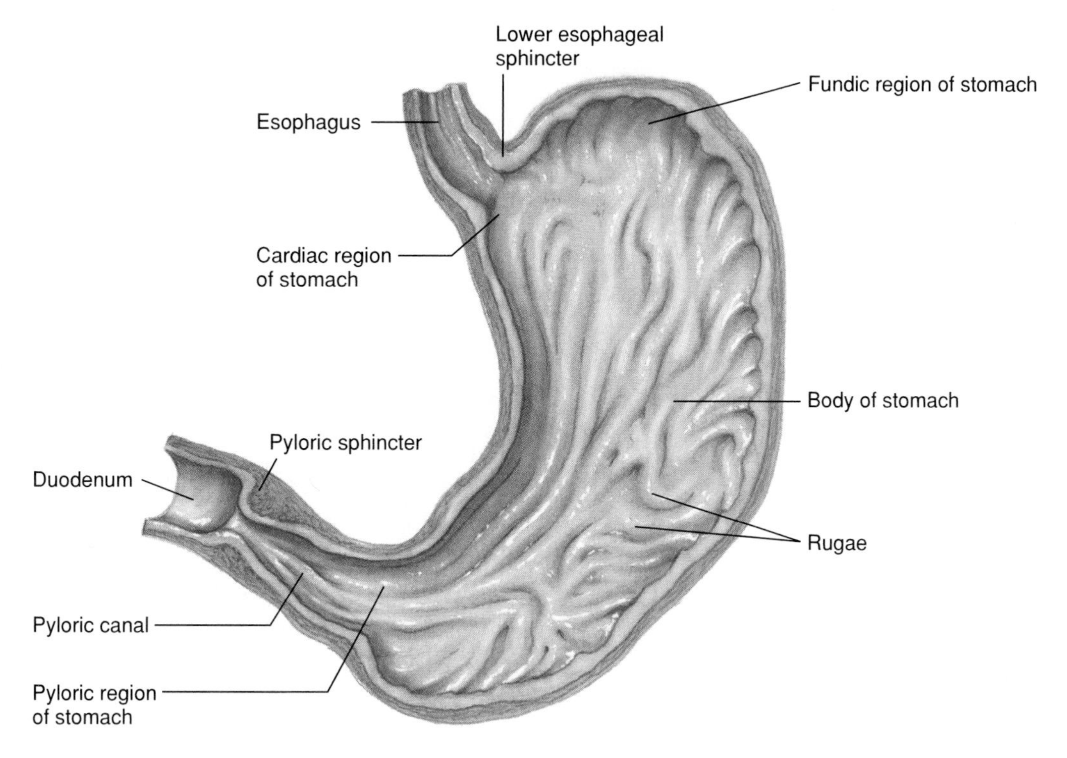

Figure 15.11

Major regions of the stomach.

Esophagus

The **esophagus**, a straight, collapsible tube about 25 centimeters long, is a food passageway from the pharynx to the stomach (see figs. 15.1 and 15.6). The esophagus begins at the base of the pharynx and descends behind the trachea, passing through the mediastinum. It penetrates the diaphragm through an opening, the *esophageal hiatus,* and is continuous with the stomach on the abdominal side of the diaphragm.

Just above where the esophagus joins the stomach, some circular smooth muscle fibers in the esophageal wall thicken, forming the *lower esophageal sphincter* (cardiac sphincter) (fig. 15.11). These fibers usually contract and close the entrance to the stomach, preventing regurgitation of stomach contents into the esophagus. When peristaltic waves reach the stomach, these muscle fibers relax and allow food to enter.

Mucous glands are scattered throughout the submucosa of the esophagus. Their secretions moisten and lubricate the tube's inner lining.

1. Describe the regions of the pharynx.
2. List the major events that occur during swallowing.
3. What is the function of the esophagus?

In a *hiatal hernia,* a portion of the stomach protrudes through a weakened area of the diaphragm, through the esophageal hiatus, and into the thorax. As a result of a hiatal hernia, regurgitation (reflux) of gastric juice into the esophagus may inflame the esophageal mucosa, causing heartburn, difficulty in swallowing, or ulceration and blood loss. In response to the destructive action of gastric juice, columnar epithelium may replace the squamous epithelium that normally lines the esophagus (see chapter 5). This condition, called *Barrett's esophagus,* increases the risk of developing esophageal cancer. ■

Stomach

The **stomach** is a J-shaped, pouchlike organ that hangs under the diaphragm in the upper left portion of the abdominal cavity and has a capacity of about 1 liter or more (figs. 15.1 and 15.11 and reference plates 4 and 5). Thick folds (rugae) of mucosal and submucosal layers mark the stomach's inner lining and disappear when the stomach wall distends. The stomach receives food from the esophagus, mixes the food with gastric juice, initiates protein digestion, does a limited amount of absorption, and moves food into the small intestine.

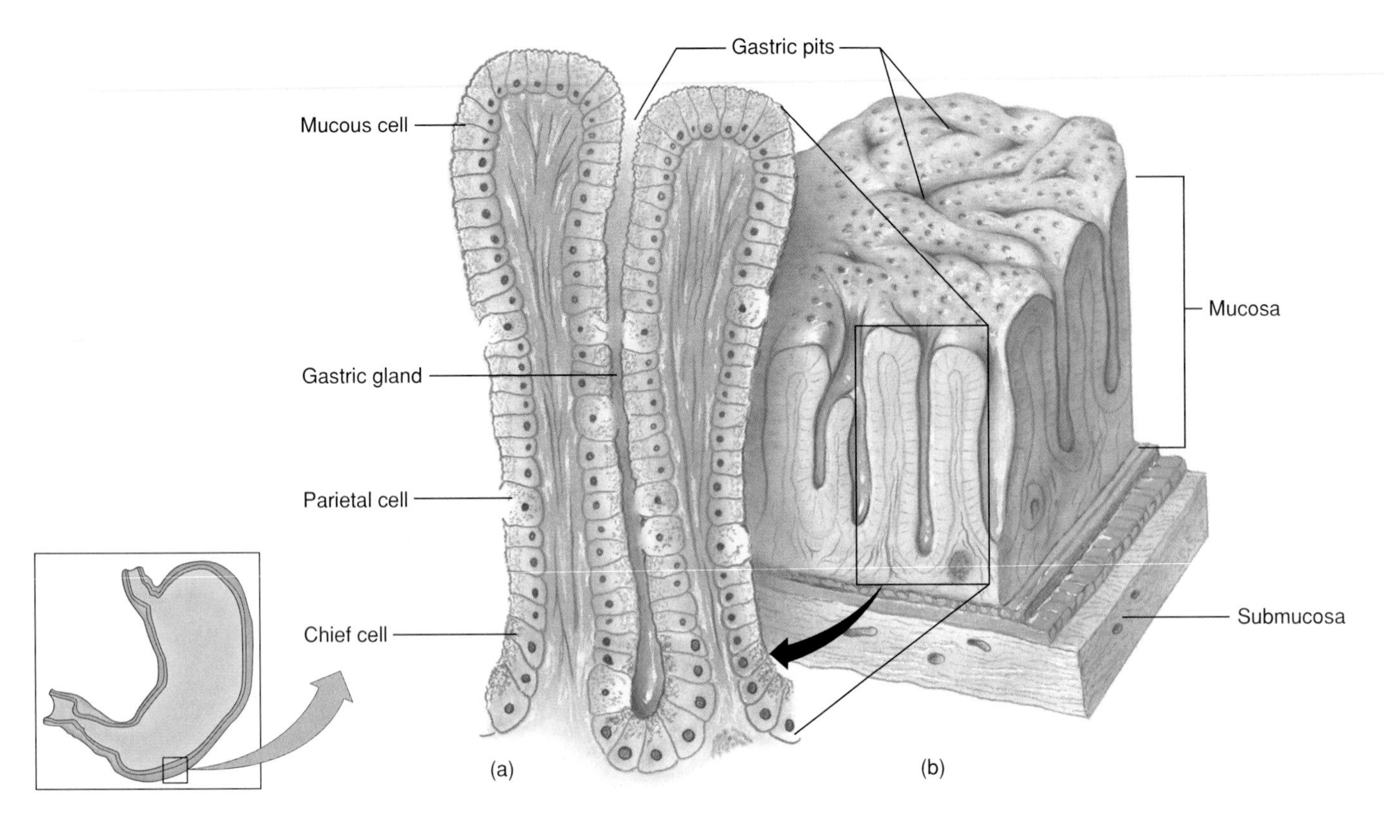

Figure 15.12

(*a*) Gastric glands include mucous cells, parietal cells, and chief cells. (*b*) Gastric pits are the openings of the gastric glands and stud the stomach mucosa.

Parts of the Stomach

The stomach is divided into the cardiac, fundic, body, and pyloric regions (fig. 15.11). The *cardiac region* is a small area near the esophageal opening. The *fundic region,* which balloons above the cardiac portion, is a temporary storage area. The dilated *body region,* which is the main part of the stomach, lies between the fundic and pyloric portions. The *pyloric region* narrows and becomes the *pyloric canal* as it approaches the small intestine.

At the end of the pyloric canal, the muscular wall thickens, forming a powerful circular muscle, the **pyloric sphincter** (pylorus). This muscle is a valve that controls gastric emptying.

Gastric Secretions

The mucous membrane that forms the stomach's inner lining is relatively thick, and many small openings called *gastric pits* stud its surface. Gastric pits are at the ends of tubular **gastric glands** (figs. 15.12 and 15.13).

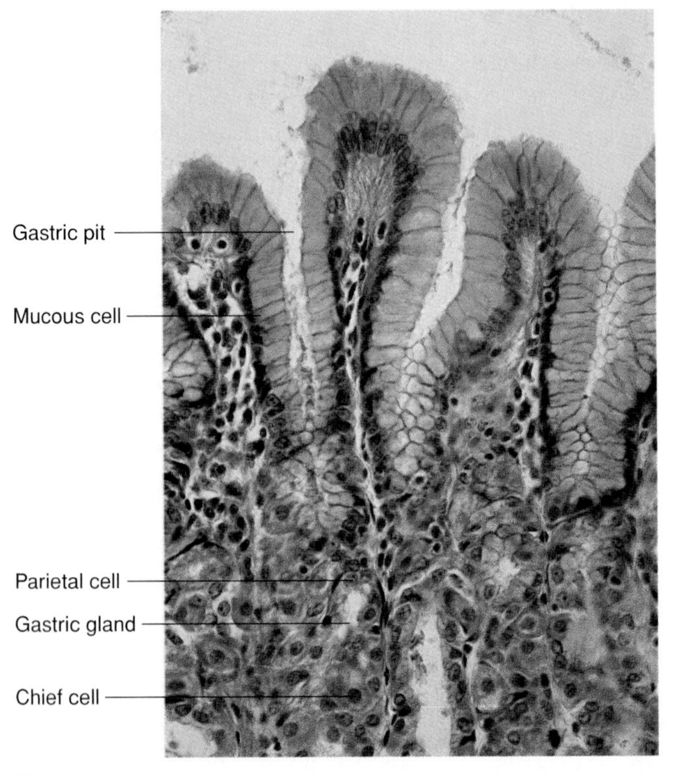

Figure 15.13

Light micrograph of cells associated with the gastric glands (50×).

Table 15.2

Major Components of Gastric Juice

Component	Source	Function
Pepsinogen	Chief cells of the gastric glands	Inactive form of pepsin
Pepsin	Formed from pepsinogen in the presence of hydrochloric acid	A protein-splitting enzyme that digests nearly all types of protein
Hydrochloric acid	Parietal cells of the gastric glands	Provides the acidic environment needed for the conversion of pepsinogen into pepsin and for the action of pepsin
Mucus	Goblet cells and mucous glands	Provides a viscous, alkaline protective layer on the inside stomach wall
Intrinsic factor	Parietal cells of the gastric glands	Aids in vitamin B_{12} absorption

Gastric glands generally contain three types of secretory cells. *Mucous cells* (goblet cells) occur in the necks of the glands, near the openings of the gastric pits. *Chief cells* and *parietal cells* are in the deeper parts of the glands. Chief cells secrete digestive enzymes, and parietal cells release hydrochloric acid. The products of the mucous cells, chief cells, and parietal cells form **gastric juice.**

Of the several digestive enzymes in gastric juice, **pepsin** is by far the most important. The chief cells secrete it as the inactive enzyme precursor **pepsinogen.** When pepsinogen contacts the hydrochloric acid from the parietal cells, part of it is snipped off rapidly, leaving pepsin. Pepsin begins the digestion of nearly all types of dietary protein. This enzyme is most active in an acidic environment, which the hydrochloric acid in gastric juice provides.

The mucous cells of the gastric glands secrete abundant thin mucus. In addition, the cells of the mucous membrane associated with the stomach's inner lining and between the gastric glands, release a more viscous and alkaline secretion, which coats the inside stomach wall. This coating normally prevents pepsin from digesting proteins in the stomach lining itself.

Another component of gastric juice is **intrinsic factor,** which the parietal cells secrete. Intrinsic factor is needed for vitamin B_{12} absorption from the small intestine.

Table 15.2 summarizes the components of gastric juice.

1. What do chief cells and parietal cells secrete?
2. Which is the most important digestive enzyme in gastric juice?
3. Why does the stomach not digest itself?

Regulation of Gastric Secretions

Gastric juice is produced continuously, but the rate varies considerably and is controlled both neurally and hormonally. When a person tastes, smells, or even sees pleasant food, or when food enters the stomach, parasympathetic impulses on the vagus nerves stimulate acetylcholine (Ach) release from nerve endings. This Ach release causes gastric glands to secrete large amounts of gastric juice, which is rich in hydrochloric acid and pepsinogen. These parasympathetic impulses also stimulate certain stomach cells to release the peptide hormone **gastrin,** which increases the secretory activity of gastric glands (fig. 15.14).

Food moving into the small intestine inhibits gastric juice secretion due to sympathetic nerve impulses that acid in the upper part of the small intestine triggers. Also, proteins and fats in this region of the intestine cause the intestinal wall to release the peptide hormone **cholecystokinin.** This hormonal action decreases gastric motility as the small intestine fills with food.

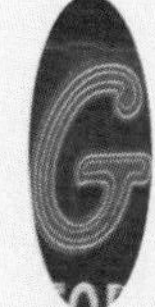

Gastrin stimulates cell growth in the mucosa of the stomach and intestines, except where gastrin is produced. This helps repair mucosal cells damaged by normal stomach function, disease, or medical treatments. ■

Gastric Absorption

Gastric enzymes begin breaking down proteins, but the stomach wall is not well-adapted to absorb digestive products. The stomach absorbs only small quantities of water and certain salts, as well as alcohol, and some lipid-soluble drugs.

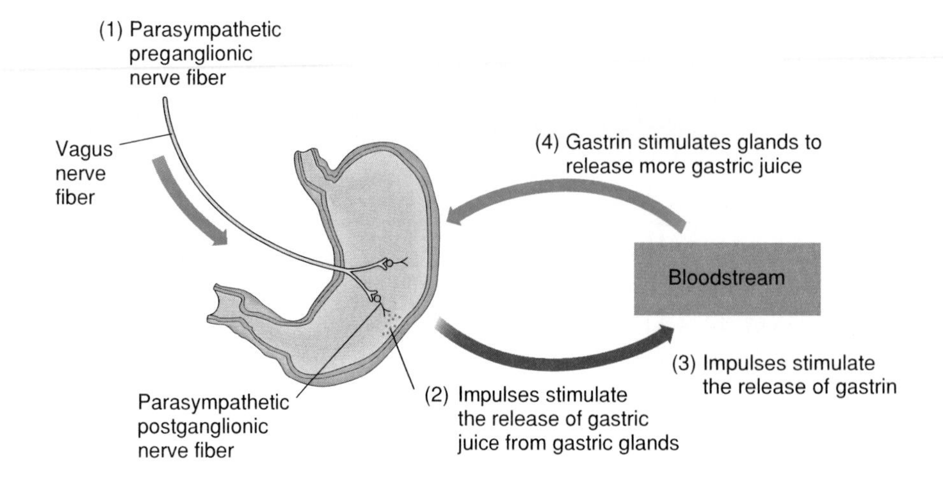

Figure 15.14

Parasympathetic nerve impulses that stimulate the release of gastric juice and gastrin partially regulate gastric juice secretion.

An *ulcer* is an open sore in the mucous membrane resulting from localized tissue breakdown. Peptic ulcers occur in the digestive tract. Types of peptic ulcers are gastric ulcers in the stomach, and duodenal ulcers in the region of the small intestine nearest the stomach.

For many years, both types of ulcers were attributed to stress and treated with medications to decrease stomach acid secretion. In 1982, two Australian researchers boldly suggested that stomach infection by the bacterium *Helicobacter pylori* causes gastric ulcers. When the medical community did not believe them, one of the researchers swallowed the bacteria to demonstrate its effects—and soon developed stomach pain. Still, it was twelve years before U.S. government physicians advised colleagues to treat gastric ulcers as an infection in people with evidence of *Helicobacter pylori.* Today, a short course of antibiotics, often combined with acid-lowering drugs, can cure a gastric ulcer. ■

1. What controls gastric juice secretion?
2. What is the function of cholecystokinin?
3. What substances can the stomach absorb?

Mixing and Emptying Actions

After a meal, the mixing movements of the stomach wall help produce a semifluid paste of food particles and gastric juice called **chyme.** Peristaltic waves push chyme toward the pyloric region of the stomach, and as chyme accumulates near the pyloric sphincter, this muscle begins to relax. This allows stomach contractions to push chyme a little at a time into the small intestine.

The rate at which the stomach empties depends on chyme's fluidity and the type of food present. For example, liquids usually pass through the stomach quite rapidly, but solids remain until they are well mixed with gastric juice. Fatty foods may remain in the stomach 3–6 hours; foods high in proteins move through more quickly; and carbohydrates usually pass through more rapidly than either fats or proteins.

As chyme enters the duodenum (the first portion of the small intestine), accessory organs add their secretions. These organs include the pancreas, liver, and gallbladder.

Vomiting results from a complex reflex that empties the stomach through the esophagus, pharynx, and mouth. Irritation or distension in the stomach or intestines can trigger vomiting. Sensory impulses travel from the site of stimulation to the *vomiting center* in the medulla oblongata, and several motor responses follow. These include taking a deep breath, raising the soft palate and thus closing the nasal cavity, closing the opening to the trachea (glottis), relaxing the circular muscle fibers at the base of the esophagus, contracting the diaphragm so it presses downward over the stomach, and contracting the abdominal wall muscles so that pressure inside the abdominal cavity increases. As a result, the stomach is squeezed from all sides, forcing its contents upward and out. ■

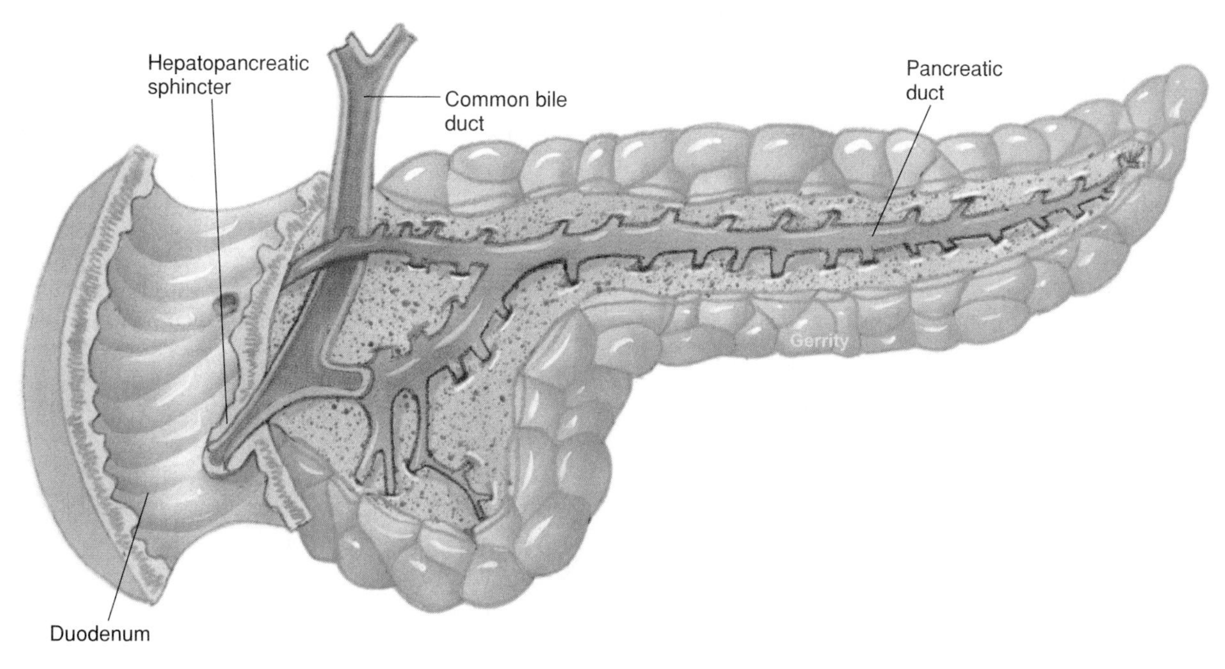

Figure 15.15

The pancreas is closely associated with the duodenum.

1. How is chyme produced?
2. What factors influence how quickly chyme leaves the stomach?

Pancreas

The **pancreas**, discussed as an endocrine gland in chapter 11, also has an exocrine function—secretion of a digestive juice called **pancreatic juice.**

Structure of the Pancreas

The pancreas is closely associated with the small intestine. It extends horizontally across the posterior abdominal wall in the C-shaped curve of the duodenum (figs. 15.1 and 15.15).

The cells that produce pancreatic juice, called *pancreatic acinar cells,* make up the bulk of the pancreas. These cells cluster around tiny tubes, into which they release their secretions. The smaller tubes unite to form larger ones, which, in turn, give rise to a *pancreatic duct* extending the length of the pancreas. The pancreatic duct usually connects with the duodenum at the same place where the bile duct from the liver and gallbladder joins the duodenum although other connections may be present (fig. 15.15). A hepatopancreatic sphincter controls the movement of pancreatic juices into the duodenum.

Pancreatic Juice

Pancreatic juice contains enzymes that digest carbohydrates, fats, nucleic acids, and proteins. The carbohydrate-digesting enzyme, **pancreatic amylase,** splits molecules of starch or glycogen into double sugars (disaccharides). The fat-digesting enzyme, **pancreatic lipase,** breaks triglyceride molecules into fatty acids and glycerol. Pancreatic juice also contains two **nucleases,** which are enzymes that break nucleic acid molecules into nucleotides.

The protein-splitting (proteolytic) enzymes are **trypsin, chymotrypsin,** and **carboxypeptidase.** These enzymes break the bonds between particular combinations of amino acids in proteins. Because no single enzyme can split all the possible amino acid combinations, complete digestion of protein molecules requires several enzymes.

The proteolytic enzymes are stored within tiny structures in cells called *zymogen granules.* These enzymes, like gastric pepsin, are secreted in inactive forms. After they reach the small intestine, other enzymes activate them. For example, pancreatic cells release inactive *trypsinogen,* which is activated to trypsin when it contacts the enzyme **enterokinase,** which the mucosa of the small intestine secretes.

A painful condition called *acute pancreatitis* results from blockage of pancreatic juice release. Trypsinogen, activated as pancreatic juice builds up, digests parts of the pancreas. Alcoholism, gallstones, certain infections, traumatic injuries, or the side effects of some drugs can cause pancreatitis. ■

In cystic fibrosis, abnormal chloride channels in cells in various tissues cause water to be drawn into the cells from interstitial spaces. This dries out secretions in the lungs and pancreas, leaving a very sticky mucus that impairs the functioning of these organs. When the pancreas is plugged with mucus, its secretions, containing digestive enzymes, cannot reach the duodenum. Individuals with cystic fibrosis must take digestive enzyme supplements—usually as a powder mixed with a soft food such as applesauce—to prevent malnutrition. ■

Regulation of Pancreatic Secretion

As with gastric and small intestine secretions, the nervous and endocrine systems regulate release of pancreatic juice. For example, when parasympathetic impulses stimulate gastric juice secretion, other parasympathetic impulses stimulate the pancreas to release digestive enzymes. Also, as acidic chyme enters the duodenum, the duodenal mucous membrane releases the peptide hormone **secretin** into the bloodstream (fig. 15.16). This hormone stimulates secretion of pancreatic juice that has a high concentration of bicarbonate ions. These ions neutralize the acid of chyme and provide a favorable environment for digestive enzymes in the intestine.

Proteins and fats in chyme within the duodenum also stimulate the intestinal wall to release cholecystokinin. Like secretin, cholecystokinin travels via the bloodstream to the pancreas. Pancreatic juice secreted in response to cholecystokinin has a high concentration of digestive enzymes.

1. List the enzymes in pancreatic juice.
2. What are the functions of the enzymes in pancreatic juice?
3. What regulates secretion of pancreatic juice?

Liver

The **liver** is in the upper right quadrant of the abdominal cavity, just below the diaphragm. It is partially surrounded by the ribs and extends from the level of the fifth intercostal space to the lower margin of the ribs. The reddish brown liver is well supplied with blood vessels (see fig. 15.1 and reference plates 4 and 5).

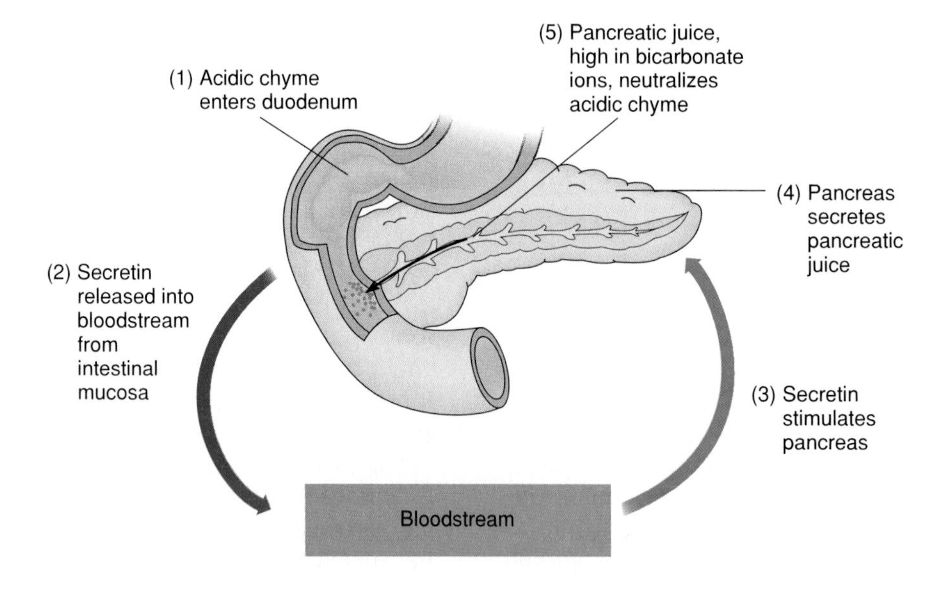

Figure 15.16

Acidic chyme entering the duodenum from the stomach stimulates release of secretin, which, in turn, stimulates release of pancreatic juice.

Liver Structure

A fibrous capsule encloses the liver, and connective tissue divides the organ into a large *right lobe* and a smaller *left lobe* (fig. 15.17). Each lobe is separated into many tiny **hepatic lobules,** which are the liver's functional units (figs. 15.18 and 15.19). A lobule consists of many hepatic cells radiating outward from a *central vein.* Vascular channels called **hepatic sinusoids** separate platelike groups of these cells from each other. Blood from the digestive tract, which is carried in the portal vein (see chapter 13), brings newly absorbed nutrients into the sinusoids and nourishes the hepatic cells.

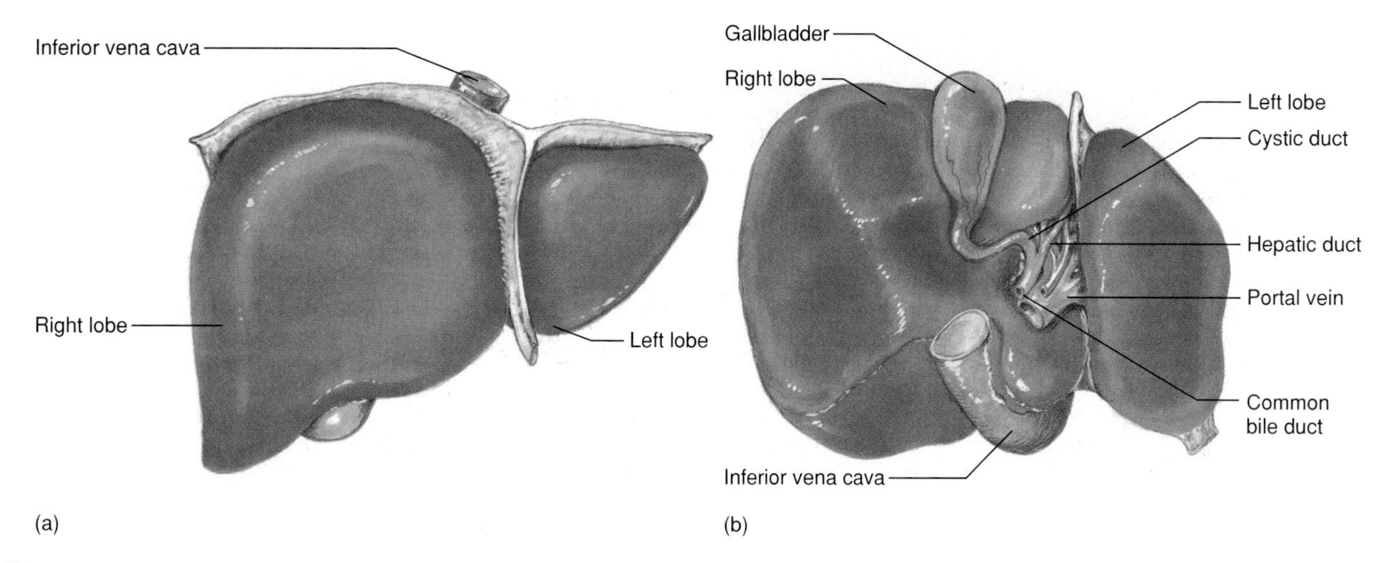

Figure 15.17

The lobes of the liver as viewed from (*a*) the front and (*b*) below.

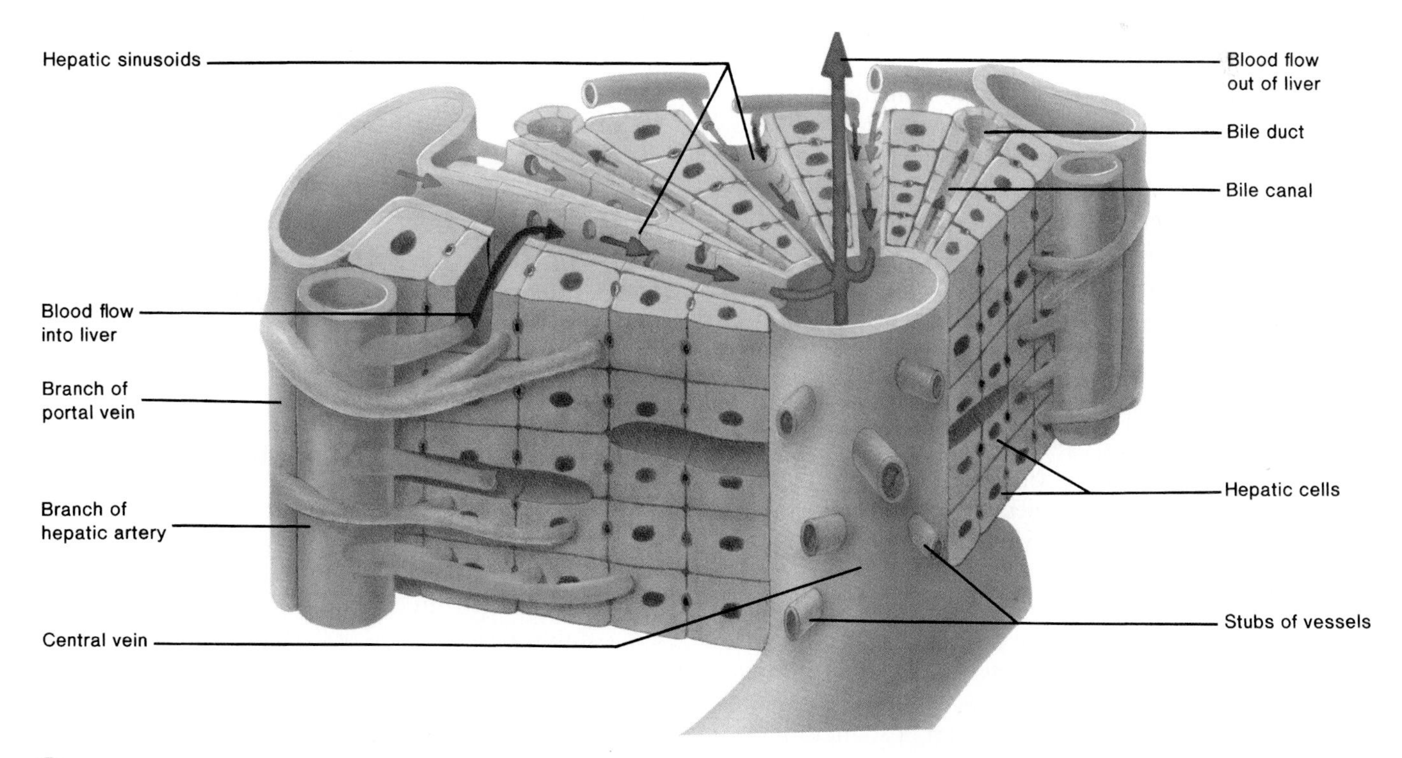

Figure 15.18

Cross section of a hepatic lobule, showing the paths of blood and bile through the lobule.

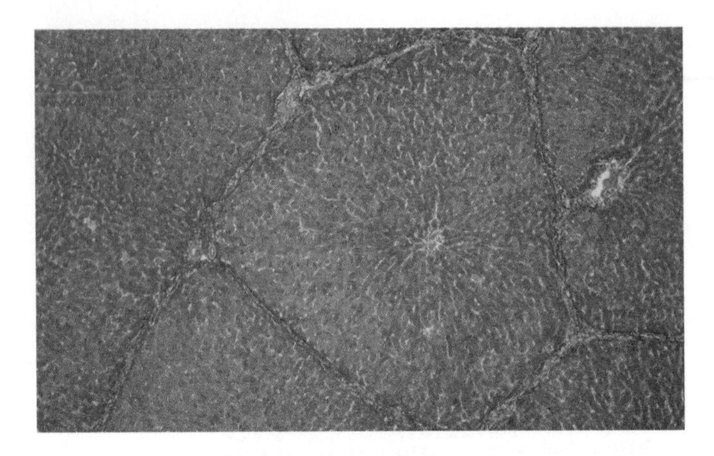

Figure 15.19
Light micrograph of a cross section of hepatic lobules (160×).

Large phagocytic macrophages called *Kupffer cells* are fixed to the inner linings of the hepatic sinusoids. They remove bacteria or other foreign particles that enter the blood in the portal vein through the intestinal wall. Blood passes from these sinusoids into the central veins of the hepatic lobules and exits the liver.

Within the hepatic lobules are many fine *bile canals,* which receive secretions from hepatic cells. The canals of neighboring lobules unite to form larger ducts and then converge to become the **hepatic ducts.** These ducts merge, in turn, to form the **common hepatic duct.**

Liver Functions

The liver has many important metabolic activities. Recall from chapter 11 that the liver plays a key role in carbohydrate metabolism by helping maintain the normal concentration of blood glucose. Liver cells responding to hormones such as insulin and glucagon decrease blood glucose by polymerizing glucose to glycogen, and increase blood glucose by breaking down glycogen to glucose or by changing noncarbohydrates into glucose.

The liver's effects on lipid metabolism include oxidizing fatty acids at an especially high rate (see chapter 4); synthesizing lipoproteins, phospholipids, and cholesterol; and changing portions of carbohydrate and protein molecules into fat molecules. Blood transports fats synthesized in the liver to adipose tissue for storage.

The most vital liver functions concern protein metabolism. They include deaminating amino acids; forming urea (see chapter 4); synthesizing plasma proteins, such as clotting factors (see chapter 12); and changing certain amino acids to other amino acids.

The liver also stores many substances, including glycogen, iron, and vitamins A, D, and B_{12}. In addition, macrophages in the liver help destroy damaged red blood cells and phagocytize foreign antigens. The liver also removes toxic substances such as alcohol from blood (detoxification) and secretes bile.

Many of these liver functions are not directly related to the digestive system and are discussed in other chapters. Bile secretion, however, is important to digestion and is explained later in this chapter.

Table 15.3 summarizes the major functions of the liver.

Table 15.3 Major Functions of the Liver

General Function	Specific Function
Carbohydrate metabolism	Polymerizes glucose to glycogen, breaks down glycogen to glucose, and changes noncarbohydrates to glucose
Lipid metabolism	Oxidizes fatty acids; synthesizes lipoproteins, phospholipids, and cholesterol; changes portions of carbohydrate and protein molecules into fats
Protein metabolism	Deaminates amino acids; forms urea; synthesizes plasma proteins; changes certain amino acids to other amino acids
Storage	Stores glycogen, iron, and vitamins A, D, and B_{12}
Blood filtering	Removes damaged red blood cells and foreign substances by phagocytosis
Detoxification	Removes toxins from blood
Secretion	Secretes bile

1. Describe the location of the liver.
2. Describe a hepatic lobule.
3. Review liver functions.

Composition of Bile

Bile is a yellowish green liquid that hepatic cells secrete continuously. In addition to water, bile contains *bile salts, bile pigments* (bilirubin and biliverdin), *cholesterol,* and *electrolytes.* Of these, bile salts are the most abundant and are the only bile substances with a digestive function. Bile pigments are products of red blood cell breakdown and are normally excreted in the bile (see chapter 12).

TOPIC OF INTEREST Hepatitis

Hepatitis is an inflammation of the liver. It has several causes, but the various types have similar symptoms.

For the first few days, hepatitis may resemble the flu, producing mild headache, low fever, fatigue, lack of appetite, nausea and vomiting, and perhaps, stiff joints. Between days 3 and 7, more distinctive symptoms appear; a rash, pain in the upper right quadrant of the abdomen, dark and foamy urine, and pale feces. About this time, the skin and sclera of the eyes begin to turn yellow from accumulating bile pigments (a form of jaundice). Great fatigue may continue for two or three weeks, and then gradually, the person begins to feel better.

This is hepatitis in its most common, least dangerous acute guise. About half a million people develop hepatitis in the United States each year, and 6,000 die. In a rare form called *fulminant hepatitis,* symptoms occur suddenly and severely, along with altered behavior and personality. Medical attention is necessary to prevent kidney or liver failure, or coma. Hepatitis that persists for more than six months is termed chronic. As many as 300 million people worldwide are hepatitis carriers. They do not have symptoms but can infect others. Five percent of carriers develop liver cancer.

Rarely, hepatitis results from alcoholism, autoimmunity, or use of certain drugs. Usually, one of five types of viruses, designated A through E in order of their discovery, cause hepatitis. The route of infection and the detection of a surface molecule specific to a particular virus type identify which virus type is causing a particular case of hepatitis. The viral types are distinguished as follows:

Hepatitis A This virus spreads by contact with food or objects contaminated with virus-containing feces. In day-care centers, it spreads through diaper changing. An outbreak affecting children in several states was traced to contaminated strawberries in school lunches. The course of hepatitis A is short and mild.

Hepatitis B This form of the illness spreads by contact with virus-containing body fluids, such as blood, saliva, or semen. It may be transmitted by blood transfusions, hypodermic needles, or sexual activity.

Hepatitis C The hepatitis C virus is believed to be responsible for about half of all cases of hepatitis. This virus is primarily transmitted in blood—from sharing razors or needles, from pregnant woman to fetus, or in blood transfusions or use of blood products. As many as 60% of individuals infected with the hepatitis C virus suffer chronic symptoms.

Hepatitis D This form of hepatitis occurs in people already infected with the hepatitis B virus. It is blood borne, and associated with blood transfusions and intravenous drug use. About 20% of individuals infected with this virus die.

Hepatitis E The hepatitis E virus is usually transmitted in water contaminated with feces in developing nations—not to residents, who are immune, but most often to visitors.

Because a virus usually causes hepatitis, antibiotic drugs, which are effective against bacteria, are not helpful. Usually, the person must just wait out the symptoms. Hepatitis C, however, sometimes responds to a form of interferon, an immune system biochemical.

Jaundice turns the skin and eye whites yellow. The distinctive skin color reflects buildup of bile pigments. The condition can have several causes. In *obstructive jaundice,* bile ducts are blocked. In *hepatocellular jaundice,* the liver is diseased. In *hemolytic jaundice,* red blood cells are destroyed too rapidly. ■

Gallbladder

The **gallbladder** is a pear-shaped sac in a depression on the liver's inferior surface. It connects to the **cystic duct,** which, in turn, joins the common hepatic duct (see figs. 15.1 and 15.21). The gallbladder is lined with epithelial cells and has a strong, muscular layer in its wall. It stores bile between meals, reabsorbs water to concentrate bile, and releases bile into the small intestine.

Cholesterol in bile may precipitate and form crystals called *gallstones* under certain conditions. Gallstones entering the common bile duct may block bile flow into the small intestine and cause considerable pain. A surgical procedure called *cholecystectomy* removes the gallbladder when gallstones are obstructive. The surgery is often performed with a laser, which shortens recovery time (fig. 15.20). ■

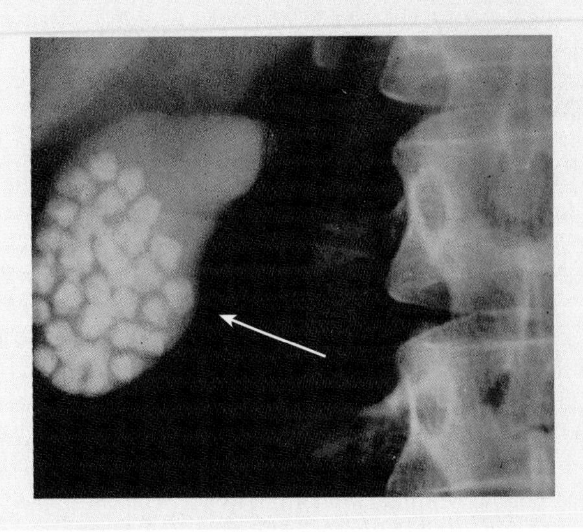

Figure 15.20

X-ray film of a gallbladder with gallstones (*arrow*).

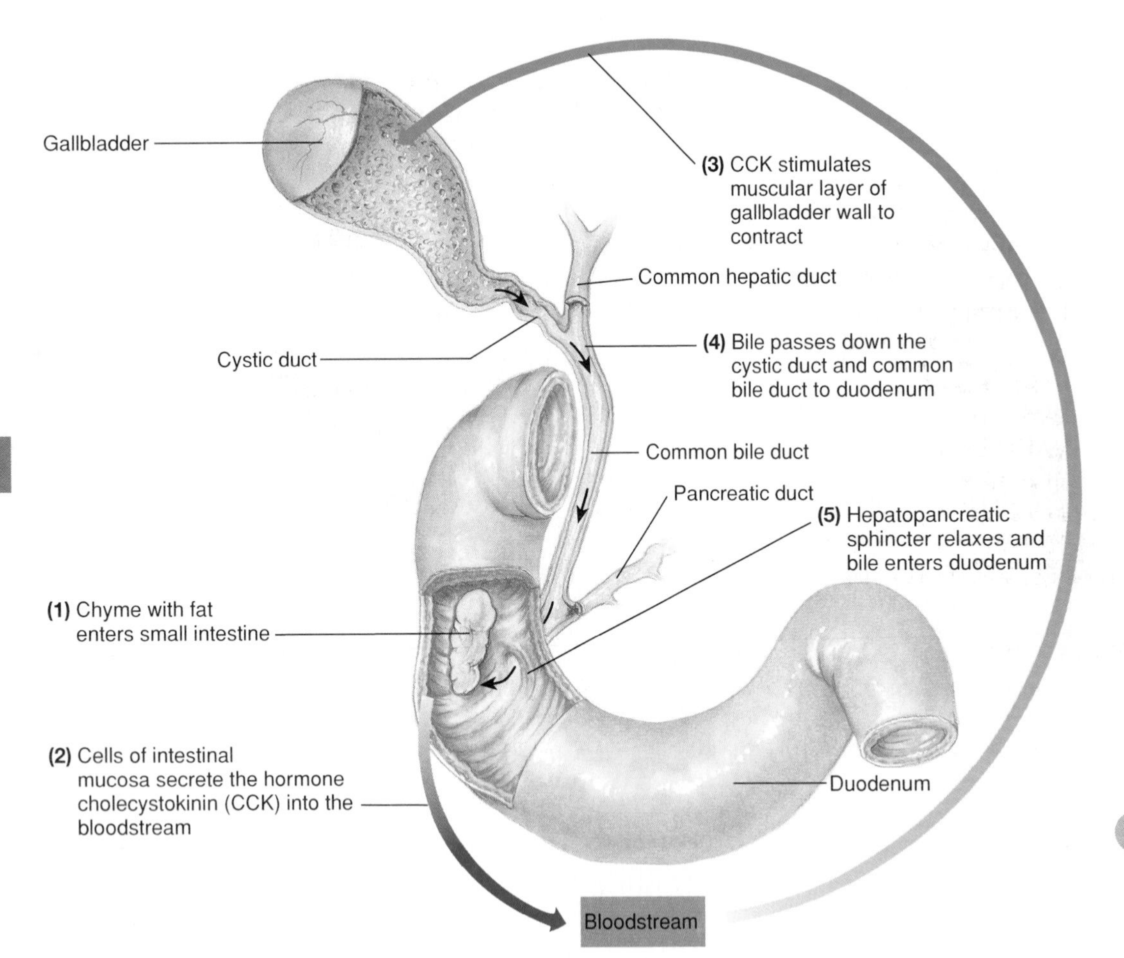

Figure 15.21

Fatty chyme entering the duodenum stimulates the gallbladder to release bile.

The common hepatic and cystic ducts join to form the common bile duct. It leads to the duodenum (see fig. 15.15), where the *hepatopancreatic sphincter* guards its exit. This sphincter normally remains contracted, so that bile collects in the common bile duct and backs up into the cystic duct. When this happens, bile flows into the gallbladder, where it is stored.

Regulation of Bile Release

Normally, bile does not enter the duodenum until cholecystokinin stimulates the gallbladder to contract. The intestinal mucosa releases this hormone in response to proteins and fats in the small intestine (recall its action to stimulate pancreatic enzyme secretion).

Table 15.4 Hormones of the Digestive Tract

Hormone	Source	Function
Gastrin	Gastric cells, in response to food	Causes gastric glands to increase their secretory activity
Cholecystokinin	Intestinal wall cells, in response to proteins and fats in the small intestine	Causes gastric glands to decrease their secretory activity and inhibits gastric motility; stimulates pancreas to secrete fluid with a high digestive enzyme concentration; stimulates gallbladder to contract and release bile
Secretin	Cells in the duodenal wall, in response to acidic chyme entering the small intestine	Stimulates pancreas to secrete fluid with a high bicarbonate ion concentration

The hepatopancreatic sphincter usually remains contracted until a peristaltic wave in the duodenal wall approaches it. Then, the sphincter relaxes, and bile squirts into the small intestine (fig. 15.21).

Table 15.4 summarizes the hormones that help control digestive functions.

Functions of Bile Salts

Bile salts aid digestive enzymes. Bile salts affect *fat globules* (clumped molecules of fats) much like a soap or detergent would affect them. That is, bile salts break fat globules into smaller droplets, an action called **emulsification.** Emulsification greatly increases the total surface area of the fatty substance, and the tiny droplets mix with water. Fat-splitting enzymes (lipases) can then digest fat molecules more effectively.

Bile salts also enhance absorption of fatty acids, cholesterol, and the fat-soluble vitamins A, D, E, and K. Lack of bile salts results in poor lipid absorption and vitamin deficiencies.

1. Explain how bile originates.
2. Describe the function of the gallbladder.
3. How is bile secretion regulated?
4. How do bile salts function in digestion?

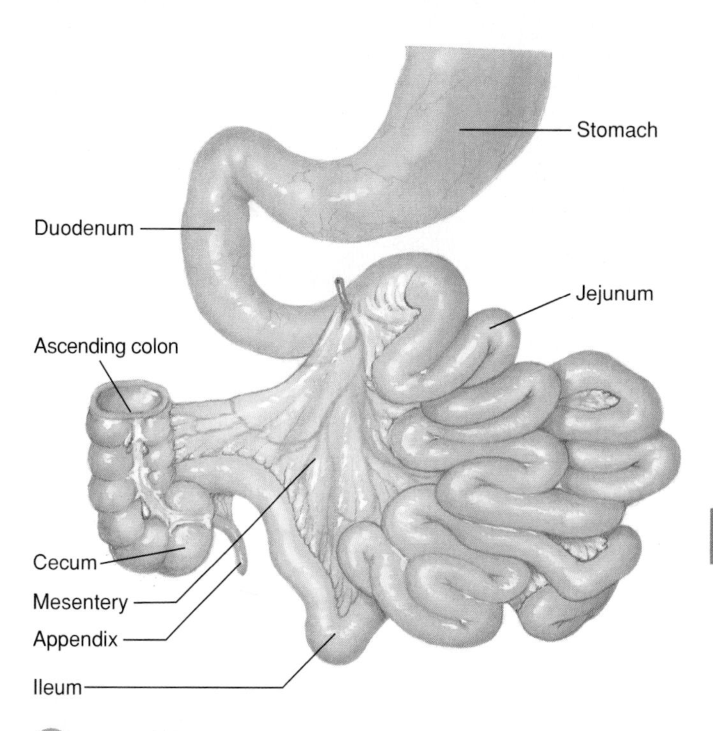

Figure 15.22

The three parts of the small intestine are the duodenum, the jejunum, and the ileum.

Small Intestine

The **small intestine** is a tubular organ that extends from the pyloric sphincter to the beginning of the large intestine. With its many loops and coils, it fills much of the abdominal cavity (see fig. 15.1 and reference plates 4 and 5).

The small intestine receives secretions from the pancreas and liver. It also completes digestion of the nutrients in chyme, absorbs the products of digestion, and transports the residues to the large intestine.

Parts of the Small Intestine

The small intestine consists of three portions: the duodenum, the jejunum, and the ileum (figs. 15.22 and 15.23). The duodenum, which is about 25 centimeters long and 5 centimeters in diameter, lies behind the parietal peritoneum and is the most fixed portion of

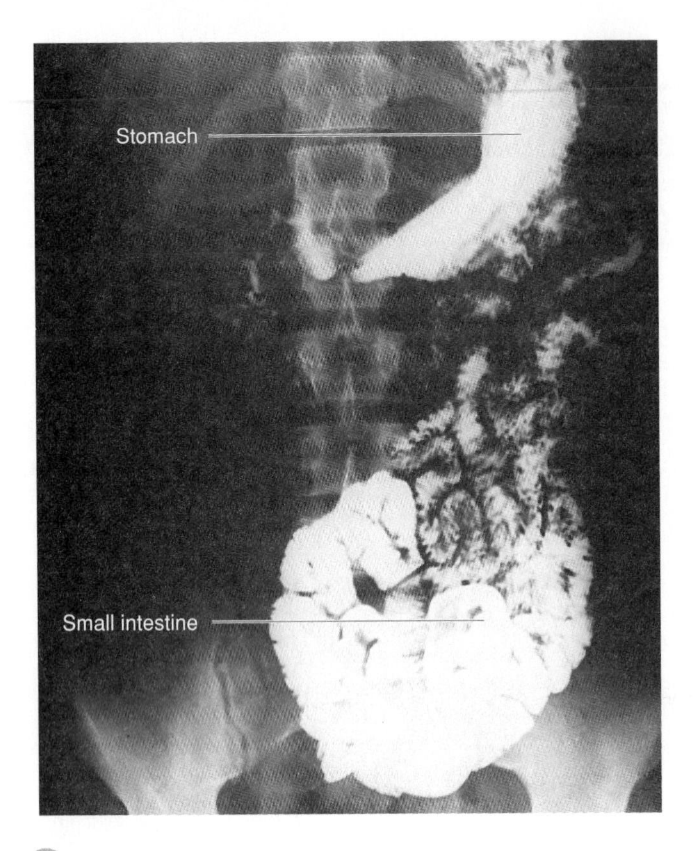

Figure 15.23

X-ray film showing a normal small intestine containing a radiopaque substance that the patient ingested.

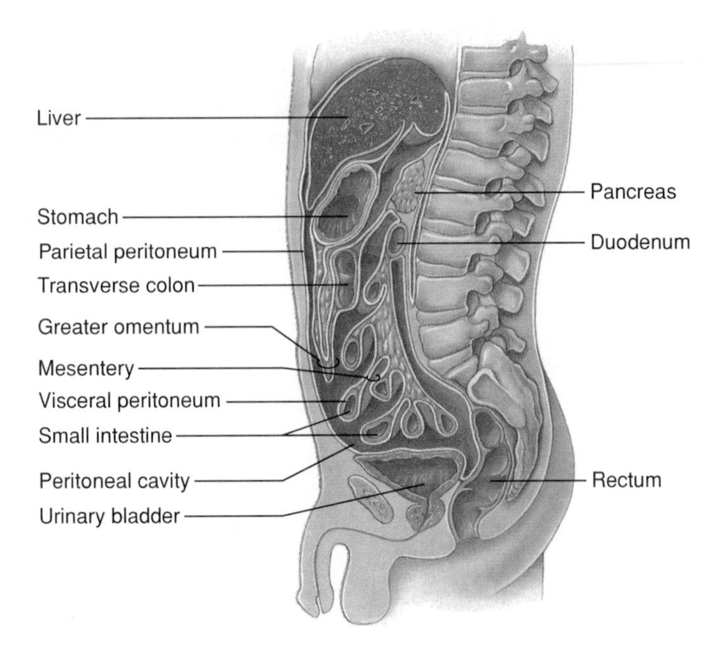

Figure 15.24

Mesentery formed by folds of the peritoneal membrane suspends portions of the small intestine from the posterior abdominal wall.

the small intestine. It follows a C-shaped path as it passes in front of the right kidney and the upper three lumbar vertebrae.

The remainder of the small intestine is mobile and lies free in the peritoneal cavity. The proximal two-fifths of this portion is the **jejunum**, and the remainder is the **ileum.** A double-layered fold of peritoneal membrane called **mesentery** suspends these portions from the posterior abdominal wall (figs. 15.22 and 15.24). The mesentery supports the blood vessels, nerves, and lymphatic vessels that supply the intestinal wall. The jejunum and ileum are not distinctly separate parts; however, the diameter of the jejunum is greater and its wall thicker, more vascularized, and more active than that of the ileum.

A filmy, double-layered fold of peritoneal membrane called the *greater omentum* drapes like an apron from the stomach over the transverse colon and the folds of the small intestine. If infections occur in the wall of the alimentary canal, cells from the omentum may adhere to the inflamed region and help wall it off so that the infection is less likely to enter the peritoneal cavity (fig. 15.24).

Structure of the Small Intestine Wall

Throughout its length, the inner wall of the small intestine appears velvety due to many tiny projections of mucous membrane called **intestinal villi** (figs. 15.3, 15.25, and 15.26). These structures are densest in the duodenum and the proximal portion of the jejunum. They project into the lumen of the alimentary canal, contacting the intestinal contents. Villi greatly increase the surface area of the intestinal lining, aiding absorption of digestive products.

Each villus consists of a layer of simple columnar epithelium and a core of connective tissue containing blood capillaries, a lymphatic capillary called a *lacteal,* and nerve fibers. Blood and lymph capillaries carry away absorbed nutrients, and nerve fibers transmit impulses to stimulate or inhibit villus activities. Between the bases of adjacent villi are tubular **intestinal glands** that extend downward into the mucous membrane (figs. 15.3, 15.25, and 15.26).

The epithelial cells that form the lining of the small intestine are continually replaced. New cells form within the intestinal glands by mitosis and migrate outward onto the villus surface. When the migrating cells reach the tip of the villus, they are shed. This *cellular turnover* renews the small intestine's epithelial lining every three to six days. ■

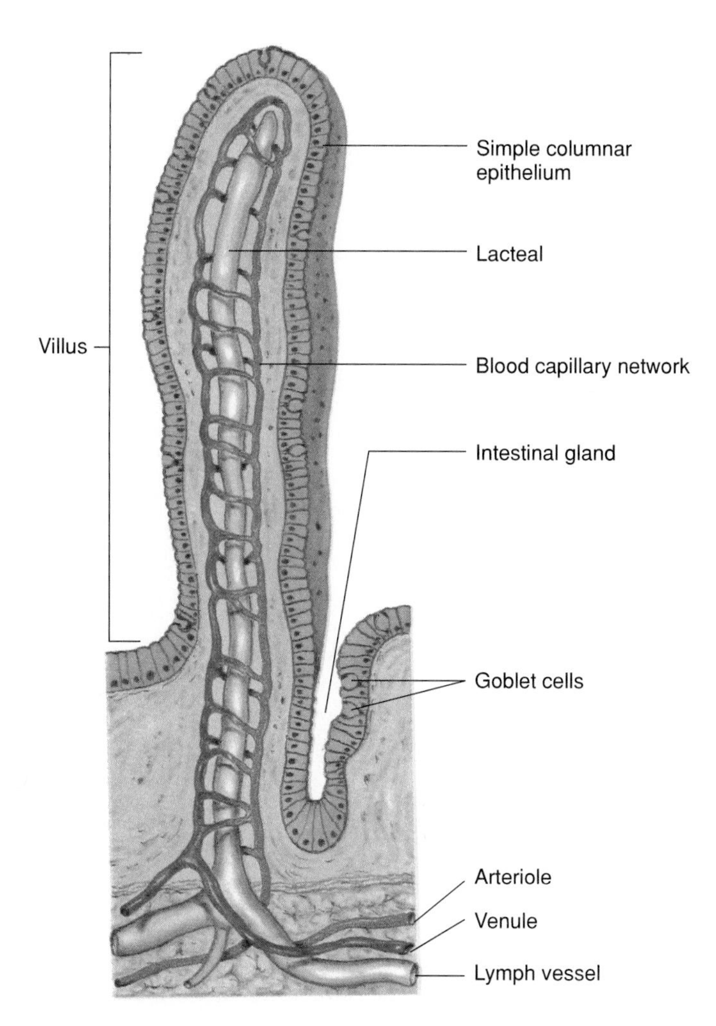

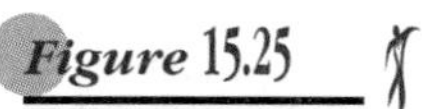
Figure 15.25

Structure of a single intestinal villus.

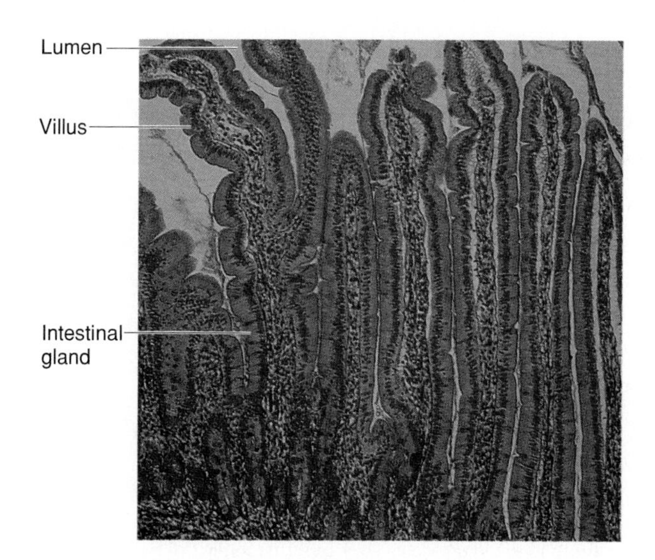

Figure 15.26

Light micrograph of intestinal villi in the wall of the duodenum (100×).

Secretions of the Small Intestine

Mucus-secreting goblet cells are abundant throughout the mucosa of the small intestine. In addition, many specialized *mucus-secreting glands* in the submucosa within the proximal duodenum secrete large quantities of thick, alkaline mucus in response to certain stimuli.

The intestinal glands at the bases of the villi secrete large amounts of a watery fluid. The villi's rapid reabsorption of this fluid provides a vehicle for moving digestive products into the villi. The fluid the intestinal glands secrete has a nearly neutral pH (6.5–7.5), and it lacks digestive enzymes. The epithelial cells of the intestinal mucosa, however, have digestive enzymes embedded in the membranes of their microvilli on their luminal surfaces. These enzymes break down food molecules just before absorption. The enzymes include **peptidases**, which split peptides into amino acids; **sucrase**, **maltase**, and **lactase**, which split the double sugars (disaccharides) sucrose, maltose, and lactose into the simple sugars (monosaccharides) glucose, fructose, and galactose; and **intestinal lipase**, which splits fats into fatty acids and glycerol.

Table 15.5 summarizes the sources and actions of the major digestive enzymes.

Many adults do not produce sufficient lactase to adequately digest lactose, or milk sugar. In this *lactose intolerance,* lactose remains undigested, increasing osmotic pressure of the intestinal contents and drawing water into the intestines. At the same time, intestinal bacteria metabolize undigested sugar, producing organic acids and gases. The overall result of lactose intolerance is bloating, intestinal cramps, and diarrhea. To avoid these unpleasant symptoms, people with lactose intolerance can take lactase in pill form before eating dairy products. Infants with lactose intolerance can drink formula based on soybeans rather than milk. ■

Regulation of Small Intestine Secretions

Goblet cells and intestinal glands secrete their products when chyme provides both chemical and mechanical stimulation. Distension of the intestinal wall activates the nerve plexuses within the wall and stimulates parasympathetic reflexes that also trigger release of small intestine secretions.

Table 15.5

Summary of the Major Digestive Enzymes

Enzyme	Source	Digestive Action
Salivary enzyme		
Amylase	Salivary glands	Begins carbohydrate digestion by breaking down starch and glycogen to disaccharides
Gastric enzyme		
Pepsin	Gastric glands	Begins protein digestion
Pancreatic enzymes		
Amylase	Pancreas	Breaks down starch and glycogen into disaccharides
Lipase	Pancreas	Breaks down fats into fatty acids and glycerol
Proteolytic enzymes	Pancreas	Breaks down proteins or partially digested proteins into peptides
(a) Trypsin		
(b) Chymotrypsin		
(c) Carboxypeptidase		
Nucleases	Pancreas	Breaks down nucleic acids into nucleotides
Intestinal enzymes		
Peptidase	Mucosal cells	Breaks down peptides into amino acids
Sucrase, maltase, lactase	Mucosal cells	Breaks down disaccharides into monosaccharides
Lipase	Mucosal cells	Breaks down fats into fatty acids and glycerol
Enterokinase	Mucosal cells	Converts trypsinogen into trypsin

1. Describe the parts of the small intestine.
2. What is the function of an intestinal villus?
3. What is the function of the intestinal glands?
4. List types of intestinal digestive enzymes.

Absorption in the Small Intestine

Villi greatly increase the surface area of the intestinal mucosa, making the small intestine the most important absorbing organ of the alimentary canal. So effective is the small intestine in absorbing digestive products, water, and electrolytes that very little absorbable material reaches its distal end.

Carbohydrate digestion begins in the mouth with the activity of salivary amylase, and enzymes from the intestinal mucosa and pancreas complete the process in the small intestine. Villi absorb the resulting monosaccharides, which enter blood capillaries. Simple sugars are absorbed by active transport or facilitated diffusion (see chapter 3).

Pepsin activity begins protein digestion in the stomach, and enzymes from the intestinal mucosa and the pancreas complete digestion in the small intestine. During this process, large protein molecules are broken down into amino acids, which are then actively transported into the villi and carried away by the blood.

Enzymes from the intestinal mucosa and pancreas digest fat molecules almost entirely. The resulting fatty acids and glycerol molecules diffuse into villi epithelial cells. The endoplasmic reticula of the cells use the fatty acids to resynthesize fat molecules similar to those digested. These fats are encased in protein to form *chylomicrons,* which make their way to the lacteals of the villi. Lymph in the lacteals and other lymphatic vessels carries chylomicrons to the bloodstream (see chapter 14) (fig. 15.27). Some fatty acids with relatively short carbon chains may be absorbed directly into the blood capillary of a villus without being changed back into fat.

The intestinal villi absorb electrolytes and water in addition to the products of carbohydrate, protein, and fat digestion. Table 15.6 summarizes the absorption process.

1. Which substances resulting from digestion of carbohydrate, protein, and fat molecules does the small intestine absorb?
2. Describe how fatty acids are absorbed and transported.

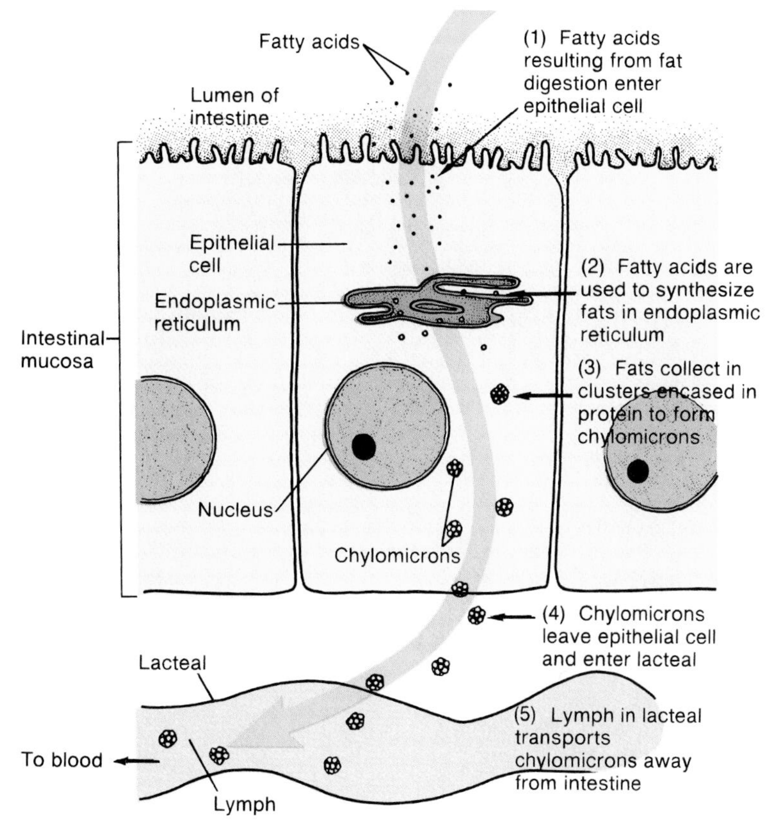

Figure 15.27

Fatty acid absorption has several steps.

Table 15.6

Intestinal Absorption of Nutrients

Nutrient	Absorption Mechanism	Means of Transport
Monosaccharides	Facilitated diffusion and active transport	Blood in capillaries
Amino acids	Active transport	Blood in capillaries
Fatty acids and glycerol	Facilitated diffusion of glycerol; diffusion of fatty acids into cells	Lymph in lacteals
	(a) Most fatty acids are resynthesized into fats and incorporated in chylomicrons for transport	
	(b) Some fatty acids with relatively short carbon chains are transported without being changed back into fats	Blood in capillaries
Electrolytes	Diffusion and active transport	Blood in capillaries
Water	Osmosis	Blood in capillaries

In *malabsorption,* the small intestine digests, but does not absorb, some nutrients. Causes of malabsorption include surgical removal of a portion of the small intestine, obstruction of lymphatic vessels due to a tumor, or interference with the production and release of bile as a result of liver disease.

Another cause of malabsorption is a reaction to *gluten,* found in certain grains, especially wheat and rye. This condition is called *celiac disease.* Microvilli are damaged, and in severe cases, villi may be destroyed. Both of these effects reduce the absorptive surface of the small intestine, preventing absorption of some nutrients. Symptoms of malabsorption include diarrhea, weight loss, weakness, vitamin deficiencies, anemia, and bone demineralization. ■

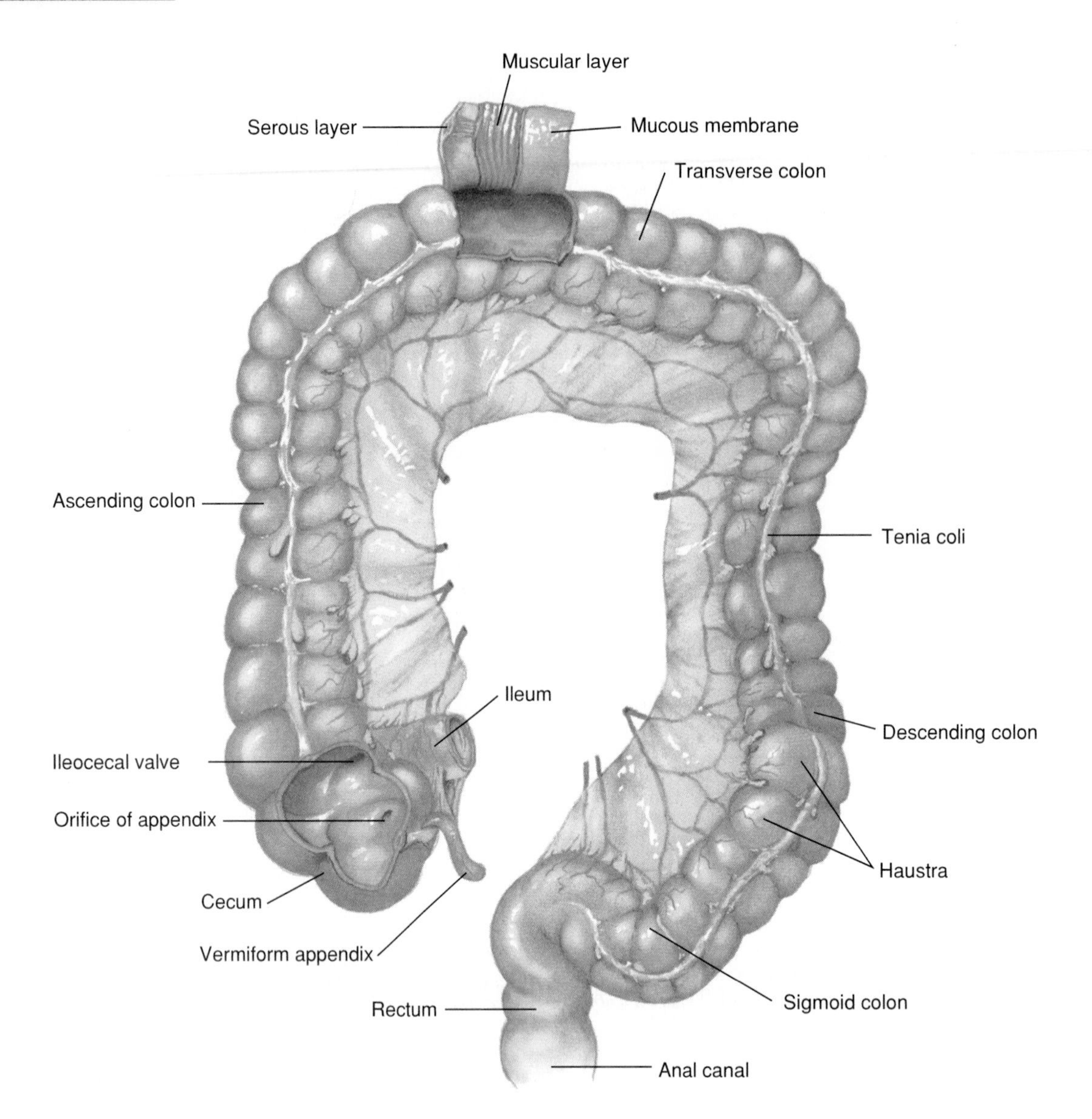

Figure 15.28

Parts of the large intestine (anterior view).

Movements of the Small Intestine

Like the stomach, the small intestine carries on mixing movements and peristalsis. The mixing movements include small, periodic, ringlike contractions that cut chyme into segments and move it back and forth.

Weak peristaltic waves propel chyme short distances through the small intestine. Consequently, chyme moves relatively slowly through the small intestine, taking from 3–10 hours to travel its length.

If the small intestine wall becomes overdistended or irritated, a strong *peristaltic rush* may pass along the organ's entire length. This movement sweeps the contents of the small intestine into the large intestine so quickly that water, nutrients, and electrolytes that would normally be absorbed are not. The result is *diarrhea,* a condition in which defecation becomes more frequent and stools are watery. Prolonged diarrhea causes imbalances in water and electrolyte concentrations.

At the distal end of the small intestine, a sphincter muscle called the **ileocecal valve** joins the small intestine's ileum to the large intestine's cecum (fig. 15.28). Normally, this sphincter remains constricted, preventing the contents of the small intestine from entering the large intestine, and the contents of the large intestine from backing up into the ileum. After a meal, however, a gastroileal reflex increases peristalsis in the ileum and relaxes the sphincter, forcing some of the contents of the small intestine into the cecum.

1. Describe the movements of the small intestine.
2. What stimulus relaxes the ileocecal valve?

Large Intestine

The **large intestine** is so named because its diameter is greater than that of the small intestine. This portion of the alimentary canal is about 1.5 meters long, and it begins in the lower right side of the abdominal cavity, where the ileum joins the cecum. From there, the large intestine extends upward on the right side, crosses obliquely to the left, and descends into the pelvis. At its distal end, it opens to the outside of the body as the anus (see fig. 15.1).

The large intestine reabsorbs water and electrolytes from chyme remaining in the alimentary canal. It also forms and stores feces.

Parts of the Large Intestine

The large intestine consists of the cecum, colon, rectum, and anal canal (figs. 15.28 and 15.29 and reference plates 4 and 5). The **cecum,** at the beginning of the large intestine, is a dilated, pouchlike structure that hangs slightly below the ileocecal opening. Projecting downward from it is a narrow tube with a closed end called the **vermiform appendix.** The human appendix has no known digestive function. However, it contains lymphatic tissue.

In *appendicitis,* the appendix becomes inflamed and infected. Surgery is required to prevent the appendix from rupturing. If it does break open, the contents of the large intestine may enter the abdominal cavity and cause a serious infection of the peritoneum called *peritonitis.* ■

The **colon** is divided into four portions—the ascending, transverse, descending, and sigmoid colons. The **ascending colon** begins at the cecum and travels upward against the posterior abdominal wall to a point just below the liver. There, it turns sharply to the left and becomes the **transverse colon.** The transverse colon is the longest and the most movable part of the large intestine. It is suspended by a fold of peritoneum and sags in the middle below the stomach. As the transverse colon approaches the spleen, it turns abruptly downward and becomes the **descending colon.** At the brim of the pelvis, the descending colon makes an S-shaped curve called the **sigmoid colon** and then becomes the rectum.

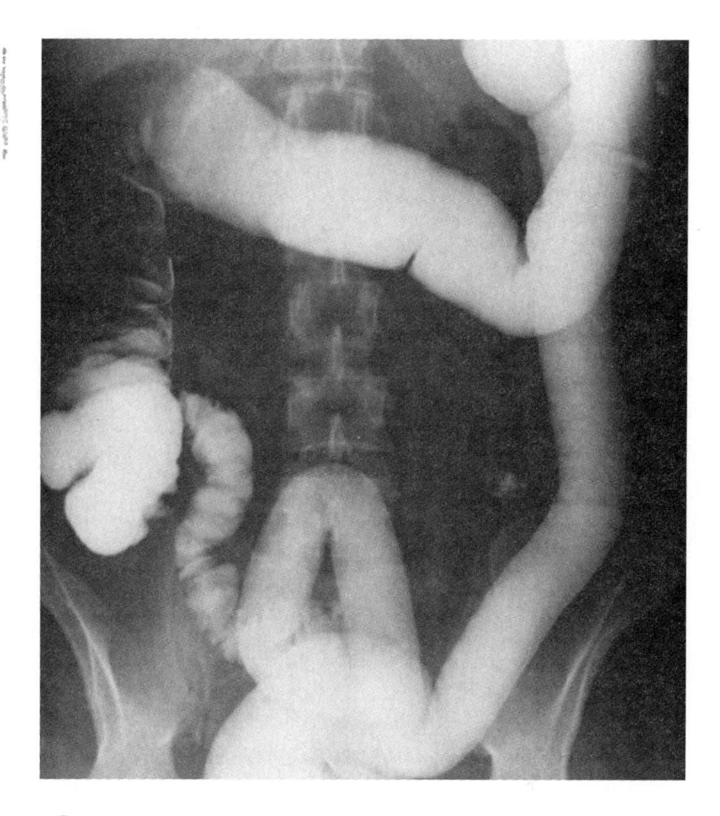

Figure 15.29

X-ray film showing the large intestine containing a radiopaque substance that the patient ingested.

The **rectum** lies next to the sacrum and generally follows its curvature. The peritoneum firmly attaches the rectum to the sacrum, and the rectum ends about 5 centimeters below the tip of the coccyx, where it becomes the anal canal (see fig. 15.28).

The last 2.5–4.0 centimeters of the large intestine form the **anal canal** (fig. 15.30). The mucous membrane in the canal folds into a series of six to eight longitudinal *anal columns.* At its distal end, the canal opens to the outside as the **anus.** Two sphincter muscles guard the anus—an *internal anal sphincter muscle* composed of smooth muscle under involuntary control and an *external anal sphincter muscle* composed of skeletal muscle under voluntary control.

1. What is the general function of the large intestine?
2. Describe the parts of the large intestine.

Rectum
Levator ani muscle
Anal canal
Anal columns
Internal anal sphincter
External anal sphincter
Anus

Figure 15.30

The rectum and anal canal are at the distal end of the alimentary canal.

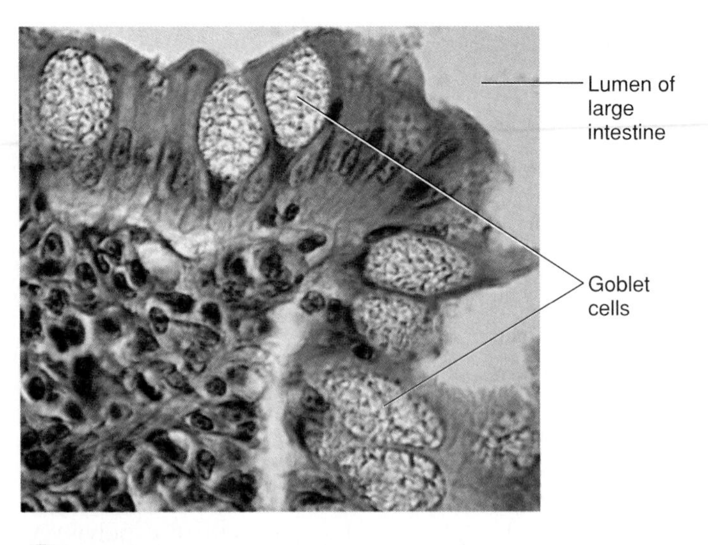

Figure 15.31

Light micrograph of the large intestine mucosa (250×).

Hemorrhoids are, literally, a pain in the rear. Enlarged and inflamed branches of the rectal vein in the anal columns cause intense itching, sharp pain, and sometimes, bright red bleeding. The hemorrhoids may be internal (which do not produce symptoms) or bulge out of the anus. Causes of hemorrhoids include anything that puts prolonged pressure on the delicate rectal tissue, including obesity, pregnancy, constipation, diarrhea, and liver disease.

Eating more fiber-rich foods and drinking lots of water can usually prevent or cure hemorrhoids. Warm soaks in the tub, cold packs, and careful wiping of painful areas also help, as do external creams and ointments. Surgery—with a scalpel or a laser—can remove severe hemorrhoids. ■

Structure of the Large Intestine Wall

The wall of the large intestine is composed of the same types of tissues as other parts of the alimentary canal but has some unique features. The large intestine wall, for example, lacks the villi characteristic of the small intestine. Also, the layer of longitudinal muscle fibers does not cover the large intestine wall uniformly. Instead, the fibers form three distinct bands (teniae coli) that extend the entire length of the colon (see fig. 15.28). These bands exert tension lengthwise on the wall, creating a series of pouches (haustra).

Functions of the Large Intestine

Unlike the small intestine, which secretes digestive enzymes and absorbs digestive products, the large intestine has little or no digestive function. The mucous membrane that forms the large intestine's inner lining, however, contains many tubular glands. Structurally, these glands are similar to those of the small intestine, but they are composed almost entirely of goblet cells. Consequently, mucus is the large intestine's only significant secretion (fig. 15.31).

Mucus secreted into the large intestine protects the intestinal wall against the abrasive action of the materials passing through it. Mucus also binds particles of fecal matter, and its alkalinity helps control the pH of the large intestine contents.

Chyme entering the large intestine contains materials that the small intestine did not digest or absorb. It also contains water, electrolytes, mucus, and bacteria. The large intestine normally absorbs water and electrolytes in the proximal half of the tube. Substances that remain in the tube become feces and are stored for a time in the distal portion of the large intestine.

1. How does the structure of the large intestine differ from that of the small intestine?
2. What substances does the large intestine absorb?

Movements of the Large Intestine

The movements of the large intestine—mixing and peristalsis—are similar to those of the small intestine, although more sluggish. Also, peristaltic waves of the large intestine happen only two or three times each day. These waves produce *mass movements* in which a large section of the intestinal wall constricts vigorously, forcing the intestinal contents toward the rectum. Mass movements usually follow a meal, as a result of the gastrocolic reflex initiated in the small intestine. Irritations of the intestinal mucosa also can trigger such movements. For instance, a person with an inflamed colon (colitis) may experience frequent mass movements.

A person usually can initiate a *defecation reflex* by holding a deep breath and contracting the abdominal wall muscles. This action increases internal abdominal pressure and forces feces into the rectum. As the rectum fills, its wall distends, triggering the defecation reflex. This stimulates peristaltic waves in the descending colon, and the internal anal sphincter relaxes. At the same time, other reflexes involving the sacral region of the spinal cord strengthen the peristaltic waves, lower the diaphragm, close the glottis, and contract the abdominal wall muscles. These actions further increase internal abdominal pressure and squeeze the rectum. The external anal sphincter is signaled to relax, and feces are forced to the outside. Contracting the external anal sphincter allows voluntary inhibition of defecation.

Taking a deep breath, closing the glottis, and forcibly contracting the abdominal wall muscles is called the *Valsalva maneuver*. This action increases internal abdominal pressure and aids defecation. The Valsalva maneuver also increases internal thoracic pressure and may interfere with the return of blood to the heart. For this reason, the Valsalva maneuver may be hazardous to persons with heart disease. ■

Feces

Feces are composed of materials that were not digested or absorbed, plus water, electrolytes, mucus, and bacteria. Usually, feces are about 75% water, and their color derives from bile pigments that bacterial action has altered somewhat. Feces' pungent odor results from a variety of compounds bacteria produce.

1. How does peristalsis in the large intestine differ from peristalsis in the small intestine?
2. List the major events of defecation.
3. Describe the composition of feces.

Nutrition and Nutrients

Nutrition is the process by which the body takes in and utilizes necessary food substances. These food substances, or **nutrients,** include carbohydrates, lipids, proteins, vitamins, minerals, and water (see chapters 2 and 4). Foods provide nutrients, and digestion breaks them down to sizes that can be absorbed and transported in the bloodstream. Nutrients that human cells cannot synthesize, such as certain amino acids, are called **essential nutrients.**

Carbohydrates

Carbohydrates are organic compounds used primarily to supply energy for cellular processes.

Carbohydrate Sources

Carbohydrates are ingested in a variety of forms, including starch from grains and vegetables; glycogen from meats; disaccharides from cane sugar, beet sugar, and molasses; and monosaccharides from honey and fruits. Digestion breaks down complex carbohydrates into monosaccharides, which are small enough to be absorbed.

Cellulose is a complex carbohydrate that is abundant in food—it gives celery its crunch and lettuce its crispness. Humans cannot digest cellulose, so most of it passes through the alimentary canal largely unchanged. Thus, cellulose provides bulk (also called fiber or roughage) against which the muscular wall of the digestive system can push, facilitating the movement of food.

Carbohydrate Utilization

The monosaccharides absorbed from the digestive tract include *fructose, galactose,* and *glucose.* Liver enzymes change fructose and galactose into glucose, which is the carbohydrate form most commonly oxidized for cellular fuel.

Some excess glucose is polymerized to *glycogen* and stored in the liver and muscles. When required to supply energy, glucose can be mobilized rapidly from glycogen. Only a certain amount of glycogen can be stored, however, and excess glucose is usually changed into fat and stored in adipose tissue.

Cells also use carbohydrates as starting materials for such vital biochemicals as the five-carbon sugars *ribose* and *deoxyribose,* required for production of the nucleic acids RNA and DNA. Carbohydrates are also required for the disaccharide *lactose* (milk sugar), synthesized when the breasts are actively secreting milk.

Many cells can also obtain energy by oxidizing fatty acids. Some cells, however, such as neurons, normally depend on a continuous supply of glucose for survival. Even a temporary decrease in the glucose supply can seriously impair nervous system function. Consequently, the body requires a minimum amount of carbohydrates. If foods do not provide an adequate carbohydrate supply, the liver may change some noncarbohydrates, such as amino acids from proteins, into glucose. Thus, the requirement for glucose has physiological priority over the requirement to synthesize proteins from available amino acids.

Carbohydrate Requirements

Because carbohydrates are the primary fuel source for cellular processes, the requirement for carbohydrates varies with individual energy expenditure. Physically active individuals require more fuel than sedentary ones. The minimal carbohydrate requirement in the human diet is unknown, but a daily carbohydrate intake of at least 125–175 grams is necessary to spare protein (that is, to avoid protein breakdown) and to avoid metabolic disorders resulting from excess fat utilization.

1. List several common sources of carbohydrates.
2. Explain the importance of cellulose in the diet.
3. Explain why the requirement for glucose has priority over protein synthesis.

Lipids

Lipids are organic compounds that include fats, oils, and fatlike substances (see chapter 2). They supply energy for cellular processes and for building structures, such as cell membranes. Lipids include fats, phospholipids, and cholesterol. The most common dietary lipids are the fats called *triglycerides*.

Lipid Sources

Triglycerides are found in plant- and animal-based foods. Saturated fats (which should comprise no more than 10% of the diet) are found mainly in foods of animal origin, such as meats, eggs, milk, and lard. Unsaturated fats are in seeds, nuts, and plant oils, such as corn oil, peanut oil, and olive oil.

Cholesterol is abundant in liver and egg yolk, and to a lesser extent, in whole milk, butter, cheese, and meats. It is generally not present in foods of plant origin.

Lipid Utilization

Digestion breaks down triglycerides into fatty acids and glycerol. After being absorbed, these products are transported in lymph and blood to tissues. Liver and adipose tissues control lipid metabolism.

The liver can change fatty acids from one form to another, but it cannot synthesize certain fatty acids, such as *linoleic acid*. This **essential fatty acid** is required for phospholipid synthesis, which, in turn, is necessary for cell membrane formation and the transport of circulating lipids. Good sources of linoleic acid include corn oil, cottonseed oil, and soy oil.

The liver uses free fatty acids to synthesize triglycerides, phospholipids, and lipoproteins that may then be released into the blood. Thus, the liver regulates circulating lipids. It also controls the total amount of cholesterol in the body by synthesizing cholesterol and releasing it into the blood, or by removing cholesterol from the blood and excreting it into the bile. The liver also uses cholesterol to produce bile salts.

Cholesterol is not an energy source. It provides structural material for cell and organelle membranes, however, and furnishes starting materials for the synthesis of certain sex hormones and adrenal hormones.

Adipose tissue stores excess triglycerides. If blood lipid concentration drops (in response to fasting, for example), some of these triglycerides are hydrolyzed into free fatty acids and glycerol, and then released into the bloodstream.

Lipid Requirements

The amounts and types of fats required for health are unknown. Because linoleic acid is an essential fatty acid and to prevent deficiency conditions from developing, nutritionists recommend that formula-fed infants receive 3% of the energy intake in the form of linoleic acid. A typical adult diet of a variety of foods usually provides adequate fats. Fat intake must also supply required amounts of fat-soluble vitamins.

Before 1993, "lite" and similar markings on prepared food packages could, and did, mean almost anything: A bottle of vegetable oil labeled "lite" referred to the product's color; a "lite" cheesecake referred to its texture! Now, "lite," or any similar spelling, has a distinct meaning: The product must have a third fewer calories than the "real thing" or half the fat calories.

Be wary of claims that a food product is "99% fat-free." This usually refers to percentage by weight—not calories, which is what counts. A creamy concoction that is 99% fat-free may be largely air and water, and in that form, fat comprises very little of it. But when the air is compressed and the water absorbed, as happens in the stomach, the fat percentage may skyrocket. ■

1. Which fatty acid is an essential nutrient?
2. What is the liver's role in lipid utilization?
3. What is the function of cholesterol?

Proteins

Proteins are polymers of amino acids with a wide variety of functions. Proteins include enzymes that control metabolic rates, clotting factors, the keratin of skin and hair, elastin and collagen of connective tissue, plasma proteins that regulate water balance, the muscle components actin and myosin, certain hormones, and the antibodies that protect against infection. Amino acids are also potential sources of energy.

Protein Sources

Foods rich in proteins include meats, fish, poultry, cheese, nuts, milk, eggs, and cereals. Legumes, including beans and peas, contain lesser amounts. Digestion breaks down proteins into amino acids—smaller molecules that intestinal tissues absorb and blood transports.

The cells of an adult can synthesize all but eight required amino acids, and the cells of a child can produce all but ten. Amino acids that the body can synthesize are termed nonessential; those that it cannot synthesize are **essential amino acids.** This term refers only to dietary intake since all amino acids are needed for normal protein synthesis.

All of the amino acids must be present in the body at the same time for growth and tissue repair to occur. In other words, if one essential amino acid is missing from the diet, protein cannot be synthesized. This is because many proteins include all twenty amino acids.

Table 15.7 Amino Acids Found in Foods

Alanine	Leucine (e)
Arginine (ch)	Lysine (e)
Asparagine	Methionine (e)
Aspartic acid	Phenylalanine (e)
Cysteine	Proline
Glutamic acid	Serine
Glutamine	Threonine (e)
Glycine	Tryptophan (e)
Histidine (ch)	Tyrosine
Hydroxyproline	Valine (e)
Isoleucine (e)	

Human cells cannot synthesize eight essential amino acids (e), which must be provided in the diet. Two additional amino acids (ch) are essential in growing children.

Table 15.7 lists the amino acids in foods and indicates those that are essential.

Proteins are classified as complete or incomplete on the basis of the amino acid types they provide. **Complete proteins**, which include those available in milk, meats, and eggs, contain adequate amounts of the essential amino acids. **Incomplete proteins**, such as *zein* in corn, which has too little of the essential amino acids tryptophan and lysine, are unable by themselves to maintain human tissues or to support normal growth and development.

A protein called *gliadin* in wheat is an example of a **partially complete protein.** It does not contain enough lysine to promote growth, but it contains enough to maintain life.

Plant proteins typically contain too little of one or more essential amino acids to provide adequate nutrition. Combining appropriate plant foods can provide a diversity of dietary amino acids. For example, beans are low in methionine but have enough lysine. Rice lacks lysine but has enough methionine. A meal of beans and rice provides enough of both types of amino acids. ■

1. Which foods are rich sources of proteins?
2. Why are some amino acids called essential?
3. Distinguish between complete and incomplete proteins.

Protein Requirements

Proteins supply essential amino acids and also provide nitrogen and other elements for the synthesis of nonessential amino acids and certain nonprotein nitrogenous substances. Consequently, the amount of protein individuals require varies according to body size, metabolic rate, and nitrogen requirements.

For an average adult, nutritionists recommend a daily protein intake of about 0.8 grams per kilogram of body weight. For a pregnant woman, the recommendation increases an additional 30 grams of protein per day. Similarly, a nursing mother requires an extra 20 grams of protein per day to maintain a high level of milk production.

A food's potential energy can be expressed in calories, which are units of heat. A **calorie** is defined as the amount of heat required to raise the temperature of a gram of water by 1° Celsius. The calorie used to measure food energy, however, is 1,000 times greater. This larger calorie (Cal) is called a kilocalorie, but nutritional studies commonly refer to it simply as a calorie. As a result of cellular oxidation, 1 gram of carbohydrate or 1 gram of protein yields about 4.1 calories, but 1 gram of fat yields 9.5 calories. ■

1. What are the physiological functions of proteins?
2. How much protein is recommended for an adult diet?

Vitamins

Vitamins are organic compounds (other than carbohydrates, lipids, and proteins) required in small amounts for normal metabolic processes, but which body cells cannot synthesize in adequate amounts. Thus, they are essential nutrients that foods must supply.

Vitamins can be classified on the basis of solubility because some are soluble in fats (or fat solvents) and others are soluble in water. *Fat-soluble vitamins* include vitamins A, D, E, and K; *water-soluble vitamins* include the B vitamins and vitamin C.

Fat-Soluble Vitamins

Because fat-soluble vitamins dissolve in fats, they associate with lipids and are influenced by the same factors that affect lipid absorption. For example, bile salts in the intestine promote absorption of these vitamins. As a group, fat-soluble vitamins are stored in moderate quantities within various tissues. They resist the effects of heat; therefore, cooking and food processing usually do not destroy them.

Table 15.8 lists the characteristics, functions, sources, and recommended dietary allowances (RDA) for adults of the fat-soluble vitamins.

1. What are vitamins?
2. How do bile salts affect absorption of fat-soluble vitamins?

Water-Soluble Vitamins

The water-soluble vitamins include the B vitamins and vitamin C. The **B vitamins** are several compounds that are essential for normal cellular metabolism. They help oxidize carbohydrates, lipids, and proteins. Since the B vitamins often occur together in foods, they are usually referred to as the *vitamin B complex*. Members of this group differ chemically and functionally. Cooking and food processing destroy some of them.

Vitamin C (ascorbic acid) is one of the least stable vitamins and is fairly widespread in plant foods. It is necessary for collagen production, conversion of folacin to folinic acid, and the metabolism of certain amino acids. Vitamin C also promotes iron absorption and synthesis of certain hormones from cholesterol.

Table 15.9 lists the characteristics, functions, sources, and RDAs for adults of the water-soluble vitamins.

1. Name the water-soluble vitamins.
2. What is the vitamin B complex?
3. Distinguish between fat-soluble and water-soluble vitamins.

Table 15.8

Fat-Soluble Vitamins

Vitamin	Characteristics	Functions	Sources and RDA* for Adults
Vitamin A	Occurs in several forms; synthesized from carotenes; stored in liver; stable in heat, acids, and alkalis; unstable in light	Necessary for synthesis of visual pigments, mucoproteins, and mucopolysaccharides; for normal development of bones and teeth; and for maintenance of epithelial cells	Liver, fish, whole milk, butter, eggs, leafy green vegetables, yellow and orange vegetables and fruits 4,000–5,000 IU†
Vitamin D	A group of steroids; resistant to heat, oxidation, acids, and alkalis; stored in liver, skin, brain, spleen, and bones	Promotes absorption of calcium and phosphorus; promotes development of teeth and bones	Produced in skin exposed to ultraviolet light; in milk, egg yolk, fish liver oils, fortified foods 400 IU
Vitamin E	A group of compounds; resistant to heat and visible light; unstable in presence of oxygen and ultraviolet light; stored in muscles and adipose tissue	An antioxidant; prevents oxidation of vitamin A and polyunsaturated fatty acids; may help maintain stability of cell membranes	Oils from cereal seeds, salad oils, margarine, shortenings, fruits, nuts, and vegetables 30 IU
Vitamin K	Occurs in several forms; resistant to heat, but destroyed by acids, alkalis, and light; stored in liver	Needed for synthesis of prothrombin, which functions in blood clotting	Leafy green vegetables, egg yolk, pork liver, soy oil, tomatoes, cauliflower 55–70 μg

*RDA = recommended dietary allowance.
†IU = international unit.

Minerals

Dietary **minerals** are elements other than carbon that are essential in human metabolism. Plants usually extract minerals from soil, and humans obtain minerals from plant foods or from animals that have eaten plants.

Characteristics of Minerals

Minerals are responsible for about 4% of body weight and are most concentrated in the bones and teeth. Minerals are usually incorporated into organic molecules. For example, phosphorus is found in phospholipids, iron in hemoglobin, and iodine in thyroxine. Some minerals are part of inorganic compounds, such as the calcium phosphate of bone. Other minerals are free ions, such as sodium, chloride, and calcium ions in blood.

Minerals comprise parts of the structural materials in all body cells. They are also portions of enzyme molecules, contribute to the osmotic pressure of body fluids, and play vital roles in nerve impulse conduction, muscle fiber contraction, blood coagulation, and maintaining the pH of body fluids.

Major Minerals

The minerals *calcium* and *phosphorus* account for nearly 75% by weight of the mineral elements in the body; thus, they are **major minerals.** Other major minerals, each of which accounts for 0.05% or more of the body weight, include potassium, sulfur, sodium, chlorine, and magnesium.

Table 15.10 lists the distribution, functions, sources, and RDAs for adults of major minerals.

1. How are minerals obtained?
2. What are the major functions of minerals?

Table 15.9

Water-Soluble Vitamins

Vitamin	Characteristics	Functions	Sources and RDA* for Adults
Thiamine (vitamin B_1)	Destroyed by heat and oxygen, especially in alkaline environment	Part of coenzyme required to oxidize carbohydrates; coenzyme required for ribose synthesis	Lean meats, liver, eggs, whole-grain cereals, leafy green vegetables, legumes 1.5 mg
Riboflavin (vitamin B_2)	Stable to heat, acids, and oxidation; destroyed by bases and ultraviolet light	Part of enzymes and coenzymes required to oxidize glucose and fatty acids and for cellular growth	Meats, dairy products, leafy green vegetables, whole-grain cereals 1.7 mg
Niacin (nicotinic acid)	Stable to heat, acids, and alkalis; converted to niacinamide by cells; synthesized from tryptophan	Part of coenzymes required to oxidize glucose and to synthesize proteins, fats, and nucleic acids	Liver, lean meats, peanuts, legumes 20 mg
Vitamin B_6	Group of three compounds; stable to heat and acids; destroyed by oxidation, bases, and ultraviolet light	Coenzyme required to synthesize proteins and certain amino acids, to convert tryptophan to niacin, to produce antibodies, and to synthesize nucleic acids	Liver, meats, bananas, avocados, beans, peanuts, whole-grain cereals, egg yolk 2 mg
Pantothenic acid	Destroyed by heat, acids, and bases	Part of coenzyme A required to oxidize carbohydrates and fats	Meats, whole-grain cereals, legumes, milk, fruits, vegetables 10 mg
Cyanocobalamin (vitamin B_{12})	Complex, cobalt-containing compound; stable to heat; inactivated by light, strong acids, and strong bases; absorption regulated by intrinsic factor from gastric glands; stored in liver	Part of coenzyme required to synthesize nucleic acids and to metabolize carbohydrates; plays role in myelin synthesis	Liver, meats, milk, cheese, eggs 3–6 μg
Folacin (folic acid)	Occurs in several forms; destroyed by oxidation in acid environment or by heat in alkaline environment; stored in liver, where it is converted into folinic acid	Coenzyme required for metabolism of certain amino acids and for DNA synthesis; promotes red blood cell production	Liver, leafy green vegetables, whole-grain cereals, legumes 0.4 mg
Biotin	Stable to heat, acids, and light; destroyed by oxidation and bases	Coenzyme required to metabolize amino acids and fatty acids, and to synthesize nucleic acids	Liver, egg yolk, nuts, legumes, mushrooms 0.3 mg
Ascorbic acid (vitamin C)	Similar to monosaccharides; stable in acids but destroyed by oxidation, heat, light, and bases	Required to produce collagen, to convert folacin to folinic acid, and to metabolize certain amino acids; promotes absorption of iron and synthesis of hormones from cholesterol	Citrus fruits, tomatoes, potatoes, leafy green vegetables 60 mg

*RDA = recommended dietary allowance.

Table 15.10 Major Minerals

Mineral	Distribution	Functions	Sources and RDA* for Adults
Calcium (Ca)	Mostly in the inorganic salts of bones and teeth	Structure of bones and teeth; essential for nerve impulse conduction, muscle fiber contraction, and blood coagulation; increases permeability of cell membranes; activates certain enzymes	Milk, milk products, leafy green vegetables 800 mg
Phosphorus (P)	Mostly in the inorganic salts of bones and teeth	Structure of bones and teeth; component in nearly all metabolic reactions; constituent of nucleic acids, many proteins, some enzymes, and some vitamins; occurs in cell membrane, ATP, and phosphates of body fluids	Meats, cheese, nuts, whole-grain cereals, milk, legumes 800 mg
Potassium (K)	Widely distributed; tends to be concentrated inside cells	Helps maintain intracellular osmotic pressure and regulate pH; promotes metabolism; needed for nerve impulse conduction and muscle fiber contraction	Avocados, dried apricots, meats, nuts, potatoes, bananas 2,500 mg
Sulfur (S)	Widely distributed; abundant in skin, hair, and nails	Essential part of various amino acids, thiamine, insulin, biotin, and mucopolysaccharides	Meats, milk, eggs, legumes None established
Sodium (Na)	Widely distributed; large proportion occurs in extracellular fluids and bound to inorganic salts of bone	Helps maintain osmotic pressure of extracellular fluids and regulate water movement; needed for conduction of nerve impulses and contraction of muscle fibers; aids in regulation of pH and in transport of substances across cell membranes	Table salt, cured ham, sauerkraut, cheese, graham crackers 2,500 mg
Chlorine (Cl)	Closely associated with sodium (as chloride); most highly concentrated in cerebrospinal fluid and gastric juice	Helps maintain osmotic pressure of extracellular fluids, regulate pH, and maintain electrolyte balance; essential in formation of hydrochloric acid; aids transport of carbon dioxide by red blood cells	Same as for sodium None established
Magnesium (Mg)	Abundant in bones	Needed in metabolic reactions that occur in mitochondria and are associated with ATP production; plays role in the breakdown of ATP to ADP	Milk, dairy products, legumes, nuts, leafy green vegetables 300–350 mg

*RDA = recommended dietary allowance.

Table 15.11

Trace Elements

Trace Element	Distribution	Functions	Sources and RDA* for Adults
Iron (Fe)	Primarily in blood; stored in liver, spleen, and bone marrow	Part of hemoglobin molecule; catalyzes vitamin A formation; incorporated into a number of enzymes	Liver, lean meats, dried apricots, raisins, enriched whole-grain cereals, legumes, molasses 10–18 mg
Manganese (Mn)	Most concentrated in liver, kidneys, and pancreas	Occurs in enzymes needed for fatty acid and cholesterol synthesis, urea formation, and normal functioning of the nervous system	Nuts, legumes, whole-grain cereals, leafy green vegetables, fruits 2.5–5 mg
Copper (Cu)	Mostly highly concentrated in liver, heart, and brain	Essential for hemoglobin synthesis, bone development, melanin production, and myelin formation	Liver, oysters, crabmeat, nuts, whole-grain cereals, legumes 2–3 mg
Iodine (I)	Concentrated in thyroid gland	Essential component for synthesis of thyroid hormones	Food content varies with soil content in different geographic regions; iodized table salt 0.15 mg
Cobalt (Co)	Widely distributed	Component of cyanocobalamin; needed for synthesis of several enzymes	Liver, lean meats, milk None established
Zinc (Zn)	Most concentrated in liver, kidneys, and brain	Constituent of several enzymes involved in digestion, respiration, bone metabolism, liver metabolism; necessary for normal wound healing and maintaining skin integrity	Meats, cereals, legumes, nuts, vegetables 15 mg
Fluorine (F)	Primarily in bones and teeth	Component of tooth structure	Fluoridated water 1.5–4.0 mg
Selenium (Se)	Concentrated in liver and kidneys	Occurs in enzymes	Lean meats, fish, cereals 0.05–2.00 mg
Chromium (Cr)	Widely distributed	Essential for use of carbohydrates	Liver, lean meats, wine 0.05–2.00 mg

*RDA = recommended dietary allowance.

Trace Elements

Trace elements are essential minerals found in minute amounts, each making up less than 0.005% of adult body weight. They include iron, manganese, copper, iodine, cobalt, zinc, fluorine, selenium, and chromium.

Table 15.11 lists the distribution, functions, sources, and RDAs for adults of the trace elements.

1. Distinguish between a major mineral and a trace element.
2. Name the major minerals and trace elements.

Adequate Diets

An adequate diet provides sufficient energy, essential fatty acids, essential amino acids, vitamins, and minerals to support optimal growth and to maintain and repair body tissues. Because individual requirements for nutrients vary greatly with age, sex, growth rate, amount of physical activity, and level of stress, as well as with genetic and environmental factors, designing a diet that is adequate for everyone is impossible. Nutrients are so widely distributed in foods, however, that satisfactory amounts and combinations can usually be obtained in spite of individual food preferences.

If, however, the diet lacks essential nutrients or a person fails to use available foods to best advantage, **malnutrition** results. This condition may involve *undernutrition* and include the symptoms of deficiency diseases, or it may be due to *overnutrition* arising from excess nutrient intake.

The factors leading to malnutrition vary. For example, a deficiency condition may stem from lack of availability or poor quality of food. On the other hand, malnutrition may result from overeating or taking too many vitamin supplements.

1. What is an adequate diet?
2. What factors influence individual nutrient requirements?
3. What causes malnutrition?

Clinical Terms Related to the Digestive System and Nutrition

Achalasia (ak″ah-la′ze-ah) Failure of the smooth muscle to relax at some junction in the digestive tube, such as that between the esophagus and stomach.

Achlorhydria (ah″klor-hi′dre-ah) Lack of hydrochloric acid in gastric secretions.

Anorexia nervosa (ă-nah-rek′se-ah ner vo′sah) Self-starvation.

Aphagia (ah-fa′je-ah) Inability to swallow.

Cachexia (kah-kek′se-ah) A state of chronic malnutrition and physical wasting.

Celiac disease (se′le-ak dĭ-zēz) Inability to digest or use fats and carbohydrates.

Cholecystitis (ko″le-sis-ti′tis) Inflammation of the gallbladder.

Cholelithiasis (ko″le-lĭ-thi′ah-sis) Stones in the gallbladder.

Cholestasis (ko″le-sta′sis) Blockage in bile flow from the gallbladder.

Cirrhosis (sĭ-ro′sis) A liver condition in which the hepatic cells degenerate and the surrounding connective tissues thicken.

Diverticulitis (di″ver-tik″u-li′tis) Inflammation of small pouches (diverticula) that form in the lining and wall of the colon.

Dumping syndrome (dum′ping sin′drōm) Symptoms, including diarrhea, that often occur following a gastrectomy.

Dysentery (dis′en-ter″e) An intestinal infection caused by viruses, bacteria, or protozoans that causes diarrhea and cramps.

Dyspepsia (dis-pep′se-ah) Indigestion; difficulty in digesting a meal.

Dysphagia (dis-fa′je-ah) Difficulty in swallowing.

Enteritis (en″tĕ-ri′tis) Inflammation of the intestine.

Esophagitis (e-sof″ah-ji′tis) Inflammation of the esophagus.

Gastrectomy (gas-trek′to-me) Partial or complete removal of the stomach.

Gastrostomy (gas-tros′to-me) The creation of an opening in the stomach wall through which food and liquids can be administered when swallowing is not possible.

Glossitis (glŏs-si′tis) Inflammation of the tongue.

Hyperalimentation (hi″-per-al″-ĭ-men-ta′shun) Long-term intravenous nutrition.

Ileitis (il″e-i′tis) Inflammation of the ileum.

Pharyngitis (far″in-ji′tis) Inflammation of the pharynx.

Polyphagia (pol″e-fa′je-ah) Overeating.

Pyloric stenosis (pi-lor′ik stĕ-no′sis) A congenital obstruction at the pyloric sphincter due to an enlarged pyloric muscle.

Pylorospasm (pi-lor′o-spazm) A spasm of the pyloric portion of the stomach or of the pyloric sphincter.

Pyorrhea (pi″o-re′ah) Inflammation of the dental periosteum with pus formation.

Stomatitis (sto″mah-ti′tis) Inflammation of the lining of the mouth.

organization

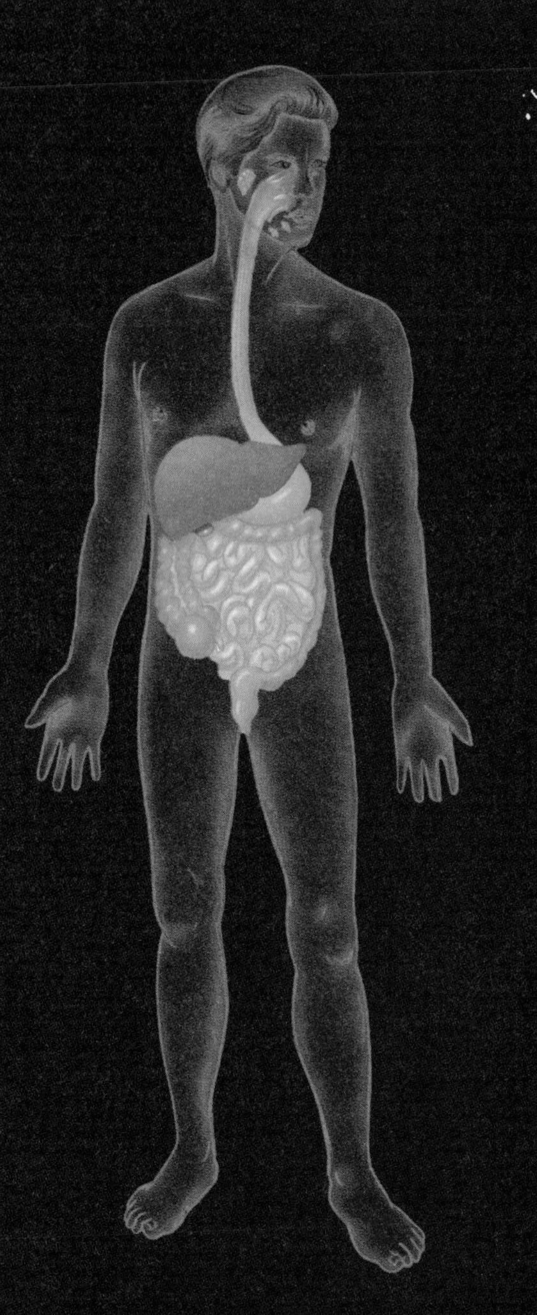

Digestive System

The digestive system ingests, digests, and absorbs nutrients for use by all body cells.

Integumentary System

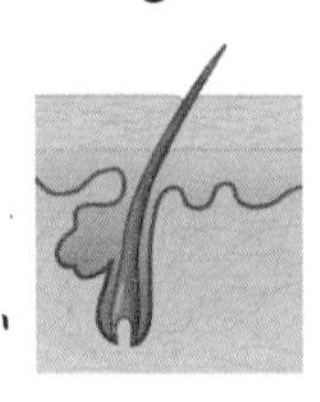

Vitamin D activated in the skin plays a role in absorption of calcium from the digestive tract.

Cardiovascular System

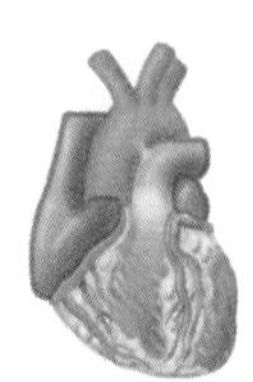

The bloodstream carries absorbed nutrients to all body cells.

Skeletal System

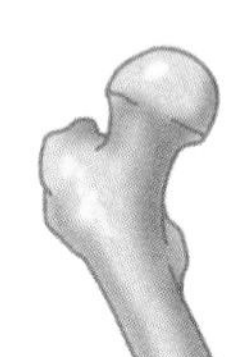

Bones are important in mastication.

Lymphatic System

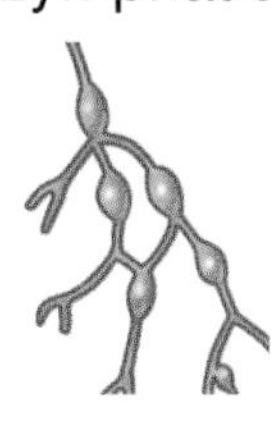

The lymphatic system plays a major role in the absorption of fats.

Muscular System

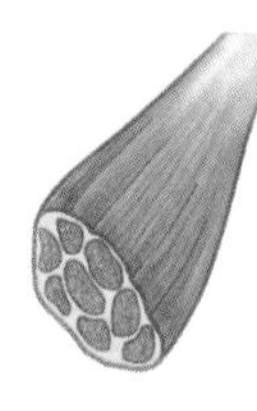

Muscles are important in mastication, swallowing, and the mixing and moving of digestion products through the gastrointestinal tract.

Respiratory System

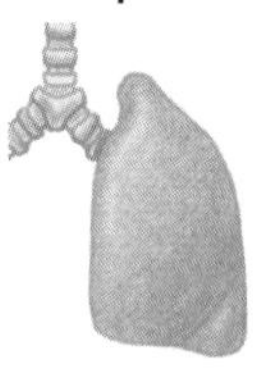

The digestive system and the respiratory system share common anatomical structures.

Nervous System

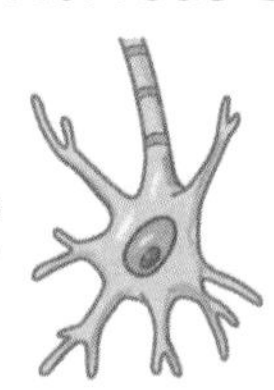

The nervous system can influence digestive system activity.

Urinary System

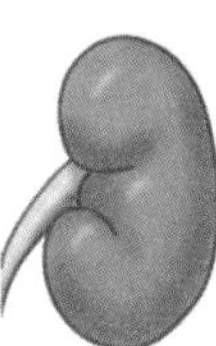

The kidneys and liver work together to activate vitamin D.

Endocrine System

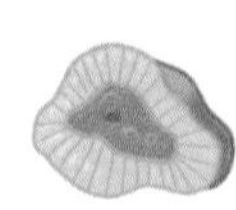

Hormones can influence digestive system activity.

Reproductive System

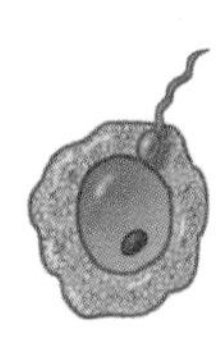

Adequate availability of nutrients, including fats, is essential for conception and normal development.

Summary Outline

Introduction (p. 409)

Digestion is the process of mechanically and chemically breaking down foods so that they can be absorbed. The digestive system consists of an alimentary canal and several accessory organs.

General Characteristics of the Alimentary Canal (p. 409)

Regions of the alimentary canal perform specific functions.

1. Structure of the wall
   The wall consists of four layers—the mucosa, submucosa, muscular layer, and serosa.
2. Movements of the tube
   Motor functions include mixing and propelling movements.

Mouth (p. 411)

The mouth receives food and begins digestion.

1. Cheeks and lips
   a. Cheeks consist of outer layers of skin, pads of fat, muscles associated with expression and chewing, and inner linings of epithelium.
   b. Lips are highly mobile and contain sensory receptors.
2. Tongue
   a. The tongue's rough surface handles food and contains taste buds.
   b. Lingual tonsils are on the root of the tongue.
3. Palate
   a. The palate includes hard and soft portions.
   b. The soft palate closes the opening to the nasal cavity during swallowing.
   c. Palatine tonsils are located on either side of the tongue in the back of the mouth.
4. Teeth
   a. There are twenty primary and thirty-two secondary teeth.
   b. Teeth mechanically break food into smaller pieces, increasing the surface area exposed to digestive actions.
   c. Each tooth consists of a crown and root, and is composed of enamel, dentin, pulp, nerves, and blood vessels.
   d. A periodontal ligament attaches a tooth to the alveolar process.

Salivary Glands (p. 415)

Salivary glands secrete saliva, which moistens food, helps bind food particles, begins chemical digestion of carbohydrates, makes taste possible, and helps cleanse the mouth.

1. Salivary secretions
   Salivary glands include serous cells that secrete digestive enzymes and mucous cells that secrete mucus.
2. Major salivary glands
   a. The parotid glands secrete saliva rich in amylase.
   b. The submandibular glands produce viscous saliva.
   c. The sublingual glands primarily secrete mucus.

Pharynx and Esophagus (p. 416)

The pharynx and esophagus are important passageways.

1. Structure of the pharynx
   The pharynx is divided into a nasopharynx, oropharynx, and laryngopharynx.
2. Swallowing mechanism
   Swallowing occurs in three stages:
   a. Food is mixed with saliva and forced into the pharynx.
   b. Involuntary reflex actions move the food into the esophagus.
   c. Peristalsis transports food to the stomach.
3. Esophagus
   a. The esophagus passes through the diaphragm and joins the stomach.
   b. Circular muscle fibers at the distal end of the esophagus help prevent regurgitation of food from the stomach.

Stomach (p. 417)

The stomach receives food, mixes it with gastric juice, carries on a limited amount of absorption, and moves food into the small intestine.

1. Parts of the stomach
   a. The stomach is divided into cardiac, fundic, body, and pyloric regions.
   b. The pyloric sphincter is a valve between the stomach and small intestine.
2. Gastric secretions
   a. Gastric glands secrete gastric juice.
   b. Gastric juice contains pepsin, hydrochloric acid, and intrinsic factor.
3. Regulation of gastric secretions
   a. Parasympathetic impulses and the hormone gastrin enhance gastric secretion.
   b. Food in the small intestine reflexly inhibits gastric secretions.
4. Gastric absorption
   The stomach wall may absorb a few substances, such as water and other small molecules.
5. Mixing and emptying actions
   a. Mixing movements help produce chyme. Peristaltic waves move chyme into the pyloric region.
   b. The muscular wall of the pyloric region regulates chyme movement into the small intestine.
   c. The rate of emptying depends on the fluidity of chyme and the type of food.

Pancreas (p. 421)

1. Structure of the pancreas
   a. The pancreas produces pancreatic juice that is secreted into a pancreatic duct.
   b. The pancreatic duct leads to the duodenum.
2. Pancreatic juice
   a. Pancreatic juice contains enzymes that can split carbohydrates, proteins, fats, and nucleic acids.
   b. Pancreatic juice has a high bicarbonate ion concentration that helps neutralize chyme and causes intestinal contents to be alkaline.

3. Hormones regulate pancreatic secretion
   a. Secretin stimulates the release of pancreatic juice with a high bicarbonate ion concentration.
   b. Cholecystokinin stimulates the release of pancreatic juice with a high concentration of digestive enzymes.

Liver (p. 422)

1. Liver structure
   a. The right and left lobes of the liver consist of hepatic lobules, the functional units of the gland.
   b. Bile canals carry bile from hepatic lobules to hepatic ducts.
2. Liver functions
   a. The liver metabolizes carbohydrates, lipids, and proteins; stores some substances; filters blood; destroys toxins; and secretes bile.
   b. Bile is the only liver secretion that directly affects digestion.
3. Composition of bile
   a. Bile contains bile salts, bile pigments, cholesterol, and electrolytes.
   b. Only the bile salts have digestive functions.
4. Gallbladder
   a. The gallbladder stores bile between meals.
   b. A sphincter muscle controls bile release from the common bile duct.
5. Regulation of bile release
   a. Cholecystokinin from the small intestine stimulates bile release.
   b. The sphincter muscle at the base of the common bile duct relaxes as a peristaltic wave in the duodenal wall approaches.
6. Functions of bile salts
   Bile salts emulsify fats and aid in the absorption of fatty acids, cholesterol, and certain vitamins.

Small Intestine (p. 427)

The small intestine receives secretions from the pancreas and liver, completes nutrient digestion, absorbs the products of digestion, and transports the residues to the large intestine.

1. Parts of the small intestine
   The small intestine consists of the duodenum, jejunum, and ileum.
2. Structure of the small intestine wall
   a. The wall is lined with villi that increase the surface area and aid in mixing and absorption.
   b. Intestinal glands are located between the villi.
3. Secretions of the small intestine
   a. Secretions include mucus and digestive enzymes.
   b. Digestive enzymes split molecules of sugars, proteins, and fats into simpler forms.
4. Regulation of small intestine secretions
   Gastric juice, chyme, and reflexes stimulated by distension of the small intestine wall stimulate small intestine secretions.
5. Absorption in the small intestine
   a. Enzymes on microvilli perform the final steps in digestion.
   b. Villi absorb monosaccharides, amino acids, fatty acids, and glycerol.
   c. Fat molecules with longer chains of carbon atoms enter the lacteals of the villi.
   d. Fatty acids with relatively short carbon chains enter blood capillaries of the villi.
6. Movements of the small intestine
   a. Movements include mixing and peristalsis.
   b. The ileocecal valve controls movement of the intestinal contents from the small intestine into the large intestine.

Large Intestine (p. 433)

The large intestine reabsorbs water and electrolytes, and forms and stores feces.

1. Parts of the large intestine
   a. The large intestine consists of the cecum, colon, rectum, and anal canal.
   b. The colon is divided into ascending, transverse, descending, and sigmoid portions.
2. Structure of the large intestine wall
   a. The large intestine wall resembles the wall in other parts of the alimentary canal.
   b. The large intestine wall has a unique layer of longitudinal muscle fibers organized in distinct bands.
3. Functions of the large intestine
   a. The large intestine has little or no digestive function.
   b. It secretes mucus.
   c. The large intestine absorbs water and electrolytes.
   d. The large intestine forms and stores feces.
4. Movements of the large intestine
   a. Movements are similar to those in the small intestine.
   b. Mass movements occur two to three times each day.
   c. A defecation reflex stimulates defecation.
5. Feces
   a. Feces consist largely of water, undigested material, electrolytes, mucus, and bacteria.
   b. The color of feces is due to bile salts that bacterial actions alter.

Nutrition and Nutrients (p. 435)

Nutrition is the process of ingestion and utilization of necessary food substances (nutrients).

Carbohydrates (p. 435)

Carbohydrates are organic compounds that primarily supply cellular energy.

1. Carbohydrate sources
   a. Starch, glycogen, disaccharides, and monosaccharides are carbohydrates.
   b. Cellulose is a polysaccharide that human enzymes cannot digest.
2. Carbohydrate utilization
   a. Oxidation releases energy from glucose.
   b. Excess glucose is stored as glycogen or converted to fat.
   c. Most carbohydrates supply energy.
   d. Some cells require a continuous supply of glucose to survive.
3. Carbohydrate requirements
   Humans survive with a wide range of carbohydrate intakes.

Lipids (p. 436)

Lipids are organic compounds that supply energy and are used to build cell structures.

1. Lipid sources
   a. Foods of plant and animal origin provide triglycerides.
   b. Foods of animal origin provide most cholesterol.
2. Lipid utilization
   a. The liver and adipose tissue control triglyceride metabolism.
   b. Linoleic acid is an essential fatty acid.
3. Lipid requirements
   a. The amounts and types of fats required for health are unknown.
   b. Fat intake must be sufficient to supply fat-soluble vitamins.

Proteins (p. 437)

Proteins are organic compounds that serve as structural materials, function as enzymes, and provide energy.

1. Protein sources
   a. Meats, dairy products, cereals, and legumes provide most proteins.
   b. Complete proteins contain adequate amounts of all the essential amino acids.
   c. Incomplete proteins lack adequate amounts of one or more essential amino acids.
2. Protein requirements
   Proteins and amino acids must supply essential amino acids and nitrogen for the synthesis of nitrogen-containing molecules.

Vitamins (p. 438)

Vitamins are organic compounds (other than carbohydrates, lipids, and proteins) that are essential for normal metabolic processes and that body cells cannot synthesize in adequate amounts.

1. Fat-soluble vitamins
   a. These include vitamins A, D, E, and K.
   b. They occur in association with lipids and are influenced by factors that affect lipid absorption.
   c. They resist the effects of heat; thus, cooking or food processing does not destroy them.
2. Water-soluble vitamins
   a. This group includes the B vitamins and vitamin C.
   b. B vitamins make up a group (the vitamin B complex) and oxidize carbohydrates, lipids, and proteins.
   c. Cooking or food processing destroys some water-soluble vitamins.

Minerals (p. 439)

Dietary minerals are elements other than carbon that are essential in human metabolism.

1. Characteristics of minerals
   a. Most minerals are in the bones and teeth.
   b. Minerals are usually incorporated into organic molecules; some occur in inorganic compounds or as free ions.
   c. They comprise structural materials, function in enzymes, and are vital in metabolic processes.
2. Major minerals
   They include calcium, phosphorus, potassium, sulfur, sodium, chlorine, and magnesium.
3. Trace elements
   They include iron, manganese, copper, iodine, cobalt, zinc, fluorine, selenium, and chromium.

Adequate Diets (p. 443)

1. An adequate diet provides sufficient energy and essential nutrients to support optimal growth, maintenance, and repair of tissues.
2. Individual requirements vary so greatly that designing a diet that is adequate for everyone is not possible.
3. Malnutrition is poor nutrition due to lack of foods or failure to make the best use of available foods.

Review Exercises

1. List and describe the locations of the major parts of the alimentary canal. (p. 409)
2. List and describe the locations of the accessory organs of the digestive system. (p. 409)
3. Name the four layers of the wall of the alimentary canal. (p. 409)
4. Distinguish between mixing movements and propelling movements. (p. 411)
5. Define *peristalsis*. (p. 411)
6. Discuss the functions of the mouth and its parts. (p. 411)
7. Distinguish between the lingual, palatine, and pharyngeal tonsils. (p. 412)
8. Compare the primary and secondary teeth. (p. 412)
9. Describe the structure of a tooth. (p. 413)
10. Explain how a tooth is anchored to its socket. (p. 413)
11. List and describe the locations of the major salivary glands. (p. 415)
12. Explain how the secretions of the salivary glands differ. (p. 415)
13. Discuss the digestive functions of saliva. (p. 415)
14. Explain the function of the esophagus. (p. 416)
15. Describe the structure of the stomach. (p. 417)
16. List the enzymes in gastric juice, and explain the function of each enzyme. (p. 419)
17. Define *cholecystokinin*. (p. 419)
18. Explain the regulation of gastric secretions. (p. 419)
19. Describe the location of the pancreas and the pancreatic duct. (p. 421)
20. List and explain the function of each enzyme found in pancreatic juice. (p. 421)
21. Explain the regulation of pancreatic secretions. (p. 422)
22. Describe liver structure. (p. 423)
23. List the major liver functions. (p. 424)
24. Describe the composition of bile. (p. 424)
25. Explain the functions of bile salts. (p. 427)
26. List and describe the locations of the parts of the small intestine. (p. 427)

27. Describe the functions of intestinal villi. (p. 428)
28. Name and explain the function of each enzyme of the intestinal mucosa. (p. 429)
29. Explain the regulation of small intestine secretions. (p. 429)
30. Summarize how each major type of digestive product is absorbed. (p. 430)
31. List and describe the locations of the parts of the large intestine. (p. 433)
32. Explain the general functions of the large intestine. (p. 434)
33. Describe the defecation reflex. (p. 435)
34. List some common sources of carbohydrates. (p. 435)
35. Summarize the importance of cellulose in the diet. (p. 435)
36. Explain why a temporary drop in glucose concentration may produce functional disorders of the nervous system. (p. 436)
37. List some common sources of lipids. (p. 436)
38. Describe the liver's role in fat metabolism. (p. 436)
39. List some common protein sources. (p. 437)
40. Distinguish between essential and nonessential amino acids. (p. 437)
41. Distinguish between complete and incomplete proteins. (p. 437)
42. Discuss the general characteristics of fat-soluble vitamins. (p. 438)
43. List the fat-soluble vitamins, and describe the major functions of each. (p. 439)
44. Discuss the general characteristics of the mineral nutrients. (p. 439)
45. List the water-soluble vitamins, and describe the major functions of each. (p. 440)
46. List the major minerals, and describe the major functions of each. (p. 441)
47. List the trace elements, and describe the major functions of each. (p. 442)
48. Define *adequate diet.* (p. 443)
49. Define *malnutrition.* (p. 443)

Critical Thinking

1. How does mechanical digestion enhance chemical digestion?
2. How can too little fat in the diet lead to a vitamin deficiency, even if a person takes vitamin supplements?
3. Why are vitamins required only in small amounts?
4. How can people consume vastly different diets, yet all obtain adequate nourishment?
5. If a patient has 95% of the stomach removed (subtotal gastrectomy) as treatment for severe ulcers or cancer, how would the digestion and absorption of foods be affected? How would the patient's eating habits have to be altered? Why?
6. Why may a person with an inflammation of the gallbladder (cholecystitis) also develop an inflammation of the pancreas (pancreatitis)?
7. Why does the blood sugar concentration of a person whose diet is relatively low in carbohydrates remain stable?
8. Examine the label information on the packages of a variety of dry breakfast cereals. Which types of cereals provide adequate sources of vitamins and minerals?

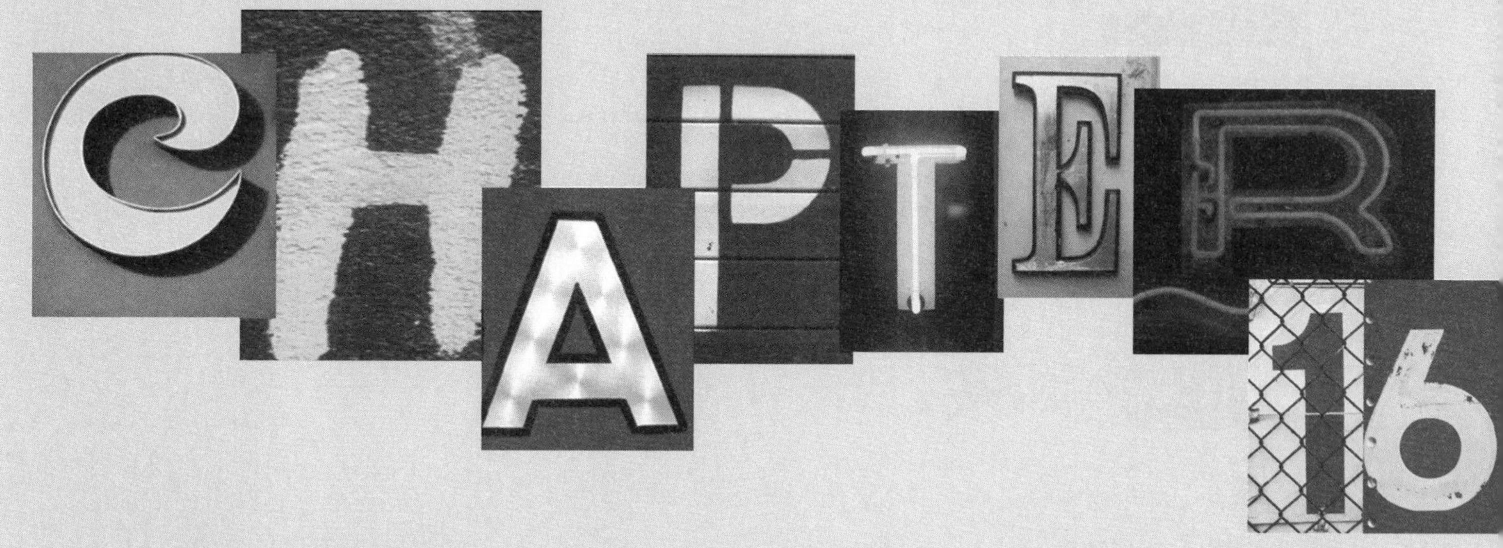

Chapter 16 Respiratory System

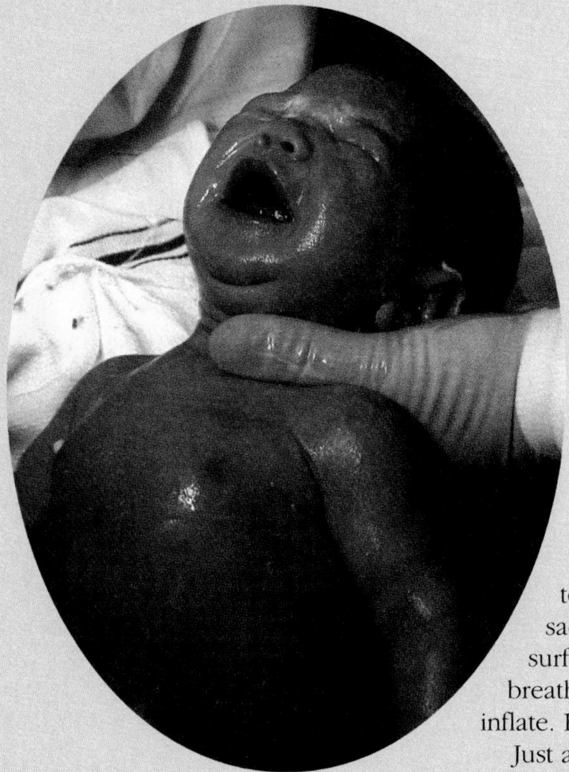

When Benjamin McClatchey was born nearly twelve weeks prematurely in 1990, he weighed only 2 pounds, 13 ounces. Like many of the 380,000 "preemies" born each year in the United States, Benjamin had respiratory distress syndrome (RDS)—today, a treatable disorder.

RDS occurs because a premature baby's lungs are too immature to produce a mixture of lipoproteins called surfactant. This substance lies between the watery inner surface of the lungs and the inhaled air, reducing surface tension and thereby enabling the millions of microscopic air sacs called alveoli to inflate. Normally, fetal lungs manufacture surfactant in the weeks preceding birth, so that when that first breath of air enters the respiratory system, the tiny air sacs can inflate. Benjamin had not had time to do this.

Just a decade ago, Benjamin might not have survived RDS. But with the help of a synthetic surfactant dripped into his lungs through an endotracheal tube and a ventilator machine designed to assist breathing in premature infants, he survived. Unlike conventional ventilators, which force air into the lungs at pressures that could damage delicate newborn lungs, the high-frequency ventilator used with preemies delivers lifesaving oxygen in tiny, gentle puffs.

Photo: We begin life with a healthy respiratory system. Exposure to pollutants and smoking can severely damage this system.

Cells require oxygen to oxidize nutrients and release energy, and must excrete the carbon dioxide that results from oxidation. Obtaining oxygen and removing carbon dioxide are the primary functions of the *respiratory system*. The respiratory organs also trap particles from incoming air, help control the temperature and water content of incoming air, produce vocal sounds, and play important roles in the sense of smell and in regulation of blood pH.

Chapter Objectives

After studying this chapter, you should be able to:

1. List the general functions of the respiratory system. (p. 451)
2. Name and describe the locations of the organs of the respiratory system. (p. 451)
3. Describe the functions of each organ of the respiratory system. (p. 451)
4. Explain the mechanisms of inspiration and expiration. (p. 457)
5. Name and define each of the lung volumes and respiratory capacities. (p. 461)
6. Locate the respiratory center, and explain how it controls normal breathing. (p. 463)
7. Discuss how various factors affect the respiratory center. (p. 464)
8. Describe the structure and function of the respiratory membrane. (p. 465)
9. Explain how air and blood exchange gases and how blood transports these gases. (p. 466)

Aids to Understanding Words

alveol- [small cavity] *alveol*us: A microscopic air sac within a lung.

bronch- [windpipe] *bronch*us: A primary branch of the trachea.

cric- [ring] *cric*oid cartilage: A ring-shaped mass of cartilage at the base of the larynx.

epi- [upon] *epi*glottis: A flaplike structure that partially covers the opening into the larynx during swallowing.

hem- [blood] *hem*oglobin: The pigment in red blood cells that transports oxygen and carbon dioxide.

Key Terms

alveolus (al-ve′o-lus); plural: **alveoli** (al-ve′o-li)

bronchial tree (brong′ke-al tre)

carbaminohemoglobin (kar-bam″ĭ-no-he″mo-glo′bin)

carbonic anhydrase (kar-bon′ik an-hi′drās)

cellular respiration (sel′u-lar res″pĭ-ra′shun)

expiration (ek″spĭ-ra′shun)

glottis (glot′is)

hemoglobin (he″mo-glo′bin)

hyperventilation (hi″per-ven″tĭ-la′shun)

inspiration (in″spĭ-ra′shun)

oxyhemoglobin (ok″se-he″mo-glo′bin)

partial pressure (par′shil presh′ur)

pleural cavity (ploo′ral kav′ĭ-te)

respiratory capacities (re-spi′rah-to″re kah-pas′ĭ-tēz)

respiratory center (re-spi′rah-to″re sen′ter)

respiratory membrane (re-spi′rah-to″re mem′brān)

respiratory volume (re-spi′rah-to″re vol′ūm)

surface tension (ser′fas ten′shun)

surfactant (ser-fak′tant)

Introduction

The respiratory system includes tubes that remove particles from (filter) incoming air and transport air into and out of the lungs. The system also includes many microscopic air sacs where gases are exchanged.

The entire process of gas exchange between the atmosphere and body cells is called **respiration.** Events of respiration include: (1) movement of air in and out of the lungs—commonly called breathing or *ventilation;* (2) gas exchange between blood and air in the lungs; (3) gas transport in blood between the lungs and body cells; and (4) gas exchange between blood and body cells. Oxygen utilization and carbon dioxide production at the cellular level is called **cellular respiration.**

Organs of the Respiratory System

The organs of the respiratory system include the nose, nasal cavity, paranasal sinuses, pharynx, larynx, trachea, bronchial tree, and lungs (fig. 16.1 and reference plates 3, 4, 5, and 6). The parts of the respiratory system can be divided into two sets, or tracts. Those organs outside the thorax constitute the *upper respiratory tract;* those within the thorax comprise the *lower respiratory tract.*

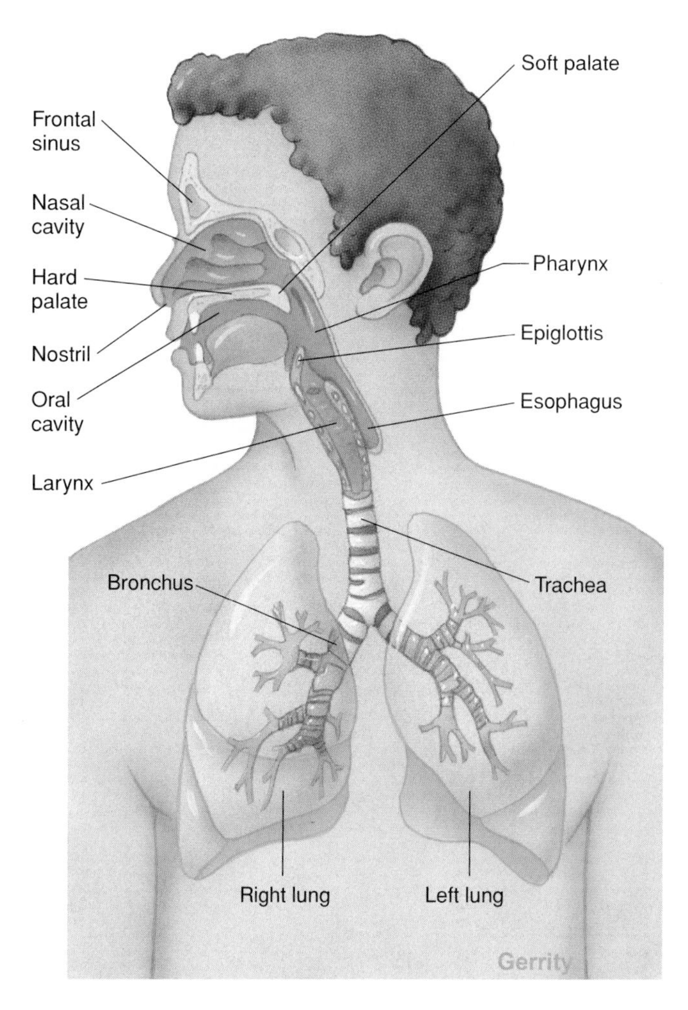

Figure 16.1

Organs of the respiratory system.

Nose

Bone and cartilage support the **nose** internally. Its two *nostrils* are openings through which air can enter and leave the nasal cavity. Many internal hairs guard the nostrils, preventing entry of large particles carried in the air.

Nasal Cavity

The **nasal cavity** is a hollow space behind the nose. The **nasal septum**, composed of bone and cartilage, divides the nasal cavity into right and left portions. **Nasal conchae** curl out from the lateral walls of the nasal cavity on each side, dividing the cavity into passageways (fig. 16.2). Nasal conchae also support the mucous membrane that lines the nasal cavity and help increase its surface area.

The mucous membrane contains pseudostratified ciliated epithelium that is rich in mucus-secreting goblet cells (see chapter 5). It also includes an extensive network of blood vessels, and as air passes over the membrane, heat leaves the blood and warms the air, adjusting the air's temperature to that of the body. In addition, incoming air moistens as water evaporates from the mucous lining. The sticky mucus the mucous membrane secretes entraps dust and other small particles entering with the air.

The nasal septum is usually straight at birth, although it sometimes bends as the result of a birth injury. It remains straight throughout early childhood, but as a person ages, the septum bends toward one side or the other. Such a *deviated septum* may obstruct the nasal cavity, making breathing difficult. ■

As the cilia of the epithelial lining move, they push a thin layer of mucus and entrapped particles toward the pharynx, where the mucus is swallowed (fig. 16.3). In the stomach, gastric juice destroys microorganisms in the mucus.

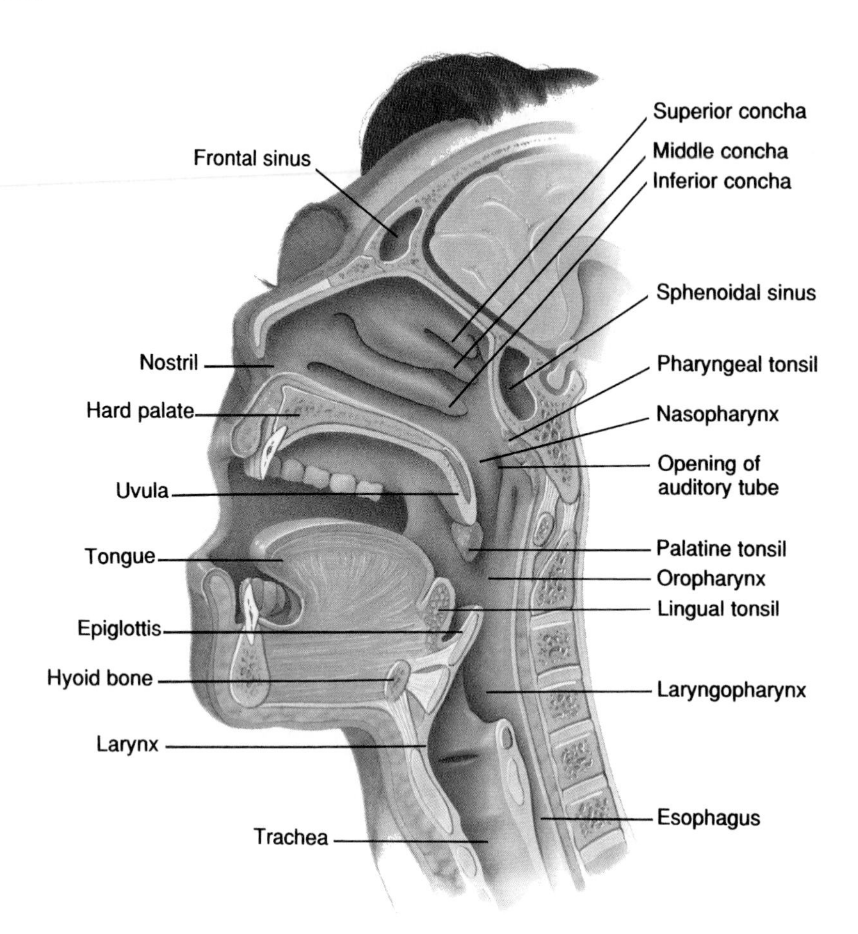

Figure 16.2

Major features of the upper respiratory tract.

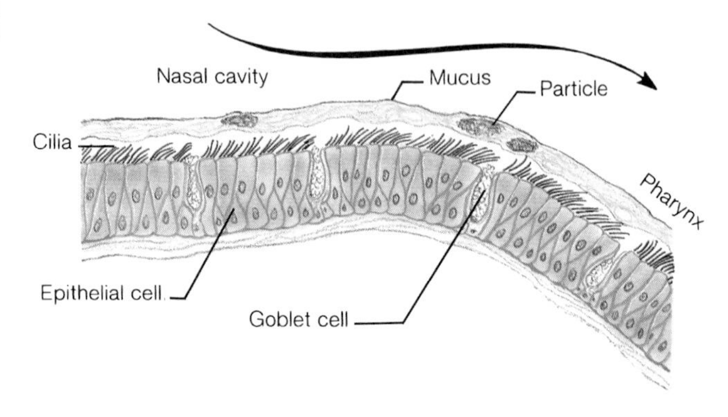

Figure 16.3

Cilia move mucus and trapped particles from the nasal cavity to the pharynx.

1. What is respiration?
2. What organs constitute the respiratory system?
3. What are the functions of the mucous membrane that lines the nasal cavity?

Paranasal Sinuses

Recall from chapter 7 that the **paranasal sinuses** are air-filled spaces located within the *maxillary, frontal, ethmoid,* and *sphenoid bones* of the skull and opening into the nasal cavity. Mucous membranes line the sinuses and are continuous with the lining of the nasal cavity. The paranasal sinuses reduce the weight of the skull and are resonant chambers that affect the quality of the voice.

A painful sinus headache can result from blocked drainage, caused by an infection or allergic reaction. ■

1. Where are the paranasal sinuses located?
2. What are the functions of the paranasal sinuses?

Pharynx

The **pharynx** (throat) is behind the oral cavity and between the nasal cavity and larynx (see figs. 16.1 and 16.2). It is a passageway for food traveling from the oral cavity to the esophagus and for air passing between the nasal cavity and larynx. It also helps produce the sounds of speech. Chapter 15 describes the subdivisions of the pharynx—nasopharynx, oropharynx, and laryngopharynx.

Larynx

The **larynx** is an enlargement in the airway at the top of the trachea and below the pharynx. It conducts air in and out of the trachea and prevents foreign objects from entering the trachea. It also houses the *vocal cords.*

The larynx is composed of a framework of muscles and cartilages bound by elastic tissue. The largest of the cartilages are the *thyroid, cricoid,* and *epiglottic cartilages* (fig. 16.4).

Inside the larynx, two pairs of horizontal folds composed of muscle tissue and connective tissue with a covering of mucous membrane extend inward from the lateral walls. The upper folds are called *false vocal cords* because they do not produce sounds (fig. 16.5*a*). Muscle fibers within these folds help close the airway during swallowing.

The lower folds of muscle tissue and elastic fibers are the *true vocal cords.* Air forced between these vocal cords causes the cords to vibrate from side to side, which generates sound waves. Changing the shapes of the pharynx and oral cavity and using the tongue and lips transform these sound waves into words.

Contracting or relaxing muscles that alter the tension on the vocal cords controls the pitch (musical tone) of a sound. Increasing tension raises pitch, and decreasing tension lowers pitch. The intensity (loudness) of a sound reflects the force of air passing through the vocal folds. Stronger blasts of air produce louder sound; weaker blasts produce softer sound.

Damage to the nerves (recurrent laryngeal nerves) that supply the laryngeal muscles can alter the quality of a person's voice. These nerves pass through the neck as parts of the vagus nerves, and they can be injured from trauma or surgery to the neck or thorax. *Nodules* or other growths on the margins of the vocal folds that interfere with the free flow of air over the folds can also cause vocal problems. Surgery can remove such lesions. ■

During normal breathing, the vocal cords are relaxed, and the opening between them, called the **glottis**, is a triangular slit (fig. 16.5). When food or liquid is swallowed, however, muscles within the false vocal cords close the glottis, which prevents food or liquid from entering the trachea.

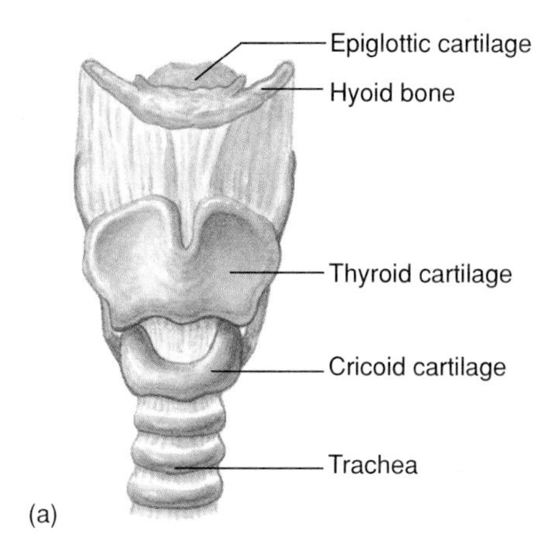

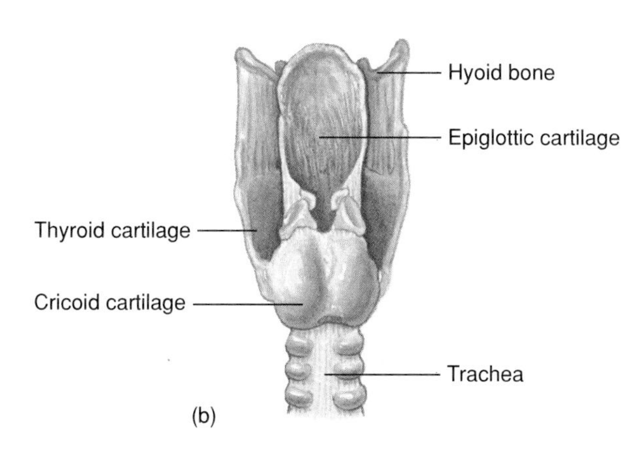

Figure 16.4

(*a*) Anterior and (*b*) posterior views of the larynx.

The epiglottic cartilage supports a flaplike structure called the **epiglottis.** This structure usually stands upright and allows air to enter the larynx. During swallowing, however, the larynx rises, and the epiglottis presses downward to partially cover the opening into the larynx. This helps prevent foods and liquids from entering the air passages (see chapter 15).

Laryngitis—hoarseness or lack of voice—occurs when the mucous membrane of the larynx becomes inflamed and swollen from an infection or an irritation from inhaled vapors, and prevents the vocal cords from vibrating as freely as before. Laryngitis is usually mild, but may be dangerous if swollen tissues obstruct the airway and interfere with breathing. Inserting a tube (endotracheal tube) into the trachea through the nose or mouth can restore the passageway until the inflammation subsides. ■

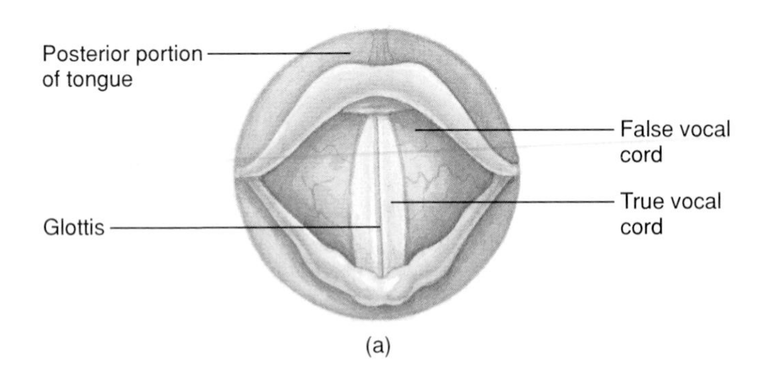

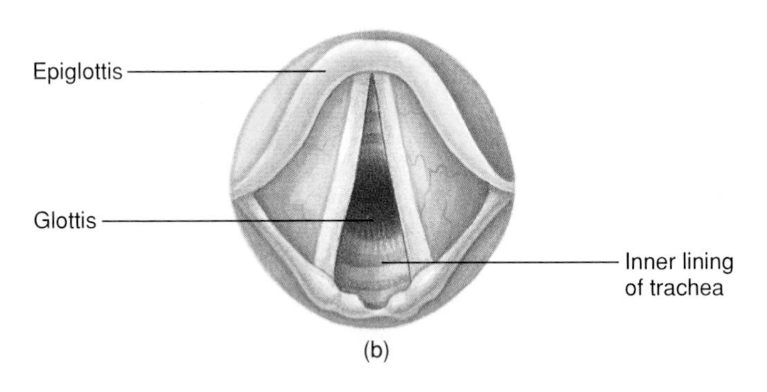

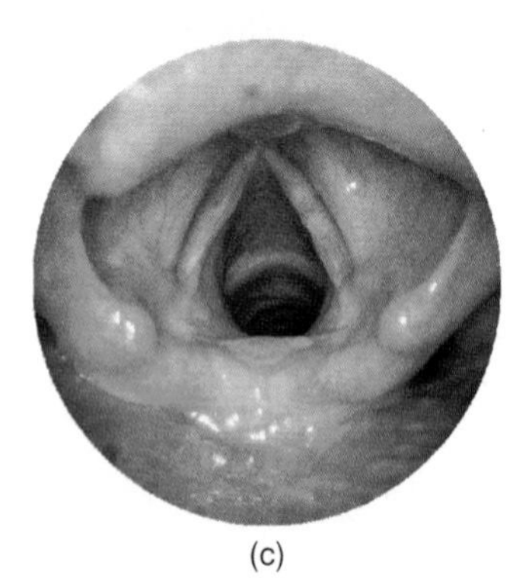

Figure 16.5

Vocal cords as viewed from above with the glottis (*a*) closed and (*b*) open. (*c*) Photograph of the glottis and vocal folds.

1. Describe the structure of the larynx.
2. How do the vocal cords produce sounds?
3. What is the function of the glottis? The epiglottis?

Trachea

The **trachea** (windpipe) is a flexible, cylindrical tube about 2.5 centimeters in diameter and 12.5 centimeters in length (fig. 16.6). It extends downward in front of the esophagus and into the thoracic cavity, where it splits into right and left bronchi.

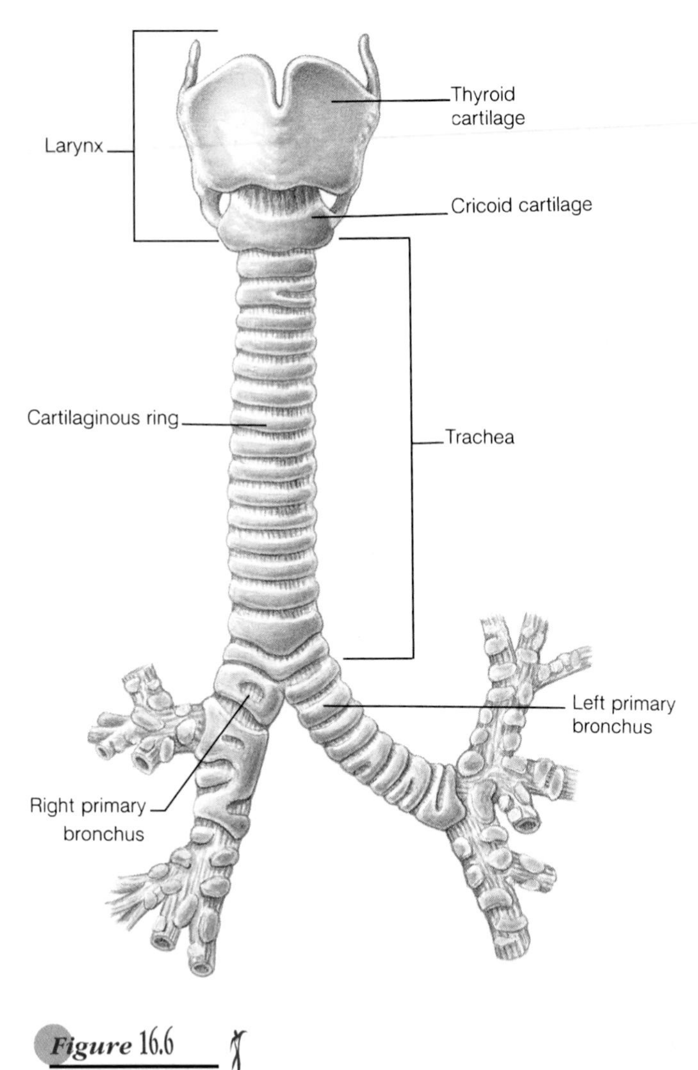

Figure 16.6

The trachea transports air between the larynx and the bronchi.

A ciliated mucous membrane with many goblet cells lines the trachea's inner wall. This membrane filters incoming air and moves entrapped particles upward into the pharynx, where the mucus can be swallowed.

Within the tracheal wall are about twenty C-shaped pieces of hyaline cartilage, one above the other. The open ends of these incomplete rings are directed posteriorly, and smooth muscle and connective tissues fill the gaps between the ends. These cartilaginous rings prevent the trachea from collapsing and blocking the airway. The soft tissues that complete the rings in the back allow the nearby esophagus to expand as food moves through it to the stomach.

Bronchial Tree

The **bronchial tree** consists of branched airways leading from the trachea to the microscopic air sacs in the lungs (fig. 16.7). Its branches begin with the right and left **primary bronchi**, which arise from the trachea at the level of the fifth thoracic vertebra.

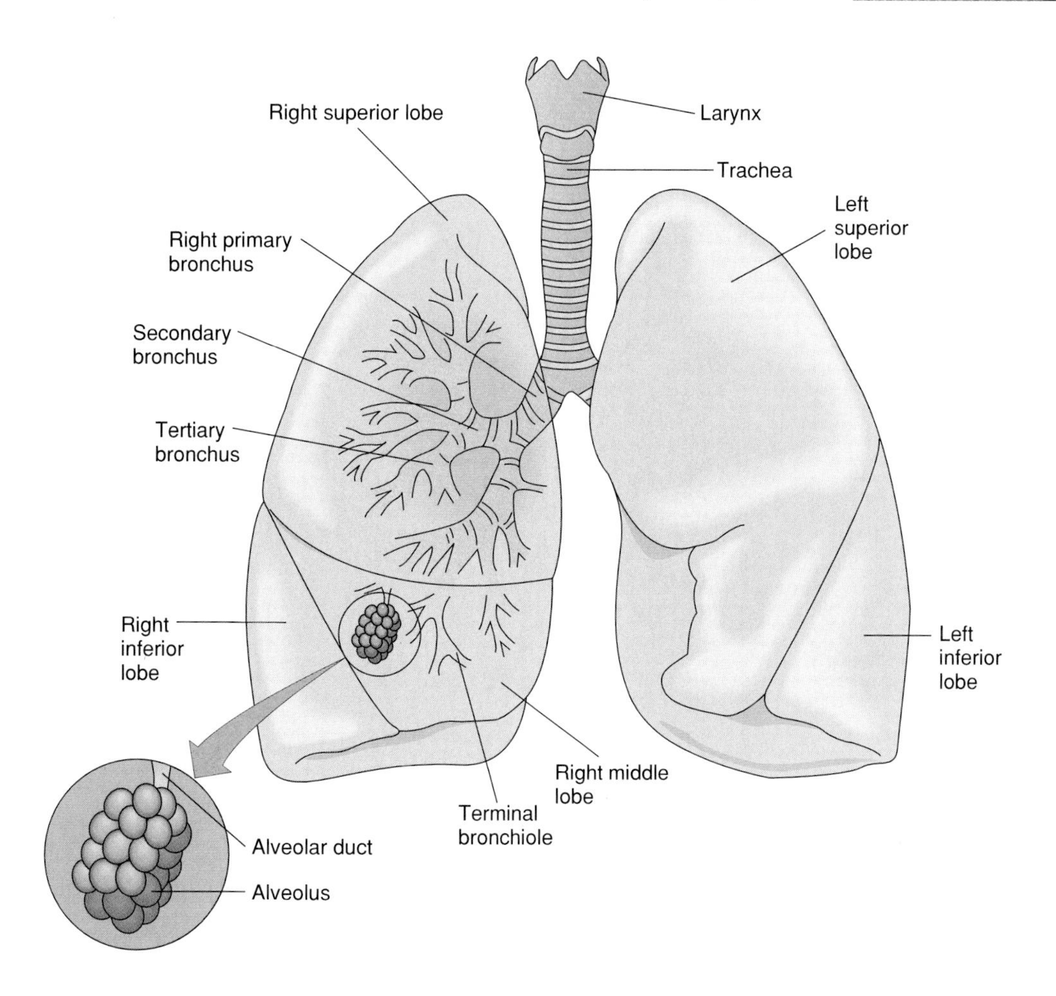

Figure 16.7

The bronchial tree consists of the passageways that connect the trachea and alveoli. The alveolar duct and alveoli are enlarged to show their location.

A short distance from its origin, each primary bronchus divides into secondary bronchi, which in turn, branch into finer and finer tubes. Among these smaller tubes are **bronchioles** that continue to divide, giving rise to very thin tubes called **alveolar ducts.** These ducts terminate in groups of microscopic air sacs called **alveoli**, which lie within capillary nets (figs. 16.8 and 16.9).

The structure of a bronchus is similar to that of the trachea, but as finer and finer branch tubes appear, the amount of cartilage in the walls decreases and finally disappears in the bronchioles. As the cartilage diminishes, a layer of smooth muscle surrounding the tube becomes more prominent. This muscular layer remains even in the smallest bronchioles, but only a few muscle fibers are in the alveolar ducts.

The branches of the bronchial tree are air passages, whose mucous membranes filter incoming air and distribute the air to alveoli throughout the lungs. The alveoli provide a large surface area of thin simple squamous epithelial cells through which gases can easily be exchanged. Oxygen diffuses through alveolar walls and enters blood in nearby capillaries, and carbon dioxide diffuses from blood through the walls and enters alveoli (fig. 16.10). An adult lung has about 300 million alveoli, providing a total surface area half the size of a tennis court.

1. What is the function of the cartilaginous rings in the tracheal wall?
2. Describe the bronchial tree.
3. Explain how gases are exchanged in the alveoli.

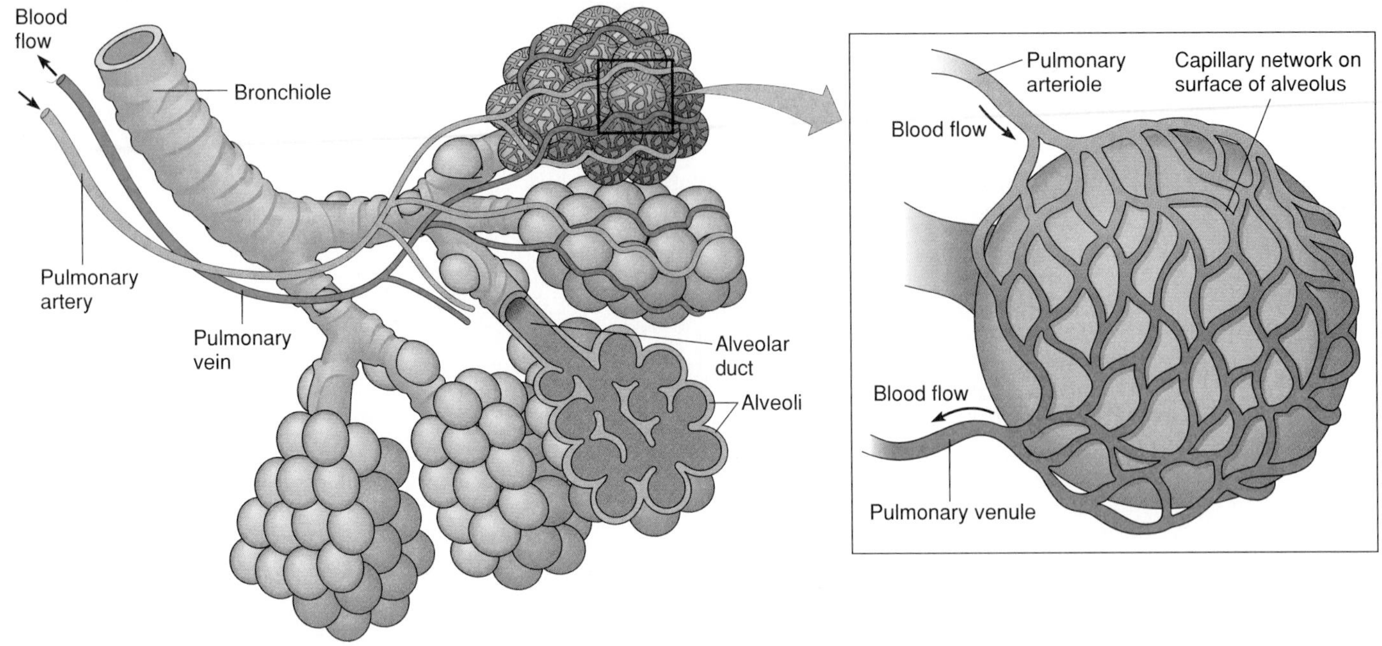

Figure 16.8

The respiratory tubes end in tiny alveoli, each of which is surrounded by a capillary network.

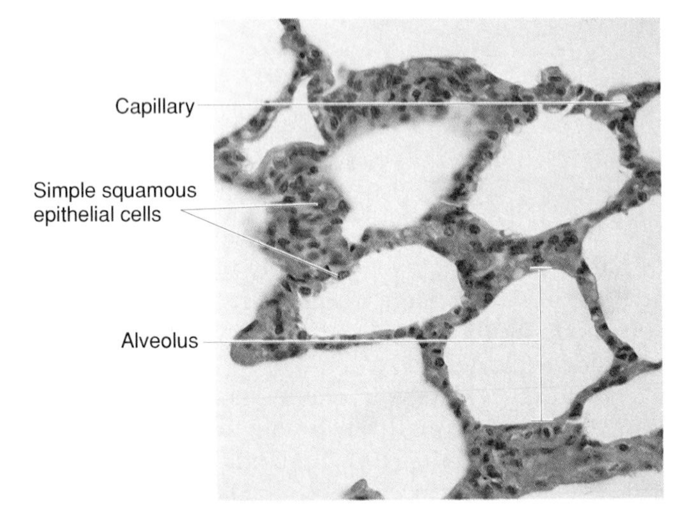

Figure 16.9

Light micrograph of alveoli (250×).

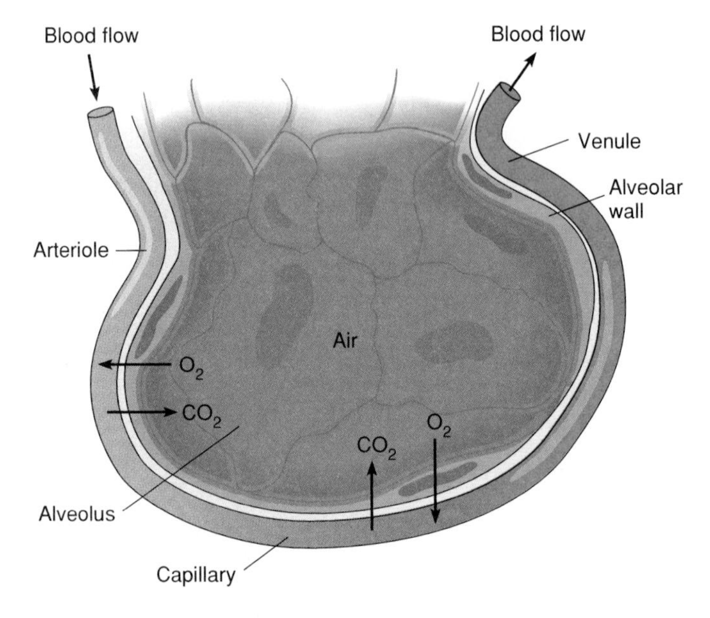

Figure 16.10

Oxygen (O_2) diffuses from air within the alveolus into the capillary, while carbon dioxide (CO_2) diffuses from blood within the capillary into the alveolus.

Table 16.1 Parts of the Respiratory System

Part	Description	Function
Nose	Part of face centered above mouth and below space between eyes	Nostrils provide entrance to nasal cavity; internal hairs begin to filter incoming air
Nasal cavity	Hollow space behind nose	Conducts air to pharynx; mucous lining filters, warms, and moistens incoming air
Paranasal sinuses	Hollow spaces in certain skull bones	Reduce weight of skull; serve as resonant chambers
Pharynx	Chamber behind oral cavity and between nasal cavity and larynx	Passageway for air moving from nasal cavity to larynx and for food moving from oral cavity to esophagus
Larynx	Enlargement at top of trachea	Passageway for air; prevents foreign objects from entering trachea; houses vocal cords
Trachea	Flexible tube that connects larynx with bronchial tree	Passageway for air; mucous lining continues to filter particles from incoming air
Bronchial tree	Branched tubes that lead from trachea to alveoli	Conducts air from trachea to alveoli; mucous lining continues to filter incoming air
Lungs	Soft, cone-shaped organs that occupy a large portion of the thoracic cavity	Contain air passages, alveoli, blood vessels, connective tissues, lymphatic vessels, and nerves of the lower respiratory tract

Lungs

The **lungs** are soft, spongy, cone-shaped organs in the thoracic cavity (see fig. 16.1 and reference plates 4 and 5). The heart and mediastinum separate the right and left lungs medially, and the diaphragm and thoracic cage enclose them.

Each lung occupies most of the thoracic space on its side. A bronchus and some large blood vessels suspend each lung in the cavity. These tubular structures enter the lung on its medial surface. A layer of serous membrane, the **visceral pleura**, firmly attaches to each lung surface and folds back to become the **parietal pleura.** The parietal pleura, in turn, forms part of the mediastinum and lines the inner wall of the thoracic cavity (see fig. 1.8).

No significant space exists between the visceral and parietal pleurae, but the potential space between them is called the **pleural cavity.** It contains a thin film of serous fluid that lubricates adjacent pleural surfaces, reducing friction as they move against one another during breathing. This fluid also helps hold the pleural membranes together, as explained in the next section.

The right lung is larger than the left one and is divided into three lobes. The left lung has two lobes (see fig. 16.1).

A major branch of the bronchial tree supplies each lobe. A lobe also has connections to blood and lymphatic vessels and lies within connective tissues. Thus, a lung includes air passages, alveoli, blood vessels, connective tissues, lymphatic vessels, and nerves.

Table 16.1 summarizes the characteristics of the major parts of the respiratory system.

1. Where are the lungs located?
2. What is the function of serous fluid within the pleural cavity?
3. What kinds of structures make up a lung?

Breathing Mechanism

Breathing, or ventilation, is the movement of air from outside the body into and out of the bronchial tree and alveoli. The actions providing these air movements are termed **inspiration** (inhalation) and **expiration** (exhalation).

Inspiration

Atmospheric pressure due to the weight of air is the force that moves air into the lungs. At sea level, this pressure is sufficient to support a column of mercury about 760 millimeters (mm) high in a tube. Thus, normal air pressure is equal to 760 mm of mercury (Hg).

TOPIC OF INTEREST

Emphysema and Lung Cancer

Emphysema is a progressive, degenerative disease that destroys alveolar walls. As a result, clusters of small air sacs merge to form larger chambers, which drastically decreases the surface area of the respiratory membrane and thereby reduces the volume of gases that can be exchanged through the membrane. Alveolar walls lose some of their elasticity, and capillary networks associated with the alveoli diminish (fig. 16A)

Loss of tissue elasticity makes it increasingly difficult for a person with emphysema to force air out of the lungs because normal expiration involves the passive elastic recoil of inflated tissues. Consequently, the person must exert abnormal muscular effort to exhale.

Emphysema may develop in response to prolonged exposure to respiratory irritants, such as those in tobacco smoke and polluted air. The disease may also result from an inherited enzyme deficiency.

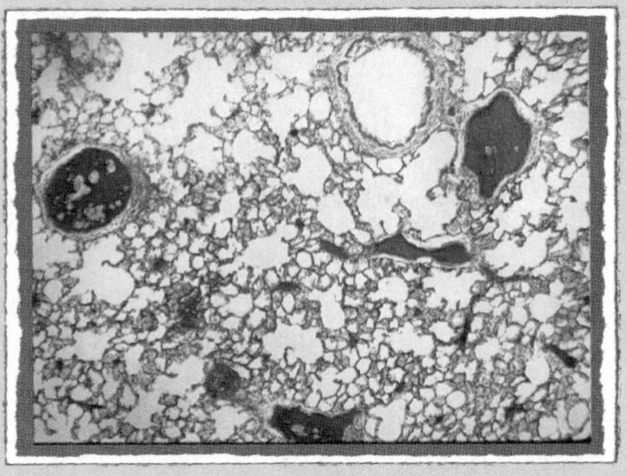

(a)

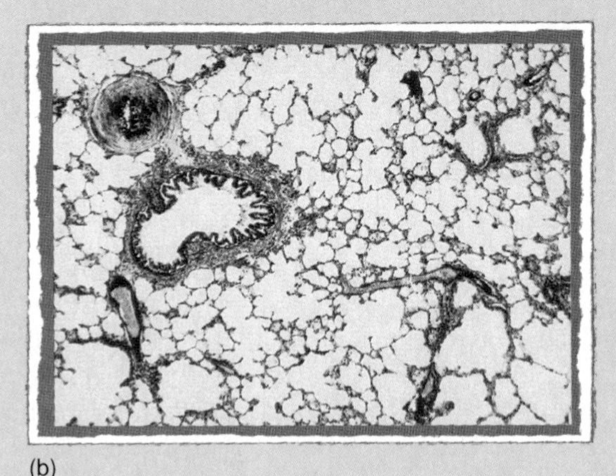

(b)

***Figure* 16A**

(*a*) Normal lung tissue. (*b*) As emphysema develops, the alveoli tend to merge, forming larger chambers.

Air pressure is exerted on all surfaces in contact with the air, and because people breathe air, the inside surfaces of their lungs also are subjected to pressure. The pressures on the inside of the lungs and alveoli and on the outside of the thoracic wall are about the same.

If the pressure inside the lungs and alveoli decreases, atmospheric pressure will push outside air into the airways. That is what happens during normal inspiration. Impulses carried on the phrenic nerves, which are associated with the cervical plexuses (see chapter 9), stimulate muscle fibers in the dome-shaped *diaphragm* below the lungs to contract. The diaphragm moves downward, the thoracic cavity enlarges, and the pressure within the alveoli falls to about 2 mm Hg below that of atmospheric pressure. In response, atmospheric pressure forces air into the airways (fig. 16.11).

While the diaphragm is contracting and moving downward, the *external (inspiratory) intercostal muscles* between the ribs may be stimulated to contract. This raises the ribs and elevates the sternum, enlarging the thoracic cavity even more. As a result, the pressure inside is reduced further and the relatively greater atmospheric pressure forces even more air into the airways.

Lung expansion in response to movements of the diaphragm and chest wall depends on movements of the pleural membranes. Only a thin film of serous fluid separates the parietal pleura on the inner wall of the thoracic cavity from the visceral pleura attached to the surface of the lungs. The water molecules in this fluid greatly attract one another, creating a force called **surface tension** that holds the moist surfaces of the pleural membranes tightly together. Any separation of

Lung cancer, like other cancers, is the uncontrolled growth of abnormal cells that rob normal cells of nutrients and oxygen, eventually crowding them out. Some cancerous growths in the lungs result secondarily from cancer cells that have spread (metastasized) from other parts of the body, such as the breasts, intestines, liver, or kidneys. Cancers that begin in the lungs are called *primary pulmonary cancers.* These may arise from epithelia, connective tissue, or blood cells. The most common form originates from epithelium and is called *bronchogenic carcinoma.* This type of cancer is a response to irritation, such as prolonged exposure to tobacco smoke (fig. 16B). Susceptibility to primary pulmonary cancers may be inherited.

Once cancer cells appear, they grow into masses that obstruct air passages and reduce gas exchange. Furthermore, bronchogenic carcinoma can spread quickly and establish secondary cancers in the lymph nodes, liver, bones, brain, or kidneys.

Lung cancer is difficult to control. Usually, it is treated with surgery, ionizing radiation, and drugs (chemotherapy). Despite treatment, however, the survival rate among lung cancer patients remains low.

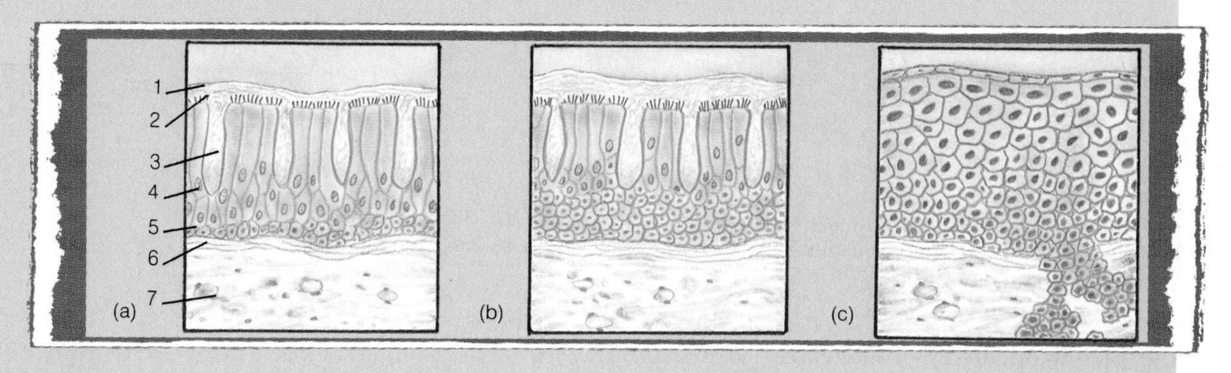

Figure 16B

Lung cancer usually starts in the lining (epithelium) of a bronchus. (*a*) The normal lining shows (4) columnar cells with (2) hairlike cilia, (3) goblet cells that secrete (1) mucus, and (5) basal cells from which new columnar cells arise. (6) A basement membrane separates the epithelial cells from (7) the underlying connective tissue. (*b*) In the first stage of lung cancer, the basal cells divide repeatedly. The goblet cells secrete excess mucus, and the cilia are less efficient in moving the heavy mucus secretion. (*c*) Continued multiplication of basal cells displaces the columnar and goblet cells. The basal cells penetrate the basement membrane and invade the deeper connective tissue.

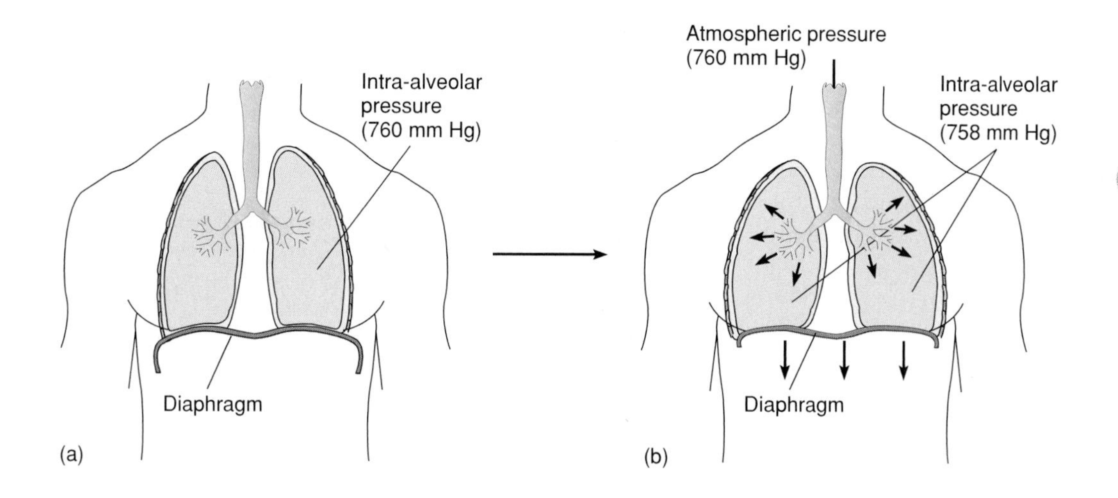

Figure 16.11

(*a*) Prior to inspiration, the intra-alveolar pressure is 760 millimeters of mercury (mm Hg). (*b*) The intra-alveolar pressure decreases to about 758 mm Hg as the thoracic cavity enlarges, and atmospheric pressure forces air into the airways.

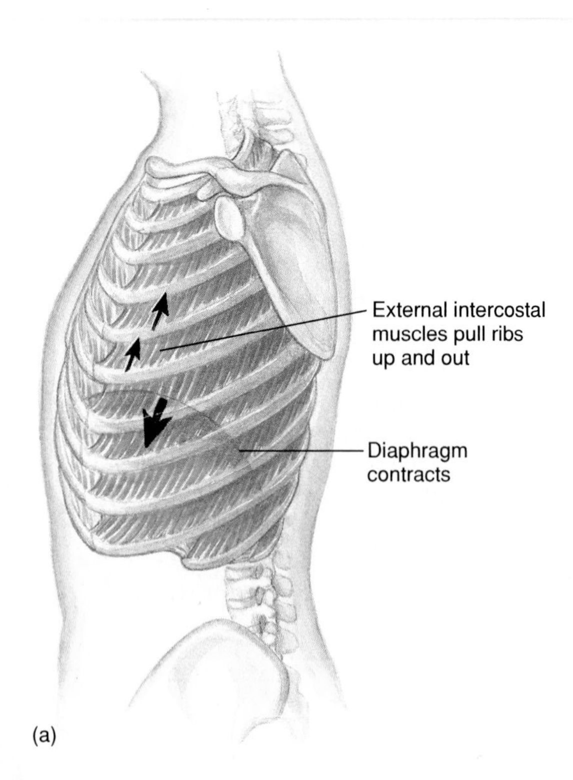

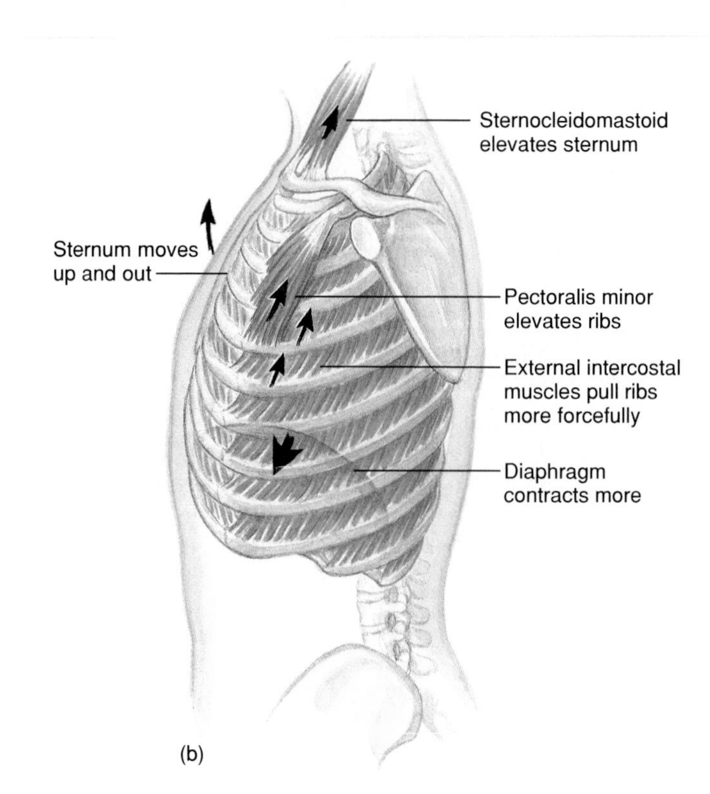

Figure 16.12

(*a*) Shape of the thoracic cavity at the end of normal inspiration. (*b*) Shape of the thoracic cavity at the end of maximal inspiration, aided by contraction of the sternocleidomastoid and pectoralis minor muscles.

the pleural membranes decreases pressure in the intrapleural space, which also tends to hold these membranes together. Consequently, when the intercostal muscles move the thoracic wall upward and outward, the parietal pleura moves, too, and the visceral pleura follows it. This helps expand the lungs in all directions.

The surface tension between the adjacent moist membranes is sufficient to collapse the alveoli, which have moist inner surfaces. Certain alveolar cells, however, synthesize a mixture of lipoproteins called **surfactant.** Surfactant, which is secreted continuously into alveolar air spaces, reduces surface tension and decreases alveoli's tendency to collapse when lung volume is low.

If a person needs to take a deeper than normal breath, the diaphragm and external intercostal muscles contract more. Additional muscles, such as the pectoralis minors and sternocleidomastoids, can also pull the thoracic cage farther upward and outward, enlarging the thoracic cavity and decreasing internal pressure (fig. 16.12).

Expiration

The forces responsible for normal expiration come from the *elastic recoil* of tissues and from surface tension. The lungs and thoracic wall contain considerable elastic tissue, and lung expansion during inspiration stretches this tissue. Also, as the diaphragm lowers, it compresses the abdominal organs beneath it. As the diaphragm and external intercostal muscles relax following inspiration, the elastic tissues cause the lungs and thoracic cage to recoil and return to their original shapes. Similarly, the abdominal organs spring back into their previous shapes, pushing the diaphragm upward (fig. 16.13*a*). At the same time, the surface tension that develops between the moist surfaces of the alveolar linings decreases the diameters of the alveoli. Each of these factors increases aveolar pressure about 1 mm Hg above atmospheric pressure, so that the air inside the lungs is forced out through respiratory passages. Thus, normal expiration is a passive process.

Because surface tension adheres the visceral and parietal pleural membranes, no actual space normally exists in the pleural cavity between them. A puncture in the thoracic wall, however, allows atmospheric air to enter the pleural cavity and create a real space between the membranes. This condition, called *pneumothorax,* may collapse the lung on the affected side because of the lung's elasticity. ■

Diaphragm

Abdominal organs recoil and press diaphragm upward

(a)

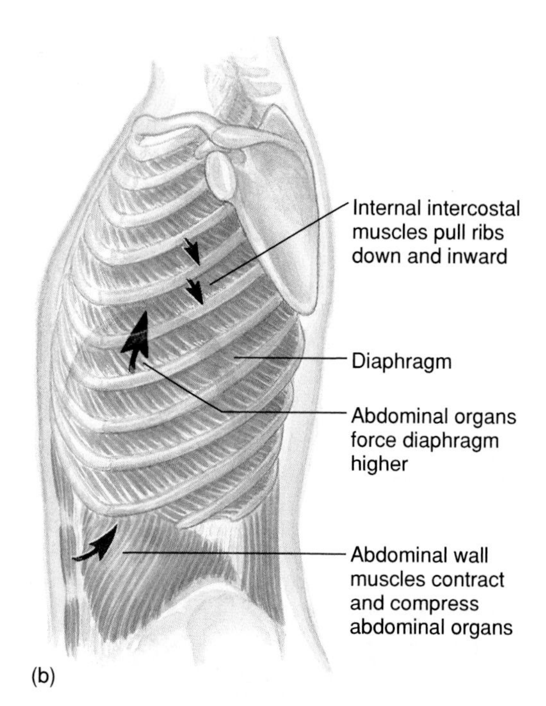

Figure 16.13

(*a*) Normal resting expiration is due to elastic recoil of lung tissues, the thoracic wall, and abdominal organs. (*b*) Contraction of abdominal wall muscles and posterior internal intercostal muscles aids maximal expiration.

If a person needs to exhale more air than normal, the posterior *internal (expiratory) intercostal muscles* can be contracted (fig. 16.13*b*). These muscles pull the ribs and sternum downward and inward, increasing the pressure in the lungs. Also, the *abdominal wall muscles,* including the external and internal obliques, transversus abdominis, and rectus abdominis, can squeeze the abdominal organs inward (see fig. 8.18). Thus, the abdominal wall muscles can increase pressure in the abdominal cavity and force the diaphragm still higher against the lungs. These actions squeeze additional air out of the lungs.

1. Describe the events in inspiration.
2. How does surface tension expand the lungs during inspiration?
3. What forces cause normal expiration?

Respiratory Air Volumes and Capacities

Different intensities in breathing move different volumes of air in or out of the lungs. *Spirometry* measures such air volumes, revealing four distinct **respiratory volumes.** The amount of air that enters the lungs during inspiration (about 500 milliliters [mL] at rest) is approximately the same amount that leaves during a normal expiration. One inspiration plus the following expiration is called a **respiratory cycle.** The volume of air that enters (or leaves) during a single respiratory cycle is termed the **tidal volume** (fig. 16.14).

During forced inspiration, air in addition to the resting tidal volume enters the lungs. This extra volume is the **inspiratory reserve volume** (complemental air), and at maximum, it equals about 3,000 mL.

During forced expiration, the lungs can expel up to about 1,100 mL of air beyond the resting tidal volume. This quantity is called the **expiratory reserve volume** (supplemental air). Even after the most forceful expiration, however, about 1,200 mL of air remains in the lungs. This is called the **residual volume.**

Residual air remains in the lungs at all times, and consequently, newly inhaled air always mixes with air already in the lungs. This prevents oxygen and carbon dioxide concentrations in the lungs from fluctuating greatly with each breath.

Combining two or more of the respiratory volumes yields four **respiratory capacities.** Combining the inspiratory reserve volume (3,000 mL) with the tidal volume (500 mL) and the expiratory reserve volume (1,100 mL) gives the **vital capacity** (4,600 mL). This is

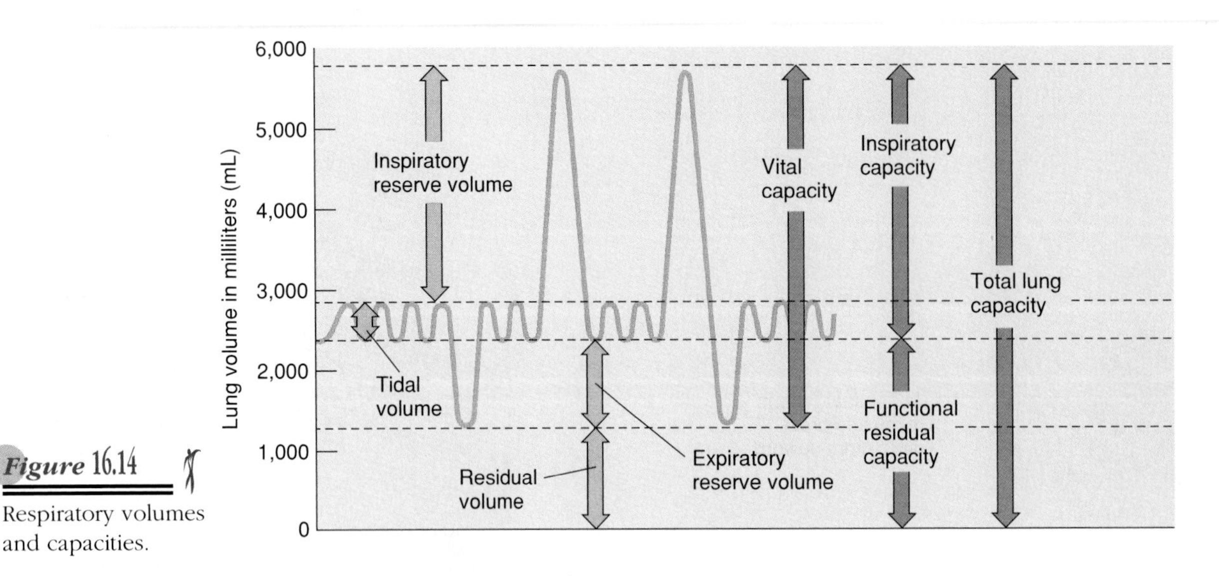

Figure 16.14

Respiratory volumes and capacities.

Table 16.2

Respiratory Air Volumes and Capacities

Name	Volume (average)	Description
Tidal volume (TV)	500 mL	Volume moved in or out of lungs during respiratory cycle
Inspiratory reserve volume (IRV)	3,000 mL	Volume that can be inhaled during forced breathing in addition to tidal volume
Expiratory reserve volume (ERV)	1,100 mL	Volume that can be exhaled during forced breathing in addition to tidal volume
Residual volume (RV)	1,200 mL	Volume that remains in lungs even after maximal expiration
Inspiratory capacity (IC)	3,500 mL	Maximum volume of air that can be inhaled following exhalation of tidal volume: IC = TV + IRV
Functional residual capacity (FRC)	2,300 mL	Volume of air that remains in the lungs following exhalation of tidal volume: FRC = ERV + RV
Vital capacity (VC)	4,600 mL	Maximum volume of air that can be exhaled after taking the deepest breath possible: VC = TV + IRV + ERV
Total lung capacity (TLC)	5,800 mL	Total volume of air that the lungs can hold: TLC = VC + RV

the maximum amount of air a person can exhale after taking the deepest breath possible.

The tidal volume (500 mL) plus the inspiratory reserve volume (3,000 mL) gives the **inspiratory capacity** (3,500 mL), which is the maximum volume of air a person can inhale following a resting expiration. Similarly, the expiratory reserve volume (1,100 mL) plus the residual volume (1,200 mL) equals the **functional residual capacity** (2,300 mL), which is the volume of air that remains in the lungs following a resting expiration.

The vital capacity plus the residual volume equals the **total lung capacity** (about 5,800 mL). This total varies with age, sex, and body size.

Some of the air that enters the respiratory tract during breathing does not reach the alveoli. This volume (about 150 mL) remains in the passageways of the trachea, bronchi, and bronchioles. Because gas is not exchanged through the walls of these passages, this air is said to occupy *anatomic dead space.*

Table 16.2 summarizes the respiratory air volumes and capacities.

An instrument called a *spirometer* measures respiratory air volumes, except residual volume, which requires a special technique. Such measurements evaluate the courses of emphysema, pneumonia, and lung cancer, conditions in which functional lung tissue is lost. These measurements may also track the progress of bronchial asthma and other diseases that obstruct air passages. ■

1. What is tidal volume?
2. Distinguish between inspiratory and expiratory reserve volumes.
3. How is vital capacity determined?
4. How is total lung capacity calculated?

Control of Breathing

Normal breathing is a rhythmic, involuntary act that continues even when a person is unconscious. The respiratory muscles, however, are under voluntary control.

Respiratory Center

Groups of neurons that comprise the **respiratory center** in the brain stem control breathing. These neurons are widely scattered throughout the pons and medulla oblongata (fig. 16.15). Two areas of the respiratory center are of special interest—the rhythmicity area of the medulla and the pneumotaxic area of the pons.

The **medullary rhythmicity area** includes two neuron groups that extend the length of the medulla oblongata. They are the dorsal respiratory group and the ventral respiratory group.

The *dorsal respiratory group* controls the basic rhythm of breathing. These neurons emit bursts of impulses that signal the diaphragm and other inspiratory muscles to contract. The impulses of each burst begin weakly, strengthen for about 2 seconds, and cease abruptly. The breathing muscles that contract in response to the impulses steadily increase the volume of air entering the lungs. The neurons remain inactive while expiration occurs passively, and then they emit another burst of inspiratory impulses, repeating the inspiration-expiration cycle.

The *ventral respiratory group* is quiet during normal breathing, but when more forceful breathing is required, these neurons generate impulses that increase inspiratory movements. Other neurons in the group activate muscles associated with forceful expiration.

The neurons in the **pneumotaxic area** continuously transmit impulses to the dorsal respiratory group and regulate the duration of inspiratory bursts originating from the dorsal group. In this way, the pneumotaxic neurons control breathing rate. More specifically, when pneumotaxic signals are strong, the inspiratory bursts are shorter, and the breathing rate increases; when pneumotaxic signals are weak, the inspiratory bursts are longer, and the breathing rate decreases (fig. 16.16).

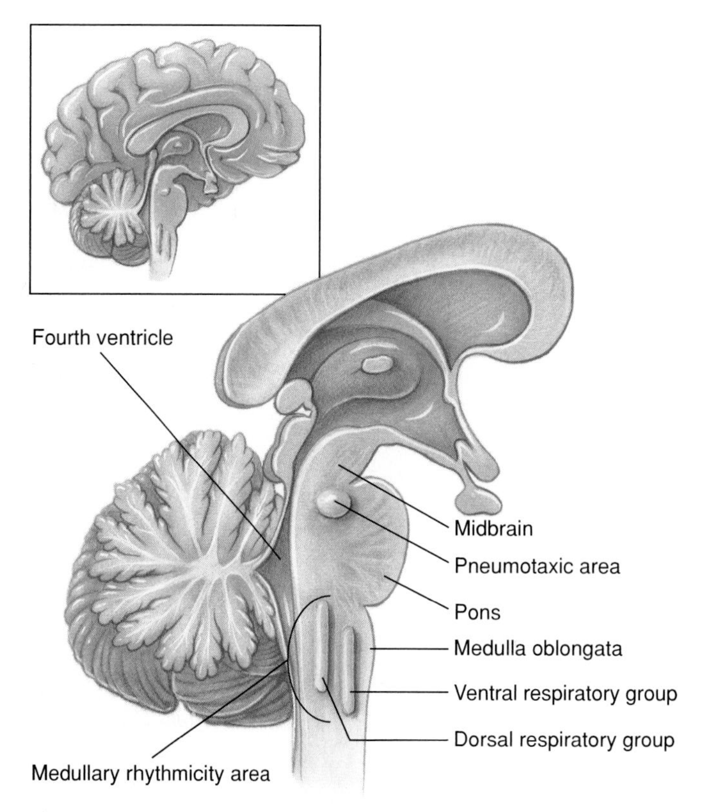

Figure 16.15

The respiratory center is in the pons and the medulla oblongata.

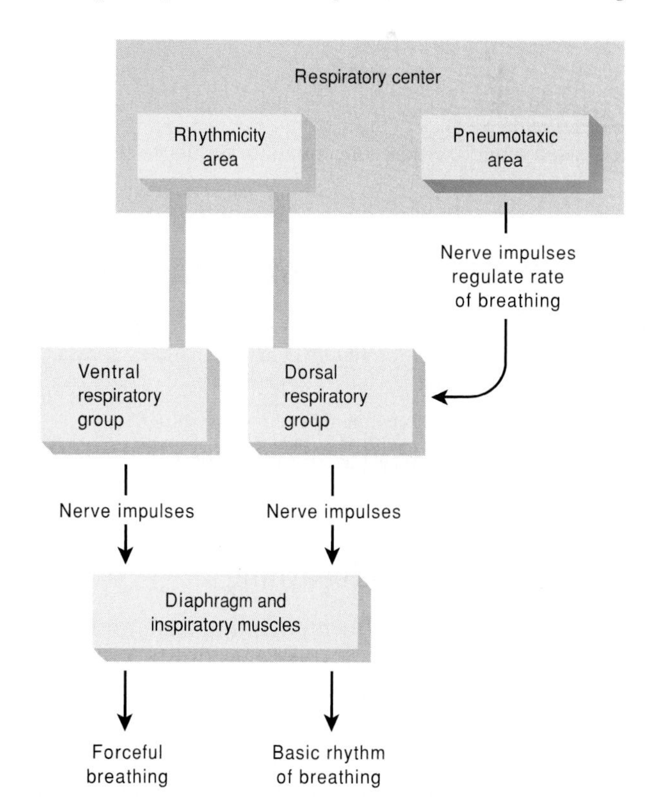

Figure 16.16

The medullary rhythmicity and pneumotaxic areas of the respiratory center control breathing.

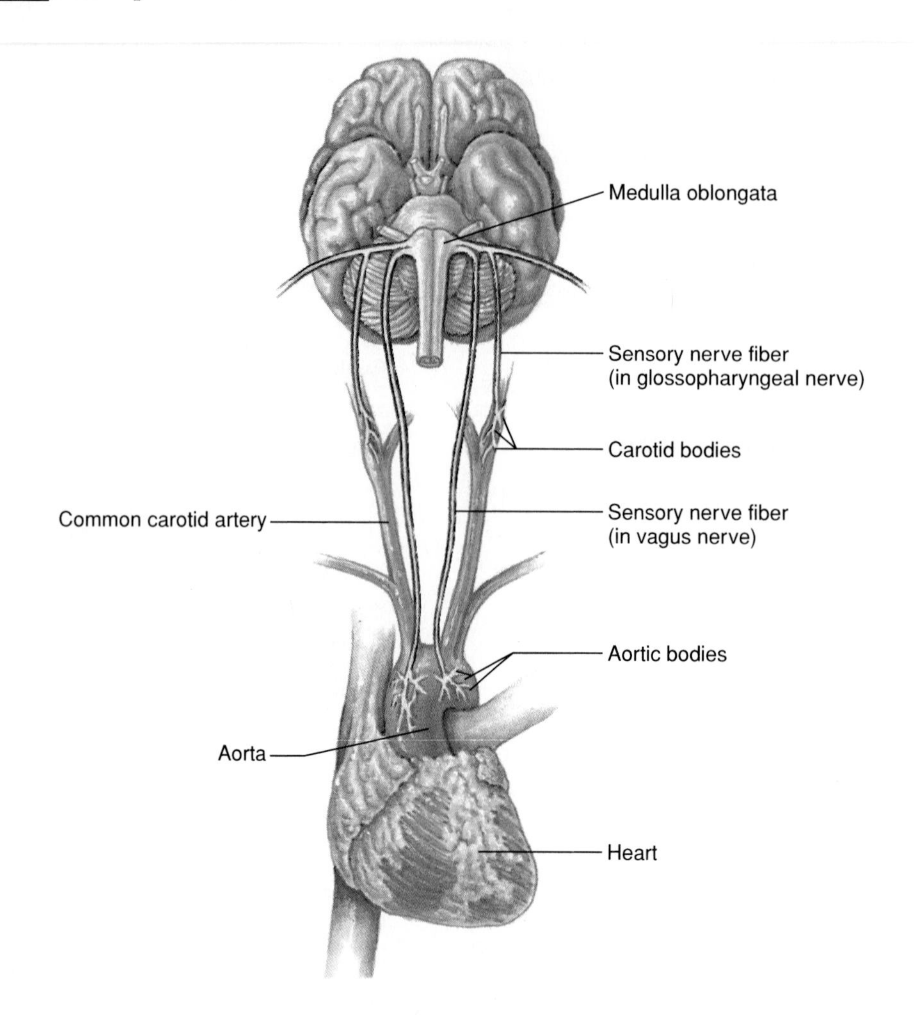

Figure 16.17

Decreased blood oxygen concentration stimulates chemoreceptors in the carotid and aortic bodies.

1. Where is the respiratory center?
2. Describe how the respiratory center maintains a normal breathing pattern.
3. Explain how the breathing pattern may change.

Factors Affecting Breathing

The respiratory center affects breathing rate and depth, and so do certain chemicals in body fluids, the degree to which lung tissues stretch, and a person's emotional state. For example, *chemosensitive areas* (central chemoreceptors) within the respiratory center, located in the ventral portion of the medulla oblongata near the origins of the vagus nerves, sense changes in blood concentrations of carbon dioxide and hydrogen ions. If either of these concentrations rises, the central chemoreceptors signal the respiratory center, and respiratory rate and tidal volume increase. As a result of the increased ventilation, more carbon dioxide is exhaled, blood concentrations of these chemicals falls, and breathing rate decreases.

Adding carbon dioxide to air can stimulate the rate and depth of breathing. Ordinary air is about 0.04% carbon dioxide. Inhaling air containing 4% carbon dioxide usually doubles breathing rate. ■

Low blood oxygen has little direct effect on the central chemoreceptors associated with the respiratory center. Instead, *peripheral chemoreceptors* in specialized structures called the *carotid* and *aortic bodies* sense changes in blood oxygen concentration. Peripheral chemoreceptors are in the walls of certain large arteries (the carotid arteries and the aorta) in the neck and thorax (fig. 16.17). Stimulated peripheral chemoreceptors transmit impulses to the respiratory center, increasing breathing rate. Blood oxygen concentration, however, must be very low to trigger this mechanism. Thus, oxygen plays only a minor role in control of normal respiration.

An *inflation reflex* helps regulate depth of breathing. This reflex occurs when stretched lung tissues stimulate stretch receptors in the visceral pleura, bronchioles, and alveoli. The sensory impulses of this reflex travel via the vagus nerves to the pneumotaxic area of the respiratory center and shorten the duration of inspiratory movements. This action prevents overinflation of the lungs during forceful breathing.

Emotional upset can alter the normal breathing pattern. Fear and pain typically increase breathing rate. Conscious control of breathing is also possible because the respiratory muscles are voluntary. A person can stop breathing altogether for a very short time.

If breathing stops, blood concentrations of carbon dioxide and hydrogen ions rise, and oxygen concentration falls. These changes (primarily the increased carbon dioxide) stimulate the respiratory center, and soon, the urge to inhale overpowers the desire to hold the breath. A person can increase breath-holding time by breathing rapidly and deeply in advance. This action, called **hyperventilation**, lowers blood carbon dioxide concentration. Following hyperventilation, it takes longer than usual for the carbon dioxide concentration to produce an overwhelming effect on the respiratory center. (**Note:** Hyperventilation should never be used to help in holding the breath while swimming because the person may lose consciousness underwater and drown.)

Sometimes, a person who is emotionally upset may hyperventilate, become dizzy, and lose consciousness. This condition is due to a lowered carbon dioxide concentration followed by a rise in pH (respiratory alkalosis), a localized vasoconstriction of cerebral arterioles, and consequently, a decreased blood flow to nearby brain cells. Interference with the oxygen supply to the brain causes fainting. ■

1. What chemical factors affect breathing?
2. Describe the inflation reflex.
3. How does hyperventilation decrease respiratory rate?

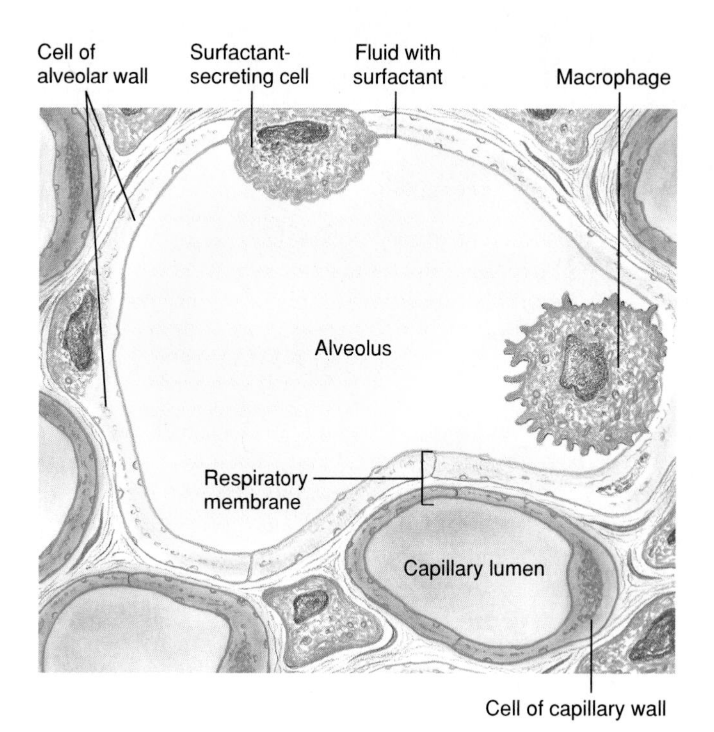

Figure 16.18

The respiratory membrane consists of the walls of the alveolus and the capillary.

Alveolar Gas Exchanges

The parts of the respiratory system discussed so far conduct air in and out of air passages. The alveoli carry on the vital process of exchanging gases between air and blood.

Alveoli

Alveoli are microscopic air sacs clustered at the distal ends of the narrowest respiratory tubes, the alveolar ducts (see fig. 16.8). Each alveolus consists of a tiny space within a thin wall that separates it from adjacent alveoli.

Respiratory Membrane

The wall of an alveolus consists of an inner lining of simple squamous epithelium and a dense network of capillaries, which are also lined with simple squamous epithelial cells. Thin, fused basement membranes separate the layers of these flattened cells, and in the spaces between the cells are elastic and collagenous fibers that support the alveolar wall. At least two thicknesses of epithelial cells and a layer of fused basement membranes separate the air in an alveolus from blood in a capillary (fig. 16.18). These layers comprise the **respiratory membrane**, across which blood and alveolar air exchange gases.

TOPIC OF INTEREST: Exercise and Breathing

Moderate to heavy physical exercise greatly increases the amount of oxygen the skeletal muscles use. For example, a young man at rest utilizes about 250 mL of oxygen per minute, but maximal exercise may require 3,600 mL of oxygen per minute.

As oxygen utilization increases, the volume of carbon dioxide produced also increases. Because decreased blood oxygen and increased blood carbon dioxide concentrations stimulate the respiratory center, exercise increases breathing rate. Studies reveal, however, that blood oxygen and carbon dioxide concentrations usually do not change during exercise. This reflects the respiratory system's effectiveness in obtaining oxygen and releasing carbon dioxide to the outside.

The cerebral cortex and sensory structures called *proprioceptors* that are associated with muscles and joints cause much of the increased breathing rate during vigorous exercise. Specifically, whenever the cerebral cortex signals skeletal muscles to contract, it also transmits stimulating impulses to the respiratory center. Muscular movements stimulate proprioceptors, triggering a *joint reflex* that travels to the respiratory center, increasing breathing rate.

When breathing rate increases during exercise, increased blood flow is also required to power skeletal muscles. Thus, physical exercise places demands on both the circulatory and respiratory systems. If either of these systems fails to keep up with cellular demands, the person begins to feel out of breath. This feeling usually reflects inability of the heart and circulatory system to move enough blood between the lungs and cells, rather than the respiratory system's inability to provide enough air.

Diffusion across the Respiratory Membrane

Recall from chapter 3 that molecules diffuse from regions where they are in higher concentration toward regions where they are in lower concentration. For gases, it is more useful to think of diffusion occurring from regions of higher pressure toward regions of lower pressure. The pressure of a gas determines the rate at which it diffuses from one region to another.

Measured by volume, ordinary air is about 78% nitrogen, 21% oxygen, and 0.04% carbon dioxide. Air has small amounts of other gases that have little or no physiological importance.

In a mixture of gases such as air, each gas accounts for a portion of the total pressure the mixture produces. The amount of pressure each gas contributes, called the **partial pressure**, is directly proportional to the concentration of the gas in the mixture. For example, because air is 21% oxygen, this gas accounts for 21% of the atmospheric pressure (21% of 760 mm Hg), or 160 mm Hg. Thus, the partial pressure of oxygen, symbolized P_{O_2}, in atmospheric air is 160 mm Hg. Similarly, the partial pressure of carbon dioxide (P_{CO_2}) in air is 0.3 mm Hg.

When a mixture of gases dissolves in blood, the resulting concentration of each gas is proportional to its partial pressure. Each gas diffuses between blood and its surroundings from areas of higher partial pressure to areas of lower partial pressure until the partial pressures in the two regions reach equilibrium. For example, the P_{CO_2} in capillary blood is 45 mm Hg, but the P_{CO_2} in alveolar air is 40 mm Hg. Because of the difference in these partial pressures, carbon dioxide diffuses from blood, where its partial pressure is higher, across the respiratory membrane and into alveolar air (fig. 16.19). When blood leaves the lungs, its P_{CO_2} is 40 mm Hg, which is the same as the P_{CO_2} of alveolar air. Similarly, the P_{O_2} of capillary blood is 40 mm Hg, but that of alveolar air is 104 mm Hg. Thus, oxygen diffuses from alveolar air into blood, and blood leaves the lungs with a P_{O_2} of 104 mm Hg. (Because of the large volume of air always in the lungs, as long as breathing continues, alveolar P_{O_2} stays relatively constant at 104 mm Hg.)

The respiratory membrane is so thin that certain chemicals other than carbon dioxide may diffuse into alveolar air and be exhaled. This is why breath analysis can reveal alcohol in blood or acetone in the breath of a person who has untreated diabetes mellitus. Breath analysis may also detect substances associated with kidney failure, certain digestive disturbances, and liver disease. ■

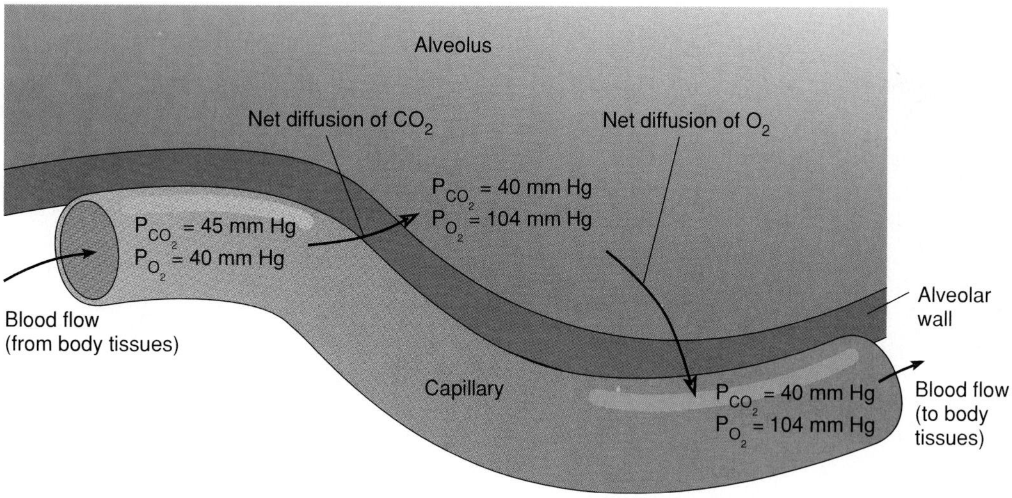

Figure 16.19

Gases are exchanged between alveolar air and capillary blood because of differences in partial pressures.

1. Describe the structure of the respiratory membrane.
2. What is the partial pressure of a gas?
3. What force moves oxygen and carbon dioxide across the respiratory membrane?

Gas Transport

Blood transports oxygen and carbon dioxide between the lungs and cells. As these gases enter blood, they dissolve in the liquid portion (plasma) or combine chemically with blood components.

Oxygen Transport

Almost all the oxygen (over 98%) that blood transports combines with the iron-containing protein *hemoglobin* in red blood cells. The remainder of the oxygen dissolves in plasma.

In the lungs, where the P_{O_2} is relatively high, oxygen dissolves in blood and combines rapidly with the iron atoms of hemoglobin, forming **oxyhemoglobin** (fig. 16.20). The chemical bonds between oxygen and hemoglobin molecules are unstable, and as the P_{O_6} decreases, oxyhemoglobin molecules release oxygen, which diffuses into nearby cells that have depleted their oxygen supplies in cellular respiration.

Several other factors affect the amount of oxygen that oxyhemoglobin releases. More oxygen is released as the blood concentration of carbon dioxide increases, as blood becomes more acidic, or as blood temperature increases. This explains why more oxygen is released to skeletal muscles during physical exercise. The increased muscular activity and oxygen utilization increase carbon dioxide concentration, decrease pH, and raise temperature. Less active cells receive proportionately less oxygen.

Carbon monoxide (CO) is a toxic gas produced in gasoline engines as a result of incomplete fuel combustion. It is also a component of tobacco smoke. Carbon monoxide is toxic because it combines with hemoglobin more effectively than does oxygen and does not dissociate readily from hemoglobin. Thus, when a person breathes carbon monoxide, less hemoglobin is available for oxygen transport, and cells soon develop oxygen deficiency. The effects of carbon monoxide on hemoglobin may explain the lower average birth weights of infants born to women who smoked cigarettes while pregnant. ■

1. How is oxygen transported from the lungs to cells?
2. What stimulates blood to release oxygen to tissues?

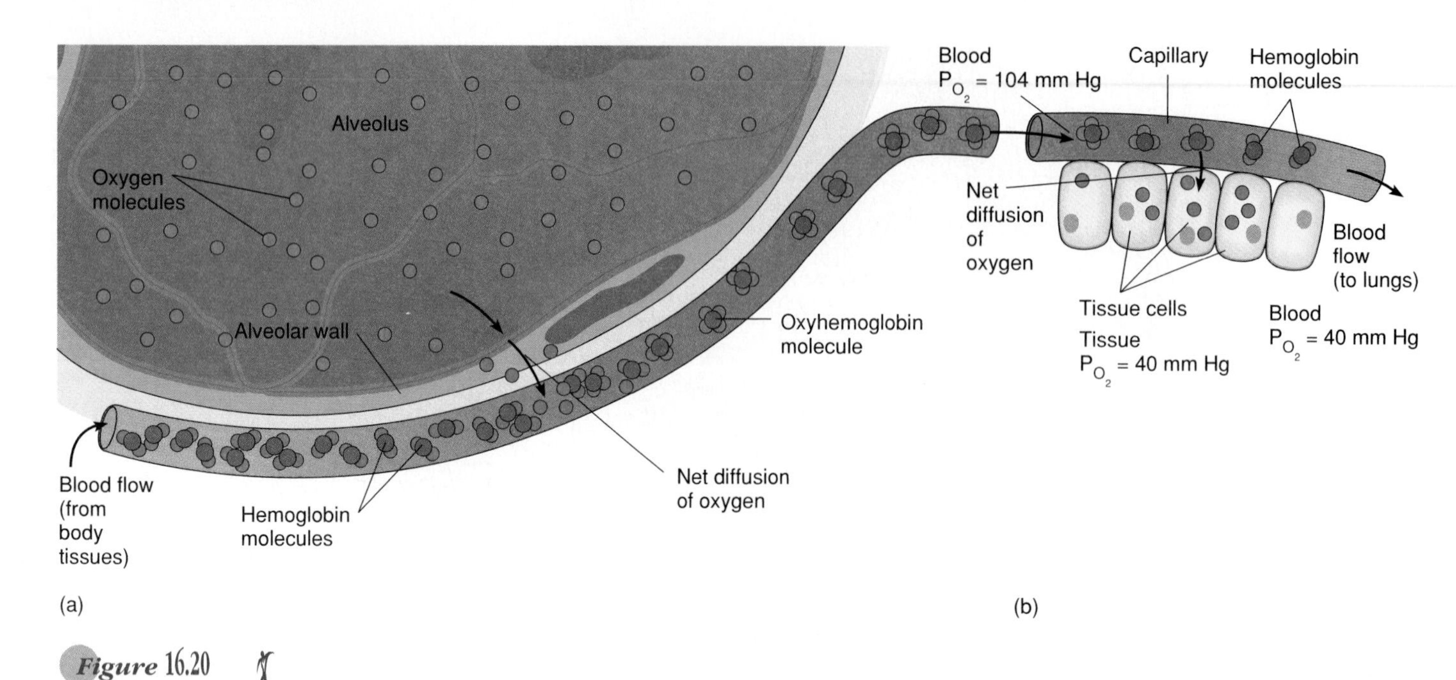

Figure 16.20

(*a*) Oxygen molecules entering blood from the alveolus bond to hemoglobin, forming oxyhemoglobin. (*b*) In the regions of the body cells, oxyhemoglobin releases oxygen. Note that much oxygen is still bound to hemoglobin at the P_{O_2} of systemic venous blood.

Carbon Dioxide Transport

Blood flowing through the capillaries of the tissues gains carbon dioxide because the tissues have a relatively high P_{CO_2}. Blood transports carbon dioxide to the lungs in one of three forms: as carbon dioxide dissolved in plasma, as part of a compound formed by bonding to hemoglobin, or as part of a bicarbonate ion (fig. 16.21).

The amount of carbon dioxide that dissolves in plasma is determined by its partial pressure. The higher the P_{CO_2} of the tissues, the more carbon dioxide will go into solution. However, only about 7% of the carbon dioxide that blood transports is in this form.

Unlike oxygen, which combines with the iron atoms of hemoglobin molecules, carbon dioxide bonds with the amino groups ($-NH_2$) of these molecules. Consequently, oxygen and carbon dioxide do not compete for binding sites, and a hemoglobin molecule can transport both gases at the same time.

Carbon dioxide combining with hemoglobin forms a loosely bound compound called **carbaminohemoglobin.** This molecule decomposes readily in regions of low P_{CO_2}, releasing its carbon dioxide. This method of transporting carbon dioxide is theoretically quite effective, but carbaminohemoglobin forms slowly. Only about 23% of the carbon dioxide that blood transports is in this form.

The most important carbon dioxide transport mechanism involves the formation of **bicarbonate ions** (HCO_3^-). Carbon dioxide reacts with water to form carbonic acid (H_2CO_3):

$$CO_2 + H_2O \rightarrow H_2CO_3$$

This reaction occurs slowly in plasma, but much of the carbon dioxide diffuses into red blood cells. These cells contain the enzyme **carbonic anhydrase,** which speeds the reaction between carbon dioxide and water. The resulting carbonic acid then dissociates, releasing hydrogen ions (H^+) and bicarbonate ions (HCO_3^-):

$$H_2CO_3 \rightarrow H^+ + HCO_3^-$$

Most of the hydrogen ions combine quickly with hemoglobin molecules and, thus, do not accumulate and greatly change blood pH. The bicarbonate ions diffuse out of red blood cells and enter the plasma. Nearly 70% of the carbon dioxide that blood transports is in this form.

When blood passes through the capillaries of the lungs, its dissolved carbon dioxide diffuses into alveoli in response to the relatively low P_{CO_2} of aveolar air (fig. 16.22). At the same time, hydrogen ions and bicarbonate ions in red blood cells recombine to form

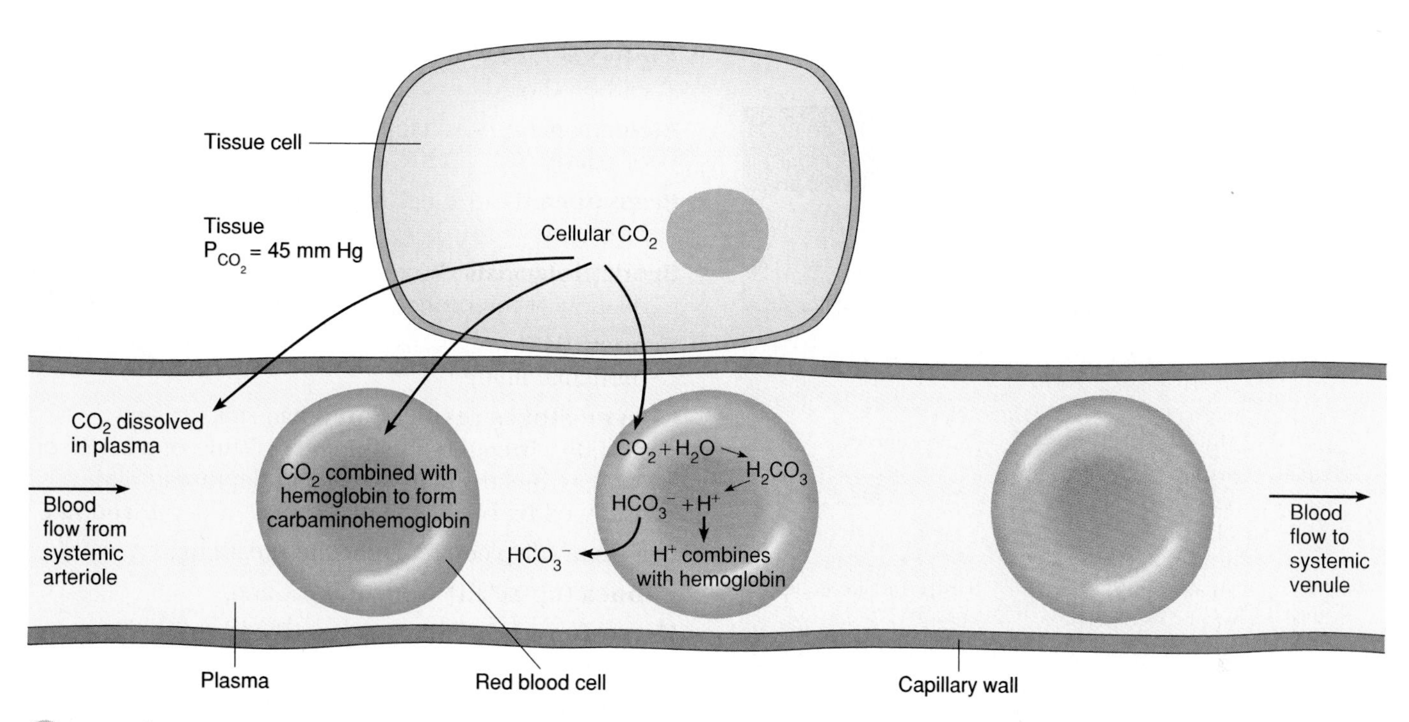

Figure 16.21

Blood transports carbon dioxide from tissue cells in a dissolved state, combined with hemoglobin, or in the form of bicarbonate ions (HCO_3^-).

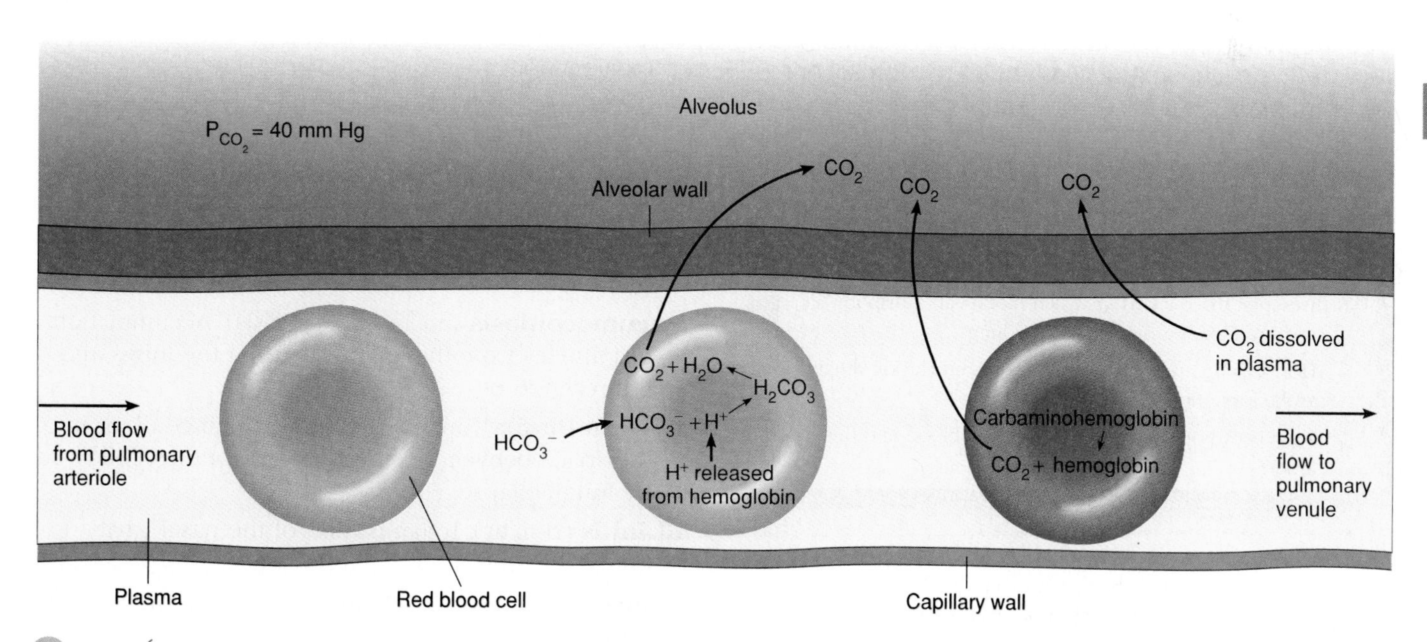

Figure 16.22

In the lungs, carbon dioxide diffuses from the plasma into alveoli.

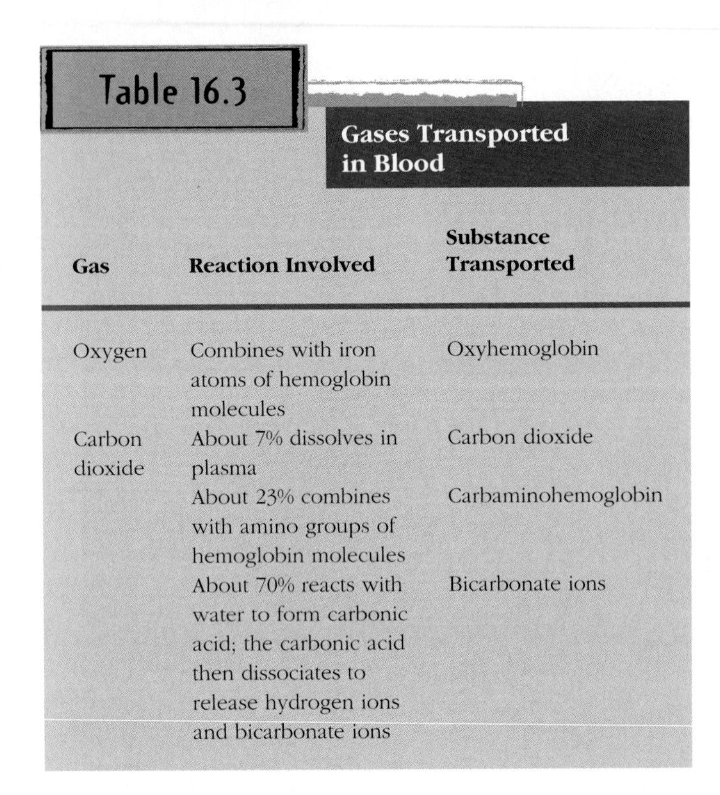

Table 16.3 Gases Transported in Blood

Gas	Reaction Involved	Substance Transported
Oxygen	Combines with iron atoms of hemoglobin molecules	Oxyhemoglobin
Carbon dioxide	About 7% dissolves in plasma	Carbon dioxide
	About 23% combines with amino groups of hemoglobin molecules	Carbaminohemoglobin
	About 70% reacts with water to form carbonic acid; the carbonic acid then dissociates to release hydrogen ions and bicarbonate ions	Bicarbonate ions

carbonic acid, and under the influence of carbonic anhydrase, the carbonic acid quickly breaks down to yield carbon dioxide and water:

$$H^+ + HCO_3^- \rightarrow H_2CO_3 \rightarrow CO_2 + H_2O$$

Carbaminohemoglobin also releases its carbon dioxide, and carbon dioxide continues to diffuse out of blood until the P_{CO_2} of blood and of alveolar air are in equilibrium.

Table 16.3 summarizes transport of blood gases.

1. Describe three forms in which blood can transport carbon dioxide from cells to the lungs.
2. How can hemoglobin carry oxygen and carbon dioxide at the same time?
3. How is carbon dioxide released from blood into the lungs?

Clinical Terms Related to the Respiratory System

Anoxia (ah-nok′se-ah) Absence or deficiency of oxygen within tissues.

Apnea (ap-ne′ah) Temporary cessation of breathing.

Asphyxia (as-fik′se-ah) Oxygen deficiency and excess carbon dioxide in blood and tissues.

Atelectasis (at″e-lek′tah-sis) Collapse of a lung or part of a lung.

Bradypnea (brad″e-ne′ah) Abnormally slow breathing.

Bronchiolectasis (brong″ke-o-lek′tah-sis) Chronic dilation of the bronchioles.

Bronchitis (brong-ki′tis) Inflammation of the bronchial lining.

Cheyne-Stokes respiration (chān stōks res″pĭ-ra′shun) Irregular breathing consisting of a series of shallow breaths that increase in depth and rate, followed by breaths that decrease in depth and rate.

Dyspnea (disp′ne-ah) Difficulty breathing.

Eupnea (up-ne′ah) Normal breathing.

Hemothorax (he″mo-tho′raks) Blood in the pleural cavity.

Hypercapnia (hi″per-kap′ne-ah) Excess carbon dioxide in blood.

Hyperoxia (hi″per-ok′se-ah) Excess oxygenation of blood.

Hyperpnea (hi″perp-ne′ah) Increase in the depth and rate of breathing.

Hyperventilation (hi″per-ven″tĭ-la′shun) Prolonged, rapid, and deep breathing.

Hypoxemia (hi″pok-se′me-ah) Deficiency in blood oxygenation.

Hypoxia (hi-pok′se-ah) Diminished availability of oxygen in tissues.

Lobar pneumonia (lo′ber nu-mo′ne-ah) Pneumonia that affects an entire lung lobe.

Pleurisy (ploo′rĭ-se) Inflammation of the pleural membranes.

Pneumoconiosis (nu″mo-ko″ne-o′sis) Accumulation of particles from the environment in the lungs and the reaction of tissues to them.

Pneumothorax (nu″mo-tho′raks) Entrance of air into the space between the pleural membranes, followed by lung collapse.

Rhinitis (ri-ni′tis) Inflammation of the nasal cavity lining.

Sinusitis (si″nŭ-si′tis) Inflammation of the sinus cavity lining.

Tachypnea (tak″ip-ne′ah) Rapid, shallow breathing.

Tracheotomy (tra″ke-ot′o-me) An incision in the trachea for exploration or for removal of a foreign object.

organization

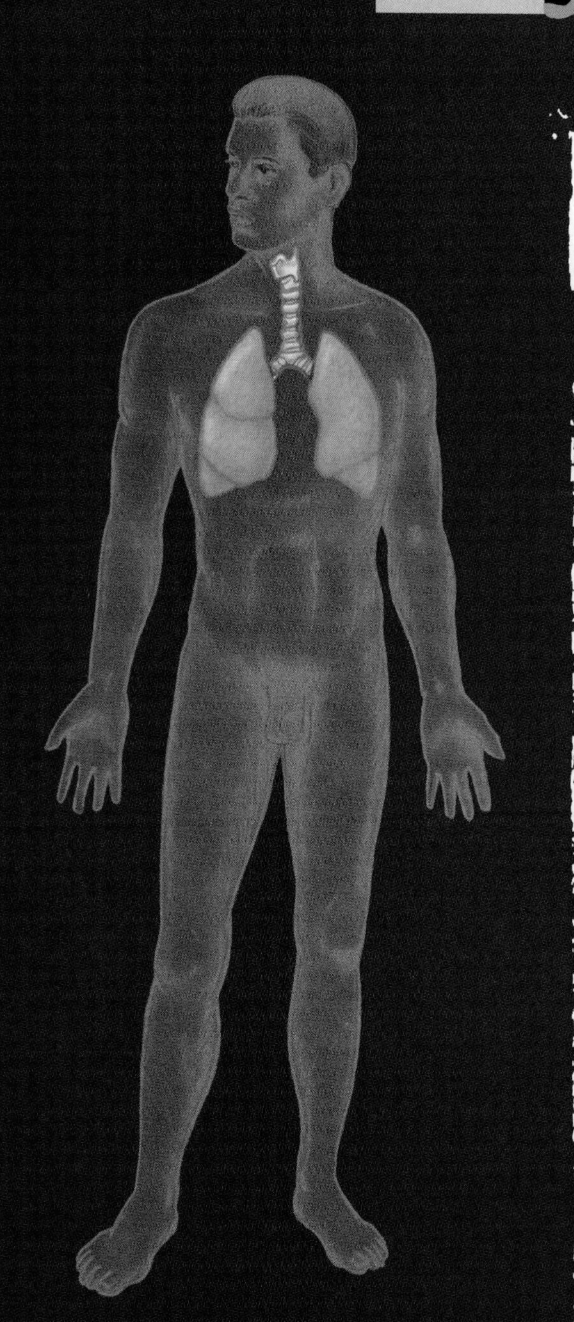

Respiratory System

The respiratory system provides oxygen for the internal environment and excretes carbon dioxide.

Integumentary System

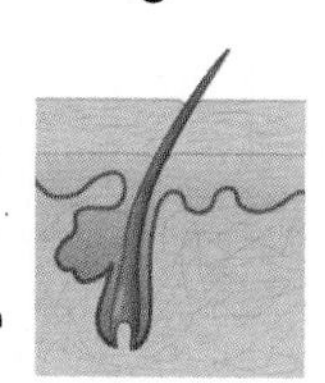

Stimulation of skin receptors may alter respiratory rate.

Cardiovascular System

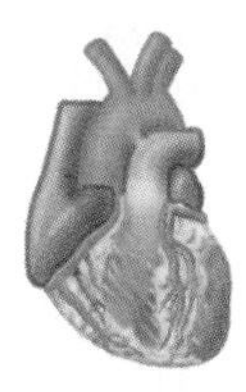

As the heart pumps blood through the lungs, the lungs oxygenate the blood and excrete carbon dioxide.

Skeletal System

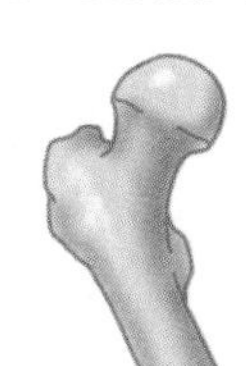

Bones provide attachments for muscles involved in breathing.

Lymphatic System

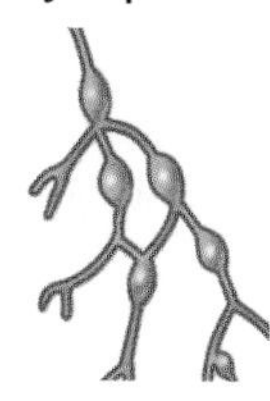

Cells of the immune system patrol the lungs and defend against infection.

Muscular System

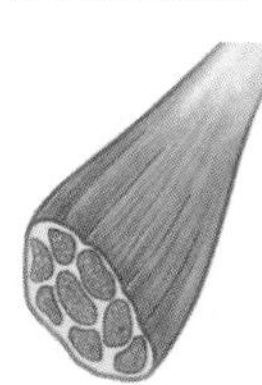

The respiratory system eliminates carbon dioxide produced by exercising muscles.

Digestive System

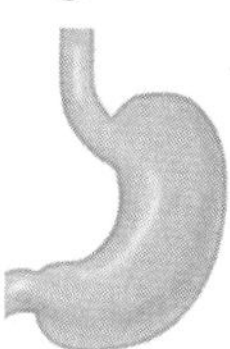

The digestive system and respiratory system share openings to the outside.

Nervous System

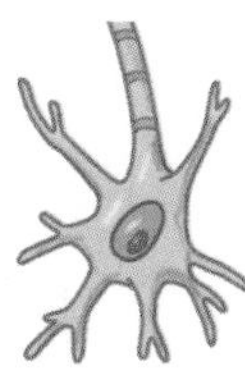

The brain controls the respiratory system. The respiratory system helps control pH of the internal environment.

Urinary System

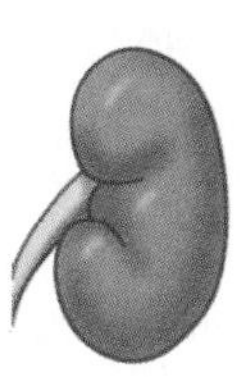

The kidneys and the respiratory system work together to maintain blood pH. The kidneys compensate for water lost through breathing.

Endocrine System

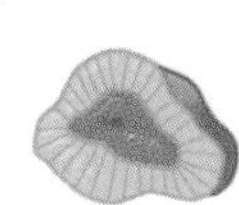

Hormonelike substances control the production of red blood cells that transport oxygen and carbon dioxide.

Reproductive System

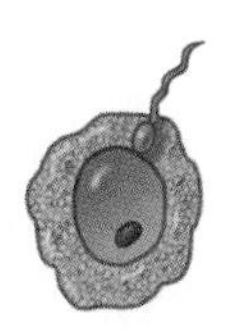

Respiration increases during sexual activity. Fetal "respiration" begins before birth.

Summary Outline

Introduction (p. 451)

The respiratory system includes tubes that remove particles from incoming air and transport air to and from the lungs and the air sacs where gases are exchanged. Respiration is the entire process of gas exchange between the atmosphere and body cells.

Organs of the Respiratory System (p. 451)

The respiratory system includes the nose, nasal cavity, paranasal sinuses, pharynx, larynx, trachea, bronchial tree, and lungs.

The upper respiratory tract includes the respiratory organs outside the thorax; the lower respiratory tract includes those respiratory organs within the thorax.

1. Nose
   a. Bone and cartilage support the nose.
   b. The nostrils are openings for air.
2. Nasal cavity
   a. Nasal conchae divide the nasal cavity into passageways and help increase the surface area of the mucous membrane.
   b. The mucous membrane filters, warms, and moistens incoming air.
   c. Ciliary action carries particles trapped in mucus to the pharynx, where they are swallowed.
3. Paranasal sinuses
   a. The paranasal sinuses are spaces in the bones of the skull that open into the nasal cavity.
   b. Mucous membrane lines sinuses.
4. Pharynx
   a. The pharynx is behind the oral cavity and between the nasal cavity and larynx.
   b. It is a passageway for air and food.
5. Larynx
   a. The larynx conducts air and helps prevent foreign objects from entering the trachea.
   b. It is composed of muscles and cartilages and is lined with mucous membrane.
   c. The larynx contains the vocal cords, which vibrate from side to side and produce sounds when air passes between them.
   d. The glottis and epiglottis help prevent foods and liquids from entering the trachea.
6. Trachea
   a. The trachea extends into the thoracic cavity in front of the esophagus.
   b. It divides into right and left bronchi.
7. Bronchial tree
   a. The bronchial tree consists of branched air passages that lead from the trachea to the air sacs.
   b. Alveoli are at the distal ends of the narrowest tubes, the alveolar ducts.
8. Lungs
   a. The mediastinum separates the left and right lungs, and the diaphragm and thoracic cage enclose them.
   b. The visceral pleura attaches to the surface of the lungs. The parietal pleura lines the thoracic cavity.
   c. Each lobe of the lungs is composed of alveoli, blood vessels, and supporting tissues.

Breathing Mechanism (p. 457)

Changes in the size of the thoracic cavity accompany inspiration and expiration.

1. Inspiration
   a. Atmospheric pressure forces air into the lungs.
   b. Inspiration occurs when the pressure inside alveoli decreases.
   c. Pressure within alveoli decreases when the diaphragm moves downward and the thoracic cage moves upward and outward.
   d. Surface tension aids lung expansion.
2. Expiration
   a. Elastic recoil of tissues and surface tension within alveoli provide the forces of expiration.
   b. Thoracic and abdominal wall muscles aid expiration.
3. Respiratory air volumes and capacities
   a. One inspiration followed by one expiration is a respiratory cycle.
   b. The amount of air that moves in (or out) during a single respiratory cycle is the tidal volume.
   c. Additional air that can be inhaled is the inspiratory reserve volume. Additional air that can be exhaled is the expiratory reserve volume.
   d. Residual volume remains in the lungs after a maximal expiration.
   e. The vital capacity is the maximum amount of air a person can exhale after taking the deepest breath possible.
   f. The inspiratory capacity is the maximum volume of air a person can inhale following exhalation of the tidal volume.
   g. The functional residual capacity is the volume of air that remains in the lungs after a person exhales the tidal volume.
   h. The total lung capacity equals the vital capacity plus the residual volume.

Control of Breathing (p. 463)

Normal breathing is rhythmic and involuntary.

1. Respiratory center
   a. The respiratory center is in the brain stem and includes portions of the pons and medulla oblongata.
   b. The medullary rhythmicity area includes two groups of neurons.
      (1) The dorsal respiratory group controls the basic rhythm of breathing.
      (2) The ventral respiratory group increases inspiratory and expiratory movements during forceful breathing.
   c. The pneumotaxic area regulates breathing rate.
2. Factors affecting breathing
   a. Chemicals, stretching of lung tissues, and emotional states affect breathing.
   b. Chemosensitive areas (central chemoreceptors) are associated with the respiratory center.
      (1) Blood concentrations of carbon dioxide and hydrogen ions influence the central chemoreceptors.
      (2) Stimulation of these receptors increases breathing rate.
   c. Peripheral chemoreceptors are in the walls of certain large arteries.
      (1) These chemoreceptors sense low oxygen concentration.
      (2) When oxygen concentration is low, breathing rate increases.

d. Overstretching lung tissues triggers an inflation reflex.
   (1) This reflex shortens the duration of inspiratory movements.
   (2) The inflation reflex prevents overinflation of the lungs during forceful breathing.

e. Hyperventilation decreases blood carbon dioxide concentration, but *this is very dangerous when done before swimming underwater.*

Alveolar Gas Exchanges (p. 465)

Gas exchange between air and blood occurs in alveoli.

1. Alveoli
   Alveoli are tiny air sacs clustered at the distal ends of alveolar ducts.
2. Respiratory membrane
   a. This membrane consists of alveolar and capillary walls.
   b. Blood and alveolar air exchange gases across this membrane.
3. Diffusion across the respiratory membrane
   a. The partial pressure of a gas is proportional to the concentration of that gas in a mixture or the concentration dissolved in a liquid.
   b. Gases diffuse from regions of higher partial pressure toward regions of lower partial pressure.
   c. Oxygen diffuses from alveolar air into blood. Carbon dioxide diffuses from blood into alveolar air.

Gas Transport (p. 467)

Blood transports gases between the lungs and cells.

1. Oxygen transport
   a. Blood mainly transports oxygen in combination with hemoglobin molecules.
   b. The resulting oxyhemoglobin is unstable and releases its oxygen in regions where the P_{O_2} is low.
   c. More oxygen is released as the blood concentration of carbon dioxide increases, as blood becomes more acidic, and as blood temperature increases.
2. Carbon dioxide transport
   a. Carbon dioxide may be carried in solution, bound to hemoglobin, or as a bicarbonate ion.
   b. Most carbon dioxide is transported in the form of bicarbonate ions.
   c. The enzyme carbonic anhydrase speeds the reaction between carbon dioxide and water to form carbonic acid.
   d. Carbonic acid dissociates to release hydrogen ions and bicarbonate ions.

Review Exercises

1. Describe the general functions of the respiratory system. (p. 451)
2. Distinguish between the upper and lower respiratory tracts. (p. 451)
3. Explain how the nose and nasal cavity filter incoming air. (p. 451)
4. Describe the locations of the major paranasal sinuses. (p. 452)
5. Distinguish between the pharynx and larynx. (p. 453)
6. Name and describe the locations of the larger cartilages of the larynx. (p. 453)
7. Distinguish between the false vocal cords and the true vocal cords. (p. 453)
8. Compare the structure of the trachea with that of the branches of the bronchial tree. (p. 454)
9. Distinguish between the visceral pleura and parietal pleura. (p. 457)
10. Explain normal inspiration and forced inspiration. (p. 457)
11. Define *surface tension,* and explain how it aids breathing. (p. 458)
12. Define *surfactant,* and explain its function. (p. 460)
13. Explain normal expiration and forced expiration. (p. 460)
14. Distinguish between the vital capacity and total lung capacity. (p. 461)
15. Describe the location of the respiratory center, and name its major components. (p. 463)
16. Describe the control of the basic rhythm of breathing. (p. 463)
17. Explain the function of the pneumotaxic area of the respiratory center. (p. 463)
18. Describe the function of the chemoreceptors in the carotid and aortic bodies. (p. 464)
19. Describe the inflation reflex. (p. 465)
20. Define *hyperventilation,* and explain how it affects the respiratory center. (p. 465)
21. Define *respiratory membrane,* and explain its function. (p. 465)
22. Explain the relationship between the partial pressure of a gas and its diffusion rate. (p. 466)
23. Summarize the gas exchanges across the respiratory membrane. (p. 466)
24. Describe how blood transports oxygen. (p. 467)
25. List three factors that increase release of blood oxygen. (p. 467)
26. Explain how the blood transports carbon dioxide. (p. 468)

Critical Thinking

1. It is below 0° F outside, but the dedicated runner bundles up and hits the road anyway. "You're crazy," shouts a neighbor. "Your lungs will freeze." Why is the well-meaning neighbor wrong?
2. Why does breathing through the mouth dry out the throat?
3. George Washington went for a walk in the freezing rain on a bleak December day in 1799. The next day, he had trouble breathing and swallowing. A doctor suggested cutting a hole in the president's throat so he could breathe, but other doctors voted him down, instead bleeding the patient, plastering his throat with bran and honey, and placing beetles on his legs to produce blisters. Soon, Washington's voice became muffled, his breathing was more labored, and he grew restless. For a short time, he seemed euphoric; then, he died. Washington had epiglottitis, in

which the epiglottis swells to ten times its normal size. How does this diagnosis explain his symptoms? Which suggested treatment might have worked?

4. Why can you not commit suicide by holding your breath?
5. When a woman is very close to delivering a baby, she may hyperventilate. Breathing into a paper bag regulates her breathing. How does this action return her breathing to normal?
6. Why were the finishing times of endurance events at the 1968 Olympics, held in 2,200-meter-high Mexico City, rather slow?
7. Why might it be dangerous for a heavy smoker to use a cough suppressant?
8. Emphysema reduces the lungs' capacity to recoil elastically. Which respiratory air volumes does emphysema affect?
9. What changes would you expect in the relative concentrations of blood oxygen in a patient who breathes rapidly and deeply for a prolonged time? Why?
10. If a person has stopped breathing and is receiving pulmonary resuscitation, would it be better to administer pure oxygen or a mixture of oxygen and carbon dioxide? Why?

Urinary System

Felicia had looked forward to summer camp all year, especially the overnight hikes. A three-day expedition in July was wonderful, but five days after returning to camp, Felicia developed severe abdominal cramps. So did seventeen other campers and two counselors, some of whom had bloody diarrhea, too. Several of the stricken campers were hospitalized, Felicia among them. While the others improved in a few days and were released, Felicia suffered from a complication, called hemolytic uremic syndrome (HUS). Her urine had turned bloody, and she also had blood abnormalities—severe anemia and lack of platelets.

Camp personnel reported the outbreak to public health officials, who quickly recognized the signs of food poisoning and traced the illness to hamburgers cooked outdoors on the trip. The burgers were served rare, the red meat not hot enough to kill a strain of *Escherichia coli* bacteria that releases a poison called shigatoxin.

Most people who eat meat tainted with *E. coli* toxin become ill, but the damage usually is restricted to the digestive tract, producing cramps and diarrhea for several days. In about 6% of affected people, mostly children, HUS develops because the bloodstream transports the toxin to the kidneys. Here, the toxin destroys cells of the microscopic tubules that normally filter proteins and blood cells from forming urine. With the tubule linings compromised, proteins and blood cells, as well as damaged kidney cells, appear in the urine.

HUS is a leading cause of acute renal (kidney) failure, killing 3–5% of affected children. Felicia was in the lucky majority. Blood clotted around the sites of her damaged kidney cells, and over a few weeks, new cells formed. Three weeks after the bloody urine began, her urine was once again clear, and she was healthy.

Photo: Cook that burger! Hemolytic uremic syndrome is a complication of infection by bacteria that produce shigatoxin. Destruction of the proximal portions of kidney tubules allows proteins and blood cells to enter urine.

Cells produce a variety of wastes, which, if they accumulate, are toxic. Body fluids, such as blood and lymph, carry wastes from the tissues that produce them, and other structures remove wastes from blood and transport them to the outside. The respiratory system removes carbon dioxide from blood, and the *urinary system* removes certain salts and nitrogenous wastes. The urinary system also helps maintain the normal concentrations of water and electrolytes within body fluids; regulates the pH and volume of body fluids; and helps control red blood cell production and blood pressure.

Chapter Objectives

After studying this chapter, you should be able to:

1. Name and list the general functions of the organs of the urinary system. (p. 477)
2. Describe the locations and structure of the kidneys. (p. 477)
3. List the functions of the kidneys. (p. 479)
4. Trace the pathway of blood through the major vessels within a kidney. (p. 479)
5. Describe a nephron, and explain the functions of its major parts. (p. 480)
6. Explain how glomerular filtrate is produced, and describe its composition. (p. 482)
7. Explain the factors that affect the rate of glomerular filtration and how this rate is regulated. (p. 483)
8. Discuss the role of tubular reabsorption in urine formation. (p. 485)
9. Define *tubular secretion,* and explain its role in urine formation. (p. 488)
10. Describe the structure of the ureters, urinary bladder, and urethra. (p. 490)
11. Explain the process and control of micturition. (p. 491)

Aids to Understanding Words

calyc- [small cup] major *calyc*es: Cuplike divisions of the renal pelvis.

cort- [covering] renal *cort*ex: The shell of tissues surrounding the inner kidney.

detrus- [to force away] *detrus*or muscle: A muscle within the bladder wall that expels urine.

glom- [little ball] *glom*erulus: A cluster of capillaries within a renal corpuscle.

mict- [to pass urine] *mict*urition: The process of expelling urine from the bladder.

nephr- [pertaining to the kidney] *nephr*on: The functional unit of a kidney.

papill- [nipple] renal *papill*ae: Small elevations that project into a renal calyx.

trigon- [triangular shape] *trigon*e: A triangular area on the internal floor of the urinary bladder.

Key Terms

afferent arteriole (af′er-ent ar-te′re-ōl)

detrusor muscle (de-truz′or mus′l)

efferent arteriole (ef′er-ent ar-te′re-ōl)

glomerular capsule (glo-mer′u-lar kap′sul)

glomerulus (glo-mer′u-lus)

juxtaglomerular apparatus (juks″tah-glo-mer′u-lar ap″ah-ra′tus)

micturition (mik″tu-rish′un)

nephron (nef′ron)

peritubular capillary (per″ĭ-tu′bu-lar kap′ĭ-ler″e)

renal corpuscle (re′nal kor′pusl)

renal cortex (re′nal kor′teks)

renal medulla (re′nal mĕ-dul′ah)

renal tubule (re′nal tu′būl)

retroperitoneal (re″tro-per″ĭ-to-ne′al)

Introduction

The urinary system consists of a pair of kidneys, which remove substances from blood, form urine, and help regulate certain metabolic processes; a pair of tubular ureters, which transport urine from the kidneys; a saclike urinary bladder, which stores urine; and a tubular urethra, which conveys urine to the outside of the body. Figure 17.1 and reference plate 6 show these organs.

Kidneys

A **kidney** is a reddish brown, bean-shaped organ with a smooth surface. An adult kidney is about 12 centimeters long, 6 centimeters wide, and 3 centimeters thick, and is enclosed in a tough, fibrous capsule (fig. 17.2).

Location of the Kidneys

The kidneys lie on either side of the vertebral column in a depression high on the posterior wall of the abdominal cavity. The upper and lower borders of the kidneys are generally at the levels of the twelfth thoracic and third lumbar vertebrae, respectively. The left kidney is usually 1.5–2.0 centimeters higher than the right one.

The kidneys are positioned **retroperitoneally**, which means they are behind the parietal peritoneum and against the deep muscles of the back. Connective tissue and masses of adipose tissue surround the kidneys and hold them in position (see fig. 1.9).

Kidney Structure

The lateral surface of each kidney is convex, but its medial side is deeply concave. The resulting medial depression leads into a hollow chamber called the **renal sinus**. The entrance to this sinus is termed the *hilum,* and through it pass blood vessels, nerves, lymphatic vessels, and the ureter (see fig. 17.1).

The superior end of the ureter expands to form a funnel-shaped sac called the **renal pelvis** inside the renal sinus. The pelvis subdivides into two or three tubes, called *major calyces* (singular, *calyx*), and they, in turn, subdivide into several *minor calyces* (fig. 17.2).

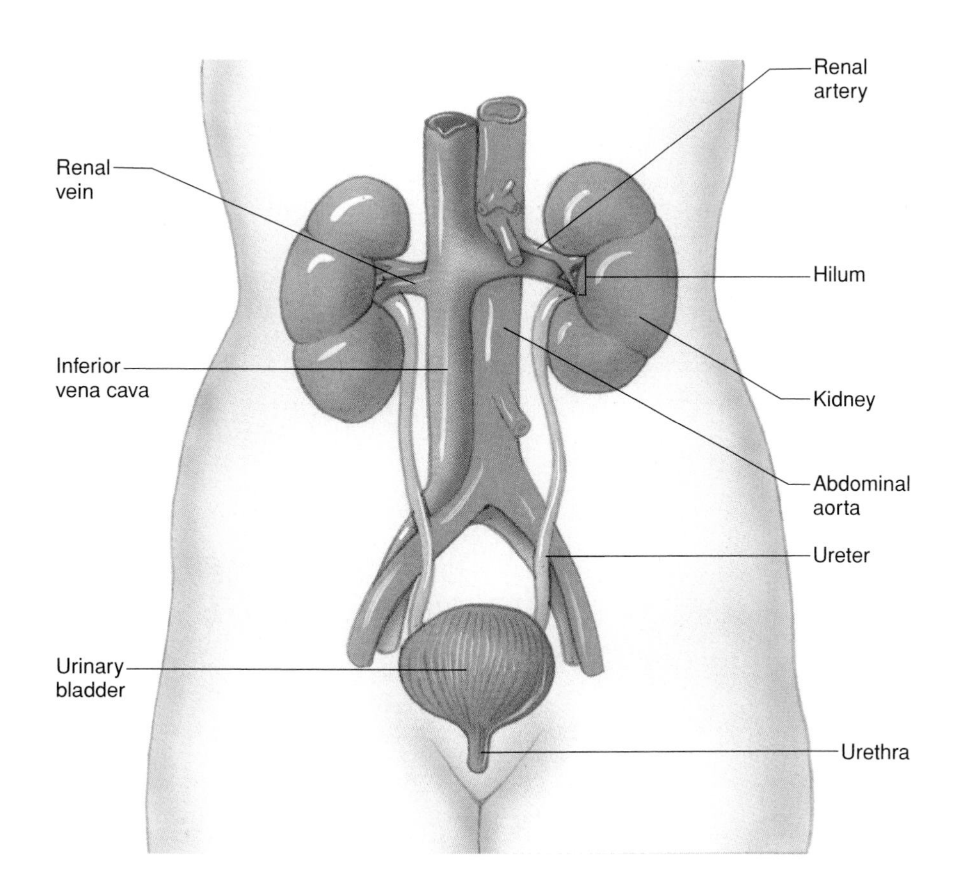

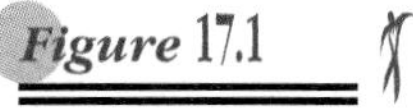
Figure 17.1

The urinary system includes the kidneys, ureters, urinary bladder, and urethra. Notice the relationship of these structures to the major blood vessels.

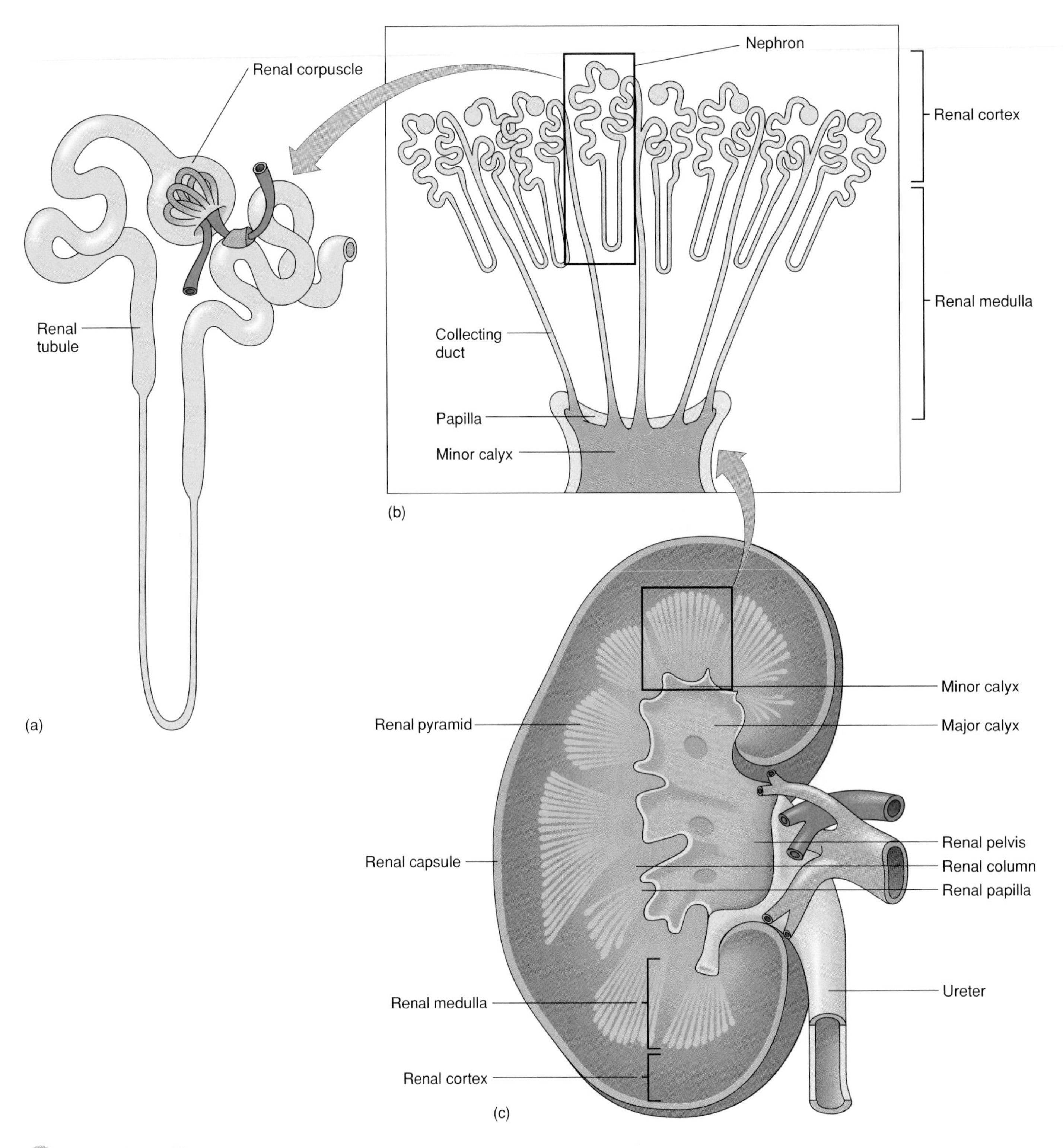

Figure 17.2

(*a*) A single nephron. (*b*) A renal pyramid containing nephrons. (*c*) Longitudinal section of a kidney.

A series of small elevations called *renal papillae* project into the renal sinus from its wall. Tiny openings that lead into a minor calyx pierce each projection.

Each kidney has two distinct regions—an inner medulla and an outer cortex. The **renal medulla** is composed of conical masses of tissue called *renal pyramids* and appears striated. The **renal cortex** forms a shell around the medulla and dips into the medulla between renal pyramids, forming *renal columns*. The granular appearance of the cortex is due to the random organization of tiny tubules associated with the **nephrons**, the kidney's functional units (fig. 17.2).

1. Where are the kidneys located?
2. Describe kidney structure.
3. Name the kidney's functional unit.

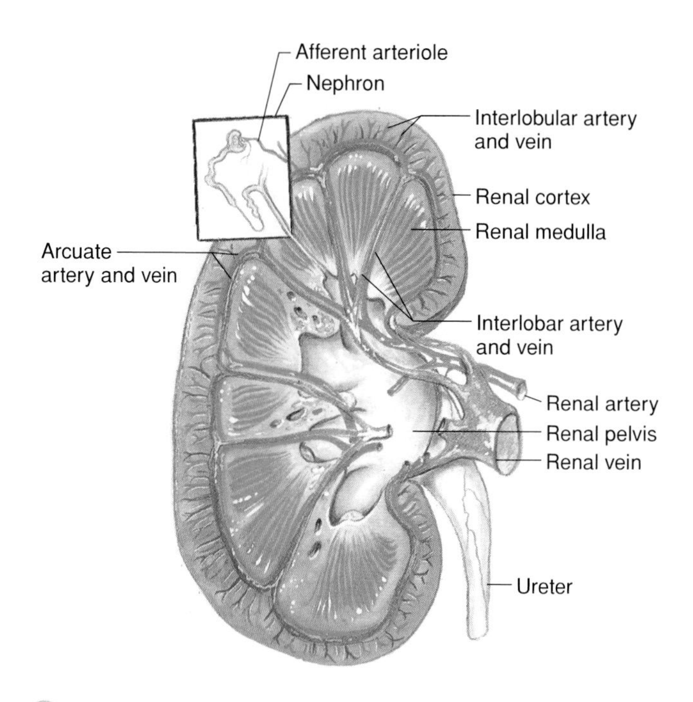

Figure 17.3

Main branches of the renal artery and renal vein.

Kidney Functions

The kidneys regulate the volume, composition, and pH of body fluids. In the process, they remove metabolic wastes from blood, combine the wastes with excess water and electrolytes to form urine, and excrete urine to the outside. The kidneys secrete the hormone *erythropoietin* (see chapter 12) to help control the rate of red blood cell formation and the enzyme *renin* to help regulate blood pressure.

Renal Blood Vessels

The **renal arteries**, which arise from the abdominal aorta, supply blood to the kidneys. These arteries transport a large volume of blood. When a person is at rest, the renal arteries usually carry 15–30% of the total cardiac output into the kidneys.

A renal artery enters a kidney through the hilum and gives off several branches, called *interlobar arteries*, which pass between the renal pyramids. At the junction between the medulla and cortex, the interlobar arteries branch to form a series of incomplete arches, the *arcuate arteries*, which, in turn, give rise to *interlobular arteries*. The final branches of the interlobular arteries, called **afferent arterioles**, lead to the nephrons (figs. 17.3 and 17.4).

Venous blood returns through a series of vessels that correspond generally to arterial pathways. The **renal vein** then joins the inferior vena cava as it courses through the abdominal cavity (see fig. 17.1).

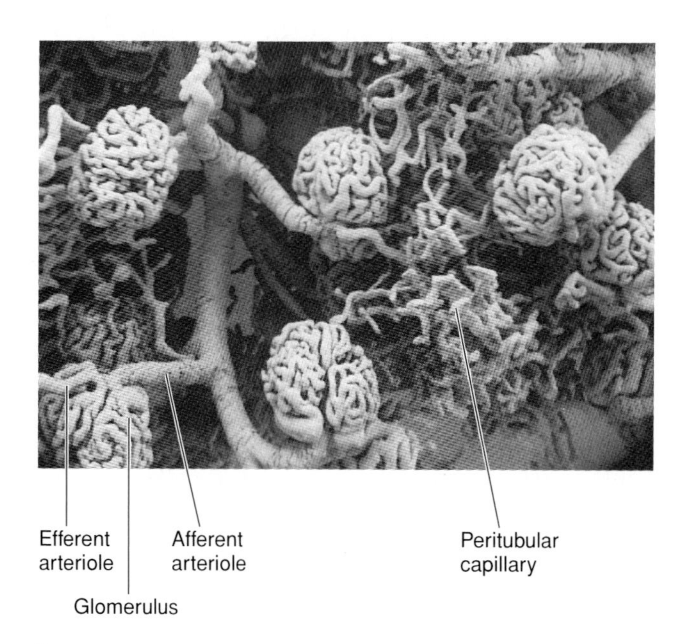

Figure 17.4

Scanning electron micrograph of a cast of the renal blood vessels associated with glomeruli (260×). From *Tissues and Organs: A Text-Atlas of Scanning Electron Microscopy*, by R. G. Kessel and R. H. Kardon, © 1979 W. H. Freeman and Company.

A kidney transplant can help patients with end-stage renal disease. In this procedure, a surgeon places a kidney from a living donor or a cadaver whose tissues are antigenically similar (histocompatible) to those of the recipient in the depression on the medial surface of the right or left ilium (iliac fossa). The surgeon then connects the renal artery and vein of the donor kidney to the recipient's iliac artery and vein, respectively, and the ureter of the donor kidney to the dome of the recipient's urinary bladder. ■

Nephrons

Nephron Structure

A kidney contains about 1 million nephrons, each consisting of a **renal corpuscle** and a **renal tubule** (see fig. 17.2*a*). Fluid flows through renal tubules on its way out of the body.

A renal corpuscle is composed of a tangled cluster of blood capillaries called a **glomerulus.** Glomerular capillaries filter fluid, the first step in urine formation. A thin-walled, saclike structure called a **glomerular capsule** surrounds the glomerulus (fig. 17.5). The glomerular capsule, an expansion at the proximal end of a renal tubule, receives the fluid the glomerulus filters. The renal tubule leads away from the glomerular capsule and becomes highly coiled. This coiled portion is called the *proximal convoluted tubule.*

The proximal convoluted tubule dips toward the renal pelvis, becoming the *descending limb of the nephron loop* (loop of Henle). The tubule then curves back toward its renal corpuscle and forms the *ascending limb of the nephron loop.* The ascending limb returns to the region of the renal corpuscle, where it becomes highly coiled again and is called the *distal convoluted tubule.*

Distal convoluted tubules from several nephrons merge in the renal cortex to form a *collecting duct,* which, in turn, passes into the renal medulla and enlarges as other collecting ducts join it. The resulting tube empties into a minor calyx through an opening in a renal papilla. Figure 17.6 summarizes the structure of a nephron and its associated blood vessels.

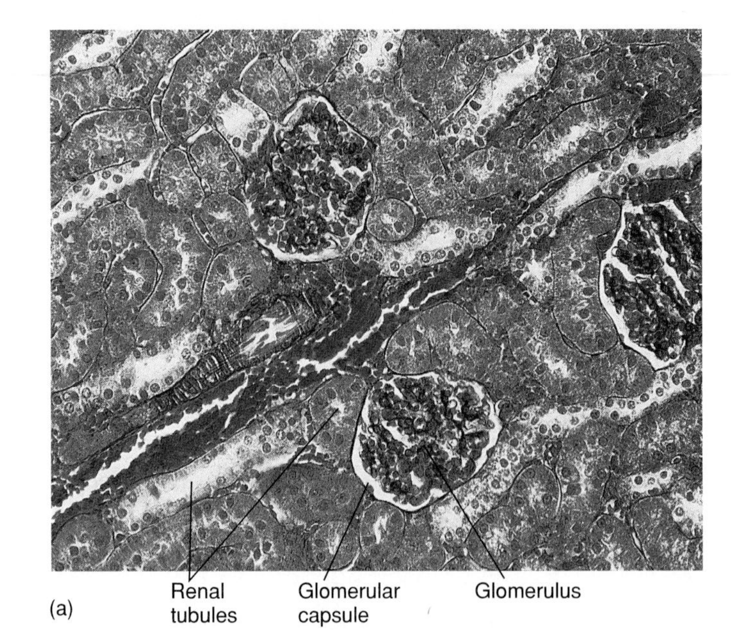

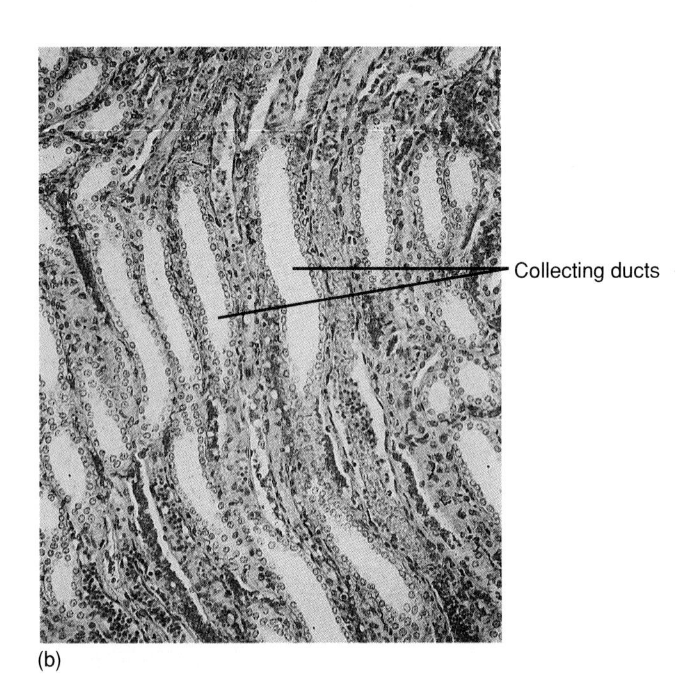

Figure 17.5

(*a*) Light micrograph of a section of the human renal cortex (300×). (*b*) Light micrograph of the renal medulla (400×).

1. List the general functions of the kidneys.
2. Trace the blood supply to the kidney.
3. Name the parts of a nephron.

Blood Supply of a Nephron

The cluster of capillaries that forms a glomerulus arises from an afferent arteriole. After passing through the glomerular capillaries, blood enters an **efferent arteriole** (rather than a venule), whose diameter is smaller

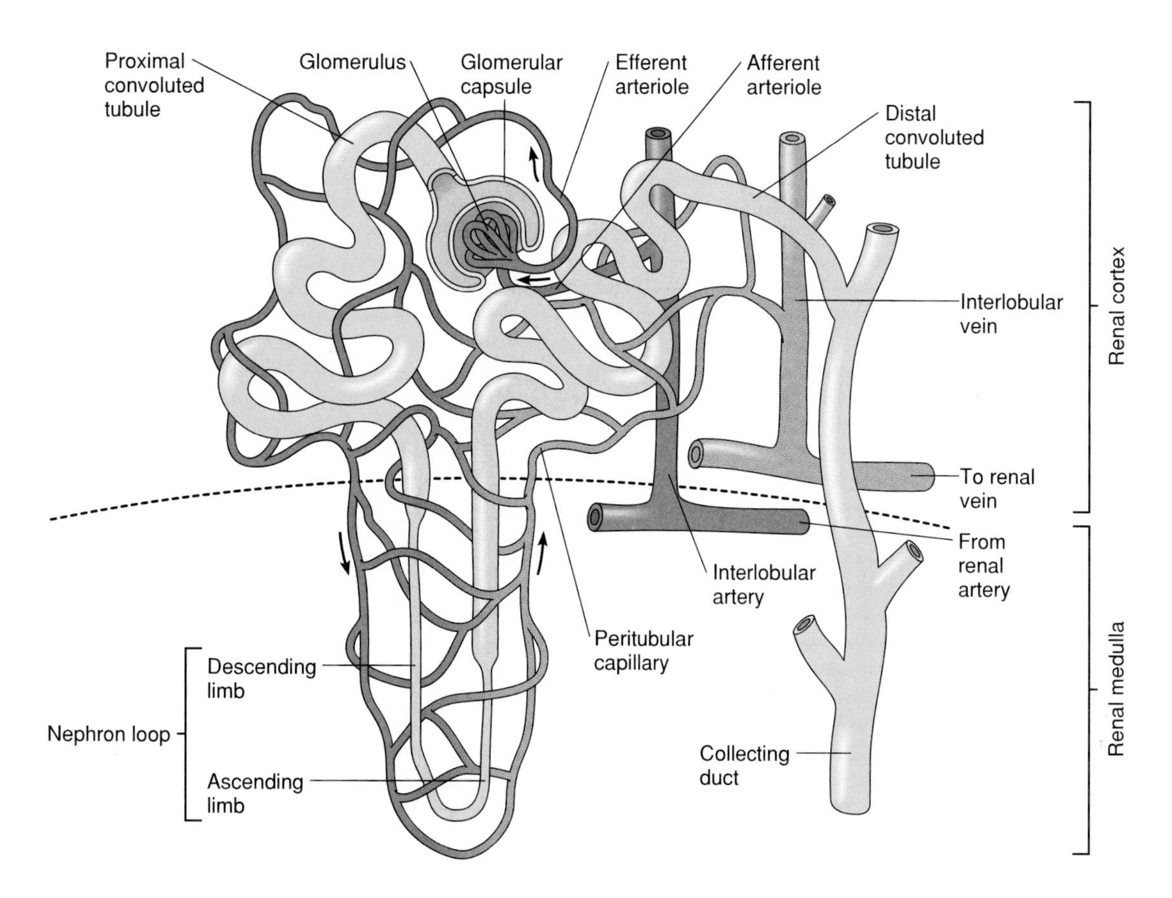

Figure 17.6

Structure of a nephron and the blood vessels associated with it.

than that of the afferent vessel (see fig. 17.4). The efferent arteriole resists blood flow to some extent. This backs up blood into the glomerulus, increasing pressure in the glomerular capillary.

The efferent arteriole branches into a complex, freely interconnecting network of capillaries, called the **peritubular capillary system,** that surrounds the renal tubule (see figs. 17.4 and 17.6). Blood in the peritubular capillary system is under low pressure. After flowing through the capillary network, the blood rejoins blood from other branches of the peritubular capillary system and enters the venous system of the kidney.

Juxtaglomerular Apparatus

Near its beginning, the distal convoluted tubule passes between and contacts afferent and efferent arterioles. At the point of contact, the epithelial cells of the distal tubule are quite narrow and densely packed. These cells comprise a structure called the *macula densa.*

Close by, in the walls of the arterioles near their attachments to the glomerulus, are some enlarged smooth muscle cells called *juxtaglomerular cells.* With cells of the macula densa, they constitute the **juxtaglomerular apparatus** (complex) (fig. 17.7). Its function in the control of renin secretion is described later in the chapter.

1. Describe the system of blood vessels that supplies a nephron.
2. What structures form the juxtaglomerular apparatus?

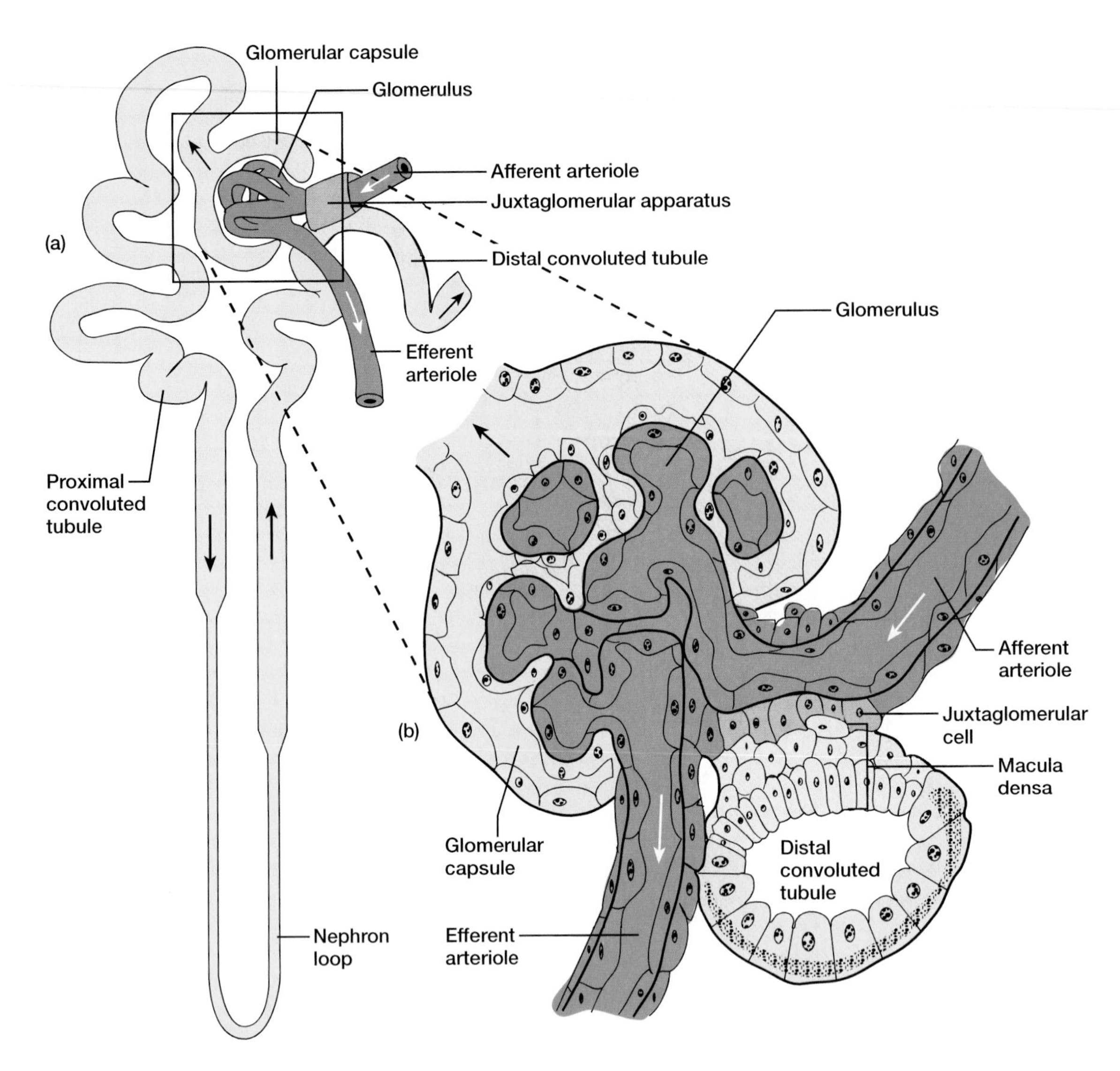

Figure 17.7

(*a*) Location of the juxtaglomerular apparatus. (*b*) Enlargement of a section of the juxtaglomerular apparatus, which consists of the macula densa and the juxtaglomerular cells. White arrows indicate direction of blood flow. Black arrows indicate flow of glomerular filtrate and tubular fluid.

Urine Formation

Nephrons regulate water and electrolyte concentrations within body fluids and remove wastes from blood. The end product of these functions is **urine**, which the body excretes. Urine contains metabolic wastes and excess water and electrolytes. Urine formation occurs through three processes: glomerular filtration, tubular reabsorption, and tubular secretion.

Glomerular Filtration

Urine formation begins when water and certain dissolved substances are filtered out of glomerular capillaries and into glomerular capsules. This filtration is similar to filtration at the arteriolar ends of other capillaries. Many tiny openings (fenestrae) in glomerular capillary walls, however, make glomerular capillaries much more permeable than capillaries in other tissues (fig. 17.8).

The glomerular capsule receives the resulting **glomerular filtrate**, which is similar in composition to the filtrate that becomes tissue fluid elsewhere in the body. That is, glomerular filtrate is mostly water and the same components as blood plasma, except for the large protein molecules.

Table 17.1 shows the relative concentrations of some substances in plasma, glomerular filtrate, and urine.

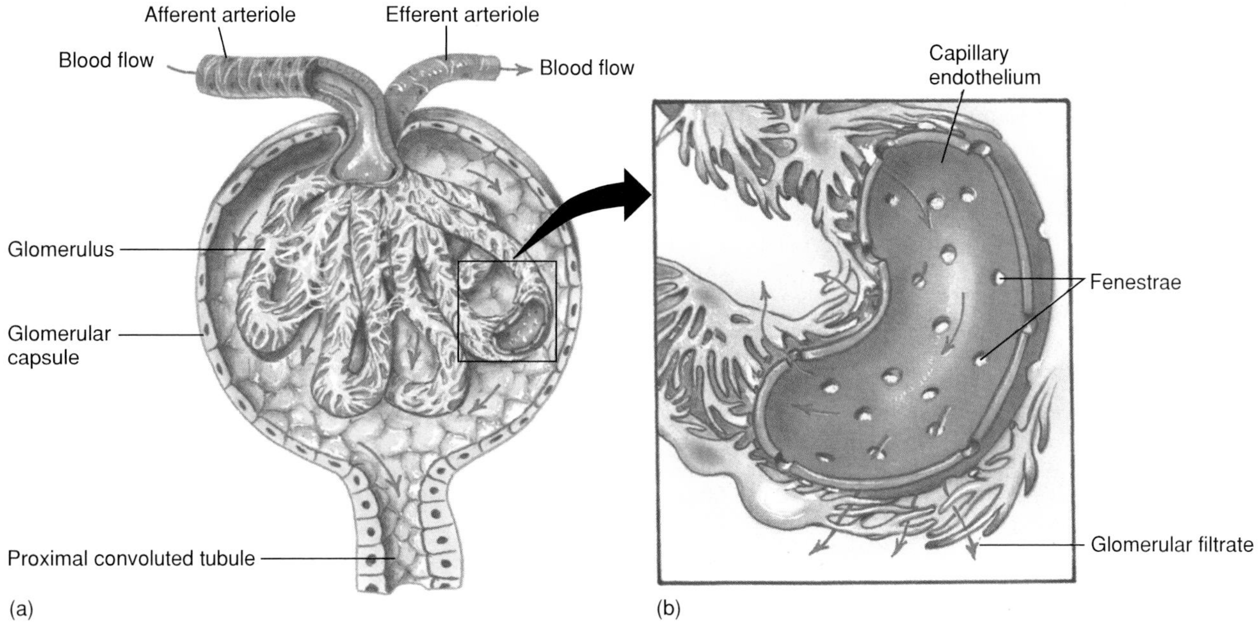

***Figure* 17.8**

(*a*) The first step in urine formation is filtration of substances out of glomerular capillaries and into the glomerular capsule. (*b*) Glomerular filtrate passes through fenestrae of the capillary endothelium.

Table 17.1

Relative Concentrations of Substances in the Plasma, Glomerular Filtrate, and Urine

Concentrations (mEq/l)

Substance	*Plasma*	*Glomerular filtrate*	*Urine*
Sodium (Na^+)	142	142	128
Potassium (K^+)	5	5	60
Calcium (Ca^{+2})	4	4	5
Magnesium (Mg^{+2})	3	3	15
Chloride (Cl^-)	103	103	134
Bicarbonate (HCO_3^-)	27	27	14
Sulfate (SO_4^{-2})	1	1	33
Phosphate (PO_4^{-3})	2	2	40

Concentrations (mg/100 ml)

Substance	*Plasma*	*Glomerular filtrate*	*Urine*
Glucose	100	100	0
Urea	26	26	1,820
Uric acid	4	4	53

Note: mEq/l = milliequivalent per liter.

Filtration Pressure

As in other capillaries, the hydrostatic pressure of blood forces substances through the glomerular capillary wall. (Recall that glomerular capillary pressure is high compared to that of other capillaries.) The osmotic pressure of plasma in the glomerulus and the hydrostatic pressure inside the glomerular capsule also influence this movement. An increase in either of these pressures opposes movement out of the capillary and, thus, reduces filtration. The net pressure forcing substances out of the glomerulus is the **filtration pressure.**

If arterial blood pressure plummets, as may occur during *shock,* glomerular hydrostatic pressure may fall below the level required for filtration. At the same time, epithelial cells of the renal tubules may not receive sufficient nutrients to maintain their high rates of metabolism. As a result, cells may die (tubular necrosis), impairing renal functions. Such changes may result in chronic renal failure. ■

Filtration Rate

The glomerular filtration rate is directly proportional to filtration pressure. Consequently, factors that affect glomerular hydrostatic pressure, glomerular plasma osmotic pressure, or hydrostatic pressure in the glomerular capsule also affect filtration rate. For example, any change in the diameters of the afferent and efferent arterioles changes glomerular hydrostatic pressure, also altering glomerular filtration rate.

The afferent arteriole, through which blood enters the glomerulus, may constrict in response to sympathetic nerve impulses. Blood flow diminishes, filtration pressure decreases, and filtration rate drops. On the other hand, if the efferent arteriole (through which blood leaves the glomerulus) constricts, blood backs up into the glomerulus, filtration pressure increases, and filtration rate rises. Vasodilation of these vessels causes opposite effects.

In capillaries, the plasma osmotic pressure that attracts water inward (see chapter 12) opposes the blood pressure that forces water and dissolved substances outward. During filtration through the capillary wall, proteins remaining in the plasma raise osmotic pressure within the glomerular capillary. When this pressure reaches a certain high level, filtration ceases. Conversely, conditions that decrease plasma osmotic pressure, such as a decrease in plasma protein concentration, increase filtration rate.

In *glomerulonephritis*, the glomerular capillaries are inflamed and become more permeable to proteins, which appear in the glomerular filtrate and in urine (proteinuria). At the same time, the protein concentration in blood plasma decreases (hypoproteinemia), and this decreases blood osmotic pressure. As a result, less tissue fluid moves into the capillaries, and edema develops. ■

The hydrostatic pressure in the glomerular capsule sometimes changes because of an obstruction, such as a stone in a ureter or an enlarged prostate gland pressing on the urethra. If this occurs, fluids back up into renal tubules and raise the hydrostatic pressure in the glomerular capsule. Because any increase in capsular pressure opposes glomerular filtration, filtration rate may decrease significantly.

In an average adult, the glomerular filtration rate for the nephrons of both kidneys is about 125 milliliters per minute, or 180,000 milliliters (180 liters, or nearly 45 gallons) in 24 hours. Only a small fraction is excreted as urine. Instead, most fluid that passes through renal tubules is reabsorbed and reenters the plasma.

1. What processes form urine?
2. What forces affect filtration pressure?
3. What factors influence the rate of glomerular filtration?

Regulation of Filtration Rate

Glomerular filtration rate usually is relatively constant. To help maintain homeostasis, however, glomerular filtration rate may increase when body fluids are in excess and may decrease when the body must conserve fluid.

Sympathetic nervous system reflexes that respond to changes in blood pressure and blood volume can alter glomerular filtration rate. If blood pressure or volume drops sufficiently, afferent arterioles vasoconstrict, decreasing glomerular filtration rate. This helps ensure that less urine forms when the body must conserve water. Conversely, vasodilation of afferent arterioles increases glomerular filtration rate to counter increased blood volume or blood pressure.

Another mechanism to control filtration rate uses the enzyme renin. Juxtaglomerular cells secrete renin (1) whenever special cells in the afferent arteriole sense a drop in blood pressure; (2) in response to sympathetic stimulation; and (3) when the macula densa (see fig. 17.7) senses decreased amounts of chloride, potassium, and sodium ions reaching the distal tubule. Once in the bloodstream, renin reacts with the plasma protein *angiotensinogen* to form *angiotensin I*. A second enzyme (*angiotensin converting enzyme,* or ACE) in the lungs and in plasma quickly changes angiotensin I to angiotensin II.

Angiotensin II has a number of actions that help maintain sodium balance, water balance, and blood pressure (fig. 17.9). Angiotensinogen II vasoconstricts the efferent arteriole, which causes blood to back up into the glomerulus, thus raising glomerular capillary hydrostatic pressure. This important action helps minimize the decrease in glomerular filtration rate when systemic blood pressure is low. Angiotensin II has a major effect on the kidneys by stimulating secretion of the adrenal hormone aldosterone, which stimulates tubular reabsorption of sodium.

The heart secretes another hormone, atrial natriuretic peptide (ANP), when blood volume increases. ANP increases sodium excretion by a number of mechanisms, including increasing glomerular filtration rate.

Elevated blood pressure (hypertension) is sometimes associated with excess release of renin, followed by increased formation of the vasoconstrictor angiotensin II. Patients with this form of high blood pressure often take a drug that prevents the formation of angiotensin II by inhibiting the action of the enzyme that converts angiotensin I into angiotensin II. Such a drug is called an *angiotensin converting enzyme inhibitor* (ACE inhibitor). ■

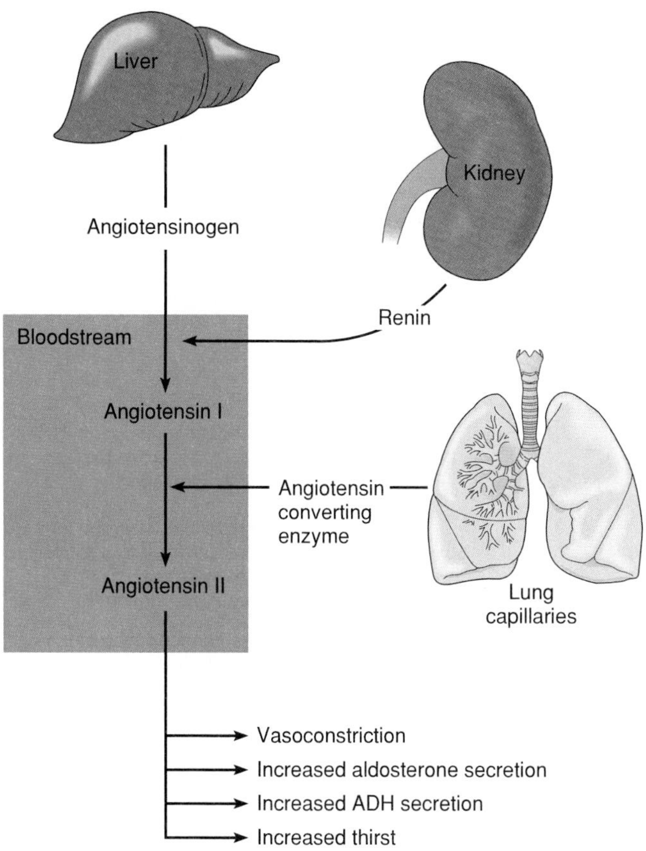

Liver
Kidney
Angiotensinogen
Renin
Bloodstream
Angiotensin I
Angiotensin converting enzyme
Angiotensin II
Lung capillaries
Vasoconstriction
Increased aldosterone secretion
Increased ADH secretion
Increased thirst

Figure 17.9

The formation of angiotensin II in the bloodstream involves several organs. Angiotensin II has multiple actions that conserve sodium and water. (ADH stands for antidiuretic hormone.)

1. What is the function of the macula densa?
2. How does renin help regulate filtration rate?

Tubular Reabsorption

Comparing the composition of glomerular filtrate entering the renal tubule with that of urine leaving the tubule reveals that the fluid changes as it passes through the tubule (see table 17.1). For example, glucose is present in glomerular filtrate but absent in urine. In contrast urea and uric acid are much more concentrated in urine than in glomerular filtrate. Such changes in fluid composition are largely the result of **tubular reabsorption,** the process that transports substances out of tubular fluid, through the epithelium of the renal tubule, and into the blood of the peritubular capillary (fig. 17.10).

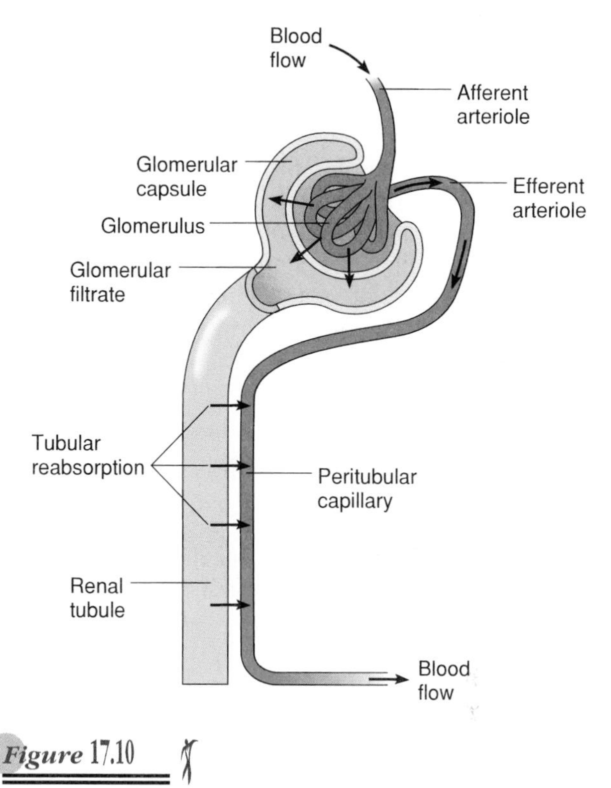

Figure 17.10

Tubular reabsorption transports substances from the renal tubule into the blood of the peritubular capillary.

Most of the plasma flowing through the kidney escapes filtration. Approximately 80% continues on through the peritubular capillaries. Peritubular capillary blood is under low pressure because it has already passed through two arterioles. Also, the wall of the peritubular capillary is more permeable than that of other capillaries. Both of these factors enhance the rate of fluid reabsorption from the renal tubule.

Tubular reabsorption occurs throughout the renal tubule, but most occurs in the proximal convoluted portion. The epithelial cells here have many microscopic projections called *microvilli* that form a "brush border" on their free surfaces. These tiny extensions greatly increase the surface area exposed to glomerular filtrate and enhance reabsorption.

Segments of the renal tubule are adapted to reabsorb specific substances, using particular modes of transport. Active transport, for example, reabsorbs glucose through the walls of the proximal tubule. Osmosis rapidly reabsorbs water through the epithelium of the proximal tubule. Portions of the distal tubule and collecting duct, however, are almost impermeable to water, a characteristic important in the regulation of urine concentration and volume, as described later in this chapter.

Active transport utilizes carrier molecules in cell membranes (see chapter 3). These carriers transport certain molecules across the membrane, release them, and then repeat the process. Such a mechanism, however, has a *limited transport capacity;* that is, it can only transport a certain number of molecules in a given time because the number of carriers is limited.

Usually, carrier molecules are able to transport all of the glucose in glomerular filtrate. When plasma glucose concentration increases to a critical level, called the *renal plasma threshold,* more glucose molecules are in the filtrate than the active transport mechanism can handle. As a result, some glucose remains in the tubular fluid and is excreted in urine.

Glucose in urine, called *glucosuria* (or *glycosuria*), may occur following intravenous administration of glucose. It may also occur in a patient with diabetes mellitus, in whom blood glucose concentration rises abnormally (see chapter 11). ■

Amino acids enter glomerular filtrate and are reabsorbed in the proximal convoluted tubule. Three different active transport mechanisms reabsorb different groups of amino acids, whose members have similar structures. Normally, only a trace of amino acids remains in urine.

Glomerular filtrate is nearly free of protein except for traces of albumin, a protein so small that pinocytosis reabsorbs it through the brush border of epithelial cells lining the proximal convoluted tubule. Proteins inside epithelial cells are broken down to amino acids, which then move into blood of the peritubular capillary.

The epithelium of the proximal convoluted tubule also reabsorbs creatine; lactic, citric, uric, and ascorbic (vitamin C) acids; and phosphate, sulfate, calcium, potassium, and sodium ions. Active transport mechanisms with limited transport capacities reabsorb these chemicals. These substances usually do not appear in urine, however, until glomerular filtrate concentration exceeds a particular substance's threshold.

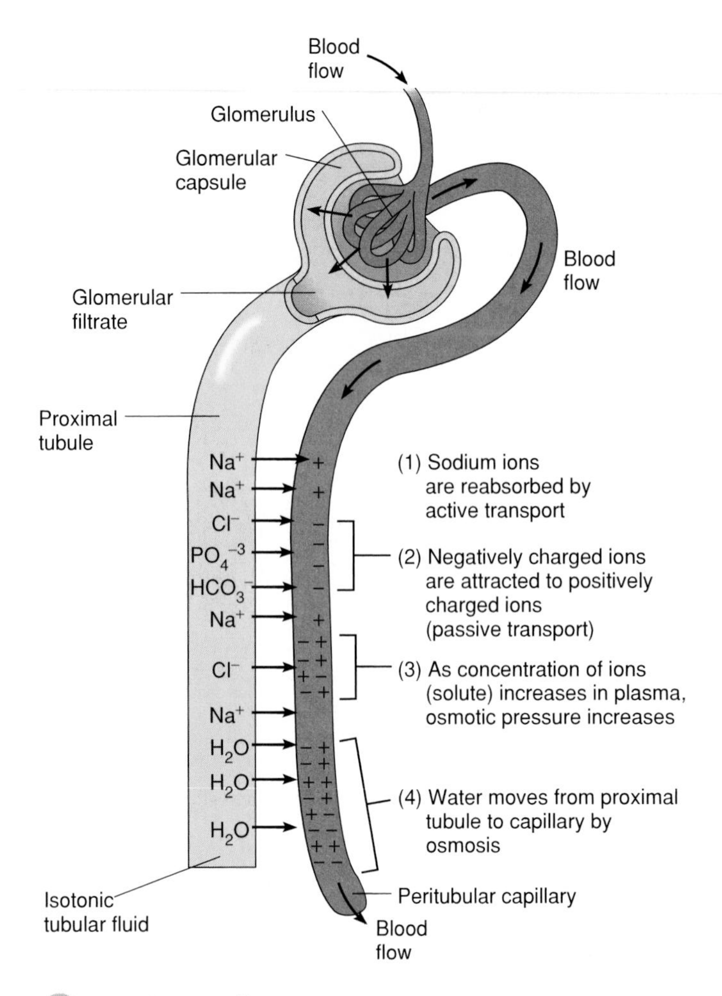

Figure 17.11

In the proximal portion of the renal tubule, osmosis reabsorbs water in response to active transport reabsorbing sodium and other solutes.

Sodium and Water Reabsorption

Substances that remain in the renal tubule become more concentrated as water is reabsorbed from the filtrate. Water reabsorption occurs passively by osmosis, primarily in the proximal convoluted tubule, and is closely associated with the active reabsorption of sodium ions (fig. 17.11). If sodium reabsorption increases, water reabsorption increases; if sodium reabsorption decreases, water reabsorption decreases also.

Active transport (the sodium pump mechanism) reabsorbs about 70% of sodium ions in the proximal segment of the renal tubule. As these positively charged ions (Na^+) move through the tubular wall, negatively charged ions, including chloride ions (Cl^-), phosphate ions (PO_4^{-3}), and bicarbonate ions (HCO_3^-) accompany them. These negatively charged ions move because of the electrochemical attraction between particles of opposite charge. This **passive transport** does not require direct expenditure of cellular energy.

As active transport moves more sodium ions into the peritubular capillary, along with various negatively charged ions, the concentration of solutes within the peritubular blood increases. Since water moves across cell membranes from regions of lesser solute concentration (hypotonic) toward regions of greater solute concentration (hypertonic), water moves by osmosis from the renal tubule into the peritubular capillary. Movement of solutes and water into the peritubular capillary greatly reduces the fluid volume within the renal tubule. The end of the proximal convoluted tubule is in osmotic equilibrium, and the remaining tubular fluid is isotonic.

1. What chemicals are normally in the glomerular filtrate but are not present in urine?
2. What mechanisms reabsorb solutes from the glomerular filtrate?
3. Describe the role of passive transport in urine formation.

Regulation of Urine Concentration and Volume

Active transport continues to reabsorb sodium ions as the tubular fluid moves through the nephron loop, the distal convoluted segment, and the collecting duct. Water is absorbed passively by osmosis in various segments of the renal tubule. As a result, almost all the sodium and water that enter the renal tubule as part of the glomerular filtrate are reabsorbed before urine is excreted.

The hormones aldosterone and ADH (antidiuretic hormone) may stimulate additional reabsorption of sodium and water, respectively. The changes in sodium and water excretion in response to these hormones are the final adjustments the kidney makes to maintain a constant internal environment.

As discussed in chapter 11, the adrenal gland secretes aldosterone in response to changes in blood concentrations of sodium and potassium ions. Aldosterone stimulates the distal tubule to reabsorb sodium and secrete potassium. Angiotensin II is another important stimulator of aldosterone secretion.

Neurons in the hypothalamus produce ADH, which the posterior pituitary releases in response to a decreasing water concentration in blood or a decrease in blood volume. When ADH reaches the kidney, it increases the permeability to water of the epithelial linings of the distal convoluted tubule and collecting duct, and water moves rapidly out of these segments by osmosis (water is reabsorbed). Consequently, urine volume falls, and urine concentrates soluble wastes and other substances in minimal water. Concentrated urine minimizes loss of body fluids when dehydration is likely.

If body fluids become too dilute, ADH secretion decreases. In the absence of ADH, the epithelial linings of the distal segment and collecting duct become less permeable to water, less water is reabsorbed, and urine is more dilute, excreting the excess water.

Table 17.2 summarizes the role of ADH in urine production.

Table 17.2 Role of ADH in Regulating Urine Concentration and Volume

1. Concentration of water in blood decreases.
2. Increase in osmotic pressure of body fluids stimulates osmoreceptors in hypothalamus in brain.
3. Hypothalamus signals posterior pituitary to release ADH.
4. Blood carries ADH to kidneys.
5. ADH causes distal convoluted tubules and collecting ducts to increase water reabsorption by osmosis.
6. Urine concentrates, and urine volume decreases.

Urea and Uric Acid Excretion

Urea is a by-product of amino acid catabolism. Consequently, its plasma concentration is directly proportional to the amount of protein in the diet. Urea enters the renal tubule by filtration. About 50% of it is reabsorbed, and the remainder is excreted in urine.

Uric acid forms from the metabolism of certain organic bases in nucleic acids. Active transport reabsorbs all the uric acid normally present in glomerular filtrate, but about 10% of the amount filtered is secreted into the renal tubule for excretion in urine.

Elevated concentrations of uric acid in the plasma cause a condition called *gout*. Because uric acid is relatively insoluble, it precipitates when in excess. In gout, crystals of uric acid deposit in joints and other tissues, where they produce inflammation and extreme pain. The joints of the great toes are most often affected, but other hand and feet joints may also be involved. Drugs that inhibit uric acid reabsorption, thus increasing its excretion, treat gout. People once thought gluttony caused gout, but the condition may be inherited. ■

1. How does the hypothalamus regulate urine concentration and volume?
2. Explain how urea and uric acid are excreted.

Tubular Secretion

In **tubular secretion,** certain substances move from the plasma in the peritubular capillary into the fluid of the renal tubule (fig. 17.12). As a result, the amount of a particular substance excreted into the urine may exceed the amount filtered from the plasma in the glomerulus.

Active transport mechanisms similar to those that function in reabsorption secrete some substances. Secretory mechanisms, however, transport substances in the opposite direction. For example, the epithelium of the proximal convoluted segment actively secretes certain organic compounds, including penicillin, creatinine, and histamine, into the tubular fluid.

Hydrogen ions are also actively secreted throughout the entire renal tubule. Secretion of hydrogen ions is important in regulating the pH of body fluids, as chapter 18 explains.

Most potassium ions in the glomerular filtrate are actively reabsorbed in the proximal convoluted tubule, but some may be secreted in the distal segment and collecting duct. During this process, active reabsorption of sodium ions from the tubular fluid produces a negative electrical charge within the tube. Because positively charged potassium ions (K^+) and hydrogen ions (H^+) are attracted to negatively charged regions, these ions move passively through the tubular epithelium and enter the tubular fluid (fig. 17.13). Active processes also secrete potassium ions.

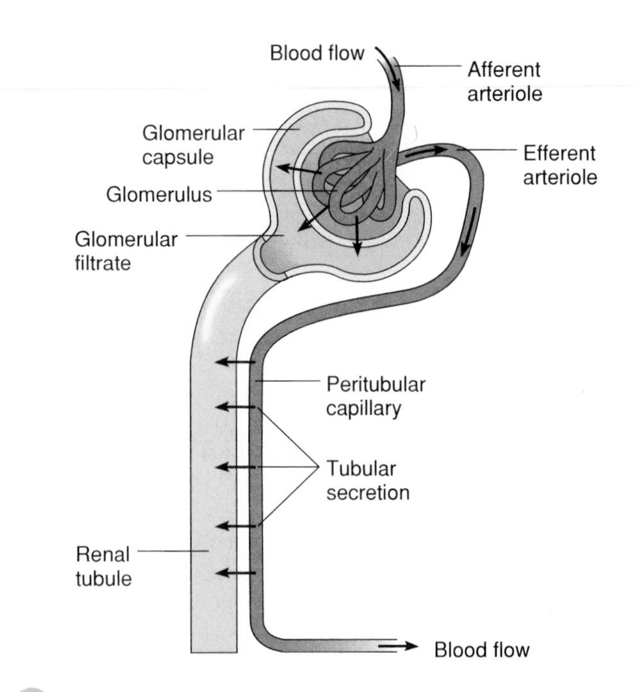

Figure 17.12

Secretory mechanisms move substances from the plasma of the peritubular capillary into the fluid of the renal tubule.

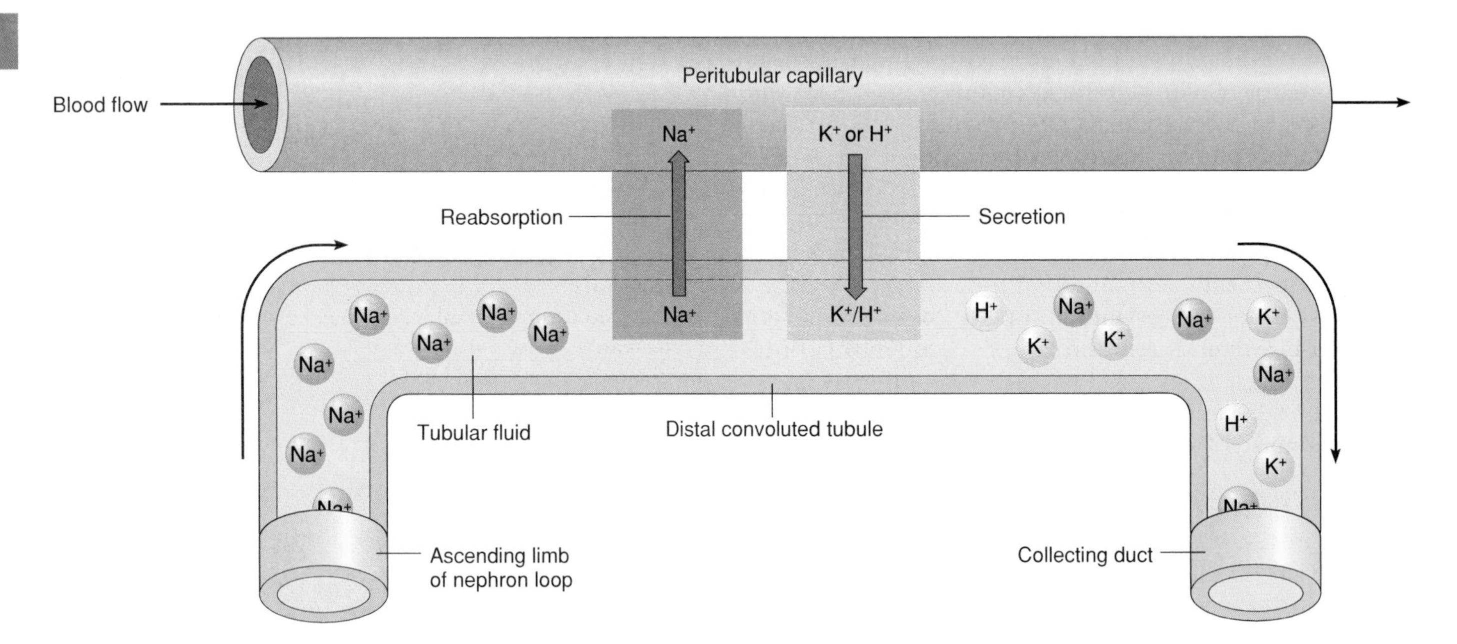

Figure 17.13

In the distal convoluted tubule, potassium ions (or hydrogen ions) may be passively secreted in response to the active reabsorption of sodium ions.

Table 17.3 Functions of Nephron Components

Part	Function
Renal corpuscle	
Glomerulus	Filtration of water and dissolved substances from plasma
Glomerular capsule	Receives glomerular filtrate
Renal tubule	
Proximal convoluted tubule	Reabsorption of glucose; amino acids; creatine; lactic, uric, citric, and ascorbic acids; phosphate, sulfate, calcium, potassium, and sodium ions by active transport Reabsorption of water by osmosis Reabsorption of chloride ions and other negatively charged ions by electrochemical attraction Active secretion of substances such as penicillin, histamine, creatinine, and hydrogen ions
Descending limb of nephron loop	Reabsorption of water by osmosis
Ascending limb of nephron loop	Reabsorption of sodium, potassium, and chloride ions by active transport
Distal convoluted tubule	Reabsorption of sodium ions by active transport Reabsorption of water by osmosis Active secretion of hydrogen ions Secretion of potassium ions both actively and by electrochemical attraction
Collecting duct	Reabsorption of water by osmosis

Table 17.3 summarizes the functions of the nephron components.

1. Define *tubular secretion*.
2. What substances are actively secreted?
3. How does sodium reabsorption affect potassium secretion?

Urine Composition

Urine composition reflects the amounts of water and solutes that the kidneys must eliminate from the body or retain in the internal environment to maintain homeostasis. The composition differs considerably from time to time because of variations in dietary intake and physical activity. Urine is about 95% water, and usually contains urea and uric acid. It may also contain a trace of amino acids and a variety of electrolytes, whose concentrations vary directly with amounts in the diet (see table 17.1).

The volume of urine produced is usually between 0.6 and 2.5 liters per day, depending on fluid intake, environmental temperature, relative humidity of the surrounding air, and the person's emotional condition, respiratory rate, and body temperature. Urine output of 50–60 milliliters per hour is normal; output of less than 30 milliliters per hour may indicate kidney failure.

Abnormal substances in urine include glucose, proteins, hemoglobin, ketones, and blood cells. The significance of such substances in urine, however, may depend on the amounts present and on other factors. For example, glucose in urine may follow a large intake of carbohydrates, proteins may appear following vigorous physical exercise, and ketones may appear after a prolonged fast. Also, some pregnant women have glucose in their urine as birth nears. ■

1. List the normal constituents of urine.
2. What factors affect urine volume?

Urine Elimination

After urine forms in the nephrons, it passes from the collecting ducts through openings in the renal papillae and enters the calyces of the kidney (see fig. 17.2). From there, it passes through the renal pelvis, and a ureter conveys it to the urinary bladder (see fig. 17.1 and reference plate 6). The urethra excretes urine to the outside.

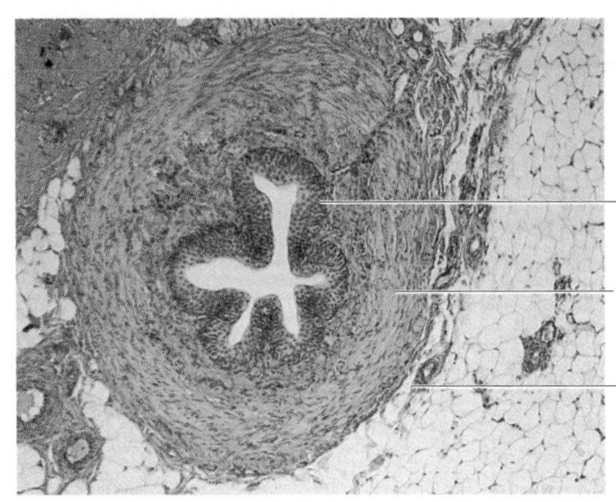

Figure 17.14

Cross section of a ureter (16×).

Ureters

Each **ureter** is a tube about 25 centimeters long that begins as the funnel-shaped renal pelvis. It extends downward behind the parietal peritoneum and runs parallel to the vertebral column. Within the pelvic cavity, each ureter courses forward and medially and joins the urinary bladder from underneath.

The ureter wall has three layers. The inner layer, or *mucous coat,* is continuous with the linings of the renal tubules and the urinary bladder. The middle layer, or *muscular coat,* consists largely of smooth muscle fibers. The outer layer, or *fibrous coat,* is connective tissue (fig. 17.14).

The muscular walls of the ureters propel the urine. Muscular peristaltic waves, originating in the renal pelvis, force urine along the length of the ureter. When a peristaltic wave reaches the urinary bladder, a jet of urine spurts into the bladder. A flaplike fold of mucous membrane covers the opening through which urine enters the bladder. This fold acts as a valve, allowing urine to enter the bladder from the ureter but preventing it from backing up.

1. Describe the structure of a ureter.
2. How is urine moved from the renal pelvis to the urinary bladder?
3. What prevents urine from backing up from the urinary bladder into the ureters?

Urinary Bladder

The **urinary bladder** is a hollow, distensible, muscular organ that stores urine and forces it into the urethra (see fig. 17.1 and reference plate 6). It is within the pelvic cavity, behind the symphysis pubis, and beneath the parietal peritoneum.

The pressure of surrounding organs alters the shape of the somewhat spherical bladder. When empty, the inner wall of the bladder forms many folds, but as the bladder fills with urine, the wall becomes smoother. At the same time, the superior surface of the bladder expands upward into a dome.

The internal floor of the bladder includes a triangular area called the *trigone,* which has an opening at each of its three angles (fig. 17.15). Posteriorly, at the base of the trigone, the openings are those of the ureters. Anteriorly, at the apex of the trigone, a short, funnel-shaped extension, called the *neck* of the bladder, contains the opening into the urethra.

The wall of the urinary bladder has four layers. The inner layer, or *mucous coat,* includes several thicknesses of transitional epithelial cells. The thickness of this tissue changes as the bladder expands and contracts. Thus, during distension, the tissue may be only two or three cells thick; during contraction, it may be five or six cells thick (see chapter 5).

The second layer of the bladder wall is the *submucous coat.* It consists of connective tissue and contains many elastic fibers.

The third layer of the bladder wall, or *muscular coat,* is composed primarily of coarse bundles of smooth muscle fibers. These bundles are interlaced in

Kidney stones, which are usually composed of uric acid, calcium oxalate, calcium phosphate, or magnesium phosphate, can form in the collecting ducts and renal pelvis. Such a stone passing into a ureter causes severe pain that begins in the region of the kidney and radiates into the abdomen, pelvis, and lower limbs. It may also cause nausea and vomiting.

About 60% of kidney stones pass from the body on their own. Other stones were once removed surgically but are now shattered with intense sound waves. In this procedure, called *extracorporeal shock-wave lithotripsy* (*ESWL*), the patient is placed in a stainless steel tub filled with water. A spark-gap electrode produces shock waves underwater, and a reflector concentrates and focuses the shock-wave energy on the stones. The resulting sandlike fragments then leave in urine. ■

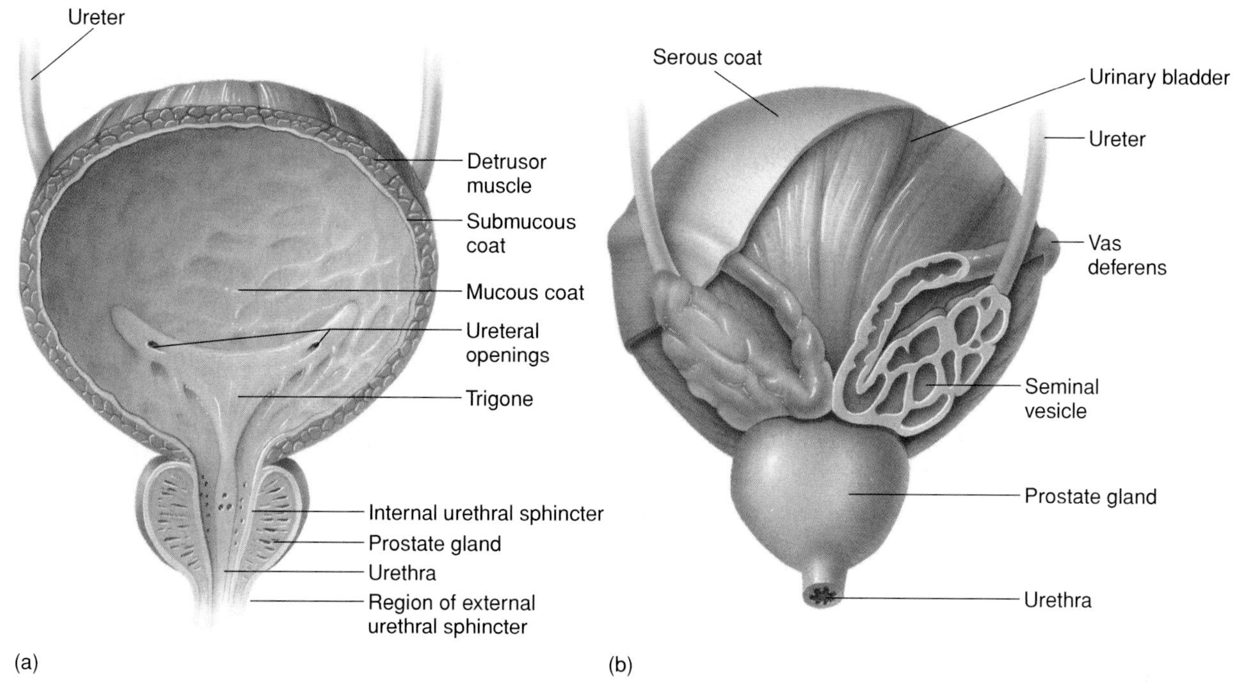

(a)

(b)

Figure 17.15

A male urinary bladder. (*a*) Coronal section. (*b*) Posterior view.

all directions and at all depths, and together, they comprise the **detrusor muscle.** The portion of the detrusor muscle that surrounds the neck of the bladder forms an *internal urethral sphincter.* Sustained contraction of this muscle prevents the bladder from emptying until pressure within the bladder increases to a certain level. The detrusor muscle is innervated with parasympathetic nerve fibers that function in the micturition reflex, discussed in the section that follows.

The outer layer of the bladder wall, or *serous coat,* consists of the parietal peritoneum. This layer is only on the bladder's upper surface. Elsewhere, the outer coat is composed of fibrous connective tissue.

Because the linings of the ureters and the urinary bladder are continuous, infectious agents, such as bacteria, may ascend from the bladder into the ureters. Inflammation of the bladder, called *cystitis,* is more common in women than men because the female urethral pathway is shorter. Inflammation of the ureter is called *ureteritis.*■

1. Describe the trigone of the urinary bladder.
2. Describe the structure of the bladder wall.
3. What kind of nerve fibers supply the detrusor muscle?

Micturition

Micturition (urination) is the process that expels urine from the urinary bladder. In micturition, the detrusor muscle contracts, as do muscles in the abdominal wall and pelvic floor. At the same time, muscles in the thoracic wall and diaphragm do not contract. Micturition also requires relaxation of the *external urethral sphincter.* This muscle, which is part of the urogenital diaphragm described in chapter 8, surrounds the urethra about 3 centimeters from the bladder and is composed of voluntary skeletal muscle tissue.

Distension of the bladder wall as it fills with urine stimulates stretch receptors, triggering the micturition reflex. The *micturition reflex center* is in the spinal cord. When sensory impulses from the stretch receptors signal the reflex center, parasympathetic motor impulses travel to the detrusor muscle, which contracts rhythmically in response. A sensation of urgency accompanies this action.

The urinary bladder may hold as much as 600 milliliters of urine before stimulating pain receptors, but the urge to urinate usually begins when it contains about 150 milliliters. As urine volume increases to 300 milliliters or more, the sensation of fullness intensifies, and contractions of the bladder wall become more powerful. When these contractions are strong enough to force the internal urethral sphincter open, another reflex signals the external urethral sphincter to relax, and the bladder can empty.

Because the external urethral sphincter is composed of skeletal muscle, it is under conscious control. Thus, the sphincter muscle ordinarily remains contracted until a person decides to urinate. Nerve centers in the brain stem and cerebral cortex that can partially inhibit the micturition reflex aid this control. When a person decides to urinate, the external urethral sphincter relaxes, and the micturition reflex is no longer inhibited. Nerve centers within the pons and the hypothalamus of the brain heighten the micturition reflex. Consequently, the detrusor muscle contracts, and urine is excreted through the urethra. Within a few moments, the neurons of the micturition reflex fatigue, the detrusor muscle relaxes, and the bladder begins to fill with urine again.

Damage to the spinal cord above the sacral region destroys voluntary control of urination. If the micturition reflex center and its sensory and motor fibers are uninjured, however, micturition may continue to occur reflexly. In this case, the bladder collects urine until its walls stretch enough to trigger a micturition reflex, and the detrusor muscle contracts in response. This condition is called an *automatic bladder.* ■

Urethra

The **urethra** is a tube that conveys urine from the urinary bladder to the outside (see fig. 17.1 and reference plate 7). Its wall is lined with mucous membrane and contains a thick layer of smooth muscle tissue, whose fibers are generally directed longitudinally. The urethral wall also contains numerous mucous glands, called *urethral glands,* which secrete mucus into the urethral canal (fig. 17.16).

1. Describe micturition.
2. How is it possible to consciously inhibit the micturition reflex?
3. Describe the structure of the urethra.

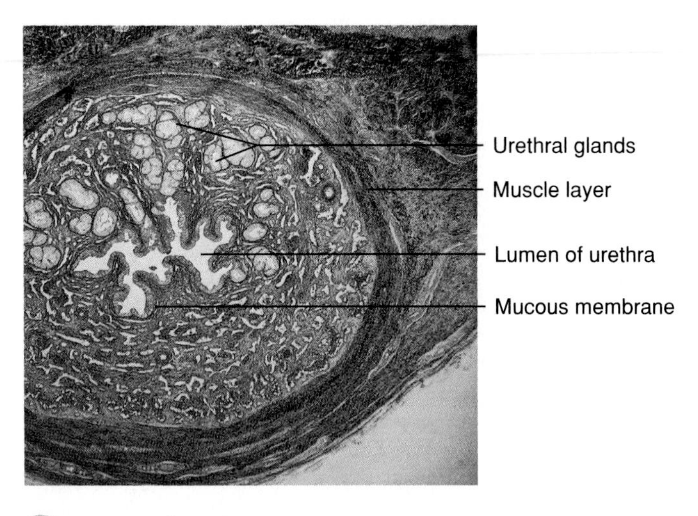

Figure 17.16

Cross section through the urethra (10×).

Clinical Terms Related to the Urinary System

Anuria (ah-nu′re-ah) Absence of urine due to failure of kidney function or to an obstruction in a urinary pathway.

Bacteriuria (bak-te″re-u′re-ah) Bacteria in urine.

Cystectomy (sis-tek′to-me) Surgical removal of the urinary bladder.

Cystitis (sis-ti′tis) Inflammation of the urinary bladder.

Cystoscope (sis′to-skōp) Instrument used to visually examine the interior of the urinary bladder.

Cystotomy (sis-tot′o-me) An incision of the urinary bladder wall.

Diuresis (di″u-re′sis) Increased urine excretion.

Diuretic (di″u-ret′ik) A substance that increases urine production.

Dysuria (dis-u′re-ah) Painful or difficult urination.

Hematuria (hem″ah-tu′re-ah) Blood in urine.

Incontinence (in-kon′ti-nens) Inability to control urination and/or defecation reflexes.

Nephrectomy (nĕ-frek′to-me) Surgical removal of a kidney.

Nephrolithiasis (nef″ro-lĭ-thi′ah-sis) Kidney stones.

Nephroptosis (nef″rop-to′sis) A movable or displaced kidney.

Oliguria (ol″ĭ-gu′re-ah) Scanty urine output.

Polyuria (pol″e-u′re-ah) Excess urine output.

Pyelolithotomy (pi″ĕ-lo-lĭ-thot′o-me) Removal of a stone from the renal pelvis.

Pyelonephritis (pi″ĕ-lo-ne-fri′tis) Inflammation of the renal pelvis.

Pyelotomy (pi″ĕ-lot′o-me) An incision into the renal pelvis.

Pyuria (pi-u′re-ah) Pus (white blood cells) in urine.

Uremia (u-re′me-ah) Accumulation in blood of substances ordinarily excreted in urine.

Ureteritis (u-re″ter-i′tis) Inflammation of the ureter.

Urethritis (u″re-thri′tis) Inflammation of the urethra.

organization

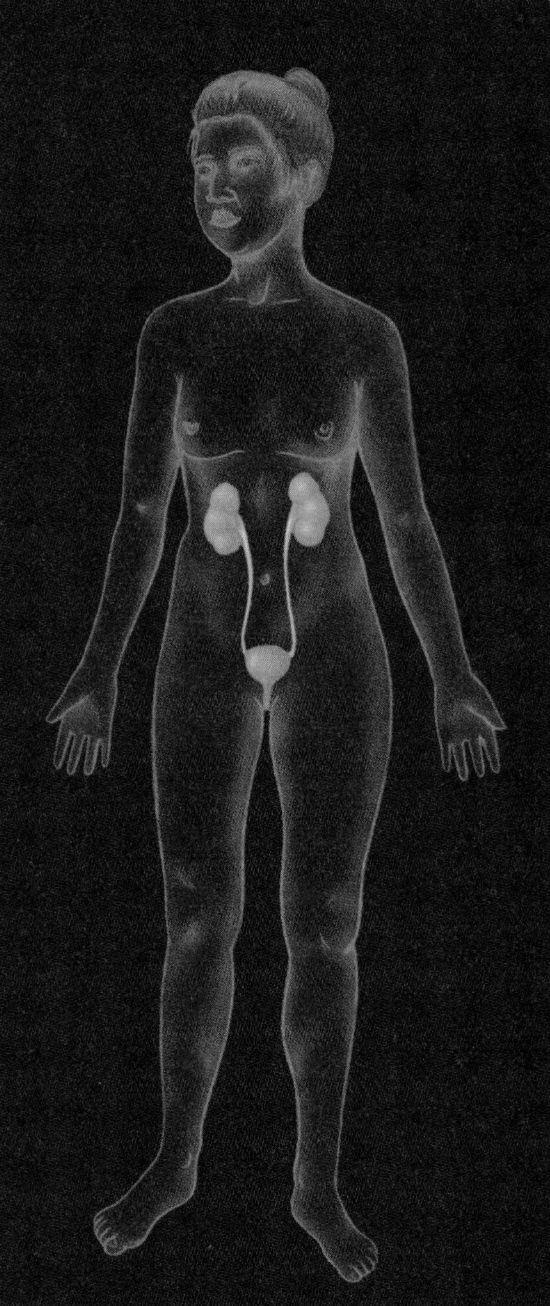

Urinary System

The urinary system controls the composition of the internal environment.

Integumentary System

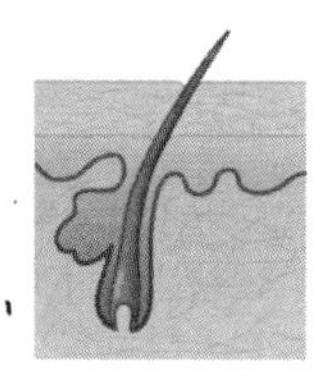

The urinary system compensates for water loss due to sweating. The kidneys and skin both play a role in vitamin D production.

Cardiovascular System

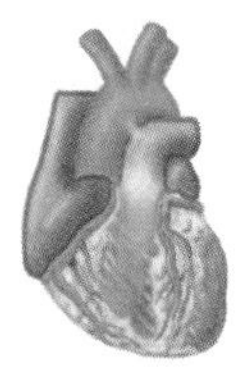

The urinary system controls blood volume. Blood volume and blood pressure play a role in determining water and solute excretion.

Skeletal System

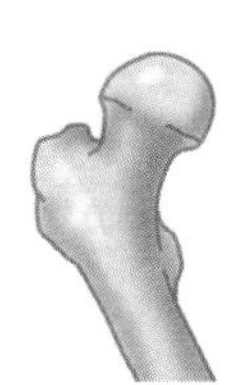

The kidneys and bone tissue work together to control plasma calcium levels.

Lymphatic System

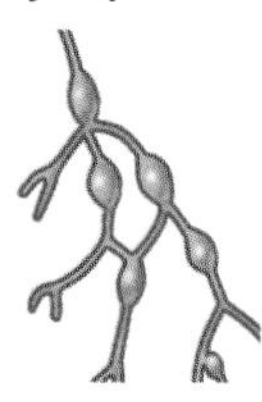

The kidneys control extracellular fluid volume and composition (including lymph).

Muscular System

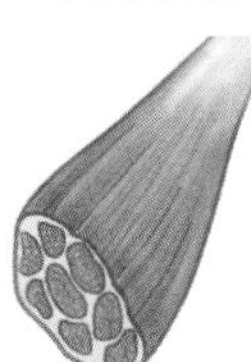

Muscle tissue controls urine elimination from the bladder.

Digestive System

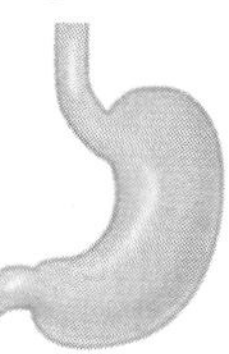

The kidneys compensate for fluids lost by the digestive system.

Nervous System

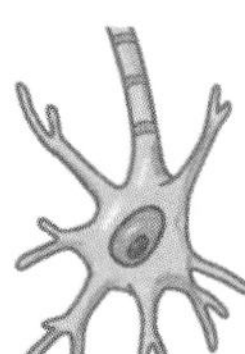

The nervous system influences urine production and elimination.

Respiratory System

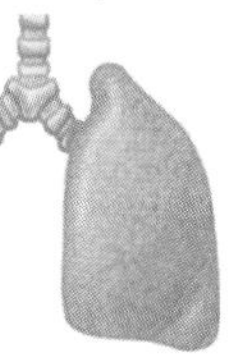

The kidneys and the lungs work together to control the pH of the internal environment.

Endocrine System

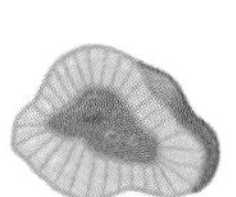

The endocrine system influences urine production.

Reproductive System

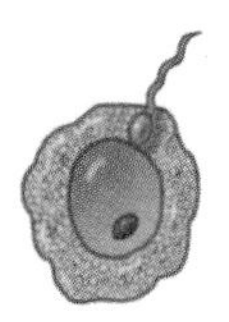

The urinary system in males shares common organs with the reproductive system. The kidneys compensate for fluids lost from the male and female reproductive systems.

Summary Outline

Introduction (p. 477)

The urinary system consists of the kidneys, ureters, urinary bladder, and urethra.

Kidneys (p. 477)

1. Location of the kidneys
   a. The kidneys are high on the posterior wall of the abdominal cavity.
   b. They are behind the parietal peritoneum.
2. Kidney structure
   a. A kidney contains a hollow renal sinus.
   b. The ureter expands into the renal pelvis.
   c. Renal papillae project into the renal sinus.
   d. Each kidney divides into a medulla and a cortex.
3. Kidney functions
   a. The kidneys remove metabolic wastes from blood and excrete them.
   b. They also help regulate red blood cell production; blood pressure; and the volume, composition, and pH of body fluids.
4. Renal blood vessels
   a. Arterial blood flows through the renal artery, interlobar arteries, arcuate arteries, interlobular arteries, afferent arterioles, glomerular capillaries, efferent arterioles, and peritubular capillaries.
   b. Venous blood returns through a series of vessels that correspond to the arterial pathways.
5. Nephrons
   a. Nephron structure
      (1) A nephron is the functional unit of the kidney.
      (2) It consists of a renal corpuscle and a renal tubule.
         (a) The corpuscle consists of a glomerulus and glomerular capsule.
         (b) Portions of the renal tubule include the proximal convoluted tubule, nephron loop (ascending and descending limbs), distal convoluted tubule, and collecting duct.
      (3) The collecting duct empties into the minor calyx of the renal pelvis.
   b. Blood supply of a nephron
      (1) The glomerular capillary receives blood from the afferent arteriole and passes it to the efferent arteriole.
      (2) The efferent arteriole gives rise to the peritubular capillary system, which surrounds the renal tubule.
   c. Juxtaglomerular apparatus
      (1) The juxtaglomerular apparatus is at the point of contact between the distal convoluted tubule and the afferent and efferent arterioles.
      (2) It consists of the macula densa and juxtaglomerular cells.

Urine Formation (p. 482)

Nephrons remove wastes from blood and regulate water and electrolyte concentrations. Urine is the end product.

1. Glomerular filtration
   a. Urine formation begins when water and dissolved materials filter out of glomerular capillaries.
   b. Glomerular capillaries are much more permeable than the capillaries in other tissues.
   c. The composition of the filtrate is similar to that of tissue fluid.
2. Filtration pressure
   a. Filtration is due mainly to hydrostatic pressure inside glomerular capillaries.
   b. The osmotic pressure of plasma and the hydrostatic pressure in the glomerular capsule also affect filtration.
   c. Filtration pressure is the net force moving material out of the glomerulus and into the glomerular capsule.
3. Filtration rate
   a. Rate of filtration varies with filtration pressure.
   b. Filtration pressure changes with the diameters of the afferent and efferent arterioles.
   c. As osmotic pressure in the glomerulus increases, filtration rate decreases.
   d. As hydrostatic pressure in a glomerular capsule increases, filtration rate decreases.
   e. The kidneys produce about 125 milliliters of glomerular fluid per minute, most of which is reabsorbed.
4. Regulation of filtration rate
   a. Glomerular filtration rate remains relatively constant, but may increase or decrease as required.
   b. Increased sympathetic nerve activity can decrease glomerular filtration rate.
   c. When the macula densa senses decreased amounts of chloride, potassium, and sodium ions in the distal tubule, it causes juxtaglomerular cells to release renin.
   d. This triggers a series of changes leading to vasoconstriction of afferent and efferent arterioles, which may affect glomerular filtration rate, and aldosterone secretion, which stimulates tubular sodium reabsorption.
5. Tubular reabsorption
   a. Substances are selectively reabsorbed from glomerular filtrate.
   b. The peritubular capillary's permeability adapts it for reabsorption.
   c. Most reabsorption occurs in the proximal tubule, where epithelial cells have microvilli.
   d. Different modes of transport reabsorb various substances in particular segments of the renal tubule.
      (1) Active transport reabsorbs glucose and amino acids.
      (2) Osmosis reabsorbs water.
   e. Active transport mechanisms have limited transport capacities.
6. Sodium and water reabsorption
   a. Substances that remain in the filtrate are concentrated as water is reabsorbed.
   b. Active transport reabsorbs sodium ions.
   c. As positively charged sodium ions move out of the filtrate, negatively charged ions follow them.
   d. Water is passively reabsorbed by osmosis.
7. Regulation of urine concentration and volume
   a. Most sodium is reabsorbed before urine is excreted.
   b. Antidiuretic hormone increases the permeability of the distal convoluted tubule and collecting duct, promoting water reabsorption.

8. Urea and uric acid excretion
   a. Diffusion passively reabsorbs urea. About 50% of the urea is excreted in urine.
   b. Active transport reabsorbs uric acid. Some uric acid is secreted into the renal tubule.
9. Tubular secretion
   a. Secretion transports substances from plasma to the tubular fluid.
   b. Various organic compounds and hydrogen ions are secreted actively.
   c. Potassium ions are secreted both actively and passively.
10. Urine composition
    a. Urine is about 95% water, and it also usually contains urea and uric acid.
    b. Urine contains varying amounts of electrolytes and may contain a trace of amino acids.
    c. Urine volume varies with fluid intake and with certain environmental factors.

Urine Elimination (p. 489)

1. Ureters
   a. The ureter extends from the kidney to the urinary bladder.
   b. Peristaltic waves in the ureter force urine to the urinary bladder.
2. Urinary bladder
   a. The urinary bladder stores urine and forces it through the urethra during micturition.
   b. The openings for the ureters and urethra are at the three angles of the trigone.
   c. A portion of the detrusor muscle forms an internal urethral sphincter.
3. Micturition
   a. Micturition expels urine.
   b. Micturition contracts the detrusor muscle and relaxes the external urethral sphincter.
   c. Micturition reflex
      (1) Distension stimulates stretch receptors in the bladder wall.
      (2) The micturition reflex center in the spinal cord sends parasympathetic motor impulses to the detrusor muscle.
      (3) As the bladder fills, its internal pressure increases, forcing the internal urethral sphincter open.
      (4) A second reflex relaxes the external urethral sphincter unless voluntary control maintains its contraction.
      (5) Nerve centers in the cerebral cortex and brain stem aid control of urination.
4. Urethra
   The urethra conveys urine from the urinary bladder to the outside.

Review Exercises

1. Name and list the general functions of the organs of the urinary system. (p. 477)
2. Describe the external and internal structure of a kidney. (p. 477)
3. List the functions of the kidneys. (p. 479)
4. Name the vessels through which blood passes as it travels from the renal artery to the renal vein. (p. 479)
5. Distinguish between a renal corpuscle and a renal tubule. (p. 480)
6. Name the parts through which fluid passes from the glomerulus to the collecting duct. (p. 480)
7. Describe the location and structure of the juxtaglomerular apparatus. (p. 481)
8. Compare the composition of glomerular filtrate with that of blood plasma. (p. 482)
9. Define *filtration pressure*. (p. 483)
10. Explain how the diameters of the afferent and efferent arterioles affect the rate of glomerular filtration. (p. 484)
11. Explain how changes in the osmotic pressure of blood plasma affect glomerular filtration rate. (p. 484)
12. Explain how the hydrostatic pressure of a glomerular capsule affects the rate of glomerular filtration. (p. 484)
13. Describe two mechanisms by which the body regulates filtration rate. (p. 484)
14. Discuss how tubular reabsorption is selective. (p. 485)
15. Explain how the peritubular capillary is adapted for reabsorption. (p. 485)
16. Explain how epithelial cells of the proximal convoluted tubule are adapted for reabsorption. (p. 485)
17. Explain why active transport mechanisms have limited transport capacities. (p. 486)
18. Define *renal plasma threshold*. (p. 486)
19. Explain how amino acids and proteins are reabsorbed. (p. 486)
20. Describe the effect of sodium reabsorption on the reabsorption of negatively charged ions. (p. 486)
21. Explain how sodium reabsorption affects water reabsorption. (p. 486)
22. Describe the function of ADH. (p. 487)
23. Compare the processes that reabsorb urea and uric acid. (p. 487)
24. Explain how potassium ions may be secreted passively. (p. 488)
25. List common constituents of urine and their sources. (p. 489)
26. List some of the factors that affect the urine volume produced daily. (p. 489)
27. Describe the structure and function of a ureter. (p. 490)
28. Explain how the muscular wall of the ureter helps move urine. (p. 490)
29. Describe the structure and location of the urinary bladder. (p. 490)
30. Define *detrusor muscle*. (p. 491)
31. Distinguish between the internal and external urethral sphincters. (p. 491)
32. Describe the micturition reflex. (p. 491)
33. Explain how the micturition reflex can be voluntarily controlled. (p. 492)

Critical Thinking

1. Image you are adrift at sea. Why will you dehydrate more quickly if you drink seawater instead of fresh water to quench your thirst?
2. Urinary tract infections frequently accompany sexually transmitted diseases. Why?
3. Would an excess or deficiency of renin be likely to cause hypertension (high blood pressure)? Cite a reason for your answer.
4. Why is protein in urine a sign of kidney damage? What structures in the kidney are probably affected?
5. Why are people following high-protein diets advised to drink large quantities of water?
6. How might very low blood pressure impair kidney function?
7. An infant is born with narrowed renal arteries. What effect will this condition have on urine volume?
8. If a patient who has had major abdominal surgery receives intravenous fluids equal to blood volume lost during surgery, would you expect urine volume to be greater or less than normal? Why?
9. If blood pressure plummets in a patient in shock as a result of a severe injury, how would you expect urine volume to change? Why?

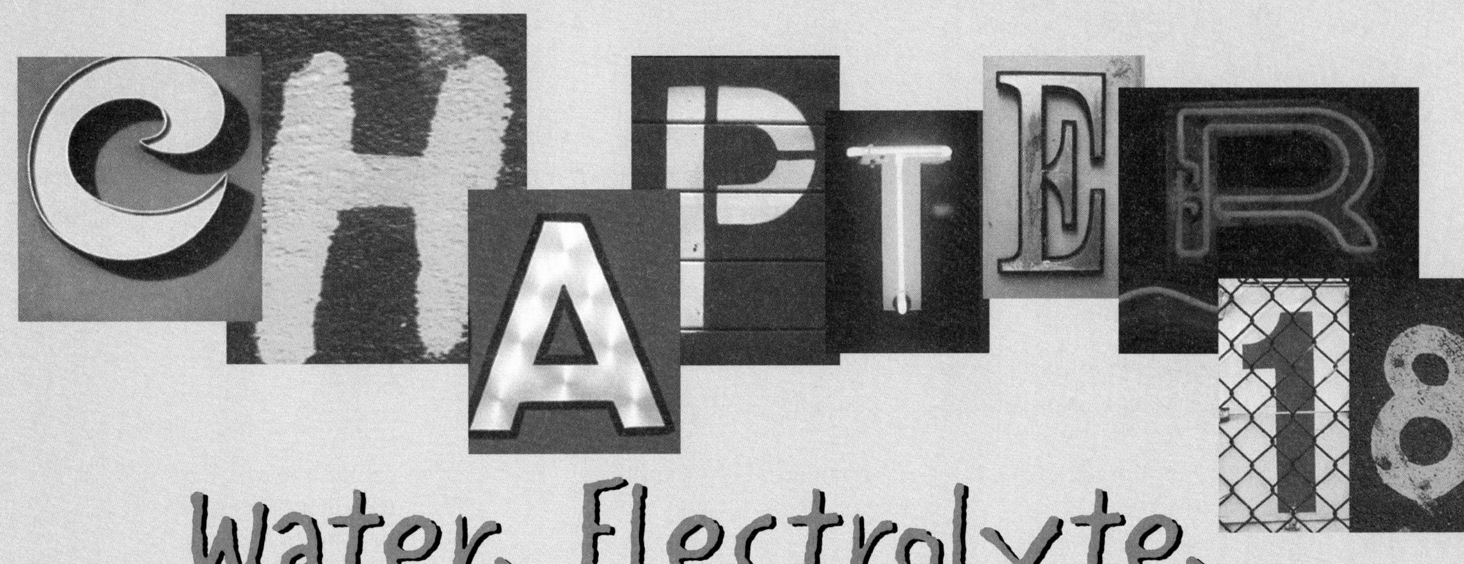

Chapter 18

Water, Electrolyte, and Acid-Base Balance

The summer of 1995 was a brutal time in Chicago. A fierce heat wave swept the city on July 12–16, with temperatures hovering between 93–104° F. The Cook County medical examiner's office worked round the clock during this time to handle 465 heat-related deaths.

Several factors point to a heat-related death:

- Body temperature at or soon after death of greater than 105° F
- Evidence that heat caused the death
- Decomposition of the corpse
- No other identifiable cause of death
- Evidence that the person was alive just before the heat wave

In the many blazing hot, un-air-conditioned apartments of Chicago, mostly elderly people fell victim. An untold number suffered severe dehydration, which may have been the direct cause of death. As body fluids became concentrated, with not enough water to form sweat to counter the rising and relentless heat, body temperatures soared. Wastes accumulating in extracellular fluids may have caused symptoms of cerebral disturbance, including mental confusion, delirium, and coma, preventing many people from seeking help.

Other communities also lost citizens to the heat wave of 1995. As a result, local governments now issue public service announcements informing people how to deal with heat and provide air-conditioned areas for people to stay. A piece of advice for all—when the mercury rises, drink! The human body cannot function for long with too little water.

Photo: Heat waves are deadly. It is important to keep cool.

Cell functions and survival depend on homeostasis—a stable internal environment. In such an environment, body cells continually receive oxygen and nutrients, and wastes are continually carried away. At the same time, the water and dissolved electrolyte concentrations and the pH of body fluids remain constant. Homeostasis requires *water, electrolyte,* and *acid-base balance.*

Chapter Objectives

After studying this chapter, you should be able to:

1. Explain water and electrolyte balance, and discuss its importance. (p. 499)
2. Describe how body fluids are distributed within compartments, how fluid composition differs between compartments, and how fluids move from one compartment to another. (p. 499)
3. List the routes by which water enters and leaves the body, and explain the regulation of water intake and output. (p. 501)
4. Explain how electrolytes enter and leave the body, and the regulation of electrolyte intake and output. (p. 503)
5. List the major sources of hydrogen ions in the body. (p. 506)
6. Distinguish between strong and weak acids and bases. (p. 507)
7. Explain how chemical buffer systems, the respiratory center, and the kidneys minimize changes in the pH of body fluids. (p. 507)

Aids to Understanding Words

de- [separation from] *de*hydration: Removal of water from cells or body fluids.

extra- [outside] *extra*cellular fluid: Fluid outside of cells.

im- [not] *im*balance: Condition in which factors are not in equilibrium.

intra- [within] *intra*cellular fluid: Fluid within cells.

neutr- [neither one nor the other] *neutr*al: A solution that is neither acidic nor basic.

Key Terms

acid (as′id)

acid-base buffer system (as′id-bās buf′er sis′tem)

base (bās)

electrolyte balance (e-lek′tro-līt bal′ans)

extracellular (ek″strah-sel′u-lar)

intracellular (in″trah-sel′u-lar)

transcellular (trans-sel′u-lar)

water balance (wot′er bal′ans)

Introduction

The term *balance* suggests a state of equilibrium. For water and **electrolytes** (molecules that release ions in water), balance means that the quantities entering the body equal the quantities leaving it. Maintaining such a balance requires mechanisms that replace lost water and electrolytes, and excrete excesses. As a result, levels of water and electrolytes in the body remain stable.

Water balance and electrolyte balance are interdependent because electrolytes dissolve in the water of body fluids. Consequently, anything that alters electrolyte concentrations necessarily alters water concentration by adding or removing solutes. Likewise, anything that changes water concentration changes electrolyte concentrations by concentrating or diluting them.

Distribution of Body Fluids

Body fluids are not uniformly distributed throughout tissues but occur in regions, or *compartments,* that contain fluids of varying compositions. Movement of water and electrolytes between these compartments is regulated so that their distribution and the composition of body fluids remain stable.

Fluid Compartments

The body of an average adult female is about 52% water by weight, and that of an average male is about 63% water. The difference between the sexes is due to females generally having more adipose tissue, which has little water, than males. Water in the body (about 40 liters), together with its dissolved electrolytes, is distributed into two major compartments—an intracellular fluid compartment and an extracellular fluid compartment.

The **intracellular fluid compartment** includes all the water and electrolytes that cell membranes enclose. In other words, intracellular fluid is fluid within cells, and in an adult, it represents about 63% by volume of total body water.

The **extracellular fluid compartment** includes all the fluid outside cells—within tissue spaces (interstitial fluid), blood vessels (plasma), and lymphatic vessels (lymph). Epithelial layers separate a specialized fraction of extracellular fluid from other extracellular fluids. This **transcellular fluid** includes *cerebrospinal fluid* of the central nervous system, *aqueous* and *vitreous humors* of the eyes, *synovial fluid* of the joints, *serous fluid* within body cavities, and fluid *secretions* of exocrine glands. The fluids of the extracellular compartment constitute about 37% by volume of total body water (fig. 18.1).

Body Fluid Composition

Extracellular fluids generally have similar compositions, including high concentrations of sodium, chloride, and bicarbonate ions, and lesser concentrations of potassium, calcium, magnesium, phosphate, and sulfate ions. The blood plasma fraction of extracellular fluid contains considerably more protein than do either interstitial fluid or lymph.

Intracellular fluid contains high concentrations of potassium, phosphate, and magnesium ions. It has a greater concentration of sulfate ions, and lesser concentrations of sodium, chloride, and bicarbonate ions than do extracellular fluids. Intracellular fluid also has a greater protein concentration than plasma. Figure 18.2 shows these relative concentrations.

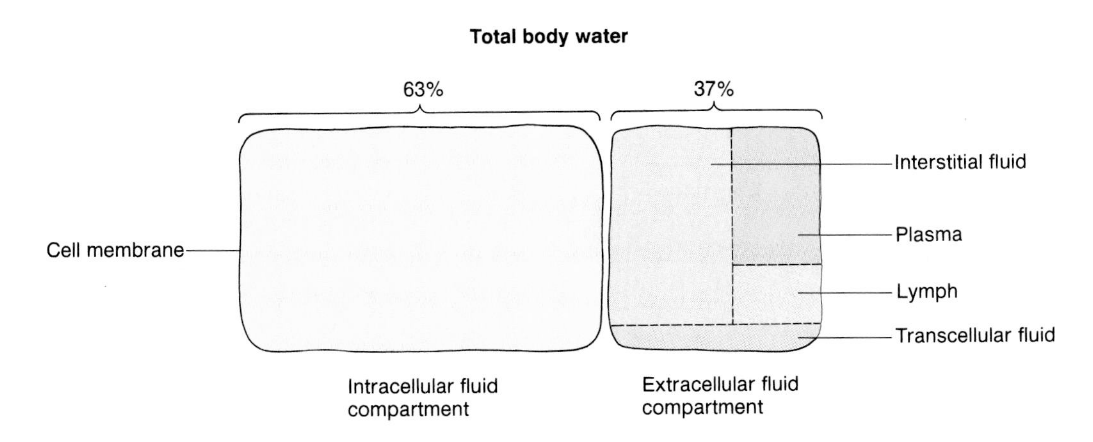

Figure 18.1

Cell membranes separate fluid in the intracellular compartment from fluids in the extracellular compartment.

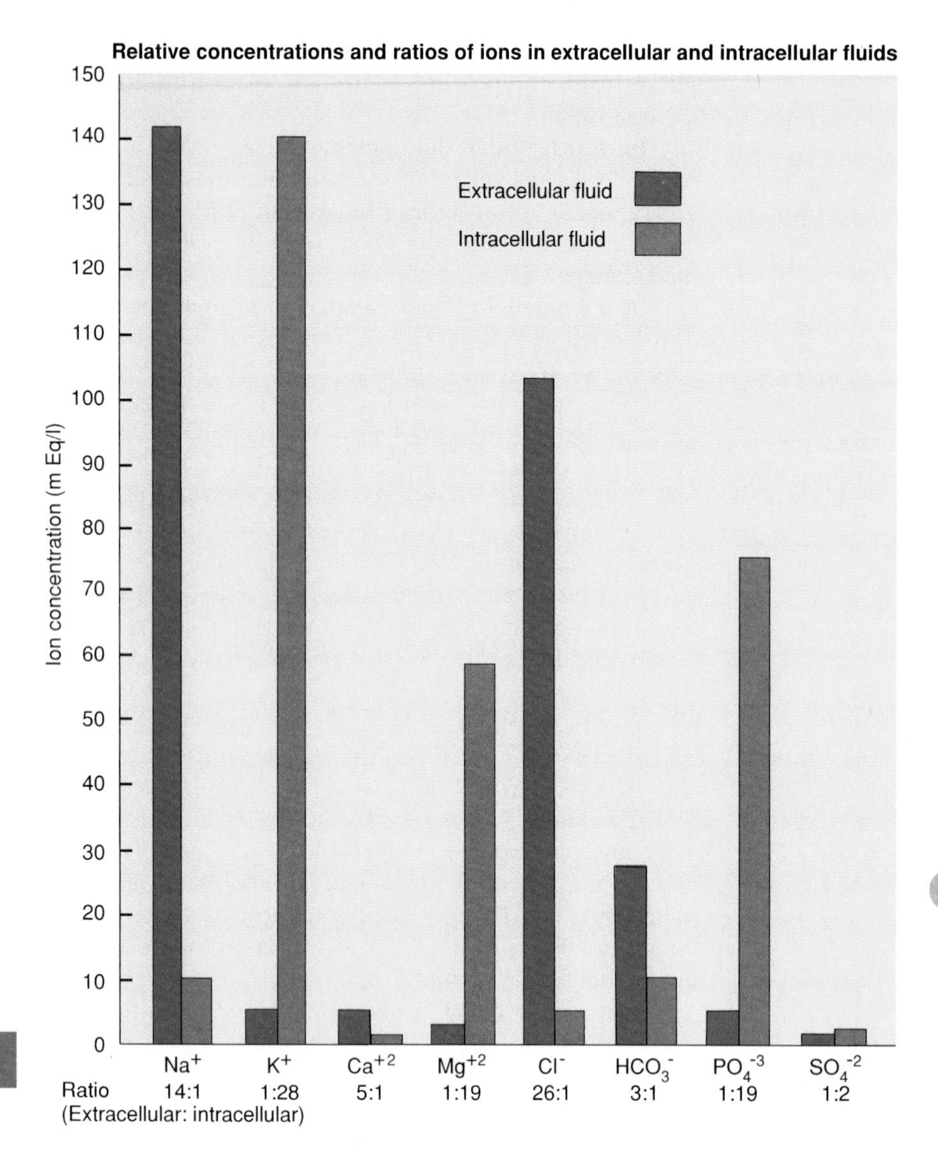

Figure 18.2

Extracellular fluids have relatively high concentrations of sodium (Na^+), chloride (Cl^-), and bicarbonate (HCO_3^-) ions. Intracellular fluid has relatively high concentrations of potassium (K^+), magnesium (Mg^{+2}), and phosphate (PO_4^{-3}) ions.

1. How are water balance and electrolyte balance interdependent?
2. Describe the normal distribution of water within the body.
3. Which electrolytes are in higher concentrations in extracellular fluids? In intracellular fluid?
4. How does protein concentration vary in different body fluids?

Movement of Fluid between Compartments

Two major factors regulate the movement of water and electrolytes from one fluid compartment to another: *hydrostatic pressure* and *osmotic pressure* (fig. 18.3). As explained in chapter 13, fluid leaves plasma at the arteriolar ends of capillaries and enters interstitial spaces because of the net outward force of hydrostatic pressure (blood pressure). Fluid returns to plasma from interstitial spaces at the venular ends of capillaries because of the net inward force of osmotic pressure. Likewise, as mentioned in chapter 14, the hydrostatic pressure that develops within interstitial spaces forces fluid in interstitial spaces into lymph capillaries. Lymph circulation returns interstitial fluid to plasma.

Pressures similarly control fluid movement between the intracellular and extracellular compartments. Because hydrostatic pressure within cells and surrounding interstitial fluid is ordinarily equal and stable, a change in osmotic pressure is the likely cause of any net fluid movement.

The sodium ion concentration in extracellular fluids is especially high. A decrease in this concentration causes net movement of water from the extracellular

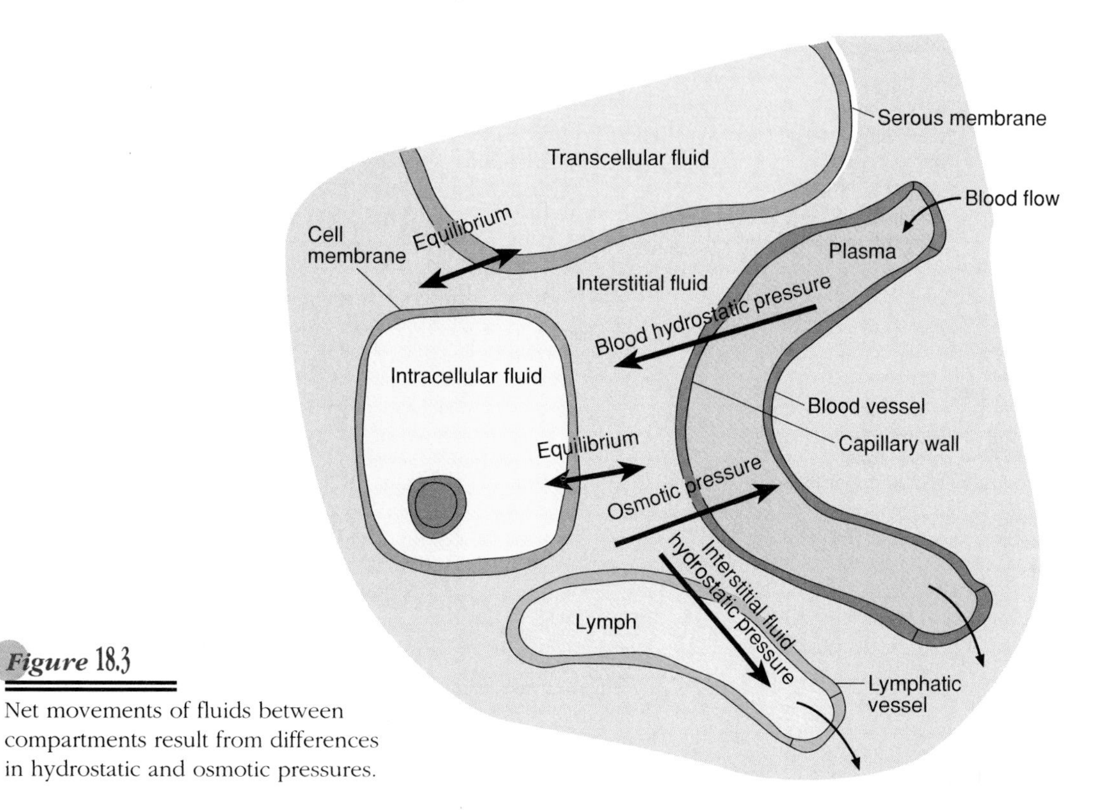

Figure 18.3

Net movements of fluids between compartments result from differences in hydrostatic and osmotic pressures.

compartment into the intracellular compartment by osmosis. As a consequence, cells swell. Conversely, if the sodium ion concentration in interstitial fluid increases, the net movement of water is outward from the intracellular compartment, and cells shrink as they lose water.

1. What factors control the movement of water and electrolytes from one fluid compartment to another?
2. How does the sodium ion concentration within body fluids affect the net movement of water between compartments?

Water Balance

Water balance exists when total water intake and total water loss are equal. Homeostatic mechanisms maintain water balance.

Water Intake

The volume of water gained each day varies from individual to individual. An average adult living in a moderate environment takes in about 2,500 milliliters. Of this amount, drinking water or beverages supply probably 60%, while moist foods provide another 30%. The remaining 10% is a by-product of the oxidative metabolism of nutrients and is called **water of metabolism** (fig. 18.4).

Regulation of Water Intake

The primary regulator of water intake is thirst. The intense feeling of thirst derives from the osmotic pressure of extracellular fluids and a *thirst center* in the hypothalamus. As the body loses water, the osmotic pressure of extracellular fluids increases. This stimulates *osmoreceptors* in the thirst center, which cause the person to feel thirsty and to seek water.

Thirst is a homeostatic mechanism, normally triggered whenever total body water decreases by as little as 1%. As a thirsty person drinks water, the act of drinking and the resulting stomach wall distension trigger nerve impulses that inhibit the thirst mechanism. Thus, drinking stops long before the swallowed water is absorbed.

1. What is water balance?
2. Where is the thirst center located?
3. What stimulates fluid intake? What inhibits it?

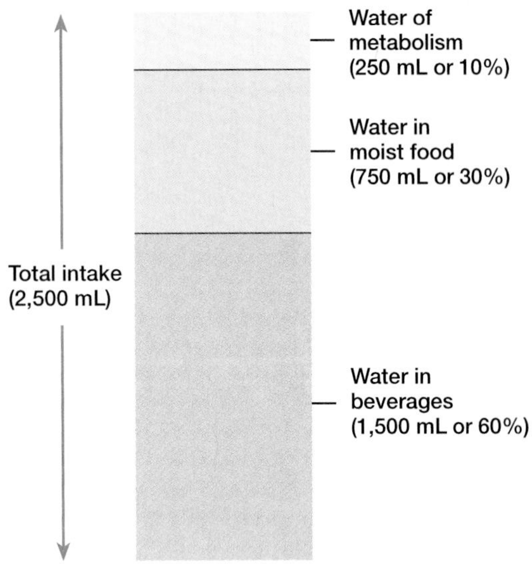

Average daily intake of water
Water of metabolism (250 mL or 10%)
Water in moist food (750 mL or 30%)
Total intake (2,500 mL)
Water in beverages (1,500 mL or 60%)

***Figure* 18.4**

Major sources of body water.

Average daily output of water
Water lost in sweat (150 mL or 6%)
Water lost in feces (150 mL or 6%)
Water lost through skin and lungs (700 mL or 28%)
Total output (2,500 mL)
Water lost in urine (1,500 mL or 60%)

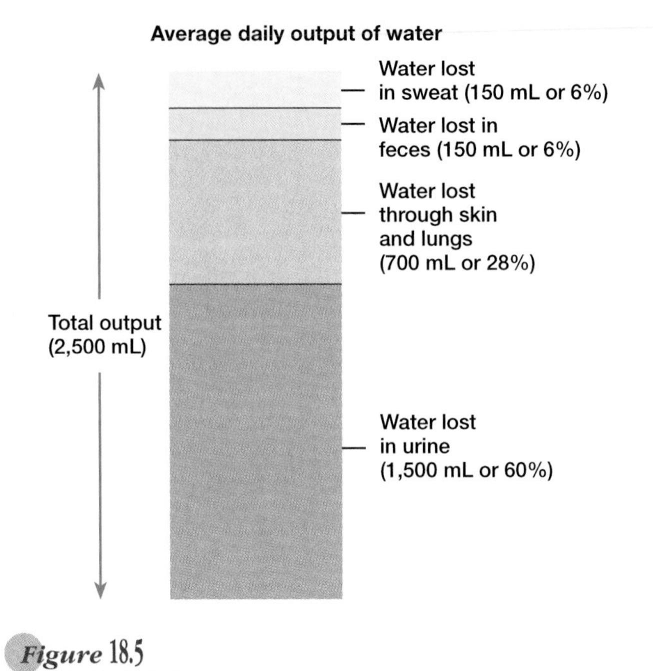

***Figure* 18.5**

Routes by which the body loses water. Urine production is most important in the regulation of water balance.

Water Output

Water normally enters the body only through the mouth, but it can be lost by a variety of routes. These include obvious losses in urine, feces, and sweat (sensible perspiration), as well as less obvious losses, such as evaporation of water from the skin (insensible perspiration) and from the lungs during breathing.

If an average adult takes in 2,500 milliliters of water each day, then 2,500 milliliters must be eliminated to maintain water balance. Of this volume, perhaps 60% is lost in urine, 6% in feces, and 6% in sweat. About 28% is lost by evaporation from the skin and lungs (fig. 18.5). These percentages vary with environmental temperature and relative humidity, and with physical exercise.

Regulation of Water Balance

The primary means of regulating water output is urine production. The distal convoluted tubules and collecting ducts of the nephrons are most important in regulating the volume of water excreted in urine. The epithelial linings of these segments of the renal tubule remain relatively impermeable to water unless antidiuretic hormone (ADH) is present. ADH increases the permeability of the distal tubule and collecting duct, thereby increasing water reabsorption and reducing urine production. In the absence of ADH, less water is reabsorbed, and more urine is produced (see chapter 17).

1. By what routes does the body lose water?
2. What role do renal tubules play in water balance regulation?

Diuretics are substances that promote urine production. A number of common substances, such as caffeine in coffee and tea, have diuretic effects, as do a variety of drugs used to reduce body fluid volume.

Diuretics produce their effects in different ways. Some, such as alcohol and certain narcotic drugs, promote urine formation by inhibiting ADH release. Other diuretics, such as caffeine, inhibit the reabsorption of sodium ions or other solutes in portions of renal tubules. As a consequence, the osmotic pressure of the tubular fluid increases, reducing osmotic reabsorption of water and increasing urine volume. ■

Electrolyte Balance

Electrolyte balance exists when the quantities of electrolytes the body gains equal those lost. Homeostatic mechanisms maintain electrolyte balance.

Electrolyte Intake

The electrolytes most important to cellular functions release sodium, potassium, calcium, magnesium, chloride, sulfate, phosphate, bicarbonate, and hydrogen ions. Foods provide most of these electrolytes, but drinking water and other beverages are also sources. Some electrolytes are by-products of metabolic reactions.

Regulation of Electrolyte Intake

Ordinarily, responding to hunger and thirst provides sufficient electrolytes. A severe electrolyte deficiency may produce a *salt craving,* a strong desire to eat salty foods.

Electrolyte Output

The body loses electrolytes by perspiring, with more lost in sweat on warmer days and during strenuous exercise. Varying amounts of electrolytes are lost in feces. Kidney function and urine production, however, vary electrolyte output to maintain balance.

1. Which electrolytes are most important to cellular functions?
2. What mechanisms ordinarily regulate electrolyte intake?
3. By what routes does the body lose electrolytes?

Regulation of Electrolyte Balance

Precise concentrations of positively charged ions, such as sodium (Na^+), potassium (K^+), and calcium (Ca^{+2}), are required for nerve impulse conduction, muscle fiber contraction, and maintenance of cell membrane potential. *Sodium ions* account for nearly 90% of positively charged ions in extracellular fluids. The kidneys and the hormone aldosterone provide the primary mechanism regulating these ions. Aldosterone, which the adrenal cortex secretes, increases sodium ion reabsorption in the distal convoluted tubules and collecting ducts of the kidneys' nephrons.

Aldosterone also regulates *potassium ions.* A rising potassium ion concentration directly stimulates the adrenal cortex to secrete aldosterone. This hormone enhances reabsorption of sodium ions, and at the same time, causes renal tubules to excrete potassium ions (fig. 18.6).

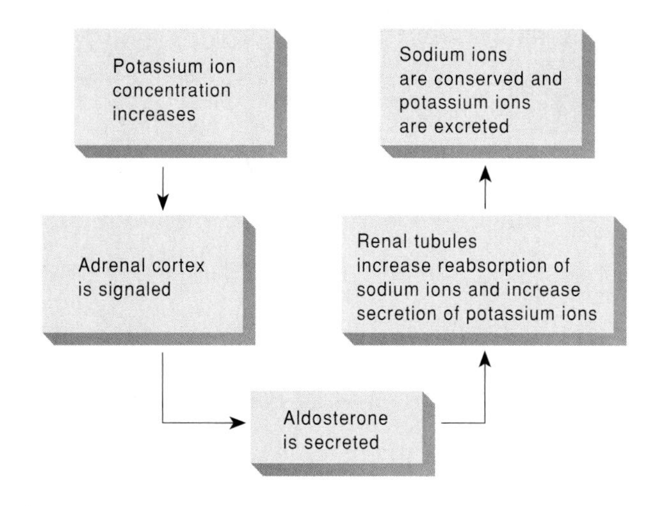

Figure 18.6

If potassium ion concentration increases, the kidneys conserve sodium ions and excrete potassium ions.

Recall from chapter 11 that the parathyroid glands, as well as calcitonin from the thyroid gland, regulate the concentration of *calcium ions* in extracellular fluids. Calcium ion concentration dropping below normal directly stimulates the parathyroids to secrete parathyroid hormone. Parathyroid hormone increases the concentrations of calcium and phosphate ions in extracellular fluids.

Generally, the regulatory mechanisms that control positively charged ions secondarily control concentrations of negatively charged ions. For example, renal tubules passively reabsorb chloride ions (Cl^-), the most abundant negatively charged ions in extracellular fluids, in response to active reabsorption of sodium ions. That is, negatively charged chloride ions are electrically attracted to positively charged sodium ions and accompany them as they are reabsorbed (see chapter 17).

Active transport mechanisms with limited transport capacities partially regulate some negatively charged ions, such as phosphate ions (PO_4^{-3}) and sulfate ions (SO_4^{-2}). Thus, if extracellular phosphate ion concentration is low, renal tubules reabsorb phosphate ions. On the other hand, if the renal plasma threshold is exceeded, excess phosphate is excreted in urine.

1. How are concentrations of sodium and potassium ions controlled?
2. How are calcium ions regulated?
3. What mechanism regulates the concentrations of most negatively charged ions?

TOPIC OF INTEREST Water Balance Disorders

Among the more common disorders involving an imbalance in the water of body fluids are dehydration, water intoxication, and edema.

Dehydration

In 1994, the world watched in horror as thousands of starving people died in the African nation of Rwanda. Lack of food did not kill most of these people, but cholera toxin that cripples the ability of intestinal lining cells to reabsorb water. The severe diarrhea that develops can kill in days, sometimes even hours.

In *dehydration,* water output exceeds water intake. This condition may develop following excess sweating or prolonged water deprivation accompanied by continued water output. In either case, as water is lost, extracellular fluid concentrates, and water leaves cells by osmosis (fig. 18A). Dehydration may also accompany prolonged vomiting or diarrhea that depletes body fluids. During dehydration, the skin and mucous membranes of the mouth feel dry, and body weight drops. Hyperthermia may develop as the body's temperature-regulating mechanism loses efficiency due to lack of water for sweat.

Because infants' kidneys are less able to conserve water than those of adults, infants are more likely to become dehydrated. Elderly people are also especially susceptible to developing water imbalances because the sensitivity of their thirst mechanism decreases with age, and physical disabilities may make it difficult for them to obtain adequate fluids.

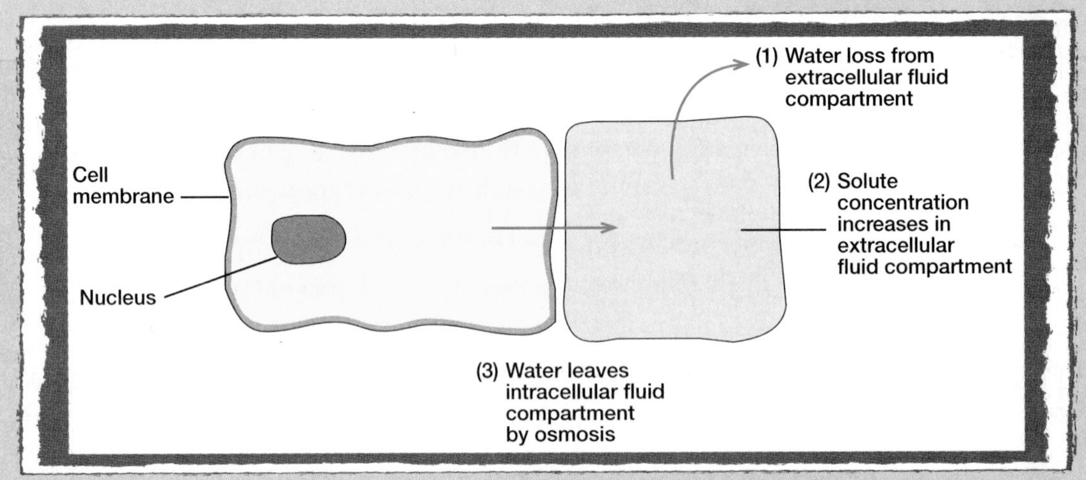

***Figure* 18A**

If excess extracellular fluids are lost, cells dehydrate by osmosis.

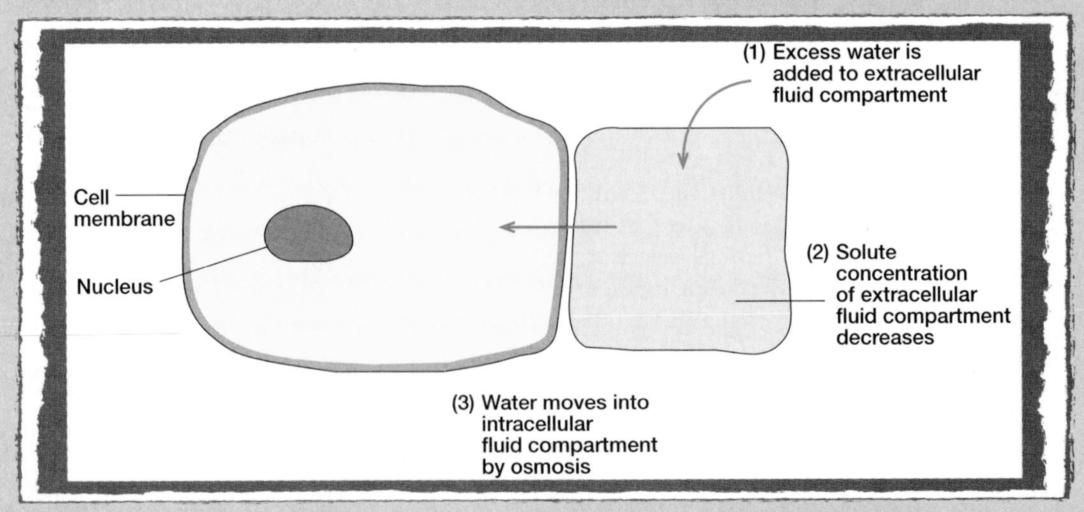

***Figure* 18B**

If excess water is added to the extracellular fluid compartment, cells gain water by osmosis.

The treatment for dehydration is to replace the lost water and electrolytes. If only water is replaced, the extracellular fluid becomes more dilute than normal, producing a condition called water intoxication.

Water Intoxication

Babies rushed to emergency rooms because they are having seizures sometimes have drunk too much water, a rare condition called *water intoxication*. This can occur when a baby under six months of age is given several bottles of water a day or very dilute infant formula. The hungry infant gobbles down the water, and soon its tissues swell with excess fluid. When the serum sodium level drops, the eyes begin to flutter, and a seizure occurs. As extracellular fluid becomes hypotonic, water enters the cells rapidly by osmosis (fig. 18B). Coma resulting from swelling brain tissues may follow unless water intake is restricted and hypertonic salt solutions given. Usually, recovery is complete within a few days.

Edema

Edema is an abnormal accumulation of extracellular fluid within interstitial spaces. A variety of factors can cause edema, including decrease in plasma protein concentration (hypoproteinemia), obstruction of lymphatic vessels, increased venous pressure, and increased capillary permeability.

Hypoproteinemia may result from liver disease that hinders plasma protein synthesis; kidney disease (glomerulonephritis) that damages glomerular capillaries, allowing proteins to enter urine; or starvation, in which amino acid intake is insufficient to support plasma protein synthesis. In each of these instances, plasma protein concentration decreases, which decreases plasma osmotic pressure, reducing the normal return of tissue fluid to the venular ends of capillaries. Tissue fluid consequently accumulates in interstitial spaces.

As discussed in chapter 14, surgery or parasitic infections of lymphatic vessels may result in *lymphatic obstructions*. Back pressure develops in the lymphatic vessels, interfering with the normal movement of tissue fluid into them. At the same time, proteins that the lymphatic circulation ordinarily removes accumulate in interstitial spaces, raising osmotic pressure of interstitial fluid. This effect attracts still more fluid into interstitial spaces.

If blood outflow from the liver into the inferior vena cava is blocked, venous pressure within the liver and portal blood vessels increases greatly. As a result, fluid with a high protein concentration is exuded from the surfaces of the liver and intestine into the peritoneal cavity. This increases the osmotic pressure of the abdominal fluid, which, in turn, attracts more water into the peritoneal cavity by osmosis. This condition, called *ascites,* distends the abdomen and is quite painful.

Edema may also result from increased capillary permeability accompanying *inflammation*. Recall that inflammation is a response to tissue damage and usually releases chemicals such as histamine from damaged cells. Histamine causes vasodilation and increased capillary permeability, so that excess fluid leaks out of capillaries and enters interstitial spaces.

Table 18A summarizes the factors that result in edema.

Table 18A

Factors Associated with Edema

Factor	Cause	Effect
Low plasma protein concentration	Liver disease and failure to synthesize proteins; kidney disease and loss of proteins in urine; lack of proteins in diet due to starvation	Plasma osmotic pressure decreases; less fluid enters venular ends of capillaries by osmosis
Obstruction of lymphatic vessels	Surgical removal of portions of lymphatic pathways; certain parasitic infections	Back pressure in lymphatic vessels interferes with movement of fluid from interstitial spaces into lymph capillaries
Increased venous pressure	Venous obstructions or faulty venous valves	Back pressure in veins interferes with return of fluid from interstitial spaces into venular ends of capillaries
Inflammation	Tissue damage	Capillaries become abnormally permeable; fluid leaks from plasma into interstitial spaces

TOPIC OF INTEREST: Sodium and Potassium Imbalances

Extracellular fluids usually have high sodium ion concentrations, and intracellular fluid usually has high potassium ion concentrations. Renal regulation of sodium is closely related to that of potassium because secretion (and excretion) of potassium accompanies active reabsorption of sodium (under the influence of aldosterone). Therefore, conditions resulting from sodium ion imbalance often also involve potassium ion imbalance.

Such disorders include:

1. *Low sodium concentration (hyponatremia)* Possible causes of sodium deficiencies include prolonged sweating, vomiting, or diarrhea; renal disease in which sodium is reabsorbed inadequately; adrenal cortex disorders in which aldosterone secretion is insufficient to promote sodium reabsorption (Addison's disease); and drinking too much water. Possible effects of hyponatremia include the development of hypotonic extracellular fluid that promotes water movement into cells by osmosis, producing symptoms of water intoxication.
2. *High sodium concentration (hypernatremia)* Possible causes of elevated sodium concentration include excess water loss by evaporation (despite decreased sweating), as may occur during high fever, and increased water loss accompanying diabetes insipidus. In one form of diabetes insipidus, ADH secretion is insufficient for renal tubules to maintain water conservation. Hypernatremia may disturb the central nervous system, causing confusion, stupor, and coma.
3. *Low potassium concentration (hypokalemia)* Possible causes of potassium deficiency include the adrenal cortex releasing excess aldosterone (Cushing's syndrome), which increases renal excretion of potassium; use of diuretic drugs that promote potassium excretion; kidney disease; and prolonged vomiting or diarrhea. Possible effects of hypokalemia include muscular weakness or paralysis, respiratory difficulty, and severe cardiac disturbances, such as atrial or ventricular arrhythmias.
4. *High potassium concentration (hyperkalemia)* Possible causes of elevated potassium concentration include renal disease, which decreases potassium excretion; use of drugs that promote renal conservation of potassium; the adrenal cortex releasing insufficient aldosterone (Addison's disease); and a shift of potassium from intracellular to extracellular fluid, a change that accompanies an increase in hydrogen ion concentration (acidosis). Possible effects of hyperkalemia include paralysis of the skeletal muscles and severe cardiac disturbances, such as cardiac arrest.

Acid-Base Balance

Recall from chapter 2 that electrolytes that ionize in water and release hydrogen ions are called **acids,** and that substances that combine with hydrogen ions are called **bases.** Maintenance of homeostasis depends on control of concentrations of acids and bases within body fluids.

Sources of Hydrogen Ions

Most of the hydrogen ions in body fluids originate as by-products of metabolic processes, although the digestive tract may directly absorb small quantities.

The major metabolic sources of hydrogen ions include (fig. 18.7):

1. **Aerobic respiration of glucose** This process produces carbon dioxide and water. Carbon dioxide diffuses out of cells and reacts with water in extracellular fluids to form *carbonic acid,* which then ionizes to release hydrogen ions and bicarbonate ions:

$$H_2CO_3 \rightarrow H^+ + HCO_3^-$$

2. **Anaerobic respiration of glucose** Anaerobically metabolized glucose produces *lactic acid,* which adds hydrogen ions to body fluids.
3. **Incomplete oxidation of fatty acids** This process produces *acidic ketone bodies,* which increase hydrogen ion concentration.
4. **Oxidation of amino acids containing sulfur** This process yields *sulfuric acid* (H_2SO_4), which ionizes to release hydrogen ions.

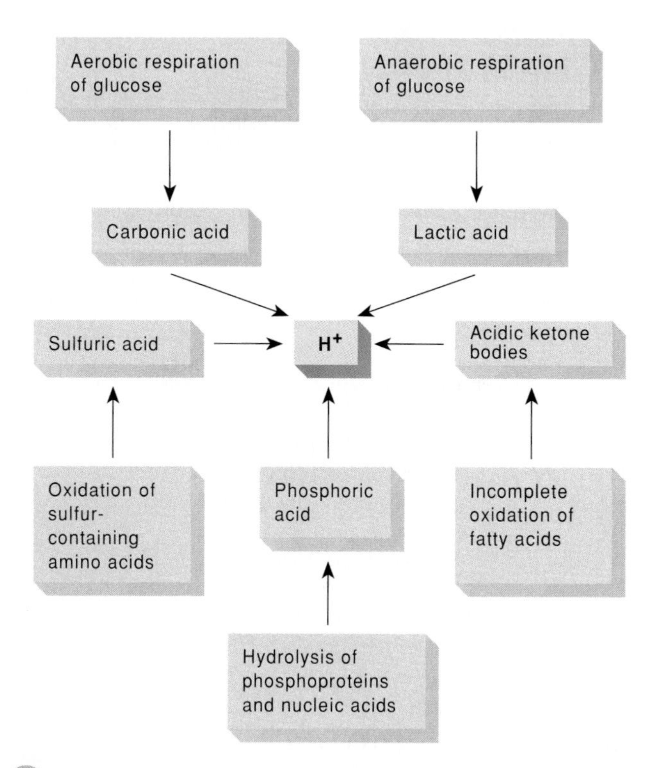

Figure 18.7

Some of the metabolic processes that provide hydrogen ions.

5. **Breakdown (hydrolysis) of phosphoproteins and nucleic acids** Phosphoproteins and nucleic acids contain phosphorus. Their oxidation produces *phosphoric acid* (H_3PO_4), which ionizes to release hydrogen ions.

The acids resulting from metabolism vary in strength. Thus, their effects on the hydrogen ion concentration of body fluids vary.

1. Distinguish between an acid and a base.
2. What are the major sources of hydrogen ions in the body?

Strengths of Acids and Bases

Acids that ionize more completely are *strong acids,* and those that ionize less completely are *weak acids.* For example, the hydrochloric acid (HCl) of gastric juice is a strong acid, but the carbonic acid (H_2CO_3) produced when carbon dioxide reacts with water is weak.

Bases are substances that, like hydroxyl ions (OH^-), combine with hydrogen ions. *Chloride ions* (Cl^-) and *bicarbonate ions* (HCO_3^-) are also bases. Chloride ions combine less readily with hydrogen ions and are *weak bases.* Bicarbonate ions, which combine more readily with hydrogen ions, are *strong bases.*

Regulation of Hydrogen Ion Concentration

Acid-base buffer systems, the respiratory center in the brain stem, and the nephrons in the kidneys regulate hydrogen ion concentration, measured by pH, in body fluids.

Acid-Base Buffer Systems

Acid-base buffer systems are in all body fluids and are usually composed of sets of two or more chemicals that combine with acids or bases when in excess. More specifically, the chemical components of a buffer system can change strong acids, which tend to release large quantities of hydrogen ions, into weak acids, which release fewer hydrogen ions. Likewise, these buffers can combine with strong bases and change them into weak bases. Such activity helps minimize pH changes in body fluids.

The three most important acid-base buffer systems in body fluids are:

1. **Bicarbonate buffer system** The bicarbonate acid-base buffer system, present in intracellular and extracellular body fluids, consists of carbonic acid (H_2CO_3) and sodium bicarbonate ($NaHCO_3$). A strong acid, like hydrochloric acid, reacts with the sodium bicarbonate. The products of the reaction are carbonic acid, which is a weaker acid, and sodium chloride. This minimizes an increase in hydrogen ion concentration in body fluids:

$$\underset{\text{(strong acid)}}{HCl} + NaHCO_3 \rightarrow \underset{\text{(weak acid)}}{H_2CO_3} + NaCl$$

On the other hand, a strong base like sodium hydroxide (NaOH) reacts with carbonic acid. The products are sodium bicarbonate ($NaHCO_3$),which is a weaker base, and water. This minimizes a shift toward a more basic (alkaline) state:

$$\underset{\text{(strong base)}}{NaOH} + H_2CO_3 \rightarrow \underset{\text{(weak base)}}{NaHCO_3} + H_2O$$

2. **Phosphate buffer system** The phosphate acid-base buffer system, also present in intracellular and extracellular body fluids, is particularly important as a regulator of hydrogen ion concentrations in the tubular fluid of the nephrons and in urine. This buffer system consists of two phosphate compounds—sodium monohydrogen phosphate (Na_2HPO_4) and sodium dihydrogen phosphate (NaH_2PO_4). Sodium monohydrogen phosphate reacts with any strong acids present to produce weaker acids. Sodium dihydrogen phosphate reacts with any strong bases present to produce weaker bases.
3. **Protein buffer system** The protein acid-base buffer system consists of the plasma proteins, such as albumins, and certain proteins within cells, including the hemoglobin of red blood cells. As described in chapter 2, proteins are chains of amino acids. Some of these amino acids have freely exposed groups of atoms called *carboxyl groups* (–COOH). When the solution pH rises, these carboxyl groups can ionize, releasing hydrogen ions:

$$-COOH \rightarrow -COO^- + H^+$$

Some amino acids within a protein molecule also contain freely exposed *amino groups* ($-NH_2$). When the solution pH falls, these amino groups can accept hydrogen ions:

$$-NH_2 + H^+ \rightarrow -NH_3^+$$

Thus, protein molecules can function as acids by releasing hydrogen ions from their carboxyl groups or as bases by accepting hydrogen ions into their amino groups. This special property allows protein molecules to operate as an acid-base buffer system, minimizing changes in pH.

Neurons are particularly sensitive to changes in the pH of body fluids. If interstitial fluid becomes more alkaline than normal (alkalosis), neurons become more excitable, and seizures may result. Conversely, acidic conditions (acidosis) depress neuron activity, reducing the level of consciousness. ■

Table 18.1 summarizes the actions of the three major buffer systems.

Table 18.1 Chemical Acid-Base Buffer Systems

Buffer System	Constituents	Actions
Bicarbonate system	Sodium bicarbonate ($NaHCO_3$)	Converts a strong acid into a weak acid
	Carbonic acid (H_2CO_3)	Converts a strong base into a weak base
Phosphate system	Sodium monohydrogen phosphate (Na_2HPO_4)	Converts a strong acid into a weak acid
	Sodium dihydrogen phosphate (NaH_2PO_4)	Converts a strong base into a weak base
Protein system (and amino acids)	$-COO^-$ group of an amino acid within a protein	Released a hydrogen ion in the presence of excess base
	$-NH_3^+$ group of an amino acid within a protein	Accepted a hydrogen ion in the presence of excess acid

1. How does a strong acid or base differ from a weak acid or base?
2. How does a chemical buffer system help regulate pH of body fluids?
3. List the major buffer systems of the body.

The Respiratory Center

The respiratory center in the brain helps regulate hydrogen ion concentration in body fluids by controlling the rate and depth of breathing (see chapter 16). Specifically, if body cells increase their carbon dioxide production, as during physical exercise, carbonic acid production increases. As carbonic acid dissociates, the

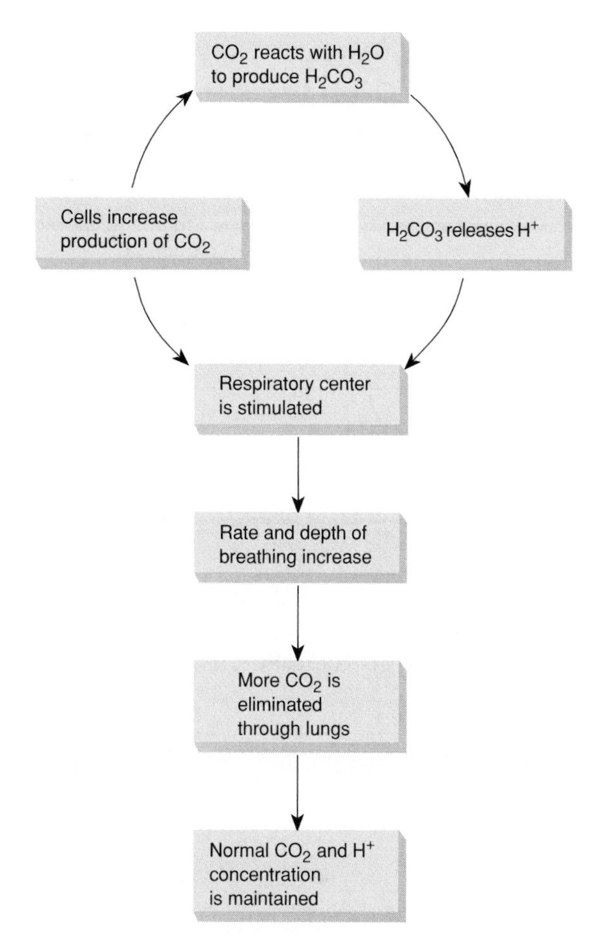

Figure 18.8

An increase in carbon dioxide elimination follows an increase in carbon dioxide production.

concentration of hydrogen ions increases, and the pH of body fluids drops. Such an increasing concentration of carbon dioxide in the central nervous system and the subsequent increase in hydrogen ion concentration in the cerebrospinal fluid stimulates chemosensitive areas within the respiratory center.

In response, the respiratory center increases the depth and rate of breathing so that the lungs excrete more carbon dioxide. This causes the hydrogen ion concentration in body fluids to return toward normal because the released carbon dioxide comes from carbonic acid (fig. 18.8):

$$H_2CO_3 \rightarrow CO_2 + H_2O$$

Conversely, if body cells are less active, production of carbon dioxide and hydrogen ions in body fluids remains relatively low. As a result, breathing rate and depth stay closer to resting levels.

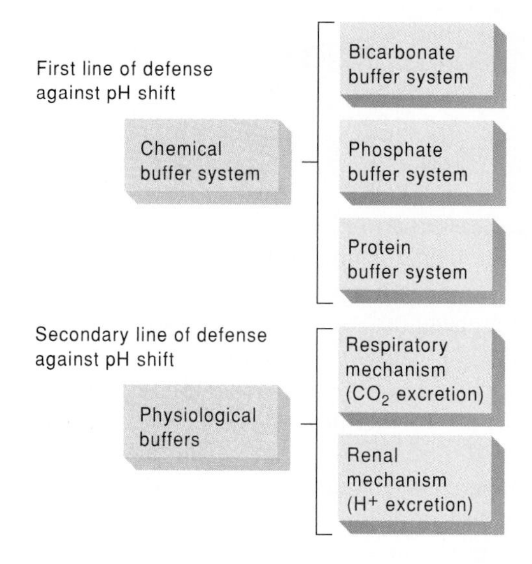

Figure 18.9

Chemical buffers act rapidly, while physiological buffers may require several minutes to several days to begin resisting a change in pH.

The Kidneys

Nephrons help regulate the hydrogen ion concentration of body fluids by excreting hydrogen ions in urine. Recall from chapter 17 that epithelial cells that line certain segments of renal tubules secrete hydrogen ions into tubular fluid.

Rates of Regulation

Regulators of hydrogen ion concentration operate at different rates (fig. 18.9). Acid-base buffers can change strong acids or bases into weak acids or bases almost immediately. For this reason, these chemical buffer systems are sometimes called the body's *first line of defense* against shifts in pH.

Physiological buffer systems, such as the respiratory and renal mechanisms, function more slowly and constitute *secondary defenses*. The respiratory mechanism may require several minutes to begin resisting a change in pH, and the renal mechanism may require one to three days to regulate a changing hydrogen ion concentration.

1. How does the respiratory system help regulate acid-base balance?
2. How do the kidneys respond to excess hydrogen ions?
3. How do the rates of chemical and physiological buffer systems differ?

Topic of Interest: Acid-Base Imbalances

Chemical and physiological buffer systems generally maintain the hydrogen ion concentration of body fluids within very narrow pH ranges. The pH of arterial blood is normally 7.35–7.45. Abnormal conditions may disturb the acid-base balance. A value below 7.35 produces *acidosis*. A pH above 7.45 produces *alkalosis*. Such shifts in the pH of body fluids can be life threatening. A person usually cannot survive if pH drops to 6.8 or rises to 8.0 for longer than a few hours (fig. 18C).

Accumulation of acids or loss of bases, both of which increase hydrogen ion concentrations of body fluids, causes acidosis. Conversely, loss of acids or accumulation of bases, and the consequent decreases in hydrogen ion concentrations, cause alkalosis (fig. 18D).

The two major types of acidosis are *respiratory acidosis* and *metabolic acidosis*. Factors that increase carbon dioxide, also increasing the concentration of carbonic acid (the respiratory acid), cause respiratory acidosis. Metabolic acidosis is due to accumulation of any other acids in the body fluids or to loss of bases, including bicarbonate ions.

Respiratory acidosis may be due to hindered pulmonary ventilation, which increases carbon dioxide. This may result from (fig. 18E):

1. Injury to the respiratory center of the brain stem, decreasing rate and depth of breathing.
2. Obstructions in air passages that interfere with air movement into alveoli.
3. Diseases that decrease gas exchanges, such as pneumonia, or those that reduce surface area of the respiratory membrane, such as emphysema.

Any of these conditions can increase the level of carbonic acid and hydrogen ions in body fluids, lowering pH. Chemical buffers, such as hemoglobin, may resist this shift in pH. At the same time, increasing concentrations of carbon dioxide and hydrogen ions stimulate the respiratory center, increasing breathing rate and depth, thereby lowering carbon dioxide concentration. Also, the kidneys may begin to excrete more hydrogen ions. Eventually, these chemical and physiological buffers return the pH of body fluids to normal. The acidosis is thus *compensated*.

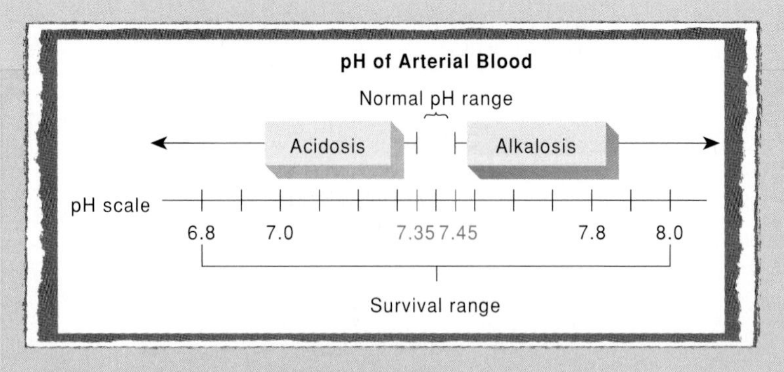

***Figure* 18C**

If the pH of arterial blood drops to 6.8 or rises to 8.0 for more than a few hours, the person usually cannot survive.

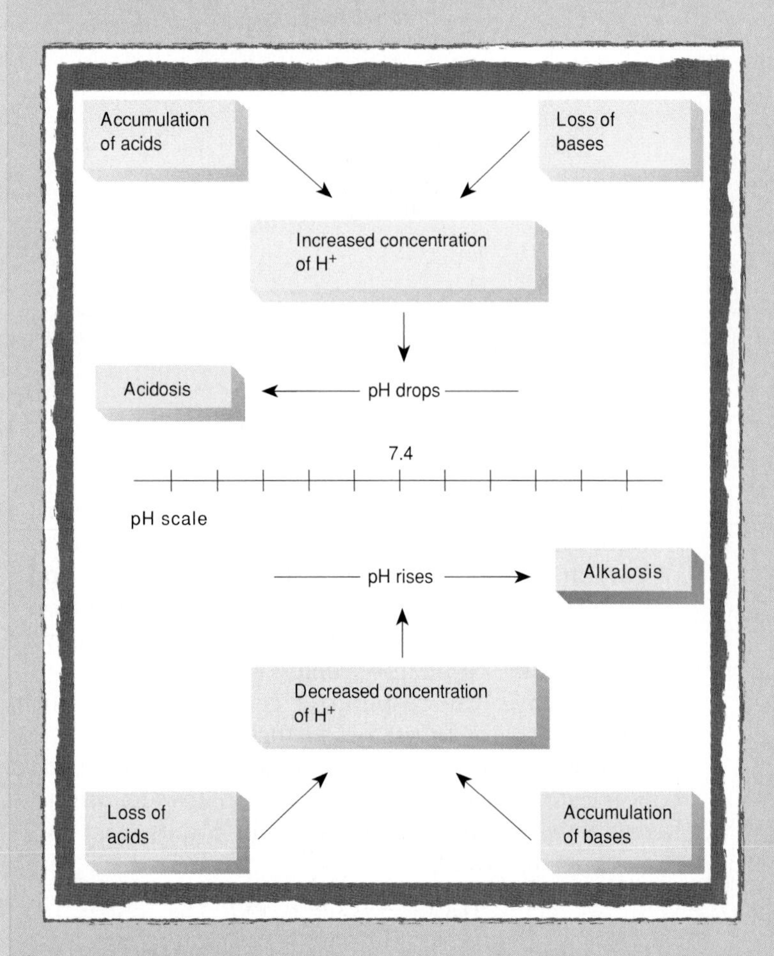

***Figure* 18D**

Acidosis results from accumulation of acids or loss of bases. Alkalosis results from loss of acids or accumulation of bases.

Symptoms of respiratory acidosis result from depression of central nervous system function and include drowsiness, disorientation, stupor, labored breathing, and cyanosis. In *uncompensated acidosis,* the person may become comatose and die.

Metabolic acidosis is due to accumulation of nonrespiratory acids or loss of bases. Factors that may lead to this condition include (fig. 18F):

1. Kidney disease that reduces glomerular filtration so that kidneys fail to excrete acids produced in metabolism (uremic acidosis).
2. Prolonged vomiting with loss of alkaline contents of the upper intestine and stomach contents. (Losing only the stomach contents produces metabolic alkalosis.)
3. Prolonged diarrhea with loss of excess alkaline intestinal secretions (especially in infants).
4. Diabetes mellitus, in which some fatty acids are changed into ketone bodies, such as *acetoacetic acid, beta-hydroxybutyric acid,* and *acetone.* Normally, these molecules are rare, and cells oxidize them as energy sources. However, if fats are utilized at an abnormally high rate, as in diabetes mellitus, ketone bodies may accumulate faster than they can be oxidized. At such times, these compounds may be excreted in urine (ketonuria). The lungs may excrete acetone, which is volatile and imparts a fruity odor to the breath. More seriously, accumulation of acetoacetic acid and beta-hydroxybutyric acid may lower pH (ketoacidosis). These acids may also combine with bicarbonate ions in urine. As a result, excess bicarbonate ions are excreted, interfering with the function of the bicarbonate acid-base buffer system.

In each case, the pH lowers. Countering this lower pH are chemical buffer systems, which accept excess hydrogen ions; the respiratory center, which increases breathing rate and depth; and the kidneys, which excrete more hydrogen ions.

The two major types of alkalosis are *respiratory alkalosis* and *metabolic alkalosis.* Excess loss of carbon dioxide and consequent loss of carbonic acid cause respiratory alkalosis. Metabolic alkalosis is due to excess loss of hydrogen ions or gain of bases.

Respiratory alkalosis develops from *hyperventilation* (described in chapter 16), producing

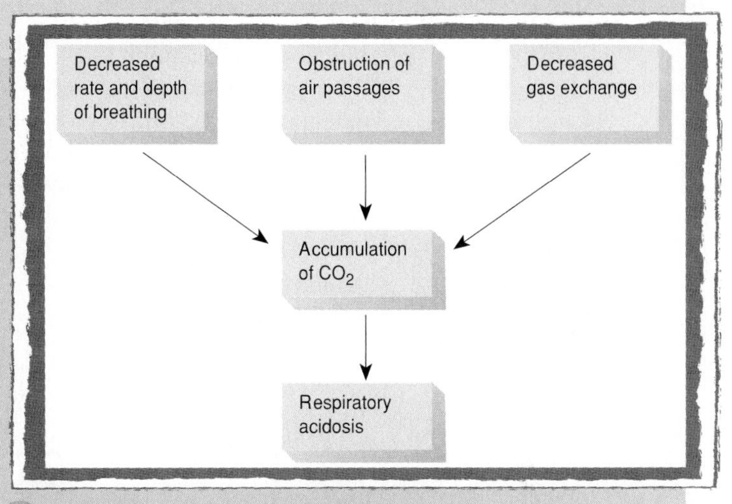

***Figure* 18E**

Some of the factors that lead to respiratory acidosis.

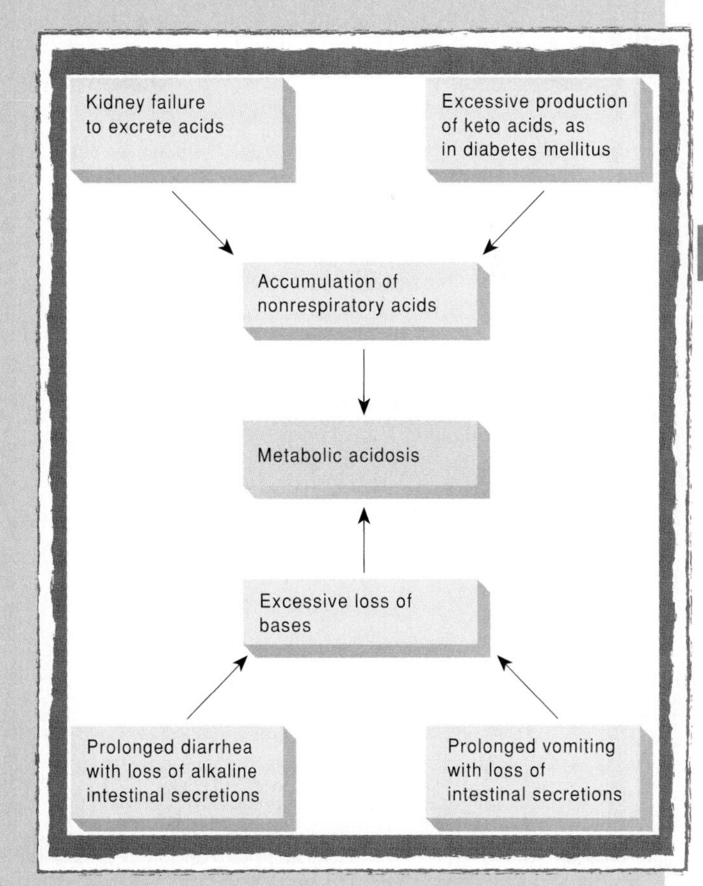

***Figure* 18F**

Some of the factors that lead to metabolic acidosis.

Continued

Acid-Base Imbalances—Continued

too great a loss of carbon dioxide and consequent decreases in carbonic acid and hydrogen ion concentrations (fig. 18G). Hyperventilation may occur in response to anxiety or may accompany fever or poisoning from salicylates, such as aspirin. At high altitudes, hyperventilation may be a response to low oxygen partial pressure. Musicians, who must provide a large volume of air when playing sustained passages, sometimes hyperventilate. In each case, rapid, deep breathing depletes carbon dioxide, and the pH of body fluids increases.

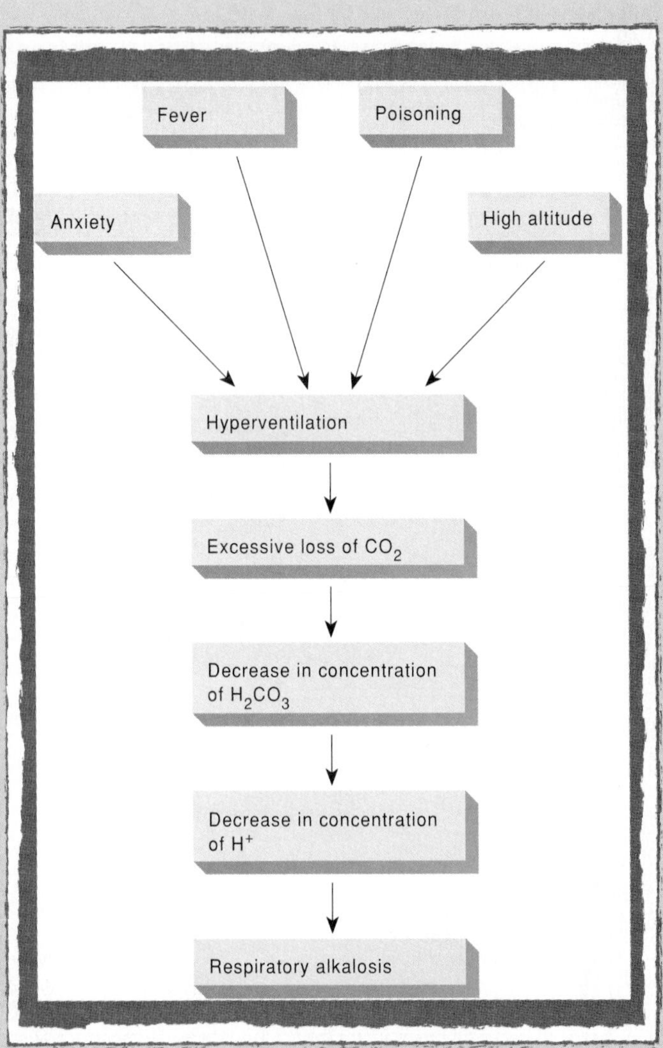

***Figure* 18G**

Some of the factors that lead to respiratory alkalosis.

Chemical buffers, such as hemoglobin, that release hydrogen ions resist the increase in pH. The lower concentrations of carbon dioxide and hydrogen ions result in less stimulation of the respiratory center. This inhibits the hyperventilation, thus reducing further carbon dioxide loss. The kidneys decrease secretion of hydrogen ions, and the urine becomes alkaline as bases are excreted.

Symptoms of respiratory alkalosis include lightheadedness, agitation, dizziness, and tingling sensations. In severe cases, peripheral nerves may spontaneously trigger impulses, and muscles may contract tetanically (see chapter 8).

Metabolic alkalosis results from a great loss of hydrogen ions or from a gain in bases, both of which increase blood pH (alkalemia) (fig. 18H). This condition may follow gastric drainage (lavage), prolonged vomiting of stomach contents, or use of certain diuretic drugs. Because gastric juice is very acidic, its loss leaves body fluids more basic. Metabolic alkalosis may also develop from ingesting too much antacid, such as sodium bicarbonate. Symptoms of metabolic alkalosis include decreasd breathing rate and depth, which, in turn, increases the blood carbon dioxide concentration.

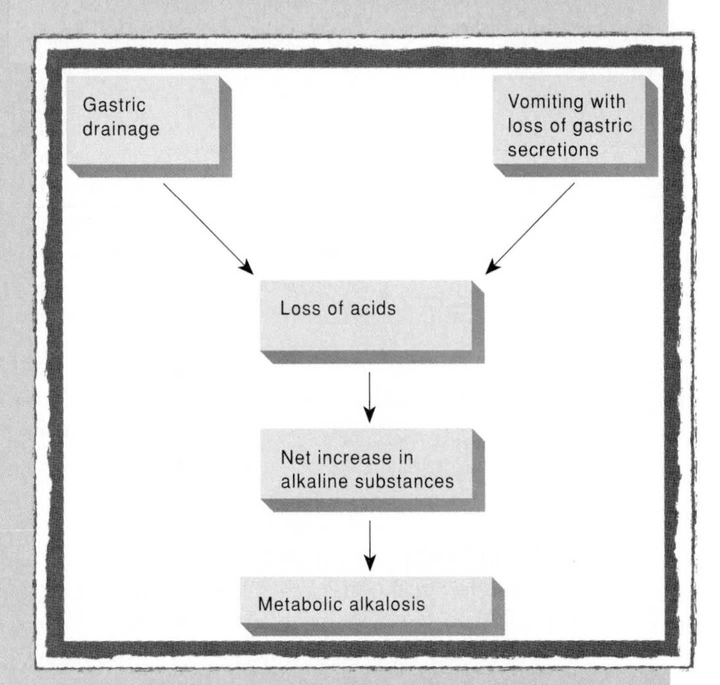

***Figure* 18H**

Some of the factors that lead to metabolic alkalosis.

Clinical Terms Related to Water and Electrolyte Balance

Acetonemia (as″ĕ-to-ne′me-ah) Abnormal amounts of acetone in blood.

Acetonuria (as″ĕ-to-nu′re-ah) Abnormal amounts of acetone in urine.

Albuminuria (al-bu″mĭ-nu′re-ah) Albumin in urine.

Anasarca (an″ah-sar′kah) Widespread accumulation of tissue fluid.

Antacid (ant-as′id) A substance that neutralizes an acid.

Anuria (ah-nu′re-ah) Absence of urine excretion.

Azotemia (az″o-te′me-ah) Accumulation of nitrogenous wastes in blood.

Diuresis (di″u-re′sis) Increased urine production.

Glycosuria (gli″ko-su′re-ah) Excess sugar in urine.

Hyperglycemia (hi″per-gli-se′me-ah) Abnormally high concentration of blood sugar.

Hyperkalemia (hi″per-kah-le′me-ah) Excess potassium in blood.

Hypernatremia (hi″per-na-tre′me-ah) Excess sodium in blood.

Hyperuricemia (hi″per-u″rĭ-se′me-ah) Excess uric acid in blood.

Hypoglycemia (hi″po-gli-se′me-ah) Abnormally low concentration of blood sugar.

Ketonuria (ke″to-nu′re-ah) Ketone bodies in urine.

Ketosis (ke″to′sis) Acidosis due to excess ketone bodies in body fluids.

Proteinuria (pro″te-ĭ-nu′re-ah) Protein in urine.

Uremia (u-re′me-ah) A toxic condition resulting from nitrogenous wastes in blood.

Summary Outline

Introduction (p. 499)

Maintenance of water and electrolyte balance requires equal quantities of these substances entering and leaving the body. Altering the water balance affects the electrolyte balance.

Distribution of Body Fluids (p. 499)

1. Fluid compartments
   a. The intracellular fluid compartment includes the fluids and electrolytes cell membranes enclose.
   b. The extracellular fluid compartment includes all the fluids and electrolytes outside cell membranes.
2. Body fluid composition
   a. Extracellular fluids have high concentrations of sodium, chloride, and bicarbonate ions with less potassium, calcium, magnesium, phosphate, and sulfate ions. Plasma contains more protein than does either interstitial fluid or lymph.
   b. Intracellular fluid contains high concentrations of potassium, phosphate, and magnesium ions. It also has a greater concentration of sulfate ions and lesser concentrations of sodium, chloride, and bicarbonate ions than do extracellular fluids.
3. Movement of fluid between compartments
   a. Hydrostatic and osmotic pressure regulate fluid movements.
      (1) Hydrostatic pressure forces fluid out of plasma, and osmotic pressure returns fluid to plasma.
      (2) Hydrostatic pressure drives fluid into lymph vessels.
      (3) Osmotic pressure regulates fluid movement in and out of cells.
   b. Sodium ion concentrations are especially important in fluid movement regulation.

Water Balance (p. 501)

1. Water intake
   a. Most water comes from consuming liquids or moist foods.
   b. Oxidative metabolism produces some water.
2. Regulation of water intake
   a. Thirst is the primary regulator of water intake.
   b. Drinking and the resulting stomach distension inhibit thirst.
3. Water output
   Water is lost in urine, feces, and sweat, and by evaporation from the skin and lungs.
4. Regulation of water balance
   The distal convoluted tubules and collecting ducts of the nephrons regulate water balance.

Electrolyte Balance (p. 503)

1. Electrolyte intake
   a. The most important electrolytes in body fluids release ions of sodium, potassium, calcium, magnesium, chloride, sulfate, phosphate, bicarbonate, and hydrogen.
   b. These ions are obtained in foods and beverages or as by-products of metabolic processes.
2. Regulation of electrolyte intake
   a. Food and drink usually provide sufficient quantities of electrolytes.
   b. A severe electrolyte deficiency may produce a salt craving.
3. Electrolyte output
   a. Electrolytes are lost through perspiration, feces, and urine.
   b. Quantities lost vary with temperature and physical exercise.
   c. Most electrolytes are lost through the kidneys.
4. Regulation of electrolyte balance
   a. Concentrations of sodium, potassium, and calcium ions in body fluids are particularly important.
   b. The adrenal cortex secretes aldosterone to regulate sodium and potassium ions.
   c. Parathyroid hormone and calcitonin regulate calcium ions.
   d. The mechanisms that control positively charged ions secondarily regulate negatively charged ions.

Acid-Base Balance (p. 506)

Acids are electrolytes that release hydrogen ions. Bases combine with hydrogen ions.

1. Sources of hydrogen ions
   a. Aerobic respiration of glucose produces carbonic acid.
   b. Anaerobic respiration of glucose produces lactic acid.
   c. Incomplete oxidation of fatty acids releases acidic ketone bodies.
   d. Oxidation of sulfur-containing amino acids produces sulfuric acid.
   e. Hydrolysis of phosphoproteins and nucleic acids produces phosphoric acid.
2. Strengths of acids and bases
   a. Acids vary in ionization extent.
      (1) Strong acids, such as hydrochloric acid, ionize more completely.
      (2) Weak acids, such as carbonic acid, ionize less completely.
   b. Bases vary in strength also.
3. Regulation of hydrogen ion concentration
   a. Acid-base buffer systems
      (1) Buffer systems change strong acids into weaker acids, or strong bases into weaker bases.
      (2) They include the bicarbonate buffer system, phosphate buffer system, and protein buffer system.
      (3) Buffer systems minimize pH changes.
   b. The respiratory center controls the rate and depth of breathing to regulate pH.
   c. Kidney nephrons secrete hydrogen ions to regulate pH.
   d. Chemical buffers act more rapidly. Physiological buffers act less rapidly.

Review Exercises

1. Explain how water balance and electrolyte balance are interdependent. (p. 499)
2. Name the body fluid compartments, and describe their locations. (p. 499)
3. Explain how extracellular and intracellular fluids differ in composition. (p. 499)
4. Describe the control of fluid movements between body fluid compartments. (p. 500)
5. Prepare a list of sources of normal water gain and loss to illustrate how water intake equals water output. (p. 501)
6. Define *water of metabolism*. (p. 501)
7. Explain the regulation of water intake. (p. 501)
8. Explain how nephrons regulate water output. (p. 502)
9. List the most important electrolytes in body fluids. (p. 503)
10. Explain the regulation of electrolyte intake. (p. 503)
11. List the routes by which electrolytes leave the body. (p. 503)
12. Explain how the adrenal cortex regulates electrolyte balance. (p. 503)
13. Describe how the parathyroid glands regulate electrolyte balance. (p. 503)
14. Distinguish between an acid and a base. (p. 506)
15. List five sources of hydrogen ions in body fluids, and an acid that originates from each. (p. 506)
16. Distinguish between a strong acid and a weak acid, and cite an example of each. (p. 507)
17. Distinguish between a strong base and a weak base, and give an example of each. (p. 507)
18. Explain how an acid-base buffer system functions. (p. 507)
19. Describe how the bicarbonate buffer system resists changes in pH. (p. 507)
20. Explain why a protein has acidic as well as basic properties. (p. 508)
21. Explain how the respiratory center regulates acid-base balance. (p. 508)
22. Explain how the kidneys regulate acid-base balance. (p. 509)
23. Distinguish between a chemical buffer system and a physiological buffer system. (p. 509)

Critical Thinking

1. A twenty-five-year-old male contracted a strain of *Escherichia coli* that produces a shigatoxin from eating an undercooked hamburger and developed diarrhea. How would this affect his blood pH, urine pH, and respiratory rate?
2. A student hyperventilates and is disoriented just before an exam. Is this student likely to be experiencing acidosis or alkalosis? How will the body compensate in an effort to maintain homeostasis?
3. A ten-year-old female is rescued from a swimming pool after several minutes of floundering in the water. What is (are) the cause(s) of the girl's acidosis? What treatment(s) will bring the body back to homeostasis?
4. A thirty-eight-year-old woman contracted *Mycoplasma* pneumonia and ran a temperature of 104° for five days. Even though the woman drank copious amounts of liquid, her blood pressure dropped to 70/50, indicating dehydration. Should the woman receive intravenous hypertonic glucose or normal saline? Why?
5. Some time ago, several newborn infants died due to sodium chloride mistakenly substituted for sugar in their formula. What symptoms would this produce? Why are infants more prone to the hazard of excess salt intake than adults?
6. An elderly, semiconscious patient is tentatively diagnosed with acidosis. What components of arterial blood should be examined to determine if the acidosis is of respiratory origin?
7. Radiation therapy may damage the mucosa of the stomach and intestines. What effect might this have on the patient's electrolyte balance?
8. If the right ventricle of a patient's heart is failing, increasing venous pressure, what changes might occur in the patient's extracellular fluid compartments?

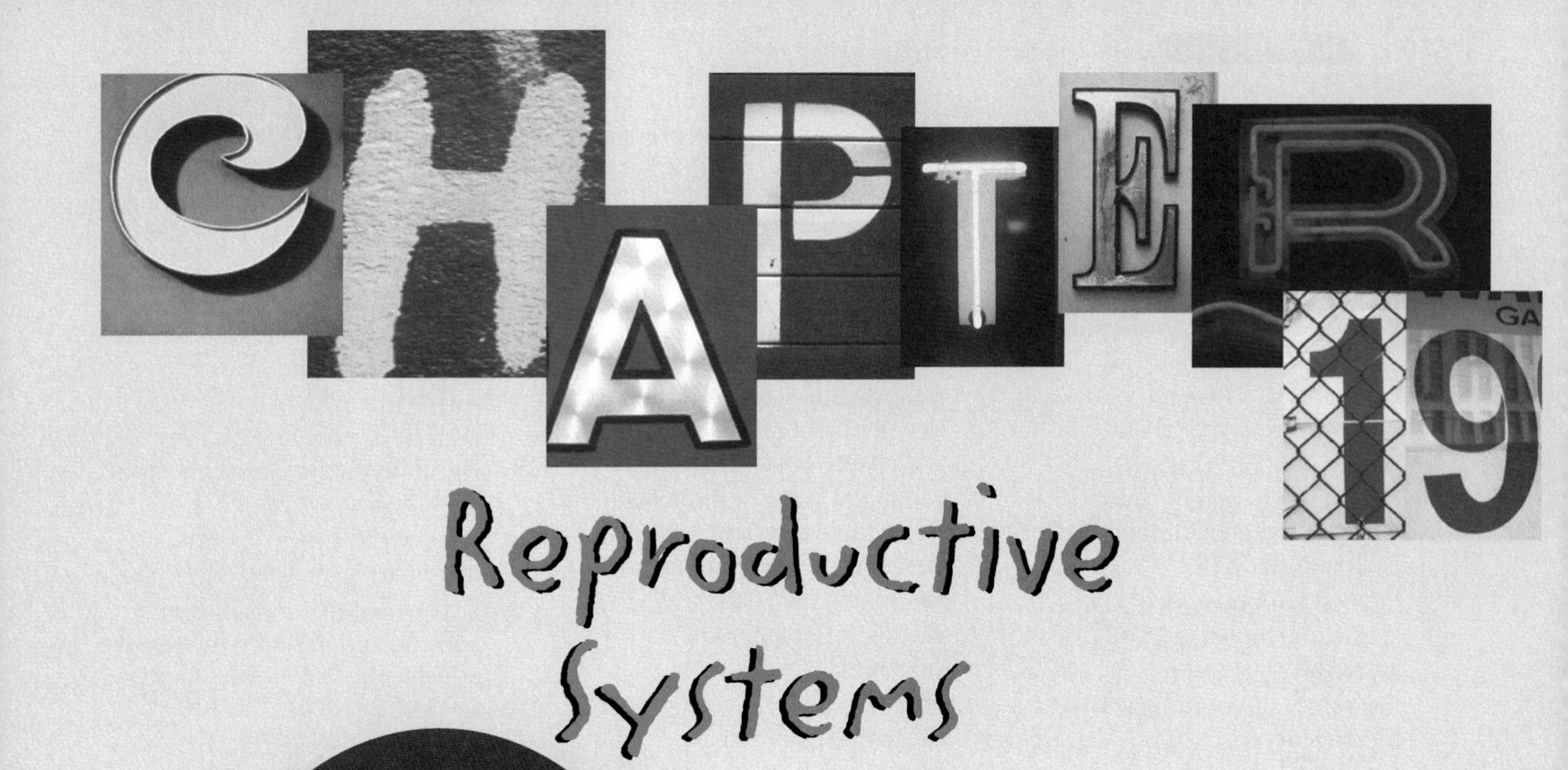

Chapter 19 Reproductive Systems

Leslie and Howard spent nearly a decade taking measures to avoid a pregnancy. It is ironic that now, when they desperately wish to become parents, pregnancy is elusive.

Leslie used birth control pills while she and Howard were in graduate school and starting their careers. Now that they are both in their early thirties, the time finally seems right to start a family. But even after ten months of trying, Leslie is not pregnant.

Fearful that one of them has a medical problem that prevents pregnancy, Leslie and Howard visit a fertility specialist, who conducts tests and determines that both of them have minor problems that time alone might treat. Leslie's irregular menstrual periods since discontinuing use of birth control pills have made it difficult for her to predict when she is fertile. The doctor suggests that she use an ovulation indicator kit to detect hormonal changes that indicate when an ovary will release an egg cell. Howard has a low sperm count.

The fertility specialist suggests that the couple continue trying to conceive for several months. If they have no luck, Howard can donate sperm samples that can be combined, concentrated, and then inserted into Leslie's reproductive tract (artificial insemination).

But this is not necessary. With the ovulation test as a guideline, Leslie is pregnant two months later. The couple is fortunate—many causes of infertility are not so easily remedied.

Photo: Too many abnormally shaped sperm cells can make a male infertile.

A new life begins with two existing lives. The male and female reproductive systems are unique in that they are not necessary for survival of the individual, but they are essential for perpetuation of the human species.

Chapter Objectives

After studying this chapter, you should be able to:

1. Name the parts of the male reproductive system, and describe the general functions of each part. (p. 517)
2. Outline the process of spermatogenesis. (p. 519)
3. Trace the path sperm cells follow from their site of formation to the outside. (p. 519)
4. Explain how hormones control the activities of male reproductive organs and the development of male secondary sex characteristics. (p. 525)
5. Name the parts of the female reproductive system, and describe the general functions of each part. (p. 526)
6. Outline the process of oogenesis. (p. 527)
7. Describe how hormones control the activities of female reproductive organs and the development of female secondary sex characteristics. (p. 532)
8. Describe the major events of the menstrual cycle. (p. 532)
9. Review the structure of the mammary glands. (p. 534)
10. List several methods of birth control, and describe the relative effectiveness of each method. (p. 538)
11. List general symptoms of sexually transmitted diseases. (p. 541)

Aids to Understanding Words

andr- [man] *andr*ogens: Male sex hormones.

ejacul- [to shoot forth] *ejacul*ation: The process of expelling semen from the male reproductive tract.

fimb- [fringe] *fimb*riae: Irregular extensions on the margin of the infundibulum of the uterine tube.

follic- [small bag] *follic*le: The ovarian structure containing an egg.

genesis- [origin] spermato*genesis*: The process of forming sperm cells.

germ- [to bud or sprout] *germ*inal epithelium: Tissue that gives rise to sex cells by special cell division.

labi- [lip] *labi*a minora: Flattened, longitudinal folds extending along the margins of the female vestibule.

mens- [month] *mens*trual cycle: The monthly female reproductive cycle.

mons- [mountain] *mons* pubis: Rounded elevation overlying the pubic symphysis in females.

puber- [adult] *puber*ty: Time when a person becomes able to reproduce.

Key Terms

androgen (an′dro-jen)
contraception (kon″trah-sep′shun)
ejaculation (e-jak″u-la′shun)
emission (e-mish′un)
estrogen (es′tro-jen)
gonadotropin (go-nad″o-trōp′in)
meiosis (mi-o′sis)
menopause (men′o-pawz)
menstrual cycle (men′stroo-al si′kl)
oogenesis (ō″o-jen′ĕ-sis)
orgasm (or′gazm)
ovulation (o″vu-la′shun)
primary follicle (pri′ma-re fol′ĭ-kl)
progesterone (pro-jes′tĕ-rōn)
puberty (pu′ber-te)
semen (se′men)
spermatogenesis (sper″mah-to-jen′ĕ-sis)
testosterone (tes-tos′tĕ-rōn)
zygote (zi′gōt)

Introduction

Male and female reproductive systems are a connected series of organs and glands that produce and nurture sex cells, and transport them to sites of fertilization. Some reproductive organs secrete hormones vital in the development and maintenance of sexual characteristics and the regulation of reproductive physiology.

Organs of the Male Reproductive System

Organs of the male reproductive system produce and maintain male sex cells, or **sperm cells**; transport these cells and supporting fluids to the outside; and secrete male sex hormones. A male's *primary sex organs* (gonads) are the two **testes** (singular, *testis*) in which sperm cells and male sex hormones form. The *accessory sex organs* of the male reproductive system are the internal and external reproductive organs (fig. 19.1 and reference plates 3 and 4).

Testes

The testes are ovoid structures about 5 centimeters long and 3 centimeters in diameter. Both testes are within the cavity of the saclike *scrotum*.

Structure of the Testes

A tough, white, fibrous capsule encloses each testis. Along the capsule's posterior border, the connective tissue thickens and extends into the testis, forming thin septa that subdivide the testis into about 250 *lobules*.

Each lobule contains one to four highly coiled, convoluted **seminiferous tubules**, each approximately 70 centimeters long uncoiled. These tubules course posteriorly and unite to form a complex network of channels. These channels give rise to several ducts that join a tube called the **epididymis** (plural, epididymides). The epididymis coils on the outer surface of the testis and continues to become the **vas deferens** (plural, *vasa deferentia*) (fig. 19.2*a*).

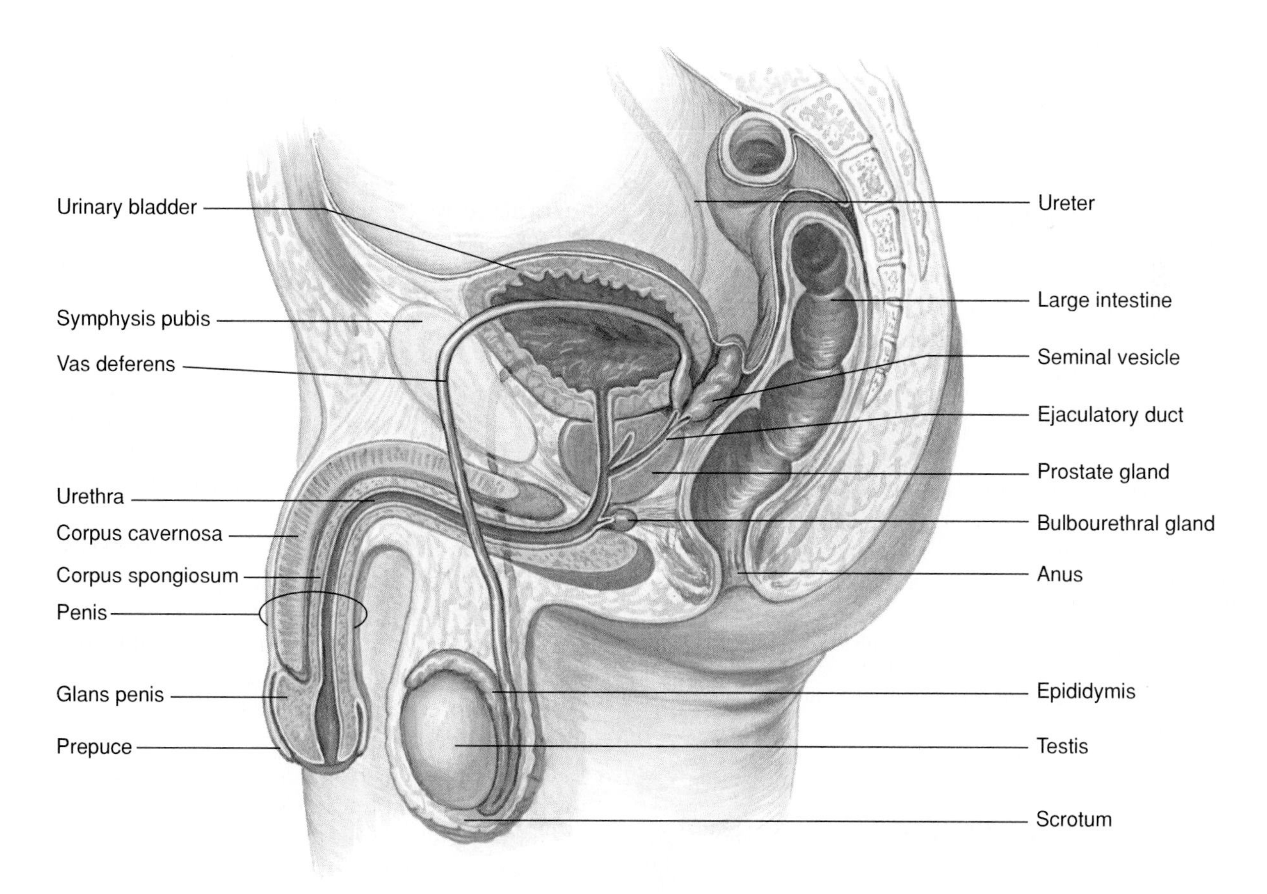

Figure 19.1

Male reproductive organs (sagittal view).

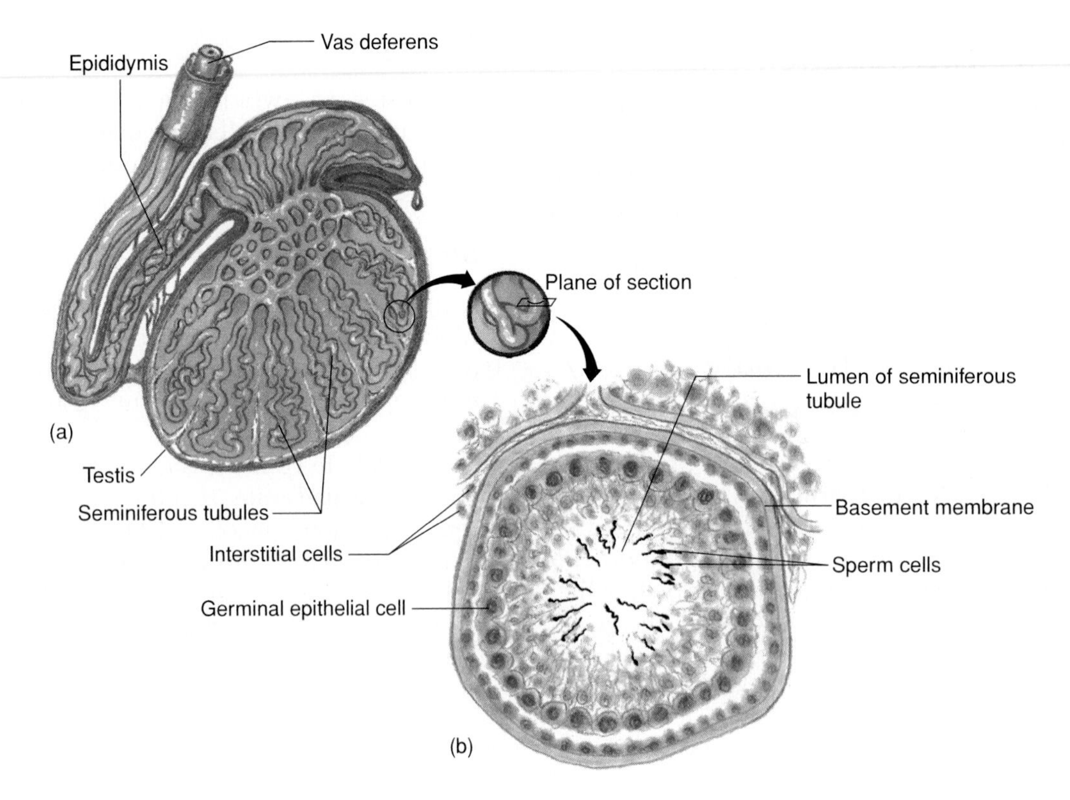

Figure 19.2

(*a*) Sagittal section of a testis. (*b*) Cross section of a seminiferous tubule.

A specialized stratified epithelium with **spermatogenic cells**, which give rise to sperm cells, lines the seminiferous tubules. Other specialized cells, called **interstitial cells**, lie in the spaces between the seminiferous tubules (fig. 19.2*b*). Interstitial cells produce and secrete male sex hormones.

The epithelial cells of the seminiferous tubules can give rise to *testicular cancer,* a common cancer in young men. In most cases, the first sign is a painless testis enlargement or a scrotal mass that attaches to a testis.

If a biopsy (tissue sample) reveals cancer cells, surgery removes the affected testis (orchiectomy). Radiation and/or chemotherapy often prevent(s) the cancer from recurring. ■

1. Describe the structure of a testis.
2. Where are sperm cells produced within the testes?
3. Which cells produce male sex hormones?

Formation of Sperm Cells

The epithelium of the seminiferous tubules consists of supporting cells and spermatogenic cells. Supporting cells support, nourish, and regulate the spermatogenic cells, which give rise to sperm cells.

Males produce sperm cells continually throughout their reproductive lives. Sperm cells collect in the lumen of each seminiferous tubule, then pass to the epididymis, where they accumulate and mature.

A mature sperm cell is a tiny, tadpole-shaped structure about 0.06 millimeters long. It consists of a flattened head, a cylindrical midpiece (body), and an elongated tail (flagellum) (figs. 19.3 and 3.8*b*).

The oval *head* of a sperm cell is composed primarily of a nucleus and contains highly compacted chromatin of twenty-three chromosomes. A small protrusion at its anterior end, called the *acrosome,* contains enzymes that help the sperm cell penetrate an egg cell during fertilization. (Chapter 20 describes this process.)

The *midpiece* of a sperm cell has a central, filamentous core and many mitochondria in a spiral. The *tail* consists of several microtubules enclosed in an extension of the cell membrane. The mitochondria provide ATP for the tail's lashing movement that propels the sperm cell through fluid.

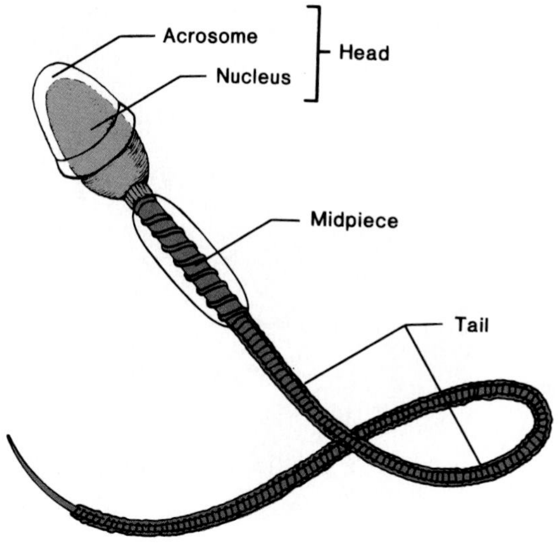

Acrosome
Nucleus
Head
Midpiece
Tail

Figure 19.3

Parts of a mature sperm cell.

Spermatogenesis

In the male embryo, spermatogenic cells are undifferentiated and are called *spermatogonia*. Each spermatogonium contains forty-six chromosomes in its nucleus, the usual number for human cells. During early adolescence, hormones stimulate spermatogonia to undergo mitosis (see chapter 3), and some of them enlarge to become *primary spermatocytes* (fig. 19.4).

During a special type of cell division called **meiosis,** each primary spermatocyte divides to form two *secondary spermatocytes*. Each of these cells, in turn, divides to form two *spermatids,* which mature into sperm cells. Meiosis also reduces the number of chromosomes in each cell by half. Consequently, for each primary spermatocyte that undergoes meiosis, four sperm cells with twenty-three chromosomes in each of their nuclei form (fig. 19.5). This process of producing sperm cells is called **spermatogenesis.**

1. Explain the function of supporting cells in the seminiferous tubules.
2. Describe the structure of a sperm cell.
3. Review the events of spermatogenesis.

Male Internal Accessory Organs

The internal accessory organs of the male reproductive system include the epididymides, vasa deferentia, ejaculatory ducts, and urethra, as well as the seminal vesicles, prostate gland, and bulbourethral glands.

Epididymis

Each epididymis is a tightly coiled, threadlike tube about 6 meters long (see figs. 19.1 and 19.2). The epididymis connects to ducts within a testis. It emerges from the top of the testis, descends along the posterior surface of the testis, and then courses upward to become the vas deferens.

Immature sperm cells reaching the epididymis are nonmotile. As rhythmic peristaltic contractions help move these cells through the epididymis, however, sperm cells mature. Following this aging process, sperm cells can move independently and fertilize egg cells.

Vas Deferens

Each **vas deferens,** also called ductus deferens, is a muscular tube about 45 centimeters long (see fig. 19.1). It passes upward along the medial side of a testis and through a passage in the lower abdominal wall (inguinal canal), enters the abdominal cavity, and ends behind the urinary bladder. Just outside the prostate gland, the vas deferens unites with the duct of a seminal vesicle to form an **ejaculatory duct,** which passes through the prostate gland and empties into the urethra.

Seminal Vesicle

A **seminal vesicle** is a convoluted, saclike structure about 5 centimeters long that attaches to the vas deferens near the base of the urinary bladder (see fig. 19.1). The glandular tissue lining the inner wall of a seminal vesicle secretes a slightly alkaline fluid. This fluid helps regulate the pH of the tubular contents as sperm cells travel to the outside. Seminal vesicle secretions also contain *fructose,* a monosaccharide that provides energy to sperm cells, and *prostaglandins* (see chapter 11) that stimulate muscular contractions within the female reproductive organs, aiding the movement of sperm cells toward the egg cell.

1. Describe the structure of the epididymis.
2. Trace the path of the vas deferens.
3. What is the function of a seminal vesicle?

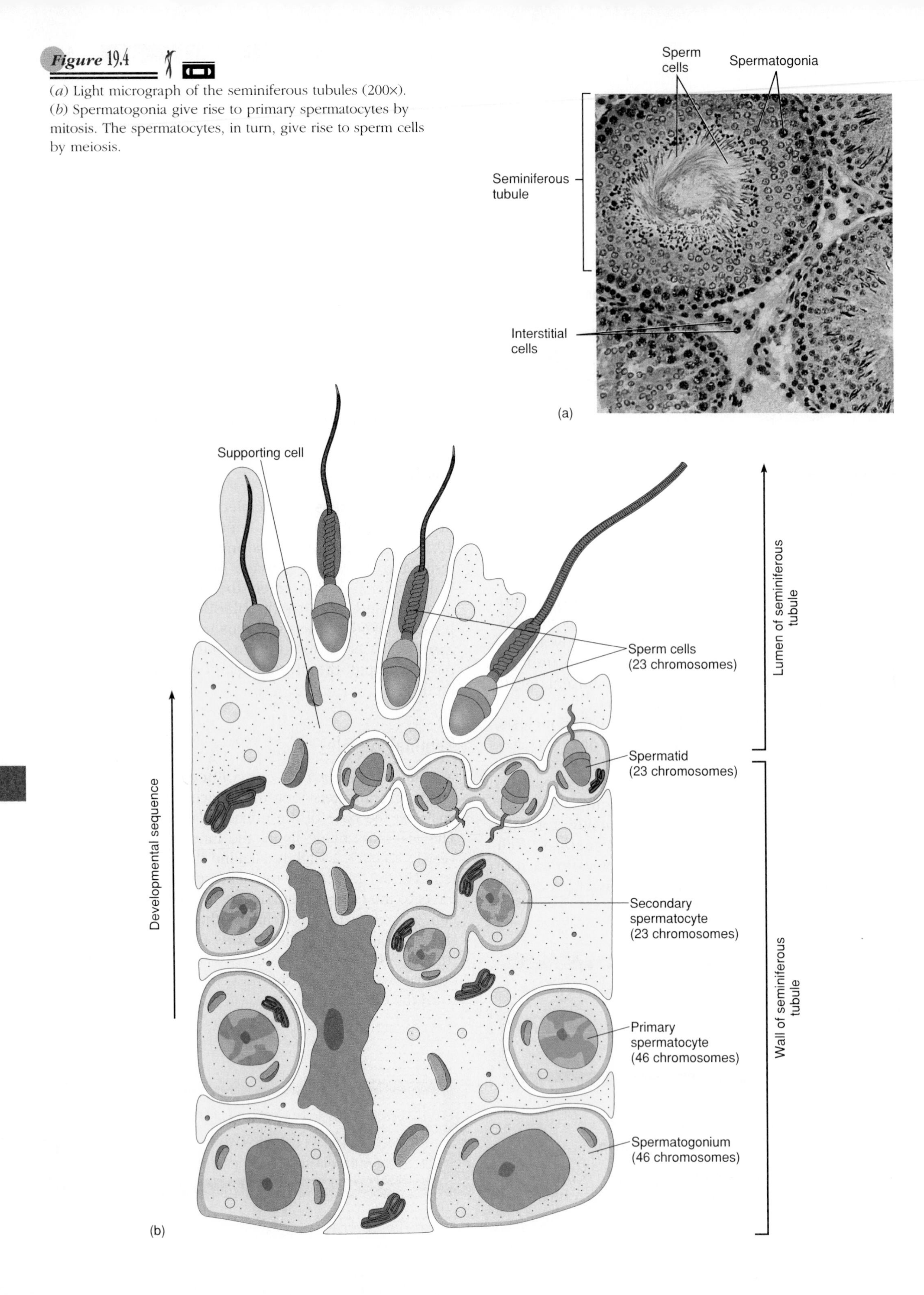

Figure 19.4

(*a*) Light micrograph of the seminiferous tubules (200×). (*b*) Spermatogonia give rise to primary spermatocytes by mitosis. The spermatocytes, in turn, give rise to sperm cells by meiosis.

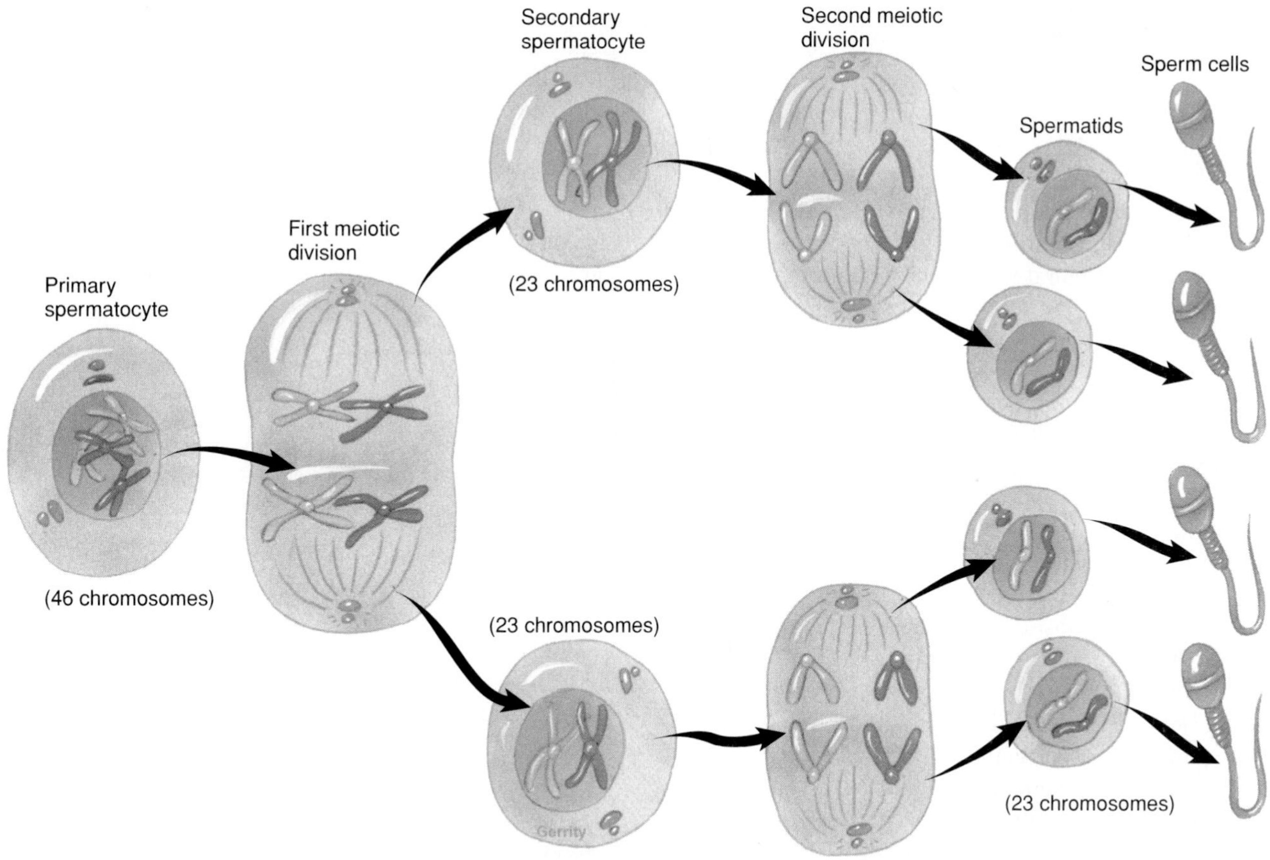

Figure 19.5

Spermatogenesis involves two successive meiotic divisions.

Prostate Gland

The **prostate gland** is a chestnut-shaped structure about 4 centimeters across and 3 centimeters thick that surrounds the beginning of the urethra, just below the urinary bladder (see fig. 19.1). It is enclosed in connective tissue and composed of many branched tubular glands, whose ducts open into the urethra.

The prostate gland secretes a thin, milky fluid with an alkaline pH. This secretion neutralizes the fluid containing sperm cells, which is acidic from metabolic wastes that stored sperm cells produce. Prostatic fluid also enhances the motility of sperm cells and helps neutralize the acidic secretions of the vagina.

Bulbourethral Glands

The two **bulbourethral glands** (Cowper's glands) are each about a centimeter in diameter and are posterior to the prostate gland within muscle fibers of the external urethral sphincter (see fig. 19.1). Bulbourethral glands have many tubes whose epithelial linings secrete a mucuslike fluid in response to sexual stimulation. This fluid lubricates the end of the penis in preparation for sexual intercourse. Females secrete most of the lubricating fluid for sexual intercourse, however.

Semen

Semen is the fluid the male urethra conveys to the outside during ejaculation. It consists of sperm cells from the testes and secretions of the seminal vesicles, prostate gland, and bulbourethral glands. Semen is slightly alkaline (pH about 7.5), and its contents include prostaglandins and nutrients.

The volume of semen released at one time varies from 2–6 milliliters. The average number of sperm cells in the fluid is about 120 million per milliliter.

Sperm cells are nonmotile while in the ducts of the testis and epididymis, but are activated as they mix with accessory gland secretions. Sperm cells cannot fertilize an egg cell, however, until they enter the female reproductive tract. Acquiring the ability to fertilize an egg cell is called *capacitation,* and it reflects weakening of sperm cells' acrosomal membranes.

Topic of Interest: Prostate Enlargement

The prostate gland is relatively small in boys, begins to grow in early adolescence, and reaches adult size several years later. Usually, the gland does not grow again until age fifty, when in half of all men, it enlarges enough to press on the urethra. This produces a feeling of pressure because the bladder cannot empty completely and leads to frequent urination. Retained urine can lead to infection and inflammation, bladder stones, or kidney disease.

Medical researchers do not know what causes prostate enlargement. Risk factors include a fatty diet, having had a vasectomy, and possibly, occupational exposure to batteries or the metal cadmium. The enlargement may be benign or cancerous. Because prostate cancer is nearly 100% treatable if detected early, men should have their prostates examined regularly.

Diagnostic tests include a rectal exam, as well as a blood test to detect prostate specific antigen (PSA), a cell surface protein normally found on prostate cells. Elevated PSA levels indicate an enlarged prostate, possibly from a benign or cancerous growth. Ultrasound may provide further information. Table 19A summarizes treatments for an enlarged prostate.

Table 19A

Medical Treatments for an Enlarged Prostate Gland
Surgical removal of prostate
Radiation
Drugs to block testosterone's growth-stimulating effect on prostate
Microwave energy delivered through a probe inserted into urethra or rectum
Balloon inserted into urethra and inflated with liquid
Tumor frozen with liquid nitrogen delivered by probe through skin
Device (stent) inserted between lobes of prostate to relieve pressure on urethra

1. Where is the prostate gland?
2. What is the function of the prostate gland's secretion?
3. What are the components of semen?

Male External Reproductive Organs

The male external reproductive organs are the scrotum, which encloses the testes, and the penis, through which the urethra passes.

Scrotum

The **scrotum** is a pouch of skin and subcutaneous tissue that hangs from the lower abdominal region behind the penis (see fig. 19.1). A medial septum subdivides the scrotum into two chambers, each of which encloses a testis. Each chamber also contains a serous membrane, which covers the testis and helps ensure that it moves smoothly within the scrotum.

Penis

The **penis** is a cylindrical organ that conveys urine and semen through the urethra to the outside (see fig. 19.1). During **erection,** it enlarges and stiffens so that it can be inserted into the vagina during sexual intercourse.

The *body,* or shaft, of the penis has three columns of erectile tissue—a pair of dorsally located *corpora cavernosa* and a single, ventral *corpus spongiosum.* A tough capsule of white fibrous connective tissue surrounds each column. Skin, a thin layer of subcutaneous tissue, and a layer of connective tissue enclose the penis.

The corpus spongiosum, through which the urethra extends, enlarges at its distal end to form a sensitive, cone-shaped **glans penis.** The glans covers the ends of the corpora cavernosa and bears the urethral opening (external urethral orifice). The skin of the glans is very thin and hairless, and contains sensory receptors for sexual stimulation. A loose fold of skin called the *prepuce* (foreskin) begins just behind the glans and extends forward to cover the glans as a sheath. The surgical procedure called *circumcision* removes the prepuce.

1. Describe the structure of the penis.
2. What is circumcision?

Erection, Orgasm, and Ejaculation

During sexual stimulation, parasympathetic nerve impulses pass from the sacral portion of the spinal cord to the arteries leading into the penis, dilating them. At the same time, the increasing pressure of arterial blood entering the vascular spaces of erectile tissue compresses the veins of the penis, reducing flow of venous blood away from the penis. Consequently, blood accumulates in erectile tissues, and the penis swells and elongates, producing an erection.

The culmination of sexual stimulation is **orgasm,** a pleasurable feeling of physiological and psychological release. Emission and ejaculation accompany male orgasm.

Emission is the movement of sperm cells from the testes and secretions from the prostate gland and seminal vesicles into the urethra, where they mix to form semen. Emission occurs in response to sympathetic nerve impulses from the spinal cord, which stimulate peristaltic contractions in the walls of the testicular ducts, epididymides, vasa deferentia, and ejaculatory ducts. At the same time, other sympathetic impulses stimulate rhythmic contractions of the seminal vesicles and prostate gland.

The urethra filling with semen stimulates sensory impulses into the sacral portion of the spinal cord. In response, the spinal cord transmits motor impulses to certain skeletal muscles at the base of the penile erectile columns, causing them to contract rhythmically. This increases the pressure within the erectile tissues and helps force semen through the urethra to the outside—a process called **ejaculation.**

The sequence of events during emission and ejaculation is coordinated so that fluid from the bulbourethral glands is expelled first. This is followed by the release of fluid from the prostate gland, the passage of sperm cells, and finally, the ejection of fluid from the seminal vesicles.

Immediately after ejaculation, sympathetic impulses constrict the arteries that supply the erectile tissue, reducing blood inflow. Smooth muscles within the walls of the vascular spaces partially contract again, and the veins of the penis carry excess blood out of these spaces. The penis gradually returns to its flaccid state.

Table 19.1 summarizes the functions of the male reproductive organs.

Table 19.1 Functions of the Male Reproductive Organs

Organ	Function
Testis	
Seminiferous tubules	Produce sperm cells
Interstitial cells	Produce and secrete male sex hormones
Epididymis	Stores sperm cells undergoing maturation; conveys sperm cells to vas deferens
Vas deferens	Conveys sperm cells to ejaculatory duct
Seminal vesicle	Secretes an alkaline fluid containing nutrients and prostaglandins that helps neutralize the acidic components of semen
Prostate gland	Secretes an alkaline fluid that helps neutralize semen's acidity and enhances sperm cell motility
Bulbourethral gland	Secretes fluid that lubricates end of penis
Scrotum	Encloses and protects testes
Penis	Conveys urine and semen to outside of body; inserted into vagina during sexual intercourse; glans penis is richly supplied with sensory nerve endings associated with feelings of pleasure during sexual stimulation

Spontaneous emissions and ejaculations are common in adolescent males during sleep. Changes in hormonal concentrations that accompany adolescent development and sexual maturation cause these *nocturnal emissions.* ■

1. What controls blood flow into penile erectile tissues?
2. Distinguish between orgasm, emission, and ejaculation.
3. Review the events associated with emission and ejaculation.

TOPIC OF INTEREST

Male Infertility

Male infertility—the inability of sperm cells to fertilize an egg cell—has several causes. If during fetal development the testes do not descend into the scrotum, the higher temperature of the abdominal cavity or inguinal canal destroys sperm cells developing in the seminiferous tubules. Certain diseases, such as mumps, may inflame the testes (orchitis) and cause infertility by destroying cells in the seminiferous tubules.

Both the quality and quantity of sperm cells are essential factors in a man's ability to father a child. If a sperm head is misshapen, if a sperm cell cannot swim, or if sperm cells are too few, completing the journey to the egg cell may be impossible.

Computer-aided sperm analysis (CASA) uses criteria for normalcy in human male seminal fluid and the sperm cells it contains. For this analysis, a man abstains from intercourse for two to three days, and then provides a sperm sample, which must be examined within the hour. The man also provides information about his reproductive history and possible exposure to toxins. The sperm sample is placed on a slide under a microscope, and a video camera sends an image to a videocassette recorder, which projects a live or digitized image. The camera also sends the image to a computer, which traces sperm cell trajectories and displays them on a monitor. Figure 19A shows a CASA of normal sperm cells, depicting different swimming patterns as they travel. Table 19B lists the components of a semen analysis.

Table 19B Semen Analysis

Characteristic	Normal Value
Volume	2–6 milliliters/ejaculate
Sperm cell density	120 million cells/milliliter
Percent motile	> 40%
Motile sperm cell density	> 8 million/milliliter
Average velocity	> 20 micrometers/second
Motility	> 8 micrometers/second
Percent abnormal morphology	> 40%
White blood cells	> 5 million/milliliter

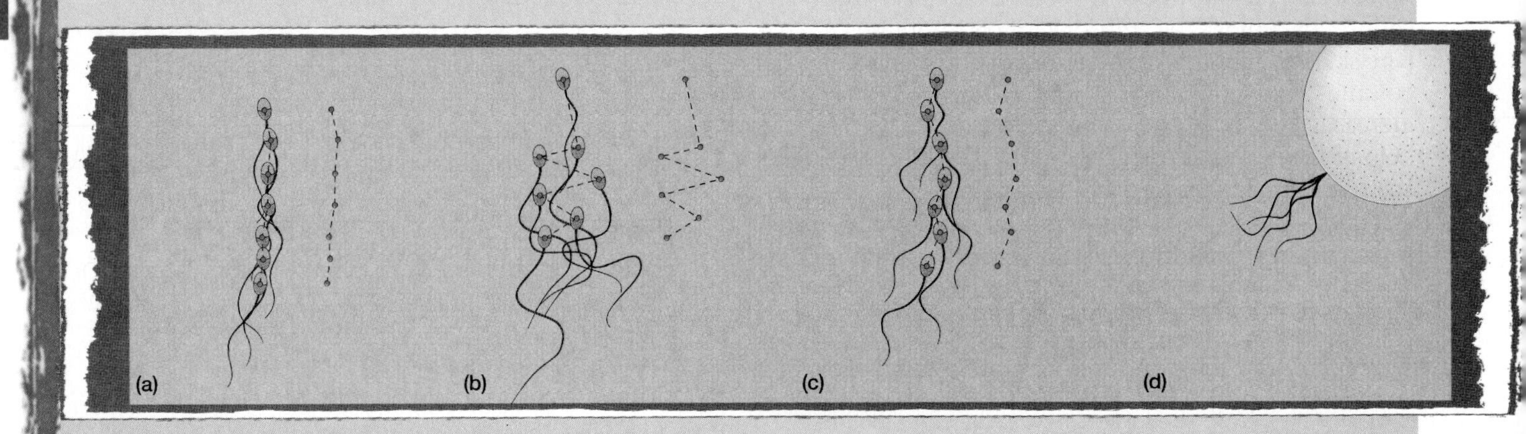

***Figure* 19A**

A computer tracks sperm cell movements. In semen, sperm cells swim in a straight line (*a*), but as they are activated by biochemicals normally found in the woman's body, their trajectories widen (*b*). The sperm cells in (*c*) are in the mucus of a woman's cervix, and the sperm cells in (*d*) are attempting to digest through the structures surrounding an egg cell.

Hormonal Control of Male Reproductive Functions

The hypothalamus, anterior pituitary gland, and testes secrete hormones that control male reproductive functions. These hormones initiate and maintain sperm cell production, and oversee the development and maintenance of male sexual characteristics.

Hypothalamic and Pituitary Hormones

Prior to ten years of age, the male body is reproductively immature. It is childlike, with spermatogenic cells undifferentiated. Then, a series of changes leads to development of a reproductively functional adult. The hypothalamus controls many of these changes.

Recall from chapter 11 that the hypothalamus secretes gonadotropin-releasing hormone (GnRH), which enters blood vessels leading to the anterior pituitary gland. In response, the anterior pituitary secretes the **gonadotropins** called *luteinizing hormone* (LH) and *follicle-stimulating hormone* (FSH). LH, which in males is also called interstitial cell-stimulating hormone (ICSH), promotes development of testicular interstitial cells, and they, in turn, secrete male sex hormones. FSH stimulates the supporting cells of the seminiferous tubules to respond to the effects of the male sex hormone **testosterone.** Then, in the presence of FSH and testosterone, these supporting cells stimulate spermatogenic cells to undergo spermatogenesis, which produces sperm cells.

Male Sex Hormones

Male sex hormones are termed **androgens.** Testicular interstitial cells produce most of them, but the adrenal cortex synthesizes small amounts (see chapter 11).

Testosterone is the most abundant androgen. It loosely attaches to plasma proteins for secretion and transport in blood.

Testosterone secretion begins during fetal development and continues for several weeks following birth; then it nearly ceases during childhood. Between the ages of thirteen and fifteen, a young man's androgen production usually increases rapidly. This phase in development, when an individual becomes reproductively functional, is **puberty.** After puberty, testosterone secretion continues throughout the life of a male.

Actions of Testosterone

During puberty, testosterone stimulates enlargement of the testes (the primary male sex characteristic) and accessory organs of the reproductive system, as well as development of *male secondary sex characteristics.* Secondary sex characteristics are special features associated with the adult male body and include the following developments:

1. The growth of body hair, particularly on the face, chest, axillary region, and pubic region, increases. Sometimes, hair growth on the scalp slows.
2. The larynx enlarges, and the vocal folds thicken, lowering the pitch of the voice.
3. The skin thickens.
4. Muscular growth increases, the shoulders broaden, and the waist narrows.
5. The bones thicken and strengthen.

Testosterone also increases the rate of cellular metabolism and red blood cell production, so that the average number of red blood cells in a cubic millimeter of blood usually is greater in males than in females. Testosterone stimulates sexual activity by affecting certain portions of the brain.

Regulation of Male Sex Hormones

The extent to which male secondary sex characteristics develop directly relates to the amount of testosterone that interstitial cells secrete. A negative feedback system involving the hypothalamus regulates the quantity of testosterone (fig. 19.6).

An increasing blood testosterone concentration inhibits the hypothalamus, and its stimulation of the anterior pituitary gland by GnRH decreases. As the pituitary's secretion of LH (ICSH) falls in response, the amount of testosterone the interstitial cells release decreases.

As blood testosterone concentration drops, the hypothalamus becomes less inhibited, and it once again stimulates the anterior pituitary to release LH. Increasing LH secretion causes interstitial cells to release more testosterone, and blood testosterone concentration increases.

Testosterone level decreases somewhat during and after the *male climacteric,* a decline in sexual function with aging. At any given age, the testosterone concentration in the male body is regulated so that it remains relatively constant.

1. Which hormone initiates the changes associated with male sexual maturity?
2. Describe several male secondary sexual characteristics.
3. List the functions of testosterone.
4. Explain the regulation of secretion of male sex hormones.

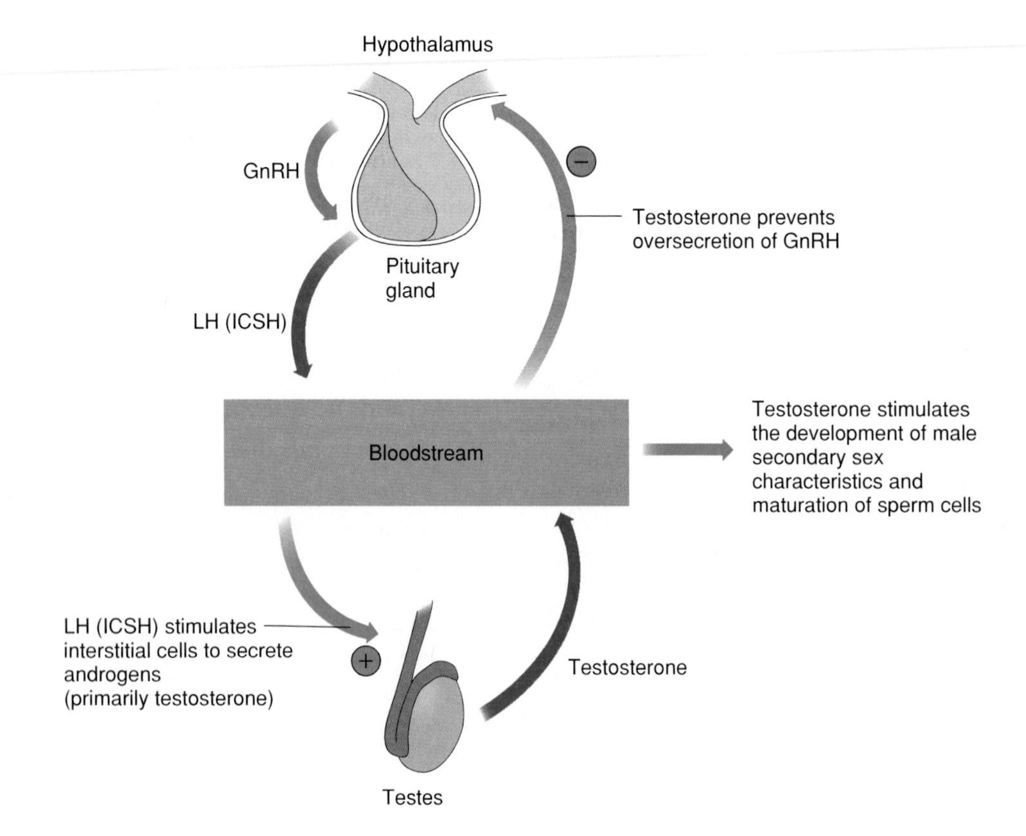

Figure 19.6

The hypothalamus controls maturation of sperm cells and development of male secondary sex characteristics. A negative feedback mechanism operating between the hypothalamus, the anterior pituitary gland, and the testes controls testosterone concentration.

Organs of the Female Reproductive System

The organs of the female reproductive system produce and maintain the female sex cells, or egg cells (ova); transport these cells to the site of fertilization; provide a favorable environment for a developing offspring; move the offspring to the outside; and produce female sex hormones. A female's *primary sex organs* (gonads) are the two ovaries, which produce the female sex cells and sex hormones. The *accessory sex organs* of the female reproductive system are the internal and external reproductive organs (fig. 19.7 and reference plates 5 and 6).

Ovaries

The two **ovaries** are solid, ovoid structures, each about 3.5 centimeters long, 2 centimeters wide, and 1 centimeter thick. The ovaries lie in shallow depressions in the lateral wall of the pelvic cavity (fig. 19.7).

Ovary Structure

Ovarian tissues are subdivided into two indistinct regions—an inner *medulla* and an outer *cortex*. The ovarian medulla is composed of loose connective tissue, and contains many blood vessels, lymphatic vessels, and nerve fibers. The ovarian cortex consists of more compact tissue and has a somewhat granular appearance due to tiny masses of cells called *ovarian follicles*.

A layer of cuboidal epithelium covers the ovary's free surface. Just beneath this epithelium is a layer of dense connective tissue.

1. What are the primary sex organs of the female?
2. Describe the structure of an ovary.

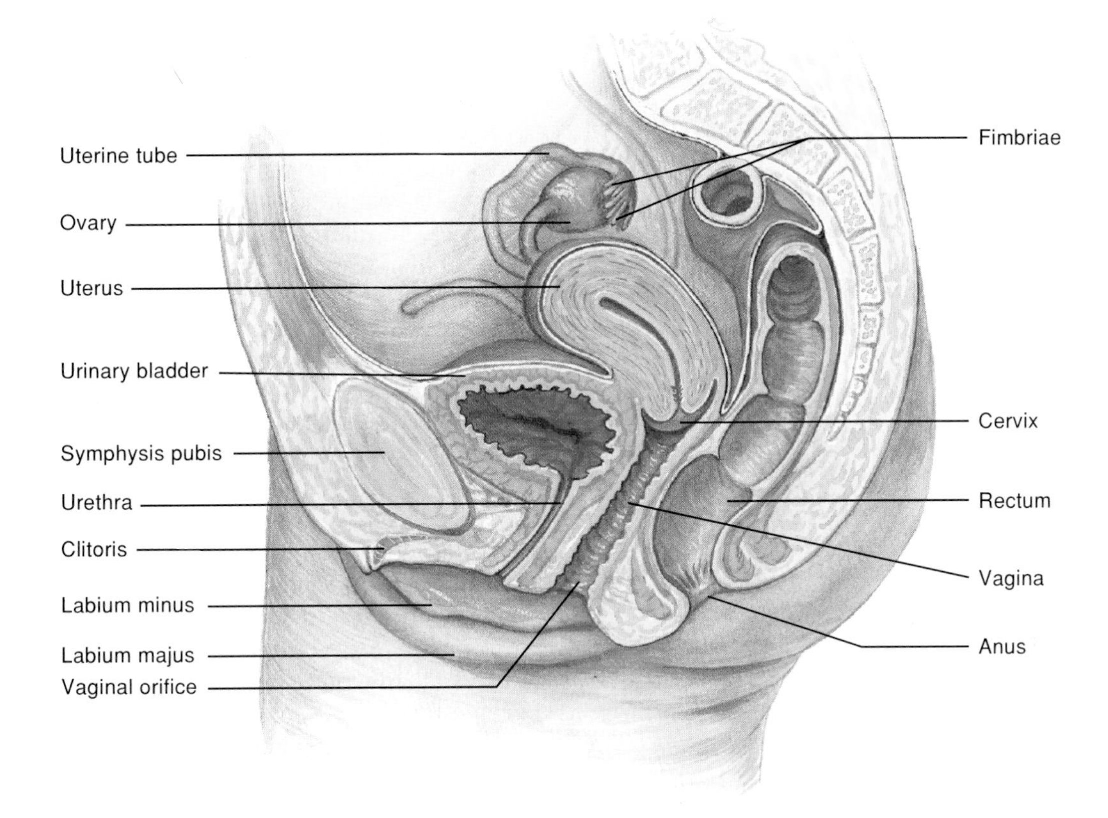

Figure 19.7

Female reproductive organs (sagittal view).

Primordial Follicles

During prenatal (before birth) development, small groups of cells in the outer region of the ovarian cortex form several million **primordial follicles.** Each of these structures consists of a single, large cell, called a *primary oocyte,* closely surrounded by epithelial cells called *follicular cells.*

Early in development, primary oocytes begin to undergo meiosis, but the process soon halts and does not continue until puberty. Once the primordial follicles appear, no new ones form. Instead, the number of oocytes in the ovary steadily declines, as many of the oocytes degenerate. Of the several million oocytes formed originally, only a million or so remain at birth, and perhaps 400,000 are present at puberty. The ovary releases fewer than 400 or 500 oocytes during a female's reproductive life.

Oogenesis

Oogenesis is the process of egg cell formation. Beginning at puberty, some primary oocytes are stimulated to continue meiosis. As in the case of sperm cells, the resulting cells have half as many chromosomes (twenty-three) in their nuclei as their parent cells.

When a primary oocyte divides, the distribution of the cytoplasm is unequal. One of the resulting cells, called a *secondary oocyte,* is large, and the other, called the *first polar body,* is small (fig. 19.8).

The large secondary oocyte represents a future egg cell that a sperm cell can fertilize. If this happens, the oocyte divides unequally to produce a tiny *second polar body* and a large fertilized egg cell, or **zygote.**

The polar bodies have no further function and soon degenerate. Their role in reproduction is to allow the egg cell to accumulate a large amount of cytoplasm, necessary to nurture a zygote and early embryo.

1. How does the timing of egg cell production differ from that of sperm cells?
2. Describe the major events of oogenesis.

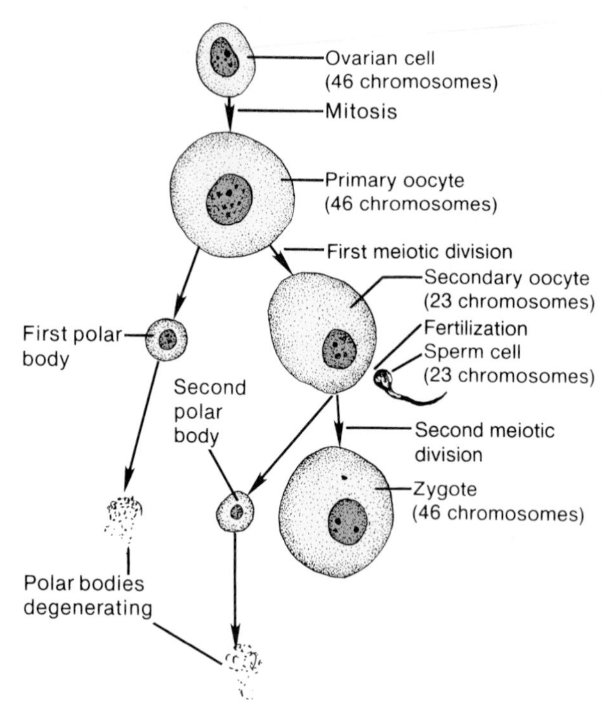

Figure 19.8

During oogenesis, a single egg cell (secondary oocyte) results from the meiosis of a primary oocyte. If the egg cell is fertilized, it generates a second polar body and becomes a zygote.

Follicle Maturation

At puberty, the anterior pituitary gland secretes increased amounts of FSH, and the ovaries enlarge in response. At the same time, some of the primordial follicles mature into **primary follicles.**

During maturation, the oocyte of a primary follicle enlarges, and surrounding follicular cells proliferate by mitosis. These follicular cells become organized into layers, and soon, a cavity appears in the cellular mass. A clear *follicular fluid* fills the cavity and bathes the oocyte. The enlarging fluid-filled cavity presses the oocyte to one side. In time, the follicle reaches a diameter of 10 millimeters or more and bulges outward on the ovary surface like a blister (fig. 19.9).

The oocyte within the mature follicle is a large, spherical cell, and a membrane (zona pellucida) and follicular cells (corona radiata) surround and enclose it. Processes from the follicular cells extend through the zona pellucida and supply the oocyte with nutrients.

As many as twenty primary follicles may begin maturing at any one time, but one follicle usually outgrows the others. Typically, only the dominant follicle fully develops, and the other follicles degenerate (fig. 19.10).

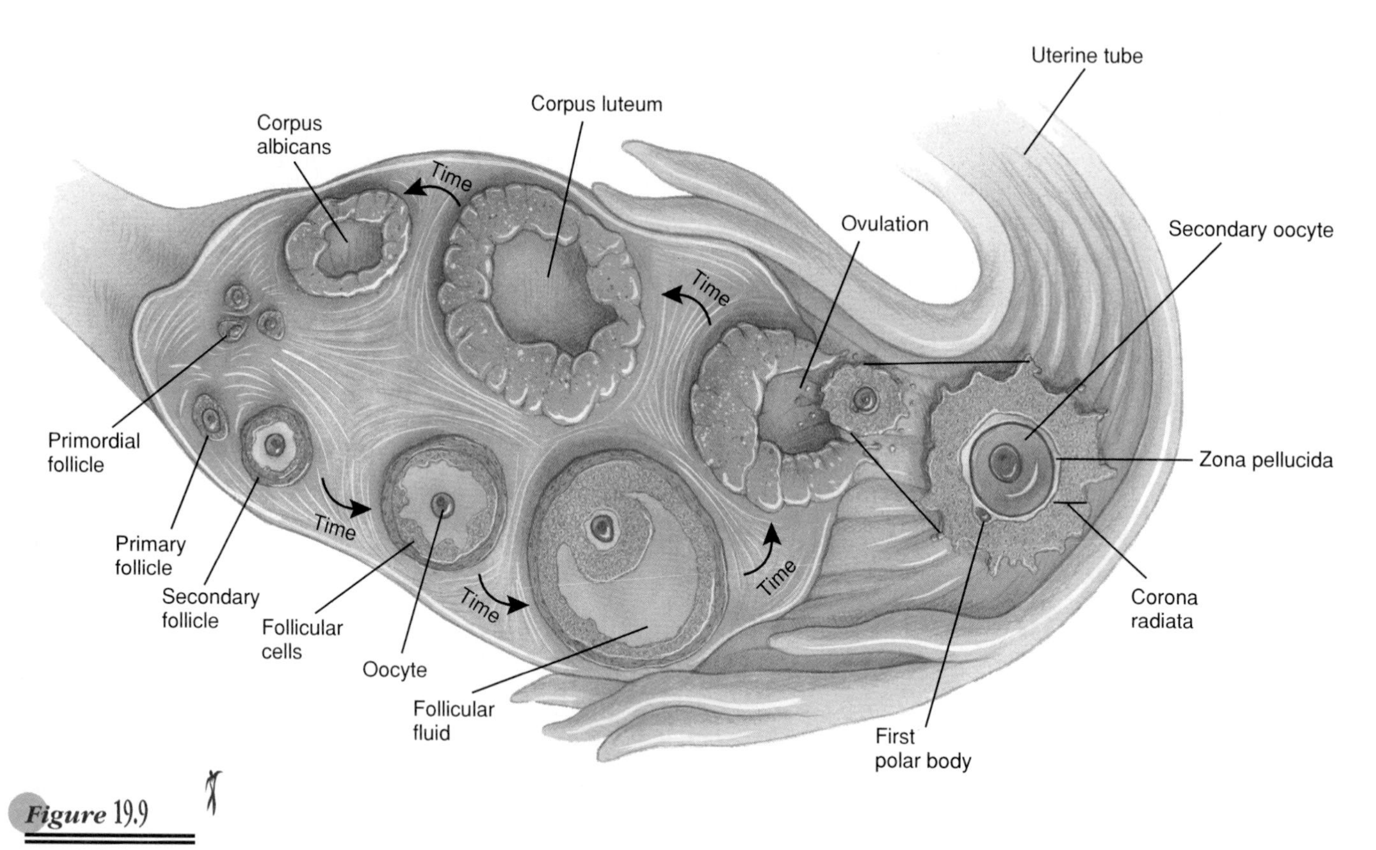

Figure 19.9

As a follicle matures, the egg cell enlarges, and follicular cells and fluid surround it. Eventually, the mature follicle ruptures, releasing the egg cell.

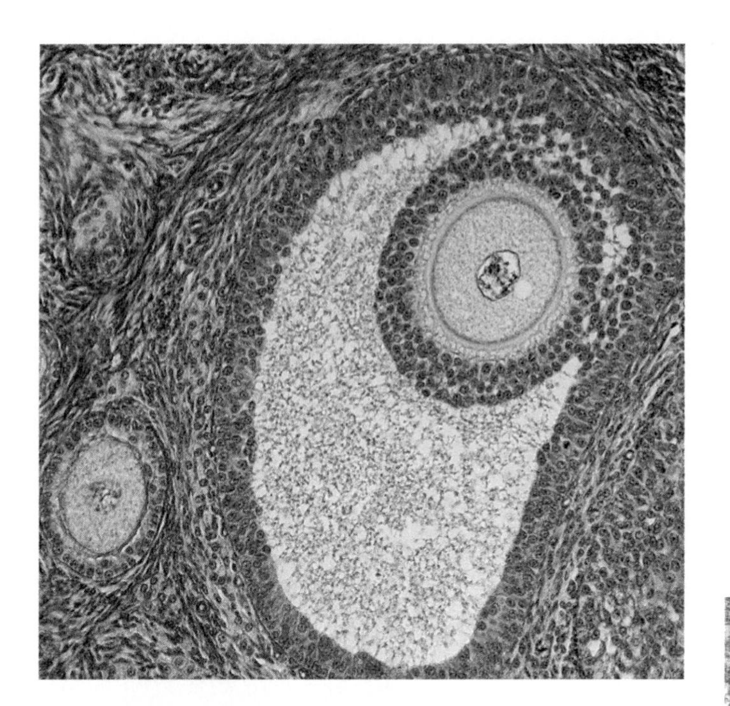

Figure 19.10
Light micrograph of a maturing follicle (250×).

Ovulation

As a follicle matures, its primary oocyte undergoes oogenesis, giving rise to a secondary oocyte and a first polar body. The process of **ovulation** releases these cells from the follicle.

Hormones from the anterior pituitary stimulate ovulation, causing the mature follicle to swell rapidly and its wall to weaken. Eventually, the wall ruptures, and follicular fluid and the secondary oocyte ooze from the ovary's surface (see fig. 19.9).

After ovulation, the secondary oocyte and one or two layers of follicular cells surrounding it are usually propelled to the opening of a nearby uterine tube (fig. 19.11). If the oocyte is not fertilized within a relatively short time, it degenerates.

1. What changes occur in a follicle and its oocyte during maturation?
2. What causes ovulation?
3. What happens to an oocyte following ovulation?

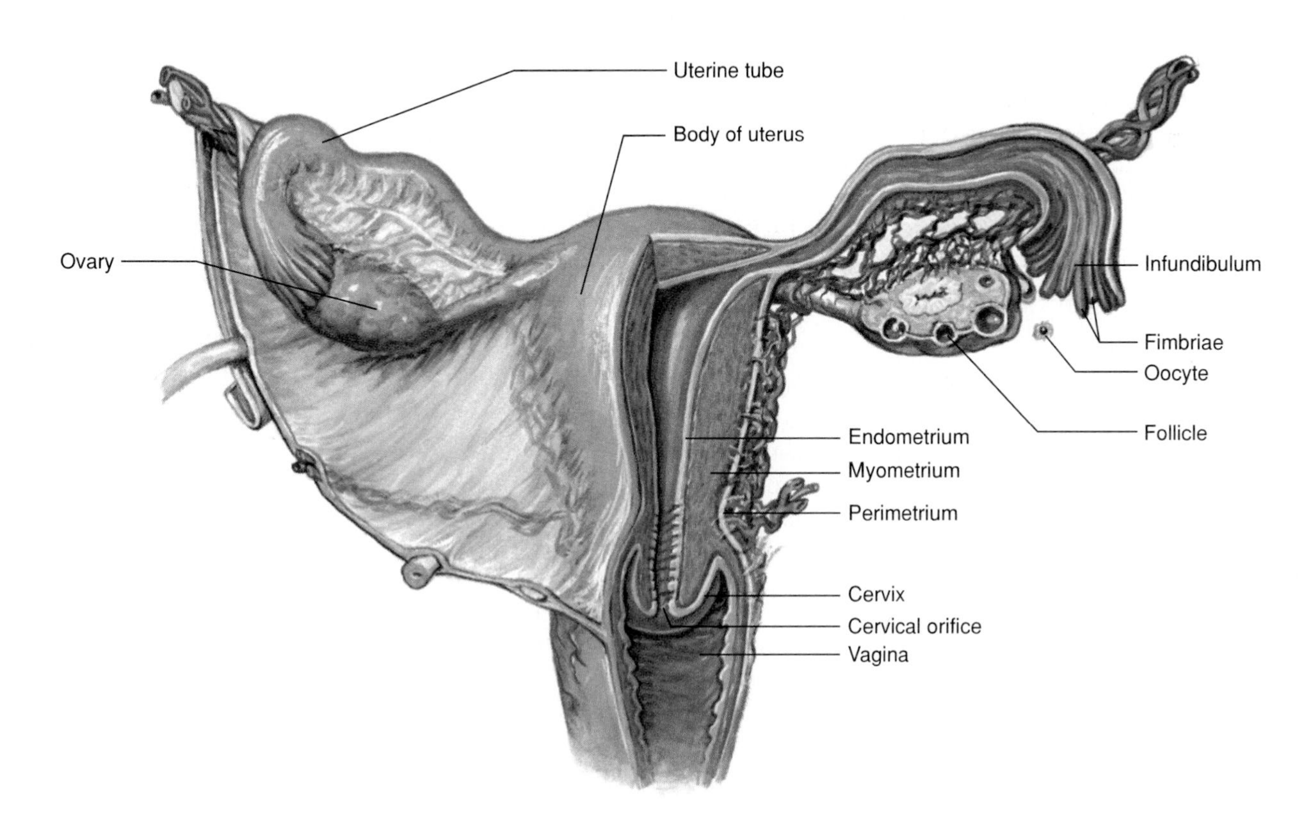

Figure 19.11
The funnel-shaped infundibulum of the uterine tube partially encircles the ovary (posterior view).

Female Internal Accessory Organs

The internal accessory organs of the female reproductive system include a pair of uterine tubes, a uterus, and a vagina.

Uterine Tubes

The **uterine tubes** (fallopian tubes or oviducts) open near the ovaries (fig. 19.11). Each tube is about 10 centimeters long and passes medially to the uterus, penetrates its wall, and opens into the uterine cavity.

Near each ovary, a uterine tube expands to form a funnel-shaped *infundibulum,* which partially encircles the ovary medially. Fingerlike extensions called *fimbriae* fringe the infundibulum margin. Although the infundibulum generally does not touch the ovary, one of the larger extensions connects directly to the ovary.

Simple columnar epithelial cells, some of which are *ciliated,* line the uterine tube. The epithelium secretes mucus, and the cilia beat toward the uterus. These actions help draw the egg cell and expelled follicular fluid into the infundibulum following ovulation. Ciliary action and peristaltic contractions of the uterine tube's muscular layer help transport the egg cell down the uterine tube.

Uterus

If the egg cell is fertilized in the uterine tube, the **uterus** receives the embryo and sustains its development. The uterus is a hollow, muscular organ shaped somewhat like an inverted pear (fig. 19.11).

Uterus size changes greatly during pregnancy. In its nonpregnant, adult state, the uterus is about 7 centimeters long, 5 centimeters wide (at its broadest point), and 2.5 centimeters in diameter. It is located medially within the anterior portion of the pelvic cavity, above the vagina, and usually bends forward over the urinary bladder.

The upper two-thirds, or *body,* of the uterus has a dome-shaped top. The uterine tubes enter the top of the uterus at its broadest part. The lower third of the uterus is called the **cervix.** This tubular part extends downward into the upper portion of the vagina. The cervix surrounds the opening called the *cervical orifice,* through which the uterus opens to the vagina.

The uterine wall is thick and has three layers. The **endometrium,** the inner mucosal layer, is covered with columnar epithelium and contains abundant tubular glands. The **myometrium,** a thick, middle, muscular layer, consists largely of bundles of smooth muscle fibers. During the monthly female reproductive cycles and during pregnancy, the endometrium and myometrium change extensively. The **perimetrium** is an outer serosal layer that covers the body of the uterus and part of the cervix (fig. 19.12).

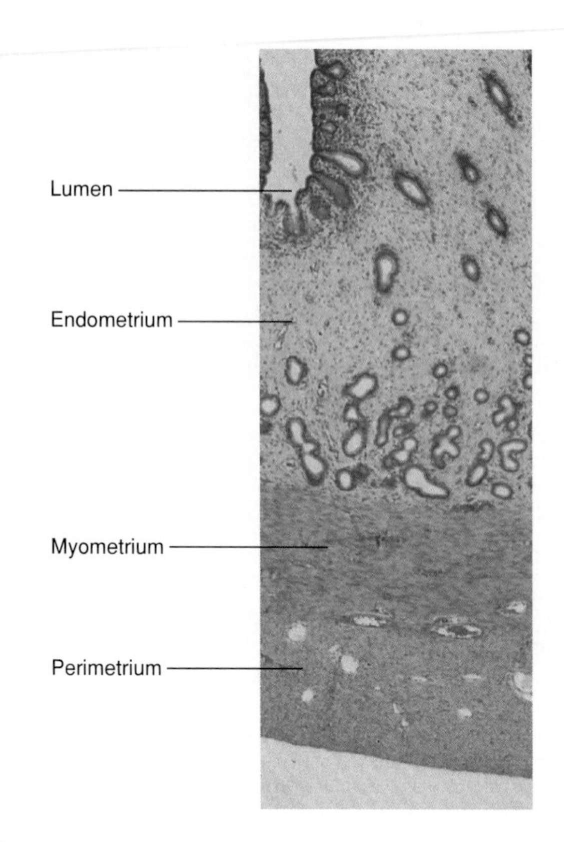

Figure 19.12

Light micrograph of the uterine wall (10.5×).

A procedure called the *Pap* (Papanicolaou) *smear test* can usually detect cancer of the cervix. A doctor or nurse scrapes off a tiny sample of cervical tissue, smears the sample on a glass slide, stains it, and sends it to a laboratory, where cytotechnologists examine it for the presence of abnormal cells. Cervical cancer detected and treated early has a high cure rate. ■

Vagina

The **vagina** is a fibromuscular tube, about 9 centimeters long, extending from the uterus to the outside (see fig. 19.7). It conveys uterine secretions, receives the erect penis during sexual intercourse, and provides an open channel for offspring during birth.

The vagina extends upward and back into the pelvic cavity. It is posterior to the urinary bladder and urethra, and anterior to the rectum, and attaches to these structures by connective tissues.

A thin membrane of connective tissue and stratified squamous epithelium called the **hymen** partially closes the *vaginal orifice*. A central opening of varying size allows uterine and vaginal secretions to pass to the outside.

The vaginal wall has three layers. The inner *mucosal layer* is stratified squamous epithelium. This layer lacks mucous glands; the mucus in the lumen of the vagina comes from uterine glands and from vestibular glands at the mouth of the vagina.

The middle *muscular layer* consists mainly of smooth muscle fibers. A thin band of striated muscle at the lower end of the vagina helps close the vaginal opening. A voluntary muscle (bulbospongiosus), however, is primarily responsible for closing this orifice.

The outer *fibrous layer* consists of dense fibrous connective tissue interlaced with elastic fibers. It attaches the vagina to surrounding organs.

1. How is an egg cell moved along a uterine tube?
2. Describe the structure of the uterus.
3. Describe the structure of the vagina.

Female External Reproductive Organs

The external accessory organs of the female reproductive system include the labia majora, labia minora, clitoris, and vestibular glands. These structures surround the openings of the urethra and vagina, and compose the **vulva** (see fig. 19.7).

Labia Majora

The **labia majora** (singular, *labium majus*) enclose and protect the other external reproductive organs. They correspond to the scrotum, and are composed of rounded folds of adipose tissue and a thin layer of smooth muscle, covered by skin.

The labia majora lie closely together. A cleft that includes the urethral and vaginal openings separates the labia longitudinally. At their anterior ends, the labia merge to form a medial, rounded elevation of adipose tissue called the *mons pubis,* which overlies the symphysis pubis (see fig. 19.7).

Labia Minora

The **labia minora** (singular, *labium minus*) are flattened, longitudinal folds between the labia majora (see fig. 19.7). They are composed of connective tissue richly supplied with blood vessels, giving a pinkish appearance. Posteriorly, the labia minora merge with the labia majora, while anteriorly, they converge to form a hoodlike covering around the clitoris.

Clitoris

The **clitoris** is a small projection at the anterior end of the vulva between the labia minora (see figure 19.7). It is usually about 2 centimeters long and 0.5 centimeters in diameter, including a portion embedded in surrounding tissues. The clitoris corresponds to the penis and has a similar structure. It is composed of two columns of erectile tissue called *corpora cavernosa*. At its anterior end, a small mass of erectile tissue forms a **glans** richly supplied with sensory nerve fibers.

Vestibule

The labia minora enclose a space called the **vestibule.** The vagina opens into the posterior portion of the vestibule, and the urethra opens in the midline, just in front of the vagina and about 2.5 centimeters behind the glans of the clitoris.

A pair of **vestibular glands**, corresponding to the bulbourethral glands, lie one on either side of the vaginal opening. Beneath the mucosa of the vestibule on either side is a mass of vascular erectile tissue called the **vestibular bulb.**

1. What is the male counterpart of the labia majora? Of the clitoris?
2. What structures are within the vestibule?

Erection, Lubrication, and Orgasm

Erectile tissues in the clitoris and around the vaginal entrance respond to sexual stimulation. Following such stimulation, parasympathetic nerve impulses pass out from the sacral portion of the spinal cord, dilating the arteries associated with the erectile tissues. As a result, blood inflow increases, and the erectile tissues swell. At the same time, the vagina expands and elongates.

If sexual stimulation is sufficiently intense, parasympathetic impulses stimulate the vestibular glands to secrete mucus into the vestibule. This moistens and lubricates the tissues surrounding the vestibule and the lower end of the vagina, facilitating insertion of the penis into the vagina.

The clitoris is abundantly supplied with sensory nerve fibers, which are especially sensitive to local stimulation. The culmination of such stimulation is orgasm.

Just prior to orgasm, the tissues of the outer third of the vagina engorge with blood and swell. This increases the friction on the penis during intercourse. Orgasm initiates a series of reflexes involving the sacral and lumbar portions of the spinal cord. In response to these reflexes, the muscles of the perineum and the

Table 19.2 Functions of the Female Reproductive Organs

Organ	Function
Ovary	Produces egg cells and female sex hormones
Uterine tube	Conveys egg cell toward uterus; site of fertilization; conveys developing embryo to uterus
Uterus	Protects and sustains embryo during pregnancy
Vagina	Conveys uterine secretions to outside of body; receives erect penis during sexual intercourse; provides open channel for offspring during birth process
Labia majora	Enclose and protect other external reproductive organs
Labia minora	Form margins of vestibule; protect openings of vagina and urethra
Clitoris	Produces feelings of pleasure during sexual stimulation due to abundant sensory nerve endings in glans
Vestibule	Space between labia minora that contains vaginal and urethral openings
Vestibular glands	Secrete fluid that moistens and lubricates vestibule

walls of the uterus and uterine tubes contract rhythmically. These contractions help transport sperm cells through the female reproductive tract toward the upper ends of the uterine tubes.

Table 19.2 summarizes the functions of the female reproductive organs.

1. What events result from parasympathetic stimulation of the female reproductive organs?
2. What changes occur in the vagina just prior to and during orgasm?

Hormonal Control of Female Reproductive Functions

The hypothalamus, anterior pituitary gland, and ovaries secrete hormones that control development and maintenance of female secondary sex characteristics, maturation of female sex cells, and changes during the monthly reproductive cycle.

Female Sex Hormones

The female body is reproductively immature until about age ten. Then, the hypothalamus begins to secrete increasing amounts of GnRH. GnRH, in turn, stimulates the anterior pituitary to release the gonadotropins FSH and LH. These hormones play primary roles in controlling female sex cell maturation and in producing female sex hormones.

Several tissues, including the ovaries, the adrenal cortices, and the placenta (during pregnancy), secrete female sex hormones belonging to two major groups—**estrogen** and **progesterone.** The ovaries are the primary source of estrogen (in a nonpregnant female). At puberty, under the influence of the anterior pituitary, the ovaries secrete increasing amounts of estrogen. Estrogen stimulates enlargement of accessory organs, including the vagina, uterus, uterine tubes, ovaries, and external reproductive structures. Estrogen also develops and maintains the *female secondary sex characteristics,* which include:

1. Development of the breasts and the ductile system of the mammary glands within the breasts.
2. Increased deposition of adipose tissue in the subcutaneous layer generally and in the breasts, thighs, and buttocks particularly.
3. Increased vascularization of the skin.

The ovaries are also the primary source of progesterone (in a nonpregnant female). This hormone promotes changes in the uterus during the female reproductive cycle, affects the mammary glands, and helps regulate the anterior pituitary's secretion of gonadotropins.

Androgen (male sex hormone) concentrations produce certain other changes in females at puberty. For example, increased hair growth in the pubic and axillary regions is due to androgen the adrenal cortices secrete. Conversely, development of the female skeletal configuration, which includes narrow shoulders and broad hips, is a consequence of low androgen concentrations.

1. What factors initiate sexual maturation in a female?
2. What is the function of estrogen?
3. What is the function of androgen in a female?

Female Reproductive Cycle

The female reproductive cycle, or **menstrual cycle,** consists of regular, recurring changes in the uterine lining, which culminate in menstrual bleeding (menses).

Such cycles usually begin around age thirteen, continue into middle age, and then cease.

Women athletes may have disturbed menstrual cycles, ranging from diminished menstrual flow (oligomenorrhea) to complete stoppage (amenorrhea). The more active an athlete, the more likely are menstrual problems. This effect results from a loss of adipose tissue and a consequent decline in estrogen, which adipose tissue synthesizes from adrenal androgens. ■

A female's first menstrual cycle (menarche) occurs after the ovaries and other organs of the reproductive control system have matured and respond to certain hormones. Then, hypothalamic secretion of GnRH stimulates the anterior pituitary to release threshold levels of FSH and LH. FSH stimulates maturation of an ovarian follicle. The follicular cells produce increasing amounts of estrogen and some progesterone. LH stimulates certain ovarian cells to secrete precursor molecules (testosterone) used to produce estrogen.

In a young female, estrogen stimulates development of secondary sex characteristics. Estrogen secreted during subsequent menstrual cycles continues development and maintenance of these traits.

Increasing estrogen concentration during the first week or so of a menstrual cycle changes the uterine lining, thickening the glandular endometrium (proliferative phase). Meanwhile, the developing follicle completes maturation, and by the fourteenth day of the cycle, the follicle appears on the ovary surface as a blisterlike bulge.

Within the follicle, the follicular cells, which surround and connect the oocyte to the inner wall, loosen. Follicular fluid accumulates rapidly.

While the follicle matured, estrogen that it secreted inhibited the anterior pituitary's release of LH, but allowed LH to be stored in the gland. Estrogen also made anterior pituitary cells more sensitive to the action of GnRH, which the hypothalamus released in rhythmic pulses about 90 minutes apart.

Near the fourteenth day of follicular development, anterior pituitary cells finally release the stored LH in response to the GnRH pulses. The resulting surge in LH concentration, which lasts about 36 hours, weakens and ruptures the bulging follicular wall, which sends the oocyte and follicular fluid from the ovary (ovulation).

Following ovulation, the remnants of the follicle within the ovary rapidly change. The space that the follicular fluid occupied fills with blood, which soon clots, and under the influence of LH, follicular cells enlarge to form a temporary glandular structure called a **corpus luteum** ("yellow body").

Follicular cells secrete progesterone during the first part of the menstrual cycle. Corpus luteum cells secrete abundant progesterone and estrogen during the last half of the cycle. Consequently, as a corpus luteum forms, blood progesterone concentration increases sharply.

Progesterone causes the uterine endometrium to become more vascular and glandular. It also stimulates uterine glands to secrete more glycogen and lipids (secretory phase). As a result, endometrial tissues fill with fluids containing nutrients and electrolytes, providing a favorable environment for embryo development.

Estrogen and progesterone inhibit the anterior pituitary's release of LH and FSH. Consequently, no other follicles are stimulated to develop when the corpus luteum is active. If a sperm cell does not fertilize the oocyte released at ovulation, however, the corpus luteum begins to degenerate on about the twenty-fourth day of the cycle. Eventually, fibrous connective tissue replaces it. The remnant of such a corpus luteum is called a *corpus albicans*.

When the corpus luteum ceases to function, estrogen and progesterone concentrations decline rapidly, and blood vessels in the endometrium constrict in response. This reduces the supply of oxygen and nutrients to the thickened uterine lining, and these lining tissues soon disintegrate and slough off. At the same time, blood escapes from damaged capillaries, creating a flow of blood and cellular debris that passes through the vagina as the *menstrual flow* (menses). This flow usually begins about the twenty-eighth day of the cycle and continues for three to five days while the estrogen concentration is low.

The menstrual flow marks the end of a menstrual cycle and the beginning of a new cycle. Figure 19.13 graphs this cycle, and table 19.3 summarizes it.

Low blood concentrations of estrogen and progesterone at the beginning of the menstrual cycle mean that the hypothalamus and anterior pituitary are no longer inhibited. Consequently, FSH and LH concentrations soon increase, stimulating a new follicle to mature. As this follicle secretes estrogen, the uterine lining undergoes repair, and the endometrium begins to thicken again.

Menopause

After puberty, menstrual cycles continue at regular intervals into the late forties or early fifties, when they become increasingly irregular. Then in a few months or years, the cycles cease altogether. This period in life is called **menopause.**

Aging of the ovaries causes menopause. After about thirty-five years of cycling, few primary follicles remain for pituitary gonadotropins to stimulate. Consequently, the follicles no longer mature, ovulation does not occur, and blood estrogen concentration plummets.

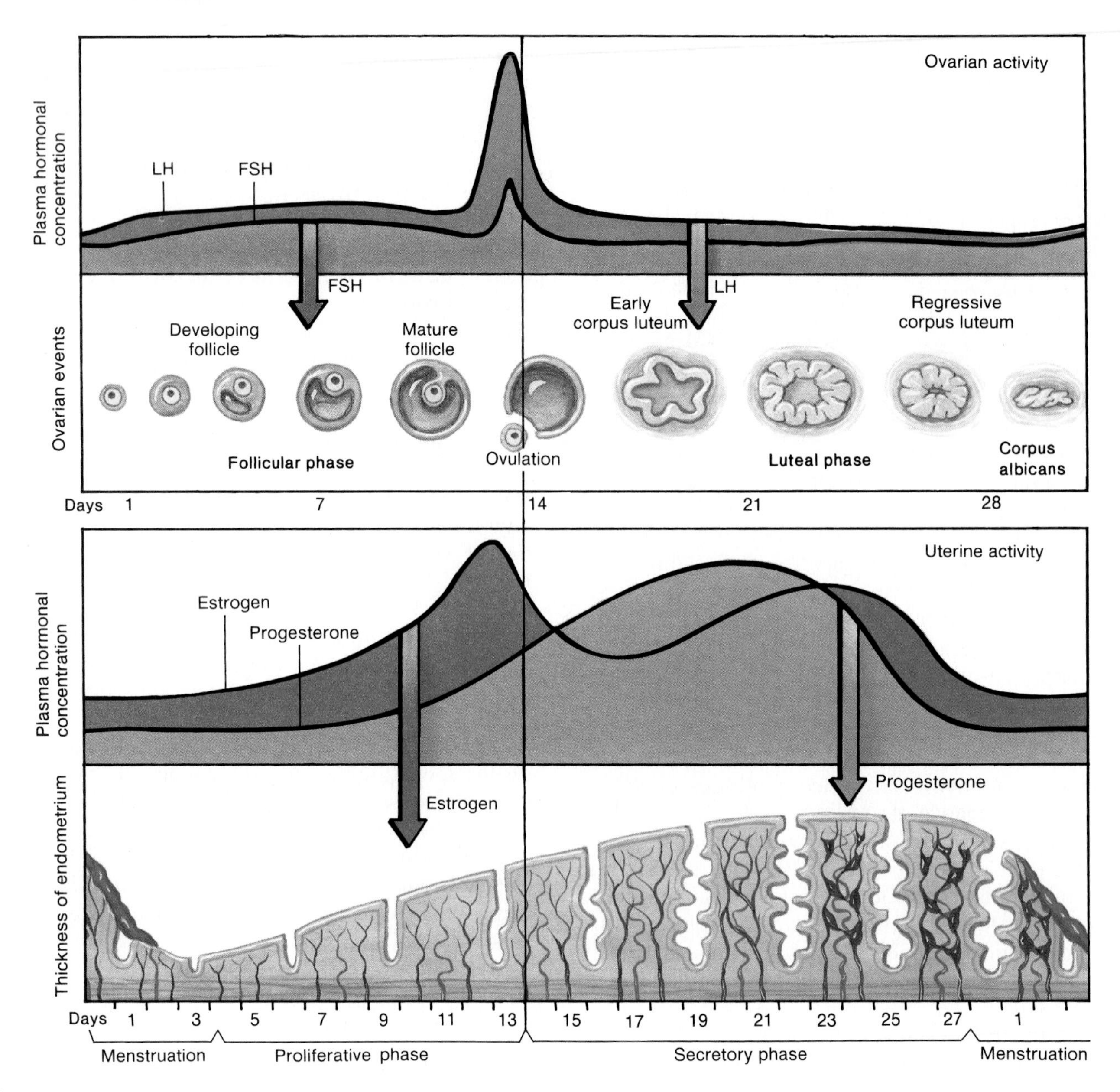

Figure 19.13

Major events in the ovarian and menstrual cycles.

Reduced estrogen concentration and lack of progesterone may change the female secondary sex characteristics. The breasts, vagina, uterus, and uterine tubes may shrink, and the pubic and axillary hair may thin.

1. Trace the events of the female menstrual cycle.
2. What causes menstrual flow?
3. What are some changes that may occur at menopause?

Mammary Glands

The **mammary glands** are accessory organs of the female reproductive system that are specialized to secrete milk following pregnancy (fig. 19.14 and reference plate 1). They are in the subcutaneous tissue of the anterior thorax within elevations called *breasts*. The breasts overlie the *pectoralis major* muscles and extend from the second to the sixth ribs and from the sternum to the axillae.

Table 19.3 Major Events in a Menstrual Cycle

1. Anterior pituitary gland secretes follicle-stimulating hormone (FSH) and luteinizing hormone (LH).
2. FSH stimulates maturation of a follicle.
3. Follicular cells produce and secrete estrogen.
   a. Estrogen maintains secondary sex traits.
   b. Estrogen causes uterine lining to thicken.
4. Anterior pituitary releases a surge of LH, which stimulates ovulation.
5. Follicular cells become corpus luteum cells, which secrete estrogen and progesterone.
   a. Estrogen continues to stimulate uterine wall development.
   b. Progesterone stimulates uterine lining to become more glandular and vascular.
   c. Estrogen and progesterone inhibit anterior pituitary from secreting LH and FSH.
6. If egg cell is not fertilized, corpus luteum degenerates and no longer secretes estrogen and progesterone.
7. As concentrations of estrogen and progesterone decline, blood vessels in uterine lining constrict.
8. Uterine lining disintegrates and sloughs off, producing menstrual flow.
9. Anterior pituitary is no longer inhibited and again secretes FSH and LH.
10. The menstrual cycle repeats.

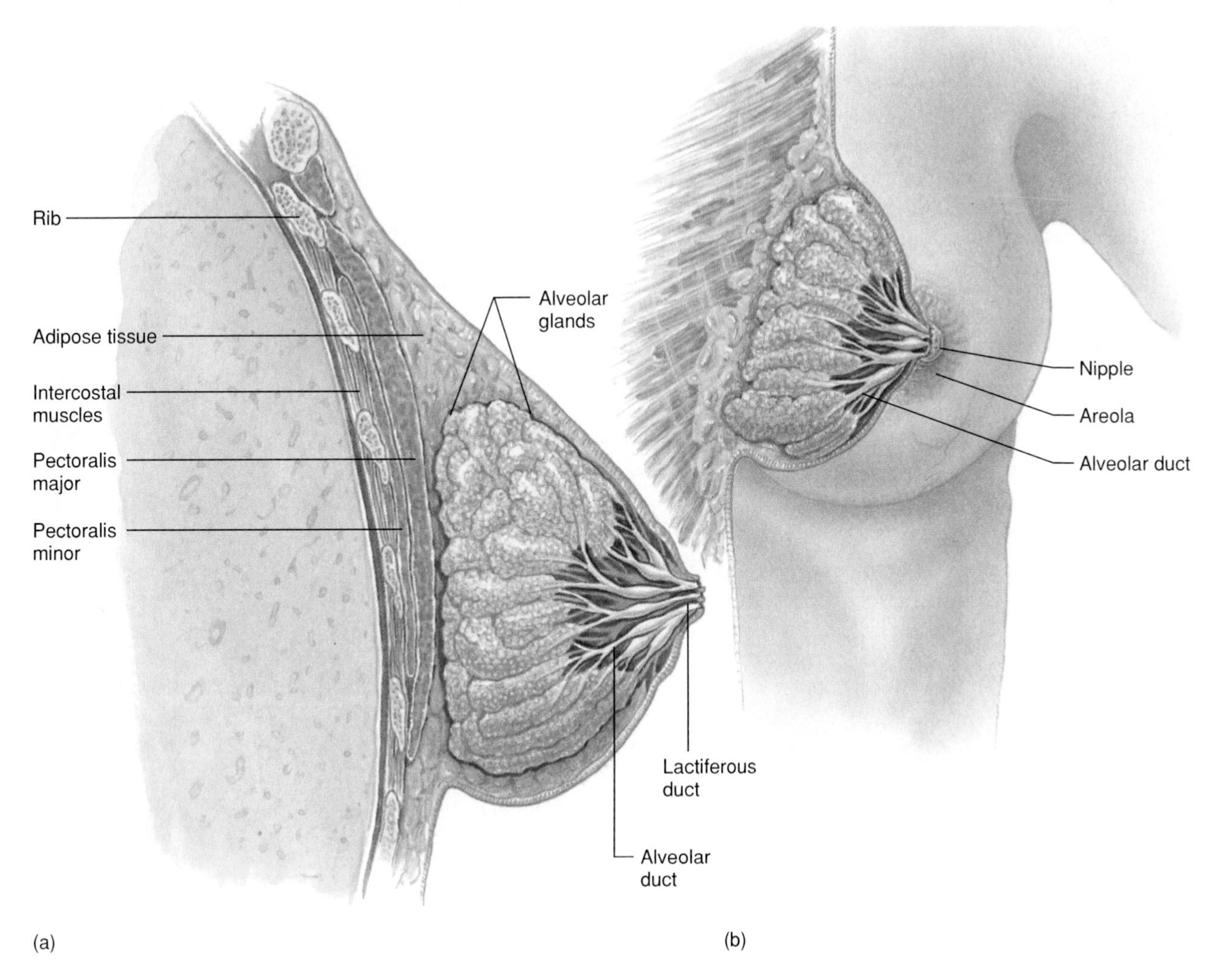

Figure 19.14

The mammary gland is located within the breast. (*a*) Sagittal section of the breast. (*b*) Anterior view.

TOPIC OF INTEREST Breast Cancer Update

Finding the Lump

Few discoveries are as terrifying to a woman as finding a lump in her breast. The woman may first notice a small area of thickening, a dimple, a change in contour, or a nipple that is flatter than usual, points in an unusual direction, or produces a discharge.

The next step is a physical exam. The doctor palpates the breast and performs a *mammogram,* an X-ray scan that pinpoints the location and approximate extent of abnormal tissue (fig. 19B). An ultrasound scan can distinguish between a cyst and a tumor. If an area is suspicious, the next step is a biopsy, in which a very thin needle samples the affected tissue.

Eighty percent of the time, a breast lump is a sign of fibrocystic breast disease, which is benign. The lump may be a fluid-filled sac of glandular tissue (a cyst) or a solid, fibrous mass of connective tissue (a fibroadenoma). Treatment includes taking vitamin E or synthetic androgens under a doctor's care, lowering caffeine intake, and examining unusual lumps further.

Treatment

If biopsied breast cells are cancerous, treatment is usually surgical. A *lumpectomy* removes a small tumor and some surrounding tissue; a *modified mastectomy* removes an entire breast; and a *radical mastectomy* removes the breast and surrounding lymph and muscle tissue. If abnormal tissue extends to nearby lymph nodes, chemotherapy and possibly radiation therapy follows surgery. Chemotherapy is also indicated if a bone scan or other test reveals that the cancer has spread.

Causes of Breast Cancer

Experimental genetic tests can tell whether a woman with a strong family history of breast cancer has a high likelihood of developing an inherited form of the illness. Ten percent of all breast cancers are inherited, and researchers are searching for the environmental triggers of the remaining 90%. One candidate is prolonged estrogen exposure. Researchers first recognized the estrogen link in the 1970s, after noting that young women whose ovaries had been removed rarely developed breast cancer. Lack of estrogen apparently protected these women against breast cancer.

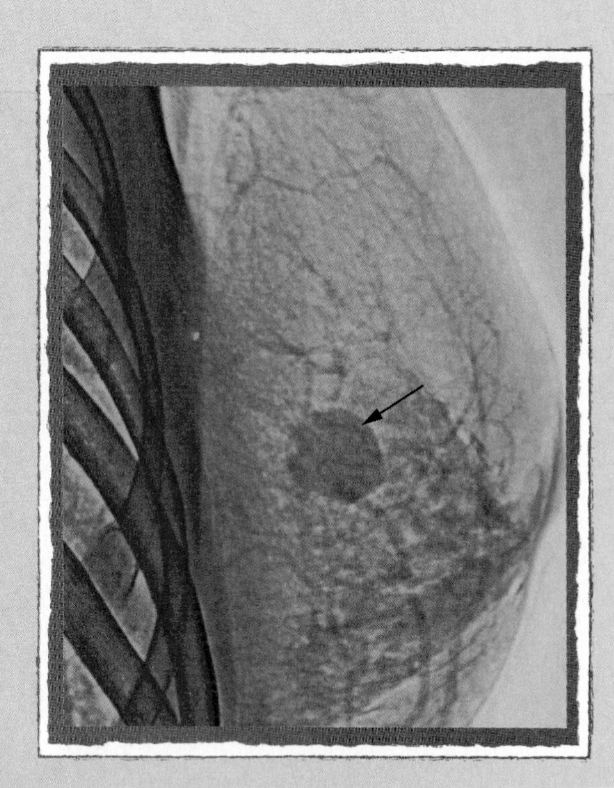

Figure 19B

Mammogram of a breast with a tumor (*arrow*).

Prolonged estrogen exposure can occur in a variety of ways:

- Early menarche (first period) and late menopause (last period).
- Exposure to pesticide residues and other pollutants. These contaminants stimulate cell division like estrogens do. Wildlife populations exposed to environmental estrogens demonstrate reproductive problems, such as infertility, undersized genitalia, and thin eggshells.
- Having no children, or having a first child after the age of thirty.
- Not breast-feeding. Women who do not breast-feed have a higher breast cancer risk than women who do.

Breast Cancer Statistics

Nearly 3 million women in the United States currently have breast cancer. People frequently misunderstand the oft-quoted figure that a woman's lifetime risk of developing breast cancer is one in eight, believing that at any given time, one in eight U.S. women has breast cancer. This figure, however, refers to the lifetime risk for women who live to be ninety-five (table 19C).

Statistics indicate that breast cancer is on the rise. The lifetime risk of one in eight was one in sixteen in the 1940s. It has increased 1% a year since then, and 4% a year since 1987. At least some of that rise is due to more wide reaching and earlier diagnosis. The illness is the second most common cancer in women after lung cancer and causes 46,000 deaths in the United States each year.

Prevention

Health agencies advise women to have baseline mammograms by the age of forty and yearly mammograms after age 40. Women with a family history of breast cancer may benefit from having earlier mammograms. A mammogram may spot a tumor two years before a woman can feel it.

Table 19.C **Breast Cancer Risk**

By age	Odds	By age	Odds
25	1 in 19,608	60	1 in 24
30	1 in 2,525	65	1 in 17
35	1 in 622	70	1 in 14
40	1 in 217	75	1 in 11
45	1 in 93	80	1 in 10
50	1 in 50	85	1 in 9
55	1 in 33	95 or older	1 in 8

A *nipple* is located near the tip of each breast at about the level of the fourth intercostal space. A circular area of pigmented skin, called the *areola,* surrounds each nipple.

A mammary gland is composed of fifteen to twenty lobes. Each lobe contains glands (alveolar glands) and an alveolar duct (lactiferous duct) that leads to the nipple and opens to the outside. Dense connective and adipose tissues separate the lobes. These tissues also support the glands and attach them to the fascia of the underlying pectoral muscles. Other connective tissue, which forms dense strands called *suspensory ligaments,* extends inward from the dermis of the breast to the fascia, helping to support the breast's weight.

The mammary glands of males and females are similar. As children reach puberty, the glands in males do not develop, whereas in females, ovarian hormones stimulate development of the glands. As a result, the alveolar glands and ducts enlarge, and fat forms deposits around and within the breasts.

Chapter 20 describes the hormonal mechanism that stimulates mammary glands to produce and secrete milk.

1. Describe the structure of a mammary gland.
2. What changes do ovarian hormones cause in mammary glands?

Birth Control

Birth control is voluntary regulation of the number of offspring produced and when they are conceived. This control requires a method of **contraception** to avoid fertilization of an egg cell following sexual intercourse (coitus) or to prevent the hollow ball of cells that will develop into an embryo (a blastocyst) from nestling (implanting) in the uterine wall.

Coitus Interruptus

Coitus interruptus is withdrawing the penis from the vagina before ejaculation, which prevents entry of sperm cells into the female reproductive tract. This method may result in pregnancy because a male may find it difficult to withdraw just prior to ejaculation. Also, small quantities of semen containing sperm cells may reach the vagina before ejaculation.

Rhythm Method

The *rhythm method* (also called timed coitus or natural family planning) requires abstinence from sexual intercourse a few days before and a few days after ovulation. The rhythm method results in a high rate of pregnancy because accurately identifying infertile times to have intercourse is difficult. Another disadvantage of the rhythm method is that it restricts spontaneity in sexual activity.

Mechanical Barriers

Mechanical barriers prevent sperm cells from entering the female reproductive tract during sexual intercourse. The *male condom* consists of a thin latex or natural membrane sheath placed over the erect penis before intercourse to prevent semen from entering the vagina upon ejaculation (fig. 19.15*a*).

A recently introduced *female condom* resembles a small plastic bag. A woman inserts it into her vagina prior to intercourse to block sperm cells from reaching the cervix.

Condoms are inexpensive and may also protect against sexually transmitted diseases. However, a male condom may decrease sensitivity of the penis during intercourse, and its use interrupts the sex act.

Another mechanical barrier is the *diaphragm,* a cup-shaped device with a flexible ring forming the rim. A woman inserts the diaphragm into the vagina so that it covers the cervix, preventing entry of sperm cells into the uterus (fig. 19.15*b*). To be effective, a diaphragm must be fitted for size by a physician, inserted properly, and used with a chemical spermicide applied to the surface adjacent to the cervix and to the rim of the diaphragm. The device must be left in position for several hours following sexual intercourse. A diaphragm can be inserted up to 6 hours prior to sexual contact.

Similar to but smaller than the diaphragm is the *cervical cap,* which adheres to the cervix by suction. A woman inserts it with her fingers before intercourse. Different cultures have used cervical caps, made of such varied substances as beeswax, lemon halves, paper, and opium poppy fibers, for centuries.

Chemical Barriers

Chemical barrier contraceptives include creams, foams, and jellies with spermicidal properties. Within the vagina, such chemicals create an unfavorable environment for sperm cells (fig. 19.15*c*).

Chemical barrier contraceptives are fairly easy to use but have a high failure rate when used alone. They are most effective when used with a diaphragm.

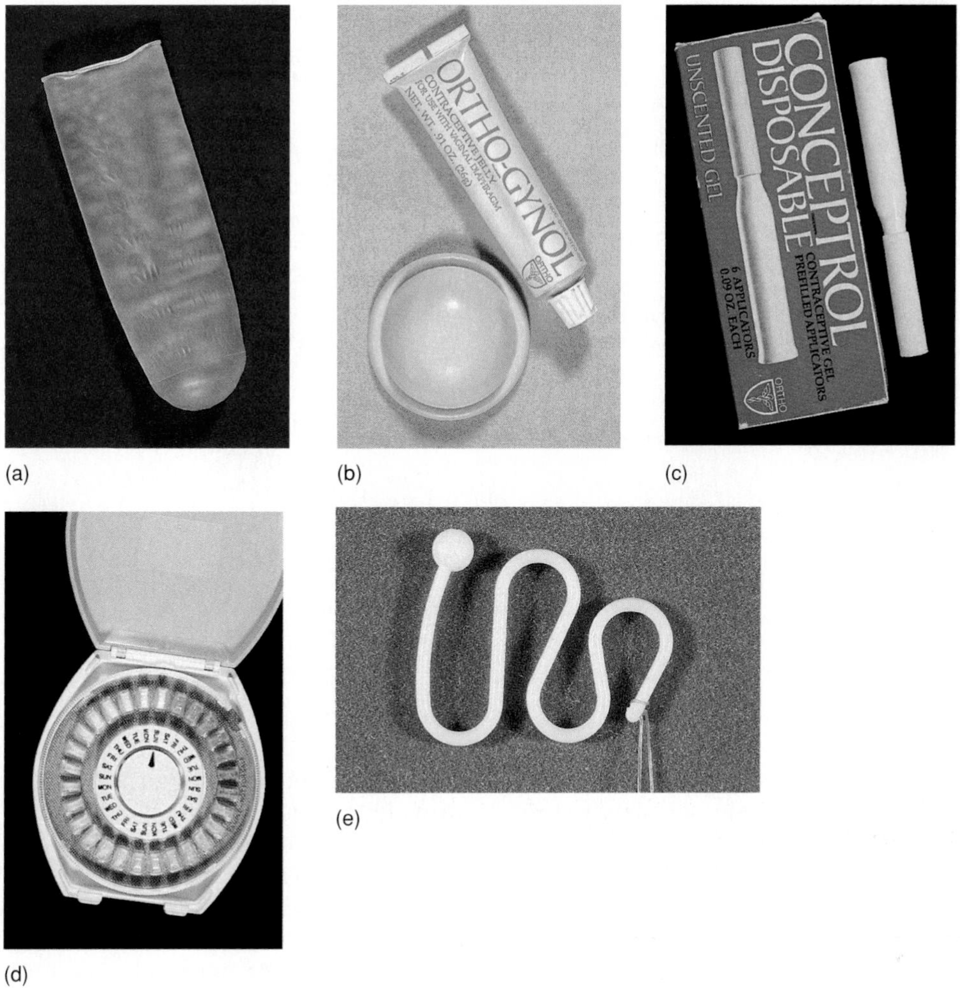

(a)

(b)

(c)

(d)

(e)

Figure 19.15

Devices and substances used for birth control. (*a*) Condom. (*b*) Diaphragm. (*c*) Spermicidal gel. (*d*) Oral contraceptives. (*e*) IUD. (Photographs are not to scale.)

Oral Contraceptives

An *oral contraceptive,* or birth control pill, contains synthetic estrogen-like and progesterone-like chemicals. In women, these drugs disrupt the normal pattern of gonadotropin secretion and prevent the LH surge that triggers ovulation. They also interfere with buildup of the uterine lining necessary for implantation (fig. 19.15*d*).

Oral contraceptives, if used correctly, prevent pregnancy nearly 100% of the time. They may cause nausea, retention of body fluids, increased skin pigmentation, and breast tenderness, however. Also, some women, particularly those over age thirty-five who smoke, may develop intravascular blood clots, liver disorders, or high blood pressure when using certain types of oral contraceptives.

Injectable Contraception

An intramuscular injection of Depo-Provera (medroxyprogesterone acetate) protects against pregnancy for three months by preventing maturation and release of a secondary oocyte. It also alters the uterine lining, making it less hospitable for a blastocyst. Depo-Provera is long-acting—it takes ten to eighteen months after the last injection for the effects to wear off. Use of Depo-Provera requires a doctor's care because potential side effects make it risky for women with certain medical conditions.

Contraceptive Implants

A *contraceptive implant* is a set of small progesterone-containing capsules or rods inserted surgically under

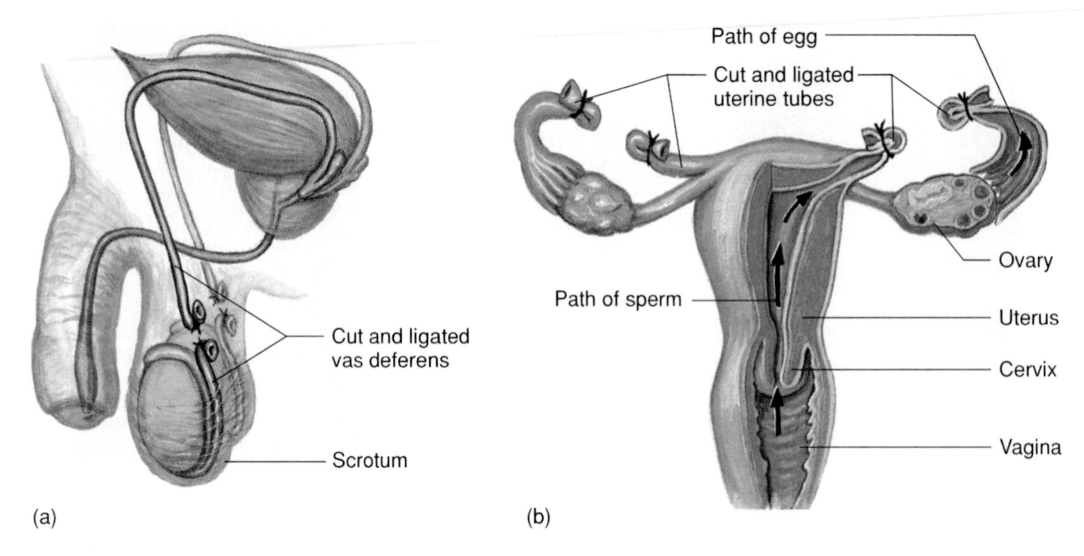

Figure 19.16

(*a*) Vasectomy cuts and ligates each vas deferens. (*b*) Tubal ligation cuts and ligates each uterine tube.

the skin of a woman's arm or scapular region. The progesterone, which the implant releases slowly, prevents ovulation in much the same way as do oral contraceptives. A contraceptive implant is effective for up to five years, and removal of the implant reverses its contraceptive action.

Intrauterine Devices

An *intrauterine device,* or *IUD,* is a small, solid object that a physician places within the uterine cavity. An IUD interferes with implantation of a blastocyst, perhaps by inflaming uterine tissues (fig. 19.15*e*).

The uterus may spontaneously expel the IUD, or the IUD may produce abdominal pain or excessive menstrual bleeding. It may also injure the uterus or produce other serious health problems. A physician should regularly check IUD placement.

Surgical Methods

Surgical methods of contraception sterilize the male or female. In the male, a physician performs a *vasectomy,* removing a small section of each vas deferens near the epididymis and tying (ligating) the cut ends of the ducts. A vasectomy is a simple operation with few side effects, although it may cause some pain for a week or two.

After a vasectomy, sperm cells cannot leave the epididymis; thus, they are not included in the semen. However, sperm cells may already be present in portions of the ducts distal to the cuts. Consequently, the sperm count may not reach zero for several weeks.

The corresponding procedure in the female is *tubal ligation.* The uterine tubes are cut and tied (ligated) so that sperm cells cannot reach an egg cell.

Neither a vasectomy nor a tubal ligation changes hormonal concentrations or sexual drives. These procedures, shown in figure 19.16, are the most reliable forms of contraception. Reversing them requires microsurgery.

1. What factors make the rhythm method less reliable than some other methods of contraception?
2. Describe two methods of contraception that use mechanical barriers.
3. How do oral contraceptives, injectable contraceptives, and contraceptive implants prevent pregnancy?

Sexually Transmitted Diseases

The twenty recognized **sexually transmitted diseases** (STDs) are often called "silent infections" because the early stages may not produce symptoms, especially in women. By the time symptoms appear, it is often too late to prevent complications or spread of the infection to sexual partners. Many STDs have similar symptoms, some of which are also seen in diseases or allergies that are not sexually related. A physician should be consulted if one or a combination of the following symptoms appears:

1. Burning sensation during urination.
2. Pain in the lower abdomen.
3. Fever or swollen glands in the neck.
4. Discharge from the vagina or penis.
5. Pain, itch, or inflammation in the genital or anal area.
6. Pain during intercourse.
7. Sores, blisters, bumps, or a rash anywhere on the body, particularly the mouth or genitals.
8. Itchy, runny eyes.

One possible complication of the STDs gonorrhea and chlamydia is **pelvic inflammatory disease,** in which bacteria enter the vagina and spread throughout the reproductive organs. The disease begins with intermittent cramps, followed by sudden fever, chills, weakness, and severe cramps. Hospitalization and intravenous antibiotics can stop the infection. The uterus and uterine tubes are often scarred.

Acquired immune deficiency syndrome (AIDS) is an STD that destroys the immune system. Infections and often cancer, diseases that the immune system usually conquers, overrun the body. The AIDS virus (human immunodeficiency virus, or HIV) passes from one person to another in body fluids such as semen, blood, and breast milk. Unprotected intercourse and using a needle containing contaminated blood are the most frequent means of transmission.

1. Why are sexually transmitted diseases sometimes difficult to diagnose?
2. What are some common symptoms of sexually transmitted diseases?

Clinical Terms Related to the Reproductive Systems

Amenorrhea (a-men″o-re′ah) Absence of menstrual flow, usually due to a disturbance in hormonal concentrations.

Conization (ko″nĭ-za′shun) Surgical removal of a cone of tissue from the cervix for examination.

Curettage (ku″rĕ-tahzh′) Surgical procedure in which the cervix is dilated and the endometrium of the uterus is scraped (commonly called D and C for dilation and curettage).

Dysmenorrhea (dis″men-ŏ-re′ah) Painful menstruation.

Endometriosis (en″do-me″tre-o′sis) Tissue similar to the inner lining of the uterus occurring within the pelvic cavity.

Endometritis (en″do-mĕ-tri′tis) Inflammation of the uterine lining.

Epididymitis (ep″ĭ-did″ĭ-mi′tis) Inflammation of the epididymis.

Hematometra (hem″ah-to-me′trah) Accumulation of menstrual blood within the uterine cavity.

Hysterectomy (his″tĕ-rek′to-me) Surgical removal of the uterus.

Mastitis (mas″ti′tis) Inflammation of a mammary gland.

Oophorectomy (o″of-o-rek′to-me) Surgical removal of an ovary.

Oophoritis (o″of-o-ri′tis) Inflammation of an ovary.

Orchiectomy (or″ke-ĕk′to-me) Surgical removal of a testis.

Orchitis (or-ki′tis) Inflammation of a testis.

Prostatectomy (pros″tah-tek′to-me) Surgical removal of a portion or all of the prostate gland.

Prostatitis (pros″tah-ti′tis) Inflammation of the prostate gland.

Salpingectomy (sal″pin-jek′to-me) Surgical removal of a uterine tube.

Vaginitis (vaj″ĭ-ni′tis) Inflammation of the vaginal lining.

Varicocele (var′ĭ-ko-sēl″) Distension of the veins within the spermatic cord.

organization

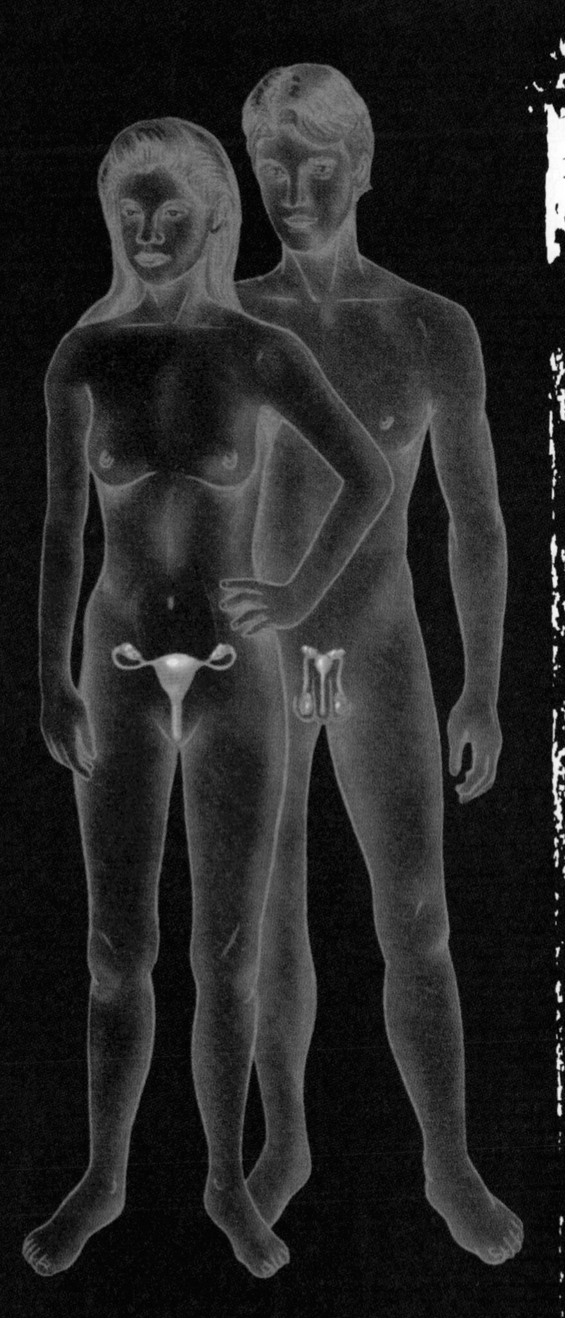

Reproductive Systems

Gamete production, fertilization, fetal development, and childbirth are essential for survival of the species.

Integumentary System

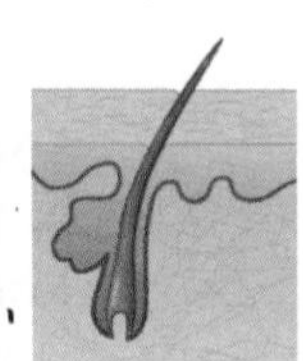

Skin sensory receptors play a role in sexual pleasure.

Cardiovascular System

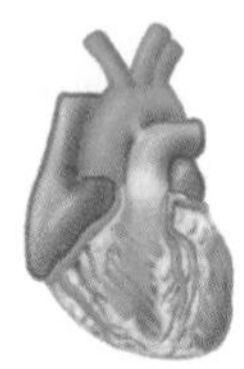

Blood pressure is necessary for the normal function of erectile tissue in the male and female.

Skeletal System

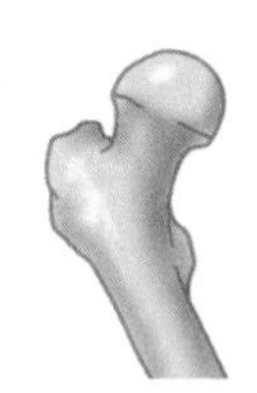

Bones can be a temporary source of calcium during lactation.

Lymphatic System

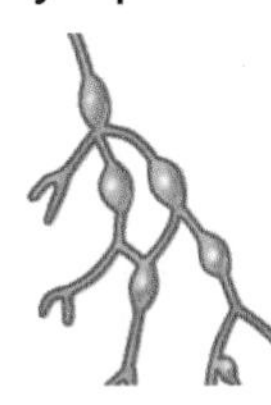

Special mechanisms inhibit the female immune system from attacking sperm as foreign invaders.

Muscular System

Skeletal, cardiac, and smooth muscles all play a role in reproductive processes and sexual activity.

Digestive System

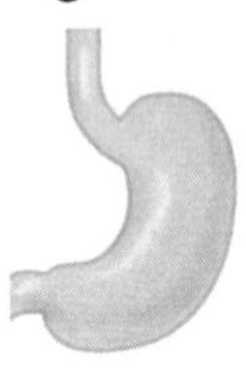

Proper nutrition is essential for the formation of normal gametes and for normal fetal development during pregnancy.

Nervous System

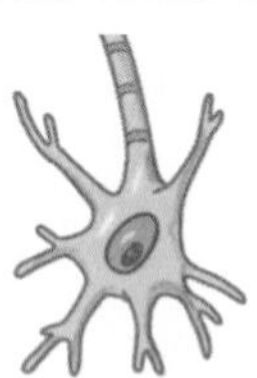

The nervous system plays a major role in sexual activity and sexual pleasure.

Respiratory System

During pregnancy, the placenta provides oxygen to the fetus and removes cabon dioxide.

Endocrine System

Hormones control the production of ova in the female and sperm in the male.

Urinary System

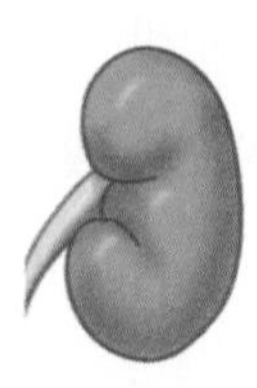

Male urinary and reproductive systems share common structures. Kidneys compensate for fluid loss from the reproductive systems. Pregnancy may cause fluid retention.

Summary Outline

Introduction (p. 517)

Reproductive organs produce sex cells and sex hormones, sustain these cells, or transport them.

Organs of the Male Reproductive System (p. 517)

The primary male sex organs are the testes, which produce sperm cells and male sex hormones. Accessory organs include the internal and external reproductive organs.

1. Testes
   a. Structure of the testes
      (1) The testes are composed of lobules separated by connective tissue and filled with the seminiferous tubules.
      (2) The epithelium lining the seminiferous tubules produces sperm cells.
      (3) The interstitial cells produce male sex hormones.
   b. Formation of sperm cells
      (1) The epithelium lining the seminiferous tubules includes supporting cells and spermatogenic cells.
         (a) Supporting cells support and nourish spermatogenic cells.
         (b) Spermatogenic cells give rise to sperm cells.
      (2) A sperm cell consists of a head, midpiece, and tail.
   c. Spermatogenesis
      (1) Spermatogonia give rise to sperm cells.
      (2) Meiosis reduces the number of chromosomes in sperm cells by half (forty-six to twenty-three).
      (3) Spermatogenesis produces four sperm cells from each primary spermatocyte.
2. Male internal accessory organs
   a. Epididymis
      (1) The epididymis is a tightly coiled tube that leads into the vas deferens.
      (2) It stores and nourishes immature sperm cells and promotes their maturation.
   b. Vas deferens
      (1) The vas deferens is a muscular tube.
      (2) It passes through the inguinal canal, enters the abdominal cavity, courses medially into the pelvic cavity, and ends behind the urinary bladder.
      (3) It fuses with the duct from the seminal vesicle to form the ejaculatory duct.
   c. Seminal vesicle
      (1) The seminal vesicle is a saclike structure attached to the vas deferens.
      (2) It secretes an alkaline fluid that contains nutrients, such as fructose, and prostaglandins.
   d. Prostate gland
      (1) The prostate gland surrounds the urethra just below the urinary bladder.
      (2) It secretes a thin, milky fluid that neutralizes the pH of semen and the acidic secretions of the vagina.
   e. Bulbourethral glands
      (1) The bulbourethral glands are two small structures beneath the prostate gland.
      (2) They secrete a fluid that lubricates the penis in preparation for sexual intercourse.
   f. Semen
      (1) Semen consists of sperm cells and secretions of the seminal vesicles, prostate gland, and bulbourethral glands.
      (2) This fluid is slightly alkaline and contains nutrients and prostaglandins.
      (3) Semen activates sperm cells, but these cells are unable to fertilize egg cells until they enter the female reproductive tract.
3. Male external reproductive organs
   a. Scrotum
      The scrotum is a pouch of skin and subcutaneous tissue that encloses the testes.
   b. Penis
      (1) The penis is specialized to become erect for insertion into the vagina during sexual intercourse.
      (2) Its body is composed of three columns of erectile tissue.
4. Erection, orgasm, and ejaculation
   a. During erection, the vascular spaces within the erectile tissue engorge with blood.
   b. Orgasm is the culmination of sexual stimulation. Emission and ejaculation accompany male orgasm.
   c. Sympathetic reflexes propel semen.

Hormonal Control of Male Reproductive Functions (p. 525)

1. Hypothalamic and pituitary hormones
   a. The male body remains reproductively immature until the hypothalamus releases gonadotropin-releasing hormone (GnRH), which stimulates the anterior pituitary gland to release gonadotropins.
   b. Follicle-stimulating hormone (FSH) stimulates spermatogenesis.
   c. Luteinizing hormone (LH), known in males as interstitial cell-stimulating hormone (ICSH), stimulates interstitial cells to produce male sex hormones.
2. Male sex hormones
   a. Male sex hormones are called androgens, with testosterone the most important.
   b. Androgen production increases rapidly at puberty.
   c. Actions of testosterone
      (1) Testosterone stimulates development of the male reproductive organs.
      (2) It also develops and maintains male secondary sex characteristics.
   d. Regulation of male sex hormones
      (1) A negative feedback mechanism regulates testosterone concentration.
         (a) A rising testosterone concentration inhibits the hypothalamus and reduces the anterior pituitary's secretion of gonadotropins.
         (b) As testosterone concentration falls, the hypothalamus signals the anterior pituitary to secrete gonadotropins.
      (2) The testosterone concentration remains stable from day to day.

Organs of the Female Reproductive System (p. 526)

The primary female sex organs are the ovaries, which produce female sex cells and sex hormones. Accessory organs are internal and external.

1. Ovaries
   a. Ovary structure
      (1) Each ovary is subdivided into a medulla and a cortex.
      (2) The medulla is composed of connective tissue, blood vessels, lymphatic vessels, and nerves.
      (3) The cortex contains ovarian follicles within cuboidal epithelium.
   b. Primordial follicles
      (1) During prenatal development, groups of cells in the ovarian cortex form millions of primordial follicles.
      (2) Each primordial follicle contains a primary oocyte and a layer of follicular cells.
      (3) The primary oocyte begins meiosis, but the process stalls until puberty.
      (4) The number of oocytes steadily declines throughout a female's life.
   c. Oogenesis
      (1) Beginning at puberty, some oocytes are stimulated to continue meiosis.
      (2) When a primary oocyte undergoes oogenesis, it gives rise to a secondary oocyte in which the original chromosome number is halved (from forty-six to twenty-three).
      (3) Fertilization of a secondary oocyte produces a zygote.
   d. Follicle maturation
      (1) At puberty, FSH initiates follicle maturation.
      (2) During maturation, the oocyte enlarges, the follicular cells multiply, and a fluid-filled cavity forms.
      (3) Usually, only one follicle at a time fully develops.
   e. Ovulation
      (1) Ovulation is the release of an oocyte from an ovary.
      (2) A rupturing follicle releases the oocyte.
      (3) After ovulation, the oocyte is drawn into the opening of the uterine tube.
2. Female internal accessory organs
   a. Uterine tubes
      (1) The end of each uterine tube expands, and its margin bears irregular extensions.
      (2) Ciliated cells that line the tube and peristaltic contractions in the wall of the tube help transport the egg cell down the uterine tube.
   b. Uterus
      (1) The uterus receives the embryo and sustains it.
      (2) The uterine wall includes the endometrium, myometrium, and perimetrium.
   c. Vagina
      (1) The vagina receives the erect penis, conveys uterine secretions to the outside, and provides an open channel for the fetus during birth.
      (2) Its wall consists of mucosal, muscular, and fibrous layers.
3. Female external reproductive organs
   a. Labia majora
      (1) The labia majora are rounded folds of adipose tissue and skin.
      (2) The upper ends form a rounded elevation over the symphysis pubis.
   b. Labia minora
      (1) The labia minora are flattened, longitudinal folds between the labia majora.
      (2) They are well supplied with blood vessels.
   c. Clitoris
      (1) The clitoris is a small projection at the anterior end of the vulva. It corresponds to the penis.
      (2) It is composed of two columns of erectile tissue.
   d. Vestibule
      (1) The vestibule is the space between the labia minora.
      (2) The vestibular glands secrete mucus into the vestibule during sexual stimulation.
4. Erection, lubrication, and orgasm
   a. During periods of sexual stimulation, the erectile tissues of the clitoris and vestibular bulbs engorge with blood and swell.
   b. The vestibular glands secrete mucus into the vestibule and vagina.
   c. During orgasm, the muscles of the perineum, uterine wall, and uterine tubes contract rhythmically.

Hormonal Control of Female Reproductive Functions (p. 532)

The hypothalamus, anterior pituitary gland, and ovaries secrete hormones that control sex cell maturation, and the development and maintenance of female secondary sex characteristics.

1. Female sex hormones
   a. A female body remains reproductively immature until about ten years of age, when gonadotropin secretion increases.
   b. The most important female sex hormones are estrogen and progesterone.
      (1) Estrogen develops and maintains most female secondary sex characteristics.
      (2) Progesterone changes the uterus.
2. Female reproductive cycle
   a. FSH initiates a menstrual cycle by stimulating follicle maturation.
   b. Maturing follicular cells secrete estrogen, which maintains the secondary sex traits and thickens the uterine lining.
   c. The anterior pituitary secreting a relatively large amount of LH triggers ovulation.
   d. Following ovulation, follicular cells give rise to the corpus luteum.
      (1) The corpus luteum secretes progesterone, which causes the uterine lining to become more vascular and glandular.
      (2) If an oocyte is not fertilized, the corpus luteum begins to degenerate.
      (3) As estrogen and progesterone concentrations decline, the uterine lining disintegrates, causing menstrual flow.
   e. During this cycle, estrogen and progesterone inhibit the release of LH and FSH. As estrogen and progesterone concentrations decline, the anterior pituitary secretes FSH and LH again, stimulating a new menstrual cycle.
3. Menopause
   a. Menopause is termination of the menstrual cycle due to aging of the ovaries.
   b. Reduced estrogen concentration and lack of progesterone may cause regressive changes in female secondary sex characteristics.

Mammary Glands (p. 534)

1. The mammary glands are in the subcutaneous tissue of the anterior thorax.
2. They are composed of lobes that contain glands and a duct.
3. Dense connective and adipose tissues separate the lobes.
4. Ovarian hormones stimulate female breast development.
   a. Alveolar glands and ducts enlarge.
   b. Fat deposits around and within the breasts.

Birth Control (p. 538)

Birth control is voluntary regulation of the number of children produced and when they are conceived. It usually involves some method of contraception.

1. Coitus interruptus
   Coitus interruptus is withdrawal of the penis from the vagina before ejaculation.
2. Rhythm method
   The rhythm method is abstinence from sexual intercourse for several days before and after ovulation.
3. Mechanical barriers
   Males and females can use condoms. Females can also use diaphragms and cervical caps.
4. Chemical barriers
   Spermicidal creams, foams, and jellies provide an unfavorable environment in the vagina for sperm survival.
5. Oral contraceptives
   Birth control pills contain synthetic estrogen-like and progesterone-like substances that disrupt a female's normal pattern of gonadotropin secretion, and prevent ovulation and the normal buildup of the uterine lining.
6. Injectable contraception
   Intramuscular injection with medroxyprogesterone acetate every three months acts similarly to oral contraceptives to prevent pregnancy.
7. Contraceptive implants
   A contraceptive implant is a set of progesterone-containing capsules or rods inserted under a woman's skin. Progesterone released from the implant prevents ovulation.
8. Intrauterine devices (IUD)
   An IUD is a solid object inserted in the uterine cavity that prevents pregnancy by interfering with implantation.
9. Surgical methods
   Vasectomies in males and tubal ligations in females are surgical sterilization procedures.

Sexually Transmitted Diseases (p. 541)

1. Sexually transmitted diseases (STDs) are passed during sexual contact and may go undetected for years.
2. The twenty recognized STDs share similar symptoms.

Review Exercises

1. List the general functions of the male reproductive system. (p. 517)
2. Distinguish between the primary and accessory male reproductive organs. (p. 517)
3. Describe the structure of a testis. (p. 517)
4. Review the process of meiosis. (p. 519)
5. Describe the epididymis, and explain its function. (p. 519)
6. Trace the path of the vas deferens from the epididymis to the ejaculatory duct. (p. 519)
7. On a diagram, locate the seminal vesicles, prostate gland, and bulbourethral glands, and describe the composition of their secretions. (p. 519)
8. Describe the composition of semen. (p. 521)
9. Describe the structure of the penis. (p. 522)
10. Explain the mechanism that produces penile erection. (p. 523)
11. Distinguish between emission and ejaculation. (p. 523)
12. Explain the mechanism of ejaculation. (p. 523)
13. Explain the role of gonadotropin-releasing hormone (GnRH) in the control of male reproductive functions. (p. 525)
14. List several male secondary sex characteristics. (p. 525)
15. Explain the regulation of testosterone concentration. (p. 525)
16. List the general functions of the female reproductive system. (p. 526)
17. Describe the structure of an ovary. (p. 526)
18. Describe how a follicle matures. (p. 528)
19. On a diagram, locate the uterine tubes, and explain their function. (p. 530)
20. Describe the structure of the uterus. (p. 530)
21. On a diagram, locate the clitoris, and describe its structure. (p. 531)
22. Explain the role of gonadotropin-releasing hormone (GnRH) in regulating female reproductive functions. (p. 532)
23. List several female secondary sex characteristics. (p. 532)
24. Define *menstrual cycle*. (p. 532)
25. Summarize the major events in a menstrual cycle. (p. 532)
26. Describe the structure of a mammary gland. (p. 534)
27. Define *contraception*. (p. 538)
28. List several methods of contraception, and explain how each prevents pregnancy. (p. 538)
29. List several symptoms of sexually transmitted diseases. (p. 541)

Critical Thinking

1. How are the human male and female reproductive tracts similar? How are the structures of the testis and ovary similar?
2. Why must the chromosome number be halved in sperm cells and egg cells?
3. Some men are unable to become fathers because their spermatids do not mature into sperm. Injection of their spermatids into their partner's secondary oocytes sometimes results in conception. A few men have fathered healthy babies this way. Why would this procedure work with spermatids, but not with primary spermatocytes?
4. *Contraception* literally means "against conception." According to this definition, is an intrauterine device a contraceptive? Why or why not?
5. Understanding the causes of infertility can be valuable in developing new birth control methods. Cite a type of contraceptive based on each of the following causes of infertility: (a) failure to ovulate due to a hormonal imbalance; (b) a large fibroid tumor that disturbs the uterine lining; (c) endometrial tissue blocking uterine tubes; (d) low sperm count (too few sperm per ejaculate).
6. How can a couple use "fertility awareness" methods to conceive a child and also to prevent pregnancy?
7. Sometimes, a sperm cell fertilizes a polar body rather than an oocyte. An embryo does not develop, and the fertilized polar body degenerates. Why is a polar body unable to support development of an embryo?
8. What changes, if any, would a male who has had one testis removed experience? A female who has had one ovary removed?
9. Does a tubal ligation cause a woman to enter menopause prematurely? Why or why not?

Chapter 20 Pregnancy, Growth, and Development

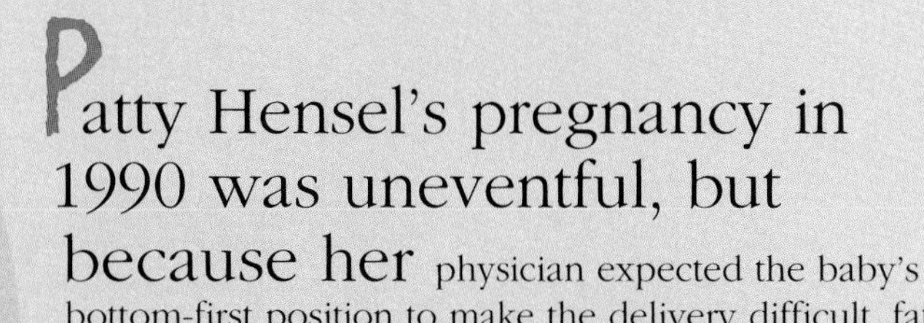

Patty Hensel's pregnancy in 1990 was uneventful, but because her physician expected the baby's bottom-first position to make the delivery difficult, father-to-be Mike was not permitted in the delivery room, and Patty was sedated. Even the experienced medical staff were not prepared for what happened next.

The baby had two heads and two necks, yet appeared to share the rest of the body, with two legs and two arms in the correct places, and a third arm between the heads. The baby was actually two individuals, Abigail and Brittany. They were conjoined twins. Each twin had her own neck, head, heart, stomach, gallbladder, and nervous system. They shared a large liver, bloodstream, and all organs below the navel, including the reproductive tract, and had three lungs and three kidneys.

Abby and Britty were strong and healthy, and their parents soon confronted the many tasks involved in caring for their daughters. Doctors suggested surgery to separate the twins. Aware that only one child would likely survive surgery, Mike and Patty Hensel chose to let their daughters be.

Conjoined twins occur in only 1 in 50,000 births, and about 40% are stillborn. Abby and Britty Hensel are rare among the rare, being joined in a manner seen only four times before. They are the result of incomplete twinning, which probably occurred during the first two weeks of gestation.

Through magazine articles and television appearances, Abby and Britty have captured many hearts. The girls have distinctive personalities, and attend school, swim, ride bikes, and play like any other children.

Photo: Abby and Britty Hensel.

A sperm cell and egg cell unite, forming a zygote, and the journey of prenatal development begins. Following thirty-eight weeks of cell division, growth, and specialization into distinctive tissues and organs, a new human being will enter the world.

Chapter Objectives

After studying this chapter, you should be able to:

1. Distinguish between growth and development. (p. 549)
2. Define *pregnancy,* and describe the process of fertilization. (p. 549)
3. Describe the major events of cleavage. (p. 550)
4. Distinguish between an embryo and a fetus. (p. 550)
5. Describe the hormonal changes in the maternal body during pregnancy. (p. 552)
6. Explain how the primary germ layers originate, and list the structures each layer produces. (p. 557)
7. Describe the major events of the embryonic stage of development. (p. 557)
8. Describe the formation and function of the placenta. (p. 557)
9. Describe the major events of the fetal stage of development. (p. 560)
10. Trace the general path of blood through the fetal circulatory system. (p. 561)
11. Describe the birth process, and explain the role of hormones in this process. (p. 564)
12. Describe the major circulatory and physiological adjustments required of the newborn. (p. 568)

Aids to Understanding Words

allant- [sausage-shaped] *allant*ois: Tubelike structure extending from the yolk sac into the connecting stalk of the embryo.

chorio- [skin] *chorio*n: Outermost membrane surrounding the fetus and other fetal membranes.

cleav- [to divide] *cleav*age: Period of development in which the zygote divides into smaller and smaller cells.

lacun- [pool] *lacun*a: Space between the chorionic villi filled with maternal blood.

morul- [mulberry] *morul*a: A solid ball of about sixteen cells that looks somewhat like a mulberry.

nat- [to be born] pre*nat*al: Period of development before birth.

troph- [nourishment] *troph*oblast: Cellular layer surrounding the inner cell mass and helping nourish it.

umbil- [navel] *umbil*ical cord: Structure connecting the fetus to the placenta.

Key Terms

amnion (am′ne-on)
chorion (ko′re-on)
cleavage (klēv′ij)
embryo (em′bre-o)
fertilization (fer″tĭ-lĭ-za′shun)
fetus (fe′tus)
neonatal (ne″o-na′tal)
placenta (plah-sen′tah)
postnatal (pōst-na′tal)
prenatal (pre-na′tal)
primary germ layers (pri′mar-e jerm la′erz)
umbilical cord (um-bil′ĭ-kal kord)
zygote (zi′gōt)

Introduction

Growth is an increase in size. In humans, growth entails an increase in cell numbers, followed by enlargement of the newly formed cells. Development, which includes growth, is the continuous process by which an individual changes from one life phase to another. These life phases include a **prenatal period** that begins with fertilization and ends at birth, and a **postnatal period** that begins at birth and ends at death.

Pregnancy

Pregnancy is the presence of a developing offspring in the uterus. It results from the union of an egg cell and a sperm cell—an event called **fertilization.**

Transport of Sex Cells

Prior to fertilization, a female ovulates an egg cell (secondary oocyte), which enters a uterine tube. During sexual intercourse, the male deposits semen containing sperm cells in the vagina near the cervix. To reach the egg cell, the sperm cells must move upward through the uterus and uterine tube. Prostaglandins in the semen stimulate lashing of sperm tails and muscular contractions of the uterus and uterine tube, which help sperm cells move. Also, high estrogen concentrations during the first part of the menstrual cycle stimulate the uterus and cervix to secrete a thin, watery fluid that promotes sperm transport and survival. Conversely, during the latter part of the cycle, when progesterone concentration is high, the female reproductive tract secretes a viscous fluid that hampers sperm transport and survival.

Sperm cells reach the upper portions of the uterine tube within an hour following sexual intercourse. Many sperm cells may reach the egg cell, but only one actually fertilizes it (fig. 20.1).

Fertilization

A sperm cell that reaches the egg cell invades the follicular cells that adhere to the egg cell's surface (corona radiata) and binds to the *zona pellucida* surrounding the egg cell membrane. The acrosome of the sperm cell releases an enzyme (hyaluronidase) that aids penetration (fig. 20.2).

Union of the egg cell and sperm cell triggers lysosome-like vesicles just beneath the egg cell membrane to release enzymes that harden the zona pellucida around the fertilized egg cell. This reduces the chance that other sperm cells will penetrate.

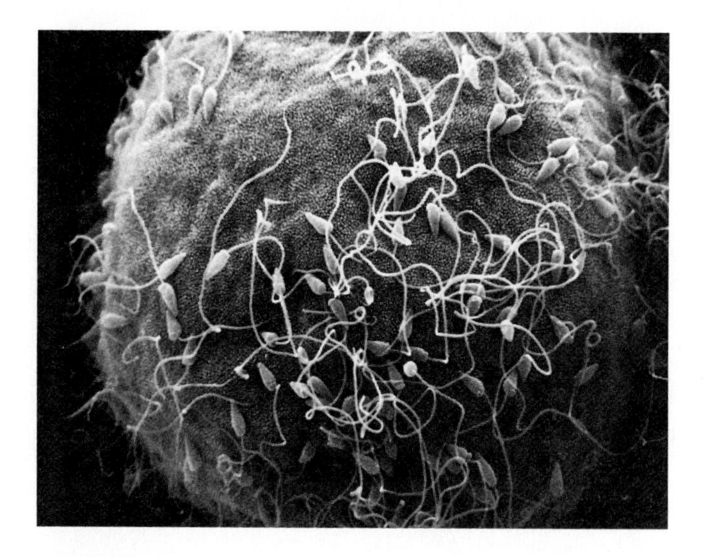

Figure 20.1

Scanning electron micrograph of sperm cells on the surface of an egg cell (1200×). Only one sperm cell may actually fertilize the egg cell.

Once a sperm cell enters the egg cell's cytoplasm, the nucleus in the sperm cell's head swells. The egg cell then divides unequally to form a large cell and a tiny second polar body, which is later expelled. Next, the nuclei of the egg cell and sperm cell unite. Their nuclear membranes disassemble, and their chromosomes mingle, completing fertilization.

Because the sperm cell and the egg cell each provides twenty-three chromosomes, the product of fertilization is a cell with forty-six chromosomes—the usual number in a human body cell (somatic cell). This cell, called a **zygote**, is the first cell of the future offspring.

The approximate time of fertilization is fourteen days after the onset of the last menstruation. The estimated time of birth is 266 days from fertilization. Most women give birth within ten to fifteen days of this calculated time. ■

1. What factors aid the movements of sperm cells through the female reproductive tract?
2. Where in the female reproductive system does fertilization normally occur?
3. List the events of fertilization.

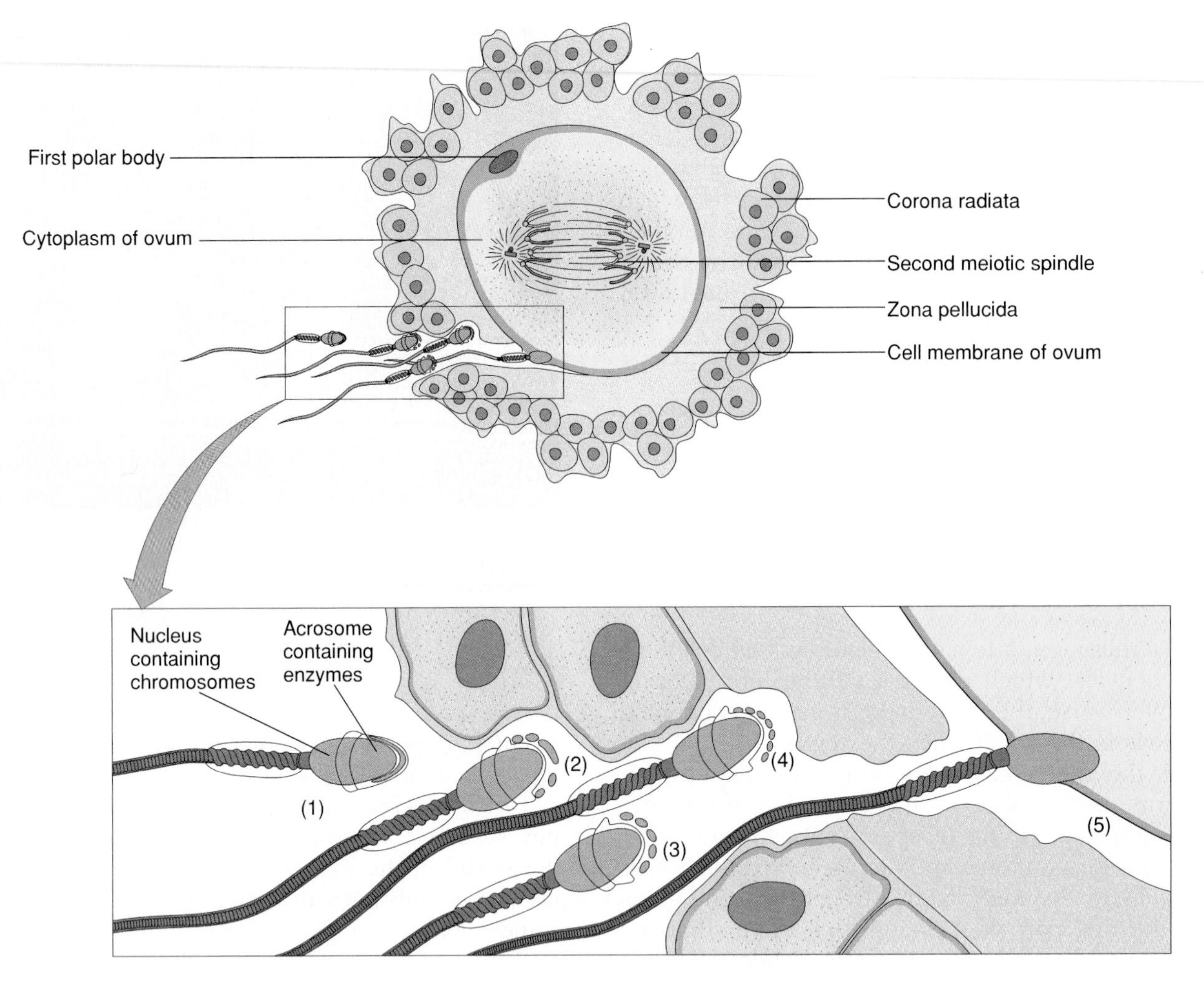

Figure 20.2

Steps in fertilization. (1) Sperm cell reaches corona radiata surrounding egg cell. (2) Acrosome of sperm cell releases protein-digesting enzyme. (3 and 4) Sperm cell penetrates zona pellucida surrounding egg cell. (5) Sperm cell's plasma membrane fuses with egg cell membrane.

Prenatal Period

Early Embryonic Development

About 30 hours after forming, the zygote undergoes *mitosis,* giving rise to two new cells (blastomeres). These cells, in turn, divide into four cells, which divide into eight cells, and so forth. These divisions take place rapidly with little time to grow. Thus, each subsequent division results in smaller and smaller cells. This phase of development is termed **cleavage** (fig. 20.3).

Meanwhile, the tiny mass of cells moves through the uterine tube to the uterine cavity. This trip takes about three days, and by then, the structure consists of a solid ball (morula) of about sixteen cells (fig. 20.4).

The morula remains free within the uterine cavity for about three days. During this stage, the zona pellucida of the original egg cell degenerates, and the structure, which now consists of a hollow ball of cells (blastocyst), begins to attach to the uterine lining. By the end of the first week of development, it superficially *implants* in the endometrium (fig. 20.5).

About the time of implantation, certain cells within the blastocyst organize into a group (inner cell mass) that will give rise to the body of the offspring. This marks the beginning of the **embryonic stage** of development. The offspring is termed an **embryo** until the end of the eighth week, by which time the basic structural form of the human body is recognizable. After the eighth week and until birth, the offspring is called a **fetus.**

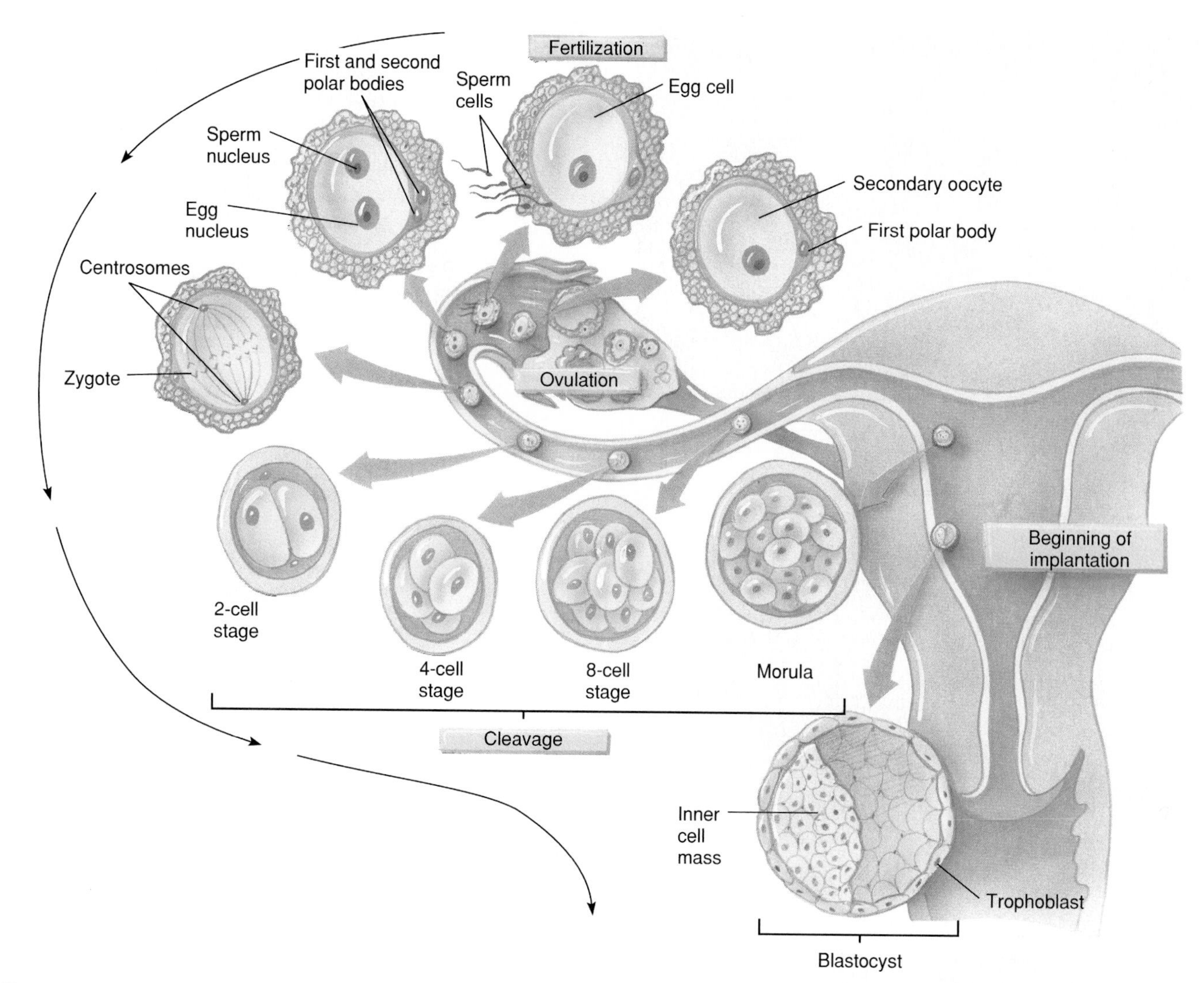

Figure 20.3

Stages in early human development.

Eventually, the cells surrounding the embryo, with cells of the endometrium, form a complex vascular structure called the **placenta.** This organ attaches the embryo to the uterine wall and exchanges nutrients, gases, and wastes between maternal blood and blood of the embryo. The placenta also secretes hormones.

1. What is cleavage?
2. What is implantation?
3. How do an embryo and a fetus differ?

When two ovarian follicles release egg cells simultaneously and both are fertilized, the resulting zygotes develop into fraternal (dizygotic) twins. Such twins are no more alike genetically than siblings. In other instances, twins develop from a single fertilized egg (monozygotic twins). This may happen if two inner cell masses form within a blastocyst and each produces an embryo. Monozygotic twins usually share a single placenta and are identical genetically. Thus, they are always the same sex and look alike. ■

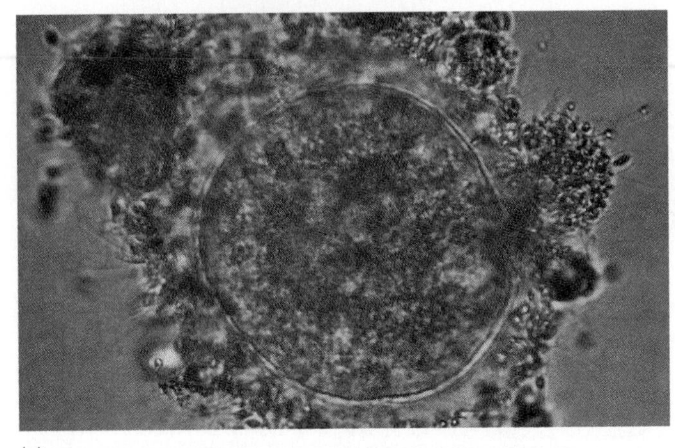

(a)

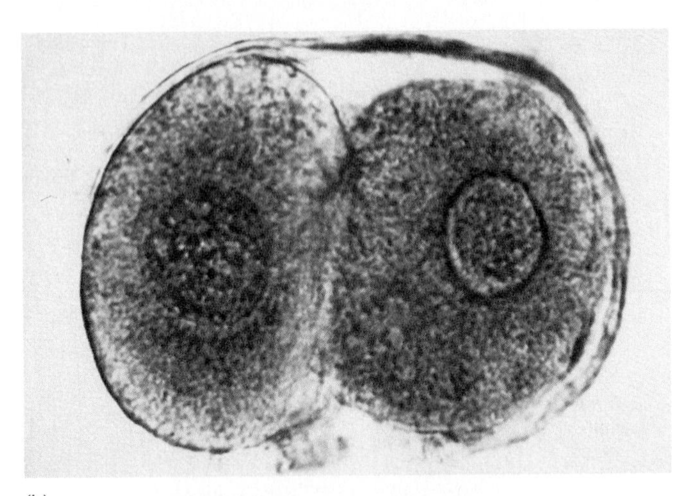

(b)

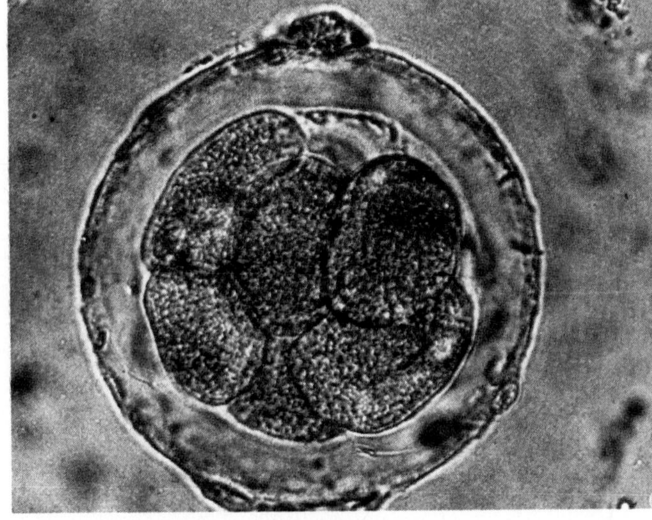

(c)

Figure 20.4

Light micrographs of (*a*) a human egg cell surrounded by follicular cells and sperm cells, (*b*) two-cell stage, (18×) and (*c*) morula.

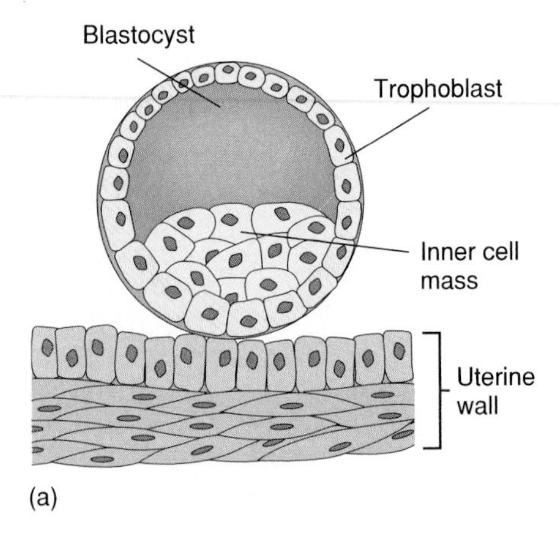

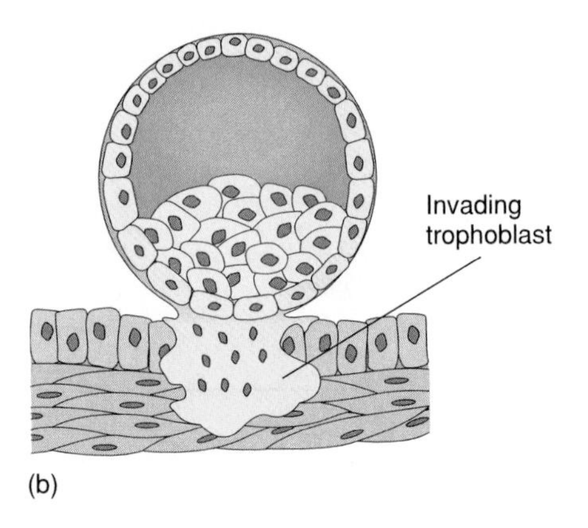

Figure 20.5

About the sixth day of development, the blastocyst (*a*) contacts the uterine wall and (*b*) begins to implant.

Hormonal Changes during Pregnancy

During a typical menstrual cycle, the corpus luteum degenerates about two weeks after ovulation. Consequently, estrogen and progesterone concentrations decline rapidly, the uterine lining breaks down, and the endometrium sloughs away as menstrual flow. If this occurs following implantation, the embryo is lost (spontaneously aborted).

The hormone **human chorionic gonadotropin** (hCG) normally prevents pregnancy loss. A layer of embryonic cells (trophoblast), which surrounds the developing embryo and later helps form the placenta, secretes hCG (fig. 20.5). This hormone, similar in function to luteinizing hormone (LH), maintains the corpus luteum, which secretes estrogen and progesterone, stimulating the uterine wall to grow and develop. At the same time, the anterior pituitary's release of follicle-stimulating hormone (FSH) and LH is inhibited, so normal menstrual cycles cease.

TOPIC OF INTEREST

Female Infertility

Infertility is the inability to conceive after a year of trying. In 90% of cases, infertility has a physical cause, and the abnormality is in the female's reproductive system 60% of the time.

A common cause of female infertility is the anterior pituitary secreting too little (hyposecreting) gonadotropic hormones, followed by failure to ovulate (anovulation). Testing the urine for *pregnanediol,* a product of progesterone metabolism, detects an anovulatory cycle. Progesterone concentration normally rises after ovulation. No increase in pregnanediol in the urine during the latter part of the menstrual cycle suggests lack of ovulation.

Fertility specialists can treat anovulation due to hyposecretion of gonadotropic hormones by administering human chorionic gonadotropin (hCG) obtained from human placentas. Another ovulation-stimulating biochemical, human menopausal gonadotropin (hMG), contains luteinizing hormone (LH) and follicle-stimulating hormone (FSH) and is obtained from urine of postmenopausal women. Either hCG or hMG may overstimulate the ovaries, however, and cause many follicles to release egg cells simultaneously, resulting in multiple births.

Another cause of female infertility is *endometriosis,* in which small pieces of the inner uterine lining (endometrium) move up through the uterine tubes during menses and implant in the abdominal cavity. Here, the tissue changes in a similar way to the uterine lining during the menstrual cycle. The abnormally located tissue breaks down at the end of the cycle but cannot be expelled. Instead, it remains in the abdominal cavity, irritating the lining (peritoneum) and causing considerable pain. This tissue also stimulates formation of fibrous tissue (fibrosis) that may encase the ovary, preventing ovulation or obstructing the uterine tubes.

Infections, such as gonorrhea, cause some women to become infertile. These infections can inflame and obstruct the uterine tubes, or stimulate production of viscous mucus that plugs the cervix and prevents sperm entry.

Finding the right treatment for a particular patient requires determining the infertility's cause. Table 20A describes diagnostic tests for female infertility.

Table 20A Tests to Assess Female Infertility

Test	What It Checks
Hormone levels	If ovulation occurs
Ultrasound	Placement and appearance of reproductive organs and structures
Postcoital test	Cervix examined soon after unprotected intercourse to see if mucus is thin enough to allow sperm through
Endometrial biopsy	Small piece of uterine lining sampled and viewed under microscope to see if it can support an embryo
Hysterosalpingogram	Dye injected into uterine tube and followed with scanner shows if tube is clear or blocked
Laparoscopy	Small, lit optical device inserted near navel to detect scar tissue blocking tubes, which could be missed in ultrasound
Laparotomy	Scar tissue in tubes removed through incision made for laparoscopy

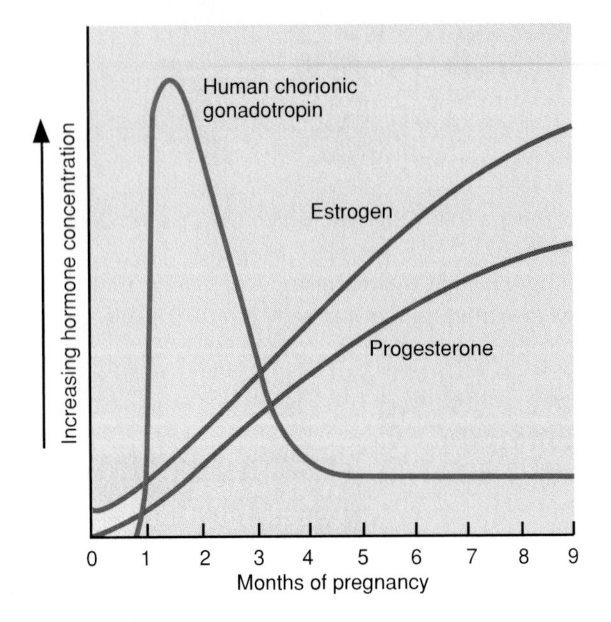

Figure 20.6

Relative concentrations of three hormones in maternal blood during pregnancy.

Secretion of hCG continues at a high level for about two months, then declines by the end of four months. The corpus luteum persists throughout pregnancy, but its function as a hormone source becomes less important after the first three months (first trimester), when the placenta secretes sufficient estrogen and progesterone (fig. 20.6).

lastocyst cells secrete hCG, which is excreted in urine, and its presence there signals pregnancy. Secretion of hCG peaks at fifty to sixty days of gestation, then falls to a much lower level for the remainder of pregnancy. ■

Placental estrogen and *placental progesterone* maintain the uterine wall during the second and third trimesters of pregnancy. The placenta also secretes a hormone called **placental lactogen** that, with placental estrogen and progesterone, stimulates breast development and prepares the mammary glands for milk secretion. Placental progesterone and a polypeptide hormone called *relaxin* from the corpus luteum inhibit the smooth muscles in the myometrium, suppressing uterine contractions until the birth process begins.

Table 20.1 Hormonal Changes during Pregnancy

1. Following implantation, embryonic cells begin to secrete human chorionic gonadotropin.
2. Human chorionic gonadotropin maintains the corpus luteum, which continues to secrete estrogen and progesterone.
3. The developing placenta secretes large quantities of estrogen and progesterone.
4. Placental estrogen and progesterone
   a. Stimulate the uterine lining to continue development.
   b. Maintain the uterine lining.
   c. Inhibit the anterior pituitary's secretion of follicle-stimulating hormone (FSH) and luteinizing hormone (LH).
   d. Stimulate development of mammary glands.
   e. Inhibit uterine contractions (progesterone).
   f. Enlarge the reproductive organs (estrogen).
5. Relaxin from the corpus luteum inhibits uterine contractions and relaxes the pelvic ligaments.
6. The placenta secretes placental lactogen that stimulates breast development.
7. Aldosterone from the adrenal cortex promotes sodium reabsorption.
8. Parathyroid hormone from the parathyroid glands helps maintain a high concentration of maternal blood calcium.

The high concentration of placental estrogen during pregnancy enlarges the vagina and external reproductive organs. Also, relaxin relaxes the ligaments joining the symphysis pubis and sacroiliac joints during the last week of pregnancy, allowing greater movement at these joints and aiding the passage of the fetus through the birth canal.

Other hormonal changes of pregnancy include increased adrenal secretion of aldosterone, which promotes renal reabsorption of sodium, leading to fluid retention. The parathyroid glands secrete parathyroid hormone, which helps maintain a high concentration of maternal blood calcium (see chapter 11).

Table 20.1 summarizes the hormonal changes of pregnancy.

1. Which hormone normally prevents pregnancy loss?
2. What is the source of the hormones that sustain the uterine wall during pregnancy?
3. What other hormonal changes occur during pregnancy?

Assisted Reproductive Technologies

Michele and Ray L'Esperance wanted children badly, but Michele's uterine tubes had been removed due to scarring. A procedure called *in vitro fertilization (IVF)* enabled the couple to have children.

First, Michele received human menopausal gonadotropin to stimulate development of ovarian follicles. When an ultrasound scan showed that the follicles had grown to a certain diameter, she received human chorionic gonadotropin to induce ovulation. Then, Michele's physician used an optical instrument called a laparoscope to examine the interior of her abdomen and take the largest oocytes from an ovary. The oocytes were incubated at 37° C in a medium buffered at pH 7.4. When the oocytes matured, they were mixed in a laboratory dish with Ray's sperm cells, which had been washed to remove inhibitory factors. Secretions from Michele's reproductive tract were added to activate the sperm.

Next, fertilized egg cells were selected and incubated in a special medium for about 60 hours. At this stage, five of the eight- to sixteen-cell balls of cells were transferred through Michele's cervix and into her uterus to increase the chances that one or two would complete development. The L'Esperances beat the odds—they had quintuplets (fig. 20A)!

Success rates for IVF vary from clinic to clinic, ranging from 0–40%, with the average about 14%. Pregnancy via IVF is expensive, costing thousands of dollars. Table 20B describes other assisted reproductive technologies.

***Figure* 20A**

In vitro fertilization worked for Michele and Ray L'Esperance. Five fertilized ova implanted in Michele's uterus are now Erica, Alexandria, Veronica, Danielle, and Raymond.

Table 20B

Assisted Reproductive Technologies

Technology	Procedure	Condition It Treats
Artificial insemination	Donated sperm cells or pooled specimens are placed near a woman's cervix.	Male infertility—lack of sperm cells or low sperm count
Surrogate mother	An oocyte fertilized in vitro is implanted in a woman other than the one who donated the oocyte. The surrogate, or "gestational mother," gives the newborn to the "genetic mother" and her partner, the sperm donor.	Female infertility—lack of a uterus
Gamete intrafallopian transfer (GIFT)	Oocytes are removed from a woman's ovary, then placed along with donated sperm cells into a uterine tube.	Female infertility—bypasses blocked uterine tube
Zygote intrafallopian transfer (ZIFT)	An oocyte fertilized in vitro is placed in a uterine tube. It travels to the uterus on its own.	Female infertility—bypasses blocked uterine tube
Embryo adoption	A woman is artificially inseminated with sperm cells from a man whose partner cannot ovulate healthy oocytes. If the woman conceives, the morula is flushed from her uterus and implanted in the uterus of the sperm donor's partner.	Female infertility—a woman has nonfunctional ovaries, but a healthy uterus

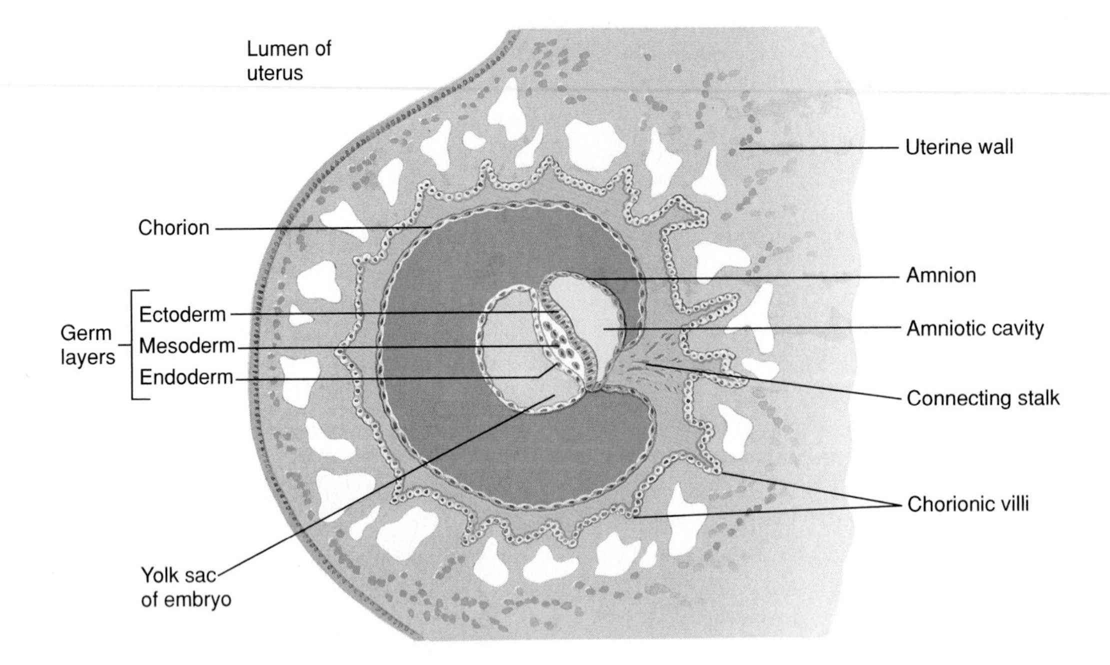

Figure 20.7

Early in the embryonic stage of development, the three primary germ layers form.

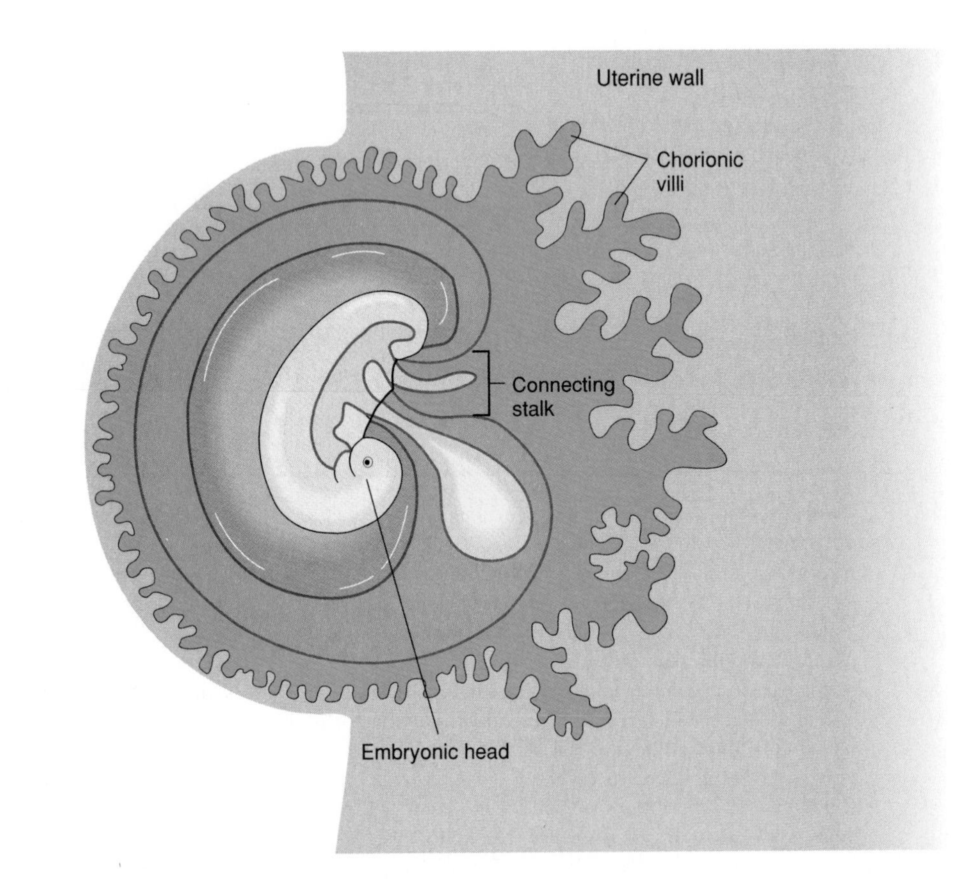

Figure 20.8

During the fourth week, the flat embryonic disk becomes cylindrical.

Embryonic Stage

The embryonic stage extends from the end of the second week through the eighth week of development. During this time, the placenta forms, the main internal organs develop, and the major external body structures appear.

Early in the embryonic stage, the cells of the inner cell mass organize into a flattened **embryonic disk** with two distinct layers—an outer *ectoderm* and an inner *endoderm*. A short time later, the ectoderm and endoderm fold, and a third layer of cells, the *mesoderm,* forms between them. All organs form from these three cell layers, called the **primary germ layers**. A *connecting stalk* attaches the embryonic disk to the developing placenta (fig. 20.7).

Ectodermal cells give rise to the nervous system, portions of special sensory organs, the epidermis, hair, nails, glands of the skin, and linings of the mouth and anal canal. Mesodermal cells form all types of muscle tissue, bone tissue, bone marrow, blood, blood vessels, lymphatic vessels, connective tissues, internal reproductive organs, kidneys, and the epithelial linings of the body cavities. Endodermal cells produce the epithelial linings of the digestive tract, respiratory tract, urinary bladder, and urethra.

As the embryo continues to implant in the uterus, slender projections grow out from the trophoblast into the surrounding endometrium. These projections, called **chorionic villi,** branch, and by the end of the fourth week, they are well formed.

Also during the fourth week of development, the flat embryonic disk transforms into a cylindrical structure (fig. 20.8). The head and jaws develop, the heart beats and forces blood through the blood vessels, and tiny buds, which will give rise to the upper and lower limbs, form.

During the fifth through the seventh weeks, as figure 20.9 shows, the head grows rapidly and becomes rounded and erect. The face, with developing eyes, nose, and mouth, becomes more humanlike. The upper and lower limbs elongate, and fingers and toes appear (fig. 20.10). By the end of the seventh week, all the main internal organs are present, and as these structures enlarge, the body takes on a humanlike appearance.

As the chorionic villi develop, embryonic blood vessels appear within them and are continuous with those passing through the connecting stalk to the body of the embryo. At the same time, irregular spaces called **lacunae** form around and between the villi. These spaces fill with maternal blood that escapes from eroded endometrial blood vessels.

A thin **placental membrane** separates embryonic blood within the capillary of a chorionic villus from maternal blood in a lacuna. Across this membrane, composed of the epithelium of the chorionic villus and the

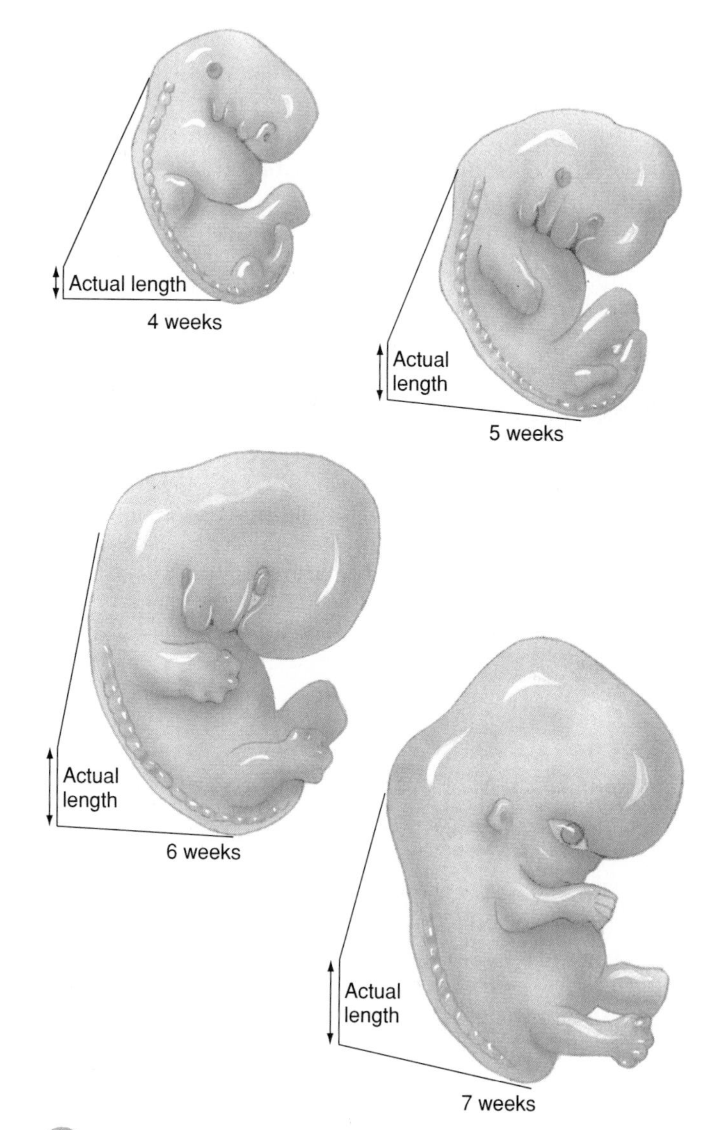

Figure 20.9

In the fifth through the seventh weeks of development, the embryonic body and face develop a humanlike appearance.

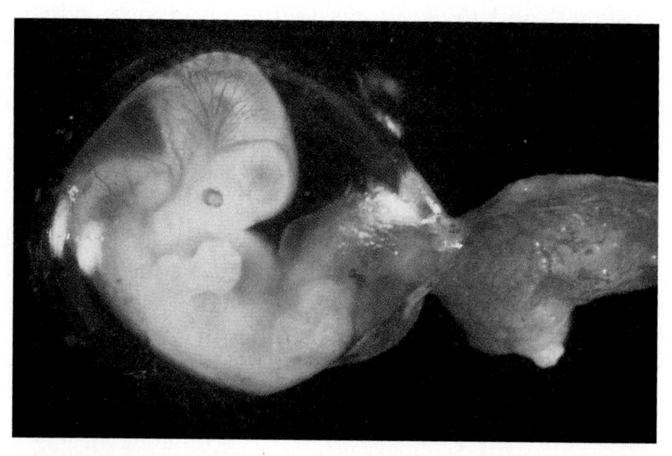

Figure 20.10

Human embryo after about six weeks of development.

epithelium of the capillary inside the villus, maternal and embryonic blood exchange substances (fig. 20.11). Oxygen and nutrients diffuse from maternal blood into the embryo's blood, and carbon dioxide and other wastes diffuse from the embryo's blood into maternal blood. Various substances also cross the placental membrane by active transport and pinocytosis.

If a pregnant woman takes an addictive drug, her newborn may suffer from withdrawal symptoms when amounts of the drug it was accustomed to as a fetus suddenly plummet after birth. Newborn addiction occurs with certain addictive drugs of abuse, such as heroin, and with some prescription drugs used to treat anxiety. It also occurs with very large doses of vitamin C. Vitamin C is not addictive, but if a fetus is accustomed to megadoses, the sudden drop in vitamin C level after birth may bring on symptoms of vitamin C deficiency. ■

1. Describe the major events of the embryonic stage of development.
2. Which tissues and structures develop from ectoderm? From mesoderm? From endoderm?
3. How do embryonic and maternal blood exchange substances?

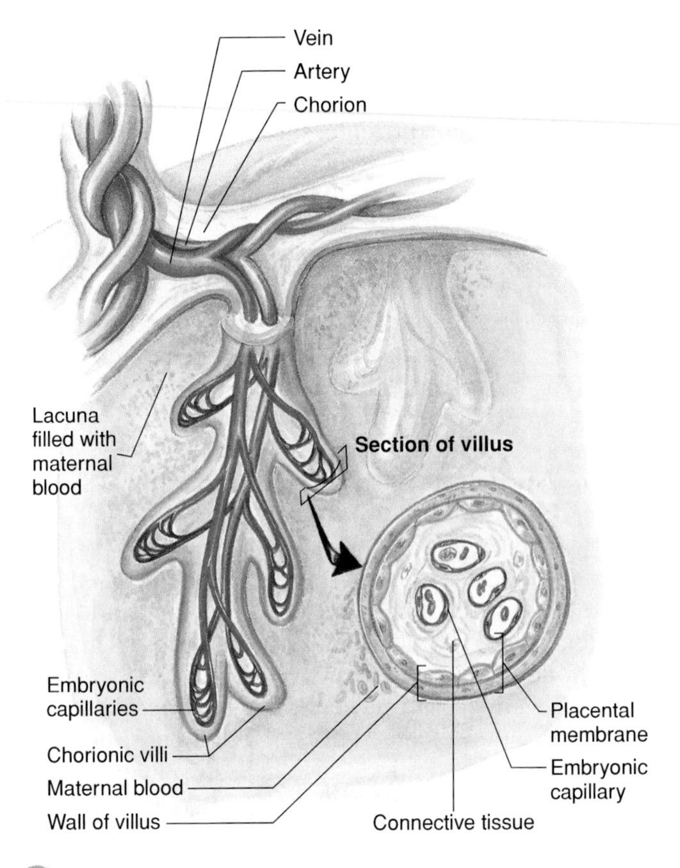

Figure 20.11

As illustrated in the section of villus (lower part of figure), the placental membrane consists of the epithelial wall of an embryonic capillary and the epithelial wall of a chorionic villus.

Until about the end of the eighth week, chorionic villi cover the entire surface of the former trophoblast, now called the **chorion.** As the embryo and the chorion surrounding it enlarge, however, only those villi in contact with the endometrium endure. The others degenerate, and the portions of the chorion to which they were attached become smooth. The disk-shaped region of the chorion still in contact with the uterine wall becomes the placenta.

The embryonic portion of the placenta is the chorion and its villi; the maternal portion is the area of the uterine wall where the villi attach (fig. 20.12). When fully formed, the placenta is a reddish brown disk about 20 centimeters long and 2.5 centimeters thick, and weighing about 0.5 kilogram.

While the placenta forms, another membrane, called the **amnion,** develops around the embryo during the second week. Its margin attaches around the edge of the embryonic disk, and **amniotic fluid** fills the space between the amnion and embryonic disk.

As the embryo becomes more cylindrical, the amnion margins fold, enclosing the embryo in the amnion and amniotic fluid. The amnion envelops the tissues on the underside of the embryo, by which the embryo attaches to the chorion and the developing placenta. In this manner, the **umbilical cord** forms (fig. 20.13).

The umbilical cord contains three blood vessels—two *umbilical arteries* and one *umbilical vein*—that transport blood between the embryo and the placenta (see fig. 20.12). The umbilical cord suspends the embryo in the *amniotic cavity.* The amniotic fluid allows the embryo to grow freely without compression from surrounding tissues and also protects the embryo from jarring movements of the woman's body.

Two other embryonic membranes form during development—the yolk sac and the allantois (fig. 20.13). The **yolk sac** forms during the second week and attaches to the underside of the embryonic disk. It forms blood cells in the early stages of development and gives rise to the cells that later become sex cells. The

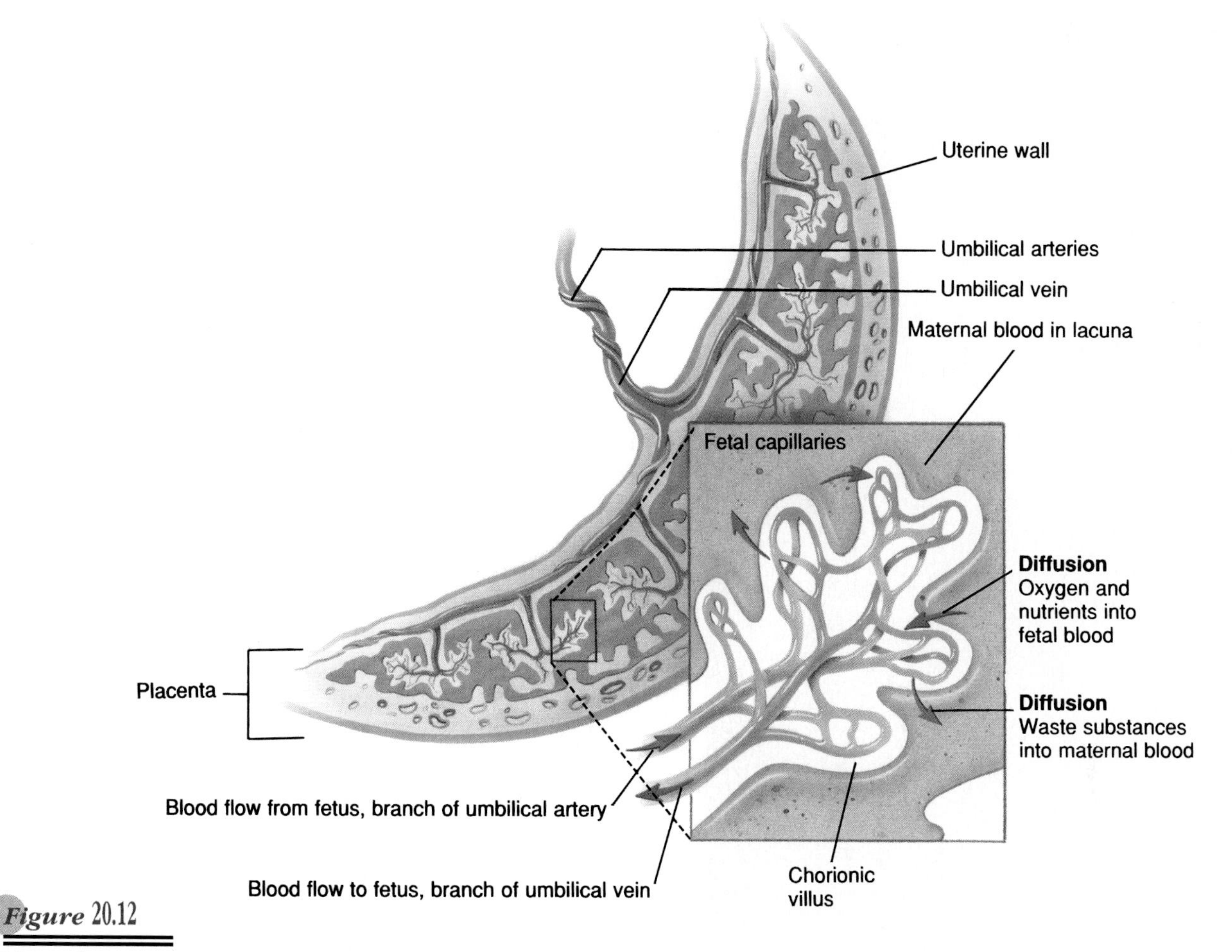

Figure 20.12

The embryonic portion of the placenta consists of the chorion and its villi. The maternal portion is part of the uterine wall to which the chorionic villi attach.

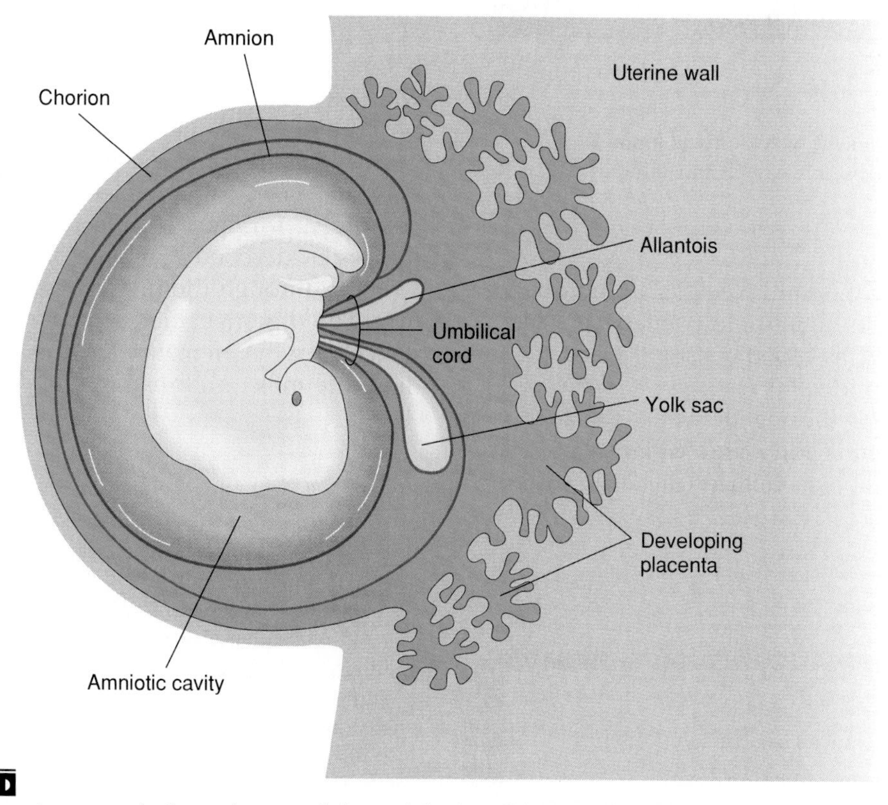

Figure 20.13

As the amnion develops, it surrounds the embryo, and the umbilical cord begins to form from structures in the connecting stalk.

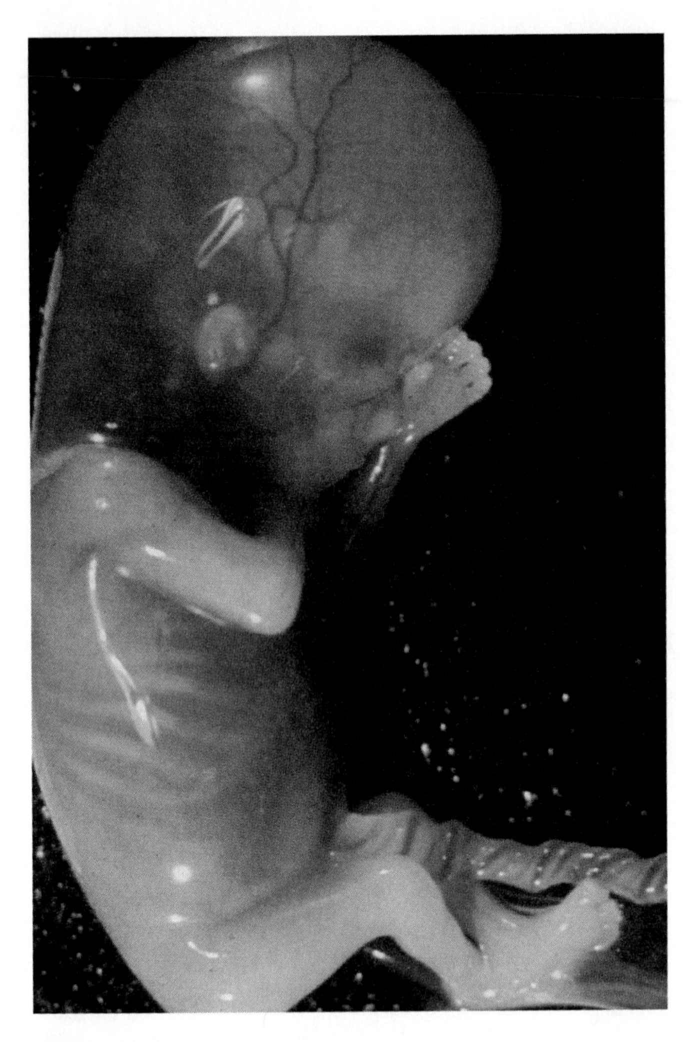

Figure 20.14

By the beginning of the eighth week of development, the embryonic body is recognizable as human (3×).

allantois forms during the third week as a tube extending from the early yolk sac into the connecting stalk of the embryo. It, too, forms blood cells and gives rise to the umbilical arteries and vein.

By the beginning of the eighth week, the embryo is usually 30 millimeters long and weighs less than 5 grams. It is recognizable as human (fig. 20.14).

1. Describe how the placenta forms.
2. What is the function of amniotic fluid?
3. What is the significance of the yolk sac?

Fetal Stage

The **fetal stage** begins at the end of the eighth week of development and lasts until birth. During this period, growth is rapid, and body proportions change considerably. At the beginning of the fetal stage, the head is disproportionately large, and the lower limbs are short. Gradually, proportions become more like those of a child.

During the third month, growth in body length accelerates, but head growth slows. The upper limbs reach the length they will maintain throughout development, and ossification centers appear in most bones. By the twelfth week, the external reproductive organs are distinguishable as male or female.

In the fourth month, the body grows rapidly and reaches a length of up to 20 centimeters. The lower limbs lengthen considerably, and the skeleton continues to ossify.

In the fifth month, growth rate decreases. The lower limbs reach their final proportions. Skeletal muscles become active, and the pregnant woman may first detect fetal movements. Hair appears on the head. Fine, downy hair and a cheesy mixture of dead epidermal cells and sebum from the sebaceous glands covers the skin.

During the sixth month, the fetal body gains substantial weight. The eyebrows and eyelashes appear. The skin is wrinkled and translucent, and blood vessels in the skin give the fetus a reddish appearance.

In the seventh month, fat is deposited in subcutaneous tissues, making the skin smoother. The eyelids, which fused during the third month, reopen. At the end of this month, the fetus is about 40 centimeters long.

In the final trimester, fetal brain cells rapidly form networks, as organs specialize and grow. A layer of fat completes formation beneath the skin. In the male, the testes descend from regions near the developing kidneys, through the inguinal canal, and into the scrotum. The digestive and respiratory systems mature last, which is why premature infants often have difficulty digesting milk and breathing.

Premature infants' survival chances increase directly with age and weight. Survival is more likely if the lungs are sufficiently developed with the thin respiratory membranes necessary for rapid exchange of oxygen and carbon dioxide, and if the lungs produce enough surfactant to reduce alveolar surface tension (see chapter 16). A fetus of less than twenty-four weeks or weighing less than 600 grams at birth seldom survives, even with intensive medical care. *Neonatology* is the medical field that deals with premature and ill newborns. ■

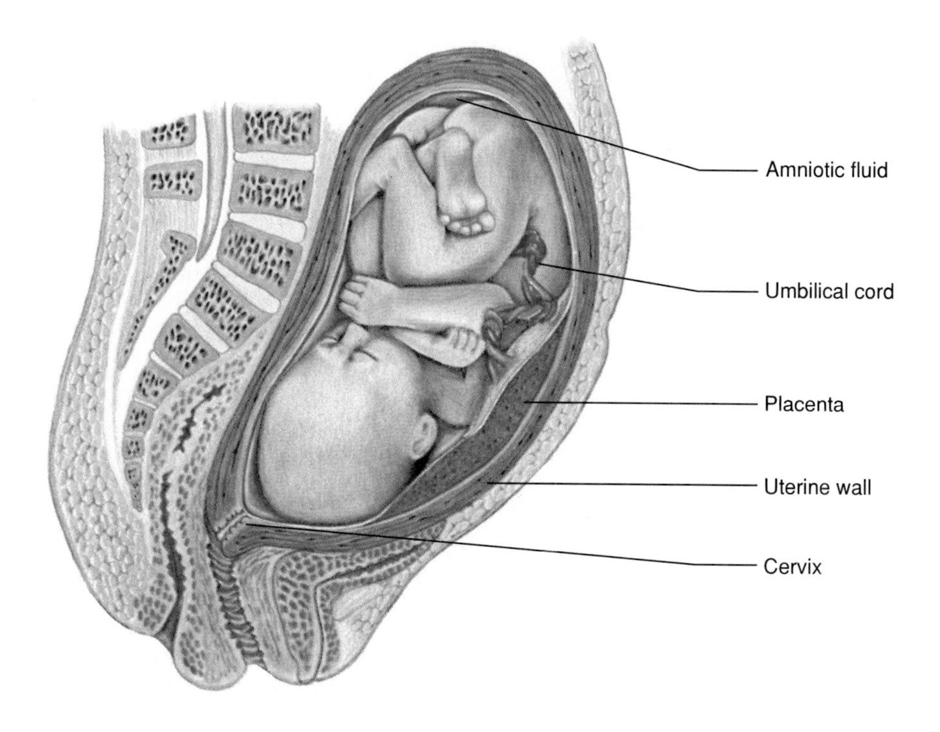

Figure 20.15

A full-term fetus usually is positioned with its head near the cervix.

At the end of the ninth month (on average 266 days), the fetus is *full term*. It is about 50 centimeters long and weighs 2.7–3.6 kilograms. The skin has lost its downy hair, but sebum and dead epidermal cells still coat it. Hair usually covers the scalp. The fingers and toes have well-developed nails. The skull bones are largely ossified. As figure 20.15 shows, the fetus is usually positioned upside down, with its head toward the cervix.

1. What major changes occur during the fetal stage of development?
2. Describe a full-term fetus.

Fetal Blood and Circulation

Throughout fetal development, the maternal blood supplies oxygen and nutrients, and carries away wastes. These substances diffuse between maternal and fetal blood across the placental membrane, and umbilical blood vessels carry them to and from the fetal body.

The fetal blood and vascular system must adapt to intrauterine existence. The concentration of oxygen-carrying hemoglobin in fetal blood is about 50% greater than in maternal blood, and fetal hemoglobin has a greater attraction for oxygen than does adult hemoglobin. At a particular oxygen partial pressure, fetal hemoglobin can carry 20–30% more oxygen than can adult hemoglobin.

In the fetal circulatory system, the umbilical vein transports blood rich in oxygen and nutrients from the placenta to the fetal body. This vein enters the body and extends along the anterior abdominal wall to the liver. About half the blood it carries passes into the liver, and the rest enters a vessel called the **ductus venosus**, which bypasses the liver.

The ductus venosus extends a short distance and joins the inferior vena cava. There, oxygenated blood from the placenta mixes with deoxygenated blood from the lower parts of the fetal body. This mixture continues through the vena cava to the right atrium.

In an adult heart, blood from the right atrium enters the right ventricle and is pumped through the pulmonary trunk and arteries to the lungs (see chapter 13). The fetal lungs, however, are nonfunctional, and blood largely bypasses them. Much of the blood from the inferior vena cava that enters the fetal right atrium is shunted directly into the left atrium through an opening in the

TOPIC OF INTEREST

Some Causes of Birth Defects

Thalidomide

The idea that the placenta protects the embryo and fetus from harmful substances was tragically disproven between 1957 and 1961, when 10,000 children in Europe were born with flippers in place of limbs. Doctors soon determined that the teratogen (the agent causing the birth defect) was the mild tranquilizer *thalidomide,* which all of the mothers of deformed infants had taken early in pregnancy, during the time of prenatal limb formation. The United States was spared a thalidomide disaster because an astute government physician noted the drug's adverse effects on monkeys in experiments, and she halted testing.

Rubella

At about the same time as the thalidomide crisis, another teratogen, a virus, was sweeping the United States. In the early 1960s, a *rubella* (German measles) epidemic caused 20,000 birth defects and 30,000 stillbirths.

Alcohol

A pregnant woman who has just one or two alcoholic drinks a day, or perhaps a large amount at a crucial time in prenatal development, risks *fetal alcohol syndrome* (FAS) in her unborn child. Animal studies show that even small amounts of alcohol can alter fetal brain chemistry. It is best to avoid drinking alcohol entirely when pregnant, or when trying to become pregnant.

A child with FAS has a small head, misshapen eyes, and a flat face and nose (fig. 20B). He or she grows slowly before and after birth. Intellect is impaired, ranging from minor learning disabilities to mental retardation.

Teens and young adults with FAS are short and have small heads. Many remain at an early grade school level of development, and often lack social and communication skills.

In the United States today, FAS is the third most common cause of mental retardation in newborns. One to three of every 1,000 infants has the syndrome—more than 40,000 each year.

Cigarettes

Chemicals in cigarette smoke stress a fetus. Carbon monoxide crosses the placenta and plugs up sites on the fetus's hemoglobin molecules that normally bind oxygen. Other chemicals in smoke prevent nutrients from reaching the fetus. Studies comparing placentas of smokers and nonsmokers show that smoke-exposed placentas lack important growth factors. The result of these assaults is poor growth before and after birth. Cigarette smoking during pregnancy is linked to spontaneous abortion, stillbirth, prematurity, and low birth weight.

Nutrients

Certain nutrients in large amounts, particularly vitamins, act in the body as drugs. The acne medication *isotretinoin* (Accutane) is a derivative of vitamin A that causes spontaneous abortions and defects of the heart, nervous system, and face. A vitamin-A–based drug used to treat psoriasis, as well as excesses of vitamin A itself, also cause birth defects because some forms of vitamin A are stored in body fat for up to three years after ingestion.

Malnutrition in a pregnant woman threatens the fetus. Obstetric records of pregnant women before, during, and after World War II link inadequate nutrition early in pregnancy to an increase in spontaneous abortions. The aborted fetuses had very little brain tissue. Poor nutrition later in pregnancy affects placenta development. The infant has a low birth weight and is at high risk for short stature, tooth decay, delayed sexual development, learning disabilities, and possibly, mental retardation.

Occupational Hazards

The workplace can be a source of teratogens. Women who work with textile dyes, lead, certain photographic chemicals, semiconductor materials, mercury, and cadmium have increased rates of spontaneous abortion and delivering children with birth defects. The male's role in environmentally caused birth defects is not well understood. Men whose jobs expose them to sustained heat, such as smelter workers, glass manufacturers, and bakers, however, may produce sperm that can fertilize an oocyte and possibly lead to spontaneous abortion or a birth defect. A virus or a toxic chemical carried in semen may also cause a birth defect.

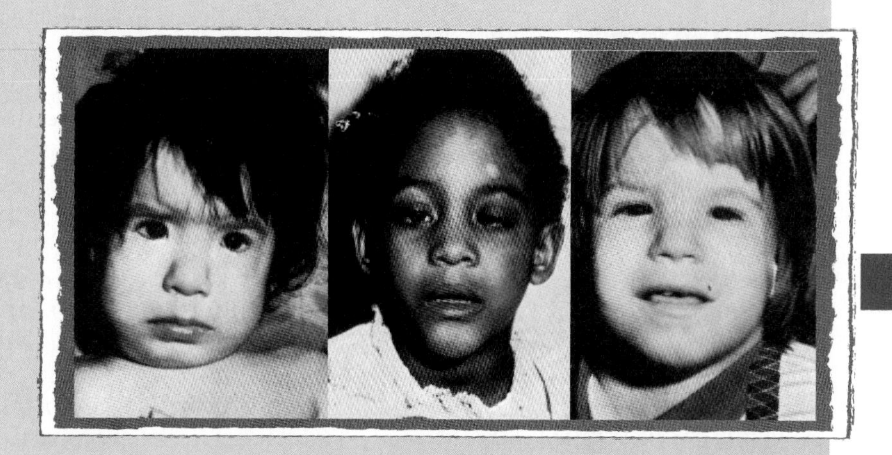

***Figure* 20B**

Fetal alcohol syndrome. Some children whose mothers drank alcohol during pregnancy have characteristic flat faces that are strikingly similar in children of different races. Women who drink excessively while pregnant have a 30–45% chance of having a child affected to some degree by prenatal exposure to alcohol. However, only 6% of exposed offspring have full-blown fetal alcohol syndrome.

atrial septum called the **foramen ovale.** Blood passes through the foramen ovale because blood pressure in the right atrium is somewhat greater than that in the left atrium. Furthermore, a small valve on the left side of the atrial septum overlies the foramen ovale and helps prevent blood from moving in the reverse direction.

The rest of the fetal blood entering the right atrium, including a large proportion of the deoxygenated blood entering from the superior vena cava, passes into the right ventricle and out through the pulmonary trunk. Only a small volume of blood enters the pulmonary circuit because the lungs are collapsed, and their blood vessels have a high resistance to blood flow. Enough blood does reach lung tissues, however, to sustain them.

Most of the blood in the pulmonary trunk bypasses the lungs by entering a fetal vessel called the **ductus arteriosus,** which connects the pulmonary trunk to the descending portion of the aortic arch. As a result of this connection, blood with a relatively low oxygen concentration, which is returning to the heart through the superior vena cava, bypasses the lungs. At the same time, it is prevented from entering the portion of the aorta that branches to the heart and brain.

The more highly oxygenated blood that enters the left atrium through the foramen ovale mixes with a small amount of deoxygenated blood returning from the pulmonary veins. This mixture moves into the left ventricle and is pumped into the aorta. Some of it reaches the myocardium through the coronary arteries, and some reaches the brain tissues through the carotid arteries.

Blood carried by the descending aorta is partially oxygenated and partially deoxygenated. Some of it is carried into the branches of the aorta that lead to the lower regions of the body. The rest passes into the umbilical arteries, which branch from the internal iliac arteries and lead to the placenta. There, the blood is reoxygenated (fig. 20.16).

Table 20.2 summarizes the major features of fetal circulation. At birth, the fetal circulatory system must adjust when the placenta ceases to function and the newborn begins to breathe.

1. Which umbilical vessel carries oxygen-rich blood to the fetus?
2. What is the function of the ductus venosus?
3. How does fetal circulation allow blood to bypass the lungs?

Birth Process

Pregnancy usually continues for thirty-eight weeks from conception, which is forty weeks from the woman's last menstrual period. Pregnancy ends with the *birth process*.

A declining progesterone concentration plays a major role in initiating birth. During pregnancy, progesterone suppresses uterine contractions. As the placenta ages, progesterone concentration within the uterus declines, which stimulates synthesis of a prostaglandin that promotes uterine contractions.

Another stimulant of the birth process is stretching of the uterine and vaginal tissues late in pregnancy. This initiates nerve impulses to the hypothalamus, which, in turn, signals the posterior pituitary gland to release the hormone **oxytocin** (see chapter 11). Oxytocin stimulates powerful uterine contractions. Combined with the greater excitability of the myometrium due to the decline in progesterone secretion, oxytocin aids *labor* in its later stages.

During labor, rhythmic, muscular contractions begin at the top of the uterus and extend down its length. Since the fetus is usually positioned head downward, labor contractions force the head against the cervix (fig. 20.17). This action stretches the cervix, which elicits a reflex that stimulates still stronger labor contractions. Thus, a *positive feedback system* operates, in which uterine contractions produce more intense uterine contractions. At the same time, cervix dilation reflexly stimulates the posterior pituitary to increase oxytocin release. As labor continues, positive feedback stimulates abdominal wall muscles to contract, which also helps force the fetus through the cervix and vagina to the outside.

The umbilical cord usually contains two arteries and one vein. A small percentage of newborns have only one umbilical artery. Since this condition is often associated with other cardiovascular disorders, the vessels within the severed cord are routinely counted following birth. ■

An infant passing through the birth canal can tear the delicate tissues between the vulva and anus (perineum). An *episiotomy* is a clean cut a physician makes in the perineal tissues to avoid a ragged tear. ■

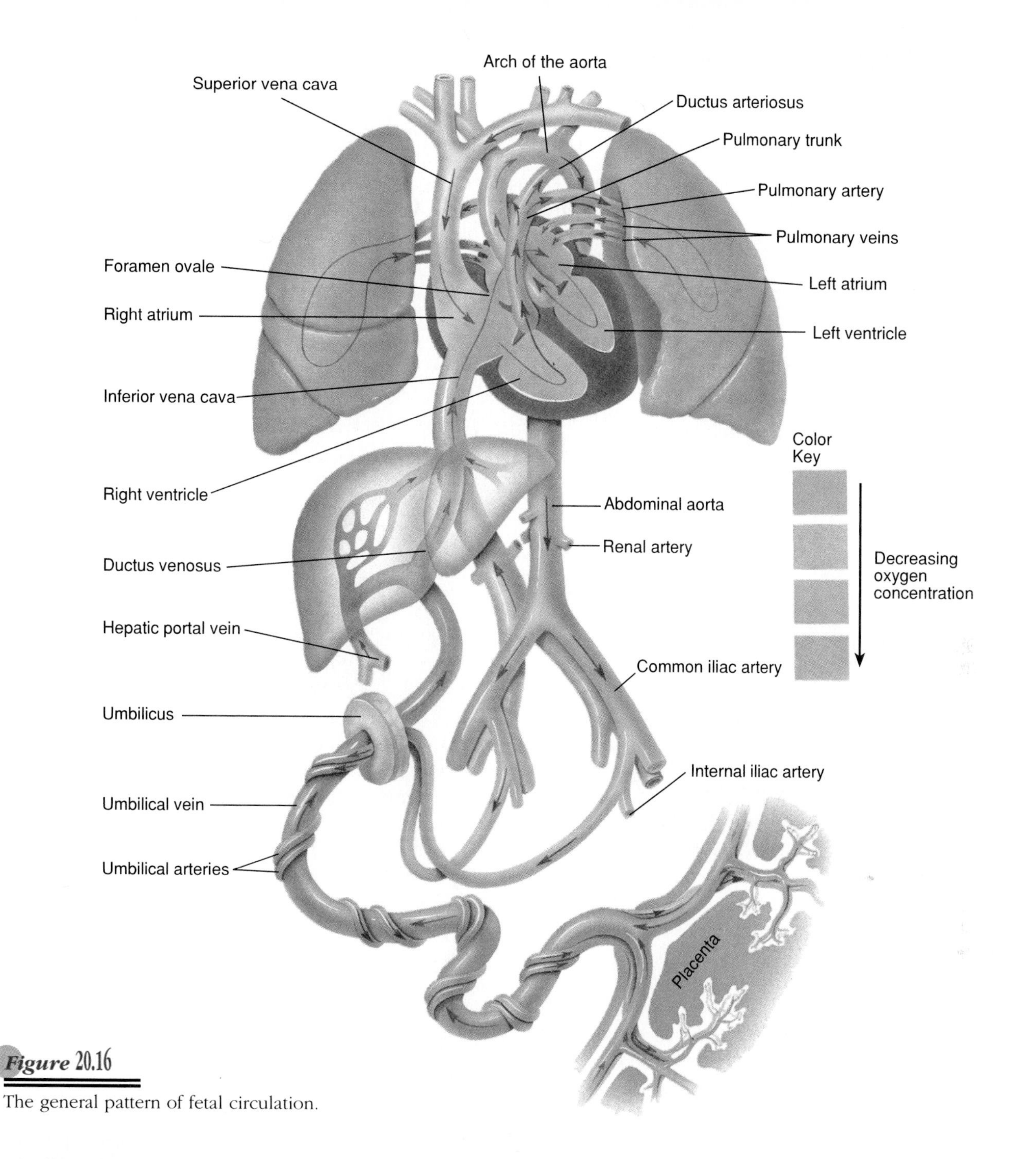

Figure 20.16

The general pattern of fetal circulation.

Table 20.2

Fetal Circulatory Adaptations

Adaptation	Function
Fetal blood	Has greater oxygen-carrying capacity than blood in an adult
Umbilical vein	Carries oxygenated blood from placenta to fetus
Ductus venosus	Conducts about half the blood from umbilical vein directly to inferior vena cava, bypassing liver
Foramen ovale	Conveys much blood entering right atrium from inferior vena cava, through atrial septum, and into left atrium, bypassing lungs
Ductus arteriosus	Conducts some blood from pulmonary trunk to aorta, bypassing lungs
Umbilical arteries	Carry blood from internal iliac arteries to placenta

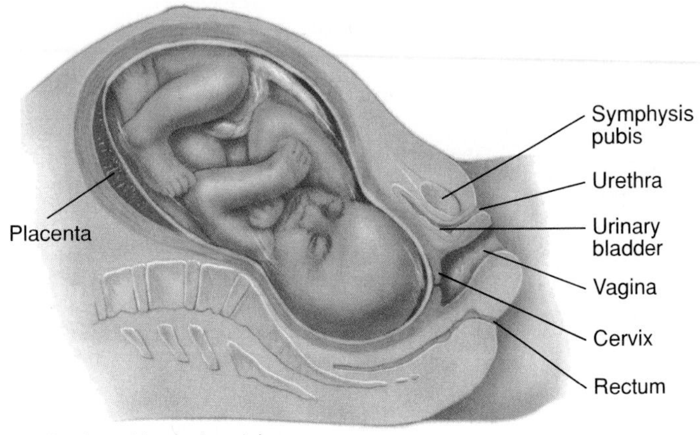

(a) Fetal position before labor

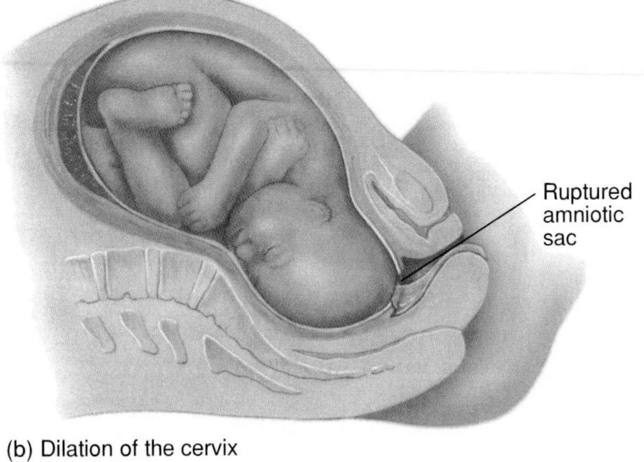

(b) Dilation of the cervix

Placenta

(c) Expulsion of the fetus

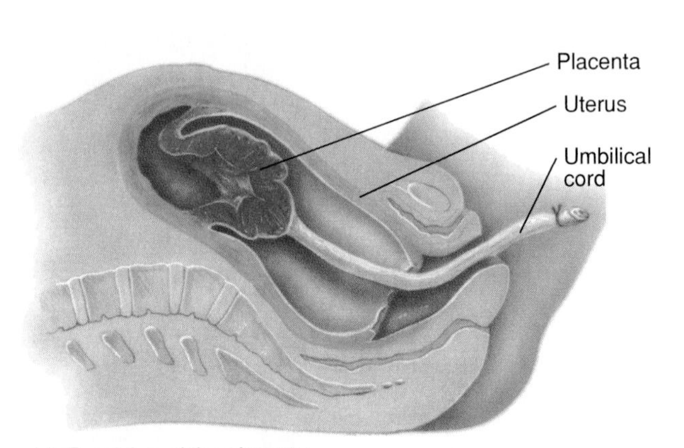

(d) Expulsion of the placenta

Figure 20.17

Stages in birth. (*a*) Fetal position before labor. (*b*) Dilation of the cervix. (*c*) Expulsion of the fetus. (*d*) Expulsion of the placenta.

Following birth of the fetus, the placenta separates from the uterine wall, and uterine contractions expel it through the birth canal. Bleeding accompanies the expelled placenta, termed the *afterbirth,* because the separation damages vascular tissues. Oxytocin stimulates continued uterine contraction, however, which compresses the bleeding vessels and minimizes blood loss.

1. Describe the role of progesterone in initiating labor.
2. Explain how dilation of the cervix affects labor.

Postnatal Period

Following birth, both mother and newborn experience physiological and structural changes.

Milk Production and Secretion

During pregnancy, placental estrogen and progesterone stimulate further development of the mammary glands. Estrogen causes the ductile systems to grow and branch, and deposits abundant fat around them. Progesterone stimulates development of the alveolar glands at the ends of the ducts. Placental lactogen also promotes these changes.

TOPIC OF INTEREST: Fetal Testing

Ultrasound

Many pregnant women undergo *ultrasound* exams. Sound waves are bounced off of the embryo or fetus, and the pattern of deflected sound waves is converted into an image. Sound waves are applied externally with a transducing device on the woman's abdomen or internally with a device inserted into the vagina.

Ultrasound is used early in pregnancy to estimate date of conception and to confirm viability. Later in the pregnancy, ultrasound can visualize major structures, the heartbeat, and even yawns, kicks, and thumb sucking. The technique can detect some types of birth defects.

Chorionic Villus Sampling

Chorionic villus sampling (CVS) examines the chromosomes in chorionic villus cells to obtain information on fetal health. Villus cells are genetically identical to fetal cells because they derive from the same fertilized egg. The test carries a risk of causing miscarriage. Thus, only women who have previously had a child with a detectable chromosome abnormality usually have the test. CVS is performed at the tenth week of gestation.

Amniocentesis

Amniocentesis is performed after the fourteenth week of gestation. A physician uses ultrasound as a guide to insert a needle into the amniotic sac and withdraw about 5 milliliters of fluid. Fetal fibroblasts in the fluid are cultured and their chromosomes checked. It takes about a week to grow these cells. A faster technique uses DNA probes to highlight specific chromosomes.

Amniocentesis carries about a 0.5% chance of causing miscarriage. Only women whose risk of having a fetus with a chromosomal anomaly equals or exceeds the risk of the procedure have amniocentesis. This includes women of any age who have had a child with a detectable chromosomal abnormality and women over age thirty-five. (Older women are more likely to produce egg cells that have extra or missing chromosomes, producing abnormal fetuses.)

Hormonal activity doubles breast size during pregnancy, and the mammary glands become capable of secreting milk. Milk is not secreted, however, because placental progesterone inhibits milk production, and placental lactogen blocks the action of *prolactin* (see chapter 11).

Following childbirth and the expulsion of the placenta, maternal blood concentrations of placental hormones decline rapidly. In two or three days, prolactin, which is no longer inhibited, stimulates the mammary glands to secrete milk. Meanwhile, the glands secrete a thin, watery fluid called *colostrum*. Colostrum is rich in proteins, but its carbohydrate and fat concentrations are lower than those of milk. Colostrum contains antibodies from the mother's immune system that protect the newborn from certain infections.

Milk ejection requires contraction of specialized *myoepithelial cells* surrounding the alveolar glands. Suckling or mechanical stimulation of the nipple or areola elicits the reflex action that controls this process. Impulses from sensory receptors within the breasts go to the hypothalamus, which signals the posterior pituitary gland to release oxytocin. Oxytocin travels in the bloodstream to the breasts and stimulates myoepithelial cells to contract. As a result, milk is ejected into a suckling infant's mouth in about 30 seconds (fig. 20.18).

As long as milk is removed from the breasts, prolactin and oxytocin release continues, and the mammary glands produce milk. If milk is not removed regularly, the hypothalamus inhibits prolactin secretion, and within about one week, the mammary glands stop producing milk.

Human milk is the best possible food for human babies. Milk of other animals contains different concentrations of nutrients than human milk.

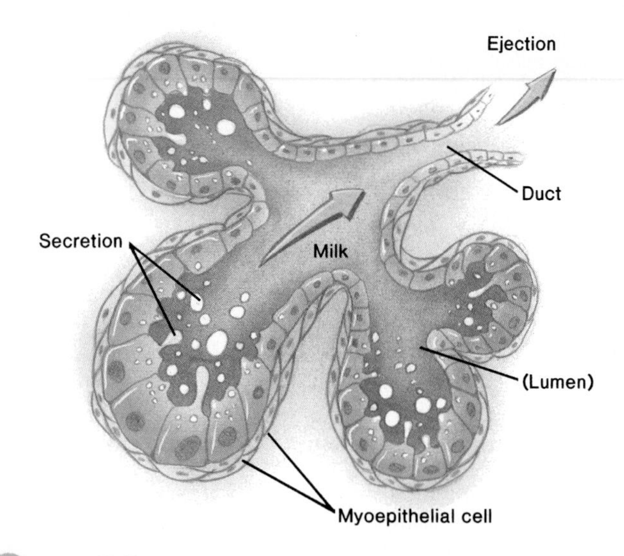

Figure 20.18

Myoepithelial cells eject milk from an alveolar gland.

1. How does pregnancy affect the mammary glands?
2. What stimulates the mammary glands to produce milk?
3. What causes milk to flow into the ductile system of a mammary gland?

Neonatal Period

The **neonatal period** begins abruptly at birth and extends to the end of the first four weeks. At birth, the newborn must make quick physiological adjustments to become self-reliant. It must respire, obtain and digest nutrients, excrete wastes, and regulate body temperature.

A newborn's most immediate requirement is to obtain oxygen and excrete carbon dioxide. The first breath must be particularly forceful because the newborn's lungs are collapsed, and its small airways offer considerable resistance to air movement. Also, surface tension adheres the moist membranes of the lungs. The lungs of a full-term fetus continuously secrete *surfactant* (see chapter 16), however, which reduces surface tension. Thus, after the first powerful breath begins to expand the lungs, breathing eases.

The newborn has a high metabolic rate, and its immature liver may be unable to supply enough glucose to support its metabolic requirements. Consequently, the newborn typically utilizes stored fat for energy.

1. Why must an infant's first breath be particularly forceful?
2. What does a newborn use for energy during its first few days of life?

A newborn's kidneys are usually unable to produce concentrated urine, so they excrete a dilute fluid. For this reason, the newborn may become dehydrated, and develop a water and electrolyte imbalance. Also, some of the newborn's homeostatic control mechanisms may not function adequately. For example, the temperature-regulating system may be unable to maintain a constant body temperature.

When the placenta ceases to function and breathing begins, the newborn's circulatory system also changes. Following birth, the umbilical vessels constrict. The umbilical arteries close first, and if the umbilical cord is not clamped or severed for a minute or so, blood continues to flow from the placenta to the newborn through the umbilical vein, adding to the newborn's blood volume. Similarly, the ductus venosus constricts shortly after birth and appears in the adult as a fibrous cord (ligamentum venosum) superficially embedded in the wall of the liver.

The foramen ovale closes from the blood pressure changes in the right and left atria as fetal vessels constrict. As blood ceases to flow from the umbilical vein into the inferior vena cava, blood pressure in the right atrium falls. Also, as the lungs expand with the first breathing movements, resistance to blood flow through the pulmonary circuit decreases, more blood enters the left atrium through the pulmonary veins, and blood pressure in the left atrium increases.

As blood pressure in the left atrium rises and that in the right atrium falls, the valve on the left side of the atrial septum closes the foramen ovale. In most individuals, this valve gradually fuses with the tissues along the margin of the foramen. In an adult, a depression called the *fossa ovalis* marks the site of the previous opening.

The ductus arteriosus, like the other fetal vessels, constricts after birth. After the ductus arteriosus closes, blood can no longer bypass the lungs by moving from the pulmonary artery directly into the aorta. In an adult, a cord called the *ligamentum arteriosum* represents the ductus arteriosus.

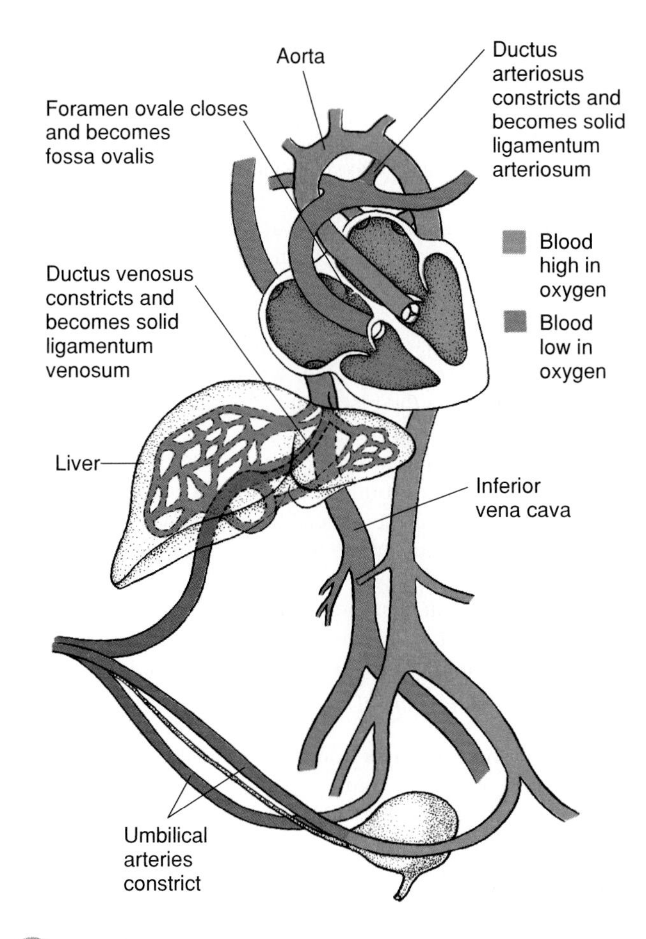

***Figure* 20.19**

Major changes in the newborn's circulatory system.

In *patent ductus arteriosus* (PDA), the ductus arteriosus fails to close completely. This condition is common in newborns whose mothers were infected with rubella virus (German measles) during the first three months of pregnancy.

After birth, the metabolic rate and oxygen consumption in neonatal tissues increase, in large part to maintain body temperature. If the ductus arteriosus remains open, the newborn's blood oxygen concentration may be too low to adequately supply tissues, including the myocardium. If PDA is not corrected surgically, the heart may fail, even though the myocardium is normal. ■

Changes in the newborn's circulatory system are gradual. Constriction of the ductus arteriosus may be functionally complete within 15 minutes, but the permanent closure of the foramen ovale may take up to a year.

Recall that fetal hemoglobin is slightly different and has a greater affinity for oxygen than the adult type. Fetal hemoglobin production falls after birth, and by the time an infant is four months old, most of the circulating hemoglobin is the adult type. Figure 20.19 illustrates circulatory changes in the newborn.

1. How do the kidneys of a newborn differ from those of an adult?
2. What changes occur in the newborn's circulatory system?

Clinical Terms Related to Pregnancy, Growth, and Development

Abruptio placentae (ab-rup′she-o plah-cen′tā) Premature separation of the placenta from the uterine wall.

Dizygotic twins (di″zi-got′ik twinz) Twins resulting from two sperm cells fertilizing two egg cells.

Hydatid mole (hi′dah-tid mōl) A type of uterine tumor that originates from placental tissue.

Hydramnios (hi-dram′ne-os) Excess amniotic fluid.

Intrauterine transfusion (in″trah-u′ter-in trans-fu′zhun) Transfusion administered by injecting blood into the fetal peritoneal cavity before birth.

Lochia (lo′ke-ah) Vaginal discharge following childbirth.

Meconium (mĕ-ko′ne-um) Anal discharge from the digestive tract of a full-term fetus or a newborn.

Monozygotic twins (mon″o-zi-got′ik twinz) Twins resulting from one sperm cell fertilizing one egg cell, which then splits.

Perinatology (per″ĭ-na-tol′o-je) Branch of medicine concerned with the fetus after twenty-five weeks of development and with the newborn for the first four weeks after birth.

Postpartum (pōst-par′tum) Occurring after birth.

Teratology (ter″ah-tol′o-je) Study of substances that cause abnormal development and congenital malformations.

Trimester (tri-mes′ter) Each third of the total period of pregnancy.

Ultrasonography (ul″trah-son-og′rah-fe) Technique to visualize the size and position of fetal structures from patterns of deflected ultrasonic waves.

Summary Outline

Introduction (p. 549)

Growth is an increase in size. Development is the process of changing from one life phase to another.

Pregnancy (p. 549)

Pregnancy is the presence of a developing offspring in the uterus.

1. Transport of sex cells
   a. A male deposits semen in the vagina during sexual intercourse.
   b. A sperm cell lashes its tail to move, and muscular contractions in the female reproductive tract help.
2. Fertilization
   a. An enzyme helps a sperm cell penetrate the zona pellucida.
   b. When a sperm cell penetrates an egg cell membrane, changes in the membrane and the zona pellucida prevent entry of additional sperm cells.
   c. Fusion of the nuclei of a sperm cell and an egg cell completes fertilization.
   d. The product of fertilization is a zygote with forty-six chromosomes.

Prenatal Period (p. 550)

1. Early embryonic development
   a. Cells undergo mitosis, giving rise to smaller and smaller cells during cleavage.
   b. The developing offspring moves down the uterine tube to the uterus, where it implants in the endometrium.
   c. The offspring is called an embryo from the second through the eighth week of development. Thereafter, it is a fetus.
   d. Eventually, embryonic and maternal cells form a placenta.
2. Hormonal changes during pregnancy
   a. Embryonic cells produce human chorionic gonadotropin (hCG), which maintains the corpus luteum.
   b. Placental tissue produces high concentrations of estrogen and progesterone.
      (1) Estrogen and progesterone maintain the uterine wall and inhibit secretion of follicle-stimulating hormone (FSH) and luteinizing hormone (LH).
      (2) Progesterone and relaxin inhibit contraction of uterine muscles.
      (3) Estrogen enlarges the vagina.
      (4) Relaxin helps relax the ligaments of the pelvic joints.
   c. Placental lactogen stimulates development of the breasts and mammary glands.
   d. During pregnancy, increased aldosterone secretion promotes retention of sodium and body fluid. Increased secretion of parathyroid hormone helps maintain a high concentration of maternal blood calcium.
3. Embryonic stage
   a. The embryonic stage extends from the end of the second week through the eighth week of development.
   b. During this stage, the placenta and main internal and external body structures develop.
   c. The cells of the inner cell mass organize into primary germ layers.
   d. The embryonic disk becomes cylindrical and attaches to the developing placenta.
   e. The placental membrane consists of the epithelium of the chorionic villi and the epithelium of the capillaries inside the villi.
      (1) Oxygen and nutrients diffuse from maternal blood across the placental membrane and into fetal blood.
      (2) Carbon dioxide and other wastes diffuse from fetal blood across the placental membrane and into maternal blood.
   f. A fluid-filled amnion develops around the embryo.
   g. The umbilical cord forms as the amnion envelops the tissues attached to the underside of the embryo.
   h. The yolk sac forms on the underside of the embryonic disk.
   i. The allantois extends from the yolk sac into the connecting stalk.
   j. By the beginning of the eighth week, the embryo is recognizable as human.
4. Fetal stage
   a. The fetal stage extends from the end of the eighth week of development until birth.
   b. Existing structures grow and mature. Only a few new parts appear.
   c. The fetus is full term at the end of the tenth month (thirty-eight weeks from conception).
5. Fetal blood and circulation
   a. Umbilical vessels carry blood between the placenta and the fetus.
   b. Fetal blood carries a greater concentration of oxygen than does maternal blood.
   c. Blood enters the fetus through the umbilical vein and partially bypasses the liver through the ductus venosus.
   d. Blood enters the right atrium and partially bypasses the lungs through the foramen ovale.
   e. Blood entering the pulmonary trunk partially bypasses the lungs through the ductus arteriosus.
   f. Blood enters the umbilical arteries from the internal iliac arteries.
6. Birth process
   a. During pregnancy, placental progesterone inhibits uterine contractions.
   b. A variety of factors promote birth.
      (1) A decreasing progesterone concentration and the release of a prostaglandin initiate the birth process.
      (2) The posterior pituitary gland releases oxytocin.
      (3) Oxytocin stimulates uterine muscles to contract, and labor begins.
   c. Following birth, placental tissues are expelled.

Postnatal Period (p. 566)

1. Milk production and secretion
   a. Following childbirth, concentrations of placental hormones decline, the action of prolactin is no longer blocked, and the mammary glands begin to secrete milk.
   b. Reflex response to mechanical stimulation of the nipple stimulates the posterior pituitary to release oxytocin, which causes the alveolar ducts to eject milk.
2. Neonatal period
   a. The neonatal period extends from birth to the end of the first four weeks.
   b. The newborn must begin to respire, obtain nutrients, excrete wastes, and regulate body temperature.
   c. The first breath must be powerful to expand the lungs.
   d. The liver is immature and unable to supply sufficient glucose, so the newborn depends primarily on stored fat for energy.
   e. A newborn's immature kidneys cannot concentrate urine well.
   f. A newborn's homeostatic mechanisms may function imperfectly, and body temperature may be unstable.
   g. The circulatory system changes when placental circulation ceases.
      (1) Umbilical vessels constrict.
      (2) The ductus venosus constricts.
      (3) A valve closes the foramen ovale as blood pressure in the right atrium falls and pressure in the left atrium rises.
      (4) The ductus arteriosus constricts.

Review Exercises

1. Define *growth* and *development*. (p. 549)
2. Define *pregnancy*. (p. 549)
3. Describe how sperm cells move within the female reproductive tract. (p. 549)
4. Describe the process of fertilization. (p. 549)
5. Define *cleavage*. (p. 550)
6. Describe the process of implantation. (p. 550)
7. Define *embryo*. (p. 550)
8. Define *fetus*. (p. 550)
9. Explain the major hormonal changes in the maternal body during pregnancy. (p. 552)
10. Explain how the primary germ layers form. (p. 557)
11. List the major body parts derived from ectoderm, mesoderm, and endoderm. (p. 557)
12. Define *placental membrane*. (p. 557)
13. Describe the formation of the placenta, and explain its functions. (p. 557)
14. Distinguish between the chorion and the amnion. (p. 558)
15. Explain the function of amniotic fluid. (p. 558)
16. Describe the formation of the umbilical cord. (p. 558)
17. Explain how the yolk sac and allantois form. (p. 558)
18. List the major changes in the fetal stage of development. (p. 560)
19. Describe a full-term fetus. (p. 561)
20. Compare the properties of fetal hemoglobin with those of adult hemoglobin. (p. 561)
21. Trace the pathway of blood from the placenta to the fetus and back to the placenta. (p. 561)
22. Discuss the events of the birth process. (p. 564)
23. Explain the roles of prolactin and oxytocin in milk production and secretion. (p. 566)
24. Explain why a newborn's first breath must be particularly forceful. (p. 568)
25. Explain why newborns tend to develop water and electrolyte imbalances. (p. 568)
26. Describe the changes in the newborn's circulatory system. (p. 568)

Critical Thinking

1. Why can twins resulting from a single fertilized egg exchange blood or receive organ transplants from each other without rejection, but twins resulting from two fertilized eggs sometimes cannot?
2. What symptoms may appear if a newborn's ductus arteriosus fails to close?
3. What kinds of studies and information are required to determine whether a man's exposure to a potential teratogen can cause birth defects years later? How would such analysis differ if a woman was exposed?
4. In Aldous Huxley's book *Brave New World*, egg cells are fertilized in vitro and develop assembly-line style. To render some of the embryos less intelligent, lab workers give them alcohol. What medical condition does this scenario invoke?
5. What technology would enable a fetus born in the fourth month to survive in a laboratory setting?
6. Toxins usually cause more severe medical problems if exposure is during the first eight weeks of pregnancy rather than during the later weeks. Why?
7. Milk from a cow has a higher percentage of protein and a lower percentage of fat than does human milk. Why do you think this is so?

Aids to Understanding Words

acetabul-, vinegar cup: *acetabul*um
adip-, fat: *adip*ose tissue
agglutin-, to glue together: *agglutin*ation
aliment-, food: *aliment*ary canal
allant-, sausage-shaped: *allant*ois
alveol-, small cavity: *alveol*us
an-, without: *an*aerobic respiration
ana-, up: *ana*bolic
andr-, man: *andr*ogens
append-, to hang something: *append*icular
ax-, axis: *ax*ial skeleton
bil-, bile: *bil*irubin
-blast, budding: osteo*blast*
brady-, slow: *brady*cardia
bronch-, windpipe: *bronch*us
calat-, something inserted: inter*calat*ed disk
calyc-, small cup: *calyc*es
cardi-, heart: peri*cardi*um
carp-, wrist: *carp*als
cata-, down: *cata*bolic
chondr-, cartilage: *chondr*ocyte
chorio-, skin: *chorio*n
choroid, skinlike: *choroid*
chym-, juice: *chym*e
-clast, broken: osteo*clast*
cleav-, to divide: *cleav*age
cochlea, snail: *cochlea*
condyl-, knob: *condyl*e
corac-, beaklike: *corac*oid process
cort-, covering: *cort*ex
cribr-, sievelike: *cribr*iform plate
cric-, ring: *cric*oid cartilage
crin-, to secrete: endo*crin*e
crist-, ridge: *crist*a galli
cut-, skin: sub*cut*aneous
cyt-, cell: *cyt*oplasm
de-, undoing: *de*amination
decidu-, falling off: *decidu*ous
dendr-, tree: *dendr*ite
derm-, skin: *derm*is
detrus-, to force away: *detrus*or muscle
di-, two: *di*saccharide
diastol-, dilation: *diastol*e
diuret-, to pass urine: *diuret*ic
dors-, back: *dors*al
ejacul-, to shoot forth: *ejacul*ation
embol-, stopper: *embol*us
endo-, within: *endo*plasmic reticulum
epi-, upon: *epi*thelial tissue
erg-, work: syn*erg*ist
erythr-, red: *erythr*ocyte
exo-, outside: *exo*crine gland
extra-, outside: *extra*cellular
fimb-, fringe: *fimb*riae
follic-, small bag: hair *follic*le
fov-, pit: *fov*ea
funi-, small cord or fiber: *funi*culus
gangli-, a swelling: *gangli*on
gastr-, stomach: *gastr*ic gland
-gen, to be produced: aller*gen*
-genesis, origin: spermato*genesis*
germ-, to bud or sprout: *germ*inal
glen-, joint socket: *glen*oid cavity
-glia, glue: neuro*glia*
glom-, little ball: *glom*erulus
glyc-, sweet: *glyc*ogen
-gram, something written: electrocardio*gram*
hema-, blood: *hema*toma
hemo-, blood: *hemo*globin
hepat-, liver: *hepat*ic duct
homeo-, same: *homeo*stasis
humor-, fluid: *humor*al
hyper-, above: *hyper*tonic
hypo-, below: *hypo*tonic
im-, (or in-), not: *im*balance
immun-, free: *immun*ity
inflamm-, to set on fire: *inflamm*ation
inter-, between: *inter*phase
intra-, inside: *intra*membranous
iris, rainbow: *iris*
iso-, equal: *iso*tonic
kerat-, horn: *kerat*in
labi-, lip: *labi*a
labyrinth, maze: *labyrinth*
lacri-, tears: *lacri*mal gland
lacun-, pool: *lacun*a
laten-, hidden: *laten*t
-lemm, rind or peel: neuri*lemm*a
leuko-, white: *leuko*cyte
lingu-, tongue: *lingu*al tonsil

lip-, fat: *lip*ids
-logy, study of: physio*logy*
-lyte, dissolvable: electro*lyte*
macro-, large: *macro*phage
macula, spot: *macula* lutea
meat-, passage: auditory *meat*us
melan-, black: *melan*in
mening-, membrane: *mening*es
mens-, month: *mens*trual
meta-, change: *meta*bolism
mict-, to pass urine: *mict*urition
mit-, thread: *mit*osis
mono-, one: *mono*saccharide
mons-, mountain: *mons* pubis
morul-, mulberry: *morul*a
moto-, moving: *mot*or
mut-, change: *mut*ation
myo-, muscle: *myo*fibril
nat-, to be born: pre*nat*al
nephr-, kidney: *nephr*on
neutr-, neither one nor the other: *neutr*al
nod-, knot: *nod*ule
nutri-, nourish: *nutri*ent
odont-, tooth: *odont*oid process
olfact-, to smell: *olfact*ory
-osis, abnormal increase in production: leukocyt*osis*
oss-, bone: *oss*eous tissue
papill-, nipple: *papill*ary muscle
para-, beside: *para*thyroid glands
pariet-, wall: *pariet*al membrane
patho-, disease: *patho*gen
pelv-, basin: *pelv*ic cavity
peri-, around: *peri*cardial membrane
phag-, to eat: *phag*ocytosis
pino-, to drink: *pino*cytosis
pleur-, rib: *pleur*al membrane
plex-, interweaving: choroid *plex*us
poie-, to make: hemoto*poie*sis
poly-, many: *poly*unsaturated
pseudo-, false: *pseudo*stratified epithelium
puber-, adult: *puber*ty
pylor-, gatekeeper: *pylor*ic sphincter
sacchar-, sugar: mono*sacchar*ide
sarco-, flesh: *sarco*plasm
scler-, hard: *scler*a
seb-, grease: *seb*aceous gland
sens-, feeling: *sens*ory neuron
-som, body: ribo*some*
squam-, scale: *squam*ous epithelium
stasis-, standing still: homeo*stasis*
strat-, layer: *strat*ified
syn-, together: *syn*thesis
systol-, contraction: *systol*e
tachy-, rapid: *tachy*cardia
tetan-, stiff: *tetan*ic
thromb-, clot: *thromb*ocyte
toc-, birth: oxy*toc*in
-tomy, cutting: ana*tomy*
trigon-, triangle: *trigon*e
troph-, well fed: muscular hyper*troph*y
-tropic, influencing: adrenocortico*tropic*
tympan-, drum: *tympan*ic membrane
umbil-, navel: *umbil*ical cord
ventr-, belly or stomach: *ventr*icle
vill-, hair: *vill*i
vitre-, glass: *vitre*ous humor
zym-, ferment: en*zym*e

Appendix B Related Web Sites

If you have problems linking to any of the web sites listed, please visit the homepage for *Hole's Essentials of Human Anatomy and Physiology,* sixth edition, by Shier et al. at www.mhhe.com/sciencemath/biology/holeessentials/ for the most current web sites related to this textbook.

Chapter 1: Introduction to Human Anatomy and Physiology

Items of General Interest

http://www.merck.com/!!s9OfD17kBs9Oj532jO/pubs/mmanual/html/sectoc.htm
This site is the table of contents for the Internet version of the Merck Manual and has fairly technical explanations for most disorders.

http://www.hhmi.org/beyondbio101
The web site from the Howard Hughes Medical Institute has numerous links to biology education sites. There is a long discussion about undergraduate teaching innovations as well as interviews with professional biologists. You can find information about careers in the Biological Sciences and how science works today.

http://physiology.med.cornell.edu/wwwvl/physioWeb.html
This is a physiology master web site with links to many physiology topics.

http://www.med.wright.edu/SOM/academic/anatomy/beyond.html
Various imaging techniques are used to display the body: cross sections, CT scans, X rays. It includes self-testing.

http://www.npac.syr.edu/projects/vishuman/VisibleHuman.html
Pictures from the Visible Human Project. This project used an actual cadaver and presented photographs and other imaging techniques of sections of the person.

http://www.scar.rad.washington.edu/RadAnatomy.html
X-ray images from various perspectives.

http://www9.biostr.washington.edu/da.html
A human anatomy atlas illustrating numerous systems.

http://www.largnet.uwo.ca/med/i-way.html
CT and ultrasounds of different body systems.

http://www.lifeart.com/
Free samples from the anatomy and physiology clipart packages come out each month. You are invited to download the images.

http://www.gdb.org/Dan/softsearch/biol-links.html
A place to look for biological software.

Chapter 2: Chemical Basis of Life

http://the-tech.mit.edu/Chemicool/
A beautifully colored periodic table where each element is linked to further tables that give more detailed information about that element.

http://members.aol.com/jeff555555/table/ptable.html
Another periodic table with links: this one loads more quickly.

http://www.virtual-pc.com/mindweb/merlin/alchemy/alchmore.html#principles
This is an introduction to an Internet college-level chemistry course. There is a small fee and the purchase of a book.

http://studentweb.tulane.edu/~riker/chempoin.html
Chemistry department of Tulane University web site has downloadable chemistry software including clipart.

http://yip5.chem.wfu.edu/yip/organic/compgraf.html#viz
Movies of models of atoms are shown.

Chapter 3: Cells

http://www.chemie.fu-berlin.de/chemistry/bio/amino-acids.html
This has a list of amino acids. Clicking on the name of the amino acid will bring up another page showing different models of it.

http://yip5.chem.wfu.edu/yip/organic/compgraf.html#viz
A linked movie at this site illustrates a rotating model of DNA. A model of an enzyme also can be viewed from a link on this site.

http://lenti.med.umn.edu/~mwd/cell_www/cell_intro.html
Some useful information about the chemical makeup of cells.

http://esg-www.mit.edu:8001/esgbio/cb/org/animal.gif
A beautifully colored picture of an animal cell but with no links.

http://lenti.med.umn.edu/~mwd/cell_www/cell_intro.html
A good picture of the cell membrane and its constituents.

http://lenti.med.umn.edu/~mwd/cell_www/chapter2/membrane.html#SEMIPERM
Additional details about the actual chemicals making up the cell membrane and a discussion of osmosis and diffusion.

http://lenti.med.umn.edu/~mwd/cell_www/chapter2/protein.html#MEMBRANE
Active transport of proteins is discussed along with linked pictures illustrating their action.

http://cancer.med.upenn.edu/resources/p53/
This page provides links to many other places with information about the p53 tumor-suppressor factor. It provides additional links related to cancer as well.

http://lenti.med.umn.edu/~mwd/cell_www/chapter2/mitochondria.html
A picture and brief description of the mitochondrion.

Chapter 4: Cellular Metabolism

http://www.ornl.gov/TechResources/Human_Genome/home.html
This is the home page of the Human Genome Project. There are links to sites related to specific conditions and chromosomes.

http://www.informatik.uni-rostock.de/HUM-MOLGEN/index.html
Use this address instead of the above for the same information. This is a mirror site. These are handy because sometimes the mirror is busy or unavailable for some other reason.

Chapter 6: Skin and the Integumentary System

http://www.medic.mie-u.ac.jp/derma/anatomy.html
This has a very clear color micrograph of a skin section with clickable areas for more details on certain layers, cells, or subjects. The site includes links to other dermatology pages.

http://www-sci.lib.uci.edu/~martindale/Medical.html#Derm
Links to many good dermatology sites but no skin information in it.

http://www.maui.net/~southsky/introto.html
A very useful page for skin cancer information. There are a number of links and detailed explanations.

http://www.pinch.com/skin/
This is a page with links to several skin disease pages and includes a search link.

http://www.hkma.com.hk/std/menuskin.htm
Very detailed information about different types of eczema.

http://www.ehap.musc.edu/Science/Silver.html
Detailed technical information about scleroderma.

http://www-medlib.med.utah/WebPath/INFEHTML/INFECIDX.html
Pictures of pathological conditions including skin parasites such as lice and mites.

Chapter 7: Skeletal System

Bone Tissue

http://weber.u.washington.edu/~dboone/key/subjects/arthritis/xxxxxxxx1_1.html
Links to Paget disease as well as other bone problems.

http://www.osteo.org/
Information on frequency of osteoporosis among groups of people, prevention, and ways to slow its effects. There are additional links to osteogenesis imperfecta, Paget disease, and other bone tissue disorders.

Skeletal Organization

http://www.cs.brown.edu/people/oa/Bin/skeleton.html
Click on the bone and hear its name pronounced!

http://www.scar.rad.washington.edu/RadAnatomy.html
X-ray images from various perspectives.

Joints

http://www.rad.upenn.edu/rundle/InteractiveKnee.html
Pictures, text, and links about this most complex of all joints.

http://www.meddean.luc.edu/develop/afang/KNEE.HTM
MRIs of the knee joint and associated structures including muscles.

http://rpisun1.mda.uth.tmc.edu/se/anatomy/
This contains a clickable human. One can examine sections of the foot, knee, and elbow joints.

http://www.rad.washington.edu/Anatomy/TMJ/TMJ.html
Various types of images, text, and links concerning temporomandibular joint disorder.

http://weber.u.washington.edu/~dboone/key/subjects/arthritis/xxxxxxxx1_1.html
An arthritis site with links to numerous other skeletal problems such as carpal tunnel syndrome and Paget disease.

http://anatomy.uams.edu/HTMLpages/anatomyhtml/pectorals.html
Lists of bones and muscles plus a skeletal outline for major joints from the University of Arkansas.

Chapter 8: Muscular System

Anatomy of the Muscular System

http://rpiwww.mdacc.tmc.edu:80/se/anatomy/arm/
The viewer can click on any of several sections through the forearm to examine the positions of the muscles.

http://www.healthink.com/muscles.html
This site provides simple pictures and a brief lay summary of functions and locations of certain major skeletal muscles such as the deltoid and the latissimus dorsi.

Muscle Physiology

http://www.megaweb.co.za/md
Duchenne muscular dystrophy with links to many sites related to this specific condition as well as other muscular dystrophies. Multiple support groups are listed.

http://www.cs.sfu.ca/css/update/vol8/8.1-muscles.html
This site explains in lay terms the effect of insulin-like growth factor (IGF) on the growth of muscle cells. There are no other links.

Chapter 9: Nervous System

Nervous Tissue

http://neuron.duke.edu/
Go to this address to obtain a downloadable neuron simulation program. Complete documentation is available.

http://www.physiol.arizona.edu/CELL/Instruct/BodyElect.html
This has some very interesting material, pictures and text, from the University of Arizona from the course entitled: "Body electric." There is information about excitable cells here.

Coverings of the Central Nervous System

http://www9.biostr.washington.edu/cgi-bin/imageform
This site has human brain atlases including some attractive pictures with vivid colors, a movie list, and three dimensional reconstructions.

http://www.nucmed.buffalo.edu/nrlgy1.htm#fdg_normal_brainx31
You can see a PET scan of normal brains with links to images with pathological conditions.

http://www.coa.uky.edu/ADReview
This web page contains a great deal of useful information about Alzheimer disease and includes numerous links to other useful sites pertaining to the topic.

http://www.merck.com/!!s9OfD17kBs9OwL2Wwl/disease/preventable/hib/
Haemophilus B is a bacterium that can cause meningitis. This site offers a brief discussion of the mode of transmission and lists those at greatest risk.

http://www.loni.ucla.edu/humandata/human.html
This site offers brain sections using various imaging techniques.

http://uta.marymt.edu/~psychol/brain.html
You can click on the cerebral cortex and get the names and brief descriptions of superficial structures. Very attractive colors.

Peripheral Nervous System

http://www.geocities.com/HotSprings/1161/
This site has a huge number of links providing information on polio (affects the anterior motor horn) and other diseases with similar symptoms.

Chapter 10: Somatic and Special Senses

http://www.blindness.org/Latest_Research.html
This site contains recent research news about macular degeneration, retinitis pigmentosum, and related disorders.

http://www.merck.com/!!s9OfD17kBs9OqO00Pa/disease/glaucoma/
Brief and simple explanations for intraocular pressure, diagnosis, and other topics related to glaucoma.

http://www.tinnitus-pjj.com/
A good explanation of tinnitus plus links to additional information and bibliography.

http://www.adworks.com/dizzy/vestib.html
This web site has many links that provide information on equilibrium disorders, including pictures of otoscopic exams and downloadable videos.

Chapter 11: Endocrine System

http://www.niddk.nih.gov/DiabetesDocs.html
Diabetes mellitus and conditions caused by diabetes, such as neuropathy. Information for patients.

http://home.ican.net/~thyroid/English/Guides.html
Multiple links to web sites concerning thyroid conditions. Pictures are available on some of the sites.

http://www.cs.sfu.ca/css/update/vol8/8.1-muscles.html
This site explains in lay terms the effect of insulin-like growth factor (IGF) on muscle cells and relates the phenomenon to aging. Additionally, a clear drawing explains the roles of receptors and of carrier molecules in this process.

Chapter 12: Blood

http://www.med.nagoya-u.ac.jp/pathy/Pictures/atlas.html
A photomicrographic atlas of blood cells and bone marrow. Some very good pictures include leukemias, thalassemia (a disorder affecting red blood cells), and anemias.

http://www.merck.com/!!s9OfD17kBs9OwL2Wwl/disease/preventable/hepb/
Brief facts on hepatitis B: symptoms, statistics, spread, and prevention.

http://cancer.med.upenn.edu/pdq.400113.html
This site contains the leukemia research report.

http://www-medlib.med.utah.edu/WebPath/INFEHTML/INFEC041.html
This site illustrates photomicrographs of blood parasites and has links to other image-containing sites.

Chapter 13: Cardiovascular System

The Heart

http://www.usc.edu/hsc/pharmacy/ced/dietchol/DietChl2.htm
A technical discussion about the relationship between free radicals and atherosclerosis (contributes to heart attacks).

http://sln2.fi.edu/biosci/heart.html
This web page has an open heart movie (requires QTV) as well as other attractive graphics, plus activities related to EKGs and X rays.

http://www.merck.com/!!s9OfD17kBs9Oma1w8T/disease/heart/
This site covers heart disease risk factors in layperson's terms.

http://www-medlib.med.utah.edu/WebPath/ATHHTML/ATHIDX.html
Photomicrographs of atheroscleroses and thromboses and pictures of an infarcted heart.

http://www/vesalius.com/story/story.html
Heart illustrations and text appear at this site.

http://www.physiol.arizona.edu/CELL/Instruct/BodyElect.html
This has some very interesting material, pictures and text, from the University of Arizona from the course entitled: "Body electric." It includes an EKG.

The Blood Vessels

http://neurosurgery.mgh.harvard.edu/vaschome.htm#AVMs
This site explains strokes, aneurysms, and other phenomena related to cerebral circulation. Many links to other medical centers contain diagrams and images of angiograms.

Chapter 14: Lymphatic System and Immunity

http://www.merck.com/!!s9OfD17kBs9PBm0Hsz/disease/hiv/
Several aspects of HIV and AIDS are covered here including an extensive presentation on opportunistic infections.

http://www.jem.org/cgi/content/full/185/1/55
Apoptosis, macrophages, CD-4, and HIV are discussed in this Journal of Experimental Medicine article.

http://www-medlib.med.utah.edu/WebPath/IMMHTML/IMM009.html
Illustrations of lupus erythematosus butterfly rash with links to related topics.

Chapter 15: Digestion and Nutrition

Digestive System

http://www.merck.com/!!s9OfD17kBs9OwL2Wwl/disease/preventable/hepa/
Brief, simple explanation of hepatitis A, vaccination, and statistics.

http://www.geocities.com/Paris/4664/introli.html
An explanation of lactose intolerance in layperson's terminology, with links to other sources of information about this condition.

Nutrition and Metabolism

http://www.blindness.org/Latest_Research.html
Numerous links to pages having information about hypoglycemia.

Chapter 16: Respiratory System

http://www.merck.com/!!s9OfD17kBs9OwL2Wwl/disease/preventable/pneu/
A brief explanation of pneumococcal pneumonia and a list of those at risk.

http://www9.biostr.washington.edu/cgi-bin/PageMaster?atlas:Thorax+ffpathIndex:Thoracic^Viscera+2
Beautiful, color, three dimensional pictures of the lungs and other viscera.

Chapter 17: Urinary System

http://www.gamewood.net/rnet/section/CME.htm
A tutorial with links to radiology and pathology images, biopsies, and case studies.

Chapter 18: Water, Electrolyte, and Acid-Base Balance

http://www3.ncbi.nlm.nih.gov/htbin-post/Omim/dispmim?261600
This site has information about PKU: cause, clinical signs, genetics, population biology, and dietary recommendations.

http://www.niddk.nih.gov/DiabetesDocs.html
Diabetes mellitus and conditions caused by diabetes, such as neuropathy; information for patients. This site is not very technical.

http://www.msud-support.org/descrip.htm
This is a non-technical site explaining the diagnosis, cause, and genetics of Maple Syrup Urine Disease.

Chapter 19: Reproductive Systems

http://www.luc.edu/depts/biology/meiosis.htm
A good diagram of meiosis. There is a link to a mitosis picture.

http://www.luc.edu/depts/biology/meiosis.html
A picture showing the steps of meiosis along with a brief explanation. There are links to other sites included.

http://cancer.med.upenn.edu/specialty/gyn_onc/ovarian/
This site provides brief pieces of information about ovarian cancer, technical data about specific types of ovarian cancer, and has several links to other pages.

http://cancer.med.upenn.edu/classroom/colp/
Basically a tutorial on cervical cancer examination. It includes effective diagrams of normal tissue and progressive stages of cancer.

http://www.menshealth.com/features/mensconf/indexg.html
Men's health issues such as potency are presented on this lively site.

Chapter 20: Pregnancy, Growth, and Development

http://www.luc.edu/depts/biology/meiosis.htm
A good diagram of meiosis, with a link to a mitosis picture.

http://hawley-lab2.ucdavis.edu/Meiosis.html
A more technical discussion of meiosis.

Glossary

A phonetic guide to pronunciation follows each glossary word. Any unmarked vowel that ends a syllable or stands alone as a syllable has the long sound. Thus, the word *play* is phonetically spelled *pla*. Any unmarked vowel followed by a consonant has the short sound. The word *tough*, for instance, is phonetically spelled *tuf*. If a long vowel appears in the middle of a syllable (followed by a consonant), it is marked with a macron (¯), the sign for a long vowel. Thus, the word *plate* is phonetically spelled *plāt*. Similarly, if a vowel stands alone or ends a syllable, but has a short sound, it is marked with a breve (˘).

A

abdomen (ab-do′men) Portion of body between diaphragm and pelvis.

abdominopelvic cavity (ab-dom″ĭ-no-pel′vik kav′ĭ-te) The space between the diaphragm and the lower portion of the trunk of the body.

abduction (ab-duk′shun) Movement of a body part away from midline.

absorption (ab-sorp′shun) Taking in of substances by cells or membranes.

accessory organs (ak-ses′o-re or′ganz) Organs that supplement functions of other organs.

accommodation (ah-kom″o-da′shun) Adjustment of lens of eye for close or distant vision.

acetylcholine (as″ĕ-til-ko′lēn) Type of neurotransmitter, which is a biochemical secreted at axon ends of many neurons; transmits nerve messages across synapses.

acetyl coenzyme A (as′ĕ-til ko-en′zīm) Intermediate compound produced during oxidation of carbohydrates and fats.

acid (as′id) Substance that ionizes in water to release hydrogen ions.

acidosis (as″ĭ-do′sis) A relative increase in acidity of body fluids.

ACTH Adrenocorticotropic hormone.

actin (ak′tin) Protein that, with myosin, contracts muscle fibers.

action potential (ak′shun po-ten′shal) Sequence of electrical changes when a nerve cell membrane is exposed to a stimulus that exceeds its threshold.

active transport (ak′tiv trans′port) Process that uses metabolic energy to move a substance across a cell membrane, usually against the concentration gradient.

adaptation (ad″ap-ta′shun) Adjustment to environmental conditions.

adduction (ah-duk′shun) Movement of body part toward midline.

adenosine diphosphate (ah-den′o-sēn di-fos′fāt) **(ADP)** Molecule produced when adenosine triphosphate loses terminal phosphate.

adenosine triphosphate (ah-den′o-sēn tri-fos′fāt) **(ATP)** Organic molecule that stores and releases energy for use in cellular processes.

adenylate cyclase (ah-den′i-lāt si′klās) Enzyme activated when certain hormones combine with receptors on cell membranes circularizing ATP to cyclic AMP.

ADH Antidiuretic hormone.

adipose tissue (ad′ĭ-pōs tish′u) Fat-storing tissue.

ADP Adenosine diphosphate.

adrenal cortex (ah-dre′nal kor′teks) Outer portion of the adrenal gland.

adrenal glands (ah-dre′nal glandz) Endocrine glands located on superior portions of kidneys.

adrenal medulla (ah-dre′nal me-dul′ah) Inner portion of adrenal gland.

adrenergic fiber (ad″ren-er′jik fi′ber) Nerve fiber that secretes norepinephrine at the axon terminal.

adrenocorticotropic hormone (ad-re″no-kor″te-ko-trōp′ik hor′mōn) **(ACTH)** Hormone that the anterior pituitary secretes to stimulate activity in adrenal cortex.

aerobic respiration (a″er-o′bik res″pĭ-ra′shun) Phase of cellular respiration that requires oxygen.

afferent arteriole (af′er-ent ar-te′re-ōl) Vessel that conveys blood to glomerulus of nephron within kidneys.

agglutination (ah-gloo″tĭ-na′shun) Clumping of blood cells in response to a reaction between an antibody and an antigen.

agranulocyte (a-gran′u-lo-sīt) Nongranular leukocyte.

albumin (al-bu′min) Plasma protein that helps regulate osmotic concentration of blood.

aldosterone (al dos′ter-ōn″) Hormone that the adrenal cortex secretes to regulate sodium and potassium ion concentrations and fluid volume.

alimentary canal (al″i-men′tar-e kah-nal′) Tubular portion of digestive tract that leads from the mouth to the anus.

alkaline (al′kah-līn) Pertaining to or having properties of a base or alkali; basic.

alkalosis (al″kah-lo′sis) A relative increase in alkalinity of body fluids.

allantois (ah-lan′to-is) Embryonic structure that forms umbilical blood vessels.

allergen (al′er-jen) Foreign substance that stimulates an allergic reaction.

all-or-none response (al′or-nun′ re-spons′) Muscle fiber or neuron responding completely when exposed to a threshold stimulus.

alveolar ducts (al-ve′o-lar dukts′) Fine tubes that carry air to air sacs of the lungs.

alveolus (al-ve′o-lus) (plural, *alveoli*) Air sac of lung; saclike structure.

amino acid (ah-me′no as′id) Small organic compound that contains an amino group ($-NH_2$) and a carboxyl group ($-COOH$); structural unit of a protein molecule.

amnion (am′ne-on) Extra-embryonic membrane that encircles a developing fetus and contains amniotic fluid.

amniotic fluid (am″ne-ot′ik floo′id) Fluid within the amniotic cavity that surrounds the developing fetus.

ampulla (am-pul′ah) Expansion at the end of each semicircular canal that contains a crista ampullaris.

amylase (am′ĭ-lās) Enzyme that hydrolyzes starch.

anabolism (ah-nab′o-liz″em) Metabolic process by which larger molecules are synthesized from smaller ones; anabolic metabolism.

anaerobic respiration (an-a″er-o′bik res″pĭ-ra′shun) Phase of cellular respiration that occurs in the absence of oxygen.

anaphase (an′ah-fāz) Stage in mitosis when duplicate chromosomes move to opposite poles of cell.

anatomy (ah-nat′o-me) Branch of science dealing with the form and structure of body parts.

androgen (an′dro-jen) Male sex hormone, such as testosterone.

antagonist (an-tag′o-nist) Muscle that opposes a prime mover.

antebrachial (an″te-bra′ke-al) Pertaining to the forearm.

antecubital (an″te-ku′bĭ-tal) Region in front of elbow joint.

anterior (an-te′re-or) Front.

anterior pituitary (an-te′re-or pĭ-tu′ĭ-tār″e) Front lobe of pituitary gland.

antibody (an′tĭ-bod″e) Protein (immunoglobulin) that B cells of the immune system produce in response to the presence of a nonself antigen; it reacts with the antigen.

antibody-mediated immunity (an′tĭ-bod″e me′de-ā-tĭd ĭ-mu′nĭ-te) Destruction of cells bearing foreign (nonself) antigens by circulating antibodies; humoral immunity.

anticodon (an″tĭ-ko′don) Three contiguous nucleotides of a transfer RNA molecule that are complementary to a specific mRNA.

antidiuretic hormone (an″tĭ-di″u-ret′ik hor′mōn) **(ADH)** Hormone that the posterior pituitary lobe releases to enhance water conservation by kidneys.

antigen (an′tĭ-jen) Chemical that stimulates cells to produce antibodies.

aorta (a-or′tah) Major systemic artery that receives blood from left ventricle.

aortic sinus (a-or′tik si′nus) Swelling in aortic wall behind each cusp of the semilunar valve.

aortic valve (a-or′tik valv) Flaplike structures in wall of aorta near its origin that prevent blood from returning to left ventricle of heart.

apocrine gland (ap′o-krin gland) Gland whose secretions contain parts of secretory cells.

aponeurosis (ap″o-nu-ro′sis) Sheetlike tendon that attaches certain muscles to other parts.

appendicular (ap″en-dik′u-lar) Pertaining to upper or lower limbs.

aqueous humor (a′kwe-us hu′mor) Watery fluid that fills the anterior cavity of the eye.

arachnoid mater (ah-rak′noid ma′ter) Delicate, weblike middle layer of meninges.

arrector pili muscle (ah-rek′tor pil′i mus′l) Smooth muscle in skin associated with a hair follicle.

arrhythmia (ah-rith′me-ah) Abnormal heart action characterized by a loss of rhythm.

arteriole (ar-te′re-ōl) Small branch of an artery that communicates with a capillary network.

artery (ar′ter-e) Vessel that transports blood away from heart.

articular cartilage (ar-tik′u-lar kar′tĭ-lij) Hyaline cartilage that covers ends of bones in synovial joints.

articulation (ar-tik″u-la′shun) Joining of structures at a joint.

ascending tracts (ah-send′ing trakts) Groups of nerve fibers in spinal cord that transmit sensory impulses upward to brain.

ascorbic acid (as-kor′bik as′id) Vitamin C; a water-soluble vitamin.

assimilation (ah-sim″ĭ-la′shun) Chemically changing absorbed substances.

association area (ah-so″se-a′shun a′re-ah) Region of the cerebral cortex controlling memory, reasoning, judgment, and emotions.

astrocyte (as′tro-sīt) Type of neuroglial cell that connects neurons to blood vessels.

atherosclerosis (ath″er-o-sklĕ-ro′sis) Abnormal accumulation of fatty substances on inner linings of arteries.

atmospheric pressure (at″mos-fĕr′ik presh′ur) Pressure exerted by the weight of air; about 760 millimeters of mercury at sea level.

atom (at′om) Smallest particle of an element that has the properties of that element.

atomic number (ah-tom′ik num′ber) Number of protons in an atom of an element.

atomic weight (ah-tom′ik wāt) Number approximately equal to number of protons plus number of neutrons in an atom of an element.

ATP Adenosine triphosphate.

ATPase Enzyme that causes ATP molecules to release the energy stored in their terminal phosphate bonds.

atrioventricular bundle (a″tre-o-ven-trik′u-lar bun′dl) **(A-V bundle)** Group of specialized fibers that conduct impulses from the atrioventricular node to ventricular muscle of heart.

atrioventricular node (a″tre-o-ven-trik′u-lar nōd) **(A-V node)** Specialized mass of cardiac muscle fibers in interatrial septum of heart; transmits cardiac impulses from sinoatrial node to A-V bundle.

atrium (a′tre-um) (plural, *atria*) Chamber of heart that receives blood from veins.

atrophy (at′ro-fe) A wasting away or decrease in size of an organ or tissue.

auditory (aw′di-to″re) Pertaining to the ear, or sense of hearing.

auditory ossicle (aw′di-to″re os′i-kl) Bone of middle ear.

auditory tube (aw′di-to″re tūb) Tube that connects middle ear cavity to pharynx; eustachian tube.

auricle (aw′ri-kl) Earlike structure; portion of heart that forms part of atrial wall.

autonomic nervous system (aw″to-nom′ik ner′vus sis′tem) Portion of nervous system that controls actions of viscera and skin.

axial (ak′se-al) Pertaining to the head, neck, and trunk.

axillary (ak′sĭ-ler″e) Pertaining to the armpit.

axon (ak′son) Nerve fiber that conducts nerve impulse away from neuron cell body.

B

baroreceptor (bar″o-re-sep′tor) Receptor that senses changes in pressure.

basal ganglion (ba′sal gang′gle-on) Mass of gray matter deep within a cerebral hemisphere of brain.

base (bās) Substance that ionizes in water to release hydroxyl ions (OH^-) or other ions that combine with hydrogen ions.

basement membrane (bās′ment mem′brān) Layer of nonliving material that anchors epithelial tissue to underlying connective tissue.

basophil (ba′so-fil) White blood cell containing cytoplasmic granules that stain with basic dye.

B cell (sel) Lymphocyte that produces and secretes antibodies to fight foreign substances in body; B lymphocyte.

beta oxidation (ba′tah ok″sĭ-da′shun) Chemical process by which fatty acids are broken down into molecules of acetyl coenzyme A.

bicuspid valve (bi-kus′pid valv) Heart valve between left atrium and left ventricle; mitral valve.

bile (bīl) Fluid secreted by liver and stored in gallbladder.

bilirubin (bil″ĭ-roo′bin) Bile pigment produced from hemoglobin breakdown.

biliverdin (bil″ĭ-ver′din) Bile pigment produced from hemoglobin breakdown.

biotin (bi′o-tin) Water-soluble vitamin; member of vitamin B complex.

blastocyst (blas′to-sist) Early stage of preembryonic development that consists of a hollow ball of cells.

brachial (bra′ke-al) Pertaining to the arm.

brain stem (brān stem) Portion of brain that includes midbrain, pons, and medulla oblongata.

Broca's area (bro′kahz a′re-ah) Region of frontal lobe that coordinates complex muscular actions of mouth, tongue, and larynx, making speech possible.

bronchial tree (brong′ke-al tre) Bronchi and their branches that carry air from trachea to alveoli of lungs.

bronchiole (brong′ke-ōl) Small branch of a bronchus within lung.

bronchus (brong′kus) (plural, *bronchi*) Branch of trachea that leads to lung.

buccal (buk′al) Pertaining to the mouth and inner lining of the cheeks.

buffer (buf′er) Substance that can react with strong acid or base to form weaker acid or base and thus resist change in pH.

buffer system (buf′er sis′tem) Sets of chemical reactions that occur in body fluids to maintain a particular pH.

bulbourethral glands (bul″bo-u-re′thral glandz) Glands that secrete viscous fluid into male urethra during sexual excitement; Cowper's glands.

bursa (bur′sah) Saclike, fluid-filled structure, lined with synovial membrane, near a joint.

C

calcitonin (kal″sĭ-to′nin) Hormone the thyroid gland secretes to help regulate blood calcium concentration.

calorie (kal′o-re) Unit to measure heat energy and the energy content of foods.

canaliculus (kan″ah-lik′u-lus) Microscopic canal that connects lacunae of bone tissue.

cancellous bone (kan′sĕ-lus bōn) Bone tissue with a latticework structure; spongy bone.

capacitation (kah-pas″i-ta′shun) Activation of sperm cell to fertilize egg cell.

capillary (kap′ĭ-ler″e) Small blood vessel that connects an arteriole and a venule.

carbohydrate (kar″bo-hi′drāt) Organic compound that contains carbon, hydrogen, and oxygen, in a 1:2:1 ratio.

carbonic anhydrase (kar-bon′ik an-hi′drās) Enzyme that catalyzes reaction between carbon dioxide and water to form carbonic acid.

carboxyhemoglobin (kar-bok″se-he″mo-glo′bin) Compound formed by union of carbon dioxide and hemoglobin.

carboxypeptidase (kar-bok″se-pep′tĭ-dās) Protein-splitting enzyme in pancreatic juice.

cardiac conduction system (kar′de-ak kon-duk′shun sis′tem) System of specialized cardiac muscle fibers that conducts cardiac impulses from S-A node into myocardium.

cardiac cycle (kar′de-ak si′kl) Series of myocardial contractions and relaxations that constitutes a complete heartbeat.

cardiac muscle (kar′de-ak mus′l) Specialized muscle tissue found only in the heart.

cardiac output (kar′de-ak owt′poot) The volume of blood per minute pumped by the heart (calculated by multiplying stroke volume in milliliters by heart rate in beats per minute).

carpals (kar′pals) Bones of wrist.

carpus (kar′pus) Wrist; wrist bones as a group.

cartilage (kar′tĭ-lij) Type of connective tissue in which cells are within lacunae and separated by a semisolid matrix.

cartilaginous joint (kar-tĭ-laj′ĭ-nus joint) A type of slightly movable joint.

catabolism (kă-tab′o-lizm) Metabolic process that breaks down large molecules into smaller ones; catabolic metabolism.

catalyst (kat′ah-list) Chemical that increases the rate of a chemical reaction but is not permanently altered by that reaction.

celiac (se′le-ak) Pertaining to the abdomen.

cell (sel) Smallest living structural and functional unit of an organism.

cell body (sel bod′e) Portion of nerve cell that includes a cytoplasmic mass and a nucleus, and from which nerve fibers extend.

cell-mediated immunity (sel me′de-ā″tid ĭ-mu′nĭ-te) Attack of T cells and their secreted products on foreign cells.

cell membrane (sel mem′brān) Selectively permeable outer boundary of a cell consisting of a phospholipid bilayer embedded with proteins; plasma membrane or cytoplasmic membrane.

cellular respiration (sel′u-lar res″pĭ-ra′shun) Cellular process that releases energy from organic compounds.

cellulose (sel′u-lōs) Polysaccharide abundant in plant tissues that human enzymes cannot break down.

cementum (se-men′tum) Bonelike material that fastens the root of a tooth into its bony socket.

central canal (sen′tral kah-nal′) Tube within spinal cord that is continuous with brain ventricles and contains cerebrospinal fluid.

central nervous system (sen′tral ner′vus sis′tem) **(CNS)** Portion of nervous system that consists of the brain and spinal cord.

centriole (sen′tre-ōl) Cellular organelle built of microtubules that organizes mitotic spindle.

centromere (sen′tro-mēr) Portion of chromosome to which spindle fibers attach during mitosis.

centrosome (sen′tro-sōm) Cellular organelle consisting of two centrioles.

cephalic (sĕ-fal′ik) Pertaining to the head.

cerebellar cortex (ser″ĕ-bel′ar kor′teks) Outer layer of cerebellum.

cerebellum (ser″ĕ-bel′um) Portion of brain that coordinates skeletal muscle movement.

cerebral cortex (ser′ĕ-bral kor′teks) Outer layer of cerebrum.

cerebral hemisphere (ser′ĕ-bral hem′ĭ-sfēr) One of the large, paired structures that constitute the cerebrum.

cerebrospinal fluid (ser″ĕ-bro-spi′nal floo′id) Fluid in ventricles of brain, subarachnoid space of meninges, and central canal of spinal cord.

cerebrum (ser′ĕ-brum) Portion of brain that occupies upper part of cranial cavity and provides higher mental functions.

cervical (ser′vĭ-kal) Pertaining to the neck as in the cervix of the uterus.

cervix (ser′viks) Narrow, inferior "neck" of uterus that leads into vagina.

chemoreceptor (ke″mo-re-sep′tor) Receptor stimulated by binding of certain chemicals.

chief cell (chēf sel) Cell of gastric gland that secretes various digestive enzymes, including pepsinogen.

cholecystokinin (ko″le-sis″to-ki′nin) Hormone the small intestine secretes to stimulate release of pancreatic juice from pancreas and bile from gallbladder.

cholesterol (ko-les′ter-ol) Lipid that body cells use to synthesize steroid hormones; excreted into bile.

cholinergic fiber (ko″lin-er′jik fi′ber) Nerve fiber that secretes acetylcholine at axon terminal.

cholinesterase (ko″lin-es′ter-ās) Enzyme that catalyzes breakdown of acetylcholine.

chondrocyte (kon′dro-sīt) Cartilage cell.

chorion (ko′re-on) Extra-embryonic membrane that forms outermost covering around fetus and contributes to formation of placenta.

chorionic villi (ko″re-on′ik vil′i) Projections that extend from outer surface of chorion and help attach embryo to uterine wall.

choroid coat (ko′roid kōt) Vascular, pigmented middle layer of wall of eye.

choroid plexus (ko′roid plek′sus) Mass of specialized capillaries that secretes cerebrospinal fluid into ventricle of brain.

chromatid (kro′mah-tid) One-half of a replicated chromosome or a single unreplicated chromosome.

chromatin (kro′mah-tin) DNA and complexed protein that condenses to form chromosomes during mitosis.

chromosome (kro′mo-sōm) Rodlike structure that condenses from chromatin in a cell's nucleus during mitosis.

chylomicron (ki″lo-mi′kron) Microscopic droplet of fat in the blood following fat digestion.

chyme (kīm) Semifluid mass of partially digested food that passes from stomach to small intestine.

chymotrypsin (ki″mo-trip′sin) Protein-splitting enzyme in pancreatic juice.

cilia (sil′e-ah) Microscopic, hairlike processes on exposed surfaces of certain epithelial cells.

ciliary body (sil′e-er″e bod′e) Structure associated with the choroid layer of eye that secretes aqueous humor and contains the ciliary muscle.

circadian rhythm (ser″kah-de′an rithm) Pattern of repeated behavior associated with cycles of night and day.

circle of Willis (ser′kl uv wil′is) Arterial ring on ventral surface of brain.

circular muscles (ser′ku-lar mus′lz) Muscles whose fibers are organized in circular patterns, usually around an opening or in the wall of a tube; sphincter muscles.

circulation (ser-ku-la′shun) Path of blood through system of vessels.

circumduction (ser″kum-duk′shun) Movement of a body part, such as a limb, so that the end traces out a cone.

cisternae (sis-ter′ne) Enlarged portions of sarcoplasmic reticulum near actin and myosin filaments of a muscle fiber.

citric acid cycle (sit′rik as′id si′kl) Series of chemical reactions that oxidizes certain molecules, releasing energy; Krebs cycle.

cleavage (klēv′ij) Early successive divisions of blastocyst cells into smaller and smaller cells.

clitoris (kli′to-ris) Small, erectile organ in anterior of vulva; corresponding to penis.

clone (klōn) Group of cells that originate from a single cell and are therefore genetically identical.

CNS Central nervous system.

coagulation (ko-ag″u-la′shun) Blood clotting.

cochlea (kok′le-ah) Portion of inner ear that contains hearing receptors.

codon (ko′don) Set of three nucleotides of a messenger RNA molecule corresponding to a particular amino acid.

coenzyme (ko-en′zīm) Nonprotein organic molecule required for the activity of a particular enzyme.

cofactor (ko′fak-tor) Small molecule or ion that must combine with an enzyme for activity.

collagen (kol′ah-jen) Protein in white fibers of connective tissues and in bone matrix.

common bile duct (kom′mon bīl dukt) Tube that transports bile from cystic duct to duodenum.

compact bone (kom-pakt′ bōn) Dense tissue in which cells are organized in osteons (Haversian systems) with no apparent spaces.

complement (kom′plĕ-ment) Group of enzymes activated by antibody combining with antigen; enhances reaction against foreign substances within body.

complete protein (kom-plēt′ pro′te-in) Protein that contains adequate amounts of essential amino acids to maintain body tissues and to promote normal growth and development.

compound (kom′pownd) Substance composed of two or more chemically bonded elements.

condyle (kon′dīl) Rounded process of a bone, usually forming a joint.

cones (kōns) Color receptors in retina of eye.

conjunctiva (kon″junk-ti′vah) Membranous covering on anterior surface of eye.

connective tissue (kŏ-nek′tiv tish′u) One of the basic types of tissue that includes bone, cartilage, and loose and fibrous connective tissue.

convergence (kon-ver′jens) Nerve impulses arriving at the same neuron.

convolution (kon″vo-lu′shun) Elevation on a structure's surface caused by infolding of structure upon itself.

cornea (kor′ne-ah) Transparent anterior portion of outer layer of eye wall.

coronary artery (kor′o-na″re ar′ter-e) Artery that supplies blood to wall of heart.

coronary sinus (kor′o-na″re si′nus) Large vessel on posterior heart surface into which cardiac veins drain.

corpus albicans (kor′pus al′bĭ-kanz) Remnant of corpus luteum in ovary; composed of fibrous connective tissue.

corpus callosum (kor′pus kah-lo′sum) Mass of white matter within brain; composed of nerve fibers connecting right and left cerebral hemispheres.

corpus luteum (kor′pus loot′e-um) Structure that forms from tissues of ruptured ovarian follicle and secretes female hormones.

cortex (kor′teks) Outer layer of an organ, such as the adrenal gland, cerebrum, or kidney.

cortisol (kor′tĭ-sol) Glucocorticoid that the adrenal cortex secretes.

costal (kos′tal) Pertaining to the ribs.

covalent bond (ko′va-lent bond) Chemical bond formed by electron sharing between atoms.

cranial (kra′ne-al) Pertaining to the cranium.

cranial nerve (kra′ne-al nerv) Nerve that arises from the brain.

creatine phosphate (kre′ah-tin fos′fāt) Muscle biochemical that stores energy.

crenation (kre-na′shun) Cell shrinkage due to contact with hypertonic solution.

crest (krest) Ridgelike projection of bone.

cricoid cartilage (kri′koid kar′tĭ-lij) Ringlike cartilage that forms lower end of larynx.

crista ampullaris (kris′tah am-pul′ar-is) Sensory organ within semicircular canal that functions in sense of dynamic equilibrium.

cubital (ku′bi-tal) Pertaining to the forearm.

cutaneous (ku-ta′ne-us) Pertaining to the skin.

cystic duct (sis′tik dukt) Tube that connects the gallbladder to the common bile duct.

cytocrine secretion (si′to-krin se-kre′shun) Transfer of melanin granules from melanocytes into epithelial cells.

cytoplasm (si′to-plazm) Contents of a cell, excluding the nucleus and cell membrane.

D

deamination (de-am″ĭ-na′shun) Removing amino groups ($-NH_2$) from amino acids.

deciduous teeth (de-sid′u-us tēth) Teeth that are shed and replaced by permanent teeth; primary teeth.

decomposition (de-kom″po-zish′un) Breakdown of molecules into simpler compounds.

defecation (def″ĕ-ka′shun) Discharge of feces from rectum through anus.

dehydration synthesis (de″hi-dra′shun sin′thĕ-sis) Anabolic process that joins small molecules; synthesis.

dendrite (den′drīt) Nerve fiber that transmits impulses toward neuron cell body.

dentin (den′tin) Bonelike substance that forms the bulk of a tooth.

deoxyhemoglobin (de-ok″se-he″mo-glo′bin) Hemoglobin to which oxygen is not bound.

deoxyribonucleic acid (de-ok′si-ri″bo-nu-kle″ik as′id) **(DNA)** Genetic material; double-stranded polymer of nucleotides, each containing a phosphate group, a nitrogenous base (adenine, thymine, guanine, or cytosine), and the sugar deoxyribose.

depolarization (de-po″lar-ĭ-za′shun) Loss of negative electrical charge from inner membrane surface.

dermis (der′mis) Thick layer of skin beneath epidermis.

descending tracts (de-send′ing trakts) Groups of nerve fibers that carry nerve impulses from brain through spinal cord.

detrusor muscle (de-trūz′or mus′l) Muscular wall of urinary bladder.

diapedesis (di″ah-pĕ-de′sis) Squeezing of leukocytes between cells of blood vessel walls.

diaphragm (di′ah-fram) Sheetlike structure composed largely of skeletal muscle and connective tissue that separates thoracic and abdominal cavities; also, a caplike contraceptive device inserted in vagina.

diaphysis (di-af′ĭ-sis) Shaft of a long bone.

diastole (di-as′tol-le) Phase of cardiac cycle when heart chamber wall relaxes.

diastolic pressure (di-a-stol′ik presh′ur) Lowest arterial blood pressure reached during diastolic phase of cardiac cycle.

diencephalon (di″en-sef′ah-lon) Portion of brain in region of third ventricle that includes thalamus and hypothalamus.

differentiation (dif″er-en″she-a′shun) Cell specialization due to differential gene expression.

diffusion (dĭ-fu′zhun) Random movement of molecules from region of higher concentration toward one of lower concentration.

digestion (di-jes′chun) Breaking down of large nutrient molecules into smaller molecules that can be absorbed; hydrolysis.

dipeptide (di-pep′tīd) Molecule composed of two joined amino acids.

disaccharide (di-sak′ah-rīd) Sugar produced by union of two monosaccharides.

distal (dis′tal) Further from the midline or origin; opposite of *proximal*.

divergence (di-ver′jens) A neuron sending impulses to multiple other neurons.

DNA Deoxyribonucleic acid.

dorsal root (dor′sal root) Sensory branch of spinal nerve by which nerve joins spinal cord.

dorsal root ganglion (dor′sal root gang′gle-on) Mass of sensory neuron cell bodies in dorsal root of spinal nerve.

dorsum (dors′um) Pertaining to the back surface of a body part.

ductus arteriosus (duk′tus ar-te″re-o′sus) Blood vessel that connects pulmonary artery and aorta in fetus.

ductus venosus (duk′tus ven-o′sus) Blood vessel that connects umbilical vein and inferior vena cava in fetus.

dura mater (du′rah ma′ter) Tough outer layer of meninges.

dynamic equilibrium (di-nam′ik e″kwĭ-lib′re-um) Maintenance of balance when head and body are suddenly moved or rotated.

E

eccrine gland (ek′rin gland) Sweat gland that maintains body temperature.

ECG Electrocardiogram.

ectoderm (ek′to-derm) Outermost primary germ layer.

edema (ĕ-de′mah) Fluid accumulation within tissue spaces.

effector (ĕ-fek′tor) Muscle or gland that responds to stimulation.

efferent arteriole (ef′er-ent ar-te′re-ol) Vessel that conducts blood away from the glomerulus of a kidney nephron.

ejaculation (e-jak″u-la′shun) Discharge of semen containing sperm cells from male urethra.

elastin (e-las′tin) Protein that comprises the yellow, elastic fibers of connective tissue.

electrocardiogram (el-lek″tro-kar′de-o-gram″) **(ECG** or **EKG)** Recording of electrical activity associated with heartbeat.

electrolyte (e-lek′tro-līt) Substance that ionizes in water solution.

electrolyte balance (e-lek′tro-līt bal′ans) Condition when the quantities of electrolytes entering the body equal those leaving it.

electron (e-lek′tron) Small, negatively charged particle that revolves around the nucleus of an atom.

element (el′ĕ-ment) Pure chemical substance with only one type of atom.

embolus (em′bo-lus) A blood clot or gas bubble that obstructs a blood vessel.

embryo (em′bre-o) Prenatal stage of development after germ layers form but before rudiments of all organs are present.

emission (e-mish′un) Movement of sperm cells from the vas deferens into the ejaculatory duct and urethra.

emulsification (e-mul″sĭ-fĭ-ka′shun) Dispersing fat globules into smaller droplets by the action of bile salts.

enamel (e-nam′el) Hard covering on the exposed surface of a tooth.

endocardium (en″do-kar′de-um) Inner lining of heart chambers.

endochondral bone (en″do-kon′dral bōn) Bone that begins as hyaline cartilage and is replaced by bone tissue.

endocrine gland (en′do-krin gland) Gland that secretes hormones directly into the blood or body fluids.

endocytosis (en″do-si-to′sis) Process by which a cell membrane envelops a substance and draws it into the cell in a vesicle.

endoderm (en′do-derm) Innermost primary germ layer.

endolymph (en′do-limf) Fluid within membranous labyrinth of inner ear.

endometrium (en″do-me′tre-um) Inner lining of uterus.

endomysium (en″do-mis′e-um) Sheath of connective tissue that surrounds each skeletal muscle fiber.

endoplasmic reticulum (en-do-plaz′mic rĕ-tik′-u-lum) Organelle composed of a system of connected membranous tubules and vesicles along which protein is synthesized.

endosteum (en-dos′te-um) Tissue lining medullary cavity within bone.

endothelium (en″do-the′le-um) Layer of epithelial cells that forms the inner lining of blood vessels and heart chambers.

energy (en′er-je) Ability to cause something to move and thus do work.

enzyme (en′zīm) Protein that catalyzes a specific biochemical reaction.

eosinophil (e″o-sin′o-fil) White blood cell containing cytoplasmic granules that stain with acidic dye.

ependyma (ĕ-pen′dĭ-mah) Membrane composed of neuroglial cells that lines brain ventricles.

epicardium (ep″ĭ-kar′de-um) Visceral portion of pericardium on surface of heart.

epicondyle (ep″ĭ-kon′dīl) Projection of bone above condyle.

epidermis (ep″ĭ-der′mis) Outer epithelial layer of skin.

epididymis (ep″ĭ-did′ĭ-mis) Highly coiled tubule that leads from the seminiferous tubules of the testis to the vas deferens.

epidural space (ep″ĭ-du′ral spās) Space between dural sheath of spinal cord and bone of vertebral canal.

epigastric region (ep″ĭ-gas′trik re′jun) Upper middle portion of abdomen.

epiglottis (ep″ĭ-glot′is) Flaplike, cartilaginous structure at back of tongue near entrance to trachea.

epimysium (ep″i-mis′e-um) Outer sheath of connective tissue surrounding a skeletal muscle.

epinephrine (ep″ĭ-nef′rin) Hormone the adrenal medulla secretes during stress.

epiphyseal disk (ep″ĭ-fiz′e-al disk) Cartilaginous layer within the long bone epiphysis that grows.

epiphysis (ĕ-pif′ĭ-sis) End of long bone.

epithelium (ep″ĭ-the′le-um) Tissue type that covers all free body surfaces.

equilibrium (e″kwĭ-lib′re-um) State of balance between two opposing forces.

erythroblast (ĕ-rith′ro-blast) Immature red blood cell.

erythrocyte (ĕ-rith′ro-sīt) Red blood cell.

erythropoiesis (ĕ-rith″ro-poi-e′sis) Red blood cell formation.

erythropoietin (ĕ-rith″ro-poi′ĕ-tin) Kidney hormone that promotes red blood cell formation.

esophagus (ĕ-sof′ah-gus) Tubular portion of digestive tract that leads from pharynx to stomach.

essential amino acid (ĕ-sen′shal ah-me′no as′id) Amino acid required for health that body cells cannot synthesize in adequate amounts.

essential fatty acid (ĕ-sen′shal fat′e as′id) Fatty acid required for health that body cells cannot synthesize in adequate amounts.

estrogen (es′tro-jen) Hormone that stimulates development of female secondary sex characteristics.

evaporation (e″vap′o-ra-shun) Changing a liquid into a gas.

eversion (e-ver′zhun) Outward turning movement of sole of foot.

exchange reaction (eks-chānj re-ak′shun) Chemical reaction in which parts of two kinds of molecules trade positions.

excretion (ek-skre′shun) Elimination of metabolic wastes.

exocrine gland (ek′so-krin gland) Gland that secretes its products into a duct or onto a body surface.

exocytosis (eks-o-si-to′sis) Transport of substances out of a cell in vesicles.

expiration (ek″spĭ-ra′shun) Expulsion of air from lungs.

extension (ek-sten′shun) Movement increasing angle between parts at joint.

extracellular (ek″strah-sel′u-lar) Outside of cells; refers to the internal environment, body fluids outside individual cells.

extrapyramidal tract (ek″strah-pĭ-ram′i-dal trakt) Nerve tracts, other than the corticospinal tracts, that transmit impulses from cerebral cortex to spinal cord.

extremity (ek-strem′ĭ-te) Limb.

F

facet (fas′et) Small, flattened surface of a bone.

facilitated diffusion (fah-sil″ĭ-tāt′ed dĭ-fu′zhun) Diffusion in which carrier molecules transport substances across membranes from a region of higher concentration to a region of lower concentration.

facilitation (fah-sil″ĭ-ta′shun) Subthreshold stimulation of a neuron makes it more responsive to further stimulation.

fascia (fash′e-ah) Sheet of fibrous connective tissue that encloses a muscle.

fat (fat) Adipose tissue; an organic molecule containing glycerol and fatty acids.

fatty acid (fat′e as′id) Building block of a fat molecule.

feces (fe′sēz) Material expelled from digestive tract during defecation.

fertilization (fer″tĭ-lĭ-za′shun) Union of egg cell and sperm cell.

fetus (fe′tus) Human embryo after eight weeks of development.

fibril (fi′bril) Tiny fiber or filament.

fibrin (fi′brin) Insoluble, fibrous protein formed from fibrinogen during blood coagulation.

fibrinogen (fi-brin′o-jen) Plasma protein converted into fibrin during blood coagulation.
fibroblast (fi′bro-blast) Cell that produces fibers in connective tissues.
fibrous joint (fi′brus joint) A kind of immovable joint.
filtration (fil-tra′shun) Movement of material across a membrane as a result of hydrostatic pressure.
fissure (fish′ur) Narrow cleft that separates parts, such as the lobes of the cerebrum.
flexion (flek′shun) Bending at a joint to decrease the angle between bones.
follicle (fol′ĭ-kl) Pouchlike depression or cavity.
follicle-stimulating hormone (fol′ĭ-kl stim′u-la″ting hor′mōn) **(FSH)** Substance that the anterior pituitary secretes to stimulate follicular development in a female or sperm cell production in a male.
follicular cells (fŏ-lik′u-lar selz) Ovarian cells that surround a developing egg cell and secrete female sex hormones.
fontanel (fon″tah-nel′) Membranous region between certain cranial bones in skull of fetus or infant.
foramen (fo-ra′men) (plural, *foramina*) Opening, usually in a bone or membrane.
foramen magnum (fo-ra′men mag′num) Opening in occipital bone of skull through which spinal cord passes.
foramen ovale (fo-ra′men o-val′e) Opening in interatrial septum of fetal heart.
formula (fōr′mu-lah) Group of symbols and numbers for expressing the composition of a compound.
fossa (fos′ah) Depression in a bone or other part.
fovea (fo′ve-ah) Tiny pit or depression.
fovea centralis (fo′ve-ah sen-tral′is) Region of retina, consisting of densely packed cones, that provides the greatest visual acuity.
frenulum (fren′u-lum) Fold of tissue that anchors and limits movement of a body part.
frontal (frun′tal) Pertaining to the forehead.
FSH Follicle-stimulating hormone.
functional syncytium (funk′shun-al sin-sish′e-um) Cells performing as a unit; those of the heart join electrically.

G

gallbladder (gawl′blad-er) Saclike organ associated with liver that stores and concentrates bile.
ganglion (gang′gle-on) (plural, *ganglia*) Mass of neuron cell bodies, usually outside central nervous system.
gastric gland (gas′trik gland) Gland within stomach wall that secretes gastric juice.
gastric juice (gas′trik jo͞os) Secretion of gastric glands within stomach.
gastrin (gas′trin) Hormone that the stomach lining secretes to stimulate gastric juice secretion.
gene (jēn) Portion of a DNA molecule that encodes information to synthesize a protein, a control sequence, or tRNA or rRNA; the unit of inheritance.
gene expression (jēn eks-presh′un) Cellular synthesis of a protein coded for by a particular gene. Not all genes are expressed by any one cell.
genetic code (jĕ-net′ik kōd) Correspondence between particular DNA nucleotide base triplets and specific amino acids.
germinal epithelium (jer′mĭ-nal ep″ĭ-the′le-um) Tissue within an ovary that gives rise to sex cells.
germ layers (jerm la′ers) Embryonic cell layers that form organs during development; ectoderm, mesoderm, and endoderm.
GH Growth hormone.
gland (gland) Group of cells that secrete a product.
globin (glo′bin) Protein portion of a hemoglobin molecule.
globulin (glob′u-lin) Type of protein in blood plasma.
glomerular capsule (glo-mer′u-lar kap′sūl) Proximal portion of renal tubule that encloses glomerulus of nephron; Bowman's capsule.
glomerulus (glo-mer′u-lus) Capillary tuft within glomerular capsule of nephron.
glottis (glot′is) Slitlike opening between true vocal folds or vocal cords.
glucagon (gloo′kah-gon) Hormone that pancreatic islets of Langerhans secrete to release glucose from storage.
glucocorticoid (gloo″ko-kor′tĭ-koid) Any one of a group of hormones that the adrenal cortex secretes to influence carbohydrate, fat, and protein metabolism.
glucose (gloo′kōs) Monosaccharide in blood that is the primary source of cellular energy.
gluteal (gloo′te-al) Pertaining to the buttocks.
glycerol (glis′er-ol) Organic compound that is a building block for fat molecules.
glycogen (gli′ko-jen) Polysaccharide that stores glucose in liver and muscles.
glycoprotein (gli″ko-pro′te-in) Compound composed of a carbohydrate combined with a protein.
goblet cell (gob′let sel) Epithelial cell specialized to secrete mucus.
Golgi apparatus (gol′je ap″ah-ra′tus) An organelle that prepares cellular products for secretion.
gonadotropin (go-nad″o-trōp′in) Hormone that stimulates activity in gonads.
granulocyte (gran′u-lo-sīt) Leukocyte with granules in its cytoplasm.
gray matter (gra mat′er) Region of central nervous system that generally lacks myelin and thus appears gray.
groin (groin) Region of body between abdomen and thighs.
growth (grōth) Process by which a structure enlarges.
growth hormone (grōth hor′mōn) **(GH)** Hormone that the anterior pituitary secretes to promote growth of the organism; somatotropin.

H

hair follicle (hār fol′ĭ-kl) Tubelike depression in skin in which a hair develops.
hapten (hap′ten) Small molecule that combines with larger molecule to form an antigenic substance.
hematocrit (he-mat′o-krit) Volume percentage of red blood cells within a sample of whole blood.
hematoma (he″mah-to′mah) Mass of coagulated blood within tissues or a body cavity.
hematopoiesis (hem″ah-to-poi-e′sis) Blood production and blood cells; hemopoiesis.
heme (hēm) Iron-containing portion of hemoglobin.
hemocytoblast (he″mo-si′to-blast) Cell that gives rise to blood cells.
hemoglobin (he″mo-glo′bin) Pigment of red blood cells that transports oxygen.
hemorrhage (hem′o-rij) Loss of blood from circulatory system; bleeding.
hemostasis (he″mo-sta′sis) Stoppage of bleeding.
hepatic (hĕ-pat′ik) Pertaining to the liver.
hepatic lobule (hĕ-pat′ik lob′ul) Functional unit of liver.
holocrine gland (ho′lo-krin gland) Gland whose secretion contains entire secretory cells.

homeostasis (ho″me-o-sta′sis) State of balance in which the body's internal environment remains in the normal range.

hormone (hor′mōn) Substance that an endocrine gland secretes and that blood or body fluids transport.

humoral immunity (hu′mor-al ĭ-mu′nĭ-te) Circulating antibodies' destruction of cells bearing foreign (nonself) antigens.

hydrogen bond (hi′dro-jen bond) Weak chemical bond between a hydrogen atom and a nitrogen or oxygen atom, sharing with another nearby nitrogen or oxygen atom.

hydrolysis (hi-drol′ĭ-sis) Enzymatically adding a water molecule to split a molecule into smaller portions.

hydrostatic pressure (hy″dro-stat′ik presh′ur) Pressure a fluid exerts, such as blood pressure.

hydroxyl ion (hi-drok′sil i′on) OH^-.

hymen (hi′men) Membranous fold of tissue that partially covers vaginal opening.

hyperplasia (hi″per-pla′ze-ah) Increased production and growth of new cells.

hypertonic (hi″per-ton′ik) Describes a solution containing a greater concentration of dissolved particles than the solution with which it is compared.

hypertrophy (hi″per′tro-fe) Enlargement of an organ or tissue.

hyperventilation (hi″per-ven″tĭ-la′shun) Abnormally deep and prolonged breathing.

hypochondriac region (hi″po-kon′dre-ak re′jun) Portion of abdomen on either side of middle or epigastric region.

hypogastric region (hi″po-gas′trik re′jun) Lower middle portion of abdomen.

hypothalamus (hi″po-thal′ah-mus) Portion of brain below thalamus and forming floor of third ventricle.

hypotonic (hi″po-ton′ik) Describes a solution containing a lesser concentration of dissolved particles than the solution (usually body fluids) to which it is compared.

I

iliac region (il′e-ak re′jun) Portion of abdomen on either side of lower middle or hypogastric region.

ilium (il′e-um) One of the bones of a coxal bone or hipbone.

immunity (ĭ-mu′nĭ-te) Resistance to effects of specific disease-causing agents.

immunoglobulin (im″u-no-glob′u-lin) Globular plasma proteins that function as antibodies.

implantation (im″plan-ta′shun) Embedding an embryo in the lining of the uterus.

impulse (im′puls) Wave of depolarization conducted along a nerve or muscle fiber.

incomplete protein (in″kom-plēt′ pro′te-in) Protein that lacks adequate amounts of essential amino acids.

inferior (in-fer′e-or) Situated below something else; pertaining to the lower surface of a part.

inflammation (in″flah-ma′shun) Tissue response to stress that causes blood vessel dilation and fluid accumulation in the affected region.

inguinal (ing′gwĭ-nal) Pertaining to the groin region.

inorganic (in″or-gan′ik) Chemical substances that lack carbon and hydrogen.

insertion (in-ser′shun) The end of a muscle attached to a movable part.

inspiration (in″spĭ-ra′shun) Breathing in; inhalation.

insula (in′su-lah) Cerebral lobe deep within the lateral sulcus.

insulin (in′su-lin) Hormone that pancreatic islets of Langerhans secrete to control carbohydrate metabolism.

integumentary (in-teg-u-men′tar-e) Pertaining to the skin and its accessory organs.

intercalated disk (in-ter″kah-lāt′ed disk) Membranous boundary between cardiac muscle cells.

intercellular (in″ter-sel′u-lar) Between cells.

intercellular fluid (in″ter-sel′u-lar floo′id) Tissue fluid between cells other than blood cells.

interneuron (in″ter-nu′ron) Neuron between a sensory neuron and a motor neuron; internuncial or association neuron.

interphase (in′ter-fāz) Period between two cell divisions when a cell is carrying on its normal functions and prepares for division.

interstitial cell (in″ter-stish′al sel) Hormone-secreting cell between seminiferous tubules of testis.

interstitial fluid (in″ter-stish′al floo′id) Same as intercellular fluid.

intervertebral disk (in″ter-ver′tĕ-bral disk) Layer of fibrocartilage between bodies of vertebrae.

intestinal gland (in-tes′tĭ-nal gland) Tubular gland at the base of a villus within the intestinal wall.

intestinal juice (in-tes′tĭ-nal jo͞os) Fluid that intestinal glands secrete, containing digestive enzymes.

intracellular (in″trah-sel′u-lar) Within cells.

intracellular fluid (in″trah-sel′u-lar floo′id) Fluid within cells.

intramembranous bone (in″trah-mem′brah-nus bōn) Bone that forms from membranelike layers of primitive connective tissue.

intrinsic factor (in-trin′sik fak′tor) Substance that gastric glands produce to promote absorption of vitamin B_{12}.

inversion (in-ver′zhun) Inward turning movement of sole of foot.

involuntary (in-vol′un-tar″e) Not consciously controlled; functions automatically.

ion (i′on) Atom or molecule with electrical charge.

ionic bond (i-on′ik bond) Chemical bond formed between two ions by transfer of electrons.

ionization (i″on-ĭ-za′shun) Dissociation into ions.

iris (i′ris) Colored, muscular portion of eye that surrounds the pupil and regulates its size.

irritability (ir″ĭ-tah-bil′ĭ-te) Ability of an organism to react to changes in its environment.

ischemia (is-ke′me-ah) Deficiency of blood in a body part.

isotonic (i″so-ton′ik) Describes a solution with the same concentration of dissolved particles as the solution (usually body fluids) with which it is compared.

isotope (i′so-tōp) Atom that has the same number of protons as other atoms of an element but has a different number of neutrons.

joint (joint) Union of two or more bones; articulation.

juxtaglomerular apparatus (juks″tah-glo-mer′u-lar ap″ah-ra′tus) Structure in arteriolar walls near the glomerulus that regulates renin secretion.

K

keratin (ker′ah-tin) Protein in epidermis, hair, and nails.

keratinization (ker″ah-tin″ĭ-za′shun) Process by which cells form fibrils of keratin and harden.

ketone body (ke′tōn bod′e) Type of compound produced during fat catabolism, including acetone, acetoacetic acid, and betahydroxybuteric acid.

kinase (ki′nās) An enzyme that activates a precursor form of another enzyme by adding a phosphate group.

Kupffer cell (koop′fer sel) Large, fixed phagocyte in liver that removes bacterial cells from blood.

L

labor (la′bor) Process of childbirth.

labyrinth (lab′ĭ-rinth) System of connecting tubes within inner ear, including cochlea, vestibule, and semicircular canals.

lacrimal gland (lak′rĭ-mal gland) Tear-secreting gland.

lactase (lak′tās) Enzyme that catalyzes breakdown of lactose into glucose and galactose.

lactation (lak-ta′shun) Production of milk by mammary glands.

lacteal (lak′te-al) Lymphatic capillary associated with a villus of the small intestine.

lactic acid (lak′tik as′id) Organic compound formed from pyruvic acid during anaerobic respiration.

lacuna (lah-ku′nah) A hollow cavity.

lamella (lah-mel′ah) Layer of matrix in bone tissue.

laryngopharynx (lah-ring″go-far′ingks) Lower portion of pharynx near opening to larynx.

larynx (lar′ingks) Structure between pharynx and trachea that houses the vocal cords.

latent period (la′tent pe′re-od) Time between application of a stimulus and beginning of a response in a muscle fiber.

lateral (lat′er-al) Pertaining to the side.

leukocyte (lu′ko-sīt) White blood cell.

lever (lev′er) Simple mechanical device consisting of a rod, fulcrum, weight, and a source of energy that is applied to some point on the rod.

LH Luteinizing hormone.

ligament (lig′ah-ment) Cord or sheet of connective tissue binding two or more bones at a joint.

limbic system (lim′bik sis′tem) Group of connected structures within the brain that produces emotional feelings.

lingual (ling′gwal) Pertaining to the tongue.

lipase (li′pās) Fat-digesting enzyme.

lipid (lip′id) Fat, oil, or fatlike compound that usually has fatty acids in its molecular structure.

lipoprotein (lip-o-pro′te-in) Complex of lipid and protein.

lumbar (lum′bar) Pertaining to the region of the loins.

lumen (lu′men) Space within a tubular structure, such as a blood vessel or intestine.

luteinizing hormone (lu′te-in-īz″ing hor′mōn) **(LH; ICSH** in males) Hormone that the anterior pituitary secretes to control formation of corpus luteum in females and testosterone secretion in males.

lymph (limf) Fluid that lymphatic vessels carry.

lymphatic pathway (lim-fat′ik path′wa) Pattern of vessels that transport lymph.

lymph node (limf nōd) Mass of lymphoid tissue.

lymphocyte (lim′fo-sīt) Type of white blood cell that provides immunity; B cell or T cell.

lysosome (li′so-sōm) Organelle that contains digestive enzymes.

M

macrophage (mak′ro-fāj) Large phagocytic cell.

macroscopic (mak″ro-skop′ik) Large enough to be seen with unaided eye.

macula lutea (mak′u-lah lu′te-ah) Yellowish depression in retina of eye that is associated with acute vision.

malignant (mah-lig′nant) The power to threaten life; cancerous.

malnutrition (mal″nu-trish′un) Physical symptoms resulting from lack of specific nutrients.

mammary (mam′ar-e) Pertaining to the breast.

marrow (mar′o) Connective tissue in spaces within bones that includes blood-forming stem cells.

mast cell (mast sel) Cell to which antibodies, formed in response to allergens, attach, bursting the cell and releasing allergy mediators, which cause symptoms.

mastication (mas″tĭ-ka′shun) Chewing movement.

matrix (ma′triks) Intercellular material of connective tissue.

matter (mat′er) Anything that has weight and occupies space.

meatus (me-a′tus) Passageway or channel, or external opening of passageway.

mechanoreceptor (mek″ah-no-re-sep′tor) Sensory receptor that senses mechanical stimulation, such as changes in pressure or tension.

medial (me′de-al) Toward or near the midline.

mediastinum (me″de-ah-sti′num) Tissues and organs of the thoracic cavity that form a septum between the lungs.

medulla (mĕ-dul′ah) Inner portion of an organ.

medulla oblongata (mĕ-dul′ah ob″long-gah′tah) Portion of the brain stem between the pons and the spinal cord.

medullary cavity (med′u-lār″e kav′ĭ-te) Cavity containing marrow within diaphysis of long bone.

megakaryocyte (meg″ah-kar′e-o-sīt) Large bone marrow cell that gives rise to blood platelets.

meiosis (mi-o′sis) Cell division that halves the genetic material, resulting in egg and sperm cells (gametes).

melanin (mel′ah-nin) Dark pigment normally found in skin and hair.

melanocyte (mel′ah-no-sīt) Melanin-producing cell.

melatonin (mel″ah-to′nin) Hormone that the pineal gland secretes.

memory cell (mem′o-re sel) B lymphocyte or T lymphocyte produced in the primary immune response that can be activated rapidly if the same antigen is encountered in the future.

meninges (mĕ-nin′jēz) (singular, *meninx*) Three membranes that cover the brain and spinal cord.

meniscus (men-is′kus) (plural, *menisci*) Fibrocartilage that separates articulating bone surfaces in the knee.

menopause (men′o-pawz) Termination of the menstrual cycle.

menstrual cycle (men′stroo-al si′kl) Reoccurring changes in the uterine lining of a woman of reproductive age due to cycling hormones.

menstruation (men″stroo-a′shun) Shedding of blood and other tissue from uterine lining at end of female reproductive cycle.

merocrine gland (mer′o-krin gland) Gland whose cells secrete a fluid without losing cytoplasm.

mesentery (mes′en-ter″e) Fold of peritoneal membrane that attaches abdominal organ to abdominal wall.

mesoderm (mez′o-derm) Middle primary germ layer.
messenger RNA (mes′in-jer) RNA that transmits information for protein's amino acid sequence from cell nucleus to cytoplasm.
metabolic rate (met″ah-bol′ic rāt) Rate at which the body's biochemistry builds up and breaks down molecules to store and release energy.
metabolism (mĕ-tab′o-lizm) All chemical changes within cells considered together.
metacarpals (met″ah-kar′pals) Bones of hand between wrist and finger bones.
metaphase (met′ah-fāz) Stage in mitosis when chromosomes align in the middle of the cell.
metatarsals (met″ah-tar′sals) Bones of foot between ankle and toe bones.
microfilament (mi″kro-fil′ah-ment) Tiny rod of actin protein in cytoplasm that provides structural support and movement.
microglia (mi-krog′le-ah) Neuroglial cells that support neurons and phagocytize.
microscopic (mi″kro-skop′ik) Too small to be seen by unaided eye.
microtubule (mi″kro-tu′būl) Minute, hollow rod of the protein tubulin.
microvilli (mi″kro-vil′i) Tiny, cylindrical processes that extend from some epithelial cell membranes and increase membrane surface area.
micturition (mik″tu-rish′un) Urination.
midbrain (mid′brān) Small region of brain stem between diencephalon and pons.
mineral (min′er-al) Element not found in organic compounds that is essential in human metabolism.
mineralocorticoid (min″er-al-o-kor′tĭ-koid) Hormones that the adrenal cortex secretes to influence electrolyte concentrations in body fluids.
mitochondrion (mi″to-kon′dre-on) (plural, *mitochondria*) Organelle housing enzymes that catalyze reactions of aerobic respiration.
mitosis (mi-to′sis) Division of a somatic cell to form two genetically identical cells.
mixed nerve (mikst nerv) Nerve that includes both sensory and motor neuron fibers.
molecular formula (mo-lek′u-lar fōr′mu-lah) Abbreviation for the number of atoms of each element in a compound.
molecule (mol′ĕ-kūl) Particle composed of two or more joined atoms.
monocyte (mon′o-sīt) Type of white blood cell that is a phagocyte.
monosaccharide (mon″o-sak′ah-rīd) Simple sugar, such as glucose or fructose.
motor area (mo′tor a′re-ah) Region of brain from which impulses to muscles or glands originate.
motor end plate (mo′tor end plāt) Specialized portion of muscle fiber membrane at neuromuscular junction.
motor nerve (mo′tor nerv) Nerve that consists of motor neuron fibers.
motor neuron (mo′tor nu′ron) Neuron that transmits impulses from central nervous system to an effector.
motor unit (mo′tor u′nit) Motor neuron and its associated muscle fibers.
mucosa (mu-ko′sah) Membrane that lines tubes and body cavities that open to outside of body; mucous membrane.
mucous cell (mu′kus sel) Glandular cell that secretes mucus.
mucous membrane (mu′kus mem′brān) Mucosa.
mucus (mu′kus) Fluid secretion of mucous cells.
muscle impulse (mus′el im′puls) Impulse that travels along sarcolemma and down transverse tubules.
muscle tissue (mus′el tish′u) Contractile tissue consisting of filaments of actin and myosin, which slide past each other, shortening cells.
mutagen (mu′tah-jen) Agent that can cause mutations.
mutation (mu-ta′shun) Change in a gene.
myelin (mi′ĕ-lin) Fatty material that forms sheathlike covering around some nerve fibers.
myocardium (mi″o-kar′de-um) Muscle tissue of heart.
myofibril (mi″o-fi′bril) Contractile fibers within muscle cells.
myoglobin (mi″o-glo′bin) Pigmented compound in muscle tissue that stores oxygen.
myogram (mi′o-gram) Recording of a muscular contraction.
myometrium (mi″o-me′tre-um) Layer of smooth muscle tissue within uterine wall.
myosin (mi′o-sin) Protein that, with actin, contracts and relaxes muscle fibers.

nasal cavity (na′zal kav′ĭ-te) Space within nose.
nasal concha (na′zal kong′kah) Shell-like bone extending out from wall of nasal cavity; a turbinate bone.
nasal septum (na′zal sep′tum) Wall of bone and cartilage that separates nasal cavity into two portions.
nasopharynx (na″zo-far′ingks) Portion of pharynx associated with nasal cavity.
negative feedback (neg′ah-tiv fēd′bak) Mechanism activated by an imbalance that corrects the imbalance.
neonatal (ne″o-na′tal) First four weeks of life.
nephron (nef′ron) Functional unit of kidney, consisting of a renal corpuscle and a renal tubule.
nerve (nerv) Bundle of nerve fibers.
nerve impulse (nerv im′puls) Electrochemical process of depolarization and repolarization along a nerve fiber.
nervous tissue (ner′vus tish′u) Neurons and neuroglia.
neurilemma (nu″rĭ-lem′ah) Sheath formed from Schwann cells on outside of some nerve fibers.
neurofibrils (nu″ro-fi′brilz) Fine, cytoplasmic threads that extend from the cell body into the processes of neurons.
neuroglial cell (nu-rog′le-ahl sel) Specialized cell of the nervous system that produces myelin, communicates between cells, and maintains the ionic environment.
neuromuscular junction (nu″ro-mus′ku-lar jungk′shun) Point of communication between a nerve cell and muscle cell.
neuron (nu′ron) Nerve cell.
neurotransmitter (nu″ro-trans′mit-er) Chemical that axon end secretes to control another neuron or an effector.
neutral (nu′tral) Neither acidic nor alkaline.
neutron (nu′tron) Electrically neutral particle in atomic nucleus.
neutrophil (nu′tro-fil) Type of phagocytic leukocyte.
niacin (ni′ah-sin) Vitamin of the B-complex group; nicotinic acid.
Nissl bodies (nis′l bod′ēz) Membranous sacs within cytoplasm of nerve cells that have ribosomes attached to their surfaces.
nonelectrolyte (non″e-lek′tro-līt) Substance that does not dissociate into ions when dissolved.

nonprotein nitrogenous substance (non-pro′te-in ni-troj′ĕ-nus sub′stans) Substance, such as urea or uric acid, that contains nitrogen but is not a protein.
norepinephrine (nor″ep-ĭ-nef′rin) Neurotransmitter released from axons of some nerve fibers.
nuclease (nu′kle-ās) Enzyme that catalyzes decomposition of nucleic acids.
nucleic acid (nu-kle′ik as′id) Substance composed of bonded nucleotides; RNA or DNA.
nucleolus (nu-kle′o-lus) (plural, *nucleoli*) Small structure within cell nucleus that contains RNA and proteins.
nucleotide (nu′kle-o-tīd″) Building block of nucleic acid molecule, consisting of a sugar, a nitrogenous base, and a phosphate group.
nucleus (nu′kle-us) (plural, *nuclei*) Cellular organelle enclosed by double-layered, porous membrane and containing DNA; dense core of atom composed of protons and neutrons.
nutrient (nu′tre-ent) Chemical that body requires from environment.

O

occipital (ok-sip′ĭ-tal) Pertaining to the lower, back portion of the head.
olfactory (ol-fak′to-re) Pertaining to the sense of smell.
olfactory nerves (ol-fak′to-re nervz) First pair of cranial nerves, which conduct impulses associated with sense of smell.
oligodendrocyte (ol″ĭ-go-den′dro-sīt) Type of neuroglial cell that forms myelin.
oocyte (o′o-sīt) Immature egg cell.
oogenesis (o″o-jen′ĕ-sis) Differentiation of an egg cell.
ophthalmic (of-thal′mik) Pertaining to the eye.
optic (op′tik) Pertaining to the eye.
optic chiasma (op′tik ki-az′mah) X-shaped structure on underside of brain; formed by partial crossing over of optic nerve fibers.
optic disk (op′tik disk) Region in the retina of eye where nerve fibers leave to become part of the optic nerve.
oral (o′ral) Pertaining to the mouth.
organ (or′gan) Structure consisting of a group of tissues with a specialized function.
organelle (or″gah-nel′) Part of a cell that performs a specialized function.
organic (or-gan′ik) Carbon-containing molecules.
organism (or′gah-nizm) Individual living thing.
orgasm (or′gaz-um) Culmination of sexual excitement.
orifice (or′ĭ-fis) Opening.
origin (or′ĭ-jin) End of muscle that attaches to immovable part.
oropharynx (o″ro-far′ingks) Portion of pharynx in posterior of oral cavity.
osmoreceptor (oz″mo-re-sep′tor) Receptor sensitive to changes in osmotic pressure of body fluids.
osmosis (oz-mo′sis) Diffusion of water through selectively permeable membrane in response to concentration gradient.
osmotic (oz-mot′ik) Pertaining to osmosis.
osmotic pressure (oz-mot′ik presh′ur) Amount of pressure required to stop osmosis; a solution's potential pressure caused by nondiffusible solute particles in the solution.
ossification (os″ĭ-fĭ-ka′shun) Bone tissue formation.
osteoblast (os′te-o-blast″) Bone-forming cell.
osteoclast (os′te-o-klast″) Cell that erodes bone.
osteocyte (os′te-o-sīt) Mature bone cell.
osteon (os′te-on) Cylinder-shaped unit containing bone cells that surround an osteonic canal; Haversian system.
osteonic canal (os′te-on-ik kah-nal′) Tiny channel in bone tissue that contains a blood vessel; Haversian canal.
otolith (o′to-lith) Small particle of calcium carbonate associated with receptors of equilibrium.
oval window (o′val win′do) Opening between stapes and inner ear.
ovarian (o-va′re-an) Pertaining to the ovary.
ovary (o′vah-re) Primary female reproductive organ; an egg-cell-producing organ.
oviduct (o′vĭ-dukt) Tube that leads from ovary to uterus; uterine tube or fallopian tube.
ovulation (o″vu-la′shun) Release of egg cell from mature ovarian follicle.
oxidation (ok″sĭ-da′shun) Process by which oxygen combines with another chemical; the removal of hydrogen or the loss of electrons; opposite of *reduction*.
oxygen debt (ok′sĭ-jen det) Amount of oxygen required after physical exercise to convert accumulated lactic acid to glucose.
oxyhemoglobin (ok″sĭ-he″mo-glo′bin) Compound formed when oxygen combines with hemoglobin.
oxytocin (ok″sĭ-to′sin) Hormone that the posterior pituitary releases to contract smooth muscles in the uterus and mammary glands.

P

pacemaker (pās′māk-er) Mass of specialized muscle tissue that controls the rhythm of heartbeat; sinoatrial node.
pain receptor (pān re″sep′tor) Sensory nerve ending associated with feeling pain.
palate (pal′at) Roof of mouth.
palatine (pal′ah-tīn) Pertaining to the palate.
palmar (pahl′mar) Pertaining to the palm of the hand.
pancreas (pan′kre-as) Glandular organ in abdominal cavity that secretes hormones and digestive enzymes.
pancreatic (pan″kre-at′ik) Pertaining to the pancreas.
pancreatic juice (pan″kre-at′ik jo͞os) Digestive secretions of the pancreas.
pantothenic acid (pan″to-then′ik as′id) Vitamin of the B-complex group.
papilla (pah-pil′ah) Tiny, nipplelike projection.
papillary muscle (pap′ĭ-ler″e mus′l) Muscle that extends inward from ventricular walls of heart and to which chordae tendineae attach.
paranasal sinus (par″ah-na′zal si-nus) Air-filled cavity in cranial bone; lined with mucous membrane and connected to nasal cavity.
parasympathetic nervous system (par″ah-sim″pah-thet′ik ner′vus sis′tem) Portion of autonomic nervous system that arises from brain and sacral region of spinal cord.
parathyroid glands (par″ah-thi′roid glandz) Small endocrine glands embedded in posterior thyroid gland.
parathyroid hormone (par″ah-thi′roid hor′mōn) **(PTH)** Hormone that parathyroid glands secrete to help regulate level of blood calcium and phosphate ions.
parietal (pah-ri′ĕ-tal) Pertaining to the wall of an organ or a cavity.
parietal cell (pah-ri′ĕ-tal sel) Cell of a gastric gland that secretes hydrochloric acid and intrinsic factor.
parietal pleura (pah-ri′ĕ-tal ploo′rah) Membrane that lines the inner wall of thoracic cavity.

parotid glands (pah-rot′id glandz) Large salivary glands on sides of face just in front and below ears.

partial pressure (par′shal presh′ur) Pressure that one gas produces in a mixture of gases.

pathogen (path′o-jen) Any disease-causing agent.

pectoral (pek′tor-al) Pertaining to the chest.

pectoral girdle (pek′tor-al ger′dl) Part of skeleton that supports and attaches to upper limbs.

pelvic (pel′vik) Pertaining to the pelvis.

pelvic girdle (pel′vik ger′dl) Portion of skeleton to which lower limbs attach.

pelvis (pel′vis) Bony ring formed by sacrum and coxal bones.

penis (pe′nis) Male external reproductive organ through which urethra passes.

pepsin (pep′sin) Protein-splitting enzyme that stomach gastric glands secrete.

pepsinogen (pep-sin′o-jen) Inactive form of pepsin.

peptide (pep′tīd) Compound of two or more bonded amino acid molecules.

peptide bond (pep′tīd bond) Bond between carboxyl group of one amino acid and amino group of another.

pericardial (per″ĭ-kar′de-al) Pertaining to the pericardium.

pericardium (per″ĭ-kar′de-um) Serous membrane that surrounds heart.

perichondrium (per″ĭ-kon′dre-um) Layer of fibrous connective tissue that encloses cartilaginous structures.

perilymph (per′ĭ-limf) Fluid in space between membranous and osseous labyrinths of inner ear.

perimetrium (per-ĭ-me′tre-um) Outer serosal layer of uterine wall.

perimysium (per″i-mis′e-um) Sheath of connective tissue that encloses bundle of skeletal muscle fibers.

perineal (per″ĭ-ne′al) Pertaining to the perineum.

perineum (per′ĭ-ne′um) Body region between scrotum or urethral opening and anus.

periodontal ligament (per″e-o-don′tal lig′ah-ment) Fibrous membrane that surrounds a tooth and attaches it to jawbone.

periosteum (per″e-os′te-um) Fibrous connective tissue covering on bone surface.

peripheral (pĕ-rif′er-al) Pertaining to parts near the surface or toward the outside.

peripheral nervous system (pĕ-rif′er-al ner′vus sis′tem) Portions of nervous system outside central nervous system.

peripheral resistance (pĕ-rif′er-al re-zis′tans) Resistance to blood flow due to friction between blood and blood vessel walls.

peristalsis (per″ĭ-stal′sis) Rhythmic waves of muscular contraction in walls of certain tubular organs.

peritoneal (per″ĭ-to-ne′al) Pertaining to the peritoneum.

peritoneal cavity (per″i-to-ne′al kav′ĭ-te) Potential space between parietal and visceral peritoneal membranes.

peritoneum (per″ĭ-to-ne′um) Serous membrane that lines abdominal cavity and encloses abdominal viscera.

peritubular capillary (per″ĭ-tu′bu-lar kap′ĭ-ler″e) Capillary that surrounds renal tubule and functions in reabsorption and secretion during urine formation.

pH The negative logarithm of the hydrogen ion concentration used to indicate the acidic or alkaline condition of a solution; values range from 0 to 14.

phagocytosis (fag″o-si-to′sis) Process by which a cell engulfs and digests solid substances.

phalanx (fa′langks) (plural, *phalanges*) Bone of a finger or toe.

pharynx (far′ingks) Portion of digestive tube between mouth and esophagus.

phospholipid (fos″fo-lip′id) Lipid that contains two fatty acid molecules and a phosphate group combined with a glycerol molecule.

photoreceptor (fo″to-re-sep′tor) Sensory receptor sensitive to light energy; rods and cones.

physiology (fiz″e-ol′o-je) Study of body functions.

pia mater (pi′ah ma′ter) Inner layer of meninges that encloses brain and spinal cord.

pineal gland (pin′e-al gland) Small structure in central part of brain.

pinocytosis (pin″o-si-to′sis) Process by which a cell engulfs droplets from its surroundings.

pituitary gland (pĭ-tu′ĭ-tār″e gland) Endocrine gland attached to base of brain consisting of anterior and posterior lobes.

placenta (plah-sen′tah) Structure that attaches the fetus to the uterine wall, providing for delivery of nutrients to and removal of wastes from the fetus.

plantar (plan′tar) Pertaining to the sole of the foot.

plasma (plaz′mah) Fluid portion of circulating blood.

plasma cell (plaz′mah sel) Antibody-producing cell that forms when activated B cells proliferate.

plasma protein (plaz′mah pro′te-in) Proteins dissolved in blood plasma.

platelet (plāt′let) Cytoplasmic fragment formed in bone marrow that helps blood clot.

pleural (ploo′ral) Pertaining to pleura or membranes surrounding the lungs.

pleural cavity (ploo′ral kav′ĭ-te) Potential space between pleural membranes.

pleural membranes (ploo′ral mem′brānz) Serous membranes that enclose the lungs and line the chest wall.

plexus (plek′sus) Network of interlaced nerves or blood vessels.

polar body (po′lar bod′e) Small, nonfunctional cell that is a product of meiosis in the female.

polarization (po″lar-ĭ-za′shun) Development of an electrical charge on a cell membrane surface due to an unequal distribution of positive and negative ions on either side of the membrane.

polypeptide (pol″e-pep′tīd) Compound formed by union of many amino acid molecules.

polysaccharide (pol″e-sak′ah-rīd) Carbohydrate composed of many joined monosaccharides.

polyunsaturated fatty acid (pol″e-un-sach′ĕ-ra-ted fat′e as′id) Fatty acid containing one or more double bonds in its carbon atom chain.

pons (ponz) Portion of brain stem above medulla oblongata and below midbrain.

popliteal (pop″lĭ-te′al) Pertaining to region behind knee.

positive feedback (poz′ĭ-tiv fēd′bak) Process by which changes cause additional similar changes, producing unstable conditions.

posterior (pos-tēr′e-or) Toward the back; the opposite of *anterior*.

posterior pituitary (pos-tēr′e-or pĭ-tu′ĭ-tār″e) Lobe of pituitary gland that secretes oxytocin and antidiuretic hormone (vasopressin).

postganglionic fiber (pōst″gang-gle-on′ik fi′ber) Autonomic nerve fiber on distal side of a ganglion.

postnatal (pōst-na′tal) After birth.

preganglionic fiber (pre″gang-gle-on′ik fi′ber) Autonomic nerve fiber on proximal side of a ganglion.

pregnancy (preg′nan-se) Condition in which a female has a developing offspring in her uterus.

prenatal (pre-na′tal) Before birth.
primary follicle (pri′ma-re fol′ĭ-kl) Primordial follicle that begins to mature in response to hormonal changes in a female at puberty.
primary reproductive organs (pri′ma-re re″pro-duk′tiv or′ganz) Sex-cell-producing parts; testes in males and ovaries in females.
prime mover (prīm moov′er) Muscle responsible for a particular body movement.
PRL Prolactin.
process (pros′es) Prominent bone projection.
progesterone (pro-jes′tě-rōn) Female hormone that corpus luteum and placenta secrete.
projection (pro-jek′shun) Process by which brain causes a sensation to seem to come from region of body being stimulated.
prolactin (pro-lak′tin) **(PRL)** Hormone that the anterior pituitary secretes to stimulate milk production in mammary glands.
pronation (pro-na′shun) Downward or backward rotation of palm of hand.
prophase (pro′fāz) Stage of mitosis when chromosomes become visible.
proprioceptor (pro″pre-o-sep′tor) Nerve ending that senses changes in muscle or tendon tension.
prostaglandins (pros″tah-glan′dins) Group of compounds with powerful, hormonelike effects.
prostate gland (pros′tāt gland) Gland surrounding male urethra below urinary bladder that adds its secretion to semen just prior to ejaculation.
protein (pro′te-in) Nitrogen-containing organic compound of joined amino acid molecules.
prothrombin (pro-throm′bin) Plasma protein that leads to formation of blood clots.
proton (pro′ton) Positively charged particle in atomic nucleus.
protraction (pro-trak′shun) Forward movement of a body part.
proximal (prok′sĭ-mal) Closer to midline or origin; opposite of *distal.*
PTH Parathyroid hormone.
puberty (pu′ber-te) Stage of development in which reproductive organs become functional.
pulmonary (pul′mo-ner″e) Pertaining to the lungs.
pulmonary circuit (pul′mo-ner″e ser′kit) System of blood vessels that carries blood between heart and lungs.
pulse (puls) Surge of blood felt through walls of arteries due to contraction of heart ventricles.
pupil (pu′pil) Opening in iris through which light enters eye.
Purkinje fibers (pur-kin′je fi′berz) Specialized muscle fibers that conduct cardiac impulses from the A-V bundle into the ventricular walls.
pyramidal cells (pĭ-ram′ĭ-dal sels) Large, pyramid-shaped neurons within the cerebral cortex.
pyruvic acid (pi-roo′vik as′id) Intermediate product of carbohydrate oxidation.

R

radioactive (ra″de-o-ak′tiv) Describes an atom that emits energy at a constant rate.
rate-limiting enzyme (rāt lim′i-ting en′zīm) An enzyme, usually present in small amounts, that controls the rate of a metabolic pathway by regulating one of its steps.
recruitment (re-kroot′ment) Increase in number of motor units activated as stimulation intensity increases.
red marrow (red mar′o) Blood-cell-forming tissue in spaces within bones.
referred pain (re-ferd′ pān) Pain that feels as if it is originating from a part other than the site being stimulated.
reflex (re′fleks) Rapid, automatic response to a stimulus.
reflex arc (re′fleks ark) Nerve pathway, consisting of a sensory neuron, interneuron, and motor neuron, that forms the structural and functional bases for a reflex.
refraction (re-frak′shun) Bending of light as it passes between media of different densities.
relaxin (re-lak′sin) Hormone from corpus luteum that inhibits uterine contractions during pregnancy.
renal (re′nal) Pertaining to the kidney.
renal corpuscle (re′nal kor′pusl) Part of nephron that consists of a glomerulus and a glomerular capsule.
renal cortex (re′nal kor′teks) Outer portion of a kidney.
renal medulla (re′nal mě-dul′ah) Inner portion of a kidney.
renal pelvis (re′nal pel′vis) Cavity in a kidney that channels urine to the ureter.
renal tubule (re′nal tu′būl) Portion of nephron that extends from renal corpuscle to collecting duct.
renin (re′nin) Enzyme that kidneys release to maintain blood pressure and blood volume.
replication (rep″lĭ-ka′shun) Reproduction of an exact copy of a DNA molecule.
reproduction (re″pro-duk′shun) Offspring formation.
resorption (re-sorp′shun) Decomposition and assimilation of a structure as a result of physiological activity.
respiration (res″pĭ-ra′shun) Cellular process that releases energy from nutrients; breathing.
respiratory capacities (re-spi′rah-to″re kah-pas′ĭ-tēz) Value obtained by adding two or more respiratory volumes.
respiratory center (re-spi′rah-to″re sen′ter) Portion of brain stem that controls breathing depth and rate.
respiratory membrane (re-spi′rah-to″re mem′brān) Membrane composed of a capillary wall and an alveolar wall through which blood and inspired air exchange gases.
response (re-spons′) Action resulting from a stimulus.
resting potential (res′ting po-ten′shal) Difference in electrical charge between inside and outside of an undisturbed nerve cell membrane.
reticular formation (rě-tik′u-lar for-ma′shun) Complex network of nerve fibers within brain stem that arouses cerebrum.
retina (ret′ĭ-nah) Inner layer of eye wall that contains visual receptors.
retinal (ret′ĭ-nal) Form of vitamin A; retinene.
retinene (ret′ĭ-nēn) Chemical precursor of rhodopsin, a visual pigment.
retraction (rě-trak′shun) Movement of a part toward the back.
retroperitoneal (ret″ro-per″i-to-ne′al) Located behind the peritoneum.
reversible reaction (re-ver′sĭ-bl re-ak′shun) Chemical reaction in which end products can change back into reactants.
rhodopsin (ro-dop′sin) Light-sensitive pigment in rods of the retina; visual purple.
riboflavin (ri″bo-fla′vin) Vitamin of B-complex group; vitamin B_2.
ribonucleic acid (ri″bo-nu-kle′ik as′id) **(RNA)** Nucleic acid whose nucleotides contain ribose.
ribose (ri′bōs) Five-carbon sugar in RNA.

ribosome (ri′bo-sōm) Organelle composed of RNA and protein that is a structural support for protein synthesis.

RNA Ribonucleic acid.

rod (rod) Type of receptor that provides colorless vision.

rotation (ro-ta′shun) Movement turning a body part on its longitudinal axis.

round window (rownd win′do) Membrane-covered opening between inner ear and middle ear.

S

sagittal (saj′ĭ-tal) Plane or section that divides a structure into right and left portions.

S-A node (nōd) Sinoatrial node.

sarcomere (sar′ko-mēr) Structural and functional unit of a myofibril.

sarcoplasmic reticulum (sar″ko-plaz′mik rĕ-tik′u-lum) Membranous network of channels and tubules within a muscle fiber, corresponding to the endoplasmic reticulum of other cells.

saturated fatty acid (sat′u-rāt″ed fat′e as′id) Fatty acid molecule without double bonds between its carbon atoms.

Schwann cell (shwahn sel) Type of neuroglial cell that surrounds a fiber of a peripheral neuron, forming the neurilemmal sheath and myelin.

sclera (skle′rah) White, fibrous outer layer of eyeball.

scrotum (skro′tum) Pouch of skin that encloses testes.

sebaceous gland (se-ba′shus gland) Skin gland that secretes sebum.

sebum (se′bum) Oily secretion of sebaceous glands.

secretin (se-kre′tin) Hormone that small intestine secretes to stimulate pancreas to release pancreatic juice.

selectively permeable (se-lek′tiv-le per′me-ah-bl) Describes membrane that allows some molecules through but not others; semipermeable.

semen (se′men) Fluid containing sperm cells and secretions discharged from male reproductive tract at ejaculation.

semicircular canal (sem″ĭ-ser′ku-lar kah-nal′) Tubular structure within inner ear that contains receptors providing sense of dynamic equilibrium.

seminiferous tubule (sem″ĭ-nif′er-us tu′būl) Tubule within testes where sperm cells form.

sensation (sen-sa′shun) Feeling resulting from the brain's interpretation of sensory nerve impulses.

sensory adaptation (sen′so-re ad″ap-ta′shun) Sensory receptors becoming less responsive after constant repeated stimulation.

sensory area (sen′so-re a′re-ah) Portion of cerebral cortex that receives and interprets sensory nerve impulses.

sensory nerve (sen′so-re nerv) Nerve composed of sensory nerve fibers.

sensory neuron (sen′so-re nu′ron) Neuron that transmits an impulse from a receptor to the central nervous system.

sensory receptor (sen′so-re re″sep′tor) Specialized structure associated with the peripheral end of a sensory neuron specific to detecting a particular sensation and triggering a nerve impulse in response, which is transmitted to the central nervous system.

serotonin (se″ro-to′nin) Vasoconstrictor that blood platelets release when blood vessels break, controlling bleeding. Also a neurotransmitter.

serous cell (ser′us sel) Glandular cell that secretes watery fluid with high enzyme content.

serous fluid (ser′us floo′id) Secretion of a serous cell.

serous membrane (ser′us mem′brān) Membrane that lines a cavity without an opening to the outside of the body.

serum (se′rum) Fluid portion of coagulated blood.

sesamoid bone (ses′ah-moid bōn) Round bone in tendons adjacent to joints.

simple sugar (sim′pl shoog′ar) Monosaccharide.

sinoatrial node (si″no-a′tre-al nōd) **(S-A node)** Specialized tissue in wall of right atrium that initiates cardiac cycles; pacemaker.

sinus (si′nus) Cavity or space in a bone or other body part.

skeletal muscle (skel′ĭ-tal mus′l) Type of muscle tissue in muscles attached to bones.

smooth muscle (smooth mus′l) Type of muscle tissue in walls of hollow viscera; visceral muscle.

solute (sol′ūt) Substance dissolved in a solution.

solution (so-lu′shun) Homogenous mixture of substances (solutes) within a dissolving medium (solvent).

solvent (sol′vent) Liquid portion of a solution in which a solute is dissolved.

somatic cell (so-mat′ik sel) Any cell of the body other than the sex cells.

somatic nervous system (so-mat′ik ner′vus sis′tem) Motor pathways of peripheral nervous system that lead to skeletal muscles.

special sense (spesh′al sens) Sense that involves receptors associated with specialized sensory organs, such as the eyes and ears.

spermatid (sper′mah-tid) Intermediate stage in sperm cell formation.

spermatocyte (sper-mat′o-sīt) Early stage in sperm cell formation.

spermatogenesis (sper″mah-to-jen′ĕ-sis) Sperm cell production.

spermatogonium (sper″mah-to-go′ne-um) Undifferentiated spermatogenic cell in the germinal epithelium of a seminiferous tubule.

sphincter (sfingk′ter) Circular muscle that closes opening or lumen of tubular structure.

spinal (spi′nal) Pertaining to the spinal cord or to the vertebral canal.

spinal cord (spi′nal kord) Portion of the central nervous system extending from the brain stem through the vertebral canal.

spinal nerve (spi′nal nerv) Nerve that arises from spinal cord.

spleen (splēn) Large organ in the upper left region of abdomen.

spongy bone (spunj′e bōn) Bone that consists of bars and plates separated by irregular spaces; cancellous bone.

squamous (skwa′mus) Flat or platelike.

starch (starch) Polysaccharide common in foods of plant origin.

static equilibrium (stat′ik e″kwĭ-lib′re-um) Maintenance of balance when the head and body are motionless.

steroid (ste′roid) Type of organic molecule including rings of carbon and hydrogen atoms.

stimulus (stim′u-lus) (plural, *stimuli*) Change in environmental conditions followed by response of an organism or a cell.

stomach (stum′ak) Digestive organ between the esophagus and small intestine.

stratified (strat′ĭ-fīd) Arranged in layers.

stratum basale (strat′tum ba′sal-e) Deepest layer of epidermis in which cells divide; stratum germinativum.

stratum corneum (stra′tum kor′ne-um) Outer, horny layer of epidermis.

stressor (stres′or) Factor capable of stimulating a stress response.

stroke volume (strōk vol′ūm) Amount of blood the ventricle discharges with each heartbeat.

structural formula (struk′cher-al for′mu-lah) Representation of the way atoms bond to form a molecule, using symbols for each element and lines to indicate chemical bonds.
subarachnoid space (sub″ah-rak′noid spās) Space within meninges between arachnoid mater and pia mater.
subcutaneous (sub″ku-ta′ne-us) Beneath the skin.
sublingual (sub-ling′gwal) Beneath the tongue.
submucosa (sub″mu-ko′sah) Layer of connective tissue underneath a mucous membrane.
substrate (sub′strāt) Substance upon which an enzyme acts.
sucrase (su′krās) Digestive enzyme that catalyzes the breakdown of sucrose.
sucrose (soo′krōs) Disaccharide; table sugar.
sulcus (sul′kus (plural, *sulci*) Shallow groove, such as that between convolutions on brain surface.
summation (sum-ma′shun) Phenomena in which the degree of change in membrane potential directly correlates with stimulation frequency.
superficial (soo″per-fish′al) Near the surface.
superior (soo-pe′re-or) Pertaining to a structure that is vertically higher than another structure.
supination (soo″pĭ-na′shun) Forearm rotation so palm faces upward when arm is outstretched.
surface tension (ser′fas ten′shun) Force due to mutual attraction of water molecules that holds moist membranes together.
surfactant (ser-fak′tant) Substance produced by the lungs that reduces surface tension within the alveoli.
suture (soo′cher) Immovable joint, such as that between flat bones of skull.
sweat gland (swet gland) Exocrine gland in skin that secretes sweat.
sympathetic nervous system (sim″pah-thet′ik ner′vus sis′tem) Portion of autonomic nervous system that arises from thoracic and lumbar regions of spinal cord.
symphysis (sim′fĭ-sis) Slightly movable joint between bones separated by a pad of fibrocartilage.
synapse (sin′aps) Junction between axon of one neuron and dendrite or cell body of another neuron.
synaptic knob (sĭ-nap′tik nob) Tiny enlargement at end of axon that secretes a neurotransmitter.
syncytium (sin-sish′e-um) Mass of merging cells.
synergist (sin′er-jist) Muscle that assists action of a prime mover.
synovial fluid (sĭ-no′ve-al floo′id) Fluid that the synovial membrane secretes.
synovial joint (sĭ-no′ve-al joint) Freely movable joint.
synovial membrane (sĭ-no′ve-al mem′brān) Membrane that forms inner lining of capsule of freely movable joint.
synthesis (sin′thĕ-sis) Building large molecules from smaller ones that join.
system (sis′tem) Group of organs coordinated to carry on a specialized function.
systemic circuit (sis-tem′ik ser′kit) Vessels that conduct blood between the heart and all body tissues except the lungs.
systole (sis′to-le) Phase of cardiac cycle when heart chamber wall contracts.
systolic pressure (sis-tol′ik presh′ur) Arterial blood pressure during systolic phase of cardiac cycle.

T

target tissue (tar′get tish′u) Specific tissue on which a hormone exerts its effect.
tarsus (tar′sus) The ankle bones.
taste bud (tāst bud) Organ containing receptors associated with sense of taste.
T cell (sel) Type of lymphocyte that interacts directly with antigen-bearing particles and causes cell-mediated immunity.
telophase (tel′o-fāz) Stage in mitosis when newly formed cells separate.
tendon (ten′don) Cordlike or bandlike mass of white fibrous connective tissue that connects a muscle to a bone.
testis (tes′tis) (plural, *testes*) Primary male reproductive organ; sperm-cell-producing organ.
testosterone (tes-tos′tĕ-rōn) Male sex hormone that interstitial cells of the testes secrete.
tetanus (tet′ah-nus) Continuous, forceful muscular contraction (tetanic contraction) without relaxation.
thalamus (thal′ah-mus) Mass of gray matter at base of cerebrum in wall of third ventricle.
thermoreceptor (ther″mo-re-sep′tor) Sensory receptor sensitive to temperature changes; heat and cold receptors.
thiamine (thi′ah-min) Vitamin B_1.
thoracic (tho-ras′ik) Pertaining to the chest.
threshold stimulus (thresh′old stim′u-lus) Stimulation level that must be exceeded to elicit nerve impulse or muscle contraction.
thrombus (throm′bus) Blood clot that remains at its formation site in blood vessel.
thymosins (thi′mo-sins) Group of peptides that the thymus gland secretes to increase production of certain types of white blood cells.
thymus (thi′mus) Glandular organ in mediastinum behind sternum and between lungs.
thyroid gland (thi′roid gland) Endocrine gland just below larynx and in front of trachea that secretes thyroid hormones.
thyroxine (thi-rok′sin) One hormone that the thyroid gland secretes.
tissue (tish′u) Group of similar cells that performs a specialized function.
trachea (tra′ke-ah) Tubular organ that leads from larynx to bronchi.
transcellular fluid (trans″sel′u-lar floo′id) Portion of the extracellular fluid, including fluid within special body cavities.
transcription (trans-krip′shun) Manufacturing a complementary RNA from DNA.
transfer RNA (trans′fer) RNA molecule that carries an amino acid to a ribosome in protein synthesis.
translation (trans-la′shun) Assembly of an amino acid chain according to the sequence of base triplets in an mRNA molecule.
transverse tubule (trans-vers′ tu′būl) Membranous channel that extends inward from a muscle fiber membrane and passes through the fiber.
tricuspid valve (tri-kus′pid valv) Heart valve between right atrium and right ventricle.
triglyceride (tri-glis′er-īd) Lipid composed of three fatty acids combined with a glycerol molecule.
triiodothyronine (tri″i-o″do-thi′ro-nēn) Type of thyroid hormone.
trochanter (tro-kan′ter) Broad process on a bone.
trochlea (trok′le-ah) Pulley-shaped structure.
tropic hormone (tro′pik hor′mōn) Hormone whose target tissue is an endocrine gland.
trypsin (trip′sin) Enzyme in pancreatic juice that breaks down protein molecules.

tubercle (tu′ber-kl) Small, rounded process on a bone.

tuberosity (tu″bĕ-ros′ĭ-te) Elevation or protuberance on a bone.

twitch (twich) Brief muscular contraction followed by relaxation.

tympanic membrane (tim-pan′ik mem′brān) Thin membrane that covers auditory canal and separates external ear from middle ear; eardrum.

U

umbilical cord (um-bil′ĭ-kal kord) Cordlike structure that connects fetus to placenta.

umbilical region (um-bil′ĭ-kal re′jun) Central portion of abdomen.

unsaturated fatty acid (un-sat′u-rāt″ed fat′e as′id) Fatty acid molecule with one or more double bonds between atoms of its carbon chain.

urea (u-re′ah) Nonprotein nitrogenous substance resulting from protein metabolism.

ureter (u-re′ter) Muscular tube that carries urine from the kidney to the urinary bladder.

urethra (u-re′thrah) Tube leading from the urinary bladder to the outside of body.

urine (u′rin) Wastes and excess water removed from the blood and excreted by the kidneys into the ureters, to the urinary bladder, and out of the body through the urethra.

uterine (u′ter-in) Pertaining to the uterus.

uterine tube (u′ter-in tūb) Tube that extends from the uterus on either side toward the ovary and transports sex cells; fallopian tube or oviduct.

uterus (u′ter-us) Hollow, muscular organ within female pelvis in which fetus develops.

utricle (u′trĭ-kl) Enlarged portion of membranous labyrinth of inner ear.

uvula (u′vu-lah) Fleshy portion of soft palate that hangs down above root of tongue.

V

vaccine (vak′sēn) Substance that contains antigens for stimulating immune response.

vagina (vah-ji′nah) Tubular organ that leads from uterus to vestibule of female reproductive tract.

Valsalva's maneuver (val-sal′vahz mah″noo′ver) Increasing intrathoracic pressure by forcing air from lungs against a closed glottis.

vascular (vas′ku-lar) Pertaining to blood vessels.

vas deferens (vas def′er-ens) (plural, *vasa deferentia*) Tube that leads from the epididymis to the urethra of the male reproductive tract.

vasoconstriction (vas″o-kon-strik′shun) Decrease in the diameter of a blood vessel.

vasodilation (vas″o-di-la′shun) Increase in the diameter of a blood vessel.

vein (vān) Vessel that carries blood toward the heart.

vena cava (ve′nah kav′ah) One of two large veins (superior and inferior) that convey deoxygenated blood to right atrium of the heart.

ventral root (ven′tral root) Motor branch of spinal nerve by which nerve attaches to spinal cord.

ventricle (ven′trĭ-kl) A cavity, such as brain ventricles filled with cerebrospinal fluid, or heart ventricles that contain blood.

venule (ven′ūl) Vessel that carries blood from capillaries to a vein.

vesicle (ves′ĭ-kl) Membranous cytoplasmic sac formed by infolding of cell membrane.

villus (vil′us) (plural, *villi*) Tiny, fingerlike projection that extends outward from lining of small intestine.

visceral (vis′er-al) Pertaining to the contents of a body cavity.

visceral peritoneum (vis′er-al per′ĭ-to-ne′um) Membrane that covers organ surfaces within abdominal cavity.

visceral pleura (vis′er-al ploo′rah) Membrane that covers lung surfaces.

viscosity (vis-kos′ĭ-te) Tendency for fluid to resist flowing due to internal friction of its molecules.

vitamin (vi′tah-min) Organic compound, other than a carbohydrate, lipid, or protein, required for normal metabolism but that the body cannot synthesize in adequate amounts.

vitreous humor (vit′re-us hu′mor) Substance between lens and retina of eye.

vocal cords (vo′kal kordz) Folds of tissue within larynx that vibrate and produce sounds.

voluntary (vol′un-tār″e) Capable of being consciously controlled.

vulva (vul′vah) External female reproductive parts that surround vaginal opening.

W

water balance (wot′er bal′ans) When the volume of water entering the body is equal to the volume leaving it.

water of metabolism (wot′er uv mĕ-tab′o-lizm) Water produced as a by-product of metabolic processes.

yellow marrow (yel′o mar′o) Fat storage tissue in certain bone cavities.

Z

zygote (zi′got) Cell produced when egg and sperm fuse; fertilized egg cell.

zymogen granule (zi-mo′jen gran′ūl) Cellular structure that stores inactive forms of protein-splitting enzymes in a pancreatic cell.

Credits

Photographs

Chapter 1

Opener: © Zig Lesczcynski/Animals Animals/Earth Scenes; **1A*a*, p.17:** © Science Photo Library/Photo Researchers, Inc.; **1A*b*, p.17:** © CNRI/SPL/Photo Researchers, Inc.;)

Chapter 2

Opener: Helen Blau; **2A, p.35:** © Mark Antman/The Image Works; **2B*a*, p. 35:** © SIU/Peter Arnold, Inc.; **2.8*a-b*, 2.16*b*, 2.17*b*:** John W. Hole Jr.

Chapter 3

Opener: © Dorothea-Zucker-Franklin, M.D. NYU Medical Center/Phototake; **3.4*a*:** © D.W. Fawcett/Photo Researchers, Inc.; **3.6:** © Bill Longcore/Photo Researchers, Inc.; **3.7*a*:** © Gordon Leedale/Biophoto Assoc.; **3.8*a*:** Courtesy of American Lung Association; **3.8*b*:** © Manfred Kage/Peter Arnold, Inc.; **3.10*b*:** © Stephen L. Wolfe; **3.19:** © Ed Reschke; **3.A:** © Dr. Tony Brain/SPL/Photo Researchers, Inc.

Chapter 4

Opener: © L.A. Times, photo by Tammy Lechner; **4.14:** © A.C. Barrington Brown/Photo Researchers, Inc.; **4A, p. 85:** Planet Art

Chapter 5

Opener: Courtesy of Organogenesis Inc.; **5.1*b*, 5.2*b*:** © Ed Reschke; **5.3*b*:** © Manfred Kage/Peter Arnold, Inc.; **5.4*b*:** © Ed Reschke; **5.5*b*:** © Fred Hossler/Visuals Unlimited; **5.6*b*, 5.7*b*:** © The McGraw-Hill Companies, Inc./Carol D. Jacobson, Ph.D., Department of Veterinary Anatomy, Iowa State University; **5.8*a*:** © Ed Reschke; **5.10:** © David M. Phillips/Visuals Unlimited; **5.11:** Courtesy Jean Paul Revel; **5.12:** © Veronica Burmeister/Visuals Unlimited; **5.13*b*, 5.14*b*, 5.15*b*, 5.16*b*, 5.17*b*:** © Ed Reschke; **5.18*b*:** © The McGraw-Hill Companies, Inc./Carol D. Jacobson, Ph.D., Department of Veterinary Anatomy, Iowa State University; **5.19*b*:** © Victor B. Eichler; **5.19*c*:** © Biophoto Associates/Photo Researchers, Inc.; **5.20*b*:** © Ed Reschke/Peter Arnold, Inc.; **5.21, 5.22*b*:** © Ed Reschke; **5.23*b*:** © Manfred Kage/Peter Arnold, Inc.; **5.24*b*:** © Ed Reschke

Chapter 6

Opener: © Sunstar/Photo Researchers, Inc.; **6.2*a*:** © Victor B. Eichler; **6.2*b*:** © M. Schliwa/Visuals Unlimited; **6.A, p. 122:** © SPL/Photo Researchers, Inc.; **6.4*b*:** © Victor B. Eichler, Ph.D.; **6.5:** © CNRI/SPL/Photo Researchers, Inc.; **6.6:** © Per H. Kjeldsen, University of Michigan

Chapter 7

Opener: © Corbis-Bettmann; **7.2:** Courtesy of John W. Hole, Jr.; **7.4:** © Biophoto Associates/Photo Researchers, Inc.; **7.21:** Courtesy of Eastman Kodak; **7.28:** © Martin Rotker

Reference Plates

RP 8-11: Courtesy of John W. Hole, Jr.

Chapter 8

Opener: AP/Wide World Press; **8.3*a*, 8.8*a*:** © H.E. Huxley; **8A, p. 206:** Gransanti/Sygma; **8.1*a-b*, p. 210:** John C. Mese

Chapter 9

Opener: AP/Wide World Photos; **9.1:** © Ed Reschke; **9.12:** © Don Fawcett/Photo Researchers, Inc.; **9.20*b*:** © Per H. Kjeldsen, University of Michigan

Chapter 10

Opener: © Jeff LePore/Photo Researchers, Inc.; **10.11:** © Fred Hossler/Visuals Unlimited; **10.23:** © Per H. Kjeldsen, University of Michigan; **10.24*a*:** © Carroll Weiss/Camera M.D. Studios; **10.26*c*:** © Frank S. Werblin

Chapter 11

Opener: UPI/Corbis-Bettmann; **11.15:** © Ed Reschke; **11A, p. 311:** Courtesy of Eli Lilly and Company Archives

Chapter 12

Opener: Corbis-Bettmann; **12.2*b*:** © Bill Longcore/Photo Researchers, Inc.; **12.6:** © Warren Rosenberg/Biological Photo Service; **12.7, 12.8, 12.9:** © Ed Reschke; **12.10, 12A*a*, p. 330:** © Ed Reschke/Peter Arnold, Inc.; **12A*b*, p. 330:** © Ed Reschke; **12.13:** © David M. Phillips/Visuals Unlimited; **12.14*a*:** © Carroll Weiss/Camera M.D. Studios; **12.14*b*:** © Camera M.D. Studios

Chapter 13

Opener: UPI/Corbis-Bettmann; **13.5:** © The McGraw-Hill Companies, Inc./Karl Ruben, photographer; **13.21:** © Don W. Fawcett, from Bloom, W. and Fawcett, D.W. "Textbook of Histology," 10/e W.B. Saunders Co., 1975 photo by T. Kuwabara; **13A, p. 360:** © Ed Reschke/Peter Arnold, Inc; **13B, p. 360:** © J.&L. Weber/Peter Arnold, Inc.; **13C, p. 360** © Alfred Pasieka/Peter Arnold, Inc.; **13.23*a*:** © Ed Reschke; **13.23*b*:** © Fred Hossler/Visuals Unlimited

Chapter 14

Opener: Courtesy of Hancock Family; **14.3:** © Ed Reschke; **14.7:** Kent Van De Graaff; **14.9*b*:** © John Cunningham/ Visuals Unlimited; **14.10*b*:** © Biophoto Associates/Photo Researchers, Inc.; **14.12:** Courtesy of Sloan-Kettering Institute for Cancer Research/Etienne de Harven, M.D. and Nina Lampen

Chapter 15

Opener: Teri Currie; **15.7:** John W. Hole, Jr. **15.13:** © Ed Reschke; **15.19:** © Victor B. Eichler; **15.23:** Armed Forces Institute of Pathology; **15.26:** © Manfred Kage/Peter Arnold, Inc.; **15.29:** © James Shaffer; **15.31:** © Ed Reschke

Chapter 16

Opener: © SIU/Photo Researchers, Inc.; **16.5*c*:** © CNRI/Phototake; **16.9:** © Dwight R. Khun; **16A*a*, p. 458:** © Victor B. Eichler; **16A*b*, p. 458:** © Victor B. Eichler

Chapter 17

Opener: © Edward Lettau/Photo Researchers, Inc.; **17.4:** From *Tissues and Organs: A Text-Atlas of Scanning Electron Microscopy* by Richard G. Kessel and

Randy H. Kardon, W.H. Freeman and Company, 1979; **17.5*a-b*,** © Manfred Kage/Peter Arnold, Inc.; **17.14:** © Per H. Kjeldsen, University of Michigan; **17.16:** © Ed Reschke

Chapter 18

Opener: © UPI/Corbis-Bettmann

Chapter 19

Opener: © Dr. Tony Brain/SPL/Photo Researchers, Inc.; **19.4*a*:** © Biophoto Associates/Photo Researchers, Inc.; **19.10:** © Ed Reschke/Peter Arnold, Inc.; **19.12:** © The McGraw-Hill Companies, Inc./Carol D. Jacobson, Ph.D., Department of Veterinary Anatomy, Iowa State University; **19B, p. 536:** Courtesy of Southern Illinois University School of Medicine; **19.15:** © MHHE/Bob Coyle, photographer

Chapter 20

Opener: © Steve Wernka/Impact Visuals; **20.1:** From M.M. Tegner and D. Epel, "Sea Urchin Sperm: Eggs Interaction Studied with the Scanning Electron Microscope," *Science* 179: 685-688. Copyright 1973 by the AAAS.; **20.4*a*:** © A. Tsiaras/Photo Researchers, Inc.; **20.4*b*:** © Omikron/Photo Researchers, Inc.; **20.4c:** © Petit Format/Nestle/Photo Researchers, Inc.; **20A, p. 535:** "People Weekly" © Taro Yamasaki, 1992; **20.10, 20.14:** © Donald Yaeger/Camera M.D. Studios; **20B, p. 563:** From Anne Pytkowiez Streissguth, *Science,* 209:353-361 (18 July 1980) © AAAS.

Line Art

Chapter 3

3.16: From Stuart Ira Fox, *Human Physiology,* 4th ed. Copyright © 1993 The McGraw-Hill Companies, Inc. All Rights Reserved. Reprinted by permission.

Chapter 6

6.1: From Kent M. Van De Graaff and Stuart Ira Fox, *Concepts of Human Anatomy and Physiology,* 4th edition. Copyright © 1995 The McGraw-Hill Companies, Inc. All Rights Reserved. Reprinted by permission.

Chapter 7

7.7, 7.8, 7.10, 7.11, 7.12, 7.13: From Kent M. Van De Graaff, *Human Anatomy,* 3d ed. Copyright © 1992 The McGraw-Hill Companies, Inc. All Rights Reserved. Reprinted by permission.

Chapter 8

8.3: From Sylvia S. Mader, *Human Biology,* 4th ed. Copyright © 1995 The McGraw-Hill Companies, Inc. All Rights Reserved. Reprinted by permission. **8.14, 8.15:** From Kent M. Van De Graaff and Stuart Ira Fox, *Concepts of Human Anatomy and Physiology,* 4th ed. Copyright © 1995 The McGraw-Hill Companies, Inc. All Rights Reserved. Reprinted by permission.

Chapter 9

9.4, 9.24: From Kent M. Van De Graaff, *Human Anatomy,* 4th ed. Copyright © 1995 The McGraw-Hill Companies, Inc. All Rights Reserved. Reprinted by permission. **9.20:** From Kent M. Van De Graaff, *Human Anatomy,* 3d ed. Copyright © 1992 The McGraw-Hill Companies, Inc. All Rights Reserved. Reprinted by permission. **9.25:** From Kent M. Van De Graaff and Stuart Ira Fox, *Concepts of Human Anatomy and Physiology,* 2d ed. Copyright © 1989 The McGraw-Hill Companies, Inc. All Rights Reserved. Reprinted by permission.

Chapter 10

10.2, 10.20: From Kent M. Van De Graaff and Stuart Ira Fox, *Concepts of Human Anatomy and Physiology,* 3d ed. Copyright © 1992 The McGraw-Hill Companies, Inc. All Rights Reserved. Reprinted by permission.
10.10*a*, 10.21: From Kent M. Van De Graaff and Stuart Ira Fox, *Concepts of Human Anatomy and Physiology,* 4th ed. Copyright © 1995 The McGraw-Hill Companies, Inc. All Rights Reserved. Reprinted by permission.
10.10*b*: From Stuart Ira Fox, *Human Physiology,* 4th ed. Copyright © 1993 The McGraw-Hill Companies, Inc. All Rights Reserved. Reprinted by permission.

Chapter 12

12.18: From Ricki Lewis, *Human Genetics.* Copyright © 1994 The McGraw-Hill Companies, Inc. All Rights Reserved. Reprinted by permission.

Chapter 13

13.28: From Kent M. Van De Graaff and Stuart Ira Fox, *Concepts of Human Anatomy and Physiology,* 4th ed. Copyright © 1995 The McGraw-Hill Companies, Inc. All Rights Reserved. Reprinted by permission.

Chapter 14

14.2: From Kent M. Van De Graaff, *Human Anatomy,* 3d ed. Copyright © 1992 The McGraw-Hill Companies, Inc. All Rights Reserved. Reprinted by permission.

Chapter 15

15.3, 15.28, 15.30: From Kent M. Van De Graaff and Stuart Ira Fox, *Concepts of Human Anatomy and Physiology,* 4th edition. Copyright © 1995 The McGraw-Hill Companies, Inc. All Rights Reserved. Reprinted by permission. **15.9:** From Kent M. Van De Graaff, *Human Anatomy,* 4th ed. Copyright © 1995 The McGraw-Hill Companies, Inc. All Rights Reserved. Reprinted by permission.

Chapter 16

16.17: From Kent M. Van De Graaff, *Human Anatomy,* 4th ed. Copyright © 1995 The McGraw-Hill Companies, Inc. All Rights Reserved. Reprinted by permission.

Chapter 17

17.7: From Kent M. Van De Graaff and Stuart Ira Fox, *Concepts of Human Anatomy and Physiology,* 4th edition. Copyright © 1995 The McGraw-Hill Companies, Inc. All Rights Reserved. Reprinted by permission.

Chapter 19

19.2: From Kent M. Van De Graaff, *Human Anatomy,* 2d ed. Copyright © 1988 The McGraw-Hill Companies, Inc. All Rights Reserved. Reprinted by permission. **19A:** From Ricki Lewis, *Human Genetics.* Copyright © 1994 The McGraw-Hill Companies, Inc. All Rights Reserved. Reprinted by permission.
19.11, 19.14: From Kent M. Van De Graaff and Stuart Ira Fox, *Concepts of Human Anatomy and Physiology,* 4th edition. Copyright © 1995 The McGraw-Hill Companies, Inc. All Rights Reserved. Reprinted by permission.

Illustrators

Kristen Wienandt: 2.3; 3.5; 4.8; 4.17
Peg Gerrity: 2Bb; 5.8 icon; 5.9 icon; 5.17 icon; 5.20 icon; 5.24 icon; 9.21; 9.22; 9.32; 10.1; 10.3; 10.9; 10.10; 10.27; 11.14; 13.6 icon; 13.8; 13.17; 13.29; 14.11; 15.1; 15.15; 16.1; 17.1; 19.5; 20.9; 11 ORGANization figures, enhancement of icons.

Index

Page numbers set in *italics* refer to figures; numbers followed by a *t* designate tables.

A

D

E

H

M

P

Q

T

U

V

W

X

Y

Z

Aids to Understanding Words

acetabul-, vinegar cup: *acetabul*um
adip-, fat: *adip*ose tissue
agglutin-, to glue together: *agglutin*ation
aliment-, food: *aliment*ary canal
allant-, sausage-shaped: *allant*ois
alveol-, small cavity: *alveol*us
an-, without: *an*aerobic respiration
ana-, up: *ana*bolic
andr-, man: *andr*ogens
append-, to hang something: *append*icular
ax-, axis: *ax*ial skeleton
bil-, bile: *bil*irubin
-blast, budding: osteo*blast*
brady-, slow: *brady*cardia
bronch-, windpipe: *bronch*us
calat-, something inserted: inter*calat*ed disk
calyc-, small cup: *calyc*es
cardi-, heart: peri*cardi*um
carp-, wrist: *carp*als
cata-, down: *cata*bolic
chondr-, cartilage: *chondr*ocyte
chorio-, skin: *chorio*n
choroid-, skinlike: *choroid*
chym-, juice: *chym*e
-clast, broken: osteo*clast*
cleav-, to divide: *cleav*age
cochlea, snail: *cochlea*
condyl-, knob: *condyl*e
corac-, beaklike: *corac*oid process
cort-, covering: *cort*ex
cribr-, sievelike: *cribr*iform plate
cric-, ring: *cric*oid cartilage
crin-, to secrete: endo*crin*e
crist-, ridge: *crist*a galli
cut-, skin: sub*cut*aneous
cyt-, cell: *cyt*oplasm
de-, undoing: *de*amination
decidu-, falling off: *decidu*ous
dendr-, tree: *dendr*ite
derm-, skin: *derm*is
detrus-, to force away: *detrus*or muscle
di-, two: *di*saccharide
diastol-, dilation: *diastol*e
diuret-, to pass urine: *diuret*ic
dors-, back: *dors*al
ejacul-, to shoot forth: *ejacul*ation
embol-, stopper: *embol*us
endo-, within: *endo*plasmic reticulum
epi-, upon: *epi*thelial tissue
erg-, work: syn*erg*ist
erythr-, red: *erythr*ocyte
exo-, outside: *exo*crine gland
extra-, outside: *extra*cellular
fimb-, fringe: *fimb*riae
follic-, small bag: hair *follic*le
fov-, pit: *fov*ea
funi-, small cord or fiber: *funi*culus
gangli-, a swelling: *gangli*on
gastr-, stomach: *gastr*ic gland
-gen, to be produced: aller*gen*
genesis-, origin: spermato*genesis*
germ-, to bud or sprout: *germ*inal
glen-, joint socket: *glen*oid cavity
-glia, glue: neuro*glia*
glom-, little ball: *glom*erulus
glyc-, sweet: *glyc*ogen
-gram, something written: electrocardio*gram*
hema-, blood: *hema*toma
hemo-, blood: *hemo*globin
hepat-, liver: *hepat*ic duct
homeo-, same: *homeo*stasis
humor-, fluid: *humor*al
hyper-, above: *hyper*tonic
hypo-, below: *hypo*tonic
im-, (or in-), not: *im*balance
immun-, free: *immun*ity
inflamm-, to set on fire: *inflamm*ation
inter-, between: *inter*phase
intra-, inside: *intra*membranous